THE PHYSICS OF SOLIDS

The Physics of Solids

J. B. Ketterson

Northwestern University, USA

OXFORD
UNIVERSITY PRESS

OXFORD
UNIVERSITY PRESS

Great Clarendon Street, Oxford, OX2 6DP,
United Kingdom

Oxford University Press is a department of the University of Oxford. It furthers the University's objective of excellence in research, scholarship, and education by publishing worldwide. Oxford is a registered trade mark of Oxford University Press in the UK and in certain other countries

First Edition published in 2016

Published in the United States of America by Oxford University Press
198 Madison Avenue, New York, NY 10016, United States of America

British Library Cataloguing in Publication Data
Data available

Library of Congress Control Number: 2015951923

ISBN 978–0–19–874290–6

Preface

It is now common to offer a one-year course in condensed matter physics in many graduate schools. Along with trying to give students a broader perspective, this allows the inclusion of topics such as fluid mechanics, liquid crystals, polymers, critical phenomena, etc. But many of these topics are now part of the emerging field of soft matter physics, and books on this subject are beginning to appear. Hence the time may have come to rein-in the subject matter to something approximating its older boundaries. The choice for the title of this book, *The Physics of Solids*, reflects the author's goal of distinguishing its contents from the broader field of condensed matter physics. That said, solid-state physics has expanded tremendously, and it easily consumes a year's study, even at the introductory graduate level.

I would like *The Physics of Solids* to be useful for the long haul. One learns to find one's way through useful books and if the coverage is sufficiently broad and deep, they can remain valuable beyond graduate school. With this in mind, the chapters in Part I are at the undergraduate level. They assume only elementary mathematics, mechanics, electricity and magnetism, and thermodynamics; for the most part the statistical and quantum physics that is needed is derived along the way. In particular, this material is at a level where it could be the solid-state physics component of a modern physics course. As more sophisticated topics are introduced in successive parts more is asked of the student, it being assumed that other components of their education (graduate quantum mechanics, electrodynamics, and statistical physics) are now available for them to apply to solid-state physics. If I have been successful in all of this, the student will be able to start at an elementary level and proceed to the point where graduate research begins, and beyond, using this book as the main source. This has the advantage of not having to intimately navigate through two or more books.

Some of the topics covered, though widely judged as important, are specialized or rather technical (an example being the theory of the interacting electron liquid); most of this material has been relegated to chapter appendices, and can be safely passed over by many readers. In addition there are two appendices which treat the lagrangian formulation of fields (utilized in variational formulations) and the second quantized or occupation number representation (typically used to discuss many-body, particularly BCS, theory), since they are often underemphasized in typical course sequences.

The book aims to cover all the material traditionally included in a solid-state physics course. This includes discussions of the mechanical, cohesive, dielectric, magnetic, transport (thermal and electrical), optical, and structural properties, using both phenomenological and quantum-microscopic approaches. For metals, the Fermi surface plays a major role and the scattering of low-lying electrons underlies low temperature electrical transport. Phonons and their interactions, including the electron–phonon interaction, yield a semi-quantitative understanding of thermal conductivity and high temperature electrical transport. Superconductivity, perhaps the most dramatic metallic phenomenon, is treated in some detail along with a discussion of its microscopic origin in the electron–phonon interaction (at least for conventional superconductors). However, the text is not limited to these more traditional, though absolutely fundamental, topics. Sections treating more recent advances are included, such as the quantum Hall effects, semiconductor quantum wells, excitonic Bose condensation, highly correlated systems, high temperature and exotic superconductivity, transport in highly disordered media, and non-crystalline solids. Because it has played such an important role in understanding the properties of solids, particularly unconventional superconductivity, we have included sections on superfluidity in the quantum liquids ^{3}He and ^{4}He, from which the quasiparticle concept, so important in the understanding of solid-state excitations, evolved. Of course the field of quantum gasses was greatly impacted by the technology developed to optically trap and cool various atoms and molecules and the resulting discoveries, which will not be addressed here. Finally, some of the tools used to examine solids are discussed, including the de Haas van Alphen effect, photoemission, X-ray/neutron diffraction, optical absorption, magnetometry, magnetic resonance, etc. Each chapter ends with problems gathered from a wide range of sources along with some (hopefully) original ones. As always, serious students will work on as many of these as they can. Recommendations for further reading are also given.

One topic, semiconductor physics, seems to have gravitated to electrical engineering and materials science departments in the United States. Conversely, it is commonly part of physics departments in Europe. This includes both transport and optical properties (along with associated devices). Since many graduates ultimately end up in the semiconductor industry, emphasis on these topics should be an important component of their physics education.

Theoretical solid-state physics through much of its history has been divided into two camps. The first camp concentrates on evolving the technology to calculate the ground state and excited state energies of solids with ever increasing accuracy, usually employing some form of Hartree mean-field theory. The development of density functional theory (DFT) has had an enormous effect on this program. Not only did it legitimize the faith in earlier mean-field-like approaches, but it heralded the need to develop ever better density functionals to accurately carry out calculations. The results obtained are sometimes called first principles calculations, although many approximations are involved. DFT uses one-electron-like states and associated band structures to calculate total energies and many other solid-state properties; these include lattice constants, compressibility and elastic constants, phonon spectra, Fermi surfaces in metals, energy gaps in insulators, and optical critical points. For those systems that have a magnetic ground state, the nature of the magnetism can often be predicted with great accuracy, e.g., the saturation magnetization of a ferromagnet or the spin structure of an antiferromagnet. Recently much progress has been made in predicting excited state properties, which were originally thought to lie beyond density functional theory; an example here is the band gap in semiconductors.

But density functional theory as currently practiced fails in so-called strongly correlated systems which, loosely speaking, are those where the time spent by electrons orbiting the nucleus of some atoms is comparable to or longer than the time characterizing their hopping between neighboring atoms; the motion of such electrons is then strongly affected by other occupied orbitals on that atom. Going beyond DFT to treat strongly correlated systems is currently one of the major challenges in solid-state physics. Computing superconducting transition temperatures involving other electronic pairing mechanisms is even more challenging.

The above-referred-to second camp seeks to predict new or explain observed phenomena, starting from some appropriate model Hamiltonian which contains parameters that are adjusted to achieve desired results or elaborate various possibilities. Perhaps the most spectacular success of this latter approach was an explanation of the nature of the superconducting state. But as theoretical solid-state physics has evolved the number of problems that can be accurately treated by the first approach continually expands. As this happens, practitioners in the second camp turn their attention to more subtle effects, usually where many-body effects are judged to play a role that cannot be smoothed over. Problems in this latter category include the quantum Hall effects, electron interactions in highly disordered materials, highly correlated electron transport, and quantum phase transitions. The "first principles" and model building approaches have their merits and shortcomings. In this book I will utilize both, depending on the problem at hand.

Quantum mechanics books typically emphasize few-electron problems: e.g., the hydrogen atom, the hydrogen molecule, and the helium atom. One must not lose site of the fact that many-electron problems as they arise in atomic and molecular physics are effectively unsolvable. This inescapable fact arises from what W. Kohn has called the *exponential wall.*[1] Take the case of an O_2 molecule, which has a total of 16 electrons. Its electrons move in a $3 \times 16 = 48$-dimensional space (ignoring spin). If one (coarsely) approximates the wave function with just ten values along all axes of this configuration space there will be 10^{48} numbers; no known computer memory technology will store these numbers and, as seen, the difficulties increase exponentially. And solids typically contain 10^{22} electrons. Hence, developing strategies to make reliable estimates of typical properties is one of the main intellectual and practical challenges of both chemistry and solid-state physics.

In conclusion I would like to thank the many generations of students who have listened to my lectures, and, as a part of their homework assignments, corrected and critiqued the notes I distributed, from which this book has evolved. I also learned from their project reports. Many people have commented on parts of the book including, M.H. Cohen, M.L. Cohen, G. Crabtree,

[1] W. Kohn, Nobel Lecture: Electronic structure of matter—wave functions and density functionals, *Rev. Mod. Phys.* 71, 1253 (1999).

Y. Eckstein, A. Feinerman, A.J. Freeman, E. Fukushima, A. Garg, G. Graviton, M. Grayson, J. Jain, D. Khmelnitskii, B. Maple, D. Mattis, I.K. Schuller, S. Shafraniuk, and P. Sievert; although they were very helpful they bear no responsibility for remaining issues of clarity or accuracy. At an earlier stage, Ms. Arlene Jackson produced Word files from illegible hand-written pages, without which the project would never have started.

J.B. Ketterson, fall 2015.

Contents

Part II Crystal Structure and its Determination

Part III Electronic Structure of Periodic Solids

Part V Lattice Dynamics

Part VI Electron Transport and Conduction Electron Dynamics

Part I

Introductory Topics

Elastic Behavior of Solids

1

The defining property of a solid is its ability to resist shape changes in the presence of external stresses, examples of the latter being a compression or shear, which we define shortly. Hence we begin our study of solid-state physics with a discussion of stresses and how a solid responds to them.

1.1 The stress tensor

A body at rest must, according to Newton's second law, have no net force acting on it. Subjecting a solid to two forces that are equal in magnitude, but act in opposite directions along some axis normal to two parallel surfaces will "squeeze" the material, but leave it at rest. Applying such equal and opposite forces tangent to two parallel surfaces (referred to as a couple) results in no linear acceleration but will cause an angular acceleration; hence such couples must be balanced for the body as a whole to be at rest.

But macroscopic forces applied to a solid must always be distributed over some region of the surface;[1] a force applied to a vanishingly small area results in a diverging local pressure which will greatly deform the surface locally, ultimately causing it to permanently (irreversibly) change its shape (as when the application of a sharp object, such as a pin, results in a hole or scratch). For this reason the concept of *stress* is introduced.

Stress is defined as *force per unit area* and will be denoted by σ. We will limit ourselves here to stresses that do not irreversibly deform the solid. Stresses are separated into two classes: *compression and shear.*[2] These are shown schematically in Fig. 1.1 for a cube where the z faces are parallel to the plane of the page. Figure 1.1(a) shows an example of a compression involving distributed forces in the $\pm$x directions applied *normal to the x faces* of the cube; it is denoted as σ_{xx}. Figure 1.1(b) shows an example of a shear stress involving distributed forces acting in the $\pm$x directions *tangent to the y faces*; it is denoted as σ_{xy}. Note that this unbalanced shear would, as mentioned, result in a net torque which would in turn result in an angular acceleration; hence for the body to be in equilibrium this shear must be counter-balanced by a second shear σ_{yx} on the x faces with forces acting in the $\pm$y directions, as shown in Fig. 1.1(c); i.e., $\sigma_{xy} = \sigma_{yx}$. It is apparent that the various stress components form a tensor[3] that we write in the form

$$\sigma = \begin{pmatrix} \sigma_{xx} & \sigma_{xy} & \sigma_{zx} \\ \sigma_{xy} & \sigma_{yy} & \sigma_{yz} \\ \sigma_{zx} & \sigma_{yz} & \sigma_{zz} \end{pmatrix}, \tag{1.1}$$

where the equilibrium constraint $\sigma_{ij} = \sigma_{ji}$ has been incorporated.

[1] Here we neglect so-called *body forces*, like gravity, which act directly on the interior of a material.

[2] In a liquid, which cannot support a shear, we must have an isotropic compression such that $\sigma_{11} = \sigma_{22} = \sigma_{33} = -P$.

[3] Under rotations σ transforms as $\sigma'_{ij} = \sum_{kl} \sigma_{kl} R_{ki} R_{lj}$, where **R** is a Euler rotation matrix.

The Physics of Solids. First Edition. J. B. Ketterson.

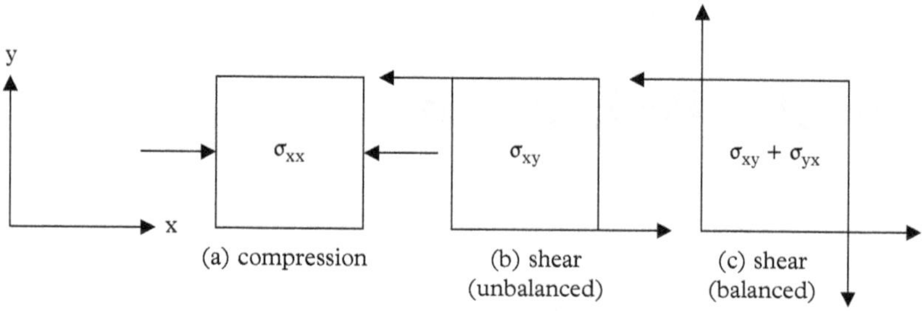

Figure 1.1 *Basic stresses acting on the faces of a cube: (a) a compression; (b) an unbalanced shear (which would result in an angular acceleration); (c) a balanced shear (achieved with a second shear acting over the remaining pair of faces).*

1.2 The strain tensor

As previously noted, the defining property of a solid is its tendency to resist changing shape in the presence of external stresses. This being the case one can picture a vector **r**, measured relative to an arbitrary origin, that specifies some point in the body (say some atom). When the solid is deformed, as a result of the application of stresses, this point (atom) will move to a new position in the body specified by a vector **r**′ that differs from the old vector by some amount **u**; i.e., $\mathbf{r}' = \mathbf{r} + \mathbf{u}$, as shown in the accompanying drawing.

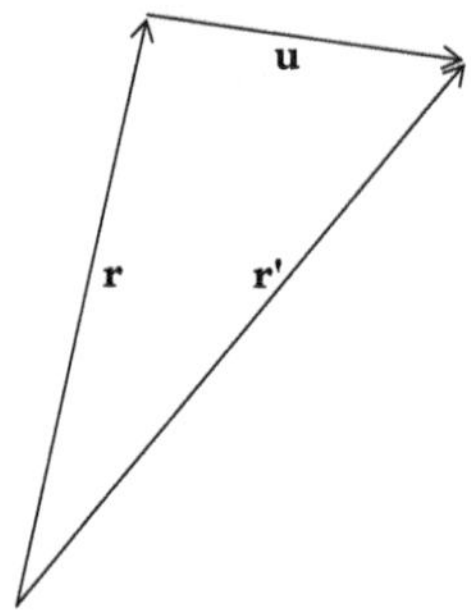

The coordinates of **r**′ are assumed to be a function of the coordinates prior to the deformation; hence $\mathbf{u} = \mathbf{u}(\mathbf{r})$. Now consider two closely spaced points separated by a distance $d\mathbf{r}$; as a result of the deformation, $d\mathbf{r}$ changes to $d\mathbf{r}'$, given by $d\mathbf{r}' = d\mathbf{r} + d\mathbf{u}$. We expand $d\mathbf{u}(\mathbf{r})$ as

$$du_i = \sum_j \frac{\partial u_i}{\partial x_j} dx_j, \tag{1.2a}$$

where x_j denotes the coordinates $j = 1$ to 3. We next factor the change du_i into symmetric and antisymmetric parts by writing

$$\begin{aligned} du_i &= \sum_j \left(\frac{1}{2}\left(\frac{\partial u_i}{\partial x_j} + \frac{\partial u_j}{\partial x_i} \right) + \frac{1}{2}\left(\frac{\partial u_i}{\partial x_j} - \frac{\partial u_j}{\partial x_i} \right) \right) dx_j \\ &\equiv \sum_j \left[\varepsilon_{ij} + \omega_{ij} \right] dx_j, \end{aligned} \tag{1.2b}$$

where in the second step introduced the symmetric and antisymmetric tensors

$$\varepsilon_{ij} \equiv \frac{1}{2}\left(\frac{\partial u_i}{\partial x_j} + \frac{\partial u_j}{\partial x_i}\right) \tag{1.3a}$$

and

$$\omega_{ij} \equiv \frac{1}{2}\left(\frac{\partial u_i}{\partial x_j} - \frac{\partial u_j}{\partial x_i}\right). \tag{1.3b}$$

Now the change in du arising strictly from the second term in (1.2b) may be written as $\boldsymbol{\omega} \times \mathbf{dr}$ where $\omega = -(\omega_{23}, \omega_{32}, \omega_{13})$; i.e., it corresponds to a *rotation* by an angle ω and hence does not represent a true deformation of the solid. The symmetric tensor ε_{ij} then corresponds to a pure deformation and is called the *strain tensor.*

1.3 Hooke's law

We start by assuming that in the absence of an external stress a body maintains its shape. We then assume that when a stress is applied the body distorts, taking up a new shape, which it maintains as long as the stress is present.[4] This distorted shape is characterized by the position-dependent vector **u** and the associated strain tensor ε_{ij}.[5] In leading order of a Taylor expansion we expect a linear relation between the applied stress and the resulting strain that we write as

$$\varepsilon_{ij} = \sum_{k,l} s_{ij;kl}\sigma_{kl}, \tag{1.4a}$$

where the $s_{ij;kl} = \partial\varepsilon_{ij}/\partial\sigma_{kl}$ are called *elastic compliance constants* and are characteristic of a given material. However, it is more common to write this relationship with the strain as the independent variable and hence we write

$$\sigma_{ij} = \sum_{k,l} c_{ij;kl}\varepsilon_{kl}, \tag{1.4b}$$

where the $c_{ij;kl} = \partial\sigma_{ij}/\partial\varepsilon_{kl}$ and are called *elastic stiffness constants*, and we henceforth confine ourselves to this description.

As written there are $3^4 = 81$ elements in the fourth-rank tensor $c_{ij;kl}$. The symmetry of the stress and strain tensors leads to the relations $c_{ij;kl} = c_{ji;kl} = c_{ij;lk} = c_{ji;lk}$.[6] As an example, the σ_{xx} component is given by

$$\begin{aligned}\sigma_{xx} = {} & c_{xx;xx}\varepsilon_{xx} + c_{xx;xy}\varepsilon_{xy} + c_{xx;xz}\varepsilon_{xz} + c_{xx;yx}\varepsilon_{yx} + c_{xx;yy}\varepsilon_{yy}\\ & + c_{xx;yz}\varepsilon_{yz} + c_{xx;zx}\varepsilon_{zx} + c_{xx;zy}\varepsilon_{zy} + c_{xx;zz}\varepsilon_{zz}.\end{aligned}$$

Applying the above symmetry relations for $c_{ij;kl}$ and ε_{ij} we may rewrite this expression as

$$\sigma_{xx} = c_{xx;xx}\varepsilon_{xx} + c_{xx;yy}\varepsilon_{yy} + c_{xx;zz}\varepsilon_{zz} + c_{xx;xy}2\varepsilon_{xy} + c_{xx;yz}2\varepsilon_{yz} + c_{xx;zx}2\varepsilon_{zx}.$$

[4] This excludes so-called viscoelastic materials of which "silly putty" is an example.

[5] A material can have an *internal strain*, as occurs in polycrystalline materials consisting of many contacting parts which mutually distort each other. We ignore such effects here.

[6] Additional symmetry considerations to be discussed shortly further reduce the number of independent elastic constants.

It is common to *rename* all the factors involved in this expression and write it in the form

$$\sigma_1 \equiv c_{11}\varepsilon_1 + c_{12}\varepsilon_2 + c_{13}\varepsilon_3 + c_{14}\varepsilon_4 + c_{15}\varepsilon_5 + c_{16}\varepsilon_6.$$

Proceeding in an analogous manner with the remaining components of the stress tensor, we can then express Eq. (1.4b) in a matrix form (note the tensor transformation properties are lost in this process). In place of the above symmetric 3×3 stress and strain tensors we have introduced *six-component* vectors which are connected by a 6×6 matrix of elastic moduli, c_{ij}, having the form[7]

$$\begin{pmatrix} \sigma_{xx} \\ \sigma_{yy} \\ \sigma_{zz} \\ \sigma_{yz} \\ \sigma_{zx} \\ \sigma_{xy} \end{pmatrix} = \begin{pmatrix} c_{11} & c_{12} & c_{13} & c_{14} & c_{15} & c_{16} \\ c_{21} & c_{22} & c_{23} & c_{24} & c_{25} & c_{26} \\ c_{31} & c_{32} & c_{33} & c_{34} & c_{35} & c_{36} \\ c_{41} & c_{42} & c_{43} & c_{44} & c_{45} & c_{46} \\ c_{51} & c_{52} & c_{53} & c_{54} & c_{55} & c_{56} \\ c_{61} & c_{62} & c_{63} & c_{64} & c_{65} & c_{66} \end{pmatrix} \begin{pmatrix} \varepsilon_{xx} \\ \varepsilon_{yy} \\ \varepsilon_{zz} \\ 2\varepsilon_{yz} \\ 2\varepsilon_{zx} \\ 2\varepsilon_{xy} \end{pmatrix}, \tag{1.5}$$

or, in a more compact form,

$$\sigma_i = \sum_{j=1}^{6} c_{ij}\varepsilon_j, \tag{1.6}$$

where we have numbered the six components of the strain vector $\boldsymbol{\varepsilon}$ as

$$(\varepsilon_{xx}, \varepsilon_{yy}, \varepsilon_{zz}, 2\varepsilon_{yz}, 2\varepsilon_{zx}, 2\varepsilon_{xy}) \to (\varepsilon_1, \varepsilon_2, \varepsilon_3, \varepsilon_4, \varepsilon_5, \varepsilon_6) \tag{1.7a}$$

and the six components of the stress vector $\boldsymbol{\sigma}$ as

$$(\sigma_{xx}, \sigma_{yy}, \sigma_{zz}, \sigma_{yz}, \sigma_{zx}, \sigma_{xy}) \to (\sigma_1, \sigma_2, \sigma_3, \sigma_4, \sigma_5, \sigma_6). \tag{1.7b}$$

Note the somewhat awkward inclusion of factors of two in the relations between the off-diagonal elements of the strain tensor ε_{ij} and the strain six-vector ε_i. The 81 elements of the tensor $c_{ij;kl}$ are now replaced by the 36 elements of the matrix c_{ij}.

1.4 The energy density

Consider a cube with edges of length L aligned parallel to the Cartesian x, y, and z axis. Assume that pairs of distributed forces F_x acting along the x direction, of the type pictured in Fig. 1.1(a,b), are applied to this object; we assume the body is thermally isolated so that any work done on it goes into raising the internal energy. These forces are distributed over the cube such that they correspond to a

[7] The matrix c_{ij} is not a tensor and hence its rotation properties must be obtained from those for $c_{ij;kl}$.

compression component, $L^2\sigma_1$, and two shear components, $L^2\sigma_5$ and $L^2\sigma_6$; this results in corresponding displacements of opposing faces normal to the x, y, and z of the cube relative to each other by the amounts $Ld\varepsilon_1$, $Ld\varepsilon_5$, and $Ld\varepsilon_6$. The resulting change in the internal energy of the cube is therefore

$$F_x du_x = L^3 (\sigma_1 d\varepsilon_1 + \sigma_5 d\varepsilon_5 + \sigma_6 d\varepsilon_6). \tag{1.8}$$

Generalizing to force couples acting in an arbitrary direction, the increase in energy per unit volume of the sample, de, is given by

$$de = \sum_{i=1}^{6} \sigma_i \, d\varepsilon_i. \tag{1.9}$$

Inserting Eq. (1.4b) and integrating (1.9), we have[8,9]

$$e = \frac{1}{2}\sum_{i,j}^{6} c_{ij}\, \varepsilon_i \varepsilon_j = \frac{1}{2}\sum_{i}^{6} \sigma_i \varepsilon_i. \tag{1.10}$$

This quadratic expression for the energy must not change on relabeling the strain variables and this in turn demands that

$$c_{ij} = c_{ji}; \tag{1.11}$$

this symmetry decreases the maximum number of independent elastic moduli c_{ij} from 36 to 21; in terms of the fourth-rank elastic constants discussed, the reduction is from 81 to 21.

1.5 Elastic moduli of cubic and hexagonal systems

By performing various symmetry operations associated with a crystal having a given symmetry, relations can be generated among the elements of the matrix that further reduce the number of independent constants, c_{ij}. Although crystal symmetry will not be introduced until later,[10] Table 1.1 lists the number of independent elastic constants for the various crystal symmetries; the number in the second column corresponds to a reduction that is possible in lower symmetry systems through a special choice of crystal axes.

The conclusions of such an analysis for two common crystal symmetries, cubic and hexagonal, are

$$\textbf{cubic: } c_{11} = c_{22} = c_{33}; c_{12} = c_{23} = c_{31}; c_{44} = c_{55} = c_{66}; \tag{1.12a}$$

$$\textbf{hexagonal: } \begin{cases} c_{11} = c_{22}; c_{12}; c_{13} = c_{23}; c_{33} \\ c_{44} = c_{55}; c_{66} = \dfrac{1}{2}(c_{11} - c_{12}); \end{cases} \tag{1.12b}$$

for a total of three and five independent parameters respectively; all other $c_{ij} = 0$. Amorphous or fully averaged polycrystalline materials are usually regarded as

[8] The tensor forms corresponding to (1.9) and (1.10) are

$$de = \sum_{i,j=1}^{3} \sigma_{ij} d\varepsilon_{ij} \tag{1.9'}$$

and

$$e = \frac{1}{2}\sum_{i,j,k,l=1}^{3} c_{ij;kl}\varepsilon_{ij}\varepsilon_{kl} = \frac{1}{2}\sum_{i,j=1}^{3} \sigma_{ij}\varepsilon_{ij} \tag{1.10'}$$

[9] In Section 16.7 we will discuss a method to obtain elastic constants from electronic structure calculations.

[10] Crystal symmetry will be discussed in Chapter 8, as will the symmetry designations in Table 1.1.

Table 1.1 *Number of independent elastic constants for various crystal symmetries.*

Symmetry	#		Symmetry	#	
Triclinic	21	18	Rhombohedral (C_3, S_6)	7	6
Monoclinic	13	12	Rhombohedral (C_{3v}, D_3, D_{3d})	6	6
Orthorhombic	9	9	Hexagonal	5	5
Tetragonal (C_4, S_4, S_{4h})	7	6	Cubic	3	3
Tetragonal ($C_{4v}, D_{2d}, D_4, D_{4h}$)	6	6	–		

Table 1.2 *Elastic compliances of some cubic crystals (at room temperature unless indicated), in units of* 10^{12} *dyne/cm.*

Crystal	c_{11}	c_{12}	c_{44}
Al	1.08	0.62	0.28
C (diamond)	10.76	1.25	5.76
Cu	1.684	1.214	0.754
Ge	1.29	0.48	0.67
Fe	2.37	1.41	1.16
K	0.046	0.037	0.026
Na (210 K)	0.055	0.042	0.049
Ni	2.50	1.60	1.185
Pb	0.48	0.41	0.14
Si	1.66	0.64	0.79
W	5.01	1.98	1.15
AgBr	0.56	0.33	0.073
AgCl	0.60	0.36	0.062
KBr	0.35	0.058	0.050
KCl	0.40	0.062	0.062
KI	0.27	0.043	0.042
NaBr	0.33	0.13	0.13
NaCl	0.486	0.127	0.128

being elastically isotropic and the three independent elastic moduli of the cubic case (the highest crystal symmetry) are then related as

$$\textbf{isotropic:}\quad c_{11};c_{12};c_{44}=\frac{1}{2}(c_{11}-c_{12}); \tag{1.13}$$

i.e., there are only two independent elastic constants. We will show this in Section 1.8. Table 1.2 gives the elastic compliances of some cubic crystals.

1.6 Distortions of a cubic crystal

We now give a few examples that relate distortions of a cubic crystal through the elastic moduli to external stress for various prescribed conditions. As a simple example, consider the application of a uniaxial stress[11] $\sigma_1 = -P$ along the x axis of a cube under the condition that no displacements can occur in the y or z direction. Here, we can imagine pushing on a very soft material constrained inside a square hole in a very hard material, as shown in Fig. 1.2; we will assume that the hard material does not yield and, in addition, the materials can slip with respect to each other under compression. We then have $\varepsilon_2 = \varepsilon_3 = 0$ and consequently Eqs. (1.6) yield only the condition

$$\sigma_1 = -P = c_{11}\varepsilon_1. \tag{1.14}$$

Note the stresses $\sigma_2 = \sigma_3 \neq 0$ and can now be determined from ε_1.

Next, imagine that the compressional stress is applied to the x face and that the cube is free to expand or contract in the y and z directions (i.e., no stresses are present in these directions). Equations (1.6) may then be written as

$$\begin{aligned} -P &= +c_{11}\varepsilon_1 + c_{12}\varepsilon_2 + c_{12}\varepsilon_3;\\ 0 &= +c_{12}\varepsilon_1 + c_{11}\varepsilon_2 + c_{12}\varepsilon_3;\\ 0 &= +c_{12}\varepsilon_1 + c_{12}\varepsilon_2 + c_{11}\varepsilon_3. \end{aligned} \tag{1.15a–c}$$

[11] The sign of a stress is defined with respect to the outward normal across the face on which it acts, so pressing on the sample corresponds to a negative stress.

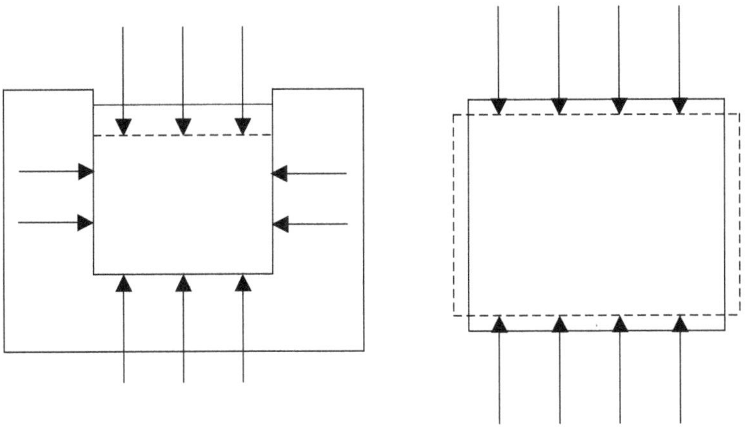

Figure 1.2 *The figure to the left shows the effect of a uniaxial stress for a sample constrained in the horizontal plane; the figure to the right shows the effect of a uniaxial compression when the sample is free to expand in the horizontal plane.*

Adding the second two equations, and noting that by symmetry $\varepsilon_2 = \varepsilon_3$, we obtain

$$\frac{\varepsilon_2}{\varepsilon_1} = -\frac{c_{12}}{(c_{11} + c_{12})}. \tag{1.16}$$

This ratio of the transverse to the longitudinal displacements is called *Poisson's ratio* and will be denoted by ν.[12] Substituting this expression into Eq. (1.15a) we obtain

$$Y \equiv \frac{P}{\varepsilon_1} = -\frac{(c_{11} + c_{12})\, c_{11} - 2c_{12}^2}{c_{11} + c_{12}}, \tag{1.17}$$

where we have introduced the *Young's modulus*, Y.

Lastly, we assume that the crystal is subjected to a hydrostatic pressure P that acts normal to all three faces of the cube,

$$\sigma_1 = \sigma_2 = \sigma_3 = -P. \tag{1.18}$$

Equations (1.6) now become

$$\begin{aligned} -P &= c_{11}\varepsilon_1 + c_{12}\varepsilon_2 + c_{12}\varepsilon_3, \\ -P &= c_{12}\varepsilon_1 + c_{11}\varepsilon_2 + c_{12}\varepsilon_3, \\ -P &= c_{12}\varepsilon_1 + c_{12}\varepsilon_2 + c_{11}\varepsilon_3. \end{aligned} \tag{1.19a–c}$$

From symmetry, $\varepsilon_1 = \varepsilon_2 = \varepsilon_3 \equiv \varepsilon$ and each cube edge, assumed to have length L, changes by an amount $dL = \varepsilon L$. The volume in the presence of the applied pressure is then $V = (L + \delta L)^3 \cong L^3 + 3L^2\delta L$ or $\frac{\delta V}{L^3} = 3\varepsilon$. The compressibility K is defined by the relation $K = -V\delta P/\delta V$ and on adding Eqs. (1.19a–c) we obtain

$$K = \frac{c_{11} + 2c_{12}}{3}. \tag{1.20}$$

Since they correspond to well-defined measurement conditions, Y, ν, and K are commonly used as alternative parameters for characterizing the elastic response of materials (cubic in this case). For the isotropic case, where we only have c_{11} and c_{12}, it is common to use two of the three parameters K, Y, and ν in place of the moduli c_{11} and c_{12}.

1.7 Elastic waves

From our previous discussion—see Eq. (1.8)—the ith component of the differential force dF_i arising from the differential surface elements dS_j is

$$dF_i = \sum_j \sigma_{ij} dS_j. \tag{1.21}$$

[12] It is also denoted by σ in some treatments, which must then not be confused with the stress tensor.

The total force component acting on the entire surface S that encloses some volume V is then

$$F_i = \int_S \sum_j \sigma_{ij} dS_j. \tag{1.22}$$

Alternatively we can write (1.22) as

$$F_i \equiv \int_V f_i dV, \tag{1.23}$$

where we have introduced the components f_i of a *force density* vector $\mathbf{f}$ (which has units of force per unit volume) or, as it is more commonly called, the *volume force*. Applying Green's theorem we have

$$f_i = \sum_j \frac{\partial \sigma_{ij}}{\partial x_j}. \tag{1.24}$$

Note that for the dynamic case where the body is locally accelerated $\mathbf{f} \neq 0$; from (1.24) we see that this arises from a position dependence of σ_{ij}.

By Newton's second law, the force must be equal to the mass times the acceleration of the volume in question; i.e.,

$$\int_V f_i(\mathbf{r}, t) dV = \int_V \rho\, \ddot{u}_i(\mathbf{r}, t) dV, \tag{1.25}$$

where ρ is the mass per unit volume (the density) and $\mathbf{u}$ is the local displacement vector introduced earlier, which now depends on time as well as the spatial coordinates. Substituting (1.24) we have

$$\int_V \rho\, \ddot{u}_i(\mathbf{r}, t) dV = \int_V \sum_j \frac{\partial \sigma_{ij}(\mathbf{r}, t)}{\partial x_j} dV. \tag{1.26}$$

Noting that this must hold for any volume, we may directly equate the integrands in the two expressions; i.e.,

$$\rho \ddot{u}_i(\mathbf{r}, t) = \sum_j \frac{\partial \sigma_{ij}(\mathbf{r}, t)}{\partial x_j}. \tag{1.27}$$

Inserting Eq. (1.4b), we obtain the equation of motion for the ith component of the displacement,

$$\begin{aligned} \rho \ddot{u}_i(\mathbf{r}, t) &= \sum_j \sum_{k,l} c_{ij;kl} \frac{\partial \varepsilon_{kl}(\mathbf{r}, t)}{\partial x_j} \\ &= \sum_j \sum_{k,l} c_{ij;kl} \frac{\partial}{\partial x_j} \frac{1}{2} \left(\frac{\partial u_k(\mathbf{r}, t)}{\partial x_l} + \frac{\partial u_l(\mathbf{r}, t)}{\partial x_k} \right). \end{aligned} \tag{1.28}$$

This equation is second order in both space and time and has wave-like solutions.

If we revert to our matrix notation, Eq. (1.27) must be written out component by component (this is a disadvantage of the matrix notation); on doing this we find that the three components are given by

$$\rho\ddot{u}_1(\mathbf{r},t) = \frac{\partial\sigma_1(\mathbf{r},t)}{\partial x} + \frac{\partial\sigma_6(\mathbf{r},t)}{\partial y} + \frac{\partial\sigma_5(\mathbf{r},t)}{\partial z},$$

$$\rho\ddot{u}_2(\mathbf{r},t) = \frac{\partial\sigma_6(\mathbf{r},t)}{\partial x} + \frac{\partial\sigma_2(\mathbf{r},t)}{\partial y} + \frac{\partial\sigma_4(\mathbf{r},t)}{\partial z},$$

$$\rho\ddot{u}_3(\mathbf{r},t) = \frac{\partial\sigma_5(\mathbf{r},t)}{\partial x} + \frac{\partial\sigma_4(\mathbf{r},t)}{\partial y} + \frac{\partial\sigma_3(\mathbf{r},t)}{\partial z}; \qquad (1.29\text{a–c})$$

the six components of the stress tensor are related to the strain tensor through Eq. (1.5). The solutions to this set of equations are in general quite complex, especially when the crystal symmetry is low, where all the components of u_i tend to couple and many elastic moduli enter. We may seek solutions in the form $\mathbf{u}(\mathbf{r},t) = \mathbf{u}^{(0)}e^{-i\omega t + i\mathbf{k}\cdot\mathbf{r}}$ and write $\mathbf{k} = k\mathbf{n}$, where $\mathbf{n}$ is a unit vector parallel to the wave vector $\mathbf{k}$, which may be parameterized in terms of the polar angles θ and φ; inserting this form into (1.29) we would obtain a set of equations having the form

$$\begin{pmatrix} M_{11}(k,\mathbf{n},c_{ij}) - \rho\omega^2 & M_{12}(k,\mathbf{n},c_{ij}) & M_{13}(k,\mathbf{n},c_{ij}) \\ M_{21}(k,\mathbf{n},c_{ij}) & M_{22}(k,\mathbf{n},c_{ij}) - \rho\omega^2 & M_{23}(k,\mathbf{n},c_{ij}) \\ M_{31}(k,\mathbf{n},c_{ij}) & M_{32}(k,\mathbf{n},c_{ij}) & M_{33}(k,\mathbf{n},c_{ij}) - \rho\omega^2 \end{pmatrix} \begin{pmatrix} u_1^{(0)} \\ u_2^{(0)} \\ u_3^{(0)} \end{pmatrix} = 0, \qquad (1.30)$$

where the M_{ij} involve products of c_{ij}, k^2, and n_in_j. The eigenvalues, $\rho\omega^2$, of this matrix for a given k and direction of propagation $\mathbf{n}$ give the frequencies, while the corresponding eigenvectors give the polarizations, telling us how the displacement $\mathbf{u}$ of the mode is related to the propagation direction $\mathbf{n}$.

1.8 Waves in a cubic crystal

We will only consider the case of sound propagation in a cubic crystal. In this case the equations of motion following from Eqs. (1.29a–c) for the three components (reverting to x, y, z, notation) take the form

$$\rho\ddot{u}_1(\mathbf{r},t) = c_{11}\frac{\partial\varepsilon_1}{\partial x} + c_{12}\left(\frac{\partial\varepsilon_2}{\partial x} + \frac{\partial\varepsilon_3}{\partial x}\right) + c_{44}\left(\frac{\partial\varepsilon_6}{\partial y} + \frac{\partial\varepsilon_5}{\partial z}\right),$$

$$\rho\ddot{u}_2(\mathbf{r},t) = c_{11}\frac{\partial\varepsilon_2}{\partial y} + c_{12}\left(\frac{\partial\varepsilon_1}{\partial y} + \frac{\partial\varepsilon_3}{\partial y}\right) + c_{44}\left(\frac{\partial\varepsilon_4}{\partial z} + \frac{\partial\varepsilon_6}{\partial x}\right),$$

$$\rho\ddot{u}_3(\mathbf{r},t) = c_{11}\frac{\partial\varepsilon_3}{\partial z} + c_{12}\left(\frac{\partial\varepsilon_2}{\partial z} + \frac{\partial\varepsilon_1}{\partial z}\right) + c_{44}\left(\frac{\partial\varepsilon_5}{\partial x} + \frac{\partial\varepsilon_4}{\partial y}\right). \qquad (1.31\text{a–c})$$

If we write out the x component of Eq. (1.31a) in terms of the components u_i using Eqs. (1.3) and (1.7a), we obtain

$$\rho\ddot{u}_1(\mathbf{r},t) = c_{11}\frac{\partial^2 u_1}{\partial x^2} + c_{12}\left(\frac{\partial^2 u_2}{\partial x \partial y} + \frac{\partial^2 u_3}{\partial x \partial z}\right)$$
$$+ c_{44}\left[\frac{\partial}{\partial y}\left(\frac{\partial u_2}{\partial x} + \frac{\partial u_1}{\partial y}\right) + \frac{\partial}{\partial z}\left(\frac{\partial u_1}{\partial z} + \frac{\partial u_3}{\partial x}\right)\right]. \quad (1.32)$$

For longitudinal modes the displacement is parallel to the propagation direction. The mode propagating parallel to the x axis has the form

$$u_1(x,t) = u_1^{(0)} e^{-i\omega t + ikx}. \quad (1.33)$$

From Eq. (1.32) we have $\rho\omega^2 = c_{11}k^2$, corresponding to a longitudinal sound velocity v_l given by

$$v_l = \frac{\omega}{k} = \sqrt{\frac{c_{11}}{\rho}}. \quad (1.34)$$

For shear modes the displacement is perpendicular to the propagation direction. A mode polarized along x and propagating parallel to y has the form $u_1(y,t) = u_1^{(0)} e^{-i\omega t + iky}$; inserting this into Eq. (1.32) gives $\rho\omega^2 = c_{44}k^2$ corresponding to a transverse sound velocity v_t,

$$v_t = \frac{\omega}{k} = \sqrt{\frac{c_{44}}{\rho}}. \quad (1.35)$$

Sound velocities in typical solids are of order 10^5 cm/sec. Note that for an arbitrary propagation direction **n**, **u** is neither parallel nor perpendicular to **k**; i.e., the wave is neither longitudinal nor transverse.

Equations (1.31a) can be rearranged as

$$\rho\ddot{u}_1(\mathbf{r},t) = (c_{11} - c_{12} - 2c_{44})\frac{\partial^2 u_1}{\partial x^2} + c_{44}\left(\frac{\partial^2 u_1}{\partial x^2} + \frac{\partial^2 u_1}{\partial y^2} + \frac{\partial^2 u_1}{\partial z^2}\right)$$
$$+ (c_{12} + c_{44})\frac{\partial}{\partial x}\left(\frac{\partial u_1}{\partial x} + \frac{\partial u_2}{\partial y} + \frac{\partial u_3}{\partial z}\right), \quad (1.36)$$

or

$$\rho\ddot{u}_1(\mathbf{r},t) = (c_{11} - c_{12} - 2c_{44})\frac{\partial^2 u_1}{\partial x^2} + c_{44}\nabla^2 u_1 + (c_{12} + c_{44})\frac{\partial}{\partial x}\nabla\cdot\mathbf{u}. \quad (1.37)$$

Note that if we invoke the elastic isotropy condition (see Eq. (1.13))

$$c_{11} - c_{12} = 2c_{44}, \quad (1.38)$$

the first term drops out. Combining the corresponding equations for $\rho\ddot{u}_2(\mathbf{r},t)$ and $\rho\ddot{u}_3(\mathbf{r},t)$ we obtain the vector equation

$$\rho\ddot{\mathbf{u}}(\mathbf{r},t) = c_{44}\nabla^2\mathbf{u} + (c_{12} + c_{44})\,\nabla(\nabla\cdot\mathbf{u}). \tag{1.39}$$

Using the vector identity $\nabla\times(\nabla\times\mathbf{u}) = \nabla(\nabla\cdot\mathbf{u}) - \nabla^2\mathbf{u}$ we can write this as

$$\rho\ddot{\mathbf{u}}(\mathbf{r},t) = (c_{12} + 2c_{44})\,\nabla^2\mathbf{u} + (c_{12} + c_{44})\,\nabla\times(\nabla\times\mathbf{u}), \tag{1.40}$$

or

$$\rho\ddot{\mathbf{u}}(\mathbf{r},t) = (c_{12} + 2c_{44})\,\nabla(\nabla\cdot\mathbf{u}) - c_{44}\nabla\times(\nabla\times\mathbf{u}). \tag{1.41}$$

These equations, which as far as the spatial derivatives are concerned only involve the ∇ operator, are invariant with respect to rotations of the coordinate system. Hence if the material satisfies the condition (1.38) it corresponds to an *isotropic* (amorphous or polycrystalline) solid. The anisotropy constant,

$$A = \frac{2c_{44}}{c_{11} - c_{12}}, \tag{1.42}$$

is a measure of the departure of a material from being isotropic. However the condition (1.38) is rarely satisfied for crystals and A can have quite large values (e.g., A = 6 for potassium).

1.9 Isotropic media

If condition (1.38) is satisfied, we have $c_{11} = c_{12} + 2c_{44}$ and the stress–strain relation may be written as

$$\sigma_{ij} = \lambda\delta_{ij}\sum_k \frac{\partial u_k}{\partial x_k} + \mu\left(\frac{\partial u_i}{\partial x_j} + \frac{\partial u_j}{\partial x_i}\right), \tag{1.43}$$

where λ and μ are called the *Lamé constants* and are related to c_{12}, c_{44}, and K by

$$c_{12} = \lambda, \tag{1.44a}$$

$$c_{44} = \mu, \tag{1.44b}$$

$$c_{11} = \lambda + 2\mu, \tag{1.44c}$$

and

$$K = \lambda + \frac{2}{3}\mu. \tag{1.44d}$$

We note that Eq. (1.43) can be written in an alternative form by adding and subtracting the average of the sum of the diagonal terms entering the second term,

$$\sigma_{ij} = \left(\lambda + \frac{2}{3}\mu\right)\delta_{ij}\sum_k \frac{\partial u_k}{\partial x_k} + \mu\left(\frac{\partial u_i}{\partial x_j} + \frac{\partial u_j}{\partial x_i} - \frac{2}{3}\sum_k \frac{\partial u_k}{\partial x_k}\right) \tag{1.45a}$$

or

$$\sigma_{ij} = K\delta_{ij}\sum_k \frac{\partial u_k}{\partial x_k} + \mu\left(\frac{\partial u_i}{\partial x_j} + \frac{\partial u_j}{\partial x_i} - \frac{2}{3}\sum_k \frac{\partial u_k}{\partial x_k}\right); \tag{1.45b}$$

written in this form the second term involves only a *pure shear* with no change in volume.

The expression for the energy density analogous to (1.10′) is

$$e = \frac{1}{2}\lambda\sum_i u_{ii}^2 + \mu\sum_{ik} u_{ik}^2. \tag{1.46}$$

The wave equation, Eq. (1.41), for this isotropic case becomes

$$\rho\ddot{u}(\mathbf{r}, t) = (\lambda + 2\mu)\,\nabla^2\mathbf{u} + (\lambda + \mu)\,\nabla \times (\nabla \times \mathbf{u}), \tag{1.47}$$

or in component form

$$\rho\ddot{u}_i - (\lambda + \mu)\frac{\partial^2 u_i}{\partial x_i^2} - \mu\sum_k \frac{\partial^2 u_k}{\partial x_i^2} = 0. \tag{1.48}$$

The plane wave solutions to (1.48) are again of two kinds (independent of direction): longitudinal (l or $\parallel$) and transverse (t or $\perp$). The resulting wave equations are

$$\nabla^2\mathbf{u}_\perp - \frac{1}{v_t^2}\frac{\partial^2\mathbf{u}_\perp}{\partial t^2} = 0 \tag{1.49}$$

and

$$\nabla^2 u_\parallel - \frac{1}{v_\ell^2}\frac{\partial^2 u_\parallel}{\partial t^2} = 0, \tag{1.50}$$

where

$$v_t^2 = \frac{\mu}{\rho} \tag{1.51a}$$

and

$$v_l^2 = \frac{\lambda + 2\mu}{\rho}. \tag{1.51b}$$

1.10 Viscosity in solids: internal friction

Conventional liquids cannot maintain a static shear stress. However they do support "dynamic" shear (and bulk) stresses that are proportional to the *rate of change of the strain, but* which relax toward zero in an irreversible manner. Based on symmetry, the most general form that such stresses can have is

$$\sigma'_{ij} = \zeta\delta_{ij}\sum_k \frac{\partial v_k}{\partial x_k} + \eta\left(\frac{\partial v_i}{\partial x_j} + \frac{\partial v_j}{\partial x_i} - \frac{2}{3}\sum_k \frac{\partial v_k}{\partial x_k}\right), \tag{1.52}$$

where the v_k are components of the velocity. Note the form is similar to Eq. (1.45) with $v_k = \dot{u}_k$ replacing u_k. The two parameters η and ζ are referred to as the first and second viscosity; the latter is also called the bulk viscosity. An analogous phenomenon occurs in solids and has the same mathematical form as Eq. (1.52); we write it as (see Eq. (1.45))

$$\sigma'_{ij} = \lambda'\delta_{ij}\sum_k \frac{\partial v_k}{\partial x_k} + \mu'\left(\frac{\partial v_i}{\partial x_j} + \frac{\partial v_j}{\partial x_i}\right) \tag{1.53a}$$

or

$$\sigma'_{ij} = K'\delta_{ij}\sum_k \frac{\partial v_k}{\partial x_k} + \mu'\left(\frac{\partial v_i}{\partial x_j} + \frac{\partial v_j}{\partial x_i} - \frac{2}{3}\sum_k \frac{\partial v_k}{\partial x_k}\right). \tag{1.53b}$$

For the case of sound waves with $u_k \propto e^{-i\omega t}$ it is natural to introduce complex elastic parameters and write[13]

$$\lambda \to \lambda - i\omega\lambda'; \quad \mu \to \mu - i\omega\mu'; \quad K \to K - i\omega K'; \tag{1.54a–c}$$

in analogy with a liquid, we could equally well write the second two coefficients as η and ζ. To describe the attenuation of sound we introduce a complex propagation constant $k \to k + i\alpha$, where α is called the *attenuation coefficient*, which is a measure of how rapidly the amplitude decays as a function of distance. With these definitions, Eq. (1.48) yields

$$-(k + i\alpha)^2(\mu - i\omega\mu') + \rho\omega^2 = 0 \tag{1.55}$$

for the shear mode and

$$-(k + i\alpha)^2\left(\lambda + 2\mu - i\omega(\lambda' + 2\mu')\right) + \rho\omega^2 = 0 \tag{1.56}$$

for the longitudinal mode. Separating the real and imaginary parts we find, to leading order in μ' and λ', the same sound velocities as given in (1.51a,b); for the attenuation coefficients we obtain

$$\alpha_t = \frac{\omega^2\mu'}{2\rho v_t^3} \tag{1.57}$$

[13] For the case of crystal elasticity we would write $c_{ij} \to c_{ij} - i\omega c'_{ij}$.

and

$$\alpha_l = \frac{\omega^2(\lambda' + 2\mu')}{2\rho v_l^3}$$
$$= \frac{\omega^2\left(K' + \frac{4}{3}\mu'\right)}{2\rho v_l^3} \qquad (1.58)$$

for the transverse and longitudinal waves respectively.[14] Note the attenuation scales as the square of the frequency.

1.11 Measuring elastic constants

The most accurate way to determine the elastic constants of a material is to measure the velocity of longitudinal and transverse sound for propagation along various crystallographic directions and use the associated expressions for the velocity to solve for the constants c_{ij}. Of course one needs enough directions and polarizations to allow an unambiguous solution, and preferably some extra combinations to provide cross-checks. The sound waves themselves are generated using piezoelectric plates,[15] one as a transmitter and the other as a receiver of the sound; the fundamental resonance frequency corresponds to when the thickness is equal to a half wavelength (the transducers can be excited on odd overtones of this frequency). A typical frequency used is 10^7Hz.

[14] A more careful analysis shows that there is an additional contribution to the attenuation for longitudinal waves. The oscillating density associated with such waves induces temperature oscillations and the associated internal heat transport leads to dissipation (see Landau and Lifshitz *Theory of Elasticity*, section 35).

[15] Piezoelectric effects will be discussed in Chapter 26.

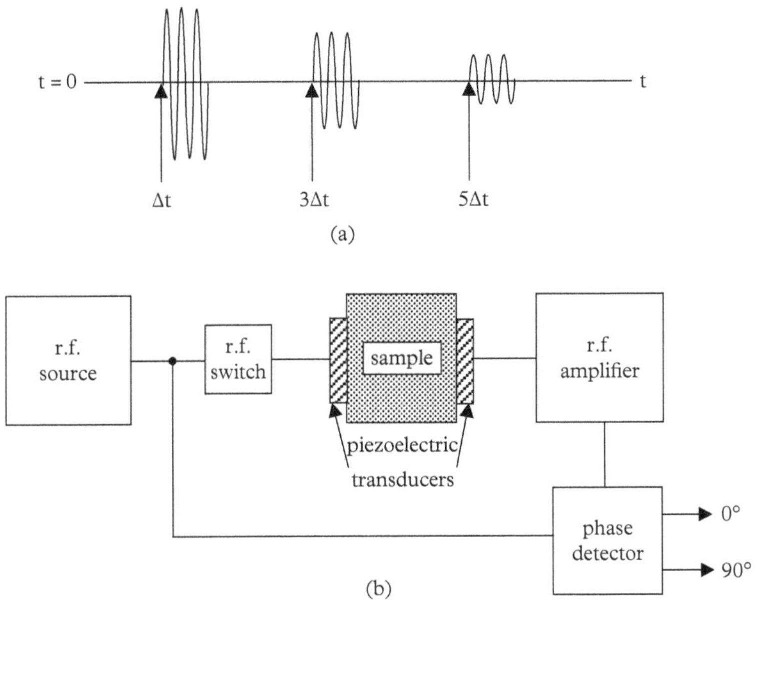

Figure 1.3 *(a) An "echo train" of acoustic pulses bouncing back and forth between two parallel surfaces of a sample. (b) A phase sensitive ultrasonic detector which resolves the arriving pulse into in-phase (0°) and out-of-phase (90°) components.*

In experiments one applies a burst of r.f. energy formed by gating the output of a continuously running r.f. generator to the transmitting transducer, thereby launching a sound wave into the attached specimen. After a *transit time* $\Delta t = d/v_S$, where d is the sample thickness and v_S is the velocity of sound, the signal arrives at the receiving transducer, at which point a portion is converted back into a (now much weaker) r.f. signal (via the inverse piezoelectric effect) which is detected by a phase sensitive detector; by measuring this arrival time of the pulse we obtain the sound velocity. But most of the sound is reflected back into the sample, and at a time $3\Delta t$ it again arrives back at the receiving transducer. The subsequent "echoes" are attenuated due to losses in the sample. This echo train is depicted schematically in Fig. 1.3(a); a schematic of the electronic components of a typical experimental set-up is shown in Fig. 1.3(b).

The use of a phase sensitive detector greatly increases the accuracy with which *changes* in the velocity with respect to some parameter can be detected. A small shift in the sound velocity generates a small shift in the phase of the arriving sound relative to the free running oscillator with which it is compared. Examples of parameters that shift the velocity are temperature, pressure, and magnetic field.

SUPPLEMENTARY READING

A good discussion of elastic properties of solids can be found in the 1955 edition of *Introduction to Solid State Physics*, C. Kittel, J. Wiley and Sons, New York.

Clear and concise, but at a more advanced level is *Theory of Elasticity*, L.D. Landau and E.M. Lifshitz, Pergamon Press (1986).

Problems

1.1. Show that the longitudinal and transverse velocities of sound in the [110] and [111] are given by

	Longitudinal	Transverse
[110]	$v_l^{[110]} = \sqrt{\dfrac{c_{11} + c_{12} + 2c_{44}}{2\rho}}$	$v_t^{[110]} = \sqrt{\dfrac{c_{11} - c_{12}}{2\rho}}$
		$v_t^{[110]} = \sqrt{\dfrac{c_{44}}{\rho}}$
[111]	$v_l^{[111]} = \sqrt{\dfrac{c_{11} + 2c_{12} + 4c_{44}}{3\rho}}$	$v_t^{[111]} = \sqrt{\dfrac{c_{11} - c_{12} + c_{44}}{3\rho}}$

1.2. Show that the stresses acting on a plane with its normal specified by the direction cosines α, β, and γ are given by

$$\sigma_{nx} = \alpha\sigma_{xx} + \beta\sigma_{xy} + \gamma\sigma_{xz}$$

$$\sigma_{ny} = \alpha\sigma_{yx} + \beta\sigma_{yy} + \gamma\sigma_{yz}.$$

$$\sigma_{nz} = \alpha\sigma_{zx} + \beta\sigma_{zy} + \gamma\sigma_{zz}$$

σ_{nx}, σ_{ny}, and σ_{nz} are the stresses acting in the x, y, and z directions with respect to the surface normal, **n**.

1.3. Show that in terms of the Lamé constants λ and μ for an isotropic body, the compression modulus, *Poisson's ratio*, and Young's modulus are given by

$$K = \left(\lambda + \frac{2}{3}\mu\right); \quad \nu = \frac{1}{2}\frac{\lambda}{\lambda + \mu} = \frac{1}{2}\frac{3K - 2\mu}{3K + \mu}; \quad Y = \frac{9K\mu}{3K + \mu}$$

and, conversely, that Young's modulus and Poisson's ratio are given in terms of λ and μ by the expressions

$$\lambda = \frac{Y\nu}{(1 + \nu)(1 - 2\nu)}, \quad \mu = \frac{Y}{2(1 + \nu)}.$$

Show further that the stress–strain relation may be written in terms of Y and ν as

$$\sigma_{ij} = \frac{Y}{1 + \nu}\left(\varepsilon_{ij} + \frac{\nu}{1 - 2\nu}\delta_{ij}\sum_k \varepsilon_{kk}\right).$$

1.4. Suppose you have cube of edge L made from an elastically isotropic material that is attached to a rigid (non-yielding) base to which a uniformly distributed force F is applied to the top surface (see figure). Calculate the displacement **u**.

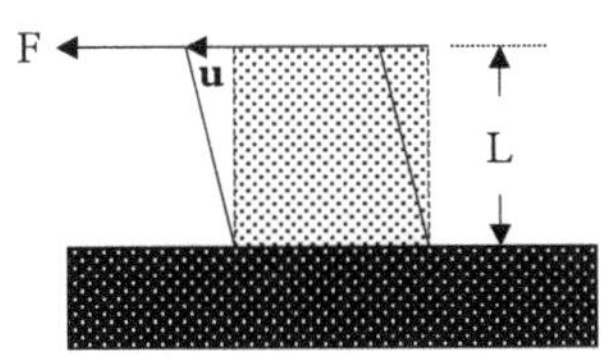

1.5. Consider a small disc (a pillbox) implanted at the interface between two elastic media as shown in the accompanying figure.

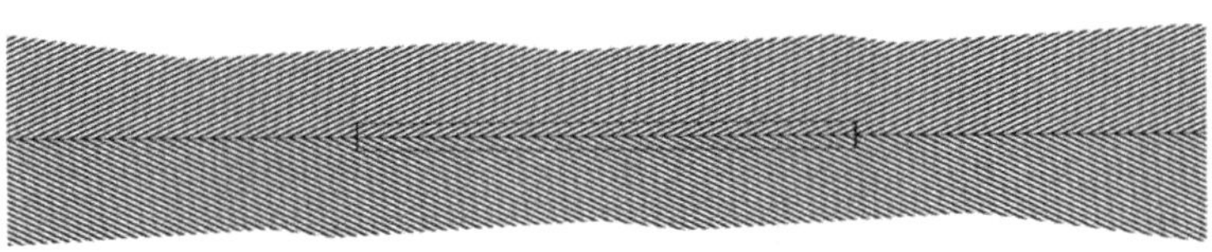

(a) Assuming the two media are in intimate contact, argue that the displacement vector, **u**, associated with some distortion, present in the two media is continuous across the boundary; this constitutes a *boundary condition.*

(b) By considering the forces acting on the pillbox, show that in the limit it is infinitesimally thin, the three components of the quantity $\sum \sigma_{ij} n_j$ must be continuous across the interface, where n is a vector perpendicular to the interface; this constitutes a second boundary condition.

1.6. Assume you have two isotropic media, 1 and 2, characterized by Lame constants, λ_1, μ_1 and λ_2, μ_2, that are in intimate contact and separated by a plane boundary located at $x = 0$.

(a) Assume there is a longitudinal wave propagating in the x direction. Using the results of Problem 1.5, calculate the amplitude of the *reflected* wave in medium 1 and *transmitted* wave in medium 2.

(b) Do the same for a shear wave polarized along y.

Electric Behavior of Insulators

2

We recall that Maxwell's theory of electromagnetism in free space involves two fields: the electric field, **E**, and the magnetic field, **H**. When material media are present two more fields are introduced: the electric displacement, **D**, and the magnetic induction, **B**, for a total of four, all of which are position and time dependent in general.[1] Here we limit ourselves to static electric behaviors associated with the two fields **E** and **D**; the two fields associated with magnetic behavior will be introduced in Chapter 5. As we will discuss, **E** and **D** are related through a third field, the polarization field, **P**, according to $\mathbf{D} = \mathbf{E} + 4\pi\mathbf{P}$ in c.g.s. units.[2] **P** arises from a distribution of microscopic electric dipoles within the material, as we will also discuss. We recall that an individual dipole consists of equal and opposite charges separated by some distance. Hence, we will begin our discussion of the electrical behavior of matter with a brief review of the fields arising from arrays of such dipoles and how they respond to external fields.

The electric field experienced by an isolated "point" particle[3] (one not embedded in some medium) is taken to be the same as the external electric field at the point where it is located; if we do not regard the particle as being extended, it is not meaningful to ask what effects that field might have on some possible charge distribution within the particle. However, if we have an *assembly* of particles in an external field, starting with atoms and proceeding to molecules and larger assemblies, we cannot avoid asking what effect the remaining particles have on the field at a specific particle. If we are to *calculate* the response of individual atoms or molecules of a material to an external field we must examine this problem in some detail.

2.1 The potential from a distribution of dipoles: the polarization

Assume we have an insulating material, one that cannot transport charge. In addition we will initially assume it is electrically neutral; i.e., there are no "excess" electrostatic charges: in the atoms or molecules making up the material the positive charges of the nuclei are canceled by the negative charge of the surrounding electrons. In the presence of an external field the infinitesimal volume elements making up the material (ultimately the individual atoms and molecules) become

[1] This book makes exclusive use of the c.g.s. system of units as opposed to the m.k.s. system. The former is primarily used in the field of condensed matter physics.

[2] Here we suppress the effects of quantum mechanics and regard the fields as arising from point particles, rather than charge distributions associated with particles described by a wave function.

[3] Electrons and nuclei will be regarded as point particles.

The Physics of Solids. First Edition. J. B. Ketterson.

"polarized," i.e., they acquire a dipole moment, as a result of displacements of the nuclei and the surrounding electrons. Certain materials, called electrets,[4] are polarized even in the absence of an externally applied field; these we encounter much later. We start by examining the macroscopic behavior associated with dipoles, as it is treated in electrostatics. A curious result we will obtain is that the field produced by the polarization depends on the *macroscopic shape* of the body.

From electrostatics, the potential, $\phi_i(\mathbf{r})$, at the point $\mathbf{r}$ due to the ith dipole that is located at the point $\mathbf{r}_i$ is

$$\phi_i(\mathbf{r}) = \mathbf{p}_i \cdot (\mathbf{r}-\mathbf{r}_i)/\left|\mathbf{r}-\mathbf{r}_i\right|^3; \tag{2.1}$$

the associated electric field $\mathbf{E}(\mathbf{r}) = -\nabla\phi(\mathbf{r})$ is then

$$\mathbf{E}_i(\mathbf{r}) = -\frac{\mathbf{p}_i}{\left|\mathbf{r}-\mathbf{r}_i\right|^3} \cdot + \frac{3}{\left|\mathbf{r}-\mathbf{r}_i\right|^5}(\mathbf{p}_i \cdot (\mathbf{r}-\mathbf{r}_i))(\mathbf{r}-\mathbf{r}_i). \tag{2.2}$$

The polarization, $\mathbf{P}$, previously introduced, is defined as the *dipole moment per unit volume*, which is the vector sum of the dipole moments $\mathbf{p}$ within some small volume element divided by the volume element itself; i.e., $\mathbf{P}$ is a *dipole density*. The *total* electrostatic potential, $\phi_P(\mathbf{r})$, arising from a *distribution of* dipoles in some region is then obtained by replacing $\mathbf{p}$ by $\mathbf{P}(\mathbf{r}')$ and integrating over the volume of the body; i.e.,

$$\begin{aligned}\phi_P(\mathbf{r}) &= \sum_i \phi_i(\mathbf{r}) = \sum_i \frac{\mathbf{p}_i \cdot (\mathbf{r}-\mathbf{r}_i)}{\left|\mathbf{r}-\mathbf{r}_i\right|^3}\\ &\cong \int d^3r'\mathbf{P}(\mathbf{r}') \cdot \frac{\mathbf{r}-\mathbf{r}'}{\left|\mathbf{r}-\mathbf{r}'\right|^3},\end{aligned} \tag{2.3}$$

which we may rewrite as

$$\phi_P(\mathbf{r}) = \int d^3r'\,\mathbf{P}(\mathbf{r}') \cdot \nabla' \frac{1}{\left|\mathbf{r}-\mathbf{r}'\right|}. \tag{2.4}$$

Writing

$$\nabla' \cdot \frac{\mathbf{P}(\mathbf{r}')}{\left|\mathbf{r}-\mathbf{r}'\right|} = \frac{1}{\left|\mathbf{r}-\mathbf{r}'\right|}\nabla' \cdot \mathbf{P}(\mathbf{r}') + \mathbf{P}(\mathbf{r}') \cdot \nabla' \frac{1}{\left|\mathbf{r}-\mathbf{r}'\right|}, \tag{2.5}$$

and using Green's theorem, (2.4) becomes

$$\begin{aligned}\phi_P(\mathbf{r}) &= \int dS' \cdot \frac{\mathbf{P}(\mathbf{r}')}{\left|\mathbf{r}-\mathbf{r}'\right|} - \int d^3r' \frac{\nabla' \cdot \mathbf{P}(\mathbf{r}')}{\left|\mathbf{r}-\mathbf{r}'\right|}\\ &= \int dS' \frac{\sigma_P}{\left|\mathbf{r}-\mathbf{r}'\right|} + \int d^3r' \frac{\rho_P(\mathbf{r}')}{\left|\mathbf{r}-\mathbf{r}'\right|};\end{aligned} \tag{2.6}$$

[4] Also called ferro- or pyro-electrics depending on their behavior with temperature, as we discuss in Chapter 25.

here we have defined the quantities

$$\sigma_P = P_n \tag{2.7}$$

and

$$\rho_P(\mathbf{r}') = -\nabla' \cdot \mathbf{P}(\mathbf{r}'), \tag{2.8}$$

which are called the surface and bulk polarization (or bound) charge densities respectively, and P_n denotes the normal component of $\mathbf{P}$. Note that if $\mathbf{P}$ is uniform the potential arises *solely* from the surface bound charge density; in the absence of a polarization $\phi(\mathbf{r})$ vanishes altogether.

The *total* dipole moment of a body, which is typically denoted as $\mathbf{p}$, is clearly

$$\mathbf{p} = \sum_i \mathbf{p}_i \cong \int d^3r' \mathbf{P}(\mathbf{r}'). \tag{2.9}$$

Now, $\nabla' \cdot \left((\mathbf{r}')_j \mathbf{P}(\mathbf{r}')\right) = (\mathbf{P}(\mathbf{r}'))_j + (\mathbf{r}')_j \nabla' \cdot \mathbf{P}(\mathbf{r}')$, where the subscript j denotes the jth Cartesian component. Using Green's theorem we can then write the total dipole moment of the body as

$$\mathbf{p} = \int \mathbf{r}' (d\mathbf{S}' \cdot \mathbf{P}(\mathbf{r}')) - \int d^3r' \, \mathbf{r}' (\nabla' \cdot \mathbf{P}(\mathbf{r}')), \tag{2.10}$$

or using (2.5a,b) we have

$$\mathbf{p} = \int dS' \mathbf{r}' \sigma_P (\mathbf{r}') + \int d^3r' \mathbf{r}' \rho_P (\mathbf{r}'). \tag{2.11}$$

Equations (2.6) and (2.11) must be interpreted carefully. If the surface of integration is placed immediately outside the body (and nowhere intersects it), where the polarization is formally zero, the surface term will vanish; however, since the polarization contains a discontinuity at the surface, ρ_P will be singular there (even if it vanishes in the bulk, as it does for a uniformly polarized body) and, on carrying the volume integration through the surface, will produce a contribution from the surface in addition to any bulk contribution (arising from a non-uniform $\mathbf{P}$). Alternatively, if we place the surface of integration immediately inside the body, where $\mathbf{P}$ is not singular, we would need to include a non-vanishing contribution from the surface-charge density when considering the *total* dipole moment of the body.

With the understanding that integrations are to be carried out over all space (with any singularities due to surfaces included) we only need to retain the second term in Eq. (2.6) and we can write

$$\phi_P(\mathbf{r}) = \int d^3r' \frac{\rho_P}{|\mathbf{r} - \mathbf{r}'|}. \tag{2.12}$$

This has the same form as for free charges. If we now include a free (or true) charge density, ρ_{free}, and write $\phi_{\text{total}} = \phi_{\text{free}} + \phi_P$, we can write the Poisson equation of electrostatics as

$$\nabla^2\phi_{\text{total}} = -4\pi\rho_{\text{total}} = -4\pi\,(\rho_P + \rho_{\text{free}}); \tag{2.13}$$

writing the total electric field as $\mathbf{E}_{\text{total}} = -\nabla\phi_{\text{total}}$, we then have

$$\nabla\cdot\mathbf{E} = 4\pi\,(\rho_P + \rho_{\text{free}}), \tag{2.14}$$

where we have dropped the subscript. Using the definition of ρ_P given by Eq. (2.8) we obtain Maxwell's first equation

$$\nabla\cdot\mathbf{D} = \rho_{\text{free}}, \tag{2.15}$$

where we have defined

$$\mathbf{D}(\mathbf{r}) \equiv \mathbf{E}(\mathbf{r}) + 4\pi\mathbf{P}(\mathbf{r}). \tag{2.16}$$

Note ρ_f is usually written simply as ρ, with the understanding that it is the free charge.

The **D** and **E** fields obey the Maxwell boundary conditions at an interface between two dielectrics (that we give without proof)[5]

$$\mathbf{n}\cdot\mathbf{D}_1 = \mathbf{n}\cdot\mathbf{D}_2 \tag{2.17}$$

and

$$\mathbf{n}\times\mathbf{E}_1 = \mathbf{n}\times\mathbf{E}_2, \tag{2.18}$$

where **n** is a unit vector perpendicular to the interface.

2.2 The local electric field

As noted at the beginning of this chapter, if we are to calculate the response of the atoms of a solid to an external electric field we must first understand what effect the surrounding medium has on this field; i.e., we must calculate the local electric field, $\mathbf{E}_{\text{loc}}$ acting on a specific molecule, allowing for the effect of the surrounding medium. Strictly speaking this requires a quantum mechanical description. However, here we will use a simplified description where we divide this local electric field into *four* contributions (as first done by H.A. Lorentz):

$$\mathbf{E}_{\text{loc}} = \mathbf{E}_0 + \mathbf{E}_1 + \mathbf{E}_2 + \mathbf{E}_3, \tag{2.19}$$

where $\mathbf{E}_0$ is the external field (which we assume to be uniform, as one would have between a pair of parallel capacitor plates in the absence of edge effects), $\mathbf{E}_1$ is the so-called *depolarization field* which is produced by *induced* charges on the surface of a sample, as previously discussed (here we assume there is no free

[5] See, e.g., J.D. Jackson: *Classical Electrodynamics*, John Wiley and Sons or Anupam Garg, *Classical Electromagnetism in a Nutshell*, Princeton University Press.

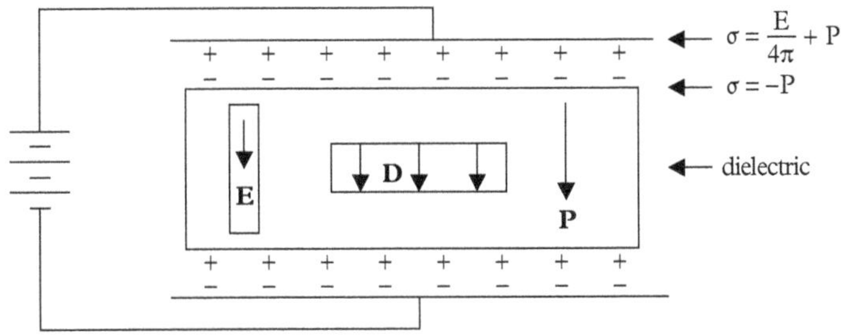

Figure 2.1 *A slab of dielectric located between two capacitor plates showing the surface charge densities on all external surfaces. Shown also are the E and D fields and the respective cavities required to measure them (consistent with the boundary conditions (2.17) and (2.18)); note neither of these fields corresponds to the electric field experienced by an atom or molecule within the dielectric.*

charge involved), $\mathbf{E}_2$ is the field induced on the *surface* of an imaginary spherical cavity within the sample (where the reason for its introduction will be discussed shortly), and $\mathbf{E}_3$ is the field produced by all the dipoles *within* that cavity.

We begin with a discussion of $\mathbf{E}_1$. To fix the geometry, consider a slab dielectric with parallel faces aligned between two parallel capacitor plates which are connected to a battery, as shown in Fig. 2.1 (the space between the dielectric and the capacitor plates has been enlarged slightly to show the separate charge densities). When the dielectric is inserted between the plates an additional surface charge $\pm\mathbf{P}$ flows onto the plates to cancel the corresponding surface charge induced on the dielectric; therefore in the chosen geometry the field corresponding to $\mathbf{E}_1$ is $4\pi\mathbf{P}$. Note that inside the long thin cavity we have a field $\mathbf{E}$, while inside the short fat cavity we have a field $\mathbf{D}$, as required by Eqs. (2.17) and (2.18); however *neither* of these corresponds to the true local field, as we will see later.

The depolarizing field $\mathbf{E}_1$ for the earlier case of parallel plates is a special case of a more general situation known from the electrostatics of dielectrics: if the dielectric has the shape of an ellipsoid with the field parallel to one of its principal axes, a solution of the associated boundary value problem shows that the field will be uniform within the body, and parallel to the external field,[6] with the depolarization field, $\mathbf{E}_1$, given by

$$E_{1i} = -N_i P_i, \qquad (2.20)$$

where N_i is called the *depolarization factor* and i denotes x, y, and z principal axes of the ellipsoid (the negative sign in (2.20) indicates that the induced field $\mathbf{E}_1$ is in the opposite direction of $\mathbf{P}$ and hence the designation depolarization field). In tensor notation we have

$$\begin{bmatrix} E_{1x} \\ E_{1y} \\ E_{1z} \end{bmatrix} = \begin{bmatrix} N_{xx} & 0 & 0 \\ 0 & N_{yy} & 0 \\ 0 & 0 & N_{zz} \end{bmatrix} \begin{bmatrix} P_x \\ P_y \\ P_z \end{bmatrix}. \qquad (2.21)$$

For the case of an ellipsoid, the N_i must be calculated explicitly; for the case of a sphere or a "degenerate" form of an ellipsoid (an infinite slab or a cylinder), we have the results given in Table 2.1. The two cases for the thin slab correspond directly to the Maxwell boundary condition.

[6] The polarization $\mathbf{P}$ is also uniform when the external field $\mathbf{E}$ is at an arbitrary angle, but $\mathbf{E}$ and $\mathbf{P}$ are then not parallel.

Table 2.1 *Depolarization factors for simple shapes*

Shape	Axis	N
Sphere	Any	$4\pi/3$
Thin slab	Normal	4π
Thin slab	In-plane	0
Long circular cylinder	Normal to axis	2π
Long circular cylinder	Parallel to axis	0

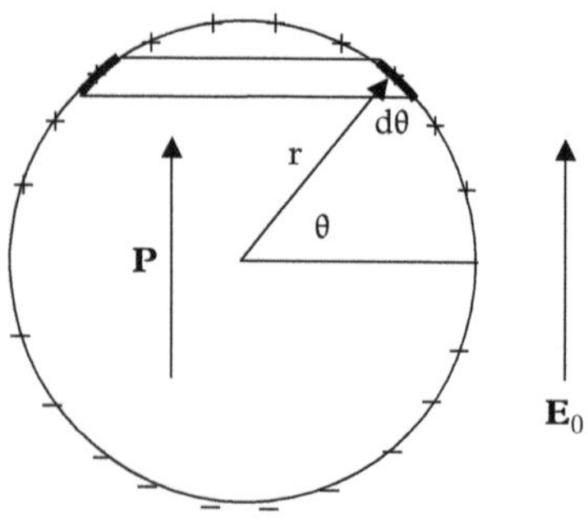

Figure 2.2 *Polarization charges distributed on the surface of a dielectric sphere in a uniform electric field, E_0.*

We now calculate the depolarization field for the case of a sphere. Figure 2.2 shows a uniformly polarized sphere on the surface of which there is a charge density, $\sigma_P = P_n = P\cos\theta$. The charge on the ring at an angle θ, having an angular width $d\theta$, is $\sigma_P(2\pi r\sin\theta)(rd\theta)$. The z component of the electric field at the center of a sphere is therefore

$$E_z = \int \frac{1}{r^2}(2\pi r^2 \sin\theta d\theta)(P\cos\theta)\cos\theta = -\frac{4\pi P}{3}. \tag{2.22}$$

(The field is uniform, as we have said previously, so it is sufficient to evaluate it at the center.) From (2.20), we then have $N = 4\pi/3$, in agreement with the entry in Table 2.1.

We now return to the question of the local field and the reason for introducing the two additional fields $\mathbf{E}_2$ and $\mathbf{E}_3$ in writing Eq. (2.19). Lorentz constructed the following trick to evaluate the local electric field acting on an atom. First he imagined a hollow cavity being created within the dielectric around the atom in question. The size of the cavity is intermediate in scale—large compared to an atom but small compared with the body itself. In making this cavity it is assumed that the polarization in the surrounding medium is not altered. It turns out to be simplest if we choose a spherical cavity, and we denote the field arising from the polarization charges on the inside of this sphere as $\mathbf{E}_2$. Lorentz then introduced a discrete sum over the dipoles within this cavity, with the exception of the dipole at the center: this field is denoted as $\mathbf{E}_3$. On physical grounds one should be able to use a cavity of any shape to calculate $\mathbf{E}_2$ as long as the field $\mathbf{E}_3$ from the dipoles within it is properly calculated.[7]

As previously noted, the $\mathbf{E}_2$ contribution to the local field arises from the surface charges on the *inside* surface of the imaginary sphere that we embedded in the dielectric: it therefore has the *opposite sign* from that on the outside of a dielectric sphere, which we just calculated in (2.22); hence we may write immediately

$$\mathbf{E}_2 = \frac{4\pi \mathbf{P}}{3}. \tag{2.23}$$

[7] We could just as well choose this cavity to be an ellipsoid, which would then alter the result for the field $\mathbf{E}_2$; the field $\mathbf{E}_3$ arising from the dipole sum would then also be different, however the sum $\mathbf{E}_2 + \mathbf{E}_3$ would remain the same.

Finally, we need to calculate the field $\mathbf{E}_3$ arising from the sum over all the dipoles within the sphere. The electric field from a single dipole was given in (2.2) as

$$\mathbf{E}(\mathbf{r}) = -\frac{\mathbf{p}}{r^3} + \frac{3(\mathbf{p}\cdot\mathbf{r})\mathbf{r}}{r^5}. \tag{2.24}$$

Placing the field point in the center of the sphere and summing over all the dipoles we would have

$$\mathbf{E} = -\sum_i \left[\frac{\mathbf{p}_i}{r_i^3} - \frac{3(\mathbf{p}_i\cdot\mathbf{r}_i)\mathbf{r}_i}{r_i^5}\right]. \tag{2.25}$$

For a cubic or isotropic material and assuming all the dipoles are aligned along z, the term $\overline{3(\mathbf{p}_i\cdot\mathbf{r}_i)\mathbf{r}} = \overline{3z^2\mathbf{n}_z}$, where $\mathbf{n}_z$ is a unit vector along the z axis. Because of the symmetry associated with our choice of a spherical cavity,

$$\sum_i x_i^2 = \sum_i y_i^2 = \sum_i z_i^2 = \frac{1}{3}\sum_i r_i^2; \tag{2.26}$$

hence the sum in Eq. (2.25) vanishes. For the case shown in Fig. 2.1 the local electric field will be given by[8]

$$\mathbf{E}_{\text{local}} = \mathbf{E} + \frac{4\pi}{3}\mathbf{P}. \tag{2.27}$$

We must emphasize that this is a model calculation only, although it is widely (and sometimes blindly) applied. This is simply because the sum over dipoles involves microscopic (quantum) considerations. In general it is of semi-quantitative value only.

2.3 The dielectric constant

In general the polarization is a function of the electric field; i.e., $\mathbf{P} = \mathbf{P}(\mathbf{E})$. For sufficiently small electric fields we may expand the jth component of $\mathbf{P}$ in a power series in $\mathbf{E}$ as

$$P_j(\mathbf{E}) = P_j(0) + \sum_{k=1}^{3} \left.\frac{\partial P_j}{\partial E_k}\right|_{E=0} E_k + \cdots, \tag{2.28}$$

where the higher-order terms can usually be neglected.[9] Under the assumption that the material is unpolarized in the absence of an external electric field, we will have $P_j(0) = 0$. It is then natural to introduce an *electric susceptibility tensor* through the definition,

$$\chi_{jk}^E = \frac{\partial P_j}{\partial E_k}; \tag{2.29}$$

[8] We note that the vanishing of $\mathbf{E}_3$ is subject to the restrictions we have assumed, and is not necessarily zero. In practice it would have to be evaluated using the density matrix calculated from a full quantum mechanical treatment.

[9] Very large dynamic electric fields can be generated with lasers; it may then be necessary to include higher order terms in the expansion. This will be discussed much later in Chapter 31.

we henceforth drop the superscript.[10] The displacement field $\mathbf{D} = \mathbf{E} + 4\pi\mathbf{P}$ is then written as

$$\begin{aligned} D_j &= E_j + 4\pi P_j \\ &= \sum_{j=1}^{3} (\delta_{jk} + 4\pi\chi_{jk}) E_k \\ &= \sum_{j=1}^{3} \varepsilon_{jk} E_k, \end{aligned} \tag{2.30}$$

where we have defined the *dielectric tensor*,

$$\varepsilon_{jk} \equiv \delta_{jk} + 4\pi\chi_{jk}. \tag{2.31}$$

For a liquid or gas, and amorphous or cubic solids, the dielectric tensor can be written as a scalar, $\varepsilon = 1 + 4\pi\chi$, and we then have $\mathbf{D} = \varepsilon\mathbf{E}$. We will usually assume that we are dealing with such materials. Table 2.2 lists the dielectric constants of some common materials.

[10] From the fact that the energy density is a quadratic function of the electric field it follows that the tensor χ^{E}_{jk} is symmetric.

Table 2.2 *Dielectric constants of various materials.*

Material	Dielectric constant	Material	Dielectric constant
Air	1.0	Paraffin	2.1
Bakelite	4.4 to 5.4	Plexiglass	2.8
Barium titanate	100–1250	Polyethylene	2.3
Carbon tetrachloride	2.17	Polystyrene	2.6
Cellulose acetate	3.3–3.9	Porcelain	5.1–5.9
Epoxy resin	3.4–3.7	Silicones	3.2–4.7
Ethyl alcohol	24	Titanium dioxide	100
Glass	7.6–8.0	Quartz	3.8
Kevlar	3.5–4.5	Rubber	2.8
Lucite	2.5	Teflon	2.1
Mica	5.4	Vinyl	2.8–4.5
Mylar	3.2	Water	76.5–80
Paper	3.0		

To connect the susceptibility to a microscopic quantity, we start by writing the dipole moment induced on the ith molecule as

$$\mathbf{p}_i = \alpha_i \mathbf{E}_{local}, \tag{2.32}$$

where E_{local} is the local electric field at the ith molecule and α_i is a parameter called the electric *polarizability*; generally it is a tensor.[11] The polarization is then

$$\mathbf{P} = \sum_i \mathbf{E}_{local} n_i \alpha_i, \tag{2.33}$$

where n_i is the number of molecules per unit volume of type i. Using (2.33) and (2.27) we then have

$$\chi = \frac{|\mathbf{P}|}{|\mathbf{E}|} = \frac{P}{E_{local} - \frac{4\pi}{3}P} \tag{2.34}$$

or

$$\chi = \frac{\sum_i n_i \alpha_i}{1 - \frac{4\pi}{3}\sum_i n_i \alpha_i}. \tag{2.35}$$

On the other hand,

$$\chi = \frac{D - E}{4\pi E} = \frac{\varepsilon - 1}{4\pi}. \tag{2.36}$$

Equating (2.34) and (2.36) we have

$$\frac{\varepsilon - 1}{4\pi} = \frac{\sum_i n_i \alpha_i}{1 - \frac{4\pi}{3}\sum_i n_i \alpha_i}, \tag{2.37}$$

from which we obtain

$$\frac{\varepsilon - 1}{\varepsilon + 2} = \frac{4\pi}{3}\sum_i n_i \alpha_i; \tag{2.38}$$

this is the *Clausius–Mossotti relation* or, with ε replaced by n^2, the Lorenz–Lorentz equation, where n is the index of refraction. For the case of a material with a single polarizability (a monatomic material), Eq. (2.38) becomes

$$\frac{M}{\rho}\frac{\varepsilon - 1}{\varepsilon + 2} = \frac{4\pi}{3}A\alpha, \tag{2.39}$$

where M is the atomic mass, ρ is the mass density, and A is Avogadro's number.

[11] We are here assuming that the molecules have no permanent dipole moments; the case where the molecules do have such a moment is treated in Section 2.4.

2.4 Orientational polarization

Solids that are composed of molecules possessing a *permanent* dipole moment can have an enhanced dielectric constant. We first consider a model in which the molecules are small and, as a result of thermal processes, can rotate about their lattice sites; the application of an electric field will cause a partial alignment of the molecules which we now calculate. The interaction energy of the dipoles with the local electric field is given by

$$U = -\mathbf{p} \cdot \mathbf{E}. \tag{2.40}$$

Using the normalized Boltzmann weight factor,[12] we may write the average value of the dipole moment as

$$\bar{p} = \frac{\int d\Omega p \cos\theta e^{-U(\theta)/k_B T}}{\int d\Omega e^{-U(\theta)/k_B T}} = \frac{\int_0^\pi \sin\theta d\theta p \cos\theta e^{pE\cos\theta/k_B T}}{\int_0^\pi \sin\theta d\theta e^{pE\cos\theta/k_B T}} = p\,\overline{\cos\theta}, \tag{2.41}$$

where $d\Omega$ is the differential element of solid angle. We may write this expression as

$$\begin{aligned}\overline{\cos\theta} &= \frac{d}{dx}\ln\left[\int_0^\pi \sin\theta d\theta e^{x\cos\theta}\right] \\ &= \frac{d}{dx}\ln\left[\frac{1}{x}\left(e^x - e^{-x}\right)\right],\end{aligned} \tag{2.42}$$

where $x = \frac{pE}{k_B T}$. Differentiating and then expanding in the small parameter x, we have

$$\bar{p} = p\left[\cot x - \frac{1}{x}\right] = \frac{p^2 E}{3k_B T} + \cdots. \tag{2.43}$$

We then have for the polarizability and polarization in the high temperature limit,

$$\alpha = \frac{p^2}{3k_B T} \tag{2.44}$$

and

$$P = n\frac{p^2 E}{3k_B T}, \tag{2.45}$$

where n is the number density of the molecules.

[12] See Eqs. (4.2) and (4.3).

We limit ourselves to the case where we have only one kind of molecule. In the limit where the polarization is small, we may write Eq. (2.37) as

$$\varepsilon = 1 + 4\pi n\alpha, \tag{2.46}$$

where we write α as

$$\alpha = \alpha_0 + \frac{p^2}{3k_BT}, \tag{2.47}$$

where α_0 denotes the ionic and electronic contributions to the polarizability. If we neglect α_0 and use Eq. (2.37) without further approximation we obtain

$$\varepsilon = 1 + \frac{4\pi np^2}{3k_B(T - T_c)}, \tag{2.48}$$

where we have defined a "transition temperature," $T_c = 4\pi np^2/9k_B$. The divergence that sets in at this temperature is referred to as the "polarization catastrophe." Actually, (2.48) becomes inapplicable well before this point, however to discuss it further would involve a discussion of critical phenomena that we will not enter into.

If the dipoles are restricted in their movements, the earlier treatment needs to be modified. As an example, suppose we have a uniaxial material where the dipoles can point in two directions, parallel and antiparallel, relative to an electric field aligned along the symmetry axis. The ratio of the number of molecules in these states is given by

$$\frac{n_1}{n_2} = e^{2pE/k_BT} = e^{2x}. \tag{2.49}$$

The fractional excess pointed parallel to the field is

$$\frac{n_1 - n_2}{n_1 + n_2} = \frac{e^x - e^{-x}}{e^x + e^{-x}} = \tanh x \cong x. \tag{2.50}$$

The polarization produced by these dipoles is then given by

$$P = npx = \frac{np^2E}{k_BT} \tag{2.51}$$

which, other than a numerical factor, agrees in form with Eq. (2.45). Note that by replacing electric dipole moment p by the magnetic dipole moment μ, Eqs. (2.44) and (2.51) are, respectively, identical to the classical Langevin and the quantum mechanical (for a spin 1/2 system) expressions for the magnetization which will be discussed in Chapter 5.

2.5 Dielectric relaxation

The motion of dipoles in response[13] to an external field will be accompanied by dissipation. Under the influence of an oscillatory electric field of the form $E = \Re E_0 e^{-i\omega t}$, we can expect the polarizability to have a Drude-like form[14]

$$\alpha = \frac{\alpha_0}{1 - i\omega\tau}, \tag{2.52}$$

where τ is a rotational relaxation time. Debye[15] modeled τ by assuming the molecules acted as spheres of radius a rotating in a fluid of viscosity η under the influence of thermal fluctuations. He obtained

$$\tau = \frac{4\pi\eta a^3}{k_B T}; \tag{2.53}$$

the times τ are typically of order 10^{-11}sec.

We can construct a model for the relaxation in a solid of the type discussed earlier, in which there are two stable orientations (uniaxial symmetry), by writing some phenomenological rate equations for the populations in the two states of the form

$$\begin{aligned} \frac{dn_1}{dt} &= -w_{12}n_1 + w_{21}n_2 \\ \frac{dn_2}{dt} &= w_{12}n_1 - w_{21}n_2 \end{aligned}, \tag{2.54}$$

where w_{12} and w_{21} are the transition rates from 1 to 2 and 2 to 1, respectively. In equilibrium, $\dot{n}_1 = \dot{n}_2 = 0$. This in turn requires that

$$\frac{n_1}{n_2} = \frac{w_{21}}{w_{12}}. \tag{2.55}$$

However, in equilibrium the two populations must satisfy Eq. (2.49); hence we may write

$$n_1 = Ae^{pE/k_B T}; n_2 = Ae^{-pE/k_B T}. \tag{2.56}$$

From Eq. (2.55) it follows that the transition rates w_{12} and w_{21} may then be written as

$$w_{12} = \frac{1}{2\tau}e^{-pE/k_B T}; w_{21} = \frac{1}{2\tau}e^{pE/k_B T}, \tag{2.57}$$

where τ is a relaxation time and the choice of the factor 2 will be apparent from the final form we obtain. In the limit $pE/k_B T \ll 1$, where we can expand the exponentials in (2.57), Equations (2.54) become

$$2\tau\frac{dn_1}{dt} = -(n_1 - n_2) + \frac{pE}{k_B T}(n_1 + n_2) \tag{2.58}$$

[13] The dielectric response will always be associated with some characteristic times, t_0, and accompanying frequencies, $\omega_0 = 1/t_0$, that measure the response to a changing electric field, which is discussed at various points in this book.

[14] The Drude model will be discussed in Chapter 3.

[15] The 1936 Nobel Prize in Chemistry was awarded to Petrus (Peter) Josephus Wilhelmus Debye "for his contributions to our knowledge of molecular structure through his investigations on dipole moments and on the diffraction of X-rays and electrons in gases."

and

$$2\tau\frac{dn_2}{dt} = (n_1 - n_2) - \frac{pE}{k_B T}(n_1 + n_2). \tag{2.59}$$

Now, $n_1 + n_2 = n$, the total number of dipoles per unit volume, so the only time dependence in the second term arises from E; if we assume $E \propto e^{-i\omega t}$, these equations have the solution

$$n_1 - n_2 = \frac{n}{1 - i\omega\tau}\frac{pE}{k_B T}. \tag{2.60}$$

The dipole moment per unit volume, the polarization P, may then be written as

$$P = \frac{p(n_1 - n_2)}{E} = \frac{n}{1 - i\omega\tau}\frac{p^2 E}{k_B T}. \tag{2.61}$$

The relaxation times in solids are typically much longer than in liquids.

SUPPLEMENTARY READING

The dielectric properties of insulators (including ferroelectrics) are discussed in various editions of C. Kittel's *Introduction to Solid State Physics*, John Wiley and Sons.

A standard reference at an advanced level is *Electrodynamics of Continuous Media*, L.D. Landau and E.M. Lifshitz, Pergamon Press, New York (1984).

The Theory of Electrons, and its Application to Light and Radiant Heat, by H.A. Lorentz, reprinted by Dover, New York (1952), is still worth reading.

Problems

2.1. Assume you are given a spherical particle with radius, a and dielectric constant, ε, located in vacuum in a constant electric field, **E**. Show that the Maxwell boundary conditions are consistent with the internal field being uniform and go on to calculate the total dipole moment of the particle.

2.2. Calculate the depolarization factor for an ellipsoid of revolution for the external field along the symmetry axis.

2.3. If the imaginary sphere used in the Lorentz model is replaced by a cube with one of its faces perpendicular to the internal polarization field, show that the electric field at the cube center is identical to that for a sphere. Start by evaluating the polarization surface charge density on the upper and lower faces of this cube.

3 Metals and the Drude–Lorentz Model

In its more modern context, the historical model of Drude and Lorentz views a material as consisting of an array of (i) fixed positive ions and free electrons, or (ii) neutral atoms, or balanced positive and negative ions atoms (as in NaCl), with no free electrons. In the first case we have a metal and in the second an insulator. Let N_i be the number of ions and N_e the number of electrons; the corresponding number densities are N_i/V and N_e/V, where V is the volume. In what follows, it will be assumed that the number of free (detached) electrons is given by

$$N_e = zN_i; \tag{3.1}$$

i.e., each ion contributes z electrons. Figure 3.1(a) shows this model schematically.

In the Drude model, the free electrons are pictured as randomly moving through the solid as a result of their thermally induced motions.[1] In the region between the atoms we might think of them as moving almost freely, but when they encounter an atom core they are scattered by the potential it sets up. This situation is pictured in Fig. 3.1(b).[2]

3.1 Ohm's law

In the presence of an electric field, **E**, a force $\mathbf{F} = -|e|\mathbf{E} \equiv e\mathbf{E}$ acts on each of the free electrons; here we are using a convention in which e is itself a negative number.[3] A force $+|e|z\mathbf{E}$ is also exerted on the ions, but it will be assumed that the interatomic forces fix their positions in a solid material. In the absence of irreversible processes, the electrons would accelerate indefinitely; however collisions with the ions themselves, other electrons, or various defect structures within the solid, result in a loss of momentum. We will model this effect by introducing a viscous drag force that we write as $-m\mathbf{v}/\tau$, where **v** is a "drift" velocity acquired as a result of the external electric field (i.e., over and above that associated with thermal motion) and τ is a "relaxation time." In a typical metal at room temperature $\tau \sim 10^{-14}$ sec, but it can be several orders of magnitude longer in very pure metals at low temperatures (~ 4 K). Combining these ideas, Newton's equation of motion is then

$$m\dot{\mathbf{v}} + \frac{m}{\tau}\mathbf{v} = e\mathbf{E}, \tag{3.2}$$

[1] Later, when we come to discuss the quantum theory of free electrons we will find that, as a result of the Pauli exclusion principle, the velocities of electrons are much higher than thermal velocities.

[2] This model is highly over-simplified and is given for historical purposes only. In a real solid, the potential arising from the ion cores and the other electrons extends throughout the unit cells of the crystal lattice. Furthermore, when the quantum theory is introduced, the distance traveled between collisions will turn out to be far larger than a lattice spacing.

[3] The convention where e is taken as positive is also commonly used in the literature.

The Physics of Solids. First Edition. J. B. Ketterson.

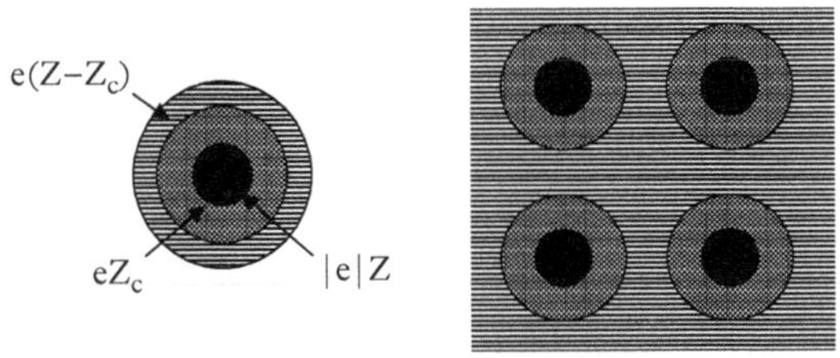

Figure 3.1(a) *On the left is depicted a semi-classical model of an atom. The total charge number of the atom, Z, has been partitioned into a nuclear contribution, +Z (in black), a core or bound contribution, Z_C, and a valence contribution, $Z - Z_C$. For a metal, z of the valence electrons will be assumed to be free to move through the solid. On the right we show an assembly of bound atoms making up a solid, with the conducting electron fraction z spread out and free to move through the solid.*

where m is the electron mass and e is again the charge of the electron. Assuming the initial drift velocity vanishes, this equation has the solution

$$v = \frac{e\tau}{m}\mathbf{E}\left(1 - e^{-t/\tau}\right). \tag{3.3}$$

Equation (3.3) has the limiting forms

$$v = \frac{et}{m}\mathbf{E} \quad (t \ll \tau) \tag{3.4a}$$

and

$$\begin{aligned} v &= \frac{e\tau}{m}\mathbf{E} \\ &= \mu\mathbf{E} \quad (t \gg \tau), \end{aligned} \tag{3.4b}$$

where $\mu \equiv e\tau/m$ is called the *mobility*. If we ignore the acceleration term (which vanishes at steady state) we arrive at (3.4b) immediately. Now the current density is defined by $\mathbf{j} = nev$, where n is the number of electrons per unit volume, N/V; hence,

$$\begin{aligned} j &= \frac{ne^2\tau}{m}\mathbf{E} \\ &= \sigma\mathbf{E}, \end{aligned} \tag{3.5a}$$

where $\sigma \equiv ne^2\tau/m$ is called the *electrical conductivity*. Equation (3.5a) forms the basis of *Ohm's law* in the Drude model for the conduction of electricity. We can also write

$$\mathbf{E} = \rho\mathbf{j}, \tag{3.5b}$$

where $\rho \equiv 1/\sigma$ is called the *resistivity*.

Table 3.1 lists the conductivities and resistivities of many of the metals at room temperature in MKS units.

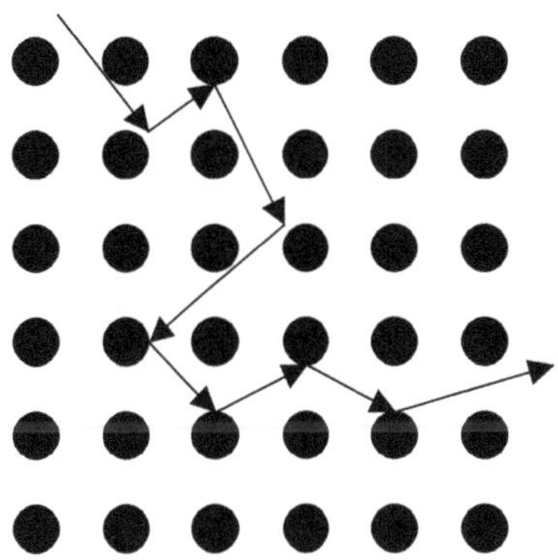

Figure 3.1(b) *We show, schematically, the motion of an electron through a periodic array of atoms under the influence of an electric field and the "collisions" encountered with the atoms along its path.*

Table 3.1 *Electrical conductivity and electrical resistivity of metal elements at* 20 °C.

	$\sigma \times 10^{-8}$ Ωm	$\rho \times 10^{7}$ / Ωm		$\sigma \times 10^{-8}$ Ωm	$\rho \times 10^{7}$ / Ωm		$\sigma \times 10^{-8}$ Ωm	$\rho \times 10^{7}$ / Ωm
Ag	6.21	1.61	In	1.14	8.75	Re	0.54	18.6
Au	4.55	2.20	Ir	1.96	5.1	Rh	2.08	4.8
Al	3.65	2.74	K	1.39	7.19	Ru	1.35	7.4
Ba	0.26	39	La	0.13	79	Sb	0.24	41.3
Be	3.08	3.25	Li	1.07	9.32	Sc	0.21	46.8
Bi	0.086	116	Lu	0.19	53	Sm	0.10	99
Ca	2.78	3.6	Mg	2.33	4.30	Sn	0.91	11.0
Ce	0.12	81	Mn	0.072	139	Sr	0.47	21.5
Cd	1.38	7.27	Mo	1.89	5.3	Ta	0.76	13.1
Co	1.72	5.8	Na	2.11	4.75	Tb	0.09	111
Cr	0.78	12.9	Nb	0.69	14.5	Tc	0.7	14
Cs	0.50	20.0	Nd	0.17	59	Th	0.66	15.2
Cu	5.88	1.70	Ni	1.43	7.0	Ti	0.23	43.1
Dy	0.11	90	Np	0.085	118	Tl	0.61	16.4
Er	0.12	81	Os	1.10	9.1	Tm	0.16	62
Eu	0.11	89	Pb	0.48	21	U	0.39	25.7
Fe	1.02	9.8	Pd	0.95	10.5	V	0.50	19.9
Ga	0.67	14.85	Pr	0.15	67	W	1.89	5.3
Gd	0.07	134	Po	0.22	46	Y	0.17	58.5
Hf	0.33	30.6	Pt	0.96	10.4	Yb	0.38	26.4
Hg	0.10	95.9	Pu	0.07	143	Zn	1.69	5.92
Ho	0.13	77.7	Rb	0.80	12.5	Zr	0.24	42.4

3.2 The Hall effect

We next examine the motion of the electrons in the simultaneous presence of a static electric and static magnetic field, where the latter is oriented perpendicular to the direction of the current flow, as shown in Fig. 3.2.

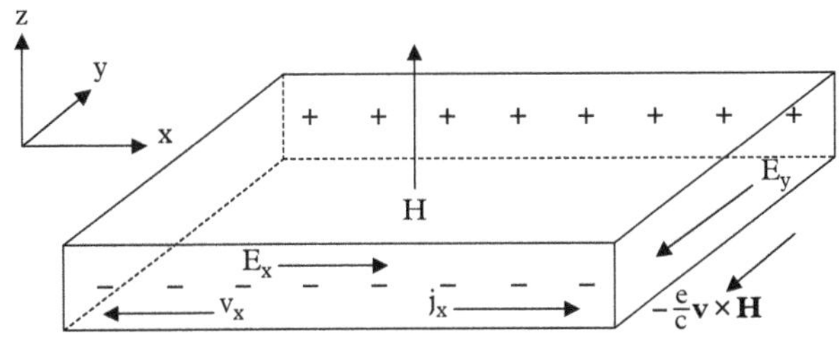

Figure 3.2 *Diagram of the geometry of the Hall effect. The electric field,* E_x, *results in a velocity,* v_x, *in the x direction. The tendency of a magnetic field aligned in the z direction to bend the electron trajectories is canceled by the field* E_y, *without which a current would arise in the y direction, which is forbidden, since motion in that direction is constrained.*

The motion of the electrons is now determined by combining the Lorentz force, $\mathbf{F} = \frac{e}{c}(\mathbf{E} + \mathbf{v} \times \mathbf{H})$, with Newton's law which gives

$$m\dot{v}_x + \frac{m}{\tau}v_x = eE_x + \frac{e}{c}v_yH, \tag{3.6a}$$

$$m\dot{v}_y + \frac{m}{\tau}v_y = eE_y - \frac{e}{c}v_xH, \tag{3.6b}$$

and

$$m\dot{v}_z + \frac{m}{\tau}v_z = 0. \tag{3.6c}$$

Here, E_x is the applied (longitudinal) field and E_y is a resultant (transverse) field, the necessity for which will become apparent. We discuss only the steady-state solution for which $\dot{\mathbf{v}} = 0$; furthermore we must have $v_z = 0$ and $v_y = 0$, since there are no "sources" or "sinks" for currents in these directions (motion in the x direction is sustained by electrons entering and leaving at external contacts). Equation (3.6b) then yields $E_y = \frac{v_x}{c}H$. But $j_x = nev_x$ and therefore

$$E_y = \frac{j_xH}{nec} = R_Hj_xH, \tag{3.7a}$$

where

$$R_H = \frac{1}{nec}. \tag{3.7b}$$

The appearance of a transverse voltage under the simultaneous presence of a transport current and a mutually perpendicular field is called the *Hall Effect.*[4] It is commonly used to measure the carrier density, n.[5] Alternatively, once n (or more phenomenologically R_H) is known, the Hall effect can be used to measure the magnetic fields. The effect is larger in semiconductors where the carrier densities are smaller.

[4] On a new action of the magnet on electric currents, Edwin Hall, *Am. J. Math.* 2, 287 (1879).

[5] The presence of multiple charge carriers complicates the analysis and the simple model used here is then no longer valid.

The presence of E_y implies the total electric field is rotated through an angle θ relative to the x axis, given by

$$\tan\theta = \frac{E_y}{E_x} = \frac{j_x H}{necE_x} = \frac{\sigma E_x H}{necE_x} = \frac{ne^2\tau H}{necm} = -\omega_c\tau, \tag{3.8}$$

where $\omega_c \equiv \frac{|e|H}{mc}$ is the so-called *cyclotron frequency*. Note that a long relaxation time (weak scattering) implies a large Hall angle.

3.3 Frequency-dependent conductivity

Rather than applying a constant electric field we may employ an oscillatory field of the form $\mathbf{E} = \Re\mathbf{E}_0 e^{-i\omega t}$, where $\Re$ denotes the real part. Equation (3.2) then becomes

$$m\dot{\mathbf{v}} + \frac{m}{\tau}\mathbf{v} = e\Re\mathbf{E}_0 e^{-i\omega t}. \tag{3.9}$$

Assuming a steady-state solution of the form $\mathbf{v} = \mathrm{Re}\mathbf{v}_0 e^{-i\omega t}$, we have

$$\mathbf{v} = \Re\frac{\frac{e\tau}{m}\mathbf{E}_0 e^{-i\omega t}}{1 - i\omega\tau}. \tag{3.10}$$

The resulting current density is

$$\mathbf{j} = \Re\frac{\sigma(0)\mathbf{E}}{1 - i\omega\tau} = \Re\sigma(\omega)\mathbf{E}, \tag{3.11}$$

where we define a frequency-dependent conductivity, $\sigma(\omega) = \sigma(0)/(1 - i\omega\tau)$. For frequencies where $\omega\tau \gg 1$, the complex conductivity becomes purely imaginary and hence the current and field are 90° out of phase, implying that the dissipation vanishes.

3.4 Dielectric constant of a metal

We next examine the dielectric constant of a metal. The dielectric constant is defined through the relation $\mathbf{D} = \varepsilon\mathbf{E} = \mathbf{E} + 4\pi\mathbf{P}$, where $\mathbf{P}$ is the polarization; thus,

$$\varepsilon = 1 + 4\pi\frac{|\mathbf{P}|}{|\mathbf{E}|}. \tag{3.12}$$

The polarization (defined as the dipole moment per unit volume) is

$$\mathbf{P} = ne\mathbf{r}, \tag{3.13}$$

where **r** is obtained by integrating **v** in Eq. (3.10); requiring the constant of integration to vanish (i.e., no spontaneous polarization) and ignoring any polarization from any bound atomic core electrons (which would contribute a background dielectric constant), we obtain

$$\varepsilon(\omega) = 1 - \frac{\frac{4\pi ne^2}{m}}{\omega^2 + \frac{i\omega}{\tau}}. \tag{3.14}$$

For $\omega \gg \frac{1}{\tau}$ we may write this as

$$\varepsilon(\omega) = 1 - \frac{\omega_p^2}{\omega^2}, \tag{3.15}$$

where we have introduced the so-called plasma frequency, defined by $\omega_p^2 \equiv 4\pi ne^2/m$.

When $\omega = \omega_p$ the dielectric constant vanishes. From the relation $\mathbf{D} = \varepsilon\mathbf{E}$, this implies that we can have an electric field, **E**, with no accompanying displacement field, **D**. Now at a boundary the normal component of **D** is continuous and hence we can have a longitudinal electric field within the metal that produces no external field. Physically, the plasma oscillations correspond to a periodic transfer of energy between the kinetic energy and the electrostatic-potential energy of the electrons. We can arrive at the plasma oscillations directly by simultaneously solving (i) the equation of charge continuity, (ii) Maxwell's second equation, and (iii) Newton's equation of motion:

$$\dot{\rho} + \nabla \cdot \mathbf{j} = 0, \tag{3.16a}$$

$$\nabla \cdot \mathbf{E} = 4\pi\rho, \tag{3.16b}$$

and

$$m\dot{\mathbf{v}} = e\mathbf{E}. \tag{3.16c}$$

Taking the time derivative of (3.16b), substituting it into (3.16a), and assuming the presence of a spatial dependence $\Re e^{i\mathbf{k}\cdot\mathbf{r}}$, we obtain

$$\frac{\partial \mathbf{E}}{\partial t} = -4\pi\mathbf{j} = -4\pi ne\mathbf{v}. \tag{3.17}$$

Taking the time derivative of (3.17) and inserting it in (3.16c) we obtain

$$\frac{\partial^2 \mathbf{E}}{\partial t^2} + \frac{4\pi ne^2}{m}\mathbf{E} = 0, \tag{3.18}$$

which has an oscillatory solution with a frequency $\omega_p = 4\pi ne^2/m$.

We will discuss plasma oscillations in greater detail in Sections 14.2.2, 19.13, 30.4, and 30.5. In a metal, $\omega_p \cong 10^{16}$ Hz or 10 eV, which is in the ultraviolet. For frequencies above ω_p the metal becomes transparent in this model.

3.5 Dielectric constant of an insulator

In an insulator the electrons may be thought of as "attached" to the ions; the simplest model would involve attachment via springs. The Drude–Lorentz equation may then be written as

$$m\left(\ddot{\mathbf{r}} + \frac{1}{\tau}\dot{\mathbf{r}} + \omega_0^2\mathbf{r}\right) = e\mathbf{E}; \tag{3.19}$$

here $\omega_0^2 = k/m$, where k is the spring constant. This equation has the solution

$$\mathbf{r} = \Re\frac{\dfrac{e}{m}}{\omega_0^2 - \omega^2 - \dfrac{i\omega}{\tau}}\mathbf{E}. \tag{3.20}$$

Using Eqs. (3.12) and (3.13) we obtain

$$\varepsilon(\omega) = 1 + \frac{\omega_p^2}{\left(\omega_0^2 - \omega^2\right) - \dfrac{i\omega}{\tau}}, \tag{3.21}$$

where we again introduce the plasma frequency, $\omega_p^2 = 4\pi ne^2/m$. Separating $\varepsilon(\omega)$ into its real and imaginary parts, $\varepsilon(\omega) = \varepsilon_1(\omega) + i\varepsilon_2(\omega)$, we have

$$\varepsilon_1 = 1 + \frac{\omega_p^2\left(\omega_0^2 - \omega^2\right)}{\left(\omega_0^2 - \omega^2\right)^2 + \dfrac{\omega^2}{\tau^2}} \tag{3.22a}$$

and

$$\varepsilon_2 = \frac{\omega_p^2\omega/\tau}{\left(\omega_0^2 - \omega^2\right)^2 + \dfrac{\omega^2}{\tau^2}}. \tag{3.22b}$$

These expressions are shown graphically in Fig. 3.3 as a function of ω/ω_0 for the case $\omega_p^2/\omega_0^2 = 2$ and $\omega_0\tau = 3$. Note $\varepsilon_1(\omega \to \infty) = 1$ and $\varepsilon_1(\omega \to 0) = 1 + \omega_p^2/\omega_0^2$; the latter is clearly always greater than unity. Also note that the dielectric constant can be negative for ω just above ω_0, and for very high frequencies (e.g., X-ray frequencies) $\varepsilon_1(\omega)$ is always less than one. For ω near ω_0 the imaginary part of ε goes through a maximum, i.e., the absorption (or attenuation) is maximal.

At optical frequencies the quantity measured experimentally is usually the index of refraction, defined as

$$n = \frac{c}{v_p}, \tag{3.23}$$

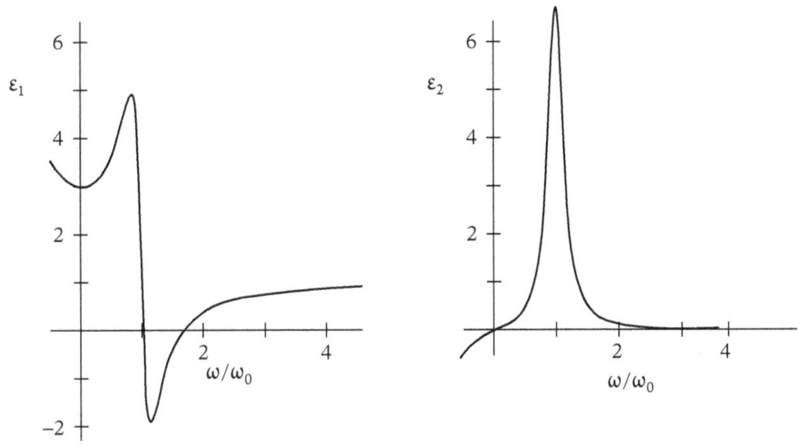

Figure 3.3 *A plot of Eqs. (3.22a) and (3.22b) as a function of* ω/ω_0 *for the case* $\omega_p^2/\omega_0^2 = 2$ *and* $\omega_0\tau = 3$.

where $v_p = c/\varepsilon^{1/2}$ is the phase velocity; i.e., $n^2 = c^2/v_p^2 = \varepsilon$. We write $n = n_1 + in_2$ and

$$n^2 = n_1^2 + 2in_1n_2 - n_2^2 = \varepsilon_1 + i\varepsilon_2, \tag{3.24}$$

which yields

$$\varepsilon_1 = n_1^2 - n_2^2 \tag{3.25a}$$

and

$$\varepsilon_2 = 2n_1n_2. \tag{3.25b}$$

Solving these equations for n_1 and n_2 we obtain

$$n_1 = \sqrt{\frac{\varepsilon_1}{2} + \frac{1}{2}\sqrt{\varepsilon_1^2 + \varepsilon_2^2}} \tag{3.26a}$$

and

$$n_2 = \frac{\varepsilon_2}{2n_1} = \frac{\varepsilon_2}{\sqrt{2\varepsilon_1 + 2\sqrt{\varepsilon_1^2 + \varepsilon_2^2}}}. \tag{3.26b}$$

From the definition (3.23) we see that the phase velocity becomes complex in the presence of an imaginary contribution to ε; we interpret this as follows. We write $v_p = \omega/K$, where $K = k + i\alpha$ is a complex propagation vector. Thus, $n = c/v_p = cK/\omega = c(k + i\alpha)/\omega$, where $k = \omega n_1/c$ and $\alpha = \omega n_2/c$. Assuming a spatial behavior of the amplitude of the form $e^{iKx} = e^{ikx-\alpha x}$, we then refer to k as the

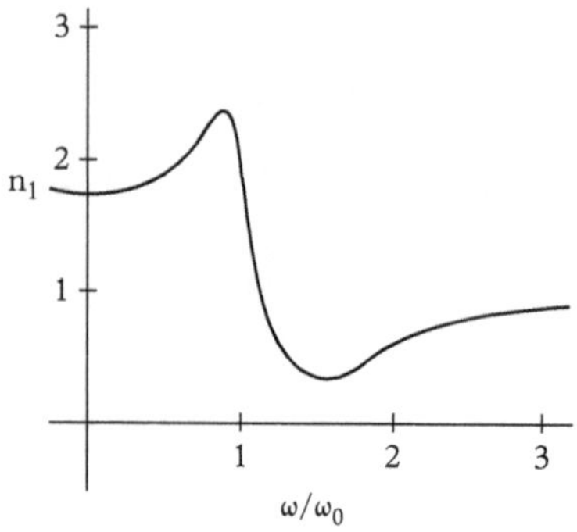

Figure 3.4 *A plot of the phase velocity associated with a plane wave propagating in a dielectric solid; the parameters are the same as in Fig. 3.3.*

propagation constant (or wave vector) and α as the attenuation. Therefore if ε (or n) is complex we have damping of the wave.

Figure 3.4 shows a plot of the real part of the index of refraction n_1 for the same parameters as used in Fig. 3.3. In addition to the phase velocity one can define a group velocity $v_g = d\omega/dk$, which, outside the vicinity of the resonance, is an approximate measure of the velocity of a pulse. Note that the group velocity and phase velocity exceed that of light for some range of frequencies near the resonance and hence neither can correspond to a physical velocity of propagation of the wave in this region. The definition of the true velocity and the nature of the propagation of the wave in this so-called anomalous region are mathematically complicated and will not be discussed here.[6]

3.6 A metal in a constant magnetic field and an oscillatory electric field

Let us return to the geometry of the Hall effect involving the application of mutually perpendicular electric and magnetic fields, only in the present case the electric field will be oscillatory. The problem is easiest to treat if we apply a circular polarized electric field that we write as

$$\mathbf{E} = E_0(\cos\omega t\hat{\mathbf{x}} + \sin\omega t\hat{\mathbf{y}}). \tag{3.27}$$

Our Drude–Lorentz equations of motion are then

$$m\left(\dot{v}_x + \frac{v_x}{\tau}\right) = \frac{e}{c}v_y H + eE_0\cos\omega t, \tag{3.28a}$$

$$m\left(\dot{v}_y + \frac{v_y}{\tau}\right) = -\frac{e}{c}v_x H + eE_0\sin\omega t, \tag{3.28b}$$

and

$$m\left(\dot{v}_z + \frac{v_z}{\tau}\right) = 0. \tag{3.28c}$$

We again restrict to the steady-state solutions where v_z vanishes and the transient motion associated with the x and y motion has died out. To solve (3.28a,b), we multiply (3.28b) by i and subtract the resulting equation from (3.28a) to obtain

$$m(\dot{v}_x - i\dot{v}_y) + \frac{m}{\tau}(v_x - iv_y) - \frac{ie}{c}(v_x - iv_y)H = eE_0e^{-i\omega t}. \tag{3.29}$$

We introduce a new variable $\tilde{v} = v_x - iv_y$ and (3.29) becomes

$$m\dot{\tilde{v}} + m\left(\frac{1}{\tau} + i\omega_c\right)\tilde{v} = eE_0e^{-i\omega t}, \tag{3.30}$$

[6] For a discussion of the true or signal velocity see *Optics*, A. Sommerfeld, Academic Press, New York, p. 114 (1954) or *Electromagnetic Theory*, J.A. Stratton, McGraw-Hill Book Co., p. 333 (1941).

where $\omega_c = \dfrac{|e|H}{mc}$ is the cyclotron resonance frequency introduced earlier. Assuming a form $\tilde{v} = \tilde{v}_0 e^{-i\omega t}$, we have

$$\tilde{v} = \frac{ie/m}{(\omega - \omega_c) + i/\tau} E_0 e^{-i\omega t}. \tag{3.31}$$

Resolving this into real and imaginary parts yields

$$v_x = \frac{\dfrac{e}{m} E_0 \left[\dfrac{1}{\tau}\cos\omega t + (\omega - \omega_c)\sin\omega t\right]}{(\omega - \omega_c)^2 + \dfrac{1}{\tau^2}} \tag{3.32a}$$

and

$$v_y = -\frac{\dfrac{e}{m} E_0 \left[-(\omega - \omega_c)\cos\omega t + \dfrac{1}{\tau}\sin\omega t\right]}{(\omega - \omega_c)^2 + \dfrac{1}{\tau^2}}. \tag{3.32b}$$

If we introduce a phase angle through the definition

$$\tan\phi = (\omega - \omega_c)\tau, \tag{3.33}$$

then we may rewrite Eq. (3.32) as

$$v_x = \frac{\dfrac{e}{m} E_0}{\left[(\omega - \omega_c)^2 + \dfrac{1}{\tau^2}\right]^{1/2}} \cos(\omega t - \phi) \tag{3.34a}$$

and

$$v_y = -\frac{\dfrac{e}{m} E_0}{\left[(\omega - \omega_c)^2 + \dfrac{1}{\tau^2}\right]^{1/2}} \sin(\omega t - \phi). \tag{3.34b}$$

In the limit where $\omega_c\tau$ is large, the velocity of the electrons becomes large when we satisfy the resonance condition $\omega = \omega_c$. This phenomenon is referred to as *cyclotron resonance* since the condition is the same as that in a cyclotron (a device used historically to accelerate charged particles, primarily protons). Precisely at the resonance, the phase angle $\phi = 0$ and the dissipation is maximal (here the velocity and the electric force are parallel); as we move from low frequency to high frequency (passing through the resonance) the phase angle changes from $\phi \cong -\pi/2$ to $\phi \cong +\pi/2$.

Later, in our discussion of the quantum mechanics of electron states in solids, we will find that in responding to external fields, electrons can behave as if they have an altered mass or an *effective mass* (which may be either greater than or less than the free electron mass; furthermore it may have either sign). Cyclotron resonance then provides a means of measuring this mass. The phenomenon can only be observed at low temperatures (~ 4 K) in a pure metal or semiconductor, where long scattering times with $\omega_c\tau \gg 1$ are then achievable. For a frequency $f = 10^{10}$Hz the resonance occurs at a field of 3.57 KG for free electrons. The experiment typically involves measuring the power absorbed by a sample in a microwave cavity at cryogenic temperatures.

3.7 Thermal conductivity in a metal

Let us denote the heat flux by **q**; this vector measures the thermal energy transport across a unit area (the surface normal of which defines the direction) in unit time. By Newton's law of cooling, this heat flux is proportional to the temperature gradient; hence,

$$\mathbf{q} = -\kappa \nabla T. \tag{3.35}$$

The proportionality constant, κ, is referred to as the *thermal conductivity*. The minus sign reflects the fact that heat flows from higher to lower temperatures. For an anisotropic system (e.g., a hexagonal crystal), κ is a tensor and we write

$$q_i = -\sum_{j=1}^{3} \kappa_{ij} \frac{\partial T}{\partial x_j}. \tag{3.36}$$

From kinetic theory, the thermal conductivity is given by

$$\kappa = \frac{1}{3} c_V \bar{v}\, l, \tag{3.37}$$

where $l = \bar{v}\tau$ is the mean free path, $\bar{v}$ is the root mean square velocity, and c_V is the heat capacity at constant volume (per unit volume). The thermal conductivity of some common metals at room temperature is given in Table 3.2, in MKS units.

It is interesting to examine the ratio of the electronic-thermal conductivity and the electrical conductivity,

$$\frac{\kappa}{\sigma} = \frac{\frac{1}{3} c_V \bar{v}^2 \tau}{\frac{ne^2\tau}{m}}. \tag{3.38}$$

Classically, the heat capacity at constant volume of the electron gas is $3nk_B/2$, where k_B is Boltzmann's constant.[7] The corresponding energy density of the gas is $nm\bar{v}^2/2 = 3nk_BT/2$, with $\bar{v}^2 = \bar{v}_x^2 + \bar{v}_y^2 + \bar{v}_z^2$; here, T is the absolute temperature

Table 3.2 *Thermal conductivity of some metals.*

Material	Thermal conductivity (W/m K)
Silver	406.0
Copper	385.0
Gold	314
Aluminum	205.0
Iron	79.5
Lead	34.7
Mercury	8.3

[7] This result will be derived in Chapter 4.

and each of the three directions contributes $nk_BT/2$ (to be derived in Chapter 4); this is known as the *equipartition* theorem. Equation (3.38) then becomes

$$\frac{\kappa}{\sigma T} = \frac{3k_B^2}{2e^2} = 1.24 \times 10^{-13} \left(\frac{\text{erg}}{\text{esu K}}\right)^2. \tag{3.39}$$

This relation is referred to as the *Wiedemann–Franz law* and the numerical factor is called the Lorentz number; some values are given in Table 3.3, in MKS units. The magnitude is close to what is observed experimentally in metals, but the agreement is accidental since quantum mechanics significantly alters the temperature dependence and magnitude of both κ and σ.

Table 3.3 *Wiedemann–Franz ratio* $L = \frac{\kappa}{\sigma T} = \frac{\pi^2 k_B^2}{3e^2} = 2.45 \times 10^{-8} \frac{W\Omega}{K^2}$.

Metal	$\kappa/\sigma T(10^{-8}\ W\Omega/K^2)$
Cu	2.23
Ag	2.31
Au	2.35
Zn	2.31
Cd	2.42
Sn	2.52
Mo	2.61
Pb	2.47
Pt	2.51

3.8 Thermoelectric effect

In the presence of a temperature gradient there will be a position dependence of the mean electronic velocity (the electrons moving more rapidly on the hotter side of the metal). In the absence of a "compensating effect" this would result in a net thermal drift velocity, v_q, of the electrons (from hot toward cold), given in kinetic theory by $v_q = l\, dv_x/dx$, where l is again the free path, $l = \bar{v}\tau$. We then have

$$v_q = (\bar{v}\tau)\frac{d\,\bar{v}_x}{dx} = \frac{1}{3}(\bar{v}\tau)\frac{d\bar{v}}{dx} = \frac{1}{3}\frac{\tau}{m}\frac{d}{dx}\left(\frac{1}{2}m\bar{v}^2\right)$$

$$= \frac{\tau}{3m}\frac{d}{dx}\left(\frac{3}{2}k_BT\right) = \frac{\tau}{3mn}c_V\frac{dT}{dx}. \tag{3.40}$$

The compensating effect referred to is the spontaneous appearance of an electric field which causes a drift velocity $v_d = \mu E$, where $\mu = e\tau/m$ is the mobility given earlier, such that $v_d + v_q = 0$; i.e.,

$$\mu \mathbf{E} + \frac{\tau}{3mn}c_V \nabla T = 0$$

or

$$\mathbf{E} = \frac{c_V}{3en}\nabla T = \frac{k_B}{2e}\nabla T. \tag{3.41}$$

The predicted value of the electric field is about 100 times larger than what is observed experimentally; this is but one of several failures of the classical theory of the electron gas. The defects of the classical theories are all remedied by the quantum theory of electrical transport, which we will take up in Chapter 19.

SUPPLEMENTARY READING

The references to the original work by Drude are: Zur Elektronentheorie der Metalle, *Annalen der Physik* **306**, 566 (1900); Zur Elektronentheorie der Metalle; II. Teil: Galvanomagnetische und thermomagnetische Effecte, *Annalen der Physik* **308**, 369 (1900). There is an excellent discussion of the Drude model in *Solid State Physics*, N.W. Ashcroft and N.D. Mermin, Holt, Rinehaert and Winston, Philadelphia (1976).

The Theory of Electrons, and its Application to Light and Radiant Heat, by H.A. Lorentz, reprinted by Dover, New York (1952), is still worth reading.

The concepts from the kinetic theory of gases used in our treatment of the thermal conductivity and the thermoelectric effect are discussed in *Thermodynamics, the Kinetic Theory of Gases, and Statistical Mechanics*, F.W. Sears and G.L. Salinger, Addison Wesley Publishing Co., Cambridge, MA, Chapters 9 and 10 (1975).

Problems

3.1. Suppose you are given a spherical metallic particle of radius a located in vacuum. Using the dielectric constant given by Eq. (3.14), calculate the frequency for which the dipole moment, **p**, is largest. This may be interpreted as the plasmon resonance frequency of the particle.

3.2. Show that at zero frequency the polarizibility of a metallic sphere of radius a is given by $\alpha = a^3$. (Note this gives the right order of magnitude for the polarizibility of an atom.) If we have n such spheres per unit volume, what condition would be required for the dielectric constant to be approximated as $\varepsilon = 1 + 4\pi na^3$?

3.3. Calculate the orbits of an electron in free space having a velocity $\mathbf{v}_0$ at $t=0$ under the influence of an electric field **E** and a magnetic field **H** directed along the x and z directions, respectively. Discuss the behavior of the possible orbits qualitatively.

3.4. Suppose you have a parallel plate capacitor where the region between the plates consists of two parallel layers, one with dielectric constant ε, zero conductivity, and thickness d, while the second has dielectric constant zero, conductivity σ, and a thickness qd. Show that the effective dielectric constant at a frequency ω is given by

$$\varepsilon^* = \frac{\varepsilon(1+q)}{1+(i\omega q/4\pi\,\sigma)}.$$

3.5. Using the Maxwell boundary conditions calculate the reflectivity of an electromagnetic wave incident normal to a bulk metallic film with a conductivity given by Eq. (3.15); at what frequency will waves begin to propagate in the metal?

3.6. Show that in the presence of a magnetic field, $\mathbf{H} = H\mathbf{n}_z$, the Drude model has a solution for a circularly polarized wave, $\mathbf{E}(t) = \mathrm{Re}\mathbf{E}_0(\mathbf{n}_x \pm i\mathbf{n}_y)e^{-i\omega t}$, where $\mathbf{n}_x, \mathbf{n}_y$, and $\mathbf{n}_z$ are unit vectors along the x, y, and z axes and that the associated current has the form

$$j = \mathrm{Re}(\mathbf{n}_x \pm i\mathbf{n}_y)\frac{\sigma(0)}{1-i(\omega \mp \omega_c)\tau}\mathbf{E}_0; \quad j_z = 0.$$

Now use this current–field relation together with Maxwell's equations to show that there is a solution for a propagating wave $\mathbf{E}(t) = \text{Re}\mathbf{E}_0(\mathbf{n}_x \pm i\mathbf{n}_y)e^{-i\omega t+ikz}$ with the dispersion equation $\omega^2 = \left(c^2/\varepsilon(\omega)\right)k^2$, where

$$\varepsilon = 1 - \frac{\omega_p^2}{\omega}\left(\frac{1}{\omega \mp \omega_c + i/\tau}\right).$$

At low frequencies for a pure metal at low temperatures satisfying the condition $\omega_c\tau \gg 1$, we have the limiting form $\varepsilon = \pm\,\omega_p^2/\omega\,\omega_c$; this leads to the dispersion relation,

$$\omega = \frac{\omega_c}{\omega_p^2}c^2k^2.$$

The resulting modes are called *helicons*. Calculate the resulting frequency for a wavelength of 1 cm at a field of 10 kG for an electron density of 10^{22}cm^{-3}. These modes also propagate for long distances (over much of the earth) in the ionosphere where they are excited (over a wide range of frequencies) by lightning flashes. Those modes propagating in the audible range can be heard (particularly in the early morning and referred to as the dawn chorus) simply by attaching a long wire to a high gain amplifier followed by earphones. They are called *whistlers* and have the property that they start at a very high frequency and then progress to very low frequencies; explain this.

4 Elementary Theories of the Thermal Properties of Solids

A fundamental thermodynamic property of a body is its heat capacity, $C \equiv dQ/dT$, where dQ is an increment of heat transferred to the body and dT is the associated temperature change. When measured at constant volume (where no work is done by the sample on its environment), the heat added goes directly into increasing the internal energy, E, and we may then write this quantity as

$$C_V = \left(\frac{\partial E(T,V)}{\partial T}\right)_V. \tag{4.1}$$

The internal energy is computed using the methods of statistical physics, which we now briefly summarize. Suppose we want to compute the average value of some quantity Q. For a mechanical system, $Q = Q(p_1 \cdots p_{3N}, q_1 \cdots q_{3N})$, where $p_i = p_1 \ldots p_{3N}$ and $q_i = q_1 \ldots q_{3N}$ are the momenta and coordinates of the 6N Hamiltonian degrees of freedom associated with N particles moving in three dimensions. Taken together, the 3N p_i and q_i make up a high-dimensional space called *phase space*.

We define $P = P(p_1 \cdots p_{3N}, q_1 \cdots q_{3N})$ to be the *probability density*, i.e., the probability of finding a particle in a volume element $d\Gamma = \prod_{i=1}^{3N} dp_i dq_i$ of our phase space; then from statistics the average of Q, which we denote as $\bar{Q}$, is given by

$$\bar{Q} = \int Q(p_1 \cdots p_{3N}, q_1 \cdots q_{3N})\, P(p_1 \cdots p_{3N}, q_1 \cdots q_{3N})d\Gamma. \tag{4.2a}$$

The most important conclusion of classical statistical physics is that the normalized probability density is given by the *Boltzmann form*

$$P(p_1 \cdots p_{3N}, q_1 \cdots q_{3N}) = \frac{e^{-\beta H(p_1 \cdots p_{3N}, q_1 \cdots q_{3N})}}{\int e^{-\beta H(p_1 \cdots p_{3N}, q_1 \cdots q_{3N})} d\Gamma}, \tag{4.2b}$$

where H is the total Hamiltonian (usually the total energy) of the system and $\beta \equiv 1/k_B T$, where T is the absolute temperature and k_B is Boltzmann's constant. It is common to introduce the so-called *partition function*, Z, according to the definition

$$Z = \int e^{-\beta H(p_1 \cdots p_{3N}, q_1 \cdots q_{3N})} d\Gamma, \tag{4.3}$$

The Physics of Solids. First Edition. J. B. Ketterson.

which is easily seen to be the normalization factor in Eq. (4.2b); hence we can write

$$\bar{Q} = \frac{1}{Z}\int Q\,(p_1 \cdots p_{3N}, q_1 \cdots q_{3N})\, e^{-\beta H(p_1 \cdots p_{3N}, q_1 \cdots q_{3N})} d\Gamma. \tag{4.4}$$

Note that if we take minus the derivative of ln Z with respect to β we obtain

$$-\frac{\partial}{\partial\beta}\ln Z = \frac{\int H(p_1 \cdots p_{3N}, q_1 \cdots q_{3N}) e^{-\beta H(p_1 \cdots p_{3N}, q_1 \cdots q_{3N})} d\Gamma}{\int e^{-\beta H(p_1 \cdots p_{3N}, q_1 \cdots q_{3N})} d\Gamma}; \tag{4.5}$$

$$= \bar{E}$$

i.e., we obtain the average total energy. Various other thermodynamic averages can be evaluated by computing appropriate derivatives of Z.

A special case of Eq. (4.4), which is the only one we will consider here, is when the particles in our system *do not interact with each other.* Then the Hamiltonian can be written as a sum,

$$H(p_1 \cdots p_{3N}, q_1 \cdots q_{3N}) = H(p_1, q_1) + H(p_2, q_2) \cdots$$

and hence

$$e^{-\beta H(p_1 \cdots p_{3N}, q_1 \cdots q_{3N})} = e^{-\beta H(p_1, q_1)} e^{-\beta H(p_2, q_2)} e^{-\beta H(p_3, q_3)} \cdots .$$

We will further assume that the quantity Q being averaged can similarly be written as a sum,

$$Q(p_1 \cdots p_{3N}, q_1 \cdots q_{3N}) = Q(p_1, q_1) + Q(p_2, q_2) \cdots .$$

The integrals in Eq. (4.4) can then be done *one at a time*; those involving a coordinate other than some specific one in $Q\,(p_1 \cdots p_{3N}, q_1 \cdots q_{3N})$ will cancel from the top and bottom of (4.4) and the total is simply the sum of the individual terms. In an isotropic three-dimensional system, the result is simply 3N times the average for a single coordinate associated with a single particle.[1]

4.1 The equipartition law for free and bound particles

As an example, we derive the equipartition law of kinetic theory for an ideal gas. If we consider a single particle in our gas, the average energy would be

$$\frac{1}{2}m\,\overline{v_i^2} = \int_{-\infty}^{+\infty} \frac{1}{2}mv_i^2 e^{-\frac{\beta}{2}mv_i^2} dv_i \Big/ \int_{-\infty}^{+\infty} e^{-\frac{\beta}{2}mv_i^2} dv_i, \tag{4.6}$$

[1] One can then write $d\Gamma = \prod_{i=1}^{3N} d\gamma_i$, where $d\gamma_i = dp_i dq_i$ and $d\gamma_i$ denotes the differential "volume" element in phase space associated with a *single* mechanical degree of freedom.

where i refers to the x, y, or z component of that particle. Carrying out the integral we obtain $m\bar{v}_i^2/2 = k_BT/2$; i.e., we have $k_BT/2$ for each translational degree of freedom.

We next consider a solid, pictured as atoms attached by springs to their neighbors. From mechanics, we know that the mechanical motion of such a system may be represented by a superposition of normal modes, the coordinates of which have equations of motion of the form $\ddot{q}_i + \omega_i^2 q_i = 0$ (the q_i being linear combinations of the coordinates of all the atoms of our solid). There are 3N such equations, where N is the number of atoms and the factor 3 accounts for motion in the x, y, and z directions; each normal mode may have its own frequency, ω_i. The Hamiltonian for each mode may be written as

$$H(q,p) = \frac{p^2}{2m} + \frac{1}{2}m\omega^2 q^2 \tag{4.7}$$

The average energy of each normal mode is

$$\bar{E} = \frac{\int dp \frac{p^2}{2m} e^{-\beta p^2/2m}}{\int dp e^{-\beta p^2/2m}} + \frac{\int dq \frac{m}{2}\omega^2 q^2 e^{-\beta m\omega^2 q^2/2}}{\int dq\, e^{-\beta m\omega^2 q^2/2}};$$

$$= \frac{1}{2}k_BT + \frac{1}{2}k_BT. \tag{4.8}$$

Hence the average energy of a normal mode is k_BT ($k_BT/2$ for the kinetic energy and $k_BT/2$ for the potential energy). This is called the *equipartition theorem* of classical statistical mechanics. The x, y, and z directions represent independent vibrational modes of the system, and hence the heat capacity of a collection of N atoms is $3Nk_B$. Most solids have a heat capacity of order $3Nk_B$ at room temperature, which is known as the *Dulong and Petit law.*

But what about the electrons? In the Drude–Lorentz model each free electron would contribute $3k_B/2$ and each bound electron $3k_B$ to the heat capacity. Since there are Z electrons per atom this would result in a total heat capacity that greatly exceeds the observed value. (This failure of the classical theory is connected with that mentioned earlier in Section 3.8.) We will account for the missing electronic contribution to the heat capacity later, but here we assume that only the atoms (or nuclei) can contribute to the room temperature heat capacity, which we will refer to as the lattice heat capacity.

4.2 The lattice heat capacity at low temperatures: the Einstein model

Figure 4.1 shows the heat capacity of diamond as a function of temperature.[2] We see that at low temperature the heat capacity goes rapidly to zero, in clear

[2] The Planck theory of radiation and the theory of specific heat, A. Einstein, *Ann. Phys.* **22**, 180 (1907).

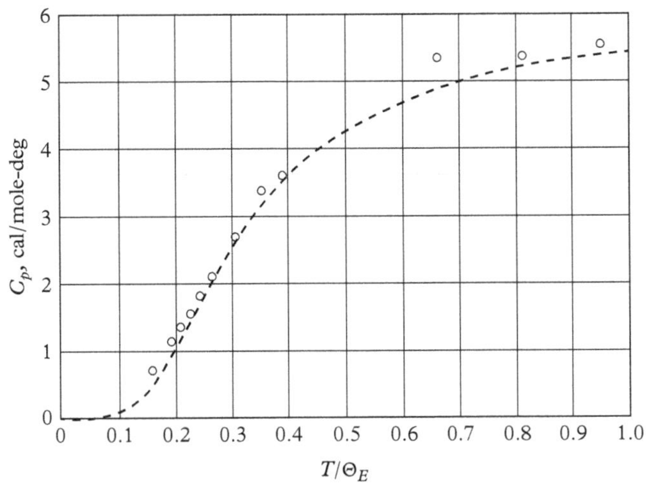

Figure 4.1 *A comparison of the heat capacity of diamond with the Einstein model using* $\Theta_E = 1320\,\mathrm{K}$. *(After A. Einstein (1907).)*

violation of the Dulong–Petit (equipartition) law. This paradox was resolved by Einstein and was one of the early applications of the quantum hypothesis (after Planck's treatment of the blackbody radiation problem). We now reconstruct the argument used in the previous section for obtaining the average energy of an oscillator, to allow for the fact that the energy levels must be *quantized* according to Planck's formula,

$$E_n = n\hbar\omega; \tag{4.9}$$

here n is an integer, ω is the angular frequency of some mode, and $\hbar = h/2\pi$, where h is Planck's constant.[3] We must now examine how to alter the statistical procedure for computing thermodynamic averages given by Eqs. (4.2a,b) to comply with quantum mechanics.

Formally, the transition to quantum mechanics is made by regarding Q as an *operator*, which we write as $\hat{Q}$. We anticipate that the *integrals* over coordinates in Eqs. (4.2) and (4.3) are then replaced by *summations* over all values of the quantum numbers associated with these operators, n;[4] we then have

$$\bar{Q} = \frac{1}{Z}\sum_n \langle n|\,\hat{Q}\,|n\rangle e^{-\beta E_n}, \tag{4.10}$$

where $\langle n|\,\hat{Q}\,|n\rangle$ is the diagonal matrix element associated with the operator $\hat{Q}$, which in the present case will be E_n. The partition function Z is given by an analogous form

$$Z = \sum_n e^{-\beta E_n}. \tag{4.11}$$

[3] Here we are neglecting the quantum zero-point energy which contributes to the total energy (which cannot be directly measured) but not to the heat capacity (which of course can be measured).

[4] Our derivation here is a historical one. The more modern derivation of the distribution function starts from an evaluation of the probability for the distribution of n identical objects (in the present case the number of quanta associated with the oscillator) among g identical boxes (the number of oscillators), there being no limit on the number of such objects in each box; this is the behavior expected for Bose particles, where we may have multiple occupancy of a given quantum state. The better derivation is identical to that used to obtain the Fermi–Dirac distribution given in Chapter 6 (where we may have only one object in each box), the sole difference being that the probability function, w, employed there—see Eq. (6.16)—is replaced by $w = g!/n!(g-n)!$.

Inserting (4.9) into (4.10) we have

$$\bar{E} = \frac{\hbar\omega\left[e^{-\beta\hbar\omega} + 2e^{-2\beta\hbar\omega} + 3e^{-3\beta\hbar\omega}\ldots\right]}{\left[1 + e^{-\beta\hbar\omega} + e^{-2\beta\hbar\omega} + e^{-3\beta\hbar\omega} + \ldots\right]}$$

$$= \hbar\omega x \frac{d}{dx} ln\left[\sum_{n=0}^{\infty} x^n\right] = -\hbar\omega x \frac{d}{dx} ln(1-x),$$

$$= \frac{\hbar\omega}{e^{\beta\hbar\omega} - 1} \qquad (4.12)$$

where we have introduced $x = \exp[-\beta\hbar\omega]$ in the intermediate steps.

Einstein's model for a solid was to replace the spectrum of normal mode frequencies associated with the coupled vibrating atoms by a *single* frequency, now referred to as the Einstein frequency, ω_E. The total thermal vibrational energy of a solid consisting of 3N oscillators is then given by

$$E = \frac{3N\hbar\omega_E}{e^{\hbar\omega_E/k_BT} - 1}. \qquad (4.13)$$

The heat capacity, $\left(\frac{\partial E}{\partial T}\right)_V$, is then

$$C_V = \frac{\left(3N\hbar^2\omega_E^2/k_BT^2\right) e^{\hbar\omega_E/k_BT}}{(e^{\hbar\omega_E/k_BT} - 1)^2}. \qquad (4.14)$$

As $T \to \infty$ we obtain

$$C_V = \frac{\left(3N\hbar^2\omega_E^2/k_BT^2\right)\left(1 + \frac{\hbar\omega_E}{k_BT} + \ldots\right)}{\left(1 + \frac{\hbar\omega_E}{k_BT} + \ldots - 1\right)^2} = 3Nk_B, \qquad (4.15)$$

thereby recovering the law of Dulong and Petit, while in the opposite limit where $T \to 0$,

$$C_V = \frac{3N\hbar^2\omega_E^2}{k_BT^2} e^{-\hbar\omega_E/k_BT}. \qquad (4.16a)$$

If we introduce a quantity having the units of temperature, $\theta_E = \hbar\omega_E/k_B$, called the Einstein temperature, we may write (4.16a) as

$$C = 3nR\left(\frac{\theta_E}{T}\right)^2 e^{-\theta_E/T}, \qquad (4.16b)$$

where $R = Ak_B$ is the gas constant; here A is Avogadro's number and n is the number of moles. The exponential dominates the low temperature behavior and involves the characteristic temperature θ_E. Figure 4.1 shows experimental data on the heat capacity of diamond; the dashed plot shows a fit to the this model of the data. The Einstein model accounts for the Dulong–Petit law at high temperature and the rapid decrease to zero at low temperatures. Experimentally, however, the low temperature heat capacity shows the power law behavior $C \sim T^3$, rather than an exponential behavior. This disagreement arises from Einstein's oversimplified model of the vibration spectrum that assumed all modes have the same frequency; we correct this defect in Section 4.3.

4.3 The Debye model

As previously discussed, the experimentally determined heat capacity decreases as T^3 rather than as $e^{-\theta_E/T}$ in an insulator at low temperatures. The origin of this disagreement lies with the Einstein assumption that the collection of all of the normal mode frequencies may be approximated by the single (Einstein) frequency, ω_E. In order to correctly calculate the heat capacity of a harmonic solid we must average over all of the normal mode frequencies according to

$$\bar{E}(T) = \int \frac{\hbar\omega}{e^{\hbar\omega/k_B T} - 1} N(\omega)d\omega; \tag{4.17}$$

here $N(\omega)$ is the *density of states*, defined as the number of normal modes having a frequency ω in an interval $d\omega$ (in a 1 cm^3 solid, consisting of $\sim 10^{22}$ atoms, the distribution of normal mode frequencies may be regarded as continuous).

We are now presented with the task of calculating $N(\omega)$. Consider the case of a hypothetical *one-dimensional* solid (or one-dimensional lattice) modeled as discrete mass points interconnected by springs and vibrating along the axis connecting the masses. We will assume the availability of some results from the mechanics of this problem (which is discussed later, in Section 16.1). If the "ends" of this solid are fixed, the time and position dependence of the displacement associated with the nth mass point for the lth mode has the standing-wave form

$$u_n^{(l)} = u_{n0} \sin\left(\frac{l\pi na}{L}\right) \sin(\omega_l t + \phi); \tag{4.18a}$$

here l is a positive integer and L is the length of the lattice. Note that for $l > l^{max} = L/a$, solutions $u_n^{(l)}$ reproduce those already occurring for $l < l^{max}$ and therefore do not represent new modes; i.e., the number of modes is limited and equal to the number of particles, as we expect for this one-dimensional system. This point will arise again later. Setting $na = x$ (the position along the string), we can write (4.18a) as[5]

$$u^{(l)}(x) = u_0 \sin(kx) \sin(\omega_l t + \phi), \tag{4.18b}$$

where we introduce the wave vector as

$$k = \frac{l\pi}{L}. \tag{4.19a}$$

[5] From Eq. (1.50), $\nabla^2 u_\parallel - v_l^{-2}\partial^2 u_\parallel/\partial t^2 = 0$, we see that (4.18b) is the solution corresponding to a longitudinal sound wave in the continuum limit with $\omega/k = v_l$.

The number of states dl between l and $l + dl$ is then

$$dl = \frac{L}{\pi}dk; \tag{4.19b}$$

here we assume that both l and dl are large (relative to unity). The density of modes (states) of our one-dimensional lattice is defined as

$$N(\omega) = \frac{dl}{d\omega} = \frac{L}{\pi}\frac{dk}{d\omega}. \tag{4.20}$$

We next consider a two-dimensional solid. Replacing Eq. (4.19a) are the conditions

$$l_x = \frac{L_x}{\pi}k_x; \quad l_y = \frac{L_y}{\pi}k_y, \tag{4.21}$$

where l_x and l_y are integers and L_x and L_y are the lengths of the edges of our two-dimensional lattice. The differential number of states, dN, in a differential region (area) of two-dimensional mode number space, $dl_x dl_y$, is

$$dN = dl_x dl_y = \frac{L_x}{\pi}\frac{L_y}{\pi}dk_x dk_y = \frac{A}{\pi^2}d^2k, \tag{4.22}$$

where $A = L_x L_y$ is the area occupied by our lattice. Clearly for three dimensions we have

$$l_x = \frac{L_x}{\pi}k_y; l_y = \frac{L_y}{\pi}k_y; l_z = \frac{L_z}{\pi}k_z. \tag{4.23}$$

The number of states in a differential cube in mode number space is then

$$dN = dl_x dl_y dl_z = \frac{V}{\pi^3}d^3k, \tag{4.24}$$

where V is the volume of the lattice. Now in the one-dimensional case, Eq. (4.18), we were restricted to positive l and hence positive k; in three dimensions we are restricted to k_x, k_y, $k_z > 0$ or, in spherical coordinates, to *1/8 of the associated sphere in k space.* If, in summing over all the states, we formally extend the integral to *all* of the k space,[6] we must divide by 8; we then have the important relation

$$dN = \frac{V}{(2\pi)^3}d^3k. \tag{4.25}$$

In what follows, we assume that our system is isotropic (note real crystal lattices are not isotropic). We may then write $k = k(\omega)$. In spherical coordinates $d^3k = k^2 dk d\Omega$. Integrating over all angles, the density of states per unit frequency in three dimensions, $N(\omega)$, is then

$$N(\omega) = \frac{V}{(2\pi)^3} \cdot 4\pi \cdot k^2 \frac{dk}{d\omega} \tag{4.26}$$

[6] If we had instead considered running waves and periodic boundary conditions (as we will do when considering running electron waves in Section 6.1), waves with positive and negative **k** will correspond to independent solutions. The distinction between standing and running waves and how it affects the counting of states, is not emphasized in most texts. Later, when we consider the Brillouin zones, it will be apparent that the formulation which includes all of **k** space is the more natural one.

and the total energy becomes

$$\bar{E} = \frac{V}{2\pi^2}\int \frac{\hbar\omega}{e^{\hbar\omega/k_BT}-1}k^2\frac{dk}{d\omega}d\omega. \tag{4.27}$$

The quantized lattice waves with phase velocities ω/k are called *phonons*, in analogy with the quantized electromagnetic oscillations, the photons. For small wave vectors (long wavelengths) the normal modes of a solid are longitudinal (ℓ) or transverse (t) sound waves, which have a velocity which is independent of frequency; i.e.,

$$\left(\frac{\omega}{k}\right)_{\ell,t} = c_{\ell,t}, \tag{4.28}$$

where $c_{\ell,t}$ are the longitudinal and transverse sound velocities (see Section 1.9). Note there are two independent transverse modes and each of the three modes (one longitudinal and two transverse) contributes a term to (4.27). For short wavelengths the velocities become frequency dependent, a phenomenon known as *dispersion*. However, in the Debye theory we ignore this complication and assume $c_{\ell,t} \neq c_{\ell,t}(\omega)$. We noted earlier that in a one-dimensional lattice there is an upper limit on the frequency $\int_0^{\omega_D} N(\omega)d\omega$ that can be supported. A similar thing happens in higher dimensions although here the highest frequency for a given mode depends on the direction in k space. As a further simplification the Debye model limits the integration to a *single* upper frequency for both the longitudinal and transverse modes such that the *total number* of modes is constrained by

$$\int_0^{\omega_D} N(\omega)d\omega = 3N, \tag{4.29}$$

where N is the total number of atoms; the maximum frequency, ω_D, is called the Debye frequency and is of order 10^{13} Hz. Explicitly including all three modes and using (4.27) and (4.28), Eq. (4.29) becomes

$$\frac{V}{2\pi^2}\left[\frac{1}{c_l^3}+\frac{2}{c_t^3}\right]\int_0^{\omega_D}\omega^2 d\omega = 3N. \tag{4.30}$$

For a simple cubic lattice, $N = V/a^3$, where a is the lattice spacing. Note the mass density, ρ, is given by MN/V, where M is the mass of an atom. Defining an average sound velocity by $\frac{3}{\bar{c}^3} \equiv \frac{1}{c_\ell^3} + \frac{2}{c_t^3}$, we have

$$\omega_D^3 = 6\pi^2\bar{c}^3\frac{N}{V}. \tag{4.31}$$

We may then write the total vibrational energy of our lattice in the Debye model as

$$\bar{E} = \frac{3V}{2\pi^2\bar{c}^3}\int_0^{\omega_D}\frac{\hbar\omega^3 d\omega}{e^{\hbar\omega/k_BT}-1}. \tag{4.32}$$

The resulting heat capacity is shown in Fig. 4.2.

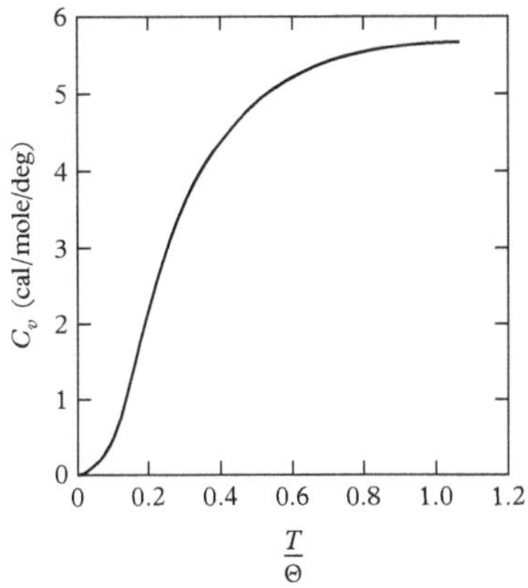

Figure 4.2 *The heat capacity according to the Debye model.*

At high temperatures we may expand the exponential in the denominator,

$$\bar{E} = \frac{3V}{2\pi^2\bar{c}^3}\int_0^{\omega_D}\frac{\hbar\omega^3 d\omega}{\left(1+\dfrac{\hbar\omega}{k_B T}+\ldots-1\right)} = \frac{3Vk_B T}{2\pi^2\bar{c}^3}\int_0^{\omega_D}\omega^2 d\omega = \frac{Vk_B T}{2\pi^2\bar{c}^3}\omega_D^3$$
$$= 3Nk_B T; \tag{4.33}$$

i.e., we again recover the law of Dulong and Petit.

In the limit of very low temperatures only low frequencies contribute to the integral and we may extend the upper limit of the integration to ∞ with negligible loss in accuracy. Introducing the variable $x = \hbar\omega/k_B T$, we may write (4.30) as

$$\bar{E} = \frac{3(k_B T)^4 V}{2\pi^2\bar{c}^3\hbar^3}\int_0^{\infty}\frac{x^3}{e^x-1}dx. \tag{4.34}$$

The integral

$$\int_0^{\infty}\frac{x^3}{e^x-1}dx = 6\zeta(4) = \sum_{n=1}^{\infty}\frac{1}{n^4} = \frac{\pi^4}{15},$$

where ζ is the zeta function.[7] We then have

$$\bar{E} = \frac{3(k_B T)^4 V}{2\pi^2\bar{c}^3\hbar^3}\frac{\pi^4}{15} = \frac{3\pi^4 N k_B}{5}\frac{T^4}{\theta_D^3}, \tag{4.35}$$

where we have defined a Debye temperature (analogous to the Einstein temperature) by $\theta_D \equiv (\hbar\bar{c}/k_B)\cdot(6\pi^2 N/V)^{1/3}$. The heat capacity is then

$$C_V = \left(\frac{\partial E}{\partial T}\right)_V = \frac{12\pi^4 nR}{5}\left(\frac{T}{\theta_D}\right)^3 \cong 234nR\left(\frac{T}{\theta_D}\right)^3, \tag{4.36}$$

where we have again introduced the gas constant. As seen from Fig. 4.3, this form agrees well with the observed heat capacity at low temperatures.

For intermediate temperatures, the details of the phonon density of states produce deviations from the Debye theory; it nonetheless serves as a reasonably accurate *interpolation formula* between the (asymptotically exact) T^3 behavior for $T \ll \theta_D$ and the classical Dulong–Petit law at high temperatures, where $T \gg \theta_D$.

A fit to the Debye model for the case of aluminum and copper is shown in Fig. 4.3.[8] Table 4.1 lists the Debye temperatures for many of the elements.

[7] Abramowitz and Stegun, *Handbook of Mathematical Functions*, Dover, New York (1972).

[8] The theory of specific warmth, P. Debye, *Ann. Phys.* **39**, 789 (1912).

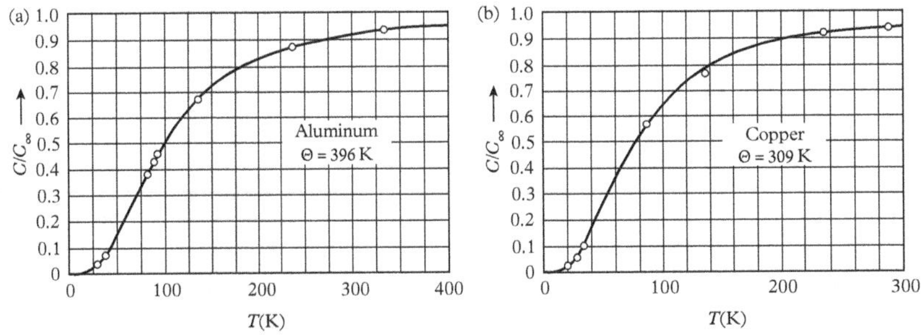

Figure 4.3 *(a) The heat capacity of aluminum fitted to the Debye model. (After P. Debye (1912)); (b) the same for the case of copper.*

Table 4.1 *Debye temperatures of some elements* (K).

Element	θ_D(K)	Element	θ_D(K)	Element	θ_D(K)
Ag	225	Hg	71.9	Rn	64
Al	428	In	108	Ru	600
Ar	92	Ir	420	Sb	211
As	282	K	91	Sc	360
Au	165	Kr	72	Se	90
Ba	110	La	142	Si	645
Be	1440	Li	344	Sn	200
Bi	119	Lu	210	Sr	147
C	2230	Mg	400	Ta	240
Ca	230	Mn	410	Te	153
Cd	209	Mo	450	Th	163
Co	445	Nb	275	Ti	420
Cr	630	Ne	75	Tl	78.5
Cs	38	Ni	450	U	207
Cu	343	Os	500	V	380
Dy	210	Pb	105	W	400
Fe	470	Pd	274	Xe	64
Ga	320	Pt	240	Y	280
Gd	200	Rb	56	Yb	120
Ge	374	Re	430	Zn	327
Hf	252	Rh	480	Zr	291

Problems

4.1. (a) Obtain as an integral the expression for the thermal energy of a two-dimensional solid based on the Debye theory. How do you define the Debye frequency and temperature?

(b) Derive the limiting forms of the heat capacity at low and high temperatures relative to the Debye temperature.

4.2. At very low temperatures what kind of excitations would you expect to be present in liquid ^{4}He (helium is the only material that does not solidify on going to absolute zero). What then would be the temperature dependence of the thermal energy and the heat capacity?

4.3. Suppose you had a lossless electrical transmission line of length L, a kind of one-dimensional electromagnetic box in which waves travel at some velocity $v \leq c$. At finite temperatures electromagnetic waves will be excited in the line.

(a) Using the Planck theory evaluate the amount of energy that would be present in an interval $\Delta\omega$ about some frequency ω when $L \gg \lambda$, where λ is the wavelength associated with ω.

(b) In equilibrium half of this energy is flowing in one direction and the remaining half in the opposite direction. What is the associated energy flux in the interval $\Delta\omega$ and its limiting form for $k_B T \gg \hbar\omega$?

(c) A transmission line has a characteristic impedance which we will assume is real and denote as R_0. Assume our transmission line is terminated by a resistor having this value. What would be the average "noise" voltage that would appear across this resistor? This is sometimes called *Johnson noise*.

Elementary Theories of Magnetism

Magnetic Ions

5

From electromagnetism, the relation between the magnetic induction, **B**, and the magnetic field, **H**, is given by $\mathbf{B} = \mathbf{H} + 4\pi\mathbf{M}$, where **M** is the magnetic moment per cm^3, or *magnetization*.[1] For those cases where there is a linear relationship between **M** and **H** we may write

$$M_i = \sum_{j=1}^{3} \chi_{ij}^{M} H_j \tag{5.1a}$$

or

$$B_i = \sum_{j=1}^{3} \mu_{ij} H_j = \sum_{j=1}^{3} \left(\delta_{ij} + 4\pi\chi_{ij}^{M}\right) H_j, \tag{5.1b}$$

where μ_{ij} and χ_{ij}^{M} are called the *magnetic permeability* and *susceptibility* tensors, respectively. For the remainder of this chapter we will drop the superscript on χ_{ij}^{M}. If the material is isotropic this relation becomes simply

$$\mathbf{B} = \mu\mathbf{H} = (1 + 4\pi\chi)\mathbf{H}. \tag{5.2}$$

With the exception of superconductors and ferromagnets, the magnetic response is generally rather weak: $\chi \approx 10^{-5}$. For this reason we may for most purposes neglect the effect of the surrounding material in determining the local value of **H**; i.e., **H** may be taken as the external field.[2]

All materials are either diamagnetic, $\chi < 0$, or paramagnetic, $\chi > 0$, hence diamagnetic materials tend to exclude flux while paramagnetic materials tend to gather flux. Superconductors and (soft) ferromagnets are extreme limits of these behaviors.[3] Furthermore, μ may involve more than one contribution, the signs of which may differ. The presence of a non-zero susceptibility depends ultimately on quantum effects—strictly classical systems cannot be magnetic,[4] although a semi-classical theory works well for many situations. We begin our discussion of the magnetic properties of matter with the semi-classical theory of diamagnetism.

[1] As in our discussion of the electrical properties in Chapter 2, we employ the c.g.s. system of units.

[2] This will not be the case when the response of the system is large, as with superconductors and ferromagnets, as we shall see later. In such cases one must account for *demagnetization effects* which are the magnetic analog of the depolarization effects discussed in Chapter 2.

[3] A so-called type I superconductor completely excludes the flux; a soft ferromagnet may have a permeability exceeding 1000.

[4] This is the so-called Bohr–van Leeuwen theorem.

The Physics of Solids. First Edition. J. B. Ketterson.

5.1 Langevin diamagnetism

In the model to be discussed here (Langevin, 1905) magnetization arises from circulating microscopic (atomic-scale) currents.[5] Classically, the magnetization of a body arises from some current distribution and is given by[6,7]

$$\mathbf{M} = \frac{1}{2c}\mathbf{r} \times \mathbf{j}; \tag{5.3a}$$

the *total* magnetic moment $\mathbf{m}$ is therefore

$$\mathbf{m} = \frac{1}{2c}\int \mathbf{r} \times \mathbf{j}\, d^3r. \tag{5.3b}$$

From classical mechanics, the equation of motion of a body (in the present case an atom or molecule) having an angular momentum $\mathbf{L}$ and subjected to an external torque $\mathbf{N}$ is $d\mathbf{L}/dt = \mathbf{N}$. If this body has a magnetic moment $\mathbf{m}$ then the torque in the presence of a magnetic field $\mathbf{H}$ is given by $\mathbf{m} \times \mathbf{H}$. Writing $\mathbf{j} = ne\mathbf{v}$ in Eq. (5.3b) and introducing the mass density, ρ, we have

$$\mathbf{m} = \frac{e}{2mc}\int \mathbf{r} \times \rho\mathbf{v}\, d^3r \tag{5.4a}$$

$$= \frac{e}{2mc}\mathbf{L}, \tag{5.4b}$$

where, in the second step, we identify $\rho\mathbf{v}$ as the momentum density and write $\mathbf{L} = \int \mathbf{r} \times \rho\,\mathbf{v}\, d^3r$. Inserting this into our equation of motion we have

$$\frac{d\mathbf{L}}{dt} = \frac{e}{2mc}\mathbf{L} \times \mathbf{B}. \tag{5.5}$$

Note that this equation is similar to the simplified gyroscope equation of motion encountered in the mechanics of rotating bodies. Since the change of $\mathbf{L}$ is here always perpendicular to $\mathbf{L}$, the magnitude of the latter must remain constant and therefore the body *precesses* at a constant gyroscopic precession frequency, given by

$$\omega_L = \frac{eB}{2mc}, \tag{5.6}$$

known as the *Larmor frequency*; this property is called *Larmor's theorem*. In practice, the external magnetic field must be applied adiabatically so that there are no electric fields induced (by Faraday's law) which could otherwise excite the atom.

Atoms with closed atomic shells have no net magnetic moment.[8,9] However, a diamagnetic (negative) moment is induced by the application of a magnetic field, which results in a diamagnetic contribution to the magnetic susceptibility of a gas, liquid, or solid composed of such atoms.[10] Using the earlier theory, let us estimate

[5] An exception is a superconductor where the currents may be regarded as macroscopic.

[6] In terms of the *total* current, Maxwell's fourth equation can be written as $\nabla \times \mathbf{B} = (4\pi/c)\mathbf{j}_{total}$, in the static limit. Writing $\mathbf{B} = \mathbf{H} + 4\pi\mathbf{M}$ and splitting the total current into a contribution $\mathbf{j}_{ext}$ arising from external sources and a microscopic magnetization current, $\mathbf{j}_{mag}$, it follows that we can write $\mathbf{j}_{mag} = c\nabla \times \mathbf{M}$, which has a solution for $\mathbf{M}$ of the form $\mathbf{M} = (1/2c)\mathbf{r} \times \mathbf{j}_{mag}$. Although this expression appears to depend on the choice of the origin, this is not the case due to the circulating nature of the microscopic currents (note angular momentum, which is written $\mathbf{L} = \mathbf{r} \times \mathbf{p}$, is also origin independent). We will henceforth drop the subscript on $\mathbf{j}_{mag}$.

[7] That such circulating currents can exist ultimately rests on quantum mechanics.

[8] Atoms with unpaired spins always have a net magnetic moment. This can yield a macroscopic paramagnetic contribution to the magnetic susceptibility that will be discussed in the next section.

[9] We ignore here any magnetic moment associated with the nucleus.

[10] Even in the presence of unfilled shells there will be a diamagnetic contribution arising from the core electrons of an atom.

the induced magnetic moment of such an atom, which we denote as $\boldsymbol{\mu}$. We write the current in Eq. (5.4) in the form $\mathbf{j} = n(\mathbf{r})e\mathbf{v}$, where now $\int n(\mathbf{r})\, d^3r = Z$, with Z being the atomic number; substituting $\mathbf{v} = \boldsymbol{\omega}_L \times \mathbf{r}$ we have

$$\boldsymbol{\mu} = \frac{e}{2c}\int n(\mathbf{r})\mathbf{r} \times (\boldsymbol{\omega}_L \times \mathbf{r})\, d^3r$$

$$= -\frac{Ze^2H}{4mc^2}\overline{r_\perp^2}\,\mathbf{n}_\mu, \tag{5.7}$$

where $\mathbf{n}_\mu$ is a unit vector parallel to $\boldsymbol{\mu}$ and

$$\overline{r_\perp^2} \equiv \frac{1}{Z}\mathbf{n}_\mu \cdot \int n(\mathbf{r})\mathbf{r} \times (\mathbf{n}_\mu \times \mathbf{r})d^3r; \tag{5.8}$$

the latter is the average of r^2 projected onto the plane perpendicular to $\boldsymbol{\mu}$. For our closed shell atom, which has spherical symmetry, we have $\overline{x^2} = \overline{y^2} = \overline{z^2} = \overline{r^2}/3$ or $\overline{r_\perp^2} = 2\overline{r^2}/3$. The magnetic susceptibility is then

$$\chi = \frac{n\mu}{H} = -\frac{nZe^2}{6mc^2}\overline{r^2}, \tag{5.9}$$

where n is the number of atoms per cm^3. One sometimes finds susceptibilities given in other units, e.g., per mole or per gram. It is then important to note that when calculating B from H using Eq. (5.2) one must always use the *susceptibility per unit volume.*

Atoms with closed atomic shells are diamagnetic. The same is true for the core contribution from ions (as in an ionic crystal). The quantity $\overline{r_\perp^2}$ may be accurately calculated using the Hartree[11] or some more refined theory and, depending on the system, can agree within a few percent. Table 5.1 lists the diamagnetic susceptibilities of some common ions.

[11] The Hartree theory is discussed in Section 15.1.

Table 5.1 *Experimental values of the diamagnetic susceptibility of ions (10^{-6} cm^3/mole).*

Ion	χ	Ion	χ	Ion	χ
F^-	–9.4	Li^+	–0.7	Mg^+	–4.3
Cl^-	–24.2	Na^+	–6.1	Ca^+	–10.7
Br^-	–34.5	K^+	–14.6	Sr^+	–18.0
I^-	–50.6	Rb^+	–22.0	Ba^+	–29.0
		Cs^+	–35.1		

5.2 Langevin paramagnetism

We next discuss what we might call the semi-classical theory of paramagnetism, also due to Langevin. Materials containing ions having unfilled d or f shells (the transition and the rare earth/actinide atoms[12]) can have *permanent* magnetic moments; i.e., they have a moment in the absence of an applied field (as opposed to the induced moments associated with diamagnetism). The s and p electrons, which usually dominate chemical bonding, are almost always paired with neighboring atoms in a solid, resulting in no net spin from these electrons.[13] According to Hund's rule, the d and f electrons align so as to maximize the total spin that is allowed by the Pauli exclusion principle. Even nickel,[14] which has the atomic configuration $1s^2 2s^2 2p^6 3d^{10}$, will be spontaneously magnetic as an ion in the solid state, since electrons are "promoted" from the d shell to the s–p shells (which lie close in energy) in order to form chemical bonds.[15]

For simplicity, we assume the spins on neighboring ions do not influence each other; i.e., they act as independent non-interacting permanent magnetic moments, $\boldsymbol{\mu}$.[16] We use Eqs. (4.2) and (4.3) to compute the thermally averaged moment. The Hamiltonian in this case is given by the usual interaction energy of the magnetic moment with the external magnetic field,

$$U = -\boldsymbol{\mu} \cdot \mathbf{H}. \tag{5.10}$$

The relevant phase space element for a single magnetic moment in this case involves the coordinates defining its orientation. Since in the absence of interactions between different moments these coordinates act independently of each other, we may write $d\Gamma = \prod_{i=1}^{N} \sin\theta_i d\theta_i d\phi_i = N \sin\theta d\theta d\phi$, where N is the total number of moments. The integral (4.2) associated with a particular moment is then

$$\begin{aligned}\bar{\mu} &= \frac{\int \mu \cos\theta\, e^{\beta\mu H \cos\theta} \sin\theta d\theta d\phi}{\int e^{\beta\mu H \cos\theta} \sin\theta d\theta d\phi} \\ &= \frac{1}{\beta}\frac{\partial}{\partial H} \ln\left[\int e^{\beta\mu H \cos\theta} \sin\theta d\theta d\phi\right] \\ &= \frac{1}{\beta}\frac{\partial}{\partial H} \ln\left[\frac{2\pi}{\beta\mu H}\left(e^{\beta\mu H} - e^{-\beta\mu H}\right)\right]\end{aligned}$$

or $\bar{\mu} = \mu L(x)$, where

$$L(x) = \coth(x) - 1/x \tag{5.11a}$$

is called the *Langevin function* and $x \equiv \mu H/k_B T$. The magnetic moment per unit volume V is therefore given by

$$M = \frac{N}{V}\mu L(x) = n\mu L(x), \tag{5.11b}$$

[12] Materials involving actinide atoms having unfilled 5f shells can also be magnetic.

[13] O_2 is an example of a paramagnetic molecule having only s and p valence electrons.

[14] Metallic nickel is an example of a ferromagnet, a material that is spontaneously magnetic in the absence of an applied field.

[15] The valence of an atom is defined as 2 × (number of unpaired spins), either in the ground state or in a "promoted" state.

[16] Such interactions can lead to ferromagnetism, antiferromagnetism, or more complex states, as discussed in Chapter 26.

where n is the number of moments per unit volume. In the high temperature limit, $x \ll 1$, we obtain

$$M(T) = \frac{n\mu^2}{3k_B T}H = \frac{C}{T}H, \tag{5.12a}$$

where $C \equiv n\mu^2/3k_B$. The magnetic susceptibility, given by $\chi = M/H$, in this limit is then

$$\chi(T) = \frac{C}{T}; \tag{5.12b}$$

Eq. (5.12b) is known as *Curie's law*. Note that the divergence implied as $T \to 0$ does not occur in practice. In fact, in the presence of a finite field $L(\mu H/k_B T) \to 1$ as $T \to 0$ corresponding to a magnetization of $n\mu$; i.e., all the spins are aligned along the applied field. However interaction effects among the individual moments become important in the low temperature limit. These effects are sometimes modeled by introducing an "effective" temperature, $T \to T + T_0$ (where T_0 may have either sign) in Eq. (5.12); this provides a more accurate description for many materials over a wider temperature range; however it always fails for temperatures approaching T_0. Note that the theory given here is essentially identical to that given earlier for non-interacting electric dipoles in Chapter 2 (see Eq. (2.45)).

5.3 Quantum theory of magnetism

The Hamiltonian prescription (see Section 18.1) for incorporating magnetic fields into quantum mechanics is to replace the kinetic energy operator associated with a single electron, $\hat{p}^2/2m$, by

$$\hat{H}_0 = \frac{1}{2m}\left(\frac{\hbar}{i}\nabla - \frac{e}{c}\mathbf{A}\right)^2. \tag{5.13}$$

The so-called *orbital* part of the magnetic interaction arises from the changes in $\hat{H}_0$ associated with $\mathbf{A}$. For a uniform magnetic field, $\mathbf{H}$, we can write the vector potential as

$$\mathbf{A} = -\frac{1}{2}\mathbf{r} \times \mathbf{H}. \tag{5.14}$$

For atoms, the additional terms generated by $\mathbf{A}$ can lead to both paramagnetic and diamagnetic contributions. For conduction electrons in metals, $\mathbf{A}$ leads to so-called Landau diamagnetism. Both of these topics will be taken up in later chapters.[17]

[17] See Sections 24.3 and 42A.

In addition to effects resulting from **A** in (5.13), we have a *direct* interaction between the external field and the electron spins, which can be obtained from the classical expression for the interaction energy of a dipole with an external field,[18]

$$U = -\hat{\mu} \cdot \mathbf{H}. \tag{5.15}$$

The spin part of our Hamiltonian is then written as

$$\hat{H}_s = g\mu_B \mathbf{H} \cdot \hat{\mathbf{S}}, \tag{5.16}$$

where $\hat{\mathbf{S}}$ is the total spin operator; $\mu_B = e\hbar/2mc$ is the Bohr magneton, and, to sufficient accuracy, we may take the electron g-factor as 2.

5.4 Quantum theory of non-interacting spins

Let us consider a collection of non-interacting spin 1/2, g = 2 particles (electrons) in a magnetic field aligned parallel or antiparallel to the z axis. Many insulators contain ions in valence states that involve unfilled d or f shells, the so-called paramagnetic ions which can have a net magnetic moment, as previously discussed. At higher temperatures the system can often be idealized as an ensemble of non-interacting spins, as we did in the semi-classical Langevin theory, which is sometimes referred to as a spin gas. This approximation is even better for nuclear spins, since the moments are of order a thousand times smaller, which results in their interaction being very small; to apply the spin 1/2 theory in that case, μ_B is replaced by a nuclear magneton and the measured nuclear g factor, which generally differs substantially from 2.

From the electron Hamiltonian (5.16) we have the energy levels

$$E = \pm g\mu_B S_z H = \pm \frac{e\hbar}{2mc} H. \tag{5.17}$$

From Eqs. (5.10) and (5.11) the average of the magnetization, M (the magnetic moment per unit volume), is

$$M = n\frac{g\mu_B}{2} \frac{e^{\frac{g\mu_B H}{2k_B T}} - e^{-\frac{g\mu_B H}{2k_B T}}}{e^{\frac{g\mu_B H}{2k_B T}} + e^{-\frac{g\mu_B H}{2k_B T}}} \tag{5.18a}$$

$$= \frac{ng\mu_B}{2} \tanh\left[\frac{g\mu_B H}{2k_B T}\right], \tag{5.18b}$$

[18] In the relativistic Dirac theory of the electron, the interaction with a magnetic field emerges naturally and one does not need to invoke the classical expression; a term identical to (5.16) emerges in the non-relativistic limit.

where n is again the number of spins per unit volume. At higher temperatures, such that $g\mu_B H/2 \ll k_B T$, we can expand the hyperbolic tangent as $\tanh(x) \approx x$, and the magnetic susceptibility $\chi = M/H$ becomes

$$\chi \cong n\frac{g^2\mu_B^2}{4k_BT}; \tag{5.19}$$

Note that we again obtain a Curie-like 1/T dependence, although the expression for C (see Eq. (5.12a)) involves a slightly different numerical factor.[19]

The above treatment can be generalized to higher spin S. For an atom with total angular momentum $\mathbf{J} = \mathbf{L} + \mathbf{S}$, the effective g-factor is given by the so-called Landé expression,

$$g_L = 1 + \frac{J(J+1) + S(S+1) - L(L+1)}{2J(J+1)}. \tag{5.20}$$

The resulting expression for the magnetization (which we do not derive, see Problem 5.1) is

$$M = ng_L J\mu_B B_J\left[\frac{g_L J\mu_B H}{k_B T}\right], \tag{5.21}$$

where

$$B_J(x) \equiv \frac{2J+1}{2J}\coth\left[\frac{2J+1}{2J}x\right] - \frac{1}{2J}\coth\left[\frac{x}{2J}\right] \tag{5.22}$$

is called the *Brillouin function*. For $x \ll 1$, the susceptibility is Curie-like and given by

$$\chi = n\frac{g_L^2 J(J+1)\mu_B^2}{3k_BT} = n\frac{p^2\mu_B^2}{3k_BT} \tag{5.23a}$$

$$= \frac{C}{T}; \tag{5.23b}$$

here,

$$C = \frac{ng_L^2 J(J+1)\mu_B^2}{3k_B} = \frac{np^2\mu_B^2}{3k_B} \tag{5.24}$$

is Curie's constant and we introduced an effective magneton number,

$$p \equiv g_L\sqrt{J(J+1)}. \tag{5.25}$$

[19] The eigenvalue of the operator $\hat{s}^2$ is $s(s+1)$, which is 3/4 for a spin 1/2 particle; allowing for this fact (5.19) becomes identical to (5.12a).

5.5 Adiabatic demagnetization

A powerful method to achieve very low temperatures is the adiabatic removal of a magnetic field from an isolated ensemble of magnetic moments (spins) associated with either electrons or nuclei; the process is referred to as *adiabatic demagnetization.*[20] To discuss this technique we write the differential change in the entropy S(H,T) as

$$\begin{aligned} dS &= \left(\frac{\partial S}{\partial H}\right)_T dH + \left(\frac{\partial S}{\partial T}\right)_H dT \\ &= \left(\frac{\partial M}{\partial T}\right)_H dH + \frac{1}{T} C_H dT \end{aligned}, \tag{5.26}$$

where in the second step we have employed the Maxwell relation, $(\partial S/\partial H)_T = (\partial M/\partial T)_H$, in the first term and introduced the heat capacity at constant magnetic field, $C_H \equiv T\,(\partial S/\partial T)_H$, in the second term.

The process of adiabatic demagnetization can be idealized as consisting of two steps. In the first step we apply a magnetic field to the sample while holding the temperature constant, resulting in an entropy change,

$$S(H,T) = S(0,T) + \int_0^H dH \left(\frac{\partial M}{\partial T}\right)_H. \tag{5.27}$$

In the absence of a magnetic field, and neglecting any interaction between the spins, the number of independent configurations, W, that N spins can adopt is $W = (2J+1)^N$; since the entropy is given by $S = k_B \ln W$, we then have

$$S(0,T) = Nk_B \ln(2J+1). \tag{5.28}$$

Note that this expression is temperature independent. In reality, this form would not set in until the thermal energy exceeded a characteristic energy, ΔE, involving the interaction energy of the magnetic dipoles making up the system; i.e., $k_B T \cong \Delta E \cong \mu^2/a_s^3$, where a_s is a spin–spin spacing. Typical temperatures are 10^{-3} K for dilute electron spin systems or 10^{-8} K for nuclei.

If very high magnetic fields satisfying the criteria $\mu H \gg k_B T$ were available most of the entropy would be removed from the system;[21] i.e., $S(\infty, T) = 0$ (here we ignore the entropy of the lattice or other thermodynamic systems in contact with the spin system). Hence the second step of the cooling process involves isolating the system and adiabatically removing the magnetic field, during which the temperature drops, with the ultimate limit set by the interaction energy among the spins mentioned earlier, below which the system orders and the entropy rapidly drops to zero.

[20] This technique was independently proposed by Giauque and Debye in 1926; it was demonstrated by Giauque and MacDougall in 1933, where a temperature of 0.53 K was reached (well below that which can be achieved with pumped ^{4}He) using the paramagnetic salt gadolinium sulphate.

[21] At such high fields the spins are aligned and the disorder, which is the source of the entropy, is removed.

In the region where Curie's law applies ($k_B T \gg \mu H$), the field-dependent entropy is given from Eqs. (5.23), with $M = \chi H$, and (5.28) by

$$\begin{aligned} S(H,T) &= Nk_B \ln(2J+1) - N\frac{g_L^2 J(J+1)\mu_B^2}{3k_B}\int_0^H dH\frac{H}{T^2}, \\ &= Nk_B \ln(2J+1) - \frac{C}{2}\frac{H^2}{T^2} \end{aligned} \tag{5.29}$$

where $C = Ng_L^2 J(J+1)\mu_B^2/3k_B$ is Curie's constant. In this limit, where spin–spin interactions are ignored and $k_B T \gg \mu H$, the entropy change ΔS scales as the ratio H^2/T^2; hence we can write

$$\frac{H_{start}}{T_{start}} = \frac{H_{final}}{T_{final}}, \tag{5.30}$$

where H_{start}, T_{start} and H_{final}, T_{final} correspond to the starting and final values of the field and temperature. It is now immediately apparent that large temperature reductions can be achieved simply by lowering the magnetic field on a pre-magnetized spin system. In practice, limits are set by various additional heat loads (e.g., the lattice entropy), irreversibility (due to demagnetizing too rapidly), heat leaking into the system, and, ultimately, ordering of the spin system. Adiabatic demagnetization of nuclei is the only known way to cool below temperatures of about 10^{-3} K.

5.6 Chemical bonding, Hund's rules, and magnetic ions

The theories we have developed for paramagnetism were based on an ensemble of non-interacting magnetic moments. One must then ask under what conditions one can have such moments in the first place, and, if so, when can they be viewed as non-interacting; we now examine these questions.

As has been noted several times, paramagnetism in solids primarily involves chemical compounds containing magnetic ions of the iron (3d) and rare-earth (4f) groups of the periodic table. To understand why such magnetic ions exist we must invoke some concepts from atomic physics and chemistry.

Figure 5.1 shows, qualitatively, the order in which atomic shells fill as one increases the charge Z of the atomic nucleus. For the same principal quantum number, an electron with higher angular momentum tends to stay further from the nucleus because of the repulsive centrifugal barrier; for this reason they do not feel the nuclear attraction as strongly and therefore lie higher in energy. The approximate placement of the levels is

$$6p > 5d > 4f > 6s > 5p > 4d > 5s > 4p > 3d > 4s > 3p > 3s > 2p > 2s > 1s.$$

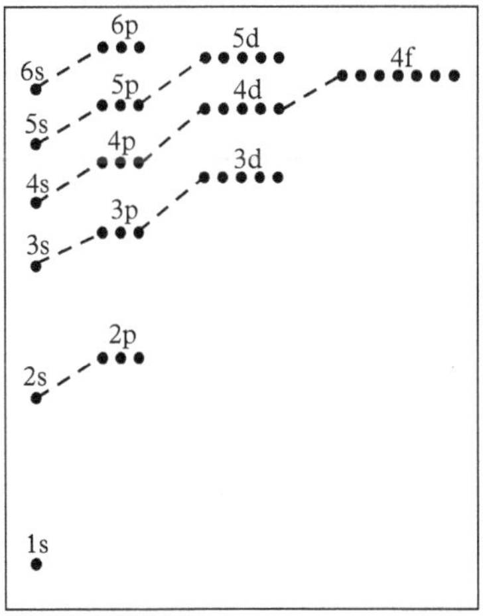

Figure 5.1 *A schematic drawing (not to scale) of the relative placement of the energy levels of atoms that approximately defines how the levels fill with increasing atomic number. The number of black dots signifies the orbital (but not the spin) degeneracy. The levels shown account for the atoms through radon.*

This energy ordering, together with the spatial extent of the associated electron states, explains much of the chemistry that is summarized by the periodic table.

In the absence of d and f electrons the valence of an atom is primarily determined by the number, n, of s and p electrons outside a filled rare-gas shell, there being $2 + 6 = 8$ such states. Those atoms with $n < 4$ are assigned a positive valence equal to n, while those atoms with $n > 4$ have a negative valence, $n - 8$. As will be discussed in Chapter 7, covalent chemical bonding, the prototypical case being carbon with a valence of four, involves the formation of *shared* electron pairs between neighboring atoms. On the other hand, ionic bonding involves the *transfer* of electrons from the cations (atoms with $n < 4$) which leaves them positively charged, to form filled s–p shells on the anions (those with $n > 4$), which leaves them negatively charged. However in both of these cases no unpaired spins remain[22] and hence there is no paramagnetism; as examples, neither sodium chloride (an ionic material) nor carbon (a covalent material) is paramagnetic.

When forming chemical compounds containing atoms with 3d and 4f electrons, the dominant bonding still involves the s and p electrons,[23] although one or more of the d or f electrons can be "promoted" to an s–p shell, resulting in *multiple valence states.* While the s and p electrons pair up to form bonds with no net moment, the "core" 3d or 4f electrons tend to maintain an atomic-like moment characteristic of the number of d or f electrons retained in that shell, as we now discuss.

For the elements of interest here, an $\mathbf{L} \cdot \mathbf{S}$ coupling model applies to the atoms where

$$\mathbf{L} = \sum_i \mathbf{l}_i \tag{5.31}$$

and

$$\mathbf{S} = \sum_i \mathbf{s}_i; \tag{5.32}$$

here, $\mathbf{l}_i$ and $\mathbf{s}_i$ are the angular momenta and spins of the individual electrons in the shell. The behavior of the vectors **L**, **S** and $\mathbf{J} = \mathbf{L} + \mathbf{S}$ are captured by the following three rules, first formulated by Hund.[24,25]

1. The atom will have the highest possible total spin **S** consistent with the Pauli exclusion principle.[26]
2. The total orbital angular momentum **L** will be that which is highest *after* rule 1 is satisfied.
3. The magnitude $|\mathbf{J}|$ of the total angular momentum **J** is $|\mathbf{L} - \mathbf{S}|$ if the shell is less than half full and $|\mathbf{L} + \mathbf{S}|$ if it is more than half full. When the shell is exactly half full we have $\mathbf{J} = \mathbf{S}; \mathbf{L} = 0$.

[22] An exception is a so-called *free radical* where lone, unpaired, electrons exist.

[23] This tendency is reduced for compounds with 4d, 5d, and 5f atoms.

[24] The first two rules have their origin in the Coulomb repulsion between electrons. If the electrons are in the same spin state the Pauli principle keeps them apart and hence the atom favors maximal spin; if the electrons adopt the maximum orbital angular momentum (consistent with having the maximum spin), this also tends to keep the electrons apart (semi-classically they then tend to rotate in the same direction and position themselves on opposite sides of the atom, whereas if they rotate in opposite directions they encounter each other twice with each rotation). Rule 3 arises from the $\mathbf{L} \cdot \mathbf{S}$ coupling, a relativistic effect that can be viewed as an interaction between the orbital and spin dipole moments.

[25] The enhanced correlation of the electrons arising from rules 1 and 2 allows the electrons, particularly in the 4f shell, to pack more tightly about the nucleus leading to a contraction in the ionic radius with increasing Z, a phenomenon known as *lanthanide contraction.*

[26] It is initially surprising that while electrons *within* an atom tend to align *parallel* to each other according to this rule, electrons involved in forming bonds *between* atoms, as previously noted, favor *antiparallel* alignment. This was first explained by Heitler and London for hydrogen. In the case of homopolar chemical bonds, formation of antiparallel singlet states maximizes the wave function overlap, and the gain in *exchange energy* outweighs the Coulomb repulsion, as discussed in texts on quantum mechanics.

When atoms with unfilled d and f shells are incorporated into chemical compounds the presence of the anions tends to isolate the moments thereby reducing their interaction.[27]

5.7 Magnetic moments of 4f and 3d ions

Here we summarize some data on the magnetic moments in chemical compounds[28] containing 3d or 4f cations. Table 5.2 lists the values of p obtained from fitting the experimental susceptibilities (measured near room temperature) to Eq. (5.23a), together with those calculated from its definition in Eq. (5.25) and the Hund value for J, for the trivalent rare-earth ions. Note that the agreement is remarkably good for such a simple model.

Table 5.3 shows the data for the 3d transition metal ions. However the agreement calculated when using $p = g_L\sqrt{J(J+1)}$ is now much poorer. The agreement is significantly better if we use $p = 2\sqrt{S(S+1)}$, i.e., if we assume the orbital angular momentum is somehow absent, a phenomenon called *quenching*. What is happening physically is that the 3d orbitals more strongly overlap the region where the chemical bonds form and hence they feel the presence of the adjacent atoms (compared to the case with the 4f electrons). The potential seen by the d electrons is then less spherically symmetric and the angular momentum quantum number then tends to lose its identity.

[27] The use of large anions (e.g., involving organic compounds), the inclusion of water of hydration, or the addition of other non-magnetic cations, will further increase the separation of the magnetic cations within the crystal and thereby reduce their interaction. Some examples are $Ce_2Mg_3(NO_3)_{12}\cdot 24H_2O$, $KCr(SO_4)_2\cdot 12H_2O$, and $NH_4Fe(SO_4)_2\cdot 12H_2O$, all of which are of historical importance with respect to cooling by adiabatic demagnetization.

[28] The idea of non-interacting atomic moments is less useful in metals; here the phenomenon of ferromagnetism commonly occurs, which is treated in Chapter 26.

[29] From C. Kittel, *Introduction to Solid State Physics*, Wiley, 5th edition, p. 442.

Table 5.2 *Effective magneton numbers of trivalent rare-earth ions.*[29]

Ion	Configuration	Spectroscopic designation	$p_{calculated}$	$p_{measured}$
Ce^{3+}	$4p^6 5s^2 4f^1$	$^3F_{5/2}$	2.54	2.4
Pr^{3+}	$4p^6 5s^2 4f^2$	$^3H_4$	3.58	3.5
Nd^{3+}	$4p^6 5s^2 4f^3$	$^4I_{9/2}$	3.62	3.5
Pm^{3+}	$4p^6 5s^2 4f^4$	$^5I_4$	2.68	–
Sm^{3+}	$4p^6 5s^2 4f^5$	$^6H_{5/2}$	0.84	1.5
Eu^{3+}	$4p^6 5s^2 4f^6$	$^7F_0$	0	3.4
Gd^{3+}	$4p^6 5s^2 4f^7$	$^8S_{7/2}$	7.94	8.0
Tb^{3+}	$4p^6 5s^2 4f^8$	$^7F_6$	9.72	9.5
Dy^{3+}	$4p^6 5s^2 4f^9$	$^6H_{15/2}$	10.63	10.6
Ho^{3+}	$4p^6 5s^2 4f^{10}$	$^5I_8$	10.60	10.4
Er^{3+}	$4p^6 5s^2 4f^{11}$	$^4I_{15/2}$	9.59	9.5
Tm^{3+}	$4p^6 5s^2 4f^{12}$	$^3H_6$	7.57	7.3
Yb^{3+}	$4p^6 5s^2 4f^{13}$	$^2F_{7/2}$	4.54	4.5

Table 5.3 *Effective magneton numbers for the 3d transition metals.*[30]

Ion	Configuration	Spectroscopic notation	$g_L\sqrt{J(J+1)}$	$2\sqrt{S(S+1)}$	$p_{measured}$
Ti^{3+}; V^{4+}	$3d^1$	$^2D_{3/2}$	1.55	1.73	1.8
V^{3+}	$3d^2$	$^3F_2$	1.63	2.83	2.8
Cr^{3+}; V^{2+}	$3d^3$	$^4F_{3/2}$	0.77	3.87	3.8
Mn^{3+}; Cr^{2+}	$3d^4$	$^5D_0$	0	4.90	4.9
Fe^{3+}; Mn^{2+}	$3d^5$	$^6S_{5/2}$	5.92	5.92	5.9
Fe^{2+}	$3d^6$	$^5D_4$	6.70	4.90	5.4
Co^{2+}	$3d^7$	$^4F_{9/2}$	6.63	3.87	4.8
Ni^{2+}	$3d^8$	$^3F_4$	5.59	2.83	3.2
Cu^{2+}	$3d^9$	$^2D_{5/2}$	3.55	1.73	1.9

FURTHER READING

The Theory of Magnetism Made Simple: An Introduction to Physical Concepts and to Some Useful mathematical methods, Daniel C. Mattis, World Scientific Publishing Company (2006).

A classic and still valuable book on magnetic response is *The Theory of Electric and Magnetic Susceptibilities*, J.H. Van Vleck, Oxford University Press (1932).

Adiabatic demagnetization is discussed in *Principles and Applications of Magnetic Cooling*, R.P. Hudson, North Holland, Amsterdam (1972).

For a discussion of adiabatic nuclear cooling, see *Experimental Principles and Methods Below 1K*, O.V. Lounasmaa, Academic Press, London (1974).

Electron Paramagnetic Resonance of Transition Ions, A. Abragam and B. Bleaney, Dover (1986).

Problems

5.1. Derive Eqs. (5.21) and (5.22).

5.2. Derive the Maxwell relation $\left(\frac{\partial S}{\partial H}\right)_T = \left(\frac{\partial M}{\partial T}\right)_H$.

5.3. Verify that the spectroscopic designation for each of the rare-earth ions is consistent with that predicted by Hund's rule.

5.4. Calculate the Landé g-factors for the rare-earth ions.

[30] From C. Kittel, *Introduction to Solid State Physics*, Wiley, 5th edition, p. 443.

The Non-interacting Fermi Gas

6

As mentioned in Chapter 4, the classical theory would result in a contribution of $3k_B$ or $3k_B/2$ to the heat capacity for each bound or free electron in a solid, which is not observed experimentally. Furthermore, and in disagreement with the Debye theory of the lattice, the specific heat of metals shows an asymptotically *linear* behavior as $T \rightarrow 0$, which is usually written $C_V = \gamma T$, where γ is a coefficient. The resolution of the electronic heat capacity problem also lies in quantum theory but, in addition, it is necessary to include the effects of the Pauli exclusion principle.[1]

The simplest model of a metal is to assume that it consists of free electrons in a box, as discussed is Section 6.3, with their mutual Coulomb repulsion "switched off." This model is not quite as bad as it initially sounds, and in some ways applies to the alkali metals. The atoms in these materials (Li, Na, K, Rb, and Cs) have low ionization energies, only weakly binding their single valence s electron to the nucleus. We then imagine that when assembled as a crystal each atom contributes one valence electron to the formation of a kind of *electron gas* or *quantum plasma*, having a number density equal to that of the atoms. On average, the nuclear charge will then compensate the free or mobile electronic charge.[2]

6.1 The quantum mechanics of non-interacting electrons in a box

The time-independent Schrödinger equation for a single particle with energy ε in three dimensions is given by[3]

$$\left[\frac{-\hbar^2}{2m}\nabla^2 + V(\mathbf{r})\right]\psi(\mathbf{r}) = \varepsilon\psi(\mathbf{r}). \tag{6.1}$$

As a simple model for electrons in a metal we will replace the effect of all the ions with a position-independent potential, $-V_0$, and, as previously noted, ignore the Coulomb repulsion between the electrons. If the electrons are very tightly bound within the metal, $|V_0|$ is large; if we take the limit $-V_0 \rightarrow -\infty$ and move the zero of energy to the bottom of this confining well, we may write

$$V(\mathbf{r}) = 0 \quad \text{(inside the box)} \tag{6.2a}$$

$$V(\mathbf{r}) = +\infty \quad \text{(outside the box)} \tag{6.2b}$$

[1] The 1945 Nobel Prize in Physics was awarded to Wolfgang Pauli "for the discovery of the Exclusion Principle, also called the Pauli principle."

[2] In Chapter 37 we will discuss another fermion system, liquid ^{3}He.

[3] We will often use the Greek letter ε for the energy of individual electrons and E for the total energy of a many-electron system, such as a solid.

The Physics of Solids. First Edition. J. B. Ketterson.

We must now calculate the energy levels associated with this potential. We write the three-dimensional wave function $\psi(\mathbf{r})$ as a product of three one-dimensional wave functions: $\psi(\mathbf{r}) = \psi_x(x)\psi_y(y)\psi_z(z)$, insert this form into (6.1), and divide the resulting equation by $\psi(\mathbf{r})$ to obtain

$$-\frac{\hbar^2}{2m}\left[\frac{1}{\psi_x}\frac{d^2\psi_x}{dx^2}+\frac{1}{\psi_y}\frac{d^2\psi_y}{dy^2}+\frac{1}{\psi_z}\frac{\partial^2\psi_z}{dz^2}\right]=\varepsilon. \tag{6.3}$$

Since the three terms in the bracket on the left are independent of each other, Eq. (6.3) has a solution only if they are separately equal to constants, k_i^2, and we write the resulting equations as

$$\frac{1}{\psi_x}\frac{d^2\psi_x}{dx^2}=-k_x^2;\ \frac{1}{\psi_y^2}\frac{d^2\psi_y}{dy^2}=-k_y^2;\ \frac{1}{\psi_z}\frac{d^2\psi_z}{dz^2}=-k_z^2, \tag{6.4a,b,c}$$

where, to satisfy (6.3),

$$\frac{\hbar^2}{2m}\left(k_x^2+k_y^2+k_z^2\right)=\varepsilon. \tag{6.5}$$

The solutions of the equations for ψ_x, ψ_y, and ψ_z are all of the same form; that for ψ_x is given by

$$\psi_x = A_x \sin k_x x + B_x \cos k_x x \tag{6.6}$$

and similarly for y and z. Our chosen form (6.2) for the potential requires that $\psi(\mathbf{r})$ vanish at the walls of our box, which for convenience we make a cube with side L. Requiring that ψ_i vanish for x, y, or $z = 0$ requires $B_i = 0$; requiring that ψ_i vanish for x, y, or $z = L$ requires $k_i = l_i\pi/L$, where the l_i are integers. Our wave function then becomes

$$\psi_{l_x l_y l_z}(x,y,z) = A\sin(l_x\pi x/L)\sin(l_y\pi y/L)\sin(l_z\pi z/L),$$

where we have combined the three constants A_i into a single constant; this latter constant is determined by the normalization condition $\int d^3r\psi^*(\mathbf{r})\psi(\mathbf{r}) = 1$. Carrying out the integral, we obtain $A = (2/L)^{3/2}$. The final expressions for the wave functions and energy levels of a particle in a cubical box are then

$$\psi_{l_x l_y l_z}(x,y,z) = \left(\frac{2}{L}\right)^{3/2}\sin\frac{l_x\pi x}{L}\sin\frac{l_y\pi y}{L}\sin\frac{l_z\pi z}{L} \tag{6.7a}$$

and

$$\varepsilon_{l_x l_y l_z} = \frac{\hbar^2}{2m}\left(\frac{\pi}{L}\right)^2\left(l_x^2+l_y^2+l_z^2\right);\ l_i = 1,2,3\ldots \tag{6.7b}$$

There is another boundary condition we can employ for our wave function which is more widely used in solid-state physics. It is called the *periodic boundary condition.* To visualize this boundary condition imagine a one-dimensional solid of length L. If we now "bend" this solid into the form of a circle and *attach the two ends*, which can be carried out in a second dimension, continuity of our wave function requires $\psi(x) = \psi(x + L)$. For the 2d case, which can be connected in a third dimension, we would form a torus (a donut), as shown in Fig. 6.1. In three dimensions this bending process involves a higher-dimensional torus. In all cases the prescription is the same and in 3d we must have

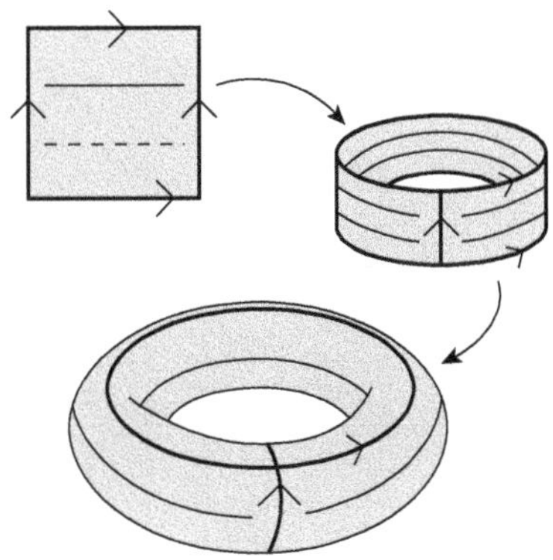

Figure 6.1 *Folding a 2-d sheet into a torus so as to achieve periodic boundary conditions.*

$$\psi(x + L, y, z) = \psi(x, y, z) \tag{6.8a}$$

$$\psi(x, y + L, z) = \psi(x, y, z) \tag{6.8b}$$

$$\psi(x, y, z + L) = \psi(x, y, z). \tag{6.8c}$$

The solutions to the Schrödinger equation may now be written as plane waves,

$$\psi(\mathbf{r}) = \frac{1}{L^{3/2}} e^{i\mathbf{k}\cdot\mathbf{r}}, \tag{6.9a}$$

where $L^{-3/2}$ is the appropriate normalization factor. The periodic boundary condition then takes the form

$$e^{i[k_x(x+L)+k_y y+k_z z]} = e^{i(k_x x+k_y y+k_z z)}$$

and similarly for the y and z coordinates; i.e., we must have $e^{ik_i L} = 1$, which requires $k_i L = 2\pi l_i$ or $k_i = (2\pi/L)\, l_i$. The expression for the energy is then

$$\varepsilon_l = \frac{\hbar^2}{2m}\left(\frac{2\pi}{L}\right)^2 \left[l_x^2 + l_y^2 + l_z^2\right] \tag{6.9b}$$

with

$$l_i = \pm 1, \pm 2 \ldots\ldots$$

Note both positive *and* negative l_i occur, since k_i may have either sign (corresponding to left and right propagating plane waves in a single dimension). The situations represented by Eqs. (6.7a) and (6.9a) correspond, as we have noted, to standing waves and running waves respectively.

The running wave solutions are usually preferred because they are naturally associated with a probability flux, **S**, defined in quantum mechanics by

$$\mathbf{S} = \frac{\hbar}{2im}\left(\psi^* \nabla \psi - \psi \nabla \psi^*\right) = \frac{\hbar \mathbf{k}}{m}. \tag{6.10}$$

The electric current density per electron, $\mathbf{j} = e\mathbf{S}$, following from Eq. (6.10) is then

$$\mathbf{j} = \frac{1}{L^3}\frac{e\hbar}{m}\mathbf{k} = \frac{e\mathbf{v}}{L^3}, \tag{6.11}$$

where $\mathbf{v}$ is its velocity. If we are discussing electrical transport associated with a given electron quantum state having quantum numbers l_x, l_y, l_z, the corresponding wave vectors entering (6.11) are $k_i = 2\pi\, l_i/L$.

At first it might seem strange that these two classes of wave functions—(6.7a) and (6.9a)—having different energy levels—(6.7b) and (6.9b)—could both form a valid basis for describing the free electron metal. We recall again that a macroscopic metallic sample has an enormous number of different quantum states, and that the quantum numbers $k_i = 2\pi\, l_i/L$ may be thought of as essentially continuous at large l_i. The fact that the energy levels involve factors of (π/L) and $(2\pi/L)$ in (6.7b) and (6.9b), and hence *different spacings* between the energy levels, is compensated for by having a *positive* l_i for the former and *positive and negative* l_i for the latter, as discussed previously in Chapter 4 in connection with counting phonon states. Assuming roughly equally occupancy of nearby energy states, we then have the same total amount of energy if we average over a volume element of quantum number space, provided it is large compared to unity. We will primarily work with the running wave solutions, and in the limit of macroscopic matter a sum over all states $\mathbf{k}$ is transformed into an integral via the prescription

$$\sum_{\mathbf{k}} = \frac{V}{(2\pi)^3}\int d^3k \tag{6.12a}$$

(see Eq. (4.25)).[4]

6.2 The non-interacting Fermi gas at absolute zero

We have yet to discuss the effect of the Pauli exclusion principle and the temperature on our free electron gas; we begin with the former effect. The Pauli formulation of the exclusion principle states that only one electron may occupy a given quantum state.[5] Since electrons have a spin angular momentum of 1/2, with projections $\sigma = \pm 1/2$, we will have a total of two spin states for each $\mathbf{k}$ state; Eq. (6.12) is then modified to

$$\sum_{\mathbf{k},\sigma} = \frac{2V}{(2\pi)^3}\int d^3k. \tag{6.12b}$$

If our box contains N electrons, then at absolute zero, where the system must be in its lowest energy or ground state, all quantum states below some energy ε_F,

[4] We can also write this sum as $\sum_l \rightarrow \int dl_x dl_y dl_z$. Writing $dl_i = (L/2\pi)dk_i$ the sum becomes $\sum_{\mathbf{k}} \rightarrow (L/2\pi)^{-3}\int dk_x dk_y dk_z$.

[5] The Dirac formulation states that the total wave function of a system of electrons must be antisymmetric, which includes but is more general than the Pauli formulation. We will not discuss the complications introduced by having an antisymmetric wave function here; we will address this issue in Section 15.1.2.

called the *Fermi energy*, will be filled and all those above will be empty. The relation between ε_F and N is fixed by the "state counting recipe" to be

$$\frac{2V}{(2\pi)^3}\int_0^{\varepsilon_F} d^3k = N. \tag{6.13a}$$

For our isotropic electron gas we may replace d^3k by $4\pi k^2 dk$. Using $\varepsilon = \hbar^2k^2/2m$, (6.13a) becomes

$$\begin{aligned} N &= \frac{2V}{(2\pi)^3}\cdot 4\pi\cdot\left(\frac{2m}{\hbar^2}\right)^{3/2}\int_0^{\varepsilon_F}\frac{1}{2}\varepsilon^{1/2}d\varepsilon \\ &= \frac{V}{3\pi^2}\left(\frac{2m}{\hbar^2}\right)^{3/2}\varepsilon_F^{3/2}. \end{aligned} \tag{6.13b}$$

Equation (6.13b) may also be rewritten as

$$N = \int_0^{\varepsilon_F} N(\varepsilon)d\varepsilon, \tag{6.13c}$$

where

$$N(\varepsilon) = \frac{V}{2\pi^2}\left(\frac{2m}{\hbar^2}\right)^{3/2}\varepsilon^{1/2} \tag{6.13d}$$

is the *number of states per unit energy*, called the density of states for short.[6] We may use (6.13b) to solve for $\varepsilon_F = \varepsilon_F(N)$, as

$$\varepsilon_F = \frac{\hbar^2}{2m}\left(3\pi^2\frac{N}{V}\right)^{2/3}. \tag{6.14a}$$

We may write ε_F as $\hbar^2k_F^2/2m$, where we have introduced a *Fermi wave vector*,

$$k_F = \left(3\pi^2\frac{N}{V}\right)^{1/3}; \tag{6.14b}$$

one may also introduce the *Fermi momentum* through $\varepsilon_F \equiv p_F^2/2m$ with $p_F = \hbar k_F$. It is useful to think of the three components of **k** as making up a new space called **k**-space (or, if we write $\mathbf{p} = \hbar\mathbf{k}$, **p**-space). The occupied states, $p < p_F$, are then separated from the unoccupied states, $p > p_F$, by a spherical surface, called the *Fermi* sphere.[7] For $T > 0$, states above the Fermi sphere will be occupied and states below will be empty, as will be discussed in the next section.

[6] $N(\varepsilon)$ for electrons is analogous to $N(\omega)$ introduced in Chapter 4 for phonons in connection with the Debye theory.

[7] In real metals this surface is generally not spherical, in which case we have a *Fermi surface*.

6.3 Fermi–Dirac statistics

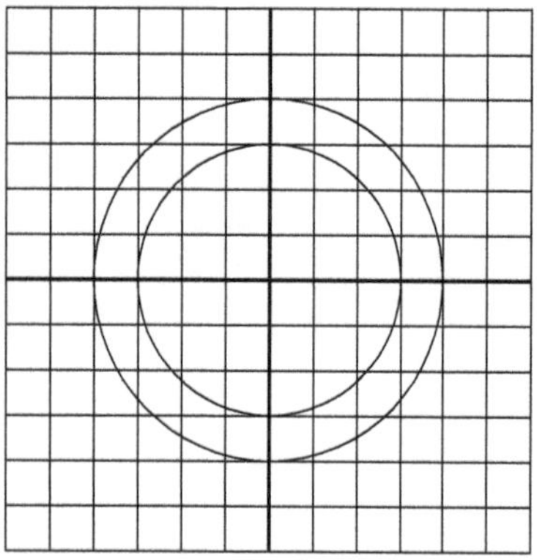

Figure 6.2 *Schematic representation of the number of states, g_E, lying between $l = 3$ and $l = 4$ in quantum number space.*

We now discuss the statistical mechanics of particles that obey the Pauli exclusion principle.[8] When $T \neq 0$ there are a large number of ways to distribute the electrons among the allowed quantum states for a given *total* energy. The strategy is to assume the system will be near a state of maximum probability density. As before, we think of the quantum numbers of the particles, l_x, l_y, l_z (or equivalently k_x, k_y, k_z), as forming some kind of space. We will denote the number of states lying between l and $l + 1$ as g_l; for large l this number is given by the volume of a "shell" in l space with an area of $4\pi l^2$ and one unit thick with $l^2 = l_x^2 + l_y^2 + l_z^2$, where, from Eq. (6.9b), $\varepsilon_l = (\hbar^2/2m) \cdot (2\pi/L)^2\, l^2$. An example of this situation is shown in Fig. 6.2 for a case in two dimensions. Here we estimate $g_l \approx 20$, corresponding to the number of states lying between the circles with radii $l = 3$ and $l = 4$; in the two-dimensional continuum limit we would have $2\pi l$, and using the mean radius gives (1/2)(3 - 4) $(2\pi) \approx 22$.

Let the number of electrons which are *actually occupying* these g_l states be n_l. We ask the following question: how many ways, w_l, are there to distribute these n_l *identical* electrons among the g_l states in such a way as to not have two electrons in the same state (which otherwise would violate the Pauli principle). Temporarily suppressing the subscript l, consider the case $g = 4$, $n = 2$, as shown in the diagram immediately below, where the states are now depicted as cells or boxes:

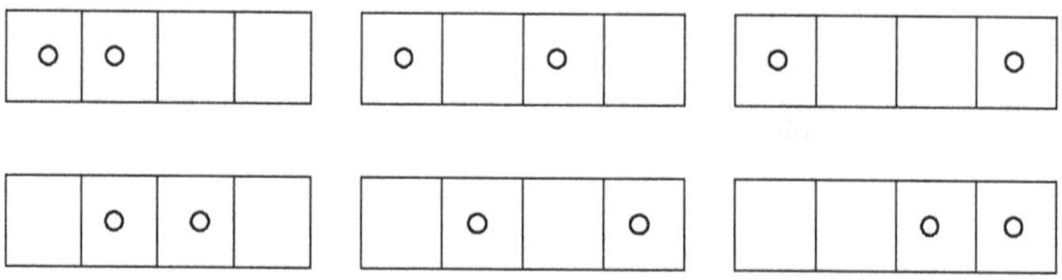

In this case, we see by directly enumerating the cases that there are six different ways to distribute the electrons. Let us develop a formula for $w = w(g,n)$. Now the number of permutations of n *distinguishable* objects among g cells, denoted as ${}_gP_n$ (and often referred to as the permutation group), is

$$ {}_gP_n = g(g-1)(g-2)\ldots\ldots(g-n+1). \tag{6.15} $$

For our $g = 4$, $n = 2$ case (with the latter represented as a circle and a cross), this expression yields 12 and the individual cases are shown in the diagram below.

[8] This topic is discussed in most texts on statistical physics; we will develop it here for the sake of completeness.

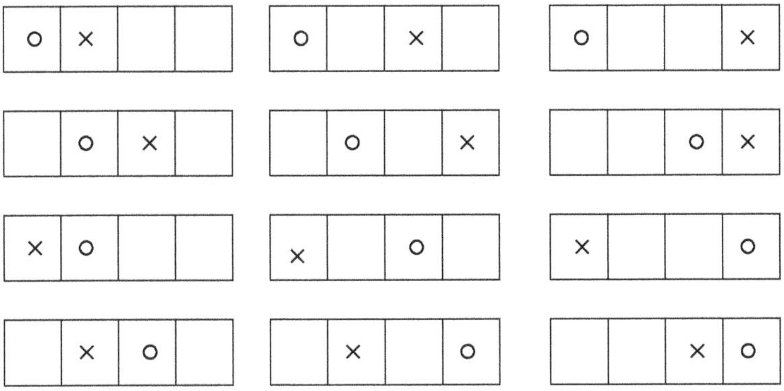

If we are to obtain the result we seek for *indistinguishable* objects, we must clearly divide by the number of permutations of the objects among themselves, n!. The resultant quantity is denoted by ${}_gC_n$ and called the *combination group*,

$$w = {}_gC_n = \frac{g(g-1)(g-2)\ldots(g-n+1)}{n!}$$
$$= \frac{g(g-1)(g-2)\ldots(g-n+1)(g-n)(g-n-1)\ldots 1}{n!(g-n)(g-n-1)\ldots 1} \quad (6.16a)$$

or

$$w(g, n) = \frac{g!}{n!(g-n)!}. \quad (6.16b)$$

Note that (6.16a) is the expression for the *binomial expansion coefficients* of

$$(1+x)^g = 1 + gx + \frac{g(g-1)}{2!}x^2 + \ldots \frac{g(g-1)(g-2)\ldots(g-n+1)}{n!}x^n + \ldots$$

and hence $w = {}_gC_n$ is often called the *binomial distribution.*

To obtain the *total* probability, W, we must form the product of the probabilities w_l for each of the energy shells $\varepsilon(l)$,

$$W = \prod_l w(g_l, n_l) \quad (6.17)$$

(note this quantity is not normalized to unity). To find the configuration(s) with highest probability we must maximize (6.17) with respect to n; i.e., we require $\partial W/\partial n_l = 0$. However this maximization must be carried out subject to *constraints*; those we choose here are that the total *number of particles is fixed* and the total *energy is fixed*; i.e.,

$$\sum_l n_l = N \quad (6.18a)$$

and

$$\sum_l n_l \varepsilon_l = E_{total}. \tag{6.18b}$$

In the absence of these constraints the n_l at different energies would be independent. It turns out to be more convenient to work with $ln W$ rather than W itself:

$$lnW = ln\left(\prod_l w\right) = \sum_l ln(w). \tag{6.19}$$

To evaluate (6.19) we use Stirling's approximation

$$ln\,(n!) \cong n\,ln(n) - n; \quad n \to \infty,$$

which yields

$$\begin{aligned} lnW = \sum_l \{&[(g_l\, ln(g_l) - g_l)] - [(n_l\, ln(n_l) - n_l)] \\ &- [(g_l - n_l)\, ln(g_l - n_l) - (g_l - n_l)]\}. \end{aligned} \tag{6.20}$$

To find the extrema of a multivariable function subject to constraints one must use Lagrange's method of undetermined multipliers. The method involves setting the sum of the derivative of the function (for which we seek an extremum) *and* the derivative of each of the constraints, multiplied by an undetermined parameter (called a multiplier) to zero; these parameters are fixed at the end of the calculation by demanding that the constraints (or equivalent ones) be satisfied. In the present case this prescription reads

$$\frac{d}{dn_l}\left[lnW + \alpha\left(N - \sum_l n_l\right) + \beta\left(E_{total} - \sum_l n_l\,\varepsilon_l\right)\right] = 0, \tag{6.21}$$

where α and β are the two multipliers. From Eqs. (6.18a,b) and (6.20) we have

$$-ln(n_l) + ln(g_l - n_l) - \alpha - \beta\varepsilon_l = 0$$

or

$$ln\left(\frac{g_l - n_l}{n_l}\right) = \alpha + \beta\varepsilon_l$$

and

$$\frac{g_l - n_l}{n_l} = e^{\alpha+\beta\varepsilon_l};$$

solving for n_l we have

$$n_l = \frac{g_l}{e^{\alpha+\beta\varepsilon_l} + 1}; \tag{6.22}$$

henceforth we drop the index l, concerning which we will expand on further.[9]

[9] In real materials, ε is not a spherically symmetric function of the magnitude l, but rather depends on l_x, l_y, l_z, or equivalently k_x, k_y, k_z; i.e., $\varepsilon = \varepsilon(\mathbf{k})$.

We must now determine the two multipliers, α and β. We redefine the first of these as $\alpha \equiv -\mu/k_B T$, where μ is another parameter, called the *chemical potential*, about which we will say more later. We may obtain β by requiring that $n(\varepsilon)$ go over into the Boltzmann distribution at high temperatures (where the particles are distributed widely in energy and hence the probability of two particles occupying the same cell is small, and with it the effects of the exclusion principle). We must then have

$$n = g e^{-\beta(\varepsilon-\mu)} \tag{6.23}$$

and we immediately identify $\beta = (k_B T)^{-1}$ as we earlier defined it.[10]

We can rewrite Eq. (6.22) as

$$n = g f(\varepsilon), \tag{6.24a}$$

where

$$f(\varepsilon) = \frac{1}{e^{(\varepsilon-\mu)/k_B T} + 1} \tag{6.24b}$$

is referred to as the *Fermi–Dirac distribution function.* The factor g in (6.24a) is usually called the *degeneracy*. If we are performing averages of quantities that only involve the energy (examples of which we take up in the next few sections), then we can take $g = N(\varepsilon)$, the density of states. However, there are cases where we must consider all quantum states separately, as numbered by l_x, l_y, l_z or equivalently k_x, k_y, k_z (together with the two spin states); then it is taken to be one and the average involves a sum over all such states.

Note that (6.24b) differs from the Planck or Bose–Einstein distribution function used for photons and phonons in that it has +1 (rather than −1) in the denominator. It turns out that all particles fall into one of two classes: fermions (following the Fermi–Dirac distribution), or bosons (following the Bose–Einstein distribution), with additional properties that are summarized in Table 6.1.

Figure (6.3) shows a plot of $f(\varepsilon)$ for the cases $T = 0$ and $T > 0$. At $T = 0$, $f(\varepsilon)$ is simply a step function (the factor $\exp[(\varepsilon-\mu)/k_B T]$ goes from 0 to ∞ as $\varepsilon-\mu$ passes from negative to positive). For $0 < k_B T \ll \mu$ the "step" in the distribution acquires a width of order $k_B T$. Physically the Fermi surface is then no longer

Table 6.1 *Particle Properties vs. spin.*

Particle class	Particle spin (in units of $\hbar$)	Wave function symmetry (on exchanging a pair of particles)
Fermi particles (fermions)	half integral	Antisymmetric
Bose particles (bosons)	Integral	Symmetric

[10] In Eq. (6.22) one might ask whether at large T one can have $\alpha(T) \equiv -\beta\mu(T) < \beta\varepsilon$, or equivalently $\varepsilon < \mu$, for a range of energies, thereby rendering the Boltzmann limit invalid for those energies. It will turn out that as the temperature increases, $\mu(T)$ goes from positive to negative and becomes increasingly so as the temperature increases, allowing (6.23) to be a valid limit for all energies.

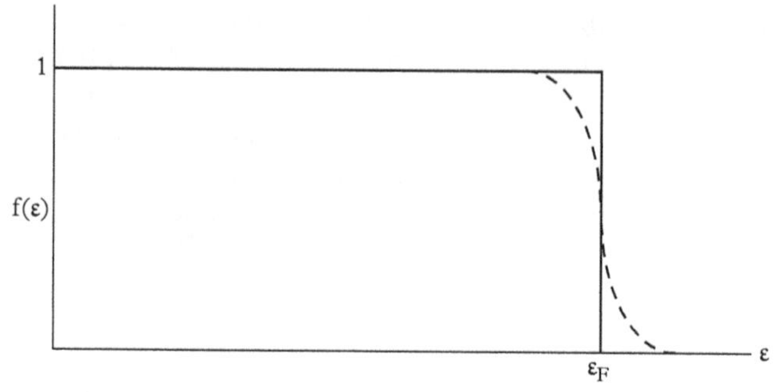

Figure 6.3 *Sketch of the Fermi–Dirac distribution function for absolute zero and some temperature* $k_BT \ll \varepsilon_F$. *The region over which the distribution is broadened is of order* k_BT.

sharp, there being unoccupied states below this surface (referred to as hole-like excitations or holes) and occupied states above (particle-like excitations or particles). From this it is clear that at $T = 0$ the chemical potential and the Fermi energy are the same.

For an arbitrary temperature the quantity μ is determined from Eq. (6.18a),

$$\int \frac{d\varepsilon\, N(\varepsilon)}{e^{(\varepsilon-\mu)/k_BT}+1} = N. \tag{6.25}$$

Evaluation of this integral yields $N = N(\mu, T)$, which may be solved, numerically if necessary, to find $\mu = \mu(N,T)$.

6.4 Evaluating integrals involving the Fermi distribution function

For an arbitrary temperature, averages involving the Fermi distribution function must usually be performed numerically. However, for some commonly occurring cases where $k_BT/\mu \ll 1$, we may obtain analytic results as an expansion in this parameter. Let $F(\varepsilon)$ be some function for which we wish to compute averages. We must then evaluate integrals, I, having the form

$$I = \int_0^\infty F(\varepsilon)f(\varepsilon)d\varepsilon. \tag{6.26}$$

As an example, $F(\varepsilon)$ could be the density of states, $N(\varepsilon)$, in Eq. (6.26). We begin our analysis by introducing a new variable, $z = (\varepsilon-\mu)/k_BT$, and rewrite (6.26) as

$$I = \int_{-\mu/k_BT}^{\infty} \frac{F(\mu + k_BTz)}{e^z+1}(k_BTdz)$$

$$= k_BT \int_0^{\mu/k_BT} \frac{F(\mu - k_BTz)}{e^{-z}+1}dz + k_BT\int_0^\infty \frac{F(\mu + k_BTz)}{e^z+1}dz.$$

We make the substitution $\frac{1}{e^{-z}+1} = 1 - \frac{1}{e^{z}+1}$ and obtain

$$I = \int_0^{\mu} F(\varepsilon)d\varepsilon - k_BT \int_0^{\mu/k_BT} \frac{F(\mu - k_BTz)}{e^z+1}dz + k_BT \int_0^{\infty} \frac{F(\mu + k_BTz)}{e^z+1}dz.$$

Since $\mu/k_BT \gg 1$ at low temperatures, we may extend the upper limit in the second integral to ∞ (this neglects exponentially small terms of the form $e^{-\mu/k_BT}$ and is referred to as an asymptotic approximation). We then have

$$I = \int_0^{\mu} F(\varepsilon)d\varepsilon + k_BT \int_0^{\infty} \frac{F(\mu + k_BTz) - F(\mu - k_BTz)}{e^z+1}dz. \tag{6.27}$$

The next step is to expand the argument of the second integral of (6.27) in a Taylor series using

$$F(\mu - k_BTz) = F(\mu) - k_BTF'(\mu)z + 1/2(k_BT)^2F''(\mu)z^2 + \ldots$$

and note that the even terms cancel in the integrand[11]. Hence we have (to fourth order)

$$I = \int_0^{\mu} F(\varepsilon)d\varepsilon + 2\,(k_BT)^2\,F'(\mu) \int_0^{\infty} \frac{zdz}{e^z+1} + \frac{1}{3}\,(k_BT)^4\,F'''(\mu) \int_0^{\infty} \frac{z^3dz}{e^z+1} + \ldots.$$

The second and third integrals in this expression are given by $\pi^2/12$ and $7\pi^4/120$, respectively, and our final expression is

$$I = \int_0^{\mu} F(\varepsilon)d\varepsilon + \frac{\pi^2}{6}(k_BT)^2F'(\mu) + \frac{7\pi^4}{360}(k_BT)^4F'''(\mu) + \ldots. \tag{6.28}$$

Using Eq. (6.28) we may compute many of the statistical averages that occur naturally, and in the next few sections we will consider some important examples.

6.5 The temperature dependence of the chemical potential

We return to Eq. (6.25; using (6.28), and restricting to second order in T, we have

$$N = \int_0^{\infty} \mathcal{N}(\varepsilon)f(\varepsilon)d\varepsilon = \int_0^{\mu(T)} \mathcal{N}(\varepsilon)d\varepsilon + \frac{\pi^2}{6}(k_BT)^2\mathcal{N}'(\mu), \tag{6.29}$$

[11] Note that if we write $F(\mu + k_BTz) = F[(\mu(1 + (k_BT/\mu)z)]$, we see that we are effectively expanding in the quantity $(k_BT/\mu)z \ll 1$. When this variable approaches zero, the term in the denominator would be $e^{\mu/k_BT} \gg 1$ and the remaining contribution to our integral is therefore exponentially small.

where $\mu(T)$ is the chemical potential. Incorporating the $T=0$ form, $\int_0^{\mu(0)} N(\varepsilon)d\varepsilon = N$, we may rewrite Eq. (6.29) as

$$\begin{aligned}\int_0^{\mu(T)} N(\varepsilon)d\varepsilon &= \int_0^{\mu(0)} N(\varepsilon)d\varepsilon + \int_{\mu(0)}^{\mu(T)} N(\varepsilon)d\varepsilon, \\ &= N + N(\mu(0))\,[\mu(T) - \mu(0)]\end{aligned} \tag{6.30}$$

where in the second term we have assumed that the density of states is constant in the small interval between $\mu(T)$ and $\mu(T=0)$.

Combining (6.29) and (6.30) we have

$$\mu(T) = \mu(0) - \frac{\pi^2}{6}(k_B T)^2 \frac{N'(\mu)}{N(\mu)}. \tag{6.31}$$

For free electrons, $N(\varepsilon) \sim \varepsilon^{1/2}$ and we may rewrite this as

$$\mu(T) = \mu(0)\left[1 - \frac{\pi^2}{12}\left(\frac{k_B T}{\mu(0)}\right)^2 + \ldots\right]. \tag{6.32}$$

6.6 Energy of an ideal Fermi gas as a function of temperature

The energy of a Fermi gas is given by

$$E_{total}(T) = \int_0^{\infty} \frac{\varepsilon N(\varepsilon)d\varepsilon}{e^{(\varepsilon-\mu)/k_B T} + 1}. \tag{6.33}$$

Using Eq. (6.28) we may write (6.33) as

$$E_{total}(T) = \int_0^{\mu(T)} \varepsilon N(\varepsilon)d\varepsilon + \frac{\pi^2}{6}(k_B T)^2 \frac{d}{d\varepsilon}[\varepsilon N(\varepsilon)]\,|_{\varepsilon=\mu} + \ldots. \tag{6.34}$$

We split the integral in (6.34) into two parts:

$$\int_0^{\mu(T)} \varepsilon N(\varepsilon)d\varepsilon = \int_0^{\mu(0)} \varepsilon N(\varepsilon)d\varepsilon + \int_{\mu(0)}^{\mu(T)} \varepsilon N(\varepsilon)d\varepsilon.$$

The first integral is simply the total energy of a Fermi gas at $T = 0$, $E_{total}(0)$. Using our expression for $N(\varepsilon)$ (given after (6.13d)), we may write it as

$$
\begin{aligned}
E_{total}(0) &= \frac{V}{2\pi^2}\left(\frac{2m}{\hbar^2}\right)^{3/2}\int_0^{\mu(0)} \varepsilon^{3/2} d\varepsilon \\
&= \frac{V}{2\pi^2}\left(\frac{2m}{\hbar^2}\right)^{3/2}\frac{2}{5}[\mu(0)]^{5/2} \\
&= \frac{3}{5}\mu(0)\frac{V}{3\pi^2}\left(\frac{2m}{\hbar^2}\right)^{3/2}[\mu(0)]^{3/2} \\
&= \frac{3}{5}N\mu(0).
\end{aligned}
\tag{6.35}
$$

Since the shift in the chemical potential is small, we may approximate the second integral as

$$\mu(0)N(\mu(0))\,[\mu(T) - \mu(0)] = -N((\mu(0))\frac{\pi^2}{12}(k_BT)^2,$$

where we have used (6.32) in obtaining the form on the right. Noting that $N(\varepsilon) \sim \varepsilon^{1/2}$, it follows that $d\,[\varepsilon N(\varepsilon)]\,/d\varepsilon = 3N(\varepsilon)/2$ and hence the second term in (6.34) is $(\pi^2/4) \cdot (k_BT)^2 \cdot N[(\mu(0))]$. Combining the above we have

$$E_{Total}(T) = E_{Total}(0) + \frac{\pi^2}{6}N(\mu(0))(k_BT)^2. \tag{6.36}$$

The heat capacity at constant volume, $C_V = (\partial E/\partial T)_V$, is

$$
\begin{aligned}
C_V(T) &= \frac{\pi^2}{3}k_B^2 N(\mu(0))T, \\
&= \gamma T
\end{aligned}
\tag{6.37a}
$$

where we have defined the coefficient

$$\gamma \equiv \frac{\pi^2}{3}k_B^2 N(\mu(0)). \tag{6.37b}$$

We see that the specific heat is proportional to temperature, as mentioned at the beginning of this chapter. This result is due to Sommerfeld. Table 6.2 lists experimental values (as determined from the measured heat capacity) and theoretical values (based on assuming the free electron mass) for those metals having one electron per atom; also listed is a ratio m^*/m, where m^* is a so-called effective mass; this ratio measures the amount by which the free electron mass m would have to be increased in order to agree with the experimental values.[12]

The next term in the expansion of the energy of an electron gas as a function of temperature goes as T^4, implying a contribution to the electronic heat capacity

[12] As will be discussed in Chapter 10, metals with two or more electrons per atom have Fermi surfaces that differ markedly from the spherical shape of the free electron model. Even with the restriction to one electron per atom, the Fermi surfaces of the noble metals and Li, although largely spherical in shape, have so-called necks that alter the basic topology. The Fermi surface of the noble metals is discussed in Chapter 20. An additional complication arising in real metals is that the interaction of electrons with the lattice can enhance the mass, as will be discussed in Chapter 21.

Table 6.2 *Sommerfeld constant* (γ) *for metals with one electron per atom (mJ/molK²).*

	$\gamma_{observed}$	$\gamma_{calculated}$	m^*/m
Li	1.63	0.749	2.18
Na	1.38	1.094	1.26
K	2.08	1.668	1.25
Rb	2.41	1.911	1.26
Cs	3.20	2.238	1.43
Cu	0.695	0.505	1.38
Ag	0.646	0.645	1.00
Au	0.729	0.642	1.14

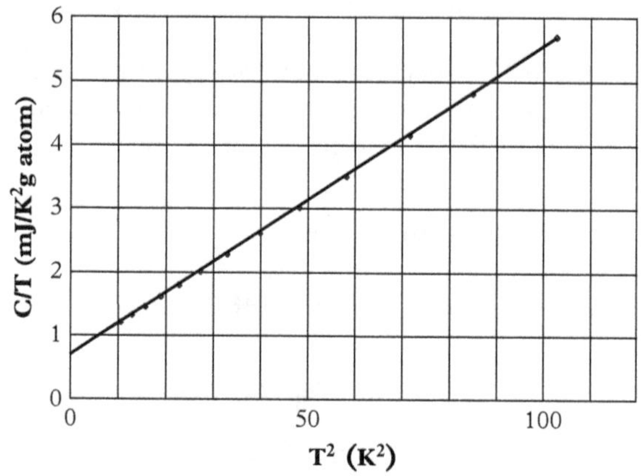

Figure 6.4 *A plot of C_P/T vs. T^2 for copper; the intercept on the vertical axis gives the coefficient γ. (After Martin (1960)).*

proportional to T^3; however it is much smaller than the leading T^3 term from the lattice specific heat and will be ignored.

We can extract the electronic and lattice contributions to the low temperature heat capacity by plotting the ratio C/T vs. T^2, as shown in Fig. 6.4 for the case of copper.[13] The intercept yields the Sommerfeld constant γ, while the coefficient of the T^3 term in the lattice heat capacity follows from the slope.

6.7 The paramagnetic susceptibility of a Fermi gas

If we apply a constant magnetic field to an electron gas, the energies of the spin up and spin down states will be shifted according to $\varepsilon_{\pm} = \varepsilon \pm \mu_B H$, where μ_B is the

[13] The specific heat of copper from 20° to 300 deg, K, D.L. Martin, *Can. J. Phys.* 38, 17 (1960).

Bohr magnetron (the magnetic moment operator is $g\mu_B\hat{s}$, which has eigenvalues $\pm\mu_B$ for $g = 2$ electrons). This effect can be incorporated by shifting the energy in the density of states by $\pm\mu_B H$. The magnetization, which is the difference of these two populations, is then

$$M(H) = \mu_B \int \left[\frac{1}{2}N(\varepsilon + \mu_B H) - \frac{1}{2}N(\varepsilon - \mu_B H)\right] f(\varepsilon)d\varepsilon, \tag{6.38}$$

where the factor 1/2 arises from our inclusion of both spin states in our earlier definition of $N(\varepsilon)$. Expanding $N(\varepsilon)$ to first order in $\mu_B H$ gives

$$M(H) = \mu_B^2 H \int d\varepsilon N'(E)f(\varepsilon). \tag{6.39}$$

At low temperature we may write $f(\varepsilon) = \theta(\mu - \varepsilon)$, where θ is the theta or step function. Integrating (6.39) by parts we have

$$\begin{aligned} M(H) &= \mu_B^2 H \int d\varepsilon N'(\varepsilon)\delta(\mu - \varepsilon) \\ &= \mu_B^2 H N\,[(\mu(0)]; \end{aligned} \tag{6.40}$$

the differential magnetic susceptibility is defined by $\chi = \partial M/\partial H$ or

$$\chi_{\text{Pauli}} = \mu_B^2 N\,[(\mu(0))], \tag{6.41a}$$

which is independent of temperature; a result due to Pauli.

There is another contribution to the magnetic susceptibility due to Landau which arises from the orbital motion of the electrons, which is given by

$$\chi_{\text{Landau}} = -\frac{1}{3}\chi_{\text{Pauli}};$$

note that this has the opposite sign to the Pauli contribution and is hence diamagnetic. The derivation of this form is somewhat more complicated and will be taken up in Chapter 20 and Appendix 42A.

6.8 Qualitative discussion of the behavior of a Fermi gas

In our earlier discussion we noted that the step in the Fermi function develops a "width" of approximately $k_B T$ at finite temperatures, implying that electrons within this interval about the Fermi surface are excited. However, the *fraction* of the electrons excited relative to the remaining electrons is only of order $k_B T/\varepsilon_F$ or T/T_F, where $T_F \equiv \varepsilon_F/k_B$. If we apply this factor to N in various classical expressions we qualitatively recover the features discussed earlier. Hence, if the classical heat capacity is $C_V = (3/2) \cdot Nk_B$, then the Fermi gas result is approximately $C_V \approx (3/2) \cdot Nk_B \cdot (T/T_F) \approx (3/2) \cdot k_B^2 \cdot (NT/\varepsilon_F)$. From Eqs. (6.13b) and (6.13d), $N/\varepsilon_F = (2/3)\mathcal{N}(\varepsilon_F)$, yielding $C_V \approx k_B^2 \mathcal{N}(\varepsilon_F)T$ versus our result, $C_V = (\pi^2/3)\, k_B^2 \mathcal{N}(\varepsilon_F)T$. Similarly, the magnetic susceptibility is classically given by $\chi = C/T$, where $C = \mu_B^2 N/3k_B$ is Curie's constant (see Section 5.2); if we multiply by T/T_F we obtain $\chi \approx (\mu_B^2/3)\cdot(N/E_F) \approx (2/9)\,\mu_B^2 \mathcal{N}(\varepsilon_F)$. In both cases we obtain the correct temperature dependence and the involvement of the density of states, although the numerical factor is correct only in order of magnitude.

Problems

6.1. Calculate the next term in the expansion given by Eq. (6.28); what would be the correction to the specific heat resulting from this term?

6.2. Suppose the kinetic energy is given by the expression

$$\varepsilon(\mathbf{k}) = \frac{\hbar^2}{2}\left[\frac{k_x^2}{m_1} + \frac{k_y^2}{m_2} + \frac{k_z^2}{m_3}\right],$$

where the m_i are parameters called *effective masses*; calculate the density of states, $\mathcal{N}(\varepsilon_F)$. What expression for the specific heat would we have in this case?

Elementary Theories of Crystal Bonding

7

7.1 Classification of bonds

A fully first principles calculation of the binding energy of a crystal is exceedingly difficult (basically impossible). Formally, the binding energy would be the lowest eigenvalue of the many-electron Schrödinger equation for all the electrons and nuclei of the crystal. And this leaves out the effects of finite temperatures. Therefore in estimating the binding energy, approximations are always introduced and the accuracy of these approximations is sometimes difficult to access. Agreement with experiment is the final test. In this section we will discuss several highly simplified models that have been used historically to discuss crystal bonding.

It is customary to divide materials into categories according to the qualitative nature of the chemical or physical bonding existing among the atoms. Table 7.1 lists various kinds of bonding.[1] The "dividing line" between these categories is not always sharp. For instance, bonding in transition metals such as tungsten, may be thought of as both metallic and covalent. In the next few sections we will discuss simple models of the Van der Waals, ionic, covalent, and metallic bonds. More accurate methods will be taken up later.

Table 7.1 *Bond classifications.*

Bond type	Example
Van der Waals	Rare gas solids (Ne, Ar, . . .), organic materials
Ionic	NaCl
Covalent	Diamond, silicon
Metallic	Sodium

7.2 The van der Waals bond: rare gas solids

For large separations, rare gas (and other) atoms interact via the attractive van der Waals force, which arises from quantum mechanical fluctuations of the dipole moments on the atoms; the potential has the form

$$V(r) = -\frac{A}{r^6}, \qquad (7.1)$$

where A is a positive constant (to be computed or obtained experimentally). As the atoms approach each other and the wave functions begin to overlap, covalent and other factors become involved and (7.1) is no longer a good representation of the potential. Rare gas atoms, which have filled atomic s and p shells, are particularly inert chemically and (7.1) forms a useful starting point in developing an interatomic potential. At short range, a repulsive force sets in, which is usually argued to arise from the Pauli principle (associated with the overlap of filled atomic

[1] Chemists also introduce the so-called hydrogen bond which we do not discuss here.

The Physics of Solids. First Edition. J. B. Ketterson.

shells). The functional form associated with this repulsion has been represented in a variety of ways. Because the next order of perturbation theory, that in lowest order yields (7.1), is repulsive and has the form $V(r) = B/r^{12}$, it is common to represent the interatomic potential by the form

$$V(r) = -\frac{A}{r^6} + \frac{B}{r^{12}} = 4\varepsilon\left[\left(\frac{\sigma}{r}\right)^{12} - \left(\frac{\sigma}{r}\right)^{6}\right], \tag{7.2}$$

which is referred to as the Lennard-Jones (or "6 – 12") potential; this potential is written in the dimensionless form $y = 4\left[x^{-12} - x^{-6}\right]$, where $x = \sigma/r$ and $y = V(x)/\varepsilon$ is plotted in Fig. 7.1. Clearly the potential changes sign for $x = 1 (r = \sigma)$, a distance which is sometimes called the *hard core radius*. The minimum occurs where

$$\frac{\partial V}{\partial r} = 0 = \frac{4\varepsilon}{\sigma}\left[-12(\sigma/r)^{13} + 6(\sigma/r)^{7}\right];$$

thus

$$(\sigma/r)^6 = \frac{1}{2},$$

corresponding to $r/\sigma = 1.122$. The potential at the minimum is therefore $-\varepsilon$, which accounts for our inclusion of the factor 4 in Eq. (7.2).

Consider now a lattice of atoms interacting via the Lennard-Jones potential. The rare gas solids crystallize in the face centered cubic structure—see

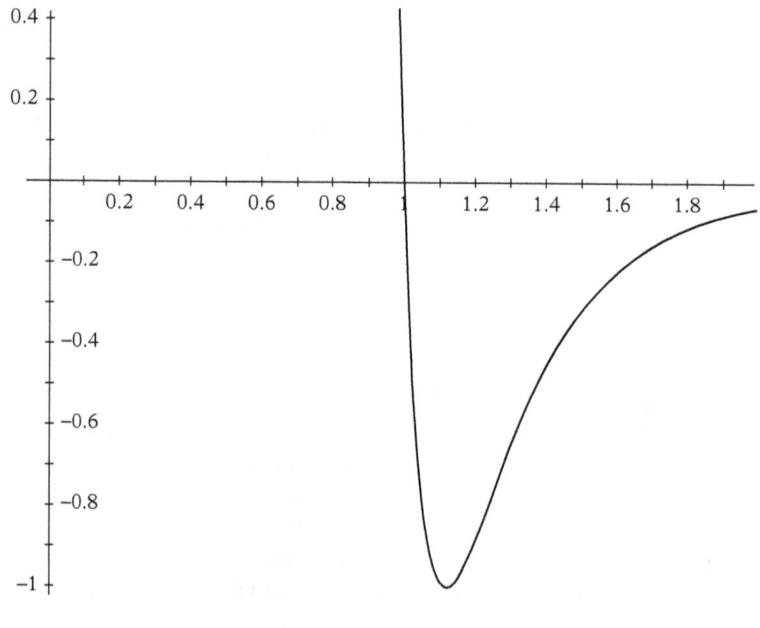

Figure 7.1 *The Lennard-Jones potential plotted in dimensionless form.*

Fig. 8.11(a). We examine the total potential energy by (i) considering a single atom interacting with the remaining atoms, (ii) multiplying by N to account for the potential felt by the totality of atoms, and (iii) dividing by two to remove the effect of double counting produced by this procedure:

$$U_{\text{total}} = \frac{1}{2} \cdot N \cdot 4\varepsilon \sum_{j \neq i} \left[\left(\frac{\sigma}{r_{ij}} \right)^{12} - \left(\frac{\sigma}{r_{ij}} \right)^{6} \right] \tag{7.3}$$

where R is the nearest-neighbor distance and the p_{ij} are the interatomic separations in units of R (i.e., $p_{12} = 1$). The sums must be evaluated numerically and for the face centered cubic lattice they are

$$\sum_{j \neq i} \left(\frac{1}{p_{ij}} \right)^{12} = 12.131, \tag{7.4a}$$

$$\sum_{j \neq i} \left(\frac{1}{p_{ij}} \right)^{6} = 14.454; \tag{7.4b}$$

note that we expect a result of order 12 since the face centered cubic lattice has 12 nearest neighbors and the range of the force is quite small.

The minimum total energy follows from setting the derivative of Eq. (7.3) to zero,

$$\frac{\partial U_{\text{total}}}{\partial R} = 0$$
$$= \frac{N}{2} 4\varepsilon \left[-\frac{12(12.13)}{R} \left(\frac{\sigma}{R} \right)^{12} + \frac{6(14.45)}{R} \left(\frac{\sigma}{R} \right)^{6} \right], \tag{7.5a}$$

which results in a value $R/\sigma = 1.09$. Table 7.2 lists experimental values of R/σ for the rare gases at $T = 0$ (where lattice vibrations, other than quantum zero-point kinetic energy, do not contribute to the energy); helium is a special case in that it is a liquid at $T = 0$, due to the comparatively large zero-point motion, and the above theory does not apply.[2]

We expect the zero-point energy to have the largest effect for the element Ne for the entries in Table. 7.2, since it is the lightest. To obtain the total energy we substitute $R/\sigma = 1.09$ into Eq. (7.3) and obtain

$$U_{\text{total}} = 2N\varepsilon \left[\frac{12.13}{(1.09)^{12}} - \frac{14.45}{(1.09)^{6}} \right] = -N(2.15)4\varepsilon. \tag{7.5b}$$

Note that we have not listed the total energies in constructing Table 7.2; this is simply because this quantity is less accurately known relative to the interatomic spacing (which can be obtained with high precision from X-ray crystallographic measurements).

Table 7.2 *Observed nearest-neighbor radii in rare gas solids.*

	Ne	Ar	Kr	Xe
R/σ	1.14	1.11	1.10	1.09

[2] Liquid helium is discussed in Chapter 36.

Another quantity that can be measured with high precision is the bulk modulus, $B \equiv -V(\partial P/\partial V)$, where P and V are the pressure and volume respectively. But $P = -(\partial E/\partial V)$ and assuming the ground state (T = 0) energy arises entirely from the potential energy, we may take $E = U_{total}$ or

$$B = V\frac{\partial^2 U_{total}}{\partial V^2}. \tag{7.6}$$

Now the face centered cubic lattice of the rare gas atoms has four atoms per unit cell. If we let the edge of the cube be a, then each atom occupies the volume $a^3/4$. The nearest-neighbor distance is $R = \left[(a/2)^2 + (a/2)^2\right]^{1/2} = a/\sqrt{2}$ or $a = \sqrt{2}R$. The volume is then given by

$$V = N \cdot \frac{1}{4} \cdot a^3 = N \cdot \frac{1}{4} \cdot \left(\sqrt{2}R\right)^3 = \frac{NR^3}{\sqrt{2}},$$

or $R^6 = 2V^2/N^2$ and $R^{12} = 4V^4/N^4$. Inserting these latter expressions into (7.3) and using (7.4a,b) yields

$$U_{total} = 2N\varepsilon\left[12{\cdot}13\frac{N^4\sigma^{12}}{4V^4} - 14.45\frac{N^2\sigma^6}{2V^2}\right] = \frac{b_{12}}{V^4} - \frac{b_6}{V^2}, \tag{7.7}$$

where $b_{12} \equiv 12{\cdot}13N^5\sigma^{12}(\varepsilon/2); b_6 \equiv 14{\cdot}45N^3\sigma^6\varepsilon$. The equilibrium volume, V_{eq}, follows from

$$\frac{\partial U_{total}}{\partial V} = 0 = -\frac{4b_{12}}{V^5} + \frac{2b_6}{V^3} \tag{7.8}$$

or

$$V_{eq} = \sqrt{\frac{2b_{12}}{b_6}}. \tag{7.9}$$

The bulk modulus follows from differentiating (7.8) a second time,

$$B = V\frac{\partial^2 U_{total}}{\partial V^2} = \frac{20b_{12}}{V^5} - \frac{6b_6}{V^3}; \tag{7.10}$$

evaluating (7.10) at the equilibrium volume (7.9) gives

$$B = \sqrt{2}\frac{b_6^{5/2}}{b_{12}^{3/2}}. \tag{7.11}$$

Data on the parameters ε and σ which describe our Lennard-Jones potential can be obtained from the virial coefficients of the gas phase (the parameters α_i in an equation of state of the form $PV/RT = 1 + \alpha_2/V^2 + \cdots$ are called virial coefficients). The experimental values of V_{eq} and B obtained with the above model are in good agreement with these data, especially for the heavier rare gas atoms where the effects of zero-point motion are smaller.

7.3 The ionic bond: alkali halide solids

Perhaps the next simplest kind of bonding (after van der Waals bonding) is the ionic bond. A typical example is NaCl, which crystallizes in the cubic rock salt structure which as shown in Fig. 7.2, here it is depicted with Cl atoms on the corners and on the face centers and Na on the edge midpoints and in the center of the cube. This structure may be viewed as two interpenetrating face centered cubic lattices spaced by 1/2 of the cube diagonal. The Na atoms give up their 3s valance electron (i.e., they ionize), leaving a $2s^2 2p^6$ neon-like shell with a net charge $+|e|$, and donate that electron to the Cl atom to form a closed argon-like $3s^2 3p^6$ shell, which then has net charge $-|e|$, the tendency to do the latter being a quantum mechanical phenomenon.[3,4] The filled argon-like shells on the Cl atoms have a significantly larger radius than the neon-like shells of the Na atoms; the former essentially form a close packed face centered cubic lattice, the latter fitting into the interstices.

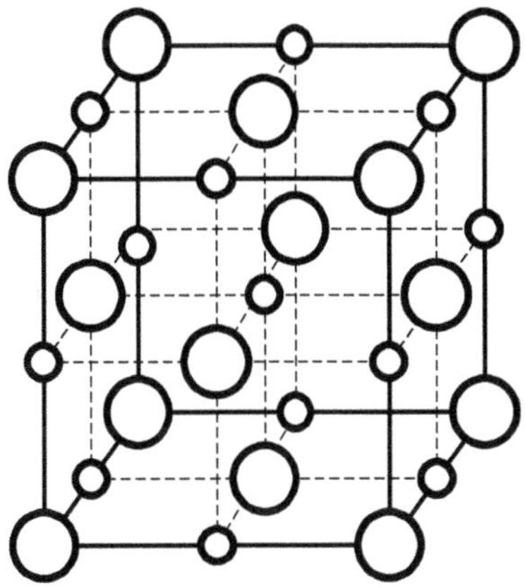

Figure 7.2 *The sodium chloride lattice.*

We will utilize the following simplified classical picture of the bonding in this and similar materials. (i) All Na–Cl atom pairs attract and all Na–Na and Cl–Cl atom pairs repel, according to the Coulomb potential $\pm q^2/r_{ij}$, where r_{ij} is the interatomic separation; and (ii) the nearest-neighbor Na–Cl pairs repel each other at short distances via a potential which we model as $\lambda e^{-r/\rho}$ where λ and ρ are parameters and r is the nearest-neighbor Na–Cl pair spacing. The total potential energy will then be written as

$$U_{total} = \sum_{i>j} \frac{\pm q^2}{r_{ij}} + Nz\lambda e^{-r/\rho}. \tag{7.12}$$

The first term in (7.12) is simply the Coulomb energy, previously discussed, while N is the total number of NaCl molecules (i.e., Na–Cl atom pairs), and z is the number of nearest neighbors (six for the NaCl lattice). The Coulomb term may be rewritten as

$$\sum_{i>j} U_{ij} = -N\frac{\alpha q^2}{R}; \tag{7.13}$$

here we have introduced the so-called *Madelung constant*, α, defined by

$$\alpha = \sum_j \frac{(\pm)}{p_{ij}}. \tag{7.14}$$

The summation in (7.14) assumes that the negative Cl ion is the origin, $r_{ij} \equiv p_{ij} R$, as before, the + applies to positive Na ions, and the – to the negative Cl ions. The equilibrium spacing follows from

$$\frac{\partial U_{total}}{\partial R} = 0 = N\left[\frac{\alpha q^2}{R^2} - \frac{z\lambda}{\rho}e^{-R/\rho}\right]$$

[3] When considering an isolated atom, the ionization energy is that energy required to remove an electron. As the nuclear charge increases and a given atomic shell fills, the ionization energies steadily increase, being maximal for the inert gases (Ne, A, Kr, ...), after which they abruptly drop as the next shell begins to fill. Electrons in the same shell lie at the same average distance from the nucleus and hence *all* feel the increasing nuclear charge as the shell fills; at the same time they correlate their motions to keep them apart, thereby minimizing their collective repulsive energy. As a result of these effects, electrons in alkali atoms (Li, Na, K, ...) are weakly bound and easily lose their outer s-electron, while halogens (F, Cl, Br, ...), which have the next to highest ionization energies, readily accept an additional electron to fill the p shell.

[4] The electron affinity is the energy gain on binding an additional electron and for Cl is 3.61 eV. This can be compared with the ionization energy of Na, 5.13 eV. Hence electron transfer between the widely separated atoms would not occur. However, when the atoms are in close proximity, the cost in Coulomb energy for the electron to transfer (rather than ionize) at distances where they bond, is smaller and there is a net energy gain on transferring the Na s-electron to fill the p shell of the Cl atom.

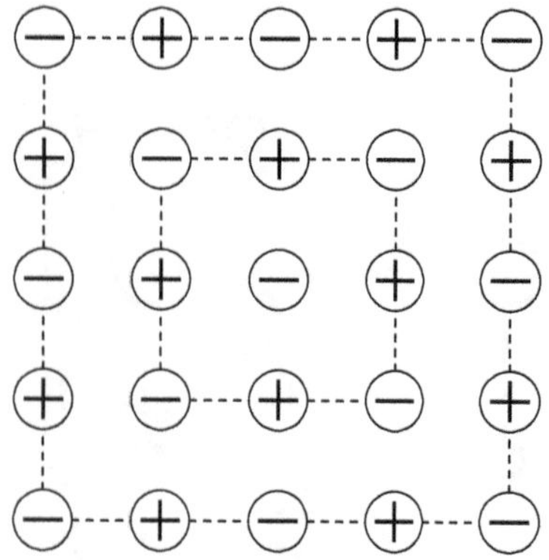

Figure 7.3 *Assigning charges to successive square "shells" in the Evjen method: atoms on the edges and corners of a shell are assigned charge fractions of 1/2 and 1/4 (with the appropriate sign) for their interior contributions, and 1/2 and 3/4 for their exterior contributions.*

or

$$R^2 e^{-R/\rho} = \frac{\alpha q^2 \rho}{z\lambda}. \tag{7.15}$$

The only remaining problem in the application of this model is calculation of the Madelung constant. The simplest case is for a hypothetical "1d" sodium chloride lattice, where Eq. (7.14) becomes

$$\begin{aligned} \alpha &= 2\left[1 - \frac{1}{2} + \frac{1}{3} - \frac{1}{4} \cdots\cdots\right] \\ &= 2\, ln2; \end{aligned} \tag{7.16}$$

here we have used the Taylor expansion of the function $ln(1 + x)$ in the second step and set $x = 1$. Note that (7.16) is only conditionally convergent; we may take this as a hint that some care will be required if we are to obtain a reasonable convergence rate in two and three dimensions.

To calculate α in two dimensions we refer the reader to Fig. 7.3. Reasonable convergence is obtained if we regard the charges as extended and divide them into parts falling in successive coordination shells, as shown in the figure. Each shell is constructed so that there is no net charge. The total charge falling in the first shell is

$$\left[-1 + 4\left(\frac{1}{2}\right) - 4\left(\frac{1}{4}\right)\right] = 0$$

and that in the second shell is

$$\left[4\left(\frac{1}{2}\right) - 4\left(\frac{3}{4}\right) - 4\left(\frac{1}{2}\right) + 8\left(\frac{1}{2}\right) - 4\left(\frac{1}{4}\right)\right] = 0.$$

The corresponding contributions to the Madelung constant are

α = (nearest shell) + (second nearest shell)

$$\alpha = \left[\frac{4\left(\frac{1}{2}\right)}{1} - \frac{4\left(\frac{1}{4}\right)}{\sqrt{2}}\right] + \left[\frac{4\left(\frac{1}{2}\right)}{1} - \frac{4\left(\frac{3}{4}\right)}{\sqrt{2}} - \frac{4\left(\frac{1}{2}\right)}{2} + \frac{8\left(\frac{1}{2}\right)}{\sqrt{5}} - \frac{4\left(\frac{1}{4}\right)}{2\sqrt{2}}\right] + \ldots \tag{7.17}$$

For the 3d case we divide the cube corners as (1/8, 7/8), the cube edges as (1/4, 3/4), and the cube faces as (1/2, 1/2). Referring to the NaCl structure shown in Fig. 7.2, the Madelung constant for the first coordination shell is

$$\alpha = \left[\frac{6\left(\frac{1}{2}\right)}{1} - \frac{12\left(\frac{1}{4}\right)}{\sqrt{2}} + \frac{8\left(\frac{1}{8}\right)}{\sqrt{3}}\right] + \ldots \tag{7.18}$$

The result for the first shell is 1.45; a similar calculation including the contribution from the second shell gives 1.75. The first few figures of the exact result are $\alpha = 1.747558$. A method used for more accurate calculations will be discussed in Appendix A of Chapter 9.

We may also calculate the bulk modulus. Note that replacing the two parameters (ε, σ) of the 6–12 potential, we now have the parameters (λ, ρ) of our repulsive potential; therefore we need two measurements to fix these parameters, which we can take to be the lattice constant and the bulk modulus, where the latter is given by the first part of Eq. (7.10). In terms of the nearest-neighbor distance, R, and the number of NaCl molecules, N, the volume is $2NR^3$. To calculate $\partial^2 U/\partial V^2$ we examine

$$\frac{\partial U}{\partial V} = \frac{\partial U}{\partial R}\frac{\partial R}{\partial V} = \frac{1}{6NR^2}\frac{\partial U}{\partial R};\ \frac{\partial^2 U}{\partial V^2} = \frac{\partial^2 U}{\partial R^2}\left(\frac{\partial R}{\partial V}\right)^2 + \frac{\partial U}{\partial R}\frac{\partial^2 R}{\partial V^2};$$

However, $\frac{\partial U}{\partial R} = 0$ at equilibrium and hence $B \equiv V\frac{\partial^2 U}{\partial V^2} = \frac{1}{18NR}\frac{\partial^2 U}{\partial R^2}$. From (7.12) and (7.13) we have

$$\frac{\partial^2 U_{\text{total}}}{\partial R^2} = N\left[\frac{z\lambda}{\rho^2}e^{-R/\rho} - \frac{2\alpha q^2}{R^3}\right]. \tag{7.19}$$

Using Eq. (7.15) we rewrite (7.19) as

$$\frac{\partial^2 U_{\text{total}}}{\partial R^2} = N\left[\frac{\alpha q^2}{\rho R} - \frac{2\alpha q^2}{R^3}\right] = \frac{N\alpha q^2}{R^3}\left[\frac{R}{\rho} - 2\right],$$

which gives the bulk modulus as

$$B = \frac{\alpha q^2}{18R^4}\left(\frac{R}{\rho} - 2\right). \tag{7.20}$$

By measuring R (using X-rays) and B (from the sound velocity) we may compute the ground state energy or, equivalently, the cohesive energy. For the material KCl one obtains $U/N = -7.26$ eV, which is to be compared with the experimental value, -7.40 eV.

7.4 The directed covalent bond: Pauling orbitals

The extreme opposite limit of the ionic bond is the non-polar covalent bond, where the electrons forming the bond are shared equally between the atoms involved. The simplest case is the binding of the gaseous hydrogen molecule, H_2, first explained by Heitler and London. As discussed in texts on quantum mechanics, the important contribution to bond formation arises from the so-called exchange integral, which has an attractive "cross term" that occurs when calculating the total energy in an approximation where the total orbital wave function is written as a symmetric combination of the atomic basis functions in a perturbation or variational calculation. This symmetric state concentrates electron density in the region *between* the H nuclei and helps to *screen* (reduce the effect of) the Coulomb repulsion between the nuclei; the spin wave function is then *antisymmetric*, so that the Pauli principle will be satisfied. These singlet pairs correspond to the *shared electrons*, or bond pairs, that were proposed by G.N. Lewis in 1916 on chemical grounds. The antisymmetric triplet contribution of the spatial wave functions, which *reduces* the density between the atoms, leads to a repulsion between the nuclei.

Atoms with multiple electrons outside a filled core (called valence electrons) allow the formation of multiple bonds having different directions in space; so-called *directed covalent bonds.*[5] This property is the basis of structural chemistry. One example is the diamond structure of the element carbon. Related materials with this crystal structure are the remaining group IV atoms in the periodic table: Si, Ge, and αSn (also called gray tin). The diamond crystal structure is shown in Fig. 7.4 and consists of atoms arranged such that the neighboring atoms of a given C atom lie at the corners of a *tetrahedron.*[6] This same local configuration applies to the methane molecule CH_4, where the H atoms also occupy the corners of a tetrahedron (with the carbon atom in the center).

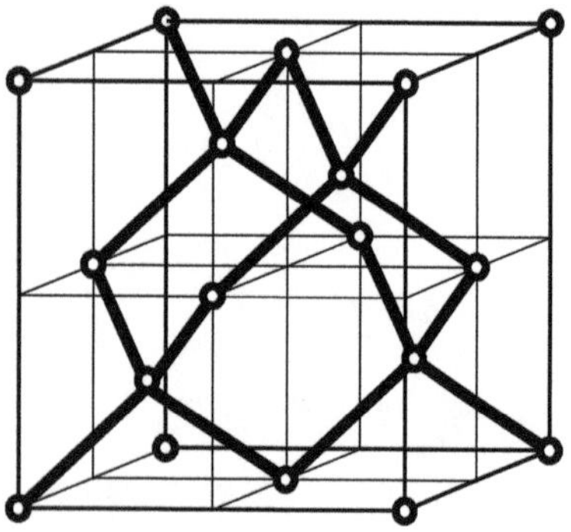

Figure 7.4 *The crystal structure of diamond, showing the tetrahedral bond arrangement.*

7.4.1 Diamond-like structures

The atomic ground state configuration of C is $1s^2 2s^2 2p^2$. The question then arises as to how the one-electron wave functions of the valence electrons can be combined to form bonds directed along the three-fold axes of a tetrahedron. We will assume that (1) the tightly bound 1s electrons do not participate in the bonding and (2) the energies of other possible configurations involving 2s and 2p electrons do not differ greatly from the ground state; i.e., we may "promote" electrons between these shells provided we more than gain the energy difference back through the "exchange integrals" of bond formation. The angular dependence of atomic s and p states (normalized over the unit sphere) can be represented by the following forms:

$$\psi_s = \sqrt{\frac{1}{4\pi}}; \qquad (7.21)$$

$$\psi_{p_z} = \sqrt{\frac{3}{4\pi}}\cos\theta; \quad \psi_{p_x} = \sqrt{\frac{3}{4\pi}}\sin\theta\cos\phi; \quad \psi_{p_y} = \sqrt{\frac{3}{4\pi}}\sin\theta\sin\phi. \qquad (7.22a\text{–}c)$$

The s states are isotropic while the p states are directed along the ±x, ±y, and ±z axes.

As a first example, let us consider the case of a single bond and find the linear combination of an s and a p orbital which maximally directs the wave function

[5] The 1954 Nobel Prize in Chemistry was awarded to Linus Carl Pauling "for his research into the nature of the chemical bond and its application to the elucidation of the structure of complex substances."

[6] Diamond may be viewed as two interpenetrating face centered cubic structures spaced 1/4 down the [111] diagonal. This crystal structure has inversion symmetry about the point along the diagonal lying midway between the two fcc sublattices. A related crystal structure, of which ZnS (the mineral zinc blende) is the prototype, does not have inversion symmetry and is hence piezoelectric. Piezoelectricity is discussed in Chapter 25.

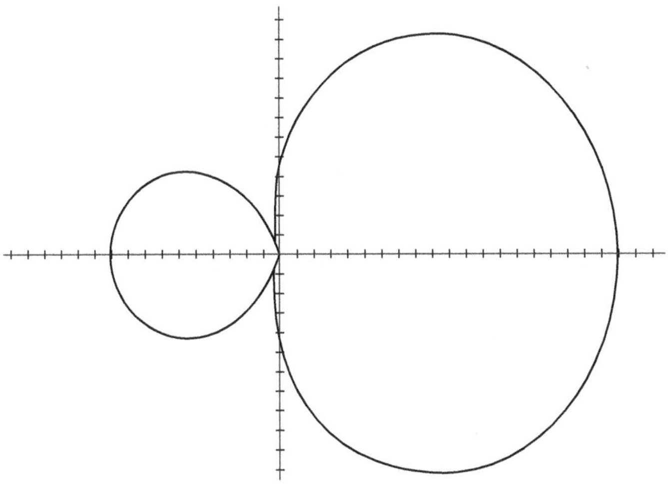

Figure 7.5 *A single sp^1 bond directed in the z direction.*

along a single axis, which we will take as the z axis. Clearly, we should use (7.22a) since it is already directed along z. The inclusion of (7.22b) or (7.22c), which are directed along x and y, would only tilt the bond off the z axis. Therefore we examine the configuration

$$\psi_1(\theta,\phi) = a\psi_s + \left(1 - a^2\right)^{1/2} \psi_{p_z}; \tag{7.23}$$

as written, this wave function is normalized for $0 \leq |a| \leq 1$. Since we are interested in a bond directed along $\hat{z}$, we maximize $\psi(\theta = 0, \phi = 0)$ as a function of the parameter a,

$$\frac{\partial\psi}{\partial a} = 1 + \frac{\sqrt{3}a}{\sqrt{1-a^2}} = 0,$$

which yields $a^2 = 1/4$ or $a = \pm 1/2$. The resultant wave function for $a = +1/2$ is

$$\psi_1(\theta,\phi) = \frac{1}{2}\psi_s + \frac{\sqrt{3}}{2}\psi_{p_z} = \frac{1}{\sqrt{4\pi}}\left(\frac{1+3\cos\theta}{2}\right); \tag{7.24}$$

the quantity in brackets has the values 2, 1/2, and −1 for $\theta = 0, = \pi/2$, and $= \pi$ respectively. For $a = -1/2$ the bond is directed in the direction $-\hat{z}$.[7] This function is graphed in Fig. 7.5. Pauling referred to the maximum of ψ as the "strength" of the bond, which is $2/\sqrt{4\pi}$ in the present case. If we had only an s or p state the values would be $1\sqrt{4\pi}$ or $\sqrt{3}/\sqrt{4\pi}$ respectively. It turns out quite generally that forming mixed or "hybrid" states, involving two or more states, allows one to "pile up" additional charge density along some bond direction.

We now construct a set of four independent orbitals from the s and the three p states which are directed from the center of a tetrahedron to its four corners, as shown schematically in Fig. 7.6; this is the case that applies to diamond.[8] A direct

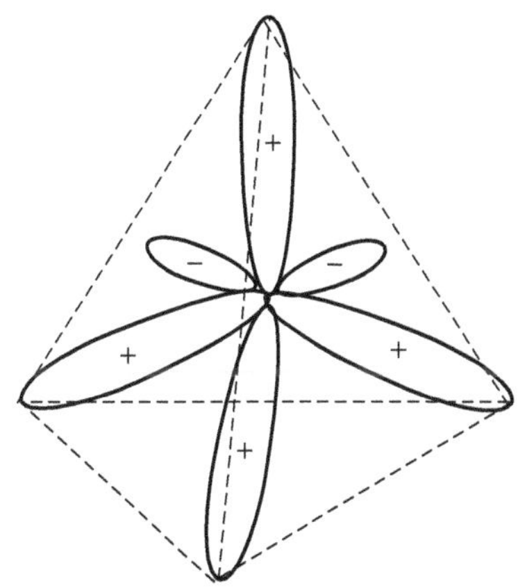

Figure 7.6 *A schematic representation of the four sp^3 bonds directed along the axes of a tetrahedron.*

[7] If the goal is to form a bond in the +z direction, then the case $a = -1/2$ would direct charge in the opposite direction and is therefore referred to as an antibonding orbital.

[8] We note that the hydrogen atoms in the methane molecule, CH_4 (which is the basis of much organic chemistry), also lie at the corners of a tetrahedron with the carbon in the center.

route to constructing these *tetrahedral orbitals* is to note that the three p orbitals can be combined and written as a vector,

$$\mathbf{R}(\theta,\phi) = \sqrt{\frac{3}{4\pi}}\left[\sin\theta\cos\phi\,\hat{\mathbf{i}} + \sin\theta\sin\phi\,\hat{\mathbf{j}} + \cos\,\theta\,\hat{\mathbf{k}}\right]$$

$$= \left[\psi_{p_x}\hat{\mathbf{i}} + \psi_{p_y}\hat{\mathbf{j}} + \psi_{p_x}\hat{\mathbf{k}}\right]. \tag{7.25}$$

A p orbital directed along an axis defined by the unit $\hat{\mathbf{n}}$ is then simply $\mathbf{R}\cdot\hat{\mathbf{n}}$. For $\hat{\mathbf{n}}$ we choose the set of four vectors from the center of a cube toward four corners that define a tetrahedron, a suitable set being

$$\hat{\mathbf{n}}_1 = \frac{1}{\sqrt{3}}\left[\hat{\mathbf{i}}+\hat{\mathbf{j}}+\hat{\mathbf{k}}\right],$$

$$\hat{\mathbf{n}}_2 = \frac{1}{\sqrt{3}}\left[\hat{\mathbf{i}}-\hat{\mathbf{j}}-\hat{\mathbf{k}}\right],$$

$$\hat{\mathbf{n}}_3 = \frac{1}{\sqrt{3}}\left[-\hat{\mathbf{i}}+\hat{\mathbf{j}}-\hat{\mathbf{k}}\right],$$

$$\hat{\mathbf{n}}_4 = \frac{1}{\sqrt{3}}\left[-\hat{\mathbf{i}}-\hat{\mathbf{j}}+\hat{\mathbf{k}}\right]. \tag{7.26a–d}$$

From Eq. (7.24) we have immediately that the four tetrahedral orbitals are

$$\psi_i = \frac{1}{2}\left[\psi_s + \sqrt{3}\,\mathbf{R}\cdot\hat{\mathbf{n}}_i\right] \tag{7.27}$$

or explicitly[9]

$$\psi_{111} = \frac{1}{2}\left[\psi_s + \psi_{p_x} + \psi_{p_y} + \psi_{p_z}\right]$$

$$\psi_{1\bar{1}\bar{1}} = \frac{1}{2}\left[\psi_s + \psi_{p_x} - \psi_{p_y} - \psi_{p_z}\right]$$

$$\psi_{\bar{1}1\bar{1}} = \frac{1}{2}\left[\psi_s - \psi_{p_x} + \psi_{p_y} - \psi_{p_z}\right]$$

$$\psi_{\bar{1}\bar{1}1} = \frac{1}{2}\left[\psi_s - \psi_{p_x} - \psi_{p_y} + \psi_{p_z}\right]. \tag{7.28a–d}$$

A desirable property of these hybrid orbitals is that they should also be orthogonal to each other; although we have not explicitly imposed this requirement (7.28a–d) satisfy this property.

These so-called sp^3 orbitals, in addition to stabilizing the crystal structures of the group four elements C, Si, Ge, and gray Sn,[10] also stabilize the tetrahedrally

[9] In forming bonds with neighboring carbon atoms, one chooses combinations where the wave functions from neighboring sites are additive (symmetric states); such orbitals are called *bonding orbitals*. One can form other (antisymmetric) combinations lying at higher energies, which are referred to as *antibonding orbitals*.

[10] Our discussion here has not clarified how the Pauling orbitals might actually be utilized to obtain the crystal binding energy. In practice, they could be used as a basis set for a quantum mechanical variational calculation based on the linear combination of atomic orbitals (LCAO) method discussed in Chapter 11.

coordinated cubic zinc blende and hexagonal wurtzite structures[11] that involve elements from the III–V, II–VI, and I–VII columns of the periodic table. Examples are listed in Table 7.3. The loss of inversion symmetry about the line separating the atoms allows a charge transfer in forming the compound. This effect increases as the constituents become more electronegative or electropositive. The resulting *ionicity* is particularly large for ZnS and CuCl, making these materials good piezoelectrics.

Table 7.3 *Some other compounds with tetrahedral bonding.*

Columns	Material
III–V	GaAs, AlAs, InSb, InP, ...
II–VI	CdS, CdTe, ZnS, ...
I–VII	CuCl, CuBr, AgBr, ...

7.4.2 Graphite

Carbon adopts another covalent crystal structure, associated with the material graphite, which consists of stacked layers of carbon atoms with a strongly bound, in-plane hexagonal structure, but with weak van der Waals, or so-called pi (see later) bonds between layers; this structure is shown in Fig. 7.7. The weak interactions between layers allow them to easily separate (cleave), a property that makes graphite function as a *solid lubricant*. The strong in-phane covalent bonds result in a very high melting temperature (only surpassed by tungsten[12]); graphite is often used as a heater element in high temperature furnaces.

In place of the four unit vectors given in Eq. (7.26), the bond directions are now given by

$$\hat{n}_1 = \hat{i},$$

$$\hat{n}_2 = -\frac{1}{2}\hat{i} + \frac{\sqrt{3}}{2}\hat{j},$$

$$\hat{n}_3 = -\frac{1}{2}\hat{i} - \frac{\sqrt{3}}{2}\hat{j},$$

$$\hat{n}_4 = \hat{k}. \qquad (7.29a\text{–}d)$$

[11] These structures will be discussed in Chapter 8.

[12] Diamond is less stable than graphite at atmospheric pressure, although once formed it will retain that phase; however at high temperatures diamond will transform into graphite.

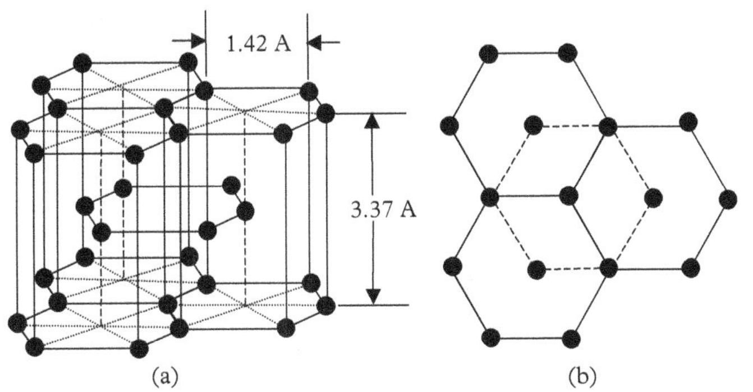

Figure 7.7 *(a) The crystal structure of graphite consists of alternating layers of hexagons with atoms in the corners but not in the center. (b) The alternating layers are displaced relative to each other such that three of the corners of the middle layer lie above the centers of the hexagons in the layer below.*

To maintain the normalization and the orthogonality the coefficient must differ from those in Eq. (7.27), and the four bonding states are now written as

$$\psi_i = \frac{1}{\sqrt{3}}\left[\psi_s + \sqrt{2}\,\mathbf{R}\cdot\hat{\mathbf{n}}_i\right], \tag{7.30}$$

or explicitly[13]

$$\psi_1 = \frac{1}{\sqrt{3}}\left[\psi_s + \sqrt{2}\psi_{p_x}\right],$$
$$\psi_2 = \frac{1}{\sqrt{3}}\left[\psi_s - \frac{\sqrt{2}}{2}\psi_{p_x} + \frac{\sqrt{6}}{2}\psi_{p_y}\right],$$
$$\psi_3 = \frac{1}{\sqrt{3}}\left[\psi_s - \frac{\sqrt{2}}{2}\psi_{p_x} - \frac{\sqrt{6}}{2}\psi_{p_y}\right],$$
$$\psi_\pi = \psi_{p_z}. \tag{7.31a–d}$$

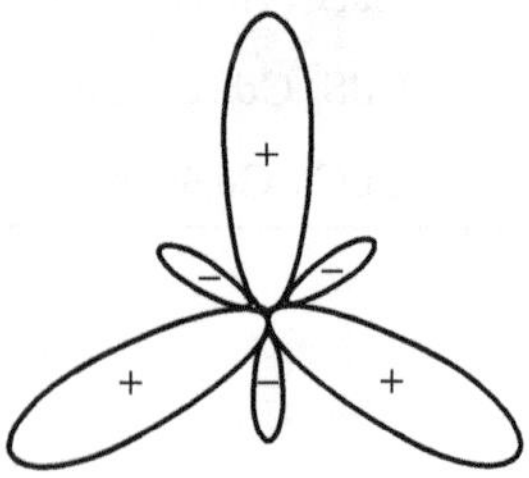

Figure 7.8 *The sp2 orbitals associated with bonding in graphite.*

The orbitals associated with (7.31a–c) are shown schematically in Fig. 7.8.

If we evaluate the bond strength for the graphite case in a manner similar to the tetrahedral case, we obtain $(1 + \sqrt{6})/\sqrt{3} = 1.991$ as opposed to 2. The three orbitals (7.31a–c), which use only two of the three p orbitals, are sometimes designed as sp^2 bonding; the fourth orbital, a p orbital which points in the z direction, forms a somewhat weaker bond which chemists refer to as a *pi bond* (π being the Greek letter for p).

Recently it has been found that carbon also adopts two other classes of structure: fullerenes and nanotubes.[14] The prototype of the first of these is the cage-like structure C_{60}, although many variants are known, C_{70} being an example. The nanotubes may be viewed as rolled graphite layers; single-walled (consisting of a single graphitic layer) and multi-walled variants exist. Furthermore, the edges of the layers can attach *with or without* a displacement along the tube axes. C_{60} crystallizes in a close-packed cubic structure with interesting properties in its own right. Because of the great variety of forms that nanotubes adopt during preparation, they have not been crystallized. However they are studied as one-dimensional wires. Single layers of graphite, when supported on some substrate, are an example of a nearly-two-dimensional system, referred to as graphene.[15]

Table 7.4 *Structures of some group V elements and II–VI compounds.*

System	Structure
Bi, Sb, As	Rhombohedral (a slight distortion of simple cubic)
PbTe, SnTe, PbS	Rock salt

Table 7.4 shows some materials in the II–VI PbTe family and the related group V semimetals. In the As column we may view the s shell as filled; the three p electrons will then form orthogonal bonds directed along the x, y, and z axes, resulting in a simple cubic structure. However this structure is rather unstable and distorts to form a rhombohedral structure.[16] In the PbTe family a charge transfer occurs which adds iconicity, but the electron shell structure then mimics the As family. The resulting p bonds and the ionicity lead to a stable rock salt structure with some covalent character. The materials are narrowband gap semiconductors.

[13] We again have antibonding orbitals associated with the in-plane directions $\phi = \pi, \pi/6$, and $-\pi/6$, along with the direction $-z$.

[14] These will be discussed in Chapter 49.

[15] Graphene will be discussed in detail in Chapter 49.

[16] The rhombohedral distortion leads to an overlap between valence and conduction bands (involving structure near the T and L points of the Brillouin zone) so that the material becomes a semimetal.

7.5 The alkali metal bond: the Wigner–Seitz model

Calculation of the cohesive energy of a metal usually involves a summation of the energies obtained from a band structure[17] and an allowance for the effects of exchange and correlation among the electrons. However, there is a simple model which works reasonably well for the alkali metals and which brings out the basic physics of metallic binding; this is the method of Wigner and Seitz, which we now discuss.

We make the following assumptions: (i) we numerically calculate the energy of the $k = 0$ state by assuming a core potential obtained from an atomic calculation and through applying a simple boundary condition to the wave function which approximates the required *crystal* boundary condition (which will be discussed in more detail shortly); (ii) the energies of the remaining states, between $0 < k \leq k_F$, are obtained by adding the free electron energy $\hbar^2k^2/2m^*$ to the $k = 0$ energy, and these energies are summed to obtain the total energy;[18] and (iii) we assume that, since each unit cell contains only one electron, the effect of exchange and correlation need not be explicitly included.

We now discuss the calculation of the energy at $k = 0$. The $k = 0$ wave function has the full translational symmetry of the crystal. It follows that the wave function derivative in a direction normal to planes that bisect the vectors joining the nearest neighbors of the atom must vanish; these planes form a polyhedron.[19] Figure 7.9 shows the corners of the body centered cubic lattice (the body centered atom being obscured, see Fig. 8.11) of the group I alkali metals and the associated polyhedron (referred to as the Wigner–Seitz cell). As an approximate boundary condition, Wigner and Seitz replaced this polyhedron with a sphere of radius r with an equivalent volume which requires

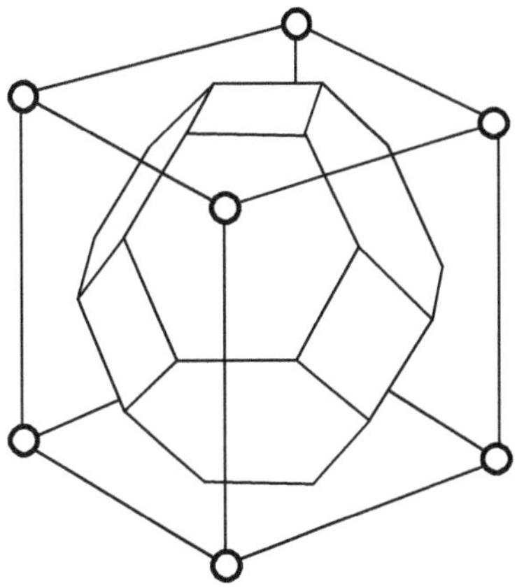

Figure 7.9 *The Wigner-Seitz polyhedron for the body centered cubic lattice.*

$$\frac{4}{3}\pi r_0^3 = \frac{1}{2}a^3; \tag{7.32}$$

the 1/2 factor compensates for the fact that there are two atoms per unit cell in the bcc structure. This leads to $r_0 = (3/8\pi)^{1/3}\,a \cong 0 \cdot 49a$. To obtain the wave function and energy at $k = 0$ one assumes a trial value of the energy and numerically integrates the radial Schrödinger equation from $r = 0$ to $r = r_0$ (the potential is obtained from a Hartree–Fock or other calculation for the free atom, and we assume that it doesn't change in the crystal). The derivative of the wave function will, in general, not vanish for $r = r_0$, and the trial energy is then changed until this condition is achieved; the smallest energy, E_0, which satisfies this condition is the desired eigenvalue (eigenvalues corresponding to core levels being excluded).

The total cohesive energy is

$$\Delta U = NE_0 + U_0 - NE_{atom}, \tag{7.33}$$

[17] We will begin a discussion of electronic band structure in Chapter 10.

[18] Note the "effective" masses differ from the free electrons as follows:

	Li	Na	K	Rb	Cs
m*/m	1.40	0.98	0.94	0.87	0.83

[19] These polyhedra are a real-space analogue of the Brillouin zones, to be discussed in Chapter 9.

where U_0 is the energy of a free Fermi gas at $T = 0$ (calculated at the electron density of the alkali metal involved) and E_{atom} is the binding energy of the s electron in the atom. Now the resulting binding energy will be a function of the unit cell dimensions, a (or equivalently r_0); i.e., $\Delta U = \Delta U(a)$. The equilibrium lattice spacing, a_0, is that value for which $\partial \Delta U(a)/\partial a = 0$.

7.6 Atomic radii

It is natural to want to assign a radius to an atom, either in isolation or embedded in a molecule or crystal. Using such radii one could then, on the basis of a "stick-and-ball" model, predict lattice spacings in simple compounds. Of course the quantum mechanical probabilistic nature of the charge density surrounding an atom precludes the existence of unambiguous atomic radii. Nonetheless, apart from their thermal and zero-point motions, atoms have rather well-defined spacings or bond lengths in crystals and molecules.[20] However, just how much of this bond spacing to assign to each atom cannot be unambiguously ascertained. Of course in a monoatomic crystal, in which all atoms have the same spacing, a bond length can be defined as half of the interatomic spacing.

In spite of these problems it turns out that useful atomic radii can be assigned by inter-comparing the spacings particular atoms have in different crystallographic or molecular settings. Tables of atomic radii have been assembled by many authors; although such radii are generally internally consistent among the atomic species fitted, one should not in general combine radii from different compilations. We will not discuss the procedures used (some of which are strictly empirical while others are semi-empirical, based on simplified quantum mechanical models for the atoms), but only list some commonly accepted atomic radii. They in general depend on a number of factors, including the *coordination* (the number and arrangement of the nearest neighbors) and the *oxidation state* (valence). Note that while we are accustomed to regarding valences as being integral, in some materials they may have non-integral values.[21] The radii further depend on the nature of the bond: metallic, covalent, ionic, or van der Waals.[22] These bond classes are qualitative; as an example, a bond might be viewed as a mixture of two limiting cases: covalent (with no charge transfer) and ionic (with complete charge transfer). But charge transfer depends to some extent on how the radii are chosen.

Table 7.5 lists the tetrahedral covalent radii applicable to semiconductors having the diamond, zinc blende/wurtzite, and chalcopyrite structures, as originally determined by Pauling. Table 7.6 gives the metallic radii for twelve-fold coordinated structures (face centered cubic or hexagonal close packed). Finally, Tables 7.7 and 7.8 give ionic radii applicable to structures which are locally six-fold (also called octahedrally) coordinated structures (as in the sodium chloride structure). All entries are in angstroms.

[20] Atomic spacings can be determined using X-ray, electron, or neutron diffraction.

[21] In so-called mixed valence compounds the valence is sometimes viewed as fluctuation between two states at a very high rate.

[22] A fourth class of bond, the so-called hydrogen bond, which is important in biological and other applications, will not be discussed here.

Table 7.5 *Four-fold coordinated (tetrahedral) covalent radii.*[23]

	Be	B	C	N	O	F
	1.06	0.88	0.77	0.70	0.66	0.64
	Mg	Al	Si	P(III)	S(II)	Cl
	1.40	1.26	1.17	1.10	1.04	0.99
Cu	Zn	Ga	Ge	As(III)	Se(II)	Br
1.35	1.31	1.26	1.22	1.18	1.14	1.11
Ag	Cd	In	Sn	Sb	Te	I
1.53	1.48	1.44	1.40	1.36	1.32	1.28
Au	Hg	Tl	Pb	Bi		
1.50	1.48	1.47	1.46	1.46		

Table 7.6 *Twelve-fold coordinated (close packed) metallic radii.*[24]

Ag	1.445	Fe	1.274	Nb	1.468	Sn	1.623
Al	1.432	Ga	1.411	Nd	1.821	Sr	2.151
Au	1.442	Gd	1.802	Ni	1.246	Ta	1.467
Ba	2.243	Hf	1.580	Os	1.353	Tb	1.782
Be	1.128	Hg	1.573	Pb	1.750	Tc	1.360
Bi	1.70	Ho	1.766	Pd	1.376	Th	1.798
Ca	1.974	In	1.663	Pm	1.810	Ti	1.462
Cd	1.568	Ir	1.357	Pr	1.828	Tl	1.716
Ce	1.825	K	2.376	Pt	1.387	Tm	1.746
Co	1.252	La	1.877	Rb	2.546	U	1.56
Cr	1.282	Li	1.562	Re	1.375	V	1.346
Cs	2.731	Lu	1.734	Rh	1.345	W	1.408
Cu	1.278	Mg	1.602	Ru	1.339	Y	1.801
Dy	1.773	Mn	1.264	Sb	1.59	Yb	1.740
Er	1.757	Mo	1.400	Sc	1.641	Zn	1.394
Eu	2.042	Na	1.911	Sm	1.802	Zr	1.602

[23] From L. Pauling, *The Nature of the Chemical Bond*, Cornell University Press, Ithaca, NY (1948).

[24] From E. Teatum, K. Gschneidner, and J. Waber, *Compilation of Calculated Data Useful in Predicting Metallurgical Behavior of the Elements in Binary Alloy Systems*, Los Alamos Scientific Laboratory (1960).

Table 7.7 *Six-fold coordinated (octahedral) cation ionic radii for various oxidation states.*[25]

Ag(+1)	1.29	Er(+3)	1.030	Mn(+3)	0.72/0.785*	Ta(+3)	0.86
Al(+3)	0.675	Eu(+2)	1.31	Mo(+3)	0.83	Tb(+3)	1.063
Au(+1)	1.51	Eu(+3)	1.087	Na(+1)	1.16	Th(+4)	1.08
Au(+3)	0.99	Fe(+2)	0.75/0.920*	Nb(+3)	0.86	Ti(+2)	1.00
Ba(+2)	1.49	Fe(+3)	0.69/0.785*	Nd(+3)	1.123	Ti(+3)	0.810
Be(+2)	0.59	Ga(+3)	0.760	Ni(+2)	0.830	Ti(+4)	0.745
Bi(+3)	1.17	Gd(+3)	1.078	Pb(+2)	1.33	Tl(+1)	1.64
Ca(+2)	1.14	Hf(+4)	0.85	Pd(+2)	1.00	Tl(+3)	1.025
Cd(+2)	1.09	Hg(+1)	1.33	Pm(+3)	1.11	Tm(+3)	1.020
Ce(+3)	1.15	Hg(+2)	1.16	Pr(+3)	1.13	U(+3)	1.165
Ce(+4)	1.01	Ho(+3)	1.041	Pt(+2)	0.94	U(+4)	1.03
Co(+2)	0.79/0.885*	In(+3)	0.940	Rb(+1)	1.66	V(+2)	0.93
Co(+3)	0.685/0.7*	Ir(+3)	0.82	Rh(+3)	0.805	V(+3)	0.780
Cr(+2)	0.87/0.94*	K(+1)	1.52	Ru(+3)	0.82	Y(+3)	1.040
Cr(+3)	0.755	La(+3)	1.172	Sb(+3)	0.90	Yb(+2)	1.16
Cs(+1)	1.81	Li(+1)	0.90	Sc(+3)	0.885	Yb(+3)	1.008
Cu(+1)	0.91	Lu(+3)	1.001	Sm(+3)	1.098	Zn(+2)	0.880
Cu(+2)	0.87	Mg(+2)	0.860	Sr(+2)	1.32	Zr(+4)	0.86
Dy(+3)	1.052	Mn(+2)	0.81/0.970*				

*High-spin/low-spin limits. The number in parenthesis is the oxidation state.

Table 7.8 *Six-fold coordinated anion ionic radii.*

Cl(−1)	1.67	O(−2)	1.26
Br(−1)	1.82	S(−2)	1.70
F(−1)	1.19	Se(−2)	1.84
I(−1)	2.06	Te(−2)	2.07

The number in parenthesis is the oxidation state.

[25] From R.D. Shannon, Revised effective ionic radii in halides and chalcogenides, *Acta Cryst.* **A32**, 751 (1976).

ADDITIONAL READING

The Nature of the Chemical Bond, L. Pauling, Cornell University Press, Ithaca, NY (1948).

Bonding and Structure, N.W. Alcock, Ellis Horwood (1990).

Structural Inorganic Chemistry, A.F. Wells, Oxford University Press, Oxford, U.K. (1986).

Problems

7.1. Demonstrate the orthonormal character of the bonds specified by Eqs. (7.28) and (7.31).

7.2. Using the covalent tetrahedral radii given in Table 7.5, calculate the lattice constant, a, of cubic Ge, and cubic GaAs; compare with experiment.

7.3. Using the six-fold coordinated cation and anion ionic radii in Tables 7.7 and 7.8, calculate the lattice constant, a, of NaCl and KCl; compare with experiment.

7.4. Using the twelve-fold coordinated metallic radii, calculate the lattice constant of a close packed cubic compound, Cu_3Au, in which Au atoms are located on the corners of a cube and Au atoms are positioned in the center of the cube faces; compare with experiment.

7.5. Write your own computer program to calculate the Madelung constants for a one-, two-, and three-dimensional NaCl like lattice. How many terms must the expansion be carried to in order to get a stable result to six decimal places?

7.6. Carry out a similar calculation as in Problem 7.5, but for the CsCl structure where the Cs atoms are in the center of a cube and the Cl atoms lie on the corners; how many terms are required for convergence to six decimal places?

7.7. As a model for the van der Waals interaction, we start by assuming the two isolated (but identical) atoms can be described as two uncoupled one-dimensional harmonic oscillators with the Hamiltonian

$$\hat{H}_0 = \frac{p_1^2}{2m} + \frac{1}{2}kx_1^2 + \frac{p_2^2}{2m} + \frac{1}{2}kx_2^2,$$

each with a resonant frequency $\omega_0 = \sqrt{\frac{k}{m}}$. Now assume that we have an interaction between the atoms (oscillators) of the (Coulomb) form

$$\hat{H}_1 = \frac{e^2}{R} + \frac{e^2}{R + x_1 - x_2} - \frac{e^2}{R + x_1} - \frac{e^2}{R - x_2},$$

where R is the separation between the origins of the two oscillators. To examine the behavior at large separations, expand this perturbation to leading order in 1/R. The resulting total Hamiltonian can be then diagonalized by a canonical transformation of the form

$$x_s = \frac{1}{\sqrt{2}}(x_1 + x_2);\ x_a = \frac{1}{\sqrt{2}}(x_1 - x_2),$$

with corresponding expressions for the momenta. Show that the frequencies of the coupled oscillators are given by

$$\omega = \sqrt{\frac{1}{m}\left(k \pm \frac{2e^2}{R^3}\right)} = \omega_0\left[1 \pm \frac{1}{2}\left(\frac{2e^2}{kR^3}\right) - \frac{1}{8}\left(\frac{2e^2}{kR^3}\right)^2 + \cdots\right],$$

which we may designate as ω_s and ω_a. Writing the zero-point energy of the system as $E = \frac{1}{2}\hbar(\omega_s + \omega_a)$, show that the energy associated with the Coulomb interaction of the oscillators is given by

$$\Delta E = -\hbar\omega_0 \frac{1}{8}\left(\frac{2e^2}{kR^3}\right)^2 = -\frac{A}{R^6}.$$

Part II

Crystal Structure and its Determination

Lattices and Crystal Structures

8

8.1 The space lattice

The lowest energy state of most pure materials at low temperatures is a crystalline solid, where the atoms or molecules making up the substance bind together in a manner which periodically repeats itself in space. Keep in mind that atoms are not points but have a finite spatial extent. In addition they execute thermal and quantum mechanical zero-point motions. A physical quantity to keep in mind is the time-averaged periodic electron density.

The basic unit of a periodic structure is called the *unit cell* and the unit cells are stacked side by side to fill the interior space of a macroscopic crystal. Depending on the material we can have multiple atoms within this unit cell. In each unit cell we designate some point (such as one corner) as an origin from which the centers of all the atoms in the unit cell are measured. These designated points in all of the unit cells form what is called a *space lattice* (or simply a lattice for short). Figure 8.1 shows a rectangular space lattice in two dimensions, along with a unit cell. The vectors **a** and **b** (and in three dimensions a third vector **c**) which lie on the edges of the unit cell are called the *primitive translation vectors*. In general the vectors **a**, **b**, and **c** are not orthogonal for the simple reason that the atoms need not arrange themselves along orthogonal lines. We may nonetheless project the primitive vectors onto three orthogonal axes if we choose by writing

$$\mathbf{a} = a_x\hat{\mathbf{x}} + a_y\hat{\mathbf{y}} + a_z\hat{\mathbf{z}}, \tag{8.1a}$$

$$\mathbf{b} = b_x\hat{\mathbf{x}} + b_y\hat{\mathbf{y}} + b_z\hat{\mathbf{z}}, \tag{8.1b}$$

$$\mathbf{c} = c_x\hat{\mathbf{x}} + c_y\hat{\mathbf{y}} + c_z\hat{\mathbf{z}}. \tag{8.1c}$$

The coordinates at any lattice point, **R**, may be written (in three dimensions) as

$$\mathbf{R}(n_1, n_2, n_3) = n_1\mathbf{a} + n_2\mathbf{b} + n_3\mathbf{c}, \tag{8.2}$$

where n_1, n_2, and n_3 are positive or negative integers. If **r** denotes a point within a unit cell, then there is another equivalent point, $\mathbf{r}'$, at

$$\mathbf{r}' = \mathbf{r} + \mathbf{R}(n_1, n_2, n_3) \tag{8.3}$$

The Physics of Solids. First Edition. J. B. Ketterson.

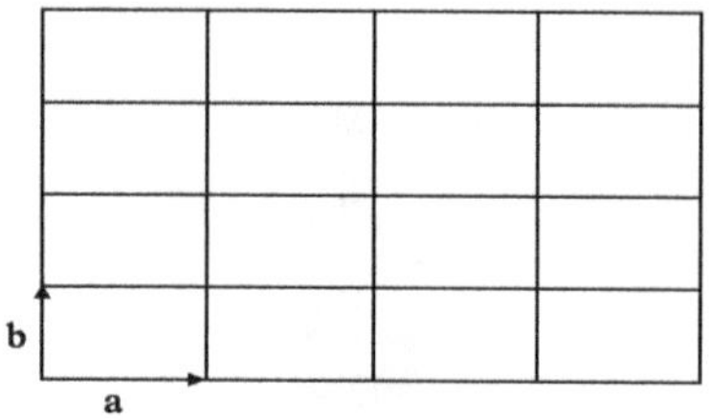

Figure 8.1 *A rectangular space lattice.*

in another unit cell designated by the numbers n_1, n_2, and n_3. We must emphasize that a space lattice is not necessarily a crystal structure.

8.2 The basis

If the crystal structure involves only one kind of atom and if, further, there is only one atom per unit cell then we may locate each atom at the origin of a unit cell. However if there are multiple atoms per unit cell we must specify their location within the unit cell, which we can do with n vectors $\mathbf{r}_i$, one for each atom, which we write as

$$\mathbf{r}_i = u_i\mathbf{a} + v_i\mathbf{b} + w_i\mathbf{c}. \tag{8.4}$$

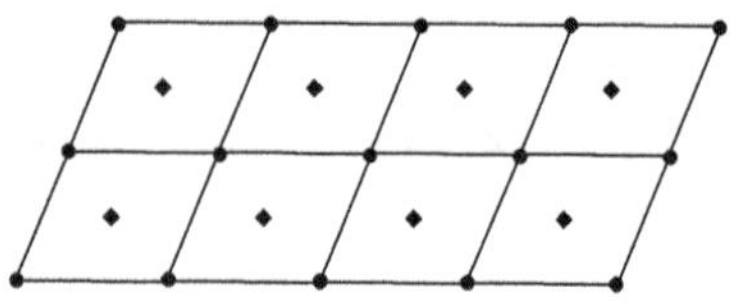

Figure 8.2 *A two-dimensional oblique crystal with two different atoms per unit cell.*

Figure 8.2 shows a crystal with two atoms per unit cell designated as ♦ and •. It is often convenient to pick one of the atoms (• in Fig. 8.2) as the origin for the unit cell (i.e., with $u = v = w = 0$ in Eq. (8.4)) and measure the positions of the other atoms relative to it. The collection of atom coordinates of Eq. (8.4) is referred to as a *basis* and together with the lattice defines a crystal structure: (lattice) + (basis) = (crystal structure).

8.3 Point groups in two dimensions

A two-dimensional (2d) crystal has translational symmetry (involving the two basic periods, **a** and **b**). It may also have additional symmetries. The first involves a discrete rotation about a third axis perpendicular to the plane of the 2d crystal. If an object (e.g., a crystal or molecule) reproduces itself on rotating through a minimum angle of $2\pi/n$ about this third axis, we say that it has *n-fold rotational symmetry*, since we may rotate the object n times (through this minimum angle), reproducing it each time, before completing a rotation of 2π and returning it to its original orientation. An n-fold axis is designated simply by writing n. If there is a line in the plane of the crystal through which it can be reflected we say it has a mirror line (a plane in three dimensions) and attach an m to the n; if there are two such lines, not related by the n-fold rotational symmetry, we attach a second m. Hence the notation, 6 mm denotes a 2d object with a six-fold axis of rotational symmetry and two independent mirror lines. This system of notation is referred to as the *International System*. The combination of the rotational symmetry together with any mirror planes is referred to as the *point group symmetry* of the object (a lattice or molecule), since the rotations are performed about a point and the mirror line runs through that point.

Combining a *space* lattice of points with a *point group symmetry* about each lattice point places restrictions on the allowed values of n; the allowed values of n are 1, 2, 3, 4, and 6, a theorem we now prove.[1]

[1] A group of materials called *quasicrystals* exists which does not have full translational symmetry, yet which has a lattice that, constructed according to a set of rules, shows a five-fold symmetric X-ray diffraction pattern.

Consider two lattice points A and B in Fig. 8.3 which are separated by a primitive translation vector, a. This figure is generated from the line segment AB by rotation through an angle $+\phi$ about the lattice point A and independently by an angle $-\phi$ about the lattice point B, where ϕ is an allowed point-group symmetry angle, $\phi = 2\phi/n$. Since these are collectively symmetry operations of the lattice as a whole, the distance A′B′ so generated must be an integral multiple of a. We write this requirement mathematically as

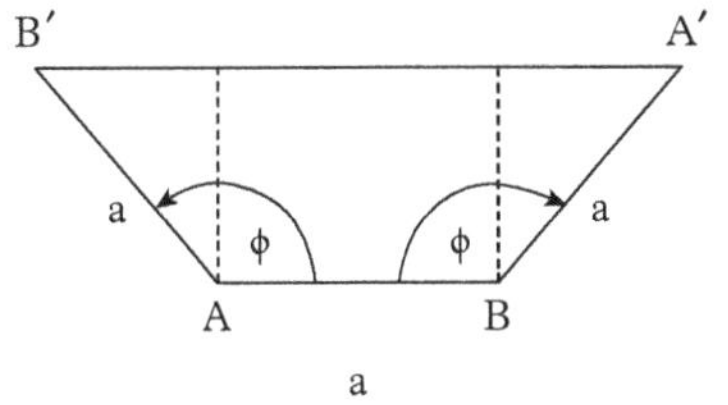

Figure 8.3 *Relating rotations to translations*

$$\begin{aligned} a + 2a\sin(\phi - \pi/2) &= a - 2a\cos\phi \\ &= pa \end{aligned} \tag{8.5}$$

where p is an integer. We therefore have $\cos\phi = (1 - p)/2$; since $|\cos\phi| \leq 1$ we can only have $p = 3, 2, 1$, or 0 corresponding to the angles $\phi = \pi, 2\pi/3, \pi/2$, or $\pi/3$; hence $n = 2, 3, 4$, or 6. When we include mirror symmetry the above restriction results in ten allowed point groups which are compatible with translational symmetry in two dimensions; these are shown schematically in Fig. 8.4. The dot(s) are used to fix the rotational symmetry (a single dot requiring a full 360° rotation to reproduce the figure). The lines passing through the center of the circle denote the reflection symmetry, and the symbol at the center also denotes

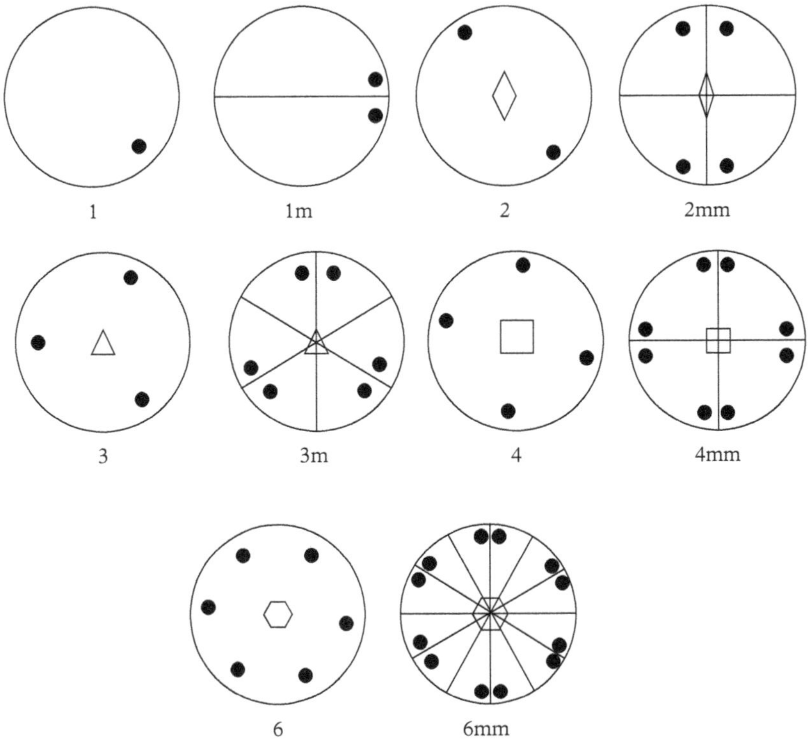

Figure 8.4 *The ten two-dimensional point groups that are compatible with translational symmetry.*

the rotational symmetry of the axis (e.g., an equilateral triangle has three-fold symmetry). Note the presence of a two-fold axis *and* a reflection line generates another reflection line at right angles to the first; this occurs for $n = 2, 4$, or 6 and is denoted by a second m (e.g., 4 mm).

8.4 Bravais lattices in two dimensions

The restriction of n to 1, 2, 3, 4, and 6 allows us to distinguish five kinds of space lattices in two dimensions; these are shown in Fig. 8.5 and are referred to as *Bravais lattices.* For the square and rectangular lattices, the angle between **a** and **b**, denoted by γ, is 90°, and they have four- and two-fold axes respectively ($n = 2$ or 4). For the hexagonal (sometimes incorrectly called triangular) case, $\gamma = 120°$ and we have a six-fold axis ($n = 6$) or, when coupled with a low symmetry basis, a three-fold axis ($n = 3$). (Note that by changing the sign of **a**, the angle γ becomes 30° and this is an equally good specification of the unit cell.) The case where $\mathbf{b}' = 2\mathbf{b} - \mathbf{a}$ is perpendicular to **a** is special since it may be viewed as a rectangular lattice with an additional lattice point in its center; and hence has a higher *two-fold* symmetry; it is referred to as a *centered rectangle* (note that a centered square lattice is not a special case since it is equivalent to a simple square lattice rotated by 45°). We see that the hexagonal case is a special case of the centered rectangle, where $\gamma = 120°$; however, it has an even higher (six-fold) symmetry. The fifth

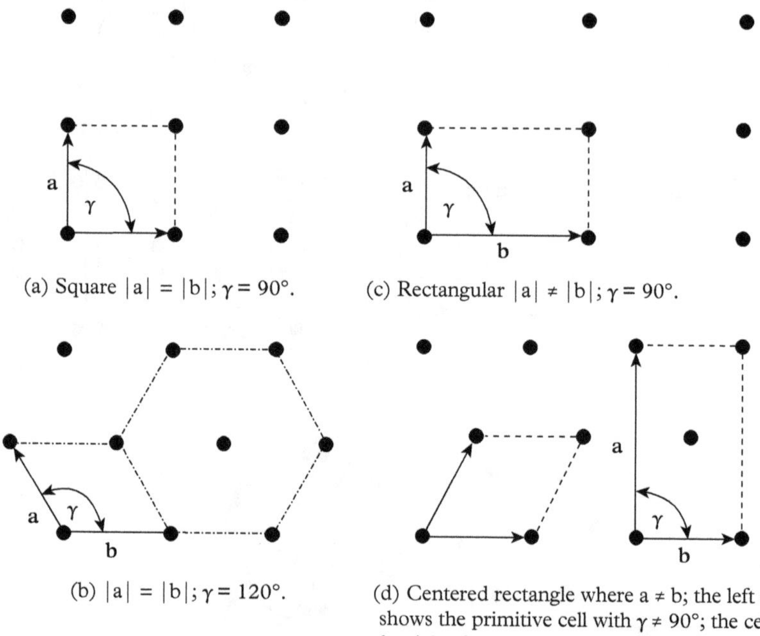

(a) Square $|a| = |b|; \gamma = 90°$.

(c) Rectangular $|a| \neq |b|; \gamma = 90°$.

(b) $|a| = |b|; \gamma = 120°$.

(d) Centered rectangle where $a \neq b$; the left side shows the primitive cell with $\gamma \neq 90°$; the cell at the right shows centered unit cell with $\gamma = 90°$.

Figure 8.5 *Two-dimensional lattices with n = 4, 6, and 2. Together with the oblique lattice they form the five two-dimensional Bravais lattices. Primitive cells are shown, completed by broken lines; the hexagon, which is not a lattice, is completed as dash-dot lines.*

Table 8.1 *The Five Two-Dimensional Bravais Lattices*

Lattice	Conventional Unit Cells	Corresponding about Lattice	Symmetry
Oblique	Parallelogram	$a \neq b, \gamma \neq 90°$	1, 2
Primitive rectangular	Rectangle	$a \neq b, \gamma = 90°$	$1m, 2mm$
Centered* rectangular	Rectangle	$a \neq b, \gamma = 90°$*	$1m, 2mm$
Square	Square	$a = b, \gamma = 90°$	$4, 4mm$
Hexagonal	60° Rhombus	$a = b, \gamma = 120°$	$3, 3m, 6, 6mm$

*$2b - a \perp a$ (see text)

Bravais lattice is the oblique lattice and corresponds to the case when $\gamma \neq 30$, 90, or 120° and the centered rectangle condition is not met; this lattice has a two-fold axis which becomes one-fold in the presence of a low symmetry basis. The characteristics of the five Bravais lattices are listed in Table 8.1 along with the point group symmetries which generate them.

8.5 Space groups in two dimensions

A Bravais lattice has a symmetry that is equal to or higher than the associated point-group symmetry. The highest symmetry point groups corresponding to the various Bravais lattices are listed in Table 8.2. The remaining point-group symmetries given in Table 8.1 involve a loss of symmetry relative to the Bravais lattice itself, resulting from the addition of a basis to the unit cell. The combination of the rotation/reflection (point-group) symmetry *and* the translational symmetry of a crystal generate what is called the *space group* (the group comprising all rotation, reflection, *and* translation operations allowed by a given crystal symmetry). Adding a *basis* in a real crystal involves giving the precise positions of the various atoms in the unit cell. However, for the purpose of specifying the space group we may simply locate objects or forms in the unit cell which accomplish a given reduction of the symmetry; an example is given in Fig. 8.6(a). Owing to the presence of the oblique triangle in the unit lattice, the rotational symmetry is reduced to $n = 1$ (even though the oblique Bravais lattice *itself* has $n = 2$ rotational symmetry). Since its construction requires only *one* object per unit cell, it is said to be *primitive*; the space group is then denoted p1 in the international notation. Suppose we located two objects inside the unit cell, as shown in Fig. 8.6(b). These objects were purposely sized and arranged so as to generate a two-fold symmetry axis. However, if we imagine the two triangles in the unit cell to be connected (by a string, so they represent a single object having two-fold symmetry), the lattice may still be thought of as primitive; the designation is then p2.

Table 8.2 *Groups that match the lattice symmetry*

System	group
Oblique	2
Rectangular	2 mm
Square	4 mm
Hexagonal	6 mm

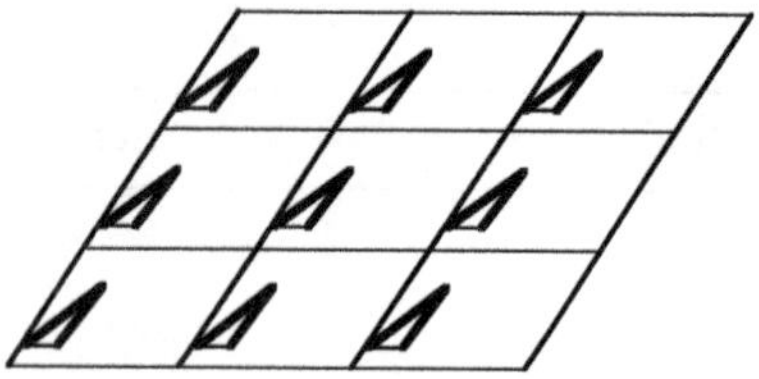

Figure 8.6(a) *Oblique lattice with a basis of one low symmetry "molecule" represented schematically by the scalene triangle.*

There are in total 17 space groups (see Fig. 8.7) in two dimensions. These are listed in Table 8.3, in the International system. The letter c denotes a "centered" structure, as discussed earlier in connection with the Bravais lattices. The letter g denotes what is called a "glide" line of symmetry, which involves a combined translation followed by a reflection (or vice versa). This operation is shown in Fig. 8.6(c).

Two-dimensional systems are of more than theoretical interest, as they do occur in nature. A good example is the so-called Langmuir monolayers—organic molecules which for chemical reasons (they are so-called ambiphylic molecules) are confined to the surface of a liquid such as water, where they are known to

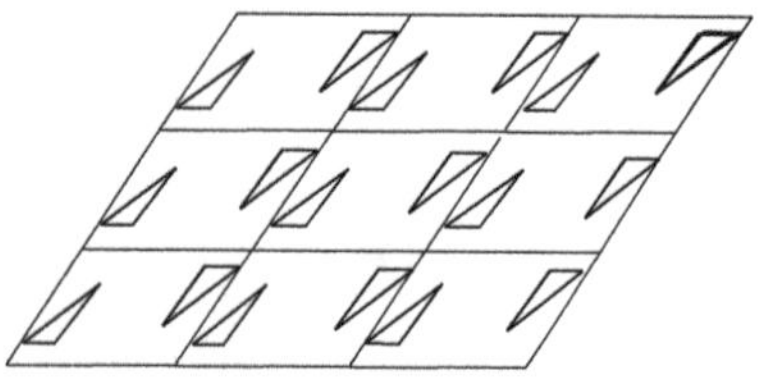

Figure 8.6(b) *Adding a second triangle to generate a two-fold axis.*

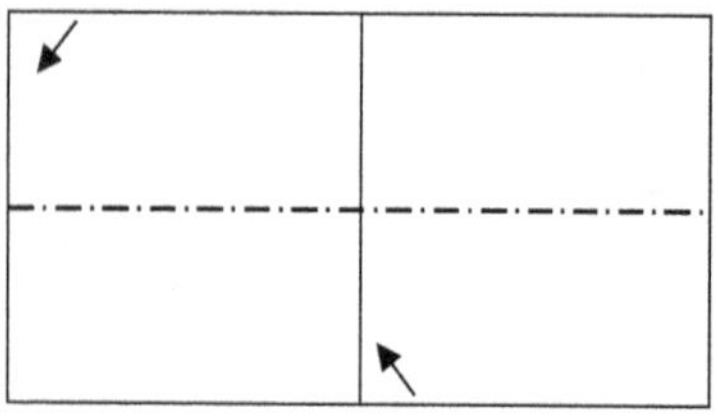

Figure 8.6(c) *The glide operation in this figure involves horizontally translating by half a period to the right and then reflecting through the dash-dot glide line.*

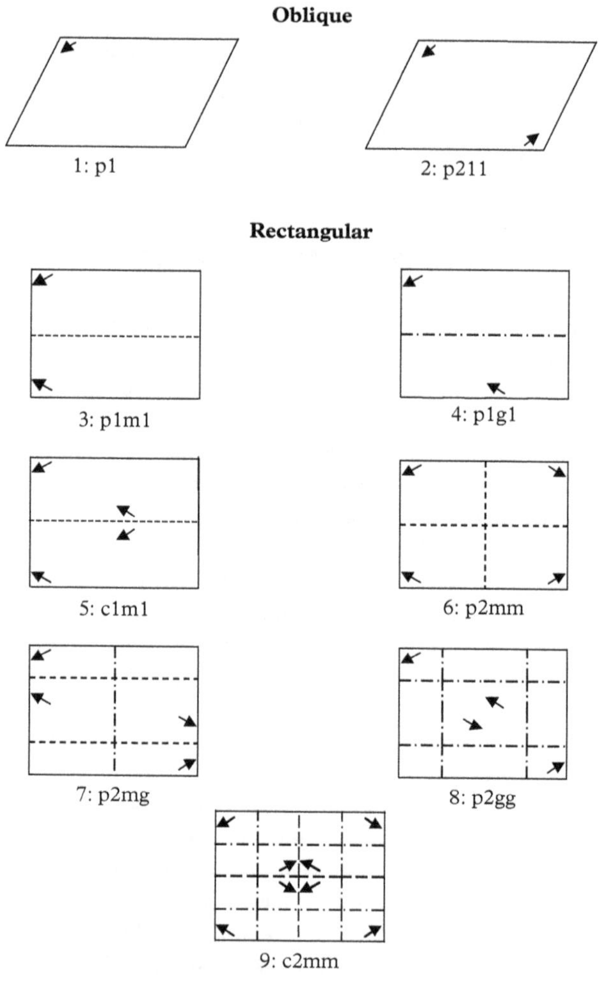

Figure 8.7 *The seventeen two-dimensional space groups.*

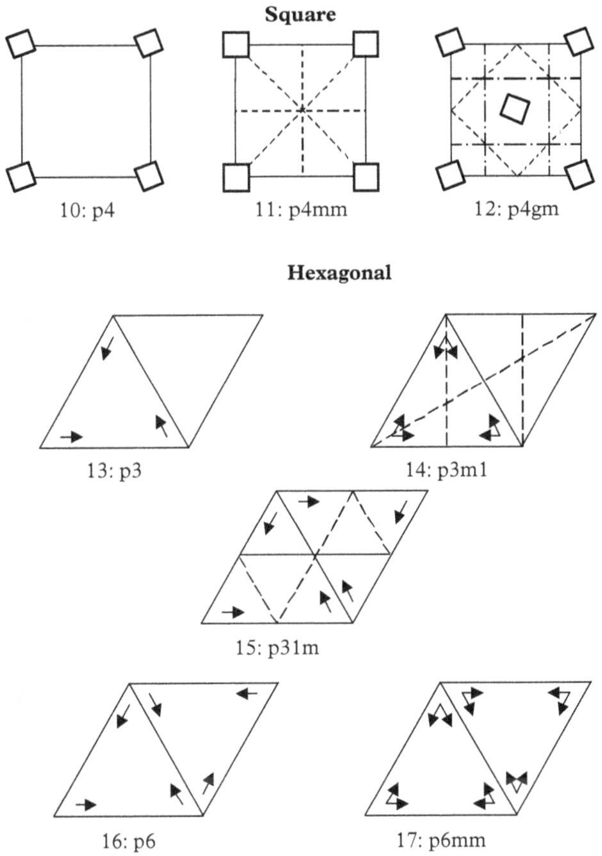

Figure 8.7 *continued*

have ordered (2d crystalline) phases. More recently, essentially 2d materials, such as graphene, have appeared (see Chapter 49) to which the model applies quite accurately.

8.6 Point groups in three dimensions

As in two dimensions, the symmetry of a three-dimensional crystal involves a combination of (i) point-group symmetry operations (about a point), (ii) translational symmetry operations, and (iii) combinations of these two symmetry operations.

We begin with an enumeration of the point-group symmetry operations applicable to crystals. In three dimensions there are two notations in common use. One is a generalization of the international notation used earlier in Section 8.3 for two dimensions and used in Fig. 8.8; the other is the so-called Schoenflies system, which is also included in Fig. 8.8.

The point-group symmetry operations involve rotations through some minimum angle of $2\pi/n$. Replacing the *lines* of reflection symmetry in two dimensions we have *planes of symmetry* or, equivalently, mirror planes, involving a reflection within the unit cell through some plane. We distinguish two kinds of mirror planes which may be associated with a symmetry axis. The first involves a vertical mirror plane *containing* the symmetry axis (i.e., the symmetry axis lies in the plane of the

Table 8.3 *The seventeen space groups.*

1	Oblique	p1
2	Oblique	p211
3	Rectangular	p1m1
4	Rectangular	p1g1
5	Rectangular	c1m1
6	Rectangular	p2mm
7	Rectangular	p2mg
8	Rectangular	p2gg
9	Rectangular	c2mm
10	Square	p4
11	Square	p4mm
12	Square	p4gm
13	hexagonal	p3
14	hexagonal	p3m1
15	hexagonal	p31m
16	hexagonal	p6
17	hexagonal	p6mm

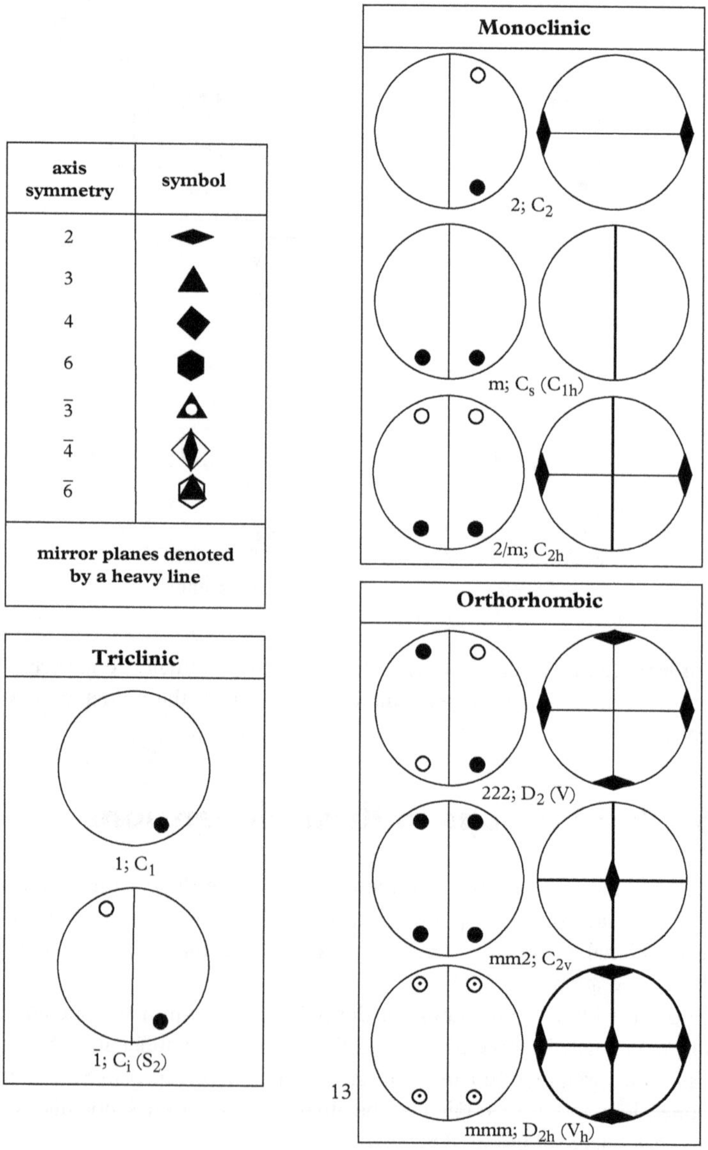

Figure 8.8 *The 32 crystallographic point groups.*

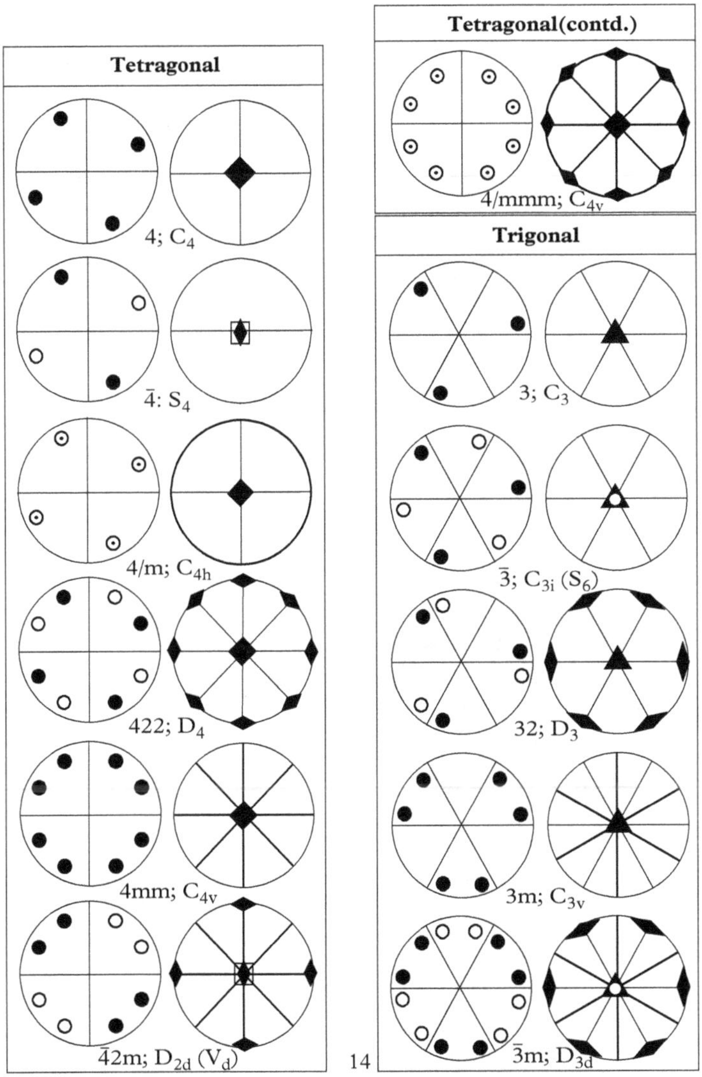

Figure 8.8 *continued*

mirror); this is the generalization to three dimensions of the symmetry line in two dimensions. The second involves a horizontal mirror plane which is *perpendicular* to the rotation axis. The combination of rotations around a primary axis and mirror planes can generate additional axes of rotational symmetry.

There are a total of 32 point groups that are consistent with translational symmetry. They are listed in Table 8.4. They are also shown as stereographic projections in Fig. 8.8, along with the various symbols used to identify the symmetry operations and with the designations given in both the international system

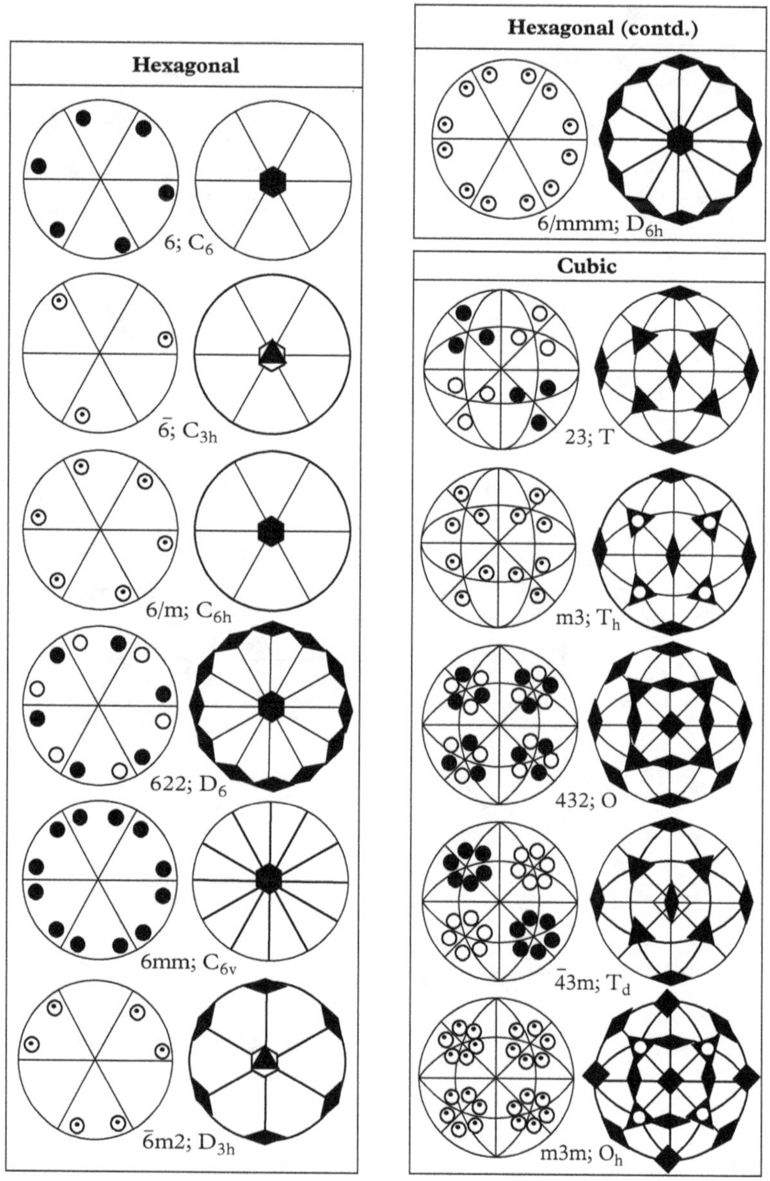

Figure 8.8 *continued*

Table 8.4 *The 32 point groups in three dimensions.*

System	International symbols (long)	International symbols (short)	Schoenflies symbols
Triclinic	1	1	C_1
	$\bar{1}$	$\bar{1}$	$C_i(S_2)$
Monoclinic	2	2	C_2
	m	m	C_s (C_{1h})
	2/m	$\frac{2}{m}$	C_{2h}
Orthorhombic	222	222	D_{2h} (V)
	mm2	mm2	C_{2v}
	mmm	$\frac{2}{m}\frac{2}{m}\frac{2}{m}$	D_{2h} (V_h)
Tetragonal	4	4	C_4
	$\bar{4}$	$\bar{4}$	S_4
	4/m	$\frac{4}{m}$	C_{4h}
	422	422	D_4
	4mm	4mm	C_{4v}
	$\bar{4}$2m	$\bar{4}$2m	$D_{2d}(V_d)$
	4/mmm	$\frac{4}{m}\frac{2}{m}\frac{2}{m}$	D_{4h}
Trigonal (Rhombohedral)	3	3	C_3
	$\bar{3}$	$\bar{3}$	C_{3i} (S_6)
	32	32	D_3
	3m	3m	C_{3v}
	$\bar{3}$m	$\bar{3}\frac{2}{m}$	D_{3d}
Hexagonal	6	6	C_6
	$\bar{6}$	$\bar{6}$	C_{3h}
	6/m	$\frac{6}{m}$	C_{6h}
	622	622	D_6
	6mm	6mm	C_{6v}
	$\bar{6}$m2	$\bar{6}$m2	D_{3h}
	6/mmm	$\frac{6}{m}\frac{6}{m}\frac{6}{m}$	D_{6h}
Cubic	23	23	T
	m3	$\frac{2}{m}\bar{3}$	T_h
	432	432	O
	$\bar{4}$3m	$\bar{4}$3m	T_d
	m3m	$\frac{4}{m}\bar{3}\frac{2}{m}$	O_h

and the Schoenflies system of Table 8.4. Note a large number of them have additional two-fold axes perpendicular to the primary symmetry axes. For those in the cubic class there are also three-fold axes or other special symmetries associated with the cube diagonals.

8.7 Bravais lattices in three dimensions

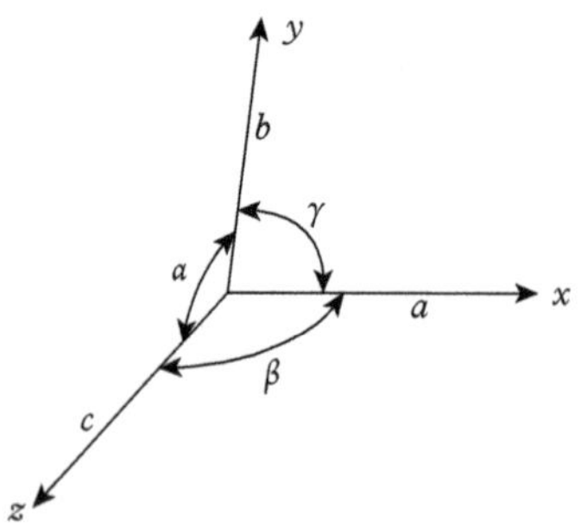

Figure 8.9 *The angles and cell dimensions describing the lattices in Table 8.5 and Fig. 8.10.*

The 32 point groups are associated with 14 Bravais lattices in three dimensions (versus the five we encountered in two dimensions). The parameters which define these lattices are given in Table 8.5 along with their designations; the angles and cell dimensions describing the lattices are shown in Fig. 8.9. The symmetries listed in the last column are those of the lattice itself. As in two dimensions, the symmetry of some particular crystal structure associated with these underlying lattices may of course be different due to additional restrictions associated with the addition of a basis. The lattices themselves are shown in Fig. 8.10. We note that for the cubic, tetragonal, orthorhombic, and monoclinic lattices there are special cases involving lattice points centered on the body diagonals (body centered or I), faces (face centered or F), or bases (base centered or C) of the lattices (which are the three-dimensional analogues of the centered rectangle Bravais lattice encountered in two dimensions); in addition, we have the primitive lattices

Table 8.5 *The 14 Bravais lattices in three dimensions.*

System	Number of lattices	Lattice symbols	Unit cell specification	Symmetry of lattice
Triclinic	1	P	$a \neq b \neq c$; α, β, γ	1, C_i
Monoclinic	2	P C	$a \neq b \neq c$; $\alpha = \gamma = 90° \neq \beta$	2/m, C_{2h}
Orthorhombic	4	P C I F	$a \neq b \neq c$; $\alpha = \beta = \gamma = 90°$	mmm, D_{2h}
Tetragonal	2	P I	$a = b \neq c$; $\alpha = \beta = \gamma = 90°$	4/mmm, D_{4h}
Cubic	3	P I C	$a = b = c$; $\alpha = \beta = \gamma = 90°$	m3m, O_h
Trigonal	1	R	$a = b = c$; $\alpha = \beta = \gamma = < 120°$; $\alpha = \beta = \gamma \neq 90°$	3m, D_{3d}
Hexagonal	1	P	$a = b \neq c$; $\alpha = \beta = 90°$; $\gamma = 120°$	6/mmm, D_{6h}

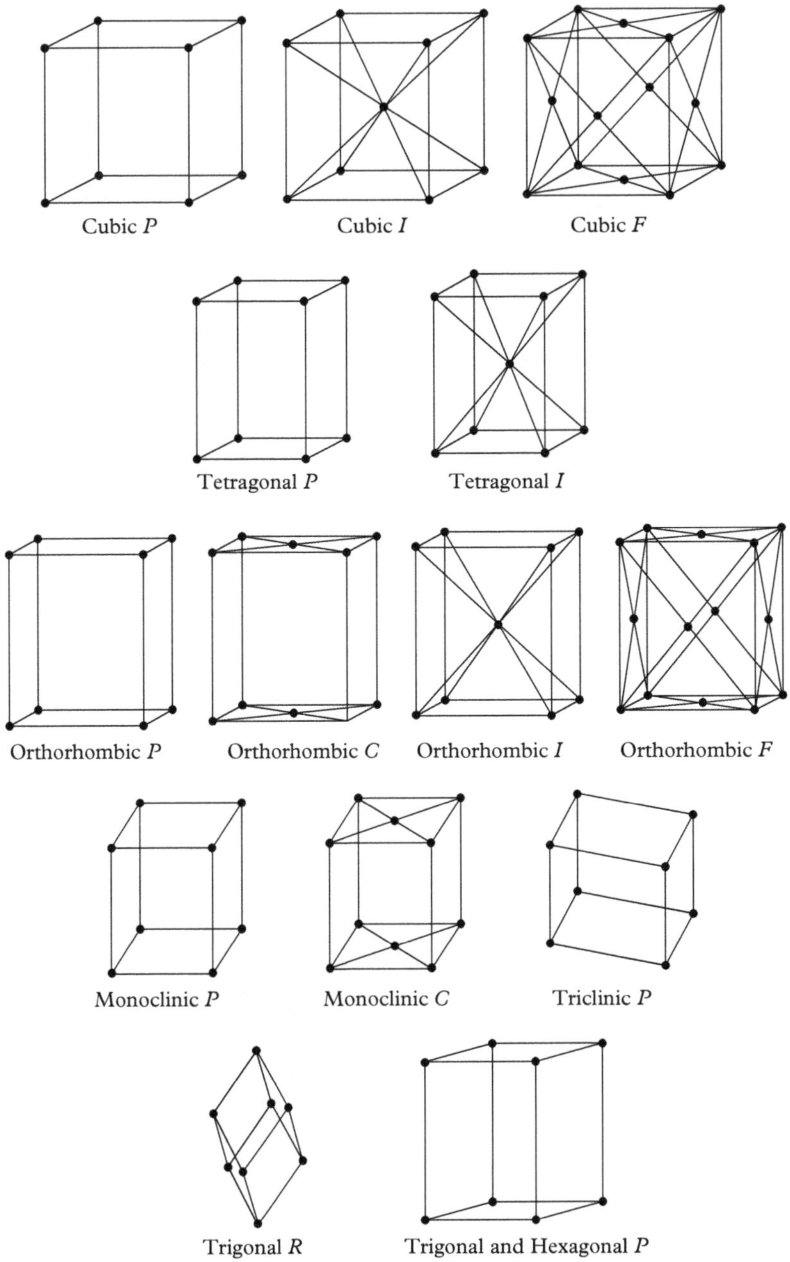

Figure 8.10 *The 14 Bravais lattices in three dimensions.*

(designated P or, for the trigonal case, R). In the face centered cubic case the unit cell is smaller than the cell drawn (since it involves only one atom per unit cell).

8.8 Crystal systems

The 14 Bravais lattices (sometimes called systems) in Table 8.5 are divided into seven classes: cubic, tetragonal, orthorhombic, monoclinic, triclinic, trigonal, and hexagonal. Although, microscopically, the Bravais lattices associated with these systems are distinguishable, for some macroscopic properties the different systems within a given class would be indistinguishable. Examples of such properties are the dielectric tensor and parts of the elastic constant tensor.

8.9 Space groups in three dimensions

As in two dimensions, the three-dimensional space groups combine the space lattices with the point-group symmetry operations and, for some space groups, additional elements. There are a total of 230 space groups which will not be specifically enumerated here. Two additional symmetry operations can also be involved: the glide plane (analogous to the two-dimensional glide line) and the screw axis; the latter involves the combination of a rotation and a translational parallel to the rotation.

8.10 Common crystal structures

In this section we describe some commonly encountered crystal structures. Polonium (Po) is the only element having the simple cubic structure (cubic P in Fig. 8.10). The next simplest structures are the face centered cubic (cubic F in Fig. 8.10 and denoted fcc) and the body centered cubic (cubic I in Fig. 8.10 and denoted bcc), which are shown in Fig. 8.11(a,b). As examples, the elements Al, Cu, Ag, Au, and Ni adopt the fcc structure, while Fe, Ta, and Nb adopt the bcc structure. In both cases the structures are also Bravais lattices. The fcc lattice is an example of what is called a close packed lattice, which we will further elaborate on. In a cubic basis there are four atoms per unit cell, while in the simplest rhombohedral Bravais lattice it will have one atom per cell. The bcc lattice is described in a cubic setting and has two atoms per unit cell.

Figure 8.12 shows two arrangements of the packing of spheres in three successive layers. The first layer corresponds to a hexagonal (triangular) array (as one has with racked pool balls). The second triangular layer rests on top of the first, but now the spheres nest into the "valleys" formed by the interstices of the first layer. The third layer may be placed into either of two *non-equivalent* sets of valleys. If the sphere centers are placed so that they fall over those of the first layer,

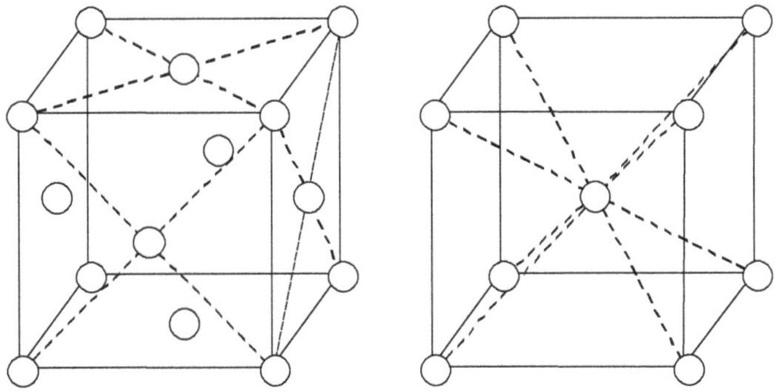

(a) (b)

Figure 8.11 *The fcc (a) and bcc (b) cubic structures; the intersection of the dashed diagonal lines helps to locate the face centered and body centered atoms respectively.*

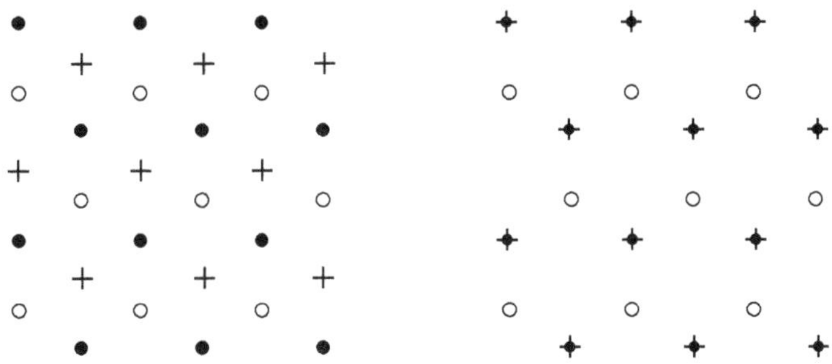

(a) (b)

● ● First-layer atoms
○ ○ Second-layer atoms
+ + Third-layer atoms

Figure 8.12 *The close packing of spheres in three successive layers. The arrangement (a) results in a face centered cubic crystal built up along the cube diagonal while the arrangement (b) results in a hexagonal close packed crystal built up along the hexagonal axis.*

and the structure is repeated so as to replicate the first two layers, we obtain a hexagonal close packed lattice. However if the spheres are placed so that they fall on the set of valleys *not* lying atop the first layer sphere centers, then we obtain, on repeating this three layer sequence, the face centered cubic lattice.

Figure 8.13 shows the atom positions in the hexagonal close packed structure, along with the underlying lattice; note there are two atoms per unit cell in the basic primitive hexagonal Bravais lattice in Fig. 8.10; i.e., it is a lattice with a basis. The elements Mg, Zn, and Cd, share this structure. Tables 8.6 and 8.7 give the lattice constants for those elements having the fcc and bcc structures. Table 8.8 gives the lattice constants for elements with the hcp structure.

We now discuss some crystals having more than one kind of atom per unit cell. The first and most commonly discussed of these is the sodium chloride structure, shown in Fig. 8.14. If the two atoms are identical, the structure becomes simple cubic. The structure may be regarded as two interpenetrating fcc lattices (one for each atom specie) displaced half a unit cell along any of the x, y, or z axes (or equivalently, half of the cube diagonal). In the actual lattice, one ion

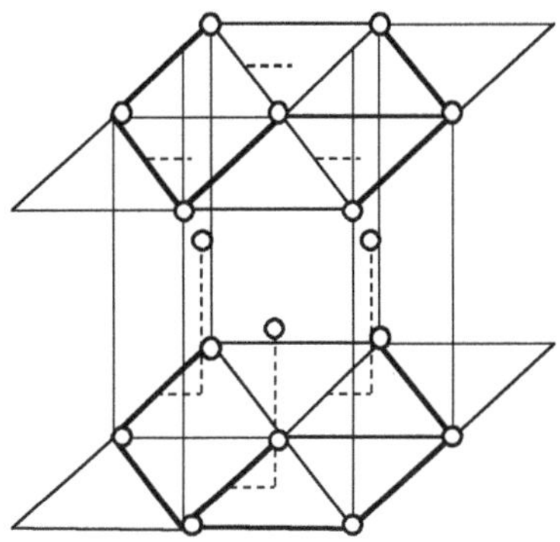

Figure 8.13 *The hexagonal close packed crystal structure and the underlying lattice.*

Table 8.6 *Lattice constants of elements having the face centered cubic structure (A).*[2]

Ar	5.26 (4.2 K)	Ir	3.84	Pt	3.92
Ag	4.09	Kr	5.72 (58 K)	δ-Pu	4.64
Al	4.05	La	5.30	Rh	3.80
Au	4.08	Ne	4.43 (4.2 K)	Sc	4.54
Ca	5.58	Ni	3.52	Sr	6.08
Ce	5.16	Pb	4.95	Th	5.08
β-Co	3.55	Pd	3.89	Xe	6.20 (58 K)
Cu	3.61	Pr	5.16	Yb	5.49

Table 8.7 *Lattice constants of elements having the body centered cubic structure (A).*

Ba	5.02	Li	3.49 (78 K)	Ta	3.31
Cr	2.88	Mo	3.15	Tl	3.88
Cs	6.05 (78 K)	Na	4.23 (5 K)	V	3.02
Fe	2.87	Nb	3.30	W	3.16
K	5.23 (5 K)	Rb	5.59 (5 K)		

Table 8.8 *Lattice constants c and a of elements having the hexagonal close packed structure.*

	a	c	c/a		a	c	c/a
Be	2.29	3.58	1.56	Os	2.74	4.32	1.58
Cd	2.98	5.62	1.89	Pr	3.67	5.92	1.61
Ce	3.65	5.96	1.63	Re	2.76	4.46	1.62
α-Co	2.51	4.07	1.62	Ru	2.70	4.28	1.59
Dy	3.59	5.65	1.57	Sc	3.31	5.27	1.59
Er	3.56	5.59	1.57	Tb	3.60	5.69	1.58
Gd	3.64	5.78	1.59	Ti	2.95	4.69	1.59
He (2 K)	3.57	5.83	1.63	Tl	3.46	5.53	1.60
Hf	3.20	5.06	1.58	Tm	3.54	5.55	1.57
Ho	3.58	5.62	1.57	Y	3.65	5.73	1.57
La	3.75	6.07	1.62	Zn	2.66	4.95	1.86
Lu	3.50	5.55	1.59	Zr	3.23	5.15	1.59
Mg	3.21	5.21	1.62				
Nd	3.66	5.90	1.61	ideal			1.63

[2] From R.W.G. Wykoff, *Crystal Structures*, Interscience, New York (1963). If not otherwise stated, values are for room temperature.

is considerably larger than the other; hence the ions in one sublattice are nearly close packed while those in the second sublattice reside in the interstices of the first. Some materials with the NaCl structure are listed in Table 8.9.

Another common system having two different atoms per unit cell is the CsCl structure, shown in Fig. 8.15. It can be viewed as two interpenetrating simple cubic structures displaced along half of the cube diagonal; it becomes bcc if the two atoms are identical. Table 8.10 gives the lattice constants of some compounds with the CsCl structure.

A third important case, with two different atoms per unit cell, is the mineral zinc blende, corresponding to the chemical compound ZnS. The structure is shown in Fig. 8.16(b) and is adopted by many important semiconducting materials such as GaN, GaAs, ZnO, InSb, etc. Perhaps the most important case is when the two atoms are identical, which corresponds to the diamond structure adopted by C, Si, Ge, and αSn, where the latter, called grey tin, differs from the common white tin form.

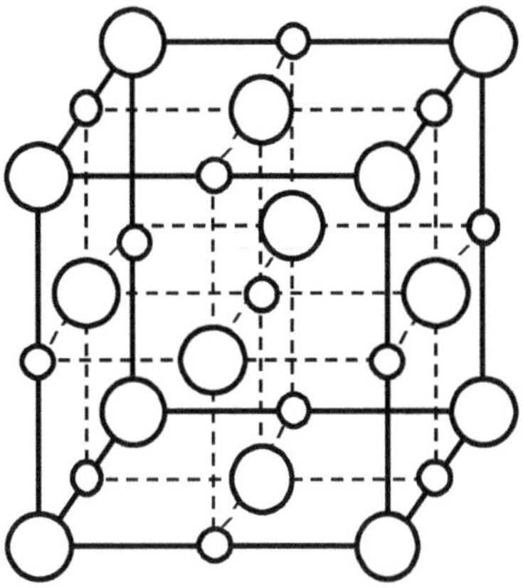

Figure 8.14 *The NaCl structure.*

Table 8.9 *Lattice constants of elements having the sodium chloride structure (A).*

LiF	4.02	RbF	5.64	CaS	5.69
LiCl	5.13	RbCl	6.58	CaSe	5.91
LiBr	5.50	RbBr	6.85	CaTe	6.34
LiI	6.00	RbI	7.34	SrO	5.16
NaF	4.62	CsF	6.01	SrS	6.02
NaCl	5.64	AgF	4.92	SrSe	6.23
NaBr	5.97	AgCl	5.55	SrTe	6.47
NaI	6.47	AgBr	5.77	BaO	5.52
KF	5.35	MgO	4.21	BaS	6.39
KCl	6.29	MgS	5.20	BaSe	6.60
KBr	6.60	MgSe	5.45	BaTe	6.99
KI	7.07	CaO	4.81		

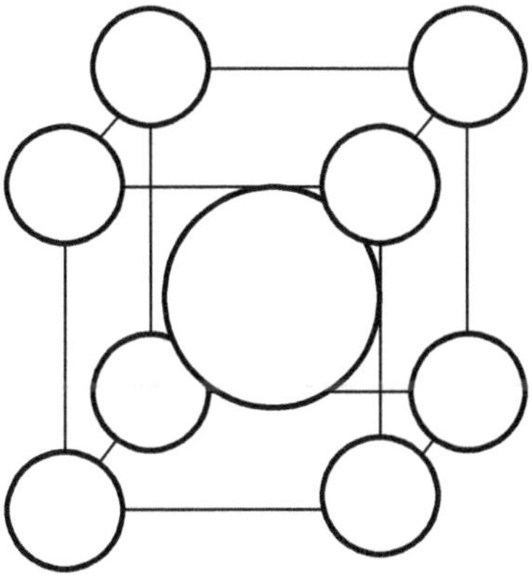

Figure 8.15 *The CsCl structure.*

Table 8.10 *Some compounds with the cesium chloride structure.*

CsCl	4.12	TlCl	3.83
CsBr	4.29	TlBr	3.97
CsI	4.57	TlI	4.20

Table 8.11 *Lattice constants of compounds having the cubic zinc blende structure (A).*

CuF	4.65	ZnS	5.41	AlSb	6.13
CuCl	5.41	ZnSe	5.67	GaP	5.45
CuBr	5.69	ZnTe	6.09	GaAs	5.65
CuI	6.04	CdS	5.82	GaSb	6.12
AgI	6.47	CdTe	6.48	InP	5.87
BeS	4.85	HgS	5.85	InAs	6.04
BeSe	5.07	HgSe	6.08	InSb	6.48
BeTe	5.54	HgTe	6.43	SiC	4.35
MnS (red)	5.60	AlP	5.45		
MnSe	5.82	AlAs	5.62		

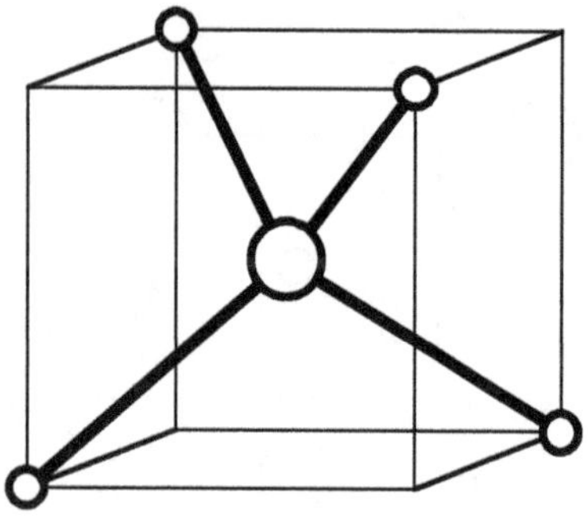

Figure 8.16(a) *The basic tetrahedrally coordinated cell.*

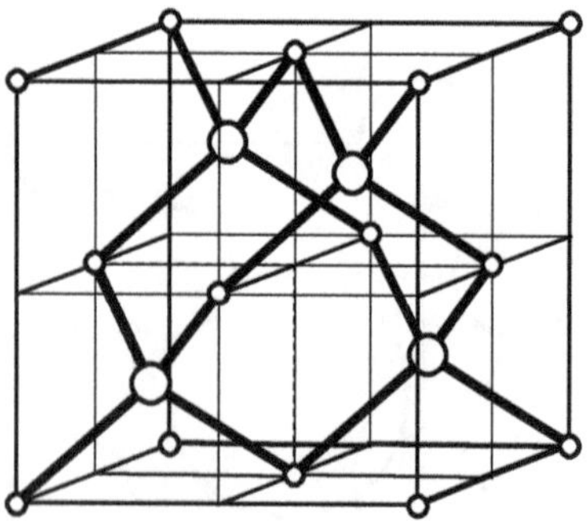

Figure 8.16(b) *The zinc blende crystal structure.*

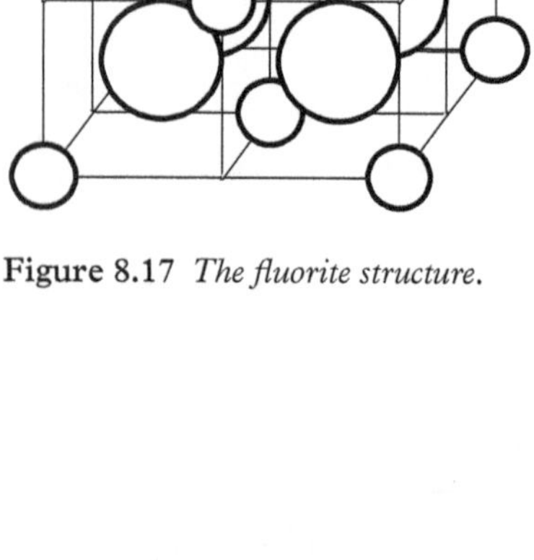

Figure 8.17 *The fluorite structure.*

The basic tetrahedrally coordinated component of the structure is shown in Fig. 8.16(a), while the unit cell with all internal, face and corner centered atoms assembled out of these basic units is shown in Fig. 8.16(b). It may be viewed as two interpenetrating fcc lattices spaced a quarter of the way down the cube diagonal. The dark lines connecting the atoms represent the chemical bonds. Those atoms not strongly bonded to other atoms within the unit cell are shown without bonds. Table 8.11 gives the lattice constants of some III–V, II–VI, and I–VIII zinc blende compounds.

Figure 8.17 shows the mineral fluorite corresponding to the compound CaF_2. This structure may be viewed as an fcc lattice of four Ca atoms containing an interior cube of eight fluorine atoms for a total of 12 atoms per cubic unit cell.

The last cubic structure we will discuss is the mineral perovskite, $KMgF_3$. This is the structure adopted by $BaTiO_3$, an important ferroelectric material. The crystal structures of the high transition temperature superconducting cuprates are also derived from this structure. Figure 8.18 shows two different ways of drawing this structure, which are related by a translation along the cube diagonal.

A common structure occurring in binary compounds is that of NiAs, which has a hexagonal unit cell. Figure 8.19 shows this structure, which consists of alternating hcp cells stacked on top of each other with each of the constituents confined to alternate layers.

A second commonly occurring hexagonal crystal structure for binary compounds is that of the mineral wurtzite, an alternate form of the mineral ZnS. Here the Zn and S sublattices are of the hcp type but displaced with respect to each other along the **c** axis; the magnitude of the displacement is such that the

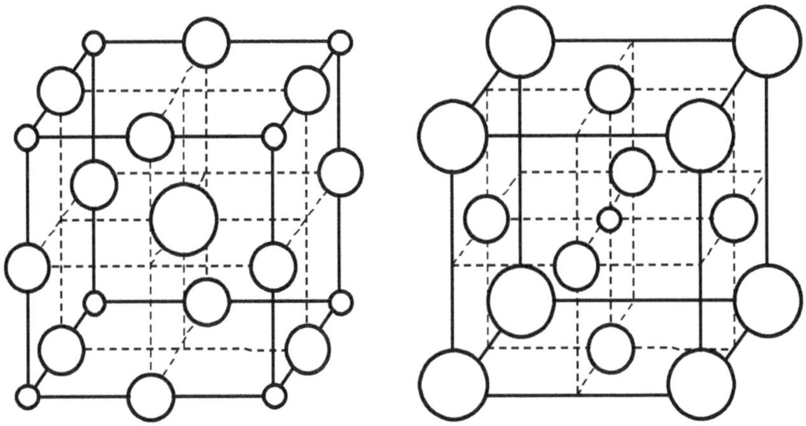

Figure 8.18 *The perovskite structure.*

atoms are *approximately* tetrahedrally coordinated although (as with the c/a ratio itself) the relative displacements vary between compounds. Among the materials adopting this structure are AgI, ZnO, CdS, CdSe, α-SiC, GaN, AlN, and BN, although in some of these cases wurtzite is not the most stable form.

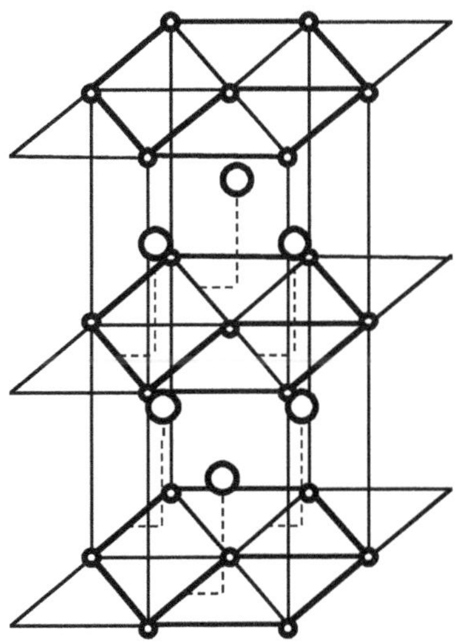

Figure 8.19 *The NiAs structure.*

8.11 Miller indices

It is common to designate a crystallographic plane by so-called Miller indices. A plane may be specified by giving the coordinates of any three non-collinear points lying in the plane. Three convenient points satisfying this criterion are those where the plane intersects the three coordinate axes of our crystal. Recall that an arbitrary lattice point is given by $\mathbf{R} = n_1\mathbf{a} + n_2\mathbf{b} + n_3\mathbf{c}$. If we let these integers have the values m, n, o, the Miller indices are defined by the following three operations:

(i) locate the intercepts of the plane with the three coordinate axes (designated m, n,o);

(ii) take the reciprocal of these numbers (which may then be fractions) $\left(\frac{1}{m}, \frac{1}{n}, \frac{1}{o}\right)$;

(iii) multiply the three numbers resulting from (ii) by the *smallest* integer which will convert them all to integers, which are then designated (h, k, ℓ); indices with a negative value are denoted with a bar (e.g., $\bar{h}$).

Figure 8.20 depicts some important planes of the cubic crystal and their associated Miller indices. Note that planes which are parallel to each other have the same indices.

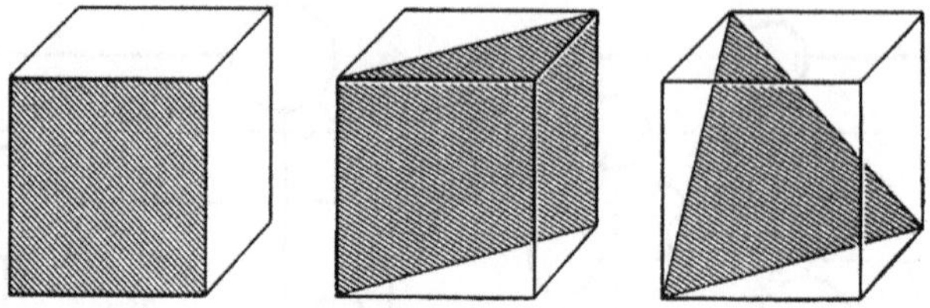

Figure 8.20 *The planes associated with the (100), (110), and (111) Miller indices of a cubic crystal.*

8.12 Wigner–Seitz polyhedra

For many purposes it is useful to surround atoms with polyhedra. If the atoms occupy sites having the full point-group symmetry of the lattice, the polyhedra will also have this symmetry; correspondingly, if the atomic sites have lower symmetry, the polyhedra will also have this lower symmetry. There are differing conventions for constructing these polyhedra. We start with the so-called *Wigner–Seitz polyhedra*, or cells, as they are also called.

Figure 8.21 shows the polyhedra obtained from the fcc lattice by constructing planes perpendicular to and intersecting the midpoints of the vectors joining the nearest 12 atoms; these atoms have the coordinates

$$\frac{a}{2\sqrt{2}}[\pm \mathbf{i} \pm \mathbf{j}], \frac{a}{2\sqrt{2}}[\pm \mathbf{j} \pm \mathbf{k}], \frac{a}{2\sqrt{2}}[\pm \mathbf{k} \pm \mathbf{i}] \tag{8.6}$$

relative to the atom at [0, 0, 0]; here the ± signs appear independently.

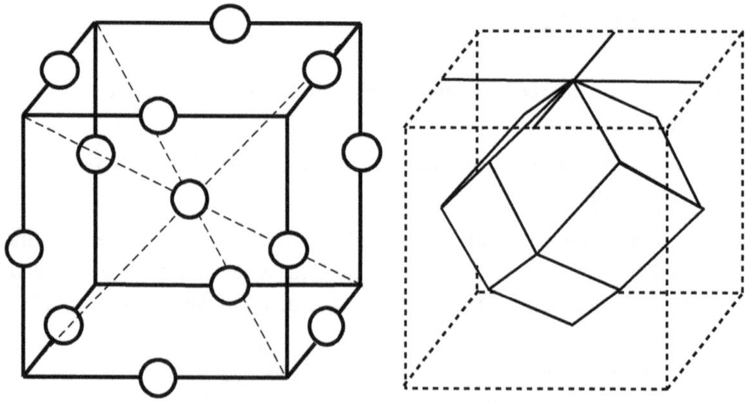

Figure 8.21 *The left side (a) shows the fcc lattice translated half a lattice constant along the x axis. The right side (b) shows the 12-sided Wigner–Seitz polyhedron constructed around the central atom from planes bisecting the lines joining the 12 nearest-neighbor atoms. Note the corners of this polyhedron touch the centers of the faces (see lines drawn on the top face). This 12-sided polyhedron is called a* rhombic dodecahedron.

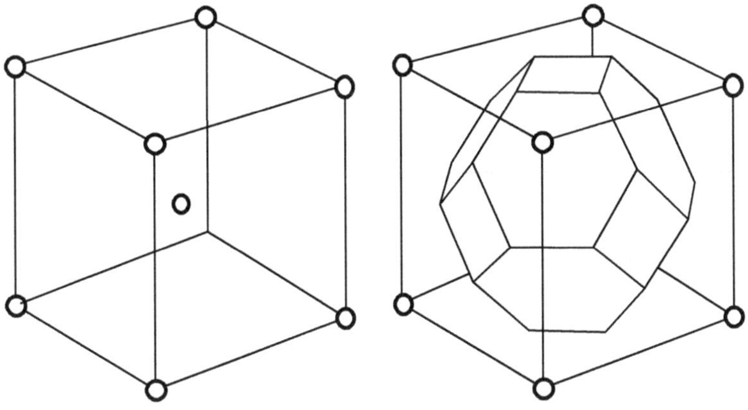

(a) (b)

Figure 8.22 *The left side (a) shows the bcc lattice while the right side (b) shows the corresponding 14-sided Wigner–Seitz cell. The centers of the square faces are located at the centers of the cube faces. This polyhedron is called a* truncated octahedron.

Figure 8.22 shows the polyhedra for the bcc lattice which are defined by the planes perpendicular to the midpoints of the following 14 vectors joining the neighboring atoms; these atoms have the coordinates

$$a[\pm \mathbf{i}],\ a[\pm \mathbf{j}],\ a[\pm \mathbf{k}],\ \frac{a}{2\sqrt{3}}[\pm \mathbf{i} \pm \mathbf{j} \pm \mathbf{k}] \tag{8.7}$$

relative to the atom at $[0, 0, 0]$. They are commonly called Wigner–Seitz cells.

8.13 Coordination polyhedra

Another class of polyhedra are the so-called *coordination polyhedra*. They are formed by connecting nearest identical atoms surrounding some contained atom

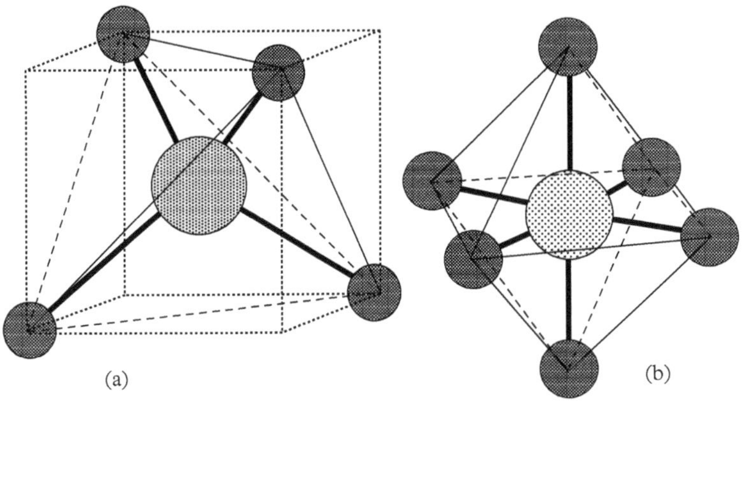

(a) (b)

Figure 8.23 *(a) Methane: a central carbon atom is surrounded by four hydrogen atoms located on the corners of a tetrahedron; the chemical bonds (heavy lines), along with the cube in which the tetrahedron is embedded, are also shown. In this case the number of faces on the polyhedron and the number of bonds coincide, and the carbon is said to be* tetrahedrally coordinated. *(b) Sulfur hexafluoride: the central sulfur atom is surrounded by six fluorine atoms; the bonds are again shown as heavy lines. The polyhedron formed by connecting the* six *fluorine atoms has* eight *faces and is called an octahedron. The sulfur is said to be* octahedrally coordinated.

in a given structure so as to generate a closed surface, the coordination polyhedra. These are particularly prevalent in the crystal chemistry literature. To discuss these we consider two different molecules: methane, CH_4, and sulfur hexafluoride, SF_6. These structures are shown in Figs. 8.23(a,b). For the case of methane, the polyhedron formed by connecting the four hydrogen atoms surrounding the carbon is a tetrahedron (which has four faces) and the carbon is then said to be *tetrahedrally coordinated.* An example of a tetrahedrally coordinated solid is diamond. For the case of sulfur hexafluoride the polyhedron formed by connecting the *six* hydrogen atoms has *eight* sides (an octahedron); the sulfur is then said to be *octahedrally coordinated.* An example of an octahedrally coordinated solid is sodium chloride: each Na atom is surrounded by six nearest-neighbor chlorines. In perovskite ($KMgF_3$, see Fig. 8.18), each of the Mg and K atoms is also surrounded by six nearest-neighbor fluorines, again forming an octahedron.[4]

In the close packed fcc structure, the polyhedron formed by connecting the 12 nearest-neighbor atoms of a given central atom is the 14-sided *cuboctahedron.* For the bcc structure the central atom is surrounded by eight atoms and the polyhedron would be a cube, which has six faces.

[4] For an extensive discussion of polyhedra as they relate to crystal chemistry, see: A.F. Wells, *Structural Inorganic Chemistry,* Oxford University Press, Oxford (1986).

X-ray Diffraction

The Reciprocal Lattice

9

The wavelengths of X-rays can be of the same order as the interatomic spacings in crystals. In optics we are familiar with the fact that a collection of equally spaced lines (a line grating) scatters (or diffracts) light at discrete angles relative to an incoming beam when the wavelength becomes smaller than the line spacing. This suggests that a similar phenomenon will occur with crystal lattices; i.e., crystal lattices can function as *three-dimensional* diffraction gratings. The utility of X-ray diffraction is that from the direction and intensity of the scattered waves we can obtain information about the atomic positions within the unit cell of a crystal and the distribution of the electron charge density.

9.1 Bragg's law

In treating the diffraction of X-rays from crystals, W.L. Bragg[1] assumed that successive parallel crystallographic planes acted as weakly reflecting "mirrors," as shown schematically in Fig. 9.1. If the reflections from successive planes add in phase we expect the intensity of the reflected X-rays to be a maximum. Recalling that the angles of incidence and reflection are identical for reflection from a plane, we find the difference in path length of the reflected beams from successive planes of a crystal, which are separated by a distance d, to be 2dsinθ (see Fig. 9.2). If this path difference is equal to a multiple of the wavelength of the incoming X-rays (and we recall there is no frequency shift induced by a simple reflection), we expect a maximum in the intensity of the X-rays scattered from the crystal. This condition,

$$n\lambda = 2d\sin\theta, \tag{9.1}$$

is known as *Bragg's law*. Clearly if we know λ and measure θ we can determined the plane spacing d (to within an integer). By examining different sets of planes (i.e., planes perpendicular to other directions in the crystal) we may deduce the primitive translation vectors associated with our lattice.

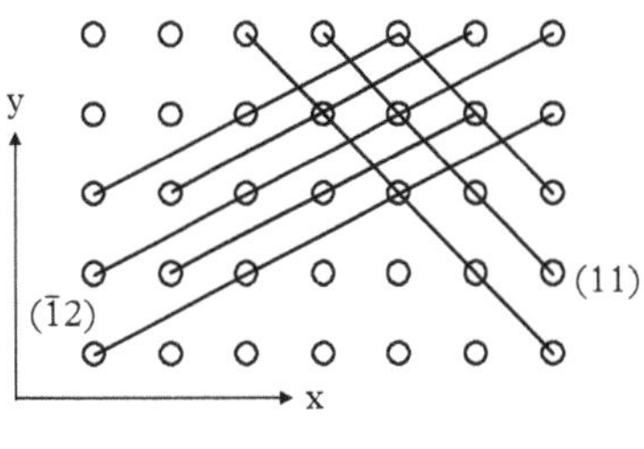

Figure 9.1 *A two-dimensional square lattices with the ($\bar{1}2$) and (11) Bragg planes indicated.*

[1] The 1915 Nobel Prize in Physics was given to William Henry Bragg and William Lawrence Bragg "for their services in the analysis of crystal structure by means of X-rays."

The Physics of Solids. First Edition. J. B. Ketterson.

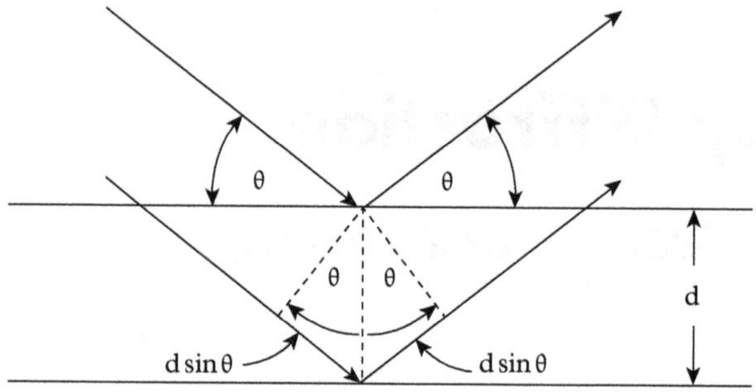

Figure 9.2 *Incoming and exiting X-rays reflecting from two successive Bragg planes; note the dashed lines are drawn parallel to the wave fronts.*

9.2 The Laue equations

Bragg's law has the virtue of providing a simple visualization of the diffraction from crystals,[2] but presents an oversimplified view of the actual diffraction process; as an example it is not immediately clear why successive atomic planes should behave as a collection of parallel, weakly reflecting, "mirrors." Clearly we require a more fundamental approach to understanding the scattering process. We begin by examining the scattering of a plane-wave X-ray beam by a single atom, as shown in Fig. 9.3.

The electric field associated with the incoming plane-wave beam accelerates the electrons in the atom causing them to radiate[3] a spherical outgoing wave at the same frequency (called Thompson scattering)[4]. If we neglect the attenuation of the incoming beam we may write the total amplitude, ϕ, arising from a particle located at $\mathbf{r} = 0$ as[5]

$$\phi(\mathbf{r}) = Ae^{i\mathbf{k}\cdot\mathbf{r}} + \phi_s(\mathbf{r}). \tag{9.2}$$

The first term in Eq. (9.2) is associated with the incoming plane wave and the second with the outgoing (scattered) spherical wave; A is an amplitude factor. We will assume $\phi_s(\mathbf{r})$ has the form of an outgoing spherical wave and write it as

$$\phi_s(\mathbf{r}) = Af(\theta, \varphi)\frac{e^{ikr}}{r}; \tag{9.3}$$

here $f(\theta, \varphi)$ represents the dependence of the amplitude of the scattered radiation on the azimuthal and polar angles θ, φ (measured relative to the direction of the incoming beam). For electromagnetic waves we would take the real part of Eq. (9.2). However particles, such as neutrons, also have wave-like properties (in the quantum theory) and in such cases Eq. (9.2) would represent the wave function of these particles. In examining the scattered radiation we will assume our observation point lies *outside* the incoming beam, allowing us to consider only the scattered radiation in our subsequent discussion.

[2] The 1914 Nobel Prize in Physics was given to Max von Laue "for his discovery of the diffraction of X-rays by crystals."

[3] The power, P, radiated by a free particle of charge q and mass m is given classically by $P = 2q^2|\ddot{\mathbf{r}}|^2/3m^2c^3$, in c.g.s. units, where $\ddot{\mathbf{r}}$ is the acceleration. Since this involves a factor m^{-2}, it is clear we may neglect any radiation from the charged nuclei.

[4] This neglects the recoil effect which for a free atom induces a downward shift in the frequency of the scattered X-ray photon; this is the basis of the *Compton effect*. Atoms in a lattice largely reflect X-rays with no frequency shift, the lattice as a whole then taking up the recoil. Inelastic processes in which electrons in atoms are raised to excited states are not coherent with the incoming beam. We note in passing that a nucleus embedded in a lattice which emits a gamma ray can undergo a certain fraction of *recoilless emissions*; this is the basis of the *Mossbauer effect* which, due to the narrow intrinsic line width of the emitted radiation, can then be exploited to carry out certain high resolution spectroscopic measurements.

[5] Although X-rays are strictly speaking vector-electromagnetic waves, we will assume the wave field may be represented as a scalar; polarization and other effects are then suppressed. We note that X-rays from a vacuum tube source are unpolarized; those from a synchrotron are highly polarized.

The scattering from a collection of atoms in a crystal having positions $\mathbf{R}_n = n_1\mathbf{a} + n_2\mathbf{b} + n_3\mathbf{c}$ is obtained by superimposing terms of the form given in Eq. (9.3).[6] Our total scattered amplitude, $\phi(\mathbf{r})_s^{\text{total}}$, is then

$$\begin{aligned}\phi(\mathbf{r})_s^{\text{total}} &= \sum_{\mathbf{R}_n} e^{i\mathbf{k}\cdot\mathbf{R}_n}\,\phi_s(\mathbf{r}-\mathbf{R}_n)\\ &= A\sum_{\mathbf{R}_n} e^{i\mathbf{k}\cdot\mathbf{R}_n}\frac{f(\hat{\mathbf{k}}')}{|\mathbf{r}-\mathbf{R}_n|}e^{ik|\mathbf{r}-\mathbf{R}_n|};\end{aligned} \tag{9.4}$$

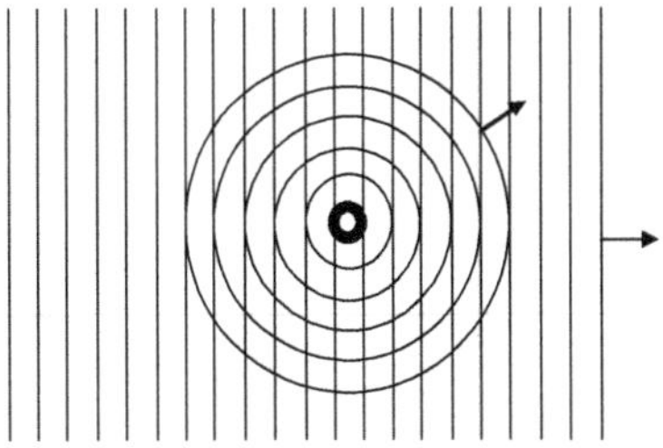

Figure 9.3 *The parallel lines show the planar wave fronts of an incoming X-ray beam, while the circles depict the outgoing spherical wave fronts resulting from Thompson scattering off the central atom.*

here $\hat{\mathbf{k}}'$ denotes a unit vector in the scattered direction (θ, φ) and the factor $e^{i\mathbf{k}\cdot\mathbf{R}_n}$ accounts for an *additional phase shift* of the incoming beam relative to our origin at $\mathbf{R}_n = 0$. We will assume our observation point, $\mathbf{r}$, is far from our crystal; i.e., $|\mathbf{r}| \gg |\mathbf{R}_n|$. Therefore, we may write

$$\begin{aligned}|\mathbf{r}-\mathbf{R}_n| &= \sqrt{r^2 - 2\mathbf{R}_n\cdot\mathbf{r} + R_n^2}\\ &\cong r\left(1 - \frac{\mathbf{R}_n\cdot\mathbf{r}}{r^2}\right)\\ &= r - \hat{\mathbf{k}}'\cdot\mathbf{R}_n = r - \frac{\mathbf{k}'\cdot\mathbf{R}_n}{k}\end{aligned} \tag{9.5}$$

where we have introduced a *scattered* wave vector, $\mathbf{k}' = k\mathbf{n} = (2\pi/\lambda)\mathbf{n}$, where $\mathbf{n}$ is a unit vector in the direction of the observation point, $\mathbf{r}$. The second term in (9.5) may be neglected in the denominator of Eq. (9.4), but must be retained in the exponent since $\mathbf{k}'\cdot\mathbf{R}_n$ is of order $2\pi\ d/\lambda$ or larger, where d is an interatomic spacing (since $d \sim \lambda$ in a diffraction experiment, the phase of the terms with different $\mathbf{R}_n$ varies rapidly, an effect which must be included in computing the total scattered amplitude). We may now rewrite Eq. (9.4) as

$$\phi_s^{\text{total}} = \frac{A}{r}e^{ikr}f(\hat{\mathbf{k}}')\sum_{\mathbf{R}_n} e^{i(\mathbf{k}-\mathbf{k}')\cdot\mathbf{R}_n}. \tag{9.6}$$

To determine the scattered amplitude we must examine the behavior of the factor $\sum_{\mathbf{R}_n} e^{-i\Delta\mathbf{k}\cdot\mathbf{R}_n}$, where $\Delta\mathbf{k} = \mathbf{k} - \mathbf{k}'$. By writing $\mathbf{R}_n = n_1\mathbf{a} + n_2\mathbf{b} + n_3\mathbf{c}$, as in Eq. (8.2), we may rewrite this in terms of the function

$$F(N_1, N_2, N_3) = \sum_{n_1=0}^{N_1-1} e^{-in_1\Delta\mathbf{k}\cdot\mathbf{a}}\sum_{n_2=0}^{N_2-1} e^{-in_2\Delta\mathbf{k}\cdot\mathbf{b}}\sum_{n_3=0}^{N_3-1} e^{-in_3\Delta\mathbf{k}\cdot\mathbf{c}}, \tag{9.7}$$

where N_i is the number of atoms along each of the three principal axes (the basis vectors) of our crystal. The total number of atoms in our crystal is then $N = N_1N_2N_3$. Consider the first sum. We define $x = \exp[-i\Delta\mathbf{k}'\cdot\mathbf{a}]$ and the

[6] At this point we will assume there is only one atom associated with each lattice site. The effect of a basis (multiple atoms per lattice unit cell) will be discussed in Section 9.8.

sum takes the form of the geometric series $\sum_{n=0}^{N-1} x^n = \frac{1-x^N}{1-x}$; our first sum then becomes

$$\sum_{n_1=0}^{N-1} e^{-in_1 \Delta \mathbf{k} \cdot \mathbf{a}} = \frac{1 - e^{-iN_1 \Delta \mathbf{k} \cdot \mathbf{a}}}{1 - e^{-i\Delta \mathbf{k} \cdot \mathbf{a}}}.$$

Since the scattered intensity is the square of the absolute magnitude of the amplitude of the wave, it involves a factor,

$$\left| \frac{1 - e^{iN_1 \Delta \mathbf{k} \cdot \mathbf{a}}}{1 - e^{i\Delta \mathbf{k} \cdot \mathbf{a}}} \right|^2 = \frac{\sin^2 \left[\frac{1}{2} N_1 \Delta \mathbf{k} \cdot \mathbf{a} \right]}{\sin^2 \left[\frac{1}{2} \Delta \mathbf{k} \cdot \mathbf{a} \right]}. \tag{9.8}$$

This expression, which occurs in optics in the theory of the diffraction grating, has a maximum value of N_1^2 when $\Delta \mathbf{k} \cdot \mathbf{a} = 2\pi m_1$, where m_1 is an integer, and drops to zero for $N_1 \Delta \mathbf{k} \cdot \mathbf{a} = 2\pi(N_1 m_1 + 1)$. The change in $\Delta \mathbf{k} \cdot \mathbf{a}$ required to pass from the maximum to the first zero is then $2\pi/N_1$, and for a solid 1 cm on a side where $N_1 \sim 10^7$ this change is very small; i.e., the peaks are very sharp. Subsequent oscillations in the intensity are of order unity (relative to N_1^2).

The *global* maximum in the intensity occurs when *all three* of the sums in Eq. (9.7) are maximal. This leads to the three conditions:

$$\Delta \mathbf{k} \cdot \mathbf{a} = 2\pi m_1; \qquad \Delta \mathbf{k} \cdot \mathbf{b} = 2\pi m_2; \qquad \Delta \mathbf{k} \cdot \mathbf{c} = 2\pi m_3; \tag{9.9}$$

where m_1, m_2, and m_3 are integers. These equations are called the *Laue equations.* In order to interpret these equations it is convenient to introduce the concept of the reciprocal lattice, which we consider next.

9.3 The reciprocal lattice

The reciprocal lattice (the reasons for this name will become apparent shortly) is defined by three basis vectors $(\mathbf{a}^*, \mathbf{b}^*, \mathbf{c}^*)$ which are related to the usual basis vectors by

$$\mathbf{a}^* = 2\pi \frac{\mathbf{b} \times \mathbf{c}}{\mathbf{a} \cdot (\mathbf{b} \times \mathbf{c})}, \quad \mathbf{b}^* = 2\pi \frac{\mathbf{c} \times \mathbf{a}}{\mathbf{a} \cdot (\mathbf{b} \times \mathbf{c})}, \quad \mathbf{c}^* = 2\pi \frac{\mathbf{a} \times \mathbf{b}}{\mathbf{a} \cdot (\mathbf{b} \times \mathbf{c})}; \tag{9.10a–c}$$

the factor of 2π is included for later convenience. From these definitions we can immediately generate the "multiplication table,"

$$\begin{array}{lll} \mathbf{a}^* \cdot \mathbf{a} = 2\pi & \mathbf{b}^* \cdot \mathbf{a} = 0 & \mathbf{c}^* \cdot \mathbf{a} = 0 \\ \mathbf{a}^* \cdot \mathbf{b} = 0 & \mathbf{b}^* \cdot \mathbf{b} = 2\pi & \mathbf{c}^* \cdot \mathbf{b} = 0 \\ \mathbf{a}^* \cdot \mathbf{c} = 0 & \mathbf{b}^* \cdot \mathbf{c} = 0 & \mathbf{c}^* \cdot \mathbf{c} = 2\pi \end{array} \tag{9.11}$$

As we did for the real space lattice we can introduce a vector (or equivalently a point) **K** in the reciprocal lattice, which is analogous to **R**, and is defined by

$$\mathbf{K} = m_1\mathbf{a}^* + m_2\mathbf{b}^* + m_3\mathbf{c}^*, \tag{9.12}$$

where (m_1, m_2, m_3) are again integers.[7] Recalling the definition of **R** from Chapter 8,

$$\mathbf{R} = n_1\mathbf{a} + n_2\mathbf{b} + n_3\mathbf{c}, \tag{9.13}$$

we see that the dot product of a reciprocal lattice vector and a real space vector is given by

$$\begin{aligned}\mathbf{R}\cdot\mathbf{K} &= (n_1m_1 + n_2m_2 + n_3m_3)2\pi \\ &= (\text{integer}) \times 2\pi.\end{aligned} \tag{9.14}$$

Equation (9.14) is analogous to the ordinary expression for the dot product of two vectors **A** and **B**: $\mathbf{A}\cdot\mathbf{B} = A_xB_x + A_yB_y + A_zB_z$.

One of the conveniences of the reciprocal lattice comes about when representing quantities in a solid which are periodic, such as the electron density. Such quantities can be represented as a three-dimensional Fourier series and must be strictly periodic under the operation $\mathbf{r} \to \mathbf{r} + \mathbf{R}$. An expansion of the form

$$\begin{aligned}\rho(\mathbf{r}) &= \sum_{m_1,m_2,m_3} \rho_{m_1m_2m_3} e^{i(m_1\mathbf{a}^*+m_2\mathbf{b}^*+m_3\mathbf{c}^*)\cdot\mathbf{r}} \\ &= \sum_{\mathbf{K}} \rho_{\mathbf{K}} e^{i\mathbf{K}\cdot\mathbf{r}}\end{aligned} \tag{9.15}$$

satisfies all the necessary requirements, where the ρ_K are the Fourier expansion coefficients. In particular, $e^{i\mathbf{K}\cdot(\mathbf{r}+\mathbf{R})} = e^{(i\mathbf{K}\cdot\mathbf{r}+2\pi n)} = e^{i\mathbf{K}\cdot\mathbf{r}}$, where from (9.14), $n = m_1n_1 + m_2n_2 + m_3n_3$, and hence is also an integer. So all terms in (9.15) are periodic under a lattice translation.

Another convenience of the reciprocal lattice is that the Laue equations (9.9) can immediately be rewritten as

$$\Delta\mathbf{k} = \mathbf{K}, \tag{9.16}$$

or, in words, the change in the scattering vector, $\mathbf{k}' - \mathbf{k}$, must be equal to some reciprocal lattice vector **K**. This is a general diffraction condition valid for X-rays, neutrons, or electrons, as we will discuss extensively in subsequent chapters.

We now show that the reciprocal lattice vector $K(m_1,m_2,m_3)$ is perpendicular to the lattice plane in **R** space having the same Miller indices. We start by considering some plane having intercepts $n_1\mathbf{a}$, $n_2\mathbf{b}$, $n_3\mathbf{c}$ on the three primitive axes,

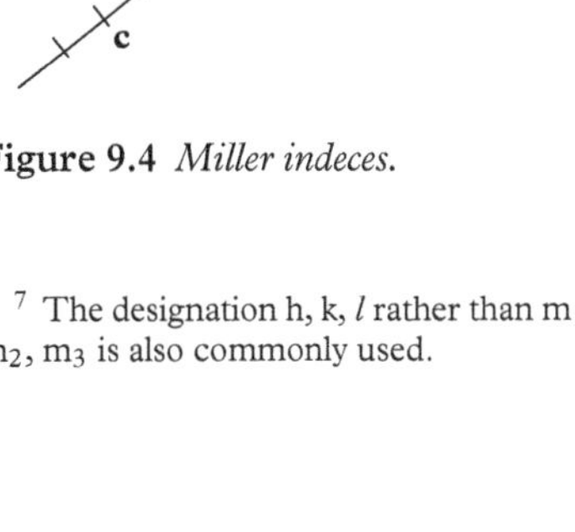

Figure 9.4 *Miller indeces.*

[7] The designation h, k, *l* rather than m_1, m_2, m_3 is also commonly used.

a, **b**, **c** (the plane $(5, 4, 3)$ is shown in Fig. 9.4). Now any two, non-collinear vectors lying in a plane define that plane. We may easily construct three such vectors by forming the combinations $n_1\mathbf{a} - n_2\mathbf{b}$, $n_2\mathbf{b} - n_3\mathbf{c}$, and $n_3\mathbf{c} - n_1\mathbf{a}$ (which form the base of the tetrahedron created by the intersection of the lattice plane with the three principal axes). If **K** is to be perpendicular to our lattice plane, its dot product with these three vectors must vanish, yielding the conditions

$$n_1m_1 - n_2m_2 = 0; \qquad n_2m_2 - n_3m_3 = 0; \qquad n_3m_3 - n_1m_1 = 0.$$

These equations are satisfied by

$$m_1 = \frac{1}{n_1},\ m_2 = \frac{1}{n_2},\ \text{and } m_3 = \frac{1}{n_3}. \tag{9.17}$$

But apart from a common integer reducing m_1, m_2, m_3 to the set of smallest integers, this is the definition of the Miller indices and hence **K** is perpendicular to this plane.

To connect the Laue equations (expressed in reciprocal space by the vector equation (9.16)) with the Bragg equation, we express the separation between neighboring parallel lattice planes in real space in terms of a reciprocal lattice vector. From the conclusion of the previous paragraph we may write the equation of a lattice plane in normal form as

$$\hat{\mathbf{K}}(m_1, m_2, m_3) \cdot \mathbf{r} = d(n_1, n_2, n_3), \tag{9.18}$$

where $\hat{\mathbf{K}} \equiv \mathbf{K}/|\mathbf{K}|$ is a unit vector parallel to **K** and d is the perpendicular distance from the origin to a plane having integer values n_1, n_2, n_3 for its intercepts (with associated vectors from the origin $n_1\mathbf{a}$, $n_2\mathbf{b}$, $n_3\mathbf{c}$). Evaluating (9.18) for each of these three intercept vectors, we obtain, using Eq. (9.11), the relations

$$\begin{aligned} 2\pi\, m_1n_1 &= |K(m_1,m_2,m_3)|\, d(n_1,n_2,n_3); \\ 2\pi\, m_2n_2 &= |K(m_1,m_2,m_3)|\, d(n_1,n_2,n_3); \\ 2\pi\, m_3n_3 &= |K(m_1,m_2,m_3)|\, d(n_1,n_2,n_3); \end{aligned} \tag{9.19}$$

i.e., the products of the integers m_in_i must be equal to the same integer, which we denote as N. If N is the *smallest* possible integer this condition is the definition of the Miller indices m_1, m_2, m_3 associated with a lattice plane defined by n_1, n_2, n_3. If we imagine a lattice plane passing through the origin (parallel to the above plane n_1, n_2, n_3) and ask for the intercepts of the *closest* plane, these intercepts are *not* the integers n_1, n_2, n_3, but rather the fractions $\frac{n_1}{N}, \frac{n_2}{N}, \frac{n_3}{N}$. Writing $d\left(\frac{n_1}{N}, \frac{n_2}{N}, \frac{n_3}{N}\right) = d_{min}$, we obtain the condition

$$|\mathbf{K}(m_1,m_2,m_3)|\, d_{min} = 2\pi, \tag{9.20}$$

where m_1, m_2, m_3 are the Miller indices of a given set of lattice planes.

9.4 Relating the Laue equations to Bragg's law

Figure 9.5 shows the geometry of the Bragg reflection process in terms of the associated wave vectors. Referring to Fig. 9.5(b), we have

$$|\Delta\mathbf{k}| = 2k\sin\theta. \tag{9.21}$$

Combining Eqs. (9.16), (9.20), and (9.21) and recalling $k = 2\pi/\lambda$ yields

$$\lambda = 2d_{min}\sin\theta. \tag{9.22a}$$

Allowing for the fact that *multiples*, n, of the shortest reciprocal vector (which is specified by the Miller indices) also satisfy the Laue conditions, we may generalize (9.22a) to

$$n\lambda = 2d\ \sin\theta, \tag{9.22b}$$

where we write d for d_{min}. This completes our task of obtaining Bragg's law from the Laue equations.

9.5 The Ewald construction

There is a geometrical way of representing the Laue diffraction conditions. The latter are given by $\Delta\mathbf{k} = \mathbf{K}$ or $\mathbf{k}' = \mathbf{K} + \mathbf{k}$. Since $|\mathbf{k}'| = |\mathbf{k}|$ and $|\mathbf{k}| = \text{const.}$ defines a sphere, we see that we will have diffraction (i.e., satisfy the Bragg condition) when a sphere in wave vector space of radius $|\mathbf{k}|$ (in reciprocal space) exists which contains K as a chord. This is referred to as the *Ewald construction.*

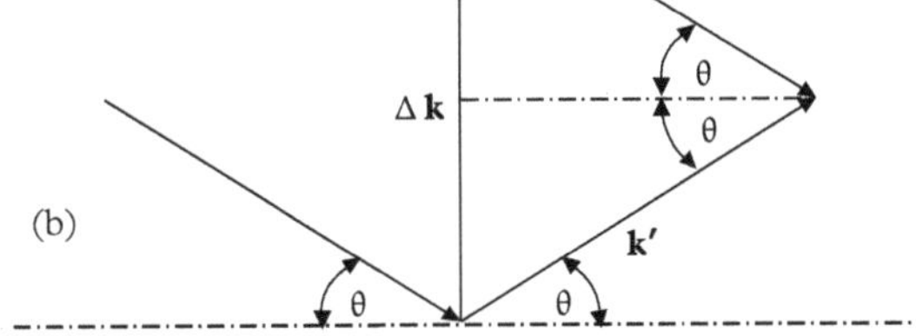

Figure 9.5 *The Bragg reflection. The upper figure (a) shows the relation of the incoming,* k, *and scattered,* k′, *wave vectors with respect to the Bragg scattering plane (having Miller indices m_1, m_2, m_3). In the lower figure (b) we have constructed the "momentum triangle" by translating the incoming wave vector.*

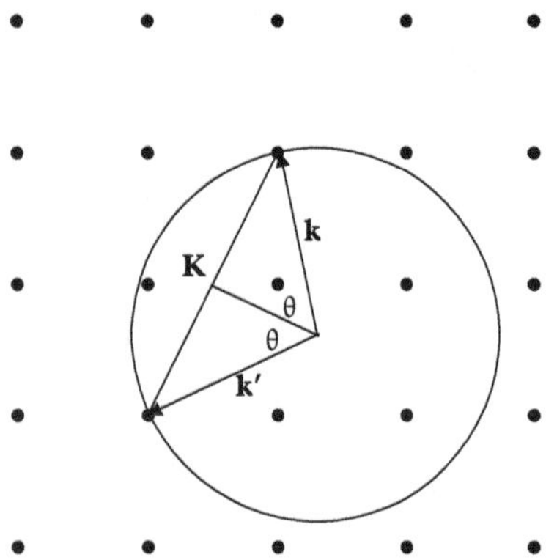

Figure 9.6 *The Ewald construction.*

This will fix the Bragg angle θ. However the Bragg condition does not fix the azimuthal angle (the angle measured from some arbitrary axis lying in the real-space Bragg plane). In the Ewald construction this arbitrariness corresponds to the freedom to rotate the (Ewald) sphere about the chord axis. This construction is shown in Fig. 9.6.

9.6 The Brillouin zone

As in the previous section, we write the Laue conditions in the form $\mathbf{k}' = \mathbf{k} + \mathbf{K}$. Squaring this relation and again noting that $|\mathbf{k}| = |\mathbf{k}'|$ we obtain

$$2\mathbf{k} \cdot \mathbf{K} = -K^2$$

or

$$\mathbf{k} \cdot \hat{\mathbf{K}} = -\frac{|\mathbf{K}|}{2}. \tag{9.23}$$

If we ask for the smallest **k** vector satisfying Eq. (9.23), it is directed antiparallel to **K** and has a length

$$k_{\min} = \frac{|\mathbf{K}|}{2}; \tag{9.24}$$

i.e., it is half the length of the associated reciprocal lattice vector. Hence any **k** vector lying on a perpendicular plane which *bisects* the reciprocal lattice vector will be diffracted.

Now each of the vectors **K** of the reciprocal lattice leads to a different plane (of the form (9.23)). Suppose we sort the reciprocal lattice vectors according to their length, each associated with its own plane. The intersections of these planes form polyhedra in reciprocal space. If we select the set of planes leading to the polyhedral having the *smallest* volume, this surface is referred to as the *first Brillouin zone*. If from the *remaining* reciprocal lattice vectors we select the set leading to the next smallest volume (having the same overall symmetry), we create the *second* Brillouin zone, and so forth. Brillouin zones play a very important role in solid-state physics, as we will see in later chapters. Figure 9.7 shows the first and second Brillouin zones for a 2d square lattice and the associated reciprocal lattice vectors.

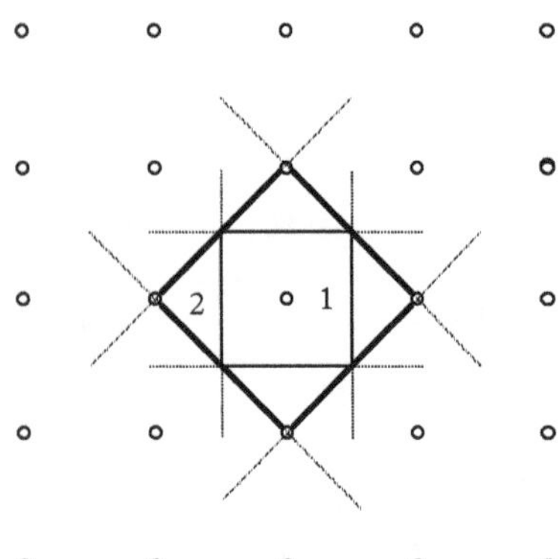

Figure 9.7 *The reciprocal lattice and the first Brillouin zone of a square real-space lattice showing the first and second Brillouin zones.*

We now summarize the rules for constructing the various Brillouin zones:

(i) pick an arbitrary reciprocal lattice point as the origin;

(ii) draw reciprocal lattice vectors from this origin to neighboring reciprocal lattice points;

(iii) bisect these vectors with planes passing normal to them;

(iv) starting with the shortest reciprocal lattice vectors, find the polyhedron formed by these planes having the smallest volume; this will be the first Brillouin zone;

(v) from the *remaining* vectors (and their associated planes) find the next-smallest-volume polyhedron (having the same overall symmetry and an added volume equal to the first zone); this surface will enclose the second Brillouin zone; etc.

9.7 The geometrical structure factor

Our derivation of the Laue equations suppresses the fact that our lattice might have a basis. It is easy to introduce a basis into the theory by considering the sum over all atoms in Eq. (9.3) as two separate sums, one over the unit cells (designed by the $\mathbf{R}_n$) and the second over the atoms within the unit cell, the position of the jth being designated by

$$\mathbf{r}_j = r_{ja}\mathbf{a} + r_{jb}\mathbf{b} + r_{jc}\mathbf{c}. \tag{9.25}$$

We write these two contributions to F in Eq. (9.7) as

$$F = \sum_{\mathbf{R}_n} e^{-i\Delta\mathbf{k}\cdot\mathbf{R}_n} \sum_j f_j\, e^{-i\Delta\mathbf{k}\cdot\mathbf{r}_j}, \tag{9.26}$$

where j runs over all atoms in the unit cell and f_j is the scattering amplitude of the jth atom; we have also inserted the Laue conditions, $\Delta\mathbf{k} = \mathbf{K}(m_1,m_2,m_3)$.

The sum involving the basis in Eq. (9.26) is referred to as the *geometrical structure factor*, or the structure factor for short, and is written as

$$S(m_1, m_2, m_3) = \sum_{j=1}^{\substack{\text{all atoms}\\ \text{in unit cell}}} f_j(\Delta\mathbf{k}(m_1, m_2, m_3))\, e^{-i\Delta\mathbf{k}(m_1,m_2,m_3)\cdot\mathbf{r}_j}. \tag{9.27}$$

Note that the structure factor may cause the diffraction intensity to vanish for special combinations of m_1, m_2, m_3 in a crystal having a basis. A well-known example is the body centered cubic lattice, where $S(m_1,\ m_2,\ m_3) = 0$ if $(m_1 + m_2 + m_3) = $ (odd integer).

9.8 The atomic scattering factor

As previously discussed, it is the individual electrons making up the atoms which scatter the X-rays, and the dipole radiation is proportional to accelerations, $\ddot{\mathbf{r}}$, where the latter may be calculated from (3.20) in the classical model of Lorentz.

In the limit where the electrons are tightly bound ($\omega_0 \gg \omega$) the acceleration follows from Eq. (3.20) as

$$\ddot{\mathbf{r}} = \frac{|e|}{m}\frac{\omega^2}{\omega_0^2}\mathbf{E}. \tag{9.28a}$$

In a semi-classical many electron atom we would have a frequency ω_{0i} for each of the electrons in various shells. In the limit where $\omega \gg \omega_0$ (free electrons) we have

$$\ddot{\mathbf{r}} = -\frac{|e|}{m}\mathbf{E}. \tag{9.28b}$$

Quantum mechanically the electrons are distributed with a density

$$n(\mathbf{r}) = \int d^3r_2 \ldots d^3r_N \,|\psi(\mathbf{r}_1 \ldots \mathbf{r}_n)|^2. \tag{9.29}$$

So we expect the scattering power within an atom to be proportional to $n(\mathbf{r})$. There will be an additional correction which is the quantum mechanical equivalent of the position-dependent spring constant (or frequency ω_{0i}) in the Lorentz model. We will not further elaborate on the problem of X-ray scattering. Instead we will simply assume that each volume element of the atom radiates with an amplitude which is proportional to $n(\mathbf{r})$ and at a phase governed by its position in the atom. The scattering amplitude of the atom will involve a sum (integral) over all of the volume elements; if we assume $n(\mathbf{r})$ is spherically symmetric, we have

$$\begin{aligned} f(\Delta\mathbf{k}) &= f_0 \int d^3r\, n(\mathbf{r}) e^{-i\Delta\mathbf{k}\cdot\mathbf{r}} \\ &= f_0 \int_0^{\infty}\int_0^{\pi}\int_0^{2\pi} r^2 dr \sin\theta d\theta d\phi\, n(r) e^{-ir\Delta k\cos\theta} \\ &= 2\pi f_0 \int_0^{\infty} r^2 n(r) dr \frac{e^{ir\Delta k} - e^{-ir\Delta k}}{ir\Delta k} \\ &= 4\pi f_0 \int_0^{\infty} n(r) \frac{\sin\Delta kr}{\Delta kr} r^2 dr, \end{aligned} \tag{9.30}$$

where the constant f_0 absorbs all remaining factors. The function $f(\Delta\mathbf{k})$ is called the *atomic scattering factor*. During Bragg diffraction it would be evaluated for $\Delta\mathbf{k} = \mathbf{K}$, where $\mathbf{K}$ is the reciprocal lattice vector associated with the diffraction.

9.9 The Debye–Waller factor

The atoms occupying the lattice (or basis) sites of a crystal are not at rest, but are in constant motion. This motion arises from two sources: (i) thermal motion and (ii) quantum mechanical zero-point motion. Intuitively, one might imagine two outcomes for the effect of this motion on the X-ray diffraction peaks: (i) the peaks diminish in amplitude, or (ii) the peaks develop a width (in addition to that associated with the diffraction function $\frac{\sin^2 Nx}{\sin^2 x}$, that is very narrow). A detailed analysis, which we will not present here, shows that the former conclusion is the correct one: the peaks, are reduced in amplitude by a multiplicative factor $e^{-2W(m_1,m_2,m_3)}$, where $W(m_1, m_2, m_3)$ is referred to as the *Debye–Waller factor*. The size of this factor increases with increasing temperature and decreases with increasing mass of the atoms. It is finite at $T = 0$ due to the zero-point motion.

9.10 Sources of X-rays

In order to study the structure of crystals using X-ray diffraction we require: (i) a source of X-rays, (ii) a material structure, and (iii) a sample geometry appropriate to this material and source. We begin with a discussion of the first of these requirements. Nowadays X-rays are obtained from two sources: an X-ray tube (the conventional source) and an electron synchrotron source.

In a tube source, electrons which are emitted thermionically (e.g., from a tungsten filament) are allowed to fall through a large potential drop (typically 50 kV) and impact a target. On entering this target the electrons are rapidly decelerated due to various inelastic processes; this results in the emission of X-rays. Such X-rays, which have a range of wavelengths, are referred to as *white* X-rays, in analogy to white light, which also contains a broad range of wavelengths. The incoming electrons also excite core electrons to unoccupied higher energy states. As these latter electrons lose energy and fall back into the unoccupied orbitals, they emit X-rays at the characteristic frequencies associated with transitions among those orbitals. The frequencies associated with these transitions are quite "sharp," being broadened only by the decay time through the energy–time uncertainty principle. Such X-rays are referred to as *monochromatic*, and their frequencies are determined by the atoms making up the target, the higher atomic number materials leading to the availability of higher characteristic emission frequencies. Among the set of discrete frequencies one may separate out a specific frequency using the Bragg diffraction process itself.

There is a limit to the amount of energy input an X-ray tube anode can sustain without rapidly destroying (or ultimately melting) the target. Greater currents can be utilized if the target rotates, thereby distributing the incoming electron beam over a wider area; voltages from 100 to 200 kV can then be utilized. Such sources are called *rotating anode* sources.

In recent years there has been a revolution in the intensity of available X-rays brought about by synchrotron accelerator sources. The electron synchrotron exploits the fact that an electron traveling in a circular orbit is being continually accelerated toward the orbit center and hence radiates. Low energy electrons would radiate at the cyclotron frequency in the usual figure-of-eight pattern associated with dipole radiation. However at relativistic electron kinetic energies the emitted radiation covers a broad band,[8] extending up to some cut-off energy and, in addition, the emitted radiation is highly directional, having the maximum intensity in a direction tangent to the beam.

In addition to its circulating path, the electron beam can be made to pass through various devices (inserted into the electron "storage" ring), known as wigglers or undulators (depending on the design), which artificially apply an additional spatially periodic acceleration (produced by a line of fixed and alternately directed magnets) to the electron beam thereby enhancing the radiation; through proper design this radiation can be highly monochromatic, even coherent.

A monochromatic X-ray beam can be extracted from a broad band synchrotron beam by again exploiting the Bragg diffraction process, typically from silicon due to the availability of highly perfect crystals. The intensities, either broad band or monochromatic, obtainable from synchrotrons greatly exceed those from tube sources by many orders of magnitude, although the expense of the instrumentation is much higher with the synchrotron.

9.11 Experimental methods to study X-ray diffraction

We now discuss some experimental geometries appropriate to both broad band and monochromatic X-ray sources; in all cases we will assume the incoming X-rays are all traveling in the same direction (referred to as a *collimated beam*). We begin with experimental techniques using a monochromatic beam.

9.11.1 The rotating crystal method

Crystalline materials may be distinguished by whether they are single crystals or polycrystals; the former is a material where the primitive crystallographic axes point in the same direction in space at all points in the material, while in the latter, boundaries exist within the material separating regions (also called grains or domains) where the primitive axes are rotated on crossing the boundary (called a grain boundary). If a monochromatic X-ray impinges on an arbitrarily oriented single crystal, the Bragg condition, $n\lambda = 2d \sin\theta$, will in general *not* be met: once the wavelength is fixed so is the Bragg angle for a given plane (described by a set of Miller indices) or order, n; other planes (with different Miller indices) will have different Bragg angles, but in general none of this discrete set of angles will match

[8] More precisely, the emitted energies involve harmonics of the cyclotron frequency $\varepsilon_n = n\hbar\omega_c$, which are very closely spaced at high energies.

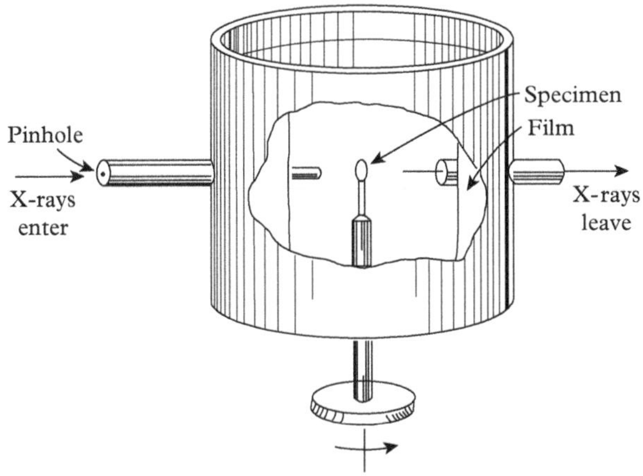

Figure 9.8 *The rotating crystal method.*

with those associated with the arbitrary orientation of the crystal (an accidental match is of course possible).

If we *rotate* the crystal about an axis other than the beam axis (an axis perpendicular to the beam axis being maximally effective), the crystal will sweep a set of Bragg planes through the Bragg angles associated with the incoming wavelength and the inter-planer d-spacings (which depend on how the crystal is oriented relative to the beam). Provided that the wavelength of the X-rays is chosen suitably, the Bragg condition will then be met for discrete values of the crystal rotation angle.

If the sample is surrounded with photographic film (usually a plane film rolled into a cylinder), the exiting or diffracted beams will produce spots on the film after development (X-rays are able to expose suitably prepared photographic film[9]). From the position of these spots and a knowledge of the placement of the film relative to the crystal, the angular positions of the Bragg plane normals and their d-spacings may be determined and with it information on the crystal structure. The relative intensity of the spots yields information about the structure factor, i.e., the placement of the atoms in the unit cell (for crystals possessing a basis). Figure 9.8 shows the geometry of the rotating crystal experiment.

9.11.2 The powder method

Another way to change the orientation of a crystal relative to the incoming beam is to grind the crystal into a powder. In so doing the resulting powdered sample presents essentially all possible orientations of the basic crystal to the incoming beam. Those crystallites (grains) satisfying the Bragg condition for some plane will diffract, while the remaining ones (which form the majority of the sample) will not. The diffracted beam will lie on a cone making an angle 2θ with the incoming beam and with a uniform intensity as a function of ϕ. The intersection of these

[9] A more efficient way to collect data is to have the diffracted beams fall on a screen containing a phosphor that emits light; this plate can then be imaged onto a charge coupled device (a CCD chip of the kind currently utilized in digital cameras) the output of which can then be transferred directly to a computer.

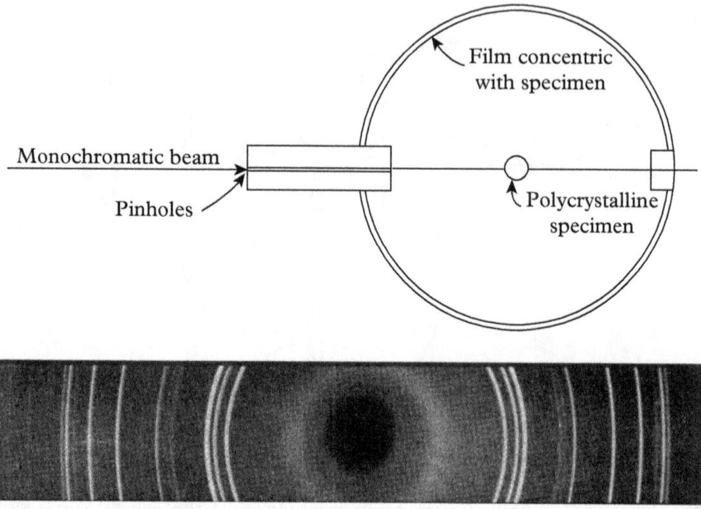

Figure 9.9 *The experimental arrangement of the powder-pattern method is shown in the upper figure; an example of a resulting film where the bright arcs correspond to the diffraction maxima.*

cones with a photographic film which is wrapped around the sample produces lines or bands on the developed film. From the placement of the film relative to the sample and the spacing of the bands, the Bragg angles can be measured and, from the known wavelength, the crystal d-spacings can be determined. As with the rotating crystal method, the relative intensity of the bands yields structure factor information. Although not necessary, if the material is finely powdered, it is customary to rotate the sample in order to obtain a better averaging over orientations. The geometry of this technique is shown in Fig. 9.9.

9.11.3 The Laue method

As discussed in connection with the rotating crystal method, we do not expect the Bragg condition to be satisfied when a monochromatic beam strikes a single crystal of an arbitrary orientation. However, if we replace the monochromatic beam with a white beam, the crystal will select those wavelengths from the incoming beam which satisfy the Bragg condition. The emerging X-rays occur at discrete angles, θ, ϕ, relative to the incoming beam. If a photographic film is placed so as to intercept several of these diffracted beams,[10] spots will appear on the film after development. From the positions of these spots and a knowledge of how the film is placed relative to the beam and sample, one can deduce the exit angles (θ, ϕ) of the diffracted X-rays. By measuring these angles we can determine the angles that the Bragg planes make with the incoming beam and from this data we may obtain information on the crystal structure and its orientation relative to the incoming beam. However since the film is sensitive to a broad range of wavelengths, information as to what wavelength was selected by the crystal is lost and with it

[10] The film (with a hole cut in it for beam passage) is usually placed perpendicular to the incoming beam; a geometry referred to as Laue back-diffraction. For small or weakly absorbing samples the film may be placed perpendicular to the beam on the far side of the sample; a geometry referred to as transmission Laue diffraction.

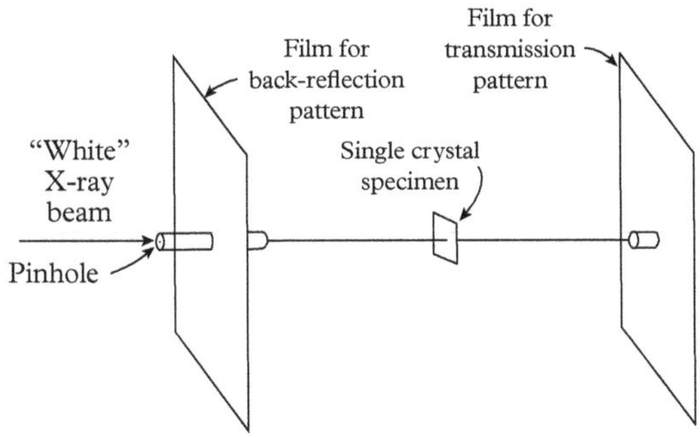

Figure 9.10 *The experimental arrangement associated with the Laue method.*

the ability to measure the Bragg plane d-spacings. Hence the Laue method is primarily used to orient crystals. Figure 9.10 shows the experimental arrangement for the Laue method.

ADDITIONAL READING

There is a brief but excellent discussion of the theory of X-ray scattering, including thermal effects and diffuse scattering (which we have not discussed) in *Electrodynamics of Continuous Media*, Chapter XVI, L.D. Landau and E.M. Lifshitz, Pergamon Press, Oxford.

Elements of X-Ray Diffraction, B.D. Cullity and S.R. Stock, Addison Wesley, Reading Massachusetts (1978).

X-ray Diffraction in Crystals, Imperfect Crystals and Amorphous Bodies, A. Guinier, Dover (1963).

Theory of X-ray Diffraction in Crystals, W.H. Zachariasen, John Wiley and Sons, Inc., (1944).

Problems

9.1. The drawing depicts a diffraction grating where the open circles are the lines of a grating viewed edge-on. What is the condition that the light be diffracted at an angle θ' if it comes in at an angle θ? What is the relation between the resulting expression and Bragg's law? Why is Bragg's law more restrictive in that we must then have $\theta = \theta'$?

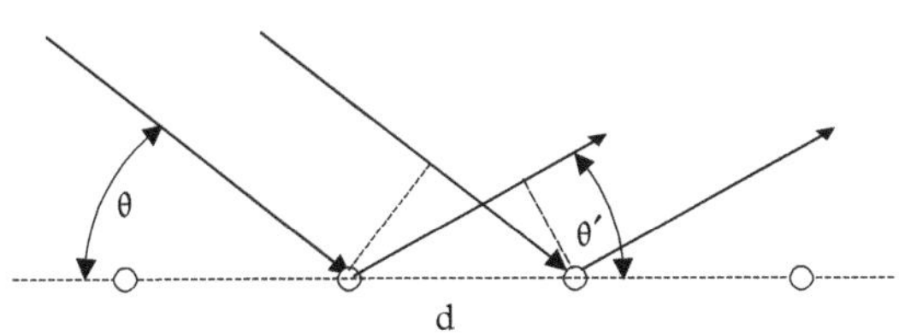

9.2. Suppose that rather than having an array of lines, as in Problem 9.1, we have an array of dots lying on a square lattice with an inter-dot spacing d. Use a Laue-like theory to calculate the angles θ' and φ' at which light with wavelength λ will be diffracted if it enters from an arbitrary direction specified by the angles θ and φ.

9.3. Find the condition(s) for the geometrical structure factor to vanish for the fcc and hcp lattices.

9.4. Suppose the number density of electrons in an atom has the form $n(\mathbf{r}) = n_0 e^{-r/r_0}$: calculate the atomic scattering factor.

Appendix 9A Evaluating lattice sums by the Ewald method

Calculation of certain solid-state properties requires repetitive evaluation of the Coulomb potential arising from a lattice of point charges.[11] We have already encountered such a sum when evaluating the Madelung constant in Section 7.3 for the NaCl structure, where we used an ad hoc, slowly convergent method. It is important to have a more rapidly convergent procedure. The formalism developed in Sections 8.1 and 8.2 to designate atom sites, together with the representation of charge densities as the Fourier expansion given by Eq. (9.15), give us the necessary mathematical tools to develop a more powerful way to evaluate sums of Coulomb potentials. The method was developed by Ewald,[12] and the basic idea is to partition the sum into real space and reciprocal space components in a manner that can be optimized so as to be rapidly convergent.

The effective charge density of the ions will here be represented as a sum of delta functions positioned at the ion sites:[13]

$$\rho(\mathbf{r}) = \sum_n \sum_j q_j \, \delta(\mathbf{r} - \mathbf{R}_n - \mathbf{r}_j), \tag{9A.1}$$

where $\mathbf{R}_n$ is the position of the nth unit cell, $\mathbf{r}_j$ is the coordinate of the jth atom in the unit cell, and q_j is the charge of that atom. The associated potential is

$$\phi(\mathbf{r}) = \sum_n \sum_j \frac{q_j}{|\mathbf{r} - \mathbf{R}_n - \mathbf{r}_j|}. \tag{9A.2}$$

The underlying idea of the Ewald method is to add and subtract auxiliary Gaussian charge densities on each atom, the effect of which is evaluated in two different ways. We write this auxiliary Gaussian charge density associated with the jth position in the unit cell (and centered on that atom) as

$$\rho_j^g(\mathbf{r}) = q_j (\alpha/\pi)^{3/2} e^{-\alpha r^2}, \tag{9A.3}$$

[11] Examples are band structure energies in the Green's function formalism and phonon spectra, to be discussed in Sections 13.2 and 16A respectively.

[12] P.P. Ewald, *Ann. Phys.* **64**, 253 (1921).

[13] It is common to combine the core electron charge with the nuclear charge, the total being that of the ion. As long as one is considering a position outside the core electron charge density, and that charge density is spherically symmetric, Gauss' law allows the core charge to be combined with the nuclear charge and concentrated at the center as an *effective ion charge*, q_j.

which has been normalized such that the integral over all space is equal to the charge q_j of the jth atom; the parameter α can be chosen to optimize the convergence. The auxiliary Gaussian charge density of the lattice *as a whole*, $\rho^{(R)}(\mathbf{r})$, is then

$$\rho^{(R)}(\mathbf{r}) = \sum_n \sum_j \rho_j^g(\mathbf{r} - \mathbf{R}_n - \mathbf{r}_j). \tag{9A.4}$$

It can also be expressed as a Fourier series,

$$\rho^{(K)}(\mathbf{r}) = \sum_{\mathbf{K}} \rho_{\mathbf{K}} e^{i\mathbf{K}\cdot\mathbf{r}}. \tag{9A.5}$$

We emphasize that $\rho^{(R)}$ and $\rho^{(K)}$ represent the same function and our labeling them differently is to identify them as sums in real space and reciprocal space respectively.

The potential associated with (9A.5) can be written as

$$\phi^{(K)}(\mathbf{r}) = \sum_{\mathbf{K}} \phi_{\mathbf{K}} e^{i\mathbf{K}\cdot\mathbf{r}} \tag{9A.6}$$

and from Poission's equation, $\nabla^2\phi = -4\pi\rho$, we have

$$\phi_{\mathbf{K}} = 4\pi \frac{\rho_{\mathbf{K}}}{K^2}. \tag{9A.7}$$

We now substitute (9A.3) into (9A.4) and equate the result to (9A.5); multiplying both sides of this equation by $e^{-i\mathbf{K}'\cdot\mathbf{r}}$, integrating over a unit cell of volume Ω, and exploiting the orthogonality for different $\mathbf{K}$ gives

$$\begin{aligned}\rho_{\mathbf{K}} &= \frac{1}{\Omega}\left(\frac{\alpha}{\pi}\right)^{3/2} \int_{\text{unit cell}} d^3r \sum_{\mathbf{R}_n} \sum_j q_j e^{-\alpha|\mathbf{r}-\mathbf{R}_n-\mathbf{r}_j|^2} e^{-i\mathbf{K}\cdot\mathbf{r}} \\ &= \left(\frac{\alpha}{\pi}\right)^{3/2} \int_{\text{all space}} d^3r \sum_j q_j e^{-\alpha|\mathbf{r}-\mathbf{r}_j|^2} e^{-i\mathbf{K}\cdot\mathbf{r}},\end{aligned}$$

where in the second step we have used the fact that by integrating only the Gaussians centered in a single cell, *but over all space*, we account for the tails from surrounding cells. Carrying out the integration by completing the square we obtain

$$\begin{aligned}\rho_{\mathbf{K}} &= \frac{1}{\Omega}\left(\sum_j q_j e^{-i\mathbf{K}\cdot\mathbf{r}_j}\right) e^{-K^2/4\alpha} \\ &= \frac{1}{\Omega} S(\mathbf{K}) e^{-K^2/4\alpha},\end{aligned} \tag{9A.8}$$

where in the second step we have introduced a *charge structure factor*, analogous to the geometric structure factor introduced in Section 9.7. Combining (9A.7) and (9A.8), Eq. (9A.6) becomes

$$\phi^{(K)}(\mathbf{r}) = \frac{4\pi}{\Omega}\sum_{\mathbf{K}}\frac{S(\mathbf{K})}{K^2}e^{-K^2/4\alpha}\,e^{i\mathbf{K}\cdot\mathbf{r}}, \tag{9A.8}$$

which is the full potential of a lattice of Gaussian charge distributions with charges equal to those of the point charge lattice given in Eq. (9A.2).

We now develop an *alternative* expression for the total potential created by the Gaussian charge densities by: (i) directly integrating the contribution from each lattice site, and (ii) summing the result over all sites. The potential $\phi_j^g(\mathbf{r})$ at a point $\mathbf{r}$ from a *single* Gaussian centered at $\mathbf{r} = 0$ is given by

$$\begin{aligned}\phi_j^g(\mathbf{r}) &= \frac{1}{r}\int_0^r \rho_j^g(r)\,d^3r' + \int_r^\infty \frac{\rho_j^g(r')}{r'}d^3r' \\ &= q_j\left(\frac{\alpha}{\pi}\right)^{3/2}\left[\frac{1}{r}\int_0^r e^{-\alpha r'^2}d^3r' + \int_r^\infty \frac{e^{-\alpha r'^2}}{r'}d^3r'\right],\end{aligned} \tag{9A.9}$$

where, since the Gaussian is isotropic, the contribution from inside a sphere of radius r follows directly from Gauss' law from the charge contained. Writing the first integral as

$$\frac{1}{r}\int_0^r e^{-\alpha r'^2}d^3r' = \frac{1}{r}\int_0^\infty e^{-\alpha r'^2}d^3r' - \frac{1}{r}\int_r^\infty e^{-\alpha r'^2}d^3r'$$

and carrying out all the integrations yields

$$\begin{aligned}\phi_j^g(\mathbf{r}) &= \frac{q_j}{r}\left[1 - 2\left(\frac{\alpha}{\pi}\right)^{1/2}\int_r^\infty e^{-\alpha r'^2}dr'\right] \\ &= \frac{q_j}{r}\left[1 - F\left(\alpha^{1/2}r\right)\right];\end{aligned} \tag{9A.10}$$

here we have introduced the function

$$F(x) = \frac{2}{\pi^{1/2}}\int_x^\infty e^{-x'^2}dx'. \tag{9A.11}$$

Equation (9A.10) can also be written in terms of the error function, $\mathrm{erf}(x) = 1 - F(x)$, where $\mathrm{erf}(x) \equiv (2/\pi^{1/2})\int_0^x e^{-x^2}dx$.[14] For the special case where $\mathbf{r} = 0$, the potential is given by

$$\begin{aligned}\phi_j^g(\mathbf{r} = 0) &= \int_0^\infty \frac{\rho_j^g(r')}{r'}d^3r' \\ &= 2q_j\left(\frac{\alpha}{\pi}\right)^{1/2}.\end{aligned} \tag{9A.12}$$

[14] *Handbook of Mathematical Functions*, M. Abramowitz and I.A. Stegun, p. 297, Dover (1972).

The *full* Gaussian potential in this representation, which we designate as $\phi^{(R)}(\mathbf{r})$, is then a sum of contributions of the form (9A.10) over all the lattice sites,

$$\begin{aligned}\phi^{(R)}(\mathbf{r}) &= \sum_{\mathbf{R}_n}\sum_{j}\frac{q_j}{|\mathbf{r}-\mathbf{R}_n-\mathbf{r}_j|}\left[1-F\left(\alpha^{1/2}|\mathbf{r}-\mathbf{R}_n-\mathbf{r}_j|\right)\right] \\ &= \phi(\mathbf{r}) - \sum_{\mathbf{R}_n}\sum_{j}\frac{q_j}{|\mathbf{r}-\mathbf{R}_n-\mathbf{r}_j|}F\left(\alpha^{1/2}|\mathbf{r}-\mathbf{R}_n-\mathbf{r}_j|\right). \end{aligned} \tag{9A.13}$$

Note that Eqs. (9A.8) and (9A.13) are different ways of expressing the *same* Gaussian lattice potential; also note that this expression explicitly contains the potential of the pure Coulomb point-charge lattice, Eq. (9A.2).

Now the object in calculating a Madelung sum is to obtain the potential at some *given* lattice site arising from the charges on all *other* site *s*. For this purpose we must leave the contribution from the chosen site, $\mathbf{n}, j$ (which is the infinite self energy), out of the sum; thus

$$\begin{aligned}\phi(\mathbf{r}=\mathbf{R}_n+\mathbf{r}_j) &= \sum_{n'}\sum_{j'}\frac{q_j}{|\mathbf{R}_n-\mathbf{R}_{n'}+\mathbf{r}_j-\mathbf{r}_{j'}|}(1-\delta_{nn'}\delta_{jj'}) \\ &\equiv \phi_{nj}.\end{aligned} \tag{9A.14}$$

The next step is to *add and subtract* the corresponding Gaussian potentials evaluated at the chosen site, but expressed as sums over real space lattice sites and reciprocal space lattice vectors respectively, as obtained through Eqs. (9A.12) and (9A.8). Choosing the origin such that $\mathbf{R}_n + \mathbf{r}_j = 0$, we then have

$$\begin{aligned}\phi_{nj} &= \phi_{nj} - \phi_{nj}^{(R)} + \phi_{nj}^{(K)} \\ &= \sum_{n}\sum_{j}\frac{q_j}{|\mathbf{R}_n+\mathbf{r}_j|}F\left(\alpha^{1/2}|\mathbf{R}_n+\mathbf{r}_j|\right) + \frac{4\pi}{\Omega}\sum_{\mathbf{K}}\frac{S(\mathbf{K})}{K^2}e^{-K^2/4\alpha} - 2q_j\left(\frac{\alpha}{\pi}\right)^{1/2},\end{aligned} \tag{9A.15}$$

where in the last term we have used (9A.12) to remove the contribution of the chosen lattice site from Eq. (9A.8); note that the pure Coulomb terms *cancel*. By choosing a suitable value of α the overall convergence of Eq. (9A.15) can be optimized.

Part III

Electronic Structure of Periodic Solids

Electrons in a Periodic Solid

10

10.1 The crystal potential

In Chapter 6 we introduced a model of a metal in which the mutual interaction between the conduction electrons and their interactions with atomic core electrons and the nucleus were neglected. This allowed us to understand some properties associated with the Pauli principle and quantum degeneracy. However, a host of other solid-state properties cannot be understood with such a simple model. For example, it tells us nothing about the origin of the differences between metals and insulators or the cohesive energy (which is responsible for binding the atoms into a solid in the first place). As a first step in improving on the free electron model of solids, we will assume that the interaction of the conduction (also called valance) electrons of the material with other conduction/valence electrons and the atomic cores, together with the nuclei, can be "averaged over" and approximated by a so-called one-body or one-electron potential energy, $V(\mathbf{r})$. In doing this, direct electron–electron interactions, as well as the effects of the quantum mechanical exchange among electrons, is replaced by an approximate, model-dependent interaction (often chosen to be a function of the electron density alone), which we will discuss in Chapter 15.

We do not initially know the one-electron potential energy $V(\mathbf{r})$, but as a starting approximation we can utilize the potential energies of the individual atoms, $U(\mathbf{r})$.[1] If we sum the contribution from each atom (for now we will assume our crystal has only one atom per unit cell), the potential energy arising from all the atoms will be

$$V(\mathbf{r}) = \sum_{R} U(\mathbf{r} - \mathbf{R}), \tag{10.1}$$

where $\mathbf{R}$ is the set of all real-space lattice vectors.[2] For the case of an infinite solid, $V(\mathbf{r})$ is periodic; i.e.,

$$V(\mathbf{r} + \mathbf{R}) = V(\mathbf{r}). \tag{10.2}$$

Figure 10.2 shows the potential of a one-dimensional system with ten atoms, each of which contributes a screened Coulomb potential.

[1] As an example see F. Herman and S. Skillman, *Atomic Structure Calculations*, Prentice-Hall, Englewood Cliffs (1963).

[2] In order for this sum to be convergent in a monatomic solid, $U(\mathbf{r})$ cannot have a long-range $1/r$ behavior; i.e., it cannot be the potential associated with a sum over ions and must be a so-called fully screened potential (that implicitly involves the other conduction electrons), or be charge compensated in some way.

The Physics of Solids. First Edition. J. B. Ketterson.

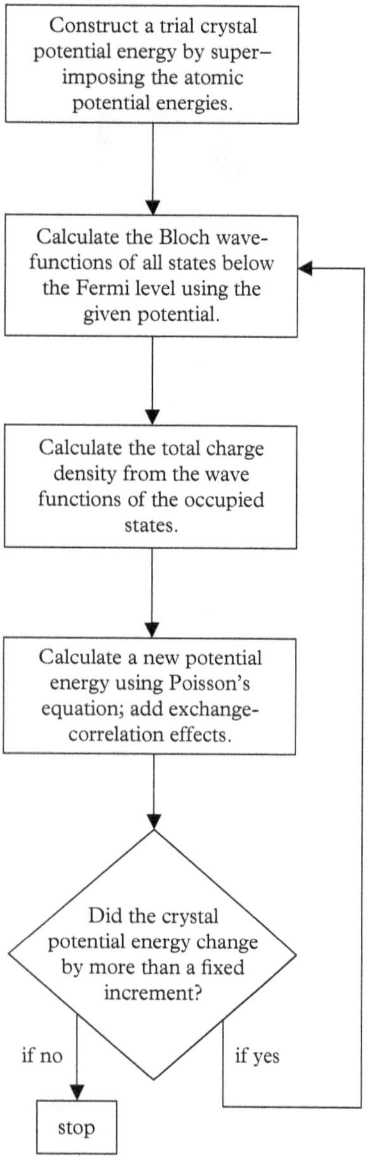

Figure 10.1 *Flow diagram for solving the one-electron Schrödinger equation in a crystal.*

For some purposes it is more convenient to write this same function as a three-dimensional Fourier series,

$$V(\mathbf{r}) = \sum_{\mathbf{K}} V_{\mathbf{K}} e^{i\mathbf{K}\cdot\mathbf{r}}, \tag{10.3}$$

where **K** is the set of all reciprocal lattice vectors (see discussion surrounding Eq. (9.15)); the $V_{\mathbf{K}}$ are then the Fourier expansion coefficients. Since the contribution from the nuclei is divergent, we expect that convergence problems will arise, which we will discuss later.

We are now faced with the problem of solving the Schrödinger equation for the motion of the conduction/valance electrons in the presence of $V(\mathbf{r})$; this will be taken up next. Of course, the process of forming the solid will modify the conduction/valance and core electron states and with it $V(\mathbf{r})$. This effect is accounted for by requiring that the charge density resulting from the crystal wave functions, calculated from $V(\mathbf{r})$, actually "produces" $V(\mathbf{r})$ through Poisson's equation; an iterative calculation is required to achieve this property, which is referred to as *self-consistency*. This process is illustrated schematically in Fig. 10.1 for the ground state of a crystal ($T = 0$).

10.2 Wave function of an electron in a periodic potential

We now establish the general form of the eigenfunctions of an electron in a periodic solid. As a result of the periodicity of our one-body potential function, $V(\mathbf{r})$, the charge density, $e|\psi(\mathbf{r})|^2$, associated with a one-electron eigenfunction, $\psi(\mathbf{r})$, must also be periodic; i.e.,

$$|\psi(\mathbf{r})|^2 = |\psi(\mathbf{r}+\mathbf{R})|^2. \tag{10.4}$$

Therefore, $\psi(\mathbf{r})$ and $\psi(\mathbf{r} + \mathbf{R})$ can differ by at most a complex constant, $C_{\mathbf{R}}$, with modulus unity; i.e., $|C_{\mathbf{R}}|^2 = 1$. We denote the constants associated with the *primitive* translations parallel to the **a**, **b**, and **c** axis as C_a, C_b, and C_c. We now assume periodic boundary conditions for translations involving N_a, N_b, and N_c displacements along the three primitive axes; for the **a** axis we must then have

$$C_a^{N_a} = 1 \tag{10.5}$$

and similarly for the **b** and **c** axes. If the extension of our crystal in the **a** direction is L_a then $N_a = L_a/|\mathbf{a}|$ (with corresponding forms for N_b and N_c). The C_a must therefore be one of the N_a^{th} *roots of unity* that are given by $\exp[i2\pi m_a/N_a]$, where $m_a = 1, 2 \ldots N_a - 1$, which satisfies (10.5) for each m_a since

$$[\exp(2\pi i m_a/N_a)]^{N_a} = \exp(2\pi i m_a) = 1.$$

We can utilize this property to examine the effect of n_a translations along the $\mathbf{a}$ axis, which generates a factor $(C_a)^{n_a}$. Writing $R_a \equiv n_a|\mathbf{a}|$ and $k_a = 2\pi m_a/N_a|\mathbf{a}|$, it then follows that we can express C_a as

$$C_a = e^{iR_a k_a}. \tag{10.6}$$

In the limit $N_a \to \infty$, k_a becomes a (countable) continuous variable and hence the phase of the exponential evolves continuously (in that sense). If we now write $\mathbf{k}_a = m_a\mathbf{a}^*/N_a$, where $\mathbf{a}^*$ is the corresponding basis vector of the reciprocal lattice (see Eq. (9.10a)), and include the contributions from the two remaining crystal axes, we may write our complex constant as

$$e^{i(k_aR_a+k_bR_b+k_cR_c)} = e^{i\mathbf{k}\cdot\mathbf{R}},$$

or equivalently

$$\psi(\mathbf{r}+\mathbf{R}) = e^{i\mathbf{k}\cdot\mathbf{R}}\psi(\mathbf{r}). \tag{10.7}$$

This form must apply at an arbitrary point $\mathbf{r}$ in the lattice.

Let us consider a particular wave function (unnormalized) having the form

$$\psi_{\mathbf{k},n}(\mathbf{r}) = e^{i\mathbf{k}\cdot\mathbf{r}}u_{\mathbf{k},n}(\mathbf{r}), \tag{10.8a}$$

where $\mathbf{k}$ and n are, respectively, continuous and discrete quantum numbers separately denoting the states of the crystal and

$$u_{\mathbf{k}n}(\mathbf{r}) = u_{\mathbf{k}n}(\mathbf{r}+\mathbf{R}) \tag{10.8b}$$

is *strictly* periodic; i.e., it contains *no position-dependent phase factor*. The form (10.8a) then satisfies (10.7) since

$$\begin{aligned}\psi_{\mathbf{k},n}(\mathbf{r}+\mathbf{R}) &= e^{i\mathbf{k}\cdot(\mathbf{r}+\mathbf{R})}u_{\mathbf{k},n}(\mathbf{r}+\mathbf{R}) = e^{i\mathbf{k}\cdot\mathbf{R}}\left(e^{i\mathbf{k}\cdot\mathbf{r}}u_{\mathbf{k},n}(\mathbf{r})\right)\\ &= e^{i\mathbf{k}\cdot\mathbf{R}}\psi_{\mathbf{k},n}(\mathbf{r});\end{aligned}$$

hence it can be taken as a general form for the eigenfunctions with respect to the quantum number $\mathbf{k}$. Our inclusion of a subscript $\mathbf{k}$ on $u_{\mathbf{k},n}(\mathbf{r})$ allows for a dependence of this periodic part on the wave vector $\mathbf{k}$; the index n is called the *band index* and its meaning will become clearer as we proceed.[3] The wave function (10.8) is referred to as a *Bloch function* (after F. Bloch (1929)); the form (10.8) was studied earlier in a mathematical context by Floquet (1883) and (10.8) is also known as the Bloch–Floquet theorem.

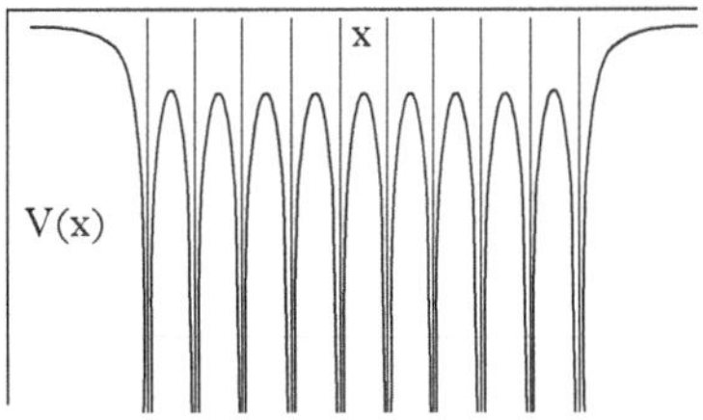

Figure 10.2 *A one-dimensional potential formed from a row of ten atoms each contributing a screened Coulomb potential.*

10.3 Properties of Bloch functions

We now examine some important properties that are associated with the Bloch periodicity condition and the resulting form of the crystal wave functions.

[3] It is often useful to think of the band index n of the Bloch wave functions as arising from the dependence of the isolated atomic wave functions on their angular momentum quantum numbers, *l*, m, and the principal quantum number, n.

10.3.1 Orthonormality

Like ordinary plane waves, $e^{i\mathbf{k}\cdot\mathbf{r}}$, the Bloch waves, $\psi_{\mathbf{k},n}(\mathbf{r})$, do not decay at large distances, and therefore cannot be normalized in the conventional way by an integral over all space. Given their periodic nature (apart from a position-dependent phase), we will normalize them to an integral over the unit cell volume, V_c, as

$$\int_{V_c} d^3r \, |\psi_{\mathbf{k},n}(\mathbf{r})|^2 = 1. \tag{10.9}$$

From the general property that non-degenerate eigenfunctions are orthogonal, we have

$$\int d^3r \, \psi^*_{\mathbf{k}',n'}(\mathbf{r})\psi_{\mathbf{k},n}(\mathbf{r}) = 0 \tag{10.10}$$

for $\mathbf{k} \neq \mathbf{k}', n \neq n'$, where the integral is over *all space*. For $\mathbf{k} = \mathbf{k}'$, (10.9) and (10.10) yield

$$\int_{V_c} d^3r \, u^*_{\mathbf{k},n'}(\mathbf{r})u_{\mathbf{k},n}(\mathbf{r}) = \delta_{nn'}. \tag{10.11}$$

10.3.2 Periodicity in k-space; energy bands

An important property of the (continuous) quantum number $\mathbf{k}$ is that it is *multi-valued*. To see this we consider the effect of a translation of the form $\mathbf{k} \to \mathbf{k} + \mathbf{K}$ on the crystal wave functions, where $\mathbf{K}$ is again a reciprocal lattice vector. From (10.7) we have

$$\begin{aligned} \psi(\mathbf{r}+\mathbf{R}) &= e^{i\mathbf{k}\cdot\mathbf{R}}\psi(\mathbf{r}) \\ &= e^{i(\mathbf{k}+\mathbf{K})\cdot\mathbf{R}}\psi(\mathbf{r}), \end{aligned} \tag{10.12}$$

where we have used the property $\mathbf{K}\cdot\mathbf{R} = 2\pi n$ in the second step, where n is an integer. Thus translation of $\mathbf{k}$ by a reciprocal lattice vector $\mathbf{K}$ results in a wave function characterized by the *same* value of the quantum number $\mathbf{k}$. Hence our crystal wave functions must be periodic in $\mathbf{K}$. Alternatively we can say that $\mathbf{k}$ is multi-valued.

Similar statements hold for the eigen energies: $E_n(\mathbf{k}) = E_n(\mathbf{k}+\mathbf{K})$ and we may therefore write

$$E_n(\mathbf{k}) = \sum_{\mathbf{R}} E_{n\mathbf{R}} e^{i\mathbf{k}\cdot\mathbf{R}}, \tag{10.13}$$

where $E_{n\mathbf{R}} = E^*_{n,-\mathbf{R}}$ since $E_n(\mathbf{k})$ is real. For these reasons we often *restrict* $\mathbf{k}$ *to lie in the first Brillouin zone* such that $|\mathbf{k}| < |\mathbf{K}|/2$, where $\mathbf{K}$ is the set of the shortest reciprocal lattice vectors which define the first Brillouin zone. This is called the *reduced zone scheme*. Since (10.13) also includes the index n, we have the property that the allowed energies in a crystal fall into *bands* within which the energies are a continuous function of the reduced wave vector, **k**.

The function $E_n(\mathbf{k})$, in addition to being periodic in **K**, must be symmetric under all crystal symmetry operations involving rotations and reflections. In addition it satisfies the relation

$$E_n(\mathbf{k}) = E_n(-\mathbf{k}), \tag{10.14}$$

even when the crystal itself does not have inversion. When time reversal symmetry is present the complex conjugates of the Bloch states must correspond to the same energy. But these states would be multiplied by a factor $e^{-i\mathbf{k}\cdot\mathbf{R}}$ on translation, and therefore correspond to wave vector $-\mathbf{k}$.[4]

10.3.3 The momentum operator and quasi-momentum

We now examine more carefully the role of the quantum number **k**; $\mathbf{p} = \hbar\mathbf{k}$ is often called the *quasi-momentum*, since it is not equal to the true momentum as we now show. We begin by calculating the expectation value of the momentum operator $\hat{\mathbf{p}} = (\hbar/i)\nabla$ within a single band:

$$\begin{aligned}\langle\hat{\mathbf{p}}\rangle &= \int d^3r\,\psi^*_{\mathbf{k},n}(\mathbf{r})\left(\frac{\hbar}{i}\nabla\right)\psi_{\mathbf{k},n}(\mathbf{r})\\ &= \hbar\mathbf{k}\int d^3r\,|\psi_{\mathbf{k},n}(\mathbf{r})|^2 + \frac{\hbar}{i}\int d^3r\,u^*_{\mathbf{k},n}(\mathbf{r})\nabla u_{\mathbf{k},n}(\mathbf{r}).\end{aligned} \tag{10.15}$$

We can expand $u_{\mathbf{k},n}(\mathbf{r})$ as

$$u_{\mathbf{k},n}(\mathbf{r}) = \sum_{\mathbf{K}} a_{\mathbf{K}}(\mathbf{k},n)e^{i\mathbf{K}\cdot\mathbf{r}}, \tag{10.16}$$

which follows from the periodicity property (10.8b). Inserting this in (10.15), we have

$$\langle\hat{\mathbf{p}}\rangle = \hbar\mathbf{k} + \sum_{\mathbf{K},\mathbf{K}'}\hbar\mathbf{K}a_{\mathbf{K}}a^*_{\mathbf{K}'}\int d^3r e^{i(\mathbf{K}-\mathbf{K}')\cdot\mathbf{r}};$$

the integral is $\delta_{\mathbf{K}\mathbf{K}'}$ and therefore we have

$$\langle\hat{\mathbf{p}}\rangle = \hbar\mathbf{k} + \hbar\sum_{\mathbf{K}}|a_{\mathbf{K}}|^2\mathbf{K}. \tag{10.17}$$

[4] In the presence of spin–orbit coupling, time reversal also involves a spin flip. When the crystal lacks inversion symmetry, as with all zinc blende and wurtzite structures, the bands $E_{n\mathbf{k}\alpha}$, where $\alpha = 1, 2$ denote the quantum numbers associated with the two spin states, have the properties

$$E_{n,\mathbf{k},1} = E_{n,-\mathbf{k},2} \neq E_{n,-\mathbf{k},1}.$$

For a more thorough discussion see: *Theory of Brillouin Zones and Electronic States in Crystals*, H. Jones, North Holland, Amsterdam, p. 262 (1962). A magnetic field breaks time reversal symmetry; however it also destroys the translational invariance from which the Bloch states arise.

More generally, however, $\hat{\mathbf{p}}$ is an off-diagonal operator with respect to the band index, n,

$$\langle \psi_{kn'} | \hat{\mathbf{p}} | \psi_{kn} \rangle = \hbar \mathbf{k} \delta_{n'n} + \left\langle u_{k,n'} \left| \frac{\hbar}{i} \nabla \right| u_{k,n} \right\rangle$$
$$= \hbar \mathbf{k} \delta_{n'n} + \mathbf{P}_{n'n} \tag{10.18a}$$

where

$$\mathbf{P}_{n'n} = \left\langle u_{k,n'} \left| \frac{\hbar}{i} \nabla \right| u_{k.n} \right\rangle$$
$$= \hbar \sum_{\mathbf{K}} \mathbf{K} a^*_{\mathbf{K}}(n', \mathbf{k}) a_{\mathbf{K}}(n, \mathbf{k}). \tag{10.18b}$$

Even more generally, we would calculate $\langle \psi_{k',n'} | (\hbar/i) \nabla | \psi_{k,n} \rangle$; however a calculation shows that it is diagonal in $\mathbf{k}$. We therefore conclude that in a state with given quasi-momentum $\hbar\mathbf{k}$, the true momentum can have any of the values $\hbar(\mathbf{k} + \mathbf{K})$, with a probability $|a_{\mathbf{K}}|^2$. The more quickly convergent the series (10.16), the less the contribution of the higher momentum values. For the case u = constant (free electrons), $\hbar\mathbf{k} = \mathbf{p}$.

10.3.4 The position operator

In order to introduce a position operator we must consider an *arbitrary* crystal wave function, $\psi(\mathbf{r})$, which we expand in terms of our basis functions, the Bloch functions, as

$$\psi(\mathbf{r}) = \sum_n \int d^3k c_{k,n} \psi_{k,n}(\mathbf{r}), \tag{10.19}$$

where the sum is over all bands. To define the position operator, $\hat{\mathbf{r}}$, we examine the identity

$$\mathbf{r}\,\psi(\mathbf{r}) = \sum_n \int d^3k c_{k,n}\, \mathbf{r}\, \psi_{k,n}(\mathbf{r})$$
$$= \sum_n \int d^3k \left[-i \frac{\partial \psi_{k,n}(\mathbf{r})}{\partial \mathbf{k}} + i e^{i\mathbf{k}\cdot\mathbf{r}} \frac{\partial u_{k,n}(\mathbf{r})}{\partial \mathbf{k}} \right] c_{k,n}. \tag{10.20}$$

Since $u_{k,n}(\mathbf{r})$ is a periodic function of $\mathbf{r}$, $\partial u_{k,n}(\mathbf{r})/\partial \mathbf{k}$ is also a periodic in $\mathbf{r}$. For some specific $\mathbf{k}$ the $u_{k,n}(\mathbf{r})$ form a complete set in which to expand $\partial u_{k,n}(\mathbf{r})/\partial \mathbf{k}$; we write this expansion as

$$\frac{\partial u_{kn}(\mathbf{r})}{\partial \mathbf{k}} = -i \sum_{n'} \mathbf{R}_{kn,kn'} u_{kn'}, \tag{10.21}$$

where the vector expansion coefficients are given by

$$\mathbf{R}_{\mathrm{kn,kn'}} = \mathrm{i}\left\langle \mathrm{u_{k,n}} \left| \frac{\partial}{\partial \mathbf{k}} \right| \mathrm{u_{k,n'}} \right\rangle \tag{10.22}$$

(do not confuse the operator $\hat{\mathbf{R}}$ with the real-space lattice vectors, $\mathbf{R}$). We integrate the first term in (10.20) by parts (where the integral over the boundary vanishes due to periodic boundary conditions) and substitute (10.22) in the second term, yielding

$$\mathbf{r}\,\psi(\mathbf{r}) = \sum_{\mathrm{n}} \int \mathrm{d}^3\mathrm{k} \left[\mathrm{i}\frac{\partial \mathrm{c_{kn}}}{\partial \mathbf{k}} \psi_{\mathrm{k,n}}(\mathbf{r}) + \mathrm{e}^{\mathrm{i}\mathbf{k}\cdot\mathbf{r}} \sum_{\mathrm{n'}} \mathbf{R}_{\mathrm{kn,kn'}} \mathrm{u_{k,n'}}(\mathbf{r}) \mathrm{c_{kn}} \right].$$

In the second term we write $\psi_{\mathrm{k,n'}} = \mathrm{e}^{\mathrm{i}\mathbf{k}\cdot\mathbf{r}}\mathrm{u_{k,n'}}$ and interchange the dummy indices n and n′; we then have

$$\mathbf{r}\,\psi(\mathbf{r}) = \sum_{\mathrm{n}} \int \mathrm{d}^3\mathrm{k} \left[\mathrm{i}\frac{\partial \mathrm{c_{kn}}}{\partial \mathbf{k}} + \sum_{\mathrm{n'}} \mathbf{R}_{\mathrm{kn',kn}} \mathrm{c_{kn'}} \right] \psi_{\mathrm{k,n}}(\mathbf{r}). \tag{10.23a}$$

We can also write (10.23a) as

$$\mathbf{r}\,\psi(\mathbf{r}) = \sum_{\mathrm{n}} \int \mathrm{d}^3\mathrm{k} \sum_{\mathrm{n'}} \left[\mathrm{i}\delta_{\mathrm{nn'}} \frac{\partial}{\partial \mathbf{k}} + \mathbf{R}_{\mathrm{kn',kn}} \right] \mathrm{c_{kn'}} \psi_{\mathrm{k,n}}(\mathbf{r}). \tag{10.23b}$$

Equation (10.23b) now allows us to identify the position operator as

$$\hat{\mathbf{r}} = \mathrm{i}\frac{\partial}{\partial \mathbf{k}} + \hat{\mathbf{R}}, \tag{10.24}$$

where the matrix elements of the operator $\hat{\mathbf{R}}$ in the Bloch basis are given by (10.22). The coefficients $\mathrm{c_{kn}}$ are the *momentum-space/band-index representation* of $\psi(\mathbf{r})$, and $\hat{\mathbf{r}}$ operates on them. These coefficients are the analogue for a periodic system of what the Fourier transform of an arbitrary wave function $\psi(\mathbf{r})$ is in a translationally invariant system where

$$\psi(\mathbf{r}) = \int \mathrm{d}^3\mathrm{k}\,\mathrm{c}(\mathbf{k})\mathrm{e}^{\mathrm{i}\mathbf{k}\cdot\mathbf{r}}. \tag{10.25}$$

The equation that parallels (10.23) is then

$$\mathbf{r}\,\psi(\mathbf{r}) = \int \mathrm{d}^3\mathrm{k}\,\mathrm{c}(\mathbf{k}) \left(-\mathrm{i}\frac{\partial}{\partial \mathbf{k}}\right) \mathrm{e}^{\mathrm{i}\mathbf{k}\cdot\mathbf{r}} = \int \mathrm{d}^3\mathrm{k}\,\mathrm{e}^{\mathrm{i}\mathbf{k}\cdot\mathbf{r}} \left(\mathrm{i}\frac{\partial \mathrm{c}(\mathbf{k})}{\partial \mathbf{k}}\right), \tag{10.26}$$

and for this case the position operator has the simpler form

$$\hat{\mathbf{r}} = \mathrm{i}\frac{\partial}{\partial \mathbf{k}}. \tag{10.27}$$

10.3.5 The velocity operator

The velocity operator follows from the Heisenberg equation of motion,

$$\dot{\hat{\mathbf{r}}} = \frac{i}{\hbar}\left[\hat{H}\hat{\mathbf{r}} - \hat{\mathbf{r}}\hat{H}\right]. \tag{10.28}$$

Inserting Eq. (12.24) for $\hat{\mathbf{r}}$, Eq. (10.28) becomes

$$\dot{\hat{\mathbf{r}}} = \frac{i}{\hbar}\left(\frac{\partial}{\partial \mathbf{k}}\hat{H} - \hat{H}\frac{\partial}{\partial \mathbf{k}}\right) + \dot{\hat{\mathbf{R}}}. \tag{10.29}$$

In the $\mathbf{k}$, n representation, where $\hat{H}$ is diagonal with eigenvalues $E_{\mathbf{k},n}$, the first term is given by $\partial E_{\mathbf{k},n}/\hbar\partial\mathbf{k}$ and is diagonal in n. On the other hand, the matrix elements associated with the operator $\dot{\hat{\mathbf{R}}}$ in this representation are

$$\left(\dot{\hat{\mathbf{R}}}\right)_{kn',kn} = \frac{i}{\hbar}\left(E_{k,n'} - E_{k,n}\right)\mathbf{R}_{kn',kn}, \tag{10.30}$$

which vanishes for $n = n'$. If we limit ourselves to the expectation value of $\langle\dot{\mathbf{r}}\rangle = \langle\mathbf{v}\rangle$, which involves only the diagonal elements in n, we have

$$\langle\mathbf{v}\rangle = \frac{1}{\hbar}\frac{\partial E_{\mathbf{k},n}}{\partial \mathbf{k}}; \tag{10.31}$$

this is the analogue of the classical Hamiltonian equation, $\mathbf{v} = \partial H/\partial \mathbf{p}$.

10.3.6 Effect of an external force

Suppose that a constant external force, $\mathbf{F}$, is applied to the electrons in our periodic solid, which can be represented by a contribution $V(\mathbf{r}) = -\mathbf{F}\cdot\mathbf{r}$ to the classical Hamiltonian, H. The corresponding operator is then $\hat{V} = -\mathbf{F}\cdot\hat{\mathbf{r}}$. The resulting rate of change of the quasi-momentum operator, $\dot{\hat{\mathbf{k}}}$, follows from the Heisenberg equation of motion, $\dot{\hat{\mathbf{k}}} = (i/\hbar)\left[\hat{H}, \hat{\mathbf{k}}\right]$. In the Bloch representation where $\hat{\mathbf{k}}$ itself is diagonal and $\hat{\mathbf{r}}$ is given by Eq. (10.24), we then have

$$\begin{aligned}\dot{\hat{\mathbf{k}}} &= -\frac{i}{\hbar}\left[(\mathbf{F}\cdot\hat{\mathbf{r}})\mathbf{k} - \mathbf{k}(\mathbf{F}\cdot\hat{\mathbf{r}})\right] \\ &= \frac{1}{\hbar}\left[\left(\mathbf{F}\cdot\frac{\partial}{\partial \mathbf{k}}\right)\mathbf{k} - \mathbf{k}\left(\mathbf{F}\cdot\frac{\partial}{\partial \mathbf{k}}\right)\right] - \frac{i}{\hbar}\left[(\mathbf{F}\cdot\hat{\mathbf{R}})\cdot\mathbf{k} - \mathbf{k}(\mathbf{F}\cdot\hat{\mathbf{R}})\right].\end{aligned} \tag{10.32}$$

The operator $\hat{\mathbf{R}}$ commutes with $\mathbf{k}$ and hence the second term vanishes, and on carrying out the differentiation in the first term, we obtain

$$\mathbf{F} = \hbar\dot{\hat{\mathbf{k}}}; \tag{10.33}$$

this equation is analogous to Newton's equation of motion.[5] We can equally well write this as

$$\hbar\dot{\mathbf{k}} = -\frac{\partial V(\mathbf{r})}{\partial \mathbf{r}} = -\frac{\partial H}{\partial \mathbf{r}}, \qquad (10.34)$$

analogous to Hamilton's second equation. The effect of a magnetic field is more subtle and will be addressed in Chapter 18.

10.3.7 Scattering among Bloch states

Let us consider the behavior of two electrons in a periodic potential. When these electrons are close to each other their wave function, in the presence of an interaction, can only be written in the non-separable form, $\psi(\mathbf{r}_1, \mathbf{r}_2)$. This wave function would be altered by a factor $e^{-i\mathbf{k}\cdot\mathbf{R}}$ on translation, where $\mathbf{k}$ is interpreted as the quasi-momentum of the combined system. When the electrons are well separated the same wave function can be written as a product and hence the same displacement would involve a factor, $e^{-i\mathbf{k}_1\cdot\mathbf{R}} e^{-i\mathbf{k}_2\cdot\mathbf{R}}$, where $\mathbf{k}_1$ and $\mathbf{k}_2$ are the quasi-momenta of the individual electrons. Since the displacement must result in the same phase factor for both cases we must have

$$\mathbf{k} = \mathbf{k}_1 + \mathbf{k}_2 + \mathbf{K}; \qquad (10.35)$$

Note that the inclusion of $\mathbf{K}$ does not alter the phase and hence, in general, must be added. From this consideration it follows that in a collision between two electrons the sum of the quasi-momenta before and after the collision is conserved *only to within a reciprocal lattice vector*; i.e.,

$$\mathbf{k}_1 + \mathbf{k}_2 = \mathbf{k}'_1 + \mathbf{k}'_2 + \mathbf{K}. \qquad (10.36)$$

Therefore, provided that the initial momenta are large enough, scattering may involve a reciprocal lattice vector; such processes are called electron *umklapp processes*.

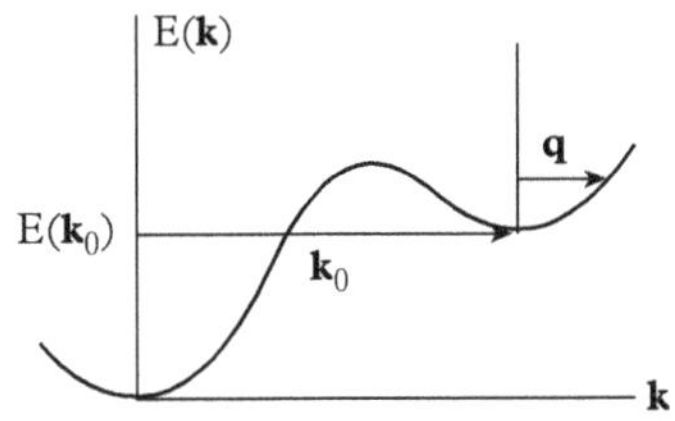

Figure 10.3 *The behavior of* $E(\mathbf{k})$ *in the vicinity of a non-degenerate point* $\mathbf{k}_0$ *having positive curvature, where* $\nabla E(\mathbf{k} = \mathbf{k}_0) = 0$.

10.4 Dispersion near a band edge

Near a point in $\mathbf{k}$ space, that we denote as $\mathbf{k}_0$, where the function $E(\mathbf{k})$ for some band n has a maximum, a minimum, or a saddle point, the linear terms in a Taylor expansion about this point vanish (see Fig. 10.3). Here we will only treat the case where this point does not involve degenerate bands. The expansion then takes the form

$$E(\mathbf{k}) = E(\mathbf{k}_0) + \frac{1}{2}(\mathbf{k}-\mathbf{k}_0)\cdot\left(\frac{\partial^2 E}{\partial \mathbf{k}\partial \mathbf{k}}\right)_{\mathbf{k}=\mathbf{k}_0}\cdot(\mathbf{k}-\mathbf{k}_0) + \cdots. \qquad (10.37a)$$

[5] According to Eq. (10.33), in the presence of an applied electric field $\mathbf{E}$, and in the absence of any scattering, the wave vector of a single electron in a Bloch band will increase *linearly in time* as $\mathbf{k}(t) = \mathbf{k}(t=0) + e\mathbf{E}t$. Let us suppose we have a single electron in a tight binding band in one dimension, where the energy is given by Eq. (11.46). The resulting velocity is $v = d\varepsilon/d(\hbar k) = -(2h_1 a/\hbar)\sin(k(t)a)$. So starting from some initial value, k will advance until it reaches the edge of the first Brillouin zone where it will be diffracted to the opposite side of the zone, and where the velocity has the *opposite sign*. Hence the motion in real space is *oscillatory*. This phenomenon is referred to as a *Bloch oscillation*. Electron scattering suppresses such oscillations in all but some unusual situations.

It is natural to introduce a tensor quantity, called the effective mass tensor, $\mathbf{m}^*$, the reciprocal or inverse of which is defined by

$$(\mathbf{m}^*)^{-1} = \frac{1}{\hbar^2}\left(\frac{\partial^2 E}{\partial \mathbf{k}\partial \mathbf{k}}\right)_{\mathbf{k}=\mathbf{k}_0} \tag{10.37b}$$

(which for free electrons is clearly the scalar m^{-1}), and write the expansion as

$$E(\mathbf{k}) = E(\mathbf{k}_0) + \frac{\hbar^2}{2}\mathbf{q}\cdot\left(\frac{1}{\mathbf{m}^*}\right)\cdot\mathbf{q} + \cdots, \tag{10.38}$$

where we have defined $\mathbf{q} = \mathbf{k} - \mathbf{k}_0$. From the equality of cross derivatives of a continuous function, (10.37a) is a symmetric tensor and its six independent elements can be expressed in terms of three principal values and three Euler angles, describing its orientation relative to the principal axes of the crystal.

We now develop an expression for the effective mass tensor and, in the process, introduce the so-called $\mathbf{k}\cdot\mathbf{p}$ method.[6] The Schrödinger equation for the eigenstate $\psi_{\mathbf{k},n} = \psi_{\mathbf{k}_0+\mathbf{q},n} = e^{i(\mathbf{k}_0+\mathbf{q}).\mathbf{r}}u_{\mathbf{k}_0+\mathbf{q},n}(\mathbf{r})$ is

$$\left(\frac{\hat{p}^2}{2m} + V(\mathbf{r})\right)\psi_{\mathbf{k}_0+\mathbf{q},n} = E_n(\mathbf{k}_0+\mathbf{q})\psi_{\mathbf{k}_0+\mathbf{q},n}. \tag{10.39}$$

Writing $\hat{p}^2\psi_{\mathbf{k}_0+\mathbf{q},n} = e^{i(\mathbf{k}_0+\mathbf{q})\cdot\mathbf{r}}(\hat{\mathbf{p}} + \hbar\mathbf{k}_0 + \hbar\mathbf{q})^2 u_{\mathbf{k}_0+\mathbf{q},n}(\mathbf{r})$, and expanding $\hat{H}$, E, and $u_{\mathbf{k}_0+\mathbf{q},n}(\mathbf{r})$ in terms of a small parameter λ, we can express (10.39) as

$$\begin{aligned}&\left(\hat{H}_0 + \lambda\hat{H}^{(1)} + \lambda^2 H^{(2)}\right)\left(u^{(0)}(\mathbf{r}) + \lambda u^{(1)}(\mathbf{r}) + \lambda^2 u^{(2)}(\mathbf{r}) + \cdots\right)\\ &\quad = \left(E^{(0)} + \lambda E^{(1)} + \lambda^2 E^{(2)}\ldots.\right)\left(u^{(0)}(\mathbf{r}) + \lambda u^{(1)}(\mathbf{r}) + \lambda^2 u^{(2)}(\mathbf{r}) + \cdots\right),\end{aligned} \tag{10.40}$$

where

$$\hat{H}_0 = \frac{1}{2m}(\hat{\mathbf{p}} + \hbar\mathbf{k}_0)^2 + V(\mathbf{r}), \tag{10.41a}$$

$$\hat{H}^{(1)} = \frac{\hbar}{m}(\hat{\mathbf{p}} + \hbar\mathbf{k}_0)\cdot\mathbf{q}, \tag{10.41b}$$

$$\hat{H}^{(2)} = \frac{\hbar^2}{2m}q^2, \tag{10.41c}$$

and we write $u^{(0)}(\mathbf{r}) = u_{\mathbf{k}_0,n}(\mathbf{r})$ and $E^{(0)} = E_n(\mathbf{k}_0)$. We expand the corrections to the periodic parts of the wave functions as

$$u^{(1,2)}_{\mathbf{k}_0+\mathbf{q},n}(\mathbf{r}) = \sum_m a^{(1,2)}_{nm}u_{\mathbf{k}_0,m}(\mathbf{r}),$$

[6] The method extends to cover a case where degeneracy is present, in Section 12.1.

and apply the procedures used in the standard development of second-order perturbation theory (noting that the linear term in the energy vanishes by assumption); we then find, on collecting the second-order terms,

$$E_n^{(2)}(\mathbf{k}_0 + \mathbf{q}) = \frac{\hbar^2}{2m}q^2 + \left(\frac{\hbar}{m}\right)^2 \sum_{m \neq n} \frac{(\mathbf{q}\cdot(\hat{\mathbf{p}})_{nm})((\hat{\mathbf{p}})_{mn}\cdot\mathbf{q})}{E_n(\mathbf{k}_0) - E_m(\mathbf{k}_0)}, \tag{10.42a}$$

where we have defined the matrix elements,

$$(\hat{\mathbf{p}})_{nm} = \langle u_{\mathbf{k}_0,n}|\hat{\mathbf{p}} + \hbar\mathbf{k}_0|u_{\mathbf{k}_0,m}\rangle = \langle \psi_{\mathbf{k}_0,n}|\hat{\mathbf{p}}|\psi_{\mathbf{k}_0,m}\rangle . \tag{10.42b}$$

On rearranging the dummy indices associated with the dot products (to make the symmetric character of the tensor explicit), and using the definition (10.38), we find the reciprocal effective mass tensor as

$$\left(\frac{1}{m^*}\right)_{\alpha\beta} = \left(\frac{1}{m}\right)\delta_{\alpha\beta} + \left(\frac{1}{m}\right)^2 \sum_{m \neq n} \frac{(\hat{\mathbf{p}})_{nm,\alpha}(\hat{\mathbf{p}})_{mn,\beta} + (\hat{\mathbf{p}})_{nm,\beta}(\hat{\mathbf{p}})_{mn,\alpha}}{E_m(\mathbf{k}_0) - E_n(\mathbf{k}_0)}. \tag{10.43a}$$

As an approximation (or as a parameterization) it is often assumed that the band extrema located at energy E_n is separated by a small energy, E_0, from a single higher-band state at $E_m = E_n + E_0$ (at the same $\mathbf{k}_0$) and that all other levels lie considerably higher in energy (or, if nearby have matrix elements which vanish due to a selection rule, such as parity); we then have a *single* matrix element to either evaluate or use as a parameter. For the case where the states m and n are s-like and p-like (which commonly occurs in semiconductors) the matrix element has the symmetry of a vector, which we denote as $\langle u_{\mathbf{k}_0,n}|\, p_\alpha + \hbar k_{0\alpha}\, |u_{\mathbf{k}_0,m}\rangle \equiv iP_\alpha(\mathbf{k}_0)$ (where we have dropped the subscripts m, n), and Eq. (10.43a) can then be written as

$$\left(\frac{1}{m^*}\right)_{\alpha\beta} = \left(\frac{1}{m}\right)\delta_{\alpha\beta} + \left(\frac{1}{m}\right)^2 \frac{P_\alpha P_\beta^* + P_\beta P_\alpha^*}{E_0}. \tag{10.43b}$$

In Chapter 11 we will treat in some detail the important case of tetrahedrally coordinated semiconductors where the energy bands are degenerate at $\mathbf{k} = 0$.

When discussing the behavior of energy bands in the vicinity of a point where $\partial E_n/\partial \mathbf{k} = 0$, we can also encounter the case where $E_n(\mathbf{k})$ passes through a local *maximum*, an example of which occurs in Fig. 10.2, in which case the principal values of the tensor are negative. Finally, one can have a mixed case where both positive and negative values occur, a three-dimensional *saddle point*.

10.5 Electrons and holes

The current carried by a *filled* band must be identically zero: for each filled state with wave vector $\mathbf{k}$ and group velocity $v_{\mathbf{k}}$ there is a corresponding state with wave vector $-\mathbf{k}$ and the opposite group velocity, even in the presence of an external field,[7] which can otherwise alter the occupation of states within an *unfilled* band. Mathematically, we may write this requirement as

$$\mathbf{j} = \frac{e}{L^3} \sum_{\text{all states}} \mathbf{v_k} = e\frac{2}{(2\pi)^3} \int_{\text{band}} d^3k\, \mathbf{v_k} = 0. \tag{10.44}$$

Now assume that the band is only partially filled, for which case we write the current as

$$\mathbf{j} = \frac{e}{L^3} \sum_{\substack{\text{occupied}\\ \text{states}}} \mathbf{v_k}. \tag{10.45}$$

Applying (10.44) it follows that

$$\mathbf{j} = -\frac{e}{L^3} \sum_{\substack{\text{unoccupied}\\ \text{states}}} \mathbf{v_k}. \tag{10.46}$$

If we introduce the distribution, $n(\mathbf{k})$, giving the probability that a given state is occupied, then (10.45) can be written

$$\mathbf{j} = e\frac{2}{(2\pi)^3} \int d^3k \mathbf{v_k} n(\mathbf{k}) \tag{10.47}$$

and (10.46) becomes

$$\mathbf{j} = -e\frac{2}{(2\pi)^3} \int_{\text{band}} d^3k \mathbf{v_k} \,(1 - n(\mathbf{k})). \tag{10.48}$$

The factor $1 - n(\mathbf{k})$ may be thought of as the "occupation number for empty states"; the latter are usually called *holes* and with this interpretation enter with the *opposite charge* (holes behave like positive electrons). These currents would vanish for a system in equilibrium and in the absence of an external field;[8] however, the presence of an external field alters the distribution function $n(\mathbf{k})$ such that a non-vanishing current is present.

There is a second context in which the terms electrons and holes appear in solid-state physics. Here we think of forming an excitation in the filled Fermi sea of a metal by adding an electron above the Fermi level *or removing* one from below. The resulting excited states are then called electrons or holes respectively. Because of the Coulomb (or other) interactions in the system, the resulting states are *not eigenstates* of the whole system; they will "decay," ultimately sharing the initial energy needed to create them with the rest of the electrons, as well as deformations of the ions in the system. Excitations in which a particle is added or

[7] Here we neglect tunneling between bands, the so-called Zener breakdown effect.

[8] We exclude superconductors here.

removed from the system are classed as *one-particle excitations*; for electrons they inevitably change the overall charge of the system and hence involve the electron work function. Note that adding or removing an electron very slowly (adiabatically) such that the system remains in equilibrium results in an energy change that corresponds to the thermodynamic chemical potential, μ.

We can also have a case where an electron from below the Fermi sea is not removed, but is placed above the Fermi level, an example being when the system absorbs a photon (but does not eject an electron). We have then created an electron–hole pair. Note these two entities can interact with each other.[9] In particular, the energy of the system is *not* simply the sum of the isolated electron and hole; however, the charge neutrality of the system is preserved in this way. Of course we can also add or remove two electrons, which would upset the charge neutrality. All three of these processes are classed as *two-particle excitations*.

At any finite temperature, multiparticle excitations are present and as an approximation one can cautiously assume that they may be represented as a sum of single-particle excitations, where the number of holes and electrons remains essentially equal to preserve charge neutrality. However as the temperature increases this approximation becomes progressively worse. In addition to the effects of temperature, various "collective" excitations involving the system as a whole exist in the presence of external forces.

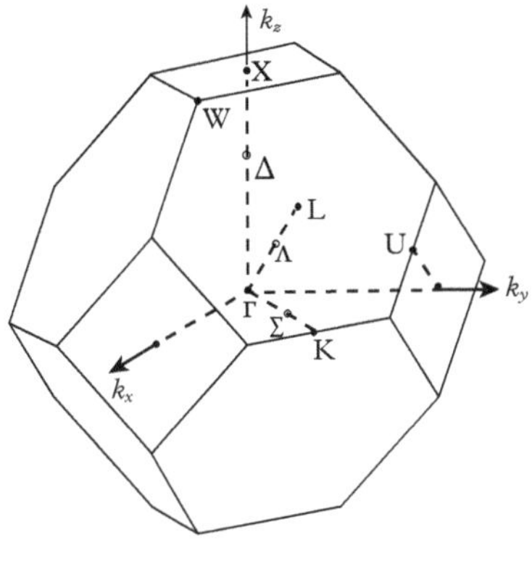

Figure 10.4 *The Brillouin zone of the face centered cubic lattice showing the symmetry points and lines.*

10.6 Brillouin zones

Brillouin zones (BZ) were introduced in Section 9.6 along with the rules for constructing them. The BZ plays a very fundamental role in the electronic structure of solids. In Figs. 10.4, 10.5, and 10.6 we show three-dimensional renderings of the BZ for the face centered cubic (fcc), body centered cubic (bcc), and hexagonal close packed (hcp) structures. In addition, certain points and axes of the BZ having a high symmetry are indicated by Greek and Roman letters that are universally used in solid-state physics for their designation. In all three cases, the point Γ at the zone center has the full point-group symmetry of the lattice.

For the fcc case the zone is comprised of six square faces and eight hexagonal faces. We note in particular the points X and L located at the center of the square and hexagonal faces, respectively. The lines Δ(connecting Γ and X), Λ(Γ to Λ), and Σ(Γ to K), have four-fold, three-fold, and two-fold symmetry respectively.

For the bcc case, the zone is comprised of 12 rhombuses having two-fold symmetry and perpendicular to the 12 two-fold Σ axes that they intersect at the points N. The eight three-fold Λ axes intersect the zone corners at points P (where three diamonds meet), and the six four-fold Δ axes intersect zone corners at points H (where four diamonds meet).

The hcp zone is enclosed by two hexagonal faces and six rectangular faces. We identify the two points Γ and A, both of which have the full hexagonal point-group symmetry. The Δ line joining these points has the hexagonal six-fold rotational

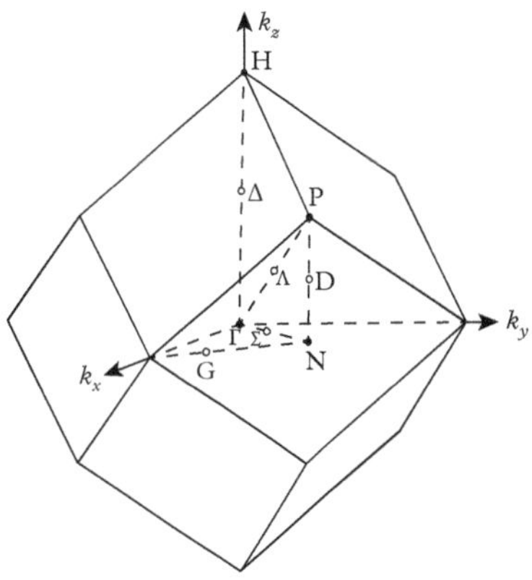

Figure 10.5 *The Brillouin zone of the body centered cubic lattice showing the symmetry points and axes.*

[9] In semiconductors this can lead to the formation of bound states called excitons, which will be discussed in Section 30.2.

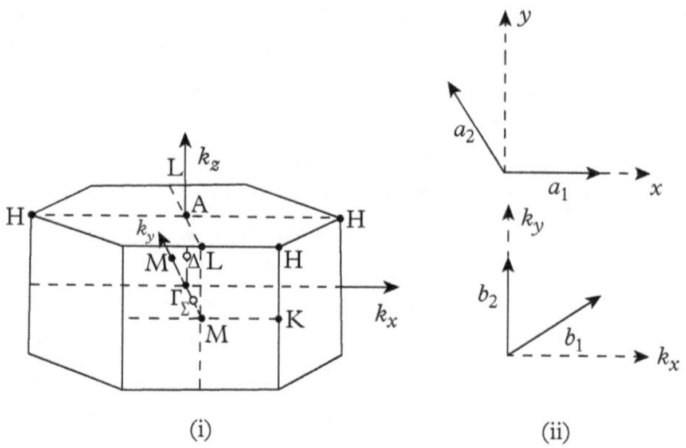

Figure 10.6 *(i) Brillouin zone for the hexagonal close packed lattice showing the symmetry points and axes. (ii) The relation between the primitive lattice vectors of the space lattice and the reciprocal lattice.*

symmetry. The lines defined by the points Γ - M (Σ) and A - L (six each) and the lines Γ - K and A - H (six each) are all two-fold axes.

10.7 Wannier functions

Intuitively (and discussed in detail in Chapter 11), we expect the Bloch functions to evolve into plane waves (i.e., $u_{kn}(\mathbf{r}) \rightarrow 1$) in the limit when the crystal potential vanishes. However our solid is built out of atoms and the crystal wave function will change very rapidly with position, especially near the atomic cores, in a distinctly non-plane-wave-like fashion. Although s and p electrons tend to spread out away from the cores, d and f electrons often remain highly localized, retaining atomic like characteristics. Rather than being nearly constant, $u_{k=0,n}(\mathbf{r})$ will be highly localized around each atom.

Let us try to examine this qualitative consideration more precisely. With Eq. (10.16) we established that the Bloch functions were *periodic in* **k** with period **K**. Hence, for some arbitrary band index n and position **r**, we can express this periodicity as a *Fourier series* in which we sum over the real-space lattice vectors, **R**, as

$$\psi_{kn}(\mathbf{r}) = \sum_{\mathbf{R}} w_{n\mathbf{R}}(\mathbf{r}) e^{i\mathbf{k}\cdot\mathbf{R}}; \tag{10.49}$$

here the "coefficients," $w_{n\mathbf{R}}(\mathbf{r})$, are themselves a new set of continuous functions of **r**, one for each band n. Using the usual procedure for finding the coefficients in a Fourier series (taken as periodic in **k**, rather than **r**), we have

$$w_{n\mathbf{R}}(\mathbf{r}) = \frac{1}{V_{BZ}} \int d^3k\, \psi_{kn}(\mathbf{r}) e^{-i\mathbf{k}\cdot\mathbf{R}}, \tag{10.50}$$

where the integral is over the first Brillouin zone and V_{BZ} is the zone volume in **k**-space.[10] We now translate the origin in (10.50) by a real-space lattice vector $\mathbf{R}'$, obtaining

$$\begin{aligned} w_{n\mathbf{R}}(\mathbf{r}-\mathbf{R}') &= \frac{1}{V_{BZ}}\int d^3k\, \psi_{\mathbf{k}n}(\mathbf{r}-\mathbf{R}')e^{-i\mathbf{k}\cdot\mathbf{R}} \\ &= \frac{1}{V_{BZ}}\int d^3k\, \psi_{\mathbf{k}n}(\mathbf{r})e^{-i\mathbf{k}\cdot\mathbf{R}'}e^{-i\mathbf{k}\cdot\mathbf{R}}, \end{aligned}$$

where we have used Eq. (10.7) in the second step. Setting $\mathbf{R} = 0$, dropping the prime on $\mathbf{R}'$, and comparing with (10.49), it follows that the expansion coefficients are a function of $\mathbf{r} - \mathbf{R}$; hence we may equally well write the expansion (10.49) as

$$\psi_{\mathbf{k}n}(\mathbf{r}) = \sum_{\mathbf{R}} w_n(\mathbf{r}-\mathbf{R})e^{i\mathbf{k}\cdot\mathbf{R}}. \tag{10.51}$$

The functions $w_n(\mathbf{r} - \mathbf{R})$ are called *Wannier functions* (after G. Wannier[11]); in terms of $\mathbf{R}$ and n, they form a complete, orthonormal set. Although formally they must be obtained from (10.50), they tend to be localized at the $\mathbf{R}^{\text{th}}$ lattice site and to some extent mimic the atomic wave functions; for this reason they are useful for some applications. Note they are *not* eigenfunctions of the crystal Hamiltonian since, as defined by (10.50), they are super-positions of the Bloch states.

10.8 The Kronig-Penney model

There is an exactly solvable solution to the Schrödinger equation, due to Kronig and Penney (1930), for the wave functions and energy levels of a particle moving in one dimension in the periodic rectangular potential shown in Fig.10.7. We seek a solution of the form

$$\psi(x) = c_1\, e^{i\kappa_1 x} + c_2\, e^{-i\kappa_1 x};\ 0 < x < a \tag{10.52}$$

and

$$\psi(x) = c_3\, e^{i\kappa_2 x} + c_4\, e^{-i\kappa_2 x};\ -b < x < 0. \tag{10.53}$$

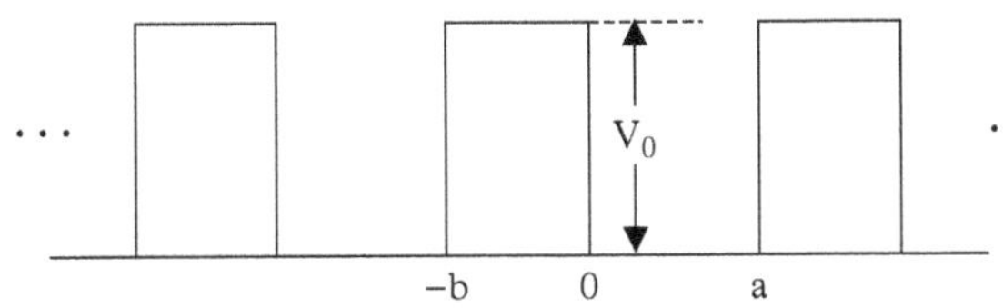

Figure 10.7 *The potential for the one-dimensional Kronig and Penney band structure model.*

[10] The prefactor may also be written as $V_{BZ}^{-1} = V/(2\pi)^3$, where V is the volume of the real-space unit cell.

[11] The structure of electronic excitation levels in insulating crystals, G.H. Wannier, *Phys. Rev.* 52, 191 (1937).

Here,

$$\kappa_1 = \sqrt{\frac{2mE}{\hbar^2}} \quad \text{and} \quad \kappa_2 = \sqrt{\frac{2m(E-V_0)}{\hbar^2}}; \tag{10.54}$$

note that κ_2 will be purely imaginary for $E < V_0$.

According to the Bloch condition, for $a < x < 2a + b$, the wave function will involve an additional factor $e^{ik(a+b)}$, $a + b$ being the period of the lattice; with this requirement the wave function within the adjacent barrier is given by

$$\psi(x) = c_3 e^{ik(a+b)} \left(e^{i\kappa_2 x} + c_4 e^{-i\kappa_2 x}\right); \quad a < x < a + b. \tag{10.55}$$

Matching the wave functions at $x = 0$ gives

$$c_1 + c_2 = c_3 + c_4, \tag{10.56}$$

while matching at $x = a$ gives

$$c_1 e^{i\kappa_1 a} + c_2 e^{-i\kappa_1 a} = e^{ik(a+b)} \left(c_3 e^{i\kappa_2 a} + c_4 e^{-i\kappa_2 a}\right). \tag{10.57}$$

Matching $d\psi/dx$ at these same points gives

$$\kappa_1 (c_1 - c_2) = \kappa_2 (c_3 - c_4) \tag{10.58}$$

and

$$\kappa_1 \left(c_1 e^{i\kappa_1 a} - c_2 e^{-i\kappa_1 a}\right) = \kappa_2 \left(c_3 e^{ik(a+b)} e^{i\kappa_2 a} - c_4 e^{ik(a+b)} e^{-i\kappa_2 a}\right). \tag{10.59}$$

Setting the determinant of the coefficients of c_1, c_2, c_3, and c_4 to zero in the latter four equations gives, after some rearrangement,

$$\cos k(a + b) = \cos(\kappa_1 a)\cos(\kappa_2 b) - \frac{1}{2}\left(\frac{\kappa_1}{\kappa_2} + \frac{\kappa_2}{\kappa_1}\right) \sin(\kappa_1 a)\sin(\kappa_2 b). \tag{10.60}$$

The solution of this transcendental equation gives the energy, $E(k)$.

Historically, the Kronig–Penney model was just a textbook problem. However, with the experimental fabrication of so-called artificial superlattices involving repeated layers of a combination like $GaAs/Ga_xAl_{1-x}As$, the model can in some cases accurately describe the motion of electrons in the direction perpendicular to the layers, as will be discussed in Chapter 47.

ADDITIONAL READING

For a discussion of early history of the quantum theory of electrons in metals, see: The development of the quantum-mechanical electron theory of metals: 1928–1933, L. Hoddeson, G. Baym, and M. Eckert, *Rev. Mod. Phys.* **59**, 287 (1987). Also interesting in this connection is, Recollections of early solid state physics, R.E. Peierls, *Proc. R. Soc. Lond. A* **371**, 28 (1980).

Symmetry properties of solids are discussed in the following books: *Theory of Brillouin Zones and Electronic States in Crystals*, H. Jones, North Holland, Amsterdam (1962); *Symmetry Principles in Solid State and Molecular Physics*, Melvin Lax, John Wiley and Sons, USA (1974); *Symmetry in the Solid State*, R.S. Knox and A. Gold, W.A. Benjamin, New York (1964). Crystal symmetry is also discussed in Chapter XIII of *Statistical Physics*, Part I, L.D. Landau and E.M. Lifshitz, Pergamon Press, Oxford (1980). Various properties of Bloch states are discussed in *Statistical Physics*, Part II, E.M. Lifshitz and L.P. Pitaevskii, Pergamon Press, Oxford (1980), section 55.

Problems

10.1. Show that the operator $\hat{p}$ is diagonal with respect to the quasi-momentum, $\mathbf{k}$.

10.2. You are given the periodic potential

$$V(x) = \alpha \sum_{n=-\infty}^{n=+\infty} \delta(x - na).$$

Find the equation analogous to Eq. (10.60) of a particle of mass m moving in this potential by taking the limit of the rectangular wave potential that corresponds to a delta function potential.

10.3. Obtain the same result as in Problem 10.2 but now using the proper boundary condition for the behavior of a Schrödinger wave function on penetrating a delta function potential.

10.4. Calculate the wave velocity $v(k)$ for a particle moving in the potential given in Problem 10.2.

10.5. Fill in the missing steps in the derivation of Eq. (10.43a).

10.6. Show that the Wannier functions form a complete orthonormal set in the "quantum numbers" $\mathbf{R}$ and n.

11 The Nearly Free Electron, OPW, Pseudopotential, and Tight Binding Methods

In this chapter we begin the development of techniques that can be used to calculate the wave functions and energy levels of Bloch electrons propagating in periodic solids, starting with the *nearly free electron* (NFE) approximation. We then treat the more powerful orthogonalized plane wave (OPW) and pseudopotential methods. Lastly we discuss the tight binding or linear combination of atomic orbital (LCAO) method. Two other approaches, the KKR and APW methods, will be taken up in Chapter 13.

11.1 The nearly free electron model

Let us suppose that the periodic potential, written in the form of Eq. (10.3), is rapidly convergent; we may then truncate the expansion after a few terms. We will also assume that the coefficients retained are small relative to characteristic energies of interest. This nearly free electron model is rather successful in describing properties of the so-called simple metals, those in which the valence states of the atoms involve only s and p electrons, such as Mg or Al.

11.1.1 The empty lattice in one dimension

In free space the unnormalized wave functions are simply the plane waves, $e^{i\mathbf{k}\cdot\mathbf{r}}$, where **k** may have any value. To begin with, it is informative to examine the form Eq. (10.8) takes in the limit of an "empty lattice" in which the "ghost" of a lattice periodicity is introduced; i.e., we partition a large box into periodic unit cells but there are no atoms in those cells. To make the plane waves conform with the Bloch form given by Eqs. (10.8a,b) we examine the behavior of such states as a function of **k** in reciprocal space (see Section 9.3), where from Eq. (10.13) the associated energies are periodic in the reciprocal lattice vector **K**. The periodicity of $u_{\mathbf{k}n}(\mathbf{r})$ permits us to write it as a Fourier series in real space, as in Eq. (10.16).

We first examine the one-dimensional case with periodicity a, where

$$u_{kn}(x) = \sum_l a_l^{(n)} e^{\frac{2\pi i l x}{a}}. \tag{11.1}$$

The Physics of Solids. First Edition. J. B. Ketterson.

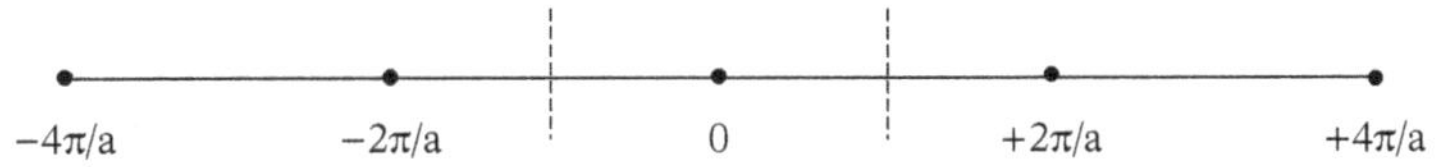

Figure 11.1 *The points denote sites of the one-dimensional reciprocal lattice. The dashed lines mark the "edges" of the first Brillouin zone at* $\pm\pi/a$.

Consider the following partitioning of our one-dimensional k space:

$$\ldots -\frac{3\pi}{a} < k < -\frac{\pi}{a};\ -\frac{\pi}{a} < k < \frac{\pi}{a};\ \frac{\pi}{a} < k < \frac{3\pi}{a};\ \frac{3\pi}{a} < k < \frac{5\pi}{a} \ldots. \tag{11.2a}$$

By subtracting the appropriate number reciprocal lattice vectors, $2\pi n/a$, and restricting the range of k to $-\pi/a < k < \pi/a$, we may write the wave vector in the various zones as

$$\ldots;\ k;\ k \pm \frac{2\pi}{a};\ k \pm \frac{4\pi}{a};\ \ldots. \tag{11.2b}$$

The restricted range of k corresponds to the *first Brillouin zone* of our 1d lattice; if we add a primitive vector of the reciprocal lattice, $2\pi/a$, we reproduce (or repeat) the first Brillouin zone, as shown in Fig. 11.1.

We may now replace our *single* plane wave function by the following *set of* wave functions:

$$\ldots, e^{ikx},\ e^{i\left(k \pm \frac{2\pi}{a}\right)x},\ e^{i\left(k \pm \frac{4\pi}{a}\right)x},\ \ldots; \tag{11.3a}$$

the corresponding functions u_{kn}, for $n = 0, \pm 1, \pm 2 \cdots$, are then

$$\ldots 1,\ e^{\pm \frac{2\pi i x}{a}},\ e^{\pm \frac{4\pi i x}{a}}\ \ldots. \tag{11.3b}$$

In other words, for this special empty lattice case we have "one-term" Fourier expansions in Eq. (11.1), and the band index n and the expansion coefficients l are uniquely associated, $\left(a_l^{(n)} = \delta_{ln}\right)$. (In the general case we would require a separate multi-term Fourier expansion for each n.) The kinetic energy associated with these empty lattice wave functions, which are the eigenvalues since they are free electrons, are

$$E_n(k) = \frac{\hbar^2}{2m}\left(k + \frac{2\pi n}{a}\right)^2, \tag{11.4}$$

where k is restricted to the first Brillouin zone and $n = 0, \pm 1, \pm 2, \cdots$.

11.1.2 The nearly free electron model in one dimension

Having considered a lattice with no atoms, let us consider a lattice where the atoms contribute a very weak potential. We write our Bloch function as

$\psi(x) = e^{ikx}u_{kn}(x)$, where $u_{kn}(x)$ is given by (11.1), and substitute it in the one-dimensional Schrödinger equation,

$$\left[-\frac{\hbar^2}{2m}\frac{d^2}{dx^2} + V(x)\right]\psi(x) = E\psi(x), \tag{11.5}$$

which yields

$$\sum_{l=-\infty}^{+\infty}\left[\frac{\hbar^2}{2m}(k+K_l)^2 + (V-E)\right]a_l e^{i(K_l+k)x} = 0, \tag{11.6}$$

where we write $K_l = 2\pi l/a$, where a is again the lattice spacing. We multiply both sides of Eq. (11.6) by e^{-iK_nx} and integrate from 0 to a (i.e., over the "unit cell") to obtain

$$\sum_{l=-\infty}^{\infty}\left\{\left[\frac{\hbar^2}{2m}(k+K_l)^2 - E\right]\delta_{ln} + V_{ln}\right\}a_l = 0, \tag{11.7}$$

where

$$\begin{aligned} V_{ln} &\equiv \frac{1}{a}\int_0^a e^{i(K_l-K_n)x}V(x)dx \\ &= (V_{nl})^*; \end{aligned} \tag{11.8}$$

where the second step follows from the fact that V(x) is real. Note that V_{ln} depends only on the difference, $n - l$. Equation (11.7) represents a set of linear homogeneous equations in the quantities a_l and the condition for there to be a solution is that the determinant of the coefficients vanishes; i.e.,

$$\left|\left[\frac{\hbar^2}{2m}(k+K_l)^2 - E\right]\delta_{ln} + V_{ln}\right| = 0. \tag{11.9}$$

If all terms in the expansion of the potential are retained, Eq. (11.9) represents a formal solution to the one-dimensional problem posed.

The nearly free electron approximation assumes

$$|V_{ln}| \ll E - V_{00}, \tag{11.10}$$

and in this case we might expect $a_l \ll a_0$ for $l > 0$. We first discuss the case where there is no degeneracy (the meaning of this term will emerge shortly). In the zeroth approximation we retain only the $l = n = 0$ term in Eq. (11.7), which yields

$$E = \frac{\hbar^2k^2}{2m} + V_{00}; \tag{11.11}$$

a_0 follows from the normalization which we may take to have the form

$$\int_0^a |u(x)|^2 dx = 1 \quad \text{or} \quad a_0 = \frac{1}{a^{1/2}}.$$

In the next approximation we neglect products of the form $a_l V_{ln}$, when l or n, but not both, is equal to zero. Equation (11.7) then yields

$$\left[\frac{\hbar^2}{2m}(k + K_n)^2 - E\right] a_n + V_{0n} a_0 = 0; \tag{11.12}$$

or setting V_{00} to zero in (11.1) (and with it all V_{ll}), we have

$$\begin{aligned} a_n &= -\frac{V_{0n} a_0}{\frac{\hbar^2}{2m}\left[(k + K_n)^2 - k^2\right]} \\ &= \frac{2m}{\hbar^2} V_{0n} \frac{a_0}{k^2 - (K_n + k)^2}. \end{aligned} \tag{11.13}$$

Our Bloch function is then

$$\psi(x) = e^{ikx}\left[1 + \frac{2m}{\hbar^2}\sum_l \frac{V_{0l}}{k^2 - (k + K_l)^2} e^{iK_l x}\right] a_0. \tag{11.14}$$

Equations (11.11) and (11.14) give the energies and wave functions of our one-dimensional solid in the NFE approximation.

We now discuss the degenerate case where k lies near a Brillouin zone edge (these are points in one dimension). It is seen from Eq. (11.14) that the expansion coefficient of the term l = n diverges when $k^2 - (k + K_n)^2 = 0$. We note that this is the condition for one-dimensional Bragg scattering and occurs when $k = -K_n/2 = -n\pi/a$. Bragg scattering leads to reflection and when this occurs we expect the wave function to contain comparable contributions from both the initial (incoming) and the reflected wave. Since this is contradictory to our earlier assumption that $a_0 \gg a_n$ for the n value causing the divergence, our perturbation treatment breaks down. The degenerate approximation involves keeping both a_0 and a_n in zeroth order. Equation (11.12) then consists of the two equations:

$$\left[\frac{\hbar^2}{2m}k^2 - E\right] a_0 + V_{0n} a_n = 0 \tag{11.15a}$$

$$V_{n0} a_0 + \left[\frac{\hbar^2}{2m}(k + K_n)^2 - E\right] a_n = 0, \tag{11.15b}$$

and Eq. (11.9) becomes

$$\begin{vmatrix} \frac{\hbar^2 k^2}{2m} - E & V_{0n} \\ V_{n0} & \frac{\hbar^2}{2m}(k + K_n)^2 - E \end{vmatrix} = 0. \tag{11.16}$$

On taking the determinant (and using the Hermiticity property given in (11.8)), we obtain

$$E = \frac{1}{2}\left\{\frac{\hbar^2}{2m}\left[k^2 + (k + K_n)^2\right] \pm \sqrt{\left[\frac{\hbar^2}{2m}\left(k^2 - (k + K_n)^2\right)\right]^2 + 4\,|V_{0n}|^2}\right\}. \tag{11.17}$$

For $V_{0n} = 0$ we have two solutions

$$E = \frac{\hbar^2 k^2}{2m}; \qquad E = \frac{\hbar^2}{2m}(k + K_n)^2.$$

These are the "empty lattice" solutions discussed earlier, in connection with Eq. (11.4). When $k^2 = (K_n + k)^2$, which as noted corresponds to $k = -K_n/2$, these two solutions for the energy become identical (which is the definition of degeneracy in quantum mechanics); when $V_{0n} \neq 0$ the degeneracy is lifted, or equivalently split, by an amount $\Delta E = 2\,|V_{on}|$. Note that the electron states are now separated into a set of "bands," which we number by n, which in this case is also the Bloch band index n introduced earlier,

$$0 < E < \frac{\hbar^2}{2m}\left(\frac{\pi}{a}\right)^2 - |V_{01}|; \qquad n = 0$$

$$\frac{\hbar^2}{2m}\left(\frac{\pi}{a}\right)^2 + |V_{01}| < E < \frac{\hbar^2}{2m}\left(\frac{2\pi}{a}\right)^2 - |V_{02}|; \qquad n = 1$$

$$\frac{\hbar^2}{2m}\left(\frac{2\pi}{a}\right)^2 + |V_{02}| < E < \frac{\hbar^2}{2m}\left(\frac{3\pi}{a}\right)^2 - |V_{03}|; \qquad n = 2$$

$$\vdots$$

$$\frac{\hbar^2}{2m}\left(\frac{n\pi}{a}\right)^2 + |V_{0n}|^2 < E < \frac{\hbar^2}{2m}\left(\frac{(n+1)\pi}{a}\right)^2 - \left|V_{0,n+1}\right|. \; n = n.$$

In this limit we retain only a single Fourier coefficient, V_{0n}, and assume all others can be ignored at the degeneracy point associated with a particular n. As previously noted for the present case, the Bloch band index, n, and index of the Fourier coefficient are the same. However this is not true in more general cases.

For $k = \pi/a$, the wave functions obtained by solving (11.15) are

$$\psi_0 = \left(\frac{2}{a}\right)^{1/2} \cos\frac{\pi}{a}x, \; \psi_1 = \left(\frac{2}{a}\right)^{1/2} \sin\frac{\pi}{a}x. \tag{11.18}$$

These two forms interact *minimally and maximally* respectively with the perturbing potential. The Fourier coefficient $|V_{02}|$ leads to a splitting when $k = 2\pi/a$, and so on. The first and second energy bands in the extended zone scheme, where we do not translate k back into the first Brillion zone, are shown in Fig. 11.2. Here we have used units such that a and $\hbar^2/2m$ are unity and hence ka and the free electron energy at the first Brillouin zone point are π and π^2 respectively.

As previously discussed, the region $-\pi/a < k < \pi/a$ is the first Brillouin zone. The "volume" it occupies in k space is then $2\pi/a$. The number of carriers in d dimensions is given by $(2/(2\pi)^d)L^dV_k$, where the 2 accounts for spin, L^d is the real space volume, and V_k is the occupied volume in k space; hence for one dimension we have $2 \cdot (L/2\pi) \cdot (2\pi/a) = 2L/a$. Since L/a is the number of atoms, we see that the *first Brillouin zone will accommodate two electrons per atom.* One can draw a significant conclusion from this fact: if our one-dimensional solid consists of atoms having two conduction electrons, thus filling the first Brillouin zone, an energy gap of $2|V_{01}|$ will separate the first excited state from the ground state. Hence we expect to have an insulator at T = 0 and an (initially) exponential behavior of the conductivity $\sigma \approx \exp(-2|V_{10}|/k_BT)$ for $T > 0$. Similarly, if we have four electrons per atom, all states in the second Brillouin zone will be filled and separated by a gap $2|V_{02}|$ from the first excited state. Hence even valence atoms (those having an even number of conduction or valance electrons) will be insulators, while those with an odd number of valance electrons, leading to a half filled zone and hence no gap, will be metals. We cannot immediately apply this conclusion to a three-dimensional solid, as will be discussed, due to an overlap of bands in different directions. It turns out that even valence atoms may be metals or insulators, but that all insulators (with rare exceptions) still have an even number of electrons per atom.

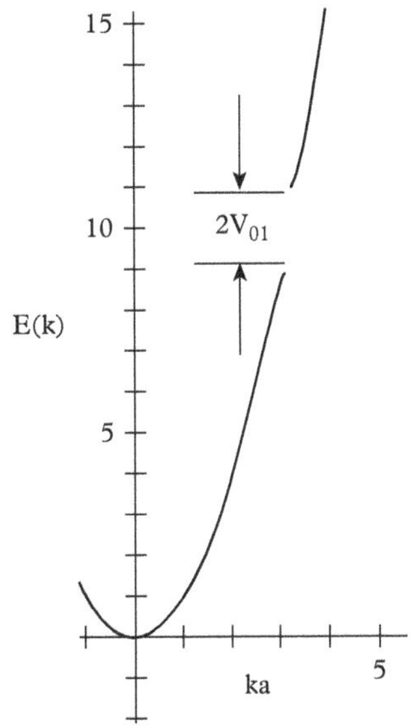

Figure 11.2 *The first and second energy bands in the extended zone scheme.*

Let us re-examine the structure of Eq. (11.6). In our perturbation theory treatment we choose the term $l = 0$ for our zeroth order form or, said differently, for our origin in reciprocal (k) space. The structure of Eq. (11.6) is invariant under the transformation $k \to k + K_n$, where K_n is an arbitrary reciprocal lattice vector, which yields

$$\sum_{l=-\infty}^{l=+\infty} \left[\frac{\hbar^2}{2m} (k + K_l + K_n)^2 + (V - E) \right] a_l e^{i(K_l + K_n + k)x} = 0. \tag{11.19}$$

If we write $l' = l + n$ and $K_{l'} = K_l + K_n$, we have

$$\sum_{l'} \left[\frac{\hbar^2}{2m} (k + K_{l'})^2 + (V - E) \right] a_{l'} e^{i(K_{l'} + k)x} = 0 \tag{11.20}$$

which is identical to (11.6), given that the origin of the summation index is arbitrary since we sum over all integers. Therefore, *any reciprocal lattice point can be used as the origin*, and the energy band structure must be invariant to the

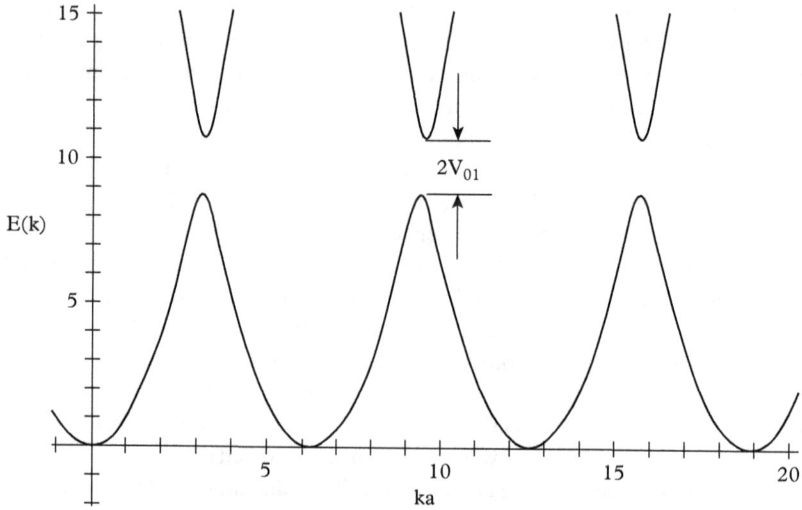

Figure 11.3 *The first and second energy bands in the repeated zone scheme.*

transformation $k \to k + K_n$ in some sense (which we will clarify). Figure 11.3 shows the same band structure as Fig. 11.2, only with successive points of the reciprocal lattice used as origins. The unperturbed bands are free electron parabolas centered at the points $2\pi n/a$ with $n = 0, 1 \cdots$. Note that the bands repeat periodically with a discrete lattice translation in reciprocal space, as argued in Eq. (10.17); this representation is called the *repeated zone scheme.*

11.1.3 The nearly free electron model in two dimensions

We now extend our discussion to cover the two-dimensional case, for which the constant energy curves can be conveniently plotted. Repeating the steps leading to Eq. (11.7) with Bloch functions of the form (10.8a,b) and with $u_{kn}(\mathbf{r})$ given by (10.11) yields

$$\sum_{\mathbf{K}} \left\{ \left[\frac{\hbar^2}{2m} (\mathbf{k} + \mathbf{K})^2 - E \right] \delta_{\mathbf{K}'\mathbf{K}} + V_{\mathbf{K}'\mathbf{K}} \right\} a_{\mathbf{K}} = 0 \tag{11.21}$$

with

$$V_{\mathbf{K}'\mathbf{K}} = \frac{1}{A_c} \int d^3r\, e^{i(\mathbf{K}-\mathbf{K}')\cdot\mathbf{r}} V(\mathbf{r}) = V^*_{\mathbf{K}\mathbf{K}'}, \tag{11.22}$$

where A_c is the area of the unit cell in real space. Equations (11.21) and (11.22) also apply to the three-dimensional case where the cell volume V_c replaces A_c in Eq. (11.22).

The simplest case is that of the square lattice, shown in Fig. 11.4. The basic translation vectors of the real and reciprocal space lattices are given by

$$\mathbf{R}_{n_1 n_2} = a(n_1\hat{\mathbf{x}} + n_2\hat{\mathbf{y}}), \tag{11.23a}$$

and

$$\mathbf{K}_{m_1 m_2} = \frac{2\pi}{a}(m_1\hat{\mathbf{x}} + m_2\hat{\mathbf{y}}). \tag{11.23b}$$

Our Bloch wave function then has the form

$$\psi(x, y) = e^{ik_x x + ik_y y} u_{kn}(x, y), \tag{11.24a}$$

where

$$u_{kn}(x, y) = \sum_{m_1, m_2} a^{(n)}_{m_1 m_2} e^{\frac{2\pi i}{a}(m_1 x + m_2 y)}. \tag{11.24b}$$

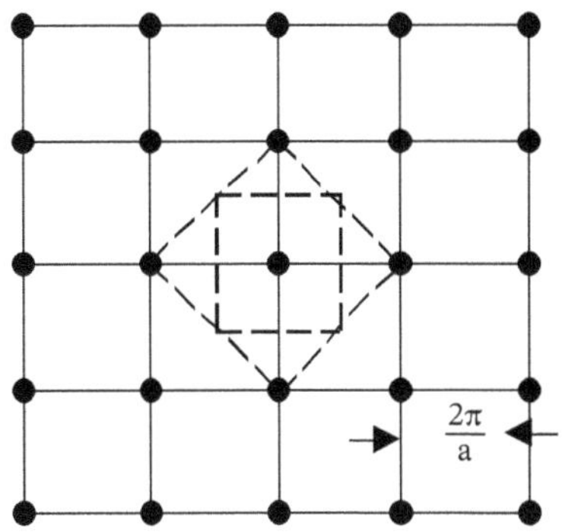

Figure 11.4 *The two-dimensional square lattice in reciprocal space; the dashed lines show the first and second Brillouin zones.*

If the $V_{\mathbf{K}'\mathbf{K}}$ are small, we may continue to neglect them in lowest order except for when $\mathbf{k}$ is at or near a Brillouin zone boundary, where Bragg diffraction requires the inclusion of more than one plane wave and we must evaluate the determinant of the coefficient in Eq. (11.21) and solve for the energies (which are the eigenvalues). If we are near the *edge* of the 2d square Brillouin zone we may continue to approximate $u_{kn}(x, y)$ with only two terms. As an example, consider the $m_1 = -1$, $m_2 = 0$ case. The 2×2 determinant following from (11.21) when $k_x \sim \pi/a$ is

$$\begin{vmatrix} \frac{\hbar^2}{2m}\left(k_x^2 + k_y^2\right) - E & V_{0-1} \\ V_{-10} & \frac{\hbar^2}{2m}\left[\left(k_x - \frac{2\pi}{a}\right)^2 + k_y^2\right] - E \end{vmatrix} = 0,$$

which has the solution

$$E = \frac{\hbar^2 k_y^2}{2m} + \frac{1}{2}\left\{\frac{\hbar^2}{2m}\left[k_x^2 + \left(k_x - \frac{2\pi}{a}\right)^2\right] \pm \sqrt{\left\{\frac{\hbar^2}{2m}\left[k_x^2 - \left(k_x - \frac{2\pi}{a}\right)^2\right]\right\}^2 - 4|V_{10}|^2}\right\}, \tag{11.25}$$

(where we have used the property $V_{0-1} = V_{10} = (V_{01})^*$). This solution gives a reasonable description unless we simultaneously involve a component k_y which lies near a different Brillouin zone edge, say $k_y = +\pi/a$; i.e., when $\mathbf{k}$ lies near the *corner* of the zone. We must then include four terms in our expansion of $u_{kn}(x, y)$,

$$u_{kn} = a_0 + a_{10} e^{\frac{2\pi i x}{a}} + a_{01} e^{\frac{2\pi i y}{a}} + a_{11} e^{\left(\frac{2\pi i x}{a} + \frac{2\pi i y}{a}\right)}, \tag{11.26}$$

which involves the Fourier components V_{00}, V_{10}, V_{01}, and V_{11} (note $V_{10} = V_{01}$ for a square lattice). The associated determinant would now be 4×4, the solution for which would be carried out using a computer.

Once a computer is invoked one can include many reciprocal lattice vectors; in particular all those reciprocal lattice vectors having the same length which are related by symmetry. Starting from the shortest vectors, for the square lattice these are

$$
\begin{aligned}
&V_{10} : \frac{2\pi}{a}\hat{\mathbf{x}},\ -\frac{2\pi}{a}\mathbf{x},\ \frac{2\pi}{a}\hat{\mathbf{y}},\ -\frac{2\pi}{a}\hat{\mathbf{y}};\ |\mathbf{K}| = \frac{2\pi}{a} \\
&V_{11} : \frac{2\pi}{a}(\hat{\mathbf{x}}+\hat{\mathbf{y}}),\ \frac{2\pi}{a}(-\hat{\mathbf{x}}+\hat{\mathbf{y}}),\ \frac{2\pi}{a}(\hat{\mathbf{x}}-\hat{\mathbf{y}}),\ \frac{2\pi}{a}(-\hat{\mathbf{x}}-\hat{\mathbf{y}});\ |\mathbf{K}| = \frac{2\sqrt{2}\pi}{a} \\
&V_{20} : \frac{4\pi}{a}\hat{\mathbf{x}},\ -\frac{4\pi}{a}\hat{\mathbf{x}},\ \frac{4\pi}{a}\hat{\mathbf{y}},\ -\frac{4\pi}{a}\hat{\mathbf{y}};\ |\mathbf{K}| = \frac{4\pi}{a} \\
&V_{12} : \frac{2\pi}{a}(2\hat{\mathbf{x}}+\hat{\mathbf{y}}),\ \frac{2\pi}{a}(-2\hat{\mathbf{x}}+\hat{\mathbf{y}}),\ \frac{2\pi}{a}(2\hat{\mathbf{x}}-\hat{\mathbf{y}}),\ \frac{2\pi}{a}(-2\hat{\mathbf{x}}-\hat{\mathbf{y}}); \\
&\qquad \frac{2\pi}{a}(\hat{\mathbf{x}}+2\hat{\mathbf{y}}),\ \frac{2\pi}{a}(-\hat{\mathbf{x}}+2\mathbf{y}),\ \frac{2\pi}{a}(\hat{\mathbf{x}}-2\hat{\mathbf{y}}),\ \frac{2\pi}{a}(-\hat{\mathbf{x}}-2\hat{\mathbf{y}});\ |\mathbf{K}| = \frac{2\sqrt{5}\pi}{a} \\
&\qquad \vdots
\end{aligned}
\tag{11.27}
$$

Summing over all the reciprocal lattice vectors of a given length and symmetry is referred to as summing over the "star," since the set of vectors forms an object which resembles the rays of a star. The stars of the vectors (11.27) are shown in Fig. 11.5. The size of a determinant involved in determining the 2d band structure for the stars shown in (11.27) would increase as: $1 + 4 + 4 + 4 + 8\ldots$; including the first four stars results in a 21×21 determinant. When we sum over the star our treatment will not be restricted to a specific face or line of the zone as, for example, was the discussion surrounding Eq. (11.26) which would be valid only for the positive quadrant ($k_x > 0$, $k_y > 0$).

To represent the functional behavior of $E = E(k_x, k_y)$ in two dimensions we can plot curves in the (k_x, k_y) plane for different values of E. For free electrons these are simply concentric circles with a radius $k = (2mE)^{1/2}/\hbar$. For nearly

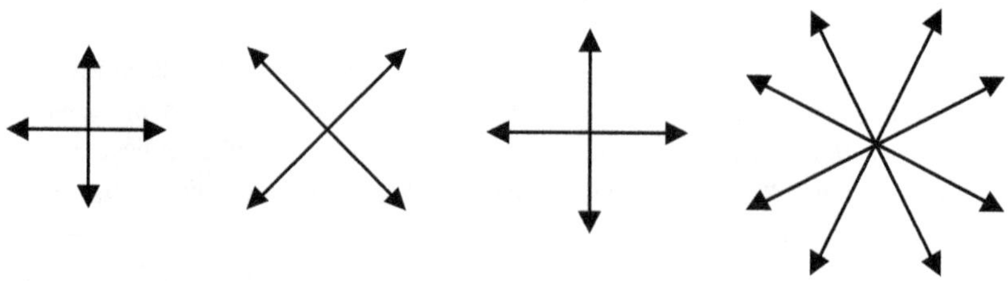

Figure 11.5 *Stars of the 2d square lattice (not to scale).*

free electrons when $E \ll (\hbar^2/2m) \cdot (\pi/2)^2$, the constant energy curves are essentially circles. However, as k increases toward the first Brillouin zone value, π/a, the constant energy surfaces are no longer circular. In particular the perimeter of the constant energy circle is interrupted in such a way that the component of the electron group velocity, v_g, normal to the Brillouin zone line *vanishes*. This is simply because Bragg diffraction results in a standing wave in the direction normal to the zone edge and a standing wave carries no current, or equivalently has no velocity in that direction. Therefore all the Fermi curves satisfies the requirement

$$v_g = \frac{1}{\hbar}\left(\frac{\partial E}{\partial \mathbf{k}}\right)_{\perp B.Z} = 0. \tag{11.28}$$

(Note (11.25) also satisfies this property in that $\partial E(k_x = \pi/a)/\partial k_x = 0$, as it must.)

We now restrict to the case where the Fourier components of the lattice potential are very small and the constant energy curves only distort (to satisfy Eq. (11.28)) very close to the Brillouin zone. However, this causes the circular curves to "break up" into *disconnected* curves (basically arcs except near the BZ). Figure 11.6 shows two different energies for a 2d square lattice. Since points in **k** space related by a reciprocal lattice vector are identical, we may *translate* sections of the constant energy curves and "reconnect" them. Figure 11.7 shows the reconnected pieces that arise from Fig. 11.6(b). The curve in Fig. 11.7(a) is assembled from pairs of opposing sides of the zone while that in 11.7(b) is formed from the four opposing corners.

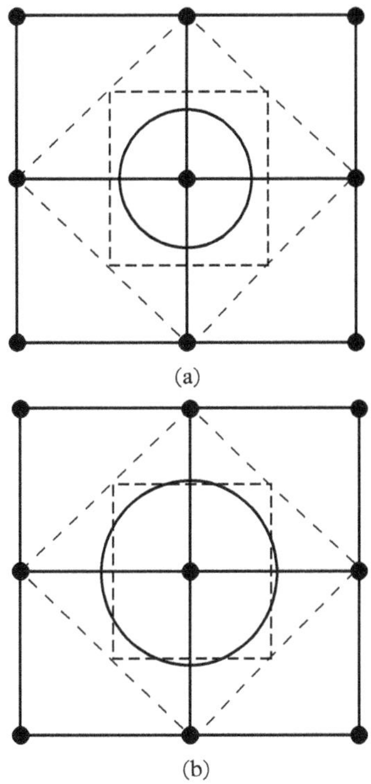

Figure 11.6 *Circles of constant energy for: (a)* $k < \pi/a$, *(b)* $k > \pi/a$.

If the chosen energy corresponds to the Fermi energy, then the region inside the circles shown in Figs. 11.6(a,b) involve *occupied* electron states, while the states outside the circle are *unoccupied* states; these two regions are commonly classified as *electrons* and *holes*. Therefore, Fig. 11.7(a) corresponds to an *electron* Fermi surface in the second Brillouin zone, while Fig. 11.7(b) is a *hole* surface in the first zone.

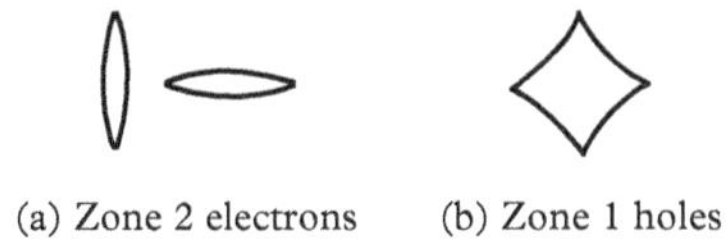

Figure 11.7 *Electron and hole surfaces following from the construction in Fig. 11.6.*

There is an alternative (but equivalent) way to construct the constant energy surface in the nearly free electron model which does not require translating sections of the surface (in the extended zone scheme) by reciprocal lattice vectors and piecing them together.[1] We start by recalling that the energy band structure is invariant to the transformation $\mathbf{k} \to \mathbf{k} + \mathbf{K}$. Hence we may draw our free electron circle about *all* lattice sites. Figure 11.8 shows this construction for the energies depicted in Fig. 11.6. Rather than having to translate sections of the curve, the complete curve is now generated, in place, by the intersections of various circles from neighboring zones (Harrison (1960)). The curves are also repeated in every zone (in accord with the translational symmetry requirements). This representation of the constant energy surfaces, shown in Fig. 11.8, is referred to as the *repeated zone* representation, as opposed to that of Fig. 11.7 where the surfaces are constructed in the *extended zone* scheme.

[1] Electronic structure of polyvalent metals, W.A. Harrison, *Phys. Rev.* **118**, 1190 (1960).

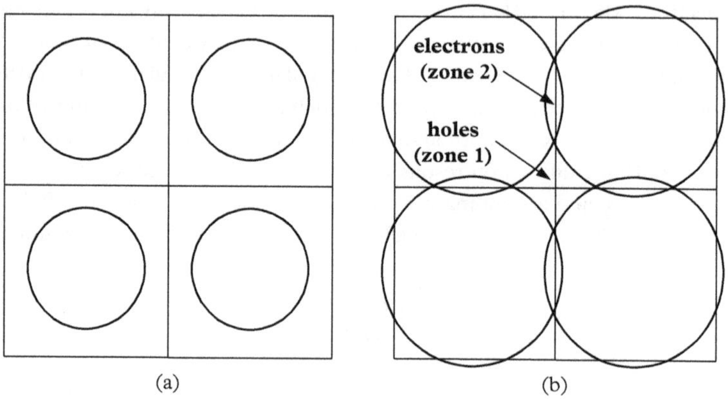

Figure 11.8 *The two-dimensional Fermi circle drawn in the repeated zone scheme.*

11.1.4 The nearly free electron model in three dimensions

The techniques just discussed for two dimensions extend naturally to three dimensions; the circles then become spheres. Zone faces, it will be recalled, are defined as the planes that bisect the reciprocal lattice vectors (drawn from some common origin); those associated with the three-dimensional star of the shortest vector make up the first zone, those of the second shortest star the second zone, and so forth. Any arc of the circle formed by the intersection of spheres drawn about two reciprocal lattice points defines a plane which bisects the reciprocal lattice vector joining those lattice points; i.e., it defines a plane of the Brillouin zone generated by the star of that vector. It follows that in the nearly free electron model, constant energy surfaces in **k** space may be constructed by drawing spheres with a radius $k = (2mE)^{1/2}/\hbar$ about all reciprocal lattice points. Gaps (or separations) in **k** space will arise whenever two spheres intersect (due to Bragg diffraction from the zone face associated with this intersection, as in the two-dimensional case); Eq. (11.28) continues to hold. Points within these surfaces may be assigned a band index n according to how many spheres enclose them and all states with this same **k** but smaller n are also occupied. This prescription defines so-called electron surfaces. In some cases, it is more convenient to visualize the unoccupied volume outside a given electron surface (in the repeated zone scheme) as a hole surface associated with the next smallest band index (as discussed in Section 10.5).

To construct the *Fermi surface* in the nearly free electron (NFE) approximation, we apply the given rules for a free electron sphere having a radius given by $k_F = (3\pi^2 n)^{1/3}$, where n is the conduction electron number density. For those metals for which the NFE approximation provides a good description we may calculate n as

$$n = \frac{(\text{valance}) \times (\#\ \text{atoms per unit cell})}{(\text{real space volume of unit cell})}. \tag{11.29}$$

Equation (11.29) assumes that all electrons which can participate in metallic bonding will contribute to the electron gas in the metallic state. Metals with multiple valance states, such as transition metals, generally cannot be described by the free electron model. Examples for which the NFE approximation provides a good description include most metals *not* having unfilled d or f shells (noble metals form an exception due to the perturbation induced by d levels lying close to the Fermi level). In such cases the relatively extended s and p orbitals interact minimally with the closed shell cores of the atoms, as we will explain more fully in Section 11.2.

Polyvalent metals for which the NFE model is particularly successful include Mg, Ca, and Al; the first adopts the hexagonal close packed structure (hcp) and the latter two, the face centered cubic (fcc) structure. Mg and Ca, which have valence 2, contribute two electrons per atom while Al with valence 3 contributes three electrons. Figure 11.9 shows the valence 2 hcp free electron Fermi surface, which corresponds approximately to Mg.[2]

It turns out that the geometrical structure factor (see Section 9.7, but now for electrons) vanishes for electron states on the hexagonal face of the Brillouin zone. However, inclusion of the spin–orbit interaction (which is a relativistic effect) largely restores a splitting of the energy bands on this surface. Figure 11.9 assumes the degeneracy *is* lifted on this zone face.[3] Figure 11.10 shows the free electron surfaces for valence 2 to 4 cases, corresponding approximately to Ca, Al, and Pb.

[2] Fermi surface of magnesium. I. Magnetoacoustic attenuation, J.B. Ketterson and R.W. Stark, *Phys. Rev.* **156**, 748 (1967).

[3] Breakdown is permitted along A-L symmetry line; see: Spin-orbit coupling in the band structure of magnesium and other hexagonal-close-packed metals, M.H. Cohen and L.M. Falicov, *Phys. Rev.* **130**, 92 (1960).

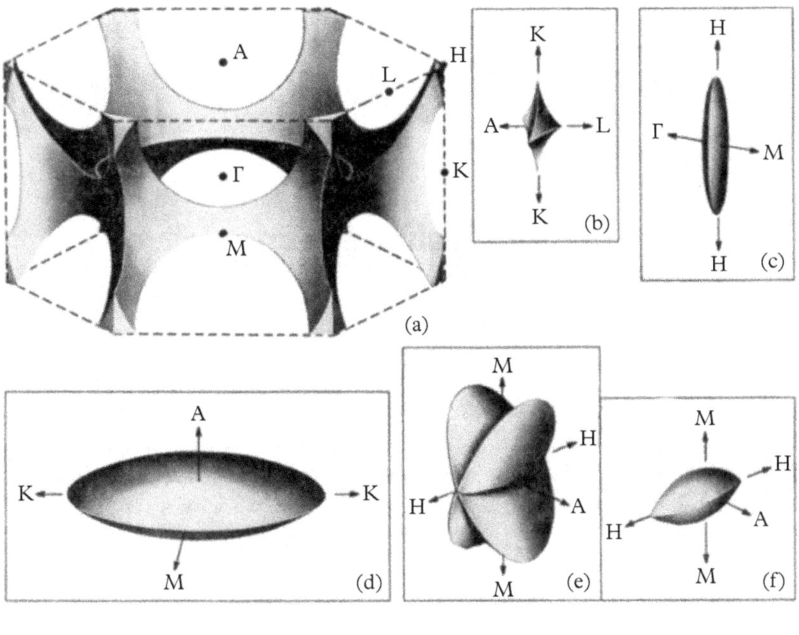

Figure 11.9 *The free electron Fermi surface generated by the intersection of the Fermi sphere with the first four Brillouin zones of the divalent metal magnesium which has the hexagonal close packed structure. The surfaces so generated have been drawn in the first or reduced zone using the Harrison prescription, as described in the text: (a) second zone holes; (b) first zone holes; (c) third zone electrons; (d) third zone electrons; (e) third zone electrons; (f) fourth zone electrons. (After Ketterson and Stark (1967).)*

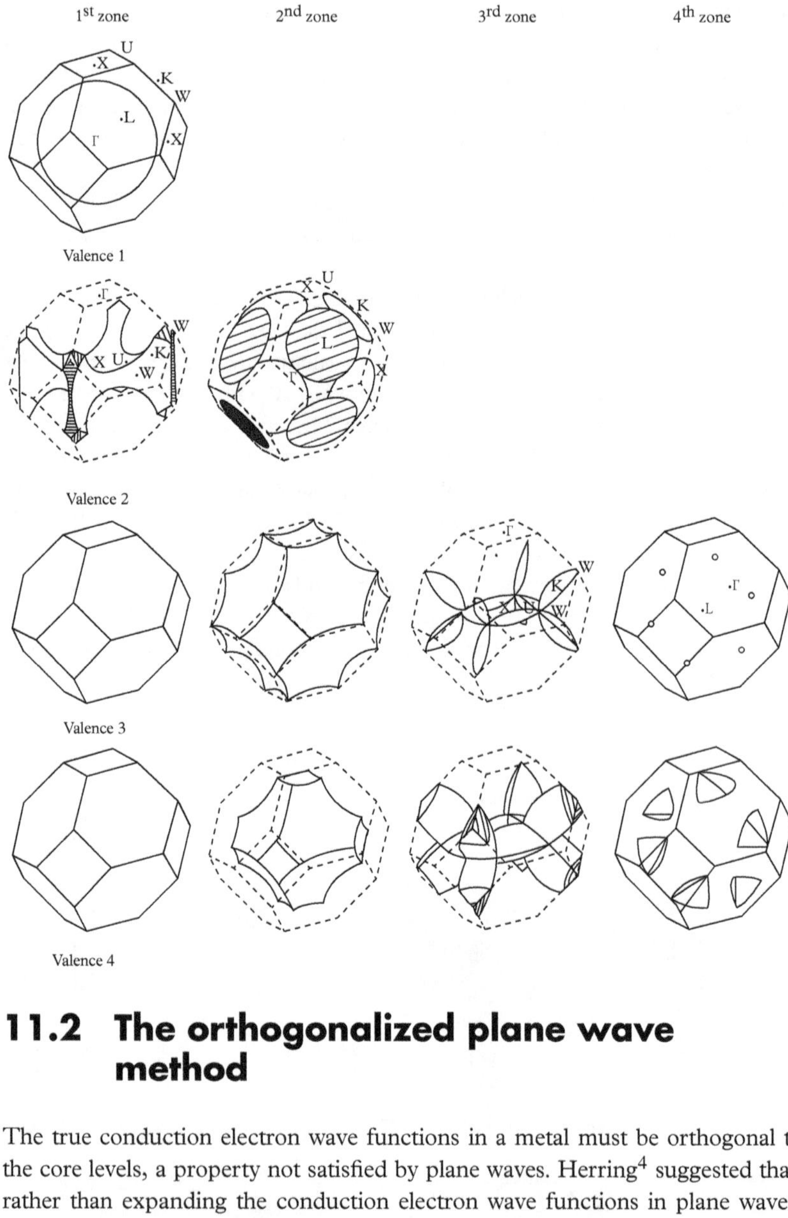

Figure 11.10 *The free electron Fermi surface for the face centered cubic structure resulting from intersections with the first four Brillouin zones, when drawn in the first Brillouin zone. The valences correspond to the number of electrons per atom extending from 1 to 4. As examples, valences 2, 3, and 4 have features in common with calcium, aluminum, and lead. (After Harrison (1960).)*

11.2 The orthogonalized plane wave method

The true conduction electron wave functions in a metal must be orthogonal to the core levels, a property not satisfied by plane waves. Herring[4] suggested that, rather than expanding the conduction electron wave functions in plane waves, a more rapidly convergent procedure is to expand in *orthogonalized plane waves* (OPWs). An OPW is that part of a plane wave that remains after its projection onto the core levels has been *subtracted off.* Suppose we have a set of occupied core levels, ϕ_i^c, with $i = 1, \ldots, n$. We define an (unnormalized) OPW, $\psi_{\mathbf{k}}$, by

$$\psi_{\mathbf{k}}(\mathbf{r}) = e^{i\mathbf{k}\cdot\mathbf{r}} - \sum_{i=1}^{n} \left\langle \phi_i^c \middle| e^{i\mathbf{k}\cdot\mathbf{r}} \right\rangle \phi_i^c(\mathbf{r}). \tag{11.30}$$

[4] A new method for calculating wave functions in crystals, C. Herring, *Phys. Rev.* 57, 1169 (1940).

Thus for sodium, which has a ground state configuration $1s^2\,2s^2\,2p^6 3s^1$, the core levels would correspond to $1s^2$, $2s^2$, and $2p^6$; i.e., we have $n = 10$ core levels. For the so-called simple metals like Na, Mg, and Al, the conduction electron wave function is usually rapidly convergent in the OPW basis. Such an expansion has the form

$$\psi_{\mathbf{k}}(\mathbf{r}) = \sum_{\mathbf{K}} a_{\mathbf{K}} \psi_{\mathbf{k+K}}(\mathbf{r}). \tag{11.31}$$

For the simple metals we need only a few OPWs to obtain a good approximation for the conduction electron wave function; throughout much of the Brillouin zone only a single OPW may give a reasonable description. However, on the basis of our earlier discussion of the nearly free electron model, we expect that in the vicinity of one of the planes in reciprocal space defining the Brillouin zone a minimum of two OPWs is required; near a BZ edge, three or more are required, depending on the number of planes intersecting; and near a corner, a still larger number are necessary.

Let us examine the one-electron Schrödinger equation in our OPW basis, which is

$$\left[-\frac{\hbar^2}{2m}\nabla^2 + V(\mathbf{r})\right] \sum_{\mathbf{K}} a_{\mathbf{K}} \psi_{\mathbf{k+K}} = E \sum_{\mathbf{K}} a_{\mathbf{K}} \psi_{\mathbf{k+K}}. \tag{11.32}$$

Substituting Eq. (11.30), we have

$$\sum_{\mathbf{K}} \left\{ \left[\frac{\hbar^2}{2m}(\mathbf{k}+\mathbf{K})^2 - E + V(\mathbf{r})\right] e^{i(\mathbf{k}+\mathbf{K})\cdot\mathbf{r}} - \left[-\frac{\hbar^2}{2m}\nabla^2 + V(\mathbf{r})\right] \sum_{i=1}^{n} \int d^3r' \phi_i^c(\mathbf{r}) \phi^{c*}(\mathbf{r}') e^{i(\mathbf{k}+\mathbf{K})\cdot\mathbf{r}'} \right\} a_{\mathbf{K}} = 0. \tag{11.33}$$

If for the core levels, $\phi_i^c(\mathbf{r})$ (which are close to the nucleus) we take $V(\mathbf{r})$ to be the atomic potential,[5] we have

$$\left[-\frac{\hbar^2}{2m}\nabla^2 + V(\mathbf{r})\right] \phi_i^c(\mathbf{r}) = E_i^c \phi_i^c(\mathbf{r}), \tag{11.34}$$

where E_i^c is the energy of the ith core level. Using (11.34) we may then rewrite the following factor in the second part of Eq. (11.33) as

$$\left[-\frac{\hbar^2}{2m}\nabla^2 + V(\mathbf{r})\right] \sum_{i=1}^{n} \int d^3r' \phi_i^c(\mathbf{r}) \phi^{c*}(\mathbf{r}') e^{i\mathbf{k}\cdot\mathbf{r}'} = \sum_{i=1}^{n} E_i^c \int \phi_i^c(\mathbf{r}')^* \phi_i^c(\mathbf{r}) e^{i\mathbf{k}\cdot\mathbf{r}'} d^3r' \equiv \hat{V}^{op} e^{i\mathbf{k}\cdot\mathbf{r}}, \tag{11.35}$$

[5] This approximation could be improved by performing the calculations self consistently.

where $\hat{V}^{op}$ is an integral operator which is "non-local"; i.e., it involves an integral over all space. Equation (11.33) then becomes

$$\sum_{\mathbf{K}}\left[\frac{\hbar^2}{2m}(\mathbf{k}+\mathbf{K})^2 - E + V(\mathbf{r}) - \hat{V}^{op}\right] a_{\mathbf{K}} e^{i(\mathbf{k}+\mathbf{K})\cdot\mathbf{r}} = 0. \tag{11.36}$$

When only s and p electrons are involved, the following qualitative behavior occurs.

(1) The effect of $\hat{V}^{op}$, or equivalently the orthogonalization, is to *add* kinetic energy, since the conduction electron wave function *oscillates rapidly* in order to remain *orthogonal* to the core electrons.

(2) This added kinetic energy (a positive quantity) has the effect of largely *canceling* the (negative) potential energy of the core; this is sometimes called the *cancellation theorem*, although this cancellation in practice is only approximate. The total "potential," $V(\mathbf{r}) - \hat{V}^{op}$, is then quite *weak*; *this is why the nearly free electron model works so well in practice for the simple metals*. The phenomenon is quite general, and also occurs in molecules (M.H. Cohen and V. Heine (1961)[6]). If we multiply Eq. (11.36) by $e^{-i\mathbf{K}'\cdot\mathbf{r}}$ and integrate over a unit cell we obtain

[6] Cancellation of kinetic and potential energy in atoms, molecules, and solids, M.H. Cohen and V. Heine, *Phys. Rev.* 122, 1821 (1961).

$$\sum_{\mathbf{K}}\left\{\left[\frac{\hbar^2}{2m}(\mathbf{k}+\mathbf{K})^2 - E\right]\delta_{\mathbf{K}\pm\mathbf{K}'} + V_{\mathbf{K}\pm\mathbf{K}'}(\mathbf{k})\right\} a_{\mathbf{K}} = 0, \tag{11.37a}$$

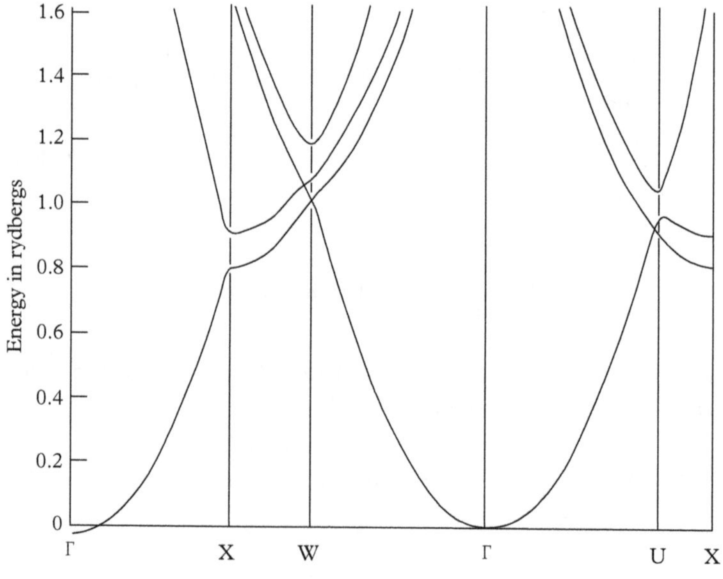

Figure 11.11 *Band structure of aluminum calculated in a four OPW approximation. Note the small band gaps at the zone boundaries, as expected for the nearly free electron-like behavior associated with this metal. (After Harrison (1960).)*

where $V_{K\pm K'}(\mathbf{k})$ depends only on the difference $\mathbf{K}\pm\mathbf{K}'$ and is given by

$$V_{\mathbf{K}}(\mathbf{k}) = \frac{1}{V_c}\int V(\mathbf{r})e^{i\mathbf{K}\cdot\mathbf{r}} - \frac{1}{V_c}\sum_{i=1}^{n}\iint e^{i(\mathbf{k}+\mathbf{K})\cdot\mathbf{r}'}E_i^c\phi_i(\mathbf{r}')^{c*}\phi_i^c(\mathbf{r})d^3rd^3r'; \tag{11.37b}$$

V_c is the volume of a unit cell. The first term in (11.37b) is independent of **k** while the second is a function of **k**. The band structure energies, $E = E(\mathbf{k})$, are then given by the eigenvalues of (11.37a) involving the non-local operator potential (11.37b).

Figure 11.11 shows the energy bands as calculated by Harrison along high symmetry directions for nearly free electron-like aluminum.[7] Note the small energy gaps at the Brillouin zone boundaries (the points L, X, U, and W).

11.3 The pseudopotential method

The difference between the NFE approximation and the OPW scheme arises from the second term in (11.37b), with the result that $V_K(\mathbf{k})$ will in general be *a function of* **k**. This comes from the *non-local* character of $\hat{V}^{op}$. In many cases it turns out that one can approximate $\hat{V}^{op}$ by some suitably chosen *local* potential which when combined with the true potential $V(\mathbf{r})$ produces an effective, overall weaker, potential referred to as the *pseudopotential* (Phillips and Kleinman (1959)[8]). This local approximation works reasonably well in many cases.

Perhaps the simplest such potential is the so-called *screened ion* form

$$V(r) = -\frac{e^2}{r};\ r > r_0 \qquad V(r) = V_0;\ r < r_0, \tag{11.38}$$

which is shown schematically in Fig. 11.12. Note that this form of the potential suppresses the strong nuclear potential for small r.

Much effort has gone into designing pseudopotentials which accurately represent the energy bands of solids, particularly in those semiconductors for which the bonding is dominated by s and p electrons.[9] One can go on to construct *non-local pseudopotentials*, in particular forms in which electrons experience different potentials depending on the angular momentum content of their wave functions. Figure 11.13 shows such a pseudopotential for the case of silicon; note the large *l*-dependent differences at small r.[10,11]

There is a vast literature on the pseudopotential method. Here we refer the reader to the book by R.M. Martin (see Further Reading section).

The pseudopotential method has been most widely applied to describe semiconductors. Figure 11.14 shows the band structure along various symmetry axes for Si and Ge, while Fig. 11.15 shows that for GaAs, all calculated using the pseudopotential method[12]. The top of the valence band (Γ_8 in Ge and GaAs, and $\Gamma_{25'}$

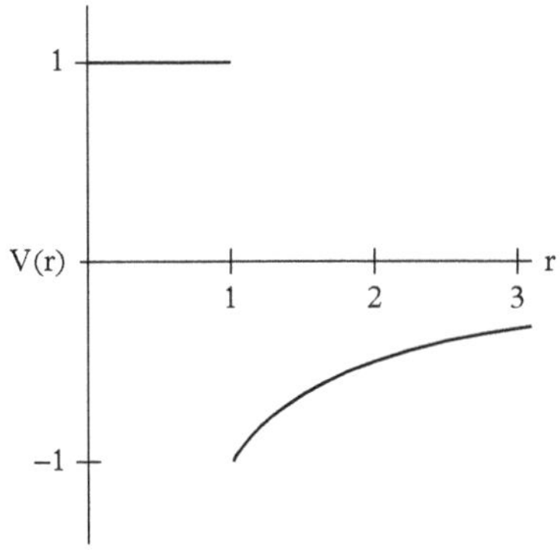

Figure 11.12 *The screened pseudo-potential.*

[7] Band Structure of Aluminum, W.A. Harrison, *Phys. Rev.* **118**, 1182 (1960).

[8] New method for calculating wave functions in crystals and molecules, J.C. Phillips and L. Kleinman, *Phys. Rev.* **116**, 287 (1959).

[9] In materials having d or f bands, the augmented plane wave or Green's function methods (to be discussed in Chapter 13) are more rapidly convergent.

[10] Theory of *ab initio* pseudopotential calculations, M.T. Yin and M.L. Cohen, *Phys. Rev. B* **25**, 7403 (1982).

[11] Norm-conserving pseudopotentials, D.R. Hamann, M. Schluter, and C. Chiang, *Phys. Rev. Lett.* **43**, 1494 (1979).

[12] Nonlocal pseudopotential calculations for the electronic structure of eleven diamond and zinc-blende semiconductors, J.R. Chelikowsky and M.L. Cohen, *Phys. Rev. B* **14**(2), 556 (1976).

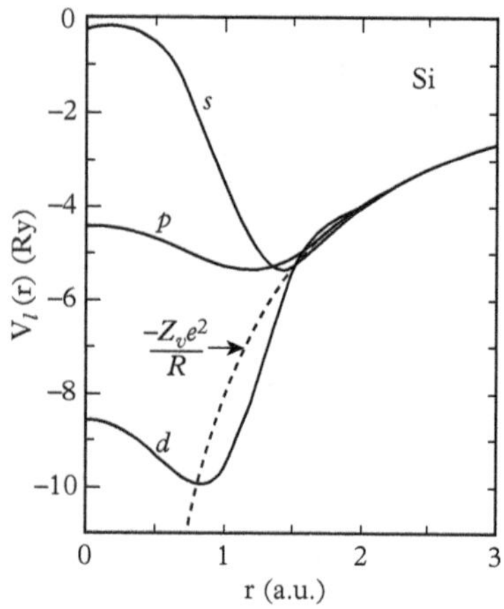

Figure 11.13 *A non-local, l-dependent, pseudopotential for the case of silicon. (After Yin and Cohen (1982).)*

in Si) lies at $\mathbf{k} = 0$ in all three materials, above which there is an energy gap separating them from the higher lying conduction band states. The structure of the valance band states, as well as formalisms to parameterize them, are discussed in Chapter 12 in some detail. Figure 11.16 shows ellipsoidal constant energy surfaces for the materials Si and Ge in the vicinity of the conduction band minima; note they are aligned along the [100], [111] axes respectively. For GaAs the conduction band minimum lies at the zone center and the constant energy surfaces are then nearly spherical for small k.

11.4 The tight binding approximation

The opposite limit of the nearly free electron model is to assume that the valance electrons are tightly bound. This approximation forms the basis of the *tight binding approximation* (TBA). It can also be used to describe the outer core electron states associated within the NFE model (rather than regarding them as part of the core potential).

The TBA assumes that the atomic wave functions represent a better starting basis then a few-plane-wave (or OPW) basis. We start our discussion of the

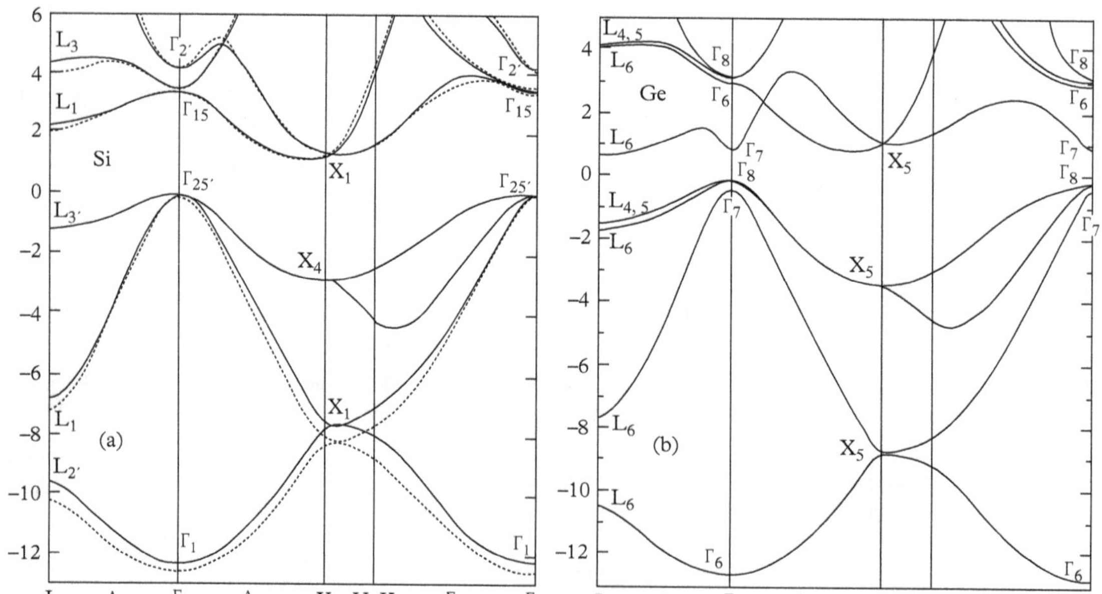

Figure 11.14 *Band structures for (a) Si and (b) Ge. The group symmetries of the states are also given. In the case of silicon, results are presented for both a non-local pseudopotential (solid line) and a local pseudopotential (dashed line). The points and lines of symmetry are as shown in Fig. 10.4. (After J. Chelikowsky and M. L. Cohen (1976).)*

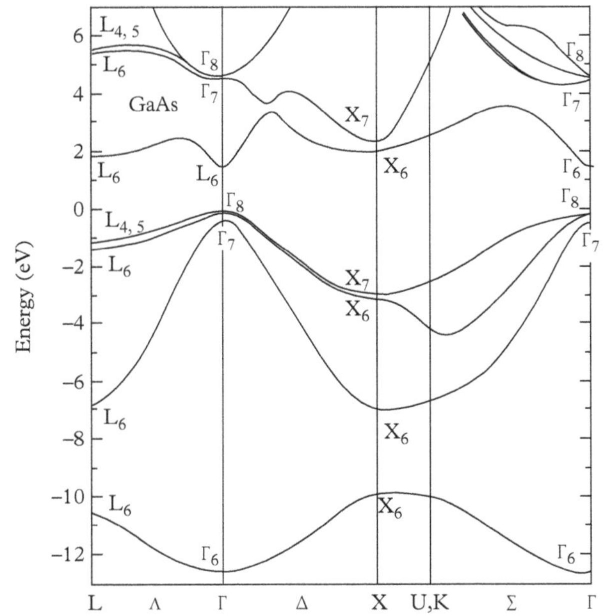

Figure 11.15 *The band structure of GaAs calculated using the pseudopotential approach. The points and lines of symmetry are as shown in Fig. 10.4. The symmetries of the states are given in the double group notation. (After J. Chelikowsky and M.L. Cohen (1976).)*

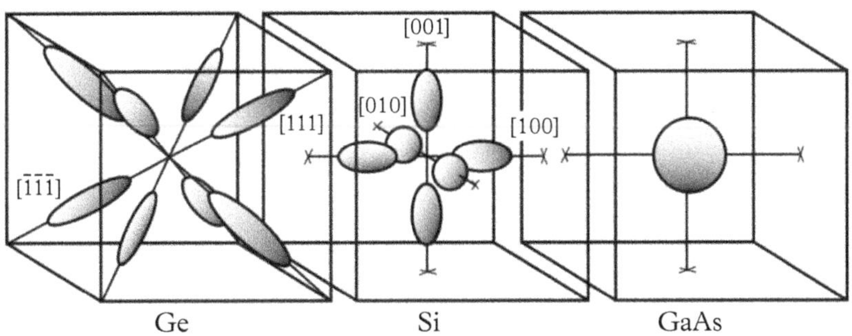

Figure 11.16 *Constant energy surfaces in the vicinity of the electron conduction band minimum for germanium, silicon, and gallium arsenide.*

TBA by writing the Schrödinger equation of the atomic valance electrons having quantum numbers *l*, m, and n as

$$\left[-\frac{\hbar^2}{2m}\nabla^2 + U(\mathbf{r})\right]\phi_{lmn}(\mathbf{r}) = E_{lmn}\phi_{lmn}(\mathbf{r}), \tag{11.39}$$

where U(**r**) is again some self-consistent (or average) potential produced by the nucleus and the core electrons, together with the averaged effect of the remaining valance (or conduction) electrons.

The TBA assumes that the conduction electron wave functions may be approximated as a *linear combination of the atomic orbitals* (LCAO).[13]

[13] We recall that an orbital is an approximate representation of a many electron system, which only depends on **r**; orbitals are also called one-electron wave functions.

This approximation is commonly used to represent the wave functions of molecules. For a solid with one atom per unit cell this takes the form

$$\psi(\mathbf{r}) = \sum_{lmn}\sum_{\mathbf{R}} a_{\mathbf{R}}^{(lmn)} \phi_{lmn}(\mathbf{r}-\mathbf{R}), \tag{11.40}$$

where **R** is the set of all real-space lattice vectors. The $a_{\mathbf{R}}^{(lmn)}$ are expansion coefficients which will be chosen to best approximate the solution to the Schrödinger equation for the crystal.

11.4.1 The tight binding approximation for s electrons

We begin with a model where we have s valance electrons with principal quantum number n (which we suppress in what immediately follows); in this case Eq. (11.40) becomes

$$\psi(\mathbf{r}) = \sum_{\mathbf{R}} a_{\mathbf{R}}\phi(\mathbf{r}-\mathbf{R}). \tag{11.41}$$

We will determine the expansion coefficients by the variational technique. This method is based on the fact that when the expectation value of the Hamiltonian is stationary with respect to the parameters controlling some trial wave function (the $a_{\mathbf{R}}$ in the present case), it represents the best solution for the energy for the chosen form. Hence we examine the quantity

$$\frac{\langle\psi(\mathbf{r})|\,\hat{H}\,|\psi(\mathbf{r})\rangle}{\langle\psi(\mathbf{r})|\,\psi(\mathbf{r})\rangle} \cong E(a_{\mathbf{R}}), \tag{11.42}$$

where $\hat{H}$ is given by

$$\hat{H} = -\frac{\hbar^2}{2m}\nabla^2 + \sum_{\mathbf{R}} U(\mathbf{r}-\mathbf{R}). \tag{11.43}$$

The inclusion of the denominator in (11.42) allows us to work with an unnormalized wave function (otherwise we must constrain the normalization using the Legrange multiplier method). The notation $E = E\,(a_{\mathbf{R}})$ denotes the fact that the approximate energy is a function of the wave function expansion coefficients in (11.41); the minimum value of E (and hence the best value for a given trial form) is obtained by solving the set of equations

$$\frac{\partial E(a_{\mathbf{R}})}{\partial a_{\mathbf{R}}^*} = 0 \tag{11.44}$$

(differentiating with respect to $a_{\mathbf{R}}$ leads to equations which are the complex conjugate of those following from (11.44)). Inserting (11.41) and (11.43) into (11.42) yields

$$E = \frac{\sum_{\mathbf{RR}'} a^*_{\mathbf{R}} a_{\mathbf{R}'} H_{\mathbf{RR}'}}{\sum_{\mathbf{RR}'} a^*_{\mathbf{R}} a_{\mathbf{R}'} J_{\mathbf{RR}'}}, \tag{11.45}$$

where we have defined

$$H_{\mathbf{RR}'} = \int d^3r\phi^*(\mathbf{r}-\mathbf{R})H\phi(\mathbf{r}-\mathbf{R}') \tag{11.46}$$

and

$$J_{\mathbf{RR}'} = -\int d^3r\phi^*(\mathbf{r}-\mathbf{R})\phi(\mathbf{r}-\mathbf{R}'); \tag{11.47}$$

the quantities $J_{\mathbf{RR}'}$ are referred to as "overlap integrals"; for the case $\mathbf{R} = \mathbf{R}'$ we have unity (assuming the atomic wave functions are normalized) but they are non-zero for $\mathbf{R} \neq \mathbf{R}'$ (the product of wave functions centered on different sites is non-zero due to their mutual "overlap"). Because of the discrete translational invariance of a crystal, we have

$$H_{\mathbf{RR}'} = H(\mathbf{R}-\mathbf{R}') \tag{11.48}$$

and

$$J_{\mathbf{RR}'} = J(\mathbf{R}-\mathbf{R}'). \tag{11.49}$$

Equation (11.46) may be rewritten as

$$\begin{aligned}
H_{\mathbf{RR}'} &= \int \phi^*_{\mathbf{R}} \left[-\frac{\hbar^2}{2m}\nabla^2 + \sum_{\mathbf{R}''} U(\mathbf{r}-\mathbf{R}'') \right] \phi_{\mathbf{R}'} d^3r \\
&= \int \phi^*_{\mathbf{R}} \left\{ -\frac{\hbar^2}{2m}\nabla^2 + U(\mathbf{r}-\mathbf{R}') + \left[\sum_{\mathbf{R}''} U(\mathbf{r}-\mathbf{R}'') - U(\mathbf{r}-\mathbf{R}') \right] \right\} \phi_{\mathbf{R}'} d^3r \\
&= \int \phi^*_{\mathbf{R}} \left\{ E_n + \left[\sum_{\mathbf{R}''} U(\mathbf{r}-\mathbf{R}'') - U(\mathbf{r}-\mathbf{R}') \right] \right\} \phi_{\mathbf{R}'} d^3r \\
&= E_n J_{\mathbf{RR}'} + h_{\mathbf{RR}'},
\end{aligned} \tag{11.50}$$

where we have used (12.1) in the third step and denoted the energy of our *atomic* s state having radial quantum number n, by E_n; the matrix $h_{\mathbf{RR}'}$ is defined by

$$\begin{aligned}
h_{\mathbf{RR}'} &= \int d^3r\phi^*_{\mathbf{R}} \left[\sum_{\mathbf{R}''} U(\mathbf{r}-\mathbf{R}'') - U(\mathbf{r}-\mathbf{R}') \right] \phi_{\mathbf{R}'} \\
&= \int d^3r\phi^*_{\mathbf{R}} \left[V(\mathbf{r}) - U(\mathbf{r}-\mathbf{R}') \right] \phi_{\mathbf{R}'},
\end{aligned} \tag{11.51}$$

where we have substituted the definition of the crystal potential (10.1) in the second step. We recall that $J_{\mathbf{RR}} = 1$ and $J_{\mathbf{RR}'}$ should converge rapidly in $\mathbf{R}-\mathbf{R}'$ if the TBA scheme is to be applicable; i.e., if the orbitals are well localized. Likewise, $h_{\mathbf{RR}'}$ should also converge rapidly and we note it contains contributions from two kinds of integrals: "two-center integrals" where $\mathbf{R} = \mathbf{R}' \neq \mathbf{R}''$ and "three-center integrals" where $\mathbf{R}$, $\mathbf{R}'$, and $\mathbf{R}''$ all differ (note the contribution from $\mathbf{R}' = \mathbf{R}''$ vanishes identically).

To obtain a stationary solution (which we will take to be the minimum) we must apply (11.44) to (11.45), which yields

$$\left(\sum_{\mathbf{RR}'} a^*_{\mathbf{R}} a_{\mathbf{R}'} J_{\mathbf{RR}'}\right)\left(\sum_{\mathbf{R}'} a_{\mathbf{R}'} H_{\mathbf{RR}'}\right) - \left(\sum_{\mathbf{R}'} a_{\mathbf{R}'} J_{\mathbf{RR}'}\right)\left(\sum_{\mathbf{RR}'} a^*_{\mathbf{R}} a_{\mathbf{R}'} H_{\mathbf{RR}'}\right) = 0.$$

Dividing by $\sum_{\mathbf{RR}'} a^*_{\mathbf{R}} a_{\mathbf{R}'} J_{\mathbf{RR}'}$ and using the definition of $E(a_{\mathbf{R}})$ given in (11.45), we have

$$\sum_{\mathbf{R}'} \left(H_{\mathbf{RR}'} - E(a_{\mathbf{R}}) J_{\mathbf{RR}'}\right) a_{\mathbf{R}'} = 0. \tag{11.52}$$

We will seek a solution of the form[14] $a_{\mathbf{R}}(\mathbf{k}) = a e^{i\mathbf{k}\cdot\mathbf{R}}$. This has the property $a_{\mathbf{R}}(\mathbf{k}+\mathbf{K}) = a_{\mathbf{R}}(\mathbf{k})$ and therefore $E(\mathbf{k}+\mathbf{K}) = E(\mathbf{k})$, which is the behavior expected for a given Bloch band (which develops from the atomic state with energy E_n). Inserting this form into (11.52), writing $E(\mathbf{k})$ in place of $E(a_{\mathbf{R}})$, and using (11.50) yields

$$\sum_{\mathbf{R}'} \left\{[E_n - E(\mathbf{k})] J_{\mathbf{RR}'} + h_{\mathbf{RR}'}\right\} e^{i\mathbf{k}\cdot\mathbf{R}'} = 0 \tag{11.53}$$

or

$$E(\mathbf{k}) = E_n + \frac{\sum_{\mathbf{R}'} h_{\mathbf{RR}'} e^{i\mathbf{k}\cdot\mathbf{R}'}}{\sum_{\mathbf{R}'} J_{\mathbf{RR}'} e^{i\mathbf{k}\cdot\mathbf{R}'}}. \tag{11.54}$$

The simplest case is that of a one-dimensional solid. We write $R = ma$, where a is the interatomic lattice spacing. If the orbitals have a width such that the overlap on adjacent sites can be limited to the nearest neighbors, we need only to retain the lattice vectors $R' - R = 0, \pm a$. Since the terms $h_{m,m\pm 1} = h_1$ contribute to (11.54) in first order[15] in the overlap, while $J_{mm} = 1$ (zeroth order), the first order correction to the atomic s-state energy, E_n, is

$$E_k = E_n + h_0 + h_1\left(e^{ika} + e^{-ika}\right) \tag{11.55a}$$

or

$$E_k = E_n + h_0 + 2h_1 \cos ka. \tag{11.55b}$$

[14] Note that the expansion coefficient, $a_{\mathbf{R}}$, cannot depend on $\mathbf{r}$. However, the form chosen comes close to mimicking the behavior of a Bloch function, as we will discuss later.

[15] This might at first seem strange and therefore requires further explanation. Now, h_0 involves the integral of $\phi^*(x)(V(x) - U(x))\phi(x)$, while h_1 involves the integral of $\phi^*(x)(V(x) - U(x \pm a))\phi(x \pm a)$. Although $\phi^*(x)$ and $\phi(x)$ are of the zeroth order, $V(x) - U(x)$ is of the first order, since this crystal potential and atomic potential tend to cancel each other within the unit cell; hence h_0 is first order. On the other hand, in the second form $V(x) - U(x \pm a)$ is zeroth order ($U(x \pm a)$ being small) while $\phi^*(x)\ \phi(x \pm a)$ is first order; hence h_1 is also first order.

Since $V(x) < U(x)$ due to the overlap from $U(x \pm na)$ (recall the overall potential is negative since the electrons are bound), h_1 and h_0 are both *negative*. If the band is half full (one electron per atom), the average energy would be

$$\overline{E(k)} = E_n + h_0 + 2\frac{1}{2\pi}h_1 a \int_0^{\pi/2a} \cos(ka)dk = E_n + h_0 + \frac{h_1}{\pi}\sin\frac{\pi}{2}$$
$$= E_n + h_0 + \frac{h_1}{\pi}.$$

On the other hand, if the band is full we have an average binding energy per electron of $\overline{E(k)} = h_0$. We see that the binding is *largest for a half full band*, but (for this case) is still present even for a full band. For the elements, tungsten has the highest melting temperature (corresponding to a large cohesive energy) and it has a half filled d band. The earlier theory is thus able to explain, qualitatively, cohesion in solids.

Note that our tight binding wave function has the form

$$\psi_{kn}(\mathbf{r}) = \sum_{\mathbf{R}} e^{i\mathbf{k}\cdot\mathbf{R}}\phi_n(\mathbf{r}-\mathbf{R}); \tag{11.56}$$

this is not precisely the Bloch form, although in translating by $\mathbf{R}$ unit cells the phase does advance by $e^{i\mathbf{k}\cdot\mathbf{R}}$, consistent with the requirement given in Eq. (10.7).[16] Since the wave function is arrived at by a variational procedure it need not have the Bloch form. However, if in place of the *atomic* wave functions we were to insert the *Wannier* functions, given by (10.50), then (11.56) would be identical to Eq. (10.51) and hence of the Bloch form.

We now discuss the three-dimensional s wave case for a face centered cubic lattice with nearest-neighbor interactions only. The set, $\{\mathbf{R}_1\}$, of the closest real-space lattice vectors to the origin ($\mathbf{R} = 0$) involves the 12 vectors:

$$\frac{a}{2}\left(\hat{i}+\hat{j}\right), \frac{a}{2}\left(-\hat{i}-\hat{j}\right), \frac{a}{2}\left(\hat{i}-\hat{j}\right), \frac{a}{2}\left(-\hat{i}-\hat{j}\right) + \left(\hat{i}\to\hat{j}, \hat{j}\to\hat{k}\right) + \left(\hat{j}\to\hat{k}, \hat{k}\to\hat{i}\right).$$

For this nearest-neighbor case, Eq. (11.54) becomes (where we again keep only J_{00})

$$E(\mathbf{k}) = E_0 + h_0 + h_1 \sum_{\mathbf{R}=\{\mathbf{R}_1\}} e^{i\mathbf{k}\cdot\mathbf{R}}$$
$$= E_0 + h_0 + h_1\Big[e^{i\frac{a}{2}(k_x+k_y)} + e^{\frac{ia}{2}(-k_x+k_y)} + e^{\frac{ia}{2}(k_x-k_y)}$$
$$+ e^{\frac{ia}{2}(-k_x-k_y)} + (k_x \to k_y, k_y \to k_z) + (k_y \to k_z, k_z \to k_x)\Big]$$

Combining the exponential factors into cosines,

$$E(\mathbf{k}) = E_0 + h_0 + 4h_1\left[\cos\left(\frac{k_x a}{2}\right)\cos\left(\frac{k_y a}{2}\right) + \cos\left(\frac{k_y a}{2}\right)\cos\left(\frac{k_z a}{2}\right)\right.$$
$$\left. + \cos\left(\frac{k_z a}{2}\right)\cos\left(\frac{k_x a}{2}\right)\right]. \tag{11.57}$$

[16] What is absent is the continual phase shift within the unit cell, characteristic of the Bloch function (10.8).

For small **k** we may expand the arguments of the cosines as

$$\cos(k_x a/2) = 1 - k_x^2 a^2/8 + \ldots \text{etc.}$$

and Eq. (11.57) becomes

$$E(\mathbf{k}) = (E_0 + h_0 + 4h_1) - h_1 a^2 \left(k_x^2 + k_y^2 + k_z^2\right). \tag{11.58}$$

For free electrons, we have $E(\mathbf{k}) = \hbar^2 k^2/2m$, and we may define the coefficient of the quadratic term as an "effective mass," m^*, by setting $-h_1 a^2 = \hbar^2/2m^*$, obtaining $m^* = -\hbar^2/2h_1 a^2$.

11.4.2 The tight binding approximation for the case $l > 0$

To extend the above theory to include the case of multiple valance electrons (s, p, d...), we must return to the wave function as it was written in Eq. (11.40). Following analogous steps we obtain the generalization of Eq. (11.52) as (where we assume we have a single radial quantum number n)

$$\sum_{l'm'}\sum_{\mathbf{R}'}\left[H_{\mathbf{RR}'}^{(lm;\,l'm')} - E J_{\mathbf{RR}'}^{(lm;\,l'm')}\right] a_{\mathbf{R}'}^{(l'm')} = 0. \tag{11.59}$$

The generalization of the variational coefficient introduced for the s wave case is $a_{\mathbf{R}}^{(lm)} = a^{(lm)} e^{i\mathbf{k}\cdot\mathbf{R}}$. The generalization of Eq. (11.52) is

$$H_{\mathbf{RR}'}^{(lm;l'm')} = E^{(l'm')} J_{\mathbf{RR}'}^{(lm;l'm')} + h_{\mathbf{RR}'}^{(lm;l'm')}, \tag{11.60}$$

where $E^{(lm)}$ are the energies of the atomic states.

If we again restrict to nearest neighbors we may neglect the overlap integrals $J_{\mathbf{RR}'}^{(lm;\,l'm')}$ for $\mathbf{R} \neq \mathbf{R}'$; assuming the original atomic states $\phi^{(lm)}$ are orthonormal, the first term in (11.60) then becomes $E^{(lm)}\delta_{\mathbf{RR}'}\delta_{ll'}\delta_{mm'}$. Equation (11.59) is then

$$\sum_{l'm'}\left[\left(E^{(l'm')} - E(\mathbf{k})\right)\delta_{ll'}\delta_{mm'} + h_0^{lm}\delta_{ll'}\delta_{mm'} + h_1^{(lm;l'm')}\right] a^{(l'm')} = 0, \tag{11.61a}$$

where

$$h_1^{(lm;l'm')} = \left\langle \phi^{(lm)}(\mathbf{r}) \right| \left[\sum_{\{\mathbf{R}\}} e^{i\mathbf{k}\cdot\mathbf{R}}\left[U(\mathbf{r}) \pm U(\mathbf{r}-\mathbf{R})\right]\right] \left| \phi^{(l'm')}(\mathbf{r}-\mathbf{R}) \right\rangle \tag{11.61b}$$

and $\{\mathbf{R}\}$ again denotes the star of the shortest (real space) lattice vector. This sum can be evaluated in a straightforward manner by exploiting the symmetry of the wave functions $\phi^{(lm)}(\mathbf{r})$. The condition that (11.61a,b) have a solution is that $E(\mathbf{k})$ satisfies the determinantal equation

$$\left|\left(E^{(l'm')} - E(\mathbf{k})\right)\delta_{ll'}\delta_{mm'} + h_0^{(lm)}\delta_{ll'}\delta_{mm'} + h_1^{(lm;l'm')}\right| = 0. \tag{11.62}$$

Rather than calculating the integrals $h_0^{(lm)}$ and $h_1^{(lm;l'm')}$ from atomic wave functions and potentials, these quantities are sometimes treated as *parameters* to fit band structures obtained by more accurate schemes. If Eq. (11.62) is used to discuss a transition metal, which has s, p, and d electrons (each with degeneracy $2l+1$), the order of the determinant would be $1 + 3 + 5 = 9$ for the case of one atom per unit cell; if spin–orbit (S–O) coupling effects were included it would increase to 18.[17]

11.4.3 The tight binding approximation for the case $l > 0$ together with a basis

In the most general case our unit cell will have a basis and Eq. (11.40) will have the form

$$\psi(\mathbf{r}) = \sum_{lmn}\sum_{\mathbf{R}}\sum_{i} a_{i\mathbf{R}}^{(lmn)}\phi_{lmn}^{(i)}(\mathbf{r} - \mathbf{R} - \mathbf{r}_i), \tag{11.63}$$

where i denotes the position vectors $\mathbf{r}_i$ of the atoms within the unit cell (see Eq. (8.4)); if these atoms differ, their wave functions $\phi_{lmn}^{(i)}$ will carry a site index. In place of (11.59) we now have the set of equations

$$\sum_{l'm'}\sum_{\mathbf{R}'}\sum_{i'}\left[H_{ii';\mathbf{RR}'}^{(lm;l'm')} - EJ_{ii';\mathbf{RR}'}^{(lm;l'm')}\right]a_{i'\mathbf{R}'}^{(l'm')} = 0, \tag{11.64}$$

where the coefficients required by the crystal periodicity condition (10.7) are now given by

$$a_{i\mathbf{R}}^{(lmn)} = a_i^{(lmn)}e^{i\mathbf{k}\cdot(\mathbf{R}+\mathbf{r}_i)} \tag{11.65}$$

and the tight binding matrix elements generalizing (11.60) are

$$H_{ii';\mathbf{RR}'}^{(lm;l'm')} = E_{i'}^{(l'm')}J_{ii';\mathbf{RR}'}^{(lm;l'm')} + h_{ii';\mathbf{RR}'}^{(lm;l'm')}. \tag{11.66}$$

We will apply this formalism to parameterize the band structures of the group four semiconductors in Section 12.3.

[17] Many of the transition metals have the bcc crystal structure which has two atoms per unit cell (see Section 11.4.3) leading to an 18 × 18 determinant; the addition of S–O coupling increases it to 36 × 36.

FURTHER READING

Quantum Theory of Solids, R.E. Peierls, Oxford University Press (1955). Chapter IV has an excellent discussion of the nearly free electron and tight binding models.

Solid State Physics, N.W. Ashcroft and N.D. Mermin, Holt, Rinehart and Winston (1976) has a thorough discussion of the nearly free electron approximation as well as Brillouin zones.

Theory of Metals and Alloys, N.F. Mott and H. Jones, Oxford University Press. An old but still useful (and quite friendly) book; Part II Section 4 discusses the nearly free electron approximation.

Pseudopotentials and other electronic structure methods are discussed in chapter 11 of *Electronic Structure; Basic Theory and Practical Methods* by R.M. Martin, Cambridge University Press (2004). Various pseudopotentials have been proposed that are optimized in some manner depending on the application. Examples can be found in: Norm-conserving pseudopotentials, D.R. Hamann, M. Schluter, and C. Chiang, *Phys. Rev. Lett.* **43**, 1494 (1979); Pseudopotentials that work: From H to Pu, G.B. Bachelet, D.R. Hamann, and M. Schluter, *Phys. Rev. B* **26**, 4199 (1982); Efficacious form for model pseudopotentials, L. Kleinman and D.M. Bylander, *Phys. Rev. Lett.* **48**, 1425 (1982); Soft self-consistent pseudopotentials in a generalized eigenvalue formalism, D. Vanderbilt, *Phys. Rev.* B **41**, 7892 (1990); and Efficient pseudopotentials for plane-wave calculations, N. Troullier and J.L. Martins, *Phys. Rev. B* 43, 1993 (1991).

Incorporating the effects of spin–orbit coupling arising from p and d electrons is discussed in Etude du couplage spin-orbite dans les metaux de transition: application au platine, J. Friedel, P. Lenglart, and G. Leman, *J. Phys. Chem. Solids* **25**, 781 (1964).

Problems

11.1. Construct the free electron (FE) Fermi curves in the first Brillouin zone for a square lattice with divalent and trivalent atoms. What is the expression for the radius of the Fermi circle in terms of both the number of electrons and the lattice constant?

11.2. Construct the first Brillouin zone for the two-dimensional hexagonal lattice. Then construct the FE Fermi curves lying in the first Brillouin zone for divalent and trivalent atoms. What is the expression for the radius of the Fermi circle for these two cases in terms of the lattice constant?

11.3. Find the eigenfunctions associated with Eq. (11.15).

11.4. Verify that the geometrical structure factor (see Section 9.7) vanishes on the AHL plane (see Fig. 10.6) of hexagonal close packed material.

11.5. Suppose you had a one-dimensional lattice of hydrogen atoms spaced uniformly by a distance a. Retaining only the on-site and nearest-neighbor overlap integrals, and approximating the potential by its value at the point of maximal overlap, determine E(k) of the lowest energy band in the tight binding approximation. What is the value of the Fermi point?

11.6. Find the eigenfunctions associated with Eq. (11.53).

11.7. Obtain the equation for E(**k**) analogous to Eq. (11.57) for the case of a simple cubic lattice.

11.8. Obtain equation (11.57) for E(**k**) for the fcc case directly from a leading order expansion of Eq. (10.13). Do the same for the simple cubic case.

11.9. In the tight binding formalism, show that if we use the Wannier functions discussed in Section 10.6 rather than atomic wave functions as the basis set, the overlap integrals will involve the kinetic energy operator rather than the potential produced by neighboring atoms.

12

The Parameterization of Band Structures: Applications to Semiconductors

Even if an accurate band structure is available, the time involved to calculate many k-points can be excessive and it is then convenient to have some sort of scheme to fit the energy bands with some reasonably small set of parameters. Here we can envisage "interpolating" between energy eigenvalues obtained from a band structure calculation, or possibly fitting experimental data, e.g., that from photoemission experiments. For materials with complicated constant energy surfaces (which cannot be represented by, say, sets of ellipsoids) the best fitting schemes are those based on a band structure formalism, e.g., the pseudopotential or the tight binding methods.[1] There are also schemes that combine the two approaches,[2,3] which we will not take up here. Readers less concerned with the details of semiconductor band structures may initially choose to ignore the material treated here.

12.1 The k·p method for diamond-like materials at the zone center

When degeneracy is present, as when treating p-like electrons at the Γ point in tetrahedrally coordinated semiconductors, the $\mathbf{k} \cdot \mathbf{p}$ theory for the dispersion near a band edge, given in Section 10.4, must be generalized. Because of its overall importance we carry out the calculations in some detail; the needed parts are obtained in the following six subsections; readers not interested in the details may proceed directly to the resulting expression for the dispersion of the hole bands, given as Eq. (12.39). To be useful for materials involving elements with higher atomic number Z, our parameterization should include the effects of spin–orbit coupling;[4] we start with a brief review of this effect.

12.1.1 Spin–orbit coupling in atoms

Spin–orbit coupling is a relativistic phenomenon arising from the property that an electron moving with velocity $\mathbf{v}$ in an electric field experiences a magnetic field in

[1] Another scheme based on energy-dependent scattering phase shifts will be discussed in a later section.

[2] Interpolation scheme for band structure of noble and transition metals: ferromagnetism and neutron diffraction in Ni, L. Hodges, H. Ehrenreich, and N.G. Lang, *Phys. Rev.* **152**, 505 (1966).

[3] Combined interpolation scheme for transition and noble metals, F.M. Mueller, *Phys. Rev.* **153**, 659 (1967).

[4] For a discussion of spin–orbit coupling, see Chapter 7 of *Theory of Brillouin Zones and Electronic States in Crystals*, H. Jones, North Holland, Amsterdam (1962).

The Physics of Solids. First Edition. J. B. Ketterson.

its rest frame. For velocities $v \ll c$, the magnetic field is given by $\mathbf{H} = -(v/c) \times \mathbf{E}$. Here the electric field arises from the electrostatic potential energy $V(\mathbf{r})$ of the nucleus and the remaining electrons; i.e., $-e\mathbf{E} = -\nabla V(\mathbf{r})$. In the so-called central field approximation one assumes that $V(\mathbf{r})$ is radially symmetric, in which case, $-e\mathbf{E} = -(\mathbf{r}/r) \cdot dV(r)/dr$. The magnetic field $\mathbf{H}$ so produced then interacts with the electron's magnetic moment, shifting its energy by $E_{SO} = -\boldsymbol{\mu} \cdot \mathbf{H}$. Writing $\hat{\boldsymbol{\mu}}$ as an operator, $\hat{\boldsymbol{\mu}} = -\mu_B g_e \hat{\boldsymbol{\sigma}}/2 = -(e\hbar/2mc)\hat{\boldsymbol{\sigma}}$ (with $\hat{\boldsymbol{\sigma}}$ the vector Pauli spin matrix), setting $g_e = 2$, including an extra factor of two in the denominator (the Thomas precession factor which we do not discuss), interpreting $\mathbf{v}$ as the operator $\hat{\mathbf{p}}/m$, and assembling these parts, yields the spin–orbit operator $\hat{H}_{SO}$ as

$$\hat{H}_{SO} = \frac{\hbar}{4m^2c^2}\left(\nabla V(\mathbf{r}) \times \hat{\mathbf{p}}\right) \cdot \hat{\boldsymbol{\sigma}}, \tag{12.1a}$$

or in the central field approximation as

$$\hat{H}_{SO} = \frac{\hbar}{4m^2c^2 r}\frac{dV(r)}{dr}\hat{\boldsymbol{\sigma}} \cdot \left(\mathbf{r} \times \hat{\mathbf{p}}\right) = \frac{\hbar^2}{4m^2c^2 r}\frac{dV(r)}{dr}\hat{\boldsymbol{l}} \cdot \hat{\boldsymbol{\sigma}}, \tag{12.1b}$$

where in (12.1b) we introduced the dimensionless angular momentum operator $\hat{\boldsymbol{l}}$. It is common at this point to define a parameter λ_{SO} and write (12.1a) as

$$\hat{H}_{SO} \equiv \frac{1}{2}\lambda_{SO}\hat{\boldsymbol{l}} \cdot \hat{\boldsymbol{\sigma}} = \lambda_{SO}\hat{\boldsymbol{l}} \cdot \hat{\mathbf{s}}. \tag{12.1c}$$

12.1.2 Including spin in atomic s and p states

In many semiconductors a representation in terms of the valence electron wave functions formed from the atomic s, $\phi_{00}(\mathbf{r})$, and the three p, $\phi_{11}(\mathbf{r})$, $\phi_{10}(\mathbf{r})$, $\phi_{1-1}(\mathbf{r})$, states (where the subscripts denote the values of l and m) is sufficient for a parameterization in the vicinity of the Fermi energy. In the isolated atoms we may write the three p states as

$$\phi_{11}(r,\theta,\phi) = R_p(r)Y_{11}(\theta,\phi),$$

$$\phi_{10}(r,\theta,\phi) = R_p(r)Y_{10}(\theta,\phi),$$

$$\phi_{1-1}(r,\theta,\phi) = R_p(r)Y_{1-1}(\theta,\phi); \tag{12.2a–c}$$

and the s state as

$$\phi_{00}(\mathbf{r}) = R_s(r); \tag{12.2d}$$

Here, $R_{s,p}(r)$ are the radial wave functions of the valence s and p orbitals and $Y_{lm}(\theta,\phi)$ are the usual spherical harmonics.

We now introduce the spin states as $\chi_{\pm 1/2}(\alpha)$, where α denotes the two components. Combining the $l = 1$ orbital states with these spin states, one can form states with total angular momentum $\mathrm{j} = 3/2$ and $\mathrm{j} = 1/2$. It is straightforward to verify that the eigenstates of the total angular momentum operator, $\hat{\mathrm{j}} = \hat{l} + \hat{\mathrm{s}}$, for the $\mathrm{j} = 3/2$ states are (where the subscripts denote j and $\mathrm{j_z}$):

$$\phi_{p,3/2,3/2}(\mathbf{r};\alpha) = \phi_{11}(\mathbf{r})\chi_{1/2}(\alpha) = \frac{1}{\sqrt{2}}\left(\phi_{p_x}(\mathbf{r}) + i\phi_{p_y}(\mathbf{r})\right)\chi_{1/2}(\alpha), \tag{12.3a}$$

$$\phi_{p,3/2,1/2}(\mathbf{r};\alpha) = \frac{1}{\sqrt{3}}\left[\phi_{11}(\mathbf{r})\chi_{-1/2}(\alpha) + \sqrt{2}\phi_{10}(\mathbf{r})\chi_{1/2}(\alpha)\right] = \frac{1}{\sqrt{3}}\left(\frac{1}{\sqrt{2}}\left(\phi_{p_x}(\mathbf{r}) + i\phi_{p_y}(\mathbf{r})\right)\chi_{-1/2}(\alpha) + \sqrt{2}\phi_{p_z}(\mathbf{r})\chi_{1/2}(\alpha)\right), \tag{12.3b}$$

$$\phi_{p,3/2,-1/2}(\mathbf{r};\alpha) = \frac{1}{\sqrt{3}}\left[\phi_{1-1}(\mathbf{r})\chi_{1/2}(\alpha) + \sqrt{2}\phi_{10}(\mathbf{r})\chi_{-1/2}(\alpha)\right] = \frac{1}{\sqrt{3}}\left(\frac{1}{\sqrt{2}}\left(\phi_{p_x}(\mathbf{r}) - i\phi_{p_y}(\mathbf{r})\right)\chi_{1/2}(\alpha) + \sqrt{2}\phi_{p_z}(\mathbf{r})\chi_{-1/2}(\alpha)\right), \tag{12.3c}$$

and

$$\phi_{p,3/2,-3/2}(\mathbf{r};\alpha) = \phi_{1-1}(\mathbf{r})\chi_{-1/2}(\alpha) = \frac{1}{\sqrt{2}}\left(\phi_{p_x}(\mathbf{r}) - i\phi_{p_y}(\mathbf{r})\right)\chi_{-1/2}(\alpha), \tag{12.3d}$$

where in the second step of (12.3a–d) we have introduced the vector p states ϕ_{p_x}, ϕ_{p_y}, and ϕ_{p_z}. The p-derived states with $\mathrm{j} = 1/2$ are

$$\phi_{p,1/2,1/2}(\mathbf{r};\alpha) = \frac{1}{\sqrt{3}}\left[\phi_{10}(\mathbf{r})\chi_{1/2}(\alpha) - \sqrt{2}\phi_{11}(\mathbf{r})\chi_{-1/2}(\alpha)\right] = \frac{1}{\sqrt{3}}\left[\phi_{p_z}(\mathbf{r})\chi_{1/2}(\alpha) - \left(\phi_{p_x}(\mathbf{r}) + i\phi_{p_y}(\mathbf{r})\right)\chi_{-1/2}(\alpha)\right], \tag{12.4a}$$

and

$$\phi_{p,1/2,-1/2}(\mathbf{r};\alpha) = \frac{1}{\sqrt{3}}\left[\phi_{10}(\mathbf{r})\chi_{-1/2}(\alpha) - \sqrt{2}\phi_{1-1}(\mathbf{r})\chi_{1/2}\right] = \frac{1}{\sqrt{3}}\left[\phi_{p_z}(\mathbf{r})\chi_{-1/2}(\alpha) - \left(\phi_{p_x}(\mathbf{r}) - i\phi_{p_y}(\mathbf{r})\right)\chi_{1/2}\right]. \tag{12.4b}$$

Finally, we can form a pair of $\mathrm{j} = 1/2$ states from the atomic s orbital $\phi_{00} \equiv \phi_s$:

$$\phi_{s,1/2,1/2}(\mathbf{r};\alpha) = \phi_s(\mathbf{r})\chi_{1/2} \tag{12.5a}$$

and

$$\phi_{s,1/2,-1/2}(\mathbf{r};\alpha) = \phi_s(\mathbf{r})\chi_{-1/2}. \tag{12.5b}$$

12.1.3 The k·p theory with spin

If we include spin–orbit coupling from Eq. (12.1a) and expand around the zone center, the $\mathbf{k}\cdot\hat{\mathbf{p}}$ Schrödinger equation associated with the Hamiltonian given in Eqs. (10.41a–c) becomes

$$\left[\frac{\hat{p}^2}{2m}+\frac{\hbar}{m}\mathbf{k}\cdot\hat{\mathbf{p}}+\frac{\hbar^2k^2}{2m}+V(\mathbf{r})+\frac{\hbar}{4mc^2}\left(\nabla V\times\hat{\mathbf{p}}\right)\cdot\hat{\sigma}\right.$$
$$\left.+\frac{\hbar^2}{4mc^2}\left(\nabla V\times\mathbf{k}\right)\cdot\hat{\sigma}\right]u_{\mathbf{k}n\nu}(\mathbf{r},\alpha)=\varepsilon_{\mathbf{k}n\nu}u_{\mathbf{k}n\nu}(\mathbf{r},\alpha), \qquad (12.6)$$

where $u_{\mathbf{k}n\nu}(\mathbf{r})$ denotes the cell periodic part of the Bloch wave function for eigenstates with wave vector $\mathbf{k}$, band index n, and spin state ν. The sixth term on the left will be neglected as it is generally small compared with the fifth.[5] As a basis in which to expand the crystal wave functions we can use the solutions to the Schrödinger equation for $\mathbf{k}=0$,

$$\left[\frac{\hat{p}^2}{2m}+V(\mathbf{r})\right]U_{n\nu}(\mathbf{r},\alpha)=\varepsilon_n U_{n\nu}(\mathbf{r},\alpha). \qquad (12.7)$$

Although spin–orbit coupling is being suppressed here we can still form solutions which are compatible with allowed crystal symmetries in the presence of spin, as we do later.

As a model for these functions in diamond/zinc blende materials, we start with $\mathbf{k}=0$ tight binding (LCAO) sums of the atomic s and p states (see Eqs. (11.63) and (11.65)) for the two atoms in the unit cell. For the s states at these sites we form the cell-periodic sums,

$$\psi_s^{(0)}(\mathbf{r})=\frac{1}{\sqrt{N}}\sum_{\mathbf{R}}\phi_s^{(0)}(\mathbf{r}-\mathbf{R}),\quad \psi_s^{(1)}(\mathbf{r})=\frac{1}{\sqrt{N}}\sum_{\mathbf{R}}\phi_s^{(1)}(\mathbf{r}-\mathbf{r}_1-\mathbf{R}), \qquad (12.8a\text{–}b)$$

where the second atom is located a quarter way along the cube diagonal at $\mathbf{r}_1=a(\mathbf{i}+\mathbf{j}+\mathbf{k})/4$. Similarly for the p states at these two sites we have

$$\begin{aligned}
\psi_{p_x}^{(0)}(\mathbf{r})&=\tfrac{1}{\sqrt{N}}\sum_{\mathbf{R}}\phi_{p_x}^{(0)}(\mathbf{r}-\mathbf{R}), & \psi_{p_x}^{(1)}(\mathbf{r})&=\tfrac{1}{\sqrt{N}}\sum_{\mathbf{R}}\phi_{p_x}^{(1)}(\mathbf{r}-\mathbf{r}_1-\mathbf{R}),\\
\psi_{p_y}^{(0)}(\mathbf{r})&=\tfrac{1}{\sqrt{N}}\sum_{\mathbf{R}}\phi_{p_y}^{(0)}(\mathbf{r}-\mathbf{R}), & \psi_{p_y}^{(1)}(\mathbf{r})&=\tfrac{1}{\sqrt{N}}\sum_{\mathbf{R}}\phi_{p_y}^{(1)}(\mathbf{r}-\mathbf{r}_1-\mathbf{R}),\\
\psi_{p_z}^{(0)}(\mathbf{r})&=\tfrac{1}{\sqrt{N}}\sum_{\mathbf{R}}\phi_{p_z}^{(0)}(\mathbf{r}-\mathbf{R}), & \psi_{p_z}^{(1)}(\mathbf{r})&=\tfrac{1}{\sqrt{N}}\sum_{\mathbf{R}}\phi_{p_z}^{(1)}(\mathbf{r}-\mathbf{r}_1-\mathbf{R}).
\end{aligned} \qquad (12.9a\text{–}f)$$

Here we will assume that our atomic-like states $\phi_i(\mathbf{r})$ are orthonormal. In order that the $U_i(\mathbf{r})$ in Eq. (12.7) themselves be normalized, we will further assume that states on neighboring sites are orthogonal; procedures to accomplish this have been developed.[6]

For s-like states we would form a linear superposition of the forms (12.8a–b), which on including independent amplitudes for each, we write as

$$\begin{aligned}
U_s(\mathbf{r})&=a_{s1}\psi_s^{(0)}(\mathbf{r})+a_{s2}\psi_s^{(1)}(\mathbf{r})\\
&=\tfrac{1}{\sqrt{N}}\sum_{\mathbf{R}}\left(a_{s1}\phi_s^{(0)}(\mathbf{r}-\mathbf{R})+a_{s2}\phi_s^{(1)}(\mathbf{r}-\mathbf{r}_1-\mathbf{R})\right).
\end{aligned} \qquad (12.10)$$

[5] Near the nucleus where the atomic momentum is large we will have $p \gg \hbar k$.

[6] On the non-orthogonality problem connected with the use of atomic wave functions in the theory of molecules and crystals, P.O. Lowdin, *J. Chem. Phys.* **18**, 366 (1950).

In place of the two constants, a_{s1} and a_{s2}, we can introduce two different constants: an overall normalization and a relative weight. For the diamond structure, which has inversion symmetry, and to which we will limit the discussion here, the relative weights must satisfy $a_{s2} = \pm a_{s1}$, and (12.10) can then have the two forms

$$U_s^{\pm}(\mathbf{r}) = \frac{1}{\sqrt{2N}} \sum_{\mathbf{R}} \left(\phi_s^{(0)}(\mathbf{r} - \mathbf{R}) \pm \phi_s^{(1)}(\mathbf{r} - \mathbf{r}_1 - \mathbf{R}) \right). \tag{12.11}$$

The s states have the property $\phi_s(-\mathbf{r}) = \phi_s(\mathbf{r})$ and it turns out that the symmetric combination has the lowest energy.

In a similar way we can construct three p-like LCAO combinations

$$U_{p_x}^{\pm}(\mathbf{r}) = \frac{1}{\sqrt{2N}} \sum_{\mathbf{R}} \left(\phi_{p_x}^{(0)}(\mathbf{r} - \mathbf{R}) \pm \phi_{p_x}^{(1)}(\mathbf{r} - \mathbf{r}_1 - \mathbf{R}) \right),$$

$$U_{p_y}^{\pm}(\mathbf{r}) = \frac{1}{\sqrt{2N}} \sum_{\mathbf{R}} \left(\phi_{p_y}^{(0)}(\mathbf{r} - \mathbf{R}) \pm \phi_{p_y}^{(1)}(\mathbf{r} - \mathbf{r}_1 - \mathbf{R}) \right),$$

and

$$U_{p_z}^{\pm}(\mathbf{r}) = \frac{1}{\sqrt{2N}} \sum_{\mathbf{R}} \left(\phi_{p_z}^{(0)}(\mathbf{r} - \mathbf{R}) \pm \phi_{p_z}^{(1)}(\mathbf{r} - \mathbf{r}_1 - \mathbf{R}) \right). \tag{12.12a–c}$$

The p states have the property $\phi_{p_i}(-\mathbf{r}) = -\phi_{p_i}(\mathbf{r})$. The lowest energy state for p states then corresponds to the *negative* sign in (11.12a–c). But before constructing additional forms we digress to discuss more generally the nature of s- and p-like states in diamond-like materials.

12.1.4 The symmetry of states in cubic crystals at the zone center

Atomic states with a given l (non-relativistically) or j (relativistically) are $2l + 1$ or $2j + 1$-fold degenerate respectively. In both cases this degeneracy results from the spherical symmetry of the atom: under an arbitrary rotation, these degenerate states transform into *linear combinations* of each other. However when the atoms are assembled into a crystal, the symmetry is lowered; for example, the only possible rotations involve the 2-, 3-, 4-, and 6-fold axes discussed in Section 8.6. Unlike atoms, where the degeneracy increases as l increases, the number of states that transform into each other in a crystal is limited. Part of the theory of crystal groups involves a determination of the number of (i.e., the degeneracy of), and the form of, the functions which transform into each other. The smallest set of such functions is called an *irreducible representation* and the minimum number of functions required is called the *dimension* of the representation; i.e., it corresponds to its degeneracy. In addition, symmetry independent sets of such functions can be present, i.e., there can be multiple irreducible representations.

The symmetry of electron states and other excitations in crystals depends on their wave vector **k**. For states with $\mathbf{k}=0$ the symmetry is highest, that of the Brillouin zone itself, and hence here they will have the highest degeneracy. For other symmetry points or axes in the Brillouin zone (see Fig. 10.2 for the fcc case of interest to us here) the degeneracy will in general be lower. The various $\mathbf{k}=0$ states are designated by the Greek letter Γ_i, where the subscript denotes the different irreducible representations. In cubic materials there are eight such representations (which as previously noted can occur multiple times); they are collected in Table 12.1. Unfortunately there are three conventions in common use which are all listed in the table; the less commonly used notations (when designating symmetries in solids) are in parenthesis. The designations $\pm$ and g/u denote states which are even (+, g) or odd (–, u) when inversion symmetry is present; for Ge, which has inversion symmetry, this designation is essential; for GaAs, which lacks this symmetry, this designation is dropped.

Table 12.1 *Designation and symmetry of the fourteen cubic* Γ*-point groups in a crystal with inversion symmetry (as in* C, Si, *and* Ge*).*

O_h designation	Symmetry
$\Gamma_1^+\,(\Gamma_1, A_g)$	1
$\Gamma_2^+\,(\Gamma_2, B_g)$	$(x^2-y^2)\,(y^2-z^2)\,(z^2-x^2)$
$\Gamma_3^+\,(\Gamma_{12}, E_g)$	$\{2z^2-x^2-y^2;\ x^2-y^2\}$
$\Gamma_4^+\,(\Gamma'_{15}, T_{1g})$	$\{xy\,(x^2-y^2);\ yz\,(y^2-z^2);\ zx\,(z^2-x^2)\}$
$\Gamma_5^+\,(\Gamma'_{25}, T_{2g})$	$\{xy;\ yz;\ zx\}$
$\Gamma_6^+\,(\Gamma_6^+)$	$\{(1/2); (-1/2)\}$
$\Gamma_7^+\,(\Gamma_7^+)$	$(x^2-y^2)\,(y^2-z^2)\,(z^2-x^2)\times\{(1/2); (-1/2)\}$
$\Gamma_8^+\,(\Gamma_8^+)$	$\{(3/2); (1/2); (-1/2); (-3/2)\}$
$\Gamma_1^-\,(\Gamma'_1, A_u)$	$xyz\,(x^2-y^2)\,(y^2-z^2)\,(z^2-x^2)$
$\Gamma_2^-\,(\Gamma'_2, B_u)$	xyz
$\Gamma_3^-\,(\Gamma'_{12}, E_u)$	$\{xyz\,(2z^2-x^2-y^2);\ xyz\,(x^2-y^2)\}$
$\Gamma_4^-\,(\Gamma_{15}, T_{1u})$	$x;\ y;\ z$
$\Gamma_5^-\,(\Gamma_{25}, T_{2u})$	$\{x\,(y^2-z^2);\ y\,(z^2-x^2);\ z\,(x^2-y^2)\}$
$\Gamma_6^-\,(\Gamma_6^-)$	$xyz\,(x^2-y^2)\,(y^2-z^2)\,(z^2-x^2)\times\{(+1/2); (-1/2)\}$
$\Gamma_7^-\,(\Gamma_7^-)$	$xyz\times\{(1/2);\ (-1/2)\}$
$\Gamma_8^-\,(\Gamma_8^-)$	$xyz\,(x^2-y^2)\,(y^2-z^2)\,(z^2-x^2)\times\{(3/2); (1/2); -(1/2); (-3/2)\}$

For boson states, which have integral spin, there are one-, two-, and three-dimensional representations in cubic crystals at Γ, and there are a total of five which are designated Γ_1 through Γ_5; they have dimensions of 1, 1, 2, 2, and 3 respectively. If the states under consideration arise from $s = 1/2$ electrons (rather than boson states associated with excitations such as phonons, photons, excitons, or magnons), then we must account for the property that a 2π rotation about some axis results in a *change of sign*. An additional symmetry operator associated with such rotations must then be included, which generates three additional representations, designated Γ_6, Γ_7, and Γ_8, with dimensions 2, 2, and 4.

Table 12.1 also lists representative functions of the spatial coordinates x, y, z that exhibit the symmetry of these eight states. When spin is present, these forms represent the *lowest order* polynomial forms (in x, y, and z) that can be written as the *product* of a *single* spin state and an orbital state. If we consider states involving several such products, then states with lower-order polynomial forms can be constructed, as we will see.[7]

In the case where there is *one atom* per unit cell we always have inversion symmetry. The four states $\phi_{p,3/2,\pm 3/2}(\mathbf{r};\alpha)$ and $\phi_{p,3/2,\pm 1/2}(\mathbf{r};\alpha)$, given by Eqs. (12.3a–d), are both odd and four-fold degenerate; they then correspond to a Γ_8^- representation. Similarly the two states $\phi_{p,1/2,\pm 1/2}(\mathbf{r};\alpha)$, given by Eqs. (12.4a,b), represent a doublet that is odd under inversion; they correspond to a Γ_7^- state. Lastly the states $\phi_{s,1/2,\pm 1/2}(\mathbf{r};\alpha)$, are two-fold degenerate and even under inversion, correspond to a Γ_6^+ representation. One can verify these assignments by performing various rotations and reflections associated with the crystal point group, being careful to properly treat their effect on the spin wave functions, and noting that the resulting coordinate permutations and sign changes are identical to the forms in the table.

Strictly speaking electron states at the zone center in cubic materials *always* correspond to Γ_6, Γ_7, and Γ_8 states, since they are formed from spin 1/2 states. However, when reporting results of a non-relativistic band structure calculation it is common to use the designations Γ_1 to Γ_5. Hence if the spin–orbit coupling is set to zero, we leave out the spin wave functions, and if we have one atom per unit cell, the three p states form a Γ_4^- representation, while the s states form a Γ_1^+ representation.

12.1.5 Basis functions for electrons in diamond-like materials at Γ

We now return to the problem of constructing the basis states $U_{n\upsilon}(\mathbf{r},\alpha)$ in diamond-like materials, where we have *two* atoms per unit cell and where we limit ourselves to s and p electrons.[8] The symmetric s-like valance band combination formed from (12.5a,b) can be shown from Table 12.1 to have Γ_6^+ symmetry,

$$U_{\Gamma_6^+,\pm 1/2}(\mathbf{r},\alpha) = U_s^+(\mathbf{r})\chi_{\pm 1/2}(\alpha), \qquad (12.13)$$

[7] As examples we note that Eqs. (12.3a–d) and (12.5a,b), which involve a *sum* of products, do not have the form of a simple product.

[8] This would not be a good approximation for a material like CuCl, where there is an appreciable contribution from d orbitals.

while the antisymmetric s-like conduction band combination has Γ_7^- symmetry,

$$U_{\Gamma_7^-,\pm 1/2}(\mathbf{r},\alpha) = U_s^-(\mathbf{r})\chi_{\pm 1/2}(\alpha). \tag{12.14}$$

On the other hand, because of the parity of the p states, the symmetric valance band combinations arise from the negative signs in (11.12 a–c). The $p_{1/2}$ states have the Γ_7^+ symmetry and are given by[9]

$$U_{\Gamma_7^+,\pm 1/2}(\mathbf{r},\alpha) = \frac{1}{\sqrt{3}}\left[U_{p_z}^-(\mathbf{r})\chi_{\pm 1/2}(\alpha) - \left(U_{p_x}^-(\mathbf{r}) \pm iU_{p_y}^-(\mathbf{r})\right)\chi_{\mp 1/2}(\alpha)\right], \tag{12.15}$$

while the antisymmetric $p_{1/2}$ conduction band combination has Γ_6^- symmetry,

$$U_{\Gamma_6^-,\pm 1/2}(\mathbf{r},\alpha) = \frac{1}{\sqrt{3}}\left[U_{p_z}^+(\mathbf{r})\chi_{\pm 1/2}(\alpha) - \left(U_{p_x}^+(\mathbf{r}) \pm iU_{p_y}^+(\mathbf{r})\right)\chi_{\mp 1/2}(\alpha)\right]. \tag{12.16}$$

Finally, the four-fold degenerate p-like j = 3/2 atomic states form the symmetric Γ_8^+ valance band combinations,

$$U_{\Gamma_8^+,\pm 3/2}(\mathbf{r},\alpha) = \frac{1}{\sqrt{2}}\left(U_{p_x}^-(\mathbf{r}) \pm iU_{p_y}^-(\mathbf{r})\right)\chi_{\pm 1/2}(\alpha), \tag{12.17}$$

$$U_{\Gamma_8^+,\pm 1/2}(\mathbf{r},\alpha) = \frac{1}{\sqrt{3}}\left(\frac{1}{\sqrt{2}}\left(U_{p_x}^-(\mathbf{r}) \pm iU_{p_y}^-(\mathbf{r})\right)\chi_{\mp 1/2}(\alpha) + \sqrt{2}U_{p_z}^-(\mathbf{r})\chi_{\pm 1/2}(\alpha)\right), \tag{12.18}$$

and the antisymmetric conduction band combinations,

$$U_{\Gamma_8^-,\pm 3/2}(\mathbf{r},\alpha) = \frac{1}{\sqrt{2}}\left(U_{p_x}^+(\mathbf{r}) \pm iU_{p_y}^+(\mathbf{r})\right)\chi_{\pm 1/2}(\alpha), \tag{12.19}$$

$$U_{\Gamma_8^-,\pm 1/2}(\mathbf{r},\alpha) = \frac{1}{\sqrt{3}}\left(\frac{1}{\sqrt{2}}\left(U_{p_x}^+(\mathbf{r}) \pm iU_{p_y}^+(\mathbf{r})\right)\chi_{\mp 1/2}(\alpha) + \sqrt{2}U_{p_z}^+(\mathbf{r})\chi_{\pm 1/2}(\alpha)\right). \tag{12.20}$$

The ± signs in (12.13) through (12.30) designate the two spin states.

We here emphasize that the combinations of the three lowest Γ^+ valence bands (see Fig. 11.14) *maximize* the charge density in the region *between* the atoms which corresponds to *bonding orbitals*, confirming the behavior discussed earlier in connection with Pauling's tetrahedral bond formation (see Section 7.4). The next three highest levels are Γ^- conduction bands and correspond to *antibonding orbitals*, which *minimize* the charge between the atoms. The relative positions of the antibonding levels varies among materials: the s states lie above the p states in silicon, but have the opposite order in germanium.

[9] Note the Γ_6^+ states are s-like while the Γ_6^- states are p-like.

12.1.6 Constructing the Hamiltonian matrix

The **k**·**p** theory considered in Section 10.4 can be applied directly to the s electron states since they are non-degenerate apart from spin. However, if we are to apply it to the six (with spin) Γ_8^+, Γ_7^+, states at the valence band edge, the theory must be generalized due the degeneracy among these states. Since we are here expanding about the zone center where $\mathbf{k} = 0$, the perturbing Hamiltonian given Eq. (10.41) has the form

$$\hat{H}' = \frac{\hbar^2 k^2}{2m} + \frac{\hbar \mathbf{k} \cdot \hat{\mathbf{p}}}{m}. \tag{12.21}$$

We can number the matrix elements of H' involving the Γ_7^+, Γ_8^+ states by n, n′. We will show that the off-diagonal matrix elements of H' between these states vanish, and therefore we will need to consider *second-order* coupling through higher conduction band states, e.g., the Γ_7^- antibonding s-like and the Γ_6^-, Γ_8^- antibonding p-like states, which we denote as m. The perturbing Hamiltonian matrix then takes the form[10]

$$H'_{nn'} \to H'_{nn'} + \sum_m \frac{H'_{nm} H'_{mn'}}{E_n^{(0)} - E_m^{(0)}}. \tag{12.22}$$

In what follows, we consider the coupling of $p_{3/2}$-like Γ_8^+ states, $U_{\Gamma^+_{8,\pm 3/2}}$ ($n = 1,4$) and $U_{\Gamma^+_{8,\pm 1/2}}$ ($n = 2, 3$), at the valence band edge, together with the split-off $p_{1/2}$-like Γ_7^+ states, $U_{\Gamma^+_{7,\pm 1/2}}$ ($n = 5, 6$) (see Fig. 11.14(b)), with the three lower-lying conduction bands involving the $s_{1/2}$-like Γ_7^- states, $U_{\Gamma^-_{7,\pm 1/2}}$, the $p_{1/2}$-like Γ_6^- states, $U_{\Gamma^-_{6,\pm 1/2}}$, and the $p_{3/2}$-like Γ_8^- states, $U_{\Gamma^-_{8,\pm 3/2}}$, $U_{\Gamma^-_{8,\pm 1/2}}$.[11] Although there are a total of 16 states, in considering the behavior of the bonding p-like electrons at the band edge, the Hamiltonian is 6×6 but with second-order corrections coupled through the eight excited states.

We are now presented with the problem of evaluating the required matrix elements. All the off-diagonal matrix elements of $\hbar^2 k^2/2m$ vanish due to orthogonality. To examine the matrix elements of $\hbar \mathbf{k} \cdot \hat{\mathbf{p}}/m$, it is helpful to use tabulated group multiplication tables.[12] Table 12.2 gives the results for the Γ representations of the O and T_d groups relevant to the diamond and zinc blende structures. Although we here examine the matrix elements one by one,[13] use of this table identifies terms that we can ignore.

The vector operator $\hat{\mathbf{p}}$ corresponds to a Γ_4^- representation. Using Table 12.2 we have

$$\Gamma_8^+ \times \Gamma_4^- = \Gamma_6^- + \Gamma_7^- + 2\Gamma_8^-. \tag{12.23a}$$

and

$$\Gamma_7^+ \times \Gamma_4^- = \Gamma_7^- + \Gamma_8^-. \tag{12.23b}$$

[10] See Section 39 of Landau and Lifshitz, *Quantum Mechanics*, Pergamon Press (1977).

[11] Coupling to lower-lying states is forbidden by the Pauli principle, since they are occupied.

[12] Such tables may be found in *Properties of the 32 Point Groups*, G.F. Koster, J.O. Dimmock, R.C. Wheeler, and H. Shatz, MIT Press, Cambridge, MA (1963).

[13] The matrix elements follow more directly using group theoretical methods; see books by Yu and Cardona, Kittel, or Snoke, referenced at the end of this section.

Table 12.2 *Multiplication table for the groups* O *and* T_d.

Γ_1	Γ_2	Γ_3	Γ_4	Γ_5	Γ_6	Γ_7	Γ_8	
Γ_1	Γ_2	Γ_3	Γ_4	Γ_5	Γ_6	Γ_7	Γ_8	Γ_1
	Γ_1	Γ_3	Γ_5	Γ_4	Γ_7	Γ_6	Γ_8	Γ_2
		$\Gamma_1+\Gamma_2+\Gamma_3$	$\Gamma_4+\Gamma_5$	$\Gamma_4+\Gamma_5$	Γ_8	Γ_8	$\Gamma_6+\Gamma_7+\Gamma_8$	Γ_3
			$\Gamma_1+\Gamma_3+\Gamma_4+\Gamma_5$	$\Gamma_2+\Gamma_3+\Gamma_4+\Gamma_5$	$\Gamma_6+\Gamma_8$	$\Gamma_7+\Gamma_8$	$\Gamma_6+\Gamma_7+2\Gamma_8$	Γ_4
				$\Gamma_1+\Gamma_3+\Gamma_4+\Gamma_5$	$\Gamma_7+\Gamma_8$	$\Gamma_6+\Gamma_8$	$\Gamma_6+\Gamma_7+2\Gamma_8$	Γ_5
					$\Gamma_1+\Gamma_4$	$\Gamma_2+\Gamma_5$	$\Gamma_3+\Gamma_4+\Gamma_5$	Γ_6
							$\Gamma_1+\Gamma_4$	Γ_7
							$\Gamma_1+\Gamma_2+\Gamma_3+2\Gamma_4+2\Gamma_5$	Γ_8

Only a Γ_1^+ representation will result in a non-zero matrix element. Since the product of any representation with itself is always Γ_1^+, we see that the product of (12.23a) with any of the three low-lying conduction band states (Γ_8^-, Γ_6^-, and Γ_7^-) will yield non-zero matrix elements. On the other hand, from (12.23b), Γ_7^+ only couples to Γ_7^- and Γ_8^-. Finally, it follows from Table 12.2 that the elements $\langle U_n^+ |\, \hbar\mathbf{k}\cdot\hat{\mathbf{p}}/m\, | U_n^+\rangle$ vanish due to parity.[14]

Up to this point we have not explicitly treated the effects of spin–orbit coupling. Here we will model this effect by simply introducing a splitting that pushes the Γ_7^+ states below the Γ_8^+ states by an amount Δ_0.

We will only calculate the H'_{11} element which now reduces to

$$H'_{11} = \frac{\hbar^2k^2}{2m} - \sum_m \left|\left\langle U_{\Gamma^+_{8,3/2}} \right| \frac{\hbar\mathbf{k}\cdot\hat{\mathbf{p}}}{m} \left| U_m \right\rangle\right|^2 \frac{1}{\Delta E_m}, \tag{12.24}$$

where $\Delta E_m \equiv E_m^{(0)} - E_n^{(0)}$. We restrict to the three lowest-lying excited states of germanium and define

$$E_0 \equiv E_{\Gamma_7^-} - E_{\Gamma_8^+} \tag{12.25a}$$

and

$$E'_0 \equiv E_{\Gamma_8^-} - E_{\Gamma_8^+}; \tag{12.25b}$$

the third energy difference, $E_{\Gamma_6^-} - E_{\Gamma_8^+}$, can be approximated by E'_0 when the spin–orbit coupling among the excited p states is suppressed.

We begin with the matrix elements of $\mathbf{k}\cdot\hat{\mathbf{p}}$ connecting $U_{\Gamma^+_{8,3/2}}$ to $U_{\Gamma^-_{7,\pm1/2}}$, which follow from (12.18) and (12.14) as

$$\left\langle U_{\Gamma^+_{8,3/2}} \right| \mathbf{k}\cdot\hat{\mathbf{p}} \left| U_{\Gamma^-_{7,1/2}} \right\rangle = \left\langle \frac{1}{\sqrt{2}}\left(U^-_{p_x}(\mathbf{r}) - iU^-_{p_y}(\mathbf{r})\right)\chi_{1/2}(\alpha) \right| \mathbf{k}\cdot\hat{\mathbf{p}} \left| U^-_s(\mathbf{r})\chi_{1/2}(\alpha) \right\rangle$$
$$= \frac{iP}{\sqrt{2}}\left(k_x - ik_y\right), \tag{12.26a}$$

[14] In zinc blende materials this term does not vanish, leading to a small linear contribution; see Section 12.1.7.

and

$$\left\langle U_{\Gamma^+_{8,3/2}} \middle| \mathbf{k}\cdot\hat{\mathbf{p}} \middle| U_{\Gamma^-_{7,-1/2}} \right\rangle = 0; \tag{12.26b}$$

here we have parameterized the momentum matrix elements as

$$\left\langle U^-_{p_x} \middle| \hat{p}_x \middle| U^-_s \right\rangle = \left\langle U^-_{p_y} \middle| \hat{p}_y \middle| U^-_s \right\rangle = \left\langle U^-_{p_z} \middle| \hat{p}_z \middle| U^-_s \right\rangle = iP. \tag{12.27}$$

The matrix elements connecting $U_{\Gamma^+_{8,3/2}}$ to the two $U_{\Gamma^-_{6,\pm1/2}}$ states are

$$\left\langle U_{\Gamma^+_{8,3/2}} \middle| \mathbf{k}\cdot\hat{\mathbf{p}} \middle| U_{\Gamma^-_{6,1/2}} \right\rangle = \frac{iQ}{\sqrt{6}}(k_y + ik_x) \tag{12.28a}$$

and

$$\left\langle U_{\Gamma^+_{8,3/2}} \middle| \mathbf{k}\cdot\hat{\mathbf{p}} \middle| U_{\Gamma^-_{6,-1/2}} \right\rangle = 0; \tag{12.28b}$$

here we have introduced a second parameter,

$$iQ \equiv \left\langle U^-_{p_x} \middle| \hat{p}_y \middle| U^+_{p_z} \right\rangle \tag{12.29}$$

which obeys cyclic permutations in x, y, and z, but changes sign if any pair is permuted. Finally, the four matrix elements connecting $U_{\Gamma^+_{8,3/2}}$ to $U_{\Gamma^-_{8,\pm3/2}}$ and $U_{\Gamma^-_{8,\pm1/2}}$ are

$$\left\langle U_{\Gamma^+_{8,3/2}} \middle| \mathbf{k}\cdot\hat{\mathbf{p}} \middle| U_{\Gamma^-_{8,3/2}} \right\rangle = Qk_z,\ \left\langle U_{\Gamma^+_{8,3/2}} \middle| \mathbf{k}\cdot\hat{\mathbf{p}} \middle| U_{\Gamma^-_{8,1/2}} \right\rangle = \frac{iQ}{\sqrt{3}}(k_y + ik_x),$$
$$\left\langle U_{\Gamma^+_{8,3/2}} \middle| \mathbf{k}\cdot\hat{\mathbf{p}} \middle| U_{\Gamma^-_{8,-1/2}} \right\rangle = 0,\ \left\langle U_{\Gamma^+_{8,3/2}} \middle| \mathbf{k}\cdot\hat{\mathbf{p}} \middle| U_{\Gamma^-_{8,-3/2}} \right\rangle = 0. \tag{12.30a–d}$$

Combining (12.24) and (12.25a,b) with (12.26) through (12.30), we obtain

$$H'_{11} = \frac{\hbar^2 k^2}{2m} + \frac{1}{2}N\left(k_x^2 + k_y^2\right) + Mk_z^2, \tag{12.31}$$

where, in accordance with existing conventions, we define

$$L \equiv -\frac{\hbar^2 P^2}{m^2 E_0},\ M \equiv -\frac{\hbar^2 Q^2}{m^2 E'_0},\ \text{and}\ N \equiv L + M. \tag{12.32}$$

Combining the nearly degenerate Γ_7^+ and Γ_8^+ levels results in a 6 × 6 matrix, the remaining elements of which are given by

$$
\begin{aligned}
&H'_{12} = -\frac{N}{\sqrt{3}}\left(k_x k_z - i k_y k_z\right); H'_{13} = -\frac{1}{2\sqrt{3}}\left[(L-M)\left(k_x^2 - k_y^2\right) - 2iNk_xk_y\right];\\
&H'_{14} = 0;\; H'_{15} = \frac{1}{\sqrt{2}}H'_{12};\; H'_{16} = -\sqrt{2}H'_{13};\\
&H'_{22} = \frac{\hbar^2k^2}{2m} + \frac{1}{3}(M+2L)k^2 - \frac{1}{2}(L-M)\left(k_x^2 + k_y^2\right);\; H'_{23} = 0;\\
&H'_{24} = H'_{13};\; H'_{25} = \frac{1}{\sqrt{2}}\left(H'_{22} - H'_{11}\right);\; H'_{26} = \sqrt{\frac{3}{2}}H'_{12};\\
&H'_{33} = H'_{22};\; H'_{34} = -H'_{12};\; H'_{35} = -H'^{*}_{26};\; H'_{36} = H'_{25};\\
&H'_{44} = H'_{11};\; H'_{45} = -\sqrt{2}H'^{*}_{13};\; H'_{46} = -H'^{*}_{15};\\
&H'_{55} = \frac{\hbar^2k^2}{2m} + \frac{1}{3}\left(2M' + L'\right)k^2 - \Delta_0;\; H'_{56} = 0;\; H'_{66} = H'_{55}. \qquad (12.33\text{a–t})
\end{aligned}
$$

To describe the split-off band we introduce the additional definitions

$$
L' = -\frac{\hbar^2P^2}{m^2\left(E_0 + \Delta_0\right)},\quad M' = -\frac{\hbar^2Q^2}{m^2\left(E'_0 + \Delta_0\right)}. \qquad (12.34\text{a,b})
$$

Here, $\Delta_0 = 3\lambda_{SO}/2$ is the spin–orbit splitting between the Γ_7^+ and Γ_8^+ states and λ_{SO} is defined by Eq. (12.1a). Note that the Γ_8^+ level lies above Γ_7^+ (consistent with the $p_{3/2}$ level lying above $p_{1/2}$ in atoms).

12.1.7 Calculating the dispersion

To find the electron dispersion curves we must diagonalize this 6 × 6 matrix. In the absence of an external magnetic field there will be three doubly-degenerate branches (due to time reversal symmetry). For values of k such that the shift in energy relative to that at the Γ point is small in comparison with the spin–orbit splitting, Δ_0, we may ignore all matrix elements coupling the Γ_7^+ and Γ_8^+ states. The electron dispersion relation for the Γ_7^+ band is then given by H'_{55}; i.e.,

$$
E(\Gamma_7^+;\mathbf{k}) = E(\Gamma_7^+;0) + \frac{\hbar^2k^2}{2m} + \frac{1}{3}\left(2M' + L'\right)k^2 - \Delta_0. \qquad (12.35)
$$

Note this band is isotropic and parabolic and, as previously stated, lies below the Γ_8^+ levels; it is sometimes referred to as the "split-off" band. Taking the second derivative with respect to k, inserting the definitions of M' and L', we obtain the effective mass of this band as

$$
\frac{m_0}{m^*_{\Gamma_7^+}} = 1 - \frac{2}{3}\left(\frac{P^2}{m_0\left(E_0 + \Delta_0\right)} + \frac{2Q^2}{m_0\left(\left(E'_0 + \Delta_0\right)\right)}\right); \qquad (12.36)
$$

here the term in parenthesis is greater than 1 and hence the band bends down, characteristic of hole-like behavior.

To obtain the dispersion relations for the remaining four-fold degenerate Γ_8^+ band in the same small k limit, we must diagonalize the 4 × 4 matrix associated with the Γ_8^+ block; i.e., we must solve the determinantal equation

$$\begin{vmatrix} H'_{11}-E' & H'_{12} & H'_{13} & 0 \\ H'^{*}_{12} & H'_{22}-E' & 0 & H'_{13} \\ H'^{*}_{13} & 0 & H'_{22}-E' & -H'_{12} \\ 0 & H'^{*}_{13} & -H'^{*}_{12} & H'_{11}-E' \end{vmatrix} = 0, \qquad (12.37)$$

leading to

$$(H'_{11}-E')\,(H'_{22}-E') = |H'_{12}|^2 + |H'_{13}|^2,$$

which has the solution[15]

$$E'_{\pm} = \frac{1}{2}(H'_{11}+H'_{22}) \pm \frac{1}{2}\sqrt{(H'_{11}+H'_{22})^2 - 4\left(H'_{11}H'_{22} - |H'_{12}|^2 - |H'_{13}|^2\right)}. \qquad (12.38)$$

Inserting the expressions for the relevant matrix elements gives (Dresselhaus, Kip and Kittel (1955)[16])

$$E'_{\pm} = Ak^2 \pm \left[B^2k^4 + C^2\left(k_x^2k_y^2 + k_y^2k_z^2 + k_z^2k_x^2\right)\right]^{1/2}, \qquad (12.39)$$

where the constants in this expression are given by

$$A = \frac{\hbar^2}{2m}\left[1 - \frac{2}{3}\left(\frac{P^2}{mE_0} + \frac{Q^2}{mE'_0}\right)\right], \qquad (12.40a)$$

$$B = \frac{\hbar^2}{2m}\frac{2}{3}\left[-\frac{P^2}{mE_0} + \frac{Q^2}{mE'_0}\right], \qquad (12.40b)$$

and

$$C = \frac{\hbar^2}{2m}\sqrt{\frac{16P^2Q^2}{3m^2E_0E'_0}}. \qquad (12.40c)$$

As defined here, A is negative since $2P^2/3m_0E_0$ and $2Q^2/3m_0E'_0$ are typically greater than 1. Furthermore, when the term involving the square root is included, the solutions $E'_{\pm}$ remain negative; therefore both of these levels correspond to holes. This is consistent with the band structures given in Figs. 11.13 and 11.14, where it is seen that the dispersion curves bend to lower energies on leaving the Γ point for both of these levels.[17] The solutions with the −

[15] Each of the roots in (12.38) is doubly degenerate; this degeneracy is lifted in zinc blende materials due to the lack of inversion symmetry, as we will discuss further.

[16] Cyclotron resonance of electrons and holes in silicon and germanium crystals, G. Dresselhaus, A.F. Kip, and C. Kittel, *Phys. Rev.* **98**, 368 (1955).

[17] If either P or Q vanishes, only one band will bend down; the heavier band will bend up. Hence it is necessary to include both contributions to obtain overall hole-like behavior and agree with the band structure calculations.

and + signs in (12.26) are referred to as *heavy and light holes*, respectively. Although only two parameters, P^2/mE_0 and Q^2/mE'_0, enter the expressions, in a more general model, A, B, and C would be independent.[18] Note the constant energy surfaces given by (12.39) are not isotropic, although they have the full cubic symmetry of the Γ point, as they must; however for any given direction the dispersion is quadratic in k. We will take up the topic of holes in group IV materials again in Section 22.5, where we discuss the Si and Ge cases.

In zinc blende materials like GaAs, which lack inversion symmetry, the pair of doubly degenerate levels predicted by Eq. (12.39) will undergo additional splitting as we leave Γ, which are initially linear in k but quickly become quadratic; the net effect is to slightly shift the band extrema from the Γ point. This effect is usually not apparent in reported band structure calculations. Figure 12.1 shows, schematically, the splitting of the Γ_8 levels as they would evolve along the [110], [100], and [111] directions.[19,20] We see that along [110] the symmetry is already low enough that the levels fully split as we leave Γ. On the other hand, we see that the Γ_7 levels remain doubly degenerate for all directions.

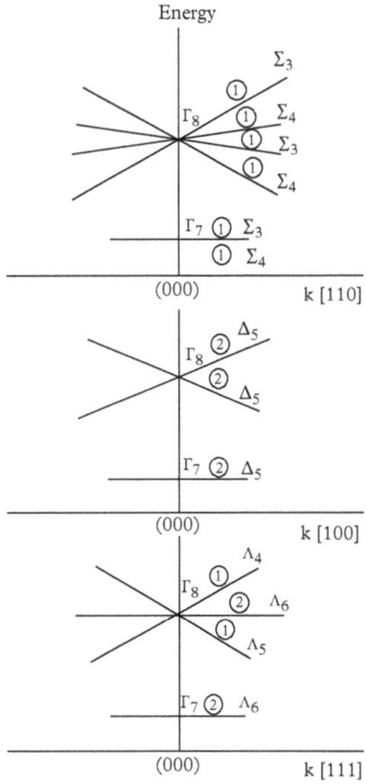

Figure 12.1 *Spin-orbit induced splittings in zinc blende materials. (After Dresselhaus (1955).)*

12.2 Inverted band structures; αSn

As we proceed down the periodic table, the s levels move lower in energy relative to the p levels in atoms and, in addition, the spin–orbit splitting of the p levels increases. As a result of these effects, band reordering can occur in solids composed of such atoms. A classic example occurs in tin, for which the low temperature alpha phase has the diamond structure. Figure 12.2 shows the band structure of αSn, which is to be compared with that of Ge shown in Fig. 11.14.[21] In Ge, the Γ_7^- level lies above the Γ_8^+ level whereas, from Fig. 12.2, it lies below Γ_8^+ (but above Γ_7^+) in αSn; i.e., E_0 in Eq. (12.25a) changes sign. In addition, Γ_7^+ has moved lower with respect to Γ_8^+ due to the increased spin–orbit coupling. Finally, we note that the dispersion of the light component of Γ_8^+ changes from negative (bending down) to positive (bending up).

To examine the behavior where the levels cross, we imagine that E_0 can be changed continuously[22] and examine the resulting band structure in the *immediate vicinity* of where it vanishes. Here the second-order coupling to states outside the Γ_8^+ group breaks down (in particular terms involving L given by Eq. (12.32)) and we will treat the problem in zeroth order by directly diagonalizing the 6 × 6 Hamiltonian in the Γ_8^+, Γ_7^- subspace. To discuss the qualitative behavior we only need to retain the linear term in $\mathbf{k} \cdot \mathbf{p}$; i.e., we drop all quadratic forms in $\mathbf{k}$, including both the diagonal contribution, $\hbar^2 k^2/2m$ and the second-order couplings through the Γ_6^- and Γ_8^-. In this approximation all off-diagonal terms in the 4 × 4 Γ_8^+ and 2 × 2 Γ_7^- subspaces vanish; only some off-diagonal terms coupling these spaces will be non-zero. To leading order in $\mathbf{k} \cdot \mathbf{p}$ our Hamiltonian in the

[18] Equation (12.39) can obtained directly using group theory methods (Luttinger (1956)); for a discussion see E.M. Lifshitz and L.P. Pitiavskii, *Statistical Physics Part II*, Section 68.

[19] Symmetry properties of the energy bands of the zinc blende structure, R.H. Parmenter, *Phys. Rev* **100**, 573 (1955)

[20] Spin orbit effects in zinc blende structures, G. Dresselhaus, *Phys. Rev.* **100**, 580 (1955).

[21] Nonlocal pseudopotential calculations for electronic-structure of 11 diamond and zinc-blende semiconductors, J.R. Chelikowsky and M.L. Cohen, *Phys. Rev. B* 14 (1976).

[22] The II–VI compound HgTe also has an inverted band structure and furthermore can be continuously alloyed with CdTe as $Hg_{1-x}Cd_xTe$, such that E_0 passes continuously through zero.

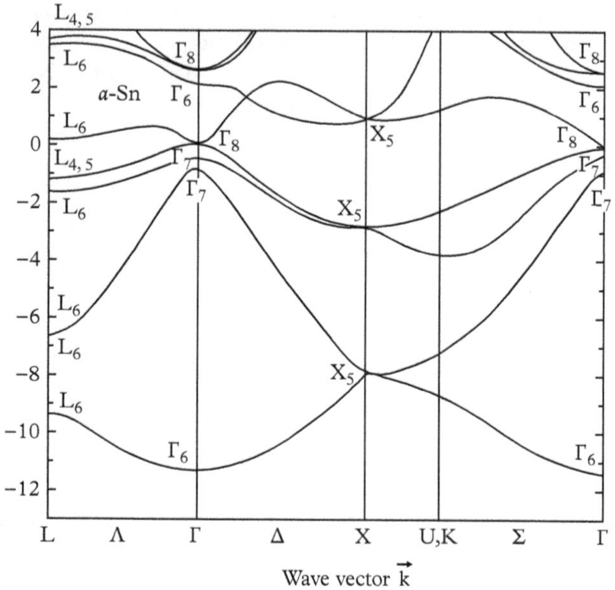

Figure 12.2 *The band structure of* αSn. *Note that when compared to* Ge, *the* Γ_7^- *level has moved below* Γ_8^+. *(After Chelikowsky and Cohen (1976).)*

6×6 space combining the Γ_7^- and Γ_8^+ levels then has the form

$$\overleftrightarrow{H} = \begin{pmatrix} 0 & 0 & 0 & 0 & H_{15} & H_{16} \\ 0 & 0 & 0 & 0 & H_{25} & H_{26} \\ 0 & 0 & 0 & 0 & H_{35} & H_{36} \\ 0 & 0 & 0 & 0 & H_{45} & H_{46} \\ H_{51} & H_{52} & H_{53} & H_{54} & E_0 & 0 \\ H_{61} & H_{62} & H_{63} & H_{64} & 0 & E_0 \end{pmatrix} \qquad (12.41)$$

where we have measured energies from Γ_8^+. The H_{15} term in (12.41) is obtained from Eq. (12.26a) and the remaining terms follow from Eqs. (12.14), (12.17), and (12.18),

$$H_{15} = \frac{i\hbar P}{\sqrt{2}m}\left(k_x - ik_y\right), \qquad H_{16} = 0.$$

$$H_{25} = \sqrt{\frac{2}{3}}\frac{i\hbar P}{m}k_z, \qquad H_{26} = \frac{i\hbar P}{\sqrt{6}m}(k_x - ik_y)$$

$$H_{35} = \frac{i\hbar P}{\sqrt{6}m}(k_x + ik_y), \qquad H_{36} = \sqrt{\frac{2}{3}}\frac{i\hbar P}{m}k_z,$$

$$H_{45} = 0, \qquad H_{46} = \frac{i\hbar P}{\sqrt{2}m}\left(k_x + ik_y\right). \qquad (12.42a–h)$$

The eigenvalues follow from

$$\left|\overleftrightarrow{H} - E\overleftrightarrow{1}\right| = \hat{0}, \qquad (12.43a)$$

which leads to the sixth-order equation

$$E^2\left[E^2(E-E_0)^2 - H_k^4\right] = 0, \tag{12.43b}$$

where we have defined

$$H_k = \sqrt{\frac{2}{3}\frac{\hbar^2 P^2}{m^2}}\,k. \tag{12.44}$$

The solutions of (12.44) are

$$E = 0, \frac{1}{2}\left[E_0 \pm \sqrt{E_0^2 + 4H_k^2}\right] \tag{12.45}$$

and come in pairs due to time reversal symmetry.

For the case $E_0 = 0$, where the Γ_8^+ and Γ_7^- levels are degenerate, Eq. (12.45) has the three solutions,

$$E(k) = 0;\ \pm\sqrt{\frac{2}{3}\frac{\hbar^2 P^2}{m^2}}\,k; \tag{12.46}$$

note that one band is flat in our approximation (corresponding to a heavy hole band) while the other two have an unusual *linear behavior* with an associated velocity $v = \sqrt{2/3}\,P/m$, as shown schematically in Fig. 12.3(a).[23]

For $E_0 > 0$ and near the band edge we have the three behaviors,

$$E(k) = 0,\ -\frac{2}{3}\frac{\hbar^2 P^2}{m^2 E_0}k^2,\ E_0 + \frac{2}{3}\frac{\hbar^2 P^2}{m^2 E_0}k^2. \tag{12.47}$$

We again have one flat band (the heavy holes) but now there are two *parabolic* bands,[24] one contacting the flat band and bending down (the light holes) and a third displaced upward by E_0 that bends up (the s-like conduction electrons). This is what we find in Ge (see Fig. 11.14(b)) and is shown schematically in Fig. 12.3(b).

The third case is when $E_0 < 0$, where we have the following three solutions:

$$E(k) = 0, +\frac{2}{3}\frac{\hbar^2 P^2}{m^2|E_0|}k^2,\ -|E_0| - \frac{2}{3}\frac{\hbar^2 P^2}{m^2|E_0|}k^2, \tag{12.48}$$

consisting of a flat heavy hole band and two parabolic bands. However, now the parabola contacting the flat band bends up while the second parabola is displaced downward by E_0 and bends down, as shown in Fig. 12.3(c). This is the so-called "inverted" behavior which is characteristic of αSn, and is shown in Fig. 12.2.

[23] We will encounter this linear behavior again in Chapter 49, where we discuss graphene, in which the band structure near the K points also has a linear behavior; these are the so-called Dirac cones.

[24] According to Eq. (12.45), the behavior becomes linear for larger k; if the quadratic terms that were earlier suppressed are included in $\overleftrightarrow{H}$, there will eventually be a crossover to quadratic behavior for still larger k.

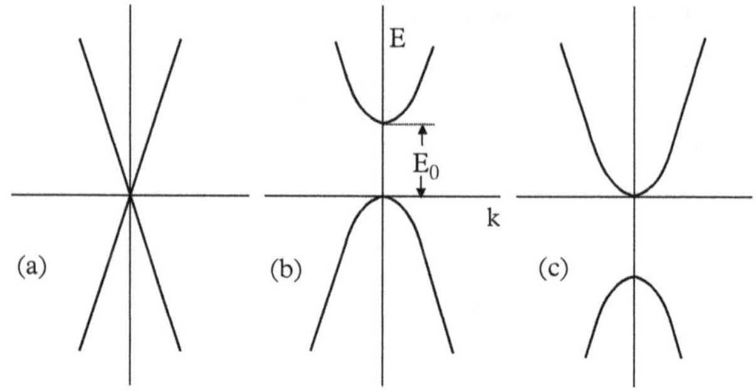

Figure 12.3 *Schematic behavior of the band structure of the hole complex in the vicinity of the* Γ *point. (a) Critical behavior at contact. (b) Normal ordering as in* Ge. *Inverted behavior as in* αSn.

12.3 The Slater–Koster method for diamond-like structures

Another powerful scheme to parameterize band structures (or represent experimental data) is to use the integrals occurring in the tight binding scheme (see Chapter 11) as fitting parameters (J.C. Slater and G.F. Koster (1954)[25]). As in our discussion of the $\mathbf{k} \cdot \mathbf{p}$ method, we limit our discussion to the diamond/zinc blende materials.

Diamond and zinc blende (ZnS) have the structures shown in Figs. 7.4 and 8.16(b) respectively. The primitive translation vectors are $\mathbf{a} = (\mathbf{i} + \mathbf{j})/\sqrt{2}$, $\mathbf{b} = (\mathbf{j}+\mathbf{k})/\sqrt{2}$, and $\mathbf{c} = (\mathbf{k}+\mathbf{i})/\sqrt{2}$ (the three vectors connecting a cube corner to the center of the three nearest faces). We recall that the diamond lattice has a basis. We place the first atom (C or Zn) at a cube corner, $\mathbf{r}_0 = 0$ and the second atom (S or another C) at $\mathbf{r}_1 = a(\mathbf{i}+\mathbf{j}+\mathbf{k})/4$, i.e., a quarter way along the cube diagonal. The tight binding s states situated at the origin, $\mathbf{r}_0 = 0$, and at the point $\mathbf{r}_1$ are given by (see Eqs. (11.63) and (11.65))

$$\psi^{(0)}_{\mathbf{k}s}(\mathbf{r}) = \sum_{\mathbf{R}} e^{i\mathbf{k}\cdot\mathbf{R}}\phi^{(0)}_{s}(\mathbf{r} - \mathbf{R}); \psi^{(1)}_{\mathbf{k}s}(\mathbf{r}) = \sum_{\mathbf{R}} e^{i\mathbf{k}\cdot(\mathbf{R}+\mathbf{r}_1)}\phi^{(1)}_{s}(\mathbf{r} - \mathbf{r}_1 - \mathbf{R}); \tag{12.49a–b}$$

which generalize (12.8a,b) to finite $\mathbf{k}$. The corresponding p states at these two sites which generalize (12.9a–d) are given by

$$\begin{aligned}
\psi^{(0)}_{\mathbf{k}p_x}(\mathbf{r}) &= \sum_{\mathbf{R}} e^{i\mathbf{k}\cdot\mathbf{R}}\phi^{(0)}_{p_x}(\mathbf{r}-\mathbf{R}); \quad \psi^{(1)}_{\mathbf{k}p_x}(\mathbf{r}) = \sum_{\mathbf{R}} e^{i\mathbf{k}\cdot(\mathbf{R}+\mathbf{r}_1)}\phi^{(1)}_{p_x}(\mathbf{r}-\mathbf{r}_1-\mathbf{R});\\
\psi^{(0)}_{\mathbf{k}p_y}(\mathbf{r}) &= \sum_{\mathbf{R}} e^{i\mathbf{k}\cdot\mathbf{R}}\phi^{(0)}_{p_y}(\mathbf{r}-\mathbf{R}); \quad \psi^{(1)}_{\mathbf{k}p_y}(\mathbf{r}) = \sum_{\mathbf{R}} e^{i\mathbf{k}\cdot(\mathbf{R}+\mathbf{r}_1)}\phi^{(1)}_{p_y}(\mathbf{r}-\mathbf{r}_1-\mathbf{R});\\
\psi^{(0)}_{\mathbf{k}p_z}(\mathbf{r}) &= \sum_{\mathbf{R}} e^{i\mathbf{k}\cdot\mathbf{R}}\phi^{(0)}_{p_z}(\mathbf{r}-\mathbf{R}); \quad \psi^{(1)}_{\mathbf{k}p_z}(\mathbf{r}) = \sum_{\mathbf{R}} e^{i\mathbf{k}\cdot(\mathbf{R}+\mathbf{r}_1)}\phi^{(1)}_{p_z}(\mathbf{r}-\mathbf{r}_1-\mathbf{R}).
\end{aligned} \tag{12.50a–f}$$

Note we must include an index j on the wave functions (and shortly on the potentials) since, for generality, we are here assuming a zinc blende structure

[25] Simplified LCAO method for the periodic potential problem, J.C. Slater and G.F. Koster, *Phys. Rev.* **94**, 1498 (1954).

where the two atoms (e.g., Zn and S) are different. The equation analogous to Eq. (11.59) is

$$\sum_{i'j'}\sum_{\mathbf{R}'}\left[H_{\mathbf{RR}'}^{(ij;i'j')} - EJ_{\mathbf{RR}'}^{(ij;i'j')}\right] a_{\mathbf{R}'}^{(i'j')} = 0, \tag{12.51}$$

where i = 0–3 denotes the four orbitals, s, p_x, p_y, and p_z, and j = 0, 1 denotes the two sites in the unit cell, $\mathbf{r}_0 = 0$ and $\mathbf{r}_1 = \frac{a}{4}(\mathbf{i}+\mathbf{j}+\mathbf{k})$, and

$$H_{\mathbf{RR}'}^{(ij;i'j')} = E^{(i'j')} J_{\mathbf{RR}'}^{(ij;i'j')} + h_{\mathbf{RR}'}^{(ij;i'j')} \tag{12.52a}$$

$$J_{\mathbf{RR}'}^{(ij;i'j')} = \left\langle \phi_i^{(j)}(\mathbf{r}-\mathbf{r}_j-\mathbf{R}) \middle| \phi_{i'}^{(j')}(\mathbf{r}-\mathbf{r}_{j'}-\mathbf{R}') \right\rangle, \tag{12.52b}$$

$$h_{\mathbf{RR}'}^{(ij;i'j')} = \left\langle \phi_i^{(j)}(\mathbf{r}-\mathbf{r}_j-\mathbf{R}) \middle| \sum_{\mathbf{R}''j''}\left[U^{(j'')}(\mathbf{r}-\mathbf{r}_j''-\mathbf{R}'') - U^{(j')}(\mathbf{r}-\mathbf{r}_{j'}-\mathbf{R}')\right] \middle| \phi_{i'}^{(j')}(\mathbf{r}-\mathbf{r}_{j'}-\mathbf{R}') \right\rangle, \tag{12.52c}$$

and

$$a_{\mathbf{R}}^{(ij)} = a^{(ij)} e^{i\mathbf{k}\cdot(\mathbf{R}+\mathbf{r}_j)}. \tag{12.52d}$$

We again restrict to the nearest neighbors where we may neglect the overlap integrals. The on-site term takes the form $h_0^{(ij;i'j')} = h_0^{(ij)}\delta_{ii'}\delta_{jj'}$ and we denote the nearest neighbor term as $h_1^{(ij;i'j')}$. Assuming the atomic orbitals on the same site are orthonormal we have $E^{(i'j')}J_{\mathbf{RR}'}^{(ij;i'j')} = E^{(ij)}\delta_{ii'}\delta_{jj'}\delta_{\mathbf{RR}'}$. Equation (12.52) then becomes

$$\sum_{i'j'}\left[\left(E^{(ij)} - E(\mathbf{k})\right)\delta_{ii'}\delta_{jj'} + h_0^{(ij)}\delta_{ii'}\delta_{jj'} + h_1^{(ij;i'j')}\right] a^{i'j'} = 0, \tag{12.53a}$$

which has a solution only if

$$\left|\left(E^{(ij)} - E(\mathbf{k})\right)\delta_{ii'}\delta_{jj'} + h_0^{(ij)}\delta_{ii'}\delta_{jj'} + h_1^{(ij;i'j')}\right| = 0. \tag{12.53b}$$

Unlike our earlier treatments, where the nearest neighbors were all on Bravis lattice sites, the nearest neighbors of our diamond lattice (which has a basis) are separated by *less than* a primitive translation vector ($\sqrt{3}a/4$ vs. $a/\sqrt{2}$). The vector positions of the nearest neighbors for our two basis sites are given by

$$\mathbf{r}_j^{(1)} = \pm\mathbf{r}_1,\ \mathbf{r}_j^{(2)} = \pm(\mathbf{r}_1-\mathbf{a}),\ \mathbf{r}_j^{(3)} = \pm(\mathbf{r}_1-\mathbf{b}),\ \mathbf{r}_j^{(4)} = \pm(\mathbf{r}_1-\mathbf{c}), \tag{12.54a}$$

or equivalently

$$\mathbf{r}_j^{(1)} = \pm\frac{a}{4}(\mathbf{i}+\mathbf{j}+\mathbf{k});\ \mathbf{r}_j^{(2)} = \pm\frac{a}{4}(-\mathbf{i}-\mathbf{j}+\mathbf{k});\ \mathbf{r}_j^{(3)} = \pm\frac{a}{4}(\mathbf{i}-\mathbf{j}-\mathbf{k});$$
$$\mathbf{r}_j^{(4)} = \pm\frac{a}{4}(-\mathbf{i}+\mathbf{j}-\mathbf{k}), \tag{12.54b}$$

where **a**, **b**, and **c** are the primitive translation vectors and the (+) and (–) signs correspond to j = 0, 1 respectively. In terms of these vectors the nearest neighbor contribution to the Hamiltonian matrix is

$$h_1^{(ij;i'j')} = \left\langle \phi_i^{(j)}(\mathbf{r}-\mathbf{r}_j) \right| \times \sum_{l=1}^{4} e^{i\mathbf{k}\cdot\mathbf{r}_{j'}^{(l)}} \left[U^{(j)}(\mathbf{r}-\mathbf{r}_j) - U^{(j')}\left(\mathbf{r}-\mathbf{r}_j+\mathbf{r}_{j'}^{(l)}\right) \right] \left| \phi_{i'}^{(j')}\left(\mathbf{r}-\mathbf{r}_j+\mathbf{r}_{j'}^{(l)}\right) \right\rangle (1-\delta_{jj'}). \tag{12.55}$$

We now examine some specific matrix elements. We start with the s - p_x matrix element,

$$h_1^{(00;11)} = \left\langle \phi_s^{(0)}(\mathbf{r}) \right| \sum_{l=1}^{4} e^{i\mathbf{k}\cdot\mathbf{r}_{j'}^{(l)}} \left[U^{(0)}(\mathbf{r}) - U^{(1)}\left(\mathbf{r}+\mathbf{r}_1^{(l)}\right) \right] \left| \phi_{p_x}^{(1)}\left(\mathbf{r}+\mathbf{r}_1^{(l)}\right) \right\rangle. \tag{12.56}$$

Defining

$$\frac{1}{4}V_{sp} \equiv \left\langle \phi_s^{(0)}(\mathbf{r}) \right| \left[U^{(0)}(\mathbf{r}) - U^{(1)}\left(\mathbf{r}+\mathbf{r}_1^{(1)}\right) \right] \left| \phi_{p_x}^{(1)}\left(\mathbf{r}+\mathbf{r}_1^{(1)}\right) \right\rangle, \tag{12.57}$$

and using the relation (which follows from the odd parity of p states)

$$\phi_{p_x}^{(1)}\left(\mathbf{r}+\mathbf{r}_1^{(1)}\right) = -\phi_{p_x}^{(1)}\left(\mathbf{r}+\mathbf{r}_1^{(2)}\right) = \phi_{p_x}^{(1)}\left(\mathbf{r}+\mathbf{r}_1^{(3)}\right) = -\phi_{p_x}^{(1)}\left(\mathbf{r}+\mathbf{r}_1^{(4)}\right). \tag{12.58a}$$

Equation (12.56) becomes

$$\begin{aligned} h_1^{(00;11)} &= \frac{V_{sp}}{4}\left[e^{i\mathbf{k}\cdot\mathbf{r}_1^{(1)}} - e^{i\mathbf{k}\cdot\mathbf{r}_1^{(2)}} + e^{i\mathbf{k}\cdot\mathbf{r}_1^{(3)}} - e^{i\mathbf{k}\cdot\mathbf{r}_1^{(4)}} \right], \\ &= V_{sp}F_1(\mathbf{k}) \end{aligned} \tag{12.59a}$$

where

$$F_1(\mathbf{k}) = -\cos\left(\frac{k_x\pi}{a}\right)\sin\left(\frac{k_y\pi}{a}\right)\sin\left(\frac{k_z\pi}{a}\right) + i\sin\left(\frac{k_x\pi}{a}\right)\cos\left(\frac{k_y\pi}{a}\right)\cos\left(\frac{k_z\pi}{a}\right). \tag{12.59b}$$

In a similar manner,

$$\phi_{p_y}^{(1)}\left(\mathbf{r}+\mathbf{r}_1^{(1)}\right) = -\phi_{p_y}^{(1)}\left(\mathbf{r}+\mathbf{r}_1^{(2)}\right) = -\phi_{p_y}^{(1)}\left(\mathbf{r}+\mathbf{r}_1^{(3)}\right) = \phi_{p_y}^{(1)}\left(\mathbf{r}+\mathbf{r}_1^{(4)}\right), \tag{12.58b}$$

$$\phi_{p_z}^{(1)}\left(\mathbf{r}+\mathbf{r}_1^{(1)}\right) = \phi_{p_z}^{(1)}\left(\mathbf{r}+\mathbf{r}_1^{(2)}\right) = -\phi_{p_z}^{(1)}\left(\mathbf{r}+\mathbf{r}_1^{(3)}\right) = -\phi_{p_z}^{(1)}\left(\mathbf{r}+\mathbf{r}_1^{(4)}\right), \tag{12.58c}$$

$$\phi_s^{(1)}\left(\mathbf{r}+\mathbf{r}_1^{(1)}\right) = \phi_s^{(1)}\left(\mathbf{r}+\mathbf{r}_1^{(2)}\right) + \phi_s^{(1)}\left(\mathbf{r}+\mathbf{r}_1^{(3)}\right) = +\phi_s^{(1)}\left(\mathbf{r}+\mathbf{r}_1^{(4)}\right), \tag{12.58d}$$

which lead to

$$
\begin{aligned}
h_1^{(00;21)} &= \frac{V_{sp}}{4}\left[e^{i\mathbf{k}\cdot\mathbf{r}_1^{(1)}} - e^{i\mathbf{k}\cdot\mathbf{r}_1^{(2)}} - e^{i\mathbf{k}\cdot\mathbf{r}_1^{(3)}} + e^{i\mathbf{k}\cdot\mathbf{r}_1^{(4)}}\right] \\
&= V_{sp}F_2(\mathbf{k}), \qquad (12.60a)
\end{aligned}
$$

$$
F_2(\mathbf{k}) = -\sin\left(\frac{k_x\pi}{a}\right)\cos\left(\frac{k_y\pi}{a}\right)\sin\left(\frac{k_z\pi}{a}\right) + i\cos\left(\frac{k_x\pi}{a}\right)\sin\left(\frac{k_y\pi}{a}\right)\cos\left(\frac{k_z\pi}{a}\right); \qquad (12.60b)
$$

$$
\begin{aligned}
h_1^{(00;31)} &= \frac{V_{sp}}{4}\left[e^{i\mathbf{k}\cdot\mathbf{r}_1^{(1)}} + e^{i\mathbf{k}\cdot\mathbf{r}_1^{(2)}} - e^{i\mathbf{k}\cdot\mathbf{r}_1^{(3)}} - e^{i\mathbf{k}\cdot\mathbf{r}_1^{(4)}}\right] \\
&= V_{sp}F_3(\mathbf{k}), \qquad (12.61a)
\end{aligned}
$$

$$
F_3(\mathbf{k}) = -\sin\left(\frac{k_x\pi}{a}\right)\sin\left(\frac{k_y\pi}{a}\right)\cos\left(\frac{k_z\pi}{a}\right) + i\cos\left(\frac{k_x\pi}{a}\right)\cos\left(\frac{k_y\pi}{a}\right)\sin\left(\frac{k_z\pi}{a}\right); \qquad (12.61b)
$$

and lastly for the s-s term,

$$
\begin{aligned}
h_1^{(00;01)} &= \frac{V_{ss}}{4}\left[e^{i\mathbf{k}\cdot\mathbf{r}_1^{(1)}} + e^{i\mathbf{k}\cdot\mathbf{r}_1^{(2)}} + e^{i\mathbf{k}\cdot\mathbf{r}_1^{(3)}} + e^{i\mathbf{k}\cdot\mathbf{r}_1^{(4)}}\right] \\
&= V_{ss}F_0(\mathbf{k}), \qquad (12.62a)
\end{aligned}
$$

$$
F_0(\mathbf{k}) = \cos\left(\frac{k_x\pi}{a}\right)\cos\left(\frac{k_y\pi}{a}\right)\cos\left(\frac{k_z\pi}{a}\right) - i\sin\left(\frac{k_x\pi}{a}\right)\sin\left(\frac{k_y\pi}{a}\right)\sin\left(\frac{k_z\pi}{a}\right). \qquad (12.62b)
$$

. In a similar way we can obtain the p-p matrix elements as

$$
\begin{aligned}
&h_1^{(10;11)} = V_{p_xp_x}F_0(\mathbf{k}),\ h_1^{(10;21)} = V_{p_xp_y}F_3(\mathbf{k}), \\
&h_1^{(20;31)} = V_{p_xp_y}F_1(\mathbf{k}),\ h_1^{(30;11)} = V_{p_xp_y}F_2(\mathbf{k}). \qquad (12.63a–d)
\end{aligned}
$$

We also need the matrix elements when the j = 0,1 states are interchanged, which follows from the even and odd parity of the s and p states, respectively. Since we will assume that our tight binding Hamiltonian matrix is used as a fitting scheme, we will absorb the diagonal elements $h_0^{(ij)}$ into the atomic energies, $E_s^0, E_s^1, E_p^0, E_p^1$).

Assembling the above we obtain the final Hamiltonian matrix as

$$
\begin{pmatrix}
E_s^{(0)}-E(\mathbf{k}) & V_{ss}F_0(\mathbf{k}) & 0 & 0 & 0 & V_{sp}F_1(\mathbf{k}) & V_{sp}F_2(\mathbf{k}) & V_{sp}F_3(\mathbf{k}) \\
V_{ss}F_0^*(\mathbf{k}) & E_s^{(1)}-E(\mathbf{k}) & -V_{sp}F_1^*(\mathbf{k}) & -V_{sp}F_2^*(\mathbf{k}) & -V_{sp}F_3^*(\mathbf{k}) & 0 & 0 & 0 \\
0 & -V_{sp}F_1(\mathbf{k}) & E_p^{(0)}-E(\mathbf{k}) & 0 & 0 & V_{p_xp_x}F_0(\mathbf{k}) & V_{p_xp_y}F_3(\mathbf{k}) & V_{p_xp_y}F_2(\mathbf{k}) \\
0 & -V_{sp}F_2(\mathbf{k}) & 0 & E_p^{(0)}-E(\mathbf{k}) & 0 & V_{p_xp_y}F_3(\mathbf{k}) & V_{p_xp_x}F_0(\mathbf{k}) & V_{p_xp_y}F_1(\mathbf{k}) \\
0 & -V_{sp}F_3(\mathbf{k}) & 0 & 0 & E_p^{(0)}-E(\mathbf{k}) & V_{p_xp_y}F_2(\mathbf{k}) & V_{p_xp_y}F_1(\mathbf{k}) & V_{p_xp_x}F_0(\mathbf{k}) \\
V_{sp}F_1^*(\mathbf{k}) & 0 & V_{p_xp_x}F_0^*(\mathbf{k}) & V_{p_xp_y}F_3^*(\mathbf{k}) & V_{p_xp_y}F_2^*(\mathbf{k}) & E_p^{(1)}-E(\mathbf{k}) & 0 & 0 \\
V_{sp}F_2^*(\mathbf{k}) & 0 & V_{p_xp_y}F_3^*(\mathbf{k}) & V_{p_xp_x}F_0^*(\mathbf{k}) & V_{p_xp_y}F_1^*(\mathbf{k}) & 0 & E_p^{(1)}-E(\mathbf{k}) & 0 \\
V_{sp}F_3^*(\mathbf{k}) & 0 & V_{p_xp_y}F_2^*(\mathbf{k}) & V_{p_xp_y}F_1^*(\mathbf{k}) & V_{p_xp_x}F_0^*(\mathbf{k}) & 0 & 0 & E_p^{(1)}-E(\mathbf{k})
\end{pmatrix}
\qquad (12.64)
$$

Table 12.3 *Tight binding parameters for* C, Si, *and* Ge *(after Chadi and Cohen).*

Element	$E_p - E_s$ (eV)	V_{ss} (eV)	V_{sp} (eV)	$V_{p_xp_x}$ (eV)	$V_{p_xp_y}$ (eV)
C	7.40	−15.2	10.2 5	3.0	8.3
Si	7.20	−8.13	5.88	3.17	7.51
Ge	8.41	−6.78	5.31	2.62	6.82

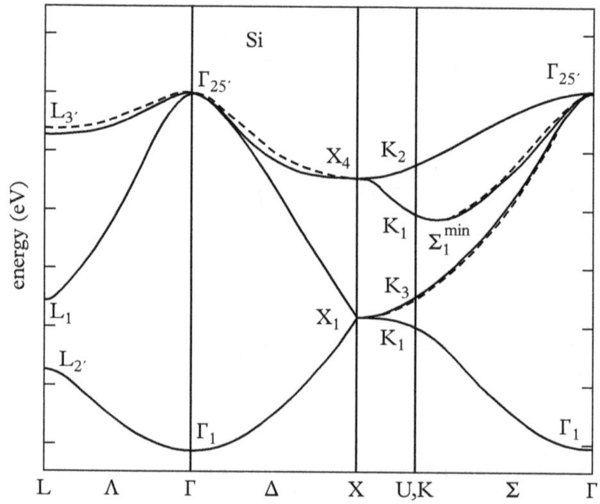

Figure 12.4 *A tight binding fit (solid line) to the band structure of* Si *as calculated via an emperical pseudo-potential approach (dashed line). (After Chadi and Cohen (1975).)*

Now the matrix (12.64) contains eight fitting parameters, $E_s^{(0)}$, $E_s^{(1)}$, $E_p^{(0)}$, $E_p^{(1)}$, V_{ss}, V_{sp}, $V_{p_xp_x}$, and $V_{p_xp_y}$. However, for the group IV materials we have $E_s^{(0)} = E_s^{(1)}$, which we write as E_s and $E_p^{(0)} = E_p^{(1)}$ written as E_p. If furthermore we work only with the relative s-p energy shift, $E_p - E_s$, then the number of parameters reduces to four. Table 12.3 shows these parameters for the group IV elements C, Si, and Ge, as obtained by Chadi and Cohen,[26] which give a good representation of the energy bands.

This tight binding fit is compared with a Si band structure obtained using an empirical pseudopotential approach in Fig. 12.4. Note the excellent agreement, demonstrating the power of the Slater–Koster scheme to represent band structures.

FURTHER READING

The multiband $\mathbf{k} \cdot \mathbf{p}$ parameterization scheme, which has been widely applied to group IV and related semiconductors, was developed by E.O. Kane in, Band

[26] Calculations of the valence bands of diamond and zincblende crystals, D.J. Chadi and M.L. Cohen, *Phys. Stat. Sol.* (b) **68**, 405 (1975).

structure of indium antimonide, *J. Chem. Phys. Solids* **1**, 249 (1957). Also see, E.O. Kane, The **k** · **p** method, in *Semiconductors and Semimetals*, Vol. 1, p. 75, R.K. Willardson and A.C. Beer, Eds., Academic Press, San Diego (1966).

Many of the important aspects of semiconductor physics are treated in *Fundamentals of Semiconductors*, P.Y. Yu and M. Cardona, Springer, Berlin (2001), after which some of our discussions are modeled. This book also gives a brief introduction to group theory.

A group theoretical discussions of holes in group IV semiconductors is given in Yu and Cardona; the topic is also discussed in *Solid State Physics*, David W. Snoke, Addison-Wesley (2009) and in *Quantum Theory of Solids*, C. Kittel, John Wiley & Sons, New York (1963).

The qualitative aspects of bonding in tetrahedrally coordinated semiconductors, going beyond the simple Pauling picture, as outlined in Section 7.4, are discussed in Chapter 3 of *Electronic Structure and the Properties of Solids*, Walter A. Harrison, W.H. Freeman & Company (1979).

Problems

12.1. Using Table 12.1 verify the group symmetries assigned to the forms given in Eqs. (12.13) through (12.20).

12.2. Calculate the remaining matrix elements given in Eqs. (12.53a–g).

12.3. Calculate the matrix elements in Eqs. (12.42a–h).

12.4. Obtain Eq. (12.44). (Hint: construct a strategy and supporting argument which obtains this result with the least algebra.)

13 Augmented Plane Wave and Green's Function Methods*

The strategies to calculate band structure developed in this chapter are widely judged to be the most powerful available. At the same time they are technically demanding relative to those treated in Chapter 11. The material is not a prerequisite for most topics treated in later chapters.

13.1 The augmented plane wave (APW) method

Near a given nucleus of a lattice of atoms, which we here assume is monatomic, the potential is large, negative, and essentially spherically symmetric. Near the surfaces of the Wigner–Seitz (WS) polyhedral enclosing the atoms[1] the potential is nearly constant (referred to as flat). This suggests that we partition the volume within the polyhedron into two regions: (i) a spherical volume centered on the atom and (ii) the remainder of the polyhedron. To a first approximation the potential may be treated as spherically symmetric within the sphere, and we denote it as $U(r)$, and constant in the remainder of the polyhedron; this approximation is referred to as the "muffin tin approximation" (a term due to Slater). The radius, r_i, of the partitioning sphere is a free parameter which may be chosen to optimize this separation, although it is common to fix it at the value corresponding to the largest inscribed sphere of the WS polyhedron. The averaging of the periodic potential required to conform to this model generally results in a discontinuity of the potential at the sphere radius. It is customary to fix the zero of the potential so that it vanishes outside the spheres.

Having partitioned the polyhedron in this way it is natural to use different basis functions to represent the wave function in the two regions. Within the sphere, a superposition of solutions of the form $R_l(E,r)Y_{lm}(\theta,\phi)$ obtained from numerically integrating the Schrödinger equation for the potential $U(r)$ is appropriate. In the region outside the spheres, plane-wave-like forms, $e^{i(\mathbf{k}+\mathbf{K})\cdot\mathbf{r}}$, satisfy both the potential free Schrödinger equation and the Bloch periodicity condition.

An augmented plane wave (APW) is defined according to the prescription

$$\phi_{\mathbf{K}}(\mathbf{k},\mathbf{r}) = e^{i(\mathbf{k}+\mathbf{K})\cdot\mathbf{r}}\theta(r-r_i) + \sum_{lm} a_{\mathbf{K}}^{(lm)}(\mathbf{k})R_l(E,r)Y_{lm}(\mathbf{n_r})\theta(r_i-r), \qquad (13.1)$$

[1] These polyhedra are formed by constructing planes perpendicular to lines connecting the nearest-neighbor atoms of our monoatomic lattice; see section 7.5.

The Physics of Solids. First Edition. J. B. Ketterson.

with the a_{lm} chosen so that ϕ_K is continuous at $r = r_i$; here $\mathbf{n_r}$ is a unit vector parallel to $\mathbf{r}$. Formally, continuity would require an infinite number of the solutions within the spheres; however in practice the expansion is limited to $l \leq l_{max}$ (leading to a small discontinuity in the basis functions), where l_{max} is of order 10.[2] Using the Rayleigh plane wave expansion,

$$e^{i\mathbf{k}\cdot\mathbf{r}} = 4\pi \sum_{l=0}^{\infty} \sum_{m=-l}^{l} i^l j_l(kr) Y^*_{lm}(\mathbf{n_k}) Y_{lm}(\mathbf{n_r}) \tag{13.2}$$

and requiring that the two forms in (13.1) match at $r = r_i$, we obtain

$$a_0^{(lm)}(\mathbf{k}) = 4\pi i^l j_l(kr_i) Y^*_{lm}(\mathbf{n_k})/R_l(E, r_i)$$

and

$$a_{\mathbf{K}}^{(lm)} = a_0^{(lm)}(\mathbf{k} + \mathbf{K}).$$

The Bloch wave may now be expanded in APWs as

$$\psi_{\mathbf{k}n}(\mathbf{r}) = \sum_{\mathbf{K}} c_{\mathbf{K}}(\mathbf{k}, n)\phi_{\mathbf{K}}(\mathbf{k}, \mathbf{r}) \tag{13.3}$$

and our task is then to determine the coefficients c_K for a specified E and $\mathbf{k}$ which most accurately represent the solution to the Schrödinger equation in the prescribed periodic potential. To accomplish this we express the one-electron Schrödinger theory as an *action principle*,[3]

$$S = \int d^3r \mathcal{L}\{\psi(\mathbf{r}), \psi^*(\mathbf{r}), \nabla\psi(\mathbf{r}), \nabla\psi(\mathbf{r})^*, V(\mathbf{r})\} \tag{13.4a}$$

where the Lagrangian density, $\mathcal{L}$, is given by

$$\mathcal{L} = \frac{\hbar^2}{2m}\nabla\psi(\mathbf{r}) \cdot \nabla\psi^*(\mathbf{r}) - (E - V(\mathbf{r}))\,|\psi(\mathbf{r})|^2. \tag{13.4b}$$

According to the action principle, we seek a minimum of S. To do this we perform a variation, δS, of S with respect to $\delta\psi^*$ and set the result equal to zero, which yields (after performing the usual integration by parts (see Appendix A Eq. (A.18))

$$\left\{\int d^3r\left[-\frac{\hbar^2}{2m}\nabla^2\psi(\mathbf{r}) + (V(\mathbf{r}) - E)\,\psi(\mathbf{r})\right] + \frac{\hbar^2}{2m}\int d^2\mathbf{r}\cdot\nabla\psi(\mathbf{r})\right\}\delta\psi^*(\mathbf{r}) = 0. \tag{13.5}$$

The next step is to insert Eq. (13.3) into (13.5), interpreting $\delta\psi^*$ as $\sum_{\mathbf{K}} \phi^*_{\mathbf{K}}\delta c^*_{\mathbf{K}}$. Treating the δc_K as independent variational parameters, Eq. (13.5) then becomes a set of linear homogeneous equations in c_K of the form

$$\sum_{\mathbf{K}} \left[H_{\mathbf{K'K}} - EO_{\mathbf{K'K}} + S_{\mathbf{K'K}}\right] c_{\mathbf{K}} = 0, \tag{13.6}$$

[2] For $r < r_i$ the potential will depart from a spherical form, particularly at larger r_i. To account for this, one can generalize the theory by expressing the angular dependence of V in terms of symmetry adapted combinations of spherical harmonics. Corrections to the potential for $r > r_i$ can be included by incorporating NFE-like Fourier components of the potential; the two corrections are referred to as *non-spherical and non-flat* respectively.

[3] The action principle is discussed in Appendix A.

where we have introduced the Hamiltonian matrix,

$$H_{\mathbf{K}'\mathbf{K}} = \int \phi^*_{\mathbf{K}'}(\mathbf{r}) \left[-\frac{\hbar^2}{2m}\nabla^2 + V(r) \right] \phi_{\mathbf{K}}(\mathbf{r}) d^3r, \tag{13.7a}$$

the "overlap" matrix,[4]

$$O_{\mathbf{K}'\mathbf{K}} = \int \phi^*_{\mathbf{K}'}(\mathbf{r})\phi_{\mathbf{K}}(\mathbf{r}) d^3r, \tag{13.7b}$$

and the matrix arising from the surface contribution,

$$S_{\mathbf{K}'\mathbf{K}} = \frac{\hbar^2}{2m} \int \phi'^*_{\mathbf{K}}(\mathbf{r}) \nabla \phi_{\mathbf{K}}(\mathbf{r}) \cdot d^2\mathbf{r}. \tag{13.7c}$$

The volume integrations in (13.7a,b) extend over the interior of the WS polyhedron and the surface integration in (13.7c) is over the inscribed sphere (of radius r_i). The contribution from the sum of (13.7a) and (13.7b) arising from inside the sphere vanishes since $R_l(r)$ satisfies the radial Schrödinger equation (having been obtained by its direct integration). Since we have fixed $V(r > r_i) = 0$, the remaining contribution becomes

$$\begin{aligned} H_{\mathbf{K}'\mathbf{K}} - EO_{\mathbf{K}'\mathbf{K}} &= \int_{|r|>r_i} d^3r \phi^*_{\mathbf{K}'}(\mathbf{r}) \left[-\frac{\hbar^2}{2m}\nabla^2 - E \right] \phi_{\mathbf{K}}(\mathbf{r}) \\ &= \int_{|r|>r_i} d^3r \left[\frac{\hbar^2}{2m}(\mathbf{k}+\mathbf{K})^2 - E \right] e^{i(\mathbf{K}-\mathbf{K}')\cdot\mathbf{r}} \\ &= \left[\frac{\hbar^2}{2m}(\mathbf{k}+\mathbf{K})^2 - E \right] \left[V_c \delta_{\mathbf{K}'\mathbf{K}} - \int_{r<r_i} e^{i(\mathbf{K}-\mathbf{K}')\cdot\mathbf{r}} d^3r \right]; \end{aligned} \tag{13.8}$$

in the last step, we have exploited the orthogonality of the functions $e^{i\mathbf{K}\cdot\mathbf{r}}$ when integrated over the (entire) unit cell volume, V_c. The integral in (13.8) is given by

$$\int_{r<r_i} e^{i\mathbf{K}\cdot\mathbf{r}} d^3r = 4\pi \int_0^{r_i} \frac{\sin Kr}{Kr} r^2 dr = 4\pi r_i^2 j_1(Kr_i)/K, \tag{13.9}$$

where j_1 is the $l = 1$ spherical Bessel function.

We must now evaluate the matrix element from the surface term. We recall that if the expansion in (l, m) is carried to high enough order, the wave function will be continuous across the sphere. Similarly, if the expansion in reciprocal lattice vectors extends to large enough $|\mathbf{K}|$, the solution will approach the exact solution (to our model potential) and hence the gradient will also be continuous. If the wave function and its slope are both continuous the surface term will vanish. We will assume that the wave function is continuous but, due to the truncation in

[4] Note that APW basis states are in general non-orthogonal.

$|\mathbf{K}|$, the derivative is discontinuous. Noting the gradient has the opposite sign on the two sides of the sphere, we may write our surface term as[5]

$$S_{\mathbf{K}'\mathbf{K}} = \frac{\hbar^2}{2m}\int e^{-i(k+K')\cdot n_r r}\left[\sum_{l,m} a_{\mathbf{K}}^{(lm)}(\mathbf{k})\frac{dR_l(\mathbf{r})}{dr}Y_{lm}(\mathbf{n_r}) - \frac{d}{dr}e^{i(k+K)\cdot n_r r}\right]_{r=r_i} r_i^2 d\Omega. \tag{13.10}$$

Expanding the plane waves with the aid of Eq. (13.2), using the addition theorem of spherical harmonics,

$$P_l\left(\mathbf{n_Q}\cdot\mathbf{n_{Q'}}\right) = \frac{4\pi}{2l+1}\sum_{m=-l}^{l} Y_{lm}^*(\mathbf{n_{Q'}})Y_{lm}(\mathbf{n_Q}) \tag{13.11}$$

and noting the orthogonality properties of spherical harmonics, (13.10) becomes

$$\begin{aligned} S_{\mathbf{K}'\mathbf{K}} &= \frac{\hbar^2}{2m}4\pi r_i^2\sum_l (2l+1)P_l\left(\mathbf{n_{Q'}}\cdot\mathbf{n_Q}\right)j_l(Q'r_i)j_l(Qr_i) \\ &\times\left[\frac{1}{R_l(r)}\frac{dR_l(r)}{dr} - \frac{1}{j_l(Qr)}\frac{dj_l(Qr)}{\partial r}\right]_{r=r_i}, \end{aligned} \tag{13.12}$$

where $\mathbf{Q} = \mathbf{K} + \mathbf{k}$ and $\mathbf{Q}' = \mathbf{K}' + \mathbf{k}$. This expression can be simplified using the identity (Abramowitz and Stegun 10.1.45)

$$\frac{\sin Rr_i}{Rr_i} = j_0(Rr_i) = \sum_{l=0}^{\infty}(2l+1)j_l(Qr_i)j_l(Q'r_i)P_l(\mathbf{n_Q}\cdot\mathbf{n_{Q'}}), \tag{13.13a}$$

where $R^2 = Q^2+Q'^2-2\mathbf{Q}\cdot\mathbf{Q}'\cos\left(\mathbf{n_Q}\cdot\mathbf{n_{Q'}}\right) = |\mathbf{Q}-\mathbf{Q}'|^2$. Differentiating (13.13a) with respect to Q we have

$$\begin{aligned} \frac{\partial j_0(Rr_i)}{\partial Q} &= -\frac{r_i}{R}\left(Q - Q'\cos\left(\mathbf{n_Q}\cdot\mathbf{n_{Q'}}\right)\right)j_1(Rr_i) \\ &= \sum_{l=0}^{\infty}(2l+1)j_l(Q'r_i)\frac{\partial j_l(Qr_i)}{\partial Q}P_l(\mathbf{n_Q}\cdot\mathbf{n_{Q'}}) \\ &= \sum_{\ell=0}^{\infty}(2l+1)j_l(Q'r_i)\frac{r_i}{Q}\frac{\partial j_\ell(Qr_i)}{\partial r}P_l(\mathbf{n_Q}\cdot\mathbf{n_{Q'}}). \end{aligned} \tag{13.13b}$$

Substituting this expression into (13.12) we have

$$\begin{aligned} S_{\mathbf{K}'\mathbf{K}} = \frac{\hbar^2}{2m}4\pi r_i^2 &\left[\sum_l (2l+1)P_l(\mathbf{n_Q}\cdot\mathbf{n_{Q'}})j_l(Q'r_i)j_l(Qr_i)\left(\frac{1}{R_l(r)}\frac{\partial R_l(r)}{\partial r}\right)_{r=r_i}\right. \\ &\left.+\frac{Q^2-\mathbf{Q}\cdot\mathbf{Q}'}{|\mathbf{Q}-\mathbf{Q}'|}j_1(|\mathbf{Q}-\mathbf{Q}'|r_i)\right]. \end{aligned} \tag{13.14}$$

[5] Note there is no contribution from the surface of the WS polyhedron since the wave function forms chosen outside the sphere automatically make this term vanish.

Combining (13.8) and (13.10), the complete expression for the APW matrix is (J.C. Slater (1937))

$$M_{\mathbf{K}'\mathbf{K}} = H_{\mathbf{K}'\mathbf{K}} - EO_{\mathbf{K}'\mathbf{K}} + S_{\mathbf{K}'\mathbf{K}}$$

$$= \left(\frac{\hbar^2}{2m}Q^2 - E\right) V_c \delta_{\mathbf{K}'\mathbf{K}} - 4\pi r_i^2 \left\{ \left(\frac{\hbar^2}{2m}\mathbf{Q}' \cdot \mathbf{Q} - E\right) \frac{j_1(|\mathbf{Q}' - \mathbf{Q}| r_i)}{|\mathbf{Q}' - \mathbf{Q}|} \right.$$

$$\left. - \frac{\hbar^2}{2m} \sum_l (2l+1) P_l(\hat{\mathbf{Q}}' \cdot \hat{\mathbf{Q}}) j_l(Q' r_i) j_l(Q r_i) L_l(r_i) \right\}, \tag{13.15a}$$

where L_l is the logarithmic derivative of the radial wave function evaluated at the muffin tin radius, r_i,

$$L_l = \frac{1}{R_l(r)} \frac{dR_l(r)}{dr}\bigg|_{r=r_i}. \tag{13.15b}$$

The condition that the system of equations

$$\sum_{\mathbf{K}} M_{\mathbf{K}'\mathbf{K}} c_{\mathbf{K}} = 0 \tag{13.16a}$$

has a solution is that

$$\mathrm{Det}|M_{\mathbf{K}'\mathbf{K}}(E, \mathbf{k})| = 0; \tag{13.16b}$$

in practice one may select an energy E, and direction $\hat{\mathbf{k}}$ and vary k until the determinant passes through zero; if no zero is found it simply means that no states exist for the chosen E and $\hat{\mathbf{k}}$. Once a solution has been found, nearby solutions may be found by regarding (13.16a) as an eigenvalue problem with eigenvalue zero (Shaw et al. (1970)),

$$\sum_{\mathbf{K}} (M_{\mathbf{K}'\mathbf{K}} - \lambda) c_{\mathbf{K}} = 0. \tag{13.17}$$

The change in λ for a change in $\mathbf{k}$ at fixed E is, using the Hellman–Feynmann theorem,

$$\left(\frac{\partial \lambda}{\partial \mathbf{k}}\right)_E = \vec{C}^* \cdot \left(\frac{\partial \overleftrightarrow{M}}{\partial \mathbf{k}}\right) \cdot \vec{C}, \tag{13.18a}$$

where $\vec{C}$ is the normalized eigenvector of the matrix $\overleftrightarrow{M}$ when $\lambda = 0$. Similarly the energy derivative of λ at fixed $\mathbf{k}$ is

$$\left(\frac{\partial \lambda}{\partial E}\right)_{\mathbf{k}} = \vec{C}^* \cdot \left(\frac{\partial \overleftrightarrow{M}}{\partial E}\right) \cdot \vec{C}. \tag{13.18b}$$

One may use these expressions to locate a new root on changing E or **k** by a small amount δE or δ**k**,

$$\lambda(E + dE) = \lambda(E) + \left(\frac{\partial \lambda}{\partial E}\right)_{\mathbf{k}} \delta E \tag{13.19a}$$

and

$$\lambda(\mathbf{k} + d\mathbf{k}) = \lambda(\mathbf{k}) + \left(\frac{\partial \lambda}{\partial \mathbf{k}}\right)_{E} \cdot \delta \mathbf{k}. \tag{13.19b}$$

Hence if we are at some value of E or **k** where λ is small but non-zero, we may change E or **k** by an amount δE or δ**k** which causes λ to vanish by solving the equation

$$\lambda + \left(\frac{\partial \lambda}{\partial E}\right)_{\mathbf{k}} \delta E = 0 \tag{13.20a}$$

or

$$\lambda + \left(\frac{\partial \lambda}{\partial \mathbf{k}}\right)_{E} \cdot \delta \mathbf{k} = 0. \tag{13.20b}$$

Finally, we may find the velocity at a point where λ = 0 by using the rules for the differentiation of implicit functions,

$$\left(\frac{\partial E}{\partial \mathbf{k}}\right)_{\lambda} = -\left(\frac{\partial \lambda}{\partial \mathbf{k}}\right)_{E} \Big/ \left(\frac{\partial \lambda}{\partial E}\right)_{\mathbf{k}}. \tag{13.21}$$

Problems arise in the application of the APW method for energies where the radial wave function has a zero at the muffin tin radius, because then the logarithmic derivative diverges (referred to as an asymptote). The term involving L_l in $M_{\mathbf{KK}'}$ then dominates, which has the effect of "uncoupling" the wave functions inside and outside the muffin tin, lowering the accuracy of the variational solution.

The band structures of the transition metals platinum[6] and niobium[7] are shown in Figs. 13.1 and 13.2; these metals have the fcc and bcc structures, respectively. The resulting Fermi surfaces will be discussed in Section 20.1.3. Both calculations employ the relativistic generalization of the augmented plane wave method (the RAPW method), which we will not discuss but which is founded on the Dirac rather than the Schrödinger equation. Certain states which are degenerate in a non-relativistic calculation are split by the spin–orbit interaction, as we discussed qualitatively in Chapter 11. We will have need of both of these band structures when we discuss their Fermi surfaces later, in Section 20.1.

13.2 The Green's function method

The last technique we will discuss for calculating wave functions and energy levels in solids is the so-called Green's function method; it is also called the scattered wave or Korringa–Kohn–Rostoker (KKR) method (J. Korringa (1947) and

[6] Fermi surfaces and effective masses in fcc transition metals, O.K. Andersen and A.R. Mackintosh, *Solid State Commun.* **6**, 285 (1968); Electronic structure of fcc transition metals Ir, Rh, Pt, and Pd, O.K. Andersen, *Phys. Rev. B* **2**, 883 (1970).

[7] Effect of non-muffin-tin terms on electronic-structure of transition-metals – niobium, N. Elyashar and D. D. Koelling, *Phys. Rev. B* **13**, 5362 (1976); *Phys. Rev. B* **15**, 3620 (1977).

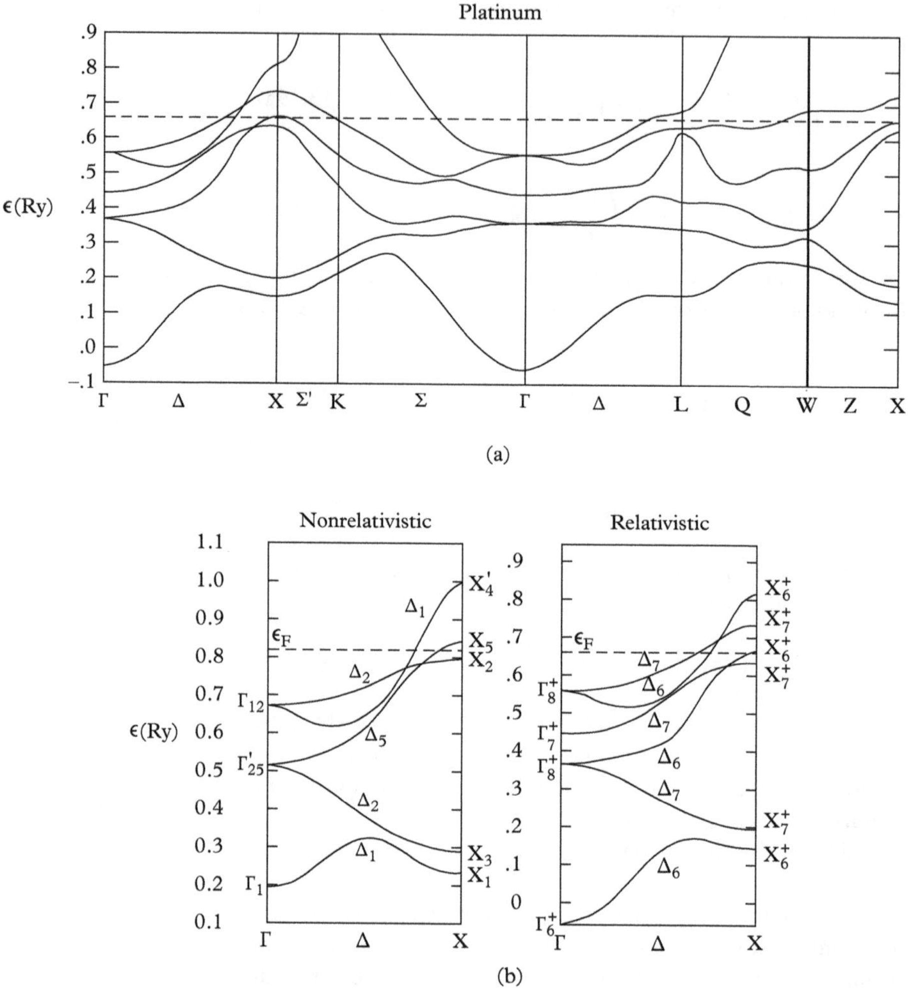

Figure 13.1 *The band structure of platinum. (a) Relativistic energy bands along the principal directions in the Brillouin zone. (b) The non-relativistic and relativistic energy bands along the [100] direction. Note that the non-relativistic Γ'_{25} level splits into a Γ_8^+ and a Γ_7^+ level. (Andersen and Mackintosh 1968.)*

W. Kohn and N. Rostoker (1954)).[8] The method also makes use of a variational principle. However, rather than using the variational principle employed for the APW method, which was based on the Schrödinger differential equation, we will employ a different form in which the Schrödinger equation is recast as an integral equation involving the empty lattice Green's function.

[8] On the calculation of the energy of a Bloch wave in a metal, *J. Korringa, Phys.*, **13**, 392 (1947); Solution of the Schrödinger equation in periodic lattices with an application to metallic lithium, W. Kohn and N. Rostoker, *Phys. Rev.* **94**, 1111 (1954).

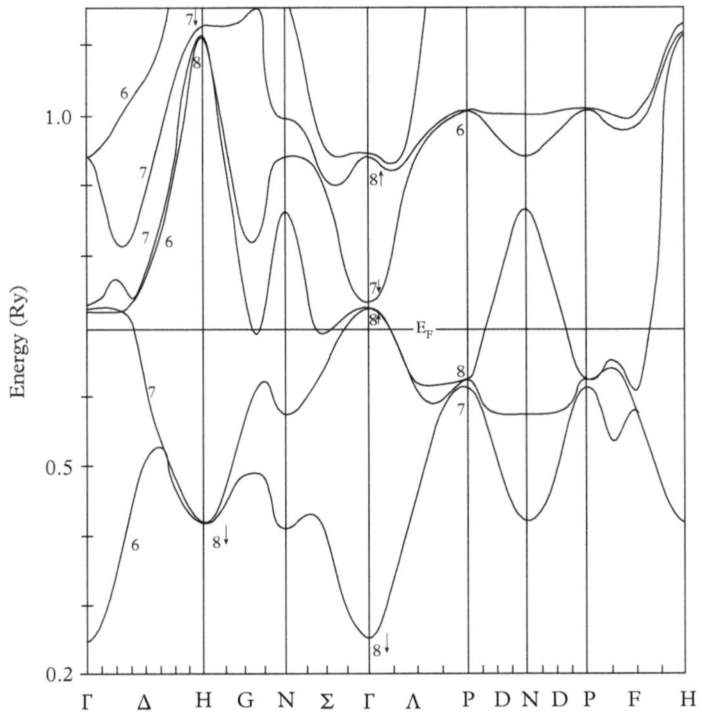

Figure 13.2 *The band structure of Nb along symmetry lines of the bcc lattice calculated with the relativistic augmented plane wave method. (Elyashar and Koelling (1977).)*

We begin by recalling the wave function of a plane wave normalized to a box of volume L^3

$$\psi_{\mathbf{k}}(\mathbf{r}) = \frac{1}{L^{3/2}} e^{i\mathbf{k}\cdot\mathbf{r}}.$$

From the completeness relation of quantum mechanics,

$$\sum_{\mathbf{k}} \psi^*_{\mathbf{k}}(\mathbf{r}')\psi_{\mathbf{k}}(\mathbf{r}) = \delta^{(3)}(\mathbf{r}-\mathbf{r}'),$$

we have

$$\frac{1}{L^3}\sum_{\mathbf{k}} e^{i\mathbf{k}\cdot(\mathbf{r}-\mathbf{r}')} = \delta^{(3)}(\mathbf{r}-\mathbf{r}'). \tag{13.22}$$

Now the usual prescription, $\sum_{\mathbf{k}} = \frac{L^3}{(2\pi)^3}\int d^3k$, of free space becomes $\frac{L^3}{(2\pi)^3}\int d^3k \sum_{\mathbf{K}}$ for a cell periodic empty lattice if we set $\mathbf{k} \to \mathbf{k} + \mathbf{K}$, where $\mathbf{K}$ is the set of all reciprocal lattice vectors and we restrict the integral to the Brillouin zone. The number of reciprocal lattice vectors is in one-to-one correspondence with the number of real space unit cells, N. Thus for $\mathbf{r} = \mathbf{r}'$, the sum (13.22) is $(N/(2\pi)^3) \int_{B.Z.} d^3k = N/V_c$, where V_c is the unit cell volume. Therefore,

$$\frac{1}{V_c}\sum_{K} e^{i(\mathbf{k}+\mathbf{K})\cdot(\mathbf{r}-\mathbf{r}')} = \delta^{(3)}(\mathbf{r}-\mathbf{r}'). \tag{13.23}$$

We define a lattice Green's function as the solution to

$$\left(-\frac{\hbar^2}{2m}\nabla^2 - E\right) G(\mathbf{r},\mathbf{r}') = \delta^{(3)}(\mathbf{r}-\mathbf{r}'), \tag{13.24}$$

where we subject $G(\mathbf{r},\mathbf{r}')$ to the Bloch periodic boundary condition,

$$G(\mathbf{r},\mathbf{r}'+\mathbf{R}) = e^{i\mathbf{k}\cdot\mathbf{R}} G(\mathbf{r},\mathbf{r}'). \tag{13.25}$$

In this way all the symmetry of the empty lattice is built into the Green's function. From Eqs. (13.23) and (13.24) it is clear that our lattice Green's function is given by

$$G(\mathbf{r},\mathbf{r}') = -\frac{1}{V_c}\sum_{K} \frac{e^{i(\mathbf{k}+\mathbf{K})\cdot(\mathbf{r}-\mathbf{r}')}}{\frac{\hbar^2}{2m}(\mathbf{k}+\mathbf{K})^2 - E}, \tag{13.26}$$

as can be verified by direct substitution into (13.24). From (13.26) it can be seen that

$$G(\mathbf{r},\mathbf{r}') = G^*(\mathbf{r}',\mathbf{r}); \tag{13.27}$$

i.e., G is Hermitian. An alternative form of the empty lattice Green's function is obtained by recalling the usual free space Green's function for the Helmholz equation,

$$(-\nabla^2 + k^2)\, G(\mathbf{r},\mathbf{r}') = \delta^{(3)}(\mathbf{r}-\mathbf{r}'),$$

which is given by

$$G(\mathbf{r},\mathbf{r}';k) = \frac{1}{4\pi}\frac{e^{ik|\mathbf{r}-\mathbf{r}'|}}{|\mathbf{r}-\mathbf{r}'|}.$$

This form is made cell periodic by replacing $\mathbf{r}'$ by $\mathbf{r}'+\mathbf{R}$, multiplying by the Bloch phase factor $e^{i\mathbf{k}\cdot\mathbf{R}}$, and summing over all $\mathbf{R}$:

$$G(\mathbf{r},\mathbf{r}') = \frac{1}{4\pi}\sum_{\mathbf{R}} e^{i\mathbf{k}\cdot\mathbf{R}} \frac{e^{i\kappa|\mathbf{r}-\mathbf{r}'-\mathbf{R}|}}{|\mathbf{r}-\mathbf{r}'-\mathbf{R}|}, \tag{13.28}$$

as may be verified by substitution into (13.24) and (13.25); to allow for the cases $E > 0$ and $E < 0$, we introduce κ, defined by

$$\kappa = +\sqrt{2mE/\hbar^2}, \quad E > 0 \tag{13.29a}$$

or

$$\kappa = +i\sqrt{-2mE/\hbar^2}, \quad E < 0. \tag{13.29b}$$

By regarding $V(\mathbf{r})\psi(\mathbf{r})$ as a "source strength" and using the superposition property, we may write the solution to the Schrödinger equation for a potential $V(\mathbf{r})$,

$$\left[-\frac{\hbar^2}{2m}\nabla^2 + V(\mathbf{r})\right]\psi(\mathbf{r}) = E\psi(\mathbf{r}),$$

as a homogeneous integral equation,

$$\psi(\mathbf{r}) = \int_{V_c} G(\mathbf{r},\mathbf{r}')\psi(\mathbf{r}')d^3r'; \tag{13.30}$$

any solution to (13.30) will satisfy the Bloch and crystal symmetry conditions since they are built into $G(\mathbf{r},\mathbf{r}')$. This integral equation is equivalent to the variational principle,

$$\delta\Lambda = 0, \tag{13.31}$$

where the Lagrangian (see Appendix A) is now given by

$$\Lambda = \int_{V_c} \psi^*(\mathbf{r})V(\mathbf{r})\psi(\mathbf{r})d^3r - \int_{V_c}\int_{V'_c} \psi^*(\mathbf{r})V(\mathbf{r})G(\mathbf{r},\mathbf{r}')V(\mathbf{r}')\psi(\mathbf{r}')d^3r d^3r'. \tag{13.32}$$

We may approximate ψ in the form

$$\psi(\mathbf{r}) = \sum_{i=1}^{N} c_i\phi_i(\mathbf{r}), \tag{13.33}$$

where the $\phi_i(\mathbf{r})$ are a set of N functions chosen on the basis that they are expected to provide a reasonably accurate representation of the true wave function. These basis functions are not required to satisfy crystal symmetry conditions, although clearly it is desirable if they do. We may then form the Hermitian "matrix elements,"

$$\Lambda_{ij} = \int_{V_c} \phi_i^*(\mathbf{r})V(\mathbf{r})\phi_j(\mathbf{r})d^3r - \int_{V_c}\int_{V'_c} \phi_i^*(\mathbf{r})V(\mathbf{r})G(\mathbf{r},\mathbf{r}')V(\mathbf{r}')\phi_j(\mathbf{r}')d^3r d^3r' \tag{13.34}$$

and write Λ as

$$\Lambda = \sum_{i=1}^{N}\sum_{j=1}^{N} c_i^* \Lambda_{ij} c_j. \qquad (13.35)$$

Minimization involves setting $\partial\Lambda/\partial c_i^* = 0$, from which we obtain the set of homogeneous equations,

$$\sum_{j=1}^{N} \Lambda_{ij} c_j = 0, \qquad (13.36)$$

which has a non-trivial solution only if

$$\mathrm{Det}|\Lambda_{ij}(E, \mathbf{k})| = 0. \qquad (13.37)$$

As with the APW method, we vary E for a fixed **k** (or vice versa) until the determinant passes through zero. Our formulation leads to only a second-order error in E or **k**.

We now restrict the discussion to potentials of the muffin tin form, introduced in the discussion of the APW method. We again choose the arbitrary additive constant of the potential so that $V = 0$ outside the muffin tin sphere having radius r_i. The volume integrals in (13.32) are then restricted to $r, r' < r_i$, and the form the wave function $\psi(\mathbf{r})$ takes for $r > r_i$ is not involved; as with the APW method, the appropriate form for the wave function for $r < r_i$ is

$$\psi(\mathbf{r}) = \sum_{l,m} c_{lm} R_{lm}(r) Y_{lm}(\mathbf{n_r}), \qquad (13.38)$$

where $R_{\ell m}(r)$ is obtained by numerically integrating the radial Schrödinger equation,

$$\left\{ -\frac{\hbar^2}{2m}\left[\frac{1}{r^2}\frac{d}{dr}r^2\frac{d}{dr} - \frac{l(l+1)}{r^2}\right] + V(r) - E \right\} R_{lm}(r) = 0, \qquad (13.39)$$

for the chosen value of E, where we require $R_l(0)$ to be finite (the normalization is arbitrary when determining $E(\mathbf{k})$). Our Lagrangian then takes the form

$$\Lambda = \frac{\hbar^2}{2m}\int_{r<r_i-2(0^+)} d^3r\psi^*(r)V(\mathbf{r})\left[\psi(\mathbf{r}) - \int_{r'<r_i-0^+} d^3r' G(\mathbf{r},\mathbf{r}')V(\mathbf{r}')\psi(\mathbf{r}')\right], \qquad (13.40)$$

where the infinitesimals introduced in the restricted region of integration are required to avoid problems with singularities for $r = r_i$. Using Schrödinger's equation to replace $V(r')\psi(\mathbf{r}')$ by $\left((\hbar^2/2m)\nabla^2 + E\right)\psi(r')$, integrating twice by parts, and using (13.24), Eq. (13.40) becomes

$$\Lambda = \left(\frac{\hbar^2}{2m}\right)^2 \times \int_{r<r_i-2(0^+)} \int r_i^2 d\Omega' \psi^*(\mathbf{r})V(\mathbf{r})\left[\psi(\mathbf{r}')\frac{\partial}{\partial r'}G(\mathbf{r},\mathbf{r}') - G(\mathbf{r},\mathbf{r}')\frac{\partial}{\partial r'}\psi(\mathbf{r}')\right]\Bigg|_{r'=r_i-0^+} . \tag{13.41}$$

Using the Schrödinger equation to now eliminate $\psi^*(\mathbf{r})V(\mathbf{r})$ and again integrating twice by parts then gives

$$\Lambda = r_i^4 \int d\Omega \times \int d\Omega' \left[\frac{\partial \psi^*(\mathbf{r})}{\partial r} - \psi^*(r)\frac{\partial}{\partial r}\right]\left[\psi(\mathbf{r}')\frac{\partial}{\partial r'}G(\mathbf{r},\mathbf{r}') - G(\mathbf{r},\mathbf{r}')\frac{\partial}{\partial r'}\psi(\mathbf{r}')\right]\Bigg|_{\substack{r=r_i-2(0^+)\\ r'=r_i-0^+}} . \tag{13.42}$$

As with the APW approach, the properties of the wave function at the sphere boundary govern the band structure.

To proceed further we require an expansion of our lattice Green's function, $G(\mathbf{r},\mathbf{r}')$, for $r, r' < r_i$ in terms of solutions to the potential-free Schrödinger equation in spherical coordinates. We write G as the sum of a part which is singular as $\mathbf{r} \to \mathbf{r}'$, $G_s(\mathbf{r},\mathbf{r}')$, and a part $G_n(\mathbf{r},\mathbf{r}')$, which is non-singular in that limit,

$$G(\mathbf{r},\mathbf{r}') = G_s(\mathbf{r},\mathbf{r}') + G_n(\mathbf{r},\mathbf{r}'). \tag{13.43}$$

G_n, in addition to including the non-singular part of G associated with the center of the unit cell chosen as the origin of $\mathbf{r}$ and $\mathbf{r}'$, also includes the additional contributions imposed by the crystal symmetry (i.e., multiple scattering effects). We first extract G_s. Writing the single cell Green's function as

$$\frac{1}{4\pi}\frac{e^{i\kappa|\mathbf{r}-\mathbf{r}'|}}{|\mathbf{r}-\mathbf{r}'|} = \frac{1}{4\pi|\mathbf{r}-\mathbf{r}'|}\left[\cos(\kappa|\mathbf{r}-\mathbf{r}'|) + i\sin(\kappa|\mathbf{r}-\mathbf{r}'|)\right]$$

and noting the singular part arises from the first term in square brackets, we have

$$G_s(\mathbf{r},\mathbf{r}') = \frac{1}{4\pi}\frac{\cos(\kappa|\mathbf{r}-\mathbf{r}'|)}{|\mathbf{r}-\mathbf{r}'|}. \tag{13.44}$$

Using an expansion analogous to (13.13) (Abramowitz and Stegun 10.1.46),

$$\begin{aligned} G_s(\mathbf{r},\mathbf{r}') &= -\frac{\kappa}{4\pi}\sum_{l=0}^{\infty}(2l+1)j_l(\kappa r)n_l(\kappa r')P_l(\mathbf{n_r}\cdot\mathbf{n_{r'}}) && (r<r') \\ &= -\kappa\sum_{l,m=0}^{\infty} j_l(\kappa r)n_l(\kappa r')Y^*_{lm}(\mathbf{n_r})Y_{lm}(\mathbf{n_{r'}}) && (r<r') \end{aligned} , \tag{13.45}$$

where we have used the addition theorem of spherical harmonics, Eq. (13.11), in the second step; here j_l and n_l are the spherical Bessel functions of the first and second kind, respectively, and for the case $r > r'$ we interchange $\mathbf{r}$ and $\mathbf{r}'$ in (13.45). The non-singular part of the Green's function, G_n, must be a solution of the homogeneous wave equation (in both $\mathbf{r}$ and $\mathbf{r}'$); it may therefore be written as a "double" expansion of the form

$$G_n(\mathbf{r},\mathbf{r}') = \sum_{l,m}\sum_{l',m'} A_{l,m;l',m'}(\kappa,\mathbf{k}) j_l(\kappa r) j_{l'}(\kappa r') Y_{lm}(\mathbf{n}_\mathbf{r}) Y^*_{l'm'}(\mathbf{n}_{\mathbf{r}'}), \tag{13.46}$$

where hermeticity requires

$$A_{l,m;l',m'} = A^*_{l',m';l,m}. \tag{13.47}$$

The full expansion of G therefore has the form

$$G(\mathbf{r},\mathbf{r}') = \sum_{l,m}\sum_{l',m'} \left[A_{l,m;l',m'} j_l(\kappa r) j_{l'}(\kappa r') - \kappa\delta_{ll'}\delta_{mm'} j_l(\kappa r) n_l(\kappa r')\right] Y_{lm}(\mathbf{n}_\mathbf{r}) Y_{l'm'}(\mathbf{n}_{\mathbf{r}'}). \tag{13.48}$$

For the case where $G(\mathbf{r},\mathbf{r}')$ is represented in the form (13.26), we obtain the coefficients $A_{l,m;l',m'}$ by expanding the plane waves $e^{i\mathbf{Q}\cdot\mathbf{r}}$ and $e^{-i\mathbf{Q}\cdot\mathbf{r}'}$ occurring in this expression (where we again write $\mathbf{Q} = \mathbf{K} + \mathbf{k}$) using Eq. (13.2) and associating them with the corresponding terms in (13.48); the result is

$$A_{lm;l'm'} = -\frac{(4\pi)^2 i^{l-l'}}{V_c j_l(\kappa r) j_{l'}(\kappa r')} \sum_{\mathbf{Q}} \frac{j_l(Qr) j_{l'}(Qr') Y_{lm}(\mathbf{n}_\mathbf{Q}) Y^*_{l'm'}(\mathbf{n}_\mathbf{Q})}{\frac{\hbar^2}{2m}Q^2 - E} - \kappa\delta_{ll'}\delta_{mm'}\frac{n_l(\kappa r')}{j_l(\kappa r')}; \qquad r < r' < r_i. \tag{13.49}$$

In spite of its appearance, (13.49) is actually independent of $\mathbf{r}$ and $\mathbf{r}'$ (as it must be since it is an expression for the expansion coefficients); however it is a function of $\mathbf{k}$ and E.

By using the fact that $G(\mathbf{r},\mathbf{r}') = G(|\mathbf{x}|)$ where $\mathbf{x} \equiv \mathbf{r} - \mathbf{r}'$, we may show that the coefficients $A_{lm;l'm'}$ are not all independent,[9]

$$G(\mathbf{x}) = -\frac{1}{V_c}\sum_{\mathbf{Q}} \frac{e^{i\mathbf{Q}\cdot\mathbf{x}}}{\frac{\hbar^2}{2m}Q^2 - E} \tag{13.50}$$

and using expressions given above we have

$$G(\mathbf{x}) = -\frac{1}{4\pi}\frac{\cos\kappa x}{x} + \sum_{L,M} D_{LM} j_L(\kappa x) Y_{LM}(\mathbf{n}_\mathbf{x}), \tag{13.51}$$

[9] Since $G(\mathbf{r},\mathbf{r}')$ is an "empty lattice" Green's function, it must still be a function of only $\mathbf{x} = \mathbf{r} - \mathbf{r}'$.

where the D_{LM} are given by

$$D_{LM} = -\frac{4\pi i^L}{V_c j_L(\kappa x)} \sum_Q \frac{j_L(Qx) Y^*_{LM}(\mathbf{n}_Q)}{\frac{\hbar^2}{2m}Q^2 - E} + \frac{\kappa}{4\pi}\delta_{LO}\delta_{MO}\cot(\kappa x), \quad (13.52)$$

with

$$D_{L-M} = (-1)^L D^*_{LM}.$$

The series in (13.51) converges slowly. However, if we write the sum occurring in (13.52) as

$$S(\eta) = \sum_Q \frac{j_L(Qx) Y^*_{LM}(\mathbf{n}_Q)}{\frac{\hbar^2}{2m}Q^2 - E} e^{(E-\frac{\hbar^2}{2m}Q^2)/\eta}, \quad (13.53)$$

Ewald has shown that $S(\eta) - S(\infty)$ approaches zero exponentially as $\eta \to \infty$; in practice the value of η required to achieve adequate convergence turns out to be rather small.[10] In actual calculations, we may evaluate the approximate but more rapidly convergent series $S(\eta)$ rather than the exact, slowly convergent series, $S(\infty)$.

Comparing (13.51) with (13.48) we obtain a relation between the A and D coefficients,

$$\sum_{L,M} D_{LM} j_L(\kappa x) Y_{LM}(\mathbf{n}_x) = \sum_{l,m}\sum_{l',m'} A_{l,m;l',m'} j_l(\kappa r) j_{l'}(\kappa r') Y_{lm}(\mathbf{n}_r) Y^*_{l'm'}(\mathbf{n}_{r'}). \quad (13.54)$$

Using the plane wave expansion (13.2) for $e^{i\kappa\cdot x}$, multiplying by $Y_{LM}(\mathbf{n}_\kappa)$, and integrating over $d\Omega_{\mathbf{n}_\kappa}$, we obtain

$$\begin{aligned} j_L(\kappa x) Y_{LM}(\mathbf{n}_x) &= \frac{1}{4\pi i^L}\int d\Omega_{\mathbf{n}_\kappa} e^{i\kappa\cdot x} Y_{LM}(\mathbf{n}_\kappa) \\ &= \frac{1}{4\pi i^L}\int d\Omega_{\mathbf{n}_\kappa} e^{-i\kappa\cdot r'} Y_{LM}(\mathbf{n}_\kappa). \end{aligned}$$

Using (13.2) to expand $e^{i\kappa\cdot r}$ and $e^{i\kappa\cdot r'}$, this expression becomes

$$\begin{aligned} j_L(\kappa x) Y_{LM}(\mathbf{n}_x) &= \frac{(4\pi)^2}{4\pi i^L}\sum_{lm}\sum_{l'm'} i^{l-l'} j_l(\kappa r) j_{l'}(\kappa r') Y_{lm}(\mathbf{n}_r) Y^*_{l'm'}(\mathbf{n}_{r'}) \\ &\quad \times \int d\Omega_{\mathbf{n}_\kappa} Y^*_{lm}(\mathbf{n}_\kappa) Y_{l'm'}(\mathbf{n}_\kappa) Y_{LM}(\mathbf{n}_\kappa) \\ &= \frac{4\pi}{i^L}\sum_{lm}\sum_{l'm'} i^{l-l'} j_l(\kappa r) j_{l'}(\kappa r') Y_{lm}(\mathbf{n}_r) Y^*_{l'm'}(\mathbf{n}_{r'}) C_{LM;lm;l'm'}, \end{aligned} \quad (13.55)$$

[10] The Ewald method is discussed in Section 9A.

where the $C_{LM;\,lm;\,l'm'}$ are Clebsh–Gordon coefficients defined by

$$C_{LM;\,lm;\,l'm'} = \int d\Omega_{\mathbf{n}_\kappa} Y^*_{lm}(\mathbf{n}_\kappa) Y_{l'm'}(\mathbf{n}_\kappa) Y_{LM}(\mathbf{n}_\kappa). \tag{13.56}$$

Substituting (13.55) into (13.54) gives the relation

$$A_{lm;l'm'} = 4\pi i^{l-l'} \sum_L \frac{1}{i^L} D_{LM} C_{LM;lm;l'm'}, \tag{13.57}$$

where, from (13.56), the coefficients, C, vanish unless $M = m - m'$; the sum on L runs from $l + l'$ to $|l - l'|$. Equations (13.48), (13.51), (13.56), and (13.57) combine to give the Green's function $G(\mathbf{r}, \mathbf{r}')$ in the desired form.

We return to the evaluation of Λ; substituting our trial function (13.38) and our Green's function in the form (13.48) into (13.42), and taking the indicated limits we obtain

$$\Lambda_{lm;l'm'} = (L_l j_l - j'_l) \left[(A_{lm;l'm'} j'_{l'} + \kappa \delta_{ll'} \delta_{mm'} n'_{l'}) - (A_{lm;l'm'} j_{l'} + \kappa \delta_{ll'} \delta_{mm'} n_{l'}) L_{l'} \right], \tag{13.58}$$

where

$$L_l = \frac{1}{R_{ll}(r)} \frac{dR_l(r)}{dr} \Big|_{r=r_i}, \tag{13.59a}$$

$$j_l = j_l(\kappa r_i), \tag{13.59b}$$

$$j'_l = \frac{d}{dr} j_l(\kappa r) \Big|_{r=r_i}, \tag{13.59c}$$

and similarly for n'_l. The condition for a minimum is

$$\mathrm{Det}|\Lambda_{lm;l'm'}| = 0, \tag{13.60}$$

where $lm;l'm'$ are arranged respectively as the rows, i, and columns, j, of a matrix. Dividing each row by $L_l j_l - j'_l$ and each column by $L_l j_{l'} - j'_{l'}$ yields the secular equation

$$\mathrm{Det}|M_{ij}| = 0, \tag{13.61a}$$

where

$$M_{ij} = A_{lm;l'm'} + \kappa \delta_{ll'} \delta_{mm'} \frac{n'_l - n_l L_l}{j'_l - j_l L_l}. \tag{13.61b}$$

We may introduce the scattering phase shifts, δ_l, according to the usual definition

$$\tan \delta_l = \frac{j'_l - j_l L_l}{n'_l - n_l L_l} \tag{13.62}$$

and rewrite (13.61) as

$$\mathrm{Det}\left|A_{lm;l'm'}(\kappa,\mathbf{k}) + \kappa\delta_{ll'}\delta_{mm'}\cot\delta_l(\kappa)\right| = 0. \qquad (13.63)$$

The procedure for finding a solution to this set of equations is the same as with the APW: one selects a **k** and varies E, or selects an E and a direction of **k** and varies its magnitude, until the determinant passes through zero. The method of taking derivatives discussed at the end of Section 13.5 can also be applied to Eq. (13.63).

In evaluating the structure constants using Eq. (13.49), care is required for energies in the vicinity of where a denominator vanishes; i.e., where $E = (\hbar^2/2m)(\mathbf{k}+\mathbf{K})^2$ (referred to as a free electron singularity). In addition the eigenvalues of (13.83) (treated as a matrix) *reorder* on crossing such a singularity.

Figure 13.3 shows the energy bands of copper, as calculated by the (non-relativistic) Green's function method.[11] Note that along Γ–X (the Δ-line) and

[11] Fermi surface and energy bands of copper, B. Segall, *Phys. Rev.* **125**, 109 (1962).

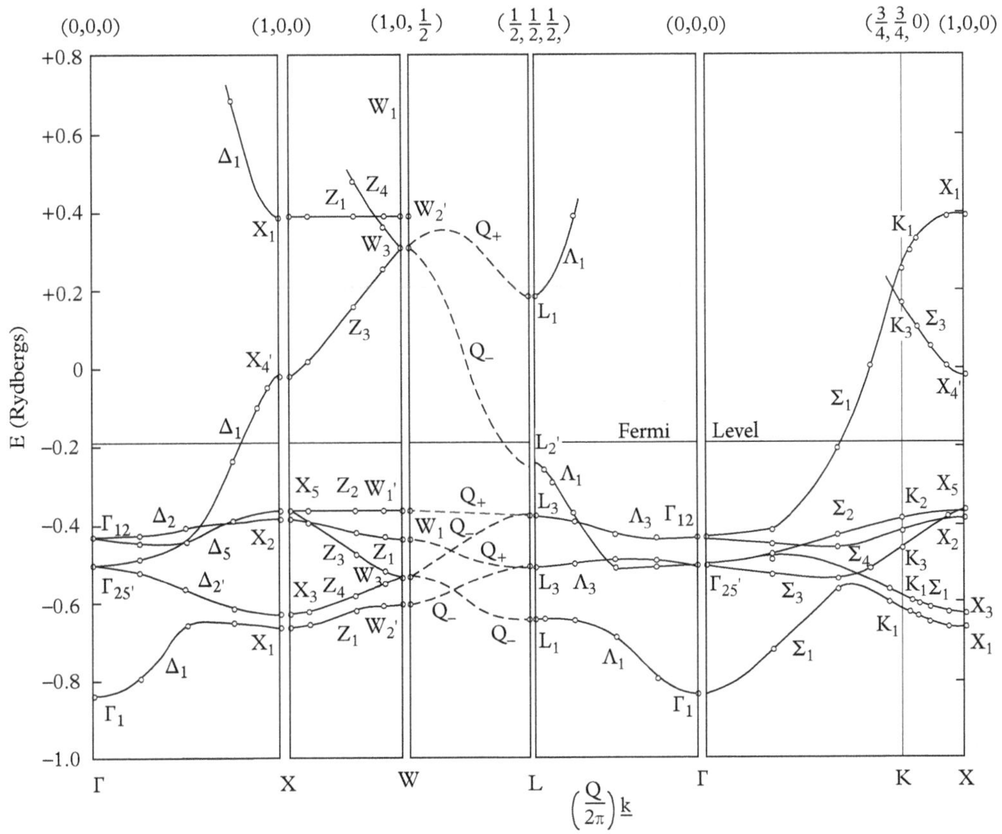

Figure 13.3 *The band structure of copper along the principal symmetry axes of the fcc Brillouin zone, as calculated by the Green's function method. (After Segall (1962).)*

Γ–K (the Σ-line) the Fermi level intersects these levels at approximately the same k-value; however along Γ–L (the Λ-line) no intersection occurs. This results in an approximately spherical Fermi surface for Cu, except for connections to the surfaces in adjacent zones along the [111] directions, which are referred to as *necks*. The volume of this surface corresponds to one electron per unit cell, in accordance with the valence +1 of Cu. We will return to the Fermi surface of copper in Section 20.1.3.

Both the APW and Green's function methods may be used to accurately parameterize the shapes of Fermi surfaces by introducing $L_l(E_F)$ or $\delta_l(E_F)$as disposable parameters (with the initial values obtained from a band structure calculation); by introducing the energy derivatives, $(dL_l/dE)_{E=E_F}$ or $(d\delta_l/dE)_{E=E_F}$ as additional parameters, one may also accurately represent the Fermi velocity (including many-body effects, as shown by Crabtree et al. (1979)). The associated methods will be discussed in Chapter 20.

13.3 More powerful approaches: linearized variational methods

There are three grounds on which the APW or KKR methods, as presented, might be criticized: (i) they are restricted to potentials of the muffin tin form, (ii) they are not relativistic, and (iii) they do not have the structure of an eigenvalue problem since one must find a zero of the determinant separately for each band rather than finding roots for all bands simultaneously through a single diagonalization of a linear (in energy), Hamiltonian-like matrix. The first objection may be removed by representing the potential for $r < r_i$ in terms of symmetry-adapted polynomials, K_{lm}, involving linear combinations of spherical harmonics, Y_{lm},

$$V(\mathbf{r}) = \sum_{l,m} V_{lm}(r)K_{lm}(\mathbf{n_r}), \tag{13.64}$$

(called "Kubic harmonics" for the case of cubic symmetry), and as a Fourier expansion in the form (13.3) for $r > r_i$ and other addressable complexities, which will not be dealt with here[12]. The second objection is important for heavier ions with strong spin–orbit coupling, but is a somewhat technical subject that will also not be discussed here.[13]

Addressing the third objection leads to the so-called linearized variational methods: the linearized augmented plane wave (LAPW), and the linearized muffin tin orbital (LMTO) methods. A hint that accurate linearized variational methods might be possible came from fitting Fermi surface data to phase shifts, δ_l, using the KKR method; it was found that the quality of such fits was extremely insensitive to the assumed value of the Fermi energy (see Shaw et al. (1970)) and it is now common to refer to the energy as the "energy parameter." The importance of this empirical result was emphasized by O. Anderson (1973). We will only discuss the LAPW method.

[12] Full-potential self-consistent linearized-augmented-plane-wave method for calculating the electronic-structure of molecules and surfaces - O_2 molecule, E. Wimmer, H. Krakauer, M. Weinert, and A. J. Freeman, *Phys. Rev. B* **24**, 864 (1981).

[13] See, e.g., Technique for relativistic spin-polarized calculations, D.D. Koelling and B.N. Harmon, *J. Phys. C-Solid State Physics* **10**, 3107 (1977).

We recall that in forming an APW basis function, the amplitude of a plane wave is matched to a sum of l-dependent solutions to the radial Schrödinger equation at the muffin tin radius, r_i; and that provided the expansion in l is carried far enough (generally $l \sim 10$), the basis function will be nearly continuous at r_i. The derivative of the wave function is not explicitly matched, but its discontinuity is minimized by the variational formulation (entering through a surface term (13.7c)). Our goal of reducing the dependence on energy of a variationally based APW-like formalism is facilitated by introducing the *energy derivatives* of the radial wave functions, $\partial R_l(r, E)/\partial E \equiv \dot{R}_l(r, E)$, as an additional set of basis functions for $r < r_i$; the extra flexibility this affords will also allow us to *match the radial derivative* of the basis functions, $\partial R_l(r, E)/\partial r \equiv R_l'(r, E)$ at $r = r_i$. The details are somewhat technical and the casual reader may skip to the last paragraph of this section where we qualitatively summarize the advantages of this method.

The radial solutions satisfy

$$\hat{H}_l R_l(r) = E R_l(r), \tag{13.65a}$$

where

$$\hat{H}_l = -\frac{\hbar^2}{2m}\left[\frac{1}{r^2}\frac{d}{dr}r^2\frac{d}{dr} - \frac{l(l+1)}{r^2}\right] + V(r); \tag{13.65b}$$

it is convenient to normalize these functions within the muffin tin sphere as

$$\int R_l^2(r) r^2 dr = 1. \tag{13.66}$$

Differentiating Eq. (13.65) with respect to E yields

$$(\hat{H}_l - E)\dot{R}_l(r, E) = R_l(r, E); \tag{13.67}$$

this is an inhomogeneous differential equation which may be numerically integrated, subject to boundary conditions to be discussed next, once $R_l(r, E)$ is known (from integrating (13.65) constrained by (13.66)). Differentiating (13.66) with respect to E yields

$$\int r^2 dr R_l(r, E)\dot{R}_l(r, E) = 0, \tag{13.68}$$

i.e., R_l and $\dot{R}_l$ are orthogonal when integrated over the muffin tin sphere. By multiplying (13.65a) by $\dot{R}_l$ and (13.67) by R_l, subtracting, integrating over the muffin tin sphere, and using Green's theorem together with the orthonormality conditions (13.68) and (13.66), we obtain

$$r_i^2\left[R_l'(r_i)\dot{R}_l(r_i) - R_l(r_i)\dot{R}_l'(r_i)\right] = 1. \tag{13.69}$$

Equations (13.68) and (13.69) define the boundary conditions allowing a definite integration of (13.67). Note that these constraints result in a normalization condition,

$$\int \dot{R}_l^2 r^2 dr = N_l, \tag{13.70}$$

where N_l is not equal to 1. In place of Eq. (13.1), our augmented basis functions are given by

$$\begin{aligned}\phi_K(\mathbf{k},\mathbf{r}) &= e^{i(\mathbf{k}+\mathbf{K})\cdot\mathbf{r}}\theta(r-r_i)\\ &+\sum_{lm}\left[a_{\mathbf{K}}^{(lm)}(\mathbf{k})R_{lm}(r,E)+b_{\mathbf{K}}^{(lm)}(\mathbf{k})\dot{R}_{lm}(r,E)\right]Y_{lm}(\mathbf{n_r})\theta(r_i-r).\end{aligned} \tag{13.71}$$

The two sets of coefficients, $a^{(lm)}$ and $b^{(lm)}$, are fixed by the requirement that the basis functions and their radial derivative are continuous (when expanded to infinite l at $r = r_i$). Using Eq. (13.2) and, separately, matching the amplitude of each spherical harmonic at $r = r_i$ yields the equations (where $\mathbf{Q} = \mathbf{k} - \mathbf{K}$)

$$a_{\mathbf{Q}}^{(lm)}R_{lm}(r_i) + b_{\mathbf{Q}}^{(lm)}\dot{R}_{lm}(r_i) = 4\pi i^l j_l(kr_i)Y_{lm}^*(\mathbf{n_Q})$$

and

$$a_{\mathbf{Q}}^{(lm)}R'_{lm}(r_i) + b_{\mathbf{Q}}^{(lm)}\dot{R}'_{lm}(r_i) = 4\pi i^l j'_l(kr_i)Y_{lm}^*(\mathbf{n_Q}).$$

Solving this set of inhomogeneous linear equations in $a^{(lm)}$ and $b^{(lm)}$, noting that the determinant of the coefficients of the left-hand side is $1/r_i^2$, from Eq. (13.69), we obtain

$$a_{\mathbf{Q}}^{(lm)} = 4\pi r_i^2 i^l Y_{lm}^*(\mathbf{n_Q})\left[j'_l(Qr_i)\dot{R}_l(r_i) - j_l(Qr_i)\dot{R}'_l(r_i)\right] \tag{13.72a}$$

and

$$b_{\mathbf{Q}}^{(lm)} = 4\pi r_i^2 i^l Y_{lm}^*(\mathbf{n_Q})\left[j_l(Qr_i)R'_l(r_i) - j'_l(Qr_i)R_l(r_i)\right]. \tag{13.72b}$$

(Note that with this choice of basis functions there is no denominator which can vanish, a possibility that led to the APW "asymptote problem" referred to earlier.) The variational wave function of the LAPW method is also of the form (13.3), with ϕ now given by (13.71). Inserting this form into the Lagrangian (13.4) yields, on integrating by parts (noting that the surface term vanishes due to the continuity of the basis functions *and* their derivatives),

$$\sum_{\mathbf{K}'}\left[H_{\mathbf{K}'\mathbf{K}} - EO_{\mathbf{K}'\mathbf{K}}\right]c_{\mathbf{K}} = 0, \tag{13.73}$$

where $H_{\mathbf{K}'\mathbf{K}}$ and $O_{\mathbf{K}'\mathbf{K}}$ are defined by (13.7a) and (13.7b). We rewrite the matrix entering (13.73) as a sum of contributions from inside and outside the muffin tin sphere,

$$(H \pm EO)_{\mathbf{K}'\mathbf{K}} = (H \pm EO)^{(\mathrm{in})}_{\mathbf{K}'\mathbf{K}} + (H \pm EO)^{(\mathrm{out})}_{\mathbf{K}'\mathbf{K}}, \qquad (13.74)$$

where, from (13.12),

$$(H - EO)^{(\mathrm{out})}_{\mathbf{K}'\mathbf{K}} = \int_{\mathrm{interstital}} d^3r e^{-(\mathbf{K}'+\mathbf{k})\cdot\mathbf{r}}(H - E)e^{i(\mathbf{K}+\mathbf{k})\cdot\mathbf{r}}, \qquad (13.75a)$$

and, using (13.15a) and (13.17),

$$(H\text{–}EO)^{(\mathrm{in})}_{\mathbf{K}'\mathbf{K}} = \sum_{ll'}\int_{\mathrm{sphere}} d^3r\left[a^{*(l'm')}_{\mathbf{Q}'}R_{l'm'} + b^{*(l'm')}_{\mathbf{Q}'}\dot{R}_{l'm'}\right]b^{(lm)}_{\mathbf{Q}'}R_{lm}Y^*_{l'm'}(\mathbf{n_r})Y_{lm}(\mathbf{n_r}). \qquad (13.75b)$$

Equation (13.75a) may be evaluated as

$$(H - E)^{(\mathrm{out})}_{\mathbf{K}'\mathbf{K}} = \left[\frac{\hbar^2}{2m}(\mathbf{K} + \mathbf{k})^2 - E\right]\Theta(|\mathbf{K}' - \mathbf{K}|) \qquad (13.76)$$

where, from (13.8) and (13.9),

$$\Theta(|\mathbf{K}' - \mathbf{K}|) = V_c\delta_{\mathbf{K}'\mathbf{K}} - 4\pi r_i^2\frac{j_1(|\mathbf{K}' - \mathbf{K}|r_i)}{|\mathbf{K}' - \mathbf{K}|} \qquad (13.77)$$

(which is V_c times the Fourier transform of the theta function). From the orthonormality relations (13.66), (13.68), and (13.70), we find for (13.75b),

$$(H - EO)^{(\mathrm{in})}_{\mathbf{K}'\mathbf{K}} = \sum_{lm}\left[a^{*(lm)}_{\mathbf{Q}'}b^{(lm)}_{\mathbf{Q}} + N_l b^{*(lm)}_{\mathbf{Q}'}b^{(lm)}_{\mathbf{Q}}\right]. \qquad (13.78)$$

We rewrite (13.72a,b) as

$$a^{(lm)}_{\mathbf{Q}} = 4\pi r_i^2 i^l Y^*_{lm}(\mathbf{n_Q})\alpha^l(Q, \mathbf{r}_i) \qquad (13.79a)$$

and

$$b^{(lm)}_{\mathbf{Q}} = 4\pi r_i^2 i^l Y^*_{lm}(\mathbf{n_Q})\beta^l(Q, \mathbf{r}_i), \qquad (13.79b)$$

where

$$\alpha^l(Q,r_i) \equiv j_l'(Qr_i)\dot{R}_l(r_i) - j_l(Qr_i)\dot{R}'_l(r_i) \qquad (13.80a)$$

and

$$\beta^l(Q, r'_i) \equiv j_l(Qr_i)R_l'(r_i) - j_l'(Qr_i)R_l(r_i). \qquad (13.80b)$$

Using these definitions we may combine (13.76) and (13.78) to write the matrices entering (13.73) as

$$O_{Q'Q} = \Theta_{K'K} + 4\pi r_i^4 \sum_l (2l+1) P_l(\mathbf{n}_{Q'} \cdot \mathbf{n}_Q) o^l_{Q'Q} \tag{13.81}$$

and

$$H_{Q',Q} = Q^2 \Theta_{Q',Q} + 4\pi r_i^4 \sum_l (2l+1) P_l(\mathbf{n}_{Q'} \cdot \mathbf{n}_Q) \left[E o^l_{Q',Q} + \alpha^l_{Q'} \beta^l_Q \right], \tag{13.82}$$

where

$$o^l_{Q',Q} \equiv \alpha^l_{Q'} \alpha^l_Q + \beta^l_{Q'} \beta^l_Q N_l \tag{13.83}$$

and we have used the addition theorem of spherical harmonics, Eq. (13.11). We may rewrite Eq. (13.82) by adding and subtracting $j_l(Q'r)R_l(r_i)\dot{R}_l(r_i)j_l(Qr_i)$ and using Eq. (13.13) to obtain the explicitly Hermitian form

$$H_{Q',Q} = Q' \cdot Q \Theta_{Q'Q} + 4\pi r_i^4 \sum_l (2l+1) P_l(\mathbf{n}_{Q'} \cdot \mathbf{n}_Q) [E o^l_{Q'Q} + \gamma^l_{Q'Q}], \tag{13.84}$$

where

$$\begin{aligned} \gamma^l_{Q'Q}(r_i) = {} & \dot{R}_l(r_i) R'_l(r_i) \left[j'_l(Q'r_i) j_l(Qr_i) + j_l(Q'r_i) j'_l(Qr_i) \right] \\ & - \left[\dot{R}'_l(r_i) R'_l(r_i) j_l(Q'r_i) j_l(Q'r_i) + \dot{R}_l(r_i) R_l(r_i) j'_l(Q'r_i) j'_l(Qr_i) \right]. \end{aligned} \tag{13.85}$$

The secular equation (13.73) constitutes the solution to the variational problem we have posed. The solution involves: (i) numerically integrating the differential equations for R_l and $\dot{R}_l$, Eqs. (13.65b) and (13.67), subject to the orthonormality conditions ((13.66), (13.68), and (13.70)); (ii) constructing the matrices $H_{Q',Q}$ and $O_{Q',Q}$ (using (13.84) and (13.81), with the definitions (13.80), (13.83), and (13.85)); and (iii) seeking an energy that results in

$$\text{Det}|H_{Q',Q} - E O_{Q',Q}| = 0. \tag{13.86}$$

At this point we are still within a variational framework; i.e., there is no obvious improvement over the usual APW formalism. However by using variational basis functions that involve both R_l and $\dot{R}_l$, an error analysis shows (O.K. Anderson (1974)) that the resulting wave functions, R_l, are accurate through *first order* in δE,

$$R_l(E + \delta E) = R_l(E) + \dot{R}_l(E)\delta E, \tag{13.87}$$

with corrections scaling as $(\delta E)^2$; this results in the leading corrections to the energy scaling as $(\delta E)^4$. This being the case, we may obtain approximate energy levels and wave functions by the following procedure: (i) We choose some appropriate reference energy parameter, E_l, which functions as a mean energy for some interval δE_l, and calculate the radial basis functions $R_l(E_l, r)$ and $\dot{R}_l(E_l, r)$. (This energy may be chosen differently for each l.) (ii) With the values of $R_l(E_l, r_i)$ and $\dot{R}_l(E_l, r_i)$ *frozen* at this energy, we evaluate the matrix elements entering the secular equation (13.73) for some chosen value of **k**. Since the only remaining energy dependence is that entering explicitly through E, Eq. (13.73) becomes a *Hamiltonian* matrix in the interval δE_l. In practice, δE_l may be chosen large enough to accurately represent most (if not all) of the filled (or partially filled) valance bands. This approximation results in an improvement by a factor $\sim 10^2$ in the speed of LAPW band structure calculations relative to the APW methods, with no significant loss in accuracy (Koelling and Arbman (1975)[14]). However, it is usually necessary to include more plane waves than in an APW calculation (the additional constraint on the local energy dependence requires a larger basis set for convergence).

[14] Use of energy derivative of radial solution in an augmented plane-wave method - application to copper, D.D. Koelling and G.O. Arbman, *J. Phys. F - Metal Physics* 5, 2041 (1975).

ADDITIONAL READING

For an interesting account of various antecedents that led to the KKR method and beyond, see: Early history of multiple scattering theory in ordered systems, *J. Korringa, Phys. Rep.* **238**, 341 (1994). Although it contains a fair amount of mathematics, one can read past that to obtain the historical flow.

The relativistic augmented plane wave method is discussed in, Augmented plane wave method, Terry Loucks, *Frontiers in Physics*, D. Pines, Ed., W.A. Benjamin, New York (1967). Also see: *Planewaves, Pseudopotentials, and the LAPW Method*, David J. Singh and Lars Nordstrom, Springer, Berlin (2005).

Although not developed in this section, the linear combination of muffin tin orbitals (LMTO) developed by O.K. Andersen is a powerful and widely used method in electronic structure calculations. It is described in the following publications: Linear methods in band theory, O.K. Andersen, *Phys. Rev. B* **12**, 3060 (1975) and, Explicit 1st principles tight-binding theory, O.K. Andersen and O. Jepsen, *Phys. Rev. Lett.* **53**, 2571 (1984).

Part IV

Electron–Electron Interaction

The Self-consistent Dielectric Function

14

14.1 The self-consistent potential method for a uniform electron system

Up to this point we have been ignoring the electron–electron interaction or assuming that it has been incorporated in some average manner. To improve on this situation we develop an approach to include such effects which is analogous to what one does in electromagnetic theory, where the effects of the environment are incorporated through a dielectric function (or in the homogeneous static limit, a dielectric constant). In what immediately follows we discuss the model of an interacting electron liquid. In order to preserve charge neutrality we assume that there is a compensating, uniform, non-deformable, positive background.[1]

Suppose we are able to apply a periodic *external* potential, $U^{(e)}(\mathbf{r}, t)$, to our electron liquid. We recall that an arbitrary space and time dependence can be represented as a Fourier expansion. Noting that any potential must be strictly real, we can write a single component of this expansion in the form

$$U^{(e)}(\mathbf{r}, t) = U^{(e)}(\mathbf{q}, \omega)\, e^{i\mathbf{q}\cdot\mathbf{r} - i\omega t + \alpha t} + U^{(e)*}(\mathbf{q}, \omega)\, e^{-i\mathbf{q}\cdot\mathbf{r} + i\omega t + \alpha t}, \tag{14.1}$$

where $\alpha \to 0+$; this latter feature allows us to apply the perturbation *adiabatically* starting at $t = -\infty$, and without this assumption integrals which will occur in our treatment would be ill-defined. We will assume the system responds *linearly*; quantities measuring the response, which we introduce shortly, may then be written as a superposition of plane wave disturbances of the form (14.1), with no mixing of the various components.

Accompanying the external potential will be an *induced* change in the electron density, $n^{(i)}(\mathbf{r}, t)$, as well as an induced (or response) potential, $U^{(i)}(\mathbf{r}, t)$. The *total* potential, $U^{(t)}(\mathbf{r}, t)$, is then

$$U^{(t)}(\mathbf{r}, t) = U^{(e)}(\mathbf{r}, t) + U^{(i)}(\mathbf{r}, t). \tag{14.2}$$

Our plan is as follows. (i) We first calculate the induced electron density using ordinary first-order, time-dependent perturbation theory. Since both the external *and* induced potentials act on the electrons, the response of the system will be

[1] A simple model for the effects of a deformable background will be discussed in Section 21.2.

The Physics of Solids. First Edition. J. B. Ketterson.

calculated with the total potential, $U^{(t)}(\mathbf{r}, t)$. (ii) We then use Poisson's equation to connect the induced charge density with the induced potential as

$$\nabla^2 U^{(i)}(\mathbf{r}, t) = -4\pi e^2 n^{(i)}(\mathbf{r}, t), \tag{14.3a}$$

which when Fourier transformed is

$$U^{(i)}(\mathbf{q}, \omega) = \frac{4\pi e^2}{q^2} n^{(i)}(\mathbf{q}, \omega). \tag{14.3b}$$

(iii) Lastly we define the dielectric constant as[2,3]

$$\frac{U^{(e)}(\mathbf{q}, \omega)}{\varepsilon(\mathbf{q}, \omega)} = U^{(t)}(\mathbf{q}, \omega), \tag{14.4a}$$

or, on combining with (14.2),

$$\varepsilon(\mathbf{q}, \omega) = 1 - \frac{U^{(i)}(\mathbf{q}, \omega)}{U^{(t)}(\mathbf{q}, \omega)}. \tag{14.4b}$$

The perturbation $U^{(t)}(\mathbf{r}, t)$ induces transitions between initial plane wave states $|\mathbf{k}\rangle$ and final states $\langle \mathbf{k}'|$. The transition amplitude of this process is governed by the following expression from first-order time-dependent perturbation theory:

$$a_{\mathbf{k}\mathbf{k}'}(t) = \frac{1}{i\hbar} \int_{-\infty}^{t} dt' \left\langle \mathbf{k}' \left| U^{(t)}(\mathbf{r}, t) \right| \mathbf{k} \right\rangle e^{-i\omega_{\mathbf{k}\mathbf{k}'} t}, \tag{14.5}$$

where $\hbar\omega_{\mathbf{k}\mathbf{k}'} = \varepsilon_{\mathbf{k}} - \varepsilon'_{\mathbf{k}}$. Writing our plane wave states (normalized over a box of volume L^3) as $|\mathbf{k}\rangle = L^{-3/2} e^{i\mathbf{k}\cdot\mathbf{r}}$, we see that the matrix element vanishes unless $\mathbf{k}' = \mathbf{k} \pm \mathbf{q}$ and therefore

$$a_{\mathbf{k},\mathbf{k}+\mathbf{q}}(t) = \frac{1}{i\hbar} \int_{-\infty}^{t} dt' U^{(t)}(\mathbf{q}, \omega) e^{-i(\omega + \omega_{\mathbf{k},\mathbf{k}+\mathbf{q}} + i\alpha)t} \tag{14.6a}$$

and

$$a_{\mathbf{k},\mathbf{k}-\mathbf{q}}(t) = \frac{1}{i\hbar} \int_{-\infty}^{t} dt' U^{(t)*}(\mathbf{q}, \omega) e^{-i(-\omega + \omega_{\mathbf{k},\mathbf{k}-\mathbf{q}} + i\alpha)t}. \tag{14.6b}$$

Integrating these expressions and writing i0 for the limiting form of $i\hbar\alpha$, we have

$$a_{\mathbf{k},\mathbf{k}+\mathbf{q}} = \frac{U^{(t)}(\mathbf{q}, \omega)}{\hbar\omega + \varepsilon_{\mathbf{k}} - \varepsilon_{\mathbf{k}+\mathbf{q}} + i0} e^{-i(\omega + \omega_{\mathbf{k},\mathbf{k}+\mathbf{q}})t} \tag{14.7a}$$

[2] In electromagnetism, the second Maxwell equation is $\nabla \cdot \mathbf{D} = 4\pi\rho^{(e)}$, where $\rho^{(e)}$ is the external charge (also called the *free or unbound* charge, see Section 2.1) in a medium and $\mathbf{D} = \varepsilon\mathbf{E} = \mathbf{E} + 4\pi\mathbf{P}$. Our external potential energy $U^{(e)}$ arises from this external charge $\rho^{(e)}$ via Poisson's equation, $\nabla^2 U^{(e)}(\mathbf{r}) = -4\pi e^2 n^{(e)}(\mathbf{r})$), and we therefore associate $U^{(e)}$ with the *displacement vector* $\mathbf{D}$ by writing $\mathbf{D} = -e^{-1}\nabla U^{(e)}$. It then follows that we should identify $U^{(t)}$ with $\mathbf{E}$, $U^{(i)}$ with $4\pi\mathbf{P}$, and write $U^{(t)} = \varepsilon^{-1} U^{(e)}$. When, $\varepsilon > 1$, $|U^{(t)}| < |U^{(e)}|$, and hence the origin of the term *screening* (as in blocking out).

[3] Our use of the symbol ε for both the dielectric constant and the energy of electron states should cause no confusion as the two may be distinguished by their functional dependencies.

and

$$a_{k,k-q} = \frac{U^{(t)*}(\mathbf{q},\omega)}{-\hbar\omega + \varepsilon_k - \varepsilon_{k-q} + i0} e^{-i(\omega+\omega_{k,k+q})t}. \tag{14.7b}$$

The perturbed time-dependent wave function is then

$$\begin{aligned}\psi_k(\mathbf{r},t) &= \psi_{\mathbf{k}}^{(0)}(\mathbf{r},t) + \psi_{\mathbf{k}}^{(1)}(\mathbf{r},t) \\ &= \frac{1}{L^{3/2}}\left[e^{i(\mathbf{k}\cdot\mathbf{r}-\omega_k t)} + a_{k,k+q}(t)e^{i((\mathbf{k}+\mathbf{q})\cdot\mathbf{r}-\omega_{k+q}t)}\right. \\ &\qquad \left. + a_{k,k-q}(t)e^{i((\mathbf{k}-\mathbf{q})\cdot\mathbf{r}-\omega_{k-q}t)}\right].\end{aligned} \tag{14.8}$$

The electron density in the host material is given by $n(\mathbf{r},t) = \sum_{\mathbf{k}} n(\mathbf{k})\,|\psi_k(\mathbf{r},t)|^2$, where we introduce the Fermi occupation factor, $n(\mathbf{k}) = n(\varepsilon_k)$, measuring whether a state $\mathbf{k}$ is actually occupied. The *induced* charge in first order is then

$$\begin{aligned}n^{(i)}(\mathbf{r},t) &= \sum_{\mathbf{k}} n(\mathbf{k})\left[\psi_{\mathbf{k}}^{(0)*}(\mathbf{r},t)\psi_{\mathbf{k}}^{(1)}(\mathbf{r},t) + \psi_{\mathbf{k}}^{(1)*}(\mathbf{r},t)\psi_{\mathbf{k}}^{(0)}(\mathbf{r},t)\right] \\ &= \frac{1}{L^3}\sum_{\mathbf{k}} n(\mathbf{k})\left[a_{k,k+q}(t)e^{i(\mathbf{q}\cdot\mathbf{r}+\omega_{k,k+q}t)}\right. \\ &\qquad \left. + a_{k,k-q}(t)e^{i(-\mathbf{q}\cdot\mathbf{r}+\omega_{k,k-q}t)} + c.c\right].\end{aligned} \tag{14.9}$$

Inserting (14.7a,b) into (14.9) gives

$$n^{(i)}(\mathbf{q},\omega) = \frac{1}{L^3}\sum_{\mathbf{k}} n(\mathbf{k})\left[\frac{1}{\hbar\omega + \varepsilon_k - \varepsilon_{k+q} + i0} + \frac{1}{-\hbar\omega + \varepsilon_k - \varepsilon_{k-q} - i0}\right]U^{(t)}(\mathbf{q},\omega). \tag{14.10}$$

Since the sum is over all $\mathbf{k}$ we may shift the origin of the second sum by $\mathbf{q}$, yielding

$$n^{(i)}(\mathbf{q},\omega) = \frac{1}{L^3}\sum_{\mathbf{k}}\left[\frac{n(\mathbf{k})}{\hbar\omega + \varepsilon_k - \varepsilon_{k+q} + i0} + \frac{n(\mathbf{k}+\mathbf{q})}{-\hbar\omega + \varepsilon_{k+q} - \varepsilon_k - i0}\right]U^{(t)}(\mathbf{q},\omega),$$

or combining terms,

$$n^{(i)}(\mathbf{q},\omega) = \frac{1}{L^3}\sum_{\mathbf{k}}\frac{n(\mathbf{k}) - n(\mathbf{k}+\mathbf{q})}{\hbar\omega + \varepsilon_k - \varepsilon_{k+q} + i0}U^{(t)}(\mathbf{q},\omega). \tag{14.11}$$

We define what is called a *polarization operator*, $\Lambda(\mathbf{q},\omega)$, as

$$n^{(i)}(\mathbf{q},\omega) \equiv \Lambda(\mathbf{q},\omega)U^{(t)}(\mathbf{q},\omega). \tag{14.12a}$$

where

$$\Lambda(\mathbf{q},\omega) \equiv \frac{1}{L^3}\sum_{\mathbf{k}}\frac{n(\mathbf{k}) - n(\mathbf{k}+\mathbf{q})}{\hbar\omega + \varepsilon_k - \varepsilon_{k+q} + i0}. \tag{14.12b}$$

Proceeding to part (ii) of our development, we next substitute (14.12a) into Poisson's equation, (14.3b), to give

$$U^{(i)}(\mathbf{q},\omega) = \frac{4\pi e^2}{q^2} n^{(i)}(\mathbf{q},\omega)$$
$$= \frac{4\pi e^2}{q^2} \Lambda(\mathbf{q},\omega) U^{(t)}(\mathbf{q},\omega). \qquad (14.13)$$

Part (iii) of our program involves replacing $U^{(i)}$ in (14.13) by $U^{(t)} - U^{(e)}$, according to Eq. (14.2), to obtain

$$U^{(e)}(\mathbf{q},\omega) = \left(1 - \frac{4\pi e^2}{q^2} \Lambda(\mathbf{q},\omega)\right) U^{(t)}(\mathbf{q},\omega);$$

on comparing this expression with the definition of $\varepsilon(\mathbf{q},\omega)$ given by (14.4) and inserting the definition of $\Lambda(\mathbf{q},\omega)$, we obtain the important result[4]

$$\varepsilon(\mathbf{q},\omega) = 1 - \frac{4\pi e^2}{q^2} \frac{1}{L^3} \sum_{\mathbf{k}} \frac{n(\mathbf{k}) - n(\mathbf{k}+\mathbf{q})}{\hbar\omega + \varepsilon_{\mathbf{k}} - \varepsilon_{\mathbf{k}+\mathbf{q}} + i0}. \qquad (14.14)$$

For the more general case where the two-body potential $U(\mathbf{q})$ is not the Coulomb potential, Eq. (14.14) becomes

$$\varepsilon(\mathbf{q},\omega) = 1 - U(\mathbf{q}) \frac{1}{L^3} \sum_{\mathbf{k}} \frac{n(\mathbf{k}) - n(\mathbf{k}+\mathbf{q})}{\hbar\omega + \varepsilon_{\mathbf{k}} - \varepsilon_{\mathbf{k}+\mathbf{q}} + i0}. \qquad (14.15)$$

We emphasize that the above treatment is limited to a translationally invariant electron liquid. In Appendix 14A we generalize the treatment to the case of a periodic electron liquid.

14.2 Applications of the dielectric function formalism

14.2.1 Static screening; the Thomas–Fermi limit

We start with the static limit; i.e., we set $\omega = 0$. If we further limit ourselves to long wavelengths we may write

$$\varepsilon(\mathbf{k}+\mathbf{q}) = \varepsilon(\mathbf{k}) + \frac{\partial \varepsilon}{\partial \mathbf{k}} \cdot \mathbf{q}$$
$$= \varepsilon(\mathbf{k}) + \hbar \mathbf{v}_F \cdot \mathbf{q} \qquad (14.16)$$

[4] Self-consistent field approach to the many-electron problem, H. Ehrenreich and M.H. Cohen, *Phys. Rev.* 115, 786 (1959); On the properties of a gas of charged particles, J. Lindhard, *K. Dansk. Vidensk. Selsk. Mat. Fys. Medd.* 28, No. 8 (1954).

and

$$n(\mathbf{k}+\mathbf{q}) = n(\mathbf{k}) + \frac{\partial n_0}{\partial \varepsilon}\frac{\partial \varepsilon}{\partial \mathbf{k}}\cdot\mathbf{q}$$
$$= n(\mathbf{k}) + \hbar\frac{\partial n_0}{\partial \varepsilon}v_F\cdot\mathbf{q}. \qquad (14.17)$$

Inserting (14.16) and (14.17) into (14.14), we obtain

$$\varepsilon(\mathbf{q}) = 1 - \frac{4\pi e^2}{q^2}\frac{1}{L^3}\sum_{\mathbf{k}}\frac{\frac{\partial n_0}{\partial \varepsilon}\frac{\partial \varepsilon}{\partial \mathbf{k}}\cdot\mathbf{q}}{\frac{\partial \varepsilon}{\partial \mathbf{k}}\cdot\mathbf{q}}$$
$$= 1 - \frac{4\pi e^2}{q^2}\frac{1}{L^3}\sum_{\mathbf{k}}\frac{\partial n_0}{\partial \varepsilon}. \qquad (14.18)$$

At absolute zero, $\partial n_0/\partial\varepsilon = -\,\delta(\mu-\varepsilon)$. Now $\sum_{\mathbf{k}}\delta(\varepsilon-\mu) = N(\mu)$, the density of states at the Fermi surface. Thus,

$$\varepsilon(\mathbf{q}) = 1 + \frac{4\pi e^2}{q^2}N(\mu) \qquad (14.19a)$$

or

$$\varepsilon(\mathbf{q}) = 1 + \frac{\kappa_{TF}^2}{q^2}, \qquad (14.19b)$$

where we have defined the *Thomas–Fermi screening wave vector*, κ_{TF}, through

$$\kappa_{TF}^2 = 4\pi e^2 N(\mu). \qquad (14.20)$$

We also introduce an associated length scale, the *Thomas–Fermi length*, $r_{TF} = \kappa_{TF}^{-1}$; for electrons at a density of $10^{22}/\text{cm}^3$ this length is about 0.6 A, which is smaller than an atomic radius and hence the effect of screening is operative at both atomic and solid-state (or chemical bond formation) scales.[5]

Static screening at $T = 0$ can be arrived at more simply through the following arguments. We start by noting that the effect of introducing a small change in the total local potential $U^{(t)}(\mathbf{r})$, which as before can be divided into external and induced components, is to cause a change of the local Fermi energy, $\delta\mu(\mathbf{r})$. Since the global chemical potential, μ, must remain constant,[6] we must have

$$\delta\mu(\mathbf{r}) + U^{(t)}(\mathbf{r}) = 0. \qquad (14.21)$$

This in turn induces a change in the local number density, $n^{(i)}(\mathbf{r})$, as

$$n^{(i)}(\mathbf{r}) = N(\mu)\delta\mu(\mathbf{r}),$$
$$= -N(\mu)U^{(t)}(\mathbf{r}), \qquad (14.22)$$

[5] Such small screening lengths imply that the small q expansion used in Eqs. (14.16) and (14.17) breaks down at short distances; this limitation will be addressed in Section 14.2.4.

[6] For a system in equilibrium with respect to the particle number, the chemical potential is a global constant.

where $N(\mu)$ is the density of states. Assuming all quantities vary in the form (14.1), we may use the Fourier transformed Poisson equation, (14.3b), to write

$$U^{(i)}(\mathbf{q}) = -\frac{4\pi e^2 N(\mu)}{q^2} U^{(t)}(\mathbf{q}). \tag{14.23}$$

Inserting (14.23) into (14.4b), we have

$$\varepsilon(\mathbf{q}) = 1 + \frac{4\pi e^2}{q^2} N(\mu) = 1 + \frac{\kappa_{TF}^2}{q^2}; \tag{14.24}$$

i.e., we obtain (14.19b).

The above approach is not limited to a degenerate charged Fermi gas; it can be adapted to a classical system: e.g., a plasma, a non-degenerate semiconductor, or an ionic solution. In these cases the relation between a change in the total local potential and the induced charge density will be governed by the Boltzmann distribution (and not the $T = 0$ Fermi distribution). The ratio between the number of particles (or particle densities) n_1 and n_2 (where n_1 denotes the group with higher energy) separated by an energy ΔE is then $n_1/n_2 = e^{-\Delta E/k_B T}$ or

$$\begin{aligned}\frac{n_1 - n_2}{n_2} &= e^{-\Delta E/k_B T} - 1\\ &\cong -\frac{\Delta E}{k_B T} \quad (\Delta E \ll k_B T).\end{aligned} \tag{14.25}$$

Identifying $n_1 - n_2 = n^{(i)}$, writing $\Delta E = U^{(t)}$, and approximating n_2 by the unperturbed density n, we obtain $n^{(i)} = -(n/k_B T)U^{(t)}$, which on substituting into (14.3b) gives $U^{(i)}(\mathbf{q}) = -(4\pi e^2 n/q^2 k_B T)U^{(t)}(\mathbf{q})$. Using (14.4b) our static dielectric function is then

$$\varepsilon(\mathbf{q}) = 1 + \frac{\kappa_{DH}^2}{q^2}, \tag{14.26}$$

where $\kappa_{DH}^2 = (4\pi n e^2/k_B T)$ and κ_{DH} is referred to as the *Debye–Huckel screening wave vector*, with which there is an associated *Debye–Huckel screening length*, $r_{DH} = \kappa_{DH}^{-1}$.

We now examine the effect that our static dielectric function has when the *external* potential results from embedding a point charge e in the media, in which case $U^{(e)}(\mathbf{r}) = e^2/r$; the Fourier transform of this potential is[7]

$$U^{(e)}(\mathbf{q}) = \frac{4\pi e^2}{q^2}. \tag{14.27}$$

Using Eq. (14.4a), the *total* potential (which we might also call the effective potential) is then $U^{(t)}(\mathbf{q}) = 4\pi e^2/\varepsilon(\mathbf{q})q^2$, which on inserting (14.19b) or (14.26) yields

$$U^{(t)}(\mathbf{q}) = \frac{4\pi e^2}{q^2 + \kappa^2}, \tag{14.28}$$

[7] This form follows immediately from (14.13) on noting that the charge density associated with a point charge is a delta function, the Fourier transform of which is 1.

where κ denotes either the Thomas–Fermi or Debye–Huckel screening wave vector. We leave it as an exercise to show that the Fourier transform of (14.28) is

$$U^{(t)}(\mathbf{r}) = \frac{e^2}{r}e^{-\kappa r}. \tag{14.29}$$

Note this Yukawa-like form has a short range, of order κ^{-1}, and one says that the charge has been "screened."

14.2.2 Plasma oscillations

We next examine the dielectric function for the case $\omega \neq 0$, but in the limit $q \ll k_F$. We first rewrite (14.14) by moving the origin of the term involving $n(\mathbf{k}+\mathbf{q})$ (see Eq. (14.10)) to give

$$\begin{aligned}\varepsilon(\mathbf{q},\omega) &= 1 - \frac{4\pi e^2}{q^2}\frac{1}{L^3}\sum_{\mathbf{k}}\frac{n(\mathbf{k}) - n(\mathbf{k}+\mathbf{q})}{\hbar\omega + \varepsilon_{\mathbf{k}} - \varepsilon_{\mathbf{k}+\mathbf{q}} - i0} \\ &= 1 + \frac{4\pi e^2}{q^2}\frac{1}{L^3}\sum_{\mathbf{k}} n(\mathbf{k})\left[\frac{1}{\varepsilon_{\mathbf{k}+\mathbf{q}} - \varepsilon_{\mathbf{k}} - \hbar\omega + i0} - \frac{1}{\varepsilon_{\mathbf{k}} - \varepsilon_{\mathbf{k}-\mathbf{q}} - \hbar\omega + i0}\right].\end{aligned} \tag{14.30}$$

Next we expand $\varepsilon_{\mathbf{k}\pm\mathbf{q}}$ as

$$\varepsilon_{\mathbf{k}\pm\mathbf{q}} = \varepsilon_{\mathbf{k}} \pm \frac{\partial\varepsilon_{\mathbf{k}}}{\partial\mathbf{k}}\cdot\mathbf{q} + \frac{1}{2}\frac{\partial^2\varepsilon_{\mathbf{k}}}{\partial\mathbf{k}^2}\mathbf{q}^2 + \cdots. \tag{14.31}$$

(For free electrons these expansions terminate with the quadratic terms.) Inserting the expansions (14.31) into the denominators in (14.30) and then expressing the resulting terms in a binomial expansion (noting the terms linear in $\mathbf{q}$ cancel) we obtain, after a little algebra,

$$\varepsilon(\mathbf{q},\omega) = 1 - \frac{4\pi e^2}{q^2}\frac{1}{L^3}\sum_{\mathbf{k}}\frac{n(\mathbf{k})}{\hbar^2\omega^2}\frac{\partial^2\varepsilon_{\mathbf{k}}}{\partial\mathbf{k}^2}\mathbf{q}^2. \tag{14.32}$$

Noting that $1/m = \partial^2\varepsilon_{\mathbf{k}}/\hbar^2\partial\mathbf{k}^2$ and $L^{-3}\sum_{\mathbf{k}} n(\mathbf{k}) = n$, where n is the number density of electrons, we have to leading order in $\mathbf{q}$,

$$\begin{aligned}\varepsilon(0,\omega) &= 1 - \frac{4\pi e^2 n}{m\omega^2} \\ &= 1 - \frac{\omega_p^2}{\omega^2},\end{aligned} \tag{14.33}$$

where $\omega_p = \left(4\pi e^2 n/m\right)^{1/2}$ is called the *plasma frequency*. This result is identical to the classical result found in Section 3.4, although for finite q the behaviors will differ.

To understand (11.47) physically, imagine we have a vacuum-metal interface. At such an interface the normal component of **D** is continuous; i.e., $D_{\perp}^{(ext)} = D_{\perp}^{(int)}$. Expressed in terms of the electric fields, we have $E_{\perp}^{(ext)} = \varepsilon E_{\perp}^{(int)}$. There is a special case which satisfies this equation: $E_{\perp}^{(ext)} = 0, \varepsilon(\omega) = 0, E_{\perp}^{(int)} \neq 0$; i.e., we can sustain internal longitudinal oscillations (in the absence of damping) when $\varepsilon(\omega) = 0$. Applying this condition to (14.33) we find that the equation $\varepsilon(\omega) = 0$ has a solution at the single frequency $\omega = \omega_p$; hence the electron gas can oscillate at this frequency, producing no external field.[8] A similar oscillatory response exists for electrons in a classical plasma with an identical frequency and hence the name plasma oscillation.

If we carry the expansion of (14.30) to the next highest order in q, we obtain

$$\varepsilon(\mathbf{q}, \omega) = 1 - \frac{4\pi e^2}{\omega^2 m} \frac{1}{L^3} \sum_{\mathbf{k}} n(\mathbf{k}) \left[1 + 2\frac{\mathbf{v} \cdot \mathbf{q}}{\omega} + 3\frac{(\mathbf{v} \cdot \mathbf{q})^2}{\omega^2} + \cdots \right], \tag{14.34}$$

where we have written $\mathbf{v} = \partial \varepsilon_{\mathbf{k}} / \hbar \partial \mathbf{k}$. Converting the summation to an integration, noting that the second term in the expansion vanishes on integrating over $d\Omega$, using

$$\frac{1}{4\pi} \int d\Omega (\mathbf{v}_F \cdot \mathbf{q})^2 = \frac{1}{3} (v_F q)^2,$$

and $n = k_F^3 / 3\pi^2$, we can write (14.34) as

$$\varepsilon(\mathbf{q}, \omega) = 1 - \frac{\omega_p^2}{\omega^2} \left[1 + \frac{3}{5} \frac{(v_F q)^2}{\omega^2} + \cdots \right]. \tag{14.35}$$

To find the plasma frequency we again set $\varepsilon(\mathbf{q}, \omega) = 0$, which yields

$$\omega = \omega_p \left[1 + \frac{3}{10} \frac{(v_F q)^2}{\omega_p^2} + \cdots \right]; \tag{14.36}$$

i.e., the frequency of the plasmon increases quadratically with the wave vector q from the value ω_p at $q = 0$.

Although ω_p occurs in expressions for the response of a plasma or an electron gas, the plasma resonance itself cannot be directly excited with transverse electromagnetic waves since the plasma oscillation is longitudinally polarized.[9] However, it can be excited by fast electrons passing through a thin metal foil; such electrons are found to exhibit an energy loss, $\Delta E = n\hbar\omega_p$, where n is an integer.

[8] In practice the dielectric constant would always have an imaginary part, as in our discussion of the Drude theory (see Eq. (3.14)). The dielectric constant then vanishes at a complex frequency given by $\omega = \omega_p - i/2\tau$ and a wave having an amplitude $e^{-i\omega t}$, once excited, would then decay in a characteristic time 2τ.

[9] In a thin conducting film (typically Ag or Au) deposited on a dielectric substrate, surface plasmons propagating at the vacuum/metal interface can be excited by light incoming from the dielectric side for a certain angle, at which there in an abrupt reduction in the amplitude of the reflected wave, a phenomenon called attenuated total reflection (ATR). This phenomenon is discussed in Section 30.5.

14.2.3 Zero sound

The analogue of plasmons can also exist in a *neutral* Fermi system; the best known example is liquid ^3He, although the fluid in a neutron star is another possible case. We can still imagine that an external potential is introduced, that couples to the particle density, although it will differ from the Coulomb form, $U(\mathbf{q},\omega) = (4\pi e^2/q^2)n(\mathbf{q},\omega)$, used in our discussion up to this point. We would then have some induced potential from which we can form a total potential, and these quantities can still be thought of as being connected by a (dimensionless) "dielectric" function, although the name is then not so appropriate. Under certain conditions, which we now derive, a new kind of mode, a sound wave called collisionless or *zero sound*, can then propagate at long wavelengths.

Assuming the limit $U(\mathbf{q} \to 0) = U_0$ exists, and using Eqs. (14.16) and (14.17), Eq. (14.15) becomes

$$\varepsilon(\mathbf{q},\omega) = 1 + U(0)\frac{1}{L^3}\sum_{\mathbf{k}} \frac{\partial n_0}{\partial \varepsilon}\frac{\mathbf{v}\cdot\mathbf{q}}{\omega - \mathbf{v}\cdot\mathbf{q}}. \tag{14.37}$$

We again seek the frequencies of any collective mode by setting $\varepsilon(\mathbf{q},\omega) = 0$. Assuming we are at absolute zero, where $\partial n_0/\partial\varepsilon = -\delta(\varepsilon - \mu)$, writing

$$\frac{1}{L^3}\sum_{\mathbf{k}} = \frac{2}{(2\pi)^3}\int d^3k = \frac{2}{(2\pi)^3}\int k^2 dk d\Omega,$$

integrating over $d\varepsilon$, introducing the density of states, $N(\mu) = k^2/\pi^2 v\hbar$, and defining the quantity $F_0 = N(\mu)\,U(0)$, we may rewrite Eq. (14.37) as

$$1 = \frac{F_0}{4\pi}\int d\Omega \frac{\mathbf{v}\cdot\mathbf{q}}{\omega - \mathbf{v}\cdot\mathbf{q}}.$$

Writing $x = \cos\theta$ and $s = \omega/kv_F$, we have

$$\frac{F_0}{2}\int_{-1}^{+1}\frac{x dx}{s - x} = 1,$$

and on carrying out the integration we obtain

$$\frac{1}{F_0} = \frac{s}{2} ln\left(\frac{s+1}{s-1}\right) - 1. \tag{14.38}$$

The quantity s introduced here is the ratio of the phase velocity, ω/q, of the collective mode to that of the Fermi velocity, v_F.

To obtain s we must solve the implicit equation (14.38). This equation has solutions only for $F_0 > 0$ which, in this simplified model, is the criterion for the existence of zero sound; i.e., in the absence of a *repulsive* (positive) interaction F_0

there is no zero sound. We examine the limiting cases of $F_0 \to 0$ and $F_0 \to \infty$. For the first of these we have

$$s = 1 + 2e^{-2/F_0}; \tag{14.39}$$

in this limit, the phase velocity of the zero sound mode approaches the Fermi velocity, v_F. In the opposite limit, $F_0 \to \infty$, $s \gg 1$ and we expand the logarithm (to the third order in s^{-1}) to obtain

$$s^2 = \frac{F_0}{3}. \tag{14.40}$$

Physically, zero sound involves two components: a displacement of the center of the Fermi distribution to correspond to the local velocity of the fluid, and a change in the shape of the Fermi surface, which is largest at the leading edge in the direction of propagation. Zero sound has been observed in liquid ^{3}He at low temperatures.[10]

14.2.4 The Kohn effect

The last property of the dielectric function we will examine, called the *Kohn effect*, is an anomaly in $\varepsilon(\mathbf{q}, 0)$ that occurs when $q = 2k_F$. We start by taking the $\omega = 0$ limit of Eq. (14.30),

$$\varepsilon(\mathbf{q}, 0) = 1 + \frac{4\pi e^2}{q^2}\frac{1}{L^3}\sum_{\mathbf{k}} n(\mathbf{k})\left[\frac{1}{\varepsilon_{\mathbf{k}+\mathbf{q}} - \varepsilon_{\mathbf{k}} + i0} + \frac{1}{\varepsilon_{\mathbf{k}-\mathbf{q}} - \varepsilon_{\mathbf{k}} - i0}\right]. \tag{14.41}$$

Writing $\varepsilon_{\mathbf{k}\pm\mathbf{q}} - \varepsilon_{\mathbf{k}} = (\hbar^2/2m)(q^2 \pm 2\mathbf{q}\cdot\mathbf{k})$ and changing from a sum to an integration, Eq. (14.41) can be written as

$$\varepsilon(\mathbf{q}, 0) = 1 + \frac{4\pi e^2}{q^2}\frac{2}{(2\pi)^3}\frac{2m}{\hbar^2}\left[F(+\mathbf{q}, +i0) + F(-\mathbf{q}, -i0)\right], \tag{14.42}$$

where

$$F(\pm q, \pm i0) = \int k^2 dk d(-\cos\theta)\, d\phi \frac{n(k)}{q^2 \pm 2qk\cos\theta \pm i0}. \tag{14.43}$$

We consider the case of the + sign in (14.43). Carrying out the integration over ϕ and noting that $n(k) = 1$ for $k < k_F$ and 0 for $k > k_F$, we have

$$F(q, i0) = 2\pi \int_0^{k_F} k^2 dk \int_{-1}^{1} dx \frac{1}{2qk\left(\frac{q}{2k} + x + i0\right)}. \tag{14.44}$$

[10] A discussion of zero sound from the perspective of Fermi liquid theory is given Section 37.2.

To evaluate this integral we make use of the operational expression

$$\frac{1}{x \mp i0} = P\left(\frac{1}{x}\right) \pm i\pi\delta(x). \tag{14.45}$$

The pole in (14.44) occurs at $x = -q/2k$. In the following expression we isolate the principal value of the integral inside the large brackets as

$$\Re F(q) = \lim_{\varepsilon\to 0} 2\pi \int_0^{k_F} dk \frac{k}{2q}\left[\int_{-1}^{(-\frac{q}{2k}-\varepsilon)} dx \frac{1}{\left(\frac{q}{2k}+x\right)} + \int_{(-\frac{q}{2k}+\varepsilon)}^{1} dx \frac{1}{\left(\frac{q}{2k}+x\right)}\right];$$

here $\Re$ denotes the real part. Evaluating the integrals in the two regions and taking the limit, we obtain

$$\Re F(q) = 2\pi \int_0^{k_F} \frac{k}{2q} dk \ln\left|\frac{1+\frac{q}{2k_F}}{1-\frac{q}{2k_F}}\right|. \tag{14.46}$$

Carrying out the remaining integration and combining the two contributions from $F(q)$ and $F(-q)$ (and noting that the imaginary parts cancel), we obtain after some algebra,

$$F(q) + F(-q) = \pi\left[\frac{4k_F^2 - q^2}{4q} \ln\left|\frac{2k_F + q}{2k_F - q}\right| + k_F\right]. \tag{14.47}$$

Our expression for $\varepsilon(\mathbf{q})$ is then

$$\varepsilon(\mathbf{q}, 0) = 1 + \frac{4\pi e^2}{q^2}\frac{2}{(2\pi)^3}\frac{2m}{\hbar^2}\frac{4\pi k_F}{2}\left[\frac{4k_F^2 - q^2}{8qk_F} \ln\left|\frac{2k_F + q}{2k_F - q}\right| + \frac{1}{2}\right]; \tag{14.48}$$

Writing $N(\mu) = k^2/\pi^2 v_F \hbar$, (14.48) becomes

$$\varepsilon(\mathbf{q}, 0) = 1 + \frac{4\pi e^2}{q^2} N(\mu)\left[\frac{4k_F^2 - q^2}{8qk_F} \ln\left|\frac{2k_F + q}{2k_F - q}\right| + \frac{1}{2}\right]. \tag{14.49}$$

We may rewrite this in the form of Eq. (14.24) as

$$\varepsilon(\mathbf{q}) = 1 + \frac{\kappa^2(q)}{q^2}, \tag{14.50}$$

where from (14.49), $\kappa^2(q)$ is

$$\kappa^2(q) = 4\pi e^2 N(\mu)\left[\frac{4k_F^2 - q^2}{8qk_F} \ln\left|\frac{2k_F + q}{2k_F - q}\right| + \frac{1}{2}\right]. \tag{14.51}$$

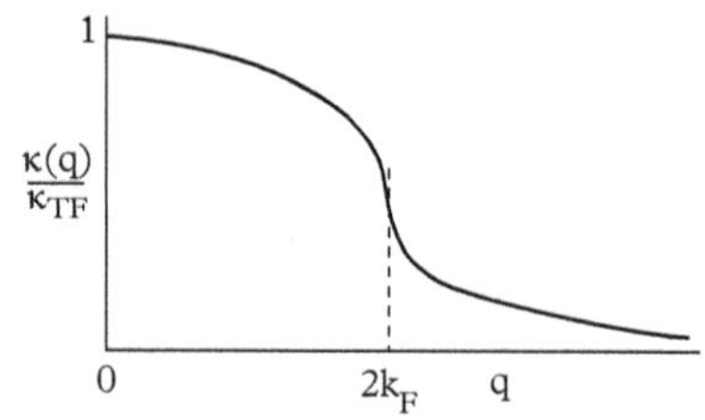

Figure 14.1 *Wave vector dependence of the Thomas-Fermi screening factor,* κ.

It is straightforward to verify that the q = 0 limit of $\kappa^2(q)$ is given by $\kappa^2 = 4\pi e^2 N(\mu)$; i.e., we recover the Thomas–Fermi result. However, at the opposite extreme we have the property $\kappa(q \to \infty) \to 0$, falling off asymptotically as q^{-2}, which is equivalent to $\varepsilon(q \to \infty) \to 1$; i.e., we recover Coulomb's law $U(r) = e^2/r$ in the limit of very small particle separations. This latter behavior is in accord with our physical intuition that screening will be suppressed at very short distances, as measured by the inter-particle spacing, which in turn fixes k_F.

The full-function $\kappa(\mathbf{q})$ is plotted in Fig. 14.1. Although not easily discernible in this plot, we note that the slope is infinite for $q = 2k_F$. Physically, we might expect some kind of anomaly when a perturbation is applied to our electron gas having a wave vector that can "connect" opposing sides of the Fermi surface where the occupation of the states changes abruptly. The singularity at $\varepsilon(\mathbf{q} = 2\mathbf{k}_F, 0)$ has a number of important experimental ramifications. It may in principle be observed directly as a small anomaly in the phonon dispersion curves, $\omega = \omega(\mathbf{q})$ for $\mathbf{q} = 2\mathbf{k}_F$. In lower dimensions the singular behavior is enhanced. For a one-dimensional system, $\varepsilon(\mathbf{q} = 2\mathbf{k}_F, 0)$ is actually divergent. Certain transition metals (such as W or Pd), having a nearly flat Fermi surface over some region of $\mathbf{k}$ space, and some polymer based organic metals, may be modeled as quasi-1d systems. In the latter case it may result in the formation of a charge density wave with a *lattice-incommensurate* wave vector, $\mathbf{q} = 2\mathbf{k}_F$. When the formalism is generalized to include the magnetic response, the coupling across flat regions of the Fermi surface can result in antiferromagnetism, as is thought to be the case for Cr, where the observed antiferromagnetism also has a lattice-incommensurate wave vector, $\mathbf{q} = 2\,\mathbf{k}_F$.

ADDITIONAL READING

One of the first many-electron systems to be treated was the atom itself;[11,12] this was done independently by L.H. Thomas and E. Fermi. The atomic case is worth some study and one of the best treatments is in *Quantum Mechanics*, L.D. Landau and E.M. Lifshitz, Pergamon Press (1981), Section 70.

Problems

14.1. Verify Eq. (14.29).

14.2. Verify Eqs. (14.34) and (14.36).

14.3. Fill in the steps leading to Eqs. (14.39) and (14.40).

14.4. Starting with Eq. (14.34), and using an appropriately normalized Boltzmann distribution, $n(\varepsilon)$, calculate the plasma frequency through terms of order $\mathbf{q}^2$.

[11] A statistical method for determining some properties of the atoms and its application to the theory of the periodic table of elements, E. Fermi, *Z. Phys.* **48**, 73 (1928).

[12] The calculation of atomic fields, L.H. Thomas, *Proc. Camb. Philos. Soc.* **23**, 542 (1927).

14.5. Evaluate the function F(q) introduced in Eq. (14.42) for the case of two dimensions and sketch the behavior of the associated screening length, $\kappa(q)$.

14.6. Do the same as in Problem 14.5, only in one dimension.

14.7. Find the leading term in the expansion of $\kappa(q)$, as given by Eq. (14.51), in a power series in q; do the same for a power series in q^{-1}.

Appendix 14A The self-consistent potential method for the periodic system

The previous discussion of the electron liquid neglected the periodicity associated with a crystalline solid and we now extend it to cover the periodic case.[13,14] The one-electron wave functions of solids have the Bloch form

$$\psi_{\mathbf{k}l} = \frac{1}{L^{3/2}} e^{i\mathbf{k}\cdot\mathbf{r}} u_{\mathbf{k}l}(\mathbf{r}), \tag{14A.1}$$

where **k** is the wave vector, which we will restrict to the first Brillouin zone, and l is the band index; the function $u_{\mathbf{k},l}(\mathbf{r})$ satisfies the crystal periodicity condition, $u_{\mathbf{k}l}(\mathbf{r}) = u_{\mathbf{k}l}(\mathbf{r}+\mathbf{R})$, where **R** is any vector of the real space lattice.

We will leave the form of the external potential $U^{(e)}(\mathbf{r}, t)$ unspecified for the moment, but since the periodic charge density of a lattice is *polarizable*, the induced electron density and self-consistent potential at some wave vector **q** will be *modulated* by the lattice periodicity and therefore takes the form

$$n^{(i)}(\mathbf{r}, t) = e^{i\mathbf{q}\cdot\mathbf{r}-i\omega t} \sum_{\mathbf{K}} n_{\mathbf{K}}^{(i)}(\mathbf{q}, \omega) e^{i\mathbf{K}\cdot\mathbf{r}} + \text{c.c.} \tag{14A.2a}$$

and

$$U^{(t)}(\mathbf{r}, t) = e^{i\mathbf{q}\cdot\mathbf{r}-i\omega t} \sum_{\mathbf{K}} U_{\mathbf{K}}^{(t)}(\mathbf{q}, \omega) e^{i\mathbf{K}\cdot\mathbf{r}} + \text{c.c.}, \tag{14A.2b}$$

where **K** denotes the vectors of the reciprocal lattice. In place of the expansion coefficients (14.5), we now have

$$a_{\mathbf{k}l,\mathbf{k}'l'} = \frac{1}{i\hbar} \int_{-\infty}^{t} dt' \left\langle \mathbf{k}'l' \right| U^{(t)}(\mathbf{r}, t) \left| \mathbf{k}l \right\rangle e^{-i\omega_{\mathbf{k}l,\mathbf{k}'l'} t}, \tag{14A.3}$$

[13] Quantum theory of dielectric constant in real solids, S.L. Adler, *Phys. Rev.* **126**, 413 (1962).

[14] Dielectric constant with local field effects included, N. Wiser, *Phys. Rev.* **129**, 62 (1963).

where $\hbar\omega_{\mathbf{k}l,\mathbf{k}'l'} = \varepsilon_{\mathbf{k}l} - \varepsilon_{\mathbf{k}'l'}$. The full matrix element entering (14A.3) is now taken between the Bloch states (14A.1); on inserting (14A.2b) it takes the form

$$\langle \mathbf{k}'l'|U^{(t)}(\mathbf{r},t)|\mathbf{k}l\rangle = \frac{1}{L^3}\int d^3r\, e^{-i\mathbf{k}'\cdot\mathbf{r}}u^*_{\mathbf{k}'l'}(\mathbf{r})\left(\sum_{\mathbf{K}} U^{(t)}_{\mathbf{K}}(\mathbf{q},\omega)e^{i(\mathbf{q}+\mathbf{K})\cdot\mathbf{r}-i\omega t} + \text{c.c.}\right)e^{i\mathbf{k}\cdot\mathbf{r}}u_{\mathbf{k}l}(\mathbf{r})$$
$$= \sum_{\mathbf{K}}\left(U^{(t)}_{\mathbf{K}}(\mathbf{q},\omega)e^{-i\omega t}M_{\mathbf{K};\mathbf{k}+\mathbf{q}l',\mathbf{k}l} + U^{(t)*}_{\mathbf{K}}(\mathbf{q},\omega)e^{+i\omega t}M_{-\mathbf{K};\mathbf{k}-\mathbf{q}l',\mathbf{k}l}\right) \tag{14A.4a}$$

where we have introduced the definition

$$M_{\mathbf{K};\mathbf{k}+\mathbf{q}l',\mathbf{k}l} \equiv \frac{1}{V_c}\int_{V_c} d^3r\, u^*_{\mathbf{k}+\mathbf{q}l'}(\mathbf{r})e^{i\mathbf{K}\cdot\mathbf{r}}u_{\mathbf{k}l}(\mathbf{r}) = \left(M_{-\mathbf{K};\mathbf{k}l,\mathbf{k}+\mathbf{q}l'}\right)^* \tag{14A.4b}$$

in which the integral is only taken over the unit cell volume, V_c. With these definitions our expansion coefficients become

$$a_{\mathbf{k}l,\mathbf{k}+\mathbf{q}l'} = \sum_{\mathbf{K}} U^{(t)}_{\mathbf{K}}(\mathbf{q},\omega)\frac{M_{\mathbf{K};\mathbf{k}+\mathbf{q}l',\mathbf{k}l}}{\hbar\omega + \varepsilon_{\mathbf{k}l} - \varepsilon_{\mathbf{k}+\mathbf{q}l'} + i0}e^{-i(\omega+\omega_{\mathbf{k}l,\mathbf{k}+\mathbf{q}l'})t} \tag{14A.5a}$$

and

$$a_{\mathbf{k}l,\mathbf{k}-\mathbf{q}l'} = \sum_{\mathbf{K}} U^{(t)*}_{\mathbf{K}}(\mathbf{q},\omega)\frac{M_{-\mathbf{K};\mathbf{k}-\mathbf{q}l',\mathbf{k}l}}{-\hbar\omega + \varepsilon_{\mathbf{k}l} - \varepsilon_{\mathbf{k}-\mathbf{q}l'} + i0}e^{-i(-\omega+\omega_{\mathbf{k}l,\mathbf{k}-\mathbf{q}l'})t}. \tag{14A.5b}$$

The induced electron density, analogous to (14.9), is given by

$$n^{(i)}(\mathbf{r},t) = \sum_{\mathbf{k},l} n_{\mathbf{k}l}\left[\psi^{(0)*}_{\mathbf{k}l}(\mathbf{r},t)\psi^{(1)}_{\mathbf{k}l}(\mathbf{r},t) + \psi^{(1)*}_{\mathbf{k}l}(\mathbf{r},t)\psi^{(0)}_{\mathbf{k}l}(\mathbf{r},t)\right],$$

where $n_{\mathbf{k},l} = n(\varepsilon_{\mathbf{k},l})$. On inserting the expansion for $\psi^{(1)}_{\mathbf{k}l}(\mathbf{r},t)$ in terms of $a_{\mathbf{k}l,\mathbf{k}+\mathbf{q}l'}$ and $a_{\mathbf{k}l,\mathbf{k}-\mathbf{q}l'}$, we obtain

$$n^{(i)}(\mathbf{r},t) = e^{i(\mathbf{q}\cdot\mathbf{r}-\omega t)}\sum_{\mathbf{K}} U^{(t)}_{\mathbf{K}}(\mathbf{q},\omega)\frac{1}{L^3}\left\{\sum_{\mathbf{k}ll'} n_{\mathbf{k}l}\left[\left(\frac{u^*_{\mathbf{k}l}u_{\mathbf{k}+\mathbf{q}l'}M_{\mathbf{K};\mathbf{k}+\mathbf{q}l',\mathbf{k}l}}{\hbar\omega + \varepsilon_{\mathbf{k}l} - \varepsilon_{\mathbf{k}+\mathbf{q}l'} + i0}\right.\right.\right.$$
$$\left.\left.\left. + \frac{u_{\mathbf{k}l}u^*_{\mathbf{k}-\mathbf{q}l'}(M_{-\mathbf{K};\mathbf{k}-\mathbf{q}l',\mathbf{k}l})^*}{-\hbar\omega + \varepsilon_{\mathbf{k}l} - \varepsilon_{\mathbf{k}-\mathbf{q}l'} - i0}\right)\right]\right\} + \text{c.c.} \tag{14A.6}$$

From (14A.2a) it follows that $n^{(i)}_{\mathbf{K}}(\mathbf{q},\omega)$ can be projected out of (14A.6) by multiplying both sides by $e^{-i\mathbf{K}'\cdot\mathbf{r}}$, integrating over all space, and using the property $\int_{V_c} d^3r e^{i(\mathbf{K}-\mathbf{K}')\cdot\mathbf{r}} = V_c\delta_{\mathbf{K}\mathbf{K}'}$, from which we obtain

$$n^{(i)}_{\mathbf{K}}(\mathbf{q},\omega) = \frac{1}{L^3}\sum_{\substack{\mathbf{k},\mathbf{K}'\\ ll'}} U^{(t)}_{\mathbf{K}'}(\mathbf{q},\omega)n_{\mathbf{k},l}\left[\frac{M_{-\mathbf{K};\mathbf{k}l',\mathbf{k}+\mathbf{q}l'}M_{\mathbf{K}';\mathbf{k}+\mathbf{q}l',\mathbf{k}l}}{\hbar\omega + \varepsilon_{\mathbf{k},l} - \varepsilon_{\mathbf{k}+\mathbf{q},l'} + i0}\right.$$
$$\left. - \frac{M_{-\mathbf{K};\mathbf{k}-\mathbf{q}l',\mathbf{k}l}(M_{-\mathbf{K}';\mathbf{k}-\mathbf{q}l',\mathbf{k}l})^*}{\hbar\omega - \varepsilon_{\mathbf{k},l} + \varepsilon_{\mathbf{k}-\mathbf{q},l'} + i0}\right].$$

Using the property $M_{\mathbf{K};\mathbf{k}+\mathbf{q}l',\mathbf{k}l} = (M_{-\mathbf{K};\mathbf{k}l,\mathbf{k}+\mathbf{q}l'})^*$ from (14A.4b), and in the second term moving the sum over $\mathbf{k}$ to one in $\mathbf{k} + \mathbf{q}$, while interchanging the labels l and l', we get

$$\begin{aligned} n_{\mathbf{K}}^{(i)}(\mathbf{q},\omega) &= \frac{1}{L^3}\sum_{\substack{\mathbf{k},\mathbf{K}'\\ ll'}} \frac{n_{\mathbf{k},l} - n_{\mathbf{k}+\mathbf{q},l'}}{\hbar\omega + \varepsilon_{\mathbf{k},l} - \varepsilon_{\mathbf{k}+\mathbf{q},l'} + i0} M^*_{\mathbf{K};\mathbf{k}+\mathbf{q}l',\mathbf{k}l} M_{\mathbf{K}';\mathbf{k}+\mathbf{q}l',\mathbf{k}l} U_{\mathbf{K}'}^{(t)}(\mathbf{q},\omega) \\ &= \sum_{\mathbf{K}'} \Lambda_{\mathbf{K}\mathbf{K}'}(\mathbf{q},\omega) U_{\mathbf{K}'}^{(t)}(\mathbf{q},\omega), \end{aligned} \tag{14A.7a}$$

where the *polarization operator* of Eq. (14.12b) now has the matrix form

$$\Lambda_{\mathbf{K}\mathbf{K}'}(\mathbf{q},\omega) = \frac{1}{L^3}\sum_{\mathbf{k},l,l'} \frac{n_{\mathbf{k},l} - n_{\mathbf{k}+\mathbf{q},l'}}{\hbar\omega + \varepsilon_{\mathbf{k},l} - \varepsilon_{\mathbf{k}+\mathbf{q},l'} + i0} M^*_{\mathbf{K};\mathbf{k}+\mathbf{q}l',\mathbf{k}l} M_{\mathbf{K}';\mathbf{k}+\mathbf{q}l',\mathbf{k}l}. \tag{14A.7b}$$

We may relate $n_{\mathbf{K}}^{(i)}(\mathbf{q},\omega)$ to the induced potential, $U_{\mathbf{K}}^{(i)}(\mathbf{q},\omega)$ by using Poisson's equation in the form (14.3b),

$$\sum_{\mathbf{K}} (\mathbf{K}+\mathbf{q})^2 U_{\mathbf{K}}^{(i)}(\mathbf{q},\omega) e^{i(\mathbf{K}+\mathbf{q})\cdot\mathbf{r}} = 4\pi e^2 \sum_{\mathbf{K}} n_{\mathbf{K}}^{(i)}(\mathbf{q},\omega) e^{i(\mathbf{K}+\mathbf{q})\cdot\mathbf{r}} \tag{14A.8}$$

for which the $\mathbf{K}^{\text{th}}$ Fourier component is

$$\begin{aligned} U_{\mathbf{K}'}^{(i)}(\mathbf{q},\omega) &= \frac{4\pi e^2}{(\mathbf{K}+\mathbf{q})^2} n_{\mathbf{K}}^{(i)}(\mathbf{q},\omega). \\ &= \frac{4\pi e^2}{(\mathbf{K}+\mathbf{q})^2} \sum_{\mathbf{K}'} \Lambda_{\mathbf{K}\mathbf{K}'}(\mathbf{q},\omega) U_{\mathbf{K}'}^{(t)}(\mathbf{q},\omega). \end{aligned} \tag{14A.9}$$

Now, by definition, $U^{(e)}(\mathbf{r},t) = U^{(t)}(\mathbf{r},t) - U^{(i)}(\mathbf{r},t)$; writing $U^{(e)}(\mathbf{r},t)$ in a form similar to (14A.2b) then leads to

$$\sum_{\mathbf{K}} U_{\mathbf{K}}^{(e)}(\mathbf{q},\omega)\, e^{i\mathbf{K}\cdot\mathbf{r}} = \sum_{\mathbf{K}} \left[U_{\mathbf{K}}^{(t)}(\mathbf{q},\omega) - U_{\mathbf{K}'}^{(i)}(\mathbf{q},\omega) \right] e^{i\mathbf{K}\cdot\mathbf{r}}.$$

Inserting (14A.9) we then have

$$\begin{aligned} U_{\mathbf{K}}^{(e)}(\mathbf{q},\omega) &= \sum_{\mathbf{K}'} \left[\delta_{\mathbf{K},\mathbf{K}'} - \frac{4\pi e^2}{(\mathbf{K}+\mathbf{q})^2} \Lambda_{\mathbf{K}\mathbf{K}'}(\mathbf{q},\omega) \right] U_{\mathbf{K}'}^{(t)}(\mathbf{q},\omega) \\ &= \sum_{\mathbf{K}} \varepsilon_{\mathbf{K}\mathbf{K}'} U_{\mathbf{K}'}^{(t)}(\mathbf{q},\omega), \end{aligned} \tag{14A.10}$$

where we have defined a frequency- and wave vector-dependent *dielectric matrix* as

$$\varepsilon_{\mathbf{K},\mathbf{K}'}(\mathbf{q},\omega) = \delta_{\mathbf{K},\mathbf{K}'} - \frac{4\pi e^2}{(\mathbf{K}+\mathbf{q})^2} \Lambda_{\mathbf{K}\mathbf{K}'}(\mathbf{q},\omega). \tag{14A.11}$$

$U^{(t)}(\mathbf{r}, t)$ then follows from (14A.10) by taking the *inverse* of this matrix,[15]

$$U_{\mathbf{k}}^{(t)}(\mathbf{q}, \omega) = \sum_{\mathbf{K}'} (\varepsilon^{-1}(\mathbf{q}, \omega))_{\mathbf{K}\mathbf{K}'} U_{\mathbf{K}'}^{(e)}(\mathbf{q}, \omega). \tag{14A.12}$$

For the case where $U^{(e)}(\mathbf{r}, t)$ is not modulated by the crystal periodicity, the sum in (14A.10) is restricted to the $\mathbf{K} = 0$ term, and (14A.12) takes the form

$$U_{\mathbf{K}}^{(t)}(\mathbf{q}, \omega) = (\varepsilon^{-1}(\mathbf{q}, \omega))_{\mathbf{K}0} U^{(e)}(\mathbf{q}, \omega). \tag{14A.13}$$

Although the total potential is modulated, when $\mathbf{q}$ is small, such that changes arising from the factor $e^{i\mathbf{q}\cdot\mathbf{r}}$ are small over a unit cell (e.g., at optical wavelengths), the *average* value of this field would correspond to its $\mathbf{K} = 0$ component; hence the macroscopic relation between the total potential and the external potential in the small $\mathbf{q}$ limit is given by

$$U^{(t)}(\mathbf{q}, \omega) = \frac{U^{(e)}(\mathbf{q}, \omega)}{\varepsilon(\mathbf{q}, \omega)}, \tag{14A.14a}$$

where

$$\varepsilon^{-1}(\mathbf{q}, \omega) = (\varepsilon^{-1}(\mathbf{q}, \omega))_{00}; \tag{14A.14b}$$

defined in this way, the so-called local field effects will be implicitly included. In this way we obtain a quantum mechanical description of the local field concept introduced in Section 2.2.

[15] Inverting the dielectric matrix is a numerically time consuming process which greatly complicates the application of this formalism.

Hartree–Fock and Density Functional Theory

15

In this chapter we continue the discussion, initiated in Chapter 14, of treating the effects resulting from the mutual Coulomb repulsion between electrons, but in more detail. As we have noted previously, the true wave function of a many-electron system, be it an atom or a macroscopic system, must depend on the coordinates of all the electrons. In the self-consistent field model one approximates the many-electron wave function as a *product* of single-particle wave functions, each depending on the coordinate of a single electron, as pioneered by Hartree for atoms; we recall that such one-particle states are called *orbitals*. The exclusion principle is then incorporated by allowing a maximum of one electron per orbital (where it is assumed that states with spin up and spin down correspond to different orbitals). In this model the effects of the other electrons are incorporated through an *effective* potential formed by *averaging* over their positions. In an atom or molecule one can further demand that the total wave function be formed from combinations of products of one-electron orbitals in such a way that it is antisymmetric when interchanging any two electrons, as is done in the Hartree–Fock theory; a determinant formed from the occupied one-electron orbitals (called a Slater determinant, after J.C. Slater) has this property and, using a wave function of this form results in an additional contribution to the effective potential that is referred to as the *exchange potential* or equivalently the *exchange interaction*. Constructing the exchange potential in a solid presents computational difficulties but model forms can still be found. Both the Hartree and the Hartree–Fock theories will be discussed in this chapter.

Many-body effects in uniform systems form an important subclass of problems.[1] As a model to understand many-electron effects in metals, much attention has been given to the problem of the electron liquid introduced in the previous chapter, although there are no physical systems that conform to the model.[2] In spite of this it has attracted large interest; however it is rather technical so we have placed the discussion in Appendix 15B.

[1] The Landau–Fermi liquid theory (to be discussed in Chapter 37) is an example of a many-body theory designed to describe a uniform neutral system such as ^{3}He or a neutron star.

[2] For many purposes, the two-dimensional electron gas that forms in quantum wells or metal-oxide-semiconductor (MOS) devices can be accurately approximated as a uniform system.

The Physics of Solids. First Edition. J. B. Ketterson.

15.1 The Hartree and Hartree–Fock approximations

15.1.1 The Hartree approximation

We now take up a discussion of the self-consistent field method as it applies to atoms and molecules. In the Hartree approximation (Hartree (1928)[3]), one *assumes* that the spatial part of the wave function can be written in the product form[4]

$$\Psi(\mathbf{r}_1, \mathbf{r}_2, \ldots\ldots \mathbf{r}_N) = \psi_{q_1}(\mathbf{r}_1)\psi_{q_2}(\mathbf{r}_2), \ldots\ldots \psi_{q_N}(\mathbf{r}_N), \tag{15.1}$$

and the spin coordinates are here being ignored. The subscripts $q_1, q_2, \cdots q_N$ denote all the quantum numbers needed to designate each of the N orbitals;[5] as a notational simplification we will here write q_i simply as i, with the understanding that the *occupied* orbitals ψ_i are numbered from $i = 1, \cdots N$.[6] The Pauli exclusion principle is incorporated simply by allowing a maximum of two electrons in each orbital (one for each spin).[7] In the ground state of an atom we would fill the levels in the order of increasing energy, starting from the lowest (1s) orbital.

For the case of a molecule containing M atoms, and when it is assumed that the positions of the nuclei are given (fixed), each of the single particle wave functions involves all the nuclear coordinates $\mathbf{R}_1, \mathbf{R}_2 \ldots \mathbf{R}_M$, denoted as $\{\mathbf{R}\}$, *as parameters*: i.e., $\psi_i(\mathbf{r}_i) \rightarrow \psi_i(\mathbf{r}_i, \{\mathbf{R}\})$. We are then, in effect, ignoring the dynamics of the nuclei, assuming the electrons adjust their orbits on a time scale much faster than a nuclear vibration period; we will consider the nuclear motions of molecules briefly in Section 15.2.2. The assumption that the electronic and nuclear motions can be treated separately is known as the *Born–Oppenheimer approximation*, which we will not further develop here (Born and Openheimer (1927)[8]).

We write the (non-relativistic) *electronic* part of the Hamiltonian, $\hat{H}^{(e)}$, for an N-electron M-atom molecule as

$$\hat{H}^{(e)}(\{\nabla\}, \{\mathbf{r}\}, \{\mathbf{R}\}) = -\frac{\hbar^2}{2m}\sum_{i=1}^{N} \nabla_i^2 - \sum_{k=1}^{M}\sum_{i=1}^{N} \frac{Z_k e^2}{|\mathbf{r}_i - \mathbf{R}_k|} + \frac{1}{2}{\sum_{i,j}^{N}}' \frac{e^2}{|\mathbf{r}_i - \mathbf{r}_j|}, \tag{15.2}$$

where $\nabla_i^2 = \frac{\partial^2}{\partial x_i^2} + \frac{\partial^2}{\partial y_i^2} + \frac{\partial^2}{\partial z_i^2}$ and the prime in the last summation denotes the exclusion of the $j = i$ terms. Of course the total energy of the molecule must include the inter-nuclear Coulomb potential, along with the nuclear kinetic energy operator, and is governed by the following Hamiltonian:

$$H^{(\text{total})} = \sum_{k=1}^{M} \frac{\hbar^2}{2M_k}\nabla_{\mathbf{R}_k}^2 + {\sum_{l,k}^{M}}' \frac{Z_k Z_l}{|\mathbf{R}_k - \mathbf{R}_l|} + \hat{H}^{(e)}(\{\nabla\}, \{\mathbf{r}\}, \{\mathbf{R}\}), \tag{15.3}$$

[3] The wave mechanics of an atom with a non-Coulomb central field Part I theory and methods, D.R. Hartree, *Proc. Camb. Phil. Soc.* **24**, 89 (1928).

[4] Although the overall wave function of a system of electrons must be strictly antisymmetric, on permuting the spatial *and* spin coordinates of any pair of electrons, this statement does not necessarily apply *separately* to the spatial and spin coordinates; i.e., they may have "mixed" symmetry representations characteristic of certain representations of the permutation group. For a discussion of this subtle issue, see: L.D. Landau and E.M. Lifshitz, *Quantum Mechanics*, Section 63.

[5] For atoms in the central field approximation, the appropriate orbital quantum numbers, q_i, are the l_i, l_{zi} and n_i; particularly simple ones are the hydrogenic-like Slater orbitals. However these quantum numbers do not apply to the atom as a whole due to the interaction between electrons.

[6] Note that our numbering the occupied orbitals in some order corresponding to $i = 1, 2 \cdots N$ does not necessarily mean that orbitals with progressively increasing quantum numbers are filled; some possible orbitals might simply be *unoccupied*, while others might contain two electrons in the same orbital (the maximum allowed by the Pauli principle). Excited states can be represented in this way.

[7] The Dirac requirement that the overall wave function be antisymmetric on interchanging both the spin and space coordinates will be discussed in Section 15.1.2.

[8] Quantum theory of molecules, M. Born and R. Oppenheimer, *Ann. Phys.* 84, 457 (1927).

where M_k is the mass of the kth nucleus. We note again that in the approximation where we ignore the dynamics of the nuclei we suppress the first term on the right of (15.3).

Using the methods of the calculus of variations (see Appendix A), we can seek an approximate solution to the many-electron Schrödinger equation using the variational form (15.1) in the Schrödinger action. To obtain a non-trivial solution we must impose the functional constraint, C_i, so that the ψ_i satisfies the N normalization conditions,[9]

$$C_i\left(\psi_i^*, \psi_i\right) = \int d^3r\, \psi_i^*(\mathbf{r})\psi_i(\mathbf{r}) = 1;\ i = 1, \cdots N. \tag{15.4}$$

For simplicity we initially restrict ourselves to atoms, where the nuclear coordinates, $\mathbf{R}_k$, do not enter. The many-electron action is

$$S = \int d^3r_1 \ldots d^3r_N\, \mathcal{L}\left[\Psi^*(\mathbf{r}_1 \ldots \mathbf{r}_N), \Psi(\mathbf{r}_1 \ldots \mathbf{r}_N), \{\nabla\Psi(\mathbf{r}_1 \ldots \mathbf{r}_N)\},\right.$$
$$\left.\{\nabla\Psi^*(\mathbf{r}_1 \ldots \mathbf{r}_N)\}\right], \tag{15.5}$$

where $\{\nabla\Psi(\mathbf{r}_1 \ldots \mathbf{r}_N)\}$ denotes the set of N gradient operators that act on the many-body wave function Ψ. The Lagrangian which yields the many-electron Schrödinger equation is given by

$$\mathcal{L} = \Psi^*(\mathbf{r}_1 \ldots \mathbf{r}_N)\left(E + \sum_k \frac{Ze^2}{r_k} - \frac{1}{2}\sum_{i,j}{}' \frac{e^2}{|\mathbf{r}_i - \mathbf{r}_j|}\right)\Psi(\mathbf{r}_1 \ldots \mathbf{r}_N)$$
$$-\frac{\hbar^2}{2m}\sum_i |\nabla_i\Psi(\mathbf{r}_1 \ldots \mathbf{r}_N)|^2. \tag{15.6}$$

For the chosen product form of our wave function, Eq. (15.1), Lagrange's equations for the individual $\psi_i(\mathbf{r})$ follow from setting the variational derivative $\delta S/\delta\psi_i^*$ to zero, which leads to the following set of equations (see Eq. (A.18) in Appendix A):

$$\frac{\partial\mathcal{L}}{\partial\psi_i^*} - \sum_{\alpha=1}^{3} \frac{\partial}{\partial x_\alpha} \frac{\partial\mathcal{L}}{\partial\left(\frac{\partial\psi_i^*}{\partial x_\alpha}\right)} = 0;\quad i = 1 \cdots N, \tag{15.7}$$

where α denotes the three components of $\mathbf{r}$.

The N constraints (15.4) are imposed by adding to the above variational derivatives, the variational derivative of each constraint multiplied by a Lagrange multiplier, λ_i:

$$\left[\frac{\delta S}{\delta\psi_i^*} + \lambda_i \frac{\delta C_i}{\delta\psi_i^*}\right]\delta\psi_i^* = 0. \tag{15.8}$$

[9] Otherwise the action is minimized by having all the individual wave functions vanish.

Performing these functional derivatives and carrying out the remaining integrations in S, making use of the normalization (15.4), yields

$$\left\{-\frac{\hbar^2}{2m}\nabla^2 - \frac{Ze^2}{r} + \lambda_i - E + \sum_{j\neq i}\int d^3r' \left[\frac{e^2}{|\mathbf{r}-\mathbf{r}'|}|\psi_j(\mathbf{r}')|^2 - \frac{\hbar^2}{2m}\psi_j^*(\mathbf{r}')\nabla'^2\psi_j(\mathbf{r}') - \frac{Ze^2}{r'}|\psi_j(\mathbf{r}')|^2\right] + \frac{e^2}{2}\sum_{k\neq i}\sum_{j\neq i}{}' \int d^3r' d^3r'' \frac{|\psi_j(\mathbf{r}')|^2\,|\psi_k(\mathbf{r}'')|^2}{|\mathbf{r}'-\mathbf{r}''|}\right\}\psi_i(\mathbf{r}) = 0, \tag{15.9}$$

where the prime in the last summation again denotes the exclusion of the j = k orbitals.[10] Now the integrals associated with the sixth, seventh, and eighth terms multiplying $\psi_i(\mathbf{r}_i)$ in Eq. (15.9) are *independent of* **r** and may therefore be treated as *constants*. We may then combine these three constants with λ_i and E to form different constants, ε_i, called *Hartree eigenvalues*, and Eq. (15.9) becomes

$$\left[-\frac{\hbar^2}{2m}\nabla^2 - \frac{Ze^2}{r} + e^2\sum_{j\neq i}\int d^3r' \frac{|\psi_j(\mathbf{r}')|^2}{|\mathbf{r}-\mathbf{r}'|}\right]\psi_i(\mathbf{r}) = \varepsilon_i\psi_i(\mathbf{r}). \tag{15.10}$$

The last term on the left-hand side of Eq. (15.10), called the *Hartree potential*, may be interpreted as an effective potential,[11]

$$V^H(\mathbf{r}) = e^2\sum_{j\neq i}^{N}\int d^3r' \frac{\psi_j^*(\mathbf{r}')\psi_j(\mathbf{r}')}{|\mathbf{r}-\mathbf{r}'|}. \tag{15.11}$$

Each of the equations (15.10) has the structure of a one-particle Schrödinger equation with eigenvalue ε_i. However they can be solved for a given orbital, ψ_i, only by assuming initial forms for each of the remaining orbitals, ψ_j. By solving all N such equations in this way we obtain a new set of functions ψ_i along with the eigen values ε_i for each i = 1 . . . N, a process referred to as an *iteration*. These new wave functions are, in turn, substituted into (15.10) to compute yet another set of ψ_i and ε_i. This process is continued until the ψ_i from successive iterations do not change (within some acceptable error). This final set of wave functions is then said to be *self consistent* (since they are consistent with themselves from one iteration to the next).

We could further constrain the wave functions by the orthonormality conditions

$$C_{ij}\left(\psi_i^*(\mathbf{r}), \psi_j(\mathbf{r})\right) = \int d^3r\psi_i^*(\mathbf{r})\psi_j(\mathbf{r})d^3r = \delta_{ij}.$$

Equation (15.7) would then become

$$\frac{\delta S}{\delta\psi_i^*} + \sum_j \lambda_{ij}\frac{\delta C_{ij}}{\delta\psi_i^*} = 0,$$

10 In performing a variation with respect to ψ_i^* it is useful to consider the terms i = j (or k) and i ≠ j (or k) separately.

11 Writing the charge density as $-|e|n_i(\mathbf{r})$, where $n_i(\mathbf{r}) = \sum_{j\neq i}\psi_j^*(\mathbf{r})\psi_j(\mathbf{r})$, we see that (15.10) is the usual expression for the electrostatic potential $\phi(\mathbf{r})$ arising from a charge distribution $-|e|n(\mathbf{r})$:

$$|e|\phi(\mathbf{r}) = e^2\int \frac{n(\mathbf{r}')}{|\mathbf{r}-\mathbf{r}'|}d^3r'.$$

where the λ_{ij} are again Lagrange multipliers. However this turns out to be unnecessary.[12]

To calculate the *total electronic energy* in the Hartree model we evaluate the expectation value of the total Hamiltonian, Eq. (15.3), with our product wave function (15.2),

$$E = \left\langle \Psi(\{\mathbf{r}_i\}) \left| \hat{H}^{(e)} \right| \Psi(\{\mathbf{r}_i\}) \right\rangle$$
$$= \int d^3r_1 \cdots d^3r_N \Psi^*(\mathbf{r}_1 \ldots \mathbf{r}_N) \sum_i \left[-\frac{\hbar^2}{2m}\nabla_i^2 - \frac{Ze^2}{r_i} + \frac{e^2}{2}\sum_j{}' \frac{1}{|\mathbf{r}_i - \mathbf{r}_j|} \right] \Psi(\mathbf{r}_1 \ldots \mathbf{r}_N).$$

Using Eq. (15.10) this equation becomes

$$E = \sum_i \varepsilon_i - \frac{e^2}{2}\sum_{i,j}{}' \int d^3r_i\, d^3r_j \frac{|\psi_i(\mathbf{r}_i)|^2\, |\psi_j(\mathbf{r}_j)|^2}{|\mathbf{r}_i - \mathbf{r}_j|}. \tag{15.12}$$

Note the total electronic energy is *not* the sum of the quantities ε_i, although those quantities approximately correspond to the ionization energies of specific orbitals, particularly for the inner shells.

For an atom with incompletely filled shells, the Hartree potential given by Eq. (15.11) is a function of the *vector* **r**. This greatly increases the complexity of applying the theory. It turns out to be a very good approximation to take a *spherical average* of this potential using the prescription

$$V^H(r) = \frac{1}{4\pi}\int d\Omega\, V^H(\mathbf{r}); \tag{15.13a}$$

this is called the *central field approximation.* It then follows that the Hartree wave functions can be written in the simple form[13]

$$\psi_i(r_i, \theta_i, \varphi_i) = R_{n_i l_i}(r_i) Y_{l_i l_{zi}}(\theta_i, \varphi_i), \tag{15.13b}$$

where (n_i, l_i, l_{zi}) denote the quantum numbers of the occupied orbitals from $1 \ldots N$. In particular, the Hartree partial differential equations then separate and we are led to a set of coupled total differential equations for the radial functions $R_{n_i l_i}(r_i)$.

The central field approximation is exact within the Hartree approximation for filled atomic shells. For unfilled shells there is a large degeneracy since, consistent with the Pauli principle, there are multiple ways to fill the l_{zi} and $s_z = \pm/2$ spin states associated with a given l. However these states actually have slightly different energies, with the lowest energy state being that given Hund's rules, and in the Hartree model the splittings arise from the non-spherical potential seen by these outer electrons.[14]

For molecules where we have $\psi_i(\mathbf{r}_i, \{\mathbf{R}_i\})$, the central field approximation fails for the valence electrons, since the atomic charge density redistributes in a spatially non-spherical manner to form the chemical bonds. Close to the atomic cores (i.e., for inner core electrons) a spherical distribution would remain a good approximation.

[12] For a proof see *Intermediate Quantum Mechanics*, H.A. Bethe and R. Jackiw, Benjamin-Cummings Publishing Co., Menlo Park, CA (1986). The matrix of coefficients λ_{ij} turns out to be Hermitian and can therefore be diagonalized and hence only diagonal Lagrange multipliers are required.

[13] Rather than treating the radial wave functions numerically, one can assume parameterized trial forms and adjust the parameters so as to give a self-consistent solution. One particularly simple set are the Slater orbitals, having the form $R_n(r) \propto r^{n-1}\exp(-\zeta r)$, where ζ involves an effective charge which accounts for screening, for which Slater gave rules; see, Atomic shielding constants, J.C. Slater, *Phys. Rev.* **36**, 57 (1930).

[14] For a discussion of the perturbation techniques used to calculate these splittings, see *Intermediate Quantum Mechanics*, H.A. Bethe and R. Jackiw, Benjamin-Cummings Publishing Co., Menlo Park, CA (1986).

15.1.2 The Hartree–Fock approximation

The set of equations (15.10) for the Hartree theory, which followed from writing the trial wave function as a simple product of one-electron orbital wave functions, can accommodate the Pauli principle by allowing a maximum of two electrons in each oribital (one for each spin). In the Hartree–Fock theory, we impose the additional Dirac requirement that the wave function changes sign on interchanging both the spatial and spin coordinates of any pair of electrons. To implement this antisymmetry we must allow each orbital i to be occupied by any of the electrons $j = 1, \ldots N$.

We start by introducing a symmetric product wave function that includes the spin coordinates, α, in the form

$$\Psi(\mathbf{r}_1\alpha_1, \mathbf{r}_2\alpha_2, \ldots\ldots \mathbf{r}_N\alpha_N) = \psi_1(\mathbf{r}_1\alpha_1)\psi_2(\mathbf{r}_2\alpha_2), \ldots\ldots \psi_N(\mathbf{r}_N\alpha_N). \quad (15.14)$$

In anticipation of antisymmetrizing this product, we write $\psi_i(\mathbf{r}_j\,\alpha_j)$, which designates a one-electron orbital with quantum numbers i associated with position *and* spin coordinates designated by j. Such a state may be further factored as

$$\psi_i(\mathbf{r}_j\alpha_j) = \psi_i(\mathbf{r}_j)\chi_i(\alpha_j). \quad (15.15a)$$

Here $\psi_i(\mathbf{r}_j)$ denotes the jth electron in the ith spatial orbital and the ith spin state; for the case of the atom in the central field approximation we may write these quantum numbers as $(n_i, l_i, l_{zi}, s_{zi})$, where $s_{zi} = \pm 1/2$. We may write these spin states as

$$\chi_{\frac{1}{2}}(\alpha_j) = \begin{pmatrix} 1 \\ 0 \end{pmatrix}; \quad \chi_{-\frac{1}{2}}(\alpha_j) = \begin{pmatrix} 0 \\ 1 \end{pmatrix}. \quad (15.15b)$$

As a trial wave function, Fock[15] used an *antisymmetric product* of one-electron wave functions which is formally generated from (15.14) by writing

$$\Psi(\mathbf{r}_1, \ldots \mathbf{r}_N; \alpha_1, \ldots \alpha_N)$$
$$= \frac{1}{\sqrt{N!}} \sum_{\nu} (-1)^{\nu} \hat{P}_{\nu} \left[\psi_1(\mathbf{r}_1)\psi_2(\mathbf{r}_2) \ldots \psi_N(\mathbf{r}_N)\chi_1(\alpha_1)\chi_2(\alpha_2) \ldots \chi_N(\alpha_N)\right], \quad (15.16)$$

where $\hat{P}_{\nu}$ is an operator which permutes the coordinates of the electrons. The sum extends over all possible N! permutations of the N electrons; these permutations are designated by the subscript ν, and the sign changes each time the coordinates $(\mathbf{r}_i, \alpha_i)$ and $(\mathbf{r}_j, \alpha_j)$ of a pair are permutated. The antisymmetric wave function represented by Eq. (15.16) is more conveniently written as a determinant (called a Slater determinate), which takes the form

[15] Approximation method for the solution of the quantum mechanical multibody problems, V. Fock, *Z. Phys.* **61**, 126 (1930).

$$\Psi(\mathbf{r}_1,\dots\mathbf{r}_N;\alpha_1,\dots\alpha_N)=\frac{1}{\sqrt{N!}}\begin{vmatrix}\psi_1(\mathbf{r}_1)\chi_1(\alpha_1) & \psi_1(\mathbf{r}_2)\chi_1(\alpha_2) & \cdots & \psi_1(\mathbf{r}_N)\chi_1(\alpha_N)\\ \psi_2(\mathbf{r}_1)\chi_2(\alpha_1) & \psi_2(\mathbf{r}_2)\chi_2(\alpha_2) & & \\ \vdots & & & \\ \psi_N(\mathbf{r}_1)\chi_N(\alpha_1) & & & \psi_N(\mathbf{r}_N)\chi_N(\alpha_N)\end{vmatrix}. \tag{15.17}$$

Let us examine two specific examples of an N = 2 case involving helium: the $1s^2$, spin-singlet ground state with $S_z = 0$ and the $1s^1 2s^1$ spin-triplet excited state with $S_z = 1$. For the first case the 1 and 2 orbitals are $\psi_{1s}(\mathbf{r}_1)\chi_\uparrow(\alpha_1)$ and $\psi_{1s}(\mathbf{r}_2)\chi_\downarrow(\alpha_2)$, which together with the two permutations (and their signs) 12 and −21 is written using (15.17) as

$$\begin{aligned}\Psi(\mathbf{r}_1,\mathbf{r}_2;\alpha_1,\alpha_2) &= \frac{1}{\sqrt{2}}\begin{vmatrix}\psi_{1s}(\mathbf{r}_1)\chi_\uparrow(\alpha_1) & \psi_{1s}(\mathbf{r}_2)\chi_\uparrow(\alpha_2)\\ \psi_{1s}(\mathbf{r}_1)\chi_\downarrow(\alpha_1) & \psi_{1s}(\mathbf{r}_2)\chi_\downarrow(\alpha_2)\end{vmatrix}\\ &= \psi_{1s}(\mathbf{r}_1)\psi_{1s}(\mathbf{r}_2)\frac{1}{\sqrt{2}}\left(\chi_\uparrow(\alpha_1)\chi_\downarrow(\alpha_2)-\chi_\downarrow(\alpha_1)\chi_\uparrow(\alpha_2)\right);\end{aligned} \tag{15.18a}$$

Note that we have a symmetric spatial form and antisymmetric spin form. For the second case we have the forms $\psi_{1s}(\mathbf{r}_1)\chi_\uparrow(\alpha_1)$ and $\psi_{2s}(\mathbf{r}_2)\chi_\uparrow(\alpha_2)$ and (15.17) gives

$$\begin{aligned}\Psi(\mathbf{r}_1,\mathbf{r}_2;\alpha_1,\alpha_2) &= \frac{1}{\sqrt{2}}\begin{vmatrix}\psi_{1s}(\mathbf{r}_1)\chi_\uparrow(\alpha_1) & \psi_{1s}(\mathbf{r}_2)\chi_\uparrow(\alpha_2)\\ \psi_{2s}(\mathbf{r}_1)\chi_\uparrow(\alpha_1) & \psi_{2s}(\mathbf{r}_2)\chi_\uparrow(\alpha_2)\end{vmatrix}\\ &= \frac{1}{\sqrt{2}}\left(\psi_{1s}(\mathbf{r}_1)\psi_{2s}(\mathbf{r}_2)-\psi_{2s}(\mathbf{r}_1)\psi_{1s}(\mathbf{r}_2)\right)\chi_\uparrow(\alpha_1)\chi_\uparrow(\alpha_2);\end{aligned} \tag{15.18b}$$

here the spin state is symmetric while the orbital state is antisymmetric. Note that for the case of the ground state the equations we would solve are identical for both the Hartree–Fock and Hartree cases.

Analogous to the Hartree case, we must now minimize the action (15.5) with respect to the $\psi_i^*(\mathbf{r})$ using Eq. (15.8), but now with a determinantal wave function of the form (15.14); in addition we must simultaneously impose the constraints, $\int d^3r\, \psi_i^*(\mathbf{r})\psi_i(\mathbf{r}) = 1$. Lumping various constant terms into the eigenvalues ε_i as before, this results in the *Hartree–Fock equations:*[16]

$$\left[-\frac{\hbar^2}{2m}\nabla^2-\frac{Ze^2}{r}+e^2\int d^3r'\sum_j{}'\frac{|\psi_j(\mathbf{r}')|^2}{|\mathbf{r}-\mathbf{r}'|}\right]\psi_i(\mathbf{r}) - e^2\sum_j{}'\delta_{s_i s_j}\int d^3r'\frac{\psi_j^*(\mathbf{r}')\psi_i(\mathbf{r})}{|\mathbf{r}-\mathbf{r}'|}\psi_j(\mathbf{r}') = \varepsilon_i\psi_i(\mathbf{r}). \tag{15.19}$$

[16] For the complete derivation see Bethe and Jackiw, ibid.

In addition to the Hartree potential, Eq. (15.11), we now have a *non-local potential*, which we can write as an integral operator of the form

$$\int d^3r' V_i^F(\mathbf{r},\mathbf{r}')\psi_i(\mathbf{r}') = \hat{V}_i^F \psi_i(\mathbf{r}), \tag{15.20a}$$

where the kernel $V_i^F(\mathbf{r},\mathbf{r}')$ is given by

$$V_i^F(\mathbf{r},\mathbf{r}') = -e^2 \sum_{j\neq i}^{N}{}' \delta_{s_i s_j} \frac{\psi_j^*(\mathbf{r}')\psi_j(\mathbf{r})}{|\mathbf{r}-\mathbf{r}'|}. \tag{15.20b}$$

We see that antisymmetrization of our wave function introduces an additional *non-local* behavior of the one-electron Schrödinger equation.

The Hartree–Fock theory yields the best solution to the many-electron Schrödinger equation that is obtainable using a trial wave function that has the form of an antisymmetrized product of one-electron wave functions. To obtain a better solution one must include variational forms that explicitly involve the coordinates of two or more electrons, e.g., $|\mathbf{r}_i - \mathbf{r}_j|$. Although the determinantal form vanishes when $\mathbf{r}_i = \mathbf{r}_j$ for parallel spin electrons, the repulsive Coulomb potential would also lead to a vanishing of the true wave function when $\mathbf{r}_i - \mathbf{r}_j = 0$ for antiparallel electrons. More generally the quasi-classical paths of all the electrons tend to adjust themselves to keep the electrons apart, while at the same time remaining close to the nucleus to take advantage of the strong attractive interaction to lower their collective energy.

The property of the many-electron wave function in which the electrons organize their motion to minimize their mutual repulsive Coulomb interaction is called *correlation*. The difference between the energy of a Hartree–Fock solution and the true (experimental) energy is called the *correlation energy*.[17] If we perform our calculations using the Hartree theory the corresponding energy difference is called the *exchange-correlation energy*. Since the parallel electrons are automatically kept far apart by the exchange interaction, there is a tendency for electrons having the same angular momentum and principal quantum numbers to adopt magnetic quantum numbers that produce the *highest* spin (consistent with the Pauli principle). This property of an atom is a component of Hund's rules, discussed earlier in Section 5.6.[18]

15.1.3 The exchange energy of a uniform electron liquid

We are now in a position to calculate the exchange energy of a uniform electron gas, which is the expectation value of the Coulomb energy,

$$\hat{V} = \frac{1}{2}\sum_{i,j}{}' \frac{e^2}{|\mathbf{r}_i - \mathbf{r}_j|}. \tag{15.21}$$

[17] Here we ignore any relativistic effects which also introduce energy shifts.

[18] For the attractive interaction among nucleons in a nucleus, a configuration of *lowest* spin is favored. This leads to the so-called odd–even effect of nuclear physics.

We write the total wave function as a product of a symmetric spin state and an antisymmetric orbital with the latter written as

$$\Psi(\mathbf{r}_1,\mathbf{r}_2,\ldots\ldots\mathbf{r}_N) = \frac{1}{\sqrt{N!}\,L^{3N/2}}\begin{vmatrix} e^{i\mathbf{k}_1\cdot\mathbf{r}_1} & e^{i\mathbf{k}_1\cdot\mathbf{r}_2} & \ldots\ldots & e^{i\mathbf{k}_1\cdot\mathbf{r}_N} \\ e^{i\mathbf{k}_2\cdot\mathbf{r}_1} & e^{i\mathbf{k}_2\cdot\mathbf{r}_2} & \ldots\ldots & e^{i\mathbf{k}_2\cdot\mathbf{r}_N} \\ \ldots\ldots & \ldots\ldots & \ldots\ldots & \ldots\ldots \\ e^{i\mathbf{k}_N\cdot\mathbf{r}_1} & e^{i\mathbf{k}_N\cdot\mathbf{r}_2} & \ldots\ldots & e^{i\mathbf{k}_N\cdot\mathbf{r}_N} \end{vmatrix}. \tag{15.22}$$

Because of the overall antisymmetry of our spatial wave function (15.22) it is sufficient to examine the expectation value of a single term in the sum (15.21), which for electrons 1 and 2 is

$$\left\langle \Psi \left| \frac{1}{2}\frac{e^2}{|\mathbf{r}_1-\mathbf{r}_2|} \right| \Psi \right\rangle = \left\langle \Psi(\mathbf{r}_1,\mathbf{r}_2,\ldots\ldots\mathbf{r}_N) \left| \frac{1}{2}\frac{e^2}{|\mathbf{r}_1-\mathbf{r}_2|} \right| \Psi(\mathbf{r}_1,\mathbf{r}_2,\ldots\ldots\mathbf{r}_N) \right\rangle$$

$$= \langle \Psi(\mathbf{r}_3,\mathbf{r}_4,\ldots\ldots\mathbf{r}_N) \,|\, \Psi(\mathbf{r}_3,\mathbf{r}_4,\ldots\ldots\mathbf{r}_N)\rangle \frac{1}{(N-1)\,(N)}\frac{1}{2}\frac{1}{L^6}\int\!\!\int d^3r_1 d^3r_2$$

$$\times \sum_{i=1,\,j=2}^{N,N} (-1)^{i+1}(-1)^j e^{-i\mathbf{k}_i\cdot\mathbf{r}_1}e^{-i\mathbf{k}_j\cdot\mathbf{r}_2}\frac{e^2}{|\mathbf{r}_1-\mathbf{r}_2|}\sum_{k=1,\,l=2}^{N,N}(-1)^{k+1}(-1)^l e^{i\mathbf{k}_k\cdot\mathbf{r}_1}e^{i\mathbf{k}_l\cdot\mathbf{r}_2}.$$

We first note that in computing the *total* Coulomb energy we must sum over all electrons in (15.21), which will produce a factor, $N(N-1)$; this cancels against the corresponding factor in the denominator. Next we rewrite the product of the four exponentials in the form $e^{-i\mathbf{k}_i\cdot\mathbf{r}_1}e^{-i\mathbf{k}_j\cdot\mathbf{r}_2}e^{i\mathbf{k}_k\cdot\mathbf{r}_1}e^{i\mathbf{k}_l\cdot\mathbf{r}_2} = e^{-i(\mathbf{k}_i-\mathbf{k}_k+\mathbf{k}_j-\mathbf{k}_l)\cdot(\mathbf{r}_1-\mathbf{r}_2)-i(\mathbf{k}_i-\mathbf{k}_k)\cdot\mathbf{r}_2}$. Integrating over d^3r_2 gives $L^3\delta_{ik}$ and that eliminates one of the sums (say i); the associated sum (over k) is then a dummy sum which gives a factor of N, the number of electrons. Carrying out the integration over $d^3\,|\mathbf{r}_1-\mathbf{r}_2|$ Fourier transforms the Coulomb potential to $V(\mathbf{k}_j-\mathbf{k}_l) = 4\pi e^2/\left|\mathbf{k}_j-\mathbf{k}_l\right|^2$ and we are left with

$$\langle\Psi|\,V\,|\Psi\rangle = \frac{N}{2L^3}\sum_l\sum_{j-l}\frac{4\pi e^2}{\left|\mathbf{k}_j-\mathbf{k}_l\right|^2}(-1)^j(-1)^l. \tag{15.23}$$

The term with $j = l$ in (15.23), which is positive and again involves a dummy sum, corresponds to the *Hartree contribution*,

$$V^H = V(\mathbf{q}=0)\,\frac{1}{2}\frac{N^2}{(2\pi)^3 L^3}; \tag{15.24a}$$

it is formally divergent, but it is cancelled by the presence of a uniform positive background which is assumed to be present.

The contribution from the double sum with $j \neq l$, which has a negative sign, constitutes the *Fock or exchange contribution.* Writing, $\mathbf{q} = \mathbf{k}_j-\mathbf{k}_l$, $\mathbf{k}_l = \mathbf{k}$ and setting $\sum_{\mathbf{k}} \to (L^3/(2\pi)^3)\int d^3k$, this part of (15.22) becomes

$$V^{EX} = -\frac{1}{2}\frac{L^3}{(2\pi)^6}\int d^3k d^3q n(\mathbf{k}-\mathbf{q})\,n(\mathbf{q})V(\mathbf{q})$$

$$= -\frac{1}{2}\frac{L^3}{(2\pi)^6}\int d^3q\frac{4\pi e^2}{q^2}\int d^3k n(\mathbf{k})n(|\mathbf{k}-\mathbf{q}|), \tag{15.24b}$$

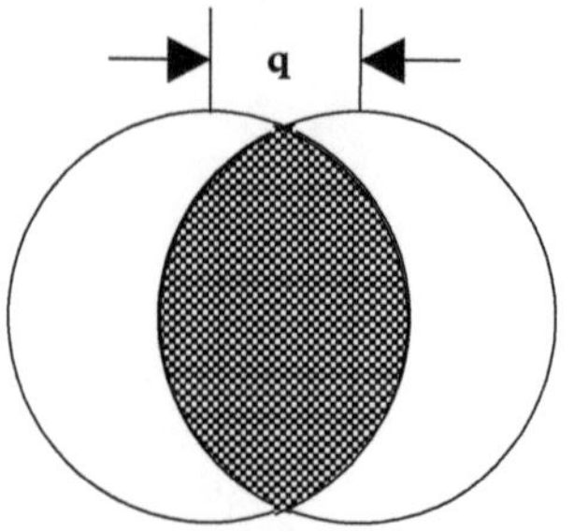

Figure 15.1 *The overlap of two Fermi spheres displaced by q.*

where n(**p**) is the Fermi distribution. The integral

$$I(q) = \int d^3k n(\mathbf{k}) n(|\mathbf{k} - \mathbf{q}|) \tag{15.25a}$$

is (see Fig. 15.1), for a given **q**, equal to the overlap volume of two spheres of radius p_F separated by a distance **q** which is given by

$$I(q) = \frac{4}{3}\pi \left[k_F^3 - \frac{3}{4} q k_F^2 + \frac{q^3}{16} \right]. \tag{15.25b}$$

Inserting this expression in (15.24b) and integrating over d^3q (with $0 < q < 2p_F$) gives (Dirac 1930)[19]

$$V^{EX} = -\frac{L^3 e^2}{4\pi^3} \left(\frac{p_F}{\hbar} \right)^4; \tag{15.26}$$

the negative sign implies that the exchange contribution lowers the energy of the system.[20] Using the expression $p_F = \hbar (3\pi^2 N/L^3)^{1/3}$ we may rewrite (15.26) as

$$\frac{V^{EX}}{N} = -\frac{3e^2}{4\pi} \left(\frac{p_F}{\hbar} \right) = -\frac{3e^2}{4\pi} \left(3\pi^2 n \right)^{1/3}, \tag{15.27}$$

where $n = N/L^3$ is the number density of electrons.

Following a suggestion of Slater (1951),[21] $V^{ex} \equiv V^{EX}/N$ is taken as an *exchange potential per electron* in an *inhomogeneous* system. Here $n = n(\mathbf{r})$ is taken to be the *local* density of electrons. This potential is then used along with the Hartree potential to perform self-consistent calculations of one-electron wave functions and energy levels of atoms, molecules, and solids. By adding a parameter such that $V^{ex} \to \alpha V^{ex}$, the resulting form has historically been used to model the effects of electron correlation; the resulting method is called the *x-α method.* Calculations by Kohn and Sham (1965) suggested that $\alpha = 2/3$.

15.2 Density functional theory (DFT)

The strategy of introducing an effective potential based on the electron density to account for exchange heralded a more general approach to electronic structure: *density functional theory* which will be developed in this and the following subsection.[22]

15.2.1 Local density theory

As discussed, Slater suggested adding the Dirac local exchange energy (per electron) to the electrostatic potential energy in the one-electron Schrödinger equation to obtain one-electron orbitals that (partly) include the effects of exchange. This model (or approximation), which incorporates an exchange potential based

[19] Note on exchange phenomena in the Thomas atom, Dirac, P.A.M., *Proc. Camb. Phil. Soc.* **26**, 376 (1930).

[20] That we obtain a negative contribution to the total energy is expected, since exchange keeps like spins farther from each other thereby lowering the Coulomb repulsion. At high densities this effect can dominate over the increase in kinetic energy resulting from an imbalance of spin populations, causing the system to become ferromagnetic; this effect was first recognized by Bloch.

[21] A simplification of the Hartree–Fock method, J.C. Slater, *Phys. Rev.* **81**, 385 (1951).

[22] The 1998 Nobel Prize in Chemistry was divided equally between Walter Kohn "for his development of the density-functional theory" and John A. Pople "for his development of computational methods in quantum chemistry."

on the uniform electron gas into a one-electron Schrödinger equation, forms the basis of a *local density approach* to electronic structure calculations.[23]

A local density calculation for an atom is performed using the following procedures. One starts with an approximate potential energy function for the atom, $e\phi(\mathbf{r})$, as might be obtained from a Thomas–Fermi or a Hartree model.[24] Using Poisson's equation we obtain $n(\mathbf{r})$ and with it the exchange potential, using Eq. (15.27). The model then assumes an effective potential given by

$$V^{eff}(\mathbf{r}) = e\phi(\mathbf{r}) + V^{ex}(\mathbf{r}). \tag{15.28}$$

Inserting (15.28) in the one-electron Schrödinger equation,

$$\left[-\frac{\hbar^2}{2m}\nabla^2 + V^{eff}(\mathbf{r})\right]\psi_i(\mathbf{r}) = \varepsilon_i\psi_i(\mathbf{r}), \tag{15.29}$$

and carrying out a numerical integration yields a set of one-electron wave functions, $\psi_i(\mathbf{r})$. This completes the initial iteration.

Further iterations are carried out by using the wave functions from the previous iteration to calculate the local electron density, $n(\mathbf{r})$, acting on the ith orbital arising from the remaining $N-1$ orbitals,[25]

$$n_i(\mathbf{r}) = \sum_{j=1}^{N}{}'\psi_j^*(\mathbf{r})\psi_j(\mathbf{r}). \tag{15.30}$$

The Coulomb potential energy is given by the usual expression

$$\begin{aligned} e\phi_i(\mathbf{r}) &= -\frac{Ze^2}{r} + V^H \\ &= -\frac{Ze^2}{r} + e^2\int d^3\mathbf{r}'\frac{n_i(\mathbf{r})}{|\mathbf{r}-\mathbf{r}'|} \end{aligned} \tag{15.31}$$

and the exchange energy is again calculated using (15.27). Forming the new (i-dependent) effective potential as

$$V_i^{eff}(\mathbf{r}) = e\phi_i(\mathbf{r}) + V^{ex}(\mathbf{r}), \tag{15.32}$$

inserting it in (15.29), and numerically integrating to obtain a new set of ψ_i completes another iteration. The process is continued until the charge densities and wave functions from successive iterations reproduce each other with sufficient accuracy.

Similar procedures may be used to calculate the orbitals of a molecule. Suppose our molecule consists of M atoms (which may have the same or different Z). The nuclear potential can then be written as

$$V^N(\mathbf{r}) = -e^2\sum_{m=1}^{M}\frac{Z_m}{|\mathbf{r}-\mathbf{R}_m|}, \tag{15.33}$$

[23] Slater's approach is therefore a hybrid of the Thomas–Fermi–Dirac method and the Schrödinger equation. The electrostatic and exchange energies enter in the same way as the Thomas–Fermi–Dirac method, but the kinetic energy is replaced by the operator $-\hbar^2\nabla^2/2m$. It is therefore initially surprising that the theory produces significantly improved numerical results.

[24] This can be obtained from the Thomas–Fermi theory; see Landau and Lifshitz, *Quantum Mechanics*, Section 70, Pergamon Press (1977).

[25] In atomic calculations one usually performs a spherical averaging of $n(\mathbf{r})$ according to $n_i(r) = \frac{1}{4\pi}\int d\Omega\, n_i(\mathbf{r})$.

where $\mathbf{R}_m$ denotes the coordinates of the mth nucleus, having charge Z_m, measured from some convenient reference point. We do not know the equilibrium coordinates $\mathbf{R}_m$ and it is necessary to assume some values (e.g., the experimental values or estimates taken from the behavior of similar molecules). We must construct an initial charge density; a reasonable choice is a superposition of the self-consistent atomic charge densities. From this initial charge density we construct the exchange potential energy and the electrostatic potential energy. Using the expression

$$-|e|\phi_i(\mathbf{r}) = -e^2\sum_{m=1}^{M}\frac{Z_m}{|\mathbf{r}-\mathbf{R}_m|} + e^2\sum_{j=1}^{N}{}' \int d^3r' \frac{\psi_j^*(\mathbf{r}')\psi_j(\mathbf{r}')}{|\mathbf{r}-\mathbf{r}'|}, \tag{15.34}$$

where the second sum runs over all orbitals on all atoms except the ith. Because of the lowered symmetry of the molecule a direct numerical integration of the Schrödinger equation may not be practical. We may then use a variational approach to solve Schrödinger's equation by introducing a set of appropriate basis functions. By appropriately summing the single particle energies we can (being careful not to double count the interaction energies) obtain the total energy of the molecule,

$$E = E(\mathbf{R}_1 \ldots \mathbf{R}_M). \tag{15.35}$$

15.2.2 Potential energy landscape and the elastic constants of molecules

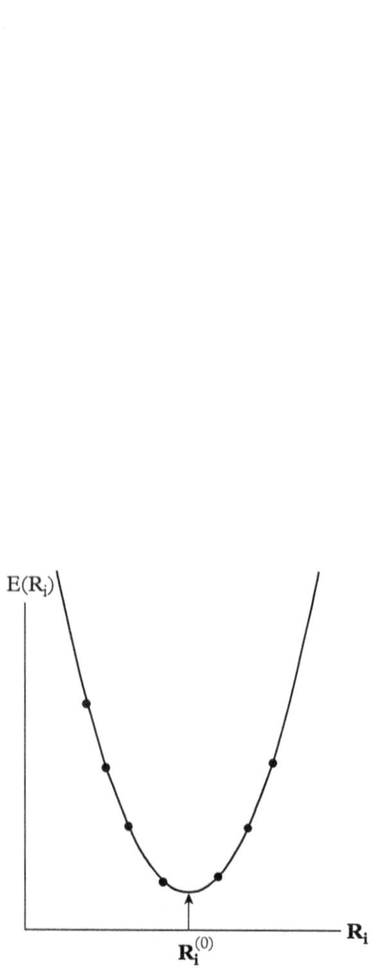

Figure 15.2 *Schematic behavior of the total energy for various values of some interatomic spacing* $\mathbf{R}_i$. *The equilibrium spacing,* $\mathbf{R}_i^{(0)}$, *can be found by interpolation with a quadratic form.*

By varying the atomic position coordinates, $\mathbf{R}_m$, until E passes through a global minimum, we obtain the ground state energy and the *equilibrium positions* of the molecules. This behavior is shown schematically for a single nuclear coordinate in Fig. 15.2. The *force constants* associated with various vibrational modes of the molecule are then obtained from

$$\kappa_{ij} = \frac{\partial^2 E}{\partial \mathbf{R}_i \partial \mathbf{R}_j}. \tag{15.36}$$

Writing the equilibrium coordinates as $\mathbf{R}_i^{(0)}$ and defining $\mathbf{U}_i = \mathbf{R}_i - \mathbf{R}_i^{(0)}$, we may write the Lagrangian for the molecular vibrations as

$$L = \sum_{i=1}^{M}\frac{1}{2}M_i\dot{U}_i^2 - \frac{1}{2}\sum_{i,j}^{M} \mathbf{U}_i \cdot \kappa_{ij} \cdot \mathbf{U}_j. \tag{15.37}$$

The normal modes and the associated frequencies are then calculated using standard procedures.

15.2.3 The Hohenberg–Kohn theorem

The discussion leading to Eq. (15.27) showed that, at least in the local density model, it is possible to write the total energy of a system as a functional of the electron density, $n(\mathbf{r})$. It is therefore natural to ask if this is true more generally; i.e., is it an exact model-independent property of a many-electron system? A theorem can be proved in this connection.[26] Suppose we are given a non-relativistic system of N spinless fermions[27] in a volume L^3 which are interacting via the two-body Coulomb potential,[28]

$$V^I = \frac{1}{2}\sum_{i,j=1}^{N}{}' \frac{e^2}{|\mathbf{r}_i - \mathbf{r}_j|}. \tag{15.38}$$

Suppose also that the ground state many-body wave function $\Psi_G(\mathbf{r}_1 \ldots\ldots \mathbf{r}_N)$ is non-degenerate. We may define a particle density, $n(\mathbf{r})$, from Ψ_G according to[29]

$$n(\mathbf{r}) = N\int |\Psi_G(\mathbf{r}, \mathbf{r}_2, \mathbf{r}_3 \ldots \mathbf{r}_N)|^2\, d^3r_2 \ldots d^3r_N; \tag{15.39a}$$

note that

$$n(\mathbf{r}) \geqq 0,$$

and, from the fact that Ψ is normalized,

$$\int d^3r\, n(\mathbf{r}) = N. \tag{15.39b}$$

Let the system be subjected to a one-body external potential, V^E, which acts on the coordinates of all the electrons and contributes a term,

$$V^E = \sum_{i=1}^{N} V(\mathbf{r}_i), \tag{15.40}$$

to the total Hamiltonian, $\hat{H}$, of the system. An important example of the "external" potential is the Coulomb potential of the nucleus (acting on the "gas" of interacting electrons surrounding it), or the potential of a collection of nuclei, as in a molecule or solid. Including the kinetic energy,

$$\hat{T} = -\frac{\hbar^2}{2m}\sum_{i=1}^{N} \nabla_i^2, \tag{15.41}$$

the total Hamiltonian is then

$$\hat{H} = \hat{T} + V^I + V^E. \tag{15.42}$$

[26] Inhomogeneous electron gas, P. Hohenberg and W. Kohn, *Phys. Rev.* **136**, 864B (1964).

[27] A relativistic treatment requires the introduction of spin wave functions, which may be suppressed in a non-relativistic treatment.

[28] The theorem may be generalized to the case of a more general potential involving two or more body forces.

[29] Note that since Ψ_G is completely antisymmetric, we may choose any $\mathbf{r}_i$ and then integrate over the remaining N–1 coordinates.

The theorem states that, under the above conditions, the electron density, $n(\mathbf{r})$, resulting from a given external potential, V^E, is *unique*. It is sometimes convenient to regard the electron density as the independent variable, in which case it then uniquely determines the external potential.[30] The proof involves a *reducto ad absurdum* argument. Let us assume that we have two external potentials V^E and V'^E leading to the *same* density $n(\mathbf{r})$ (it will turn out to not be true). The associated Hamiltonians are $\hat{H} = \hat{T} + V^I + V^E$ and $\hat{H}' = \hat{T} + V^I + V'^E$, and we will denote the corresponding many-body ground state wave functions as Ψ_G and Ψ'_G. The ground state energy, E_G, associated with $\hat{H}$ is clearly

$$E_G = \left\langle \Psi_G \left| \hat{H} \right| \Psi_G \right\rangle. \tag{15.43}$$

Now $\hat{H}'$ is assumed to differ from $\hat{H}$ (because we assume that V^E differs from V'^E). From the variational theorem of quantum mechanics we know that the expectation value of $\hat{H}$ with respect to any wave function other than Ψ_G must exceed E_G; denoting that other wave function by Ψ'_G we have

$$E_G < \left\langle \Psi'_G \left| \hat{H} \right| \Psi'_G \right\rangle = \left\langle \Psi'_G \left| \hat{H}' - V'^E + V^E \right| \Psi'_G \right\rangle. \tag{15.44}$$

Now it follows from Eqs. (15.39a,b) and (15.40) that the expectation value of V^E can be written as

$$\begin{aligned}\left\langle \Psi_G \left| V^E \right| \Psi_G \right\rangle &= \sum_{i=1}^{N} \int d^3r_1 d^3r_2 d^3r_3 \cdots d^3r_N V(\mathbf{r}_i) \left|\Psi_G(\mathbf{r}_1, \mathbf{r}_2, \mathbf{r}_3, \cdots \mathbf{r}_N)\right|^2 \\ &= N \int d^3r d^3r_2 d^3r_3 \cdots d^3r_N V(\mathbf{r}) \left|\Psi_G(\mathbf{r}, \mathbf{r}_2, \mathbf{r}_3, \cdots \mathbf{r}_N)\right|^2 \\ &= \int d^3r\, n(\mathbf{r}) V(\mathbf{r}),\end{aligned} \tag{15.45}$$

which again follows from the symmetry Ψ_G with respect to all its coordinates. We may then write (15.44) as

$$E_G < E'_G + \int d^3r \left[V(\mathbf{r}) - V'(\mathbf{r})\right] n(\mathbf{r}), \tag{15.46}$$

where here we have incorporated our assumption that both Ψ_G and Ψ'_G, although presumed different, lead through (15.39a) to the same density, $n(\mathbf{r})$. If we invoke the same procedures with the primed and unprimed quantities interchanged, we obtain the inequality

$$E'_G < E_G + \int d^3r \left[V'(\mathbf{r}) - V(\mathbf{r})\right] n(\mathbf{r}). \tag{15.47}$$

Adding (15.46) and (15.47), we obtain

$$E_G + E'_G < E'_G + E_G. \tag{15.48}$$

[30] A shift of V^E by an arbitrary constant is, of course, allowed.

Since this is impossible, our assumption must be incorrect; i.e., different external potentials V'^{E} must lead to different densities, $n'(\mathbf{r})$ or, equivalently, $n(\mathbf{r})$ is *unique* to a given V^E. We may therefore say that the total ground state energy, E_G, is a unique functional of $n(\mathbf{r})$ for a specified V^E.

The tantalizing aspect of this theorem is that it suggests that it might be possible to *actually calculate* the ground state energy by some procedure which *only* involves the function $n(\mathbf{r})$, which has *three* variables (x,y,z), rather than solving the Schrödinger equation involving the 3N coordinates of all N electrons. Let us write the ground state energy as the sum

$$E_G[n(\mathbf{r})] = F[n(\mathbf{r})] + \int V(\mathbf{r})n(\mathbf{r})d^3r, \tag{15.49}$$

where

$$F[n(\mathbf{r})] \equiv \left\langle \Psi_G \left| \hat{T} + V^I \right| \Psi_G \right\rangle. \tag{15.50}$$

We now show that the ground state energy $E_G[n(\mathbf{r})]$ has its minimum value for a given V^E when $n(\mathbf{r})$ has the form it would have for the ground state solution to the full Schrödinger equation. Suppose Ψ'_G is the ground state solution to the N-particle Schrödinger equation for an external potential $V'^E \neq V^E$. If we use this wave function to calculate the ground state energy for V^E using the variational approach, we would evaluate

$$\begin{aligned} E_G[\Psi'_G] &= \left\langle \Psi'_G \left| \hat{T} + V^I \right| \Psi'_G \right\rangle + \left\langle \Psi'_G \left| V^E \right| \Psi'_G \right\rangle \\ &= F[n'(\mathbf{r})] + \int V(\mathbf{r})\, n'(\mathbf{r}) d^3r \\ &> F[n(\mathbf{r})] + \int V(\mathbf{r})\, n(\mathbf{r}) d^3r. \end{aligned} \tag{15.51}$$

Clearly E_G would achieve its global minimum for $\Psi'_G = \Psi_G$ and hence for $n'(\mathbf{r}) = n(\mathbf{r})$. Thus the ground state energy, considered as a functional of the density $n(\mathbf{r})$, achieves its minimum value for that $n(\mathbf{r})$, which results from the solution of the associated N-body Schrödinger equation with the given V^E.[31]

Written out explicitly, the kinetic energy and potential energy terms entering $F[n(\mathbf{r})]$ are

$$\begin{aligned} \left\langle \Psi_G \left| \hat{T} \right| \Psi_G \right\rangle &= -\frac{\hbar^2}{2m} \int d^3r_1 \cdots d^3r_N\, \Psi_G^*(\mathbf{r}_1 \cdots \mathbf{r}_N) \sum_{i=1}^{N} \nabla_i^2 \Psi_G(\mathbf{r}_1 \cdots \mathbf{r}_N) \\ &= -N\frac{\hbar^2}{2m} \int d^3r \lim_{\mathbf{r}' \to \mathbf{r}} d^3r_2 \cdots dr_N^3\, \nabla^2 \Psi_G^*(\mathbf{r}', \mathbf{r}_2 \cdots \mathbf{r}_N) \Psi_G(\mathbf{r}, \mathbf{r}_2 \cdots \mathbf{r}_N) \\ &= -\frac{\hbar^2}{2m} \int d^3r \lim_{\mathbf{r}' \to \mathbf{r}} \nabla^2 \rho^{(1)}(\mathbf{r}, \mathbf{r}'), \end{aligned} \tag{15.52}$$

[31] The theorem is restricted to those $n(\mathbf{r})$ that are "V^E representable"; i.e., those $n(\mathbf{r})$ obtained from solutions to the Schrödinger equation for the given form of V^E.

where

$$\rho^{(1)}(\mathbf{r},\mathbf{r}') = N\int d^3r_2\cdots dr_N^3\,\Psi_G^*(\mathbf{r}',\mathbf{r}_2\cdots\mathbf{r}_N)\,\Psi_G(\mathbf{r},\mathbf{r}_2\cdots\mathbf{r}_N) \tag{15.53}$$

is the one-particle reduced density matrix. Similarly the potential energy term,

$$\begin{aligned}\left\langle\Psi_G\left|V^I\right|\Psi_G\right\rangle &= -\int d^3r_1\cdots d_N^3\,\Psi_G^*(\mathbf{r}_1\cdots\mathbf{r}_N)\sum_{i>j}^{N}\frac{e^2}{|\mathbf{r}_i-\mathbf{r}_j|}\Psi_G(\mathbf{r}_1\cdots\mathbf{r}_N)\\ &= -N(N-1)\int d^3r\,d^3r'\,d^3r_3\cdots dr_N^3\,\Psi_G^*(\mathbf{r},\mathbf{r}',\mathbf{r}_3\cdots\mathbf{r}_N)\frac{e^2}{|\mathbf{r}-\mathbf{r}'|}\\ &\quad\times\Psi_G(\mathbf{r},\mathbf{r}',\mathbf{r}_3\cdots\mathbf{r}_N)\\ &= -e^2\int d^3r d^3r'\frac{1}{2}\frac{\rho^{(2)}(\mathbf{r},\mathbf{r}')}{|\mathbf{r}-\mathbf{r}'|},\end{aligned} \tag{15.54}$$

where $\rho^{(2)}(\mathbf{r},\mathbf{r}')$ is defined by

$$\rho^{(2)}(\mathbf{r},\mathbf{r}') = N(N-1)\int d^3r_3\cdots dr_N^3\left|\Psi_G(\mathbf{r},\mathbf{r}',\mathbf{r}_3\cdots\mathbf{r}_N)\right|^2. \tag{15.55}$$

It is instructive to rewrite this quantity as

$$\rho^{(2)}(\mathbf{r},\mathbf{r}') = n(\mathbf{r})n(\mathbf{r}') + \rho'^{(2)}(\mathbf{r},\mathbf{r}'), \tag{15.56}$$

where

$$n(\mathbf{r}) = \rho^{(1)}(\mathbf{r},\mathbf{r}) \tag{15.57}$$

is the particle density and $\rho'^{(2)}(\mathbf{r},\mathbf{r}')$ is the non-separable part of $\rho^{(2)}(\mathbf{r},\mathbf{r}')$.

15.2.4 Application of density functional theory; Kohn–Sham theory

If the previously proved theorems are to have any utility we need to be able to construct a calculational scheme that exploits them. To implement the Kohn–Hoenberg theorem, Kohn and Sham[32] introduced a set of single-particle *Schrödinger-like* equations from which the ground state energy and density, may, in principle, be derived. In what follows we argue heuristically; for more rigorous discussions see the references in the *Additional Reading* section.

In the Kohn–Sham theory $F[n(\mathbf{r})]$ is split into *three* parts: (i) a kinetic energy that is calculated using a set of *single-particle* orbitals, $\psi_i(\mathbf{r})$, one for each particle (and now referred to as Kohn–Sham orbitals); (ii) that part of the electron interaction that is captured by the Hartree approximation; and (iii) a third term, $E^{XC}[n(\mathbf{r})]$, called the exchange-correlation energy, that absorbs the effects not included in the first two contributions. We then have

$$F[n(\mathbf{r})] = K[n(\mathbf{r})] + \frac{e^2}{2}\int d^3r'\,d^3r\frac{n(\mathbf{r})\,n(\mathbf{r}')}{|\mathbf{r}-\mathbf{r}'|} + E^{XC}[n(\mathbf{r})], \tag{15.58}$$

[32] Self-consistent equations including exchange and correlation effects, W. Kohn and L.J. Sham, *Phys. Rev.* **140**, 1133A (1965).

where $K[n(\mathbf{r})]$ has the same form as the kinetic energy in the Hartree theory

$$K[n(\mathbf{r})] = -\frac{\hbar^2}{2m}\int d^3r \sum_{i=1}^{N} \left(\nabla_i \psi_i^*(\mathbf{r})\right)\left(\nabla_i \psi_i(\mathbf{r})\right). \tag{15.59}$$

and the second term is the Hartree potential energy. The orbitals $\psi_i(\mathbf{r})$ are connected to the particle density by the requirement

$$n(\mathbf{r}) = \sum_{i=1}^{N} \psi_i^*(\mathbf{r})\psi_i(\mathbf{r}) \tag{15.60}$$

and the sum only involves occupied orbitals.

As noted, the Hartree term does not correspond to the true many-body electron–electron interaction energy due to correction term $\rho'^{(2)}(\mathbf{r},\mathbf{r}')$ in Eq. (15.56), while $K[n(\mathbf{r})]$ is not the true total kinetic energy, which cannot be represented through single-particle orbitals alone. Physically, the total Coulomb repulsion energy will be lower, since the electrons avoid each other due to both the exclusion principle and correlations in their motion. At the same time the total kinetic energy will be higher due to the additional motion involved in producing the correlations. But given that the H–K theorem tells us that the ground state energy is a functional of the density (a property both of these forms satisfy), we can formally transfer the parts that are "left out" to the third term, $E^{XC}[n(\mathbf{r})]$; then the challenge is to construct a reasonable representation for it.

To obtain the equations of motion we must minimize E_G with respect to $n(\mathbf{r})$, subject to the constraint (15.39b) on the number of particles,

$$\delta E_G[n(\mathbf{r})] = \int \delta n(\mathbf{r}) \left[\frac{\delta K[n(\mathbf{r})]}{\delta n(\mathbf{r})} + V(\mathbf{r}) + e^2 \int \frac{n(\mathbf{r}')}{|\mathbf{r}-\mathbf{r}'|} d^3r' + \frac{\delta E^{XC}[n(\mathbf{r})]}{\delta n(\mathbf{r})} - \mu\right] d^3r,$$
$$= 0 \tag{15.61}$$

where μ is the Lagrange multiplier associated with the particle number constraint and is therefore the chemical potential.[33] We combine the second, third, and fourth terms in (15.61) into an *effective* potential acting on the electrons,

$$\hat{V}^{eff}(\mathbf{r}) = V(\mathbf{r}) + e^2 \int \frac{n(\mathbf{r})}{|\mathbf{r}-\mathbf{r}'|} d^3r' + \hat{V}^{XC}(\mathbf{r}), \tag{15.62}$$

where

$$\hat{V}^{XC}(\mathbf{r}) \equiv \frac{\delta E^{XC}[n(\mathbf{r})]}{\delta n(\mathbf{r})}; \tag{15.63}$$

[33] As with the Hartree and Hartree–Fock theories we must impose the constraint that the number of particles is fixed; otherwise the energy can be trivially minimized by setting $n(\mathbf{r}) = 0$.

the latter is, in general, expected to be a *non-local* (i.e., an integral) operator involving **r** and **r**′ (as we encountered in the Hartree–Fock exchange integral). The resulting equations can be written as

$$\left[-\frac{\hbar^2}{2m}\nabla^2 + V(\mathbf{r}) + e^2 \sum_{j=1}^{N}{}' \int \frac{\psi_j^*(\mathbf{r}')\,\psi_j(\mathbf{r}')}{|\mathbf{r}-\mathbf{r}'|} d^3r' + \hat{V}^{XC}(\mathbf{r})\right] \psi_i(\mathbf{r}) = \varepsilon_i \psi_i(\mathbf{r}). \tag{15.64}$$

(As in the Hartree theory, several constants were combined here to make new constants, ε_i.) The second and third terms constitute the total electrostatic potential,

$$e\phi_i(\mathbf{r}) = V(\mathbf{r}) + e^2 \sum_{j=1}^{N}{}' \frac{\psi_j^*(\mathbf{r}')\,\psi_j(\mathbf{r}')}{|\mathbf{r}-\mathbf{r}'|} d^3r', \tag{15.65}$$

where $V(\mathbf{r})$ is given by Eq. (15.33). As was the case for the Hartree–Fock theory we can then write

$$\hat{V}^{\text{eff}}(\mathbf{r}) = e\phi(\mathbf{r}) + \hat{V}^{XC}. \tag{15.66}$$

The quantities operating on $\psi_i(\mathbf{r})$ are Hermitian and therefore the $\psi_i(\mathbf{r})$ can be constructed to be orthonormal. The ψ_i are often interpreted as one-electron orbitals in the Hartree or Hartree–Fock sense; however, strictly speaking they are only a set of functions introduced to evaluate the ground state energy.

To calculate the total energy we start by taking the expectation value of Eq. (15.64), and using the orthonormal property, we obtain

$$\begin{aligned}\sum_i \varepsilon_i &= \sum_i \langle \psi_i(\mathbf{r})| -\frac{\hbar^2}{2m}\nabla^2 + \hat{V}^{\text{eff}}(\mathbf{r})\,|\psi_i(\mathbf{r})\rangle \\ &= K + \int d^3r\, n(\mathbf{r}) \left(V(\mathbf{r}) + e^2 \int d^3r \frac{n(\mathbf{r}')}{|\mathbf{r}-\mathbf{r}'|} + V^{XC}(\mathbf{r}) \right),\end{aligned} \tag{15.67}$$

where we have restricted to a local density functional in the last step. Writing (15.49) in the form

$$E_G = K + \int d^3r\, V(\mathbf{r})n(\mathbf{r}) + \frac{e^2}{2} \int d^3r \int d^3r' \frac{n(\mathbf{r}')n(\mathbf{r})}{|\mathbf{r}-\mathbf{r}'|} + E^{XC} \tag{15.68}$$

and using (15.67) to eliminate $K + \int d^3r\, n(\mathbf{r})V(\mathbf{r})$, we find

$$E_G = \sum_i \varepsilon_i - \frac{1}{2} \int \frac{n(\mathbf{r})\, n(\mathbf{r}')}{|\mathbf{r}-\mathbf{r}'|} d^3r\, d^3r' - \int V^{XC}(\mathbf{r})n(\mathbf{r})d^3r, \tag{15.69}$$

where it is understood that the self-interaction term is omitted in the second term on the right. Thus, given some expression for the local exchange–correlation potential, we can evaluate the ground state energy of our many electron system.

One widely used form for the local exchange–correlation potential is

$$V^{XC}(\mathbf{r}) = \beta(n(\mathbf{r}))\mu^X(n(\mathbf{r})), \tag{15.70}$$

where

$$\mu^X(n) = \frac{e^2}{\pi}\left(3\pi^2 n\right)^{1/3}, \tag{15.71a}$$

$$\beta(n) = 1 + 0.7734X(n)\,ln\left(1 + \frac{1}{X(n)}\right), \tag{15.71b}$$

and

$$X = \frac{1}{21}\left(\frac{1}{4\pi n}\right)^{1/3}, \tag{15.71c}$$

where $n(\mathbf{r})$ is again the local electron density; this form was given by Hedin and Lundquist[34] and is fitted to the many-body calculations of Singwi et al.[35] for the energy of an electron gas.

The local density approximation, LDA, and a generalization called the local spin density approximation, LSDA, work surprisingly well for atoms,[36] giving good agreement between the total energies obtained from Hartree–Fock[37] and higher-order Thomas–Fermi calculations.[38] It also gives the correct overall behavior of the ionization energies (calculated as differences of total energies), at the eV level for a quantity that ranges over 25 eV across the periodic table. Finally, the LSDA yields spin polarization (Hund's rule) corrections to the total energies, where the maximum corrections occur for half-shell filling, as expected. However as usually practiced it ignores the effects of an electron interacting with itself, as was clear from the derivation of the exchange term in Section 15.2.3; methods have been developed to address this problem.[39]

Much effort has gone into trying to construct an optimum exchange correlation potential. Surprisingly most attempts to generalize to a non-local form lead to little improvement. In general, we may write the effect of $\hat{V}^{XC}(\mathbf{r})$ on $\psi(\mathbf{r})$ as an integral operator of the form

$$\hat{V}^{XC}\psi_i(\mathbf{r}) = \int V^{XC}(\mathbf{r},\mathbf{r}')\,\psi_i(\mathbf{r}')d^3r'. \tag{15.72}$$

For a further discussion of this still evolving field, the reader should consult the review article by Kohn and Vashishta.[40]

[34] Explicit local exchange-correlation potentials, L. Hedin and B.I. Lundqvist, *J. Phys.* C4, 2064 (1971).

[35] Electron correlations at metallic densities: 4, K.S. Singwi, A. Sjolander, M.P Tosi, and R.H. Land, *Phys. Rev. B* 1, 1044 (1970).

[36] Local-density-functional calculations of the energy of atoms, S. Kotochigova, Z.H. Levine, E.L. Shirley, M.D. Stiles, and C.W. Clark, *Phys. Rev. A* 55, 191 (1997).

[37] F. Herman and S. Skillman, *Atomic Structure Calculations*, Prentice-Hall, Englewood Cliffs, NJ (1963); R.D. Cowan, *The Theory of Atomic Structure and Spectra*, University of California, Berkeley, CA, p. 237 (1981).

[38] Thomas-Fermi Theory Revisited, E.H. Lieb and B. Simon, *Phys. Rev. Lett.* 31, 681 (1973). These authors develop a large Z expansion of the total energy and show that the T-F theory is exact in the limit $Z \to \infty$.

[39] Self interaction correction to density functional approximations in many electron systems, J.P. Perdew and A. Zunger, *Phys. Rev. B* 23, 5048 (1981).

[40] W. Kohn and P. Vashishta in *Theory of the Inhomogeneous Electron Gas*, S. Lundquist and N.H. March, Eds. Plenum, NY (1983). Two books on the subject are *Density Functional Theory*, R.M. Dreizler and E.K.U. Gross, Springer Verlag, NY (1990) and *Quantum Theory of the Electron Liquid*, Gabriele Giuliani and Giovanni Vignale, Cambridge University Press (2007).

15.3 Alternative strategies based on density functional theory

In the Kohn–Sham procedure, the kinetic energy and Hartree terms are separated out of the full Hamiltonian leaving an exchange correlation potential operator, $\hat{V}^{XC}$, which, in turn, reduces to a function in the local density approximation (LDA); in particular, it involves the individual K–S orbitals *only* through the condition (15.60). Within the local density approximation other strategies have been implemented which have certain advantages (see below); among these are the so-called Hartree–Fock–Kohn–Sham and the screened exchange LDA (sX-LDA), which we outline below.

15.3.1 The Hartree–Fock–Kohn–Sham (HFKS) approximation

Rather than lumping the exchange and correlation energies together, one could evaluate the Fock exchange contribution explicitly using Eq. (15.20) in terms of the K–S orbitals, leaving only the correlation energy as a functional of the density; the latter could then be evaluated in a local approximation. In the SCF electron gas model for the correlation energy, the contributions would follow from the third and higher terms on the right-hand side of Eq. (15B.41). Proceeding in this way, the HFKS Hamiltonian would take the form

$$\left[-\frac{\hbar^2}{2m}\nabla^2-\frac{Ze^2}{r}+V_H(\mathbf{r})+\hat{V}_F+\hat{V}^C(n(\mathbf{r}))\right]\psi_i(\mathbf{r})=\varepsilon_i\psi_i(\mathbf{r}). \qquad (15.73)$$

As with atoms and molecules, evaluating the exchange term adds considerable complexity; in a solid it would involve a sum over the Brillouin zone (with the singular portion separated out and evaluated analytically). One must still go on and obtain an expression for the total energy.[41]

15.3.2 The screened exchange approximation

For the electron gas, the addition of exchange causes a logarithmic singularity in the density of states (see Problem 15.4) that is not observed experimentally; by incorporating screening, as in the SCF or Thomas–Fermi models, we remove this unphysical behavior. Hence it is natural to screen the bare interaction in the HFKS model by, e.g., incorporating a factor $\exp(-\kappa_{TF}|\mathbf{r}-\mathbf{r}'|)$ in the Fock operator in Eq. (15.73); this is the basis of the sX-LDA model. The procedure must be carried out in a consistent manner such that the density-dependent part of the potential compensates for the screening in the Fock potential, and that allows an evaluation of the total energy.

[41] Various generalized Kohn–Sham schemes are discussed by A. Seidl, A. Gorling, P. Vogel, J.A. Majewski, and M. Levy, *Phys. Rev. B* 53, 3764 (1996).

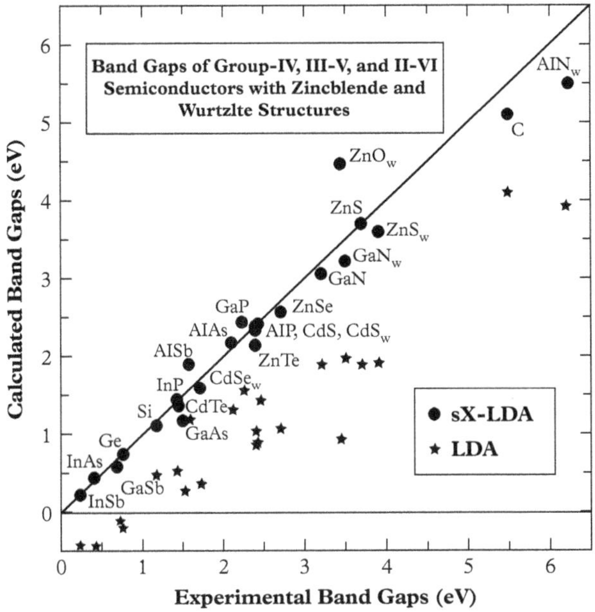

Figure 15.3 *A plot of the experimental versus calculated LDA and sX-LDA values of the band gap for semiconductors having the zinc blende and wurtzite crystal structures (after Geller et al.).*

15.3.3 Excited states

Keeping in mind that the Kohn–Sham (K–S) formalism is only applicable to calculating ground state properties, we can nonetheless examine the difference between some occupied initial state eigenvalue and an unoccupied final state eigenvalue, as a model for an electron–hole pair excitation. For semiconductors and insulators, the band gap, defined as the energy eigenvalue difference between the highest occupied state and the lowest unoccupied state,[42] significantly disagrees with experiment, even when various ground state properties (e.g., atomic separations) are in excellent agreement. It turns out that the disagreement is greatly reduced if the bands are calculated using the sX-LDA model. Figure 15.3 shows a plot of the experimental band gap versus the theoretical value, as calculated in the LDA and also the sX-LDA, by Geller et al.[43] for most of the semiconductors having the zinc-blende and wurtzite crystal structures. Note that the values arising from the screened exchange are in substantially better agreement.

The semiconductor systems mentioned are generally formed from the s and p shells of the constituent atoms. However, density functional theory and its simpler extensions, as discussed up to this point, often fails for systems containing atoms with d and f electrons having narrow bands: classes of systems exist for which metallic behavior is predicted by DFT, whereas insulators involving large excitation energies are observed experimentally. This failure is partly because strong local correlations are present in the localized orbitals of these atoms which cannot be represented with typical exchange correlation potentials. We will return to this problem in Chapter 45.

[42] For molecules, these two states are designated as the *highest occupied molecular orbital* (HOMO) and the *lowest unoccupied molecular orbital* (LUMO).

[43] Computational band-structure engineering of III–V semiconductor alloys, C.B. Geller, W. Wolf, S. Picozzi, A. Continenza, R. Asahi, W. Mannstadt, A.J. Freeman, and E. Wimmer, *Appl. Phys. Lett.* **79**, 368 (2001). Also see: Electronic structure theory for ground and excited state properties of materials, A.J. Freeman, R. Asahi, A. Continenza, and R. Wu, in *Solid State Photoemission and Related Methods*, Wiley-VCH (2003).

FURTHER READING

Many important aspects of atomic physics, including the Hartree and Hartree–Fock theories, are discussed in *Intermediate Quantum Mechanics*, H.A. Bethe and R. Jackiw, Benjamin-Cummings Publishing Co., Menlo Park, CA (1986).

The Hartree–Fock method is also discussed in *The Modern Theory of Solids,* F. Seitz, McGraw-Hill, NY (1940) and in *Quantum Mechanics,* second edition, L.I. Schiff, McGraw-Hill (1968).

For a book-length treatment of density functional theory see, *Density Functional Theory,* R.M. Dreizler and E.K.U. Gross, Springer, Berlin (1990).

Linear response in uniform systems is extensively discussed in, *The Theory of Quantum Liquids,* D. Pines and P. Nozieres, W.A. Benjamin, Inc., NY (1966).

Density functional theory is also discussed in Chapter 7 of the book *Quantum Theory of the Electron Liquid,* by Gabriele F. Giuliani and Giovanni Vignale, Cambridge University Press (2007). Another treatment is given in Theory and practice of density-functional theory, by Peter E. Blöchl, in Chapter 2 of *The LDA+DMFT Approach to Strongly Correlated Materials,* Eva Pavarini, Erik Koch, Dieter Vollhardt, and Alexander Lichtenstein (Eds.), published by Forschungszentrum Jülich GmbH Institute for Advanced Simulations ISBN 978-3-89336-734-4.

Additional references on the semiconductor band gap problem are: Good semiconductor band gaps with a modified local-density approximation, D.M. Bylander and Leonard Kleinman, *Phys. Rev. B* **41**, 7868 (1990), and Optical properties and electronic structures of semiconductors with screened-exchange LDA, R. Asahi, W. Mannstadt, and A.J. Freeman, *Phys. Rev. B* **59**, 7486 (1999).

As noted earlier, a parameterized and widely used form for the exchange–correlation potential that can be conveniently implemented for molecular and solid-state calculations is given in the paper, Explicit local exchange-correlation potentials, L. Hedin and B.I. Lundqvist, *J. Phys. C: Solid State Physics* **4**, 2064 (1971).

For a description of important historical milestones in the development of multiple scattering theory that led up to the KKR formalism see, Early history of multiple scattering theory for ordered systems, J. Korringa, *Phys. Rep.* **238**, 341 (1994).

For three accounts of the history of density functional theory, see: The beginnings and some thoughts on the future, P.C. Hohenberg, W. Kohn, and L.J. Sham, *Adv. Quantum Chem.* **21**, 7 (1990); A half century of density functional theory, A. Zangwill, *Phys. Today* (July 2015); and W. Kohn, Nobel Lecture: Electronic structure of matter—wave functions and density functionals, *Rev. Mod. Phys.* **71**, 1253 (1999).

Problems

15.1. Carry out the calculations involving Lagrange's equations leading to Eq. (15.8).

15.2. Carry out the calculations leading to (15.23), (15.25b), (15.26), and (15.27).

15.3. Show that you can write the $1s^2 2s^2$ ground state wave function for the Be atom as a product of space orbitals and spin orbitals. Can you do the same for the $1s^2 2s$ ground state of the Li atom?

15.4. Apply Eq. (15.19) to the case of a free electron gas, where the nuclear Coulomb term is replaced by a uniform positive background, and all states are taken to be plane waves. Show that this leads to an expression for the single particle energies $\varepsilon(\mathbf{k})$ of the form

$$\varepsilon(\mathbf{k}) = \frac{\hbar^2 k^2}{2m} - \int_{k'<k_F} \frac{d^3k'}{(2\pi)^3} \frac{4\pi e^2}{|\mathbf{k}-\mathbf{k}'|^2}$$
$$= \frac{\hbar^2 k^2}{2m} - \frac{2e^2}{\pi} F\left(\frac{k}{k_F}\right),$$

where

$$F(x) = \frac{1}{2} + \frac{1-x^2}{4x} \ln \frac{1+x}{1-x},$$

and that this form leads to a singularity in the heat capacity. Now use the expression for $\varepsilon(\mathbf{k})$ to evaluate the total ground state energy and show that the correction is identical to that obtained in Eq. (15.26).

15.5. Fill in the steps leading from (15.61) to (15.64).

Appendix 15A Multiple scattering theory

Whether symmetry is present or not, density functional theory (DFT) can, in principle, always be practiced within the framework of multiple scattering theory[44], as we now describe. A particular one-electron state $\psi(\mathbf{r})$ of a system of scattering centers (an array of atoms in a molecule or a cluster of atoms) can be regarded as arising from some initial incoming wave function, $\chi(\mathbf{r})$ (which is a solution of the free particle Schrödinger equation), together with a superposition of the outgoing waves, $\psi_n^{(o)}$, scattered by *all* the atoms,

$$\psi(\mathbf{r}) = \chi(\mathbf{r}) + \sum_n \psi_n^{(o)}(\mathbf{r}_n); \tag{15A.1}$$

here n denotes the individual atoms, $\mathbf{R}_n$ the position of the nth atom relative to some chosen origin, and $\mathbf{r}_n \equiv \mathbf{r}-\mathbf{R}_n$, the coordinate relative to that atom.[45] On the

[44] Some of the material in this appendix was drawn from, The modern theory of alloys, J.S. Faulkner, *Prog. Mater. Sci.* 27, 1 (1982), Pergamon Press, Great Britain.

[45] Assuming no currents are present a sum over the "incoming" waves from *all* atoms must result in a standing wave, one with no net flux.

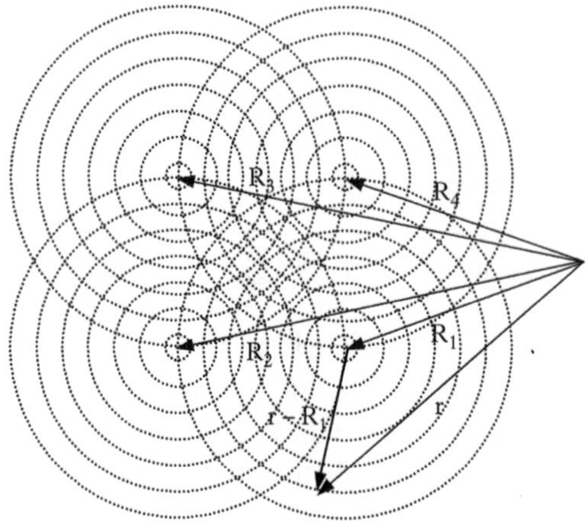

Figure 15A.1 *Outgoing spherical waves, $\psi_n^{(0)}$, from a hypothetical square molecule, along with the vectors, R_n, specifying the atom positions relative to an arbitrary origin. The incoming wave, $\psi_4^{(i)}$, on atom 4 results from outgoing waves from atoms 1, 2, and 3.*

other hand, the wave function that is incoming on a *particular* atom n, $\psi_n^{(i)}(\mathbf{r})$, results from $\chi(\mathbf{r})$ together with the outgoing waves from all the *remaining* atoms,

$$\psi_n^{(i)}(\mathbf{r}_n) = \chi(\mathbf{r}) + \sum_{n' \neq n} \psi_{n'}^{(o)}(\mathbf{r}_{n'}). \tag{15A.2}$$

A situation involving four scatterers is shown in Fig. 15A.1. This description goes back to Rayleigh[46] and Ewald[47]. For bound states (with negative energy relative to the vacuum), $\chi(\mathbf{r}) = 0$. A state of the whole system, $\psi(\mathbf{r})$, then arises from the mutual superposition of the (asymptotically decaying) outgoing waves produced by all the atoms; this is the case for a molecule.[48] For an infinite system we can also set $\chi(\mathbf{r}) = 0$ at positive energies, since there are no boundaries and hence no incoming or outgoing waves. We then have what can be pictured as a "steady-state" situation.

15A.1 Solution for a model potential

To solve these coupled equations we must relate $\psi_n^{(i)}(\mathbf{r}_n)$ and $\psi_n^{(o)}(\mathbf{r}_n)$. We will assume we can write the total potential of the system as a sum of local potentials, v_n, arising from atom n centered at position $\mathbf{R}_n$ in the form

$$V(\mathbf{r}) = \sum_n v_n(\mathbf{r}_n). \tag{15A.3}$$

where $\mathbf{r}_n = \mathbf{r} - \mathbf{R}_n$. For a binary alloy the potential on site n would be that produced by either an A or B atom: $v_n = v_A$ or v_B. Although it is not essential, given that the theory as written is linear, the potentials $v_n(\mathbf{r}_n)$ are in practice constrained to lie within non-overlapping polyhedra constructed around each atom n. The formalism is greatly simplified if the v_n are further restricted to have the muffin tin form and are furthermore spherically symmetric.

[46] J.W. Strutt, Baron Rayleigh, *Phil. Mag.* **34**, 4816 (1892).
[47] P.P. Ewald, *Ann. Phys.* IV, **49**, I (1916).
[48] Self-consistent-field Xα cluster method for polyatomic molecules and solids, J.C. Slater and K.H. Johnson, *Phys. Rev. B* 5, 844 (1972).

The outgoing wave from the site n follows from formulating the Schrödinger equation as an integral equation and iterating it, taking the incoming wave $\psi_n^{(i)}(\mathbf{r}'_n)$ on this site as given. The resulting equation is

$$\psi_n^{(o)}(\mathbf{r}_n) = \int d^3r' \int d^3r'' \Big[G^{(0)}(\mathbf{r},\mathbf{r}';\varepsilon)v_n(\mathbf{r}'_n)\delta(\mathbf{r}'-\mathbf{r}'') + G^{(0)}(\mathbf{r},\mathbf{r}';\varepsilon)v_n(\mathbf{r}'_n)\,G^{(0)}(\mathbf{r}',\mathbf{r}'';\varepsilon)v_n(\mathbf{r}''_n) + \cdots\Big]\psi_n^{(i)}(\mathbf{r}''_n), \quad (15A.4a)$$

which we rewrite as

$$\psi_n^{(o)}(\mathbf{r}_n) = \int d^3r'\,d^3r''G^{(0)}(\mathbf{r},\mathbf{r}';\varepsilon)t_n(\mathbf{r}',\mathbf{r}'';\varepsilon)\psi_n^{(i)}(\mathbf{r}''_n); \quad (15A.4b)$$

here, $G^{(0)}$ is the free particle Green's function, and in the second step we have introduced t_n, that part of the so-called T matrix which acts at the site n, as

$$t_n(\mathbf{r}',\mathbf{r}'';\varepsilon) = v_n(\mathbf{r}'_n)\Big[\delta(\mathbf{r}'-\mathbf{r}'') + G^{(0)}(\mathbf{r}',\mathbf{r}'';\varepsilon)\,v_n(\mathbf{r}''_n) + \int d^3r_1\,G^{(0)}(\mathbf{r}',\mathbf{r}_1;\varepsilon)v_n(\mathbf{r}_{1n})\,G^{(0)}(\mathbf{r}_1,\mathbf{r}'';\varepsilon)v_n(\mathbf{r}''_n) + \cdots\Big], \quad (15A.5a)$$

or, on recognizing t_n within the series,

$$t_n(\mathbf{r}',\mathbf{r}'';\varepsilon) = v_n(\mathbf{r}'_n)\delta(\mathbf{r}'-\mathbf{r}'') + v_n(\mathbf{r}'_n)\int d^3r_1G^{(0)}(\mathbf{r}',\mathbf{r}_1;\varepsilon)t_n(\mathbf{r}_1,\mathbf{r}''). \quad (15A.5b)$$

The first term in (15A.4a) is the equivalent of the Born approximation. Of course the $\psi_n^{(i)}(\mathbf{r}_n)$ at the site n in (15A.4b) actually arises from the $\psi_{n'}^{(o)}(\mathbf{r}_{n'})$ from all sites $n \neq n'$ through Eq. (15A.2) (and in the case of positive energies involves the external wave function, $\chi(\mathbf{r})$, which is here unspecified); hence these equations must be solved simultaneously for all sites n. Inserting (15A.2) into (15A.4b) yields the following set of liner homogeneous integral equations:

$$\psi_n^{(o)}(\mathbf{r}_n) = \sum_{n'\neq n}\int d^3r'\,d^3r''\,G^{(0)}(\mathbf{r},\mathbf{r}';\varepsilon)t_n(\mathbf{r}',\mathbf{r}'';\varepsilon)\psi_{n'}^{(o)}(\mathbf{r}''_{n'}). \quad (15A.6)$$

Inserting (15A.4b) into (15A.1) and continuing to take $\chi(\mathbf{r}) = 0$, the total wave function is then

$$\psi(\mathbf{r}) = \sum_n \psi_n^{(o)}(\mathbf{r}_n) = \sum_{\substack{n,n'\\ n'\neq n}}\int d^3r'd^3r''G^{(0)}(\mathbf{r},\mathbf{r}';\varepsilon)t_n(\mathbf{r}',\mathbf{r}'';\varepsilon)\psi_{n'}^{(o)}(\mathbf{r}''_{n'}). \quad (15A.7)$$

15A.2 Relating multiple scattering theory to the Kohn–Rostoker Green's function method

In Section 13.2 we discussed the Kohn-Rostoker, Green's function based method for calculating energies and wave functions in periodic solids; we now sketch how this method can be related to the multiple scattering approach used by Korringa. As assumed in Section 13.2, the calculations are greatly simplified by working in the muffin tin approximation; this also allows us to assign a unique potential to each atom. We further assume that for $\mathbf{r}$ centered on atom n, $v_n(\mathbf{r}) = v_n(r)$; i.e., the potential energy function is spherically symmetric (or averaged in some way to have this form) and vanishes for $r > r_i$, where r_i is the muffin tin radius. In the region $r > r_i$ the functions $\psi_n^{(i)}$ and $\psi_n^{(o)}$ must satisfy the potential free Schrödinger equation. Since $\psi_n^{(i)}(\mathbf{r})$ must be regular when continued within a sphere, it can be expanded in terms of spherical Bessel functions, $j_l(\kappa r)$, as

$$\psi_n^{(i)} = \sum_{l,m} a_{l,m}^{(n)} j_l(\kappa r_n) Y_{l,m}(\mathbf{n}_{\mathbf{r}_n}). \tag{15A.8}$$

On the other hand, the outgoing wave $\psi_n^{(o)}(\mathbf{r})$ must be singular at the origin of a sphere and can be expanded in outgoing spherical Hankel functions, $h_l^{(+)}$, as

$$\psi_n^{(o)} = \sum_{l,m} b_{l,m}^{(n)} h_l^{(+)}(\kappa r_n) Y_{l,m}(\mathbf{n}_{\mathbf{r}_n}). \tag{15A.9}$$

Here $\mathbf{n}_{\mathbf{r}_n}$ is a unit vector in the direction $\mathbf{r}_n$, and κ is given by (13.29). These expansions must be consistent with Eqs. (15.1) and (15A.2). This leads to two conditions relating the expansion coefficients, which we give without proof.

The first condition follows from inserting (15A.8) and (15A.9) into Eq. (15A.5) and leads to

$$b_{l,m}^{(n)} = -i\kappa\, t_l^{(n)} a_{l,m}^{(n)}; \tag{15A.10}$$

Here, $t_l^{(n)}$ is the lth component of amplitude for the scattering between incoming and outgoing spherical waves with quantum numbers l, m from a single site.[49] The $t_l^{(n)}$ can be related to scattering phase shifts as

$$\begin{aligned} t_l^{(n)} &\equiv \frac{1}{\kappa} e^{i\delta_l^{(n)}} \sin \delta_l^{(n)} \\ &= \frac{1}{2\kappa i}(e^{2i\delta_l^{(n)}} - 1); \end{aligned} \tag{15A.11}$$

here, the $\delta_l^{(n)}$ are obtained by matching the logarithmic derivative of the radial wave functions, R_l, through Eq. (13.61b), to the potential free form for $r > r_i$. The R_l themselves are obtained by integrating the radial Schrödinger equation for a given potential energy function, $v_n(r_n)$.

[49] As treated in texts on quantum mechanics, e.g., Landau and Lifshitz, the asymptotic form of the outgoing wave function for scattering of an external plane wave $\chi = e^{ikz}$ by a spherically symmetric potential v(r) has the form $\psi^{(o)}(\mathbf{r}) = f(\theta)e^{ikr}/r$, where

$$f(\theta) = \sum_{l=0}^{\infty} f_l(\theta) = \frac{1}{2ik}\sum_{l=0}^{\infty}(2l+1)(e^{2i\delta_l} - 1) \times P_l(\cos\theta),$$

which we can rewrite as

$$f(\theta) = \sum_{l=0}^{\infty} t_l (2l+1) P_l(\cos\theta).$$

The second condition, which arises from (15A.2), leads to

$$-i\kappa a^{(n)}_{l,m} = \sum_{n'\neq n} g^{(n,n')}_{l,m;l',m'} b^{(n')}_{l'm'}, \tag{15A.12}$$

where the expansion coefficients are given by

$$g^{(nn')}_{l,m;l',m'} = -4\pi i\kappa i^{l-l'} \sum_{L,M} i^{-L} C^{LM}_{l,m;l',m'} h^{(+)}_L(\kappa\,|\mathbf{R}_{nn'}|) Y_{LM}(\mathbf{n}_{\mathbf{R}_{nn'}}); \tag{15A.13}$$

here, $\mathbf{R}_{nn'} = \mathbf{R}_n - \mathbf{R}'_n$ and the $C^{LM}_{l,m;l',m'}$ are the Clebsch–Gordon coefficients given by Eq. (13.56). Substituting (15A.13) into (15A.10) yields a set of homogeneous linear equations in the coefficients $b^{(n')}_{l'm'}$. Values of κ at which the determinant of these coefficients vanishes give the allowed energies. The size of this determinant will be dictated by the total number of atoms and all the quantum numbers associated with the basis states used to represent each atom.

States are filled in the order of increasing energy until the total number of electrons in the system is accommodated (two electrons per level to allow for spin). From the wave functions within the muffin spheres one can construct a new charge density and potential energy function which, in turn, will yield new radial wave functions. This process is repeated until self-consistency is achieved. As described, the technique is applicable to a molecule (a cluster of atoms which in general will differ from each other) allowing one to calculate (within the muffin tin approximation) total energies of given configurations within DFT, and with it bond lengths and other properties, as mentioned earlier in Section 15.2.2.

Up to this point no use has been made of any available symmetries. In a periodic crystal the expansion coefficients in (15A.8) and (15A.9) must obey the Bloch symmetry conditions

$$a^{n'}_{lm} = e^{i\mathbf{k}\cdot\mathbf{R}_{nn'}} a^{n}_{lm} \tag{15A.14a}$$

and

$$b^{n'}_{lm} = e^{i\mathbf{k}\cdot\mathbf{R}_{nn'}} b^{n}_{lm}. \tag{15A.14b}$$

Applying these conditions and continuing the analysis, which we do not do, leads to a condition identical to Eq. (13.63) obtained earlier from the Green's function variational approach; in this way the Korringa multiple scattering and Kohn–Rostoker Green's function approaches can be seen to be equivalent.

Appendix 15B The uniform electron liquid

Evaluating the ground state energy of the electron liquid involves several distinct and somewhat subtle ideas. We will break the overall calculation into a series of parts, which are treated in separate subsections which follow. These parts are then assembled to obtain the final result. We outline the plan immediately below.

15B.1: We start by deriving the *Hellman–Feynman theorem*, which expresses the derivative of the ground state energy with respect to some parameter λ, in terms of the expectation value of the derivative of the Hamiltonian with respect to that same parameter (see Eq. (15B.4)). Through replacing e^2 by λe^2 we can use the Hellman–Feynman theorem to gradually "switch on" the Coulomb interaction, $\langle\psi_0| V^{(2)} |\psi_0\rangle$; integrating the expression as λ varies between 0 and 1, we obtain a formal expression for the ground state energy.

15B.2: The expectation value of $V^{(2)}$ can be related to an expectation value involving the number density operator, $\hat{n}(\mathbf{r})$, or equivalently its Fourier transform, $\hat{n}(\mathbf{q})$ (leading to Eq. (15B.11)).

15B.3: We then consider the change in the expectation value of $\hat{n}(\mathbf{q})$, or more generally $n(\mathbf{q}, \omega)$, induced by an external potential $\phi(\mathbf{q}, \omega)$ (as expressed by Eq. (15B.23)). This leads naturally to the introduction of a *generalized susceptibility*, $\chi(\mathbf{q}, \omega)$ (Eq. (15B.25)).

15B.4: The external potential $\phi(\mathbf{q}, \omega)$ is then assumed to be produced by an external charge density, $en_e(\mathbf{q}, \omega)$; these two quantities are related by Poisson's equation. The external charge can then be related to the change in the total charge density, $en(\mathbf{q}, \omega)$ (external charge plus induced charge) through a *dielectric function*, $\varepsilon(\mathbf{q}, \omega)$, which in turn involves $\chi(\mathbf{q}, \omega)$ (through Eq. (15B.29)).

15B.5: We now assemble these parts so as to express the expectation value of the Coulomb energy in terms of the imaginary part of $\varepsilon^{-1}(\mathbf{q}, \omega)$ (Eq. (13.34)). This completes the formal part of the calculation.

15B.6: As an *approximation* we can calculate the ground state energy using the self-consistent potential model for $\varepsilon(\mathbf{q}, \omega)$ given in Eq. (14.15). The resulting expressions for the real and imaginary parts obtained on integrating (14.15) are given as Eqs. (15B.37) and (15B.38). Using these resulting forms for $\varepsilon(\mathbf{q}, \omega)$ we can calculate $\text{Im}\varepsilon^{-1}(\mathbf{q}, \omega)$. Finally, to evaluate the ground state energy we must carry out the integral over λe^2; the result, first obtained by Gell-Mann and Brueckner and given as Eq. (15B.41), is then expressed as an expansion in terms of a dimensionless "length" denoted as r_s, with small values of r_s corresponding to high electron densities.

15B.1 The Hellman–Feynman theorem

We will start by deriving an expression for the ground state energy in terms of the ground state wave function. We write the Hamiltonian in the form

$$\hat{H} = \hat{H}_0 + \lambda \hat{V}^{(2)}, \tag{15B.1}$$

where $\hat{H}_0$ and $\hat{V}^{(2)}$ are the kinetic energy and Coulomb potential operators, respectively, and λ is a parameter measuring the strength or the electron–electron interaction, which we assume we can vary. If we are given the many-body ground state wave function,

$$\psi_0 = \psi_0(\mathbf{r}_1, \mathbf{r}_2, \ldots .\mathbf{r}_N), \tag{15B.2}$$

then the ground state energy is simply the expectation value of (15B.1) evaluated with this wave function. In Section 15B.5 we will obtain an expression for the expectation value of $\hat{V}^{(2)}$ in terms of the dielectric function. However, we also need to know the kinetic energy, which is clearly not given by the non-interacting value, since in the presence of the Coulomb repulsion the electrons execute complicated "motions" to avoid each other. However we can use a trick which we now describe.

The ground state energy as a function of λ is clearly given by

$$E(\lambda) = \left\langle \psi_0(\lambda) \,|\, \hat{H}(\lambda) \,|\, \psi_0(\lambda) \right\rangle; \tag{15B.3}$$

differentiating with respect to λ yields

$$\begin{aligned} \frac{\partial E}{\partial \lambda} &= \left\langle \frac{\partial \psi_0}{\partial \lambda} \left| \hat{H}(\lambda) \right| \psi_0(\lambda) \right\rangle + \left\langle \psi_0(\lambda) \left| \frac{\partial \hat{H}}{\partial \lambda} \right| \psi_0(\lambda) \right\rangle + \left\langle \psi_0(\lambda) \left| \hat{H}(\lambda) \right| \frac{\partial \psi_0(\lambda)}{\partial \lambda} \right\rangle \\ &= E(\lambda) \frac{\partial}{\partial \lambda} \langle \psi_0(\lambda) \,|\psi_0(\lambda) \rangle + \left\langle \psi_0(\lambda) \left| \frac{\partial \hat{H}}{\partial \lambda} \right| \psi_0(\lambda) \right\rangle \\ &= \left\langle \psi_0(\lambda) \left| \frac{\partial \hat{H}}{\partial \lambda} \right| \psi_0(\lambda) \right\rangle; \end{aligned} \tag{15B.4}$$

where we have assumed that $\psi_0(\lambda)$ is normalized for all λ in the last step. For our assumed linear dependence on λ we can write

$$\begin{aligned} \left\langle \psi_0(\lambda) \left| \frac{\partial \hat{H}(\lambda)}{\partial \lambda} \right| \psi_0(\lambda) \right\rangle &= \left\langle \psi_0(\lambda) \left| \hat{V}^{(2)} \right| \psi_0(\lambda) \right\rangle \\ &= \frac{1}{\lambda} \left\langle \psi_0(\lambda) \left| \lambda \hat{V}^{(2)} \right| \psi_0(\lambda) \right\rangle. \end{aligned} \tag{15B.5}$$

Note this expression is not as simple as it appears since the expectation value is a non-trivial function of λ. Integrating (15B.4) from $\lambda = 0$ to 1 and using (15B.5) we have

$$E = E(0) + \int_0^1 \frac{d\lambda}{\lambda} \left\langle \psi_0(\lambda) \,|\, \lambda \hat{V}^{(2)} \,|\, \psi_0(\lambda) \right\rangle, \tag{15B.6}$$

where E(0) is the ground state kinetic energy of a *non-interacting* Fermi gas,

$$E^{(0)} = \frac{3}{5}N\mu^{(0)}$$

with

$$\mu^{(0)} = \frac{\hbar^2}{2m}\left(3\pi^2\frac{N}{V}\right)^{2/3}$$

(evaluated at the *same density* as the final, interacting gas).[50] Equation (15B.5) is an application of the Hellman–Feynman theorem; it looks deceptively simple, but its application requires a knowledge of the matrix element of $\hat{V}^{(2)}$ with the exact ground state wave function $\psi(\lambda)$ for all λ from 0 to 1, which we pursue in the next three subsections.

15B.2 The Coulomb interaction in terms of the number operator

It turns out to be useful to derive an expression for the expectation value of the Coulomb interaction in terms of the *particle density operator*, $\hat{n}(\mathbf{r}, \{\mathbf{r}_i\})$, which is defined as

$$\hat{n}(\mathbf{r}, \{\mathbf{r}_i\}) = \sum_i \delta(\mathbf{r} - \mathbf{r}_i). \tag{15B.7}$$

Here, $\{\mathbf{r}_i\} = \mathbf{r}_1, \mathbf{r}_2, \ldots\ldots \mathbf{r}_N$, which we will often not write explicitly in what follows. The Fourier transform of (15B.7) is

$$\hat{n}(\mathbf{q}) = \int d^3r \sum_i \delta(\mathbf{r} - \mathbf{r}_i)e^{i\mathbf{q}\cdot\mathbf{r}} = \sum_i e^{i\mathbf{q}\cdot\mathbf{r}_i}, \tag{15B.8a}$$

while the inverse transform is

$$\hat{n}(\mathbf{r}) = \frac{1}{(2\pi)^3}\int d^3q\,\hat{n}(\mathbf{q})e^{-i\mathbf{q}\cdot\mathbf{r}}. \tag{15B.8b}$$

The Coulomb part of the Hamiltonian is given by

$$\hat{V}^{(2)} = \frac{1}{2}\sum_{i\neq j} V_{ij}^{(2)} = \frac{1}{2}\sum_{i\neq j}\frac{e^2}{|\mathbf{r}_i - \mathbf{r}_j|}, \tag{15B.9}$$

where the factor 1/2 avoids double counting; the required expectation value is then

$$\left\langle \psi_0 \middle| \hat{V}^{(2)} \middle| \psi_0 \right\rangle = \left\langle \psi_0 \middle| \frac{1}{2}\sum_{i\neq j}\frac{e^2}{|\mathbf{r}_i - \mathbf{r}_j|} \middle| \psi_0 \right\rangle. \tag{15B.10}$$

[50] One might think that one only needs to calculate the expectation value $\langle\psi|\hat{H}|\psi\rangle$ for $\lambda = 1$. This works for the contribution from $\hat{V}^{(2)}$. However we are then left with the non-trivial problem of calculating $\langle\psi|\hat{H}_0|\psi\rangle$. The trick of integrating over $d\lambda$ avoids this problem.

We now rewrite (15B.10) as

$$\left\langle \psi_0 \left| \hat{V}^{(2)} \right| \psi_0 \right\rangle = \langle \psi_0 | \int d^3r \int d^3r' \frac{1}{2} \frac{e^2}{|\mathbf{r}-\mathbf{r}'|} \left[\sum_{i,j} \delta(\mathbf{r}-\mathbf{r}_i)\, \delta(\mathbf{r}'-\mathbf{r}_j) \right.$$

$$\left. - \sum_i \delta(\mathbf{r}-\mathbf{r}_i)\, \delta(\mathbf{r}'-\mathbf{r}_i) \right] |\psi_0\rangle$$

$$= \langle \psi_0 | \int d^3r \int d^3r' \frac{1}{2} \frac{e^2}{|\mathbf{r}-\mathbf{r}'|} \left[\hat{n}(\mathbf{r})\hat{n}(\mathbf{r}') - \sum_i \delta(\mathbf{r}-\mathbf{r}_i)\, \delta(\mathbf{r}'-\mathbf{r}_i) \right] |\psi_0\rangle$$

Using (15B.8b) and writing $\delta(\mathbf{r}-\mathbf{r}_i) = (1/2\pi)^3 \int d^3q\, e^{-i\mathbf{q}\cdot(\mathbf{r}-\mathbf{r}_i)}$, we have

$$\left\langle \psi_0 \left| \hat{V}^{(2)} \right| \psi_0 \right\rangle = \langle \psi_0 | \int d^3r \int d^3r' \frac{1}{2} \frac{e^2}{|\mathbf{r}-\mathbf{r}'|} \frac{1}{(2\pi)^6} \int d^3q \int d^3q' \left[\hat{n}(\mathbf{q})\hat{n}(\mathbf{q}') e^{-i\mathbf{q}\cdot\mathbf{r}} e^{-i\mathbf{q}'\cdot\mathbf{r}'} \right.$$

$$\left. - \sum_i e^{-i\mathbf{q}\cdot(\mathbf{r}-\mathbf{r}_i)} e^{-i\mathbf{q}'\cdot(\mathbf{r}'-\mathbf{r}_i)} \right] |\psi_0\rangle .$$

Since the Coulomb potential is a function of $|\mathbf{r}-\mathbf{r}'|$, we rewrite the right-hand side as

$$= \langle \psi_0 | \int d^3 |\mathbf{r}-\mathbf{r}'| \int d^3r' \frac{1}{2} \frac{e^2}{|\mathbf{r}-\mathbf{r}'|} \frac{1}{(2\pi)^6} \int d^3q \int d^3q' \left[\hat{n}(\mathbf{q})\hat{n}(\mathbf{q}') \right.$$

$$\left. - \sum_i e^{i(\mathbf{q}+\mathbf{q}')\cdot\mathbf{r}_i} \right] e^{-i\mathbf{q}\cdot(\mathbf{r}-\mathbf{r}')} e^{-i(\mathbf{q}'+\mathbf{q})\cdot\mathbf{r}'} |\psi_0\rangle .$$

Carrying out the integrations over $d^3|\mathbf{r}-\mathbf{r}'|$ and d^3r', this form becomes

$$= \left\langle \psi_0 \left| \frac{1}{(2\pi)^3} \int d^3q \int d^3q' \frac{1}{2} \frac{4\pi e^2}{q^2} \delta(\mathbf{q}'+\mathbf{q}) \left[\hat{n}(\mathbf{q})\hat{n}(\mathbf{q}') - \sum_i e^{i\mathbf{q}\cdot\mathbf{r}_i} e^{i\mathbf{q}'\cdot\mathbf{r}_i} \right] \right| \psi_0 \right\rangle .$$

Integrating over d^3q', noting $\sum_i 1 = N$, and using the property $\hat{n}(-\mathbf{q}) = \hat{n}^*(\mathbf{q})$ (since $\hat{n}(\mathbf{r})$ is real), we obtain

$$\left\langle \psi_0 \left| \hat{V}^{(2)} \right| \psi_0 \right\rangle = \left\langle \psi_0 \left| \frac{1}{(2\pi)^3} \int d^3q \frac{1}{2} \frac{4\pi e^2}{q^2} \left[|\hat{n}(\mathbf{q})|^2 - N \right] \right| \psi_0 \right\rangle . \qquad (15B.11)$$

15B.3 Linear response and the response function

Equation (15B.11) involves the expectation value of the operator $\hat{n}(\mathbf{q})$, and our next task is to obtain an expression for this quantity in terms of something more physical, which will turn out to be the dielectric function. This can be achieved using so-called linear response theory. Assume we are given some external electrostatic potential, $\phi(\mathbf{r})$, which interacts with our system of N particles by way of their charge; this adds the following term to the full Hamiltonian:

$$\hat{V}(\mathbf{r}_1, \mathbf{r}_2, \ldots\ldots \mathbf{r}_N) = e\sum_{i=1}^{N}\phi(\mathbf{r}_i) = e\int d^3r\phi(\mathbf{r})\sum_{i=1}^{N}\delta(\mathbf{r}-\mathbf{r}_i)$$
$$= e\int d^3r\phi(\mathbf{r})\hat{n}(\mathbf{r}). \quad (15B.12)$$

Using the faltung theorem of Fourier analysis we can immediately write this as

$$\hat{V}(\mathbf{r}_1, \mathbf{r}_2, \ldots\ldots \mathbf{r}_N) = \frac{e}{(2\pi)^3}\int d^3q\phi(\mathbf{q})\hat{n}(-\mathbf{q}) = \frac{e}{(2\pi)^3}\int d^3q\phi(\mathbf{q})\hat{n}^*(\mathbf{q}). \quad (15B.13)$$

If we also allow ϕ to be periodic in time then we have

$$\hat{V}(\mathbf{r}_1, \mathbf{r}_2, \ldots\ldots \mathbf{r}_N, t) = \frac{e}{(2\pi)^4}\int d^3qd\omega\phi(\mathbf{q},\omega)\hat{n}^*(\mathbf{q})e^{-i\omega t}. \quad (15B.14)$$

As in Chapter 14, we will assume that the external potential is weak enough for the response of the system to be linear (and hence the term *linear response*). In this case the perturbation will not "mix" the various Fourier components in the expansion. We may then restrict (15B.14) to a single Fourier component, since the response to a general disturbance may be represented as a superposition; i.e., we write

$$\hat{V}(\mathbf{r}_1, \mathbf{r}_2, \ldots\ldots \mathbf{r}_N, \mathbf{q}, t) = e\hat{n}^*(\mathbf{r}_1, \mathbf{r}_2, \ldots\ldots \mathbf{r}_N, \mathbf{q})\phi(\mathbf{q},\omega)e^{-i\omega t} + \text{c.c.} \quad (15B.15)$$

We can calculate the response of the system to the perturbation (15B.15) using time-dependent perturbation theory. In order to make the system causal we must turn on the perturbation slowly, starting at $t = -\infty$.[51] We therefore again use the prescription

$$e^{-i\omega t} \to e^{-i\omega t+\eta t}, \quad (15B.16)$$

where η is a positive infinitesimal; this form vanishes at $t = -\infty$. Rather than using a set of plane waves as the basis states, as we did in our earlier discussion of the calculation of the self-consistent dielectric function, we assume we are given the *exact* wave functions of both the ground state as well as all the excited states; although we can *never know* these wave functions for our many-body system, this does not stop us from carrying out certain formal calculations with them. We write these time-dependent wave functions as

$$\Psi_n(\mathbf{r}_1, \mathbf{r}_2, \ldots\ldots \mathbf{r}_N, t) = \psi_n(\mathbf{r}_1, \mathbf{r}_2, \ldots\ldots \mathbf{r}_N)e^{-i\omega_n t}, \quad (15B.17)$$

[51] If the perturbation is turned on rapidly, the system will be excited into a superposition of many excited states, whereas our goal is to evolve the ground state adiabatically under the influence of the external perturbation; alternatively, if we assume that the perturbation was always present the integration leading to the coefficients $a_n(t)$ given by (15.19) is undefined at $t = -\infty$.

where $\hbar\omega_n = E_n$ is the *total* energy of the excited states, which are numbered by n (with n = 0 corresponding to the ground state). The correction to the wave function in the presence of the external perturbation can be represented by

$$\begin{aligned}\Psi^{(1)} &= \Psi^{(1)}(\mathbf{r}_1, \mathbf{r}_2, \ldots .\mathbf{r}_N, t) = \sum_n a_n(t)\Psi_n(\mathbf{r}_1, \mathbf{r}_2, \ldots ..\mathbf{r}_N, t)\\ &= \sum_n a_n(t)\psi_n(\mathbf{r}_1, \mathbf{r}_2, \ldots .\mathbf{r}_N)e^{-i\omega_n t}.\end{aligned} \tag{15B.18}$$

From time-dependent perturbation theory, the amplitudes $a_n(t)$ from our perturbation $\hat{V}(t)$ are

$$a_n(t) = \frac{1}{i\hbar}\int_{-\infty}^{t} dt \, \langle \psi_n(t)|\, \hat{V}'(t)\, |\psi_0(t)\rangle e^{i(\omega_n-\omega_0)t},$$

which on substituting Eq. (15B.15) and integrating yields

$$a_n(t) = \frac{e}{\hbar}\left[\frac{\langle \hat{n}^*(\mathbf{q})\rangle_{n0}\, \phi(\mathbf{q}, \omega)e^{(-i\omega+i\omega_{n0}+\eta)t}}{\omega - \omega_{n0} + i\eta} - \frac{\langle \hat{n}(\mathbf{q})\rangle_{n0}\, \phi^*(\mathbf{q}, \omega)e^{(i\omega+i\omega_{n0}+\eta)t}}{\omega + \omega_{n0} - i\eta}\right], \tag{15B.19}$$

where we have written $\hbar\omega_{n0} = E_n - E_0$, defined the matrix elements,

$$\langle \hat{n}(\mathbf{q})\rangle_{n0} = \langle \psi_n(\{\mathbf{r}_i\})\, |\hat{n}(\{\mathbf{r}_i\}, \mathbf{q})|\, \psi_0(\{\mathbf{r}_i\})\rangle, \tag{15B.20}$$

and set $a_0 = 1$.

We now examine the time-dependent expectation value of the density operator, the Fourier transform of which is defined as the expectation value of the associated operator, $\langle \hat{n}(\mathbf{q}, t)\rangle = \langle \Psi(\{\mathbf{r}_i\}, t)\, |\hat{n}(\{\mathbf{r}_i\}, \mathbf{q})|\, \Psi(\{\mathbf{r}_i\}, t)\rangle \cong \langle \hat{n}(\mathbf{q}, t)\rangle^{(0)} + \langle \hat{n}(\mathbf{q}, t)\rangle^{(1)}$. The term $\langle \hat{n}(\mathbf{q}, t)\rangle^{(0)}$ is just the static number density (in the absence of the perturbation), whereas $\langle \hat{n}(\mathbf{q}, t)\rangle^{(1)}$ is given by

$$\begin{aligned}\langle \hat{n}(\mathbf{q}, t)\rangle^{(1)} &= \left\langle \Psi_0(\{\mathbf{r}_i\}, t)\, |\hat{n}(\{\mathbf{r}_i\}, \mathbf{q})|\, \Psi^{(1)}(\{\mathbf{r}_i\}, t)\right\rangle\\ &\quad + \left\langle \Psi^{(1)}(\{\mathbf{r}_i\}, t)\, |\hat{n}(\{\mathbf{r}_i\}, \mathbf{q})|\, \Psi_0(\{\mathbf{r}_i\}, t)\right\rangle;\end{aligned} \tag{15B.21}$$

we will henceforth drop the superscript. We now insert (15B.18) along with the expansion coefficients (15B.19) into (15B.21); on doing this, four matrix element products occur:

$$\begin{aligned}&\langle \hat{n}(\mathbf{q})\rangle_{0n}\langle \hat{n}(\mathbf{q})\rangle_{n0} = \langle \hat{n}(\mathbf{q})\rangle_{0n}\langle \hat{n}(\mathbf{q})\rangle_{0n}^* = |\langle \hat{n}(\mathbf{q})\rangle_{0n}|^2,\\ &\langle \hat{n}(\mathbf{q})\rangle_{0n}\langle \hat{n}^*(\mathbf{q})\rangle_{n0} = \langle \hat{n}(\mathbf{q})\rangle_{0n}\langle \hat{n}^*(\mathbf{q})\rangle_{0n}^* = \langle \hat{n}(\mathbf{q})\rangle_{0n}\langle \hat{n}(-\mathbf{q})\rangle_{0n}^* = 0,\\ &\langle \hat{n}(\mathbf{q})\rangle_{n0}\langle \hat{n}^*(\mathbf{q})\rangle_{n0}^* = \langle \hat{n}(\mathbf{q})\rangle_{n0}\langle \hat{n}(-\mathbf{q})\rangle_{n0}^* = 0,\\ &\langle \hat{n}(\mathbf{q})\rangle_{n0}\langle \hat{n}(\mathbf{q})\rangle_{n0}^* = |\langle \hat{n}(\mathbf{q})\rangle_{n0}|^2.\end{aligned} \tag{15B.22a–d}$$

The second and third products vanish since, due to the translational symmetry of our system, the quantum numbers of the excited states include the momenta, and it is not possible to simultaneously couple to **q** and -**q**. Our expression for the expectation value of the number operator then reduces to

$$\left\langle \hat{n}(\mathbf{q}, t)\right\rangle = \frac{e}{\hbar}\phi(\mathbf{q}, \omega)e^{(-i\omega+\eta)t}\sum_n\left[\frac{\left|\left\langle \hat{n}(\mathbf{q})\right\rangle_{0n}\right|^2}{\omega-\omega_{n0}+i\eta} - \frac{\left|\left\langle \hat{n}(\mathbf{q})\right\rangle_{n0}\right|^2}{\omega+\omega_{n0}-i\eta}\right]. \tag{15B.23a}$$

Note that $\left|\langle\hat{n}(\mathbf{q})\rangle_{n0}\right|^2 = \langle \hat{n}(\mathbf{q})\rangle_{n0}\langle\hat{n}(\mathbf{q})\rangle^*_{n0} = \langle\hat{n}^*(\mathbf{q})\rangle^*_{0n}\langle\hat{n}^*(\mathbf{q})\rangle_{0n} = \langle\hat{n}(-\mathbf{q})\rangle^*_{0n}\langle\hat{n}(-\mathbf{q})\rangle_{0n}$; since our system has symmetry with respect to **q** and $-\mathbf{q}$, we may write $\left|\langle\hat{n}(\mathbf{q})\rangle_{n0}\right|^2 = \left|\langle\hat{n}(-\mathbf{q})\rangle_{0n}\right|^2$. Combining terms in (15B.23a), we have the alternative form

$$\left\langle \hat{n}(\mathbf{q}, \omega)\right\rangle = \frac{e}{\hbar}\phi(\mathbf{q}, \omega)\sum_n\left|\left\langle \hat{n}(\mathbf{q})\right\rangle_{0n}\right|^2\frac{2\omega_{n0}}{(\omega+i\eta)^2-\omega_{n0}^2}. \tag{15B.23b}$$

It is customary at this point to define a *response function* as

$$\chi(\mathbf{q}, \omega) = \frac{\left\langle \hat{n}(\mathbf{q}, \omega)\right\rangle}{e\phi(\mathbf{q}, \omega)}; \tag{15B.24}$$

using Eqs. (15B.23a,b), we have

$$\chi(\mathbf{q}, \omega) = \frac{1}{\hbar}\sum_n\left[\frac{\left|\left\langle \hat{n}(\mathbf{q})\right\rangle_{0n}\right|^2}{\omega-\omega_{n0}+i\eta} - \frac{\left|\left\langle \hat{n}(\mathbf{q})\right\rangle_{n0}\right|^2}{\omega+\omega_{n0}-i\eta}\right] \tag{15B.25a}$$

$$= \frac{1}{\hbar}\sum_n\left|\left\langle \hat{n}(\mathbf{q})\right\rangle_{0n}\right|^2\frac{2\omega_{n0}}{(\omega+i\eta)^2-\omega_{n0}^2}. \tag{15B.25b}$$

15B.4 Relating the response function to the dielectric function

We will now show how the number density operator is related to the *dielectric constant*. We start by writing the Fourier transformed second Maxwell equation in terms of an *external* number density, $en_e(\mathbf{q}, \omega)$; we also define a *total* charge density,

$$en_t(\mathbf{q}, \omega) = en_e(\mathbf{q}, \omega) + e\left\langle \hat{n}(\mathbf{q}, \omega)\right\rangle,$$

where $e\left\langle \hat{n}(\mathbf{q}, \omega)\right\rangle$ is the *induced* charge density, which is given by Eq. (15B.23). Through Poisson's equation these charge densities are related to **D** and **E** as

$$i\mathbf{q}\cdot\mathbf{D}(\mathbf{q}, \omega) = 4\pi en_e(\mathbf{q}, \omega) \tag{15B.26a}$$

and

$$\mathrm{i}\mathbf{q}\cdot\mathbf{E}(\mathbf{q},\omega) = 4\pi\left(\mathrm{e}n_e(\mathbf{q},\omega) + \mathrm{e}\langle\hat{n}(\mathbf{q},\omega)\rangle\right). \tag{15B.26b}$$

Using the definition $\mathbf{D}(\mathbf{q},\omega) = \varepsilon(\mathbf{q},\omega)\mathbf{E}(\mathbf{q},\omega)$, we obtain

$$\frac{1}{\varepsilon(\mathbf{q},\omega)} = 1 + \frac{\mathrm{e}\langle\hat{n}(\mathbf{q},\omega)\rangle}{\mathrm{e}n_e(\mathbf{q},\omega)}. \tag{15B.27}$$

If we write $\mathbf{D}(\mathbf{r},t) = -\nabla\phi(\mathbf{r},t)$ or $\mathbf{D}(\mathbf{q},\omega) = -\mathrm{i}\mathbf{q}\phi(\mathbf{q},\omega)$, then from (15B.26a) the potential $\phi(\mathbf{q},\omega)$ associated with the external charge density, $\mathrm{e}n_e(\mathbf{q},\omega)$, is

$$\phi(\mathbf{q},\omega) = \frac{4\pi\mathrm{e}n_e(\mathbf{q},\omega)}{q^2}. \tag{15B.28}$$

We may therefore rewrite (15B.27) as

$$\frac{1}{\varepsilon(\mathbf{q},\omega)} = 1 + \frac{4\pi\mathrm{e}^2}{q^2}\frac{\langle\hat{n}(\mathbf{q},\omega)\rangle}{\mathrm{e}\phi(\mathbf{q},\omega)} \tag{15B.29a}$$

$$= 1 + \frac{4\pi\mathrm{e}^2}{q^2}\chi(\mathbf{q},\omega), \tag{15B.29b}$$

where $\chi(\mathbf{q},\omega)$ is given by (15B.24).

15B.5 Relating the dielectric function to the ground state energy

To apply (15B.11) to calculate the ground state energy we require the matrix element

$$\left\langle\psi_0\left|\left|\hat{n}(\mathbf{q})\right|^2\right|\psi_0\right\rangle = \langle\psi_0\left|\hat{n}(\mathbf{q})\hat{n}^*(\mathbf{q})\right|\psi_0\rangle$$
$$= \sum_n\langle\psi_0\left|\hat{n}(\mathbf{q})\right|\psi_n\rangle\langle\psi_n\left|\hat{n}^*(\mathbf{q})\right|\psi_0\rangle = \sum_n\left|\langle\hat{n}(\mathbf{q})\rangle_{0n}\right|^2 \tag{15B.30}$$

where we have inserted a complete set of states between the two operators to obtain the right-hand side. We can obtain the needed sum by applying the operational formula

$$\frac{1}{x-\mathrm{i}0} = P\left(\frac{1}{x}\right) + \mathrm{i}\pi\delta(x), \tag{15B.31}$$

which is used to evaluate integrals containing this factor, and where P denotes the principal value of the associated integral, which is defined as

$$P\int_a^b dx\frac{f(x)}{x} = \lim_{\varepsilon\to 0}\left[\int_a^{-\varepsilon} dx\frac{f(x)}{x} + \int_{+\varepsilon}^b dx\frac{f(x)}{x}\right]. \tag{15B.32}$$

It then follows that

$$\mathrm{Im}\,[\chi(\mathbf{q},\omega)] = -\frac{\pi}{\hbar}\sum_n \left[\left|\langle\hat{n}(\mathbf{q})\rangle_{0n}\right|^2 \delta(\omega-\omega_{n0}) + \left|\langle\hat{n}(\mathbf{q})\rangle_{n0}\right|^2 \delta(\omega+\omega_{n0})\right]. \tag{15B.33}$$

The quantity $\sum_n |\langle n(\mathbf{q})\rangle_{0n}|^2$ is extracted simply by integrating (15B.33) over all ω.

Combining (15B.29b), (15B.30), and (15B.33), the expectation value of the Coulomb energy, Eq, (15B.11), may now be written as

$$\left\langle\psi_0\left|\hat{V}^{(2)}\right|\psi_0\right\rangle d^3q = -\frac{1}{(2\pi)^3}\int d^3q\frac{1}{2}\left[\int_{-\infty}^{\infty}\frac{d\omega}{2\pi}\mathrm{Im}\left[\frac{1}{\varepsilon(\mathbf{q},\omega)}\right] + \frac{4\pi e^2}{q^2}N\right], \tag{15B.34}$$

where we have subtracted the contribution from the background positive change. Given the function $\varepsilon(\mathbf{q},\omega)$ we can use (15B.34) and (15B.6) to calculate the ground state energy. This procedure is exact, provided we know the exact dielectric function. We can get an *approximate* value for the interaction energy by using the *self-consistent potential* expression of $\varepsilon(\mathbf{q},\omega)$ obtained in Section 14.1, as we now discuss.

15B.6 The ground state energy from the SCF dielectric function

We now sketch how the ground state energy is evaluated using Eq. (15.34) and the dielectric constant obtained in the self-consistent potential approximation. We rewrite our earlier expression for the dielectric function, Eq. (14.15), as

$$\varepsilon(\mathbf{q},\omega) = 1 - \frac{4\pi e^2}{q^2}\frac{1}{(2\pi)^3}\int d^3k\frac{n(\mathbf{k})-n(\mathbf{k}+\mathbf{q})}{\hbar\omega+\varepsilon_{\mathbf{k}}-\varepsilon_{\mathbf{k}+\mathbf{q}}-i0}. \tag{15B.35}$$

In order to evaluate the ground state energy we need

$$\mathrm{Im}\frac{1}{\varepsilon(\mathbf{q},\omega)} = \mathrm{Im}\frac{1}{\varepsilon_1(\mathbf{q},\omega)+i\varepsilon_2(\mathbf{q},\omega)} = -\frac{\varepsilon_2}{\varepsilon_1^2+\varepsilon_2^2}, \tag{15B.36}$$

where we write $\varepsilon = \varepsilon_1 + i\varepsilon_2$; i.e., we must calculate both the real and imaginary parts of $\varepsilon(\mathbf{q},\omega)$. The integral (15B.35) is evaluated using the prescription given in Eq. (15B.31) (Lindhard (1954)); the real and imaginary parts are given by

$$\varepsilon_1(\mathbf{q},\omega) = 1 + \frac{4\pi e^2}{q^2}\frac{mp_F}{2\pi^2}\left\{1 + \frac{1-\left(\omega'+\frac{1}{2}q'\right)^2}{2q'}\ln\left|\frac{1+\left(\omega'+\frac{1}{2}q'\right)}{1-\left(\omega'+\frac{1}{2}q'\right)}\right|\right.$$
$$\left. - \frac{1-\left(\omega'-\frac{1}{2}q'\right)^2}{2q'}\ln\left|\frac{1+\left(\omega'-\frac{1}{2}q'\right)}{1-\left(\omega'-\frac{1}{2}q'\right)}\right|\right\} \tag{15B.37}$$

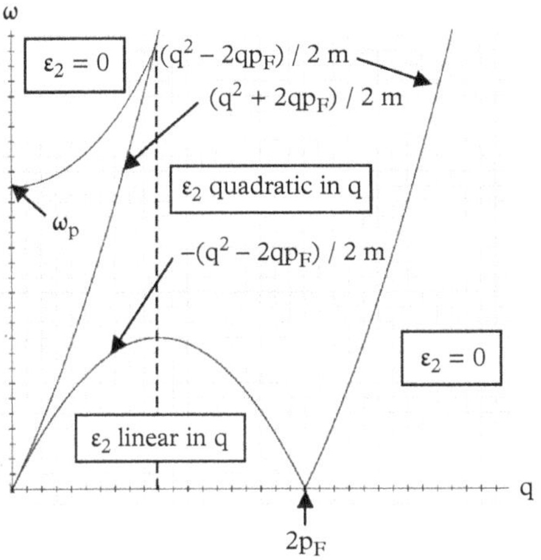

Figure 15B.1 *Plots showing the boundaries separating various regimes of the Lindhard function, as defined by Eqs. (15.37) and (15.38).*

and

$$\varepsilon_2(\mathbf{q},\omega) = \begin{cases} 0 \text{ for } \omega > \left(\frac{q^2+2qp_F}{2m}\right) \\ 0 \text{ for } q > 2p_F \text{ and } \omega < \left(\frac{q^2-2qp_F}{2m}\right) \\ \frac{2e^2m^2\omega}{q^3} \text{ for } q < 2p_F \text{ and } 0 < \omega < -\left(\frac{q^2-2qp_F}{2m}\right) \\ \frac{e^2mp_F^2}{q^3}\left[1-\left(\omega'-\frac{1}{2}q'\right)^2\right] \text{ for } 0 < -\left(\frac{q^2-2qp_F}{2m}\right) < \omega < \left(\frac{q^2+2qp_F}{2m}\right), \end{cases} \tag{15B.38}$$

where $q' \equiv \frac{q}{p_F}$ and $\omega' = \omega/\left(\frac{qp_F}{m}\right)$. The various regimes associated with (15B.37) and (15B.38) are shown in Fig. 15B.1.

Having the explicit forms for $\varepsilon(\mathbf{q},\omega)$, we can calculate the ground state energy of an electron gas in the approximation under consideration. Note that ε_1 and ε_2 both involve the "coupling constant" e^2. We can introduce a *variable* coupling constant by writing $e^2 \to \lambda e^2$, resulting in $\varepsilon_1 \to 1 + \lambda(\varepsilon_1 - 1); \varepsilon_2 \to \lambda\varepsilon_2$. We may then express the ground state energy using (15B.6), (15B.11), (15B.34), and (15B.36) as

$$E = E^{(0)} + \frac{1}{2}\int \frac{d^3q}{(2\pi)^3}\left[\int_{-\infty}^{+\infty} \frac{d\omega}{2\pi}\int_0^1 \frac{d\lambda}{\lambda}\frac{\lambda\varepsilon_2}{(1+\lambda(\varepsilon_1-1))^2+(\lambda\varepsilon_2)^2} - \frac{4\pi Ne^2}{q^2}\right]; \tag{15B.39}$$

note that the parameter λ now enters in a non-trivial manner, as opposed to how it appears to enter Eq. (15B.6). On carrying out the integration over λ we have

$$E = E^{(0)} + \frac{1}{2}\int \frac{d^3q}{(2\pi)^3}\left[\int_{-\infty}^{+\infty} \frac{d\omega}{2\pi}\tan^{-1}\frac{\varepsilon_2(\mathbf{q},\omega)}{\varepsilon_1(\mathbf{q},\omega)} - \frac{4\pi Ne^2}{q^2}\right]. \tag{15B.40}$$

We will not discuss the techniques to perform the remaining integrations (see Hubbard (1957)).[52] It is common to cast the final expression in terms of a dimensionless parameter, $r_s \equiv \frac{r_0}{a_0} = \left(\frac{3L^3}{4\pi N}\right)^{1/3} \frac{me^2}{\hbar^2}$, where $\frac{4}{3}\pi r_0^3 \equiv \frac{L^3}{N}$ and a_0 is the Bohr radius. The resulting energy per particle (in Ry) can be expressed as

$$\frac{E}{N} = \frac{e^2}{2a_0}\left[\frac{2.21}{r_s^2} - \frac{0.916}{r_s} + 0.0622 \ln r_s - 0.096 \cdots\right] \tag{15B.41}$$

(Gell-Mann and Brueckner (1957)).[53] The first term is the kinetic energy of a free Fermi gas and the second is the exchange energy[54] that we calculated in Section 15.1.3; the remaining terms are contributions from the correlation energy. The energy of an electron gas remains a topic of continuing interest.[55] For the purpose of implementing the theory in real materials via density functional theory, tables and parameterizations have been constructed.[56,57] When applying the theory to real inhomogeneous systems, further improvements would formally require terms involving the *gradient* of the density.

Problems

15B.1. Using the definition of r_s following Eq. (15B.40), the kinetic energy of a free electron gas at $T = 0$, and the expression for the exchange energy, (15.27), obtain the first two terms in the expansion given by Eq. (15B.41).

15B.2. Carry out the calculations leading to the Lindhard expressions, Eqs. (15B.37) and (15B.38).

[52] The description of collective motions in terms of many-body perturbation theory. 3. The extension of the theory to the non-uniform gas, J. Hubbard, *Proc. R. Soc.* **A244**, 199 (1958).

[53] Correlation energy of an electron gas at high density, M. Gell-Mann and K. Brueckner, *Phys.* **106**, 364 (1957).

[54] The direct Coulomb energy between electrons is of course divergent, but is canceled by the assumed uniform positive background charge.

[55] *Quantum Theory of the Electron Liquid*, G.F. Giuliani and G. Vignale, Cambridge University Press (2005).

[56] Local exchange-correlation potentials, L. Hedin, B.I. Lundqvist, and S. Lundqvist, *Solid State Commun.*, **9**, 537 (1971).

[57] Electron correlations at metallic densities-4, L. Hedin, K.S Singwi, A. Sjolander, M. Tosi, and R.H. Land, *Phys. Rev. B* **1**, 1044 (1970).

Part V

Lattice Dynamics

Harmonic Lattice Dynamics: Classical and Quantum

16

In Chapter 4 we presented some simple and historically important models of the vibrational response of atoms in solids, from which thermal properties of the lattice could be approximately calculated. In this chapter we present some more rigorous treatments. We will start with the traditional approach, which assumes the existence of a potential energy function,

$$V = V(\mathbf{U}_n^j), \tag{16.1}$$

that depends on the *displacements*, $\mathbf{U}_n^j$, of all the nuclei from their *equilibrium positions* within the unit cells;[1] here the vector index $\mathbf{n} = (n_1, n_2, n_3)$ numbers the $N = N_1N_2N_3$ unit cells of the crystal in which the atoms lie and $j = 1, 2, \ldots r$ specifies different atoms within each cell.[2] With the availability of this function, the force acting on the individual atoms can be calculated, and the classical response of the lattice then follows from Newton's law. We then go on to quantize this motion. Finally we will approach the problem from a different perspective involving the overall electronic structure; this approach is further pursued using perturbation theory in Appendix 16A of this chapter.

16.1 The 1d lattice

Because it involves a minimum of complexity, we will start with the case of a hypothetical one-dimensional solid, initially having only one atom per unit cell. Here the potential (16.1) takes the form

$$V = V(U_1, \ldots U_N). \tag{16.2}$$

We now expand (16.2) as a Taylor series in the displacements

$$\begin{aligned} V(U_1, \ldots U_N) = V_0 &+ \frac{1}{2}\sum_{n,n'}\left(\frac{\partial^2 V}{\partial U_n \partial U_{n'}}\right) U_n U_{n'} \\ &+ \frac{1}{6}\sum_{n,n',n''}\left(\frac{\partial^3 V}{\partial U_n \partial U_{n'} \partial U_{n''}}\right) U_n U_{n'} U_{n''} + \ldots \end{aligned}, \tag{16.3}$$

[1] Note that we have assumed this function does not depend explicitly on time; any time dependence of V arises from the time dependence of the coordinates $\mathbf{U}_n^j(t)$. This assumes that the electrons adjust instantaneously to the nuclear coordinates, which is the basis of the Born–Oppenheimer approximation mentioned in Sections 15.1.1 and 15.2.2. For many responses this is a good approximation since a typical plasma frequency, 10^{16} Hz, is large compared to a typical phonon frequency, 10^{13} Hz. However it fails to give the correct response of electrons in metals having energies in the same range as those of the lattice vibrations, as will be discussed in Section 21.3. Inclusion of the resulting modifications ultimately facilitates superconductivity.

[2] In the presence of a basis (see Section 8.2) the equilibrium positions of the atoms in the unit cell are given by $\mathbf{R}_n + \mathbf{r}^j$, where here we denote the lattice points

The Physics of Solids. First Edition. J. B. Ketterson.

where the linear term in U_n is absent since the first derivatives vanish at the equilibrium positions. Rather than writing out these derivatives in what follows, we introduce the definitions

$$A_{nn'} = \frac{\partial^2 V}{\partial U_n \partial U_{n'}} \tag{16.4a}$$

and

$$B_{nn'n''} = \frac{\partial^3 V}{\partial U_n \partial U_{n'} \partial U_{n''}}. \tag{16.4b}$$

Equation (16.2) then becomes

$$V(U_1, \ldots U_N) = V^{(0)} + \frac{1}{2}\sum_{n,n'} A_{nn'} U_n U_{n'} + \frac{1}{6}\sum_{n,n',n''} B_{nn'n''} U_n U_{n'} U_{n''} + \ldots. \tag{16.5}$$

From the equality of cross partial derivatives we have the conditions

$$A_{nn'} = \frac{\partial^2 V}{\partial U_n \partial U_{n'}} = \frac{\partial^2 V}{\partial U_{n'} \partial U_n} = A_{n'n};\ B_{nn'n''} = B_{n'nn''} = B_{n''n'n} = B_{nn''n'}. \tag{16.6a,b}$$

The force on the nth atom is given by $F_n = -\partial V/\partial U_n$. If we limit ourselves to a harmonic approximation where $B_{nn'n''} = 0$, we may write

$$F_n = -\sum_{n'} A_{nn'} U_{n'}. \tag{16.7}$$

From translational symmetry,

$$A_{nn'} = A_{n-n'}, \tag{16.8}$$

and from the fact that a uniform translation of all atoms, $U_n = U$, produces no net force, we have

$$\sum_{n'} A_{nn'} = 0. \tag{16.9}$$

by $\mathbf{R_n}$ (see Eq. (8.2)) and the positions within the unit cell by $\mathbf{r^j}$ (see Eq. (8.4)). Displaced atoms then have the coordinates $\mathbf{R_n} + \mathbf{r_j} + \mathbf{U^j_n}$. In one dimension we would write the position as $X_n + x^j + U^j_n$.

Using Newton's law, $F_n = M\ddot{U}_n$, where M is the atomic mass, we have

$$M\ddot{U}_n = -\sum_{n'} A_{nn'} U_{n'}. \tag{16.10}$$

Assuming a harmonic time dependence, $U_n(t) = \Re U_n e^{-i\omega t}$, where $\Re$ denotes taking the real part (which we will assume in what follows), Eq. (16.10) becomes

$$M\omega^2 U_n = \sum_{n'} A_{nn'} U_{n'}. \tag{16.11}$$

If we use periodic boundary conditions, say for a ring of atoms, there will be N–1 independent solutions of (16.11) (where we neglect the solution corresponding to a uniform rotation of the ring).

We must now seek a solution of the coupled set of equations (16.11). Since our system of equations is invariant under a lattice translation $U_n \to U_{n+1}$, any solution of (16.11) must still be a solution if $n \to n+1$. There are two possibilities: (i) we retain the original solution and (ii) we generate a new solution. In the second case we may always take linear combinations (of the old and new solutions) which behave as one of the solutions for case (i). We can therefore assume we are dealing with only a single normal mode. In analogy with our earlier discussion of tight binding Bloch states, we write the effect of the translation (which can only multiply the solution of U_n by a constant) as

$$U_{n+1} = e^{iqa} U_n, \tag{16.12}$$

where a is the lattice constant. Assuming periodic boundary conditions on traversing a cycle of our N-atom ring, we must have $(e^{iqa})^N = 1$ or $Nqa = 2\pi m$; therefore, $q = 2\pi m/Na$. By carrying out successive transformations of the form (16.12), we may write

$$U_n = U e^{iqna}, \tag{16.13}$$

where U is the amplitude at the site $n = 0$. On inserting (16.13) into (16.11) we obtain

$$M\omega^2 = \sum_{n'} A_{nn'} e^{-iqa(n-n')}. \tag{16.14}$$

Writing $n - n'$ as n and using (16.8), Eq. (16.14) can be rewritten as

$$M\omega^2 = D(q), \tag{16.15a}$$

where

$$D(q) \equiv \sum_{n} A_n\, e^{-inqa}. \tag{16.15b}$$

The simplest example of a one-dimensional lattice has nearest-neighbor interactions only. In this case, the only non-zero A_n are for $n = 0$ and $n = \pm 1$. If we write $A_0 = 2\kappa$ and use Eq. (16.9), we obtain $A_1 = A_{-1} = -\kappa$. Using these relations Eqs. (16.15a,b) become

$$M\omega^2 = 2\kappa(1 - \cos qa) = 4\kappa \sin^2\left(\frac{qa}{2}\right); \tag{16.16}$$

the corresponding equation of motion involving these A_n obtained from (16.10) is given by

$$M\ddot{U}_n = \kappa(U_{n+1} + U_{n-1} - 2U_n). \tag{16.17}$$

The parameter κ corresponds to the spring constant acting between neighboring atoms. To see this we refer to Fig. 16.1 which shows, schematically, a 1d lattice with nearest-neighbor interactions represented by such springs; here the solid vertical lines are the equilibrium positions, denoted by the real-space lattice vectors X_n, and the dashed lines show the displaced (or instantaneous) positions. The associated Newtonian equation of motion is identical to Eq. (16.17).

The relation $\omega = \omega_q$ is referred to as the *dispersion relation* and is shown in Fig. 16.2. Since $\omega/q \neq$ constant, the phase velocity $V_p \equiv \omega/q$ is not constant. It is then customary to also introduce a group velocity, $V_g \equiv d\omega/dq$, which differs substantially from V_p at large q.

As $q \to 0$ (16.16) may be expanded $\omega = \omega(\mathbf{q})$ to give $M\omega^2 = \kappa(q^2a^2 + \ldots)$ or $V_p = \omega/q = \left(\kappa a^2/M\right)^{1/2}$. The low frequency limit may be thought of as the *continuum limit* since the wavelength is then much greater than the interatomic spacing, and the discrete nature of the lattice is obscured. To make contact with the continuum theory of elasticity, we introduce the linear mass density, $\rho = M/a$, and an elastic constant, $c = \kappa a$, yielding $V_p = \sqrt{c/\rho}$ for the velocity of sound, which is the usual expression obtained in the theory of elasticity (see Chapter 1).

In the continuum limit we may write the right-hand side of Eq. (16.17) as

$$\kappa a^2 \left(\frac{U_{n+1} + U_{n-1} - 2U_n}{a^2}\right) \cong \kappa a^2 \frac{\partial^2 U_n}{\partial x^2}$$

and Eq. (16.17) becomes

$$\frac{\partial^2 U}{\partial x^2} - \frac{1}{v_p^2}\frac{\partial^2 U}{\partial t^2} = 0, \tag{16.18}$$

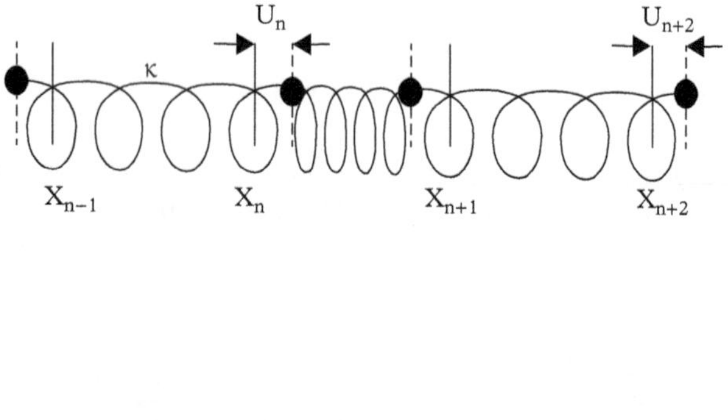

Figure 16.1 *The 1d harmonic with nearest-neighbor interactions.*

which is the one-dimensional wave equation. We note that the one-dimensional waves we are discussing here correspond to *longitudinal waves polarized along the x axis*.[3]

From (16.16), we see that there is a maximum allowed frequency in our lattice, which is given by $\omega_m = 2(\kappa/M)^{1/2}$ and occurs for $q = \pi/a$. From Eq. (16.13), $U_{n+1} = -U_n$ for this situation; i.e., neighboring atoms are moving in opposite directions. At this point we also have $V_g = 0$. We interpret these behaviors as corresponding to the presence of a *standing wave* (with no net group velocity) which results from 1d Bragg diffraction by our lattice; the value $q = \pi/a$ corresponds to the first Brillouin zone (which is a point in k-space for 1d). If we extend k beyond the value π/a the frequency starts to decrease, returning to zero at a point $q = 2\pi/a$; this is the center of the neighboring Brillouin zone and hence we see the phonon frequencies are periodic in reciprocal space. This behavior is analogous to that studied earlier for electron bands, although in the case of one atom per unit cell we have only one phonon band.

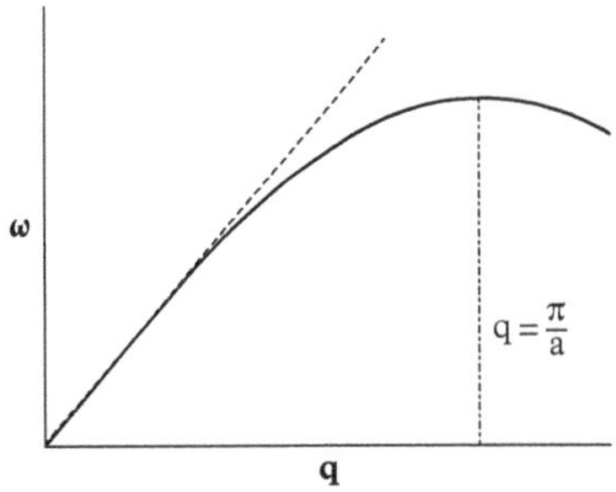

Figure 16.2 *The dispersion relation) $\omega = \omega(q)$ for a one-dimensional monatomic lattice. The dashed line shows the long-wavelength extrapolation and the dashed-dot line shows the position of the Brillouin point.*

16.2 The 1d lattice with more than one atom per unit cell

Suppose our lattice involves r different atoms per unit cell. As an example in which r = 2, we can picture a one-dimensional "NaCl lattice" made up of an alternating chain of Na and Cl atoms. So in addition to summing over unit cells, which we continue to denote by n (as in Eq. (16.5)), we must now sum over the basis of atoms in the unit cell, denoted by j. Equation (16.5) then becomes

$$V = V_0 + \frac{1}{2}\sum_{j,j'}\sum_{n,n'} A^{jj'}_{nn'} U^j_n U^{j'}_{n'} + \frac{1}{6}\sum_{j,j',j''}\sum_{n,n',n''} B^{jj'j''}_{nn'n''} U^j_n U^{j'}_{n'} U^{j''}_{n''} + \ldots \quad (16.19)$$

The force on the jth atom in the nth unit cell is $F^j_n = -\partial V/\partial U^j_n$ and Newton's law in the harmonic approximation is then

$$M_j \ddot{U}^j_n = -\sum_{n',j'} A^{jj'}_{nn'} U^{j'}_{n'}. \quad (16.20)$$

Translational symmetry again requires that

$$A^{jj'}_{nn'} = A^{jj'}_{n-n'}. \quad (16.21)$$

Equality of the cross partial derivatives, $\frac{\partial^2 V}{\partial U^j_n \partial U^{j'}_{n'}}$, yields

$$A^{jj'}_{nn'} = A^{j'j}_{n'n}. \quad (16.22)$$

[3] Transverse waves do not exist for our one-dimensional string since, in the absence of an overall tension the restoring force varies quadratically in leading order with displacement in the y or z directions.

Finally, requiring the force to vanish for a uniform translation yields

$$\sum_{n',j'} A^{jj'}_{nn'} = 0. \tag{16.23}$$

Making the substitution, $U^j_n(t) = U^j e^{-i\omega t+iqna}$, writing $n - n'$ as n, and using Eq. (16.21), Eq. (16.20) becomes

$$\begin{aligned} M_j\omega^2 U^j &= \sum_{j',n'} A^{jj'}_{nn'} U^{j'} e^{-iq(n-n')a} \\ &= \sum_{j'} D^{jj'} U^{j'}; \end{aligned} \tag{16.24}$$

here we have defined what is generally called the *dynamical matrix* as

$$D^{jj'}(qa) \equiv \sum_n A^{jj'}_n e^{-iqna}, \tag{16.25}$$

which is a generalization of Eq. (16.15b). The set of linear homogeneous equations (16.24) in U^j will have a solution only if the determinant of the coefficients vanishes,

$$\left| D^{jj'}(qa) - M_j\omega^2\delta_{jj'} \right| = 0. \tag{16.26}$$

If we have a total of r atoms per unit cell we will have r frequencies for each value of the wave vector q.

For the case of two atoms per unit cell having masses M_1 and M_2, and *nearest-cell* interactions only, $A^{jj'}_n$ is a 2×2 matrix for each n that we may write as

$$\overleftrightarrow{A}(0) = \begin{pmatrix} A^{11}_0 & A^{12}_0 \\ A^{21}_0 & A^{22}_0 \end{pmatrix}; \quad \overleftrightarrow{A}(1) = \begin{pmatrix} A^{11}_1 & A^{12}_1 \\ A^{21}_1 & A^{22}_1 \end{pmatrix}; \ldots. \tag{16.27}$$

These couplings are shown schematically in Fig. 16.3.

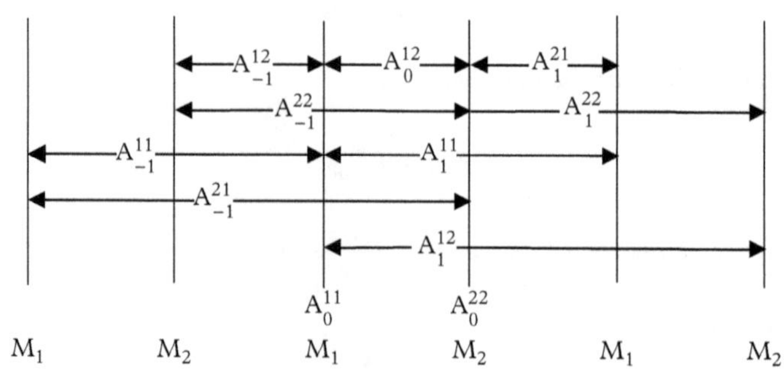

Figure 16.3 *Intra- and inter-cell elements of the* A *matrix relevant to the nth cell in the nearest-cell approximation.*

For $n = 0$ and 1, it follows from this figure that restricting interactions to *nearest neighbors* requires $A_1^{11} = A_1^{22} = A_1^{12} = 0$. Equation (16.23) gives $A_0^{21} = A_0^{12}$ and $A_1^{21} = A_{-1}^{12}$ and equation (16.22) yields the two conditions

$$j = 1: \quad A_0^{11} + A_0^{12} + A_{-1}^{12} = 0; \; A_0^{11} = -\left(A_0^{12} + A_1^{21}\right) \tag{16.28a}$$

$$j = 2: \quad A_0^{21} + A_0^{22} + A_1^{21} = 0; \; A_0^{22} = -\left(A_0^{12} + A_1^{21}\right). \tag{16.28b}$$

Using these relations the equations of motion (16.20) are

$$\omega^2 M_1 U_n^1 = -\left(A_0^{12} + A_1^{21}\right) U_n^1 + A_0^{12} U_n^2 + A_1^{21} U_{n-1}^2 \tag{16.29a}$$

and

$$\omega^2 M_2 U_n^2 = A_0^{21} U_n^1 - \left(A_0^{12} + A_1^{21}\right) U_n^2 + A_1^{21} U_{n+1}^1. \tag{16.29b}$$

If we define $A_0^{12} = -\kappa$ and $A_1^{21} = -\kappa'$, our equations of motion become

$$\omega^2 M_1 U_n^1 = (\kappa + \kappa') U_n^1 - \kappa U_n^2 - \kappa' U_{n-1}^2 \tag{16.30a}$$

and

$$\omega^2 M_2 U_n^2 = -\kappa U_n^1 + (\kappa + \kappa') U_n^2 - \kappa' U_{n+1}^1. \tag{16.30b}$$

The $\overleftrightarrow{D}$ matrix defined in (16.25) is then given by

$$\overleftrightarrow{D} = \begin{pmatrix} \kappa + \kappa' & -\kappa - \kappa' e^{iqa} \\ -\kappa - \kappa' e^{-iqa} & \kappa + \kappa' \end{pmatrix}, \tag{16.31}$$

and Eq. (16.25) is

$$\begin{vmatrix} (\kappa + \kappa') - \omega^2 M_1 & -\kappa - \kappa' e^{iqa} \\ -\kappa - \kappa' e^{-iqa} & (\kappa + \kappa') - \omega^2 M_2 \end{vmatrix} = 0. \tag{16.32}$$

Taking the determinant and solving the resultant quadratic equation yields

$$\omega^2 = \kappa \left(\frac{1}{M_1} + \frac{1}{M_2} \right) \pm \kappa \sqrt{\left(\frac{1}{M_1} + \frac{1}{M_2} \right)^2 - \frac{4}{M_1 M_2} \sin^2 qa}, \tag{16.33}$$

where we have taken $2\kappa = \kappa + \kappa'$ (note the result depends only on this combination).

For the simple case that was treated earlier it would have been more direct to introduce springs between the atoms at the outset, as shown in Fig. 16.4 and then write the equations of motion by inspection (see Problem 16.1). However this

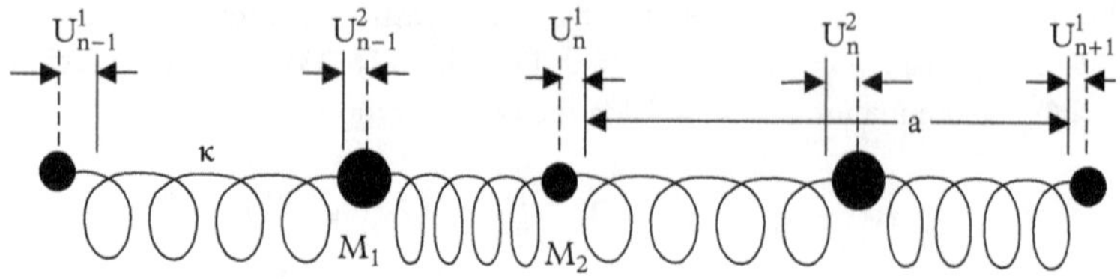

Figure 16.4 *The 1d lattice with two atoms per unit cell.*

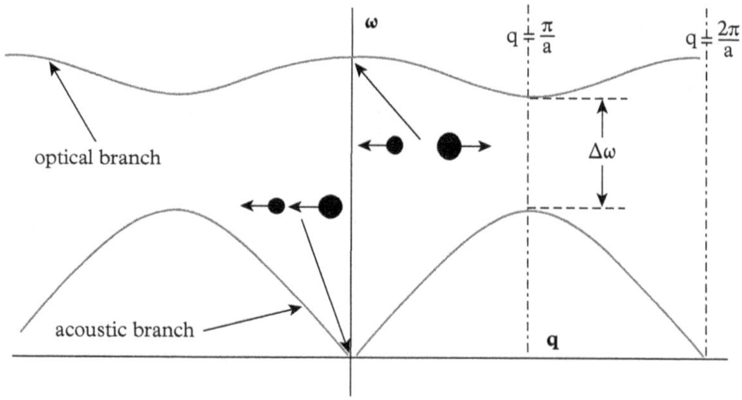

Figure 16.5 *A typical dispersion curve of a one-dimensional lattice with two atoms per unit cell. Shown also are the purely in and out of phase motions that occur for* q = 0.

would obscure how a calculations actually proceeds when using the constants $A_n^{jj'}$. Equation (16.33) is plotted in Fig. 16.5 for some arbitrarily chosen parameters.

For the case $M_1 \neq M_2$, the qa → 0 (long wavelength) limit corresponding to sound waves arises from the negative sign in (16.32),

$$\omega^2 \cong \frac{2\kappa}{M_1 + M_2}\left(\frac{qa}{2}\right)^2;$$

the corresponding phase velocity is $V_p = \left[\kappa a^2/2(M_1 + M_2)\right]^{1/2}$. The second solution as k → 0, involving the + sign in (16.32), is

$$\omega^2 = 2\kappa\left(\frac{1}{M_1} + \frac{1}{M_2}\right); \tag{16.34}$$

this corresponds to an *out of phase* motion of the two masses about their center of mass. For $q = \frac{\pi}{a}$ the two frequencies are

$$\omega^2 = \kappa\left[\left(\frac{1}{M_1} + \frac{1}{M_2}\right) \pm \left(\frac{1}{M_1} - \frac{1}{M_2}\right)\right] = \begin{cases} 2\kappa/M_1 \\ 2\kappa/M_2 \end{cases}. \tag{16.35}$$

Note we have a splitting or a gap of magnitude $\Delta\omega = 2\kappa\,|1/M_1 - 1/M_2|$, as can be seen in Fig. 16.5.

The lower branch is called the *acoustic branch*, corresponding to longitudinal sound waves as its low frequency limit. The upper branch is called the *optical mode*. If M_1 and M_2 had opposite electrical charges and an external electromagnetic field was applied, it would drive the $q = 0$ mode when its frequency was given by (16.34). This occurs for a frequency of order 10^{13} Hz in typical solids. This frequency is typically in the infrared and ionic crystals, such as NaCl, show an absorption maximum at the corresponding frequency.[4]

For the case $M_1 = M_2$ we expect to recover the one atom per unit cell case; however the lattice constant is then $a' = a/2$. For this case, Eq. (16.33) becomes

$$\omega^2 = \frac{2\kappa}{M}\left(1 \pm \cos\frac{qa}{2}\right) = \frac{2\kappa}{M}\left(1 \pm \cos qa'\right). \tag{16.36}$$

The − sign corresponds precisely to Eq. (16.16), while the + sign simply represents the extension of ω_q into the next Brillouin zone (in the repeated zone scheme).

16.3 Representing the lattice potential in 3d

The generalization to three dimensions is reasonably straightforward. Our discussion will involve the multi-atom case from the outset. The expansion of the lattice potential (16.1) then takes the form

$$\begin{aligned} V(\mathbf{U}^j_{\mathbf{n}}) = V^{(0)} &+ \frac{1}{2}\sum_{j,\mathbf{n},j',\mathbf{n}'}\left(\frac{\partial^2 V}{\partial \mathbf{U}^j_{\mathbf{n}}\partial \mathbf{U}^{j'}_{\mathbf{n}'}}\right) : \mathbf{U}^j_{\mathbf{n}}\mathbf{U}^{j'}_{\mathbf{n}'} \\ &+ \frac{1}{6}\sum_{j,\mathbf{n},j',\mathbf{n}',j'',\mathbf{n}''}\left(\frac{\partial^3 V}{\partial \mathbf{U}^j_{\mathbf{n}}\partial \mathbf{U}^{j'}_{\mathbf{n}'}\partial \mathbf{U}^{j''}_{\mathbf{n}''}}\right) \vdots\, \mathbf{U}^j_{\mathbf{n}}\mathbf{U}^{j'}_{\mathbf{n}'}\mathbf{U}^{j''}_{\mathbf{n}''} + \ldots \end{aligned}, \tag{16.37}$$

where $\mathbf{n}$ is a *vector* index denoting the unit cells of the lattice having three components n_i; and the index j denotes the r different atoms within the unit cell. Equations (16.4) and (16.5) are now written as

$$\mathbf{A}^{jj'}_{\mathbf{nn}'} = \frac{\partial^2 V}{\partial \mathbf{U}^j_{\mathbf{n}}\partial \mathbf{U}^{j'}_{\mathbf{n}'}} \tag{16.38}$$

and

$$\mathbf{B}^{jj'j''}_{\mathbf{nn}'\mathbf{n}''} = \frac{\partial^3 V}{\partial \mathbf{U}^j_{\mathbf{n}}\partial \mathbf{U}^{j'}_{\mathbf{n}'}\partial \mathbf{U}^{j''}_{\mathbf{n}''}}, \tag{16.39}$$

[4] Since light is a transverse electromagnetic wave, the propagation direction of the light would have to be normal to the axis of the string. In addition, the wavelength of light is much larger than that of the optical modes for comparable frequencies and hence resonant coupling occurs for $q \cong 0$.

where **A** and **B** are respectively second and third rank tensors. Equation (16.37) is then written as

$$V = V^{(0)} + \frac{1}{2}\sum_{j,n,j',n'} \mathbf{A}^{jj'}_{nn'} : \mathbf{U}^{j}_{n}\mathbf{U}^{j'}_{n'} + \frac{1}{6}\sum_{j,n,j',n',j'',n''} \mathbf{B}^{jj'j''}_{nn'n''} \vdots\, \mathbf{U}^{j}_{n}\mathbf{U}^{j'}_{n'}\mathbf{U}^{j''}_{n''} + \ldots . \tag{16.40}$$

From the equality of cross partial derivatives we have the conditions

$$\mathbf{A}^{jj'}_{nn'} = \frac{\partial^2 V}{\partial \mathbf{U}^{j}_{n}\partial \mathbf{U}^{j'}_{n'}} = \frac{\partial^2 V}{\partial \mathbf{U}^{j'}_{n'}\partial \mathbf{U}^{j}_{n}} = \mathbf{A}^{j'j}_{n'n} \tag{16.41}$$

and

$$\mathbf{B}^{jj'j''}_{nn'n''} = \mathbf{B}^{j'jj''}_{n'nn''} = \mathbf{B}^{jj''j'}_{nn''n'} = \mathbf{B}^{j''j'j}_{n''n'n}. \tag{16.42}$$

From translational symmetry we again have

$$\mathbf{A}^{jj'}_{nn'} = \mathbf{A}^{jj'}_{n-n'}. \tag{16.43}$$

Finally, since a uniform displacement of all atoms results in no force on the individual atoms, we have

$$\sum_{n',j''} \mathbf{A}^{jj'}_{nn'} = 0. \tag{16.44}$$

16.4 Classical theory of the harmonic lattice in 3d

In three dimensions our lattice sites are numbered by the vector index **n** with integer components (n_1,n_2,n_3); their positions are given by the real-space lattice vectors $\mathbf{R}_n$ and the atomic positions by $\mathbf{R}_n + \mathbf{r}_j + \mathbf{U}^j_n$. The force on the atom j in cell **n** is

$$\mathbf{F}^{j}_{n} = -\frac{\partial V}{\partial \mathbf{U}^{j}_{n}} = -\sum_{j',n'} \mathbf{A}^{jj'}_{nn'}\mathbf{U}^{j'}_{n'}.$$

The Newtonian equation of motion is then

$$M_j\ddot{\mathbf{U}}^{j}_{n} = -\sum_{j',n'} \mathbf{A}^{jj'}_{nn'}\cdot\mathbf{U}^{j'}_{n'}, \tag{16.45}$$

where M_j is the mass of the jth atom.

16.4.1 Normal modes and the dynamical matrix

We seek a solution in terms of propagating waves of the form

$$\mathbf{U}^{j}_{n}(t) = \mathbf{U}^{j}_{q}\, e^{i\mathbf{q}\cdot\mathbf{R}_n - i\omega t}, \tag{16.46}$$

where $\mathbf{R_n}$ denotes the real-space lattice vectors and $\mathbf{U}^j_q$ is the amplitude of the jth atom on the site $\mathbf{n} = 0$ at $t = 0$. Writing $\mathbf{A}^{jj'}_{nn'} = \mathbf{A}^{jj'}_{n-n'}$ and substituting (16.46) into (16.45) gives the set of homogenous equations

$$\sum_{j'} \left(\mathbf{D}^{jj'}(\mathbf{q}) - M_j \delta_{jj'} \omega^2 \mathbf{1}\right) \cdot \mathbf{U}^{j'}_q = 0, \tag{16.47}$$

where we define the tensor,

$$\mathbf{D}^{jj'}(\mathbf{q}) = \sum_{n'} \mathbf{A}^{jj'}_{n-n'} e^{-i\mathbf{q}\cdot(\mathbf{R_n}-\mathbf{R_{n'}})}; \tag{16.48}$$

these are the three-dimensional analogues of Eqs. (16.26) and (16.25). As before, $\mathbf{D}^{jj'}(\mathbf{q})$ is called the *dynamical matrix* which now has dimension $3r \times 3r$, where r is again the number of atoms in the unit cell. The system of equations (16.47) has a solution only if the determinant of the coefficients $\mathbf{U}^j_q$ vanishes,

$$\mathrm{Det}\left| D^{jj'}_{ii'}(\mathbf{q}) - M_j \omega^2 \delta_{ii'} \delta_{jj'} \right| = 0, \tag{16.49}$$

where i denotes the Cartesian coordinates of $\mathbf{U}^j_q$. The resulting determinant yields a polynomial equation in ω^2, the solution of which yields the normal mode eigenfrequencies, $\omega = \omega_{sq}$, for each $\mathbf{q}$; here $s = 1, \ldots 3r$, the number of degrees of freedom associated with a single unit cell. Since our system must be in equilibrium it follows that the ω_{sq} are real. The eigenvectors, $\mathbf{U}^j_{sq}$, associated with (16.47), which in general are complex, have dimension 3r.

It is easily seen that, as defined, $\mathbf{D}^{jj'}(\mathbf{q})$ is Hermitian: from Eq. (16.48) we have

$$\begin{aligned}
\left(\mathbf{D}^{jj'}(\mathbf{q})\right)^{\dagger} &= \sum_{n'} \mathbf{A}^{j'j}_{n'-n} e^{+i\mathbf{q}\cdot(\mathbf{R_{n'}}-\mathbf{R_n})} \\
&= \sum_{n'} \mathbf{A}^{jj'}_{n-n'} e^{-i\mathbf{q}\cdot(\mathbf{R_n}-\mathbf{R_{n'}})} \\
&= \mathbf{D}^{jj'}(\mathbf{q}),
\end{aligned} \tag{16.50}$$

where we have used (16.41) in the second step. Since the $\mathbf{A}^{jj'}_{nn'}$ are real, we obtain directly from (16.38) the condition

$$\left(\mathbf{D}^{jj'}(\mathbf{q})\right)^{*} = \mathbf{D}^{jj'}(-\mathbf{q}). \tag{16.51}$$

The most general solution for the motion of the lattice is a superposition of forms of the kind given in (16.46) over all possible modes $(s, \mathbf{q})$:

$$\mathbf{U}^j_n(t) = \sum_{s,q} \mathbf{U}^j_{sq} e^{i\mathbf{q}\cdot\mathbf{R_n} - i\omega_{sq} t}, \tag{16.52a}$$

which we rewrite as

$$\mathbf{U}^{j}_{n}(t) = \sum_{s,\mathbf{q}} Q_{s\mathbf{q}}(t)\mathbf{V}^{j}_{s\mathbf{q}}\, e^{i\mathbf{q}\cdot\mathbf{R}_n}, \tag{16.52b}$$

where in the second step we have introduced the *normal mode amplitudes*, $Q_{s\mathbf{q}}(t)$, into which the time dependence is absorbed, and the *normal mode vectors on site* **n**, $\mathbf{V}^{j}_{s\mathbf{q}}\, e^{i\mathbf{q}\cdot\mathbf{R}_n}$, with accompanying *polarization vectors*, $\mathbf{V}^{j}_{s\mathbf{q}}$, defining the direction of motion of the jth atom for the mode $(s, \mathbf{q})$. The complex amplitudes $Q_{s\mathbf{q}}(t)$ obey the equation

$$\ddot{Q}_{s\mathbf{q}} + \omega^2_{s\mathbf{q}} Q_{s\mathbf{q}} = 0. \tag{16.53}$$

Their magnitudes and phases would be fixed by a set of initial conditions.

16.4.2 Orthonormality and completeness relations

We now examine several properties associated with the vectors $\mathbf{V}^{j}_{s\mathbf{q}}$. To begin with, we note that because we sum over all **q** in (16.52b) we may replace **q** by −**q**; since the $\mathbf{U}^{j}_{n}$ are real it then follows that

$$\mathbf{V}^{j}_{s\mathbf{q}} = \mathbf{V}^{j*}_{s-\mathbf{q}} \tag{16.54a}$$

and

$$Q_{s\mathbf{q}} = Q^{*}_{s-\mathbf{q}}. \tag{16.54b}$$

It is of interest to examine the behavior of the vectors $\mathbf{V}^{j}_{s\mathbf{q}}$ for differing s and **q**. To do this we consider the form

$$\sum_{n}\sum_{j} M_j (\mathbf{V}^{j}_{s\mathbf{q}} e^{i\mathbf{q}\cdot\mathbf{R}_n})^{*} \cdot \mathbf{V}^{j}_{s'\mathbf{q}'}\, e^{i\mathbf{q}'\cdot\mathbf{R}_n}, \tag{16.55}$$

where the reason for our inclusion of the factor M_j will shortly become clear. The sum over **n** involves the form

$$S = \sum_{n} e^{-i\mathbf{q}\cdot\mathbf{R}_n}\, e^{i\mathbf{q}'\cdot\mathbf{R}_n}. \tag{16.56}$$

Now if we displace the lattice by an arbitrary real-space lattice vector $\mathbf{R}'$, the sum S will change according to

$$S \rightarrow e^{i(\mathbf{q}'-\mathbf{q})\cdot\mathbf{R}'}\, S.$$

In an infinite lattice, S must be invariant under such a displacement, and therefore (16.56) must vanish. Hence, $(\mathbf{V}^{j}_{sq}\, e^{i\mathbf{q}\cdot\mathbf{R_n}})^*$ and $\mathbf{V}^{j}_{s'q'}\, e^{i\mathbf{q}'\cdot\mathbf{R_n}}$ are orthogonal when summed over all unit cells. On the other hand, for the case $\mathbf{q} = \mathbf{q}'$, the sum is equal to the number of unit cells, N; therefore,

$$S = N\,\delta_{qq'}. \tag{16.57}$$

The form (16.55) then reduces to

$$N\delta_{qq'} \sum_j M_j\, \mathbf{V}^{j*}_{sq} \cdot \mathbf{V}^{j}_{s'q'}. \tag{16.58}$$

We still need to examine the behavior of (16.55) with respect to s. To do this we: (i) write $\mathbf{U}^{j'}_{q}$ as $Q_{sq}(t)\mathbf{V}^{j'}_{sq}$ in Eq. (16.47) and multiply on the left by $\mathbf{V}^{j*}_{s'q}$; (ii) do a similar substitution with the Hermitian conjugate of (16.47) for s', use (16.50), and multiply on the right by $\mathbf{V}^{j}_{sq}$; (iii) sum both expressions over j; and finally, (iv) subtract them. The result is

$$(\omega^2_{sq} - \omega^2_{s'q}) \sum_j M_j \mathbf{V}^{j*}_{s'q} \cdot \mathbf{V}^{j}_{sq} = 0. \tag{16.59}$$

This constitutes an orthogonality relation for the vectors $\mathbf{V}^{j}_{sq}$.[5] To fix the normalization for the case $s = s'$, we write

$$\sum_j M_j \mathbf{V}^{j*}_{s'q} \cdot \mathbf{V}^{j}_{sq} = M_c \delta_{ss'}, \tag{16.60a}$$

where

$$M_c = \sum_{j=1}^{r} M_j \tag{16.60b}$$

is the total mass of the r atoms in the unit cell. For the case of one atom per unit cell, the polarization vectors would satisfy

$$\mathbf{V}^{*}_{s'q} \cdot \mathbf{V}_{sq} = \delta_{ss'}\,(\text{one atom per unit cell}), \tag{16.61}$$

i.e., they are orthogonal unit vectors. Returning to the form (16.55), the normalization for what we have called the normal mode vectors then satisfies

$$\sum_{j,\,\mathbf{n}} M_j \mathbf{V}^{j*}_{s'q'} e^{-i\mathbf{q}'\cdot\mathbf{R_n}} \cdot \mathbf{V}^{j}_{sq} e^{i\mathbf{q}\cdot\mathbf{R_n}} = N M_c \delta_{qq'} \delta_{ss'} \tag{16.62}$$

when summed over all atoms in all **n** sites.

[5] If a degeneracy occurs where modes with different s have the same frequency, orthogonal states can be constructed.

We may invert (16.52b) to find the normal mode expansion coefficients, $Q_{sq}(t)$, in terms of the $\mathbf{U}_{\mathbf{n}}^j(t)$; multiplying both sides by $e^{-iq'\cdot R_n}$ and summing over j and $\mathbf{n}$, we have

$$\sum_{j,n} e^{-iq'\cdot R_n} \mathbf{U}_{\mathbf{n}}^j(t) = \sum_{j,n}\sum_{s,q} \mathbf{V}_{sq}^j \, Q_{sq}(t) e^{i(q-q')\cdot R_n}.$$

Using Eq. (16.57), taking the dot product of both sides of (16.63) with $\mathbf{V}_{s'q'}^{*j}$, carrying out the summation over s, and finally applying the orthonormality condition (16.60a) yields

$$Q_{sq}(t) = \frac{1}{NM_c} \sum_{j,n} M_j \mathbf{V}_{sq}^{*j} \cdot \mathbf{U}_{\mathbf{n}}^j(t) \, e^{-iq\cdot R_n}, \tag{16.63}$$

which can be compared with Eq. (16.52b).

A final property associated with the vectors $\mathbf{V}_{sq}^j$ is obtained by substituting (16.63) into (16.52b), which gives

$$\mathbf{U}_{\mathbf{n}}^j(t) = \frac{1}{NM_c} \sum_{s,q}\sum_{j',n'} M_{j'} \mathbf{V}_{sq}^{*j'} \cdot \mathbf{U}_{\mathbf{n}'}^{j'}(t) \, \mathbf{V}_{sq}^j \, e^{iq\cdot(R_n - R_{n'})}.$$

Since $\mathbf{U}_{\mathbf{n}}^j(t)$ is the same on both sides of this equation the remainder of it must satisfy

$$\sum_{s,q} V_{isq}^j V_{i'sq}^{*j'} e^{iq\cdot(R_n - R_{n'})} = \frac{NM_c}{M_j} \delta_{ii'} \delta_{jj'} \delta_{nn'}; \tag{16.64}$$

this is referred to as the *completeness relation.*

16.4.3 Lagrangian and Hamiltonian formulation

We can express the kinetic and potential energies of the lattice in terms of the normal coordinates, Q_{sq}. Starting with the total kinetic energy in the form

$$T = \frac{1}{2} \sum_{j,n} M_j \dot{\mathbf{U}}_{\mathbf{n}}^j \cdot \dot{\mathbf{U}}_{\mathbf{n}}^j, \tag{16.65}$$

we replace the $\dot{\mathbf{U}}_{\mathbf{n}}^j$ using (16.52b), which gives

$$T = \frac{1}{2} \sum_{j,n} M_j \sum_{s,q,s',q'} \dot{Q}_{sq} \dot{Q}_{s'q'} \mathbf{V}_{sq}^j \cdot \mathbf{V}_{s'q'}^j \, e^{i(q+q')\cdot R_n}.$$

Replacing $\mathbf{q}$ by $-\mathbf{q}$ and using (16.54a,b), together with the orthogonality relation (16.60a), we obtain

$$T = \frac{NM_c}{2} \sum_{s,q} |\dot{Q}_{sq}|^2. \tag{16.66}$$

Performing similar operations with the quadratic part of the potential energy from Eq. (16.40), and using the equations of motion as given by (16.47) and (16.48), we can write the potential energy as

$$V = \frac{NM_c}{2} \sum_{s,q} \omega_{sq}^2 |Q_{sq}|^2. \tag{16.67}$$

In terms of the normal coordinates, the Lagrangian[6] of the lattice, $T - V$, is then given by

$$L = \frac{NM_c}{2} \sum_{sq} \left[|\dot{Q}_{sq}|^2 - \omega_{sq}^2 |Q_{sq}|^2 \right]. \tag{16.68}$$

The canonical momenta[7] are given by the usual Lagrangian definition,

$$P_{sq} = \partial L / \partial \dot{Q}_{sq} = NM_c \dot{Q}_{sq}^*; \tag{16.69}$$

it then follows that the Hamiltonian, defined as $H = \sum_{s,q} P_{sq}\dot{Q}_{sq} - L$, is

$$H = \sum_{s,q} \left[\frac{|P_{sq}|^2}{2NM_c} + \frac{1}{2} NM_c \omega_{sq}^2 |Q_{sq}|^2 \right]. \tag{16.70}$$

In the limit $\omega_{sq} \to 0$, corresponding to sound waves, and for propagation along an axis of sufficiently high symmetry, all r atoms move in the same direction and in phase; one mode, Q_{1q}, will have $\mathbf{V}^j_{1q} \parallel \mathbf{q}$, while two others, Q_{2q}, Q_{3q}, propagate such that $\mathbf{V}^j_{2q}, \mathbf{V}^j_{3q} \perp \mathbf{q}$. The three modes are referred to as *longitudinal and transverse sound* and all have $\omega \propto q$. For the case of a cubic crystal the modes in this limit were discussed earlier in Section 1.8. The remaining $3r - 3$ modes have a finite frequency as $\mathbf{q} \to 0$ and are called *optical modes.*

In making the transition from a static to a dynamic lattice via Eq. (16.45), we are implicitly assuming that (in a one-electron picture) the orbitals associated with the various electronic states characterizing the ground state readjust (or deform) to "follow" the atomic displacements on a time scale which is much faster than vibrational time scales. As noted in Footnote 1, typical electronic time scales involve the plasma frequency or Fermi energy for conduction and/or valance electrons, and orbital frequencies for core electrons; these are generally on the scale 10^{15} Hz or higher, whereas the maximum lattice vibration frequencies are on the order of 10^{13} Hz. Under these conditions the electrons respond nearly adiabatically for many purposes and this approximation is referred to as the *adiabatic approximation* (it is also called the Born–Oppenheimer approximation).[8]

[6] A brief review of some components of Lagrangian mechanics is given in Appendix A.

[7] The initially surprising appearance of the factor N in P_{sq} simply reflects the fact that the canonical momentum associated with modes having a canonical velocity $\dot{Q}_{sq}$ will involve all N unit cells of the crystal.

[8] As will be discussed in Chapter 21, in the presence of the continuous excitation spectrum of the electrons, corrections to the electronic motions are required.

16.5 Quantization of lattice vibrations

The transition to quantum mechanics is accomplished by replacing $\mathbf{U}^j_n$ and $\mathbf{P}^j_n = M_j\dot{\mathbf{U}}^j_n$ by *operators* and applying the Heisenberg commutation relations for each vector component, i, atom, j, and site, n:

$$\left[\hat{U}^j_{in'}, \hat{P}^{j'}_{i'n'}\right] = i\hbar\delta_{ii'}\delta_{jj'}\delta_{nn'} \tag{16.71a}$$

or

$$\left[\hat{U}^j_{in}, M_j\dot{\hat{U}}^{j'}_{i'n'}\right] = i\hbar\delta_{ii'}\delta_{jj'}\delta_{nn'}. \tag{16.71b}$$

Using (16.52b) and (16.64), we can rewrite (16.71b) in terms of the normal coordinate operators $\hat{Q}_{s'q}$ as

$$\left[\hat{Q}_{s'q}, \dot{\hat{Q}}^*_{sq}\right] = \frac{i\hbar}{M_c N}\delta_{ss'}\delta_{qq'}; \tag{16.72a}$$

or equivalently in terms of the normal mode canonical momentum given by (16.69) as

$$\left[\hat{Q}_{s'q}, \hat{P}_{sq'}\right] = i\hbar\delta_{ss'}\delta_{qq'}. \tag{16.72b}$$

We now introduce the (dimensionless) phonon *creation* and *destruction* operators, $\hat{a}^\dagger_{sq}, \hat{a}_{sq}$, as

$$\hat{a}_{sq} = \sqrt{\frac{NM_c\omega_{sq}}{2\hbar}}\left(\hat{Q}_{sq} + \frac{i}{NM_c\omega_{sq}}\hat{P}_{sq}\right) \tag{16.73a}$$

and

$$\hat{a}^\dagger_{sq} = \sqrt{\frac{NM_c\omega_{sq}}{2\hbar}}\left(\hat{Q}_{sq} - \frac{i}{NM_c\omega_{sq}}\hat{P}_{sq}\right), \tag{16.73b}$$

which are defined such that they obey the commutation relations

$$\left[\hat{a}_{sq}, \hat{a}_{s'q'}\right] = 0, \tag{16.74a}$$

$$\left[\hat{a}^\dagger_{sq}, \hat{a}^\dagger_{s'q'}\right] = 0, \tag{16.74b}$$

and

$$\left[\hat{a}_{sq}, \hat{a}^\dagger_{s'q'}\right] = \delta_{ss'}\delta_{qq'}, \tag{16.74c}$$

which can be verified with (16.72b) and (16.73a,b). As discussed in Appendix B, and as their name implies, they create and destroy phonons in occupation number states $(s, \mathbf{q})$,

$$\hat{a}^{\dagger}_{sq}\Phi(\ldots, n_{sq}, \ldots) = \sqrt{n_{sq}+1}\,\Phi(\ldots, n_{sq}+1, \ldots) \tag{16.75a}$$

$$\hat{a}_{sq}\Phi(\ldots, n_{sq}, \ldots) = \sqrt{n_{sq}}\,\Phi(\ldots, n_{sq}-1, \ldots), \tag{16.75b}$$

where Φ is a Bose state function in the occupation number representation containing n_{sq} phonons in each of the states $s, \mathbf{q}$. In terms of the canonical variables Q_{sq} and P_{sq}, we have the inverse of (16.73a,b) as

$$\hat{Q}_{sq} = \sqrt{\frac{\hbar}{2NM_c\omega_{sq}}}\left[\hat{a}_{sq} + \hat{a}^{\dagger}_{sq}\right] \tag{16.76a}$$

and

$$\hat{P}_{sq} = -i\sqrt{\frac{\hbar NM_c\omega_{sq}}{2}}\left[\hat{a}_{sq} - \hat{a}^{\dagger}_{sq}\right]. \tag{16.76b}$$

Substituting (16.76) into the operator form of (16.70) yields

$$\hat{H} = \sum_{qs}\hbar\omega_{sq}\left(\hat{n}_{qs} + \frac{1}{2}\right), \tag{16.77}$$

where we have introduced the *number operator* as

$$\hat{n}_{qs} = \hat{a}^{\dagger}_{qs}\hat{a}_{qs}. \tag{16.78}$$

The expectation value of $\hat{n}_{sq}$ in thermal equilibrium is given by[9]

$$\langle\hat{n}_{sq}\rangle = n^{B}_{sq} = \frac{1}{e^{\hbar\omega_{sq}/k_BT} - 1}. \tag{16.79}$$

Including the zero-point motion, the average energy, $\bar{\varepsilon}_{sq}$, contributed by a mode $s, \mathbf{q}$ of a system in thermal equilibrium is

$$\bar{\varepsilon}_{sq} = \frac{\hbar\omega_{sq}}{2} + \frac{\hbar\omega_{sq}}{e^{\hbar\omega_{sq}/k_BT} - 1} \tag{16.80a}$$

or

$$\bar{\varepsilon}_{sq} = \frac{1}{2}\hbar\omega_{sq}\coth\left(\frac{\hbar\omega_{sq}}{2k_BT}\right). \tag{16.80b}$$

In the high temperature limit, each mode $(s, \mathbf{q})$ contributes an energy

$$\bar{\varepsilon}_{sq} = k_BT, \tag{16.81}$$

[9] See typical texts on Statistical Mechanics; this form can be inferred from Eq. (4.12).

which is referred to as the equipartition theorem, discussed earlier, in Chapter 4. The total energy is given by summing over the energies of all modes, which from (16.80a) is then

$$E_T = \sum_{s,q} \left[\frac{\omega_{sq}}{2} + \frac{\hbar\omega_{sq}}{e^{\hbar\omega_{sq}/k_BT} - 1} \right]. \tag{16.82}$$

At high temperature, where $\hbar\omega_{\sigma q} \ll k_BT$ for all modes, we can expand (16.82); on ignoring the zero-point energy, we obtain

$$E_T = \sum_{s,q} k_BT \left[1 - \frac{1}{2}\frac{\hbar\omega_{sq}}{k_BT} + \frac{1}{12}\left(\frac{\hbar\omega_{sq}}{k_BT}\right)^2 + \cdots \right]; \tag{16.83}$$

the heat capacity arising from the second term corresponds to the law of Dulong and Petit (see Section 4.1).

16.6 Inelastic neutron scattering

The most powerful way to study the function ω_{sq} of the phonons is with inelastic neutron scattering.[10,11] We will assume that the crystal is in the ground state and that only a single phonon is excited (multi-phonon processes are observed in practice but will not be discussed here). Hence the initial state consists of an incoming neutron with wave vector $\mathbf{k}$ and energy $\varepsilon(\mathbf{k}) = \hbar^2k^2/2M_n$, while the final state involves a neutron with wave vector $\mathbf{k}'$ and energy $\varepsilon'(\mathbf{k}) = \hbar^2k'^2/2M_n$, together with a phonon with wave vector $\mathbf{q}$ and energy $\hbar\omega_{sq}$. The equations of energy and momentum conservation are then

$$\frac{\hbar^2k^2}{2M_n} = \frac{\hbar^2k'^2}{2M_n} + \hbar\omega_{sq} \tag{16.84a}$$

and

$$\mathbf{k} = \mathbf{k}' + \mathbf{q}. \tag{16.84b}$$

Assuming the energy and direction of the incoming neutron are known, it is a simple matter to show that, by measuring the final energy and the scattering angle of the neutron, the energy and momentum of the phonon produced can be computed.

Figure 16.6 shows the phonon dispersion curves for fcc aluminum as obtained by inelastic neutron scattering for the [100] ($\Gamma \to X$) and [110] ($\Gamma \to K$) directions.[12] Note that along [100] the two transverse modes are degenerate (as we found in Chapter 1, Eq. (1.35)), while along [110] this degeneracy is lifted (see Problem 1.1).

[10] The Nobel Prize in Physics for 1994 was awarded "for pioneering contributions to the development of neutron scattering techniques for studies of condensed matter." The recipients were Bertram N. Brockhouse "for the development of neutron spectroscopy" and Clifford G. Shull "for the development of the neutron diffraction technique."

[11] Neutrons generally interact strongly with nuclei, and low energy neutrons, those with wavelengths comparable to lattice spacings respond to the collective response of a crystal.

[12] Dispersion relations for phonons in aluminum at 80 and 300 degrees K, R. Stedman and G. Nilsson, *Phys. Rev.* 145, 492 (1966).

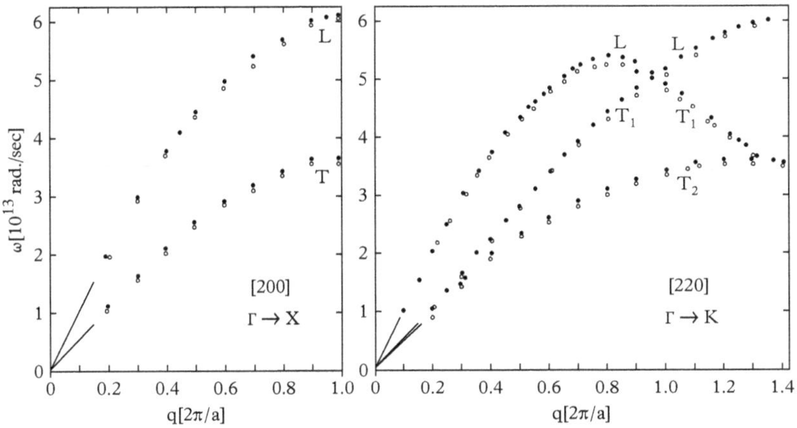

Figure 16.6 *Phonon dispersion curves for face centered cubic metallic aluminum. (a) The [200] direction; note that the two transverse branches are degenerate for this direction. (b) The [220] direction; here the transverse modes are distinct. Filled and open circles denote data at 80 K and 300 K respectively. (After R. Stedman and G. Nilsson (1966).)*

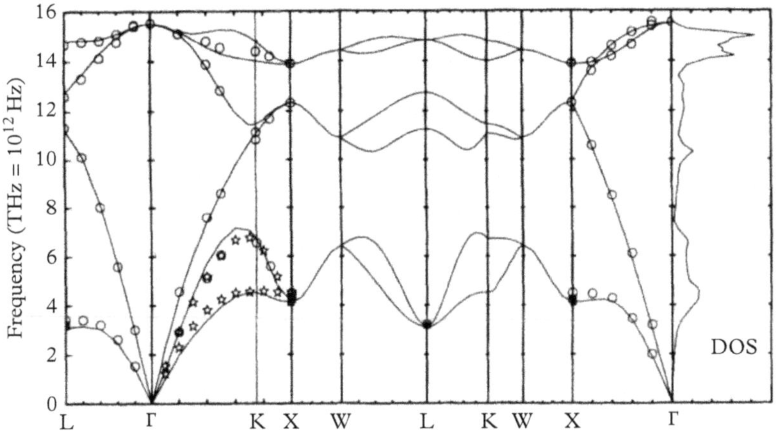

Figure 16.7 *The phonon spectrum for silicon as determined from neutron diffraction. Note the presence of optical as well as acoustical modes. The circles are from Dolling (1963) and stars from Nilsson and Nelin (1972). Solid lines are calculated results of Wei and Chou (1994). Solid squares are the frozen-phonon results of Yin and Cohen (1982). The curve at the far right shows the phonon density of states.*

Figure 16.7 shows the phonon spectrum of silicon, as determined by inelastic neutron scattering by Dolling[13] and by Nilsson and Nelin[14] along the [100] ($\Gamma \rightarrow X$), [110] ($\Gamma \rightarrow K$), and [111] ($\Gamma \rightarrow L$) directions. Since there are *two* atoms per primitive unit cell, we also have a longitudinal optical (LO) and two transverse optical (TO) modes. The smooth curves are first principles calculations.[15]

Figure 16.8 shows, schematically, a triple axis neutron spectrometer. The thermalized spectrum of neutrons in a nuclear reactor contains a sufficiently large number of neutrons with wavelengths of order crystal spacings to make it an ideal source. The spectrometer begins with a beam exiting from a port in the side of the reactor. This beam passes through a collimator to more uniquely define its direction. However this beam contains the full Maxwell spectrum of neutron wavelengths (energies). A unique wavelength is selected by exploiting Bragg's law with a crystal which minimally absorbs neutrons. The outgoing mono-directional mono-wavelength beam then falls on the crystal, the phonon spectrum of which

[13] G. Dolling, in *Inelastic Scattering of Neutrons in Solids and Liquids*, Vol. II, p. 37. Ed. S. Ekland, IAEA, Vienna (1963).

[14] G. Nilsson and G. Nelin, *Phys. Rev.* **86**, 3777 (1972).

[15] Phonon dispersions of silicon and germanium from first-principles calculations, Siqing Wei and M.Y. Chou, *Phys. Rev. B* **50**, 2221 (1994).

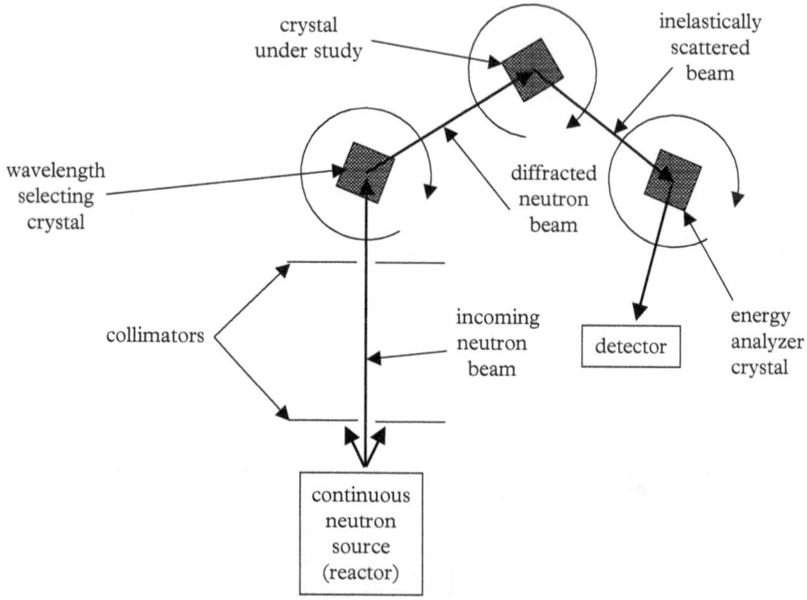

Figure 16.8 *A triple axis neutron spectrometer. Incoming thermal neutrons are collimated (to give them a unique direction) after which they pass through a crystal which selects a unique wave vector by exploiting Bragg diffraction. This beam then scatters inelastically off the crystal under investigation. The energy of this scattered beam is measured with the analyzer crystal, again exploiting Bragg's law.*

is being studied (it may be cooled if one seeks the phonon spectrum at low temperatures). The energy loss can be obtained from the direction of the emerging beam using Eqs. (16.69a,b). However, we may be interested to know the degree to which the phonon energies are broadened as a result of their lifetime. One then uses a third crystal, the analyzer crystal, to select specific energies from the emerging beam; the number of neutrons at each energy selected is then recorded using a neutron counter (the detector).

Measurements can also be made using a so-called spallation source in which a high energy proton beam incident on a heavy metal target dislodges neutrons from the nuclei, that can subsequently be cooled (thermalized) by a moderator. Here one can perform energy analysis using time-of-flight techniques.[16]

16.7 Raman scattering

One can also probe phonon spectra via inelastic light scattering, commonly referred to as Raman scattering.[17,18] With the availability of high power lasers, typically operating in the visible portion of the spectrum, this relatively weak process can now be studied systematically. Here the equations of energy and momentum conservation take the form

$$\hbar\omega_{\mathbf{k}} = \hbar\omega_{\mathbf{k}'} + \hbar\omega_{s\mathbf{q}} \tag{16.85a}$$

and

$$\mathbf{k} = \mathbf{k}' + \mathbf{q}, \tag{16.85b}$$

[16] The 1994 Nobel Prize in Physics was awarded to Bertram Brockhouse "for the development of neutron spectroscopy" and "for pioneering contributions to the development of neutron scattering techniques for studies of condensed matter."

[17] The 1930 Nobel Prize in Physics was awarded to Chandrasekhara Venkata Raman "for his work on the scattering of light and for the discovery of the effect named after him."

[18] A new radiation, C.V. Raman, *Indian J. Phys.* 2, 387 (1928); Eine neue erscheinung bei der lichtzerstreuung in krystallen, G. Landsberg and L. Mandelstam, *Naturwissenschaften* **16**, 557 (1928).

where $\omega_{\mathbf{k}}$ and $\mathbf{k}$ are the frequency and wave vector of the photon. However, unlike neutrons, which can have small energies and wavelengths comparable to lattice spacings (that are characteristic of short wavelength phonons), the wavelength of visible light is of order 10^3 times larger and hence only excitations of very long wavelength can be excited. On the other hand, the energy of the photons is much larger than the phonons and hence the frequency shifts that need to be resolved are correspondingly small. The wavelength restriction limits the region of the phonon spectrum that can be probed to very small wave vectors, and having readily observable frequency shifts (that can be resolved with typical spectrometers), largely restricts the method to the optical modes, which have finite frequencies, as $q \to 0$.

At finite temperatures the scattered photons can either lose or gain energy by emitting or absorbing an optical phonon; these processes are referred to as Stokes and anti-Stokes processes respectively.

16.8 Calculating phonon dispersion relations from first principles

We began our discussion of lattice dynamics by assuming the existence of the potential energy function, $V = V(\mathbf{U}_{\mathbf{n}}^{j})$, introduced in Eq. (16.1). Assuming that we are at absolute zero and neglecting the zero-point energy, one can interpret this function as the total energy, $E = E\,(\mathbf{U}_{\mathbf{n}}^{j})$, of the solid for some given configuration of the nuclei that are located (and frozen) at positions $\mathbf{U}_{\mathbf{n}}^{j}$. As discussed in Section 15.3.2, one would start by finding the *equilibrium* positions in the undistorted crystal by (globally) minimizing the total energy. The "spring constants" $\mathbf{A}_{\mathbf{nn}'}^{jj'}$ of Eq. (16.38a) (or equivalently the κ_{ij} of Eq. (15.36)) are then obtained by fitting a sufficiently large number of total energies for various displaced configurations so as to accurately determine the $\mathbf{A}_{\mathbf{nn}'}^{jj'}$ to some level in the expansion (nearest neighbor, next-nearest neighbors, and so on) while all the time keeping the $\mathbf{U}_{\mathbf{n}}^{j}$ small so as to stay in the harmonic regime). With these constants and the atomic masses one then can calculate the phonon frequencies, as described in Section 16.4, assuming that the crystal responds adiabatically to the atomic motions (i.e., no electronic transitions occur). This is a good approximation for an insulator with a band gap much larger than the highest phonon frequency; with certain caveats, it is also applicable to metals.[19]

The function $E = E\,(\mathbf{U}_{\mathbf{n}}^{j})$ can be calculated using some version of density functional theory. However, in applying this approach one immediately runs into a problem: the total energy of a solid is usually calculated by exploiting the full translational symmetry of the crystal, thereby gaining the enormous simplification that this allows, e.g., using Bloch functions. On the other hand, formally calculating the spring constants involves breaking this symmetry by examining various configurations of the atoms displaced from their equilibrium sites. Two techniques are commonly used to address this problem, (i) the supercell approach

[19] The one-electron states $\varepsilon(\mathbf{k})$ develop a lifetime in a region of order the Debye frequency around the Fermi surface, as will be discussed in Section 21.2.

and (ii) first-order perturbation (or equivalently linear response) theory. The first is conceptually simple, but is formally limited to certain discrete phonon wave vectors, $\mathbf{q}_i$, called frozen phonons, as we will discuss; however it rapidly becomes computationally demanding as the number of these wave vectors increases. The second is limited to harmonic lattice dynamics and yields the full phonon dispersion, but is rather formal so we have placed it in Appendix 16A of this chapter. We now outline the first approach.

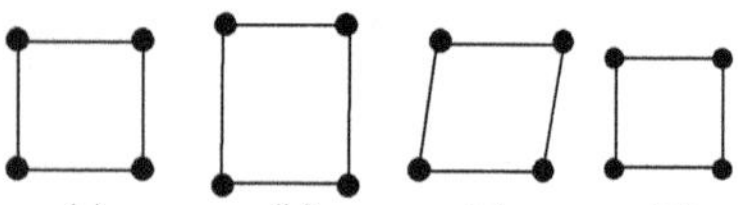

Figure 16.9 *Part (a) shows an undistorted square lattice while (b), (c), and (d) show a pure extension, a pure shear, and an isotropic compression.*

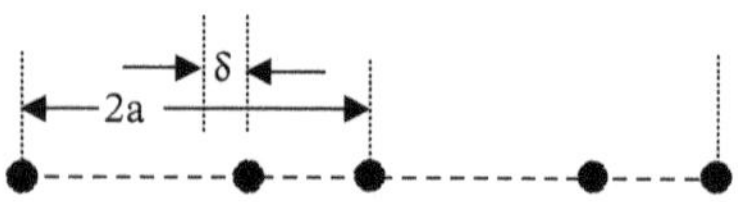

Figure 16.10 *A one-dimensional supercell showing the static displacements that would correspond to a phonon at the Brillouin zone point of the original lattice, where* $qa = \pi/2$, *but that corresponds* $q = 0$ *to of the supercell.*

16.8.1 The supercell approach

To begin with, we consider the simpler problem of calculating the elastic constants from which the long wavelength acoustic modes can be obtained, as previously discussed in Section 1.7. As an example, consider a simple square lattice in two dimensions, as shown in Fig. 16.9(a). A distortion involving a uniaxial extension of the unit cell in the y direction (while maintaining the x dimension fixed) is shown in Fig. 16.9(b); here the square lattice becomes *rectangular.* On the other hand, Fig. 16.9(c) shows a pure shear distortion in which the lattice becomes oblique. An isotropic compression, shown in Fig. 16.9(d), retains the full square symmetry. Note that the structure and its distortions are here being regarded as *periodically continued*; i.e., the space lattice symmetry of Section 8.1 is retained; however the rotational (point group) symmetry of Section 8.3 is lowered from four-fold to two-fold in Figs. 16.9(b,c). By fitting the change in the total energy to the quadratic form involving the strains given by Eq. (1.10), we can obtain the elastic constants, c_{ij}, the number of which is fixed by the symmetry of the undistorted crystal (three for the cubic lattice). The fits must be restricted to small strains, as additional terms enter for larger distortions (as parameterized by the $\mathbf{B}^{jj'j''}_{nn'n''}$ coefficients in Eq. (16.39) in the next order). Of course, the total energy calculations must be performed for lattices having the reduced symmetry associated with the distorted lattice, and one must consider at least as many independent distortions as there are elastic constants.[20]

By enlarging the number of atoms in the unit cell, the approach just described for calculating elastic constants can be extended to the case of a *non-uniform* distortion, as one has when certain short wavelength phonons propagate in the original lattice; the resulting cell is then called a *supercell.*[21,22] We have already indirectly encountered an example of a supercell when we discussed the case of a one-dimensional lattice with two atoms per unit cell for the special case when the masses of the two atoms are equal, for which the dispersion relation was given in Eq. (16.36). Here, the $qa = \pi/2$ mode at the Brillouin zone point of the original lattice becomes the $k = 0$ optical mode of the supercell. By performing independent total energy calculations as a function of (i) the lattice constant a for the uniform lattice and (ii) the internal displacement parameter δ, shown in Fig. 16.10 at fixed 2a, we would obtain two data sets that can then be fitted to quadratic forms; the curvatures associated with the two independent data sets can be represented by a combination of nearest-neighbor

[20] For a simple monatomic crystal one could use the elastic constants themselves to calculate some minimal set of spring constants (see Problem 16.3), and from the known mass of the atom in the unit cell estimate the dispersion for all $\mathbf{q}$. But qualitatively we cannot expect this to describe short wavelength phonons, which are sensitive short-range interactions whereas elastic constants involve short- and long-range interactions.

[21] *Ab initio* force constants of GaAs: a new approach to calculation of phonons and dielectric properties, K. Kunc and R.M. Martin, *Phys Rev. Lett.* **48**, 406 (1982).

[22] *Ab inito* calculation of the phonon dispersion relations: Application to Si, M.T. Yin and M.L. Cohen, *Phys. Rev. B* **25**, 4317 (1982).

and next-nearest-neighbor spring constants, and the corresponding parameters, $A_{nn'}$; in this way one can construct a potential energy landscape analogous to that discussed in Section 15.2.2 for a molecule. One could then solve the associated Newtonian equations of motion and obtain the phonon dispersion relation based on a two-parameter model. One would obtain the dispersion for all q in this way, but it is expected to best represent the two wave vectors which are compatible with the supercell, $q = 0$ and $qa = \pi$; these are examples of the frozen phonons, q_i. To improve on the calculation we can make the supercell even larger and perform additional total energy calculations representing independent displacement combinations within this larger unit cell, thereby allowing the examination of additional frozen phonon wavevectors, which in turn allow the introduction of additional constants, $A_{nn'}$.

Extending the earlier ideas to higher dimensions is straightforward. As an example, Fig. 16.11 shows a doubling of the lattice constant of the square lattice in the y direction. Figure 16.11(a) corresponds to a longitudinal wave, while Fig. 16.11(b) describes a shear wave, both "propagating" parallel to y. As with the calculation of elastic constants, the fit to the energy versus distortion for various configurations of atoms in the unit cell must be limited to small displacements, as nonlinear effects enter at larger displacements.[23]

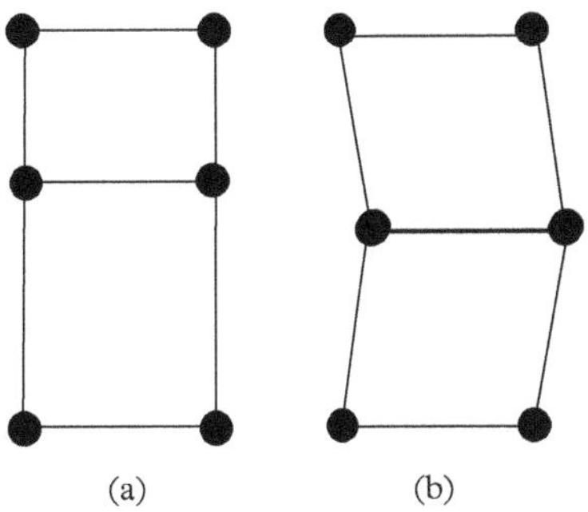

Figure 16.11 *Parts (a) and (b) can be associated with a longitudinal wave propagating in the y direction, while (b) can be associated with a shear wave, polarized in the x direction.*

The same ideas extend naturally to three dimensions: for a sufficiently large supercell, and by examining independent distortions of all atoms within that unit cell (including those in the original cell), potential energy landscapes, and with them a set of parameters, $A^{jj'}_{nn'}$, can be obtained. With these one can construct the dynamical matrix defined by Eq. (16.48), and solve for the phonon frequencies using Eq. (16.47).

We will not enter into the details of calculating frozen-phonon frequencies in real materials, but rather limit ourselves to presenting some results for the two cases for which experimental results were presented earlier: Si and Al. One of the original applications of the frozen-phonon method was to Si by Yin and Cohen (1982);[24] Figure 16.12 shows their results along the (100) ($\Gamma - X$) direction, together with the experimental results of Nilsson and Nelin.[25] Since Si has two atoms in the basic Bravais lattice, there will be longitudinal (LO) and transverse optical (TO) branches, along with the usual longitudinal (LA) and transverse optical (TA) branches. Note that the two TA and the two TO modes are expected to be separately degenerate by symmetry for the propagation along this direction. Note also that the LA and LO branches are degenerate at the X point, whereas the TO and TA are not; finally the TA and LA branches both approach zero at Γ, as expected.

Having represented a frozen-phonon calculation in terms of spring constants, one can calculate the phonon frequencies at any point in the Brillouin zone. One can then go on to calculate the phonon density of states,

$$N(\omega) = \sum_{q} \delta\left(\omega(\mathbf{q}) - \omega\right), \tag{16.86}$$

[23] Fitting the nonlinear terms, e.g., the coefficients $\mathbf{B}_{nn'n''}$ in Eq. (16.39), would allow density functional based calculations of quantities like the thermal expansion, thermal conductivity, and corrections to the heat capacity at high temperatures, as will be discussed in Chapter 17.

[24] Theory of lattice-dynamical properties of solids - application to Si and Ge, Y.T. Yin and M.L. Cohen, *Phys. Rev. B* **26**, 3259 (1982).

[25] *Ab initio* calculation of the phonon dispersion relation: Application to Si, M.T. Yin and M.L. Cohen, *Phys. Rev. B* **25**, 4317 (1982); Study of homology between silicon and germanium by thermal-neutron spectrometry, G. Nilsson and G. Nelin, *Phys. Rev. B* **6**, 3777 (1972).

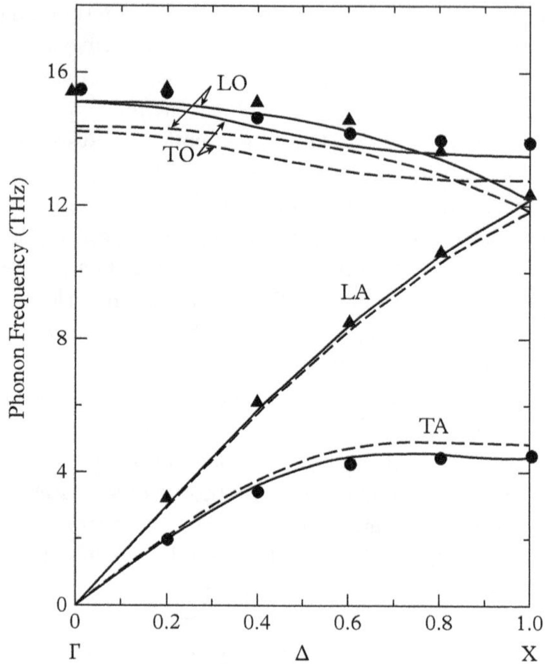

Figure 16.12 *The LA, TA, LO, and TO modes for silicon along the Γ − X direction. The solid and dashed lines are the calculated results of Yin and Cohen (involving different assumptions) using the frozen-phonon method. The points are the experimental results of Nilsson and Nelin (1972).*

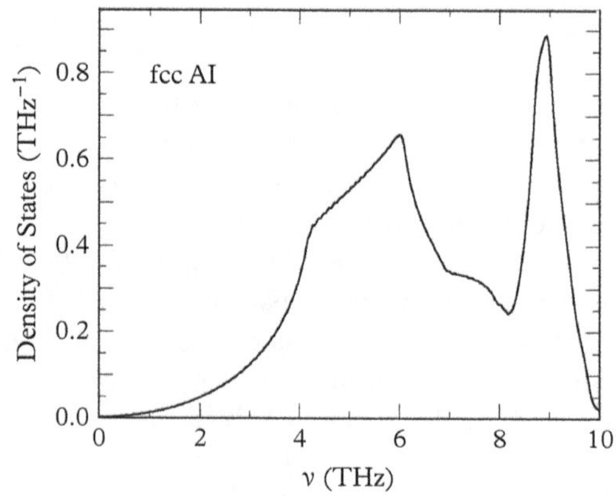

Figure 16.13 *The phonon density of states of aluminum as calculated using the frozen phonon method. (Courtesy of M. Weinert.)*

where in practice the delta function would be binned. As an example, Fig. 16.13 shows phonon density of states for the case of aluminum. Note the presence of singularities.[26]

With the availability of this function one can go on to calculate the total thermal energy, $\bar{E}(T)$, given by Eq. (4.17) and with it the phonon heat capacity, C(T) (in the harmonic approximation).

[26] These are the so-called van Hove singularities which will be discussed in connection with electronic spectra in Section 29.2.

The second approach to calculating phonon spectra, previously mentioned, uses perturbation theory and is mathematically more complex. We outline this method in Appendix 16A.

ADDITIONAL READING

An excellent discussion of lattice dynamics is given in *Quantum Theory of Solids*, R. Peierls, Oxford University Press, Oxford (1956); Peierls was a pioneer in this field. Another classic reference is *Dynamical Theory of Crystal Lattices*, M. Born and K. Huang, Clarendon Press, Oxford University Press, Oxford (1954). Valuable, but fast moving, is the treatment in *Quantum Theory of the Solid State*, J. Callaway, Academic Press, London (1974).

At a more elementary level are the treatments in *Solid State Physics*, N. Ashcroft and D. Mermin, Saunders College, PA (1975) and *Principles of the Theory of Solids*, J.M. Ziman, Cambridge University Press (1964).

Problems

16.1. Derive Eq. (16.33), but instead of exploiting properties of the force parameters, A_n, as we did in the text, write the Newtonian equations of motion for neighboring masses, M_1 and M_2, assuming that they are connected by a spring with a spring constant, κ.

16.2. Fill in the missing steps leading to (16.74a–c). Also show that the transformation to the $\hat{a}$ and $\hat{a}^\dagger$ diagonalizes the Hamiltonian.

16.3. You are given a two-dimensional square lattice. Assume that there are springs connecting the nearest neighbors and the next nearest (along the square diagonal) neighbors with spring constants κ and κ' respectively. Calculate the two-dimensional compressibility, Young's modulus, and Poisson's ratio (see Chapter 1 for the definitions of these quantities).

16.4. As in Problem 16.3, assume that there are springs connecting the atoms of a simple cubic lattice along the [100], [110], and [111] with spring constants κ, κ', and κ''; calculate the elastic constants c_{11}, c_{12}, and c_{44}.

16.5. For the lattice described in Problem 16.3, calculate the longitudinal and shear (with the displacement in the plane of the square) sound velocities for propagation in the (10) and (11) directions.

Appendix 16A Calculating the phonon spectrum for all wave vectors

Rather than directly evaluating the total energy function, $E = E(\mathbf{R}_1, \ldots \mathbf{R}_N)$, for various supercell configurations with their associated commensurate wave vectors and polarizations, all values of $\mathbf{q}$ can be calculated by using techniques based on perturbation theory.[27,28,29,30] Although the methods used are similar, our discussion here follows that of Giannozzi et al.[31] To keep things as simple as possible we will assume that there is only one atom per unit cell. This is easily generalized to cover the multi-atom case.[32]

The basic idea is that the displacements of the nuclei can be viewed as an "external" one-electron potential, $V^{(ext)}(\mathbf{r}, \{\mathbf{R}\})$, that acts on all the electrons; here we have written the set of displacements, $\mathbf{R}_1, \mathbf{R}_2, \cdots \mathbf{R}_N$ as $\{\mathbf{R}\}$. However, such a direct approach would be computationally demanding since all electrons, core and valence, would then have to be treated on an equal footing; more importantly, small displacements of the nuclei have a profound effect on tightly bound core electrons. Physically, however, the tightly bound core electrons in most materials can accurately be assumed to move *rigidly* with the nuclei, the resulting ions each contributing an effective potential $\phi^{(ion)}(\mathbf{r})$ and carrying some effective charge, Z^*. The total "external" potential acting on the remaining (valence) electrons then has the form

$$V^{(ion)}(\mathbf{r}, \{\mathbf{R}\}) = e\sum_n \phi^{(ion)}(\mathbf{r} - \mathbf{R}_n), \tag{16A.1a}$$

which we rewrite as

$$V^{(ion)}(\mathbf{r}, \{\mathbf{R}\}) = e\sum_n \phi(\mathbf{r} - \mathbf{R}_n^{(0)} - \mathbf{U}_n), \tag{16A.1b}$$

where

$$\mathbf{R}_n = \mathbf{R}_n^{(0)} + \mathbf{U}_n; \tag{16A.2}$$

Here, $\mathbf{R}_n^{(0)}$ denotes the equilibrium nuclear positions and $\mathbf{U}_n$ the displacements from equilibrium. The change in the total energy of the system resulting from the displacements can be written as

$$\Delta E^{(total)} = \Delta E^{(elec)} + \Delta E^{(ion)}, \tag{16A.3}$$

where $\Delta E^{(elec)}$ is the change in energy of the valence electrons and $\Delta E^{(ion)}$ is the change in the energy arising from ion–ion Coulomb interactions, $E^{(ion)}$, where the latter is given by[33]

$$E^{(ion)} = Z^{*2}e^2 \sum_{nn'} \frac{1}{|\mathbf{R}_n - \mathbf{R}_{n'}|}. \tag{16A.4}$$

[27] Electron-phonon interaction and phonon dispersion relations using the *augmented-plane-wave* method, S.K. Sinha, *Phys. Rev.* **169**, 477 (1968).

[28] Quantum theory of lattice dynamics: 4, P.D. De Cicco and F.A. Johnson, *Proc. R. Soc. Lond. A* **310**, 111 (1969).

[29] Electronic contribution to lattice dynamics in insulating crystals, L.J. Sham, *Phys. Rev.* **188**, 1431 (1969).

[30] Microscopic theory of force constants in the adiabatic approximation, R.M. Pick, M.H. Cohen, and R.M. Martin, *Phys. Rev.* **1**, 910 (1970).

[31] *Ab initio* calculation of phonon dispersions in semiconductors, P. Giannozzi, S. Gironcoli, P. Pavone, and S. Baroni, *Phys. Rev. B* **43**, 7231 (1991).

[32] For a general review see, Phonons and related crystal properties from density-functional perturbation theory, Stefano Baroni, Stefano de Gironcoli, and Andrea Dal Corso, *Rev. Mod. Phys.* **73**, 515 (2001).

[33] Since the cores are assumed to be rigid and spherically symmetric, even when displaced, there will be no multi-pole interactions between them.

In the Born–Oppenheimer approximation, the electrons are assumed to respond adiabatically to changes in the nuclear coordinates and that is generally a good approximation for insulators, to which our discussion here is limited; this is referred to as the adiabatic approximation. As in Eq. (16.37) we will expand $\Delta E^{(\mathrm{total})}$ to second order in the displacements,

$$\Delta E = \sum_{n} \frac{\partial E}{\partial \mathbf{R}_n} \cdot \mathbf{U}_n + \frac{1}{2} \sum_{nn'} \frac{\partial^2 E}{\partial \mathbf{R}_n \partial \mathbf{R}_{n'}} : \mathbf{U}_n \mathbf{U}_{n'} \cdots , \tag{16A.5}$$

where all derivatives are evaluated at $\mathbf{R}_n = \mathbf{R}_n^{(0)}$. The first term will vanish for a system in equilibrium. The atomic force constants enter through the second term which, in the same notation as Eq. (16.38), we write as

$$\mathbf{A}_{nn'} = \frac{\partial^2 E}{\partial \mathbf{R}_n \partial \mathbf{R}_{n'}}. \tag{16A.6}$$

However, having split the energy shift into two contributions through Eq. (16A.3), we now have two contributions to the force constants,

$$\mathbf{A}_{nn'} = \mathbf{A}_{nn'}^{(\mathrm{elec})} + \mathbf{A}_{nn'}^{(\mathrm{ion})}. \tag{16A.7}$$

We first discuss the evaluation of $\mathbf{A}_{nn'}^{(\mathrm{ion})}$. The sum in (16A.6) can be evaluated using the Ewald method;[34] although it is divergent for our monatomic lattice case, only the derivatives with respect to $\mathbf{R}_n$, which are finite, enter $\mathbf{A}_{nn'}^{(\mathrm{ion})}$. As discussed in Section 16.4, we will seek a description in terms of normal modes of the form (16.46), which in turn involves the dynamical matrix $\mathbf{D}(\mathbf{q})$, as defined by Eq. (16.48). For the ionic contribution, and after a lengthy calculation, Giannozzi et al. obtained (when applied to the monatomic case),

$$\begin{aligned} \mathbf{D}^{(\mathrm{ion})}(\mathbf{q}) = {} & \frac{4\pi e^2 Z^{*2}}{\Omega} \sum_{\mathbf{Q},\mathbf{Q}+\mathbf{q}\neq 0} \frac{e^{-(\mathbf{q}+\mathbf{Q})^2/4\alpha}}{-(\mathbf{q}+\mathbf{Q})^2} (\mathbf{q}+\mathbf{Q})(\mathbf{q}+\mathbf{Q}) \\ & - \frac{2\pi e^2 Z^{*2}}{\Omega} \sum_{\mathbf{Q}\neq 0} e^{-Q^2/4\alpha} \mathbf{Q}\mathbf{Q}, \end{aligned} \tag{16A.8}$$

where Ω is the volume of a unit cell and α is a factor chosen for optimal convergence.

We now turn to the electronic contribution to the elastic constants. In the full Hamiltonian, $V^{(\mathrm{ion})}(\mathbf{r}, \{\mathbf{R}\})$ is the only term involving the parameters $\mathbf{R}_n$. The derivative of the total electronic energy with respect to a particular displacement is then given by the Hellman–Feynman theorem as (see Section 15A.1)

$$\begin{aligned} \frac{\partial E^{(\mathrm{elec})}}{\partial \mathbf{R}_n} &= \int d^3r_1 d^3r_2 \cdots d^3r_N \Psi^*(\{\mathbf{r}\};\{\mathbf{R}\}) \frac{\partial \hat{H}(\{\mathbf{r}_i\};\{\mathbf{R}\})}{\partial \mathbf{R}_n} \Psi(\{\mathbf{r}\};\{\mathbf{R}\}), \\ &= \int d^3r_1 d^3r_2 \cdots d^3r_N \Psi^*(\{\mathbf{r}\};\{\mathbf{R}\}) \sum_i \frac{\partial V^{(\mathrm{ion})}(\mathbf{r}_i;\{\mathbf{R}\})}{\partial \mathbf{R}_n} \Psi(\{\mathbf{r}\};\{\mathbf{R}\}) \end{aligned} \tag{16A.9}$$

[34] The Ewald technique is discussed in Appendix 9A of Chapter 9. For the case of a monatomic lattice the total Ewald energy is given by $E^{\mathrm{ion}} = \frac{NZ^2e^2}{2} \sum_{n} \frac{1}{R_n} \left(1 - \mathrm{erf}(\alpha^{1/2}|\mathbf{R}_n|)\right) - NZ^2e^2\left(\frac{2\alpha}{\pi}\right)^{1/2} + \frac{4\pi NZ^2e^2}{2\Omega}\left[\sum_{\mathbf{Q}\neq 0} \frac{e^{-Q^2/4\alpha}}{Q^2} - \frac{1}{4\alpha}\right]$; for the multi-atom case see Giannozzi et al.

where $\Psi(\{\mathbf{r}_i\};\{\mathbf{R}\})$ is the many body wave function of the electrons and $\{\mathbf{r}\}$ denotes the set of all electron coordinates, $\mathbf{r}_1, \mathbf{r}_2, \cdots \mathbf{r}_N$. Because of the symmetry of the total wave function with respect to the electron coordinates, each of the integrals associated with the summation is identical and so,

$$\frac{\partial E^{(elec)}}{\partial \mathbf{R}_n} = \int d^3r\, n(\mathbf{r}) \frac{\partial V^{(ion)}(\mathbf{r};\{\mathbf{R}\})}{\partial \mathbf{R}_n}, \tag{16A.10a}$$

where

$$n(\mathbf{r};\{\mathbf{R}\}) = \int d^3r_2 \cdots d^3r_N \Psi^*(\mathbf{r}, \mathbf{r}_2, \cdots \mathbf{r}_N;\{\mathbf{R}\})\, \Psi(\mathbf{r}, \mathbf{r}_2, \cdots \mathbf{r}_N;\{\mathbf{R}\})\,; \tag{16A.10b}$$

Here, $n(\mathbf{r})$ is the electron number density. Performing a Taylor expansion of the integrand of (16A.10a) to first order in the displacements $\mathbf{U}_n$ yields

$$\frac{\partial E^{(elec)}}{\partial \mathbf{R}_n} = \int d^3r \left\{ n_0(\mathbf{r};\{\mathbf{R}\}) \frac{\partial V^{(ion)}(\mathbf{r};\{\mathbf{R}\})}{\partial \mathbf{R}_n} + \left[\sum_{n'} \frac{\partial n(\mathbf{r};\{\mathbf{R}\})}{\partial \mathbf{R}_{n'}} \frac{\partial V^{(ion)}(\mathbf{r};\{\mathbf{R}\})}{\partial \mathbf{R}_n} \right.\right.$$
$$\left.\left. + n_0(\mathbf{r};\{\mathbf{R}\}) \sum_{n'} \frac{\partial^2 V^{(ion)}(\mathbf{r};\{\mathbf{R}\})}{\partial \mathbf{R}_{n'} \partial \mathbf{R}_n} \right] \cdot \mathbf{U}_{n'} \right\}. \tag{16A.11}$$

Integrating Eq. (16A.11), we obtain the change in the total electronic energy to second order in the $\mathbf{U}_n$ as[35]

$$\Delta E^{(elec)}(\{\mathbf{R}\}) = \int d^3r \left\{ \sum_n n_0(\mathbf{r};\{\mathbf{R}\}) \frac{\partial V^{(ion)}(\mathbf{r};\{\mathbf{R}\})}{\partial \mathbf{R}_n} \cdot \mathbf{U}_n \right.$$
$$\left. + \frac{1}{2} \sum_{nn'} \left[\frac{\partial n(\mathbf{r};\{\mathbf{R}\})}{\partial \mathbf{R}_{n'}} \frac{\partial V^{(ion)}(\mathbf{r};\{\mathbf{R}\})}{\partial \mathbf{R}_n} + n_0(\mathbf{r};\{\mathbf{R}\}) \frac{\partial^2 V^{(ion)}(\mathbf{r};\{\mathbf{R}\})}{\partial \mathbf{R}_{n'} \partial \mathbf{R}_n} \right] : \mathbf{U}_n \mathbf{U}_{n'} \right\}. \tag{16A.12}$$

The electronic contribution to the force constant tensor appearing in Eq. (16.40) arises from the second term and is therefore given by

$$\mathbf{A}^{(elec)}_{nn'} = \frac{\partial^2 E^{(elec)}}{\partial \mathbf{R}_n \partial \mathbf{R}_{n'}}$$
$$= \int d^3r \left[\frac{\partial n(\mathbf{r};\{\mathbf{R}\})}{\partial \mathbf{R}_{n'}} \frac{\partial V^{(ion)}(\mathbf{r};\{\mathbf{R}\})}{\partial \mathbf{R}_n} + n_0(\mathbf{r};\{\mathbf{R}\}) \frac{\partial^2 V^{(ion)}(\mathbf{r};\{\mathbf{R}\})}{\partial \mathbf{R}_{n'} \partial \mathbf{R}_n} \right]. \tag{16A.13}$$

35 Density-functional approach to nonlinear-response coefficients of solids, X. Gonze and J.P. Vigneron, *Phys. Rev. B* **39**, 13120 (1989).

The derivatives of $V^{(ion)}$ with respect to $\mathbf{R_n}$ can be related to derivatives of the potential ϕ with respect to normal modes, $\mathbf{U_q}$, through a Taylor expansion as

$$V^{(ion)}(\mathbf{r}, \{\mathbf{R}\}) = e\sum_{n}\left[\phi(\mathbf{r}-\mathbf{R}_n^{(0)}) + \sum_{n'q}\frac{\partial\phi(\mathbf{r}-\mathbf{R}_n^{(0)})}{\partial\mathbf{R}_{n'}^{(0)}}\cdot\mathbf{U_q}e^{-i\mathbf{q}\cdot\mathbf{R}_{n'}} + \frac{1}{2}\sum_{n'n''qq'}\frac{\partial\phi^2(\mathbf{r}-\mathbf{R}_n^{(0)})}{\partial\mathbf{R}_{n'}^{(0)}\partial\mathbf{R}_{n''}^{(0)}} : \mathbf{U_q}\mathbf{U_{q'}}e^{-i(\mathbf{q}\cdot\mathbf{R}_{n'}+\mathbf{q}'\cdot\mathbf{R}_{n''})}\ldots\right]. \tag{16A.14}$$

Hence,

$$\frac{\partial V^{(ion)}(\mathbf{r})}{\partial\mathbf{U_q}} = e\sum_{n}\frac{\partial\phi(\mathbf{r}-\mathbf{R}_n^{(0)})}{\partial\mathbf{R}_n^{(0)}}e^{-i\mathbf{q}\cdot\mathbf{R_n}}, \tag{16A.15}$$

where the inverse transform is given by

$$\frac{\partial V^{(ion)}}{\partial\mathbf{R_n}} = \frac{1}{N}\sum_{q}\frac{\partial V^{(ion)}(\mathbf{r})}{\partial\mathbf{U_q}}e^{i\mathbf{q}\cdot\mathbf{R_n}}, \tag{16A.16}$$

which can be evaluated numerically; here N is the number of unit cells. The remaining derivatives in (16A.14) can be handled in a similar manner. Rather than carry all the terms in Eq. (16A.13), here we write it simply as $\mathbf{A}_{nn'}$; in terms of its Fourier transform it takes the form

$$\begin{aligned}\mathbf{A}_{nn'} &= \frac{1}{N^2}\sum_{q''q'}\mathbf{A}_{q''q'}\,e^{i(\mathbf{q}''\cdot\mathbf{R_n}+\mathbf{q}'\cdot\mathbf{R}_{n'})}\\ &= \frac{1}{N^2}\sum_{q'q''}\mathbf{A}_{q''q'}\,e^{i\mathbf{q}''\cdot(\mathbf{R_n}-\mathbf{R}_{n'})+i(\mathbf{q}''+\mathbf{q}')\cdot\mathbf{R}_{n'}}\\ &= \frac{1}{N}\sum_{q'}\mathbf{A}_{-q'q'}\,e^{i\mathbf{q}'\cdot(\mathbf{R_n}-\mathbf{R}_{n'})},\end{aligned} \tag{16A.17}$$

where in the last step we impose the translational invariance requirement that $\mathbf{A}_{nn'} = \mathbf{A}_{n-n'}$. The dynamical matrix $\mathbf{D}(\mathbf{q})$ is given by

$$\begin{aligned}\mathbf{D}(\mathbf{q}) &= \sum_{n'}\mathbf{A}_{nn'}\,e^{-i\mathbf{q}\cdot(\mathbf{R_n}-\mathbf{R}_{n'})}\\ &= \frac{1}{N}\sum_{n'q'}\mathbf{A}_{-q'q'}\,e^{i\mathbf{q}'\cdot(\mathbf{R_n}-\mathbf{R}_{n'})}e^{-i\mathbf{q}\cdot(\mathbf{R_n}-\mathbf{R}_{n'})}\\ &= \mathbf{A}_{-q'q'}.\end{aligned} \tag{16A.18}$$

Applying this transformation to Eq. (16A.13), we have

$$D^{(elec)}(\mathbf{q}) = \int d^3r \left[\left(\frac{\partial n(\mathbf{r})}{\partial \mathbf{U_q}} \right)^* \frac{\partial V^{(ion)}(\mathbf{r})}{\partial \mathbf{U_q}} + n_0(\mathbf{r}) \frac{\partial^2 V^{(ion)}(\mathbf{r})}{\partial \mathbf{U_q^*} \partial \mathbf{U_q}} \right]. \quad (16A.19)$$

Returning to Eq. (16A.11), the term involving $\partial n(\mathbf{r}) / \partial \mathbf{U_q}$ requires that we evaluate the change in electron density, $\Delta n(\mathbf{r})$, resulting from the displacements, $\mathbf{U_q}$, which can be done using first-order perturbation theory for the change in the Bloch wave functions, $\psi^{(1)}_{\mathbf{k}v\sigma}$; here v denotes the band index of the *occupied* (valence) orbitals. Now in a one-electron theory based on the Kohn–Sham equations these states are calculated from the *total* self-consistent potential (see Eq. (15.57)),

$$\left[-\frac{\hbar^2}{2m} \nabla^2 + V^{(total)}(r) \right] \psi_{\mathbf{k}v\sigma}(r) = \varepsilon_{\mathbf{k}v\sigma} \psi_{\mathbf{k}v\sigma}(r), \quad (16A.20)$$

where

$$V^{(total)}(r) = V^{(ion)}(r) + e^2 \sum_{\mathbf{k}v\sigma} \int \frac{\psi^*_{\mathbf{k}v\sigma}(\mathbf{r}')\psi_{\mathbf{k}v\sigma}(\mathbf{r})}{|\mathbf{r}-\mathbf{r}'|} d^3r' + \hat{V}^{XC}(r), \quad (16A.21)$$

and we sum over the spatial and spin quantum numbers of all the occupied orbitals. Note that the changes in $\Delta n(\mathbf{r}) = (\partial n(\mathbf{r})/\partial \mathbf{U_q}) \cdot \mathbf{U_q}$ and $\Delta V(\mathbf{r}) = (\partial V^{(total)}(\mathbf{r})/\partial \mathbf{U_q}) \cdot \mathbf{U_q}$ must be calculated self-consistently from (16A.20) and (16A.21). In the local density approximation, where we write the operator $\hat{V}^{XC}(\mathbf{r})$ as the function, $V^{XC}(n(\mathbf{r}))$, $\Delta V^{(total)}(r)$ will be governed by

$$\Delta V^{(total)}(\mathbf{r}) = \Delta V^{(ion)}(\mathbf{r}) + e^2 \int \frac{d^3r' \Delta n(\mathbf{r}')}{|\mathbf{r}-\mathbf{r}'|} + \frac{\partial V^{XC}(\mathbf{r})}{\partial n(\mathbf{r})} \Delta n(\mathbf{r}). \quad (16A.22)$$

With this quantity we can then calculate the shift in the electron density, $\Delta n(\mathbf{r})$, using first-order perturbation theory as[36]

$$\Delta n(\mathbf{r}) = \frac{1}{N} \sum_{vk\sigma} \left[\psi^{(0)*}_{vk\sigma}(\mathbf{r}) \psi^{(1)*}_{vk\sigma}(\mathbf{r}) + \psi^{(1)*}_{vk\sigma}(\mathbf{r}) \psi^{(0)}_{vk\sigma}(\mathbf{r}) \right]$$

or

$$\Delta n(\mathbf{r}) = \frac{4}{N} \sum_{vk} \psi^*_{vk}(\mathbf{r}) \sum_{ck'} \frac{\langle \psi^*_{ck'} | \Delta V | \psi_{vk} \rangle}{\varepsilon_{vk} - \varepsilon_{ck'}} \psi_{ck'}(\mathbf{r}) + \text{c.c}, \quad (16A.23)$$

where $\mathbf{k}$ and υ denote occupied states in the first Brillouin zone; note that due to the Pauli principle we only sum over the *unoccupied* (conduction) intermediate states, which are numbered by c and $\mathbf{k}'$.

[36] The Fourier components of the induced electron density change resulting from an external and a self-consistent potential were discussed earlier, in Section 14A (see Eqs. (14A.7a)).

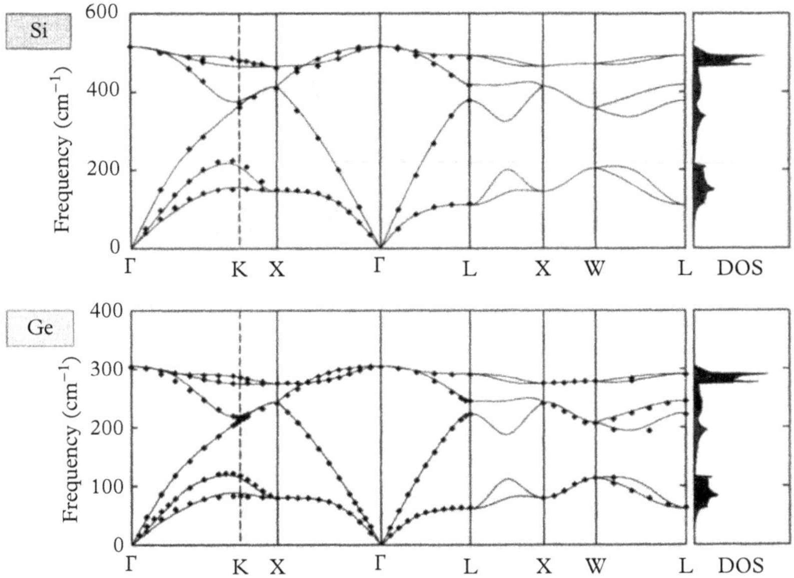

Figure 16A.1 *Calculated phonon dispersion curves and the densities of states of the elemental semiconductors,* Si *and* Ge. *(After Giannozzi et. al. (1991).) The experimental neutron scattering data are denoted by diamonds. (After Dolling (1963) and Nilsson and Nelin (1971, 1972).)*

Equation (16A.23) allows us to, at least formally, connect $\partial n(\mathbf{r})/\partial \mathbf{U}_\mathbf{q}$ to $\partial V^{(ion)}(\mathbf{r})/\partial \mathbf{U}_\mathbf{q}$, and hence evaluate $D^{(elec)}(\mathbf{q})$ using Eq. (16A.19). When combined with $\mathbf{D}^{(ion)}(\mathbf{q})$ from Eq. (16A.8) we would then have the full $\mathbf{D}(\mathbf{q})$ and be able to solve for $\omega = \omega(\mathbf{q})$, which was our goal. At this point a number of formal tricks are introduced in carrying out the actual calculations and we refer the reader to Giannozzi et al. for the details. As an example of the methodology, we show $\omega(\mathbf{q})$ for Si and Ge in Fig. 16A.1, along with the inelastic neutron scattering data.[37,38,39]

[37] G. Dolling, in *Inelastic Scattering of Neutrons in Solids and Liquids*, Vol. II, p 37. Ed. S. Ekland, IAEA, Vienna (1963).

[38] Phonon dispersion relations in Ge at 80 degrees, K. G. Nilsson and G. Nelin, *Phys. Rev. B* 3, 364 (1971).

[39] Study of homology between silicon and germanium by thermal-neutron spectrometry, G. Nilsson and G. Nelin, *Phys. Rev. B* **6**, 3777 (1972).

17 Thermal Expansion, Phonon–Phonon Interactions, and Heat Transport

The method that involves expanding the potential given by Eq. (16.1) in Chapter 16 provides a general strategy for treating lattice dynamics. However, there it was limited to the harmonic approximation and in this chapter we treat a number of applications where the effects of anharmonicity enter.[1]

In the same way that collisions among molecules are essential for establishing thermal equilibrium in a gas, so interactions among lattice vibrations are required to equilibrate the mechanical excitations in solids; these interactions arise from the anharmonic terms in the expansion of (16.1). In addition to driving the system toward equilibrium, anharmonicity also results in thermal expansion, a linear term in the high-temperature specific heat (i.e., in addition to the constant Dulong–Petit term), and finite thermal conductivity. It will turn out that we must include both the cubic and quartic terms in discussing some of these phenomena. Our treatment here will be limited to insulators, since the interaction with conduction electrons produces additional effects which will be addressed later, in Chapter 21.

17.1 Thermal expansion; the Gruneisen model

We start by expanding the Helmholz free energy to second order in the shift of the volume from its equilibrium value V_0,

$$F(T,V) = F(T,V_0) + \left(\frac{\partial F(T,V)}{\partial V}\right)_{V=V_0}(V-V_0) + \frac{1}{2}\left(\frac{\partial F^2(T,V)}{\partial V^2}\right)_{V=V_0}(V-V_0)^2 + \ldots,$$

which we can recast as

$$F(T,V) \equiv F(T,V_0) + F_1(T)\left(\frac{(V-V_0)}{V_0}\right) + \frac{1}{2}F_2(T)\left(\frac{(V-V_0)}{V_0}\right)^2 \ldots, \quad (17.1)$$

[1] Our approach will follow that of Peierls, who was a pioneer in the area of lattice dynamics.

The Physics of Solids. First Edition. J. B. Ketterson.

where F_1 and F_2 are extensive quantities. Now the pressure is given by $P = -\partial F(T,V)/\partial V$ which, in equilibrium, must vanish (here we will neglect the vapor pressure of the solid itself); we then have

$$\frac{V - V_0}{V_0} = -\frac{F_1(T)}{F_2(T)} \equiv \varepsilon(T), \tag{17.2a}$$

where we have introduced the *dilatation*, $\varepsilon(T)$, which, as indicated, is temperature dependent. The compressibility is defined by $K = -V\partial P/\partial V$ and hence

$$F_2 = KV. \tag{17.2b}$$

Since thermodynamic stability requires that $K > 0$, the denominator in (17.2a) is positive definite. In practice, K varies little with temperature and in any case is dominated by the harmonic terms; it may therefore be taken as a constant. The thermal expansion is then dominated by F_1 and arises from *anharmonic terms*. It is therefore small at low temperatures. In any case the fractional change in the volume is much less than unity.

To calculate F_1, which contributes in first order to the volume change, we would need to study the effect of a uniform rescaling of all the lattice vectors, $\mathbf{R_n}$, from their equilibrium positions.[2] For simplicity we will assume a cubic crystal with one atom per unit cell; the thermal expansion is then isotropic and $\mathbf{R_n} \to \mathbf{R_n}(1 + \varepsilon/3)$. With these assumptions the coordinates $\mathbf{U_n}$ measuring the displacement of the atoms from their zero temperature equilibrium positions can be written as

$$\mathbf{U_n} = \frac{1}{3}\varepsilon\mathbf{R_n} + \mathbf{U'_n}, \tag{17.3}$$

where the first term arises from the uniform thermal expansion and the second from the dynamic displacements relative to these new equilibrium positions. Note that since the $\mathbf{R_n}$ increase without limit for large $\mathbf{n}$, the $\mathbf{U_n}$ do likewise; i.e., they become much larger than the lattice spacing, a. On the other hand we will generally have $|\mathbf{U'_n}| \ll a$. Through the compressibility, the first term results in a global quadratic shift of the energy by an amount $KL^3\varepsilon^2/2$ (where we now write the volume as L^3 to avoid confusion with the potential energy). We may therefore rewrite the lattice potential (16.37) as

$$V = V^{(0)} + \frac{1}{2}KL^3\varepsilon^2 + \frac{1}{2}\sum_{nn'}\mathbf{A'_{nn'}} : \mathbf{U'_n}\mathbf{U'_{n'}} + \frac{1}{6}\sum_{nn'n''}\mathbf{B'_{nn'n''}} \vdots \mathbf{U'_n}\mathbf{U'_{n'}}\mathbf{U'_{n''}} + \ldots \tag{17.4a}$$

where[3]

$$\mathbf{A'_{nn'}} = \mathbf{A_{nn'}} + \frac{1}{3}\varepsilon\sum_{n''}\mathbf{B_{nn'n''}}\mathbf{R_{n''}}. \tag{17.4b}$$

[2] Here we will assume one atom per unit cell; this avoids carrying the index, j, as well as complications arising from a repositioning of atoms within the unit cell with temperature. If the positions of atoms in the unit cell are constrained by symmetry, as in the NaCl or ZnO lattices, no such repositioning occurs.

[3] Note that the second term is independent of the origin since a uniform shift by some $\mathbf{R_0}$ would produce a term $\mathbf{R_0}\sum_{n''}\mathbf{B_{nn'n''}}$, which vanishes in a manner analogous to (16.44).

We see that the structure of the problem is similar to that at T = 0, except that we now have (i), a term linear in ε that enters through the modified parameters, $\mathbf{A}'_{nn'}$ in Eq. (17.4b) and (ii) the quadratic term in ε^2. Hence (17.4a,b) is expected to lead to a free energy having the form of (17.1). We emphasize that the parameters $\mathbf{A}'_{nn'}$ are *shifted* from their zero temperature values by a term that is first order in ε and, importantly, involves the anharmonic terms.

The last term in (17.4a), involving the $\mathbf{B}'_{nn'n''}$, can be neglected. One might think that the second term in (17.4b) would diverge since it is summed over all $\mathbf{R}_n$, where the latter diverges; however this does not occur because, similar to Eq. (16.9), the sum

$$\sum_n B_{nn'n''} = 0; \tag{17.5}$$

i.e., we have a sum involving the product of two terms. When summed separately, one of these diverges (the $\mathbf{R}_n$ sum) while the other vanishes (the $\mathbf{B}_{nn'n''}$ sum); the sum of the product can then be finite.

To proceed further would require a knowledge of the anharmonic terms and in any case would be very complex. To extract the qualitative behavior we will therefore resort to a simpler calculation, involving the Debye model. To apply this model we will need to express F in terms of the energy E.

The free energy is defined as

$$F(T,V) = E(T,V) - TS(T,V), \tag{17.6a}$$

where

$$S = \int \frac{C_V}{T} dT = \int dT \frac{1}{T}\frac{dE}{dT}; \tag{17.6b}$$

here the total energy is given by (16.81) and V is held constant. To obtain E from F we divide (17.6a) by T, differentiate, and use (17.6b), obtaining

$$\begin{aligned} \frac{d}{dT}\frac{F}{T} &= \frac{d}{dT}\frac{E}{T} - \frac{1}{T}\frac{dE}{dT} \\ &= -\frac{E}{T^2} \end{aligned}$$

or

$$E = -T^2 \frac{d}{dT}\left(\frac{F}{T}\right). \tag{17.7}$$

We are now in a position to apply the Debye model. The required anharmonicity will be modeled through an assumed ε dependence of the Debye temperature, θ_D. We start by changing the variable of integration in Eq. (4.33) to $x = \theta_D/T$, where $\theta_D = \hbar\omega_D/k_B$ and ω_D is given by (4.31), with the result

$$E(T) = 9Nk_BT\left(\frac{T}{\theta_D}\right)^3 \int_0^{\theta_D/T} dx \frac{x^3}{e^x - 1}; \tag{17.8a}$$

i.e., we can write

$$E(T) = Tf\left(\frac{T}{\theta_D}\right), \tag{17.8b}$$

where f on the right-hand side is a universal function of the variable T/θ_D. Using (17.7) and (17.8b) we can now write F(T) in the Debye model as

$$F(T) = -T\int dT\frac{E(T)}{T^2} = -T\int dT\frac{f(T/\theta_D)}{T}. \tag{17.9}$$

To obtain F_1 in Eq. (17.1) we differentiate (17.9) with respect to ε,

$$\begin{aligned} F_1 = \frac{\partial F}{\partial \varepsilon} &= -T\int \frac{dT}{T}\frac{\partial f(T/\theta_D)}{\partial \varepsilon} = -T\int \frac{dT}{T}\frac{\partial f(T/\theta_D)}{\partial (T/\theta_D)}\frac{\partial (T/\theta_D)}{\partial \varepsilon} \\ &= T\frac{1}{\theta_D}\frac{\partial \theta_D}{\partial \varepsilon}\int d\left(\frac{T}{\theta_D}\right)\frac{\partial f(T/\theta_D)}{\partial (T/\theta_D)} = T\frac{1}{\theta_D}\frac{\partial \theta_D}{\partial \varepsilon} f\left(\frac{T}{\theta_D}\right); \end{aligned}$$

substituting (17.8b) then gives

$$F_1 = E(T)\frac{1}{\theta_D}\frac{\partial \theta_D}{\partial \varepsilon}. \tag{17.10}$$

Using the relation $F_2 = KV$ from (17.2b), we can now rewrite Eq. (17.2a) as

$$\varepsilon = -\frac{F_1(T, V_0)}{KV} = -\frac{E(T)}{KV}\frac{1}{\theta_D}\frac{\partial \theta_D}{\partial \varepsilon}; \tag{17.11a}$$

note that in this model the *thermal expansion is proportional to the energy density*, $e(T) = E(T)/V$, a model due to Grüneisen. Defining the proportionality constant, $\gamma \equiv -(1/\Theta)\partial\Theta/\partial\varepsilon$, called the Grüneisen constant, we can write (17.11a) as

$$\varepsilon = \gamma\frac{e(T)}{K}. \tag{17.11b}$$

At temperatures $T \ll \theta_D$, where the Debye theory becomes asymptotically exact, $e(T) \propto T^4$, while for $T \gg \theta_D$, where the law of Dulong and Petit applies, $e(T) \propto T$, with corresponding behaviors of ε.

17.2 The linear term in the high temperature specific heat

Experimentally, the specific heat at temperatures well above the Debye temperature, rather than being strictly constant as implied by the law of Dulong and Petit, contains a small contribution which is linear in the temperature. As in the previous section, this behavior stems from the higher-order terms in the expansion of the potential energy, beyond the quadratic (or harmonic) terms.

17.2.1 The heat capacity of a single anharmonic oscillator

For orientation purposes we first consider the behavior of a single oscillator of mass M that is governed by the Hamiltonian

$$H(p, x) = \frac{p^2}{2M} + V(x). \tag{17.12}$$

It will turn out that in order to obtain a consistent expansion we will need to consider both the cubic and quartic corrections to the potential V(x), which we write as

$$V(x) = \frac{1}{2}ax^2 + bx^3 + cx^4. \tag{17.13}$$

The expectation value of the energy follows from Eq. (4.5), which we can write as

$$\bar{E} = \frac{\int dpdxH(p, x)e^{-\beta H(p,x)}}{\int dpdxe^{-\beta H(p,x)}} = -\frac{\partial}{\partial\beta}\ln[Z(T)], \tag{17.14}$$

where

$$Z(T) = \int dpdx\, e^{-\beta H(p,x)} \tag{17.15}$$

is the partition function (see Eq. (4.3)). The kinetic energy contribution is a simple Gaussian in p and carrying out this integration we are left with

$$Z(T) = (2\pi M/\beta)^{1/2} \int dxe^{-\beta\left(\frac{1}{2}ax^2+bx^3+cx^4\right)}. \tag{17.16}$$

We will assume the temperature is such that the behavior is dominated by the harmonic contribution to the potential, and we expand the exponent in terms of the cubic and quartic forms as

$$Z(T) = (2\pi M/\beta)^{1/2} \int dxe^{-\beta ax^2/2}\left(1 - \beta bx^3 + \frac{1}{2}\beta^2b^2x^6 - \beta cx^4\right). \tag{17.17}$$

Note we have retained the first *two* terms in the expansion of the cubic term in the exponential; the first of these is odd and hence vanishes. Carrying out the remaining two integrations in (17.17), which are again Gaussian-like forms, we get

$$Z(T) = 2\pi\,(M/a)^{1/2}\,\beta^{-1}\left[1+\beta^{-1}\left(\frac{15}{2}\frac{b^2}{a^3}-\frac{3c}{a^2}\right)\right]. \qquad (17.18)$$

Using Eq. (17.14), the energy E(T) through terms of order T^2 is[4]

$$E(T) = k_BT + \left(\frac{15}{2}\frac{b^2}{a^3}-\frac{3c}{a^2}\right)k_B^2T^2. \qquad (17.19)$$

At this point it is apparent that carrying the additional factor in the expansion of the cubic term in the exponential was essential to obtain a consistent result through order T^2 in the energy. Differentiating (17.19) with respect to T we obtain the heat capacity as

$$C_V(T) = k_B + k_B^2\left(15\frac{b^2}{a^3}-6\frac{c}{a^2}\right)T, \qquad (17.20)$$

in agreement with the aforementioned experimental behavior.

17.2.2 The linear term in the heat capacity of a lattice

We now discuss the case of the anharmonic crystal lattice. All calculations will be done in terms of the normal coordinates, Q_{sq}, of the harmonic Hamiltonian introduced through Eq. (16.52b). The lattice potential energy may be written schematically as

$$V = V^{(2)} + V^{(3)} + V^{(4)}, \qquad (17.21)$$

where $V^{(3)}$ and $V^{(4)}$ are cubic and quartic contributions, respectively. As with the single oscillator case, we will find that both lead to corrections to the total energy of order T^2. In addition they lead to interactions among phonons which we will take up in Section 17.3.

The partition function is given by

$$Z(T) = \int d\Gamma e^{-\beta\left(H^{(0)}+V^{(3)}+V^{(4)}\right)}, \qquad (17.22)$$

[4] An alternative way to obtain this result is to form thermal averages of all pairings of the coordinates in the cubic and quartic contributions to the potential as derived from the quadratic average, $a\overline{x^2}/2 = k_BT/2$ or $\overline{x^2} = k_BT/a$. There are six such pairings for x^6 in the cubic contribution and three for x^4 in the quartic. This is the approach that will be applied also for the crystal.

where $d\Gamma$ is the volume element in the full phase space of the crystal formed from the normal coordinates Q_{sq} and P_{sq}. As we did to obtain Eq. (17.17), we can expand the exponential in the partition function as

$$\int d\Gamma e^{-\beta H} = \int d\Gamma e^{-\beta\left(E^{(kin)}+V^{(2)}+V^{(3)}+V^{(4)}\right)}$$
$$\cong \int d\Gamma e^{-\beta\left(E^{(kin)}+V^{(2)}\right)}\left[1-\beta V^{(3)}-\beta V^{(4)}+\frac{1}{2}\left(\beta V^{(3)}\right)^2\right]. \quad (17.23)$$

The term involving $-\beta V^{(3)}$ in the expansion again vanishes on integrating over $d\Gamma$ as it did in the single oscillator case; in addition, we find that the contribution to the energy from the remaining terms is, in leading order, simply their average with respect to the non-interacting Boltzmann distribution. Hence we write

$$E(T) = E_0(T) + E_q(T) + E_c(T)$$
$$= E_0(T) - \overline{\beta V^{(4)}} + \frac{1}{2}\overline{\left(\beta V^{(3)}\right)^2}, \quad (17.24)$$

where E_0 is the harmonic contribution governed by $H^{(0)}$ and leads directly to the Dulong–Petit law.

Rather than evaluating the partition function directly as we did with the single oscillator case, the contributions from the quartic and cubic terms in (17.24) will be evaluated by (i) assuming *statistical independence* of the different normal modes and (ii) assigning a mean square value of paired amplitudes of the *same* mode according to the equipartition theorem (16.80), half of which comes from the potential energy given in Eq. (16.67), which together give

$$\frac{1}{2}NM_c\omega_{sq}^2\,|Q_{sq}|^2 = \frac{1}{2}k_BT$$

for each mode, or equivalently

$$\overline{|Q_{sq}|^2} = \frac{1}{NM_c\beta\omega_{sq}^2}. \quad (17.25)$$

We begin with the contribution from the last term in (17.24). The form of $V^{(3)}$ can be taken from Eq. (16.40). However, in order to use (17.25) we must rewrite it in terms of the normal coordinates using Eqs. (16.52a,b), analogous to what we did for $V^{(2)}$ in Eq. (16.67). This yields

$$V^{(3)} = \frac{1}{6}\sum_{ss's''}\sum_{qq'q''} B_{qs,q's',q''s''}Q_{sq}Q_{s'q'}Q_{s''q''}, \quad (17.26)$$

where we have introduce the transformed coefficients

$$B_{sq,s'q',s''q''} = \sum_{jj'j''}\sum_{nn'n''} \mathbf{B}^{jj'j''}_{nn'n''} \vdots\, \mathbf{V}_{sq}\mathbf{V}_{s'q'}\mathbf{V}_{s''q''}\, e^{i(\mathbf{q}\cdot\mathbf{R}_n+\mathbf{q}'\cdot\mathbf{R}_{n'}+\mathbf{q}''\cdot\mathbf{R}_{n''})} \quad (17.27)$$

and $\mathbf{V}_{\mathrm{sq}}$ is the polarization vector for state (s, $\mathbf{q}$), introduced in Eq. (16.52b). If in Eq. (17.27) we displace all the atoms by a real-space lattice vector $\mathbf{R}$, then it will be multiplied by a factor $e^{i(\mathbf{q}\cdot\mathbf{R}+\mathbf{q}'\cdot\mathbf{R}+\mathbf{q}''\cdot\mathbf{R})}$; since the total potential energy must be unaffected by such a transformation, it follows that

$$B_{\mathrm{sq},\,\mathrm{s'q'},\,\mathrm{s''q''}} = 0, \tag{17.28}$$

unless $\mathbf{q}+\mathbf{q}'+\mathbf{q}'' = \mathbf{Q}$, where $\mathbf{Q}$ is a reciprocal lattice vector. Therefore if two of the $\mathbf{q}$ are given then the third is fixed within a reciprocal lattice vector, since we always restrict the $\mathbf{q}$ to lie in the first Brillouin zone. If all three of the $\mathbf{q}$ are small enough then we will have $\mathbf{Q} = 0$; if not, $\mathbf{Q}$ will be that reciprocal lattice vector which keeps all $\mathbf{q}$ in the first zone (which will be unique). We introduce a *generalized Kronecker delta function* which accounts for $\mathbf{Q} \neq 0$ (Umklapp) processes,

$$\Delta_{\mathrm{qq'q''}} = \begin{cases} 0 \text{ if } \mathbf{q}+\mathbf{q}'+\mathbf{q}''-\mathbf{Q} \neq 0; \\ 1 \text{ if } \mathbf{q}+\mathbf{q}'+\mathbf{q}''-\mathbf{Q} = 0. \end{cases} \tag{17.29}$$

We can then build the condition (17.28) into Eq. (17.27) and write

$$B_{\mathrm{sq},\,\mathrm{s'q'},\,\mathrm{s''q''}} \to B_{\mathrm{sq},\,\mathrm{s'q'},\,\mathrm{s''q''}}\Delta_{\mathrm{qq'q''}}. \tag{17.30}$$

Incorporating (17.30) we can now write (17.26) as

$$V^{(3)} = \frac{1}{6}\sum_{\mathrm{qq'q''}}\sum_{\mathrm{ss's''}} B_{\mathrm{sq},\,\mathrm{s'q'},\,\mathrm{s''q''}}\, Q_{\mathrm{sq}}Q_{\mathrm{s'q'}}Q_{\mathrm{s''q''}}\Delta_{\mathrm{qq'q''}}. \tag{17.31}$$

The coefficients $B_{\mathrm{sq},\,\mathrm{s'q'},\,\mathrm{s''q''}}$ have another property which we will need shortly. From the requirement (17.5), we conclude that for small $\mathbf{q}$ the coefficients $B_{\mathrm{sq},\,\mathrm{s'q'},\,\mathrm{s''q''}}$ must behave linearly in each of the variables q, q′, and q″. This conclusion is consistent with the theory of elasticity discussed in Chapter 1. The $\mathbf{q} \to 0$ limit corresponds to a quasi-uniform distortion, and hence the third-order elastic energy must correspond to a third-order form in the strain tensor, given by Eq. (1.3), where the latter is clearly proportional to $\mathbf{q}$ in this limit. If all three of the wave vectors approach zero we have the requirement

$$\lim_{\substack{\mathrm{q,q',q''}\\ \to 0}} B_{\mathrm{sq},\,\mathrm{s'q'},\,\mathrm{s''q''}} \propto \mathrm{qq'q''}. \tag{17.32}$$

We are now in a position to examine the term $\overline{\left(\beta V^{(3)}\right)^2}$ in Eq. (17.23), which produces the term Z_c in (17.24). Although the potential is overall strictly real, it is convenient to form the product of the potential with its complex conjugate wherein products of the Q_{qs} and their complex conjugates occur and which have the form

$$Q^*_{\mathrm{s_1q_1}}Q^*_{\mathrm{s_2q_2}}Q^*_{\mathrm{s_3q_3}}Q_{\mathrm{s_4q_4}}Q_{\mathrm{s_5q_5}}Q_{\mathrm{s_6q_6}}. \tag{17.33}$$

Because of the symmetry given by Eq. (17.27) we may permute the first three Q_{qs} among themselves, and similarly with the last three. To calculate averages of various terms in (17.33) we form pairs and apply Eq. (17.25). Since these averages must be real, two classes of terms need to be considered (together with permutations): (i) we pair a Q_{qs} with a Q_{-qs} and (ii) we pair a Q_{qs} with a Q^*_{qs}:

$$q_1 = -q_2; q_4 = -q_5; q_3 = q_6; s_1 = s_2; s_4 = s_5; s_3 = s_6; \tag{17.34a}$$

$$q_1 = q_4; q_2 = q_5; q_3 = q_6; s_1 = s_4; s_2 = s_5; s_3 = s_6. \tag{17.34b}$$

For case (i) the constraint (17.30) requires that $\mathbf{q}_3 = \mathbf{q}_6 = 0$ and using the property (17.32) it follows that such terms will vanish. Therefore, we need only consider the terms arising from class (ii), of which there are 3! permutations. Using Eq. (17.25) we then obtain the average[5]

$$\frac{\beta^2}{2}\overline{\left(V^{(3)}\right)^2} = \frac{\beta^2}{2}\frac{3!}{N^3M_c^3\beta^3}\sum_{ss's''}\sum_{qq'q''}\frac{\left|\frac{1}{6}B_{sq,s'q',s''q''}\right|^2}{\omega_{sq}^2\omega_{s'q'}^2\omega_{s''q''}^2}\Delta_{qq'q''}$$
$$= \frac{1}{12(NM_c)^3\beta}\sum_{ss's''}\sum_{qq'q''}\frac{\left|B_{sq,s'q',s''q''}\right|^2}{\omega_{sq}^2\omega_{s'q'}^2\omega_{s''q''}^2}\Delta_{qq'q''}. \tag{17.35}$$

The resulting contribution from the cubic term in the potential to the energy is then

$$E_c(T) = \frac{(k_BT)^2}{12(NM_c)^3}\sum_{ss's''}\sum_{qq'q''}\frac{\left|B_{sq,s'q',s''q''}\right|^2}{\omega_{sq}^2\omega_{s'q'}^2\omega_{s''q''}^2}. \tag{17.36}$$

As a result of the Δ function, the summation is restricted to two $\mathbf{q}$ variables and it is then of order N^2, resulting in an overall N^{-1} behavior of (17.36). The T^2 behavior of (17.36) results in a term proportional to T in the heat capacity.

Following Peierls, we write

$$B_{sq,s'q',s''q''} = N\omega_{sq}\omega_{s'q'}\omega_{s''q''}b_{sq,s'q',s''q''}\Delta_{qq'q''}, \tag{17.37}$$

where, due to the property (17.32), $b_{sq,s'q',s''q''}$ is a more slowly varying function of the wave vectors; Eq. (17.36) now becomes

$$E_c = \frac{(k_BT)^2}{12(NM_c)^3}\sum_{ss's''}\sum_{qq'}\left|b_{sq,s'q',s''}\right|^2. \tag{17.38}$$

We must also examine the quartic term, $V^{(4)}$. The form analogous to (17.31) is

$$V^{(4)} = \frac{1}{24}\sum_{qq'q''q'''}\sum_{ss's''s'''}C_{sq,s'q',s''q'',s'''q'''}Q_{sq}Q_{s'q'}Q_{s''q''}Q_{s'''q'''}. \tag{17.39}$$

[5] The term in which all $\mathbf{q}$ are the same in this sum is neglected since it makes a contribution N^{-1} relative to that retained in (17.40). Also, from (17.31) it follows that (17.35) is well behaved in the limit $\omega_q \rightarrow 0$.

As before, only pairs survive statistical averaging in (17.39) and there are three independent pairings of the four lattice displacements, with the result

$$-\beta\overline{V^{(4)}} = -3\beta\frac{1}{24N^2M_c^2\beta^2}\sum_{ss'}\sum_{qq'}\frac{C_{sq,sq,s'q',s'q'}}{\omega_{sq}^2\omega_{s'q'}^2}; \tag{17.40}$$

discrete translational symmetry again requires that

$$C_{sq,sq,s'q',s'q'} \to C_{sq,sq,s'q',s'q'}\Delta_{qq'q''q'''}.$$

The resulting contribution from the quartic term is then

$$E_q(T) = -\frac{(k_BT)^2}{8NM_c}\sum_{ss'}\sum_{qq'}\frac{C_{sq,sq,s'q',s'q'}}{\omega_{sq}^2\omega_{s'q'}^2}. \tag{17.41}$$

Introducing a function analogous to that defined in Eq. (17.37),

$$C_{sq,s'q',s''q'',s'''q'''} = N\omega_{sq}\omega_{s'q'}\omega_{s''q''}\omega_{s'''q'''}c_{sq,s'q',s''q'',s'''q'''}\Delta_{qq'q''q'''}, \tag{17.42}$$

the quartic contribution to the energy takes the form

$$E_q(T) = -\frac{(k_BT)^2}{8NM_c}\sum_{ss'}\sum_{qq'}c_{sq,s-q,s'-q',s'}. \tag{17.43}$$

This too produces a contribution to the heat capacity proportional to T and is of the same order as that from (17.38).

17.3 Phonon–phonon interactions

The terms in $\hat{V}^{(3)}$ and $\hat{V}^{(4)}$ lead to a coupling or equivalently a scattering among the phonons. Here we examine phonon–phonon scattering processes involving $\hat{V}^{(3)}$ and we limit ourselves to the case of one atom per unit cell. We only consider the fully quantum mechanical case since many of the interesting processes involve low temperatures, where this formulation is essential. We start by rewriting Eq. (17.31) as an operator, $\hat{V}^{(3)}$, with the three displacement operators, $\hat{Q}_{sq}$, replaced by (16.75a),

$$\hat{V}^{(3)} = \left[\frac{\hbar}{2NM}\right]^{3/2}\frac{1}{6}\sum_{ss's''}\sum_{qq'q''}B_{sq\,s'q'\,s''q''}\Delta_{qq'q''}$$
$$\times\frac{\left[\hat{a}_{sq}+\hat{a}_{sq}^{\dagger}\right]\left[\hat{a}_{s'q'}+\hat{a}_{s'q'}^{\dagger}\right]\left[\hat{a}_{s''q''}+\hat{a}_{s''q''}^{\dagger}\right]}{\sqrt{\omega_{sq}\omega_{s'q'}\omega_{s''q''}}}. \tag{17.44}$$

The matrix elements associated with the operators $\hat{a}_{sq}$ and $\hat{a}^{\dagger}_{sq}$ are given by (see Appendix B)

$$\hat{a}_{sq}\phi\left(\ldots, n_{sq}, \ldots\right) = \sqrt{n_{sq}}\,\phi(\ldots, n_{sq} - 1, \ldots) \tag{17.45a}$$

and

$$\hat{a}^{\dagger}_{sq}\phi\left(\cdots, n_{sq}, \cdots\right) = \sqrt{n_{sq} + 1}\,\phi(\cdots, n_{sq} + 1, \cdots), \tag{17.45b}$$

where $\phi\left(n_{s_1,q_1}, n_{s_2,q_2}, \cdots n_{s_N,q_N}\right)$ is a many-body Bose state function representing a possible state of the system of phonons.

The transition rate is given by the usual equation from time-dependent perturbation theory,

$$w = \frac{2\pi}{\hbar}\left|\langle f|\, V^{(3)}\, |i\rangle\right|^2 \delta\left(E_f - E_i\right). \tag{17.46}$$

The matrix element $\langle f|\, V^{(3)}\, |i\rangle$ in Eq. (17.46) contains various combinations and orderings of the $\hat{a}$ and $\hat{a}^{\dagger}$ operators. Conservation of energy forbids processes associated with $\hat{a}_{sq}\hat{a}_{s'q'}\hat{a}_{s''q''}$ and $\hat{a}^{\dagger}_{sq}\hat{a}^{\dagger}_{s'q'}\hat{a}^{\dagger}_{s''q''}$, as they would require the spontaneous loss or emergence of the energy of the three phonons involved. There remain the two classes of transitions shown in Fig. 17.1(a,b); together with their 3! different orderings, they involve the following operator products:[6]

(i) $\hat{a}_{sq}\hat{a}_{s'q'}\hat{a}^{\dagger}_{s''q''}$, where two initial-state phonons combine to form one final-state phonon;

(ii) $\hat{a}_{sq}\hat{a}^{\dagger}_{s'q'}\hat{a}^{\dagger}_{s''q''}$, where an initial-state phonon decays into two final-state phonons.

For case (i), the energy delta function in (17.46) leads immediately to the constraint

$$\omega_{sq} + \omega_{s'q'} - \omega_{s''q''} = 0, \tag{17.47a}$$

whereas the generalized Kronecker delta function in (17.44) leads to the condition

$$\mathbf{q} + \mathbf{q}' - \mathbf{q}'' = \mathbf{Q}. \tag{17.47b}$$

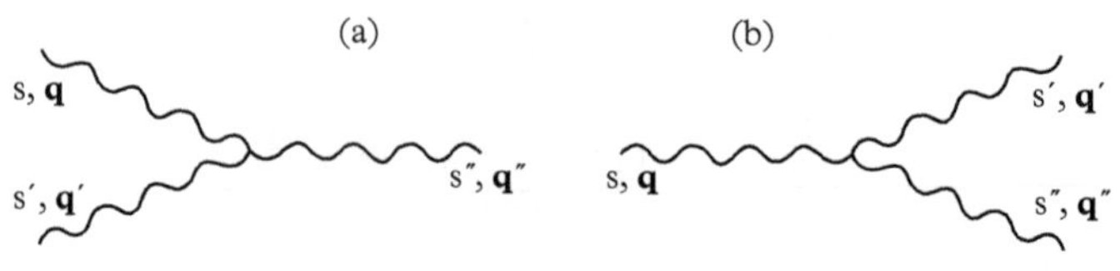

Figure 17.1 *(a) Transitions in which two initial state phonons,* s, **q** *and* s′, **q**′, *combine to form a final state phonon,* s″, **q**″; *(b) transitions in which an initial state phonon,* s, **q**, *decays into two final state phonons,* s′, **q**′ *and* s″, **q**″.

[6] Forms where two of the phonons have the same quantum numbers may be ignored in comparison with the remaining terms.

For case (ii), we have the conditions

$$\omega_{sq} - \omega_{s'q'} - \omega_{s''q''} = 0 \tag{17.48a}$$

and

$$\mathbf{q} - \mathbf{q}' - \mathbf{q}'' = \mathbf{Q}. \tag{17.48b}$$

For case (i), we have the transition probability per unit time

$$\begin{aligned} W_{sq,s'q' \to s''q''} = \frac{\pi\hbar}{4M^3N^3} & \frac{\left|B_{sq\,s'q'\,s''q''}\right|^2 n_{sq} n_{s'q'} \left(n_{s''q''} + 1\right)}{\omega_{sq}\omega_{s'q'}\omega_{s''q''}} \\ & \times \Delta_{-q,-q',q''} \delta\left(-\omega_{sq} - \omega_{s'q'} + \omega_{s''q''}\right), \end{aligned} \tag{17.49a}$$

while for case (ii) we have

$$\begin{aligned} W_{sq \to s'q',s''q''} = \frac{\pi\hbar}{4M^3N^3} & \frac{\left|B_{sq\,s'q'\,s''q''}\right|^2 n_{sq} \left(n_{s'q'} + 1\right) \left(n_{s''q''} + 1\right)}{\omega_{sq}\omega_{s'q'}\omega_{s''q''}} \\ & \times \Delta_{-q,q',q''} \delta\left(-\omega_{sq} + \omega_{s'q'} + \omega_{s''q''}\right), \end{aligned} \tag{17.49b}$$

where we have converted the δ function to frequency units.

At low frequencies, ω and q are related by $q = \omega/V_s$, where V_s is the velocity of sound associated for mode s. It is a fairly general property of the phonon dispersion curves that $\omega_{sq} < V_s q$ with increasing q (see Fig. 16.2); however the low frequency form can still be used to estimate the wave vector associated with a given frequency. We identify a characteristic temperature, $T_Q = \hbar V_s Q_c / k_B$, where $Q_c = \pi/a$ is a characteristic scale for the reciprocal lattice vectors and a is a typical unit cell spacing. For temperatures significantly above T_Q, the large wave vectors of the thermal phonons will require $\mathbf{Q} \neq 0$ in Eq. (17.48b); i.e., the sum of the wave vectors will result in a value lying outside the first Brillouin zone and require that the relevant **Q** be included in condition (17.48), in order that all **q** vectors lie within the first Brillouin zone. This can be viewed as a *non-conservation* of wave vector (or equivalently momentum). This property is an unavoidable characteristic of our crystal lattice and arises from the lack of full translational invariance of our system (which allows only discrete translations by real- space lattice vectors).[7]

For temperatures significantly less than T_Q, the wave vectors $q \cong \omega/V_s$ are small relative to Q_c. Under this condition, (17.48) is satisfied with $\mathbf{Q} = 0$ and we have conservation of the quasi-wave vector, **q**. In order of magnitude the fraction of Umklapp processes will fall off as $e^{-\gamma\theta_D/T}$ (where γ is a numerical factor less than one) since T_Q is of the same order as the Debye temperature, θ_D.

The combined effects of conservation of frequency and conservation of wave vector place important restrictions on the allowed scattering processes. Under the condition $\omega_{sq} < V_s q$ (referred to as a normal dispersion) it is not possible to simultaneously satisfy conservation of wave vector and frequency for the decay of a

[7] We should not be too concerned about this behavior. In a solid, where the atoms are localized on lattice sites, there can be no momentum transport when the solid as a whole is at rest; in a mechanical system, momentum transport must be accompanied by mass transport

mode into two final state phonons having the *same* s and this process is forbidden. When $\omega_{sq} > V_s q$, the emerging phonons propagate at a small angle with respect to each other and when $\omega_{sq} = V_s q$, the initial and final phonons propagate parallel to each other. However when $\omega_{sq} < V_s q$ it is still possible for a slower (largely transverse) mode ($s = 2, 3$) to decay into two longitudinal modes ($s = 1$) or a longitudinal and a transverse mode. The allowed processes described can also proceed in the reverse direction, in which two initial-state phonons combine to form a single final-state phonon.

Figure 17.2 shows some allowed three-phonon processes in one dimension, where two phonons combine to make a final-state phonon; here we picture propagation along a high symmetry direction where the two transverse phonons are

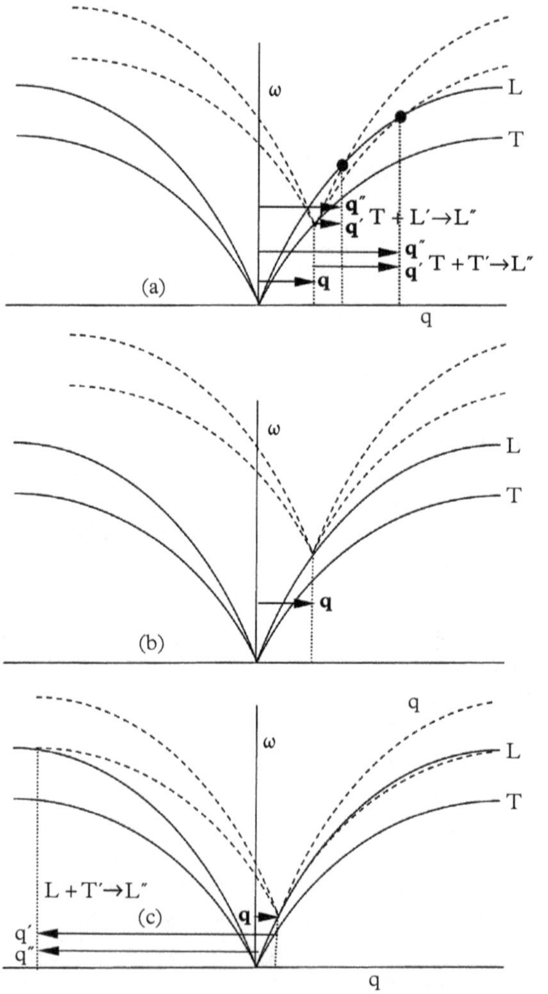

Figure 17.2 *Some three-particle processes involving phonons with wave vectors q, q′ and q″ propagating in a high symmetry direction where the two transverse modes are degenerate. The dispersion relations for the two initial states, q, and q′, shown as the full and dashed lines respectively, are combined by translating the latter with respect to the former. The following cases are shown: (a)* T+L′ → L″ *and* T+T′ → L″*; (b) a regime of a forbidden processes involving two a longitudinal phonon; (c) a process* L + T′ → L″. *(After R. Peierls (1955).)*

degenerate. One adds the wave vector of the second phonon to the first, simply displacing the ω vs. q curves for the second phonon along those for the first.

Allowed processes in an arbitrary direction require additional analysis and we will simply list the results of such an analysis from Peierls. If we order the modes according to their long-wavelength velocities, $V_1 > V_2 > V_3$ (for s = 1, 2, 3), then one obtains the following selection rules:

allowed	[3] + [3] → [2]	[3] + [3] → [1]	[3] + [2] → [2]
	[3] + [2] → [1]	[3] + [1] → [1]	[2] + [1] → [1]
forbidden	[1] + [1] → [3]	[1] + [1] → [2]	—

Note that a population consisting of only the [1] phonons has no decay channels to other modes in the presence of normal dispersion.[8]

17.4 The phonon Boltzmann transport equation

We will discuss the Boltzmann transport equation[9] in detail in Chapter 19, when we treat electron transport in metals; here we limit our discussion to that needed to treat thermal transport. The basic idea for phonons is that we generalize the equilibrium distribution function, n(T), so that, instead of depending on the temperature and frequency (energy) of the excitations (actually on the ratio $\hbar\omega_{sq}/k_B T$ in equilibrium), it instead depends on the mode s, its wave vector, $\mathbf{q}$ (momentum), the position, $\mathbf{r}$, and (more generally) the time (which are the usual Hamiltonian coordinates of a mechanical system); hence we write $n = n(s, \mathbf{q}, \mathbf{r}, t)$. For the case of thermal transport the change in the distribution function is induced through boundary conditions, by attaching different (typically opposing) surfaces of the specimen to thermal reservoirs held at different temperatures; sound waves are also launched and received at opposing surfaces when studying their velocity and attenuation, as discussed in Chapter 1. The specification of a temperature is strictly valid only for a system in thermal equilibrium, which in the present case would imply excitations distributed in frequency according to the Bose distribution; however, it is generally a good approximation to introduce a position-dependent temperature, provided that it does not vary rapidly (on a scale measured by a thermal phonon mean free path).

At this point we only consider changes in the distribution function that result from collisions among the excitations. It is therefore natural to construct a generalized *rate equation*, called the *Boltzmann equation*, by equating the *total* time derivative of $n(s, \mathbf{q}, \mathbf{r}, t)$ to a function, which we denote as I(n) and refer to as the *collision integral*; I(n) measures the overall rate that excitations s, $\mathbf{q}$ are created and

[8] One can go on to calculate the attenuation of a sound wave arising from the various three phonon processes at low temperatures; the resulting attenuation is proportional ωT^4 as first shown in: Sound absorption in solids, L.D Landau and Y.B Rumer, Phys. Zs. Sovjet, 1937.

[9] Our discussion here follows that of R. Peierls in *Quantum Theory of Solids*, Chapters 1 and 2, Oxford University Press (1955); it is based on Peierls' original work: On the kinetic theory of thermal conduction in crystals, R. Peierls, *Ann. Phys.* 3, 1055 (1929), which is reprinted in *Selected Scientific Papers of Sir Rudolf Peierls: With Commentary*, World Scientific, Singapore (1997).

destroyed, through three-particle collisions, or are scattered such that their momentum and polarization change, as a result of four-particle interactions. Hence we write

$$\frac{\partial n(s,\mathbf{q},\mathbf{r},t)}{\partial t} + \frac{\partial n(s,\mathbf{q},\mathbf{r},t)}{\partial \mathbf{r}} \cdot \dot{\mathbf{r}} + \frac{\partial n(s,\mathbf{q},\mathbf{r},t)}{\partial \mathbf{q}} \cdot \dot{\mathbf{q}} = I\left[n(s,\mathbf{q},\mathbf{r},t)\right]. \quad (17.50)$$

In what follows, we consider only time-independent cases. In the present context the third term on the left vanishes since it would require an external force capable of "accelerating" (increasing the wave vector of) our phonon excitations. We write the remaining term as $\nabla n\,(s,\mathbf{q},\mathbf{r}) \cdot \mathbf{V}^{(g)}_{sq}$, which is referred to as the drift term, where $\mathbf{V}^{(g)}_{sq} = \partial\omega_{sq}/\partial\mathbf{q}$ is the *group velocity* (not to be confused with the eigenvectors of the dynamical matrix). Our Boltzmann equation is then

$$\nabla n\,(s,\mathbf{q},\mathbf{r}) \cdot \mathbf{V}^{(g)}_{s,q} = I\left[n(s,\mathbf{q},\mathbf{r})\right]. \quad (17.51)$$

In constructing the drift term we are implicitly assuming a superposition of phonons with a narrow range of wave vectors, δq, to form wave packets which then propagate with $\mathbf{V}^{(g)}$. From the position–wave vector uncertainty principle we must have $\delta q\delta x \geq 1$. Since the largest that δx can be is of order a phonon mean free path, l, after which the mode is altered through scattering, the smallest δq is of order l^{-1}. In equilibrium, phonon occupation numbers change significantly over energies of order k_BT. The energy of a wave packet is then reasonably well defined if

$$k_BT > \delta E = \hbar\frac{\partial\omega}{\partial q}\delta q = \hbar V^{(g)}\delta q; \quad (17.52)$$

we then have

$$\delta l > \frac{1}{\delta q} > \frac{\hbar V^{(g)}}{k_BT} \approx \frac{\theta_D}{T}a, \quad (17.53)$$

where in the last step we have approximated the group velocity by the sound velocity and used the definition of the Debye temperature (see Chapter 4) with $(N/V)^{-1/3} \approx a$, where a is a characteristic atomic dimension. Hence, the temperature should not vary significantly over a length of order $a\theta_D/T$. Finally, we note that from the frequency–time uncertainty principle, $\delta\omega\delta t \geq 1$, the best precision to which the frequency can be defined is governed by $\delta\omega \geq \tau^{-1}$, where τ is a typical scattering time which we can define as $\tau = l/V^{(g)}$; we can then relax the total frequency conservation condition to[10]

$$\omega_{sq} + \omega_{s'q'} + \omega_{s''q''} \geq \tau^{-1}. \quad (17.54)$$

To construct $I(n(s,\mathbf{q},\mathbf{r})))$ we must include all processes in which:

(i) phonons are scattered *out of* state $(s,\mathbf{q})$;

(ii) phonons are scattered *in to* state $(s,\mathbf{q})$.

[10] To a very limited extent this permits collinear three-phonon processes within the same mode, in contrast to our above selection rules, which however still hold for the majority of the phonons.

Processes in category (i), which contribute to a loss rate, will enter with a negative sign. Since the initial and final states are overall constrained by the selection rules, we only need to sum over one pair of variables, which we take as $s'\mathbf{q}'$. The two classes of transitions given by Eqs. (17.49a) and (17.49b) both involve scattering *out of* the state $(s, \mathbf{q})$. However, processes in which $s'\mathbf{q}'$ and $s''\mathbf{q}''$ occur together in an *initial or final state* are not to be differentiated, which we account for by including a factor 1/2. Assembling these components and taking the limit where $\mathbf{q}$ is continuous, the collision integral will be given by

$$I_{sq} = \frac{V}{(2\pi)^3} \sum_{s's''} \int d^3q \, \Big\{ \big[w_{s''q'' \to sq,\, s'q'} - w_{sq,\, s'q' \to s''q''} \big] + \frac{1}{2} \sum_{s's''} \big[w_{s'q',\, s''q'' \to sq} - w_{sq \to s'q',\, s''q''} \big] \Big\}, \tag{17.55}$$

where the transition probabilities in this expression are given by Eqs. (17.49a,b).

If we write the *equilibrium* Bose distribution as $n^{(0)}(x) = (e^x - 1)^{-1}$, where $x = \omega_{sq}/k_B T$, then it is easy to show that

$$n^{(0)}(x)\left[n^{(0)}(x') + 1\right]\left[n^{(0)}(x'') + 1\right] = \left[n^{(0)}(x) + 1\right] n^{(0)}(x') n^{(0)}(x'') e^{x - x' - x''}; \tag{17.56}$$

since the exponential factor vanishes due to the delta function in (17.46), it follows that $I(n^{(0)})$ vanishes, as it must, since there is no net rate of change of the occupancy in thermal equilibrium. This is referred to as the *principle of detailed balancing*, which is a physical statement of the fact that in equilibrium the overall rate of transitions into and out of a given state must be identical.

With the understanding that $n^{(0)} = n^{(0)}\left(\omega_{sq}/k_B T(\mathbf{r})\right)$, which we write as $n^{(0)}(s, \mathbf{q})$, we will seek a solution of (17.50) of the form

$$n(s, \mathbf{q}, \mathbf{r}) = n^{(0)}(s, \mathbf{q}) + n^{(1)}(s, \mathbf{q}, \mathbf{r}), \tag{17.57}$$

where $n^{(1)}(s, \mathbf{q}, \mathbf{r})$ represents a small *shift* from the equilibrium Bose form. The leading order term on the left-hand side of (17.51) will then arise from $T(\mathbf{r})$ in $n^{(0)}(s, \mathbf{q})$ while the term arising from $n^{(1)}$ contributes in the next order; the left-hand side of (17.51) then becomes

$$\frac{\partial n^{(0)}}{\partial T} \nabla T \cdot \mathbf{V}^{(g)}_{sq} = n^{(0)}\left[n^{(0)} + 1\right] \frac{\omega_{sq}}{k_B T^2} \nabla T \cdot \mathbf{V}^{(g)}_{sq}. \tag{17.58}$$

As a result of (17.56), the leading term on the right involves $n^{(1)}(s, \mathbf{q}, \mathbf{r})$. Given the structure of (17.58) it is convenient to write $n^{(1)}(s, \mathbf{q}, \mathbf{r})$ in the alternative form,

$$n^{(1)}(s, \mathbf{q}, \mathbf{r}) = n^{(0)}\left[n^{(0)} + 1\right] g(s, \mathbf{q}, \mathbf{r}). \tag{17.59}$$

Inserting this form into the collision integral (17.55), using the property (17.56), and combining the result with (17.58), we obtain an *integro-differential equation* which governs the non-equilibrium distribution function, $n(s, \mathbf{q}, \mathbf{r})$, and ultimately the transport of heat, in our monatomic insulating crystal as

$$n^{(0)}\left(n^{(0)}+1\right)\frac{\hbar\omega_{sq}}{k_B T^2}\nabla T\cdot \mathbf{V}^{(g)}_{sq}$$
$$=\frac{\hbar}{32\pi^2\rho}\int d^3q'\left\{\sum_{ss'}\left|b_{sq\,s'q'\,s''}\right|^2\omega\,\omega'\omega''\,\delta\left(\omega+\omega'-\omega''\right)(n+1)(n'+1)n''(g''-g-g')\right.$$
$$\left.+\frac{1}{2}\sum_{ss''}\left|b_{sq\,s'q'\,s'''}\right|^2\omega\,\omega'\omega''\delta\left(\omega-\omega'-\omega'''\right)(n+1)n'n'''(g'+g'''-g)\right\}, \tag{17.60}$$

where to be concise we have written $n'=n^{(0)}(\hbar\omega_{s'q'}/k_BT)$, etc., redefined the coupling constants according to Eq. (17.37), and substituted the definition of the mass density, $\rho\equiv MN/L^3$.

The solution of Eq. (17.60) in a real material is horrendously difficult.[11] To begin with we require the exact phonon dispersion relations, which as described in Chapter 16, is now possible, say, using the supercell approach. But beyond that one needs the three-phonon coupling constants, $b_{sq\,s'q'\,s''q''}$. However, even with their availability, we have ignored the four-phonon processes involving the coefficients $c_{sq,\,s'q',\,s''q'',\,s'''q'''}$. Both sets of coefficients could also, in principle, be calculated using the supercell approach. The parent coefficients, $\mathbf{B}_{nn'n''}$ and $\mathbf{C}_{nn'n''n'''}$, are expected to diminish rapidly with increasing atomic separations and reasonable estimates might arise by restricting to the nearest-neighbor contributions.

17.5 The thermal conductivity in dielectric materials

In spite of the fact that the atoms of a crystal are restricted to lattice sites heat transport can still occur.[12,13] Intuitively, the energy flux in a crystal, $\mathbf{J}_\varepsilon$ (the energy transported per unit area per unit time) consists of the energy carried by each excitation, $\hbar\omega_{sq}$, multiplied by its group velocity, $\mathbf{V}^{(g)}_{sq}$, and summed over all the excitations $(s, \mathbf{q})$ in each unit of volume, where the latter is governed by the phonon distribution function, $n(s, \mathbf{q}, \mathbf{r})$. Given that there is no energy transport in equilibrium, only $n^{(1)}(s, \mathbf{q}, \mathbf{r})$ is involved and hence

$$\mathbf{J}_\varepsilon=\frac{1}{(2\pi)^3}\int d^3q\sum_s\hbar\omega_{sq}\mathbf{V}^{(g)}_{sq}n^{(1)}(s,\mathbf{q},\mathbf{r}). \tag{17.61}$$

Experimentally, the heat transport is proportional to the temperature gradient, ∇T, and is written as

$$\mathbf{J}_\varepsilon=K(T)\nabla T, \tag{17.62}$$

[11] In the high temperature regime, where quantum effects can be ignored, and in systems where the interatomic forces can be modeled by pair-wise potentials, the transport of heat can be calculated using the techniques of *molecular dynamics*, which we will not describe here.

[12] Theory of thermal conductivity of solids at low temperatures, P. Carruthers, *Rev. Mod. Phys.* 33, 92 (1961).

[13] Thermal conductivity and lattice vibrational modes, P.G. Klemens in *Solid State Physics*, Ed. F. Seitz and D. 'Turnbull, Academic Press Inc., NY (1958).

where the coefficient K(T) is the *thermal conductivity*, which in general is temperature dependent. Using similar arguments, the "momentum flux" would be

$$J_q = \frac{1}{(2\pi)^3}\sum_s \int d^3q\,\hbar\mathbf{q}\,n^{(1)}(s,\mathbf{q}). \tag{17.63}$$

Now in the absence of various interactions a flux of phonons with momentum per unit area J_q would, once set in motion, experience no drag and persist indefinitely; the energy flux they carry would also persist and the thermal conductivity would effectively be infinite. Interactions arise from three sources: (i) anharmonic terms in the expansion of the lattice potential, as considered above; (ii) phonon scattering from crystalline defects, either impurities (including different isotopes[14,15] of the chemical constituents) or structural faults;[16] and (iii) scattering from the boundaries of a finite specimen.[17] Concerning mechanism (i), the cubic and quartic terms permit Umklapp processes in which case the interacting particles can lose a momentum $\hbar\mathbf{Q}$ to the lattice in collisions; this results in a finite thermal conductivity which however increases at low temperatures, due to the falling Umklapp rate, such that $K_U(T) \propto e^{\theta_D/T}$. In the absence of Umklapp processes, collisions among the phonons would result in the phonon gas relaxing into equilibrium, but in a reference frame moving with a *finite average velocity*. This results in an infinite thermal conductivity.

The exponentially increasing conductivity at low temperatures is ultimately limited by processes (ii) and (iii). As a simple model we will use the expression for the thermal conductivity obtained in the kinetic theory of gases,

$$K = \frac{1}{3}CV_s\lambda, \tag{17.64a}$$

or equivalently

$$K = \frac{1}{3}CV_s^2\tau. \tag{17.64b}$$

Here, C is the phonon heat capacity, proportional to T^3 in the low temperature Debye limit; V_s is an average of the sound velocity; λ is a mean free path; and $\tau = \lambda/V_s$ is a relaxation time. In the absence of Umklapp processes, λ is qualitatively controlled by the defect density and, again drawing on the kinetic theory of gases, is given by $\lambda = 1/\sigma n_i$, where σ is a scattering cross section and n_i is the impurity/defect number density. If σ is taken to be independent of the thermal phonon frequencies (the average value of which falls with temperature), Eq. (17.64) predicts a thermal conductivity that falls as T^3 at low temperatures. In a chemically and isotopically pure, structurally perfect, single crystal, the only thing that then limits the mean free path is scattering from the crystal boundaries, and in this case the mean free path is of order d, a sample dimension; assuming that boundary scattering is independent of the average phonon frequency, and contains a *momentum altering* (non-specular) component, we would again obtain $K \propto T^3$. In experiments this is achieved by roughening the sample surfaces.

[14] On the thermal conductivity of dielectrics at temperatures lower than that of Debye, I. Pomeranchuk, *J. Phys.-USSR* **6**, 237 (1942).

[15] The scattering of low-frequency lattice waves by static imperfections, P.G. Klemens, *Proc. Phys. Soc. Lond. A* **68**, 1113 (1955).

[16] The thermal conductivity of dielectric solids at low temperatures – theoretical, P.G. Klemens, *Proc. R. Soc. Lond. A* **208**, 108 (1951).

[17] Note on the conduction of heat in crystals, H.B.G. Casimir, *Physica* **5**, 495 (1938).

When more than one mechanism operates in parallel, their rates will be additive and hence we can write the relaxation time as

$$\frac{1}{\tau_t} = \frac{1}{\tau_u} + \frac{1}{\tau_b} + \frac{1}{\tau_s}, \tag{17.65}$$

where τ_t, τ_u, τ_b, τ_s are the total, Umklapp, boundary, and structural contributions, with $\tau_u \propto e^{\theta_D/T}$ and $\tau_b, \tau_s \cong$ constant.

Figure 17.3 shows a log–log plot of the temperature dependence of the thermal conductivity in isotopically pure lithium fluoride.[18] Note that for temperatures above approximately 20 K, the conductivity falls rapidly; this essentially

[18] Effect of boundaries and isotopes on the thermal conductivity of LiF, P.D. Thacher, *Phys. Rev.* **156**, 975 (1967).

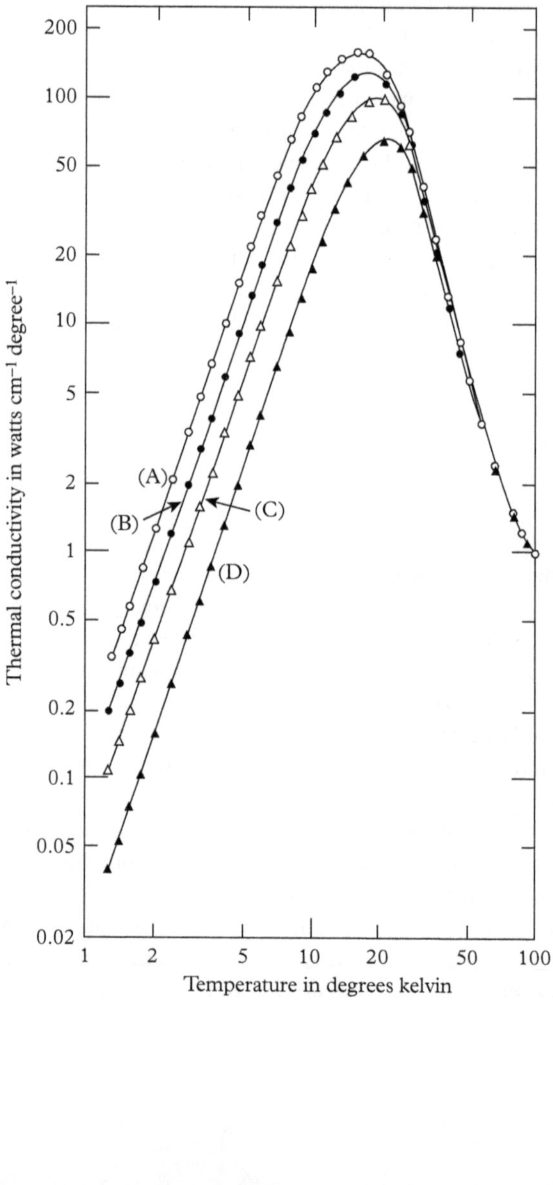

Figure 17.3 *The thermal conductivity of isotopically pure* LiF *as a function of boundary width:* (A) *7.25 mm;* (B) *4.00 mm;* (C) *2.14 mm;* (D) *1.06 mm. (After Thacher (1967).)*

exponential decrease is attributed to the increase in the fraction of Umklapp scattering events. Below 20 K the relaxation rate approaches a constant being controlled by structural and boundary effects. The conductivity then falls approximately as T^3, in proportion to the number of phonons as measured by the specific heat. Most importantly, as seen in this figure, the conductivity in this low temperature regime depends sensitively on the sample width; the larger the width, the larger the thermal conductivity. Note the surfaces were sandblasted to achieve non-specular reflections of the phonons.

17.6 Second sound

We have argued that in the absence of scattering mechanisms (i)–(iii), phonons with momentum per unit area J_q would simply drift through the solid unimpeded. We have seen that at low temperatures the Umklapp processes are suppressed; however overall momentum conserving three- and four-phonon processes remain and they can bring the phonon gas into *local equilibrium* and in the absence of mechanisms (ii) and (iii) we can then picture a phonon gas drifting in the solid that is in equilibrium in a moving reference frame, much as a gas drifts in a large diameter pipe. Now, a gas will support a sound wave, wherein the density and local velocity oscillate out of phase both spatially and temporally, so it is natural to ask if a *phonon gas* will support a similar mode. However an increase in the local phonon gas density corresponds, physically, to an increase in the local *temperature*; hence such a mode represents an *oscillating temperature wave*. To distinguish such a mode from an ordinary sound wave, the new mode is termed *second sound.*[19,20] A similar mode was first proposed and observed in superfluid ^{4}He and will be discussed in Section 36.5.

Ordinary sound waves in a gas propagate at a velocity $V_{sound} = V_{rms}/\sqrt{3}$, where V_{rms} is the root mean square thermal velocity. Hence the velocity of second sound, V_{ss}, is expected to be $V_{ss} = \overline{V_s}/\sqrt{3}$, where $\overline{V_s}$ is some average of the three sound velocities in the solid. For the mode to exist, the phonon gas must be able to come into local equilibrium in a time short compared to the period of the sound, $\tau_{pp} \ll 1/\omega$, where τ_{pp} is a phonon-phonon scattering time and ω is the angular frequency of the sound wave. At the same time we must have $1/\omega \ll \tau_u$, where τ_u is a typical Umklapp scattering time. Hence the frequency must lie in the following "window," $\tau_{pp} \ll 1/\omega \ll \tau_u$. The mode is ultimately damped by phonon viscosity (arising from τ_{pp}), structural defects, and residual Umklapp processes.

A material satisfying these conditions is solid ^{4}He. Figure 17.4 shows the measurements of Ackerman, Bertman, Fairbank, and Guyer,[21] supporting the existence of second sound. Since the velocity of this mode is essentially non-dispersive, it can be launched into the solid as a pulse using a thin carbon film heater. The temperature dependence of an identical film was used as the detector. Because of a finite recovery time of the receiving resistor following the pulse arrival, the time derivative of the temperature shift, $d\delta T/dt$ is plotted.

[19] That second sound could exist in a solid appears to have been first proposed in, The velocity of 2nd sound in liquid helium near the absolute zero, J.C. Ward and J. Wilks, *Phil. Mag.* **42**, 314 (1951).

[20] For a detailed discussion of the theory see: First and second sound in crystals, W. Gotze and K.H. Michel, *Phys. Rev.* **156**, 963 (1967).

[21] Second sound in solid helium, C.C. Ackerman, B. Bertman, H.A. Fairbank, and R.A. Guyer, *Phys. Rev. Lett.* **16**, 789 (1966).

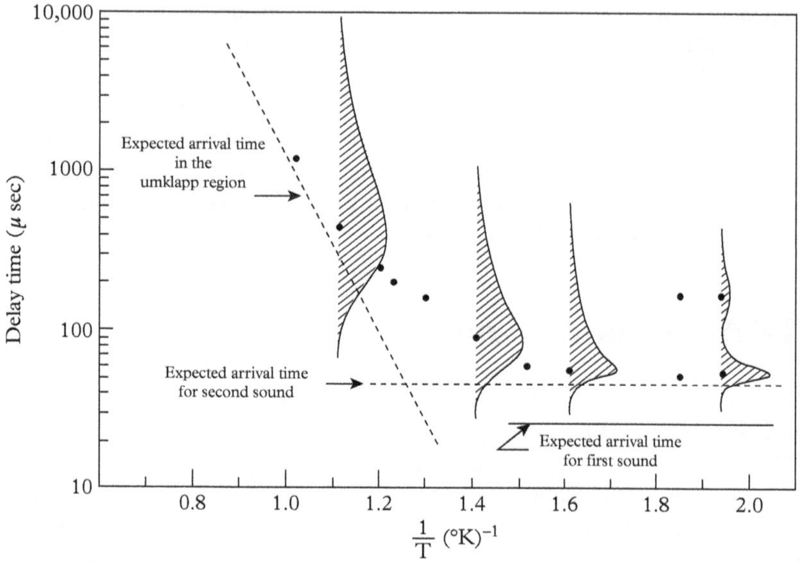

Figure 17.4 *The shapes and delay times of heat pulses as a function of the reciprocal of the temperature. Note that as the temperature falls the pulse shape evolves from broad to sharp; in addition, the pulse propagates more slowly and ultimately approaches that expected for second sound (the horizontal dashed line) at lower temperatures. (After Ackerman, Bertman, Fairbank, and Guyer (1966).)*

ADDITIONAL READING

An excellent discussion of thermal conductivity in crystals, from which much of the material in this chapter was drawn, is given in *Quantum Theory of Solids*, R. Peierls, Oxford University Press, London (1955). It is strongly recommended that Peierls be read in parallel with our treatment here. Peierls was an early pioneer in lattice dynamics.

At a simpler level there is the discussion in *Solid State Physics*, N.W. Ashcroft and N.D. Mermin, Saunders College, PA (1976).

The quadratic and higher force constants can be obtained from model interatomic forces from which one can go on to solve the phonon Boltzmann equation for the thermal conductivity; an example being silicon: Lattice thermal conductivity of silicon from empirical interatomic potentials, D.A. Broido, A. Ward, and N. Mingo, *Phys. Rev. B* 72, 014308 (2005). First principles calculations of the force constants obtained from density functional theory lead to better agreement: Intrinsic lattice thermal conductivity of semiconductors from first principles, D.A. Broido, M. Malorny, G. Birner, Natalio Mingo, and D.A. Stewart, *Appl. Phys. Lett.* **91**, 231922 (2007).

Part VI

Electron Transport and Conduction Electron Dynamics

Motion of Electrons and Holes in External Electric and Magnetic Fields

18

In this chapter we will examine the dynamics of electrons and holes in conducting solids in the presence of electric and magnetic fields. Our starting point will be a non-relativistic one-electron Hamiltonian of electrons in a periodic potential,

$$\hat{H} = \frac{\hat{p}^2}{2m} + V(\mathbf{r}), \tag{18.1}$$

where $V(\mathbf{r}) = V(\mathbf{r} + \mathbf{R})$ and $\hat{\mathbf{p}} = (\hbar/i)\nabla$ is the momentum operator.

18.1 Incorporating external electromagnetic fields; gauge invariance

In Hamiltonian mechanics, both classical and quantum, the only way to incorporate the effects of electric and magnetic fields is through the scalar and vector potentials, $\phi(\mathbf{r})$ and $\mathbf{A}(\mathbf{r})$.[1] In quantum mechanics these quantities are entered into the Hamiltonian operator (18.1) by adding the external electrostatic potential energy, $e\phi(\mathbf{r})$, to $V(\mathbf{r})$ and by identifying $\hat{\mathbf{p}}$ as[2]

$$\hat{\mathbf{p}} = \hat{\boldsymbol{\pi}} - \frac{e}{c}\mathbf{A}(\mathbf{r}), \tag{18.2}$$

where we now write $\hat{\boldsymbol{\pi}} = (\hbar/i)\nabla$; for an electron $e = -|e|$ in our convention. In general the perturbations introduced by external potentials will destroy the discrete translational invariance of a periodic lattice, and then the one-electron wave functions no longer have the Bloch form. No general solution exists, even for the simplest case of an electron in a periodic potential subject to a constant magnetic field;[3] the problem is always treated in an approximate (though usually quite sufficient) manner. We will limit ourselves to the case of weak fields.

The limiting case is that of vanishing fields; this corresponds to constant vector and scalar potentials, $\mathbf{A}_0$ and ϕ_0. The addition of a constant ϕ_0 to the potential simply changes the zero of energy: $E_n(\mathbf{k}) \to E_n(\mathbf{k}) + e\phi_0$. On the other hand, the magnetic field is given by $\mathbf{H} = \nabla \times \mathbf{A}$, and is invariant to a gauge transformation

[1] Good discussions of this formalism can be found in Section 16 of *Classical Theory of Fields* and Section 111 of *Quantum Mechanics*, L.D. Landau and E.M. Lifshitz, Pergamon Press.

[2] Hamiltonian mechanics is formulated in terms of particle momenta and coordinates. In order for Hamilton's equations to have the *same form* in the presence of magnetic fields it is necessary to introduce a so-called *canonical momentum*, $\boldsymbol{\pi} \equiv \mathbf{p} + (e/c)\mathbf{A}$, where $\mathbf{p} = m\mathbf{v}$ is the ordinary *mechanical* momentum. A static magnetic field does no work on a particle, so the kinetic energy T is then constant and given by $(1/2)mv^2$, the same as the Hamiltonian in the absence of $\mathbf{A}$; combining these properties it then follows that we must have $T = (1/2)mv^2 = (\boldsymbol{\pi} - (e/c)\mathbf{A}(\mathbf{r}))^2/2m$. In transitioning to quantum mechanics one makes $\boldsymbol{\pi}$ an operator according to the prescription, $\hat{\boldsymbol{\pi}} = (\hbar/i)\nabla$.

[3] Special cases exist which we do not discuss here. For a discussion see: Bloch electrons in a uniform magnetic field, E. Brown, *Phys. Rev.* **134**, A1038 (1964); Magnetic translation group, Y. Zak, *Phys. Rev.* **134**, A1602 (1964).

The Physics of Solids. First Edition. J. B. Ketterson.

$\mathbf{A} \to \mathbf{A} + \nabla\chi(\mathbf{r})$, where $\chi(\mathbf{r})$ is any continuous function. Classically, a change of gauge does not affect charge particle motion; however in quantum mechanics it will alter the *phase* of the of the particle's wave function, as we now demonstrate.[4]

First we consider the effect that a unitary transformation, $\hat{U}$, has on the momentum operator $\hat{p} = (\hbar/i)\nabla - (e/c)\mathbf{A}(\mathbf{r})$,

$$\hat{p} \to \hat{U}\hat{p}\hat{U}^{\dagger} = \hat{U}\left(\frac{\hbar}{i}\nabla - \frac{e}{c}\mathbf{A}(\mathbf{r})\right)\hat{U}^{\dagger} = \hat{U}\hat{U}^{\dagger}\left(\frac{\hbar}{i}\nabla - \frac{e}{c}\mathbf{A}(\mathbf{r})\right) + \hat{U}\frac{\hbar}{i}\nabla\hat{U}^{\dagger}$$
$$= \hat{p} + \frac{\hbar}{i}\hat{U}\nabla\hat{U}^{\dagger}, \qquad (18.3)$$

where we have used the property $\hat{U}\hat{U}^{\dagger} = 1$ in the second step. Now consider the effect of a gauge transformation, $\mathbf{A}(\mathbf{r}) \to \mathbf{A}(\mathbf{r}) + \nabla\chi(\mathbf{r})$, on the same operator,

$$p \to \frac{\hbar}{i}\nabla - \frac{e}{c}\mathbf{A}(\mathbf{r}) - \frac{e}{c}\nabla\chi = p - \frac{e}{c}\nabla\chi. \qquad (18.4)$$

These changes will produce the same effect if

$$\frac{\hbar}{i}\hat{U}\nabla\hat{U}^{\dagger} = -\frac{e}{c}\nabla\chi,$$

which has the solution $\hat{U}^{\dagger} = e^{-i\varphi(\mathbf{r})}$, where $\nabla\varphi(\mathbf{r}) = (e/\hbar c)\nabla\chi(\mathbf{r})$; on ignoring the integration constant we have

$$U(\mathbf{r}) = e^{(e/\hbar c)\chi(\mathbf{r}))}. \qquad (18.5)$$

Stated concisely, the gauge invariance of classical electromagnetic theory generalizes into invariance under a *position-dependent* phase in quantum mechanics. This is universally regarded as a superior perspective.

18.2 Effect of external magnetic and electric fields on Bloch functions

We start by considering the effect of a constant vector potential, $\mathbf{A}_0$; this is equivalent to a gauge function, $\chi = \mathbf{A}_0 \cdot \mathbf{r}$. According to Eq. (18.5), introducing $\mathbf{A}_0$ is equivalent to a phase shift $(e/\hbar c)\,\mathbf{A}_0 \cdot \mathbf{r}$ in the wave function of a charged particle acted upon by that vector potential. Thus in the presence of a constant vector potential $\mathbf{A}_0$, the phase of a Bloch wave function $\psi_{\mathbf{k},n} = e^{i\mathbf{k}\cdot\mathbf{r}}u_{\mathbf{k}n}(\mathbf{r})$ will be altered:

$$\psi_{\mathbf{k},\mathbf{A}_0,n} = e^{i(\mathbf{k}+(e/\hbar c)\mathbf{A}_0)\cdot\mathbf{r}}u_{\mathbf{k}n}(\mathbf{r}). \qquad (18.6a)$$

[4] In the so-called Ahronov–Bohm effect, an observable quantum interference of electrons occurs in situations where there is a position-dependent vector potential but no accompanying magnetic field. An example involves the motion of an electron outside an infinite current-carrying solenoid, where the vector potential is nonzero, but no magnetic field exists.

or equivalently

$$\psi_{\mathbf{k},\mathbf{A}_0,n}(\mathbf{r}) = e^{i\boldsymbol{\kappa}\cdot\mathbf{r}}u_{\mathbf{k}}(\mathbf{r}), \tag{18.6b}$$

where we have introduced a *crystal (or canonical) wave vector*, $\boldsymbol{\kappa}$, through the relation

$$\mathbf{k} = \boldsymbol{\kappa} - \frac{e}{\hbar c}\mathbf{A}_0. \tag{18.7}$$

Since $\boldsymbol{\kappa}$ measures the change in phase under a translation, its designation as crystal wave vector is natural.

For the case when a weak magnetic field is present, $\mathbf{A}$ is not constant, but can vary slowly in space on the scale of a lattice constant. As a "zeroth-order" approximation for the wave function we could replace $\mathbf{A}_0$ by $\mathbf{A}(\mathbf{r})$ in Eq. (18.6a),

$$\psi_{\boldsymbol{\kappa},\mathbf{A},n}(\mathbf{r}) = e^{i\boldsymbol{\kappa}\cdot\mathbf{r}}u_{\boldsymbol{\kappa}-(e/\hbar c)\mathbf{A},n}(\mathbf{r}). \tag{18.8}$$

We note that for a given $\boldsymbol{\kappa}$ the function $u(\mathbf{r})$ is now no longer strictly periodic due to the position dependence of $\mathbf{A}(\mathbf{r})$; such states could formally be expanded in terms of Bloch states ($\mathbf{k}$, n) with $\mathbf{A} = 0$, although real transitions do not occur (however, see Section 18.6).

A way to construct an approximate theory for electrons in a crystal when a magnetic field is present is to convert the Bloch energy $E_n(\mathbf{k})$ into a Hamiltonian[5] by replacing $\mathbf{k}$ by an operator $\hat{\mathbf{k}}$, and to ignore the off-diagonal matrix elements. In an r-space representation we can write this operator as[6]

$$\mathbf{k} \to \hat{\boldsymbol{\kappa}} - \frac{e}{\hbar c}\mathbf{A}(\mathbf{r}) = -i\frac{\partial}{\partial \mathbf{r}} - \frac{e}{\hbar c}\mathbf{A}(\mathbf{r}). \tag{18.9}$$

After picking some gauge such as $\mathbf{A}(\mathbf{r}) = (1/2)\,\mathbf{H}\times\mathbf{r}$, a direct calculation of the commutator $[\hat{k}_x, \hat{k}_y]$ yields

$$\hat{k}_x\hat{k}_y - \hat{k}_y\hat{k}_x = i\frac{e}{\hbar c}H_z \cdots. \tag{18.10}$$

After this substitution our r-space Hamiltonian is formally given by[7]

$$\hat{H}_{nn}(\hat{\mathbf{k}}) = \hat{E}_n\left(-i\nabla - \frac{e}{\hbar c}\mathbf{A}(\mathbf{r})\right). \tag{18.11}$$

To interpret this expression we could utilize Eq. (10.13), which in operator form in r-space is written as

$$\begin{aligned}\hat{H}_{nn}(\mathbf{k}) &= \sum_{\mathbf{R}} E_{n\mathbf{R}}e^{i\hat{\mathbf{k}}\cdot\mathbf{R}} \\ &= \sum_{\mathbf{R}} E_{n\mathbf{R}}\exp\left[\left(-i\nabla - \frac{e}{\hbar c}\mathbf{A}(\mathbf{r})\right)\cdot\mathbf{R}\right].\end{aligned} \tag{18.12}$$

[5] This approach was first applied by R.E. Peierls in, Zur theoric des diamagnetismus von leitungselektronen, I, *Z. Phys.* **80**, 763 (1933). The theory was further developed in Single band motion of conduction electrons in a uniform magnetic field, P.G. Harper, *Proc. Phys. Soc. A* **68**, 879 (1955).

[6] We could equally well carry out the subsequent analysis in a k-space representation (see Section 10.3.4) and write $\mathbf{k} \to \boldsymbol{\kappa} - (e/\hbar c)\hat{\mathbf{A}}(\hat{\mathbf{r}}) = \boldsymbol{\kappa} - (e/\hbar c)\hat{\mathbf{A}}\,(i\partial/\partial\boldsymbol{\kappa})$. This approach is discussed in *Statistical Physics*, Part II, E.M. Lifshitz and L.P. Pitaevskii, Sections 56 and 57, Pergamon Press (1980).

[7] Note that ambiguities can arise from the ordering of the components of the operators when expanding the Hamiltonian. However, due to the commutation relation (18.10) their effect is of higher order in the magnetic field and can be ignored in the weak field limit. The off-diagonal elements of the Hamiltonian, which we are ignoring, are expected to be small on the scale of a/r_c, where a is a typical lattice constant and r_c is the cyclotron radius.

As an example consider a simple cubic crystal. If we limit ourselves to the first two terms in the expansion (10.13) we obtain

$$E_n(\mathbf{k}) = E_{n000} + 2E_{n\,100}\left(\cos k_x a + \cos k_y a + \cos k_z a\right). \tag{18.13a}$$

If we further restrict to the region near the Γ point, and write $E'_{n0} = E_{n000} + 2E_{n100}$ and $m^* = \hbar^2/2E_{n\,100}a^2$, the resulting Hamiltonian operator is

$$\hat{H}_{nn} = E'_{n0} + \frac{\hbar^2}{2m^*}\left(-i\nabla - \frac{e}{\hbar c}\mathbf{A}(\mathbf{r})\right)^2, \tag{18.13b}$$

which has the same form as that for free electrons.[8]

Given the Hamiltonian (18.13b) we can calculate the time derivatives of **r** and $\hat{\boldsymbol{\kappa}}$ using the Heisenberg equations of motion:

$$\dot{\mathbf{r}} = \frac{i}{\hbar}\left[\hat{H}\left(-i\nabla - \frac{e}{\hbar c}\mathbf{A}(\mathbf{r})\right), \mathbf{r}\right] \tag{18.14a}$$

and

$$\dot{\hat{\boldsymbol{\kappa}}} = \frac{i}{\hbar}\left[\hat{H}\left(-i\nabla - \frac{e}{\hbar c}\mathbf{A}(\mathbf{r})\right), \boldsymbol{\kappa}\right], \tag{18.14b}$$

where $\hat{\boldsymbol{\kappa}}$ and **r** satisfy the commutation relations, $\left[r_i, \hat{\kappa}_j\right] = \left[r_i, -i\,\partial/\partial r_j\right] = i\delta_{ij}$.

Equations (18.12a,b) can be evaluated as follows. (i) We can start with Eq. (10.13) wherein the Bloch energies $E_n(\mathbf{k})$ are expanded as a Fourier series in the real-space lattice vectors **R**. (ii) Each of the exponential terms in this expansion is in turn expanded in a power series in $\mathbf{k}\cdot\mathbf{R}$ in the usual way, followed by the substitution $\mathbf{k} = \boldsymbol{\kappa} - (e/\hbar c)\mathbf{A}(\mathbf{r})$. (iii) We then make the substitution $\boldsymbol{\kappa} \to -i\partial/\partial\mathbf{r}$. (iv) The commutators are then calculated term by term using the canonical commutation rule. (v) The surviving terms are then regrouped to form exponentials. Equations (18.14a,b) can then be expressed as

$$\dot{\hat{\mathbf{r}}} = \frac{\partial\hat{H}}{\partial\hat{\boldsymbol{\kappa}}} \tag{18.15a}$$

and

$$\dot{\hat{\boldsymbol{\kappa}}} = -\frac{\partial\hat{H}}{\partial\mathbf{r}}, \tag{18.15b}$$

where

$$\frac{\partial\hat{H}}{\partial\hat{\boldsymbol{\kappa}}} = \left[\frac{\partial}{\partial\boldsymbol{\kappa}}E\left(\boldsymbol{\kappa} - \frac{e}{\hbar c}\mathbf{A}(\mathbf{r})\right)\right]_{\boldsymbol{\kappa}\to -i\frac{\partial}{\partial\mathbf{r}}}, \tag{18.16}$$

and similarly for $\partial\hat{H}/\partial\mathbf{r}$; Eqs. (18.15a,b) are Hamilton's equations in operator form.

[8] We will solve for the wave functions and energy levels associated with a 2d version of this Hamiltonian in Section 48.1.3.

We must also consider the case when an external electric field, $\mathbf{E}(\mathbf{r})$, is present. In the same spirit as incorporating a magnetic field, we can write

$$\hat{H}_{nn} = E_n(\hat{\mathbf{k}}) + e\phi(\mathbf{r}), \tag{18.17}$$

where $\mathbf{E}(\mathbf{r}) = -\nabla\phi(\mathbf{r})$, and we again restrict to the diagonal elements in the band index where $\hat{\kappa} = -i\partial/\partial\mathbf{r}$. Alternatively, by writing

$$\mathbf{E}(\mathbf{r}, t) = -\nabla\phi + \frac{1}{c}\frac{\partial \mathbf{A}}{\partial t}, \tag{18.18}$$

we can introduce an electric field via a time-dependent vector potential in these equations. As with the case of the magnetic field, such approximations are valid only in the weak field case; in particular a "breakdown" in which an electron makes a transition between bands (but now with *different* energies) can occur in high electric fields (referred to as Zener breakdown, see Appendix 18A).

18.3 Quasiclassical equation of motion

In the quasiclassical limit the equations of motion adopt their classical form,

$$\hbar\dot{\kappa} = -\frac{\partial H}{\partial \mathbf{r}} \tag{18.19a}$$

and

$$v = \frac{\partial H}{\hbar\partial\kappa}, \tag{18.19b}$$

where we take $H = E_n(\mathbf{k}) = E_n(\kappa - (e/\hbar c)\mathbf{A}(\mathbf{r}))$. Combining (18.19a,b), we can write

$$\dot{\kappa} = \sum_i \frac{\partial H}{\partial k_i}\frac{\partial k_i}{\partial \mathbf{r}} = \frac{e}{\hbar c}\sum_i v_i \nabla A_i(\mathbf{r}). \tag{18.20a}$$

We rewrite the left side of (18.20a) as

$$\begin{aligned} \dot{\kappa} &= \dot{\mathbf{k}} + \frac{e}{\hbar c}\dot{\mathbf{A}}(\mathbf{r}) \\ &= \dot{\mathbf{k}} + \frac{e}{\hbar c}(\mathbf{v}\cdot\nabla)\mathbf{A}(\mathbf{r}), \end{aligned} \tag{18.20b}$$

where the second term in the second step appears even though $\mathbf{A}$ has no explicit time dependence, since we are taking a total (also called "hydrodynamic") time derivative. Combining (18.20a,b) and using the vector identity

$$\sum_i v_i \nabla A_i - (\mathbf{v}\cdot\nabla)\mathbf{A} = \mathbf{v}\times(\nabla\times\mathbf{A}) = \mathbf{v}\times\mathbf{H},$$

we obtain

$$\hbar\dot{\mathbf{k}} = \frac{e}{c}\mathbf{v} \times \mathbf{H} \tag{18.21a}$$

with

$$\mathbf{v} = \frac{\partial E}{\hbar \partial \mathbf{k}}, \tag{18.21b}$$

which is a Lorentz-like equation of motion, with **k** being the *kinetic* (rather than the *canonical*) momentum.[9] If **A** has an explicit time dependence, the total time derivative entering Eq. (18.20) will contain an additional term on the right side, which via Eq. (18.18) results in an electric field contribution, and Eq. (18.21a) becomes

$$\hbar\dot{\mathbf{k}} = e\left(\mathbf{E} + \frac{1}{c}\mathbf{v} \times \mathbf{H}\right). \tag{18.22}$$

18.4 Orbits in a magnetic field

If we take the scalar product of both sides of Eq. (18.21a) with the velocity **v** and use (18.21b), we obtain

$$\frac{\partial E(\mathbf{k})}{\hbar \partial \mathbf{k}} \cdot \hbar\dot{\mathbf{k}} = \frac{\partial E(\mathbf{k})}{\partial t} = 0. \tag{18.23a}$$

Similarly, taking the dot product with **H** (assumed time independent), we obtain

$$\mathbf{H} \cdot \dot{\mathbf{k}} = \frac{\partial(\mathbf{H} \cdot \mathbf{k})}{\partial t} = 0. \tag{18.23b}$$

Together, Eqs. (18.23a,b) tell us that each particle traverses an "orbit" defined by the intersection of some surface of constant energy (to which the particle motion is constrained by the first equation) with a plane defined by the equation

$$\mathbf{k} \cdot \mathbf{H} = k_H. \tag{18.24}$$

Depending on the shape of the constant energy surfaces we may have two types of orbits. If the condition (18.24) leads to a curve that closes on itself, we say that we have a *closed orbit*, as shown in Fig. 18.1(a). If, on the other hand, the curve makes contact with the Brillouin zone in the manner shown in Fig. 18.1(b), we have an *open orbit*. As drawn, an electron would, on average, advance, depending on which side of the Fermi surface it is located, in the $-k_x$ or $+k_x$ direction in wave vector space, while drifting in the +y or −y direction in real space respectively.

[9] One sometimes regards the canonical (or total) momentum, $\boldsymbol{\pi} = \hbar\boldsymbol{\kappa}$, as consisting of two parts: a kinetic (or mechanical) momentum, $\mathbf{p} = \hbar\mathbf{k}$, and an electromagnetic momentum, $(e/c)\mathbf{A}$. Then, $\boldsymbol{\pi} = \mathbf{p} + (e/c)\mathbf{A}$.

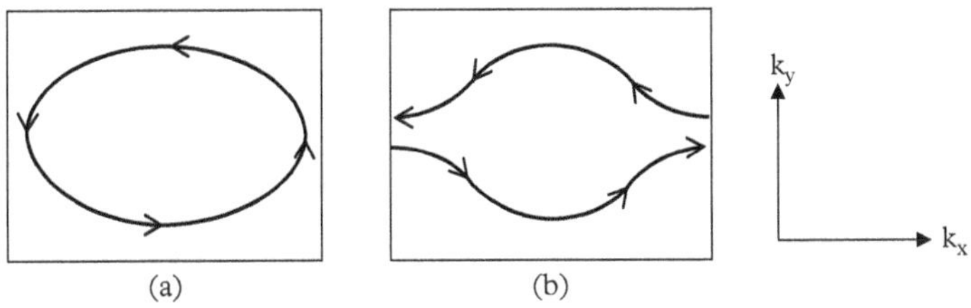

Figure 18.1 *The heavy lines depict a constant energy orbit in a plane perpendicular to a magnetic field, with the arrows showing the direction of motion (which depends on the sign of the carrier); (a) shows a closed orbit, while (b) shows an open orbit.*

We now evaluate the time required to completely traverse a closed orbit, referred to as the cyclotron period. This period is clearly given by (see Fig. 18.2)

$$
\begin{aligned}
t &= \oint dt = \oint \frac{dk_{//}}{dk_{//}/dt} \\
&= \oint \frac{\hbar dk_{//}}{\left|\frac{e}{c}\mathbf{v}_\perp \times \mathbf{H}\right|} = \frac{c\hbar^2}{|e|H}\oint \frac{dk_{//}}{\partial E/\partial k_\perp} = \frac{c\hbar^2}{|e|H}\frac{\partial}{\partial E}\oint dk_{//}dk_\perp \\
&= \frac{c\hbar^2}{|e|H}\frac{\partial A_k}{\partial E};
\end{aligned}
\tag{18.25}
$$

here, A_k is the cross-sectional area of the orbit in k-space and

$$
\frac{\partial E}{\hbar \partial \mathbf{k}_\perp} = \mathbf{v}_\perp, \tag{18.26}
$$

with $\mathbf{v}_\perp = \mathbf{v} - v_H \mathbf{n}_H$, $v_H = \mathbf{v} \cdot \mathbf{n}_H$, and $\mathbf{n}_H = \mathbf{H}/|\mathbf{H}|$. In analogy with the angular frequency of a free electron in a magnetic field, we define a *cyclotron mass*, m_c, as

$$
\omega_c = \frac{|e|H}{m_c c}. \tag{18.27}
$$

Using $\omega_c = \frac{2\pi}{t}$, it then follows from Eq. (18.25) that m_c is given by

$$
m_c = \frac{\hbar^2}{2\pi}\frac{\partial A_k}{\partial E}. \tag{18.28a}
$$

For free electrons, where $E = \hbar^2 k^2/2m$ and $A_k = \pi k^2$, Eq. (18.28) yields the free electron mass, m.

The cyclotron mass may be measured experimentally in the following way. If a metallic sample is placed in a static magnetic field, the electrons in the metal will execute cyclotron orbits (see Fig. 18.3). If an oscillating electric field (typically enhanced by a microwave cavity) is applied, we might expect electrons to resonantly absorb energy when the microwave frequency, ω_μ, matches the cyclotron frequency, ω_c. However, the microwaves only penetrate a distance of order a skin depth, δ, whereas the orbit diameter is generally much larger. Suppose our metal has a smooth free surface and that the magnetic field is aligned parallel with it. In this case the electrons receive a "kick" from the microwave field *only* when

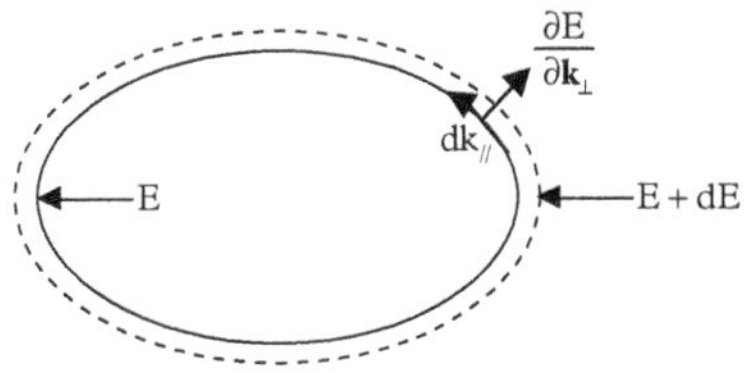

Figure 18.2 *Constant energy surfaces at* E *(solid curve) and* E + dE *(dashed curve). Also shown are the differential line element along the orbit* $dk_{//}$ *and the direction of the in-plane gradient,* $\hbar v = dE/dk_\perp$.

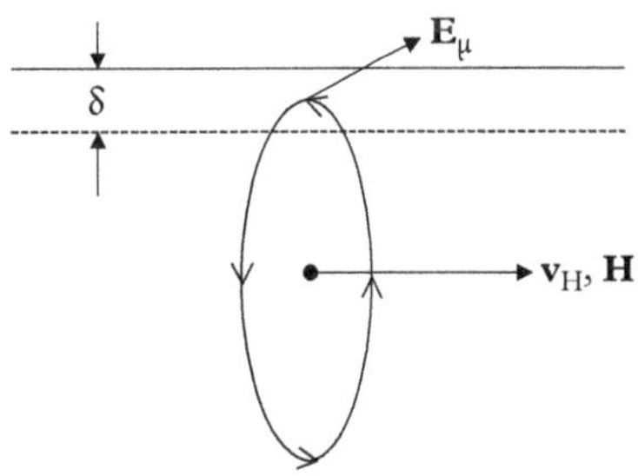

Figure 18.3 *The geometry of the Azbel-Kaner effect. With a mutually perpendicular radio frequency or microwave electric field,* E_μ, *and a static magnetic field,* H, *both of which are parallel to the surface, groups of electrons will receive a "kick" each time they enter the skin depth. If* $\omega_\mu = n\omega_c$, *the electrons will resonantly absorb energy.*

they are inside the skin depth. However, the microwave field can go through *multiple* periods, t_μ, during the time the electron goes through a single cyclotron period, t_c. Under this condition we can have resonances whenever $nt_\mu = t_c$, or equivalently $\omega_\mu = n\omega_c$. This phenomenon is called the Azbel–Kaner effect, after its discoverers.

In principle, the orbits of all electrons on the Fermi surface will participate, and in general the cyclotron frequency is different for each orbit (i.e., it is a function of k_H). In practice, electrons having a velocity component, v_H, will drift into and out of the surface (or the magnetic field can be tipped slightly causing such electrons to scatter from the, usually atomically rough, surface) and their contribution will be suppressed. Thus the resonance determines the cyclotron frequency of so-called *extremal orbits* for which $v_H = 0$. The quantity v_H may be calculated from the rules for differentiation of an implicit function. Suppose we are given the function $A_k(k_H, E)$; then,

$$v_H = \left(\frac{\partial E}{\hbar \partial k_H}\right)_{A_k} = -\frac{1}{\hbar}\left(\frac{\partial E}{\partial A_k}\right)_{k_H}\left(\frac{\partial A_k}{\partial k_H}\right)_E = -\frac{\hbar}{2\pi m_c}\left(\frac{\partial A_k}{\partial k_H}\right)_E, \tag{18.28b}$$

which clearly vanishes for an extremal orbit. The experiment must be performed in very pure metals at low temperatures to avoid electron scattering, which broadens and ultimately destroys the resonance (see Section 3.5).

18.5 Orbit quantization in a magnetic field

It is natural to expect the motion transverse to a magnetic field involving closed orbits to be quantized. The quantization of the transverse motion of free electrons in a magnetic field is treated in most texts on quantum mechanics and will not be reproduced here. However, one can essentially write down the answer based on a simple argument. Classically, the motion consists of circular orbits the center of which drifts along the magnetic field. The circular motion can be viewed as two simple harmonic oscillators 90° out of phase. Quantizing this motion we have

$$E_n = \hbar\omega_c\left(n + \frac{1}{2}\right) \tag{18.29}$$

(the zero-point energy is $\hbar\omega_c/2$, rather than $\hbar\omega_c$, as for a two-dimensional harmonic oscillator, since the phase relationship between the two oscillators is fixed; i.e., they behave as a single oscillator).[10] The drift part of the kinetic energy along **H** is $E_H = \hbar^2 k_H^2/2m$; thus,

$$\begin{aligned} E &= E_n + E_H \\ &= \left(n + \frac{1}{2}\right)\hbar\omega_c + \frac{\hbar^2 k_H^2}{2m}. \end{aligned} \tag{18.30}$$

These levels are called Landau levels.

[10] A discussion of the quantization of free-electron orbits in a magnetic field in two dimensions is given in Section 48.1.2.

For the quantization of a general constant energy surface, $E = E(\mathbf{k})$, we will use the Bohr–Sommerfeld rule, since the quantum numbers encountered in practice are usually large and hence a quasiclassical treatment is adequate. The appropriate canonical variables involve the component of $\boldsymbol{\kappa}$ and $\mathbf{r}$ tangent to the orbit and perpendicular to the magnetic field, which we will denote by $\kappa_{//}$ and $r_{//}$ (see Fig. 18.4). The quantization condition can then be written as

$$\oint \hbar\kappa_{//} d r_{//} = \oint (\hbar\mathbf{k} + \frac{e}{c}\mathbf{A}(\mathbf{r})) \cdot d\mathbf{r}_{//} = (n + \gamma)h, \qquad (18.31)$$

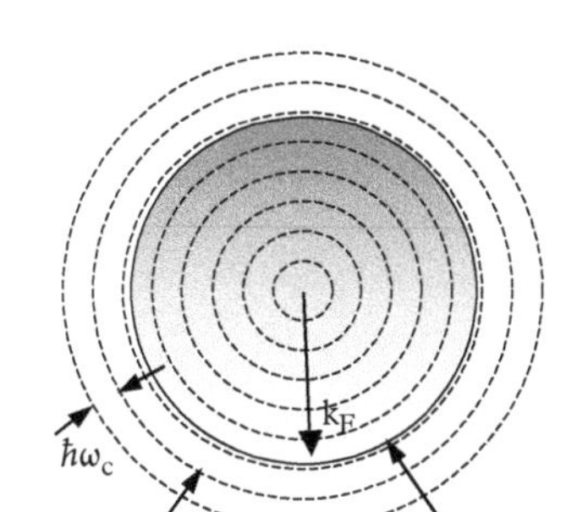

Figure 18.4 *Nested quantized energy levels in relation to the free electron sphere as viewed along the magnetic field direction. As the field increases, and with it the Landau level spacing, successive levels pass through the Fermi energy resulting in an oscillatory behavior of the thermodynamic and transport properties in the reciprocal of the magnetic field.*

where $\mathbf{k}$ represents the in-plane piece of the quasi-momentum and γ is a phase factor. Integrating the Lorentz–Newton equation of motion (18.22) with respect to time gives

$$\hbar\mathbf{k} = \frac{e}{c}\mathbf{r} \times \mathbf{H}; \qquad (18.32)$$

from (18.32) we see that the coordinate in $\mathbf{k}$-space is related to that in $\mathbf{r}$-space by a factor $eH/c\hbar$, but *rotated* by 90° in the plane of the orbit. If we now write the vector potential in the gauge $\mathbf{A}(\mathbf{r}) = \frac{1}{2}\mathbf{H} \times \mathbf{r}$, the left side of Eq. (18.31) becomes

$$\oint d\mathbf{r}_{//} \cdot \left[\frac{e}{c}\mathbf{r} \times \mathbf{H} + \frac{1}{2}\frac{e}{c}\mathbf{H} \times \mathbf{r}\right] = \frac{eH}{c}\left|\frac{1}{2}\int \mathbf{r}\times d\mathbf{r}\right| = \frac{eH}{c}A_r, \qquad (18.33)$$

where A_r is the area in $\mathbf{r}$-space. Combining (18.31) and (18.32) gives the Onsager–Lifshitz quantization rule,

$$A_r = \frac{ch}{eH}(n + \gamma); \qquad (18.34a)$$

applying Eq. (18.32) twice we may *scale* the areas in real space to those in $\mathbf{k}$-space, and (18.34a) becomes

$$A_k = \frac{2\pi eH}{c\hbar}(n + \gamma). \qquad (18.34b)$$

Note that in general $A_k = A_k(k_H)$.

We now discuss, qualitatively, the deHaas van Alphen effect, which is an oscillatory behavior of the magnetization (and other thermodynamic and transport properties) that is observed in pure metals at low temperatures. Associated with each quantized area $A(H, n, k_H)$ is an energy level $E(H,n,k_H)$. For a given H and n there is a value of k_H, which we will denote as k_H^c, where $E(H,k_H^c,n) = E_F$, in which E_F is again the Fermi energy. As the field increases at fixed n, the area $A(H,n,k_H^c)$ associated with this level increases and therefore k_H^c changes (for a spherical Fermi surface k_H^c moves toward zero). When k_H^c reaches a value where the area as a function of k_H is maximal (thereby satisfying the condition $\partial A(k_H)/\partial k_H = 0$) and can no longer increase (at the equator for a sphere), the solution for k_H^c of that level ceases to exist; the level *passes above the Fermi energy* and empties. The singular behavior associated with this phenomenon causes the various properties cited above to oscillate. A different case arises if the Fermi surface has a value of k_H where $A(H,n,k_H^c)$ passes through a minimum (where again $\partial A(k_H)/\partial k_H = 0$ vanishes); a local solution for k_H^c with increasing field then abruptly appears (rather disappearing) and the singular behavior also produces oscillations.

Having argued that only the extremal orbits where $\partial A(k_H)/\partial k_H = 0$ will contribute to the oscillatory magnetization, we need to form an expression for the period of these oscillations. Using Eq. (18.34b), consider the reciprocals of the field values at which two successive levels pass through the Fermi surface,

$$\frac{1}{H_{n+1}} = \frac{2\pi e}{c\hbar A_k}(n + 1 + \gamma) \tag{18.35a}$$

and

$$\frac{1}{H_n} = \frac{2\pi e}{c\hbar A_k}(n + \gamma)\,. \tag{18.35b}$$

Subtracting, we obtain the de Haas–van Alphen (dHvA) period, $\Delta\,(1/H)$, as

$$\frac{1}{H_{n+1}} - \frac{1}{H_n} \equiv \Delta\left(\frac{1}{H}\right) = \frac{2\pi e}{c\hbar A_k}; \tag{18.36a}$$

note the oscillations are periodic in the *reciprocal* of the magnetic field. One also introduces a (dHvA) frequency as

$$F = \frac{c\hbar A_k}{2\pi e}; \tag{18.36b}$$

thus the larger the extremal cross-sectional area, the more rapid are the oscillations with reciprocal magnetic field.

The energy separation between successive energy levels is $\hbar\omega_c$, and to avoid occupation of neighboring levels above the Fermi energy we require that $\hbar\omega_c \gtrapprox k_B T$, i.e., low temperatures (typically of order 1 K). In addition, electrons must complete cyclotron orbits requiring $\omega_c \tau \gtrapprox 1$, where τ is the electron scattering time. We will discuss the dHvA effect more rigorously in Section 20.1.

18.6 Magnetic breakdown

In some metals different sheets of the Fermi surface, associated with different bands, nearly contact each other at certain points in the Brillouin zone; this is shown schematically in Fig. 18.5. Locally we may model these surfaces by the parabolic forms

$$k_x = \pm\left(\frac{k_0}{2} + \frac{1}{2R_k}k_y^2\right), \tag{18.37}$$

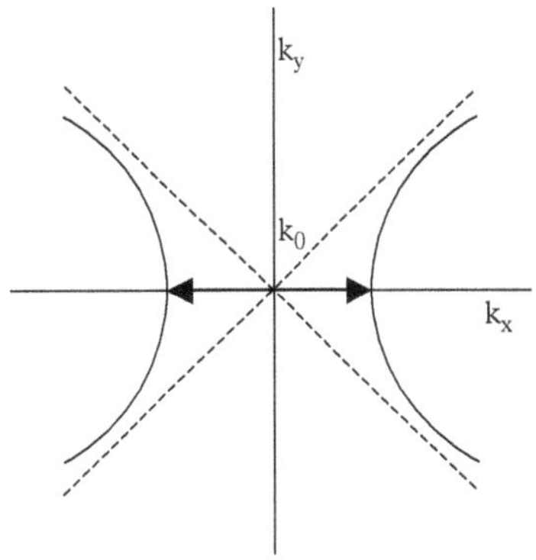

Figure 18.5 *Two near-contacting sheets of the Fermi surface.*

where R_k is the radius of curvature in k-space at the turning point of the orbit, which is typically of the same order as k_0. Such close contact can arise from a particularly weak crystal potential (as in some free-electron-like metals such as Mg) or as a result of spin–orbit coupling lifting what would otherwise be a degeneracy. An electron traversing an orbit in a magnetic field on one of these sheets can then, on encountering the near degeneracy, *tunnel* from that sheet to the adjacent sheet.[11]

We now derive a criterion for avoiding such tunneling. Our approach will be to exploit an analogy between Eq. (18.10) and the usual one-dimensional Heisenberg commutation relation, $[\hat{p}, \hat{q}] = i\hbar$; comparing these expressions, we can write $\hat{q} = \hat{k}_x\hbar c/|e|H_z$ and $\hat{p} = \hbar\hat{k}_y$. A classical, one-dimensional problem, yielding the same parabolic behavior in p and q space that (18A.20) has in k_x and k_y space, is the motion of a particle in a constant force field F, for which the Hamiltonian is $H = p^2/2m - Fq$; the associated trajectory in p–q space, which is the analogue of (18.37), has the form

$$q = \frac{p^2}{2mF} - \frac{E}{F}, \tag{18.38}$$

where E is the energy, and the radius of curvature at the turning point is $R = mF$. The Schrödinger equation associated with this Hamiltonian is

$$\left(-\frac{\hbar^2}{2m}\frac{d^2}{dq^2} - Fq\right)\psi(q) = E\psi(q). \tag{18.39}$$

Since we are interested in tunneling, we seek a solution of (18.39) in the classically forbidden region; by direct substitution one can verify that in the limit of large negative q the form

$$\psi(q) \propto q^{-1/4}e^{-(2/3)\sqrt{2mF/\hbar^2}|q|^{3/2}} \tag{18.40}$$

is an asymptotic solution to (18.39). Hence a characteristic length over which the wave function decays is then $q_c = (\hbar^2/mF)^{1/3} = (\hbar^2/R)^{1/3}$; the corresponding decay length in k-space is $k_c = (|e|H/\hbar c)q_c = (|e|H/\hbar c)\cdot(\hbar^2/R)^{1/3}$. Writing

[11] Magnetic breakdown in crystals, L.M. Falicov and M.H. Cohen, *Phys. Rev. Lett.* 7, 231 (1961).

$R = (eH\hbar/c)R_k \approx (eH\hbar/c)k_0$, we then have $k_c = (|e|\,H/\hbar c)^{2/3}(k_0)^{-1/3}$. The condition for vanishing tunneling is $k_0 \gg k_c$, which gives

$$(k_0)^2 \gg (|e|\,H/\hbar c). \tag{18.41}$$

ADDITIONAL READING

The theory of Bloch electrons in external fields is discussed in *Statistical Physics Part II*, E.M. Lifshitz and L.P. Pitaevskii, Chapter VI. Pergamon Press, NY (1980).

Perhaps the best discussion on electron dynamics is A.B. Pippard, *Dynamics of Conduction Electrons*, Documents in Modern Physics Theories, Gordon & Beach (1965). For an extended discussion of magnetic breakdown see: Magnetic Breakdown in Metals, R.W. Stark and L.M. Falicov, *Progress in Low Temperature Physics*, Vol. 5, chapter VI, Elsevier (1967).

Problems

18.1. Suppose the energy bands near the Fermi surface have the form

$$E = \frac{\hbar^2}{2m}\left(\alpha_{11}k_x^2 + \alpha_{22}k_y^2 + \alpha_{33}k_x^2\right),$$

where α_{11}, α_{22}, and α_{33} are parameters. Now assume we have a magnetic field **H** whose vector direction is specified by the usual angles θ and φ.

(a) Calculate the extremal cross-sectional area perpendicular to **H** as a function of θ and φ at an energy $E = E_F$.

(b) Calculate the cyclotron frequency by solving the Newton–Lorentz equation of motion (18.22); show that your result agrees with Eqs. (18.27) and (18.28a).

18.2. The Hamiltonian entering the Schrödinger equation in the presence of an electric field may be obtained either by introducing a scalar potential, $V(\mathbf{r}) = -e\mathbf{E}\cdot\mathbf{r}$ with $\mathbf{E} = -\nabla V(\mathbf{r}, t)$, or a vector potential, $\mathbf{A} = -c\mathbf{E}t$ with $\mathbf{E} = -\frac{1}{c}\frac{\partial \mathbf{A}}{\partial t}$. Show that with the latter approach the associated time-dependent Schrödinger wave equation in one dimension is

$$i\hbar\frac{\partial\Psi(x,t)}{\partial t} = -\frac{\hbar^2}{2m}\left(\frac{\partial}{\partial x} + \frac{ieE}{\hbar}t\right)^2\Psi(x,t)$$

and that it has a plane-wave-like solution of the form

$$\Psi(x,t) = e^{ik_0x}e^{-\frac{i}{\hbar}\int_0^t \varepsilon_{k(t')}dt'},$$

where $\varepsilon_{k(t)} = \frac{\hbar^2k^2(t)}{2m}$ and $k(t) = k_0 + \frac{eE}{\hbar}t$.

18.3. Derive the commutation relation given in Eq. (18.10).

18.4. Picture a point in k-space near a Brillouin zone face where two free-electron spheres centered in neighboring zones intersect, as shown in the adjoining figure.

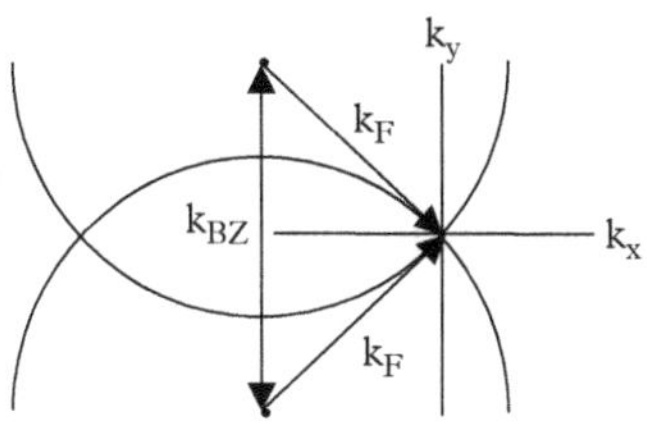

(a) Show that in a coordinate system centered at one of the intersection points of the two spheres, we can *locally* approximate the energy band structure associated with the two spheres as

$$E(\mathbf{k}) = E_F + |v_x|\, k_x + |v_y|\, k_y;$$
$$E(\mathbf{k}) = E_F + |v_x|\, k_x - |v_y|\, k_y,$$

where v_x and v_y are the components of the Fermi velocity at the intersection point and $E(\mathbf{k}) = E_F$ corresponds to being on one of the two surfaces. At this point we have a picture that corresponds to the dotted lines in Fig. 18.5.

(b) Find expressions for the components v_x and v_y of the Fermi velocity as a function of k_F and k_{BZ}.

(c) Now assume that there is a Fourier component of the underlying lattice potential associated with the Brillouin zone face we are considering. Show that to lowest order in k_x and k_y the equation for the two sheets of the Fermi surface adopts the form given in Eq. (18.37), and find expressions for R_k and k_0.

18.5. Starting with Eq. (18.13a), express the Hamiltonian $\hat{H}_{nn}$ in the presence of a vector potential $\mathbf{A}(\mathbf{r})$ for the tight-binding, simple cubic case in terms of an effective mass tensor for $\mathbf{k}$ in the vicinity of the following three points on the Brillouin zone surface: (a), $(\pi/a, 0, 0)$; (b), $(\pi/a, \pi/a, 0)$, and (c), $(\pi/a, \pi/a, \pi/a)$.

Appendix 18A Zener breakdown

In treating electrical transport properties (to be discussed in Chapters 19 and 22) one usually assumes that in the presence of an electric field, **E**, scattering from impurities constrains the electron distribution function to deviate only slightly from its equilibrium form; i.e., the electron wave vectors obey a Boltzmann distribution in a semiconductor and lie close to the Fermi surface in a metal. However, in very strong fields electrons can be very far from equilibrium and even tunnel between adjacent energy bands, particularly when there is another state nearby in k-space which is separated by only a small energy barrier. This latter effect is called *Zener breakdown* (C. Zener, 1932). Because of its great technical importance (it is the basis of laser diodes and so-called Zener diodes (See Sections 23.4.3 and 23.4.5) we will treat this process in some detail.

In the absence of electron scattering, truly free electrons would be continually accelerated by an electric field; i.e., their wave vector would increase linearly with time according to

$$\mathbf{k}(t) = \left(\mathbf{k}_0 + \frac{e}{\hbar}\mathbf{E}t\right), \tag{18A.1}$$

where $\mathbf{k}_0$ is the electron wave vector at t = 0. The associated time-dependent Schrödinger wave equation would be

$$i\hbar\frac{\partial\psi}{\partial t} = -\frac{\hbar^2}{2m}\nabla^2\psi - eEx\psi, \tag{18A.2}$$

where we have assumed $\mathbf{E}$ is parallel to x. Equation (18A.2) can be solved exactly. For simplicity, in what follows, we restrict our discussion to one dimension. We seek a plane-wave-like solution with k(t) given by (18A.1) of the form

$$\psi\,(x) = a\,(t)\,e^{i\left[\left(k_0+\frac{eE}{\hbar}t\right)x\right]}; \tag{18A.3}$$

inserting this form into (18A.2) yields

$$i\hbar\frac{da(t)}{dt} = \frac{\hbar^2}{2m}\left(k_0 + \frac{eEt}{\hbar}\right)^2 a(t) \tag{18A.4a}$$

or

$$i\hbar\frac{da(t)}{dt} = \varepsilon_{k(t)}a(t), \tag{18A.4b}$$

where $\varepsilon_{k(t)} \equiv \frac{\hbar^2}{2m}\left(k_0 + \frac{eEt}{\hbar}\right)^2$.

To illustrate the effect of a lattice we consider a one-dimensional, nearly-free-electron model with only a single Fourier component, V_K, of the periodic potential. The Hamiltonian of such a system is

$$\hat{H} = -\frac{\hbar^2}{2m}\nabla^2\psi - eEx + V\left(e^{iKx} + e^{-iKx}\right). \tag{18A.5}$$

We will restrict ourselves to motion near the Brillouin zone point, $k_{BZ} = +K/2$, where we approximate the wave function as a superposition of the a plane wave with wave vector k(t) and a second plane wave with wave vector k(t) – K, with which it is degenerate at $k = k_{BZ}$; i.e.,

$$\Psi(x,t) = a(t)\frac{1}{\sqrt{a}}e^{i\left(k+\frac{eE}{\hbar}t\right)x} + b(t)\frac{1}{\sqrt{a}}e^{i\left(k+\frac{eE}{\hbar}t-K\right)x}, \tag{18A.6}$$

where we have normalized the basis functions to a unit cell dimension, a; we rewrite this expression as

$$\Psi(x,t) = a(t)\,\psi_{k(t)}(x) + b(t)\psi_{k(t)-K}(x). \tag{18A.7}$$

Inserting this form in the time-dependent Schrödinger equation, $i\hbar\dot{\Psi}(x,t) = \hat{H}\Psi(x,t)$, with the Hamiltonian (18A.6), and taking matrix elements with respect to $\psi_{k(t)}$ and $\psi_{k(t)-K}$, we obtain a set of equations for the coefficients a (t) and b(t),

$$i\hbar\begin{bmatrix}\dot{a}(t)\\ \dot{b}(t)\end{bmatrix} = \begin{bmatrix}\varepsilon_{k(t)} & V_K\\ V_K^* & \varepsilon_{k(t)-K}\end{bmatrix}\begin{bmatrix}a(t)\\ b(t)\end{bmatrix}. \tag{18A.8}$$

As a first step in integrating these equations we write

$$a(t) = \alpha(t)e^{-\frac{i}{\hbar}\int \varepsilon_{k(t')}dt'} \tag{18A.9a}$$

and

$$b(t) = \beta(t)e^{-\frac{i}{\hbar}\int \varepsilon_{k(t')-K}dt'} \tag{18A.9b}$$

and substitute these forms into (18A.8), obtaining the pair of equations

$$i\hbar\frac{\partial\alpha(t)}{\partial t} = V_K\beta(t)e^{-\frac{i}{\hbar}\int\left[\varepsilon_{k(t')-K}-\varepsilon_{k(t')}\right]dt'} \tag{18A.10a}$$

and

$$i\hbar\frac{\partial\beta(t)}{\partial t} = V_K^*\alpha(t)e^{+\frac{i}{\hbar}\int\left[\varepsilon_{k(t')-K}-\varepsilon_{k(t')}\right]dt'}. \tag{18A.10b}$$

Differentiating (18A.10b) with respect to time gives

$$i\hbar\frac{\partial^2\beta(t)}{\partial t^2} = V_k^* e^{+\frac{i}{\hbar}\int\left[\varepsilon_{k(t')-K}-\varepsilon_{k(t')}\right]dt'}\left[\frac{\partial\alpha(t)}{\partial t} + \frac{i}{\hbar}\left(\varepsilon_{k(t)-K} - \varepsilon_{k(t)}\right)\alpha(t)\right],$$

and inserting $\frac{\partial\alpha(t)}{\partial t}$ from (18A.10a) and α(t) from (18A.10b), we get

$$\frac{\partial^2\beta(t)}{\partial t^2} + i\omega(t)\frac{\partial\beta(t)}{\partial t} + \omega_K^2\beta(t) = 0, \tag{18A.11}$$

where

$$\omega_K^2 = \frac{|V_k|^2}{\hbar^2} \tag{18A.12a}$$

and

$$\begin{aligned}\omega(t) &\equiv \frac{1}{\hbar}\left(\varepsilon_{k(t)} - \varepsilon_{k(t)-K}\right) \\ &= \frac{\hbar}{2m}\left(-K^2 + 2Kk_0 + 2K\frac{eEt}{\hbar}\right).\end{aligned} \tag{18A.12b}$$

To proceed with our discussion we must make some assumptions. First we fix the origin of time by setting $k_0 = k_{BZ}$; i.e., at $t = 0$ the electron is at the zone point. Now suppose that at some initial time $t_i \ll 0$ the system is in the state $\psi_{k(t)}(x)$, where $k(t) \ll k_{BZ}$; then $|\alpha(t_i)|^2 \cong 1$ and $|\beta(t_i)| = 0$. As time advances, the electron is accelerated to the vicinity of the Brillouin zone point, where it either Bragg reflects to the point $k(t) = -K$, or tunnels through the energy barrier $2|V_K|$. Therefore at some later time, $t_f \gg 0$, the electron will have either adiabatically transformed into the plane wave state $\psi_{k(t)-K}(x)$ with probability $|\beta(t_f)|^2$ (no tunneling), or remained in the plane wave state $\psi_{k(t)}(x)$ with probability $|\alpha(t_f)|^2 = 1 - |\beta(t_f)|^2$ (tunneling or breakdown).[12] To evaluate these probabilities we must find the appropriate asymptotic solutions of Eq. (18A.11) as $t_i = -\infty$ and $t_f = +\infty$. We first transform this equation according to

$$\beta(t) = e^{-i\int dt'\dot{\omega}(t')}U(t),$$

whereupon it becomes

$$\frac{\partial^2 U(t)}{\partial t^2} + \left(\omega_K^2 - \frac{i\dot{\omega}}{2} + \frac{\dot{\omega}^2}{2}t^2\right)U(t) = 0, \tag{18A.13a}$$

where we write

$$\dot{\omega} = \frac{d\omega(t)}{dt}. \tag{18A.13b}$$

By redefining the parameters in this equation as

$$z \equiv \dot{\omega}^{1/2}e^{-i\pi/4}t;\ n \equiv \frac{i\omega_K^2}{\dot{\omega}}, \tag{18A.14}$$

we obtain the more familiar Weber–Whittaker equation,

$$\frac{\partial^2 U(z)}{\partial z^2} + \left(n + \frac{1}{2} - \frac{1}{4}z^2\right)U(z) = 0. \tag{18A.15}$$

[12] We recall that the *band states* in the present case correspond to symmetric and antisymmetric linear combinations of $\psi_k(x)$ and $\psi_{k-K}(x)$; remaining in the state $\psi_k(x)$ in passing from $k < k_{BZ}$ to $k > k_{BZ}$ then corresponds to an inter-band transition.

As discussed, we seek a solution to this equation satisfying the boundary conditions

$$\alpha(t \to -\infty) = 1; |\beta(t \to -\infty)| = 0. \tag{18A.16a,b}$$

One class of solutions to (18A.15) is the so-called Whittaker functions $D_n(z)$;[13] however, these forms do not vanish for $t \to -\infty$. Now Eq. (18A.15) is invariant under the simultaneous substitutions $z \to \pm iz$ and $n \to -n-1$ and the functions so generated, $D_{-n-1}(\pm iz)$, form a second class of solutions which can be chosen to have the correct behavior; hence we write

$$U(z) = AD_{-n-1}(\pm iz), \tag{18A.17}$$

where the constant A is fixed by (18A.16a). The appropriate asymptotic behavior of the $D_{-n-1}(\pm iz)$ functions is given by[14]

$$D_{-n-1}(i\,|z|\,e^{-3\pi i/4})_{\,|z|\to\infty} = e^{(n+1)\pi i/4}e^{i|z|^2/4}\,|z|^{-n-1}; (t \to -\infty\,;\dot{\omega} > 0) \tag{18A.18a}$$

and

$$D_{-n-1}(i\,|z|\,e^{3\pi i/4})_{\,|z|\to\infty} = e^{3(n+1)\pi i/4}e^{-i|z|^2/4}\,|z|^{-n-1} + \frac{\sqrt{2\pi}}{\Gamma(n+1)}e^{n\pi i/4}e^{i|z|^2/4}\,|z|^n\,; (t \to +\infty; \dot{\omega} > 0). \tag{18A.18b}$$

Inserting the form (18A.17), with D_{-n-1} given by (18A.18a), into Eq. (18A.10a), integrating, and requiring (18A.16a), gives

$$|A| = \gamma^{1/2}e^{-\pi\gamma/4}; \ \gamma \equiv \frac{\omega_K^2}{\dot{\omega}}. \tag{18A.19}$$

With the availability of $|A|$ and using (18A.18b) we find

$$\begin{aligned}|\beta(t \to +\infty)|^2 &= \frac{2\pi\gamma}{\Gamma(i\gamma+1)\Gamma(-i\gamma+1)} = 2e^{-\pi\gamma}\sinh(\pi\gamma)\\ &= 1 - e^{-2\pi\gamma}\\ &= 1 - P,\end{aligned}$$

where we have used the identity[15] $\Gamma(1+iy)\Gamma(1-iy) = \frac{\pi y}{\sinh(\pi y)}$; as we will see, P is the breakdown probability, which we can rewrite as

$$\begin{aligned}P &= |\alpha(t_f \to +\infty)|^2 = e^{-2\pi\gamma}\\ &= \exp\left(-|V_K|^2\left[\hbar\frac{d}{dt}\left(\varepsilon_{k(t)} - \varepsilon_{k(t)-K}\right)\right]^{-1}\right)\\ &= \exp\left(-\frac{m\,|V_K|^2}{\hbar^2 KeE}\right).\end{aligned} \tag{18A.20}$$

[13] These functions are described in *Modern Analysis*, E.T. Whittaker and G.N. Watson, pp. 347–349, Cambridge University Press, Cambridge, UK, (1952).

[14] Slightly different forms apply for the case $\dot{\omega} < 0$.

[15] M. Abramowitz and I. Stegun, *Handbook of Mathematical Functions*, p. 264. Dover, NY (1972).

Hence, when $P \ll 1$ (small electric fields), $|\beta(t \to +\infty)|^2 \cong 1$; i.e., there is a very high probability for the electron to adiabatically transform into the Bragg reflected wave. Were there no electron scattering (which limits the acceleration) the electron would simply oscillate back and forth in the Brillouin zone between $+k_{BZ}$ and $-k_{BZ}$, a phenomena referred to as a *Bloch oscillation.*

On the other hand, the probability of breakdown increases with increasing electric field and for fields of order 10^7 volts/cm, breakdown begins to dominate for an energy gap of order a volt. Such high fields cannot readily be achieved in bulk materials but can occur in semiconductor p–n junctions.

We note in passing that the exponent in P is proportional to E^{-1} and hence cannot be obtained by a perturbation expansion in the E.

ADDITIONAL READING

The discussion of Zener breakdown in Appendix 18A follows the original treatment by C. Zener, *Proc. R. Soc. Lond. A* **137**, 696 (1932). There the problem studied was an analogous one in which transitions occur between two states which become nearly degenerate as a function of inter-particle separation during a binary collision. The problem is also discussed in L.D. Landau and E.M. Lifshitz, *Quantum Mechanics*, Pergammon Press, under the topic *Predissociation.* The calculation is summarized (and commented upon) in *Concepts in Solids*, P.W. Anderson and W.A. Benjamin, NY (1963). A more heuristic derivation is given by J. Ziman in *Principles of the Theory of Solids*, p. 163. Cambridge University Press, Cambridge, UK, (1964).

Electronic Transport Properties Governed by Static Scattering Centers

19

There are several mechanisms by which electrons in a crystal lattice may be scattered from one Bloch state to another. Among these are (i) electron–impurity, (ii) electron–phonon, and (iii) electron–electron scattering; in this section we restrict ourselves to electron–impurity scattering. We are using the term impurity loosely. It may be a "point" defect, examples being (i) a foreign atom, (ii) a missing host atom (called a vacancy), or (iii) a host atom that is located at a site other than its usual crystallographic site, such as a site inbetween other atoms (called an interstitial defect), or a lattice site normally occupied by a different atom (referred to as an anti-site defect). In addition one may consider larger structures such as (i) a surface, (ii) an interface between one part of a crystal and a second part which is not simply a translation of the first (referred to as a grain boundary), (iii) stacking faults (out of sequence stackings of successive crystallographic planes), or (iv) various kinds of dislocations. All of these electron scattering mechanisms lead to the phenomenon of *residual resistance*, a non-vanishing electrical resistivity at absolute zero. Here we will restrict ourselves to point defects that are rigidly bound to the rest of the lattice and which have no internal electronic modes that can be excited at low temperatures. In this case the scattering of a conduction electron will be *elastic*; i.e., there is no energy exchange between a conduction electron and the scattering center. The scattered electron wave vector $\mathbf{k}$ changes but it remains on the Fermi surface, including the possibility of transitioning to a different sheet.

19.1 The semi-classical approach: the distribution function

In this section we will employ a semi-classical description of the transport of electrical and thermal currents in metals, based on the Boltzmann equation, where the latter will be obtained shortly. This mode of description involves a *distribution function*, $f(n, \mathbf{k}, \mathbf{r}, t)$, which measures the number of electrons with band index n, wave vector $\mathbf{k}$, and position $\mathbf{r}$, that reside in a unit volume of the wave vector-position phase space at time t; in what immediately follows we will restrict

The Physics of Solids. First Edition. J. B. Ketterson.

ourselves to a single band and drop the band index; it will usually be straightforward to add a summation over the bands in the expressions that follow. In equilibrium, f is given by the Fermi function

$$f_0(\varepsilon) = \frac{1}{e^{(\varepsilon-\mu)/k_B T} + 1}, \tag{19.1}$$

where here we have used ε for the energy.

A sense in which a distribution function based description is semi-classical is that a simultaneous specification of the position and momentum of an electron contradicts the uncertainty principle ($\Delta k_x \Delta x \geq 1$ etc.), and hence such a description is permissible only for situations in which the distribution function does not vary too rapidly in space and time. It will also be semi-classical in that electron interference effects will not appear; electron wave functions will enter only indirectly. Rigorous, quantum mechanical approaches require a more mathematically sophisticated apparatus. We will give some simplified descriptions later, in Chapter 43.

The total number of particles, N, the number density, n(**r**,t), and the electron current density, **j**(**r**,t), are given in terms of the distribution function by the following expressions which follow from the usual prescription for integrating over phase space and where the factor 2 arises from the inclusion of electron spin,

$$N = \frac{2}{(2\pi)^3} \int d^3k d^3r\, f(\mathbf{k}, \mathbf{r}, t), \tag{19.2}$$

$$n(\mathbf{r}, t) = \frac{2}{(2\pi)^3} \int d^3k\, f(\mathbf{k}, \mathbf{r}, t), \tag{19.3}$$

and

$$\mathbf{j}(\mathbf{r}, t) = \frac{2e}{(2\pi)^3} \int d^3k\, \mathbf{v}(\mathbf{k}) f(\mathbf{k}, \mathbf{r}, t). \tag{19.4}$$

19.2 The Boltzmann equation

To make use of these expressions we need an equation to describe the time evolution of f, which is called the Boltzmann equation. We start by noting that in the absence of scattering, the quantity $f(\mathbf{k}, \mathbf{r}, t)d\gamma(t)$, which measures the number of particles in a differential volume element of phase space $d\gamma(t) = d^3r d^3k$ at time t, remains constant. At a time t + dt, the point in phase space has moved to

$$\mathbf{r}' = \mathbf{r} + \dot{\mathbf{r}}dt, \qquad \mathbf{k}' = \mathbf{k} + \dot{\mathbf{k}}dt \tag{19.5a,b}$$

and we must have

$$f(\mathbf{k}, \mathbf{r}, t)\, d\gamma(t) = f\big(\mathbf{k} + \dot{\mathbf{k}}dt, \mathbf{r} + \dot{\mathbf{r}}dt, t + dt\big)\, d\gamma(t + dt). \tag{19.6}$$

Now $d\gamma(t + dt) = Jd\gamma(t)$, where J is the Jacobian associated with the change of the coordinate system in going from t to t + dt. Although it is true in general, we only demonstrate for a two-dimensional phase space that J = 1 to first order in dt. Using Hamilton's equations, $\frac{\partial H}{\hbar\partial k} = \dot{x}$ and $\frac{\partial H}{\partial x} = -\hbar\dot{k}$, we compute the Jacobian as

$$\begin{bmatrix} \dfrac{\partial x'}{\partial x} & \dfrac{\partial x'}{\partial k} \\ \dfrac{\partial k'}{\partial x} & \dfrac{\partial k'}{\partial k} \end{bmatrix} = \begin{bmatrix} 1 + \dfrac{\partial^2 H}{\hbar\partial x\partial k}dt & \dfrac{\partial^2 H}{\hbar\partial^2 k}dt \\ -\dfrac{\partial^2 H}{\hbar\partial^2 x}dt & 1 - \dfrac{\partial^2 H}{\hbar\partial k\partial x}dt \end{bmatrix}, \tag{19.7}$$

which clearly vanishes to first order in dt.

Expanding (19.6) to first order and canceling the phase space volume element $d\gamma(t)$ from both sides, we have

$$\frac{\partial f}{\partial t} + \frac{\partial f}{\partial \mathbf{r}} \cdot \mathbf{v} + \frac{\partial f}{\partial \mathbf{k}} \cdot \dot{\mathbf{k}} = 0; \qquad \text{(collisionless form)} \tag{19.8a}$$

we may also write the left-hand side as a total derivative, $df(\mathbf{k}, \mathbf{r}, t)/dt$. However, (19.8a) is incomplete as it stands in that it does not allow for particle scattering *which causes particles to leave or enter* our phase space volume element $d\gamma(t)$; we therefore insert a so-called *collision integral* on the right-hand side, about which we will have more to say. In addition we use Newton's law to write $\hbar\dot{\mathbf{k}} = \mathbf{F}$, where $\mathbf{F}$ is a one-body force. The effect of a Hartree-like self-consistent field caused by the other particles may also be included in $\mathbf{F}$, as will be discussed later. With these two alterations Eq. (19.9a) takes the form

$$\frac{\partial f}{\partial t} + \frac{\partial f}{\partial \mathbf{r}} \cdot \mathbf{v} + \frac{\partial f}{\hbar\partial \mathbf{k}} \cdot \mathbf{F} = I\left(f(\mathbf{k}, \mathbf{r}, t)\right); \tag{19.8b}$$

Equation (19.8b) is called the *Boltzmann or kinetic equation.*

19.2.1 Evaluating the collision integral

From the so-called golden rule of quantum mechanics, the differential rate, dw, for elastic (energy-conserving) transitions between single particle states $\mathbf{k}$ and $\mathbf{k}'$ is

$$dw\left(\mathbf{k}(\varepsilon) \to \mathbf{k}'(\varepsilon')\right) = \frac{2\pi}{\hbar}\left|M_{\mathbf{kk}'}\right|\delta(\varepsilon - \varepsilon'), \tag{19.9a}$$

where $M_{\mathbf{kk}'}$ is the matrix element connecting the states $\mathbf{k}$ and $\mathbf{k}'$. The volume element d^3k' centered on $\mathbf{k}'$ contains $(2/(2\pi)^3)d^3k' = (2/(2\pi)^3\hbar\left|v'\right|)dS_{k'}d\varepsilon'$ states, where $\mathbf{v}$ is the velocity which is directed perpendicular to the Fermi surface area element dS'_k. The transition rate into this volume element would then be

$$\begin{aligned} \frac{2}{(2\pi)^3}d^3k'\,dw &= \frac{1}{2\pi^2\hbar^2\left|\mathbf{v}'\right|}\left|M_{\mathbf{kk}'}\right|^2\delta(\varepsilon - \varepsilon')dS_{k'}d\varepsilon' \\ &= W(\mathbf{k} \to \mathbf{k}')dS_{k'}d\varepsilon', \end{aligned} \tag{19.9b}$$

where $W(\mathbf{k} \to \mathbf{k}') \equiv (2\pi^2\hbar^2 |v'|)^{-1} |M_{\mathbf{kk}'}|^2 \delta(\varepsilon - \varepsilon')$. To obtain the rate of change of the distribution function for electrons, we must multiply this transition rate by the probability of an electron *occupying* the state **k** and (to satisfy the Pauli principle) the probability of finding the final state **k**′ *empty*, which yields $-f(\mathbf{k}, \mathbf{r}, t)(1 - f(\mathbf{k}', \mathbf{r}', t))W(\mathbf{k} \to \mathbf{k}')dS_{\mathbf{k}'}d\varepsilon'$, where the - sign implies that electrons are being scattered *out of* the state **k**. We must also account for the fact that electrons are simultaneously being scattered *into* the state **k** from all of the other occupied electron states, resulting in a contribution, $+f(\mathbf{k}', \mathbf{r}, t)(1 - f(\mathbf{k}, \mathbf{r}, t))W(\mathbf{k}' \to \mathbf{k})dS_{\mathbf{k}'}d\varepsilon'$. Finally we must integrate over $dS_{\mathbf{k}'}d\varepsilon'$. Our final result is therefore

$$I[f(\mathbf{k}, \mathbf{r}, t)] = -\int dS_{\mathbf{k}'}d\varepsilon'[f(\mathbf{k}, \mathbf{r}, t)(1 - f(\mathbf{k}', \mathbf{r}', t))W(\mathbf{k} \to \mathbf{k}') - f(\mathbf{k}', \mathbf{r}, t)(1 - f(\mathbf{k}, \mathbf{r}, t))W(\mathbf{k}' \to \mathbf{k})]. \tag{19.10a}$$

In equilibrium we expect I to vanish. Here the distribution function is only a function of energy, i.e., $f(\mathbf{k}, \mathbf{r}, t) = f(\varepsilon)$, and the integrand in (19.10a) will then have the form

$$[W(\mathbf{k}, \mathbf{k}')f(\varepsilon')(1 - f(\varepsilon)) - W(\mathbf{k}', \mathbf{k})f(\varepsilon)(1 - f(\varepsilon'))].$$

Note that the delta function in Eq. (19.9a) limits us to elastic collisions; i.e., $\varepsilon = \varepsilon'$, and the states **k** and **k**′ are on the same energy "shell." Therefore for the collision integral to vanish in equilibrium we must have $W(\mathbf{k}, \mathbf{k}') = W(\mathbf{k}', \mathbf{k})$. This requirement is called *detailed balancing* or *the principle of microscopic reversibility.* With this assumption (19.10a) becomes

$$I[f] = -\int dS_{\mathbf{k}'}d\varepsilon'[f(\mathbf{k}, \mathbf{r}, t) - f(\mathbf{k}', \mathbf{r}, t)]W(\mathbf{k}, \mathbf{k}'). \tag{19.10b}$$

19.2.2 The linearized Boltzmann equation

It is common to write f in the form

$$f = f_0 + \delta f, \tag{19.11}$$

where δf is a typically small change from the equilibrium form (19.1). To leading order, δf is proportional to the external driving force **F** in (19.8b); to terms linear in **F** this equation then becomes

$$\frac{\partial \delta f}{\partial t} + \frac{\partial \delta f}{\partial \mathbf{r}} \cdot \mathbf{v} + \frac{\partial f_0}{\hbar \partial \mathbf{k}} \cdot \mathbf{F} = I, \tag{19.12}$$

where $\partial f_0/\hbar\partial \mathbf{k} = (\partial f_0/\partial \varepsilon)\,\mathbf{v}$. At absolute zero we may write the Fermi function as $f_0(\varepsilon) = \theta(\mu - \varepsilon)$ (where $\theta(x) = 1$ for $x > 0$ and $\theta(x) = 0$ for $x < 0$) and

$$\frac{\partial f_0(\varepsilon)}{\partial \varepsilon} = -\delta(\mu - \varepsilon). \tag{19.13}$$

This is a good approximation for metals over a wide range of temperatures (recall the Fermi temperature is of order 10^4K), but is not generally applicable to semiconductors.

In the presence of a temperature gradient, a better starting approximation than a global equilibrium Fermi distribution is a "local equilibrium" Fermi function of the form

$$f_0(\varepsilon, \mathbf{r}) = \frac{1}{e^{(\varepsilon - \mu(\mathbf{r}))/k_B T(\mathbf{r})} + 1}; \tag{19.14}$$

we then have an additional contribution,

$$\frac{\partial f_0(\varepsilon, \mathbf{r})}{\partial \mathbf{r}} \cdot \mathbf{v} = \frac{\partial f_0}{\partial \varepsilon}\left[-\frac{\varepsilon - \mu}{T}\nabla T - \nabla \mu\right] \cdot \mathbf{v}, \tag{19.15}$$

arising from the second term in (19.8b) of the kinetic equation.

19.2.3 The relaxation time approximation

Calculating the function $W(\mathbf{k}, \mathbf{k}')$ is difficult to do in practice; even if the host wave functions are known, it is far more difficult to calculate the potential difference between the host and the impurity which actually causes the scattering. Rather than deal with the full complexity of the collision integral, it is common to introduce an approximation from which we may obtain useful expressions for transport coefficients such as the electrical conductivity. We approximate the collision integral by the form

$$I(f) = -\frac{f - f_0}{\tau(\mathbf{k})}. \tag{19.16}$$

One may further approximate this form, which we will commonly do in what follows, by assuming $\tau \neq \tau(\mathbf{k})$; i.e., τ is isotropic. As a simple example let us assume that f differs uniformly from its equilibrium form (i.e., is uniform in space) and furthermore that any forces that resulted in this situation are removed at $t = 0$. The subsequent evolution of f would then be governed by the equation

$$\frac{\partial f}{\partial t} = -\frac{\delta f}{\tau} = -\frac{f - f_0}{\tau}, \tag{19.17}$$

which has the solution

$$\delta f(t) = \delta f(0) e^{-t/\tau}; \tag{19.18}$$

i.e., it relaxes exponentially to the equilibrium form with a characteristic time, τ.

19.3 Electrical conductivity

Assume we have a uniform temperature and apply a constant electric field. In a time of order τ (typically 10^{-13} sec. at room temperature) the system will come into dynamic equilibrium or, as it is also called, a *steady-state* condition. We then have $\partial f/\partial t = 0$ and with the expectation that $\delta f \propto E$ the kinetic equation linear in E becomes

$$\frac{\partial f_0}{\partial \varepsilon}\mathbf{v}\cdot e\mathbf{E} = -\frac{\delta f}{\tau} \tag{19.19a}$$

or

$$\delta f = -e\tau\mathbf{v}\cdot\mathbf{E}\frac{\partial f_0}{\partial \varepsilon}. \tag{19.19b}$$

Inserting this form in the expression for the electrical current, (19.4), we have

$$\mathrm{j} = -\frac{2e^2\tau}{(2\pi)^3}\int d^3k\mathbf{v}(\mathbf{v}\cdot\mathbf{E})\frac{\partial f_0}{\partial \varepsilon}, \tag{19.20}$$

which we rewrite as

$$j_i = \sigma_{ij}E_j, \tag{19.21}$$

where

$$\sigma_{ij} = -\frac{2e^2\tau}{(2\pi)^3}\int d^3k v_i v_j \frac{\partial f_0}{\partial \varepsilon}. \tag{19.22}$$

We now confine ourselves to the case of a metal and employ the approximation (19.13). As in our discussion of the relaxation time we rewrite $d^3k = dSd\varepsilon/\hbar|\mathbf{v}|$. Carrying out the integral over the energy, we obtain an expression for the electrical conductivity of a metal,[1]

$$\sigma_{ij} = \frac{2e^2\tau}{(2\pi)^3\hbar}\int dS\frac{v_i v_j}{|\mathbf{v}|}. \tag{19.23}$$

For a cubic metal, where the elements of the conductivity tensor are equal and denoted by σ, we write

$$\sigma = \sigma_{zz} = \frac{2e^2\tau}{(2\pi)^3\hbar}\int dS\frac{v_z^2}{|\mathbf{v}|};$$

however,

$$\sigma_{xx} + \sigma_{yy} + \sigma_{zz} = 3\sigma = \frac{2e^2\tau}{(2\pi)^3\hbar}\int dS\frac{v_x^2 + v_y^2 + v_z^2}{|\mathbf{v}|},$$

[1] If information is available on $\tau(\mathbf{k})$ it may be left under the integral sign in Eq. (19.23).

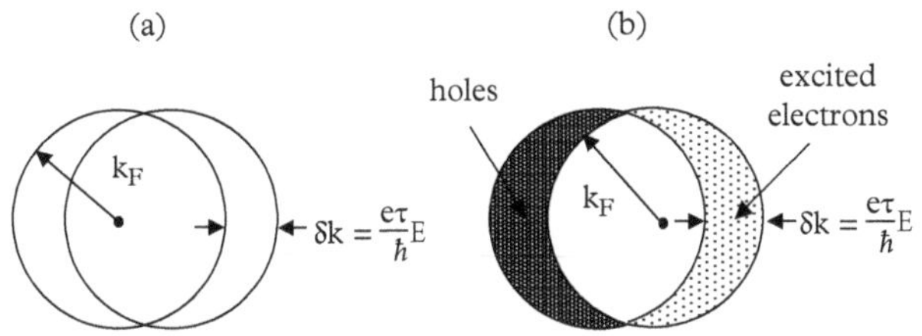

Figure 19.1 *(a) The equilibrium and the displaced Fermi sphere; (b) excited electrons and holes.*

or

$$\sigma = \frac{2e^2\tau}{3(2\pi)^3\hbar}\int dS\,|\mathbf{v}|. \quad \text{(cubic system)} \tag{19.24}$$

For free electrons, $\int dS\,|\mathbf{v}| = 4\pi k_F^2 v_F$ and $(2/(2\pi)^3)\cdot(4\pi k_F^3/3) = n$; hence $\sigma = ne^2\tau/m$, as in the Drude theory.

We can view the conduction process in two ways, as illustrated in Fig. 19.1: (i) we may think of the whole Fermi sphere as being displaced by a small amount, $\delta k = (e\tau/\hbar)E$, with all carriers participating and moving with a drift velocity, $v_D = \mu E = (e\tau/m)E \ll v_F$, or (ii) we view it as a small crescent of excited electrons on the leading edge, and an equal number of holes (empty states) on the trailing edge of the Fermi surface, both of which move much faster, essentially at the Fermi velocity. The two are related by an integration by parts.

19.4 Transport in the presence of electric fields and temperature gradients

We now generalize the previous discussion to the case where an electric field and a temperature gradient are present simultaneously. We will assume an "equilibrium" distribution of the form (19.14) with $\partial f_0(\varepsilon, \mathbf{r})/\partial \mathbf{r}$, given by (19.15). Our Boltzmann equation, (19.8), in the relaxation time approximation, (19.16), then becomes

$$\delta f(\mathbf{k}) = -\tau\left[-\frac{\varepsilon(\mathbf{k})-\mu}{T}\nabla T - \nabla\mu + \mathbf{F}\right]\cdot \mathbf{v}\frac{\partial f_0}{\partial \varepsilon}; \tag{19.25}$$

the electric current then follows as

$$\mathbf{j}(\mathbf{r}) = \frac{2e\tau}{(2\pi)^3}\int d^3k\mathbf{v}\left[\mathbf{v}\cdot\left(-\frac{\varepsilon(\mathbf{k})-\mu}{T}\nabla T - \nabla\mu + e\mathbf{E}\right)\right]\left(-\frac{\partial f_0}{\partial \varepsilon}\right). \tag{19.26}$$

In addition we will assume that a heat current, $\mathbf{J}$, is present. To incorporate this effect we appeal to the first law of thermodynamics in differential form

$$dE = dQ - PdV + \mu dN, \tag{19.27}$$

where we will assume a constant volume (i.e., we neglect thermal expansion). From this expression it is apparent that the heat flux is the difference between the energy flux and the product of the chemical potential and the particle number flux; i.e.,

$$\mathbf{J}(\mathbf{r}) = \frac{2}{(2\pi)^3} \int d^3k\, \mathbf{v}\, (\varepsilon(\mathbf{k}) - \mu)\, \delta f(\mathbf{k}),$$

which on using (19.25) becomes

$$\mathbf{J}(\mathbf{r}) = \frac{2\tau}{(2\pi)^3} \int d^3k \mathbf{v} (\varepsilon(\mathbf{k}) - \mu) \left[\mathbf{v} \cdot \left(-\frac{\varepsilon(\mathbf{k}) - \mu}{T} \nabla T - \nabla\mu + e\mathbf{E}\right)\right] \left(-\frac{\partial f_0}{\partial \varepsilon}\right). \tag{19.28}$$

In practice we do not measure the *microscopic* electric field, only the net *electromotive force*, where the latter combines the two "generalized" forces that lead to mass (and with it charge) transport: the forces arising from the internal electric field, **E**, and the flux arising from the gradient of the chemical potential, $-\Delta\mu$. Thus in place of $e\mathbf{E} - \nabla\mu$ we simply write $e\mathbf{E}$; the integral of the *effective* electric field **E** across some current path traversing the sample, which we designate as the electric potential V, is what one would measure with an external voltmeter.

We may now write Eqs. (19.26) and (19.28) in the form

$$\mathbf{j} = \mathbf{K}_{11} \cdot e\mathbf{E} + \mathbf{K}_{12} \cdot \left(-\frac{\nabla T}{T}\right) \tag{19.29a}$$

and

$$\mathbf{J} = \mathbf{K}_{21} \cdot e\mathbf{E} + \mathbf{K}_{22} \cdot \left(-\frac{\nabla T}{T}\right), \tag{19.29b}$$

which can be combined into a single matrix equation as

$$\begin{bmatrix} \mathbf{j} \\ \mathbf{J} \end{bmatrix} = \begin{bmatrix} \mathbf{K}_{11} & \mathbf{K}_{12} \\ \mathbf{K}_{21} & \mathbf{K}_{22} \end{bmatrix} \cdot \begin{bmatrix} e\mathbf{E} \\ -\left(\frac{\nabla T}{E}\right) \end{bmatrix}. \tag{19.30}$$

The tensor elements $\mathbf{K}_{ij}$ are then given by

$$(\mathbf{K}_{11})_{ij} = \frac{2\tau e}{(2\pi)^3 \hbar} \int d\varepsilon dS \frac{v_i v_j}{|\mathbf{v}|} \left(-\frac{\partial f_0}{\partial \varepsilon}\right), \tag{19.31a}$$

$$(\mathbf{K}_{12})_{ij} = \frac{2\tau e}{(2\pi)^3 \hbar} \int d\varepsilon dS \frac{v_i v_j}{|\mathbf{v}|} (\varepsilon(\mathbf{k}) - \mu) \left(-\frac{\partial f_0}{\partial \varepsilon}\right), \tag{19.31b}$$

$$(\mathbf{K}_{21})_{ij} = \frac{1}{e} (\mathbf{K}_{12})_{ij}, \tag{19.31c}$$

and

$$(\mathbf{K}_{22})_{ij} = \frac{2\tau}{(2\pi)^3 \hbar} \int d\varepsilon dS \frac{v_i v_j}{|\mathbf{v}|} (\varepsilon(\mathbf{k}) - \mu)^2 \left(-\frac{\partial f_0}{\partial \varepsilon}\right), \tag{19.31d}$$

where we have rewritten the integrals over d^3k as integrals over the energy and the Fermi surface; clearly the conductivity is given by $\sigma_{ij} = e\,(\mathbf{K}_{11})_{ij}$.

19.5 Thermal conductivity

Since there are now two "thermodynamic forces," $e\mathbf{E}$ and $-\nabla T/T$, the values of various measured fluxes depend on how the experiment is arranged. Suppose we establish a thermal gradient via insulating reservoirs, so as to ensure there is no electric current flowing in the sample; i.e., $\mathbf{j}(\mathbf{r}) = 0$. From (19.29a) we then have

$$e\mathbf{E} = (\mathbf{K}_{11})^{-1} \cdot \mathbf{K}_{12} \cdot \frac{\nabla T}{T}. \tag{19.32}$$

Inserting this result in (19.29b) we can write

$$J_i = \kappa_{ij} \frac{\partial T}{\partial x_j}, \tag{19.33}$$

where κ_{ij} is the *thermal conductivity tensor* which is given by

$$\kappa = \frac{1}{T}\mathbf{K}_{22} + \mathbf{K}_{21} \cdot (\mathbf{K}_{11})^{-1} \cdot \mathbf{K}_{12}. \tag{19.34}$$

The second term is negligible for metals, and from (19.31d) we are left with

$$\kappa_{ij} = \frac{2\tau}{(2\pi)^3 \hbar T} \int d\varepsilon dS \frac{v_i v_j}{|\mathbf{v}|} (\varepsilon(\mathbf{k}) - \mu)^2 \left(-\frac{\partial f_0}{\partial \varepsilon} \right). \tag{19.35}$$

Equation (19.35) vanishes if we approximate $-\partial f_0/\partial \varepsilon$ as a delta function, so we must include the leading correction. Carrying out an integration by parts we have

$$\kappa_{ij} = \frac{2\tau}{(2\pi)^3 \hbar T} \int d\varepsilon dS \frac{v_i v_j}{|\mathbf{v}|} f_0 \frac{\partial}{\partial \varepsilon} (\varepsilon(\mathbf{k}) - \mu)^2. \tag{19.36}$$

If in Eq. (6.28) of Section 6.4 we make the identification

$$F(\varepsilon) = \frac{\partial}{\partial \varepsilon} \left(\frac{v_i v_j}{|\mathbf{v}|} (\varepsilon(\mathbf{k}) - \mu)^2 \right)$$

and note that the first term in the expansion vanishes (since it arises from the delta function approximation), we are left with

$$\kappa_{ij} = \frac{4\pi^2 \tau k_B^2 T}{6\,(2\pi)^3 \hbar} \int dS \frac{v_i v_j}{|\mathbf{v}|}. \tag{19.37}$$

Comparing this result with Eq. (19.23) for the electrical conductivity we have

$$\kappa_{ij} = \frac{\pi^2}{3} \frac{k_B^2 T}{e^2} \sigma_{ij}. \tag{19.38}$$

This is the *Weidemann–Franz law*, obtained earlier from the classical Drude theory as Eq. (3.39), which, fortuitously, has a similar form. It is valid more generally in the residual resistance regime, and is not restricted to the relaxation time approximation, provided the scattering is elastic. In particular it does not hold for the case of phonon scattering.

19.6 Thermoelectric effects

The cross terms, $\mathbf{K}_{12}$ and $\mathbf{K}_{21}$, lead to so-called thermoelectric effects. Here we will discuss two, the Seebeck effect, which is an electric field produced by a temperature gradient, and the Peltier effect, which is a heat current produced by an electric current. We start with the former.

19.6.1 The Seebeck effect

We begin by slightly rewriting Eq. (19.32) as

$$\mathbf{E} = \frac{1}{eT}(\mathbf{K}_{11})^{-1} \cdot \mathbf{K}_{12} \cdot \nabla T = \mathbf{Q} \cdot \nabla T, \tag{19.39a}$$

where

$$\mathbf{Q} \equiv \frac{1}{eT}(\mathbf{K}_{11})^{-1} \cdot \mathbf{K}_{12}. \tag{19.39b}$$

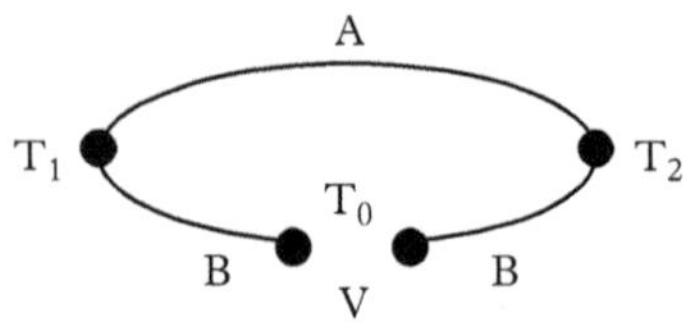

Figure 19.2 *The Seebeck effect.*

Q is called the *Seebeck* coefficient (it is also called the *thermoelectric power*).

It is interesting to consider the circuit shown in Fig. 19.2, where A and B denote wires made from *different* conductors. Integrating the Seebeck electric field around the circuit we obtain the potential difference as

$$V = \int_{T_0}^{T_1} Q_B \frac{dT}{dl} dl + \int_{T_1}^{T_2} Q_A \frac{dT}{dl} dl + \int_{T_2}^{T_0} Q_B \frac{dT}{dl} dl$$

or

$$V = \int_{T_1}^{T_2} dT\,(Q_B - Q_A). \tag{19.40}$$

Note that V is (i) independent of the path and (ii) independent of the temperature of the measuring point. These properties make the Seebeck effect a useful way to measure temperature. Once Q(T) has been calibrated and the materials employed standardized, the technique may be used as a secondary temperature standard. The point T_1 may be taken as the unknown temperature and the point T_2 is usually chosen to be 0° and is maintained by an ice bath.

19.6.2 The Peltier effect

Let us express the heat current, **J**, as a function of the electric current, **j**. From (19.29a) we have

$$e\mathbf{E} = (\mathbf{K}_{11})^{-1}\mathbf{j} + \left(\mathbf{K}_{11}{}^{-1} \cdot \mathbf{K}_{12}\right) \cdot \left(\frac{\nabla T}{T}\right);$$

inserting this in (19.29b) gives

$$\mathbf{J} = \left(\mathbf{K}_{21} \cdot \mathbf{K}_{11}{}^{-1}\right) \cdot \mathbf{j} + \left[\mathbf{K}_{22} - \left(\mathbf{K}_{21} \cdot \mathbf{K}_{11}{}^{-1} \cdot \mathbf{K}_{12}\right)\right] \cdot \left(-\frac{\nabla T}{T}\right). \tag{19.41}$$

Consider the circuit shown in Fig. 19.3 which again involves two different materials (A,B). In addition, we have two thermal reservoirs to receive and transmit thermal currents, together with a battery to supply an electrical current. Where the battery is placed in the circuit is not important for our purposes here. For simplicity we assume both reservoirs are at the same temperature. Furthermore, we must recognize that any Joule heating is *second order* in $\mathbf{j}$, whereas the thermoelectric effects are *first order*. Hence Joule heating may be neglected in first order. In steady state we must have $\mathbf{J}_{in} = \mathbf{J}_{out}$. We may then write Eq. (19.41) in the form

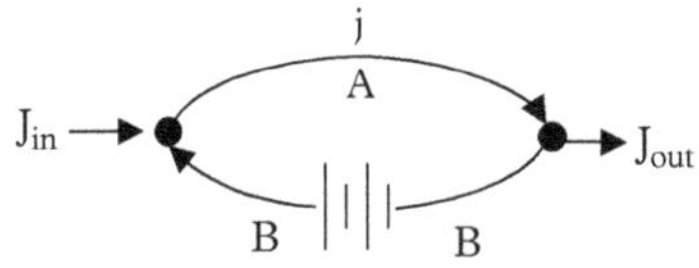

Figure 19.3 *The Peltier effect.*

$$\mathbf{J} = \mathbf{\Pi} \cdot \mathbf{j}, \tag{19.42a}$$

where

$$\mathbf{\Pi} \equiv \mathbf{K}_{21} \cdot \mathbf{K}_{11}{}^{-1}, \tag{19.42b}$$

with the latter called the *Peltier coefficient.* Since the current flows in the *opposite* direction in the two materials, we have

$$\mathbf{J}_{in} = \mathbf{J}_A + \mathbf{J}_B = (\mathbf{\Pi}_A - \mathbf{\Pi}_B) \cdot \mathbf{j} \tag{19.43}$$

(i.e., the heat currents add but the electrical currents are in opposite directions). Note that we may either extract or apply heat, depending on the sign of $\mathbf{j}$; clearly the former is of greater interest and is utilized to make *thermoelectric coolers*. Note also that the effect involves the *difference* of the Peltier coefficients and would obviously vanish for identical materials A and B. In practice one employs materials having Peltier coefficients with the *opposite* sign so that A and B both extract heat. The Peltier and Seebeck coefficients are related. Since $\mathbf{Q} = \frac{1}{eT}\mathbf{K}_{11}^{-1} \cdot \mathbf{K}_{12}, \Pi = \mathbf{K}_{21} \cdot \mathbf{K}_{11}^{-1}$, and $\mathbf{K}_{21} = \frac{1}{e}\mathbf{K}_{12}$, the two coefficients are related as $\Pi = T\mathbf{Q}$. Measurements of σ, κ, and either Π or $\mathbf{Q}$ determine all of the elements of the transport matrix $\mathbf{K}_{ij}$.[2]

19.7 Calculating thermoelectric coefficients

To evaluate either $\mathbf{Q}$ or $\mathbf{\Pi}$ we need to evaluate K_{12}, which is given by Eq. (19.31b). We limit ourselves to the low temperature behavior of a metal and again employ (19.36). In what follows it will be important to allow the scattering time to be momentum (or more simply energy) dependent; i.e., $\tau = \tau(\mathbf{k})$. We introduce the mean free path, $l(\mathbf{k})$, through the definition $l(\mathbf{k}) = |\mathbf{v}(\mathbf{k})|\tau(\mathbf{k})$. As in our discussion of the thermal conductivity coefficients we do an integration by parts but this time making the identification

$$F(\varepsilon) = \frac{\partial}{\partial \varepsilon}\left(l(\mathbf{k})(v_i v_j/v^2)(\varepsilon(\mathbf{k}) - \mu)\right),$$

[2] The relation between the off-diagonal transport coefficients is an example of a more general property, the Onsager reciprocity principle.

which gives

$$(\mathbf{K}_{12})_{ij} = \frac{2\pi^2}{3}\frac{ek_B^2T^2}{(2\pi)^3\hbar}\left[\int dS \frac{\partial}{\partial\varepsilon}\left(\frac{l(\mathbf{k})v_iv_j}{v^2}\right)\right]_{\varepsilon=\mu}. \tag{19.44}$$

Comparing (19.44) with our earlier expression (19.23) for the electrical conductivity we see that

$$(\mathbf{K}_{12})_{ij} = \frac{\pi^2k_B^2T^2}{3e}\left[\frac{\partial\sigma_{ij}}{\partial\varepsilon}\right]_{\varepsilon=\mu}. \tag{19.45}$$

For the case of cubic metals, where the tensor is diagonal, we have

$$K_{12} = \frac{e\pi^2k_B^2T^2}{9(2\pi)^3\hbar}\left[\int dS\frac{\partial l}{\partial\varepsilon}\right]_{\varepsilon=\mu}. \tag{19.46}$$

Using (19.39b), (19.42b), and (19.45), together with $\sigma = eK_{11}$, we then have

$$\Pi = \frac{\pi^2k_B^2T^2}{3e}\frac{\partial}{\partial\varepsilon}(\ln\sigma) \tag{19.47a}$$

and

$$Q = \frac{\pi^2k_B^2T}{3e}\frac{\partial}{\partial\varepsilon}(\ln\sigma). \tag{19.47b}$$

We now discuss Eqs. (19.47a,b) qualitatively (nonrigorously) and write $\sigma(\varepsilon) = en(\varepsilon)\mu(\varepsilon)$, where μ is now the mobility. We initially assume $\mu \neq \mu(\varepsilon)$, then $\partial\ln(\sigma)/\partial\varepsilon = (1/n(\varepsilon))\cdot(\partial n(\varepsilon)/\partial\varepsilon)$; using $\partial n(\varepsilon)/\partial\varepsilon = N(\varepsilon)$, the density of states, (19.47a) becomes $\Pi = (\pi^2k_B^2T^2/3e)\cdot(N(\varepsilon)/n(\varepsilon))$. But the heat capacity per electron is given by (see Eq. (6.37a)) $c = (\pi^2k_B^2T/3)\cdot(N(\varepsilon)/n(\varepsilon))$ and therefore $\Pi = cT/e$. Now the energy of each electron is approximately cT and therefore the Peltier effect contains a term proportional to the thermal energy per electron. If we also include the contribution from the energy dependence of the mean free path, we have

$$\Pi = \frac{\pi^2}{3}k_B^2T^2\left[\frac{N(\varepsilon)}{n(\varepsilon)} + \frac{\partial\ln\mu}{\partial\varepsilon}\right].$$

If $\tau(\varepsilon)$ in μ increases with energy then each electron will carry more energy.

These formulas resulted from an expansion in T/T_F and are therefore valid only for metals at low temperatures with $T/T_F \ll 1$. In particular they are not valid for semiconductors. In practice, however, the larger values of the thermopower occur in semiconductors. By exploiting materials with electron and hole dominant carriers for A and B respectively, both paths will contribute to the heat transport.

19.8 Magnetoresistance and the Hall effect

We next extend our discussions to include the effects of a magnetic field. Using the relaxation time approximation, (19.16), and writing $f = f_0 + \delta f$, the steady-state Boltzmann equation becomes

$$\frac{\partial f_0}{\partial \varepsilon}\mathbf{v}\cdot\left[e\mathbf{E} + e\frac{\mathbf{v}}{c}\times\mathbf{H}\right] + \frac{1}{\hbar}\frac{\partial \delta f}{\partial \mathbf{k}}\cdot\left[e\mathbf{E} + e\frac{\mathbf{v}}{c}\times\mathbf{H}\right] = -\frac{\delta f}{\tau}, \tag{19.48}$$

where we have now written the full Lorentz force,

$$\mathbf{F} = e\mathbf{E} + e\frac{\mathbf{v}}{c}\times\mathbf{H}. \tag{19.49}$$

The left side of (19.48) has four terms. The second term vanishes directly and the third is of order E^2 (since $\delta f \propto E$) and will be neglected; however, the fourth is of order EH and must be retained. We then have

$$-\frac{\partial f_0}{\partial \varepsilon}\mathbf{v}\cdot\mathbf{E} = \frac{\delta f}{e\tau} + \frac{1}{\hbar}\frac{\partial \delta f}{\partial \mathbf{k}}\cdot\frac{\mathbf{v}}{c}\times\mathbf{H}. \tag{19.50}$$

In what follows we restrict the discussion to the case of a free-electron-like material with

$$\mathbf{v} = \frac{\hbar\mathbf{k}}{m^*}. \tag{19.51}$$

Given Eq. (19.19b) it is natural to seek a solution of the form

$$\delta f = -e\tau\frac{\partial f_0}{\partial \varepsilon}\mathbf{v}\cdot\mathbf{X}, \tag{19.52}$$

where $\mathbf{X}$ is to be determined and is assumed independent of $\mathbf{k}$; note that for $\mathbf{H} = 0$ we have $\mathbf{X} = \mathbf{E}$, consistent with our assumption. Inserting this form in (19.50), and noting $\partial v_i/\partial k_j = \hbar\delta_{ij}/m^*$, gives

$$\mathbf{v}\cdot\mathbf{E} = \mathbf{v}\cdot\mathbf{X} + \frac{e\tau}{m^* c}\mathbf{X}\cdot(\mathbf{v}\times\mathbf{H})$$

permuting the triple scalar product and noting this equation must be satisfied for all $\mathbf{v}$ (recall $\mathbf{X} \neq \mathbf{X}(\mathbf{k})$) we have

$$\mathbf{E} = \mathbf{X} + \frac{e\tau}{m^* c}\mathbf{H}\times\mathbf{X}. \tag{19.53}$$

The electric current is then given by

$$\mathbf{j} = \frac{2e\tau}{(2\pi)^3}\int d^3k\,\mathbf{v}(e\tau\mathbf{v}\cdot\mathbf{X})\left(-\frac{\partial f_0}{\partial \varepsilon}\right). \tag{19.54}$$

This equation has the same structure as Eq. (19.20) and for our free-electron-like assumption where **v** is isotropic it can be written as

$$\mathbf{j} = \sigma \mathbf{X}. \tag{19.55a}$$

or

$$\mathbf{X} = \rho \mathbf{j}, \tag{19.55b}$$

where $\rho = \frac{1}{\sigma} = \frac{m^*}{ne^2\tau}$. Substituting (19.55b) into (19.53) we get

$$\mathbf{E} = \rho \mathbf{j} + \frac{\rho e \tau}{m^* c} \mathbf{H} \times \mathbf{j}, \tag{19.56}$$

which reduces to Ohm's law for $\mathbf{H} = 0$. Note that for an arbitrary direction of **H** there will be components of the electric field both parallel and perpendicular to j. For the case $\mathbf{j} \perp \mathbf{H}$ we have $E_{//} = \rho j$ and $E_{\perp} = (\rho e \tau / m^* c) H j$. The perpendicular voltage constitutes the *Hall effect*, discussed in Chapter 3 and we write it in the form

$$E_{\perp} = R_H j H, \tag{19.57}$$

where $R_H = \rho e \tau / m^* c$; for our free electron like case, $\rho = m^*/ne^2\tau$; thus $R_H = 1/nec$ as we obtained classically in Section 3.2. This relation appears to hold rigorously for monovalent metals. Since n then follows from a measurement of the lattice constant, a precision measurement gives a relation between E, H, and the fundamental constants, e and c.

Note that $E_{//}$ is *unaffected* by the magnetic field in this model. In the presence of anisotropy of v or τ over the Fermi surface, the longitudinal electric field does show a magnetic field dependence, which is referred to as *magnetoresistance*.

19.9 A two-band model

In order to obtain a finite magnetoresistance we need a more sophisticated model. In what follows we will outline calculations involving two spherical Fermi surfaces having carrier densities, effective masses, and relaxation times n_1, n_2; m_1, m_2; τ_1, τ_2; respectively. Such models can be extended to collections of ellipsoids described by effective mass tensors; relaxation time tensors can also be introduced. Many semiconductors and some semimetals such as Bi can to some extent be described in this way. For metals such models are of limited value in practice, since the Fermi surface is usually highly anisotropic, as are the relaxation times. Multi-carrier models of this kind can be used to represent data in these more complex systems, but the resulting parameters have only limited meaning.

We seek a solution to the problem posed by writing

$$\delta f_{1,2} = -e\tau_{1,2}\frac{\partial f_0}{\partial \varepsilon}\mathbf{v}_{1,2} \cdot \mathbf{X}_{1,2} \tag{19.58}$$

as in Eq. (19.52) for the two carrier groups where $\mathbf{X}_1$ and $\mathbf{X}_2$ satisfy equations analogous to (19.53) of the form

$$\mathbf{E} = \mathbf{X}_1 + \beta_1 \mathbf{H} \times \mathbf{X}_1; \ \mathbf{E} = \mathbf{X}_2 + \beta_2 \mathbf{H} \times \mathbf{X}_2, \tag{19.59a,b}$$

with $\beta_{1,2} = e_{1,2}\tau_{1,2}/m_{1,2}c$, and we have included a subscript on the charges to allow for the possibility of opposite signs (i.e., both electrons and holes). Corresponding to Eq. (19.56) we have

$$\mathbf{E} = \frac{\mathbf{j}_1}{\sigma_{01}} + \frac{\beta_1}{\sigma_{01}}\mathbf{H} \times \mathbf{j}_1 \tag{19.60a}$$

and

$$\mathbf{E} = \frac{\mathbf{j}_2}{\sigma_{02}} + \frac{\beta_2}{\sigma_{02}}\mathbf{H} \times \mathbf{j}_2 \tag{19.60b}$$

where $\sigma_{01} = n_{1,2}e^2\tau_{1,2}/m_{1,2}$. Next we solve (19.60a,b) to obtain $\mathbf{j}_1$ and $\mathbf{j}_2$ as[3]

$$\mathbf{j}_1 = \frac{\sigma_{01}\mathbf{E}}{1+\beta_1^2 H^2} - \frac{\sigma_{01}\beta_1}{1+\beta_1^2 H^2}\mathbf{H} \times \mathbf{E} \tag{19.61a}$$

and

$$\mathbf{j}_2 = \frac{\sigma_{02}\mathbf{E}}{1+\beta_2^2 H^2} - \frac{\sigma_{02}\beta_2}{1+\beta_2^2 H^2}\mathbf{H} \times \mathbf{E}. \tag{19.61b}$$

Writing the total current as $\mathbf{j} = \mathbf{j}_1 + \mathbf{j}_2$, we can combine (19.60a,b) to obtain

$$\mathbf{j} = (\sigma_1(H) + \sigma_2(H))\,\mathbf{E} - (\sigma_1(H)\beta_1 + \sigma_2(H)\beta_2)\,\mathbf{H} \times \mathbf{E} \tag{19.62}$$

where for compactness we introduced field-dependent individual conductivities as

$$\sigma_{1,2}(H) = \frac{\sigma_{01,2}}{1+\beta_{1,2}^2 H^2}. \tag{19.63}$$

Restricting to the case $\mathbf{H}\perp\mathbf{j}$ with $\mathbf{j} = j\hat{\mathbf{x}}$ and $\mathbf{H} = H\hat{\mathbf{z}}$ we now resolve the electric field in Eq. (19.62) into components according to $\mathbf{E} = E_{//}\hat{\mathbf{x}} + E_{\perp}\hat{\mathbf{y}}$ and on doing so we obtain the two equations

$$j = (\sigma_1(H) + \sigma_2(H))\,E_{//} + (\beta_1\sigma_1(H) + \beta_2\sigma_2(H))\,HE_{\perp} \tag{19.64a}$$

$$0 = (\sigma_1(H) + \sigma_2(H))\,E_{\perp} - (\beta_1\sigma_1(H) + \beta_2\sigma_2(H))\,HE_{//}. \tag{19.64b}$$

[3] If $\mathbf{a} = \mathbf{b} + \mathbf{c} \times \mathbf{b}$ then $\mathbf{b} = (\mathbf{a} - \mathbf{c} \times \mathbf{a})/(1 + c^2)$, as can be verified by direct substitution of one into the other.

Finally we solve (19.64a,b) for $E_\perp$ and $E_{//}$ obtaining

$$E_\perp = \frac{(\beta_1\sigma_1(H) + \beta_2\sigma_2(H))}{(\sigma_1(H) + \sigma_2(H))} HE_{//} \tag{19.65a}$$

and

$$E_{//} = \frac{(\sigma_1(H) + \sigma_2(H))}{\left[(\sigma_1(H) + \sigma_2(H))^2 + (\beta_1\sigma_1(H) + \beta_2\sigma_2(H))^2 H^2\right]} j. \tag{19.65b}$$

We can rewrite (19.65a,b) in the form

$$E_{//} = \rho(H)j \tag{19.66a}$$

and

$$E_\perp = R_H(H)Hj \tag{19.66b}$$

where $\rho(H)$ is the field-dependent resistivity and R_H is the Hall coefficient.

The Hall coefficient has the following limiting behaviors:

$$R_H(H \to 0) = \frac{(\sigma_{01}\beta_1 + \sigma_{02}\beta_2)}{(\sigma_{01} + \sigma_{02})^2} \tag{19.67a}$$

and

$$R(H \to \infty) = \frac{1}{\frac{\sigma_{01}}{\beta_1} + \frac{\sigma_{02}}{\beta_2}}. \tag{19.67b}$$

Note R_H can no longer be used to rigorously extract the carrier density although if one carrier strongly dominates it can give reasonable estimates of its concentration (and is widely used that way in semiconductors containing two carrier types).

The field-dependent resistivity has the limiting forms

$$\rho(H \to 0) = \frac{1}{\sigma_{01} + \sigma_{02}} + \mathbf{O}H^2 \tag{19.68a}$$

and

$$\rho(H \to \infty) = \frac{\frac{\sigma_{01}}{\beta_1^2} + \frac{\sigma_{02}}{\beta_2^2}}{\left(\frac{\sigma_{01}}{\beta_1} + \frac{\sigma_{02}}{\beta_2}\right)^2} + \mathbf{O}H^{-2}. \tag{19.68b}$$

This latter behavior in which the resistivity saturates at high magnetic fields is a general phenomenon for metals with *closed* Fermi surfaces. However, if the Fermi surface supports *open orbits*, the resistance associated with such field directions does not saturate.[4]

[4] See book by Pippard cited in the Additional Reading at the end of this chapter.

19.10 Calculation of the relaxation time

For the simple case of an isotropic, free-electron-like metal we will derive an expression for the electronic scattering time that enters the electrical conductivity, which we now designate as τ_{tr}. With the understanding that the collision integral vanishes in equilibrium we can rewrite (19.10b) as

$$I(f) = \int d^3k' \left[\delta f(\mathbf{k}') - \delta f(\mathbf{k})\right] W(\mathbf{k}, \mathbf{k}'). \tag{19.69}$$

Taking I(f) from (19.16b) on the left side of this expression and inserting δf from Eq. (19.19b) on both sides of the resulting expression then give

$$\frac{\mathbf{v} \cdot e\mathbf{E}}{\tau_{tr}} \left(-\frac{\partial f_0}{\partial \varepsilon}\right) = -e \int d\varepsilon' d\Omega \mathbf{E} \cdot \left[\mathbf{v}(\mathbf{k}') \left(-\frac{\partial f_0}{\partial \varepsilon'}\right) - \mathbf{v}(\mathbf{k}) \left(-\frac{\partial f_0}{\partial \varepsilon}\right)\right] W\left(\varepsilon', \psi\right), \tag{19.70}$$

where ψ is the angle between $\mathbf{k}$ and $\mathbf{k}'$ and W is defined by Eq. (19.9b). Noting that W contains the factor $\delta(\varepsilon - \varepsilon')$, we may cancel the factors involving the energy derivative of f_0 with the understanding that $\mathbf{k}$ and $\mathbf{k}'$ are then confined to the Fermi surface on both sides of (19.70), which is then written as

$$\frac{\mathbf{v} \cdot \mathbf{E}}{\tau_{tr}} = -\int d\varepsilon' d\Omega' \, \mathbf{E} \cdot \left[\mathbf{v}(\mathbf{k}') - \mathbf{v}(\mathbf{k})\right] W\left(\varepsilon', \psi\right)$$

or

$$\frac{\cos\theta}{\tau_{tr}} = 2\pi \int d\varepsilon' d\left(\cos\psi\right) \left[\cos\theta - \cos(\theta + \psi)\right] W\left(\varepsilon', \psi\right). \tag{19.71}$$

Writing $\cos(\theta + \psi) = \cos\theta\cos\psi - \sin\theta\sin\psi$, noting the second term vanishes on integrating over $d\psi$, and canceling $\cos\theta$ throughout, our final result is

$$\frac{1}{\tau_{tr}} = 2\pi \int d\varepsilon' d\left(\cos\psi\right) \left[1 - \cos\psi\right] W\left(\varepsilon', \psi\right). \tag{19.72}$$

Note that the factor $1 - \cos\psi$ strongly weights large angle scattering events and the corresponding scattering time, τ_{tr}, is then referred to as the *transport scattering time*, as opposed to the *total scattering time*, which we would write as

$$\frac{1}{\tau} = 2\pi \int d\varepsilon' \, d\left(\cos\psi\right) W\left(\psi, \varepsilon'\right). \tag{19.73}$$

We can define a transport mean free path, l_{tr}, and a transport scattering cross-section, σ_{tr} (not to be confused with the conductivity), from kinetic theory through the relations $l_{tr} = v_F\tau_{tr}$, and $l_{tr} = 1/\sigma_{tr}n_i$, where n_i is the number density of the impurities. The transport cross-section is related to the differential scattering cross-section (denoted as $\sigma(\psi)$) by

$$\sigma_{tr} = 2\pi \int d(\cos\psi)\,[1-\cos\psi]\sigma(\psi). \tag{19.74}$$

In the Born approximation, $\sigma(\psi)$ is given by

$$\sigma(\psi) = \left(\frac{2m}{\hbar^2}\right)^2 |V(\mathbf{q})|^2, \tag{19.75}$$

where $V(\mathbf{q})$ is the Fourier transform of the scattering potential; here $\mathbf{q}$ is the momentum transfer given by $q = 2k_F \sin\frac{\psi}{2}$.

For an impurity with a charge that differs from the host lattice, a screened Coulomb potential of the form

$$V(\mathbf{r}) = \frac{ze^2}{r} e^{-\kappa_{TF} r} \tag{19.76}$$

provides a reasonably good model; here z is an effective nuclear charge (e.g., +1 for the case of a Zn atom in Cu) and κ_{TF} is the Thomas Fermi screening length, discussed in Section 14.2.1, following Eq. (14.34)). The Fourier transform of (19.76) is given by

$$V(\mathbf{q}) = \frac{ze^2}{q^2 + \kappa_{TF}^2}. \tag{19.77}$$

We then have

$$\sigma_{tr} = 2\pi \left(\frac{2m}{\hbar^2}\right)^2 \int d(\cos\psi)\,[1-\cos\psi] \frac{z^2 e^4}{\left(\left(2k_F \sin\frac{\psi}{2}\right)^2 + \kappa_{TF}^2\right)^2}. \tag{19.78}$$

We will not carry out this integral, but only note that it is of order $\sigma_{tr} = \pi r_{tr}^2$, where $r_{tr} = 2mze^2/\hbar^2\kappa_{TF}^2$.

In the case of a real metal with a Fermi surface consisting of a single anisotropic sheet, the solution of (19.70) would require $\tau = \tau(\mathbf{k})$; we would then write $W(\varepsilon', \psi) \to W(\varepsilon'; \mathbf{n}, \mathbf{n}')$, where $\mathbf{n}, \mathbf{n}'$ denote unit vectors defining the position on the Fermi surface (which we here assume to be single-valued with respect to Fermi radii measured from the zone center). If the surface contains multiple sheets, the distribution function itself must have a band index $f(\mathbf{k}) \to f_n(\mathbf{k})$ and the collision integral will have *inter- and intra-band contributions.*

19.11 Limitations of the semi-classical theory

In addition to the limitations associated with over simplified models of the electronic structure and the scattering processes itself, there is a more fundamental limitation of the earlier theory, based on the use of the Boltzmann equation itself. We now discuss these limitations, as originally formulated by Peierls.

The scattering rate introduced in the collision integral (19.10) is based on the so-called golden rule of first-order time-dependent perturbation theory. In Section 10.1 we introduced the crystal potential for a perfectly periodic, static crystal, through the expression

$$V(\mathbf{r}) = \sum_{\mathbf{R}} U(\mathbf{r}-\mathbf{R}). \tag{10.1}$$

If we have a single-point defect contributing a potential $\mathbf{U}'(\mathbf{r}-\mathbf{R}')$ on site $\mathbf{R}'$ and there is no distortion of the potential associated with the surrounding atoms (i.e., no "relaxation" of their position or strength), then the new crystal potential is given by

$$\begin{aligned} V'(\mathbf{r}) &= \mathbf{U}'(\mathbf{r}-\mathbf{R}') - U(\mathbf{r}-\mathbf{R}') + \sum_{\mathbf{R}'} U(\mathbf{r}-\mathbf{R}) \\ &= \mathbf{u}(\mathbf{r}-\mathbf{R}') + V(\mathbf{r}), \end{aligned} \tag{19.79}$$

where $\mathbf{u}(\mathbf{r}-\mathbf{R}') \equiv \mathbf{U}'(\mathbf{r}-\mathbf{R}') - U(\mathbf{r}-\mathbf{R}')$; i.e., the potential of the lattice is altered by the function $\mathbf{u}(\mathbf{r}-\mathbf{R}')$. If the atoms surrounding the impurity alter their positions, our chosen form can still be used, only now the potential $\mathbf{u}(\mathbf{r})$ also includes the distortions arising from the surrounding host atoms, and would therefore be expected to have a greater spatial extent.

In the leading order of perturbation theory the electrons are scattered by the potential $\mathbf{u}(\mathbf{r})$ only "once"; this is what was assumed in calculating the matrix element $M_{\mathbf{kk}'}$ in (19.9a). Formally, the effects of multiple scattering from the same impurity are incorporated by replacing $M_{\mathbf{kk}'}$ by the so-called *scattering amplitude*, $F_{\mathbf{kk}'}$, as discussed in most quantum mechanics textbooks; $F_{\mathbf{kk}'}$ follows from $M_{\mathbf{kk}'}$ by solving an integral equation involving both functions, and for our present purposes we can simply assume that this substitution has been made.

Now let us assume that we have a number of impurities centered at a collection of sites $\mathbf{R}'$. We may write the total potential of our crystal as

$$\begin{aligned} V'(\mathbf{r}) &= \sum_{\mathbf{R}'}^{N_i} \mathbf{u}(\mathbf{r}-\mathbf{R}') + V(\mathbf{r}) \\ &= v(\mathbf{r}) + V(\mathbf{r}), \end{aligned} \tag{19.80}$$

where N_i is the total number of impurities. In evaluating the matrix element associated with a *distribution of* impurity scattering centers a number of additional problems enter. A natural first assumption is:

> Assumption 1: The number of scattering centers is small enough that the overall electronic structure of the host crystal is not significantly altered; i.e., we may continue to use the Bloch states obtained for the perfect crystal.

We next examine an assumption built into the golden rule itself, that the scattering process takes place over an infinite time. In first-order time-dependent perturbation theory one writes the total time-dependent wave function of the system as

$$\psi(\mathbf{r},t) = \sum_{\mathbf{k},n} a_{\mathbf{k},n}(t)\, \psi_{\mathbf{k},n}(\mathbf{r}) e^{-i\varepsilon_{\mathbf{k},n}t/\hbar}, \tag{19.81}$$

where we have included the band index suppressed earlier. Substituting (19.81) into the time-dependent Schrödinger equation one obtains the usual first-order differential equation for the expansion coefficients $a_{\mathbf{k},n}(t)$ as

$$i\hbar\dot{a}_{\mathbf{k},n}(t) = \sum_{\mathbf{k}',n'} \langle\mathbf{k},n|\, v(\mathbf{r})\, |\mathbf{k}',n'\rangle\, e^{i\left(\varepsilon_{\mathbf{k},n}-\varepsilon_{\mathbf{k}',n'}\right)t/\hbar} a_{\mathbf{k}',n'}. \tag{19.82}$$

Assuming the electron to be in state $\mathbf{k}', n'$ at $t = 0$, we integrate this equation and obtain

$$a_{\mathbf{k},n}(t) = -\sum_{\mathbf{k}',n'} \langle\mathbf{k},n|\, v(\mathbf{r})\, |\mathbf{k}',n'\rangle\, a_{\mathbf{k}',n'}(0) \frac{e^{i\left(\varepsilon_{\mathbf{k},n}-\varepsilon_{\mathbf{k}',n'}\right)t/\hbar} - 1}{\varepsilon_{\mathbf{k},n} - \varepsilon_{\mathbf{k}',n'}}. \tag{19.83}$$

The probability that it will be in state $\mathbf{k}, n$ at time t is therefore

$$\begin{aligned} \left|a_{\mathbf{k},n}(t)\right|^2 &= \left|\langle\mathbf{k},n|\, v(\mathbf{r})\, |\mathbf{k}',n'\rangle\right|^2 2\frac{1-\cos\left[\left(\varepsilon_{\mathbf{k}',n'}-\varepsilon_{\mathbf{k},n}\right)t/\hbar\right]}{\left(\varepsilon_{\mathbf{k}',n'}-\varepsilon_{\mathbf{k},n}\right)^2} \\ &\equiv \left|\langle\mathbf{k},n|\, v(\mathbf{r})\, |\mathbf{k}',n'\rangle\right|^2 D(t). \end{aligned} \tag{19.84}$$

One can introduce a transition probability, $w \equiv \left|a_{\mathbf{k},n}(t)\right|^2/t$, take the limit $t \to \infty$, and note that

$$\lim_{t\to\infty} \frac{D(t)}{t} = \frac{2\pi}{\hbar}\delta\left(\varepsilon_{\mathbf{k}',n'} - \varepsilon_{\mathbf{k},n}\right), \tag{19.85a}$$

yielding the golden rule in the form

$$w = \frac{2\pi}{\hbar}\left|\langle\mathbf{k},n|\, v(\mathbf{r})\, |\mathbf{k}',n'\rangle\right|^2 \delta\left(\varepsilon_{\mathbf{k}',n'} - \varepsilon_{\mathbf{k},n}\right). \tag{19.85b}$$

However taking this limit is not always justified, for reasons that we now examine.

Recall that we assumed that the electron was initially in state $\mathbf{k}, n$. However, if there were more than one initial state in Eq. (19.83), cross terms, resulting in interferences, would have occurred in the final state probability. In general, we expect the initial state to involve a superposition of many states, since the perturbation $v(\mathbf{r})$ is not generally turned on at $t = 0$, but has been acting for an extended time. As an example, if the potential is *repulsive* there must be, in the steady state, correlations among electrons in various states that result in a *reduction* in the total electron density in the vicinity of the impurity; i.e., the usual running Bloch electron waves superimpose so as to acquire a standing wave character in the vicinity of the impurity, implying special phase relationships among them. Neglecting such correlations involves a second assumption:

> Assumption 2: The nature of the perturbing potential is such that, if the electrons are distributed in the unperturbed Bloch states, then statistical averages (involving, say, transport coefficients) are unaffected by any local electron correlations near impurities.

One requirement that must be satisfied if this assumption is to be correct is that the impurities should be distributed randomly (periodically ordered impurities implies a different crystal structure). In addition, we must assume that the impurities are sufficient in number to produce well-defined averages.

If the initial state consists of a superposition of two or more Bloch states in the form $\left[a_{\mathbf{k}',n'}\left|\mathbf{k}',n'\right\rangle + a_{\mathbf{k}'',n''}\left|\mathbf{k}'',n''\right\rangle\right]$, then in forming the square of the matrix element entering (19.84) we will have diagonal terms of the form

$$\left|a_{\mathbf{k}',n'}\right|^2 \sum_{\mathbf{R}',\mathbf{R}''} \left|\langle\mathbf{k},n|\,\mathbf{u}(\mathbf{r})\left|\mathbf{k}',n'\right\rangle\right|^2 \left[e^{i(\mathbf{k}'-\mathbf{k})\cdot(\mathbf{R}'-\mathbf{R}'')}\right], \tag{19.86a}$$

and cross terms of the form

$$a^*_{\mathbf{k}'',n''}a_{\mathbf{k}',n'} \sum_{\mathbf{R}',\mathbf{R}''} \left\langle\mathbf{k}'',n''\right|\mathbf{u}(\mathbf{r})\left|\mathbf{k},n\right\rangle \langle\mathbf{k},n|\,\mathbf{u}(\mathbf{r})\left|\mathbf{k}',n'\right\rangle e^{i(\mathbf{k}-\mathbf{k}'')\cdot\mathbf{R}'-i(\mathbf{k}-\mathbf{k}')\cdot\mathbf{R}''}. \tag{19.86b}$$

Now the sums over $\mathbf{R}'$ and $\mathbf{R}''$ in (19.86a) will in general be oscillatory and uncorrelated (assuming the scattering centers are far enough apart) and should average to zero; an exception is the case when $\mathbf{R}' = \mathbf{R}''$. For electrons in the same band this term will not vanish and produces a contribution proportional to the number of scatters, N_i, which is the desired behavior: doubling the number of scatters doubles the total effective scattering (and with it, for example, the resistivity). If the electrons are in different bands they will not contribute since they have different energy and are not connected in an elastic scattering process. The terms of the form (19.86b) vanish on averaging over $\mathbf{R}''$ unless $\mathbf{k}' = \mathbf{k}''$; i.e., only the diagonal terms survive.

Collecting the above, we have the time dependence of the distribution function,

$$f(\mathbf{k}, n, t) - f(\mathbf{k}, n, 0) = \left|a_{\mathbf{k},n}(t)\right|^2 - \left|a_{\mathbf{k},n}(0)\right|^2$$
$$= \sum_{\mathbf{k}',n'} \left|\langle \mathbf{k}, n|\, v(\mathbf{r})\, |\mathbf{k}', n'\rangle\right|^2 \left[f(\mathbf{k}', n) - f(\mathbf{k}, n)\right] D(t). \quad (19.87)$$

If we extend the time t to infinity, divide both sides by t, and use (19.84), we obtain

$$\dot{f}(\mathbf{k}, n) = \frac{2\pi}{\hbar} \sum_{\mathbf{k}',n'} \left|\langle \mathbf{k}, n|\, v(\mathbf{r})\, |\mathbf{k}', n'\rangle\right|^2 \left[f(\mathbf{k}', n') - f(\mathbf{k}, n)\right], \quad (19.88)$$

which corresponds to (19.10b).

When the scattering potential $u(\mathbf{r})$ is approximately confined to a unit cell of dimension a, then for a metal where the Fermi wave vector $k_F \cong a^{-1}$, we have the situation $k_F a \cong 1$; under this condition most of the scattering events will be through fairly wide angles and it is reasonable to average the final states over the Fermi surface and introduce a "local" scattering time, $\tau(\mathbf{k})$.

At low temperatures the distribution function varies rapidly over an energy interval $k_B T$ about the Fermi energy. In order that the energies in this region should be well defined we must have

$$k_B T \gg \frac{\hbar}{t}; \quad (19.89)$$

i.e., the time t in the function D(t) must be extended to times satisfying this inequality. On the other hand, physically, we cannot extend the time beyond a relaxation time τ because subsequent collisions then occur. We are then left with the result that our semi-classical transport theory requires:

Assumption 3:
$$k_B T \gg \frac{\hbar}{\tau}. \quad (19.90)$$

When this inequality is not satisfied, the true distribution function must have a greater width than the semi-classical distribution function. This does not imply that energy is not conserved, since we are using the unperturbed states to describe our scattering problem, and are ignoring the perturbing potential itself. The total energy is always conserved, but if another scattering event occurs *before* the states being scattered escape to infinity (where their energies do not involve the perturbation for an elastic collision), then the energy is shifted by the potential $v(\mathbf{r})$, effectively spreading out the distribution.

In recent years it was recognized that when there is a high density of scattering sites and the temperature is low, individual electrons with amplitudes involving time reversed paths that return to a given scattering site can *interfere with themselves*, an effect which we have neglected. Therefore we have:

Assumption 4: Individual electrons do not undego self interference.

Modern Green's function based transport theories formally incorporate some of the effects described, but developing the methodology utilized lies beyond the scope of this book. Hence we have here limited ourselves to a qualitative discussion. However, electron interference effects associated with Assumption 4 will be treated qualitatively in Section 43.4.

19.12 Time-dependent external fields; the Chambers path integral method

Up to this point our discussions have been limited to time-independent problems. Here we develop a technique for solving the Boltzmann equation in the presence of *time-dependent* external fields. We will continue to use the relaxation time approximation for the collision integral, but with the following generalization. As was discussed in connection with Eq. (19.14), when the temperature and chemical potential are inhomogeneous, a better starting point is to assume a *local-equilibrium form* $f_0 = f_0(\varepsilon, T(\mathbf{r}), \mu(\mathbf{r}))$. If time-dependent fields are present, and if in addition the lattice is in motion, as one has when a sound wave is present (a kind of external field resulting in a local lattice velocity $\dot{\mathbf{U}}$ and, in the case of a longitudinal wave a change in the local density), then the local equilibrium distribution function depends on both space *and* time; in addition, it can reside in a *moving reference frame.* It is then natural to approximate the collision integral as

$$I[f] = -\frac{f(\mathbf{v}, \mathbf{r}, t) - f_0\left(\mathbf{v} - \dot{\mathbf{U}}, T(\mathbf{r}, t), \mu(\mathbf{r}, t)\right)}{\tau(\mathbf{k})}, \tag{19.91}$$

where

$$f_0\left(\mathbf{v} - \dot{\mathbf{U}}, T(\mathbf{r}, t), \mu(\mathbf{r}, t)\right) = \left[\exp\left(\frac{\frac{1}{2}m(\mathbf{v} - \dot{\mathbf{U}})^2 - \mu(\mathbf{r}, t)}{k_B T(\mathbf{r}, t)}\right) + 1\right]^{-1}; \tag{19.92}$$

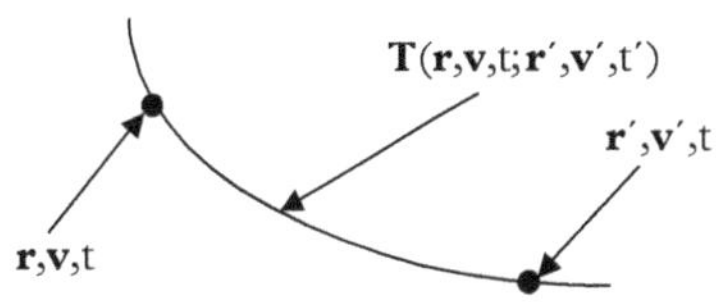

for convenience, we have used $\mathbf{v}$ rather than $\mathbf{k}$ in the distribution function.[5,6]

The essence of the Chambers method is to note that the "present" distribution function, $f(\mathbf{v}, \mathbf{r}, t)$, must arise from the product of two processes: In the first of these electrons characterized by an equilibrium distribution function $f_0(\mathbf{v}', \mathbf{r}', t')$ are scattered at a rate $f_0(\mathbf{v}', \mathbf{r}', t')/\tau$ onto a *trajectory,* $\mathbf{T}(\mathbf{r}, \mathbf{v}, t; \mathbf{r}', \mathbf{v}', t')$, that carries them from the earlier time, t', to the present time, t. In the second process,

[5] The kinetic formulation of conduction problems, R.G. Chambers, *Proc. Phys. Soc. Lond.* **A65**, 458 (1952).

[6] The effect of a moving reference frame as it affects sound propagation will be discussed in Section 21.1.

electrons on this trajectory decay between t′ and t, at a rate $e^{-(t-t')/\tau(\mathbf{k})}$. Integrating this *net* rate over all earlier times then gives

$$f(\mathbf{v}, \mathbf{r}, t) = \int_{-\infty}^{t} dt' \frac{f_0(\mathbf{v}', \mathbf{r}', t')}{\tau(\mathbf{k})} e^{-(t-t')/\tau(\mathbf{k})}. \tag{19.93}$$

The external fields enter the problem by fixing the trajectory itself, which evolves according to the Lorentz force, $\mathbf{F} = e\mathbf{E} + (e/c)\,\mathbf{v} \times \mathbf{H}$; i.e., we must solve Newton's equations for the position $\mathbf{r}$ and velocity $\mathbf{v}$ of the particle at time t with the initial conditions that it had position $\mathbf{r}'$ and velocity $\mathbf{v}'$ at an earlier time t′.

19.13 Collective excitations

Under appropriate conditions the electrons in a metal can undergo collective oscillations which, loosely speaking, are resonant modes of the electron liquid in which a macroscopic fraction of the electrons participate. Among the better known collective modes are the plasma oscillations in metals, zero-sound in ^{3}He (both of which were discussed in Chapter 14), helicons (or whistlers), magneto plasmon waves (Alven waves), and spin waves. Here we discuss the plasma oscillations.

Consider a uniform electron gas, the total charge of which is canceled by a rigid positive background continuum. If the electron gas is disturbed from equilibrium, resulting in a number density $n(\mathbf{r}, t) = n_0 + \delta n(\mathbf{r}, t)$, it will set up an electric field,

$$\mathbf{E}(\mathbf{r}, t) = -\nabla\phi(\mathbf{r}, t), \tag{19.94}$$

where $\phi(\mathbf{r}, t)$ is the electrostatic potential which can be obtained from Poisson's equation

$$\nabla^2\phi(\mathbf{r}, t) = -4\pi e\delta n(\mathbf{r}, t); \tag{19.95}$$

here $\delta n(\mathbf{r}, t)$ follows from the change in the distribution function,

$$\delta n(\mathbf{r}, t) = \int d^3k\,\delta f(\mathbf{k}, \mathbf{r}, t). \tag{19.96}$$

We will obtain $\delta f(\mathbf{k}, \mathbf{r}, t)$ from the linearized *collisionless* Boltzmann equation,[7]

$$\frac{\partial \delta f}{\partial t} + \frac{\partial \delta f}{\partial \mathbf{r}} \cdot \mathbf{v} + \frac{\partial f_0}{\partial \mathbf{k}} \cdot \dot{\mathbf{k}} = 0, \tag{19.97}$$

with $\hbar\dot{\mathbf{k}} = e\mathbf{E}$. By ignoring any scattering among the charged particles we are implicitly assuming that the dominant effect of distorting the charge distribution is to produce a *coherent* collective electric field via Poisson's equation. The theory is made self-consistent by solving equations (19.94) through (19.97)

[7] The frequency of the plasma mode is of the order of 10^{16}Hz, whereas the collision frequency at room temperature is of order 10^{13}Hz; hence ignoring collisions is amply justified.

simultaneously. Assuming that $\phi(\mathbf{r}, t) = \phi_q e^{-i\omega t + i\mathbf{q}\cdot\mathbf{r}}$ and similarly for the remaining position and time-dependent functions (19.94), (19.95), and (19.97) become

$$\mathbf{E}_q = -i\mathbf{q}\phi_q, \tag{19.98a}$$

$$\phi_q = \frac{4\pi e}{q^2} n_q, \tag{19.98b}$$

and

$$\delta f_q = -i\frac{\partial f_0}{\partial \varepsilon}\frac{e\mathbf{E}_q \cdot \mathbf{v}}{(\omega - \mathbf{q}\cdot\mathbf{v})}, \tag{19.98c}$$

respectively. Combining these equations and using (19.96) we obtain the condition

$$\frac{4\pi e^2}{q^2}\frac{2}{(2\pi)^3}\int d^3k \frac{\partial f_0}{\partial \varepsilon}\frac{qv\cos\theta}{(\omega - qv\cos\theta)} = 1. \tag{19.99a}$$

We seek a solution in the long wavelength (small q) limit. Expanding the denominator we have

$$-\frac{4\pi e^2}{q^2}\frac{2}{(2\pi)^3}\int d^3k \frac{\partial f_0}{\partial \varepsilon}\left[\frac{qv\cos\theta}{\omega} + \left(\frac{qv\cos\theta}{\omega}\right)^2 \cdots\right] = 1. \tag{19.99b}$$

The first term is odd in θ and vanishes on integrating over angles. What remains has a solution in which δf is isotropic; using the zero temperature form of $\partial f_0/\partial\varepsilon$, writing $k^2 dk = (k^2/\hbar v)\, d\varepsilon$ and $v = \hbar k/m$, and integrating over the energy we are left with

$$\begin{aligned}\omega^2 &= 4\pi e^2 \frac{k_F^3}{2\pi^2 m}\int d(\cos\theta)\cos^2\theta \\ &= \frac{4\pi e^2 n}{m},\end{aligned} \tag{19.100}$$

where n is the electron number density. This result is the same as the classical Drude result of Section 3.4, as well as that found from our analysis of the dielectric function in Section 14.2.2. The next non-vanishing term in the expansion involves $(qv\cos\theta/\omega)^4$ and yields the dispersion (see discussion in Section 14.2.2).

ADDITIONAL READING

An extensive discussion of transport properties is given in: *Electrons and Phonons: The Theory of Transport Phenomena in Solids*, J.M. Ziman, Clarendon Press, Oxford (1960).

Solid State Physics by N. Ashcroft and N.D. Mermin, Holt, Reinhart, and Winston, PA (1976), has an excellent discussion at an introductory level.

For a concise treatment of transport phenomena see: *Principles of the Theory of Solids* by J. Ziman, Cambridge University Press (1964). Our approach in this chapter was partly modeled after that of Ziman.

Chapter 12 of *Quantum Theory of Solids* by C. Kittel, John Wiley & Sons (1963) is devoted to various aspects of magnetoresistance.

Magneto resistance and other aspects of electron dynamics, particularly as they relate to Fermi surface topology, are treated in *Magnetoresistance in Metals* by A.B. Pippard, Cambridge Studies in Low Temperature Physics (2009).

Another book devoted to electron transport is: *Physics of Electronic Conduction in Solids*, F.J. Blatt, McGraw-Hill (1968).

The discussion of electrical transport in *Quantum Theory of Solids*, by R.E. Peierls, Oxford (1955), contains much insight; many of the fundamental problems with the early quantum theories of electrical transport, which lay dormant for many years, are clearly identified here.

Chapters VIII, IX, and X of *The Theory of Metals*, by A.H. Wilson, Cambridge University Press (1958), have detailed treatments of traditional models of electrical and thermal transport as parameterized by effective masses, carrier densities, and relaxation times.

Chapter IX of *Physical Kinetics*, by E.M. Lifshitz and L.P. Pitaevskii, Pergamon Press (1981), has an excellent discussion of electrical and thermal transport in metals.

A general equation for electrical conduction and other transport properties, along with the underlying theory, is given in, Statistical-mechanical theory of irreversible processes .1. general theory and simple applications to magnetic and conduction problems, R. Kubo, *J. Phys. Soc. Jpn.* **12**, 570 (1957).

Problems

19.1. Verify the statements made in the last paragraph of Section 19.3.

19.2. Assume that you have a time-dependent internal electric field in a metal, involving a single Fourier component of the form

$$\mathbf{E}(\mathbf{r}, t) = \text{Re}\left[\mathbf{E}(\mathbf{q}, \omega)e^{i\omega t - i\mathbf{q}\cdot\mathbf{r}}\right].$$

Even in the absence of a temperature gradient one will then have a non-uniform electron number density, $n = n_{eq} + \delta n(\mathbf{r}, t)$, of the form

$$\delta n(\mathbf{r}, t) = \text{Re}\left[\delta n(\mathbf{q}, \omega)e^{i\omega t - i\mathbf{q}\cdot\mathbf{r}}\right],$$

where n_{eq} is the equilibrium number density.

(a) Using the Boltzmann equation in the relaxation time approximation (under the assumption scattering does not affect the local electron density), show that the change in the distribution function has the form

$$\delta f(\mathbf{q},\omega) = \frac{\tau^{-1}\delta\mu(\mathbf{q},\omega) - e\mathbf{v}(\mathbf{k})\cdot\mathbf{E}(\mathbf{q},\omega)}{\tau^{-1} - i(\omega - \mathbf{q}\cdot\mathbf{v}(\mathbf{k}))}\left(-\frac{\partial f}{\partial \varepsilon}\right),$$

where

$$\delta\mu(\mathbf{q},\omega) = \frac{\partial\mu}{\partial n_{eq}}\delta n(\mathbf{q},\omega).$$

(b) Show that this form satisfies the equation of continuity,

$$\nabla\cdot\mathbf{j}(\mathbf{q},\omega) + \frac{\partial\rho(\mathbf{q},\omega)}{\partial t}.$$

19.3. In Problem 19.2 we found that $\delta f(\mathbf{q},\omega)$ depended on two independent variables, $\delta\mu(\mathbf{q},\omega)$ and $\mathbf{E}(\mathbf{q},\omega)$. Under the assumption that the number density remains constant, show that the frequency-dependent electrical conductivity is given by

$$\sigma(\mathbf{q},\omega) = \frac{2e^2}{(2\pi)^3}\int d^3k\frac{\mathbf{v}(\mathbf{k})\mathbf{v}(\mathbf{k})}{\tau^{-1} - i(\omega - \mathbf{q}\cdot\mathbf{v}(\mathbf{k}))}\left(-\frac{\partial f}{\partial \varepsilon}\right).$$

19.4. You are given an isotropic metal in which a plane wave mode with frequency ω and wave vector q is propagating parallel to an external magnetic field $\mathbf{H}$. By applying Maxwell's equations show that the components of the electric field and the current parallel and perpendicular to $\mathbf{H}$ are related by

$$E_{//} = \frac{4\pi}{i\omega}j_{//}, \qquad \mathbf{E}_{\perp} = \frac{4\pi i\omega}{c^2q^2 - \omega^2}\mathbf{j}_{\perp}.$$

Now assume that $\mathbf{j}_{\perp}$ and $\mathbf{E}_{\perp}$ are related by Eq. (19.56). Show that a circularly polarized wave can propagate parallel to the magnetic field, and find the relation between ω and q (in the limit $\omega/q \ll c$). What is the condition that this wave will propagate?

19.5. From Eqs. (19.91) and (19.92) show that the correction to the distribution function in lowest order may be written,

$$\delta f = -\frac{\partial f_0}{\partial \varepsilon}\int_{-\infty}^{t} dt'\left[v'\cdot(-e\mathbf{E}' + m\dot{\mathbf{U}}'/\tau) + \frac{2\varepsilon_F\delta n'}{3n_0\tau}\right]e^{-(t-t')/\tau},$$

where $\delta n'$, $\mathbf{E}'$, and $\dot{\mathbf{U}}'$ are all proportional to $e^{i\mathbf{q}\cdot\mathbf{r}'-i\omega t'}$, and $\mathbf{r}'$ and $\mathbf{v}'$ lie on the trajectory (path) traveled in the absence of the background velocity $\dot{\mathbf{U}}$ (caused by a sound wave passing through the metal). (Hint: the field $\mathbf{E}$ enters via the energy gained from the electric field along the trajectory.)

20 Measuring the Electronic Energy Spectrum On and Off the Fermi Surface

In this chapter we will describe various methods that have been used to measure the energy spectrum of electrons in both metals and insulators. For metals, a property of paramount importance is the detailed shape of the Fermi surface and the accompanying Fermi velocity, since together they affect transport and other phenomena. Methods to measure these quantities include various magneto-acoustic effects (e.g., the geometric resonance phenomena), the Gantmakher and Sondheimer size effects, the Azbel–Kaner cyclotron resonance, the anomalous skin effect, the high-field magneto-resistance (as it relates to open versus closed orbits), and various quantum oscillations, the most important of which is the de Hass van Alphen effect.

Many of these techniques do not measure the Fermi wave vector or velocity directly, but rather various averages. Therefore, their utility depends on whether these averages can be *de-convoluted*, a process we will refer to as *inversion*. Also important is the degree to which they involve Fermi surface quantities directly, and are not masked by, say, relaxation times. Hence electrical transport properties can be measured with great accuracy, but since they also involve electron scattering (and in most metals many bands), they are of limited utility for determining Fermi surface properties.

Perhaps more important is to have tools to measure the energy spectrum *away* from the Fermi surface. Here one must contend with finite lifetime effects which set limits on the accuracy with which this can be carried out. The most powerful technique for determining the energy spectrum at energies *below* the Fermi surface is angle-resolved photo-emission spectroscopy (ARPES) which, with the emergence of synchrotron sources, has become increasingly important. Also of growing importance is the inverse of photo-emission spectroscopy (IPS), a special case of which is termed bremsstrahlung isochromat spectroscopy (BIS). Here the roles of the incoming photon and the final state electron are reversed: electrons of known energy (and angle) impinge on the surface and (in the BIS variant) the energy spectrum of the emitted photons is measured. This technique probes the spectrum for energies *above* the Fermi energy.

The Physics of Solids. First Edition. J. B. Ketterson.

20.1 Determining the Fermi surface

We will begin by discussing the de Hass van Alphen (dHvA) effect. Although it is the oldest technique, it has turned out to be the most powerful, for reasons we will elaborate on. We will then give brief descriptions of the magneto-acoustic and Gantmakher techniques, which are now of historical interest only (cyclotron resonance was briefly discussed in Chapter 18).

The dHvA effect is a periodic variation of the magnetization with the reciprocal of the internal magnetic field, which is observed in pure metals at low temperatures. Its interpretation is based on the Lifshitz–Onsager quantization of the orbits of electrons in a magnetic field, discussed earlier in Section 18.5. Because of its importance we will derive the Lifshitz–Kosevich expression for the temperature and field dependence of the magnetization in its entirety.[1] Those not interested in the details can proceed directly to the expression for the magnetization, Eq. (20.18), and the discussion that follows this.

20.1.1 The Lifshitz–Kosevich equation for the dHvA effect

The derivation is most direct if we assume that the electrons in the metal are in contact with an external system that holds the chemical potential constant. If the Fermi surface of the metal consists of several sheets, we may picture the electrons as passing among these sheets so as to keep the overall chemical potential approximately constant. If however we have a single sheet then, strictly speaking, it is the number of electrons rather than the chemical potential which remains constant. The effects of these differences are usually small in practice.

Under the above circumstance it is most convenient to work with the thermodynamic potential Ω, which is defined, for a magnetic system, as $\Omega = F - G = -MH$, where F and G are the Helmholtz and Gibbs free energies respectively. We write $G = \mu N$, where N is the total number of electrons. Using the expression $dF = -SdT - MdH + \mu dN$, we obtain

$$d\Omega = -SdT - MdH - Nd\mu \tag{20.1}$$

or

$$M = -\left(\frac{\partial \Omega}{\partial H}\right)_{T,\mu}; \tag{20.2}$$

i.e., it is evaluated at constant μ, as we have discussed. For a Fermi system at zero magnetic field, the Ω potential is given by

$$\Omega = -\frac{2k_B T}{(2\pi)^3}\int d^3k \ln\left[1 + e^{(\mu-\varepsilon_n)/k_B T}\right]; \tag{20.3}$$

[1] Theory of magnetic susceptibility in metals at low temperatures, I.M. Lifshitz and A.M. Kosevich, *Sov. Phys. JETP-USSR* 2, 636 (1956).

to verify this one only has to differentiate with respect to μ, use Eq. (20.1), and note that N is given by

$$N = \frac{2}{(2\pi)^3}\int d^3k \frac{1}{e^{(\varepsilon_n - \mu)/k_B T} + 1}. \tag{20.4}$$

In the presence of a magnetic field the motion transverse to the field is quantized, with the area of the quasi-classical orbits in k-space being given by

$$A_k = \frac{2\pi eH}{c\hbar}(n + \gamma). \tag{18.34b}$$

The area of an annulus between n + 1 and n is given by $2\pi eH/c\hbar$; when H is small we can picture this as a differential, dA_k, assuming $dA_k \ll A_k$. In order to approximately conserve the number of phase space states when we go over to discrete states in the presence of a magnetic field, we set

$$\frac{2L^3}{(2\pi)^3}\int d^3k = 2L^3 \int \frac{dA(k_z)}{(2\pi)^2}\frac{dk_z}{2\pi} = \frac{eHL^3}{c\,(2\pi\hbar)}\sum_{\sigma_z}\sum_{n}\int \frac{dk_z}{2\pi}, \tag{20.5}$$

where we explicitly sum over the two spin states in the last term. The Ω potential is then given by

$$\Omega(T, H, \mu) = -\frac{eHL^3 k_B T}{c\hbar\,(2\pi)^2}\sum_{\sigma_z}\sum_{n}\int dk_z \ln\left[1 + e^{(\mu - \varepsilon(n, k_z, \sigma_z)/k_B T}\right], \tag{20.6}$$

where $\varepsilon(n, k_z, \sigma_z) = \varepsilon(n, k_z) + g\mu_B \sigma_z H$ and g is the electron g-factor.[2] We henceforth write $\mu - g\mu_B\sigma_z H = \tilde{\mu}$. To evaluate the sum over n in this expression we make use of the Poisson summation formula[3]

$$\sum_{n=1}^{\infty} f(n) = -\frac{1}{2}f(0) + \int_0^{\infty} f(n)dn + 2\Re\sum_{l=1}^{\infty}\int_0^{\infty} f(x)e^{2\pi i l n}dn. \tag{20.7}$$

Applying this expression to Eq. (20.6), we have

$$\Omega(T, H, \mu) = \Omega_{mono}(T, H, \mu) + \Omega_{osc.}(T, H, \mu), \tag{20.8}$$

where $\Omega_{mono}(T, H, \mu)$ is monotonic in H and will be ignored and

$$\Omega_{osc.}(T, H, \mu) = -\frac{eHL^3 k_B T}{c\hbar\,(2\pi)^2}\sum_{\sigma_z}\int_0^{\infty} dn \int_{k_z^{min}}^{k_z^{max}} dk_z$$

$$\times\, 2\Re\sum_{l=1}^{\infty} e^{2\pi i l n}\ln\left[1 + \exp\left(\frac{(\tilde{\mu} - \varepsilon(n, k_z))}{k_B T}\right)\right] \tag{20.9a}$$

[2] For lighter elements we would have g = 2; for heavier elements g can differ significantly from this value.

[3] This expression follows immediately from recognizing that the coefficients of a Fourier series expansion of a periodic δ function are unity; i.e., $\sum_{n=-\infty}^{n=+\infty}\delta(x-n) = \sum_{l=-\infty}^{l=+\infty} e^{2\pi i l x}$.

$$= -2\Re \frac{eHL^3 k_B T}{c\hbar (2\pi)^2} \sum_{\sigma_z} \sum_{l=1}^{\infty} \int_0^{\infty} d\varepsilon \ln\left[1 + \exp\left(\frac{(\tilde{\mu} - \varepsilon)}{k_B T}\right)\right]$$

$$\times \int_{k_z^{min}}^{k_z^{max}} \frac{dn(\varepsilon, k_z)}{d\varepsilon} e^{2\pi i l n(\varepsilon, k_z)} dk_z, \tag{20.9b}$$

where we have switched from an integration over dn (a continuous variable after introducing Eq (20.7)) to one over $d\varepsilon$, and we write $n = n(\varepsilon, k_z)$ in place of $\varepsilon = \varepsilon(n, k_z)$; from (18.34b) it follows that $(c\hbar/2\pi eH)A_k - \gamma = n$. The integrand in the integral over dk_z involves a rapidly oscillating function that averages to zero, except in the vicinity of k_{ex}, where $n = n(\varepsilon, k_z)$ is presumed to have an extremum (a maximum or minimum) with respect to k_z; we can then expand this function about k_{ex} as

$$n(\varepsilon, k_z) = n(\varepsilon, k_{ex}) + \frac{1}{2} \left.\frac{\partial^2 n(\varepsilon, k_z)}{\partial k_z^2}\right|_{k_z = k_{ex}} (k_z - k_{ex})^2 \tag{20.10}$$

and this integral becomes

$$\int_{k_z^{min}}^{k_z^{max}} \frac{dn(\varepsilon, k_z)}{d\varepsilon} e^{2\pi i l n(\varepsilon, k_z)} dk_z = \frac{dn(\varepsilon, k_{ex})}{d\varepsilon} e^{2\pi i l n(\varepsilon, k_{ex})} \int_{k_z^{min}}^{k_z^{max}} e^{\pi i l n'' (k_z - k_{ex})^2} dk_z, \tag{20.11}$$

where we have written $n'' = \partial^2 n(\varepsilon, k_z)/\partial k_z^2\big|_{k_z = k_{ex}}$ and we have taken the slowly varying part from under the integral. Writing $z = e^{\pm i\pi/4}(k_z - k_{ex})$ (where the + and – signs are used for $n'' > 0$ and $n'' < 0$ respectively), we have

$$\int_{k_z^{min}}^{k_z^{max}} e^{i\pi l n'' (k_z - k_{ex})^2} dk_z = e^{\pm i\pi/4} \int_{-\infty}^{+\infty} e^{-\pi l n'' z^2} dz = e^{\pm i\pi/4} (l n'')^{-1/2}, \tag{20.12}$$

where we have extended the limits to $\pm\infty$. If there are several extrema then we must sum (20.12) over each of them. Combining (21.11) and (21.12), Eq. (20.9) becomes

$$\Omega_{osc.}(T, H, \mu) = -2\Re \frac{eHL^3 k_B T}{c\hbar (2\pi)^2} \sum_{\sigma_z} \sum_{l=1}^{\infty} \int_0^{\infty} d\varepsilon \ln\left[1 + \exp\left(\frac{(\tilde{\mu} - \varepsilon)}{k_B T}\right)\right]$$

$$\times \frac{dn(\varepsilon, k_{ex})}{d\varepsilon} e^{2\pi i l n(\varepsilon, k_{ex})} e^{\pm i\pi/4} (l n'')^{-1/2}. \tag{20.13}$$

Integrating by parts, we have

$$k_B T \int_0^\infty d\varepsilon \ln\left[1 + \exp\left(\frac{(\tilde{\mu} - \varepsilon)}{k_B T}\right)\right] e^{2\pi i l n(\varepsilon, k_{ex})} \frac{dn(\varepsilon, k_{ex})}{d\varepsilon} d\varepsilon$$

$$= \frac{1}{2\pi i l} \int_0^\infty d\varepsilon \frac{e^{2\pi i l n(\varepsilon, k_{ex})}}{e^{(\varepsilon - \tilde{\mu})/k_B T} + 1}$$

and (20.13) becomes

$$\Omega(T, H, \mu) = -2\Re \frac{eHL^3}{(2\pi)^2 \hbar} \frac{n''}{2\pi i} \sum_{\sigma_z} \sum_l l^{-3/2} e^{\pm i\pi/4} \int_0^\infty d\varepsilon \frac{e^{2\pi i l n(\varepsilon, k_{ex})}}{e^{(\varepsilon - \tilde{\mu})/k_B T} + 1} \tag{20.14}$$

The only significant part of this integral comes from the immediate vicinity of the Fermi energy, where we can make the expansion $n(\varepsilon, k_{ex}) = n(\mu, k_{ex}) + n'(\varepsilon, k_{ex})\big|_{\varepsilon = \tilde{\mu}} (\varepsilon - \tilde{\mu})$; thus,

$$\int_0^\infty d\varepsilon \frac{e^{2\pi i l n(\varepsilon, k_{ex})}}{e^{(\varepsilon - \tilde{\mu})/k_B T} + 1} = e^{2\pi i l n(\mu, k_{ex})} \int_0^\infty d\varepsilon \frac{e^{2\pi i l n'(\varepsilon - \tilde{\mu})}}{e^{(\varepsilon - \tilde{\mu})/k_B T} + 1}.$$

Now $\int_{-\infty}^{\infty} dy \frac{e^{i\alpha y}}{e^y + 1} = -\frac{i\pi}{i \sinh(\alpha\pi)}$ and (20.14) becomes

$$\Omega(T, H, \mu) = -2\Re \frac{eHL^3 k_B T}{(2\pi)^2 \hbar c} \frac{\sqrt{n''}}{2} \sum_{\sigma_z} \sum_l l^{-3/2} e^{\pm i\pi/4} \frac{e^{2\pi i l n(\tilde{\mu}, k_{ex})}}{\sinh\left(2\pi^2 l n' k_B T\right)}. \tag{20.15}$$

Recalling that $(c\hbar/2\pi eH)A_k - \gamma = n$, we have

$$n' = \frac{c\hbar}{2\pi eH} \frac{\partial A(k_z)}{\partial \varepsilon} = \frac{\hbar c m^*}{eH} \tag{20.16a}$$

and

$$\left(\frac{\partial^2 n}{\partial k_z^2}\right)^{1/2} = \left(\frac{\hbar c}{2\pi eH}\right)^{1/2} \left(\frac{\partial^2 A_k(\mu, k_{ex})}{\partial k_z^2}\right)^{1/2}. \tag{20.16b}$$

Expanding $n(\tilde{\mu}, k_{ex}) = n(\mu, k_{ex}) \pm n'(\mu, k_{ex}) g\mu_B \sigma_z H$, summing over the two spin states, and taking the real part of the resulting expression, we obtain

$$\Omega(T, H, \mu) = \frac{L^3 k_B T}{2\pi^{3/2}} \left(\frac{eH}{\hbar c}\right)^{3/2} \sum_i \left(\frac{\partial^2 A_i(k_z)}{\partial k_z^2}\right)^{-1/2} \sum_l \frac{1}{l^{5/2}}$$
$$\times \left[\sinh \frac{2\pi^2 l k_B T c m_i^*}{e\hbar H}\right]^{-1} \cos\left(\frac{l A_i(k_z)\hbar c}{eH} \pm \frac{\pi}{4}\right) \cos\left(\frac{\pi l m^*}{m_s}\right), \tag{20.17}$$

where the sum over i accounts for the possibility of *multiple* extremal area orbits on the Fermi surface (for the given field direction) and we have introduced a spin mass defined through the relation $g/2 = m_0/m_s$.

The magnetization follows from Eq. (20.2). In taking the derivative with respect to H we will assume that the temperature is high enough and/or the magnetic fields small enough that we retain only the $l = 1$ term in (20.17); also when $(\hbar c/eH)A_i(k_z) \gg 2\pi$, the changes in the magnetization are dominated by the oscillatory term and we may ignore the monotonic field dependencies (which produce a diamagnetic contribution, see Appendix 42A). The resulting oscillatory contribution to the magnetization is

$$M_{osc}(T,H) = -\frac{2^{1/2}L^3k_BT}{\pi^{3/2}}\left(\frac{e}{\hbar cH}\right)^{1/2}\sum_i A_i(k_z)\left(\frac{\partial^2 A_i(k_z)}{\partial k_z^2}\right)^{-1/2}$$
$$\times\frac{\exp\left(\frac{-2\pi cm^*}{e\tau H}\right)}{\sinh\frac{2\pi^2 k_BTcm_i^*}{e\hbar H}}\sin\left(\frac{A_i(k_z)\hbar c}{eH}\pm\frac{\pi}{4}\right)\cos\left(\frac{\pi m^*}{m_s}\right), \quad (20.18)$$

where we have added a factor, $\exp(-2\pi cm^*/e\tau H)$, which accounts for broadening of the Landau levels due to electron scattering (and was introduced independently by Dingle and Robinson); it has the effect of decreasing the amplitude of the oscillations.

Equation (20.18) contains the following five quantities relating to the Fermi surface. In the order in which they are generally studied, the three most important quantities measured are: the extremal area, A_i, which follows from the period of the oscillations with the reciprocal of the magnetic field; the effective mass, m^*, which is determined from the temperature dependence of the relative amplitude; and the relaxation time, τ, which (once the effective mass is measured) follows from the magnetic field dependence of the amplitude of the oscillations. Determining the spin mass (or equivalently the g-factor), m_s, requires careful measurements of the relative phase of the $l = 1$ and $l = 2$ terms in (20.17).[4] We note that there is a special situation when $m^*/m_s = n$, an integer, and the last factor in (20.18) vanishes. Lastly we have the "curvature" factor, $(\partial^2 A_i(k_z)/\partial k_z^2)^{-1/2}$. However, once the shape of the Fermi surface is known this quantity can be calculated directly with greater accuracy than it can be measured (it requires a calibrated measurement of the amplitude of the oscillations, which is difficult experimentally).

20.1.2 Experimental techniques to measure the dHvA effect

Measurements of the magnetization by purely static means requires (i) placing a sample in an inhomogeneous magnetic field and measuring the resulting force (the Faraday magnetometer), or (ii) placing a sample in a homogeneous field but orienting it in a direction other than a symmetry axis (such that the magnetization

[4] A nonlinear effect arising from the difference between B and H inside the metal also contributes to the $l = 2$ term; however it turns out the two contributions can be separated.

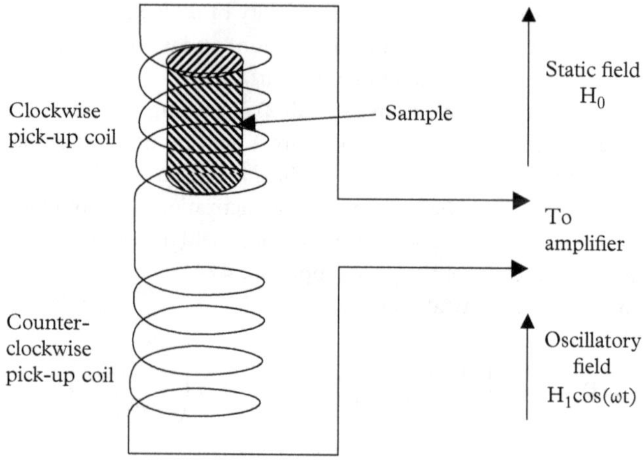

Figure 20.1 *The experimental arrangement used to detect the de-Haas-van-Alphen effect using the field modulation technique.*

is inclined relative to the applied field) and measuring the resulting torque (the torsion magnetometer).

Both of these methods are still used, but various dynamic methods are now more popular. The dynamic methods involve either moving a magnetized sample in and out of a pickup coil and measuring the induced voltage resulting from Faraday's law (the Foner magnetometer), or applying an oscillatory magnetic field, $H_1\cos(\omega t)$, in addition to the (generally much larger) static magnetic field, H_0 (the field modulation technique).[5] If the sample is placed in one of a pair of balanced pickup coils, as shown in Fig. 20.1, the signal obtained will involve oscillatory components resulting from a Taylor series expansion of the magnetization:

$$M(H_0 + H_1\cos(\omega t)) = M(H_0) + \left(\frac{dM}{dH}\right)_{H=H_0} H_1\cos(\omega t) + \frac{1}{2}\left(\frac{d^2M}{dH^2}\right)_{H=H_0} (H_1\cos(\omega t))^2 + \cdots; \qquad (20.19)$$

recalling $\cos^2(\omega t) = (1+\cos(2\omega t))/2$, we see that the signal at the second harmonic will be proportional to the second derivative of the magnetization with magnetic field.[6] Since it is the oscillatory term that varies the most rapidly we can again ignore the derivatives of all the monotonic terms in (20.18); hence the signal at the first harmonic is proportional to the susceptibility,

$$\chi(T,H) = \frac{\partial M_{osc}(T,H)}{\partial H} = \text{const.} \times \frac{T}{H^{5/2}}\sum_i A_i^2(k_z)\left(\frac{\partial^2 A_i(k_z)}{\partial k_z^2}\right)^{-1/2} \times \frac{\exp\left(\frac{-2\pi c m^*}{e\tau H}\right)}{\sinh\frac{2\pi^2 k_B T c m_i^*}{e\hbar H}} \sin\left(\frac{A_i(k_z)\hbar c}{eH} \pm \frac{\pi}{4}\right)\cos\left(\frac{\pi m^*}{m_s}\right) \qquad (20.20)$$

[5] For a detailed discussion see: Theory and technology for measuring de Haas-van Alphen type spectra in metals, R.W. Stark and L.R. Windmiller, *Cryogenics* **8**, 272 (1968); Techniques and instrumentation for measuring the de Hass-van Alphen effect in metals, L.R. Windmiller and J.B. Ketterson, *Rev. Sci. Instr.* **39**, 1672 (1968).

[6] Note the correspondence between the order of the derivative and the harmonic number breaks down beyond the second harmonic.

and the signal at the second harmonic will be proportional to $\partial\chi/\partial H$. Note that the signals involving larger cross-sectional areas, A_i (which are generally weaker as they tend to have higher effective mass), are enhanced by working at higher harmonics. The most sensitive way to determine the *anisotropy* of $A_i(\theta, \varphi)$ is firstly to determine the dHvA frequency along some initial direction (usually chosen to be a symmetry axis) by sweeping the magnetic field, and then, at constant field, rotate the sample *direction* relative to the field direction and count the number of oscillations as a function of angle to obtain the change in the frequency with angle.

20.1.3 Inversion of dHvA data and parameterization of the Fermi surface

In all cases the quantities measured in a dHvA experiment involve averages of "point" Fermi surface properties around a cyclotron orbit. Hence, if the technique is to achieve its full power, one needs techniques to "invert" these averages in order to determine the point properties. Such techniques have been developed to invert the orbital areas, cyclotron masses, relaxation times, g-factors, and changes in the areas with pressure/stress to obtain the Fermi radius, Fermi velocity, local relaxation time, local g-factor, and the change in the Fermi radius with pressure/stress. We will discuss here only some methods that have been used to determine the Fermi radius and Fermi velocity.

To perform these inversions we require a parameterization of the Fermi surface and its gradient with respect to energy in terms of two sets of functions. In developing a parametrization we must keep in mind that the space groups of crystals involve both point-rotation and translation (and occasionally mixed) operations.

Lifshitz and Pogorelov (1955) showed that, given the extremal-area function, $A(\theta,\phi)$, for a closed Fermi surface centered on a point having inversion symmetry, it was formally possible to invert this function to yield the Fermi radius function, $k(\theta,\phi)$, provided that the latter was single valued.[7] Their scheme was awkward to implement in practice; however a scheme proposed later by Mueller (1966), which we will discuss in some detail, is straightforward to apply.[8] We begin with a formal expression for the extremal area,

$$A(\theta, \phi) = \frac{1}{2} \int d\Omega' \left[k\left(\theta', \phi'\right)\right]^2 \delta\left(\mathbf{n}\left(\theta', \phi'\right) \cdot \mathbf{n}\left(\theta, \phi\right)\right) \tag{20.21}$$

where $\mathbf{n}(\theta, \phi)$ is a unit vector specified by the polar angles θ and ϕ. We now assume that $A(\theta, \phi)$ and $k^2(\theta, \phi)$ can be expanded in spherical harmonics as

$$A(\theta, \phi) = \sum_l a_{lm} Y_{lm}(\theta, \phi) \tag{20.22a}$$

and

$$k^2(\theta, \phi) = \sum_{l,m} b_{lm} Y_{lm}(\theta, \phi). \tag{20.22b}$$

[7] On the determination of Fermi surfaces and electron velocities in metals by the oscillation of magnetic susceptibility, I.M. Lifshitz and A.V. Pogorelov, *Dokl. Akad. Nauk SSSR* **96**, 1143 (1954).

[8] New inversion scheme for obtaining Fermi-surface radii from de Haas-van Alphen areas, F.M. Mueller, *Phys. Rev.* **148**, 636 (1966).

Next we expand the delta function in plane waves, and apply the Rayleigh expansion,[9]

$$e^{i\mathbf{k}\cdot\mathbf{r}} = 4\pi \sum_{l=0}^{\infty} \sum_{m=-l}^{l} i^l j_l(kr) Y^*_{lm}(\hat{\mathbf{k}}) Y_{lm}(\hat{\mathbf{r}})$$

to each of the plane waves to obtain

$$\begin{aligned} \delta(\mathbf{n}(\theta', \phi') \cdot \mathbf{n}(\theta, \phi)) &= \frac{2}{2\pi} \int_{-\infty}^{+\infty} dq e^{iq\mathbf{n}(\theta',\phi')\cdot\mathbf{n}(\theta,\phi)} \\ &= 2 \sum_{l,m} i^l Y_{lm}(\theta, \phi) Y_{lm}(\theta', \phi') I_l \end{aligned} \tag{20.23}$$

where the integral I_l in (20.23) is given by

$$I_l = \int_{-\infty}^{+\infty} j_l(q) dq.$$

Inserting the integral representation of the spherical Bessel function[10] in terms of Legendre polynomials

$$j_l(q) = \frac{(-i)^l}{2} \int_{-1}^{1} e^{iq\mu} P_l(\mu) d\mu,$$

we have

$$I_l = \int_{-\infty}^{+\infty} dq \frac{1}{2i^l} \int_{-1}^{+1} e^{i\mu q} P_l(\mu)\, d\mu. \tag{20.24}$$

Substituting (20.24) into (20.23), we have

$$\delta(\mathbf{n}(\theta', \phi') \cdot \mathbf{n}(\theta, \phi)) = \sum_{l,m} \left[\int_{-\infty}^{+\infty} dq \int_{-1}^{+1} e^{i\mu q} P_l(\mu) d\mu \right] Y_{lm}(\theta, \phi) Y_{lm}(\theta', \phi'),$$

where we have reversed the order of the integrations. Focusing on the term in the square brackets and carrying out the integration over dq yields $2\pi\delta(q)$, and the remaining integration over $d\mu$ produces $P_l(0)$; we are then left with

$$\delta(\mathbf{n}(\theta', \phi') \cdot \mathbf{n}(\theta, \phi)) = \sum_{l,m} 2\pi P_l(0) Y^*_{lm}(\theta, \phi) Y_{lm}(\theta', \phi'). \tag{20.25}$$

[9] Discussed in most texts on quantum mechanics in connection with scattering theory.

[10] See expression 10.1.14. in Handbook of Mathematical functions, M. Abramowitz and I. Stegun, Dover Publications, New York (1972).

Returning to Eq. (20.21), we insert (20.22a) on the left side and (20.22b) along with (20.24) on the right side. If we now exploit the orthonormality of the spherical harmonics, we obtain the simple result

$$a_{lm} = \pi P_l(0) b_{lm}. \tag{20.26}$$

Hence, by fitting the angular dependence of $A(\theta, \phi)$ to a series of the form (20.22a) we immediately have, using (20.26), the corresponding series (20.22b) for the square of the Fermi radius. In practice, one forms linear combinations of the $Y_{lm}(\theta, \phi)$ that transform in accordance with the point group symmetry of the Fermi surface in question. This is relatively straightforward, except for the cubic case which requires special care.[11]

We next discuss an extension of the above technique whereby one can invert the combined extremal area data together with the cyclotron mass data to obtain the Fermi velocity.[12] We start by expanding the energy derivative of the extremal area function, which is related to the effective mass by $m^* = (2\pi)^{-1}\partial A(\theta, \phi)/\partial\varepsilon$, as

$$\frac{\partial A(\theta, \phi)}{\partial \varepsilon} = \sum_{l,m} a'(\varepsilon)_{lm} Y_{lm}(\theta, \phi) \tag{20.27a}$$

where we have regarded the expansion coefficients in (20.22a) as energy dependent, giving $a'(\varepsilon)_{lm} = da(\varepsilon)_{lm}/d\varepsilon$. In the same way we write $b(\varepsilon)_{lm}$, and differentiate the expansion (20.22b) as

$$\frac{\partial k^2(\theta, \phi)}{\partial \varepsilon} = 2k\frac{\partial k}{\partial \varepsilon} = \sum_{l,m} b'_{lm} Y_{lm}(\theta, \phi), \tag{20.27b}$$

where $b'_{lm} \equiv db_{lm}/d\varepsilon$ and $\hbar\partial k(\theta, \phi)/\partial\varepsilon = v^{-1}(\theta, \phi)$ is the $\hat{\mathbf{k}}$ component of the Fermi velocity. Differentiating (20.26), we have $a'_{lm} = \pi P_l(0) b'_{lm}$. Now $k(\theta, \phi)$ is available from the inversion of the area data via Eq. (20.22b) and hence we can obtain $v(\theta, \phi)$ by combining (20.22a) and (20.27b). Writing the Fermi velocity as

$$v_F = \left(\frac{\partial \varepsilon}{\partial k}\right)_{\theta,\phi} \hat{k} + \frac{1}{k}\left(\frac{\partial \varepsilon}{\partial \theta}\right)\hat{\theta} + \frac{1}{k\sin\theta}\left(\frac{\partial \varepsilon}{\partial \phi}\right)\hat{\phi}, \tag{20.28}$$

we may evaluate the θ and ϕ components using the relations

$$\left(\frac{\partial \varepsilon}{\partial \theta}\right)_{k,\phi} = -\frac{\left(\frac{\partial k}{\partial \theta}\right)_{\varepsilon,\phi}}{\left(\frac{\partial k}{\partial \varepsilon}\right)_{\theta,\phi}}; \quad \left(\frac{\partial \varepsilon}{\partial \phi}\right)_{k,\theta} = -\frac{\left(\frac{\partial k}{\partial \phi}\right)_{\varepsilon,\theta}}{\left(\frac{\partial k}{\partial \varepsilon}\right)_{\theta,\phi}}; \tag{20.29a,b}$$

$\partial k/\partial\theta$ and $\partial k/\partial\phi$ can be evaluated from $k(\theta, \phi)$.

[11] Inversion of cubic de Haas-van Alphen data with an application to palladium, F.M. Mueller and M.G. Priestley, *Phys. Rev.* **148**, 638 (1966).

[12] Fermi velocity and Fermi radius in platinum, J.B. Ketterson, L.R. Windmiller, S. Hornfeld, and F. Mueller, *Solid State Commun.* **6**, 851 (1968).

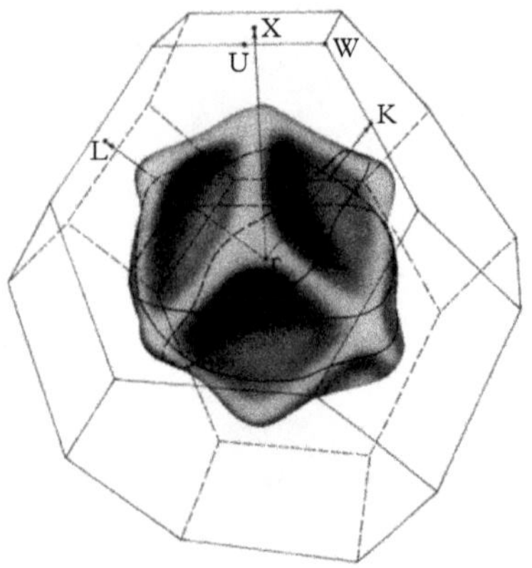

Figure 20.2 *The Γ-centered electron Fermi surface of platinum along with the fcc Brillouin zone. Shown also are the extremal orbits for the magnetic field along the [100], [110], and [111] directions. (After Windmiller et al. (1969).)*

We now show some results from the application of the above scheme to invert dHvA data for the electron surface of the transition metal platinum. We refer the reader to the band structure for Pt, shown earlier in Fig. 13.1. Note that the highest level at Γ when extended toward the Brillouin zone in the Γ-X [100], Γ-K [110], and Γ-L [111] directions intersects the Fermi level implying the presence of a closed electron surface centered at Γ. This surface is shown in Fig. 20.2.[13]

Figure 20.3(a) shows the results of the measurements of the extremal cross-sectional areas of the Γ-centered sheet of the Pt Fermi surface. Because of the cubic symmetry of Pt it is only necessary to show the data in 1/48th of the unit sphere defined by the restriction x > y > z, and this data is displayed as contours of constant area. Figure 20.3(b) shows the contours of constant radius resulting from the application of the Mueller inversion scheme.[14]

For open or multi-valued surfaces the restrictions of the Lifshitz–Pogorelov theorem are not met and an alternative representation is required. One useful representation is a Fourier series (see Eq. (10.13)) of the form[15]

$$\varepsilon(\mathbf{k}) = \sum_{\mathbf{R}} C_{\mathbf{R}} e^{i\mathbf{k}\cdot\mathbf{R}}, \tag{20.30}$$

where we sum over the set of all real-space lattice vectors and the parameters $C_{\mathbf{R}}$ can be used to fit the band structure in the vicinity of the Fermi surface. The lattice vectors sort in to "stars" which are the real space analogue of the K-space stars shown in Fig. 11.5. The sets of vectors making up these stars have the same length (although under some circumstances there are inequivalent stars which nonetheless have equal lengths). In terms of the stars Eq. (20.30) takes the form

$$\varepsilon(\mathbf{k}) = \sum_{j} C_j S_j(\mathbf{k}), \tag{20.31a}$$

[13] de Haas van Alphen effect in platinum, J.B. Ketterson and L.R. Windmiller, *Phys. Rev. B* 2, 4813 (1970).

[14] Experimental determination of Fermi radius velocity and g factor in Pd and Pt, L.R. Windmiller, J.B. Ketterson and S. Hornfeldt, *J. Appl Phys.* **40** 1291 (1969).

[15] Note that although these functions are orthogonal when integrated over the Brillouin zone, they are not orthogonal when integrated over the Fermi surface.

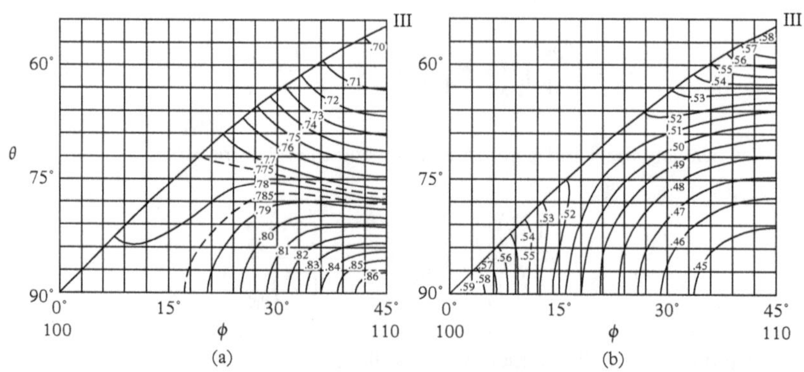

Figure 20.3 *(a) The contours of constant area of the electron Fermi surface in platinum. (b) The contours of constant radius as derived from the data using the Mueller inversion scheme. (After Windmiller et al. (1969).)*

where

$$S_j(\mathbf{k}) = \sum_{\mathbf{R}}^{\text{star}} e^{i\mathbf{k}\cdot\mathbf{R}}. \tag{20.31b}$$

If we only want to represent the Fermi surface itself then (20.31) takes the form

$$\varepsilon_F = \sum_j C_j S_j(\mathbf{k}) \tag{20.32}$$

However we can combine ε_F with C_0, the first (constant) term in the Fourier series, and then divide all coefficients by the resulting C_0 yielding the form

$$F(\mathbf{k}) = 1 + \sum_{j>0} C_j S_j(\mathbf{k}) = 0. \tag{20.33}$$

An expansion of this form was used to represent the Fermi surface of noble metals;[16,17] in the specific form as written in Eq. (20.33) it was applied to the large electron and hole surfaces of Pt,[18] as well as the Fermi surface of Au.[19] Let us assume that we have a starting set of C_j, say a set obtained by fitting the band structure near ε_F. We can then adjust these coefficients to better represent a selected set of N dHvA areas. We first calculate these areas from

$$A_i^c = \frac{1}{2}\int_0^{2\pi} k_\perp^2 \, d\theta, \tag{20.34}$$

where we write the Fermi radius in the form $\mathbf{k} = \mathbf{k}_\perp + k_{//}\mathbf{n}_H$, where $\mathbf{k}_\perp$ measures the radius in the plane of the orbit, $\mathbf{n}_H$ is a unit vector along the magnetic field direction, and $k_{//}$ measures the displacement of the orbit center along the magnetic field. Sometimes it is more convenient to measure $k_{//}$ from a point other than the Brillouin zone center, Γ, in which case it is necessary to account for this origin shift. Since (20.25) is an implicit function of $\mathbf{k}$ (rather than explicit, as is (20.21)), it must be solved iteratively using Newton's "director corrector" method. Assuming we have some starting value for $\mathbf{k}_\perp(\theta)$ for some angle θ in the plane of the orbit, we can calculate $F(\mathbf{k})$, which will not vanish since the point is presumed to be somewhat off the Fermi surface. We then carry out a Taylor expansion of $F(\mathbf{k})$ in the form

$$F(\mathbf{k} + \delta\mathbf{k}_\perp) = F(\mathbf{k}) + \nabla F(\mathbf{k}) \cdot \delta\mathbf{k}_\perp + \cdots . \tag{20.35}$$

Assuming $\mathbf{k} + \delta\mathbf{k}_\perp$ is on the Fermi surface, i.e., $F(\mathbf{k} + \delta\mathbf{k}_\perp) = 0$, we can solve for (20.35) for $\delta\mathbf{k}_\perp$ in a linear or quadratic approximation, find the change $\delta\mathbf{k}_\perp$ that brings $\mathbf{k}_\perp$ closer to the Fermi surface, and iterate this process until $|\delta\mathbf{k}_\perp|$ falls below some preset value; note this procedure will keep the orbit center, $k_{//}$, constant. Having found a radius vector in the plane of the orbit at the angle θ,

[16] Fermi surfaces of copper, silver and gold. 2. Calculation of Fermi surfaces, D.J. Roaf, *Phil. Trans. R. Soc. Lond* **A255**, 135 (1962).

[17] Fermi surfaces of noble metals, M.R. Halse, *Phil. Trans. R. Soc. Lond* **A265** 507 (1969).

[18] Fourier-series representation of Pt Fermi surface, J.B. Ketterson, F.M Mueller and L.R. Windmiller, *Phys. Rev.* **186**, 656 (1969).

[19] Measurement and inversion of de Haas-van Alphen data in gold, B. Bosacchi, J.B. Ketterson, and L.R. Windmiller, *Phys. Rev.* **B15**, 1197 (1971).

we next want a starting value for the radius at an angle $\theta + \delta\theta$. We can again use the form (20.35), but rather than an incremental vector $\delta\mathbf{k}_\perp$ perpendicular to the Fermi surface we now want an incremental vector $\delta\mathbf{k}_t$ tangent to the orbit, which is given by $\delta\mathbf{k}_t = \mathbf{k}_\perp \times \mathbf{n}_H \delta\theta$. After moving to this point we again iterate back to the Fermi surface, as before. We continue this process at each angle, θ, on a chosen grid in the plane or the orbit, thereby allowing us to calculate the area of the orbit via (20.34). $\nabla F(\mathbf{k})$ and $\nabla\nabla F(\mathbf{k})$ follow trivially from, and can be calculated in parallel with, (20.33) as

$$\nabla F(\mathbf{k}) = 1 + \sum_{j>0} C_j \nabla S_j(\mathbf{k}) \tag{20.36}$$

where

$$\nabla S_j(\mathbf{k}) = i \sum_{\mathbf{R}}^{\text{star}} \mathbf{R} e^{i\mathbf{k}\cdot\mathbf{R}}. \tag{20.37}$$

Repeating this process for each of the dHvA areas in our chosen set, we can then evaluate the mean square deviation of these calculated areas with respect to their measured counterparts, A_i^e, as

$$\Delta^2 = \frac{1}{N-1} \sum_{i=1}^{N} \left(\frac{A_i^c - A_i^e}{A_i^e} \right)^2. \tag{20.38}$$

Next we minimize $\Delta^2(C_j)$ with respect to each of the C_j in a suitably truncated expansion (20.33); to do this we expand Δ^2 in terms of the changes in the expansion coefficients δC_j, which in turn requires expanding the extremal areas in terms of the δC_j; i.e.,

$$A_i^c (C_j + \delta C_j) = A_i (C_j) + \sum_j \frac{\partial A_i}{\partial C_j} \delta C_j + \cdots .$$

The required quantities for a linear least squares fit are evaluated using the expression

$$\frac{\partial A_i^c}{\partial C_j} = -\int_0^{2\pi} d\theta S_j (\mathbf{k}) \frac{k_\perp^2}{\nabla F(\mathbf{k}) \cdot \mathbf{k}_\perp}. \tag{20.39}$$

Solving the resulting set of linear equations for the δC_j and iterating until $\Delta^2(C_j)$ is minimal, we obtain a representation of the Fermi surface in terms of the C_j.

A representation of the Fermi velocity can be obtained by regarding the Fourier expansion coefficients as energy dependent; i.e., $C_j = C_j(\varepsilon)$. The Fermi velocity involves the behavior in the immediate vicinity of the Fermi surface and we can represent the Fermi velocity by introducing a second set of coefficients, $C'_j \equiv \partial C_j/\partial\varepsilon$. Again using the rules for the differentiation of implicit functions, we have

$$v_F = \frac{1}{\hbar}\frac{\partial\varepsilon}{\partial \mathbf{k}} = -\frac{1}{\hbar}\frac{\nabla F(\mathbf{k})}{\partial F/\partial\varepsilon} = -\frac{1}{\hbar}\frac{\nabla F(\mathbf{k})}{\sum_j C'_j S_j(\mathbf{k})}. \tag{20.40a}$$

We can then evaluate the cyclotron mass, $m^* = (\hbar^2/2\pi)\ \partial A_i^c/\partial\varepsilon$, where

$$\frac{\partial A_i^c}{\partial\varepsilon} = \int_0^{2\pi} d\theta \frac{k_\perp^2}{k_\perp \cdot \nabla E(\mathbf{k})}, \tag{20.40b}$$

where $\nabla E(\mathbf{k})$ follows from (20.40a).

Once the Fermi radii and the Fermi velocity are available, one can calculate the total number of carriers, $n(\varepsilon_F)$, and the density of states, $\partial n(\varepsilon)/\partial\varepsilon|_{\varepsilon_F}$, from the usual expressions

$$n(\varepsilon_F) = \frac{2}{(2\pi)^3}\int d\Omega k^2(\theta,\phi); \quad \left.\frac{\partial n(\varepsilon)}{\partial\varepsilon}\right|_{\varepsilon_F} = \frac{2}{(2\pi)^3}\int \frac{k^2(\theta,\phi)}{\mathbf{k}\cdot\nabla E(\mathbf{k})} d\Omega. \tag{20.41}$$

For a metal like copper the number of carriers must correspond to one per atom. The density of states may be compared with that obtained from a measurement of the specific heat. One can go on to calculate other quantities such as $\partial A_i^c/\partial k_{//}$ (to ensure that the evaluated orbit is an extremal, which requires that this quantity vanishes), $\partial^2 A_i^c/\partial k_{//}^2$ (which occurs in the expression for the dHvA amplitude), or the Gaussian curvature.

The above scheme has been applied to fit the Fermi surface and Fermi velocity of gold using the experimental dHvA extremal areas and masses,[20,21] along with the cyclotron resonance masses.[22] All three of the noble metals have a Fermi surface that is essentially the same in two respects: (i) the Fermi surface is multiply connected along "necks" with their central axes along the [111] directions, and (ii) the remainder of the Fermi surface corresponds approximately to a sphere with a radius expected for a material with one electron per atom (corresponding to the valence 1 of the noble metals). Note that this shape is consistent with the band structure for copper, shown earlier in Fig. 13.3; the Fermi surface for copper is shown in Fig. 20.4.

There are four classes of dHvA orbits in Au. The first and largest are the *belly* (B) orbits, shown as the white bands that orbit the roughly spherical regions of the surface. The smallest cross sections correspond to the *necks* (N) that connect the surfaces in neighboring zones. The two remaining orbits are the *dog bone* orbit, seen in the [110] view in the right panel of Fig. 20.4 (which traverses four necks and four belly sections), which has two-fold symmetry, and the *rosette*, that appears when viewing along [100] (and also involves four necks and four belly sections), which has four-fold symmetry.

[20] Low-field de Haas-van Alphen effect in gold, A.S. Joseph, A.C. Thorson, and F.A. Blum, *Phys. Rev.* **140**, A2046 (1965).

[21] Fermi surfaces of copper, silver and gold. 1. de Haas-van Alphen effect, D. Shoenberg, *Phil. Trans. R. Soc.* **A255**, 85 (1962).

[22] Azbel-Kaner cyclotron resonance in gold, D.N. Langenberg and S.M. Marcus, *Phys. Rev.* **136**, A1383 (1964).

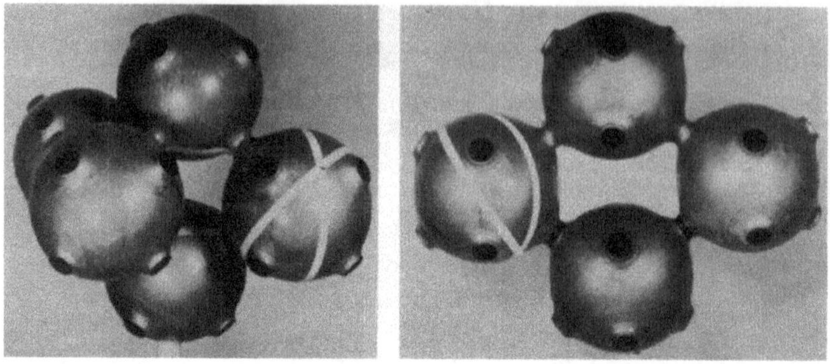

Figure 20.4 *Two views of the Fermi surface of Cu in the repeated zone scheme. Note the Fermi surface is largely spherical in shape with the exception of "necks" along the [111] directions, where contact to the surface in neighboring zones occurs. The white "bands" on the surfaces identify various extremal orbits that are observed in de Haas-van Alphen studies. (After D. Shoenberg (1962).)*

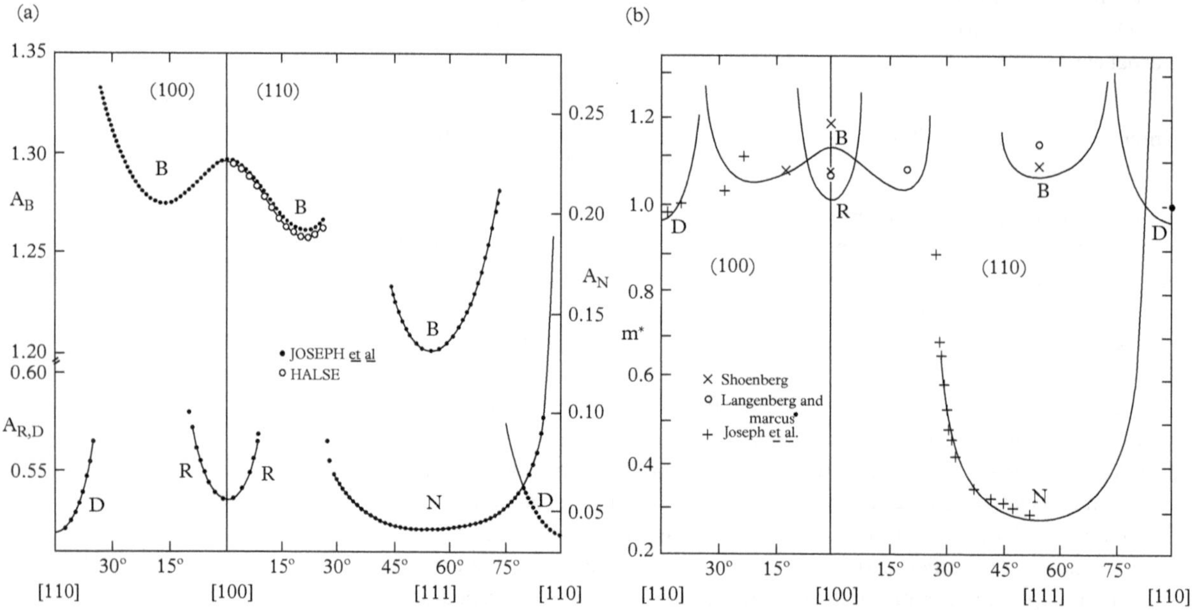

Figure 20.5 *(a) The measured extremal cross-sectional areas of the Fermi surface of Au. The smooth curve is the fit to the data by Bosacchi et al. (1971); shown also are the data of Joseph et al. (1965) (dots), and Halse (1969) (open circles). (b) The effective masses of extremal orbits on the Fermi surface of Au, together with the fit; the various points shown are the data of Shoenberg (1962), Joseph et al. (1965), and Langenberg and Marcus (1964).*

Figures 20.5(a) and 20.5(b) show the measured cross-sectional areas and effective masses of gold, as well as the results of the fits. The actual parameters C_j and C'_j are not of interest here. The Fermi radii deduced from these fits are shown in Fig. 20.6.

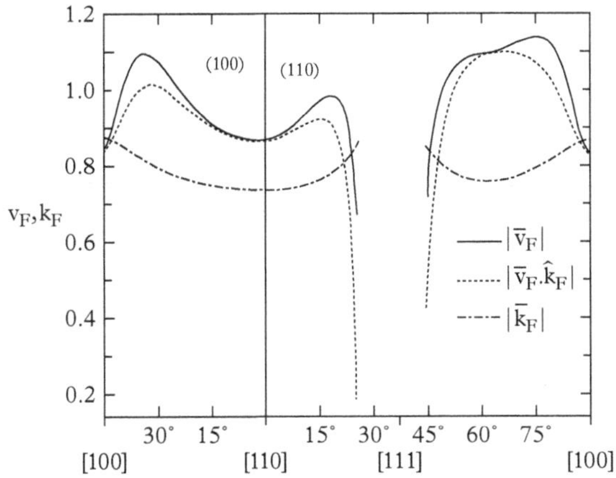

Figure 20.6 *Fermi radii and Fermi velocities in Au, as determined by the fit to the extremal areas and cyclotron masses. (After Bosacchi et al. (1971).)*

20.1.4 Band structure based parameterization of the Fermi surface

Here we discuss the inversion of dHvA data by making use of a parameterized band structure. One could use any of the approaches we have discussed earlier (nearly free electron, pseudo potential, tight binding, augmented plane wave, or Korringa–Kohn–Rostoker, KKR). We will give an example of the last of these, the KKR technique. The parameters that occur in this formalism are the energy-dependent scattering-phase shifts, $\delta_l(\varepsilon)$, from the muffin tin potential. The values of these quantities at the Fermi energy, $\delta_l(\varepsilon_F)$, determine the Fermi radius; these parameters together with their energy derivatives, $\delta_l'(\varepsilon_F) \equiv \partial\delta_l(\varepsilon)/\partial\varepsilon|_{\varepsilon=\varepsilon_F}$, determine the Fermi velocity. Efficient procedures to find these parameters were developed by Shaw et al. (1972)[23] and applied to the noble metals, which we now discuss.

When an eigenvalue, λ_i, of the KKR determinant associated with a given band vanishes, we have a solution corresponding to some energy ε and wave vector $\mathbf{k}$ of that band. If the eigenvalue does not vanish but is small, then we are near a solution and may use Newton's method to change ε or $\mathbf{k}$ to move closer to a solution; i.e.,

$$\lambda_i(\varepsilon + \delta\varepsilon, \mathbf{k}) = \lambda_i(\varepsilon) + \frac{\partial\lambda_i}{\partial\varepsilon}\delta\varepsilon \tag{20.42a}$$

and

$$\lambda_i(\varepsilon, \mathbf{k} + \delta\mathbf{k}) = \lambda_i(\varepsilon, \mathbf{k}) + \frac{\partial\lambda_i}{\partial\mathbf{k}} \cdot \delta\mathbf{k}. \tag{20.42b}$$

Since $\lambda_i(\varepsilon)$ is presumed to be a KKR eigenvalue, it vanishes and then the energy needed to move us on to the band $\varepsilon(\mathbf{k})$ is $\delta\varepsilon(\mathbf{k}) = -\lambda_i(\varepsilon, \mathbf{k}+\delta\varepsilon)\,(\partial\lambda_i/\partial\varepsilon)^{-1}$. To trace orbits and fit dHvA areas we can use $\partial\lambda_i/\partial\mathbf{k}$ in the same way we used $\partial F(\mathbf{k})/\partial\mathbf{k}$ in

[23] Inversion of Fermi-surface data using partial-wave phase-shifts and their derivatives: an application to noble metals, J.C. Shaw, J.B. Ketterson, and L.R. Windmiller, *Phys. Rev.* **10**, 3894 (1972).

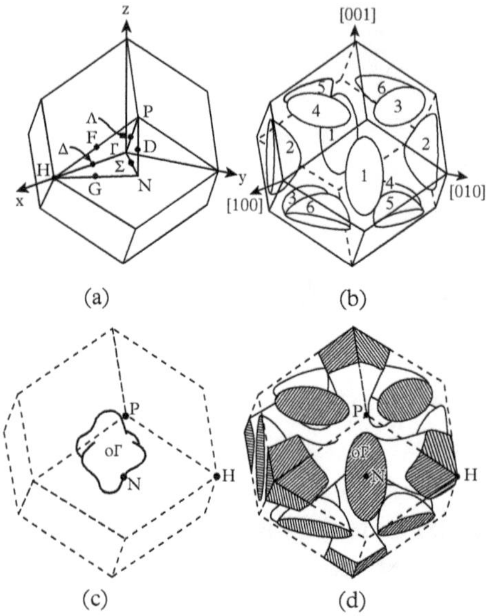

Figure 20.7 *The Fermi surface of Nb. (a) The bcc Brillouin zone, together with designed symmetry points and lines. (b) N-centered "ellipsoids." (c) The Γ-centered "octahedron." (d) The open "jungle gym." (After L.F. Mattheiss (1970).)*

the Fourier series scheme; here the parameters that are being fitted are the KKR phase shifts, $\delta_l(\varepsilon_F)$, which then hold for *all* sheets of the Fermi surface simultaneously. Derivatives of λ_i are evaluated using the Hellman–Feynman theorem and by treating the KKR determinant as a matrix for which the eigen vectors can be found (see discussion surrounding Eq. (13.18)). Both ε and $\mathbf{k}$ enter the KKR structure constants; ε also enters the phase shifts. The Fermi velocity again follows through implicit differentiating as

$$\mathbf{v} = \frac{1}{\hbar}\left(\frac{\partial\varepsilon}{\partial\mathbf{k}}\right)_\lambda = -\frac{1}{\hbar}\frac{\left(\frac{\partial\lambda}{\partial\varepsilon}\right)_\mathbf{k}}{\left(\frac{\partial\lambda}{\partial\mathbf{k}}\right)_\varepsilon}. \tag{20.43}$$

The Fermi velocity can be evaluated by introducing the energy derivatives of the phase shifts, $\delta_l'(\varepsilon_F)$, as an additional set of parameters; one can then fit the effective masses (again on all sheets of the Fermi surface) to obtain these parameters.

The number of parameters needed to fit the Fermi surface and its energy gradient is by far the smallest, if we employ a band-structure based fitting scheme like the one just described. However, the scheme used must be tailored to the crystal structure at hand, including parameters associated with all atoms in the unit cell.

We will end our discussion of the inversion of dHvA data with an application of the KKR based formalism to niobium (Crabtree et. al. (1979, 1987)).[24,25] The relativistic band structure of Nb was shown earlier, in Fig. 13.2 and the Fermi surface as given by Mattheiss (1970)[26] is depicted in Fig. 20.7. There are three sheets of the Fermi surface which are referred to as the *ellipsoids* (panel (b)) the *octahedron* (panel (c)) and the *jungle gym* (panel (d)); all three sheets are hole-like as drawn. These three surfaces support a variety of hole-like and electron-like

[24] Anisotropic many-body effects in the quasi-particle velocity of Nb, G.W Crabtree, D.H. Dye, D.P. Kirim, D.D. Koelling, and J.B. Ketterson, *Phys. Rev. Lett.* **42**, 390 (1979).

[25] Anisotropy of the Fermi-surface, Fermi velocity, many-body enhancement, and superconducting energy gap in Nb, G.W. Crabtree, D.H. Dye, D.P. Kirim, S.A. Campbell, and J.B. Ketterson, *Phys. Rev.* **35**, 1728 (1987).

[26] Electronic structure of niobium and tantalum, L.F. Mattheiss, *Phys. Rev.* **B1**, 373 (1970).

extremal dHvA orbits, which we will not discuss. The shape of the Fermi surface obtained by fitting the dHvA areas to the muffin tin phase shifts, $\delta_l(\varepsilon_F)$, is shown as a "house" diagram in Fig. 20.8; shown also are the symmetry points of the bcc Brillouin zone. The angular dependence of the Fermi velocity obtained by fitting the dHvA effective masses using the energy derivatives of the phase shifts, $\delta_l'(\varepsilon_F)$, in addition to the $\delta_l(\varepsilon_F)$ as parameters, is shown in Fig. 20.9 (again the precise values obtained for these parameters are not of interest here). Shown also are the Fermi velocities obtained from a later APW band structure by Elyashar and Koelling (1976, 1977).[27] Note that the band velocities (obtained from the band structure calculation), $\mathbf{v}_{bs}$, are always larger than the so-called quasiparticle or true velocities, $\mathbf{v}_{qp}$. This arises from a many-body effect caused by the electron–phonon interaction, which will be discussed in Chapter 21. It is customary to introduce a so-called mass enhancement factor (an increase in mass corresponds to a reduction in the velocity) through the definition

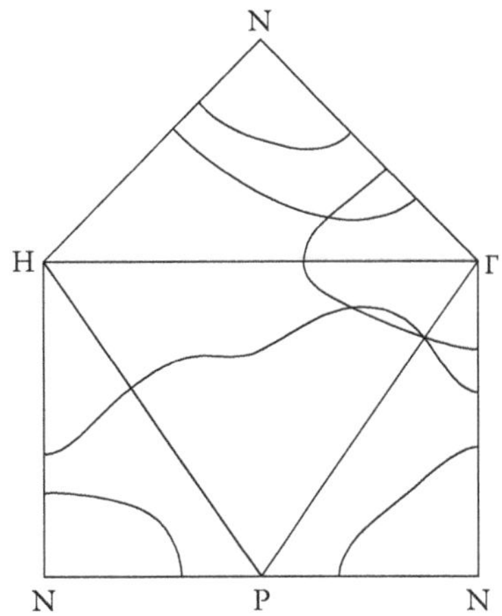

Figure 20.8 *A "house" cross section of the Fermi surface of Nb, as obtained by inverting the measured dHvA extremal areas. The main part of the house shows the cross section in the (110) plane, while the attic shows the (100) plane. Also identified are the symmetry points of the bcc Brillouin zone. (After Crabtree et al. (1987).)*

$$\lambda(\mathbf{k}) \equiv \frac{|\mathbf{v}_{bs}|}{|\mathbf{v}_{qp}|} - 1; \qquad (20.44)$$

this quantity is also plotted in Fig. 20.9. As a result of such fits the Fermi surface properties of Nb are known with rather high precision. Niobium has attracted interest due to the fact that it is the element with the highest superconducting transition.

20.1.5 Other methods to obtain Fermi surface information

As noted at the beginning of this chapter there are additional methods for obtaining specific information about the Fermi surface. In what follows immediately we discuss two techniques that yield "calipers" of the Fermi surface in directions perpendicular to an applied magnetic field.[28,29]

The *geometric resonance* (also called the magnetoacoustic effect) is an oscillatory behavior of the ultrasonic attenuation with magnetic field for ultrasound propagating perpendicular to the field. As an example we picture an elliptical orbit in k-space perpendicular to the magnetic field (which is directed out of the page), as shown in Fig. 20.10. The parallel lines show crests of a sound wave having a wave vector $q = 2\pi/\lambda$, propagating to the right. We might expect that the interaction with the sound wave is maximal when the distance between the turning points in the orbit, d_{ext}, matches an integral number of sound wavelengths; i.e., $d_{ext} = \lambda(n + \gamma)$. We recall from the Lorentz equation that dimensions in k-space are related to those in real space by a rotation of $\pi/2$ about the magnetic field and a scale factor of $eH/\hbar c$; hence

$$k_{ext} = \frac{eH}{\hbar c} d_{ext} = \frac{eH}{\hbar c} n(\lambda + \gamma). \qquad (20.45)$$

[27] Effect of non-muffin-tin terms on electronic-structure of transition-metals – niobium, N. Elyashar and D.D. Koelling, *Phys. Rev.* B13, 5362 (1976); *Phys. Rev.* B15, 3620 (1977).

[28] A proposal for determining the Fermi surface by magneto-acoustic resonance, A.B. Pippard, *Phil. Mag.* **2**, 1147 (1957).

[29] Electron resonances with ultrasonic waves in copper, R.W. Morse, H.V. Bohm, and J.D. Gavenda, *Phys. Rev.* **109**, 1394 (1958).

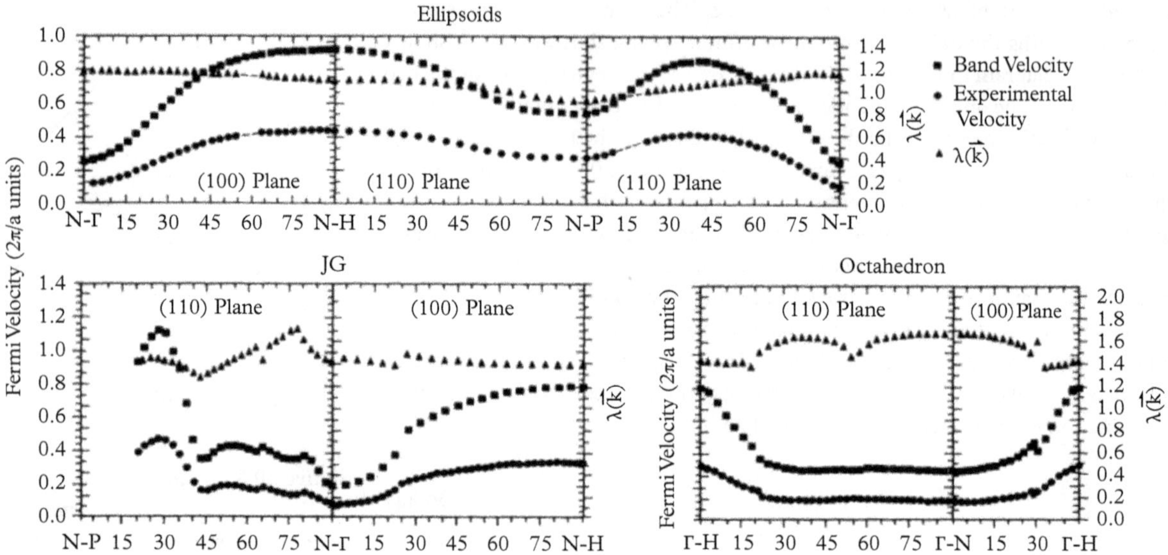

Figure 20.9 *The angular dependence of the Fermi velocity on the three sheets of the Nb Fermi surface for the (100) and (110) planes; both the band structure and quasiparticle (experimental) velocities are shown. The angular dependence of the many-body mass enhancement factor* λ(k) *is also plotted. (After Crabtree et al. (1987).)*

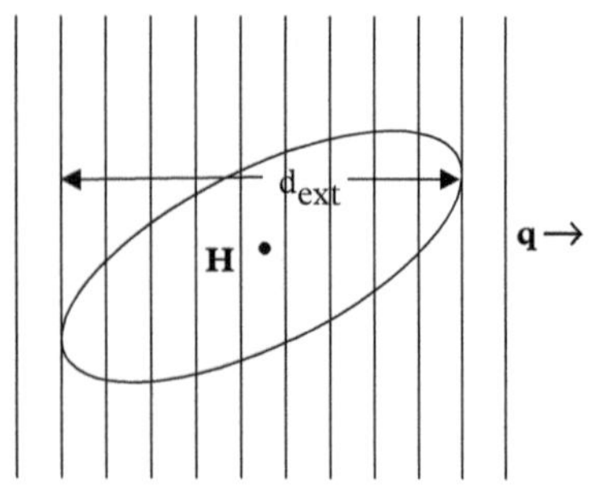

Figure 20.10 *An orbit of an electron in real space in the presence of a sound wound wave for a case where the extremal dimension matches an integral number of sound wavelengths.*

Analogous to the dHvA effect, it then follows that as the field increases, and the orbit diameter shrinks, we will have an oscillatory variation of the attenuation with a period

$$\Delta\left(\frac{1}{H}\right) = \frac{1}{H_n} - \frac{1}{H_{n+1}} = \frac{e\lambda}{\hbar c k_{ext}} = \frac{eV_s}{\hbar c k_{ext} f}, \tag{20.46}$$

where f is the sound frequency and V_s is the velocity of sound. The oscillations cease when the field is such that the orbit diameter (in real space) is less than the sound wavelength. The effect occurs for both longitudinal and transverse sound waves. As with the dHvA effect, cancellation effects emphasize that caliper which is extremal with respect to k_H, the wave vector along the magnetic field direction. Each field direction yields a different caliper (or set of calipers if the Fermi surface supports multiple extremal calipers). Although not involving it explicitly, the temperature must be low enough that the electron travels through several sound wavelengths before scattering; i.e., the mean free path l must be such that $l \gg \lambda$ or equivalently $ql \gg 1$. Clearly this criterion is favored by using high sound frequencies and pure metals.

The second technique for obtaining Fermi surfaces calipers is the Gantmakher R.F. size effect.[30] Consider the situation shown in Fig. 20.11, where we have a

[30] Dimensional effect in a metal in multiples of a certain magnetic field, V.F. Gantmakher. *Sov. Phys.-JETP* **16**, 247 (1963).

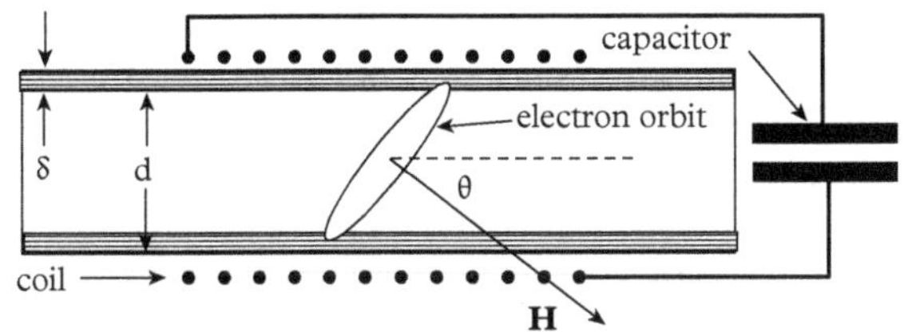

Figure 20.11 *Schematic of the r.f. size effect.*

thin slab of metal of thickness d surrounded by a coil with inductance L, which is completed by a capacitor, C, such that we have a resonant circuit (operating in the MHz range) with a resonant frequency, $\omega = \sqrt{1/LC}$.[31] The R.F. frequency is not critical here, as its primary function is for establishing a field that is confined to a region of order the skin depth, δ, that is simultaneously much less than the thickness, d (no temporal resonance is involved). If the magnetic field is such that the projection of the orbit diameter, d_{ext}, satisfies

$$d_{ext}\cos\theta = d, \tag{20.47}$$

we expect both an increase in absorption and a shift in frequency (the latter usually being more accurately determined). Measuring this effect as a function of field angle yields calipers, $d_{ext}(\theta)$, associated with the orbits. Only calipers on extremal orbits as a function of k_H contribute, as all others have a drift velocity along the field and will not return to the skin depth region on successive passages around the orbit, and hence will not effectively sample the R.F. field. As with the magneto-acoustic effect, we obtain the Fermi surface caliper of the orbit rotated by $\pi/2$ and scaled such that $k_{ext} = (eH/\hbar c)d_{ext}$. We again need metals sufficiently pure and temperatures low enough that $l \gg d$. The sample surfaces must also be carefully prepared to minimize boundary scattering.

20.2 Measuring the electronic excitation spectrum: ARPES and IPS

Methods of determining the single-particle excitation spectrum, $\varepsilon_n(\mathbf{k})$, both theoretically and experimentally, and at energies well off the Fermi energy, are of fundamental importance in solid-state physics. Angle-resolved photoelectron spectroscopy (ARPES)[32] has emerged as the most powerful experimental tool for probing the electronic structure of a material, and metals in particular, over a wide range of energies. As discussed earlier in this section, studies of the dHvA effect are capable of unsurpassed precision; however this technique can only yield the shape of the Fermi surface and the associated Fermi velocity.

In an ARPES experiment one measures the properties of electrons ejected from the surface of a metal, following the absorption of an X-ray photon. The technique was made practical by the widespread availability of the directed,

[31] Self-sustaining oscillations will occur if an appropriately d.c. biased tunnel diode is inserted in the circuit.

[32] Optical absorption, which will be discussed in Section 29.1, is limited to determining critical points in the joint density of states.

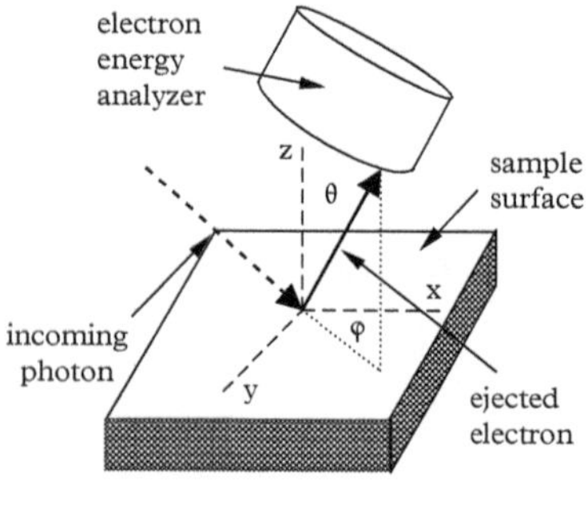

Figure 20.12 *ARPES geometry showing the incoming photon, and the ejected electron with polar angle θ and azimuthal and angles φ.*

high intensity, mono-energetic photons from modern synchrotron sources.[33] The experimental set-up is shown schematically in Fig. 20.12. After first being collimated, electrons ejected at angles θ and ϕ are passed through an electrostatic energy analyzer, after which they are collected by a position sensitive detector. What one obtains is the number of electrons ejected per unit time with energy ε_f and angles θ and ϕ for a given incoming photon energy, $h\nu$.

Since the resulting electrons must pass through the surface of the material under study, which can contain environmental contaminants or oxide layers, it is often necessary to prepare that surface *in situ*. For some crystalline materials, this can be accomplished by cleaving in ultra high vacuum and maintaining that vacuum prior to the measurements, ideally in the environment in which the experiments are actually carried out.

20.2.1 Analysis of ARPES data

Analysis of the energy spectrum of the ejected electrons assumes conservation of energy and momentum. From the first of these conditions we have

$$h\nu = \varepsilon_f - \varepsilon_i - \phi; \tag{20.48}$$

here, $h\nu$ is the energy of the incoming photon, $\varepsilon_i = \varepsilon_n(\mathbf{k})$ is the initial energy of a bound electron in the state $\psi_{\mathbf{k}n}$, $\varepsilon_f = \hbar^2 k_f^2/2m$ is the final energy of the free (ejected) electron having momentum $\mathbf{k}_f$, and ϕ is the work function. If the surface is atomically flat, the resulting in-plane translational invariance of the overall system suggests that it is reasonable to assume that the in-plane component of the momentum is conserved; i.e.,

$$\hbar \mathbf{k}_{\parallel} = \hbar \mathbf{k}_{f\parallel} \tag{20.49}$$

with

$$\hbar\,|\mathbf{k}_{f\parallel}| = \sqrt{2m\varepsilon_f}\,\sin\theta, \tag{20.50}$$

where θ is the angle of the trajectory of the emitted electron measured relative to the direction normal to the sample surface. However, to solve for $\varepsilon_n(\mathbf{k})$ we also require the perpendicular component, $\mathbf{k}_{f\perp}$. What is done in practice is to assume that *inside the crystal* we can use the free-particle-like form

$$\varepsilon_f = \hbar^2 k_f^2/2m - V_0, \tag{20.51}$$

where V_0 is the band offset relative to the vacuum, which includes the work function ϕ; this assumption is based on the property that electrons in higher Bloch bands are generally free electron like since their kinetic energy dominates. Conserving this component of the momentum yields

$$\hbar k_\perp = \sqrt{2m(\varepsilon_f \cos^2\theta + V_0)}. \tag{20.52}$$

[33] At energies of order a few electron volts, pulsed lasers, and harmonics generated from them, can be used; this approach allows the application of time-of-flight spectroscopy of the final state energies.

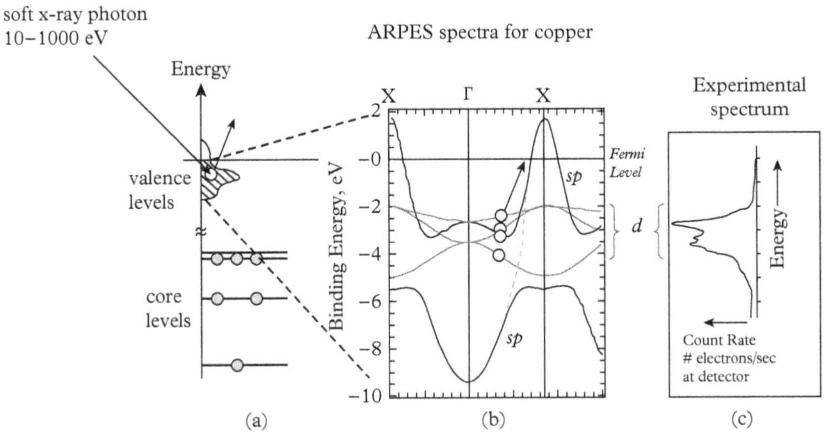

Figure 20.13 *(a) The 1s, 2s–2p, and 3s–3p atomic core levels, together with the broad 3d-4s Bloch bands of metallic copper. The circles show excited states resulting from the removal of electrons by absorption of an X-ray photon. The emissions from the core levels are referred to as X-ray photo-emission spectroscopy (XPS). (b) The band structure of copper in the 3d-4s range of energies along the Γ-X direction showing the broad s-p bands (with the free-electron parabola shown as the light dashed line) through which the narrower 3d bands pass and hybridize. The circles again represent excited states, but now where electrons in Bloch states with a* specific *wave vector have been removed by absorption of an X-ray photon and selected by the analyzer. (c) The actual count rate from the analyzer. From the four bands occupied at this wave vector, three peaks are clearly discernible. (Courtesy E. Rotenberg, Advanced Light Source, Lawrence Berkeley National Laboratory.)*

The quantity V_0 is determined by measuring the energy of the ejected electrons as a function of the incident photon energy when $\theta = 0$, to determine the threshold energy for ejection of an electron from a given Bloch band. Since there is generally a multiplicity of bands this offset must be determined for each band, n. Note that the value of $\mathbf{k}_{f\perp}$ differs inside (where it is given by Eq. (20.52)) and outside the crystal (where V_0 vanishes); i.e., the angle θ changes on leaving the crystal, reminiscent of Snell's law in optics.[34] The data will consist of a series of peaks in the count rate of ejected electrons at energies ε_f as a function of the angles θ and φ or the incoming photon energy $h\nu$; from the position of these peaks, and the determination V_0, the above equations can be solved for $\varepsilon_n(\mathbf{k})$ for the corresponding Bloch wave vector, $\mathbf{k}$.

Figure 20.13(a) shows the overall energy level structure for copper, involving the core levels together with the density of states in the vicinity of the Fermi energy. Figure 20.13(c) shows the ARPES spectrum for a chosen wave vector along the Γ-X direction in the range of energies where both the 3d and 4s–4p bands contribute. Note that all the excited states (holes in an otherwise filled Fermi sea) have a finite width due to the decay of these states by rapid filling by electrons from the surrounding Fermi sea; in addition, there will always be some instrumental broadening due to finite resolution in any of the variables ε_f, θ, and ϕ of the detector or the incoming photon frequency ν. Electrons emitted at higher energies will lie closer to the Fermi level. The Fermi wave vector itself for a given direction would correspond to that energy at which the count rate falls to zero (allowing for instrument resolution), there then being no electrons to excite for energies above ε_F. Figure 20.14 shows the behavior of E versus k along the Γ - X direction formed by superimposing energy scans like those shown in Fig. 20.13(c) for a set of discrete values of k.

[34] This also implies that there is an angle for total internal reflection where the final state electron is totally reflected back into the crystal at the surface.

20.2.2 Modeling finite lifetime effects below the Fermi energy

At absolute zero in a perfect crystal, only the (single particle) Bloch electrons at the Fermi energy will have long lifetimes.[35] Because of many-body electron–electron interactions, *all* Bloch states, with energies $\varepsilon_n(\mathbf{k}) \neq \varepsilon_F$ will decay into other Bloch states and therefore acquire a finite "width," which initially grows as $(\varepsilon_n(\mathbf{k}) - \varepsilon_F)^2$. We can write this width as $\hbar/\tau$ (from the uncertainty principle), where τ is a characteristic time that parameterizes that decay. In addition to decay into other Bloch states, processes involving the decay into a Bloch state and a phonon are also allowed off the Fermi surface.[36] For most purposes this "broadening" of a given band n can be fitted with a Lorentzian-like spectral function of the form[37]

$$A_n(\mathbf{k}, \varepsilon) = \frac{1}{\pi} \frac{\hbar/2\tau(\mathbf{k}, \varepsilon_n)}{(\varepsilon - \varepsilon_n(\mathbf{k}) - \delta\varepsilon(\mathbf{k}, \varepsilon_n))^2 + (\hbar/2\tau(\mathbf{k}, \varepsilon_n))^2}, \tag{20.53}$$

which is normalized such that

$$\int_{-\infty}^{\infty} d\varepsilon \, A_n(\mathbf{k}, \varepsilon) = 1; \tag{20.54}$$

the quantity $\delta\varepsilon_n(\mathbf{k}, \varepsilon)$, which always accompanies $\tau_n(\mathbf{k}, \varepsilon)$, is called the *level shift*. Although there is no simple way of determining $\delta\varepsilon_n(\mathbf{k}, \varepsilon)$ from the data, $\tau_n(\mathbf{k}, \varepsilon)$ can be determined by fitting the widths of the peaks observed in the ARPES spectra.[38]

An important remaining question is what determines the ejection rate of electrons for a given incoming flux of photons. Qualitatively, this rate is governed by the standard expression from time-dependent perturbation theory (the so-called golden rule), according to which

$$w = \frac{2\pi}{\hbar} \left|\hat{H}_{fi}\right|^2 \rho(\varepsilon_i, \varepsilon_f). \tag{20.55}$$

Here, $\hat{H}_{fi} = \langle \psi_f | \hat{H}' | \psi_i \rangle$ is the matrix element connecting the initial and final states of the electron,[39] $\hat{H}' = (e/mc)\mathbf{A} \cdot \hat{\mathbf{p}}$ is the perturbation caused by vector potential $\mathbf{A}(\nu)$ associated with the (classical) X-ray field, $\psi_i = \psi_{kn}$ is the Bloch initial state, ψ_f is the nearly free electron like final state (inside the crystal), and $\rho(\varepsilon)$ is the overall density of states, which is proportional to $A_n(\mathbf{k}, \varepsilon)$.[40] Note that the matrix elements $\hat{H}_{fi}$ can strongly depend on $\mathbf{k}$ and n due to the varying admixture of s-, p-, and d-like states in $\psi_{\mathbf{k}n}$.

The results from a photo-emission experiment for a Bloch band $\varepsilon_n(\mathbf{k})$ can be represented in various ways. For example, we may choose to plot the position of the maximum of the function $A_n(\mathbf{k}, \varepsilon)$ as a function of ε in $\mathbf{k}$-space for some fixed n. This would be the best representation of a constant energy surface

[35] This is the basic assumption of Landau's Fermi liquid theory, which will be discussed in Section 37.1.

[36] These processes will be discussed in Section 21.4.

[37] Representation in terms of a spectral function assumes that the electron is removed from the system before the resulting hole relaxes; this is called the impulse approximation. Formally the quantities entering $A_n(\mathbf{k}, \varepsilon)$ are related to the so-called many-body self energy, $\Sigma(\mathbf{k}, \varepsilon)$, by the expressions: $\hbar/2\tau(\mathbf{k}, \varepsilon) = \text{Im}\ \Sigma(\mathbf{k}, \varepsilon)$ and $\delta\varepsilon(\mathbf{k}, \varepsilon) = \text{Re}\Sigma(\mathbf{k}, \varepsilon)$.

[38] When static impurity scattering is present, Bloch states can acquire a finite width, $\hbar/\tau$, associated with the scattering to other single particle states *on* a constant energy surface; here, τ is a momentum scattering time.

[39] The final state here is that of the electron *inside* the material; that electron must still propagate *to and through* the surface, which is assumed to occur ballistically, together with refraction at the surface, as discussed above. However in practice the excited electron can generate plasmons and other excitations during its passage through the material, which results in additional spectral features.

[40] The quantity $\rho(\varepsilon)$ is referred to as the joint *density of states* and would include a factor, $f(\varepsilon_i)(1 - f(\varepsilon_f))$, accounting for the fact that the initial state must be occupied and the final state unoccupied.

associated with the single particle Bloch band $\varepsilon_n(\mathbf{k})$. For the case $\varepsilon_n(\mathbf{k}) = \varepsilon_F$, the resulting surface is the Fermi surface, broadened by transport of the final state into the vacuum, as well as the inherent instrumental resolution. Alternatively one could fix the direction of **k** and then plot the magnitude k as a function of the maximum of $A_n(\mathbf{k}, \varepsilon)$, which would yield the best representation of the Bloch band $\varepsilon_n(k)$ for that direction. Finally, we could fix **k** and n and plot $A_n(\mathbf{k}, \varepsilon)$ versus ε to obtain the lifetime, $\tau_n(\mathbf{k}, \varepsilon)$, for that state; Fig. 20.13(c) shows such a plot for copper for the chosen **k**, the four states lying in a region within the s–p–d complex.[41]

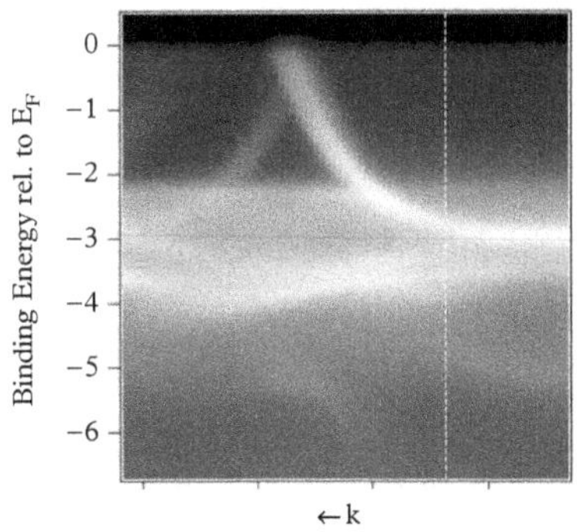

Figure 20.14 *An E versus k scan for copper along the Γ-X direction formed by superimposing energy scans for discrete values of k. The dashed line shows the value of k for which the energy scan in Fig. 20.13(c) was performed. On close examination one can identify spectrally broadened trajectories that, approximately, correspond to most of the bands shown in Fig. 20.13(b). (Courtesy E. Rotenberg, Advanced Light Source, Lawrence Berkeley National Laboratory.)*

20.2.3 Inverse photo-emission spectroscopy

Let us examine what happens if the roles of the electron and photon in a photo-emission experiment are interchanged. Equation (20.48) still applies, only now the initial state is an electron entering from the vacuum side with kinetic energy $\varepsilon_i = \hbar^2 k_i^2/2m$, and in a direction specified by the angles, θ and φ. As with ARPES, it is common to assume that on entering the crystal the in-plane momentum, $\hbar\mathbf{k}_{f\parallel}$, is conserved and that the electron is refracted as a result of the work function; the electron is then in a high energy Bloch state (one lying above the vacuum level); in this energy range the states are essentially free electron like. However at these large energies (relative to the Fermi energy), the incoming electron will excite other electrons from below the Fermi level and in the process will acquire a width, which ultimately limits the energy resolution. What is important here is that some (small) fraction of these initial states will decay by the emission of a single (typically ultraviolet) photon to lower lying unoccupied Bloch states. That photon carries negligible momentum and if we again assume conservation of in-plane momentum, $\hbar\mathbf{k}_\parallel$, between the initial and final electron Bloch states, we can probe the band structure for energies *above* ε_F in a wave-vector-specific manner. However, there are generally a large number of bands at these higher energies each of which can contribute photons of different energy to the outgoing flux. Hence it is necessary to spectroscopically examine the energies of the emitted photons if they are to be assigned to a particular band. Since the number of emitted photons is small, having an efficient, energy sensitive, detector is essential.

The data available from angle resolved IPES complements ARPES: the former allows probing of the Bloch bands lying above ε_F, whereas ARPES probes the bands below ε_F. If the incoming electrons are spin polarized the technique can be used to measure the exchange splitting in ferromagnetic materials. Of course the resolution is ultimately set by finite lifetime effects; associated with this is the fact that, since the incoming electrons penetrate only a short distance into the sample, one is really just sampling the first few atomic planes. For this reason many IPES studies often focus on probing surface states, particularly those associated with adsorbed atoms and molecules, and their effect on the work function. Figure 20.15 shows a schematic diagram of an IPES experimental set-up.

[41] Examples of such plots for the case of the two-dimensional material graphene will be given in Section 49.2.

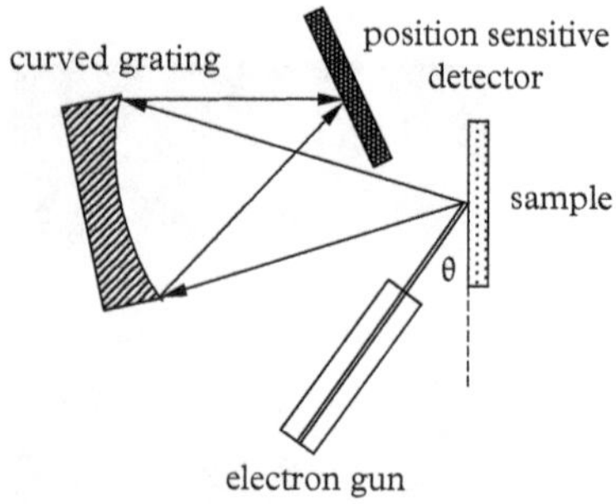

Figure 20.15 *Schematic of an angle-resolved inverse photo-electron spectrometer. Since the photons carry negligible momentum they can be collected over a wide range of solid angles, thereby increasing the total count rate without lowering the energy resolution. The combination of a curved grating and a position sensitive detector yields energy resolution and, simultaneously, the ability to acquire data over a wide range of photon energies.*

ADDITIONAL READING

Magnetic Oscillations in Metals, D. Shoenberg, Cambridge University Press, Cambridge, UK (1984). This book contains a rather complete discussion of the de Haas–van Alphen effect; Shoenberg was a pioneer in this field.

The geometric resonance (also called the magneto-acoustic) effect is discussed in the free-electron model by M.H. Cohen, M.J. Harrison, and W.A. Harrison, *Phys. Rev.* **117**, 937 (1960); a generalization to an arbitrary Fermi surface was given by A.B. Pippard, *Proc. R. Soc.* **A237**, 165 (1960).

Discussions of the theory underlying a host of techniques used for measuring Fermi surface properties, and many other topics in metals physics, are given in *Fundamentals of the Theory of Metals*, by A.A. Abrikosov, North Holland, Amsterdam (1988). Abrikosov generally starts his discussions of some phenomenon with an analysis of the physics involved, along with order of magnitude estimates of the parameters which govern it.

Some of the topics treated in this section are covered in two books by A.B. Pippard: *Dynamics of Conduction Electrons*, in the series Documents on Modern Physics, Gordon and Breach (Taylor and Francis Group), NY (1964) and *Magnetoresistance in Metals*, in the series Cambridge Studies in Low Temperature Physics, Cambridge University Press, Cambridge, UK (1989). Pippard is widely known for his work on these topics.

Angle-resolved photo-electron spectroscopy (ARPES) and inverse photo-electron spectroscopy are discussed in *Photoelectron Spectroscopy (IPS): Principles and Applications*, Stefan Hüfner, Springer (2013).

For a review of inverse photo-electron spectroscopy see: Inverse photoemission, N.V. Smith, *Rep. Prog. Phys.* **51**, 1227 (1988).

The Interacting System of Metallic Electrons and Phonons

21

Up to this point we have considered the electronic excitations and the lattice vibrations (phonons) of a solid as arising separately. In the present chapter we will examine their mutual interaction, based in part on some simple models.

High frequency sound waves (typically in the MHz range) are widely used as a probe of material properties, including the elastic response, as discussed in Section 1.11.[1] Various mechanisms contribute to the attenuation of sound waves including scattering by lattice defects and thermal phonons. To some extent these effects can be modeled by introducing an imaginary part to the elastic compliances (see Section 1.10). In a metal we must also consider the interaction with conduction electrons, particularly at low temperatures where relevant interaction lengths and times become long. For the case of free electrons the interaction with long wavelength sound waves can be developed in some detail. Here sound is taken to be a classical wave propagating in the charged background created by the ions, with the electrons treated as a degenerate gas of free electrons, the distribution function of which follows from solving the Boltzmann equation. The details are somewhat complicated and are developed in Appendix 21A. There we will find that the sound waves transfer energy to the conduction electrons, and we obtain expressions for the energy loss rate. This phenomenon forms a topic in and of itself and the remainder of our discussion is not built on it.

Lattice vibrations ultimately arise from the combined effects of electron–electron, ion–ion, and electron–ion Coulomb interactions. Using a simple model, based on a dielectric function approach that treats the positive ions classically, two collective oscillations emerge: longitudinal phonons and electron plasma oscillations. The model can also be used to construct a quantum mechanical form for the electron–phonon interaction. One result of this interaction is that zero-point phonons have an important effect on the energy and lifetime of electrons near the Fermi surface.

We will end this chapter with a more formal quantum mechanical discussion, of the electron–phonon interaction, including the effect of thermally excited phonons on the electrical resistance.

[1] Here the reader can consult various volumes of the series *Physical Acoustics*, W.P. Mason and R.N. Thurston (Eds.). Academic Press, NY.

The Physics of Solids. First Edition. J. B. Ketterson.

21.1 The dielectric function in the presence of a deformable positive background

Here we develop a simple model for an interacting system of classically moving ions and a degenerate electron gas. The "bare" Coulomb interaction between the electrons separated by a distance r is

$$V(\mathbf{r}) = \frac{e^2}{r}, \tag{21.1}$$

with its Fourier transform given by

$$V(\mathbf{q}) = \frac{4\pi e^2}{q^2}. \tag{21.2}$$

The effect of electron–electron interactions may be incorporated by introducing a dielectric function, $\varepsilon(\mathbf{q}, \omega)$ as was discussed in Chapter 14. Although the bare Coulomb interaction is frequency independent (reflecting the fact that it is instantaneous for most solid-state time scales), a frequency dependence is introduced when account is taken of the fact that the background of positive ions and the electron gas have response times of order ω_{ip}^{-1} and ω_{ep}^{-1} respectively; here $\omega_{ip} = \left(4\pi n_i e^2/M\right)^{1/2}$ is the ion-plasma frequency ($\sim 10^{13}\text{s}^{-1}$), $\omega_{ep} = \left(4\pi n_e e^2/m\right)^{1/2}$ is the electron-plasma frequency ($\sim 10^{16}\text{s}^{-1}$), n_i and n_e are ion and electron densities, and M and m are the ion and electron masses, respectively. Our *effective* electron–electron interaction will be written as

$$V(\mathbf{q}, \omega) = \frac{4\pi e^2}{\varepsilon(\mathbf{q}, \omega) q^2}, \tag{21.3}$$

where $\varepsilon(\mathbf{q}, \omega)$ contains effects due to both the ions and the background electrons.

We will use a self-consistent potential approach to calculate $\varepsilon(\mathbf{q}, \omega)$, as in Chapter 14; since we are now adding the response of the positive background, we will repeat parts of the earlier discussion. Assume that we have a plane wave electric field, which we here write as

$$\mathbf{E}(\mathbf{r}, t) = \Re\, \mathbf{E}_0(\mathbf{q}, \omega) e^{-i\omega t + i\mathbf{q}\cdot\mathbf{r}}, \tag{21.4}$$

propagating in our electron–ion system; accompanying this wave will be an electric displacement wave,

$$\mathbf{D}(\mathbf{r}, t) = \Re\, \mathbf{D}_0(\mathbf{q}, \omega)\, e^{-i\omega t + i\mathbf{q}\cdot\mathbf{r}}. \tag{21.5}$$

The dielectric function will be defined as[2]

$$\varepsilon(\mathbf{q}, \omega) = \frac{\mathbf{D}(\mathbf{q}, \omega)}{\mathbf{E}(\mathbf{q}, \omega)}. \tag{21.6}$$

[2] In what follows, all mechanisms leading to a damping of the waves are ignored and so the dielectric constant will be a real function of $\mathbf{q}$ and ω.

In order to allow us to consider a response with arbitrary ω and $\mathbf{q}$ (as opposed to some natural mode of the system having a specific dispersion behavior, $\omega = \omega(\mathbf{q})$), we imagine that a time- and space-dependent external charge density with amplitude $\rho_{ext}(\omega,\mathbf{q})$ (not involving the metal's own electrons and ions) is introduced into the system. From electrostatics, the electric field is related to the "total" charge density as

$$\nabla \cdot \mathbf{E} = 4\pi\rho_{total}; \tag{21.7}$$

here

$$\rho_{total} = \rho_e + \rho_i + \rho_{ext}, \tag{21.8}$$

where ρ_e and ρ_i are the electron-gas and background-ion charge densities, which together form the "bound" charge density of electrostatics (that arising from the media themselves). On the other hand, $\mathbf{D}$ excludes the bound charge and only involves external "free" charge, ρ_{ext}; thus,

$$\nabla \cdot \mathbf{D} = 4\pi\rho_{ext} \tag{21.9}$$

and combining Eqs. (21.6)–(21.8) we obtain the dielectric constant as

$$\varepsilon = \frac{\rho_{ext}}{\rho_e + \rho_i + \rho_{ext}}. \tag{21.10}$$

The ions will be regarded as classical; i.e., their motion is governed by Newton's law,

$$M\dot{\mathbf{V}}_i = z|e|\mathbf{E}, \tag{21.11a}$$

where $\mathbf{V}_i$ is the velocity of the ion and z is its effective charge; at long wavelengths we can regard the positive background as a continuum (called jellium) and rewrite this equation as

$$\rho\dot{\mathbf{V}} = \rho_i\mathbf{E}, \tag{21.11b}$$

where $\mathbf{V}$ is the jellium velocity, $\rho = N_i M/L^3$ is the *mass* density, and $\rho_i = N_i z\,|e|\,/L^3$ is the ion *charge* density. We can rewrite (21.11a) in terms of the ion current density and the ion number density, $n_i = N_i/L^3$, as

$$\mathbf{j}_i = n_i z\,|e|\,\mathbf{V}_i = \rho_i\mathbf{V};$$

combining with (21.11b) gives

$$\frac{d\mathbf{j}_i}{dt} = \frac{n_i e^2}{M}\mathbf{E}. \tag{21.12}$$

The accompanying ionic charge density is calculated using the equation of continuity,

$$\frac{\partial \rho_i}{\partial t} + \nabla \cdot \mathbf{j}_i = 0. \tag{21.13}$$

Taking the divergence of (21.12) and using (21.13) yields

$$\frac{\partial^2 \rho_i}{\partial t^2} + \frac{n_i e^2}{M} \nabla \cdot \mathbf{E} = 0. \tag{21.14}$$

Using Eq. (21.7) we have

$$\frac{\partial^2 \rho_i}{\partial t^2} = -\omega_{ip}^2 \, \rho_{total} \tag{21.15}$$

or

$$\omega^2 \rho_i = \omega_{ip}^2 \, \rho_{total}. \tag{21.16}$$

We now turn to the response of the electron system. Since $\omega_{ep} \gg \omega_{ip}$ the electrons will follow the ions adiabatically ($(\mu/\hbar)$ is also a characteristic frequency of the electron gas but it too greatly exceeds ω_{ip}). We will therefore calculate the electron response in the Thomas–Fermi model (see Section 14.2.1) in which one assumes that the local electron density is related to the local chemical potential $\mu = \varepsilon_F$ through the free-Fermi-gas expression

$$\mu - |e|\,\phi = \frac{\hbar^2}{2m}\left(3\pi^2 n_e\right)^{2/3}, \tag{21.17}$$

where μ is the global chemical potential and $-|e|\phi$ is a shift due to the local electrostatic potential, ϕ. Solving (21.17) for n_e we have

$$n_e = \frac{1}{3\pi^2}\left[\frac{2m}{\hbar^2}(\mu - |e|\,\phi)\right]^{3/2}. \tag{21.18}$$

Now the change in the electron density, δn_e, is much less than n_e and hence we can expand (21.18) as

$$n_e^{(0)} + \delta n_e \cong \frac{1}{3\pi^2}\left(\frac{2m\mu}{\hbar^2}\right)^{3/2}\left[1 - \frac{3}{2}\frac{|e|\,\phi}{\mu}\right];$$

Thus,

$$\delta n_e = -n_e^{(0)} \cdot \frac{3}{2}\frac{|e|\,\phi}{\mu},$$

where $n_e^{(0)} \equiv (1/3\pi^2) \cdot (2m\mu/\hbar^2)^{3/2}$. Noting that $\rho_e = -|e|\delta n_e$ we have

$$\rho_e = -\frac{3}{2}\frac{n_e^{(0)} e^2 \phi}{\mu}. \tag{21.19}$$

From Poisson's equation,

$$\nabla^2 \phi = -4\pi\rho_{total}, \tag{21.20}$$

and the electrostatic potential associated with our plane wave distortion is therefore

$$\phi = \frac{4\pi\rho_{total}}{q^2}, \tag{21.21}$$

which on inserting into (21.19) gives

$$\begin{aligned}\rho_e &= -\frac{6\pi n_e^{(0)} e^2}{\mu q^2}\rho_{total} \\ &= -\frac{\kappa_{TF}^2}{q^2}\rho_{total},\end{aligned} \tag{21.22}$$

where κ_{TF} is the Thomas–Fermi wave vector (which is the reciprocal of the Thomas–Fermi screening length), given by $\kappa_{TF} = (6\pi n_e^{(0)} e^2/\mu)^{1/2}$. Writing $\rho_{ext} = \rho_{total} - \rho_i - \rho_e$ and using (21.16) and (21.22), we have

$$\begin{aligned}\varepsilon(\mathbf{q},\omega) &= \frac{\rho_{total} - \rho_i - \rho_e}{\rho_{total}} \\ &= 1 - \frac{\omega_{ip}^2}{\omega^2} + \frac{\kappa_{TF}^2}{q^2} \\ &= \frac{\omega^2\left(q^2 + \kappa_{TF}^2\right) - \omega_{ip}^2 q^2}{\omega^2 q^2}\end{aligned} \tag{21.23}$$

or

$$\varepsilon(q,\omega)^{-1} = \frac{\omega^2 q^2}{\omega^2(q^2 + \kappa_{TF}^2) - \omega_{ip}^2 q^2}. \tag{21.24}$$

Inserting (21.24) into (21.3) we obtain an effective electron–electron interaction of the form

$$V(\mathbf{q},\omega) = \frac{4\pi e^2}{q^2 + \kappa_{TF}^2 - \dfrac{\omega_{ip}^2}{\omega^2} q^2}. \tag{21.25}$$

21.2 Limiting forms; longitudinal sound

We have two limiting forms for (21.25). When $\omega \gg \omega_{ip}$

$$V(\mathbf{q}, \omega) = \frac{4\pi e^2}{q^2 + \kappa_{TF}^2} \tag{21.26}$$

and on Fourier transforming to **r** space we have

$$V(r) = \frac{e^2}{r} e^{-i\kappa_{TF} r}; \tag{21.27}$$

we see that when **r** is large V(**r**) falls off exponentially, an effect discussed in Chapter 14 and referred to as screening. Thus in the high frequency limit, $\omega \gg \omega_{ip}$, the electron–electron interaction is repulsive, as we would expect.

We next consider frequencies $\omega \ll \omega_{ip}$, but for wave vectors of order a^{-1} where a is an interatomic spacing (or equivalently wave vectors of order the Brillouin zone dimension). If $\omega_{ip}^2/\omega^2 > (q^2 + \kappa_{TF}^2)/q^2$, we see that (21.25) is *attractive* (but ultimately vanishing) and this behavior explicitly involves the ions through ω_{ip}, which we may call an electron–phonon (electron–ion) interaction.

Let us now examine the low frequency *collective modes* of our electron–ion system. The collective modes follow from the equation[3]

$$\varepsilon(\mathbf{q}, \omega) = 0. \tag{21.28}$$

From Eq. (21.23) we obtain

$$\omega_q^2 = \frac{\omega_{ip}^2 q^2}{q^2 + \kappa_{TF}^2}. \tag{21.29}$$

The behavior of this expression is shown in Fig. 21.1.

[3] We can then sustain an internal electric field while **D** = 0. Since the normal component of **D** must be continuous across a boundary, a longitudinal mode can then propagate perpendicular to the boundary and produce no external field.

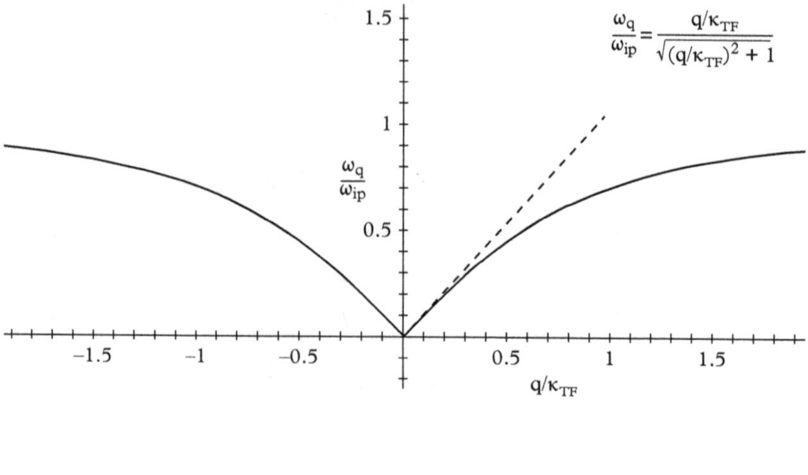

Figure 21.1 *The solid line is calculated from Eq. (21.29); the dashed line shows the slope at* q = 0.

For $q \ll \kappa_{TF}$ (which is reasonably well satisfied throughout the Brillouin zone, since $\kappa_{TF} > q_{BZ} \sim \pi/a$, where a is the lattice constant) we obtain the linear relation (due to Bohm and Staver (1950))

$$\omega_q = \frac{\omega_{ip}}{\kappa_{TF}} q. \tag{21.30}$$

A linear ω versus q relation is expected for the (longitudinal) *phonons* at small $q(\ll q_{BZ})$. The associated long wavelength velocity, $\omega_{ip}/\kappa_{TF} \sim 10^5$ cm/ sec, is typical of the sound velocity in solids. Hence, in this jellium model for the response of the ions, phonons emerge as *collective oscillations of the combined electron–ion system.*

21.3 Phonon mediated electron–electron interactions

We now examine Eq. (21.24) in more detail. We rewrite our reciprocal dielectric function in terms of ω_q in the following form:

$$\varepsilon^{-1}(\mathbf{q},\omega) = \frac{q^2}{q^2 + \kappa_{TF}^2}\left[1 + \frac{\omega_q^2}{\omega^2 - \omega_q^2}\right] \tag{21.31}$$

(Frohlich (1950); Bardeen and Pines (1955)). The associated effective potential may then be rewritten as

$$\begin{aligned} V(\omega,\mathbf{q}) &= \frac{4\pi e^2}{q^2 + \kappa_{TF}^2} + \frac{4\pi e^2}{q^2 + \kappa_{TF}^2} \cdot \frac{\omega_q^2}{\omega^2 - \omega_q^2} \\ &= V^{\text{coulomb}} + V^{(\text{el-ph})}. \end{aligned} \tag{21.32}$$

We interpret the first term in this expression as a *direct* (but screened) repulsive electron–electron interaction, while the second term is interpreted as an indirect, *phonon-induced* interaction; when we have $\omega < \omega_q$ for a given $\mathbf{q}$, the second term, and the overall form, is attractive.[4] Since the highest density of modes is for large phonon wave vectors ($q \cong q_{BZ}$), the interaction is attractive, and largest in magnitude, for frequencies of order the Debye frequency. As will be discussed later, in Chapter 34, this attractive interaction is responsible for the phenomenon of superconductivity.

It is interesting to examine how our effective interaction (21.32) alters the excitation spectrum for the electron gas.[5] To start with we construct a definition of the excitation energy based on the change of the total energy on adding a single electron with momentum $\mathbf{k}_0$ and spin σ_0 to the system. To do this we introduce a change in the distribution function by writing $n_\sigma(\mathbf{k}) \to n_\sigma(\mathbf{k}) + \delta n_\sigma(\mathbf{k})$. Now the total number of particles is given by

$$N = \frac{L^3}{(2\pi)^3}\sum_\sigma \int d^3k\, n_\sigma(\mathbf{k}) \tag{21.33}$$

[4] Curiously, when $\omega = 0$, we have $V(\omega = 0, \mathbf{q}) = 0$; i.e., the ions completely screen the Coulomb interaction. We can infer from this that there is a delicate balance between the Coulomb repulsion and the lattice response. Since superconductivity arises from the attractive interaction, this can be taken as a signal that accurately calculating superconducting transition temperatures will be difficult.

[5] Interaction between electrons and lattice vibrations in a normal metal, A.B. Migdal, *Sov. Phys. JETP-USSR* 7, 996 (1958).

and therefore the change in the distribution function required such that $N \rightarrow N+1$ when adding the particle with $\mathbf{k}_0 > \mathbf{k}_F$ must have the form

$$\delta n_{\sigma,\sigma_0}(\mathbf{k},\mathbf{k}_0) = \frac{(2\pi)^3}{L^3}\delta(\mathbf{k}-\mathbf{k}_0)\delta_{\sigma\sigma_0}; \tag{21.34}$$

for $\mathbf{k} < \mathbf{k}_F$, we interpret this expression as adding a hole.

For a non-interacting system the total energy is given by

$$E = \frac{L^3}{(2\pi)^3}\sum_\sigma \int d^3k\, \varepsilon_\sigma(\mathbf{k}) n_\sigma(\mathbf{k}) \tag{21.35}$$

and, using (21.34), the change in total energy on adding the electron with quantum numbers $\mathbf{k}_0, \sigma_0$ is just given by $\varepsilon_{\sigma_0}(\mathbf{k}_0)$. More formally we may write this as a *functional derivative*,

$$\varepsilon_\sigma(\mathbf{k}) = \frac{\delta E}{\delta n_\sigma(\mathbf{k})}. \tag{21.36}$$

Recalling the contribution of the exchange energy term in the Hartree–Fock theory of an electron gas, as given by Eq. (15.24b), we can write the total energy as

$$E = \frac{L^3}{(2\pi)^3}\sum_\sigma \int d^3k\, \varepsilon_\sigma(\mathbf{k}) n_\sigma(\mathbf{k}) - \frac{1}{2}\frac{L^3}{(2\pi)^6}\int d^3k' d^3k \left(\frac{4\pi e^2}{|\mathbf{k}-\mathbf{k}'|^2}\right) n(\mathbf{k})n(\mathbf{k}'), \tag{21.37}$$

where we have shifted the origin in the second term by writing $\mathbf{k}-\mathbf{q} = \mathbf{k}'$ in the integration over d^3q.[6] Applying our functional derivative definition of the excitation energies to Eq. (21.37) yields

$$\varepsilon(\mathbf{k}) = \varepsilon^{(0)}(\mathbf{k}) + \Delta\varepsilon(\mathbf{k}), \tag{21.38a}$$

where

$$\Delta\varepsilon(\mathbf{k}) = -\int \frac{d^3k'}{(2\pi)^3}\frac{4\pi e^2}{|\mathbf{k}-\mathbf{k}'|^2} n(\mathbf{k}'). \tag{21.38b}$$

Here, $\varepsilon^{(0)}(\mathbf{k}) = \hbar^2k^2/2m$ and we note that (21.38b) coincides with the last term on the left of Eq. (15.19) for the case of an electron gas (recall the Hartree term is cancelled by the positive background in our jellium model). Integrating (21.38b) (see the discussion leading up to Eq. (14.47)) one obtains

$$\Delta\varepsilon(\mathbf{k}) = \frac{2e^2k_F}{\pi} F\left(\frac{k}{k_F}\right), \tag{21.39a}$$

[6] The general form taken here by the exchange contribution to the total energy can be taken as a starting point for a more rigorous formulation of many-body effects known as Landau Fermi liquid theory (which will be taken up in greater detail in Chapter 37 in connection with the properties of normal liquid ^{3}He).

where

$$F(x) = \frac{1}{2} + \frac{1-x^2}{4x}\ln\frac{1+x}{1-x}. \tag{21.39b}$$

We note that the Fermi velocity, $v_F = \partial\varepsilon(\mathbf{k})/\hbar\partial\mathbf{k}$, has a logarithmic infinity for $k = k_F$; the heat capacity, which is proportional to $(v_F)^{-1}$, would then be proportional to T/lnT, as first shown by Bardeen,[7] which is in disagreement with both experiment and the Sommerfeld theory.

As a *model* for the effects of electron–electron and electron–ion screening on the single particle energies, we will utilize the jellium form, Eq. (21.32), in place of the bare potential with $\mathbf{q} = \mathbf{k} - \mathbf{k}'$ and $\hbar\omega = \varepsilon_\mathbf{k} - \varepsilon_{\mathbf{k}'}$. With these substitutions (21.38b) becomes

$$\Delta\varepsilon_\mathbf{k} = -\int \frac{d^3k'}{(2\pi)^3} n(\mathbf{k}')\left[\frac{4\pi e^2}{|\mathbf{k}-\mathbf{k}'|^2 + \kappa_{TF}^2} + \frac{4\pi e^2}{|\mathbf{k}-\mathbf{k}'|^2 + \kappa_{TF}^2}\frac{\hbar^2\omega_{\mathbf{k}-\mathbf{k}'}^2}{(\varepsilon_\mathbf{k}-\varepsilon_{\mathbf{k}'})^2 - \hbar^2\omega_{\mathbf{k}-\mathbf{k}'}^2}\right]. \tag{21.40}$$

Formally, (21.40) is an integral equation since the energy $\varepsilon_{\mathbf{k}'}$ under the integral sign is given by (21.38a), but the effect of $\Delta\varepsilon_\mathbf{k}$ can be neglected in leading order. The first term in (21.40) is the Thomas–Fermi correction to the electron energies which we can write as

$$\Delta\varepsilon_\mathbf{k}^{(TF)} = -\int_{k'<k_F} \frac{d^3k'}{(2\pi)^3}\frac{4\pi e^2}{|\mathbf{k}-\mathbf{k}'|^2 + \kappa_{TF}^2} n(\mathbf{k}'); \tag{21.41}$$

since we generally have $\kappa_{TF} > k_F$ this effect varies slowly with $\mathbf{k}$. Note the logarithmic singularity encountered with the Hartree–Fock theory, previously discussed, is now absent. In a band structure calculation the correction arising from (21.41) would normally not be applied but would be assumed to be included through some version of density functional theory.

The second term in (21.40) would cancel the first as $\omega \to 0$ but goes to zero for $\varepsilon_\mathbf{k} - \varepsilon_{\mathbf{k}'} \gg \hbar\omega_{\mathbf{k}-\mathbf{k}'}$. Since the phonon density of states increases as ω^2, but is bounded (in a monatomic periodic solid) by its value at the Brillouin zone edge, we argue that: (i) the weighted contribution from the region where the cancellation occurs is small and can be ignored, and (ii) the contribution will be maximal where the denominator in the second term of (21.40) satisfies

$$\begin{aligned}(\varepsilon_\mathbf{k}-\varepsilon_{\mathbf{k}'})^2 - \hbar^2\omega_{\mathbf{k}-\mathbf{k}'}^2 &= (\varepsilon_\mathbf{k}-\varepsilon_{\mathbf{k}'} + \hbar\omega_{\mathbf{k}-\mathbf{k}'})(\varepsilon_\mathbf{k}-\varepsilon_{\mathbf{k}'} - \hbar\omega_{\mathbf{k}-\mathbf{k}'})\\ &\cong 2\hbar\omega_{\mathbf{k}-\mathbf{k}'}(\varepsilon_\mathbf{k}-\varepsilon_{\mathbf{k}'} - \hbar\omega_{\mathbf{k}-\mathbf{k}'}) \cong 0,\end{aligned} \tag{21.42}$$

and will extend over a range of order ω_D around the Fermi surface. We can then write the second term in (21.40) as

$$\Delta\varepsilon_\mathbf{k}^{ep} \cong -\int \frac{dS_{\mathbf{k}'}d\varepsilon_{\mathbf{k}'}}{(2\pi)^3|\partial\varepsilon_{\mathbf{k}'}/\partial\mathbf{k}'|}\frac{4\pi e^2}{|\mathbf{k}-\mathbf{k}'|^2 + \kappa_{TF}^2}\frac{1}{2}\frac{\hbar\omega_{\mathbf{k}-\mathbf{k}'}}{\varepsilon_\mathbf{k}-\varepsilon_{\mathbf{k}'} - \hbar\omega_{\mathbf{k}-\mathbf{k}'} + i0}n(\varepsilon_{\mathbf{k}'}), \tag{21.43}$$

[7] J. Bardeen, *Phys. Rev.* 50, 1098 (1936).

where we rewrote the integration as one over energy and the associated constant energy surfaces, and included an infinitesimal i0 in the denominator (which would appear naturally in a more precisely formulated theory). Integrating over $d\varepsilon_{k'}$, ignoring slowly varying quantities, and retaining only the principal value contribution yields

$$\Delta\varepsilon_{\mathbf{k}}^{ep} \cong -\int \frac{dS_{\mathbf{k}'}}{(2\pi)^3 |\partial\varepsilon_{\mathbf{k}'}/\partial\mathbf{k}'|} \frac{4\pi e^2}{|\mathbf{k}-\mathbf{k}'|^2+\kappa_{TF}^2} \frac{\hbar\omega_{\mathbf{k}-\mathbf{k}'}}{2} \ln\left|\frac{\varepsilon_F - \varepsilon_{\mathbf{k}} - \hbar\omega_{\mathbf{k}-\mathbf{k}'}}{\varepsilon_F - \varepsilon_{\mathbf{k}} + \hbar\omega_{\mathbf{k}-\mathbf{k}'}}\right|. \tag{21.44}$$

Note that this contribution vanishes for $\varepsilon_{\mathbf{k}} = \varepsilon_F$; i.e., the electron–phonon interaction does not alter the shape of the Fermi surface. Using the expansion $\ln(1 + x) = x$, for energies such that $|\varepsilon_F - \varepsilon_{\mathbf{k}}| \ll \hbar\omega_{\mathbf{k}-\mathbf{k}'}$, Eq. (21.44) may be expanded as

$$\Delta\varepsilon_{\mathbf{k}}^{ep} \cong -(\varepsilon_{\mathbf{k}} - \varepsilon_F) \int \frac{dS_{\mathbf{k}'}}{(2\pi)^3 |\partial\varepsilon_{\mathbf{k}'}/\partial\mathbf{k}'|} \frac{4\pi e^2}{|\mathbf{k}-\mathbf{k}'|^2+\kappa_{TF}^2}. \tag{21.45}$$

Hence the Fermi velocity, $\frac{\partial}{\hbar\partial\mathbf{k}}\left(\varepsilon_{\mathbf{k}}^{(0)} + \Delta\varepsilon_{\mathbf{k}}^{ep}\right)$, will be reduced by a factor $(1 - \lambda)$ where

$$\lambda = \int \frac{dS_{\mathbf{k}'}}{(2\pi)^3 |\partial\varepsilon_{\mathbf{k}'}/\partial\mathbf{k}'|} \frac{4\pi e^2}{|\mathbf{k}-\mathbf{k}'|^2+\kappa_{TF}^2}; \tag{21.46}$$

to the same accuracy, a reduction in the Fermi velocity enhances the density of states, $N(\varepsilon) \rightarrow (1 + \lambda)N(\varepsilon)$, and with it the low temperature electronic heat capacity.

The imaginary infinitesimal would result in an imaginary contribution to $\Delta\varepsilon_{\mathbf{k}}^{ep}$, corresponding to a *decay of the excitations with a characteristic lifetime*, $\tau_{ep} = \hbar/\text{Im}(-\Delta\varepsilon_{\mathbf{k}}^{ep})$. This decay rate vanishes at the Fermi surface and further analysis shows

$$-\text{Im}\Delta\varepsilon_{\mathbf{k}}^{ep} \sim \begin{cases} |\varepsilon - \varepsilon_F|^3/(\hbar\omega_D)^2; & |\varepsilon - \varepsilon_F| \ll \hbar\omega_D \\ \hbar\omega_D; & \hbar\omega_D \geq |\varepsilon - \varepsilon_F| \end{cases}. \tag{21.47}$$

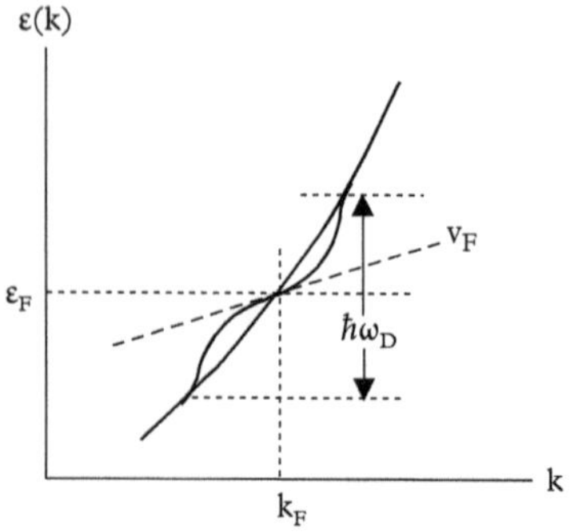

Figure 21.2 *A portion of the curve* ε(k) *in the vicinity of* ε_F *showing the change in slope produced by the electron–phonon interaction.*

In summary, we have the following: Since the frequencies ω_q cut off above a value of order the Debye frequency, ω_D, we then have only the static Thomas–Fermi form of $V(\mathbf{q})$ which varies slowly. For energy differences of order the Debye frequency, where the denominator in (21.43) is smallest, the shift will be largest. The shift vanishes for $k = k_F$ so the shape of the Fermi surface is *unaffected* by the phonon contribution; these behaviors are shown schematically in Fig. 21.2. Note that the slope, corresponding to the Fermi velocity, is *smallest* at k_F, and it is usually written in the form

$$\frac{\partial\varepsilon}{\hbar\partial\mathbf{k}} = v_F = \frac{v_F^{(0)}}{1 + \lambda(\mathbf{k})}. \tag{21.48}$$

Since the heat capacity involves an integral over the Fermi surface of $|v_F|^{-1}$, we see that there will be an extra contribution involving an average of $\lambda(\mathbf{k})$ over the surface. The enhancement is of order 20% in simple metals, but is considerably larger in certain transition metals where other mechanisms (one being so-called paramagnons) contribute to the enhancement (see discussion in Chapter 19). Finally, a more careful treatment, the results of which were summarized in Eq. (21.47), shows that the electrons off the Fermi surface, but within an interval $\hbar\omega_D$, acquire a finite lifetime due to phonon emission.

By working backward we can use (21.32) to construct a model expression for the phonon induced electron–electron interaction.[8] To do this we start by writing the correction to the total ground state energy arising from the electron–phonon interaction in second-order perturbation theory. Now the intermediate state consists of an electron with momentum $\mathbf{k}'$ and energy $\varepsilon_{k'}$, together with a phonon with momentum $\mathbf{q}$ and energy $\hbar\omega_q$.[9] Hence the perturbation in the energy has the form

$$\Delta E = \sum_{kk'} \frac{\langle \mathbf{k}|\, V^{ep}\, |\mathbf{k}'\rangle \langle \mathbf{k}'|\, V^{ep}\, |\mathbf{k}\rangle}{\varepsilon_k - (\varepsilon_{k'} - \hbar\omega_{k-k'})} n_k(1-n_{k'})$$
$$= \sum_{kk'} \frac{|g_{kk'}|^2}{\varepsilon_k - \varepsilon_{k'} + \hbar\omega_{k-k'}} n_k(1-n_{k'}), \qquad (21.49)$$

where we have denoted the matrix element by $g_{kk'}$ in the second step and required the intermediate state to be empty. Earlier, we defined the excitation energy, through Eq. (21.36), as the *first* functional derivative of the total energy. From the structure of Eq. (21.37) and our delta function form for δn (that changes the particle number by one), we are led to define the *effective* interaction through the *second* functional derivative of the total energy as

$$V(\mathbf{k},\mathbf{k}') = \frac{1}{L^3}\frac{\delta^2 E}{\delta n^2} = \frac{1}{L^3}\frac{\delta\varepsilon}{\delta n}. \qquad (21.50)$$

Applying this recipe to (21.40) gives the jellium result, while applying it to Eq. (21.49) gives

$$V^{ep}_{kk'} = |g_{kk'}|^2 \frac{2\hbar\omega(\mathbf{k}-\mathbf{k}')}{(\varepsilon_k - \varepsilon_{k'})^2 + \hbar^2\omega^2(\mathbf{k}-\mathbf{k}')}. \qquad (21.51)$$

If we compare (21.51) with (21.32), we obtain the phonon contribution to the effective potential if we make the association

$$|g_{kk'}|^2 = \frac{4\pi e^2}{|\mathbf{k}-\mathbf{k}'|^2 + \kappa_{TF}^2}\frac{\hbar\omega_{k-k'}}{2}. \qquad (21.52)$$

[8] The argument we will use, which is heuristic, follows that given by Ashcroft and Mermin, who fully acknowledge its character.

[9] Note that the intermediate state would actually involve a phonon with semi-classical amplitude, $e^{i\mathbf{q}\cdot\mathbf{r}}$, resulting in a matrix element with plane wave electron states that is proportional to

$$\langle \mathbf{k}', \mathbf{q}|\, V^{ep}\, |\mathbf{k}\rangle \propto L^{-3} \int d^3r\, e^{-i(\mathbf{k}'+\mathbf{q})\cdot\mathbf{r}} e^{\mathbf{k}\cdot\mathbf{r}} = \Delta_{\mathbf{k}-\mathbf{k}'-\mathbf{q}};$$

hence the intermediate state phonon has energy $\hbar\omega(\mathbf{k}-\mathbf{k}')$ and the total energy of the intermediate state is $\varepsilon_{k'} + \hbar\omega(\mathbf{k}-\mathbf{k}')$.

At long wavelengths we can ignore the factor $|\mathbf{k} - \mathbf{k}'|^2$ in the denominator; recalling $\kappa_{TF}^2 = \frac{6\pi ne^2}{\varepsilon_F}$, we have

$$|g_{kk'}|^2 = \frac{\hbar\varepsilon_F}{3}\omega_{k-k'} \qquad (|\mathbf{k}-\mathbf{k}'| \ll \kappa_{TF}). \tag{21.53}$$

Note that the coupling vanishes as $|\mathbf{k} - \mathbf{k}'|$, which affects the temperature dependence of the electrical resistivity, as we will discuss further.

21.4 Formal theory of the electron–phonon matrix element

In this section we outline the microscopic theory of the interaction of electrons and phonons. We begin by assuming the availability of an appropriate one-electron potential function for the distorted lattice, having the form

$$V = V(\mathbf{r}, \mathbf{r}_1 \ldots\ldots \mathbf{r}_N), \tag{21.54}$$

where $\mathbf{r}$ is the coordinate of an electron and the remaining coordinates, $\mathbf{r}_1 \ldots\ldots \mathbf{r}_N$, denote the positions of the nuclei; the latter are related to the equilibrium positions, $\mathbf{R}_n$, (see Eq. 8.2) by $\mathbf{r}_n = \mathbf{R}_n + \mathbf{U}_n$. The task of obtaining the function $V(\mathbf{r}, \mathbf{r}_1 \ldots\ldots \mathbf{r}_N)$ is formidable.[10] As a model this function could be constructed by superimposing the atomic potentials in a manner analogous to Eq. (10.1),

$$V = \sum_{\mathbf{r}_n} U(\mathbf{r} - \mathbf{r}_n) \tag{21.55}$$

(where we must distinguish the scalar atomic potentials, U, from the vector atomic displacements, **U**); this procedure would *not* include changes induced by the requirements of self consistency.

The perturbation that couples electrons and phonons is the potential resulting from the displacement of ions from their equilibrium sites $\mathbf{R}_n$ by amounts $\mathbf{U}_n$. It can be obtained by expanding the potential (21.54) (or the model form (21.55)) as a power series in $\mathbf{U}_n$. Limiting ourselves to terms linear in $\mathbf{U}_n$ this perturbation is

$$V_{ep}(\mathbf{r}) = V(\mathbf{r}, \{\mathbf{r}_n\}) - V(\mathbf{r}, \{\mathbf{R}_n\}) = \sum_n \left(\frac{\partial V\mathbf{r}, \{\mathbf{r}_n\}}{\partial \mathbf{r}_n}\right)_{\mathbf{R}_n} \cdot \mathbf{U}_n + \ldots, \tag{21.56}$$

where $\{\mathbf{r}_n\}$ denotes the set of all $\mathbf{r}_n$. The scattering between Bloch states, ψ_k, is governed by the matrix element

$$M_{k'n';kn} = \int d^3r\psi^*_{k'n'} V_{ep}\psi_{kn} \tag{21.57}$$

[10] One procedure would be to parameterize the potential from self-consistent supercell band structure calculations which yield information on (21.54) for certain configurations of the nuclei (see Chapter 36).

(where we must distinguish between a scalar band index, n, and a vector site index, **n**). If we confine ourselves for the moment to a single phonon, $\mathbf{U}_{\mathbf{n}}$ (given by Eqs. (16.46) and (16.52a,b)), (21.57) becomes

$$M^{(\mathbf{q})}_{\mathbf{k}'n';\mathbf{k}n} = \int d^3r\,\psi^*_{\mathbf{k}'n'}(\mathbf{r})\left[\sum_{\mathbf{n}} U(\mathbf{q})\,e^{i\mathbf{q}\cdot\mathbf{R}_{\mathbf{n}}}\boldsymbol{\varepsilon}(\sigma,\mathbf{q})\cdot\frac{\partial V(\mathbf{r})}{\partial \mathbf{r}_{\mathbf{n}}}\right]\psi_{\mathbf{k}n}(\mathbf{r}). \tag{21.58}$$

In evaluating this matrix element we translate all Bloch states, $\psi_{\mathbf{k}n}(\mathbf{r})$, to the site **n** yielding $e^{i\mathbf{k}\cdot\mathbf{R}_{\mathbf{n}}}\psi_{\mathbf{k}n}(\mathbf{r}-\mathbf{R}_{\mathbf{n}})$; Eq. (21.58) then contains the overall phase factor, $e^{i(\mathbf{k}-\mathbf{k}'+\mathbf{q})\cdot\mathbf{R}_{\mathbf{n}}}$. From our previous discussion,

$$\sum_{\mathbf{n}} e^{i(\mathbf{k}-\mathbf{k}'+\mathbf{q})\cdot\mathbf{R}_{\mathbf{n}}} = N\delta_{\mathbf{k}'+\mathbf{K},\,\mathbf{k}+\mathbf{q}},$$

where **K** is any reciprocal lattice vector.

To obtain the total matrix element for scattering from **k** to **k**′ we must sum over all phonon states. We also make the transition from a semi-classical treatment (where the phonon field is treated as an external potential) to a quantized phonon field by interpreting the displacement, $\mathbf{U}(\mathbf{q})$, as an operator, $\hat{\mathbf{U}}(\mathbf{q})$, as discussed in Section 16.5. The transitions for phonon absorption and emission involve the following phonon matrix elements:

$$\langle n(\mathbf{q})-1\,|\,\hat{a}(\mathbf{q})\,|\,n(\mathbf{q})\rangle = [n(\mathbf{q})]^{1/2} \qquad \text{(absorption)}$$

and

$$\langle n(\mathbf{q})+1\,|\,\hat{a}^{\dagger}(\mathbf{q})\,|\,n(\mathbf{q})\rangle = [n(\mathbf{q})+1]^{1/2} \qquad \text{(emission)},$$

where we have included the argument **q** in the phonon number $n(\mathbf{q})$ to distinguish it from the Bloch index n and the site index **n**. The total matrix element for *phonon emission* is then

$$M^{(\mathbf{q})}_{\mathbf{k}'n';\mathbf{k}n} = \left[\frac{\hbar(n(\mathbf{q})+1)}{2NM\omega(\boldsymbol{\sigma},\mathbf{q})}\right]^{1/2} I(\mathbf{k}'n';\mathbf{k}n), \tag{21.59}$$

where we have defined the ($\mathbf{R}_{\mathbf{n}}$ independent) integral,

$$I(\mathbf{k}'n',\mathbf{k}n) = N\int d^3r\,\psi^*_{\mathbf{k}'n'}(\mathbf{r}-\mathbf{R}_{\mathbf{n}})\boldsymbol{\varepsilon}(\sigma)\cdot\frac{\partial V(\mathbf{r}-\mathbf{R}_{\mathbf{n}})}{\partial \mathbf{r}_{\mathbf{n}}}\psi_{\mathbf{k}n}(\mathbf{r}-\mathbf{R}_{\mathbf{n}}) \tag{21.60}$$

and

$$\mathbf{k}' = \mathbf{k}+\mathbf{q}-\mathbf{K}. \tag{21.61a}$$

In evaluating a transition rate, a δ function appears involving the energies of the electron states and the phonon, which requires that

$$\varepsilon_n(\mathbf{k}) = \varepsilon_{n'}(\mathbf{k}') + \hbar\omega(\sigma,\mathbf{q}). \tag{21.61b}$$

The reverse process, involving *phonon absorption*, can also occur where

$$\mathbf{k}' = \mathbf{k} - \mathbf{q} - \mathbf{K} \tag{21.62a}$$

and

$$\varepsilon_n(\mathbf{k}) = \varepsilon_{n'}(\mathbf{k}') - \hbar\omega(\sigma, \mathbf{q}), \tag{21.62b}$$

with the accompanying matrix element

$$M^{(\mathbf{q})}_{\mathbf{k}'n';\,\mathbf{k}n} = \left[\frac{\hbar n(\mathbf{q})}{2NM\omega(\sigma,\mathbf{q})}\right]^{1/2} I(\mathbf{k}'n';\,\mathbf{k}n). \tag{21.63}$$

The above theory is due to F. Bloch (1928).

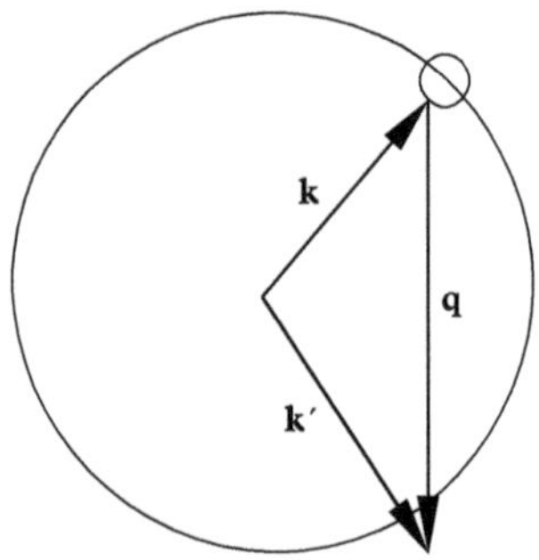

Figure 21.3 *An electron with initial energy, $\varepsilon_{\mathbf{k}} < \varepsilon_F$, that absorbs a phonon with energy, $\hbar\omega_{\mathbf{q}}$, resulting in a final state, $\varepsilon'_{\mathbf{k}} > \varepsilon_F$, but which is displaced around the Fermi surface by a (possibly large) phonon wave vector,* $\mathbf{q}$. *The small circle depicts, schematically, the range of available phonon energies. When $T \gg \Theta_D$, where Θ_D is the Debye temperature, the radius is approximately $\hbar\omega_D$; when $T \ll \Theta_D$, the range of available energies is of the order of or less than k_BT with associated wave vectors $k_BT/\hbar V_s$, where V_s is a sound velocity.*

21.5 The phonon contribution to the low temperature electrical resistance

At finite temperatures, electrons can emit or absorb (real as opposed to virtual) phonons in so-called three-particle processes involving initial and final electron states and the gain or loss of a phonon.[11] The conservation of momentum and energy conditions associated with these two processes were given previously in Eqs. (21.61a,b) and (21.62a,b). At low temperatures, where the phonon wave vectors are small, and hence $\mathbf{q} \ll \mathbf{K}$, Umklapp processes become negligible. Imposing the conservation of energy condition restricts the three-dimensional space of phonon wave vectors to a surface on which they must lie. In addition, the allowed phonon frequencies, $\omega_{\mathbf{q}}$, are bounded by ω_D and, in addition, their energies are generally much less than the Fermi energy; i.e., $\varepsilon_{\mathbf{k}} \cong \varepsilon_{\mathbf{k}'} \cong \varepsilon_F$. But as the temperature increases, $\mathbf{q}$ can be large (potentially of order $2k_F$, corresponding to electron back scattering); i.e., although phonons do not shift electron energies significantly, they can scatter them through wide angles. This situation is shown for a spherical Fermi surface in Fig. 21.3.

At high temperatures, where $T \gg \Theta_D$, with Θ_D the Debye temperature, the phonon Bose distribution can then be approximated by

$$n(\mathbf{q}) = \frac{1}{e^{\hbar\omega_{\mathbf{q}}/k_BT} - 1} \cong \frac{k_BT}{\hbar\omega_{\mathbf{q}}}; \tag{21.64}$$

essentially all the 3N vibrational modes of the lattice are excited and this expression gives the number of phonons in each of the modes. These phonons scatter electrons all around the Fermi surface thereby generating electrical resistance. Being qualitatively proportional to the number of phonons, the resistivity will then be proportional to T; i.e.,

$$\rho \propto T;\ T \gg \Theta_D. \tag{21.65}$$

[11] Four-particle processes, in which an existing phonon inelastically scatters from an electron, altering the direction (and energy) of the latter, are also possible but enter only in higher order.

Alternatively one can use an Einstein picture where all atoms vibrate with a characteristic frequency, ω_E. From the equipartition theorem the potential energy will be

$$(1/2)M\omega_E^2\overline{|\mathbf{r}|^2} = (3/2)k_BT. \tag{21.66}$$

The scattering cross section would be of order $\sigma = \pi\overline{|\mathbf{r}|^2}$, and hence proportional to T, as would the resistivity.

At low temperatures where $T \ll \Theta_D$, we have only long wavelength phonons. From the Debye theory the number of these is of order T^3 and hence we expect a total scattering rate proportional to T^3. However the scattering rate entering electrical transport also involves a loss in forward momentum (current) that in turn involves the factor $1 - \cos\theta$ (see Eq. (19.72)); at low temperature we can approximate this factor as $\theta^2/2$, where $\theta \approx q/k_F \approx k_BT/\hbar V_s k_F$. We then expect an electrical resistivity of the form $\rho = \rho_0 + \rho_1$, where $\rho_0 = m/ne^2\tau_{imp}$ arises from static scattering processes (impurities, etc.) and ρ_1 arises from phonons and combines the above two factors (T^3 and T^2); i.e.,[12]

$$\rho_1 \propto T^5;\ T \ll \Theta_D, \tag{21.67}$$

a behavior first predicted by F. Bloch.

The above discussion has been qualitative. To quantitatively calculate the temperature dependence of various transport effects we must explicitly include the electron–phonon interaction, e.g., as treated in Section 21.4. As an example, if we suppress the band indices we can write the collision integral describing the rate of change of the electron distribution function f(**k**) as (see Section 19.2.1)

$$\begin{aligned}\left(\frac{\partial f}{\partial t}\right)_{col} &= -\int \frac{d^3k'}{(2\pi)^3}\left[f(\mathbf{k})\left(1-f(\mathbf{k}')\right)w^{e-p}_{\mathbf{k}\to\mathbf{k}'} - f(\mathbf{k}')\left(1-f(\mathbf{k})\right)w^{e-p}_{\mathbf{k}'\to\mathbf{k}}\right] \\ &= -\int \frac{d^3k'}{(2\pi)^3}\left(f(\mathbf{k}) - f(\mathbf{k}')\right)w^{e-p}_{\mathbf{k}\to\mathbf{k}'},\end{aligned} \tag{21.68}$$

where $w^{e-p}_{\mathbf{k}\to\mathbf{k}'}$ is the total electron–phonon scattering rate from state **k** to **k**′ and we have assumed detailed balancing in the second step (which assumes equilibrium in the phonon system); note f(**k**) and f(**k**′) now correspond to different energies with the energy gain or loss supplied by the phonon bath. The required transition rate is given by

$$w^{e-p}_{\mathbf{k}\to\mathbf{k}'} = \int \frac{d^3q}{(2\pi)^3}\left(w_{\mathbf{k}\to\mathbf{k}-\mathbf{q}} + w_{\mathbf{k}\to\mathbf{k}+\mathbf{q}}\right), \tag{21.69}$$

where $w_{\mathbf{k}\to\mathbf{k}-\mathbf{q}}$ and $w_{\mathbf{k}\to\mathbf{k}+\mathbf{q}}$ are the golden rule transition rates involving the matrix elements for phonon emission (where the phonon number n(**q**) increases

[12] In very pure metals and at very low temperatures the resistivity will contain a contribution, $\rho_{ee} \propto T^2$, arising from electron–electron scattering. The origin of this temperature dependence will be discussed in Chapter 37 in connection with liquid ^{3}He, where it is a dominant effect.

by one) and absorption (where n(q) decreases by one); using Eqs. (21.69) and (21.63) we can write these as

$$w_{\mathbf{k}\to\mathbf{k}-\mathbf{q}} = \frac{2\pi}{\hbar}\frac{\hbar(n(\mathbf{q})+1)}{2NM\omega_\mathbf{q}}\left|I(\mathbf{k}-\mathbf{q};\mathbf{k})^2\right|\delta(\varepsilon_\mathbf{k}-\varepsilon_{\mathbf{k}-\mathbf{q}}-\hbar\omega_\mathbf{q}) \tag{21.70a}$$

and

$$w_{\mathbf{k}\to\mathbf{k}+\mathbf{q}} = \frac{2\pi}{\hbar}\frac{\hbar n(q)}{2NM\omega_\mathbf{q}}\left|I(\mathbf{k}+\mathbf{q};\mathbf{k})^2\right|\delta(\varepsilon_\mathbf{k}-\varepsilon_{\mathbf{k}+\mathbf{q}}+\hbar\omega_\mathbf{q}), \tag{21.70b}$$

where I is given by Eq. (21.60). Were we to be evaluating a transport scattering time associated with an electron current density, a factor $(1-\cos\theta)$ would enter, as discussed in Section 19.10 and incorporated above in obtaining the Bloch law.

ADDITIONAL READING

The discussions in Sections 21.1 and 21.2 are similar to those given by P.G. de Gennes in Chapter 4 of *Superconductivity in Metals and Alloys*, W.A. Benjamin (1964), and N. Ashcroft and N.D. Mermin in Chapter 26 of *Solid State Physics*, Holt, Reinhart and Winston (1976). A more rigorous discussion using Green's function methods is given by E.M. Lifshitz and L.P. Pitaevskii in Sections 64 and 65 of *Statistical Physics Part II*, Pergamon (1969).

Problem

21.1. Verify Eq. (21.44).

Appendix 21A The attenuation of sound by conduction electrons

For simplicity we will assume a free electron gas at absolute zero. We further assume that the sound is propagating in an elastically isotropic material (see Section 1.9) where the displacement associated with the sound wave, $\mathbf{u} = \mathbf{u}_0 e^{iqz-i\omega t}$, can be separated into pure longitudinal or transverse modes, where $\mathbf{u}_0 = u_{0\parallel}\hat{\mathbf{z}}$ or $\mathbf{u}_0 = \mathbf{u}_{0\perp}$ respectively, with associated sound velocities, v_l or v_t. We will discuss two mechanisms for the interaction of conduction electrons with sound waves. The first mechanism involves the self-consistent electric field generated by the moving ions; the second mechanism is discussed below.

To treat the first mechanism we regard the lattice as consisting of ions with a charge $z|e|$. There will then be an associated *ion* current

$$\mathbf{j}^{(i)} = z|e|n^{(i)}\dot{\mathbf{u}}, \tag{21A.1}$$

where $n^{(i)}$ is the number density of ions having velocity $\dot{\mathbf{u}}$. From overall charge neutrality we may set $zn^{(i)} = n_0$, where n_0 is the electron number density. This ion current combines with the *induced electron* current, $\mathbf{j}^{(e)}$, to produce a *total* current, $\mathbf{j}^{(t)}$,

$$\mathbf{j}^{(t)} = \mathbf{j}^{(i)} + \mathbf{j}^{(e)} \tag{21A.2}$$

and on inserting (21A.1) gives

$$\mathbf{j}^{(t)} = n_0\,|e|\,\dot{\mathbf{u}} + \mathbf{j}^{(e)}. \tag{21A.3}$$

The total current is related to the internal *electric field* accompanying the sound wave, by the Maxwell equations; eliminating the magnetic field by combining the second, third, and fourth Maxwell equations yields the generalized wave equation

$$\nabla^2\mathbf{E} - \nabla(\nabla\cdot\mathbf{E}) = \frac{4\pi\mu}{c^2}\frac{\partial\mathbf{j}}{\partial t} + \frac{\varepsilon\mu}{c^2}\frac{\partial^2\mathbf{E}}{\partial t^2}; \tag{21A.4}$$

here, $\mathbf{E}$ must also have the form $\mathbf{E} = \mathbf{E}_0 e^{iqz-i\omega t}$, with $\mathbf{E}_0 = \mathbf{E}_{0z}\,\hat{\mathbf{z}}$ or $\mathbf{E}_0 = \mathbf{E}_{0\perp}$ for a longitudinal or transverse wave. Inserting these forms into the wave equation we obtain

$$E_z = -\frac{4\pi i}{\varepsilon\omega} j_z^{(t)} \tag{21A.5}$$

and

$$\mathbf{E}_\perp = \frac{-\dfrac{4\pi i\mu\omega}{c^2}}{\dfrac{\varepsilon\mu\omega^2}{c^2} - q^2}\,\mathbf{j}_\perp^{(t)} = -\frac{\dfrac{4\pi i}{\varepsilon\omega}\dfrac{v_t^2}{v_p^2}}{\dfrac{v_t^2}{v_p^2} - 1}\,\mathbf{j}_\perp^{(t)}, \tag{21A.6}$$

where $v_p^2 = c^2/\varepsilon\mu$ is the phase velocity of light and $v_t = \omega/q$ is the transverse sound velocity introduced above; noting $v_p \gg v_t$, Eq. (21A.6) simplifies to

$$\mathbf{E}_\perp \cong \frac{4\pi i}{\omega\,\varepsilon}\frac{v_t^2}{v_p^2}\mathbf{j}_\perp^{(t)}. \tag{21A.7}$$

The response of the electron distribution function, $f(\mathbf{p}, \mathbf{r}, t)$, to the sound wave, which we will use below to calculate the attenuation of sound, is governed by the Boltzmann equation (see Eq. (19.8b))

$$\frac{\partial f}{\partial t} + \frac{\partial f}{\partial \mathbf{r}}\cdot\dot{\mathbf{r}} + \frac{\partial f}{\partial \mathbf{p}}\cdot\dot{\mathbf{p}} = I(f). \tag{21A.8}$$

The term on the right accounts for the collisions and will be modeled using Eq. (19.16),

$$I(f) = -\frac{f(\mathbf{r}, \mathbf{p}, t) - f_0(\varepsilon)}{\tau}, \tag{21A.9}$$

where $f_0 = 1/[\exp((\varepsilon - \varepsilon_F(n))/k_B T) + 1]$. It is here that the second mechanism for the interaction between the electrons and the ions enters. In applying Eq. (21A.9) we will assume that the non-equilibrium distribution function relaxes toward a state of *local equilibrium* involving a lattice that is both *strained and moving with a velocity* $\dot{\mathbf{u}}$ *relative to the lab frame.*[13] The first effect shifts the Fermi level due to a change in local electron density, n' according to

$$\frac{\partial f_0}{\partial \varepsilon_F}\frac{\partial \varepsilon_F}{\partial n}n' = -\frac{\partial f_0}{\partial \varepsilon}\frac{2\varepsilon_F}{3n_0}n'. \tag{21A.10}$$

Note this is non-zero only for a longitudinal wave. To account for the second effect we examine the kinetic energy in both the rest and the local frames,

$$\varepsilon_{\text{local}} = (m/2)(\mathbf{v} - \mathbf{u})^2 \cong \varepsilon_0 - \mathbf{p} \cdot \mathbf{u}. \tag{21A.11}$$

Combining (21A.10) and (21A.11), we have

$$f_0(\varepsilon) = f_0(\varepsilon_0) + \frac{\partial f_0}{\partial \varepsilon}\left(-\frac{2\varepsilon_F}{3n_0}n' - \mathbf{p} \cdot \dot{\mathbf{u}}\right). \tag{21A.12}$$

The non-equilibrium distribution function will be written as the sum of the equilibrium distribution function, $f_0(\varepsilon_0)$, and a small distortion,

$$\begin{aligned} f(r, \mathbf{p}, t) &= f_0(\varepsilon_0) + f_1(\mathbf{p}, \mathbf{r}, t) \\ &= f_0(\varepsilon_0) + \frac{\partial f_0}{\partial \varepsilon}\Phi(\mathbf{p}, \mathbf{r}.t), \end{aligned} \tag{21A.13}$$

where f_1 and Φ are first order in $\mathbf{u}$, $\mathbf{E}$, and n'. Writing $\partial f_0/\partial \mathbf{p} = \mathbf{v}\partial f_0/\partial \varepsilon$, $\dot{\mathbf{p}} = -|e|\mathbf{E}$, and assuming our plane wave form, Eq. (21A.8) becomes[14]

$$(-i\omega + iqv\cos\theta)\,\Phi - |e|\mathbf{E} \cdot \mathbf{v} = -\frac{\Phi}{\tau} - \frac{2\varepsilon_F}{3\tau}\frac{n'}{n_0} - \frac{m}{\tau}\mathbf{v} \cdot \dot{\mathbf{u}};$$

solving for Φ we have

$$\Phi = \frac{-\frac{2\varepsilon_F}{3}\frac{n'}{n_0} + |e|\tau\left(\mathbf{E} - \frac{m\dot{\mathbf{u}}}{|e|\tau}\right) \cdot \mathbf{v}}{(1 - i\omega\tau + iqv\tau\cos\theta)}. \tag{21A.14}$$

[13] This is referred to as the *collision drag effect*.

[14] To better model the force arising in a real metal, we could write $\dot{\mathbf{p}} = -\partial\varepsilon/\partial\mathbf{r} - |e|\mathbf{E}$ and introduce momentum-dependent longitudinal and transverse *deformation potential parameters*, $\lambda_l(\mathbf{p})$ and $\lambda_t(\mathbf{p})$, and write

$$\frac{\partial \varepsilon}{\partial \mathbf{r}} = \begin{cases} i\lambda_l q u_l; & \mathbf{q} \parallel \mathbf{u} \\ i\lambda_t q u_t; & \mathbf{q} \perp \mathbf{u} \end{cases}.$$

To make contact with (21A.2) and (21A.3) we require the electron current given by

$$\begin{aligned} \mathbf{j}^{(e)}(\mathbf{r},t) &= -\frac{2|e|}{(2\pi\hbar)^3}\int \mathbf{v}f(\mathbf{r},\mathbf{p},t)d^3p \\ &= -\frac{2|e|}{(2\pi\hbar)^3}\int \frac{\partial f_0(\varepsilon)}{\partial \varepsilon}\mathbf{v}\Phi(\mathbf{r},\mathbf{p},t)d^3p \\ &= \mathbf{j}_{\mathbf{E}}^{(e)}(\mathbf{r},t) + \mathbf{j}_{\dot{\mathbf{u}}}^{(e)}(\mathbf{r},t) + \mathbf{j}_{n'}^{(e)}(\mathbf{r},t), \end{aligned} \qquad (21A.15)$$

where $\mathbf{j}_{\mathbf{E}}^{(e)}, \mathbf{j}_{\dot{\mathbf{u}}}^{(e)}$, and $\mathbf{j}_{n'}^{(e)}$ are the contributions to $\mathbf{j}^{(e)}$ arising from the variations, $\mathbf{E}$, $\dot{\mathbf{u}}$, and n' in Φ; these variations are not independent of each other and part of our task is to relate them. The integrals associated with the first two currents have the same functional form and can be combined as

$$\mathbf{j}_{\mathbf{E}}^{(e)}(\mathbf{r},t) + \mathbf{j}_{\dot{\mathbf{u}}}^{(e)}(\mathbf{r},t) = \sigma \cdot \left(\mathbf{E}(\mathbf{r},t) - \frac{m\,\dot{\mathbf{u}}(\mathbf{r},t)}{|e|\tau}\right), \qquad (21A.16)$$

where $\boldsymbol{\sigma}$ is a (diagonal) frequency- and wave-vector-dependent conductivity tensor,

$$\sigma_{ij}(\mathbf{q},\omega) = -\frac{2e^2\tau}{(2\pi\hbar)^3}\int \frac{\partial f_0}{\partial \varepsilon}\frac{v_i v_j}{1 - i\omega\tau + iqv\tau\cos\theta}d^3p. \qquad (21A.17)$$

We write the last contribution, $\mathbf{j}_{n'}^{(e)}$, as

$$\mathbf{j}_{n'}^{(e)}(\mathbf{r},t) = J(\mathbf{q},\omega)\frac{n'(\mathbf{r},t)}{n_0}\mathbf{z}, \qquad (21A.18)$$

where $\mathbf{z}$ is a unit vector parallel to $\mathbf{q}$ and

$$J(\mathbf{q},\omega) = \frac{|e|\varepsilon_F}{6\pi^3\hbar^3}\int \frac{\partial f_0(\varepsilon)}{\partial \varepsilon}\frac{v\cos\theta}{1 - i\omega\tau + iqv\tau\cos\theta}d^3p, \qquad (21A.19)$$

which has units of current density. Unlike the components of $\boldsymbol{\sigma}$, the quantity $J \to 0$ in the long wavelength limit.

The quantity $n'(\mathbf{r},t)$ is related to the electron current through the equation of continuity,

$$-|e|\,\dot{n}'(\mathbf{r},t) + \nabla \cdot \mathbf{j}^{(e)}(\mathbf{r},t) = 0 \qquad (21A.20)$$

or

$$\frac{n'(\mathbf{r},t)}{n_0} = -\frac{\mathbf{q}\cdot\mathbf{j}^{(e)}(\mathbf{r},t)}{\omega\,|e|\,n_0} = -\frac{1}{|e|\,n_0 v_l}j_z^{(e)}(\mathbf{r},t), \qquad (21A.21)$$

where $j_z^{(e)}$ is the component of the electron current along the propagation direction and v_l the velocity of longitudinal sound. Equation (21A.18) now becomes

$$\mathbf{j}_{n'}^{(e)}(\mathbf{r}, t) = -J(\mathbf{q}, \omega)\frac{1}{|e|\, n_0 v_l} j_z^{(e)} \mathbf{z}. \tag{21A.22}$$

By combining (21A.21), (21A.5), and (21A.7) we can eliminate the variables $n'(\mathbf{r}, t)/n_0$ and $\mathbf{E}(\mathbf{r}, t)$ in favor of $\mathbf{u}(\mathbf{r}, t)$.

Writing $\partial f_0/\partial \varepsilon = -\delta(\varepsilon - \varepsilon_F)$ and integrating over p, we obtain σ_{zz} as[15]

$$\sigma_{zz}(\mathbf{q}, \omega) = \frac{3}{4\pi}\frac{n_0 e^2 \tau}{m}\int \frac{\cos^2\theta d\Omega}{1 - i\omega\tau + iqv_F\tau\cos\theta}$$
$$= \frac{\sigma_0}{1 - i\omega\tau}\frac{3}{a^3}\left[a - \tan^{-1} a\right]; \tag{21A.23}$$

and σ_{xx} as

$$\sigma_{xx}(\mathbf{q}, \omega) = \frac{3}{4\pi}\frac{n_0 e^2 \tau}{m}\int \frac{\sin^2\theta\cos^2\phi d\Omega}{1 - i\omega\tau + iqv_F\tau\cos\theta}$$
$$= \frac{\sigma_0}{1 - i\omega\tau}\frac{3}{2a^3}\left[(1 + a^2)\tan^{-1} a - a\right]; \tag{21A.24}$$

Here, $n_0 = (2/(2\pi\hbar)^3) \cdot (4\pi p_F^3/3)$ is the equilibrium electron number density, $\sigma_0 = n_0 e^2 \tau/m$, and we have introduced the dimensionless parameter

$$a = \frac{ql}{1 - i\omega\tau}, \tag{21A.25}$$

where l is the mean free path, $l = v_F\tau$. Performing a similar integration for J gives

$$J(\mathbf{q}, \omega) = -\frac{|e|m\varepsilon_F^2}{3\pi^3\hbar^3}\int \frac{\cos\theta d\Omega}{1 - i\omega\tau + iqv_F\tau\cos\theta}$$
$$= i\frac{4|e|m\varepsilon_F^2}{3\pi^2\hbar^3(1 - i\omega\tau)}\frac{1}{a^2}(a - \tan^{-1} a). \tag{21A.26}$$

On comparing (21A.24) and (21A.26), we find that

$$J = \frac{iamv_F}{3|e|\tau}\sigma_{zz}. \tag{21A.27}$$

Combining (21A.16), (21A.22), and (21A.27), the parallel component of the electron current in Eq. (21A.15) becomes

$$j_z^{(e)} = \frac{\sigma_{zz}}{1 + \dfrac{iav_F}{3v_l}\dfrac{\sigma_{zz}}{\sigma_0}}\left(E_z - \frac{m}{|e|\tau}\dot{u}_z\right)$$
$$\equiv \sigma'_{zz}\left(E_z - \frac{m}{|e|\tau}\dot{u}_z\right); \tag{21A.28}$$

[15] In what follows it is easiest to develop the integrals that arise in terms of logarithmic forms and then use the identity, $\tan^{-1} a = \frac{i}{2}\ln\frac{1 - ia}{1 + ia}$.

the perpendicular component in Eq. (21A.15) is simply

$$j_\perp^{(e)} = \sigma_{xx}\left[\mathbf{E}_\perp - \frac{m}{|e|\,\tau}\dot{u}_\perp\right]. \tag{21A.29}$$

We now assemble the above parts to calculate the rate at which energy in the sound wave is lost to the electron gas. One contribution involves the average rate of increase of the kinetic energy, K, of the electron gas, which is given by

$$\begin{aligned}\frac{dK}{dt} &= \overline{\int d^3p \frac{\partial f_0}{\partial \varepsilon}\Phi \frac{d}{dt}\left(\frac{1}{2}mv^2\right)} = \overline{\int d^3p \frac{\partial f_0}{\partial \varepsilon}\Phi\, \mathbf{v}\cdot\dot{\mathbf{p}}} = \overline{-|e|\mathbf{E}\cdot\int d^3p \frac{\partial f_0}{\partial \varepsilon}\Phi\, \mathbf{v}} \\ &= \frac{1}{2}\Re\left(j^{(e)*}\cdot\mathbf{E}\right),\end{aligned} \tag{21A.30}$$

where in the last step we have used the usual procedure for taking time averages of oscillating quadratic forms (with $\Re$ denoting the real part); this contribution we recognize as the ohmic heating. However there is an additional contribution that is more subtle. An electron gas in equilibrium with a uniformly moving lattice with velocity $\dot{\mathbf{u}}$ would have an average velocity, $\bar{\mathbf{v}} = \dot{\mathbf{u}}$; the momentum resulting from a departure from this condition would relax at a Drude-like rate, $-m(\bar{\mathbf{v}} - \dot{\mathbf{u}})/\tau$, which manifests itself as a force per electron on the moving lattice. The average rate that this process does work on the lattice is then $-\Re m n_0 \mathbf{u}^* \cdot (\bar{\mathbf{v}} - \dot{\mathbf{u}})/2\tau$; the minus sign implies that, on the average, this process returns energy to the sound wave. The total rate at which the sound wave dissipates energy is then

$$\dot{Q} = \frac{1}{2}\Re\left(j^{(e)}\cdot\mathbf{E}^* - mn_0\dot{\mathbf{u}}^*\cdot(\bar{\mathbf{v}}-\dot{\mathbf{u}})/\tau\right). \tag{21A.31}$$

In what follows, we will treat longitudinal and transverse waves separately.

21A.1 Longitudinal waves

To evaluate $\dot{Q}$ for a longitudinal wave we return to (21A.3), and eliminate $j_z^{(t)}$ in favor of E_z using (21A.28) or (21A.29) to obtain

$$j_z^{(e)} = -\frac{\varepsilon\omega}{4\pi i}E_z - n_0|e|\,\dot{u}_z. \tag{21A.32}$$

Setting (21A.32) equal to (21A.28), we obtain E_z as

$$\begin{aligned}E_z &= \frac{\dfrac{\sigma'_{zz}m}{|e|\,\tau} - n_0|e|}{\sigma'_{zz} + \dfrac{\varepsilon\omega}{4\pi i}}\dot{u}_z \\ &\cong \left(\frac{m}{|e|\,\tau} - \frac{n_0|e|}{\sigma'_{zz}}\right)\dot{u}_z,\end{aligned} \tag{21A.33}$$

where in the second step, we take $\varepsilon\omega/4\pi i\sigma'_{zz} \approx \varepsilon\omega/4\pi i\sigma_0 = \omega/i\omega_p^2\tau \ll 1$, which is valid for metals where $\omega_p \gg \tau^{-1} \gg \omega$; characteristic values here are $\omega_p \approx 10^{16}$, $\tau^{-1} \approx 10^{10}$(in a pure metal at low temperatures), and $\omega \leq 10^9$ (typical for sound experiments). Inserting (21A.33) into (21A.32) we have

$$\begin{aligned} j_z^{(e)} &= -\frac{\varepsilon\omega}{4\pi i}\left(\frac{m}{|e|\,\tau} - \frac{n_0\,|e|}{\sigma'_{zz}}\right)\dot{u}_z - n_0\,|e|\,\dot{u}_z \\ &\cong -n_0\,|e|\,\dot{u}_z, \end{aligned} \tag{21A.34}$$

or, equivalently, $\dot{u}_z \cong \bar{v}_z$, implying that j_z^t essentially vanishes; a result of strong screening.

Inserting (21A.33) and (21A.34) into (21A.31), we obtain the rate of energy dissipation per unit volume for longitudinal waves as

$$\begin{aligned} \dot{Q} &= \frac{1}{2}\Re\left(\frac{n_0^2\,e^2}{\sigma'_{zz}} - \frac{mn_0}{\tau}\right)|\dot{u}_z|^2 \\ &= \frac{1}{2}\,|\dot{u}_z|^2\,\frac{mn_0}{\tau}\Re\left(\frac{\sigma_0}{\sigma'_{zz}} - 1\right). \end{aligned} \tag{21A.35}$$

From the definition of σ'_{zz} through Eqs. (21A.28) and (21A.23), and in the limit $\omega\tau \ll 1$, we have[16]

$$\Re\frac{\sigma_0}{\sigma'_{zz}} \cong \Re\frac{1 + \dfrac{iav_F}{3v_l}\dfrac{3(1+i\omega\tau)}{a^3}\left[a - \tan^{-1}a\right]}{\dfrac{3}{a^3}\left[a - \tan^{-1}a\right]} = \frac{a^2\tan^{-1}a}{3\left[a - \tan^{-1}a\right]}$$

and hence,

$$\begin{aligned} \dot{Q} &= \frac{1}{2}\Re\left(\frac{n_0^2\,e^2}{\sigma'_{zz}} + \frac{mn_0}{\tau}\right)|\dot{u}_z|^2 \\ &= \frac{1}{2}\,|\dot{u}_z|^2\,\frac{n_0 m}{\tau}\left[\frac{a^2\tan^{-1}a}{3\left(a - \tan^{-1}a\right)} - 1\right]. \end{aligned} \tag{21A.36}$$

The energy flux of the sound wave is $\frac{1}{2}\rho\,|\dot{u}|^2\,v_l$, where ρ is the mass density, and the rate of energy loss per unit length, α, is therefore $\dot{Q}/(\frac{1}{2}\rho\,|\dot{u}|^2\,v_l)$ or

$$\alpha = \frac{n_0 m}{\rho v_l \tau}\left[\frac{(ql)^2\tan^{-1}ql}{3\left(ql - \tan^{-1}ql\right)} - 1\right]. \tag{21A.37}$$

(A.B. Pippard (1955)). Expanding (21A.37) in the limit $ql \ll 1$ gives

$$\begin{aligned} \alpha &= \frac{4}{15}\frac{n_0 m v_F^2\tau}{\rho v_l^3}\omega^2 \\ &= \frac{4}{3}\frac{\eta}{\rho v_l^3}\omega^2 \end{aligned} \tag{21A.38}$$

[16] The factor v_F/v_l in the second term in the numerator requires us to retain the leading correction in $\omega\tau$ entering σ_{zz}.

where, in the second step, we rewrite the expression to conform with the usual hydrodynamic result for the attenuation of sound in a liquid (see Eq. (1.58)) and we have introduced the *viscosity coefficient*, η, of a Fermi gas in the form

$$\eta = \frac{2}{5} n_0 \varepsilon_F \tau; \tag{21A.39}$$

note that the attenuation is proportional to $\omega^2\tau$ in this limit. In the opposite limit where $ql \gg 1$, but where we continue to have $\omega\tau \ll 1$, (21A.37) becomes

$$\alpha = \frac{\pi\, n_0 m v_F}{6\rho v_l^2}\omega; \tag{21A.40}$$

here we see that the attenuation scales as ω rather than $\omega^2\tau$.

21A.2 Transverse waves

The calculation of the attenuation coefficient of transverse waves proceeds in a similar manner. From (21A.3) and (21A.7) we obtain

$$\mathbf{j}_\perp^{(e)} = \frac{\omega\,\varepsilon}{4\pi\, i}\frac{v_p^2}{v_t^2}\mathbf{E}_\perp - n_0\,|e|\,\dot{\mathbf{u}}_\perp. \tag{21A.41}$$

Noting that $j_{n'}^{(e)}(\mathbf{r}, t) = 0$ for transverse waves and equating (21A.41) to (21A.16) gives

$$\mathbf{E}_\perp = \frac{n_0|e| - \dfrac{m\sigma_{xx}}{|e|\,\tau}}{\dfrac{\omega\,\varepsilon}{4\pi\, i}\dfrac{v_p^2}{v_t^2} - \sigma_{xx}}\dot{\mathbf{u}}_\perp$$

$$\cong -\left(\frac{n_0|e|}{\sigma_{xx}} + \frac{m}{|e|\,\tau}\right)\dot{\mathbf{u}}_\perp, \tag{21A.42}$$

where in the second step we have neglected the first term in the denominator;[17] substituting this approximate form in (21A.41) then gives

$$\mathbf{j}_\perp^{(e)} \cong -\frac{\omega\,\varepsilon}{4\pi\, i}\frac{v_p^2}{v_t^2}\left(\frac{n_0|e|}{\sigma_{xx}} - \frac{m}{|e|\,\tau}\right)\mathbf{u}_\perp - n_0\,|e|\,\dot{\mathbf{u}}_\perp$$

$$\cong -n_0\,|e|\,\dot{\mathbf{u}}_\perp, \tag{21A.43}$$

although the terms neglected in the second step can be important at very high sound frequencies. Substituting (21A.42) and (21A.43) into (21A.31) gives

$$\dot{Q} = \frac{1}{2}\,|\dot{u}_z|^2\,\frac{mn_0}{\tau}\Re\left(\frac{\sigma_0}{\sigma_{xx}} - 1\right); \tag{21A.44}$$

[17] Based on the parameters given above when discussing longitudinal waves, it will be several orders of magnitude smaller.

taking σ_{xx} from Eq. (21A.24), the corresponding energy attenuation per unit length is then

$$\alpha = \frac{n_0 m}{\rho v_t \tau}\left[\frac{2a^3}{3\left((1+a^2)\tan^{-1}a - a\right)} - 1\right]. \tag{21A.45}$$

Expanding in the limit of small ql gives

$$\alpha = \frac{n_0 m v_F^2 \tau}{5\rho v_t^3}\omega^2, \tag{21A.46}$$

again scaling as $\omega^2\tau$, while for $1 \ll ql \ll \omega\tau$, we obtain

$$\alpha = \frac{4 n_0 m v_F}{3\pi \rho v_t^2}\omega, \tag{21A.47}$$

which scales as ω. We see that, apart from the coefficients, the behaviors are the same for longitudinal and transverse waves.

ADDITIONAL READING

Ultrasonic attenuation in the free electron model and in the absence of an external magnetic field is discussed by A.B. Pippard, *Phil. Mag.* **46**, 1104 (1955). A deformation potential was included in the treatments by A.I. Akheizer, M.I. Kaganov, and G.I. Liubarskii, *Sov. Phys. JETP* **5**, 685 (1957) and A.B. Pippard, *Proc. R. Soc.* **A257**, 165 (1960). For a review and additional references, see: J.A. Rayne and C.K. Jones, Ultrasonic attenuation in normal metals and superconductors: Fermi-surface effects, *Physical Acoustics*, Vol. VII, W.P. Mason and R.N. Thurston (Eds.), Academic Press, NY (1970). The topic is also treated in chapter 17 of *Quantum Theory of Solids*, C. Kittel, John Wiley & Sons, NY (1963) and chapter 12 of *Fundamentals of the Theory of Metals*, A.A. Abrikosov, North Holland, Amsterdam (1988). The effects of an external magnetic field and, in particular, the emergence of the magneto-acoustic geometric resonance phenomenon, discussed qualitatively in Section 20.1.5, is derived in a free electron model by M.H. Cohen, M.J. Harrison, and W.A. Harrison, *Phys. Rev.* **117**, 937 (1960).

Part VII

Semiconductors

Homogeneous Semiconductors

22

22.1 Model band structures

A semiconductor can be regarded as an insulator with a narrow band gap, ε_g. However the term is applied rather loosely and some materials with large band gaps, such as diamond or AlN, are also called semiconductors. Most of the materials with tetrahedral coordination, including the diamond, zinc blende, wurtzite, and related structures, are termed semiconductors.

Semiconductors may be divided into two classes: those for which the points in **k**-space corresponding to the highest occupied *valence band* state, ε_v, and the lowest unoccupied *conduction band* state, ε_c, lie directly above each other (a *direct band gap* material); and those for which these two points are separated in **k**-space by some amount $\mathbf{k}_0$ (an *indirect band gap* material); these two cases are shown in Fig. 22.1(a,b). For simplicity, the behavior of the $\varepsilon(\mathbf{k})$ curves has been assumed to be parabolic in shape. This will only be true in the immediate vicinity of an extremum. The presence of a degeneracy in the band structure further complicates the structure of the constant energy surfaces near an extremum, as discussed earlier in Section 12.1, and later in Section 22.5.

At absolute zero all the states associated with the valence bands are filled (and hence their name, as these electrons contribute to the cohesive energy of the material), while those of the conduction bands are empty; the situation for the valence band is depicted schematically in Fig. 22.2(a). Figure 22.2(b) shows the same band with one electron removed. In a charge neutral system this electron would lie in a conduction band. The empty state is referred to as a *hole*. In the presence of an applied electric field, transitions will occur from the lower lying (occupied) electron state into the empty hole state; i.e., the hole will move lower in energy. However, since the hole represents the absence of an electron (having negative energy), this downward motion of the hole represents an increase in energy of the system. In steady state, where the energy gained from an external potential drop is lost to the rest of the system through scattering, the hole would drift at constant velocity. As discussed in Section 10.5, a completely filled band carries no current, and therefore the holes contribute a current of the opposite sign—they behave like positively charged electrons.

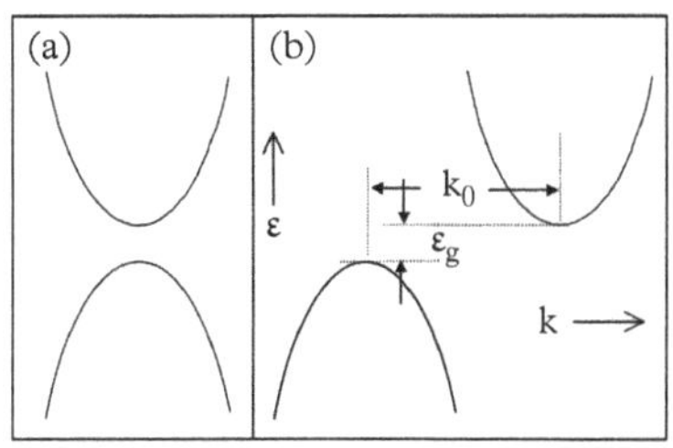

Figure 22.1 *A schematic representation of a direct-gap (a), and an indirect-gap (b) semiconductor.*

The Physics of Solids. First Edition. J. B. Ketterson.

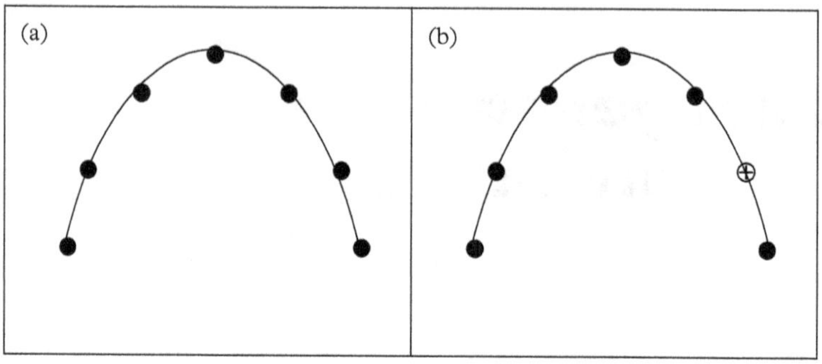

Figure 22.2 *(a) Representation of a filled band; (b) representation of the same band with one empty state, which responds as a hole.*

22.2 Parabolic bands

In the previously mentioned parabolic band approximation, we may write the **k**-dependent energies of states in the conduction and valence bands as

$$\varepsilon_e = \varepsilon_c + \frac{\hbar^2}{2m}\sum_{i,j}\alpha_{ij}k_ik_j \tag{22.1}$$

and

$$\varepsilon_h = \varepsilon_v - \frac{\hbar^2}{2m}\sum_{i,j}\beta_{ij}k_ik_j, \tag{22.2}$$

where for convenience we have assumed the band minima are at $k = 0$. The quantities α_{ij} and β_{ij} are elements of dimensionless *reciprocal* mass tensors for the electrons and holes respectively. Being proportional to second derivatives in a Taylor expansion of the energy they are symmetric tensors containing at most six independent parameters. The associated constant energy surfaces are ellipsoids. The tensors may be diagonalized and in general would have three unequal principal elements. The remaining parameters relate to three Euler angles specifying the orientation of the ellipsoid. For an arbitrary point in the Brillouin zone, all six parameters would be required to fully specify the band. In addition, we require a total number of ellipsoids such that the point-group symmetry of the crystal is satisfied.

Rather than the reciprocal effective mass tensors, we can equally well introduce effective mass tensors for the electrons and holes defined through the relations

$$m_{ij}^{(e)} = m\left(\overset{\leftrightarrow}{\alpha}{}^{-1}\right)_{ij} \tag{22.3a}$$

and

$$m_{ij}^{(h)} = m\left(\overset{\leftrightarrow}{\beta}{}^{-1}\right)_{ij}. \tag{22.3b}$$

22.3 Carrier densities and densities of states

The numbers of electrons and holes are given by

$$n_e = \int d\varepsilon_e \frac{g_e(\varepsilon_e)}{e^{(\varepsilon_e - \mu)/k_B T} + 1} \tag{22.4a}$$

and

$$n_h = \int d\varepsilon_h\, g_h(\varepsilon_h) \left[1 - \frac{1}{e^{(\varepsilon_h - \mu)/k_B T} + 1}\right], \tag{22.4b}$$

where $g_e(\varepsilon_e)$, and $g_h(\varepsilon_h)$ are the densities of states for electrons and holes respectively, and the factor $\left[1 - (e^{(\varepsilon_h - \mu)/k_B T} + 1)^{-1}\right]$ is the occupation number for holes. In order that we have charge neutrality, it will turn out that we must have $\varepsilon_e > \mu > \varepsilon_h$. In many applications we can assume that the temperature is low enough that both $\varepsilon_e - \mu$ and $\mu - \varepsilon_h$ are sufficiently larger than $k_B T$ such that we are in the Boltzmann limit;[1] this is often a good approximation at room temperatures for materials of practical importance like Si and Ge. In this limit, (22.4a,b) become

$$n_e = \int d\varepsilon g_e(\varepsilon_e) e^{-(\varepsilon_e - \mu)/k_B T} \tag{22.5a}$$

and

$$n_h = \int d\varepsilon g_h(\varepsilon_h) e^{-(\mu - \varepsilon_h)/k_B T}. \tag{22.5b}$$

By performing a principal axis transformation the energies of the electrons and holes can be written

$$\varepsilon_e = \varepsilon_c + \frac{\hbar^2}{2m}\left[\alpha_1 k_1^2 + \alpha_2 k_2^2 + \alpha_3 k_3^2\right] \tag{22.6a}$$

and

$$\varepsilon_h = \varepsilon_v - \frac{\hbar^2}{2m}\left[\beta_1 k_1^2 + \beta_2 k_2^2 + \beta_3 k_3^2\right]. \tag{22.6b}$$

We next scale the principal axes of the electrons such that $k_i' = \alpha_i^{1/2} k_i$, and similarly for the holes. Then,

$$\int d^3k = (\alpha_1 \alpha_2 \alpha_3)^{-1/2} \int d^3k' \tag{22.7}$$

and

$$\varepsilon_e = \varepsilon_c + \frac{\hbar^2}{2m} k'^2. \tag{22.8}$$

[1] It is usually sufficient to have $\varepsilon_e - \mu$ and $\mu - \varepsilon_h > 3k_B T$.

The density of states for our parabolic band is then

$$g_e(\varepsilon_e) = (\alpha_1\alpha_2\alpha_3)^{-1/2}\frac{2}{(2\pi)^3}4\pi k'^2\frac{dk'}{d\varepsilon}$$
$$= \frac{1}{2\pi^2}\left(\frac{2m_e}{\hbar^2}\right)^{3/2}(\varepsilon_e - \varepsilon_c)^{1/2}, \tag{22.9a}$$

where we have defined a *density of states effective mass* as $m_e = (\alpha_1\alpha_2\alpha_3)^{-1/3}\,m$; similarly,

$$g_h(\varepsilon_h) = \frac{1}{2\pi^2}\left(\frac{2m_h}{\hbar^2}\right)^{3/2}(\varepsilon_v - \varepsilon_h)^{1/2}, \tag{22.9b}$$

with $m_h = (\beta_1\beta_2\beta_3)^{-1/3}\,m$. The number of electrons in the Boltzmann limit is then

$$n_e = \frac{1}{2\pi^2}\left(\frac{2m_e}{\hbar^2}\right)^{3/2}\int_{\varepsilon_c}^{\infty} d\varepsilon_e\,(\varepsilon_e - \varepsilon_c)^{1/2}\,e^{-(\varepsilon_e-\mu)/k_BT}$$
$$= \frac{1}{2\pi^2}\left(\frac{2m_ek_BT}{\hbar^2}\right)^{3/2} e^{-(\varepsilon_c-\mu)/k_BT}\,2\int_0^{\infty} dx x^2 e^{-x^2};$$

the integral is given by $\sqrt{\pi}/4$ and our final expression for the electron number density is

$$n_e = \frac{1}{4}\left(\frac{2m_ek_BT}{\pi\hbar^2}\right)^{3/2} e^{-(\varepsilon_c-\mu)/k_BT}. \tag{22.10a}$$

The corresponding expression for the holes is

$$n_h = \frac{1}{4}\left(\frac{2m_hk_BT}{\pi\hbar^2}\right)^{3/2} e^{-(\mu-\varepsilon_v)/k_BT}. \tag{22.10b}$$

We note again that (22.10a,b) must be multiplied by the number of band extrema.

The product of the electron and hole concentrations is given by

$$n_en_h = n_i^2, \tag{22.11a}$$

where

$$n_i^2 = \frac{1}{16}\left(\frac{2k_BT}{\pi\hbar^2}\right)^3 (m_em_h)^{3/2}\,e^{-\varepsilon_g/k_BT}. \tag{22.11b}$$

Note that n_i is independent of the chemical potential, μ. Equation (22.11a) is known as the *law of mass action*.

For the *intrinsic* case when we have charge neutrality,[2] the chemical potential, μ_i, is determined from the condition of $n_e = n_h$, which from (22.10a,b) gives

$$\left(\frac{m_e}{m_h}\right)^{3/2} = e^{-(2\mu_i - \varepsilon_c - \varepsilon_v)/k_B T}$$

or

$$\mu_i = \frac{1}{2}\left(\varepsilon_c + \varepsilon_v + \frac{3}{2}k_B T\, ln\left(\frac{m_h}{m_e}\right)\right)$$
$$= \varepsilon_v + \frac{\varepsilon_g}{2} + \frac{3}{4}k_B T\, ln\left(\frac{m_h}{m_e}\right). \tag{22.12}$$

For the special case of $m_e = m_h$, the chemical potential lies midway in the gap; since we have assumed $k_B T \ll \varepsilon_g$ and the mass ratio enters only logarithmically, the last term in (22.12) can be regarded as a small correction.

22.4 Band properties in some specific materials

Table 22.1 shows band gaps for some typical semiconductors. The band extrema of the holes in these materials lie at the Γ point of the Brillouin zone and will be discussed shortly (αSn, also called gray tin, presents some subtleties which were discussed earlier in Section 12.2). For Si the electron band extrema lie along the six Γ–X directions; in the approximation that the constant energy surfaces are ellipsoids there is rotational symmetry about these axes. Choosing one of the Γ–X (see Fig. 11.14(a)) directions as the z axis, the energy dispersion relation is then given by

$$\varepsilon_e = \varepsilon_c + \frac{\hbar^2}{2m}\left[\alpha_\perp\left(k_x^2 + k_y^2\right) + \alpha_\parallel k_z^2\right] \tag{22.13a}$$

or

$$\varepsilon_e = \varepsilon_c + \frac{\hbar^2}{2}\left[\frac{1}{m_\perp}\left(k_x^2 + k_y^2\right) + \frac{1}{m_\parallel}k_z^2\right]. \tag{22.13b}$$

In germanium the electron band extrema lie at the L points of the Brillouin zone (see Fig. 11.14(b)). Although there are eight Γ – L directions, there are only four ellipsoidal constant energy surfaces, since one half of the ellipsoid lies in the neighboring zone.[4] If we designate the Γ – L direction as the z axis then, since these axes have three-fold symmetry, we must have constant energy surfaces that are ellipsoids of revolution in our parabolic band approximation. Hence they are still described by the expressions (22.13a,b), although, as discussed, the z axes and the number of ellipsoids differ in the two cases.

Table 22.2 lists the longitudinal, $m_{e\parallel}$, and transverse, $m_{e\perp}$, effective masses, along with the density of states mass $m_e = (m_{e\perp}^2 m_{e\parallel})^{1/3}$ for germanium and silicon (in units of the free electron mass).

Table 22.1 *Semiconductor band gaps.*[3]

Material	Energy gap (eV)	
	0 K	300 K
Si	1.17	1.11
Ge	0.74	0.66
αSn	0	-
InSb	0.23	0.17
InAs	0.43	0.36
InP	1.42	1.27
GaP	2.32	2.25
GaAs	1.52	1.43
GaSb	0.81	0.68
CdSe	1.84	1.74
CdTe	1.61	1.44
ZnO	3.44	3.2
ZnS	3.91	3.6

Table 22.2 *Electron effective mass parameters in* Si *and* Ge.

electrons	$m_{e\parallel}$	$m_{e\perp}$	m_e
Ge	1.58	0.082	0.22
Si	0.9163	0.1905	0.3216

[2] Intrinsic semiconductors are those materials that do not contain ionizable impurities, which can contribute mobile electrons.

[3] From C. Kittel, *Introduction to Solid State Physics,* 6th Ed., p. 185. John Wiley & Sons, NY (1986).

[4] Through translation using a reciprocal lattice vector this second half falls inside the original zone near the L point, related to the first by inversion symmetry.

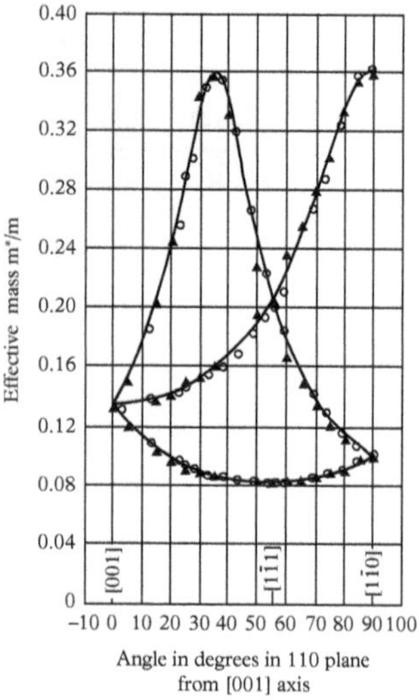

Figure 22.3 *The effective mass of electrons in* Ge *as obtained from cyclotron resonance experiments. The long axis of the ellipsoids lies along the [111] directions and hence the smallest mass is associated with this direction. For* H *in a (110) plane there are three separate branches, one associated with the principal ellipsoid and two others associated with the remaining axes. (After Dresselhaus, Kip, and Kittel (1955).)*

The effective mass parameters entering the expressions can be measured by the method of cyclotron resonance (introduced in Section 3.6). As the name implies, the method involves observing the absorption of electromagnetic radiation when its frequency matches the orbital frequency of electrons in the semiconductor spiraling around an external d.c. magnetic field. For magnetic fields of order kilogauss, the resonant frequencies are of order 10^{10} Hz, which is in the microwave region. For maximum sensitivity the microwave source, with frequency ω, is coupled to and resonated with an electromagnetic cavity in which a semiconductor sample has been placed. The frequency of the orbiting electrons, called the cyclotron frequency and designated ω_c, is then swept by varying the magnetic field through the resonant condition, $\omega = \omega_c(H,\theta)$, where a reduction in the cavity Q factor is observed.

The frequency of the resonance is found by solving the Newton–Lorentz equation of motion:

$$\hbar \dot{\mathbf{k}} = \frac{e}{c}\, \mathbf{v} \times \mathbf{H}, \tag{22.14}$$

with $\mathbf{v} = \frac{1}{\hbar}\frac{\partial \varepsilon}{\partial \mathbf{k}}$. An expression for ω_c valid for orbits arising from an arbitrary band structure was given earlier in Eq. (18.2a). It is relatively straightforward to solve (22.14) for the simpler case of the ellipsoidal surfaces described by (22.13a,b) with the result, $\omega_c = eH/m(\theta)^*c$, with

$$\frac{1}{m(\theta)^{*2}} = \frac{\cos^2\theta}{m_\perp^2} + \frac{\sin^2\theta}{m_\perp m_\parallel} \tag{22.15}$$

and where θ is the angle measured from the axis with rotational symmetry.[5] This axis lies parallel to [111] in Ge and parallel to [100] in Si (see Fig. 11.16(a,b)).

The criterion for observing cyclotron resonance (as with any resonance) is $\omega\tau \gg 1$, where τ is an appropriate electron scattering time. Hence very pure materials are required to reduce the scattering. Low temperatures are also necessary to reduce the electron–phonon scattering. For the case of a semiconductor this latter requirement also reduces the carrier concentration which, in turn, reduces the amplitude of the resonance. One way to circumvent this problem is to excite carriers by applying light to the sample, which generates electron–hole pairs. Figure 22.3 shows the measured cyclotron masses in Ge for the magnetic field in a [110] plane.[6]

22.5 Holes in Si and Ge

The band structure of Ge was shown earlier in Fig. 11.14(b). The hole bands of Si, Ge, and Sn, and many other tetrahedrally coordinated semiconductors are "derived" from the atomic p states, and the extremum is located at the Γ point of the Brillouin zone. In the presence of spin–orbit coupling the three-fold degenerate p states (six-fold with spin) are split at the Γ point into

[5] We recall that for the Azbel–Kaner cyclotron resonance observed in metals, a combination of the electrons' passage through the skin depth and a tipping of the applied magnetic field relative to the surface restricted the observed orbits to extremal sections. In a semiconductor all orbits contribute. For a parabolic band with ellipsoidal constant energy surfaces, all orbits for a given field direction have the same cyclotron frequency. For the hole bands in Si, Ge, and many other semiconductors, this is not the case and it leads to broadened resonances.

[6] Cyclotron resonance of electrons and holes in silicon and germanium crystals, G, Dresselhaus, A.F. Kip, and C. Kittel, *Phys. Rev.* **98**, 368 (1955).

a higher-energy, four-fold degenerate, $p_{3/2}$-like (Γ_8^+) state and a lower-energy, two-fold degenerate, $p_{1/2}$-like (Γ_7^+) state. On leaving the Γ point in any direction the four-fold degenerate states split into two doubly degenerate (due to the spin degeneracy) states. The resulting band structure associated with these three, two-fold degenerate, states is shown schematically in Fig. 22.4. The upper two pairs of states correspond to a group-theoretical two-dimensional E representation. As discussed in Section 12.1.7 (see Eq. (12.39)), the band structure in the parabolic approximation for such a state is described by

$$\varepsilon_h = \varepsilon_v - Ak^2 \pm \sqrt{B^2k^4 + C^2\left(k_x^2k_y^2 + k_y^2k_z^2 + k_z^2k_x^2\right)}. \tag{22.16a}$$

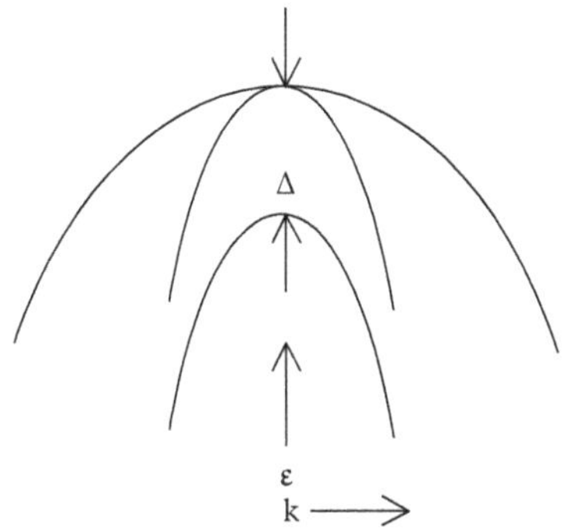

Figure 22.4 *Structure of the spin-orbit split. hole band in* Ge *and* Si.

As written here, A is positive and the positive and negative signs in (22.16a) correspond to two different bands that are degenerate at Γ. Although still corresponding to parabolic bands, in that the energy increases proportional to the magnitude k^2, they have a non-trivial (non-spherical) angular dependence, where the behavior for diamond-like structures is shown in Fig. 22.5; they are referred to as heavy (+) and light (–) holes respectively. The form for the $p_{1/2}$-like band is

$$\varepsilon_h = \varepsilon_v - \Delta - Ak^2, \tag{22.16b}$$

where Δ is a spin–orbit splitting parameter. Here the constant energy surfaces are spherical.

Returning to the surface described by Eq, (22.16a), for **k** lying along the <100>, <110>, and <111> axes the coefficient of the parabola is $A+B$, $A + \sqrt{B^2 + C^2/4}$, and $A + \sqrt{B^2 + C^2/3}$. This angular dependence complicates the interpretation of cyclotron resonance experiments on holes, which will not be discussed here. To calculate a mass appropriate to the density of states, we average Eq. (22.16a) over all angles which yields

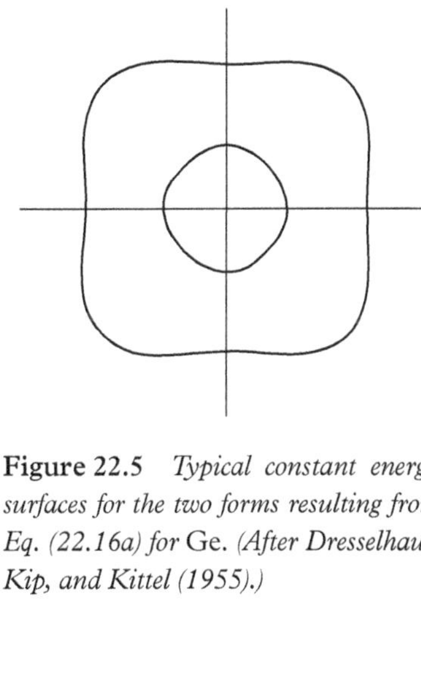

Figure 22.5 *Typical constant energy surfaces for the two forms resulting from Eq. (22.16a) for* Ge. *(After Dresselhaus, Kip, and Kittel (1955).)*

$$\begin{aligned}\varepsilon_h &= \varepsilon_v - Ak^2 \pm \frac{k^2}{4\pi}\int \sin\theta\, d\theta\, d\varphi \sqrt{B^2 + C^2(\sin^2\theta\cos\varphi\sin\varphi + \cos^2\theta)} \\ &= \varepsilon_v - Ak^2 \pm Bk^2\left(1 + \frac{2|C|^2}{15B^2}\right).\end{aligned} \tag{22.17}$$

In Table 22.3, we list the parameters A, B, and C for Ge and Si, along with the density of states masses associated with the light (L) and heavy (H) holes.

Table 22.3 *Hole mass parameters in* Ge *and* Si.

	A	B	C	m_{hH}	m_{hL}
Ge	13.38	8.48	13.15	0.347	0.042
Si	4.28	0.75	4.85	0.523	0.154

22.6 Impurity states

We limit our discussion here to substitutional impurities (those occupying a host lattice site) in the tetrahedrally coordinated crystal structures of the group IV elements, Si and Ge, involving atoms from the column III elements Al, Ga, and In, which are called *acceptors*, and the column V elements P, As, and Sb, which are called *donors*; together they are often referred to as *dopants*. The group V elements have one more valence electron than required to form the four nearest-neighbor tetrahedral covalent bonds of the diamond structure. The fifth electron will be bound to the impurity (at low temperatures), but *only weakly* since it is not involved in bond formation. It does feel the one remaining unit of nuclear charge (corresponding to the one uncompensated proton), but (qualitatively) the resulting Coulomb potential will be diminished by the dielectric constant of the group IV host, which tends to be large. We define the energy of the donor with its bound electron as ε_d which, being bound, lies below ε_c. The state ε_d is doubly degenerate due to the spin of the unpaired electron. In general there can be multiple bound states;[7] here, for simplicity, we assume there is only a single state. To be ionized the energy of this bound electron must be raised to ε_c and in so doing it will leave behind a positively charged donor site.

In the case of a group III impurity, only three electrons are available so the atom "steals" an electron from the surrounding neighborhood of the host in order to complete the fourth tetrahedral bond; this fourth electron is highly localized (at the scale of a bond length) and indistinguishable from the other three valence electrons. The resulting positive charge density left behind in the neighborhood is represented as a loosely bound hole, the overall acceptor atom then being neutral. Analogous to the donor atom, we model the acceptor as a positive hole bound by a *negative* Coulomb potential (arising from the added electron completing the local bonding), the strength of which is diminished by the host dielectric constant.

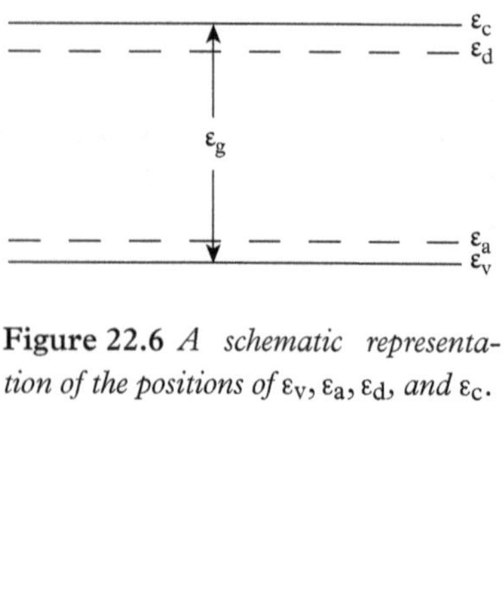

Figure 22.6 *A schematic representation of the positions of* ε_v, ε_a, ε_d, *and* ε_c.

Ionization of the bound hole involves exciting an electron from the valence band to annihilate the bound hole, leaving a singly negatively charged acceptor impurity behind and a mobile hole in the valance band. If we write the energy of the acceptor with the bound hole as ε_a, then to fill this hole and create a mobile hole in the valence band we must raise the energy of an electron in the valence band to ε_a; i.e., $\varepsilon_a > \varepsilon_v$. Ionization of this hole involves exciting an electron from the valence band to annihilate the trapped hole, leaving a singly negatively charged acceptor impurity behind. If we write the energy of the acceptor with the bound hole as ε_a, then to fill this hole and create a mobile hole in the valence band we must raise the energy of an electron in the valence band to ε_a; i.e., $\varepsilon_a > \varepsilon_v$.

Figure 22.6 shows schematically the positions of the valence band edge, the acceptor energy levels, the donor energy levels, and the conduction band edge, ε_v, ε_a, ε_d, and ε_c, respectively; the acceptor and donor levels are shown as short lines depicting the energy level of the individual atoms at some position in the lattice.

We may estimate the donor binding energies using the Bohr model, with the dielectric constant of free space replaced by that of the host, and the electron

[7] In addition, higher levels that would be degenerate under the assumption of a spherical potential are in practice split by the local tetrahedral symmetry of the diamond lattice.

Table 22.4 *Calculated and measured binding energies* $\varepsilon_c - \varepsilon_d$ (meV) *for various group V donors in* Si *and* Ge.

Measured	P	As	Sb	Calculated
Si	45	44	39	20
Ge	12	12.7	9.6	6

Table 22.5 *Measured binding energies* $\varepsilon_a - \varepsilon_v$(meV) *for various group III acceptors in* Si *and* Ge.

Measured	B	Al	Ga	In
Si	45	57	65	16
Ge	10.4	10.2	10.8	11.2

mass replaced by some appropriately averaged effective mass (recall it can be anisotropic[8]); for the donors we write this as

$$\varepsilon_c - \varepsilon_d = \frac{e^4 m_e}{2\varepsilon^2 \hbar^2}. \tag{22.18}$$

For Ge and Si, appropriate parameters are $\varepsilon = 15.8$ and 11.7 for the dielectric constants and $m_e = 0.1m$ and $0.2m$ for the masses, respectively. The calculated and measured values of the binding energies of the donors are given in Table 22.4. The complicated nature of the holes, which typically arise from degenerate bands, makes the Bohr model problematic and therefore we give only the measured values for the acceptors, which are shown in Table 22.5. Figure 22.7(a) shows, again

[8] For a discussion of the case where the effective mass is anisotropic, see Section 4.2.3. of Cardona and Yu.

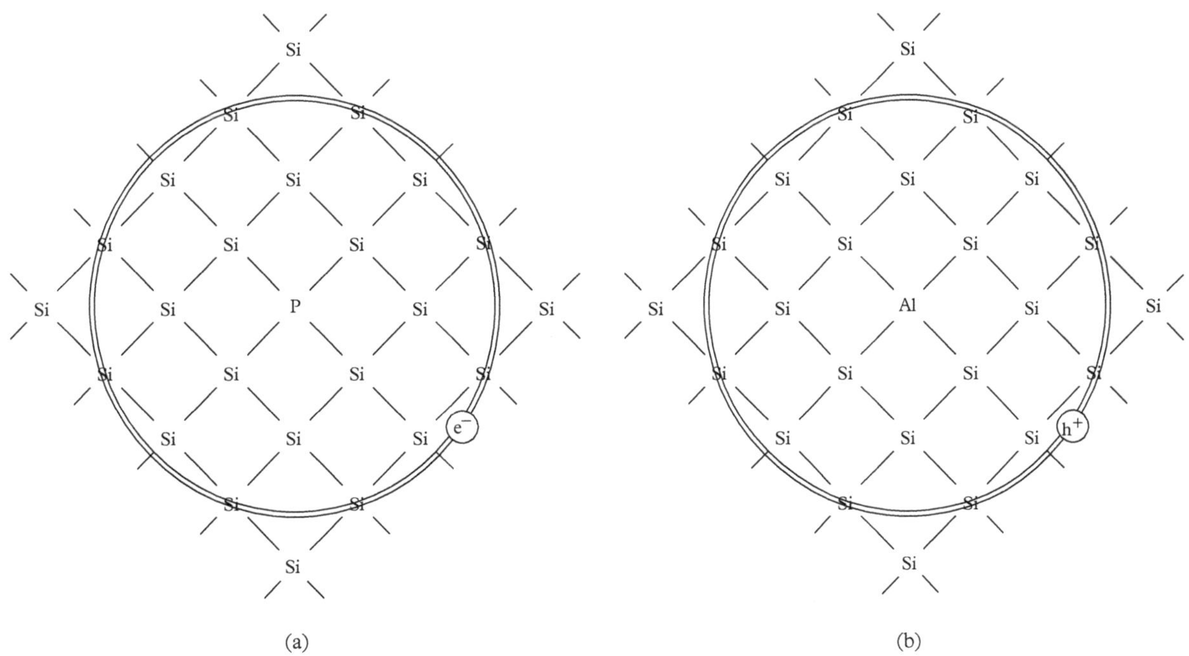

Figure 22.7 *A two-dimensional depiction of the four bonds linking the silicon atoms together in the diamond structure with (a) a phosphorous donor atom together with the Bohr orbit of the bound electron, and (b) an aluminum acceptor atom with the Bohr orbit of the bound hole.*

schematically, an embedded P donor atom in a Si lattice, while Fig. 22.7(b) shows the case for an embedded Al acceptor. Also shown are the Bohr radius (not to scale) of the corresponding orbiting electron and hole.

22.7 Calculating the occupancy of donors and acceptors

In this section we obtain expressions for the occupation probabilities of the donor and acceptor levels, which is essential if we are to know the concentration of mobile electrons and holes, which we examine in Section 22.8. To do this we must *generalize* the procedure described by Eqs. (4.2a,b) for taking statistical averages, $\bar{Q}$, of some quantity, Q, to the case of a quantum system where $\hat{Q}$ is an operator and where we have discrete states rather than continuous Hamiltonian coordinates; we then perform sums over many-body energy levels, n, rather than integrals over phase space coordinates, q_i, p_i. We also allow for a *variable number of particles*, N, which generates a second summation. The prescription then becomes

$$\bar{Q} = \sum_{N,n} Q_{N,n} P_{N,n}, \tag{22.19}$$

where the Gibbs probability weighting function is

$$P_{N,n} = \frac{e^{-\beta(E_{N,n} - \mu N)}}{\sum_{N,n} e^{-\beta(E_{N,n} - \mu N)}}; \tag{22.20}$$

here the $Q_{N,n}$ are the diagonal matrix elements of the operator $\hat{Q}$ with the many-body quantum states of the system with energies $E_{N,n}$ having numbers of particles N and quantum numbers n and $\beta = 1/k_B T$. For a *non-interacting* system we would have $E_{N,n} = N\varepsilon_n$, where ε_n are the single particle energies.

Here we take $\hat{Q}$ to be the number operator, $\hat{N}$, which has eigenvalues N, and from Eq. (22.19) the average number of particles occupying some system in the approximation where they do not interact is given by

$$\bar{N} = \frac{\sum_N N \sum_n e^{-\beta(\varepsilon_n - \mu)N}}{\sum_{N,n} e^{-\beta(\varepsilon_n - \mu)N}}. \tag{22.21}$$

We now apply this formula to calculate the fraction of an electron occupying a *single* donor site. We assume there is only one bound state, ε_d, per donor which may be either unoccupied (ionized), where N= 0, or occupied by a spin up *or* a spin down electron having energies $\varepsilon_{d\uparrow} = \varepsilon_{d\downarrow} = \varepsilon_d$ each, corresponding to N = 1.[9] For our single donor atom the sums in Eq. (22.21) then become

$$\bar{N}_d = \frac{0 + e^{-\beta(\varepsilon_{d\uparrow}-\mu)} + e^{-\beta(\varepsilon_{d\downarrow}-\mu)}}{1 + e^{-\beta(\varepsilon_{d\uparrow}-\mu)} + e^{-\beta(\varepsilon_{d\downarrow}-\mu)}} = \frac{2e^{-\beta(\varepsilon_d-\mu)}}{1 + 2e^{-\beta(\varepsilon_d-\mu)}}$$
$$= \frac{1}{1 + \frac{1}{2}e^{+\beta(\varepsilon_d-\mu)}}. \tag{22.22}$$

Assuming each donor atom acts independently, we can increase the total number of donor atoms from one to N_{0d} while continuing to denote their average occupation as $\bar{N}_d$. We also introduce a corresponding donor number density, $n_{0d} = N_{0d}/V$ and an average occupation number density, $\bar{n}_d = \bar{N}_d/V$, where V is the volume of the sample. Thus the number density of donors with a bound electron is

$$\bar{n}_d = \frac{n_{0d}}{1 + \frac{1}{2}e^{+\beta(\varepsilon_d-\mu)}}; \tag{22.23a}$$

the number density of *ionized* (and positively charged) donors is then

$$n_d^+ = n_{0d} - \bar{n}_d. \tag{22.23b}$$

These expressions hold irrespective of whether we also have acceptor atoms; the influence of one on the other enters only through their effect on the chemical potential, μ.

Analogous quantities $N_{0a}, n_{0a}, \bar{N}_a$, and $\bar{n}_a$ may be defined for the acceptors, although there are important differences in how we proceed. As previously discussed, when the crystal forms the acceptor acquires an electron from the surrounding medium allowing it to complete the tetrahedral bonding with the four nearest neighbors; this results in a *strongly localized* negative charge density which is concentrated in a region of order a bond length about the acceptor atom. The charge extracted to form this bond in turn results in a *weakly localized* electron deficiency (a positive charge density) in the surrounding medium, which we identify as a hole which is bound to the acceptor at T = 0; this positive charge density persists over a range of order the Bohr radius of the bound hole, which is safely larger than an inter-atom spacing. Here we have a maximum occupation of one hole per acceptor site which may have either spin; hence the ground state of our acceptor is doubly degenerate. At zero temperatures this hole is bound to the acceptor with an energy $\varepsilon_a - \varepsilon_v$, as discussed earlier; as the temperature increases the hole tends to ionize.[10]

From the point of view of the electrons, the acceptor system has two relevant states[11]: (i) two degenerate, neutral, one-electron states consisting of a negatively charged donor atom core along with a bound spin up or spin down hole; (ii) a singly degenerate *two-electron state* lying higher in energy by ε_a, corresponding to

[9] Because of the large coulomb repulsion, donors cannot bind two electrons which would correspond to an N = 2 many-body state.

[10] For this ionization to occur the valence band needs to lie nearby in energy (on a scale accessible with the temperature). Impurities with large values of ε_a (referred to as "deep levels") tend not ionize under normal conditions. A corresponding story holds for the donors.

[11] A third state, in which there are no added electrons on the acceptor, involves removing the strongly (chemically) bound electron that was captured from the environment to form the fourth covalent bond. That state lies much higher in energy and hence has a much smaller Boltzmann occupation factor. In the hole picture this state would correspond to two bound holes with the associated cost in coulomb repulsion.

the acceptor being ionized by drawing an electron from the valence band. If we measure our energies from ε_v, Eq. (22.21) then becomes

$$\bar{N} = \frac{e^{\beta\mu} + e^{\beta\mu} + 2e^{-\beta(\varepsilon_a - 2\mu)}}{e^{\beta\mu} + e^{+\beta\mu} + e^{-\beta(\varepsilon_a - 2\mu)}} = \frac{1 + e^{\beta(\mu - \varepsilon_a)}}{1 + \frac{1}{2}e^{\beta(\mu - \varepsilon_a)}}. \tag{22.24}$$

The mean number of holes $\bar{N}_a$ at the site is the difference between the maximum electron occupancy, 2, and $\bar{N}$; i.e., $\bar{N}_a = 2 - \bar{N}$ or

$$\bar{N}_a = \frac{1}{\frac{1}{2}e^{\beta(\mu - \varepsilon_a)} + 1}; \tag{22.25}$$

extending to multiple acceptor sites and converting to acceptor number densities, we obtain

$$\bar{n}_a = \frac{N_{0a}}{\frac{1}{2}e^{\beta(\mu - \varepsilon_a)} + 1}. \tag{22.26a}$$

The number density of *ionized* (and negatively charged) acceptors is then

$$n_a^- = n_{0a} - \bar{n}_a. \tag{22.26b}$$

Similar to what was previously noted, (22.23a,b) hold irrespective of whether donors are present, their effect entering only through μ.

Donors with bound electrons and acceptors with bound holes are charge neutral; on the other hand the concentrations n_d^+ and n_a^- that are ionized contribute positive and negative charge densities respectively. Combining these immobile ionized *charge* densities with the *number* densities of mobile electrons and holes, n_e and n_h, overall charge neutrality then requires

$$n_e - n_h = n_d^+ - n_a^-; \tag{22.27}$$

this expression fixes the chemical potential, $\mu(T)$. Here n_e and n_h are calculated with the same expressions as in the absence of impurities, Eqs. (22.10a,b), since that calculation assumes nothing about the presence or absence of donors/acceptors (although the chemical potential involved will differ from the intrinsic value). Solving (22.30) for $\mu(T)$ must be done numerically; however approximate expressions can be obtained for various regimes, which we now examine.

22.8 Carrier concentrations in the presence of donors and acceptors

In the presence of ionized impurities there will in general be an imbalance in the *mobile* charge density, $\Delta n = n_e - n_h$, although overall charge neutrality is maintained. Writing $n_e = \Delta n + n_h$ and using the mass action law, Eq. (22.11a) (which does not involve the chemical potential and hence holds in general), we have $n_i^2 = n_e n_h = (\Delta n + n_h) n_h$, which we rewrite as $n_h^2 + \Delta n n_h - n_i^2 = 0$. This equation has the solution

$$n_h = -\frac{\Delta n}{2} + \frac{1}{2}\sqrt{(\Delta n)^2 + 4n_i^2}, \tag{22.28a}$$

where we choose the positive root of the radical to ensure $n_h > 0$. Solving for n_e yields

$$\begin{aligned} n_e &= n_h + \Delta n. \\ &= +\frac{\Delta n}{2} + \frac{1}{2}\sqrt{(\Delta n)^2 + 4n_i^2}. \end{aligned} \tag{22.28b}$$

In the presence of donors and/or acceptors any changes in n_e and n_h from their intrinsic values will be accompanied by a shift of the chemical potential from its intrinsic value, where we write the latter as μ_i. We may then write the electron and hole concentrations in the form

$$n_e = n_i e^{\beta(\mu - \mu_i)} \tag{22.29a}$$

and

$$n_h = n_i e^{-\beta(\mu - \mu_i)}; \tag{22.29b}$$

combining (22.29a,b) with the definition $\Delta n = n_e - n_h$, yields

$$\frac{\Delta n}{n_i} = 2 \sinh\left[\beta\left(\mu - \mu_i\right)\right]. \tag{22.30}$$

22.8.1 High temperatures and low doping: the intrinsic regime

In the absence of donors and acceptors, charge neutrality requires $n_e = n_h = n_i$, where $n_i \propto e^{-\varepsilon_g/k_B T}$ is given by (22.11b). If the number of impurities is low and simultaneously the temperature is high, we expect that carrier densities resulting from *direct excitation* of electrons and holes across the energy gap will dominate, regardless of whether impurities are present. This is referred to as the *intrinsic regime*, since then the carrier densities closely match those of the pure (intrinsic) material.

22.8.2 Intermediate temperatures and small charge imbalance: $\mu \cong \mu_i$

In examining the effects of impurities on mobile charge densities more generally, we begin by examining a temperature regime where the chemical potential is only slightly shifted from its intrinsic value; we also assume

$$\varepsilon_d - \mu \gg k_B T \tag{22.31a}$$

and

$$\mu - \varepsilon_a \gg k_B T. \tag{22.31b}$$

It will turn out that the temperature cannot be too low, nor the *imbalance* in the carrier concentrations to large. Although it may initially appear contradictory, we will assume that *only the uncompensated charge is ionized*, which will be justified by our final result.[12] With this latter assumption we can then write $\Delta n = n_e - n_h \cong n_{0d} - n_{0a}$ and using (22.30), we have

$$\frac{n_{0d} - n_{0a}}{n_i} = 2\sinh\left[\beta\left(\mu - \mu_i\right)\right]. \tag{22.32}$$

From this equation we see that if the temperature is *high* enough, or the imbalance in the impurity concentrations small enough, so that $n_i \gg n_{0d} - n_{0a}$, then $\beta\Delta\mu$ must be small, where we write $\Delta\mu = \mu - \mu_i$. If $\Delta\mu$ is also small our assumptions (21.31a,b) are valid, since ε_d and ε_a usually lie close to ε_c and ε_v respectively and μ_i lies near the center of the gap: typically, $n_{0d} - n_{0a}$ would have to be several orders of magnitude greater than n_i for μ to differ significantly from μ_i.

Let us expand Eqs. (22.28a,b) as a power series in Δn,

$$\begin{pmatrix} n_e \\ n_h \end{pmatrix} = \pm\frac{\Delta n}{2} + n_i\sqrt{1 + \left(\frac{\Delta n}{2n_i}\right)^2} = n_i \pm \frac{\Delta n}{2} + \frac{1}{8}\frac{(\Delta n)^2}{n_i}\cdots;$$

neglecting the quadratic term (which assumes a small charge imbalance) and continuing to assume $\Delta n = n_{0d} - n_{0a}$, we have in our $\Delta n \ll n_i$ limit,

$$\begin{pmatrix} n_e \\ n_h \end{pmatrix} = n_i \pm \frac{n_{0d} - n_{0a}}{2}. \tag{22.33}$$

So we see that the shift in the intrinsic charge density only involves $n_{0d} - n_{0a}$, supporting the assumption that only the unbalanced charge density is ionized.[13]

22.8.3 Low temperatures

We have just seen that there is a tendency for only the imbalanced impurity concentration to be ionized. This being the case, if $n_{0d} > n_{0a}$ we would have a density $n_{0d} - n_{0a}$ of uncompensated donors which are ionized with their otherwise weakly bound electrons in the conduction band. With these assumptions and using (22.23a),

$$n_{0d} - n_{0a} = \bar{n}_d = \frac{n_{0d}}{\frac{1}{2}e^{\beta(\varepsilon_d - \mu)} + 1},$$

[12] It is natural to ask why a material would be doped with both donors and acceptors. Other than limits on just how high a purity can be achieved in a given material, a reason for simultaneously having both donors and acceptors is that an initially prepared material may purposely be doped with only one type of impurity, say donors, thereby making electrons the dominant carrier. However, it is often desirable to change the sign of the dominant carrier *in some region* of that material; this is commonly done by inter-diffusing or implanting a larger concentration of impurities having the opposite sign, say acceptors, thereby making holes the dominant carrier in that region.

[13] Numerical simulations confirm that over a wide range of temperatures and dopant concentrations there is a strong tendency for only the excess dopants to ionize.

from which we obtain

$$e^{-\beta(\varepsilon_d-\mu)} = \frac{n_{0d} - n_{0a}}{2n_{0a}}. \tag{22.34}$$

As $T \to 0$, $\beta \to \infty$, and therefore $\mu \to \varepsilon_d$ from below, but with our assumption in a way that keeps the right-hand side constant. However, at still lower temperatures the uncompensated donor density will itself start to fall as these atoms begin to bind electrons, as will be discussed shortly. A similar situation holds for the case $n_{0a} > n_{0d}$; then as $T \to 0$ and $\beta \to \infty$, $\mu \to \varepsilon_a$ previously.

We now examine the limit where the temperature is low enough or the charge imbalance is large enough that μ has moved close to the conduction band edge (strongly n-type with $n_{0d} \gg n_{0a}$) or to the valence band edge (strongly p-type with $n_{0d} \ll n_{0a}$), but that all the uncompensated impurities are still ionized.

Case $n_{0d} \gg n_{0a}$, $\mu \to \varepsilon_c$:

$$n_e = n_{0d} - n_{0a} \tag{22.35a}$$

and

$$n_h = \frac{n_i^2}{n_e} = \frac{n_i^2}{n_{0d} - n_{0a}}. \tag{22.35b}$$

Case $n_{0d} \ll n_{0a}$, $\mu \to \varepsilon_v$:

$$n_h = n_{0a} - n_{0d} \tag{22.36a}$$

and

$$n_e = \frac{n_i^2}{n_h} = \frac{n_i^2}{n_{0a} - n_{0d}}. \tag{22.36b}$$

22.8.4 Very low temperatures

At $T = 0$ all donor and acceptor levels are occupied; there is no free charge and the system is an insulator.[14] Hence our assumption that uncompensated impurities are still ionized breaks down. Let us examine the limiting behavior as $T \to 0$. We start by rewriting (22.10a,b) in the form

$$n_e = n_{0e} e^{-\beta(\varepsilon_c - \mu)} \tag{22.37a}$$

and

$$n_h = n_{0h} e^{-\beta(\mu - \varepsilon_v)}, \tag{22.37b}$$

[14] Here we are excluding the highly doped degenerate cases.

where

$$n_{0e} = \frac{1}{4}\left(\frac{2m_e k_B T}{\pi\hbar^2}\right)^{3/2} \tag{22.38a}$$

and

$$n_{0h} = \frac{1}{4}\left(\frac{2m_h k_B T}{\pi\hbar^2}\right)^{3/2}. \tag{22.38b}$$

We will only consider the case where there are no acceptors, $n_{0a} = 0$. We expect μ to lie in the vicinity of ε_c and ε_d and that only electrons will be involved in the charge transport. We fix the chemical potential via charge neutrality; the only relevant charge densities in this regime are the ionized donors and the conduction electrons and therefore,

$$n_e = n_d^+ = n_{0d} - \bar{n}_d. \tag{22.39}$$

Taking $\bar{n}_d$ from (22.23a) and n_e from (22.37a) gives

$$n_{0e}e^{-\beta(\varepsilon_c-\mu)} = n_{0d} - \frac{n_{0d}}{\frac{1}{2}e^{\beta(\varepsilon_d-\mu)}+1} = \frac{n_{0d}}{2e^{-\beta(\varepsilon_d-\mu)}+1},$$

which yields the following quadratic in $e^{\beta\mu}$:

$$-n_{0d}e^{+\beta\varepsilon_c} + n_{0e}e^{\beta\mu} + 2n_{0e}e^{-\beta\varepsilon_d}e^{2\beta\mu} = 0$$

having the (positive) solution

$$e^{\beta\mu} = \frac{-n_{0e} + \sqrt{n_{0e}^2 + 8n_{0e}n_{0d}e^{\beta(\varepsilon_c-\varepsilon_d)}}}{4n_{0e}e^{-\beta\varepsilon_d}}. \tag{22.40}$$

In the low temperature limit where $\beta \to \infty$, this expression becomes

$$e^{\beta\mu} = \frac{\sqrt{2}}{2}\sqrt{\frac{n_{0d}}{n_{0e}}}e^{\beta(\varepsilon_c-\varepsilon_d)/2},$$

which has the solution

$$\mu = \frac{\varepsilon_c+\varepsilon_d}{2} + \frac{1}{\beta}\ln\left[\frac{\sqrt{2}}{2}\sqrt{\frac{n_{0d}}{n_{0e}}}\right]. \tag{22.41}$$

Inserting (22.41) in Eq. (22.37a) yields n_e as

$$n_e = \frac{1}{\sqrt{2}} (n_{0d} n_{0e})^{1/2} e^{-\beta(\varepsilon_c - \varepsilon_d)/2}. \qquad (22.42)$$

Numerically, $n_{0e} \approx 10^{18}$ for $m_e \cong 0.1 m_0$ and $T = 300$ K. This relatively large value for n_{0e}, the weak $n_{0d}^{1/2}$ dependence of (22.42), and the small energy difference in the exponent tend to make impurities ionize at relatively low temperatures. (Recall that $\varepsilon_c - \varepsilon_d \cong 0.01$eV for Ge, and that room temperature corresponds to 0.026eV.)

Note that at $T = 0$, μ lies midway between ε_c and ε_d and n_e vanishes; i.e., we have an insulator, as previously noted.

22.9 Impurity band conduction

The radii of the Bohr orbits, a_{0d}, and a_{0a}, are rather large due to the screening of the Coulomb potential through the rather large host dielectric constants. On the other hand, the spacing between donor or acceptor impurities is approximately $(n_{0d})^{-1/3}$ or $(n_{0a})^{-1/3}$. When these spacings become comparable to the respective Bohr orbit dimensions, the trapped electrons or holes can quantum mechanically tunnel from impurity to impurity, resulting in a non-vanishing conductivity at low temperatures. This effect is sometimes called impurity band conduction.

At very high doping levels the semiconductor becomes a *low carrier density metal*, and at low temperatures the system becomes degenerate with the electron or hole Fermi level lying *above* the conduction or *below* the valence band edge, respectively. There are useful applications of this regime, as will be discussed in Chapter 23. In addition the nature of the conduction process itself becomes interesting. We will not go into these topics here.

22.10 Summary of the qualitative behavior on doping

In the preceding sections we have semi-quantitatively analyzed the behavior of the electron and hole carrier concentrations with the doping level and temperature. For easy reference we here summarize these results:

- For high temperatures and low doping we recover intrinsic behavior where $n_e = n_h = n_i \propto e^{-\varepsilon_g/k_B T}$.
- For a wide range of intermediate temperatures we have a region where, for $n_{0d} > n_{0a}$, *essentially all acceptor states are occupied and all unbalanced donor states are empty*, and similarly for $n_{0a} > n_{0d}$; this persists even when the chemical potential is close to its equilibrium value.

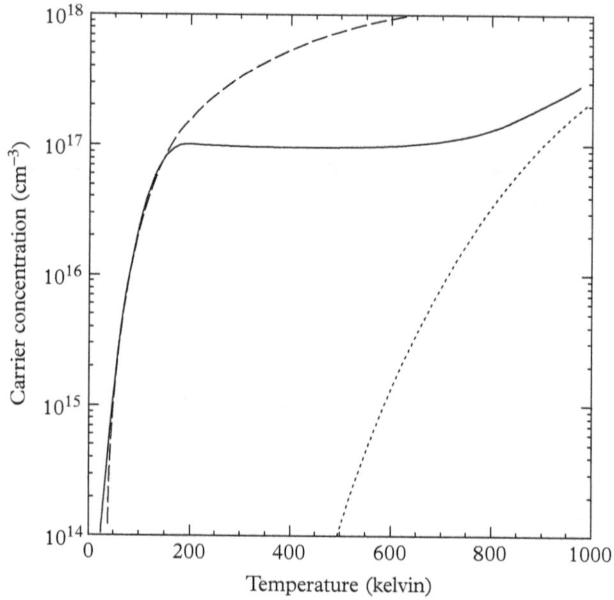

Figure 22.8 *The solid line shows the electron density in silicon for a donor density of 10^{17} cm^{-3}. At intermediate temperatures essentially all the donors are ionized and the concentration is given by the donor density. At high temperatures one enters the intrinsic regime (shown as the dotted line) where the concentration is governed by excitation across the gap. At low temperatures the concentration falls rapidly (shown as the dashed line) as the electrons reattach to the donors. (Courtesy of D. K. Ferry)*

- If the charge imbalance is very large, for $n_{0d} \gg n_{0a}$ we will have $\mu \to \varepsilon_c$; correspondingly for $n_{0a} \gg n_{0d}$, we will have $\mu \to \varepsilon_v$.
- At $T = 0$ all donor and acceptor levels are occupied and provided the orbitals do not overlap there is *no mobile charge*.
- If n_{0d} is very large, such that the Bohr orbitals overlap, and $T \to 0$, we can have $\mu > \varepsilon_c$ and the system will be a degenerate electron system.

Figure 22.8 shows the temperature dependence of the density of electrons in silicon for a doping level of 10^{17}cm^{-3}. Here we encounter three of the regimes discussed previously. At high temperatures the concentration is dominated by direct excitation across the energy gap. As the temperature falls a broad intermediate regime is encountered where the concentration arises from nearly complete ionization of the donors. Finally, at low temperatures the electrons freeze out on the donor atoms and the electron density rapidly falls.

ADDITIONAL READING

The sections in Ashcroft and Mermin on semiconductor physics are quite clear and have formed a model for the discussion given here.

An excellent all-round book on semiconductor physics is *Fundamentals of Semiconductor Physics*, Peter Y. Yu and Manuel Cardona, 3rd edition, Springer, Berlin (2001). In particular techniques to probe bulk band structure and surface states using visible/ultraviolet light and various photo-electron spectroscopies are well described. Also recommended is *Semiconductors*, David K. Ferry, Macmillan (1991) and *Electronic Materials and Devices*, D.K. Ferry and J.P. Bird, Academic Press, San Diego (2001).

Problems

22.1. Obtain Eq. (22.15).

22.2. Obtain Eq. (22.17).

23 Inhomogeneous Semiconductors

23.1 The p–n junction in equilibrium

A device called a diode, which has the property that it has asymmetric characteristics for current flow in opposite directions (and is hence used to rectify an a.c. or r.f. current), is made by forming a junction between a p-type material (dominated by acceptor levels) and an n-type material (dominated by donor levels). Figures 23.1 and 23.2 show, schematically, the position dependence of the various energies characterizing the p- and n-type sides of the device "before contact" and "after contact." After contact (Fig. 23.2) the materials come into chemical equilibrium by the exchange of electrons until their chemical potentials equilibrate.[1] For this to happen a position-dependent electrostatic potential develops which shifts the position of the conduction and valence band edges according to

$$\begin{aligned} \varepsilon_C &\rightarrow \varepsilon_C - e\phi(x) \\ \varepsilon_V &\rightarrow \varepsilon_V - e\phi(x). \end{aligned} \tag{23.1}$$

The absolute positions of the conduction and valance band energies now become a function of the position x (we restrict ourselves to one-dimensional devices); this phenomenon is called *band bending*. The electron and hole densities and the degree of ionization of the donors and acceptors also become position dependent; however other properties like the band gap or the effective masses are unaltered. We will locate the junction, where the p-type and n-type materials are joined, at $x = 0$. In the vicinity of the junction $\phi(x)$ varies in a way which we will model. The difference, $\Delta\phi = \phi(+\infty) - \phi(-\infty)$, is called the *contact potential* and is dependent on the doping levels of the bulk materials.

We now construct a simple model for the behavior of carrier densities and the potential $\phi(x)$ near the junction. In place of Eqs. (22.10a,b), as rewritten in the form of Eq.'s (22.37a,b) and (22.38a,b), we have from (23.1)

$$n_e(x) = n_{0e}(T)e^{-\beta(\varepsilon_c - e\phi(x) - \mu)} \tag{23.2a}$$

and

$$n_h(x) = n_{0h}(T)e^{-\beta(\mu - \varepsilon_v + e\phi(x))}. \tag{23.2b}$$

[1] This is proved in Section 39.1; see Eq. (39.3).

The Physics of Solids. First Edition. J. B. Ketterson.

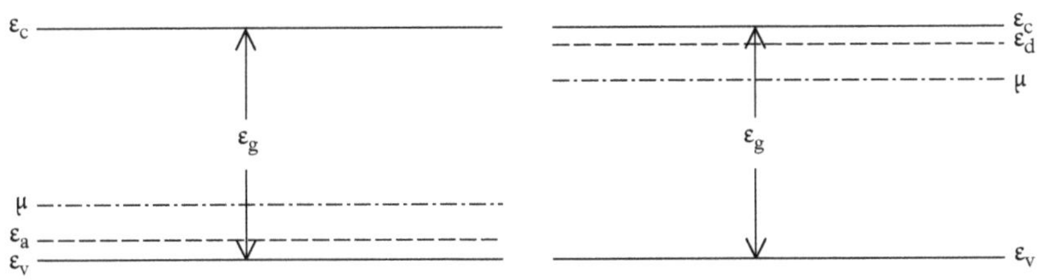

Figure 23.1 *Schematic of a p-n junction showing the position of the Fermi levels before contact.*

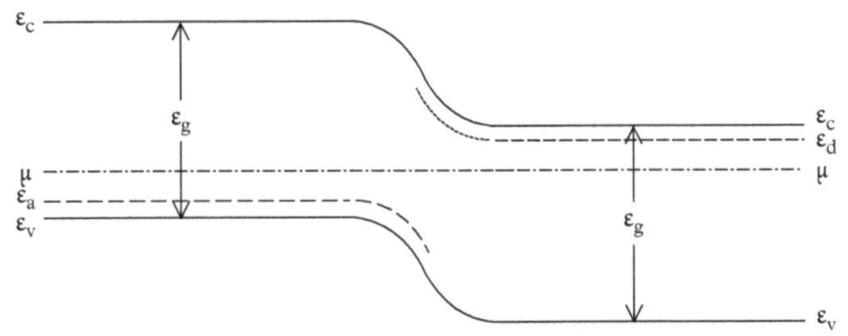

Figure 23.2 *Schematic of a p-n junction showing the position of the Fermi level after contact and the resulting band bending.*

We will assume (as is usually the case in practice) that far from the junction we are in a regime where the uncompensated impurities are *fully ionized*; to further simplify the discussion we will assume that the p and n materials contain *only* acceptors and donors respectively. Ignoring the minority carrier populations (which are small in this approximation) far from the junction we then have

$$n_{0d} = n_e = n_{0e}(T)e^{-\beta(\varepsilon_c - e\phi(+\infty) - \mu)} \tag{23.3a}$$

and

$$n_{0a} = n_h = n_{0h}(T)e^{-\beta(\mu - \varepsilon_v + e\phi(-\infty))}. \tag{23.3b}$$

Note that μ is the global chemical potential here and hence does not vary with position. Solving (23.3a) and (23.3b) for μ and equating the two expressions yields

$$\Delta\phi = \phi(+\infty) - \phi(-\infty) = \varepsilon_g + k_B T \ln\left[\frac{n_{0d}n_{0a}}{n_{0e}(T)n_{0h}(T)}\right], \tag{23.4}$$

where we have used the usual definition, $\varepsilon_c - \varepsilon_v = \varepsilon_g$.

In the region of the junction, $\phi(x)$ depends strongly on x. As one proceeds from left to right in Fig. 23.2, the band edges move such that μ lies closer to the valence band edge, ε_v, in the p-type material to being closer to the conduction band edge, ε_c, in the n-type material. At the junction (x = 0) the carrier density would be minimal since μ then lies essentially in the center of the gap; the immediate region of the junction, where the charge density is reduced, is called the *depletion zone*.

To obtain an accurate description we would have to find for the charge density at each value of the coordinate x (which is temperature and impurity concentration dependent) for some assumed form for $e\phi(x)$; we would then put the resulting total local charge density, $\rho(x)$, (bound and mobile) into Poisson's equation:

$$\frac{d^2\phi}{dx^2} = -\frac{4\pi}{\varepsilon}\rho(x). \tag{23.5}$$

Here, ε is the host dielectric constant (to be distinguished various energies) and $\rho(x)$ is the total charge density. The new potential function, $\phi(x)$, would then be used to recalculate $\rho(x)$. This process would be repeated until convergence (called self consistency) was achieved.

Rather than proceeding in this way, we solve for the potential using a model charge density which is sufficient for obtaining the qualitative behavior. We make the following assumptions: (i) The mobile charge density is zero in a depletion region that extends distances d_p and d_n into the p-type and n-type materials respectively; i.e.,

$$n_e(x) = n_h(x) = 0; -d_p \le x \le +d_n. \tag{23.6}$$

(ii) Outside this region we assume the total charge density is zero (complete compensation as implied by Eq. (22.27)). We then have the following for the charge density:

$$\rho = -e\begin{bmatrix} 0 \\ -n_{0a} \\ +n_{0d} \\ 0 \end{bmatrix} \begin{matrix} x < d_p \\ -d_p < x < 0 \\ 0 < x < +d_n \\ x > d_n \end{matrix}. \tag{23.7}$$

With this charge density we can integrate Poisson's equation with the following two boundary conditions:

$$\phi(x = -d_p) = \phi(-\infty) \tag{23.8a}$$

and

$$\phi(x = +d_n) = \phi(+\infty). \tag{23.8b}$$

Furthermore, we require $\phi(x)$ and $\phi'(x)$ to be continuous (corresponding to the absence of charged interfaces and dipole layers respectively). In the depletion layer we have

$$\phi''(x) = \frac{4\pi e}{\varepsilon}\begin{bmatrix} n_{0a} \\ -n_{0d} \end{bmatrix};$$

integrating once with respect to distance gives

$$\phi'(x) = \frac{4\pi e}{\varepsilon}\begin{bmatrix} n_{0a} \\ -n_{0d} \end{bmatrix} x + c_1.$$

and a second integration gives

$$\phi(x) = \frac{4\pi e}{\varepsilon}\begin{bmatrix} n_{0a} \\ -n_{0d} \end{bmatrix} \frac{x^2}{2} + c_1 x + c_2. \tag{23.9}$$

Requiring continuity of $\phi(x)$ and $\phi'(x)$ at $x = -d_p$ and $x = +d_n$ yields

$$\phi(x) = \phi(-\infty); x < -d_p \tag{23.10a}$$

$$\phi(x) = \frac{2\pi e n_{0a}}{\varepsilon}(x + d_p)^2 + \phi(-\infty); -d_p < x < 0 \tag{23.10b}$$

$$\phi(x) = -\frac{2\pi e n_{0d}}{\varepsilon}(x - d_n)^2 + \phi(+\infty); 0 < x < +d_n \tag{23.10c}$$

$$\phi(x) = \phi(+\infty); d_n < x. \tag{23.10d}$$

Requiring continuity of (23.10b) and (23.10c) at $x = 0$ gives

$$\Delta\phi = \frac{2\pi e}{\varepsilon}\left[n_{0d}d_n^2 + n_{0a}d_p^2\right], \tag{23.11}$$

while requiring continuity of their derivatives at $x = 0$ gives

$$n_{0a}d_p = n_{0d}d_n. \tag{23.12}$$

Equations (23.11) and (23.12) can be solved to obtain d_p and d_n yielding

$$d_{n,p} = \left[\frac{(n_{0a}/n_{0d})^{+1,-1}}{n_{0d} + n_{0a}} \frac{\varepsilon\Delta\phi}{2\pi e}\right]^{1/2}; \tag{23.13a}$$

this can be written in the more convenient form

$$d_{n,p} = 33\left[\frac{(n_{0a}/n_{0d})^{+1,-1}}{10^{-18}(n_{0d} + n_{0a})}(\varepsilon e\Delta\phi)_{eV}\right]^{1/2} \text{A}. \tag{23.13b}$$

Numerically, for energy gaps of order 0.1 eV and impurity concentrations in the range 10^{14} to 10^{18} per cubic cm, the $d_{p,n}$ are of order 10^4A to 10^2A, and the accompanying electric fields are in the range 10^5 to 10^7 volts/meter. The mobile carrier densities, charge densities, and the resulting electrostatic potential in the vicinity of the depletion zone are shown schematically in Fig. 23.3 (a), (b), (c), respectively.

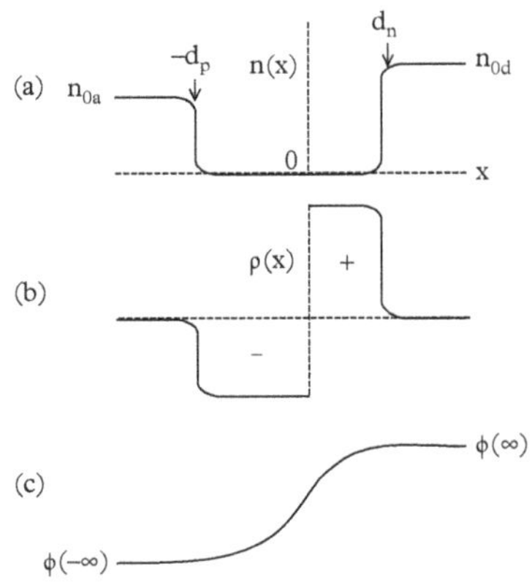

Figure 23.3 *The fully depleted model for the position dependence of carrier density (a), the dopant charge density (b), and the electrostatic potential (c), within the depletion zone.*

23.2 The p–n junction in the presence of an external electric field

Suppose we apply a voltage, V, to a p–n junction. We define this voltage as positive if it raises the potential of the p side of the junction; thus in the presence of V,

$$\Delta\phi \to (\Delta\phi)_0 - V. \tag{23.14}$$

Because of the low density of carriers in the depletion zone, this region has a very high resistance, in spite of the fact that it is very thin. Since $d_{n,p}$ scale as $(\Delta\phi)^{1/2}$, we have

$$d_{n,p}(V) = d_{n,p}(0)\left[\frac{(\Delta\phi)_0 - V}{(\Delta\phi)_0}\right]^{1/2} = d_{n,p}(0)\left[1 - \frac{V}{(\Delta\phi)_0}\right]^{1/2}. \tag{23.15}$$

Next we write the currents through the junction (which we denote as scalars, consistent with our one-dimensional treatment) in the form

$$j_t = j_e + j_h; \tag{23.16}$$

j_t, j_e, and j_h denote the total, electron, and hole currents.

At this point we will assume that the junction current can be divided into *generation* and *recombination* contributions, with the effects of an external potential incorporated into the recombination current, as we will discuss shortly. For the holes we would then write

$$j_h = j_h^g + j_h^r, \tag{23.17}$$

and similarly for the electrons. The *hole generation current* arises from thermally excited holes in the *n-type* material in the vicinity of the depletion layer; this phenomenon is to be expected since the Fermi level is moving closer to the valence band edge when the depletion zone region is entered, and hence holes will be generated spontaneously. Essentially all such holes, though small in number, are swept to the p side of the junction by the contact potential, $\Delta\phi$, and therefore they make an important contribution (note holes rise when encountering the rising valance band edge). This current is approximately constant for small applied potentials, V.

The *hole recombination current* arises from holes in the *p-type* material which have an energy *greater than* $e\Delta\phi$ and thus some fraction of them can pass across the junction to the n side where they recombine with electrons; we expect this contribution to be proportional to $e^{-\beta e\Delta\phi}$. A corresponding description holds for the electrons.

The energy diagram of an unbiased junction is shown in Fig. 23.4. Those holes with energies above the barrier height are those below the dotted line on

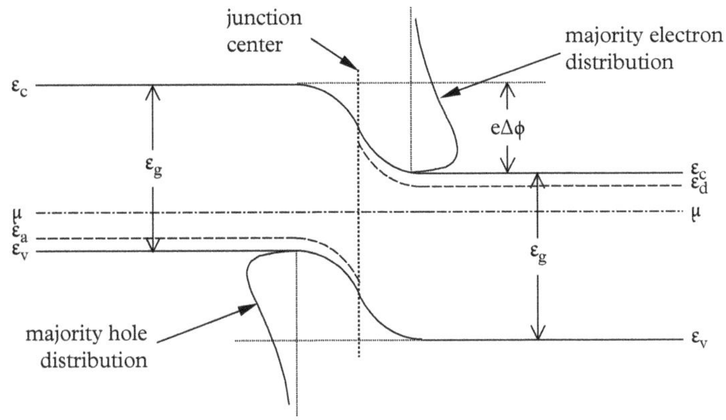

Figure 23.4 *Schematic of a p–n junction showing the position dependence of the band edges and the impurity levels. Any applied voltage will appear only across the depletion zone. Also shown is the Boltzmann distribution for majority holes on the left side and the majority electrons on the right side; these "nose-like" features should be interpreted as projecting perpendicular to the plane of the page.*

the left side of the figure, where the number of carriers per unit energy (which from Eqs. (23.3a,b) has the (scaled) functional form $x^{1/2}e^{-x}$) is plotted. Note that most of the external voltage appears across the junction and not in the remaining part of the device. As will be stressed again later, the carriers move via *diffusion* in a region surrounding the depletion region and feel the external potential only within the depletion zone.

In equilibrium the total hole current must vanish; i.e.,

$$j_h^g = -j_h^r(0). \tag{23.18}$$

In the presence of a potential we must then have, from (23.2) and (23.14),

$$j_h^g(V) = -j_h^r(0)e^{\beta eV}. \tag{23.19}$$

The total hole current may therefore be written as

$$j_h(V) = j_h^g(0)\left(1 - e^{\beta eV}\right). \tag{23.20}$$

Since the hole generation current moves to the left in the figure, we must have $j_h^g = -\left|j_h^g\right|$; thus,

$$j_h(V) = \left|j_h^g(0)\right|\left(e^{\beta eV} - 1\right). \tag{23.21}$$

Including a corresponding current for the electrons,

$$j_e(V) = \left|j_e^g(0)\right|\left(e^{\beta eV} - 1\right), \tag{23.22}$$

we have for the total junction current

$$\begin{aligned} j_t(V) &= \left[\left|j_e^g(0)\right| + \left|j_h^g(0)\right|\right]\left(e^{\beta eV} - 1\right) \\ &= j_0\left(e^{\beta eV} - 1\right); \end{aligned} \tag{23.23}$$

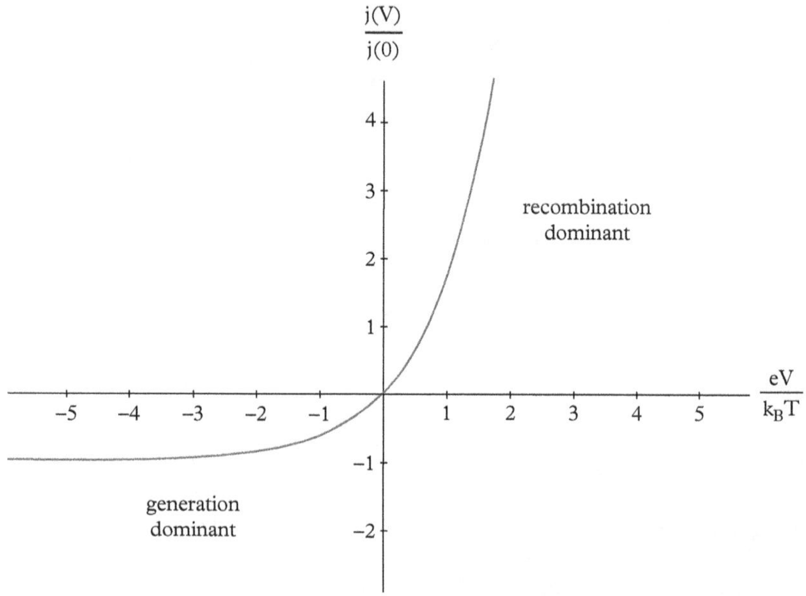

Figure 23.5 *A plot of Eq. (23.23) in the dimensionless form,* $j(V)/j(0)$ *versus* eV/k_BT.

the latter is called the diode equation, and Fig. 23.5 shows a plot of the form $e^x - 1$. Note that the function (23.23) is highly asymmetric, the current approaching a limiting value, j_0, for negative voltages and diverging exponentially for large positive voltages. This property is used to suppress one direction of the current flow of an alternating current only letting, say, the positive-going half-cycles pass, which results in a direct current component on the average; this process is called rectification, and will be discussed in Section 23.4.1.

23.3 A more detailed theory of the p–n junction

We now construct a more detailed theory of current transport in a p–n junction, both in equilibrium and under an applied voltage. Carriers are, in general, transported by two mechanisms: drift in the presence of a potential gradient, and diffusion in the presence of a concentration gradient. We define a number current density[2], **J**, and an electrical current density, **j**. Of course, $\mathbf{j}_{e,h} = \mp|e|\mathbf{J}_{e,h}$, where the − and + signs apply to electrons and holes, respectively. We write the combined drift and diffusion effects as

$$\mathbf{J}_{e,h} = n_{e,h}\mu_{e,h}\mathbf{E} - D_{e,h}\nabla n_{e,h}, \tag{23.24}$$

with $\mu_{e,h} = \mp|e|\tau^{col}_{e,h}/m_{e,h}$ being the electron and hole mobilities, with $\tau^{col}_{e,h}$ the characteristic scattering times, and $D_{e,h}$ the corresponding diffusion constants.

[2] Although we use the common symbol **J** for the number current density here and the heat current density in Section 19.5, they are different physical quantities.

The first and second terms in Eq. (23.24) are the drift and diffusion contributions respectively. The mobility and diffusion constant are related as we now show. Consider the case of a semiconducting bar of material containing carriers which, to make the discussion specific, we temporarily take to be holes. Let there be an electric field to the right which for the case of holes would result in a drift in the same direction. (Note we assume that no electrodes are present which would allow charge to enter or leave the system.) The drift would continue for some microscopic time until a density gradient built up in the gas of holes such that, in equilibrium, the drift and diffusion contributions canceled each other; i.e.,

$$\mathbf{J}_{e,h} = n_{e,h}\mu_{e,h}\mathbf{E} - D_{e,h}\nabla n_{e,h} = 0. \tag{23.25}$$

As in our previous discussions involving semiconductor transport, we will assume that Boltzmann statistics apply to the hole gas. The ratio of the two carrier densities at a point $\mathbf{r}$ and $\mathbf{r} + \mathbf{dr}$ would then be

$$\frac{n_{e,h}(\mathbf{r}+\mathbf{dr})}{n_{e,h}(\mathbf{r})} = \frac{e^{\pm\beta|e|\phi(\mathbf{r}+\mathbf{dr})}}{e^{\pm\beta|e|\phi(\mathbf{r})}} \cong 1 \pm \beta|e|\,[\phi(\mathbf{r}+\mathbf{dr}) - \phi(\mathbf{r})]$$

or

$$\nabla n_{e,h}(\mathbf{r}) = \pm n_{e,h}\beta|e|\nabla\phi(\mathbf{r}) = \mp n_{e,h}\beta|e|\mathbf{E}. \tag{23.26}$$

Inserting this expression into Eq. (23.25), we obtain

$$D_{e,h} = \frac{k_B T \mu_{e,h}}{|e|}, \tag{23.27}$$

which is called the Einstein relation.

In addition to the phenomenological transport equations for the electrons and holes, Eq. (23.24), we have the conservation laws relating to the particle numbers. If the numbers of electrons and holes were separately conserved we would have the equations

$$\frac{\partial n_{e,h}}{\partial t} + \nabla \cdot \mathbf{J}_{e,h} = 0. \tag{23.28}$$

For short times (23.27) will be obeyed. However, over some characteristic time electrons and holes can (i) recombine, or (ii) be spontaneously generated; we then rewrite (23.28) in the form

$$\frac{\partial n_{e,h}}{\partial t} + \nabla \cdot \mathbf{J}_{e,h} = s_{e,h}, \tag{23.29}$$

where $s_{e,h}$ is a so-called source-sink (generation–recombination) function. We introduce the following simple model to describe this phenomenon of carrier generation/recombination. We model the function $s_{e,h}$ by the simple relaxation form

$$s_{e,h} = -\frac{\left(n_{e,h} - n_{e,h}^{(0)}\right)}{\tau_{e,h}}, \tag{23.30}$$

where $\tau_{e,h}$ is a generation/recombination time (not to be confused with the scattering time, $\tau_{e,h}^{col}$, entering the mobility) and $n_{e,h}^{(0)}$ is the local equilibrium value of the respective carrier density.

In a spatially homogeneous (but temporally inhomogeneous) situation we would write (23.29) as

$$\frac{\partial n_{e,h}}{\partial t} = -\frac{\delta n_{e,h}}{\tau_{e,h}}, \tag{23.31}$$

where we write $\delta n_{e,h} = \left(n_{e,h} - n_{e,h}^{(0)}\right)$; this equation has the solution

$$\delta n_{e,h}(t) = \delta n_{e,h}(0)e^{-t/\tau_{e,h}}. \tag{23.32}$$

By construction, when the carrier density is disturbed from its local equilibrium value, $n_{e,h}^{(0)}$, it relaxes exponentially in a characteristic time, $\tau_{e,h}$, to that value. Note that if $n_{e,h} > n_{e,h}^{(0)}$, the carriers will recombine until $n_{e,h} = n_{e,h}^{(0)}$; hence it is natural to identify $-n_{e,h}/\tau_{e,h}$ as a recombination rate and $+n_{e,h}^{(0)}/\tau_{e,h}$ as a generation rate.

Let us for the moment assume that there are no electric fields present and that the particles move solely by diffusion. From Eqs. (23.24), (23.29), and (23.30) we obtain for our one-dimensional junction in steady state ($\partial n_{e,h}/\partial t = 0$):

$$D_{e,h}\frac{d^2 n_{e,h}}{dx^2} = -\frac{\left(n_{e,h} - n_{e,h}^{(0)}\right)}{\tau_{e,h}}. \tag{23.33}$$

The solution to this equation involves exponentials of the form $\exp\left(\pm x/L_{e,p}\right)$ where we have introduced characteristic carrier generation/recombination lengths, $L_{e,h} = (D_{e.h}\tau_{e,h})^{1/2}$. Writing $\mu_{e,h} = \mp|e|\tau_{e,h}^{col}/m_{e,h}$, $D_{e.h} = k_B T\mu_{e,h}/(\mp|e|)$, $m\bar{v}_{e,h}^2/2 = 3k_B T/2$, and $\ell_{e,h}\bar{v}_{e,h} = \tau_{e,h}^{col}$, we obtain

$$L_{e,h} = \ell_{e,h}\left(\frac{\tau_{e,h}}{3\tau_{e,h}^{col}}\right)^{1/2} = \ell_{e,h}\left(\frac{1}{3}N_{e,h}\right)^{1/2}, \tag{23.34}$$

where $N_{e,h}$ is the number of collisions the particle suffers in a generation/recombination time. This form of motion is referred to as random walk. Numerically, $L_{e,h}$ is much greater than $d_{e,p}$ and introduces another characteristic region, called the *diffusion* region, into the story. We identify the following regions of a p–n junction (in reality, of course, the boundaries are not sharp): (i) bulk p material (far from the junction), $x < -L_p$; (ii) the p-diffusion zone, $-L_p \leq x \leq -d_p$; (iii) the depletion zone, $-d_p < x < +d_n$; (iv) the n-diffusion zone, $+d_n < x < +L_n$; and (v) bulk n material, $L_n < x$. The nature of the charge transport in these five regions is as follows:

(i) $x < -L_p$. In this region there are no significant density gradients. Charge is carried almost entirely by a hole drift current in response to a small electric field (arising from the bulk resistance of the p-type material).

(ii) $-L_p \leq x \leq -d_p$. Here the hole current involves both a drift component, and a diffusion current from the density gradient. The minority electrons contribute a negligible drift current but, due to a large density gradient, contribute a diffusion current.

(iii) $-d_p < x < +d_n$. Large density gradients and junction electric fields result in large diffusion and drift currents for both electrons and holes, all of which are large compared to the total current.

(iv) $+d_n < x < +L_n$. We have the opposite of region (ii); i.e., we have large drift and diffusion currents for electrons and a diffusion but a negligible drift current for holes.

(v) $L_n < x$. Here only the electron drift current is significant.

We will not discuss how one might solve the coupled equations involving carrier transport, Boltzmann thermal occupation, and junction fields in a self-consistent way, but will again resort to a model solution. We start by assuming that charge transport is almost instantaneous in the depletion zone and therefore the electron and hole currents are *separately* conserved in moving from, $-d_p$ to $+d_n$. According to (ii) and (iv) above, the drift current of the minority carriers (electrons on the p-side and holes on the n-side) is negligible; we may therefore calculate their contributions to the total current using the diffusion equation alone. We will evaluate these currents at the depletion zone boundaries where

$$J_e(-d_p) = -D_e \left.\frac{dn_e}{dx}\right|_{x=-d_p} \tag{23.35a}$$

and

$$J_h(d_e) = -D_h \left.\frac{dn_h}{dx}\right|_{x=d_e}. \tag{23.35b}$$

We write the solution to (23.33) for electrons and holes in the p and n diffusion zones as

$$n_e(x) = n_e(-\infty) + \left(n_e(-d_p) - n_e(-\infty)\right) e^{-(x+d_p)/L_e} \tag{23.36a}$$

and

$$n_h(x) = n_h(+\infty) + (n_h(d_n) - n_h(+\infty))\, e^{-(x-d_n)/L_h}. \tag{23.36b}$$

Clearly the two constants of integration are $n_e(-\infty)$ and $n_e(-d_p)$ for the first equation and $n_h(+\infty)$ and $n_h(d_n)$ for the second. From the mass action law, Eqs. (22.11a,b), $n_e(-\infty) = n_i^2/n_{0a}$ and $n_h(+\infty) = n_i^2/n_{0d}$. Substituting (23.36a,b) into (23.35a,b) gives

$$J_e(-d_p) = -\frac{D_n}{L_n}\left(n_e(-d_p) - \frac{n_i^2}{n_{0a}}\right) \tag{23.37a}$$

and

$$J_h(d_e) = \frac{D_h}{L_p}\left(n_h(d_n) - \frac{n_i^2}{n_{0d}}\right). \tag{23.37b}$$

The total electric current is then

$$j = \frac{|e|D_e}{L_e}\left(n_e(-d_p) - \frac{n_i^2}{n_{0a}}\right) + \frac{|e|D_h}{L_p}\left(n_h(d_n) - \frac{n_i^2}{n_{0d}}\right). \tag{23.38}$$

To complete the calculation we need $n_e(-d_p)$ and $n_h(d_n)$. These two densities are related to the corresponding densities at the opposite sides of the depletion zone by the Boltzmann factor; thus,

$$n_e(-d_p) = n_e(d_n)e^{-|e|\Delta\phi/k_BT} = \left[n_e(d_n)e^{-|e|\Delta\phi^{(0)}/k_BT}\right]e^{+|e|V/k_BT} \tag{23.39a}$$

and

$$n_h(d_n) = n_h(-d_p)e^{-|e|\Delta\phi/k_BT} = \left[n_h(-d_p)e^{-|e|\Delta\phi^{(0)}/k_BT}\right]e^{+|e|V/k_BT}. \tag{23.39b}$$

To sufficient accuracy we may write $n_e(d_n) = n_{0a}$, $n_h(-d_p) = n_{0d}$, and (23.39a,b) become

$$n_e(-d_p) = \left[n_{0d}e^{-|e|\Delta\phi^{(0)}/k_BT}\right]e^{+|e|V/k_BT} \tag{23.40a}$$

and

$$n_h(d_n) = \left[n_{0a}e^{-|e|\Delta\phi^{(0)}/k_BT}\right]e^{+|e|V/k_BT}. \tag{23.40b}$$

In equilibrium the current (23.39) must vanish, so if our treatment is to be consistent the terms in square brackets in (23.40a,b) must be set equal to n_i^2/n_{0a}

and n_i^2/n_{0d} respectively, which is consistent with the mass action law. Our final expression for the current is therefore (consistent with Eq. (23.23))

$$j = |e|n_i^2\left(\frac{D_e}{L_e n_{0a}} + \frac{D_h}{L_p n_{0d}}\right)\left(e^{+|e|V/k_BT} - 1\right) = j_0\left(e^{+|e|V/k_BT} - 1\right). \quad (23.41)$$

We can now identify the electron and hole generation currents as

$$J_e = \frac{n_i^2 D_e}{L_e n_{0a}} \quad (23.42a)$$

and

$$J_h = \frac{n_i^2 D_h}{L_p n_{0d}}, \quad (23.42b)$$

respectively. Physically, we may understand these results in the following way. We expect that the minority carriers within a diffusion length will encounter the depletion zone boundary during their lifetime and, on so doing, will feel the large junction field and be swept through the junction. The flux of such particles is

$$\left(n_{e,h}L_{e,h}\right)/\tau_{e,h} = \left(n_{e,h}D_{e,h}\right)/L_{e,h},$$

in agreement with (23.42a,b).

23.4 A survey of semiconductor devices

The physical principles that were discussed in Chapter 22 and previous parts of the present one are sufficient to qualitatively describe the operation of a number of semiconductor devices. The simplest of these is the diode.

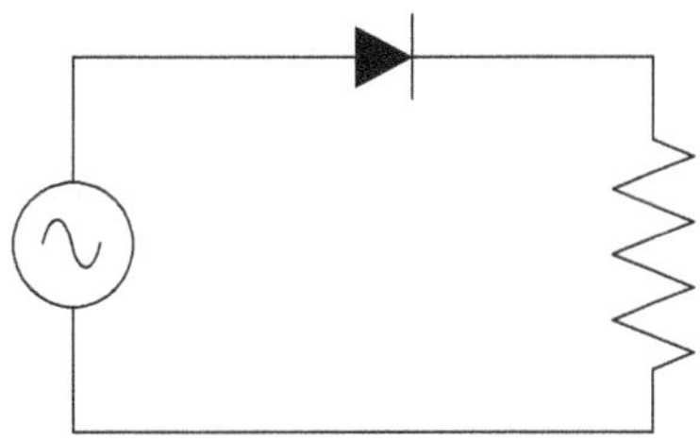

Figure 23.6 *A simple rectifier circuit.*

23.4.1 The diode rectifier

The p–n junction is used to convert an alternating current (a.c.) or radio frequency (r.f.) current into a signal, having, on the average, a direct current (d.c.) component. Consider the circuit in Fig. 23.6, consisting of an a.c. generator, a resistor, and a p-n junction, the latter referred to as a diode; these three circuit elements are denoted by the following symbols:

We idealize the diode as having zero resistance in the forward direction and infinite resistance in the reverse direction. This being the case, when the voltage from the generator is positive, a current will flow through the diode and the resistor and a voltage will develop across the latter; on the other hand, when the generator output is negative no current will flow and hence no voltage develops across the resistor. This situation is depicted in Fig. 23.7.

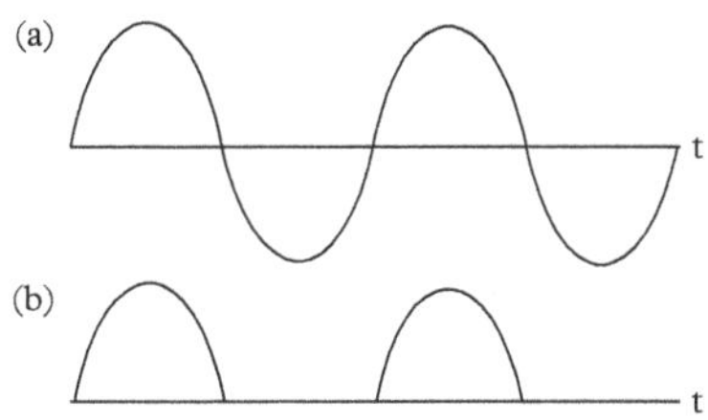

Figure 23.7 *(a) The output voltage of the generator. (b) The voltage across the resistor for a perfect diode.*

23.4.2 The p–n–p junction transistor

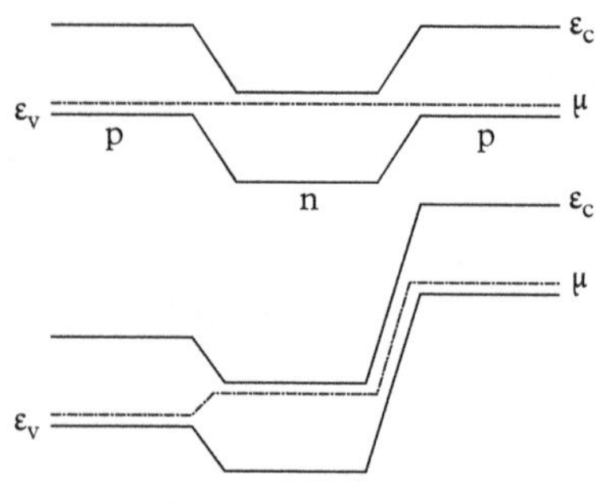

Figure 23.8 *The unbiased (upper) and biased (lower) p-n p transistor.*

We begin with the junction transistor.[3] Figure 23.8 shows the position dependence of the conduction band, valance band, and the chemical potential, without (upper) and with (lower) an external potential, for the case of the p–n–p transistor (we do not show the position of the donor and acceptor levels since now these details are no longer necessary in our description). This device consists of three successive layers termed the emitter (e), base (b), and collector (c), involving p-, n-, and p-doped semiconductors, respectively. These electrodes are separated by two junctions, a p–n (the left or first) and an n–p (the right or second). The key point, as we will now discuss, is that the thickness of the central n-layer is made small compared to a hole diffusion length. Any excess holes here can then diffuse across this n region where they encounter the depletion zone of the second junction and can fall through an electric field, which is substantially enhanced over the equilibrium contact field as a result of the back biasing of this junction. The excess of holes is created by forward biasing the first junction, which results in an exponentially increasing recombination current in the n-type region. The effective impedance of the first junction, involving the ratio of the small external potential required for forward biasing and the large recombination current, is relatively small. In the absence of injected holes the current flowing in the back-biased second junction would largely consist of the constant but small electron/hole generation current associated with this junction. The effective impedance of this junction, given by the ratio of the large back bias to the small generation current, is relatively high. The excess hole current injected at low impedance when the first junction is forward biased, and which subsequently flows through the large electric field of the second junction, results in a substantial power gain.

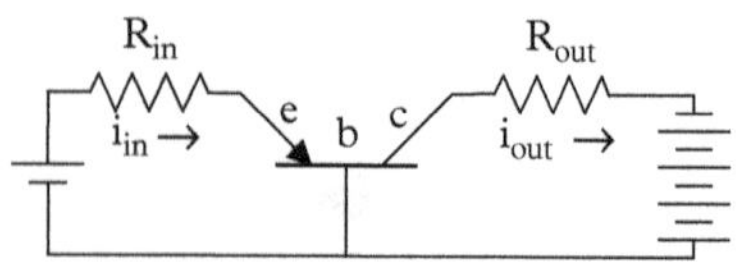

Figure 23.9 *A p-n-p transistor in a grounded base configuration.*

For historical reasons the three regions of the device itself are referred to as the emitter, base, and collector and in circuit diagrams the emitter/base and base/collector junctions are depicted as follows:

Consider the circuit shown in Fig. 23.9. We define the power gain as

$$\text{power gain} = \frac{i_{out}^2 R_{out}}{i_{in}^2 R_{in}}. \tag{23.43}$$

If no current is lost due to recombination of excess holes in traversing the base region, then $i_{in} = i_{out}$. We define as a figure of merit, $\alpha = i_{out}/i_{in}$, with $\alpha = 1$ representing the ideal case. Although we have a power gain there is no current gain.

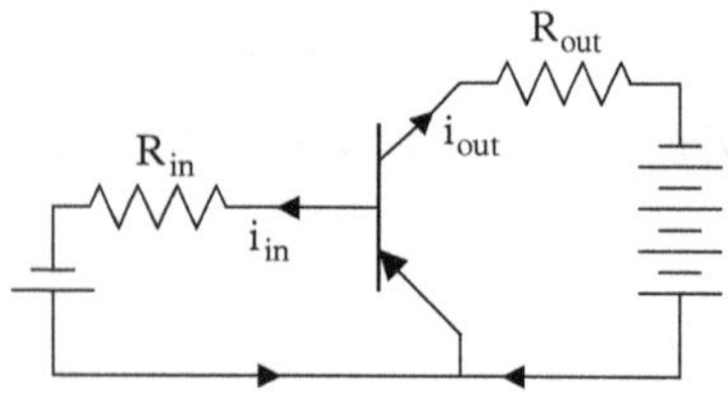

Figure 23.10 *A p n p transistor in a grounded emitter configuration*

An alternative configuration, shown in Fig. 23.10, permits both a voltage and current gain. Here we have

$$\alpha\, i_{emitter} = i_{collector} = i_{out}, \tag{23.44}$$

[3] The 1956 Nobel Prize in Physics was awarded to John Bardeen, Walter Houser Brattain, and William Bradford Shockley "for their researches on semiconductors and their discovery of the transistor effect". The first transistor, the point contact transistor, is now only of historical interest and will not be discussed.

but

$$i_{emitter} = \frac{i_{out}}{\alpha} = i_{in} + i_{out}.$$

Therefore,

$$i_{out} = \frac{\alpha}{(1-\alpha)} i_{in} \equiv \beta i_{in}, \tag{23.45}$$

where we have defined a quantity,

$$\beta = \frac{\alpha}{(1-\alpha)}; \tag{23.46}$$

a value of $\alpha = 0.95$ is typical, leading to $\beta \cong 20$. Note that in the grounded emitter configuration we have both a power and current gain. The n–p–n transistor works in an identical fashion with all circuit potentials and carrier signs reversed.

We note in passing that junction transistors and related devices do not have to be formed from identical (though differently doped) materials. Special properties can emerge from so called hetrostructure devices consisting of different materials. From a different perspective, and of enormous importance in modern electronics, is the single crystal "chip" (typically silicon) on which is patterned and integrated arrays of transistors organized to perform specific functions.[4]

23.4.3 The tunnel diode

We next consider a p-n junction that is so heavily doped that the materials become metallic, with the Fermi level *lying in the valence and conduction bands* for the p-type and n-type materials, respectively. Associated with this doping is a reduction in the width of the depletion zone, and at very high concentrations it is decreased to the point where the carriers rapidly tunnel, quantum mechanically with no energy change, through the potential barrier represented by the energy gap.[5,6]

The situation at $T = 0$ is shown schematically in Fig. 23.11 (here we assumed the doping is such that the Fermi levels measured relative to the valence and conduction bands in the p and n materials respectively are identical). If we *forward bias* this device we encounter three regimes separated by two special points, shown as B and C in Fig. 23.12. For applied voltages between zero and point B, the number of states into which the electrons can tunnel *increases* resulting in an approximately linear increase of the current. At point B the current goes through a maximum, decreasing thereafter (due to a decrease in the number of states available to tunnel into) up to point C, where it vanishes in this simple model. Above C there are no tunneling states and the current would vanish, other factors being neglected. For $T > 0$, the tail of the exponential associated with the electron–hole recombination current results in an exponential increase

[4] The Nobel Prize in Physics in 2000 was awarded "for basic work on information and communication technology". It was shared by Zhores I. Alferov and Herbert Kroemer "for developing semiconductor heterostructures used in high-speed- and opto-electronics" and Jack S. Kilby "for his part in the invention of the integrated circuit"

[5] In our previous discussions electrons and holes, due to the fact that they are in different bands, *retained their identity* when climbing through the junction contact potential.

[6] The Nobel Prize in Physics 1973 was divided, one half jointly to Leo Esaki and Ivar Giaever "for their experimental discoveries regarding tunneling phenomena in semiconductors and superconductors, respectively" and the other half to Brian David Josephson "for his theoretical predictions of the properties of a supercurrent through a tunnel barrier, in particular those phenomena which are generally known as the Josephson effects".

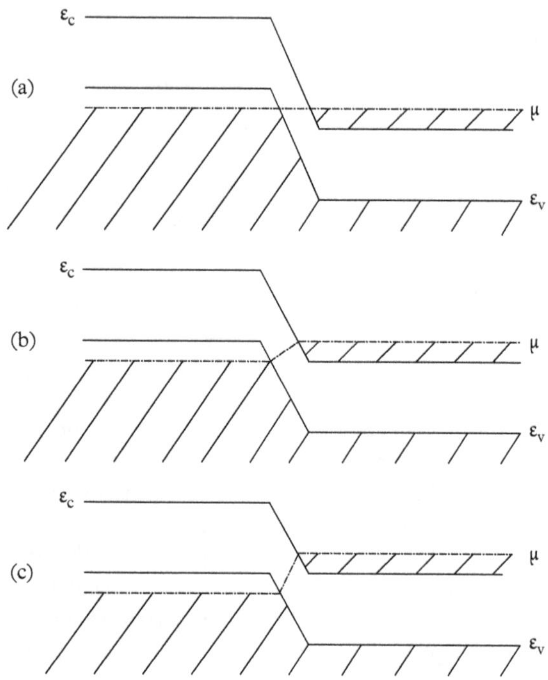

Figure 23.11 *The band alignment for a tunnel diode when: (a) no voltage is applied; (b) a small voltage is applied and states are available for electrons to tunnel into; (c) a large voltage is applied and no states are available for electrons to tunnel into, causing the tunneling current vanish.*

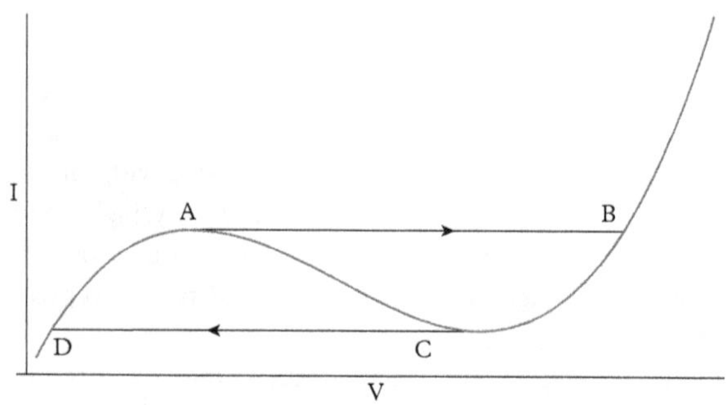

Figure 23.12 *The* I - V *characteristics of a tunnel junction when driven by a constant current source for increasing and decreasing currents, as indicated by the directions of the arrows.*

in the current. The behavior in the three regimes is summarized qualitatively in Fig. 23.12. (We consider only positive currents.) When driven by a constant current generator the current would increase to point A where it would jump to point B and after that continue along that rising curve. On returning, the current would drop to point C and then jump to D and decrease thereafter. In the region between A and B, the differential resistance, dV/dI, is negative; a property that can be used to make an oscillator. Tunnel diodes are also used for pulse shaping.

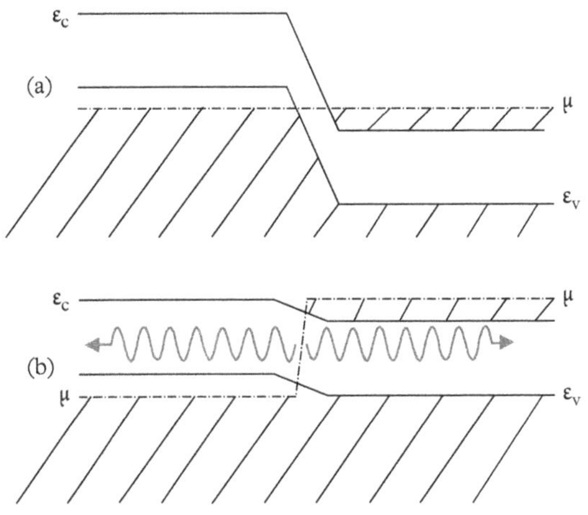

Figure 23.13 *The junction laser: (a) in equilibrium and (b) forward biased, so ε_C on the left matches μ on the right and ε_V on the right matches μ on the left.*

23.4.4 The junction laser

The junction laser may be regarded as an extreme case of a *forward biased* diode, to the point where (at T = 0) $\varepsilon_C(p) = \mu(n)$, and $\varepsilon_V(n) = \mu(p)$ (see Fig. 23.13). Electrons from the n side may then flow freely to the p side and vice versa for the holes. (At higher temperatures significant population inversion would occur at a lower voltage via the tails of the Boltzmann distribution.) Provided that we have a *direct band gap* semiconductor, such as GaAs, the electrons and holes will rapidly recombine and emit light quanta. We then have a light-emitting diode. With appropriate mirrors to provide feedback we then have conditions that can produce lasing.

Diode lasers now cover a broad spectral range extending from the infrared into the blue-violet regime. The availability of the shorter wavelengths, when combined with fluorescent emitters and other strategies, yields white light sources with far higher efficiencies than those available from incandescent lighting.[7]

23.4.5 The Zener diode

Consider an intermediately doped p–n junction that is doped to the point where the Fermi levels in the p and n regions lie within the valence and conduction bands respectively, as shown in Fig. 23.14(a), but where tunneling does not occur in the absence of a bias. When reverse biased by small voltages the resistance would rise quickly as with a conventional diode. However, above some relatively sharp threshold, electrons from the hole band tunnel directly into the electron band. Recall that the thickness of the depletion zone is given by

$$d_{n,p}(V) = d_{n,p}(0)\left[1 - \frac{V}{(\phi)_0}\right]^{1/2}; \qquad (23.47)$$

[7] The 2014 Nobel prize in physics was awarded to Isamu Akasak, Hiroshi Amano and Shuji Nakamura, "for the invention of efficient blue light-emitting diodes, which has enabled bright and energy-saving white light sources".

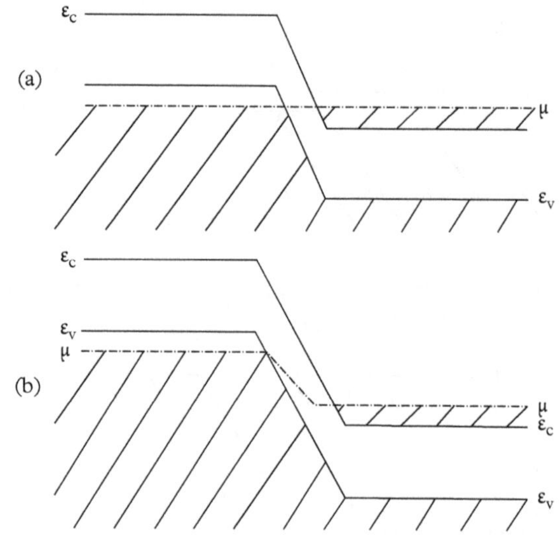

Figure 23.14 *The Zener diode: (a) in equilibrium and (b) reverse biased to the point where the electrons in the hole band tunnel into the electron band.*

i.e., as the voltage increases the size of $d_{n,p}(V)$ shrinks. The tunneling current can be calculated in the WKB approximation, where qualitatively we have

$$j \cong en_{e,h}\bar{v}T, \tag{23.48}$$

where $\bar{v}$ is the average of the Fermi or thermal velocity (depending on the doping and temperature) perpendicular to the junction, $n_{e,h}$ are the carrier concentrations, and T is the transmission probability, which is given approximately by

$$T = \exp\left[-2\kappa_\perp\left(d_n(V) + d_p(V)\right)\right], \tag{23.49a}$$

where

$$\kappa_\perp = \left(\frac{2m_{e,h}E_g}{\hbar^2}\right)^{1/2}, \tag{23.49b}$$

with $m_{e,h}$ an average mass.

Since, as previously stated, the onset voltage is rather sharp, the devices are used as *voltage references or regulators.*

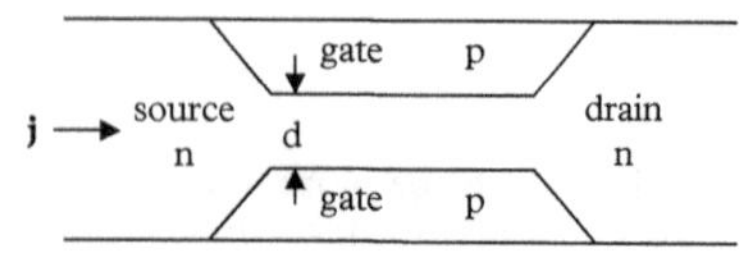

Figure 23.15 *A schematic of the planar field effect transistor (FET).*

23.4.6 The field effect transistor

The transistor may be regarded as a one-dimensional device involving the direction perpendicular to the junctions. We now consider the two-dimensional device shown in Fig. 23.15. We assume that only the surface of a Si wafer has been doped p- and n-type in the regions shown in the figure. The side from which the current j flows, on the left of the narrow region of width d separating the two p-type electrodes, is referred to as the *source* while the region to the right is called the *drain*. The two p-type electrodes are called gates. The key point is that the distance d separating the two p-type electrodes is comparable to the depletion zone width. Generally the p-type electrode is heavily doped resulting in a small $d_p(0)$, so d is dominated by d_n. Suppose

$$d > 2d_n(0), \tag{23.50a}$$

where $d_n(V_D)$ is given by Eq. (23.47) and V_D is a *positive* voltage applied to the drain; we assume the source is at zero potential and that initially no voltage is applied to the gates. As V_D increases from zero, the source–drain current initially increases from zero, but simultaneously $d_n(V_D)$ increases; ultimately we reach a drain voltage where the condition

$$d \cong 2d_n(V_D) \tag{23.50b}$$

is met. Near this point the current saturates, a phenomena referred to as "pinch off," at a value governed by the hole-generation current in the depletion region surrounding the gates. The whole process is two-dimensional in character and the local drain–gate potential difference is largest near the gate edges *closest to the drain* where the pinch is concentrated. If now a *negative* voltage, V_G, is applied to the gates (the same for both), which is the sense that increases the potential - drop to the n-type region, pinch-off will occur at a lower source–drain potential; it will also occur over a more extended region between the gates, resulting in a lower overall saturation current. This situation is depicted schematically in Fig. 23.16.

Note that only a small current flows from the gates to the n-type region since we are in a back biased condition. Yet we are controlling a relative large current flowing through the source train potential. Hence there is an enormous power gain.

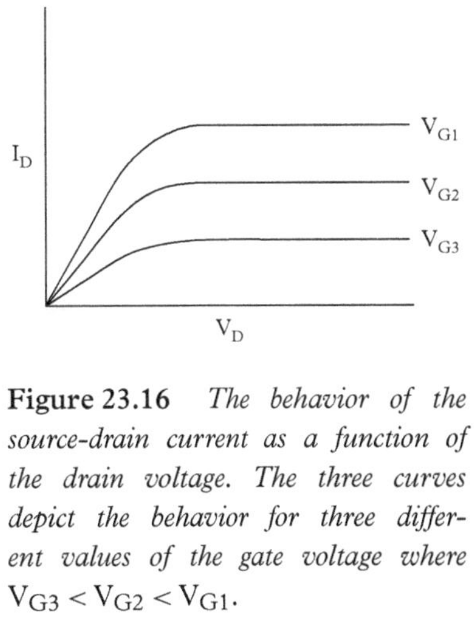

Figure 23.16 *The behavior of the source-drain current as a function of the drain voltage. The three curves depict the behavior for three different values of the gate voltage where* $V_{G3} < V_{G2} < V_{G1}$.

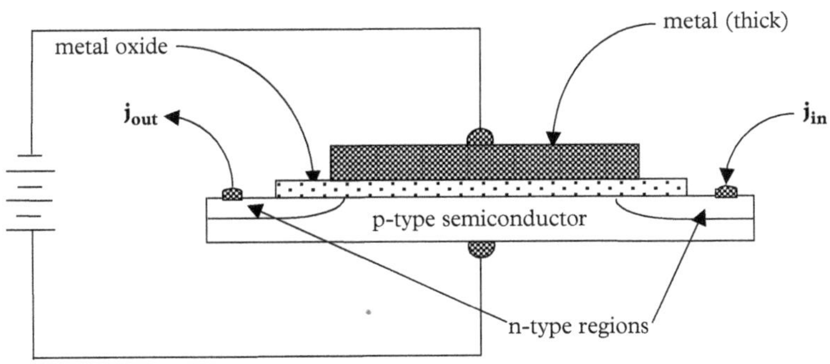

Figure 23.17 *A schematic of a metal-oxide semiconductor field-effect transistor (MOSFET)*

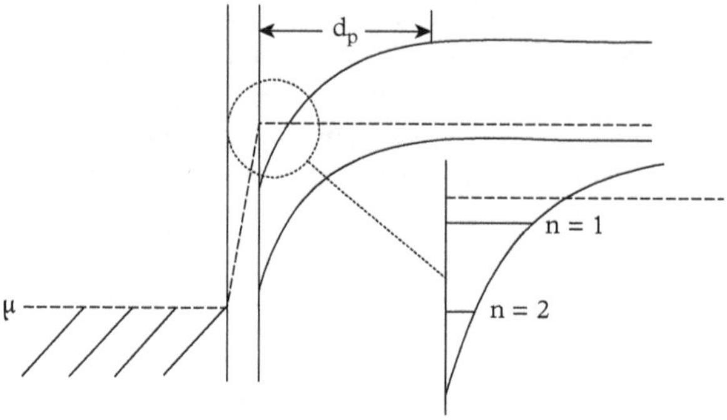

Figure 23.18 *The position of the band edges near a metal-oxide-semiconductor interface in the presence of a bias potential. Note that for sufficient bias, surface states near the interface are occupied.*

23.4.7 The MOSFET

A schematic of the MOSFET (metal-oxide–semiconductor field effect transistor) is shown in Fig. 23.17. The two n-type regions on each side of the central region are restricted to the surface; they are used to make electrical contact to the central region. Figure 23.18 shows the position dependence of the electron and hole band edges in the vicinity of the metal-oxide–semiconductor interface region in the p-type material, in the presence of a bias potential applied to the metal electrode. Note that under the influence of this bias potential, the positions of the conduction and valence band edges change continuously from their bulk values over a distance governed by the depletion length; near the interface this shift can be large enough to cause the conduction band edge to fall *below the Fermi level* and in this region *electron states are occupied.*

As a model for the potential for electrons near the surface one can assume the form

$$V(z) = \begin{cases} +\infty & z < 0 \\ |e|\, E\, z & z > 0 \end{cases}, \tag{23.51}$$

where E is the electric field. The associated Schrödinger equation is

$$\frac{d^2\psi}{dz^2} + \frac{2m}{\hbar^2}(\varepsilon + F\, z) = 0, \tag{23.52}$$

and we have introduced the force through $F = |e|E$. Defining the dimensionless variable,

$$x = -\left(z + \frac{\varepsilon}{F}\right)\left(\frac{2mF}{\hbar^2}\right)^{1/3}, \tag{23.53}$$

our Schrödinger equation becomes

$$\frac{d^2\psi}{dx^2} - x\psi = 0. \tag{23.54}$$

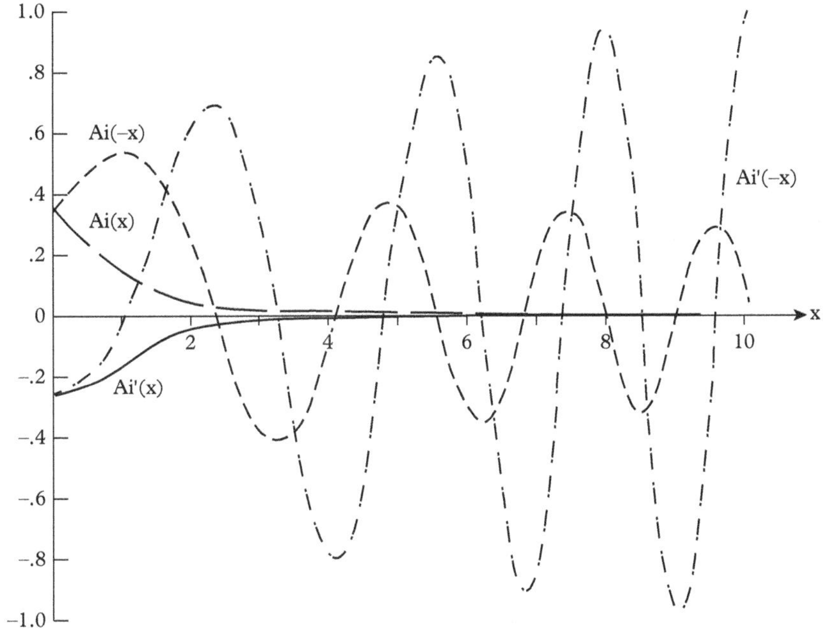

Figure 23.19 *The Airy function.*

The solution to this equation is the *Airy function,* Ai(x), which is plotted in Fig. 23.19.

The energy levels follow from the requirement that the wave function should vanish at z = 0. The values of x for the first two zeros of Ai(x) (Abromawitz and Stegun, 1972) are at x = −2.34 and x = −4.09. From (23.53) we then have for the first two levels

$$\varepsilon_1 = 2.34\left(\frac{\hbar^2 F^2}{2m}\right)^{1/3} \quad \text{and } \varepsilon_2 = 4.09\left(\frac{\hbar^2 F^2}{2m}\right)^{1/3}. \tag{23.55}$$

In addition to the binding energy at the interface we have the kinetic energy for motion parallel to the interface, so the total energy associated with a level with binding energy ε_n is

$$\varepsilon_n = \varepsilon_{n0} + \frac{\hbar^2 k_{//}^2}{m_{//}} \tag{23.56}$$

at T = 0, states will accumulate at the interface so as to satisfy $\varepsilon_n = \mu$; i.e., we will have a two-dimensional electron gas. The physics of this two-dimensional electron gas is of interest in its own right and we will return to a related system when we discuss the quantum Hall effect later (see Chapter 48).

At low applied electric fields, and in the presence of the two n-type electrodes on each side of the metal electrode, shown in Fig. 24.17, the device can be viewed as two back-to-back p–n junctions and as such there will be no overall conduction in either direction. When a field is applied to the metal electrode it induces an n-type region in the p-type material at the metal oxide interface; this provides a conduction path between the two n-type electrodes—i.e., the device can then act as a field effect transistor or switch.

ADDITIONAL READING

Our discussion of semiconductor devices has been very brief. Parts of it parallel that given in *Solid State Physics*, N.W. Ashcroft and N.D. Mermin, Holt, Rinehaert and Winston, PA (1976).

Texts on semiconductor devices include: *Device Electronics for Integrated Circuits*, Richard S. Muller, Theodore I. Kamins, and Mansun Chan, John Wiley & Sons, NY (2003); *Semiconductor Device Fundamentals*, Robert F. Pierret, Prentice Hall, Upper Saddle River, New Jersey (1995); *Solid State Electronic Devices*, B.G. Streetman and S. Banerjee, Prentice Hall, Upper Saddle River, New Jersey (2000); and *Physics of Semiconductor Devices*, S.M. Sze, Wiley-Interscience, New York, NY (1981).

A historically important and still valuable book is *Electrons and Holes in Semiconductors, with Applications to Transistor Electronics*, William Shockley, D. Van Nostrand, Canada (1950).

The first transistor was reported in The transistor: A semi-conductor triode, J. Bardeen and W. Brattain, *Phys. Rev.* **74**, 230 (1948). Junction devices were described in The theory of P-N junctions in semiconductors and P-N junction transistors, William Shockley, *Bell System Technical Journal* **28**, 435 (1949).

For a discussion of the associated history see: The discovery of the point-contact transistor, Lillian Hoddeson, *Historical Studies in the Physical Sciences* **12**, 41 (1981a) and *Crystal Fire: The Invention of the Transistor and the Birth of the Information Age.* Michael Riordan and Lillian Hoddeson, Norton, NY (1997).

Part VIII

Electric and Magnetic Properties of Insulators

Electric and Magnetic Susceptibilities

24

In Section 2.3 we introduced the electric susceptibility, χ^E, as a measure of the polarization induced by a weak electric field, **E**. Microscopically, an electric field will exert a force on all the charged constituents of a material. This force results in a displacement of various charge centers that, in equilibrium and in an insulator, is balanced by forces induced by the displacements themselves; at finite frequencies, ω, one must also account for the inertia (mass) associated with the accelerated nuclei and electrons making up the system, ultimately in a manner consistent with quantum mechanics. Such charge displacements in the body produce *dipole moments*. As discussed in Chapter 2, to account for the effect of these dipoles, in an average sense, one introduces a *polarization*, $\mathbf{P}(\omega)$ (and an associated displacement vector $\mathbf{D}(\omega)$), defined as the *dipole moment per unit volume*. Carrying out this averaging in a consistent manner yields quantities that enter the macroscopic electrodynamics of Maxwell. Such an approach assumes that the required averages are well defined which, in turn, assumes that the wavelength of the external electric field is large compared to distances over which the materials properties of the body vary, e.g., as fixed by interatomic spacings.

If the frequency, ω, is low and the fields are not too high we may picture the nuclei together with the tightly bound core electrons as having some effective charge, $\pm ze$, that displaces as a whole under the action of a force, $\mathbf{F} = \pm ze\mathbf{E}$.[1] This results in an *ionic polarization*. As noted before, in static equilibrium this force would be compensated by a force arising from displacing the atoms from their equilibrium positions (where the force on each atom vanishes). The resulting forces can be pictured as arising from stretched springs acting between atoms, where in the simplest model they only connect neighboring atoms. We will apply this model to a simple ionic solid, NaCl.

We must also discuss the dynamic behavior of ionic polarization. Qualitatively the ions, characterized by some (reduced) mass M and bound to their neighbors by springs with a constant κ, are expected to have a resonant response with a characteristic frequency, $\omega_0 \approx (\kappa/M)^{1/2}$; in the vicinity of such a resonance we expect large amplitudes and hence a large dynamic polarization, $\mathbf{P}(\omega)$. However at very high frequencies the inertial effects will dominate, and, asymptotically, the displacements **U** must fall as $\mathbf{U}(\omega \to \infty) \to \mp ze\mathbf{E}/M\omega^2$ and the polarization approaches zero as ω^{-2}.

[1] Here we will suppress effects arising from local field corrections.

The Physics of Solids. First Edition. J. B. Ketterson.

The electric field also produces a force on all the electrons. The strong nuclear electric field tightly constrains the motion of the core electrons (as we noted before they tend to move with the nucleus at low frequencies); their orbitals are nonetheless distorted by the external field resulting in a core contribution to the *electronic polarization.* On the other hand the valance electrons, being more loosely bound, are more easily polarized and make the dominant contribution to the electronic component of the induced polarization. As the frequency increases, inertial effects again enter. It is not accurate to represent the field-induced distortions of electron orbitals via springs (as we did in Section 3.5), although some qualitative features are captured by such a model. Electronic polarization is best treated quantum mechanically, as we do in Section 24.2 for the case of atoms and molecules, but the analogue of resonances, arising, e.g., at a frequency, $\hbar\omega = \varepsilon_n - \varepsilon_m$, corresponding to a transition between two atomic levels n and m, do occur where $\mathbf{P}(\omega)$ is greatly enhanced. However, as ω increases the response of electrons in various quantum states will again pass over into an inertially dominated regime where their displacements, $\mathbf{r}(\omega \to \infty) \to |e|\ \mathbf{E}/m\omega^2$; since the electrons are at least 2000 times lighter, their associated frequency regimes are correspondingly higher. At the very highest frequencies (in the X-ray range) even the core electrons enter the inertial regime.

24.1 The static ionic polarizibility

In an ionic crystal such as sodium chloride, the application of an external electric field will cause the positive and negative ions to displace in opposite directions inducing a polarization. We can estimate this polarization from the frequency of the resonance that is observed when an infrared electromagnetic wave is degenerate with the optical phonon mode. (We will consider this resonance in more detail when we discuss polaritons in Chapter 30.) For the case of the one-dimensional NaCl-like lattice considered in Section 16.2, we found a resonant frequency, ω_0, given by

$$\omega_0^2 = 2\kappa\left(\frac{1}{M_1} + \frac{1}{M_2}\right), \tag{24.1}$$

where M_1 and M_2 are the masses of the two ions and κ is the constant associated with the spring connecting the ions. Since the ion masses are known we can obtain κ by measuring ω_0.

In the presence of a *static* external electric field, E, producing a force, $\pm eE$, the displacements of the positive, U^1, and negative, U^2, ions are related as[2]

$$+eE = +2\kappa U^1 - 2\kappa U^2 \tag{24.2a}$$

and

$$-eE = -2\kappa U^1 + 2\kappa U^2. \tag{24.2b}$$

[2] The required equations follow from Eqs. (16.30a,b) by replacing the inertial force $\omega^2 M_{1,2} U_n^{1,2}$ by $\pm eE$, noting that $U_n^1 = U_{n\pm1}^1$ and $U_n^2 = U_{n\pm1}^2$ and writing $(\kappa + \kappa') \to 2\kappa$.

Subtracting (24.2a,b) we obtain

$$eE = 2\kappa(U^1 - U^2). \tag{24.3}$$

The polarizability is then

$$P = ne(U^1 - U^2) = n\frac{e^2E}{2\kappa}, \tag{24.4}$$

where $n = 1/2(a/2)^3$ is the number of sodium chloride molecules per unit (three-dimensional) volume and a is the lattice constant. Using Eq. (24.1) we can rewrite Eq. (24.4) as

$$P = n\frac{e^2E}{\omega_0^2}\left(\frac{1}{M_1} + \frac{1}{M_2}\right), \tag{24.5a}$$

corresponding to an ionic contribution to the electric susceptibility of

$$\chi^{\text{ionic}} = n\frac{e^2}{\omega_0^2}\left(\frac{1}{M_1} + \frac{1}{M_2}\right). \tag{24.5b}$$

The change in the dielectric constant in going from $\omega \ll \omega_0$ (where Eq. (24.4) applies) to $\omega \gg \omega_0$ (the inertial regime) is therefore

$$\Delta\varepsilon = 4\pi n\frac{e^2}{\omega_0^2}\left(\frac{1}{M_1} + \frac{1}{M_2}\right); \tag{24.6}$$

for sodium chloride this gives $\Delta\varepsilon = 2.7$, whereas the experimental value is 3.7.

24.2 Electronic polarizibility of atoms and molecules

In Chapter 3 we derived an expression for the dielectric constant of bound electrons based on the classical Drude–Lorentz theory. We now discuss the quantum theory of the dielectric constant. We begin by recalling from Chapter 18 the Hamiltonian of an electron in the presence of an electromagnetic field, as described by a vector potential, $\mathbf{A}(\mathbf{r}, t)$. Combining Eqs. (18.1) and (18.2a), the Hamiltonian for an N electron system is

$$\hat{H} = \frac{1}{2m}\sum_{a=1}^{N}\left[\frac{\hbar}{i}\nabla_a - \frac{e}{c}\mathbf{A}(\mathbf{r}_a, t)\right]^2 + V(\mathbf{r}_1 \ldots \mathbf{r}_N) \tag{24.7a}$$

$$= \hat{H}_0 + \hat{H}_1, \tag{24.7b}$$

where

$$\hat{H}_0 = -\frac{\hbar^2}{2m}\sum_{a=1}^{N}\nabla_a^2 + V(\mathbf{r}_1 \ldots \mathbf{r}_N) \tag{24.8a}$$

and

$$\hat{H}_1 = \frac{i\hbar e}{2mc}\sum_{a=1}^{N}\left[\nabla\cdot\mathbf{A}(\mathbf{r}_a,t) + \mathbf{A}(\mathbf{r}_a,t)\cdot\nabla\right] + \sum_{a=1}^{N}\frac{e^2}{2mc^2}\mathbf{A}(\mathbf{r}_a,t)^2. \tag{24.8b}$$

In what follows we will retain only the linear terms in the vector potential[3] and work in the London gauge where $\nabla\cdot\mathbf{A} = 0$, in which case (24.8b) becomes

$$\hat{H}_1 = \frac{i\hbar e}{mc}\sum_{a=1}^{N}\mathbf{A}(\mathbf{r}_a,t)\cdot\nabla_a = -\sum_{a=1}^{N}\frac{e}{mc}\mathbf{A}(\mathbf{r}_a,t)\cdot\hat{\mathbf{p}}_a, \tag{24.9}$$

where $\hat{\mathbf{p}}$ is the momentum operator. A harmonic electromagnetic wave may be written in the form

$$\mathbf{E}(\mathbf{r},t) = \frac{1}{2}\left[\mathbf{E}(\mathbf{r})e^{-i\omega t} + \mathbf{E}(\mathbf{r})^* e^{i\omega t}\right]. \tag{24.10}$$

The associated vector potential

$$\mathbf{A}(\mathbf{r},t) = \frac{1}{2}\left[\mathbf{A}(\mathbf{r})e^{-i\omega t} + \mathbf{A}(\mathbf{r})^* e^{i\omega t}\right], \tag{24.11a}$$

is related to the electric field through $\mathbf{E} = -\frac{1}{c}\frac{\partial\mathbf{A}}{\partial t}$, or

$$\mathbf{A}(\mathbf{r},t) = \frac{c}{2i\omega}\left[\mathbf{E}(\mathbf{r})e^{-i\omega t} - \mathbf{E}(\mathbf{r})^* e^{i\omega t}\right]. \tag{24.11b}$$

To make subsequent expressions clearly defined we will assume the exponential forms in these expressions are replaced according to the prescription

$$e^{\pm i\omega t} \rightarrow e^{\pm i\omega t + \delta t}. \tag{24.12}$$

The reason for this substitution is that it allows us to turn on the perturbation adiabatically. The fact that this form would lead to a divergence for times far in the future is not relevant since the subsequent integrations are carried out only up to the present time.[4]

We will assume that the many-body wave function can be written in a Hartree-like one-electron product form:

$$\psi(\mathbf{r}_1, \mathbf{r}_2, \cdots \mathbf{r}_N) = \psi_{n_1}(\mathbf{r}_1)\psi_{n_2}(\mathbf{r}_2)\cdots\ \psi_{n_a}(\mathbf{r}_a)\cdots\psi_{n_N}(\mathbf{r}_N),$$

[3] The term $\sum_{a=1}^{N}(e^2/2mc^2)\,\mathbf{A}(\mathbf{r}_a,t)^2$ leads to the so-called ponderomotive energy which affects the band gap and other properties of semiconductors under high-intensity illumination.

[4] A similar form was introduced in our discussion of the dielectric function in Chapter 14 (see Eq. 14.1).

with quantum numbers $n_1, n_2 \cdots n_a \cdots n_N$ (where for an atom n_a they would include the n, l, m, and s quantum numbers of that orbital) filled in accord with the Pauli principle. Our strategy will be to calculate the time-dependent corrections to each of the orbitals caused by the external electromagnetic field and use the resulting orbital to calculate the induced dipole moment of that orbital; the total moment is then taken as a sum over the moments of the individual orbitals. In addition we will assume that these orbitals are localized on the molecules or atoms making up the material; i.e., they are not delocalized as are Bloch orbitals. Hence this model is most suitable for a material bound via van der Waals forces, e.g., a rare gas or a molecular solid. However it is also useful for qualitatively describing insulators.

We denote the occupied orbitals by an index $n = 1, 2, \cdots N$ that is in one-to-one correspondence with the orbital quantum numbers $n_1, n_2 \cdots n_N$. The time-dependent wave function of the nth orbital can be expanded as

$$\Psi_n(\mathbf{r},t) = \Psi_n^{(0)}(\mathbf{r},t) + \Psi_n^{(1)}(\mathbf{r},t) + \Psi_n^{(2)}(\mathbf{r},t) + \cdots, \tag{24.13}$$

where $\Psi_n^{(1)}, \Psi_n^{(2)}$ represent successively higher-order corrections, arising from the perturbation, $\hat{H}_1$, to the zeroth-order wave functions, $\Psi_n^{(0)}(\mathbf{r},t)$. The $\Psi_n^{(i)}(\mathbf{r},t)$ are expanded in terms of the zeroth-order time-dependent eigenfunctions, $\Psi_m^{(0)}(\mathbf{r},t) = \psi_m^{(0)}(\mathbf{r})e^{-i\omega_m t}$, as

$$\Psi_n^{(i)}(\mathbf{r},t) = \sum_m a_{mn}^{(i)}(t)\psi_m^{(0)}(\mathbf{r})e^{-i\omega_m t}, \tag{24.14}$$

where the $\psi_m^{(0)}(\mathbf{r})$ are unoccupied orbitals associated with $\hat{H}_0$. We will carry out the calculation to terms linear in the electromagnetic field, which only requires the term $\Psi_n^{(1)}(\mathbf{r},t)$, for which the expansion coefficients are calculated from

$$i\hbar\dot{a}_{ln}^{(1)}(t) = \left\langle \psi_l^*(\mathbf{r}) \left| \hat{H}_1 \right| \psi_n(\mathbf{r}) \right\rangle e^{i\omega_{ln}t}, \tag{24.15}$$

where $\hat{H}_1$ is given by (24.9) and $\omega_{ln} = \omega_l - \omega_n$. Integrating (24.15) using the time dependence (24.11), we obtain

$$a_{ln}^{(1)}(t) = \frac{e}{2\hbar mc}\left[\frac{\langle\psi_l|\,\mathbf{A}(\mathbf{r})\cdot\hat{\mathbf{p}}\,|\psi_n\rangle\, e^{-i\omega t}}{\omega_{ln} - \omega - i\delta} + \frac{\langle\psi_n|\,\mathbf{A}(\mathbf{r})\cdot\hat{\mathbf{p}}\,|\psi_l\rangle^*\, e^{i\omega t}}{\omega_{ln} + \omega - i\delta}\right] e^{i\omega_{ln}t}. \tag{24.16}$$

We rewrite the matrix element entering this expression as

$$\langle\psi_l|\,\mathbf{A}(\mathbf{r})\cdot\hat{\mathbf{p}}\,|\psi_n\rangle = \sum_m \langle\psi_l|\,\mathbf{A}(\mathbf{r})\,|\psi_m\rangle\cdot\langle\psi_m|\,\hat{\mathbf{p}}\,|\psi_n\rangle. \tag{24.17}$$

Writing $\hat{\mathbf{p}} = m\dot{\hat{\mathbf{r}}}$, using the Heisenberg equation of motion, $\dot{\hat{\mathbf{r}}} = \frac{i}{\hbar}\left[\hat{H}\hat{\mathbf{r}} - \hat{\mathbf{r}}\hat{H}\right]$, and letting $\hat{H}$ act to the left and right respectively, we get

$$(\mathbf{p})_{ln} = \frac{im}{\hbar}\left[E_l - E_n\right](\mathbf{r})_{ln} = im\omega_{ln}(\mathbf{r})_{ln}; \tag{24.18}$$

$(\mathbf{r})_{ln} = \langle\psi_l|\,\mathbf{r}\,|\psi_n\rangle$ is referred to as the *dipole matrix element.*

We now write $\mathbf{A}(\mathbf{r})$ in the plane wave form, $\mathbf{A}(\mathbf{r}) = \mathbf{A}_0 e^{i\mathbf{k}\cdot\mathbf{r}}$, with a corresponding $\mathbf{E}_0 = \frac{i\omega}{c}\mathbf{A}_0$. Since k for light is of order $10^4\mathrm{cm}^{-1}$, and our atomic-like localized orbitals have a spatial extent of only a few Angstroms, it is usually a good approximation to expand the exponential and retain the first non-vanishing term; i.e.,

$$\langle\psi_l\,|\mathbf{A}(\mathbf{r})|\,\psi_n\rangle = \langle\psi_l|\,\mathbf{A}_0 e^{i\mathbf{k}\cdot\mathbf{r}}\,|\psi_n\rangle = \langle\psi_l|\,\mathbf{A}_0\left(1 + i\mathbf{k}\cdot\mathbf{r} - \frac{1}{2}(\mathbf{k}\cdot\mathbf{r})^2 + \cdots\right)|\psi_n\rangle. \tag{24.19}$$

On taking matrix elements we have

$$\left(e^{i\mathbf{k}\cdot\mathbf{r}}\right)_{ln} = \delta_{ln} + i\mathbf{k}\cdot(\mathbf{r})_{ln} - \frac{1}{2}\sum_m \mathbf{k}\cdot(\mathbf{r})_{lm}(\mathbf{r})_{mn}\cdot\mathbf{k} + \cdots. \tag{24.20}$$

The second term in this expansion also involves the dipole matrix element, but, given that this form has already appeared in (24.18), its inclusion actually involves the next highest order.[5] With these simplifications, the coefficients $a_{ln}^{(1)}(t)$ in the dipole approximation are

$$a_{ln}^{(1)}(t) = \frac{e}{2\hbar}\frac{\omega_{ln}}{\omega}\left[\frac{(\mathbf{r})_{ln}\cdot\mathbf{E}_0 e^{-i\omega t}}{\omega_{ln}-\omega-i\delta} - \frac{(\mathbf{r})^*_{nl}\cdot\mathbf{E}^*_0 e^{+i\omega t}}{\omega_{ln}+\omega-i\delta}\right]e^{i\omega_{ln}t}. \tag{24.21}$$

We now have all of the components necessary to construct the induced dipole moment, $\mathbf{p}_n(t)$), associated with the state n:

$$\begin{aligned}\mathbf{p}_n(t) &= \langle\Psi_n(\mathbf{r},t)\,|e\mathbf{r}|\,\Psi_n(\mathbf{r},t)\rangle\\ &= \left\langle\Psi_n^{(0)}(\mathbf{r},t)\,|e\mathbf{r}|\,\Psi_n^{(0)}(\mathbf{r},t)\right\rangle + \left\langle\Psi_n^{(1)}(\mathbf{r},t)\,|e\mathbf{r}|\,\Psi_n^{(0)}(\mathbf{r},t)\right\rangle\\ &\quad + \left\langle\Psi_n^{(0)}(\mathbf{r},t)\,|e\mathbf{r}|\,\Psi_n^{(1)}(\mathbf{r},t)\right\rangle + \left\langle\Psi_n^{(2)}(\mathbf{r},t)\,|e\mathbf{r}|\,\Psi_n^{(0)}(\mathbf{r},t)\right\rangle\\ &\quad + 2\left\langle\Psi_n^{(1)}(\mathbf{r},t)\,|e\mathbf{r}|\,\Psi_n^{(1)}(\mathbf{r},t)\right\rangle + \left\langle\Psi_n^{(0)}(\mathbf{r},t)\,|e\mathbf{r}|\,\Psi_n^{(2)}(\mathbf{r},t)\right\rangle + \cdots.\end{aligned} \tag{24.22}$$

The leading term is the spontaneous polarization that we assume vanishes here (for a molecule this is often not the case). The first-order contribution is[6]

$$\mathbf{p}_n^{(1)}(t) = \int d^3r\left[\sum_l a_{ln}^{(1)*}(t)\psi_l^*(\mathbf{r})e\mathbf{r}\psi_n(\mathbf{r})e^{i\omega_{ln}t} + \sum_l a_{ln}^{(1)}(t)\psi_n^*(\mathbf{r})e\mathbf{r}\psi_l(\mathbf{r})e^{i\omega_{ln}t}\right]. \tag{24.23}$$

Inserting (24.21), the ith component of $\mathbf{p}_n^{(1)}(t)$ becomes

$$\begin{aligned}p_{ni}^{(1)}(t) = \frac{e^2}{2\hbar}\sum_l\sum_j\frac{\omega_{ln}}{\omega}&\left\{\left[\frac{(r_i)_{nl}(r_j)_{ln}}{\omega_{ln}-\omega-i\delta} - \frac{(r_j)_{nl}(r_i)_{ln}}{\omega_{ln}+\omega-i\delta}\right]E_{0j}e^{-i\omega t}\right.\\ &\left.+\left[\frac{(r_i)^*_{nl}(r_j)^*_{ln}}{\omega_{ln}-\omega+i\delta} - \frac{(r_j)^*_{nl}(r_i)^*_{ln}}{\omega_{ln}+\omega+i\delta}\right]E^*_{0j}e^{+i\omega t}\right\},\end{aligned} \tag{24.24}$$

[5] The third term can be factored into symmetric (electric quadrupole) and anti-symmetric (magnetic dipole) terms as

$$\begin{aligned}\langle r_ir_j\rangle_{ln} &= \frac{1}{2}\langle r_ir_j + r_jr_i\rangle_{ln} + \frac{1}{2}\langle r_ir_j - r_jr_i\rangle_{ln}\\ &= \langle S_{ij}\rangle_{ln} + \langle A_{ij}\rangle_{ln}.\end{aligned}$$

The electric quadrupole and magnetic dipole matrix elements are defined as

$$\langle Q_{ij}\rangle_{ln} = e\langle S_{ij}\rangle_{ln};\ \langle M_{ij}\rangle_{ln} = -ie\omega\langle A_{ij}\rangle_{ln}.$$

[6] Nonlinear optical phenomena may be treated by including corrections to the polarization associated with higher-order terms in (24.24).

where we have used the property $(r_i)_{nl} = (r_i)^*_{ln}$. In the absence of a magnetic field the wave functions may be chosen as real, in which case, $(r_i)_{nl} = (r_i)_{ln}$. Equation (24.24) may then be rewritten as

$$p^{(1)}_{ni}(t) = \frac{e^2}{2\hbar}\sum_l\sum_j \frac{\omega_{ln}}{\omega}(r_i)_{nl}(r_j)_{nl}\left\{\left[\frac{1}{\omega_{ln}-\omega-i\delta}-\frac{1}{\omega_{ln}+\omega-i\delta}\right]E_{0j}e^{-i\omega t}+\text{c.c.}\right\} \tag{24.25}$$

If we write the dipole moment in a form

$$\mathbf{p}_n(t) = \mathbf{p}_{n0}e^{-i\omega t} + \mathbf{p}^*_{n0}e^{+i\omega t}, \tag{24.26}$$

we can relate it to the corresponding amplitude $\mathbf{E}_0$ through *a polarizibility tensor* $\boldsymbol{\alpha}_n(\omega)$ arising from the state n as

$$\mathbf{p}_{n0} = \boldsymbol{\alpha}_n(\omega)\cdot\mathbf{E}_0. \tag{24.27}$$

We can then write the *total* electronic polarizibility of the whole atom or molecule as

$$\alpha_{ij}(\omega) = \frac{e^2}{\hbar}\sum_{n,l}\frac{\omega_{ln}}{\omega}(r_i)_{nl}(r_j)_{nl}\left[\frac{1}{\omega_{ln}-\omega-i\delta}-\frac{1}{\omega_{ln}+\omega-i\delta}\right]. \tag{24.28}$$

Here, the sum on n is over the *occupied* ground state orbitals and, to account for the Pauli principle, the sum on l is over the *unoccupied* levels.[7] If we combine the terms in (24.28) over a single denominator, suppress the infinitesimals, and restrict to the case of cubic or isotropic symmetry, we have

$$\alpha(\omega) = \frac{e^2}{m}\sum_{n,l}\frac{f_{nl}}{\omega^2_{ln}-\omega^2}, \tag{24.29a}$$

where we have defined an *oscillator strength* by

$$f_{nl} = \frac{2m}{\hbar}\omega_{ln}(r_i)^2_{nl}. \tag{24.29b}$$

The *static* polarizibility, $\boldsymbol{\alpha}(0)$, would correspond to the $\omega = 0$ limit of (24.29a).

In our earlier classical model of a material made up of atoms with a single bound electron with resonant frequency ω_0, the atomic polarizibility in the absence of damping will be given by (see Eq. 3.20)

$$\alpha(\omega) = \frac{e^2}{m}\frac{1}{\omega_0^2-\omega^2}. \tag{24.30}$$

The quantum theory makes contact with the classical theory if we assume a *single* oscillator and a corresponding oscillator strength of $f = 1$.

[7] To account for finite temperature effects, appropriate Boltzmann occupation factors would be introduced.

We recall that the Maxwell polarization vector **P** is defined as the dipole moment per unit volume; if we have a single species and N/L^3 such species per unit volume, we can introduce the *electric susceptibility* (see Eqs. (2.28) and (2.29)) through the definition, $\mathbf{P} = \chi(\omega) \cdot \mathbf{E}$, where

$$\begin{aligned}\chi(\omega) &= \frac{N}{L^3}\alpha(\omega) \\ &= \frac{N}{L^3}\frac{e^2}{m}\sum_{n,l}\frac{f_{nl}}{\omega_{ln}^2 - \omega^2}\end{aligned} \tag{24.31a}$$

In so doing we would be ignoring the Lorentz *local field effect*; for the quasistatic limit this correction could be included using the model discussed in Section 2.2. The dielectric constant relating **D** to **E** is given by $\varepsilon(\omega)$ which on the basis of (24.31a) would be

$$\varepsilon(\omega) = 1 + 4\pi\frac{N}{L^3}\frac{e^2}{m}\sum_{n,l}\frac{f_{nl}}{\omega_{ln}^2 - \omega^2}. \tag{24.31b}$$

To calculate the electronic susceptibility of a periodic crystal where the interatomic interactions are strong we must use the Bloch wave functions, which are extended, rather than the localized atomic-like wave functions we have been assuming here. A formal expression for the longitudinal dielectric function (arising from an external potential) was given in Section 14A; it can be extended to the case of an external vector potential (to treat transverse electromagnetic waves); the local field corrections there emerge naturally as a part of the formalism.[8] Other aspects of the electromagnetic response of insulators will be discussed in Chapter 29 and 30.

This treatment will be extended in Sections 31.1 and 31.2 to include the nonlinear response of atoms and molecules in strong electric fields. Such responses have become important with the advent of pulsed lasers with very high intensities.[9]

24.3 The magnetic polarizibility of atoms and molecules

In Chapter 5 we obtained Langevin's classical expressions for the diamagnetic susceptibility of atoms, Eq. (5.9), and the paramagnetic susceptibility of noninteracting magnetic moments, Eqs. (5.11) and (5.12a). In connection with the latter, we also gave the corresponding Brillouin form, Eqs. (5.21) and (5.22), for the quantum mechanical case. In the present section we discuss the quantum theory of the magnetic response of a gas of atoms or molecules and discuss its applicability to solids. The magnetic response is typically smaller by a factor of

[8] Quantum theory of the dielectric constant in real solids, S.L. Adler, *Phys. Rev.* **126**, 413 (1962).

[9] For a discussion, see *Nonlinear Optics*, N. Bloombergen, Addison-Wesley (1965).

order 10^{-5} than the electric response and one can then neglect local field corrections.[10] Our treatment will be limited to the static ($\omega = 0$) response, however the effects of temperature will be included from the outset.

Using the same prescription as in the previous section, where we replace $\hat{\mathbf{p}}$ by $\hat{\mathbf{p}} - (e/c)\mathbf{A}$, and including the interaction energy of the electron spin with the external magnetic field, $g\mu_B\, \hat{\mathbf{s}}\cdot\mathbf{H}_0$, we can write the Hamiltonian of an N-electron atom or molecule as

$$\hat{H} = \hat{H}_0 + \sum_{a=1}^{N}\left[\frac{e}{2mc}\left(\hat{\mathbf{p}}_a\cdot\mathbf{A}(\mathbf{r}_a) + \mathbf{A}(\mathbf{r}_a)\cdot\hat{\mathbf{p}}_a\right) + \frac{e^2}{2mc^2}\mathbf{A}^2(\mathbf{r}_a) + g\mu_B\hat{\mathbf{s}}_a\cdot\mathbf{H}_0\right], \tag{24.32}$$

where $\mathbf{r}_a$ are the coordinates of all the N electrons measured from some convenient origin (the nucleus for an atom). We assume a constant external field $\mathbf{H}_0$ and, as noted before, neglect any screening of this field by the motions of the electrons themselves. We adopt the London gauge where $\nabla_a\cdot\mathbf{A}(\mathbf{r}_a) = 0$ and write $\mathbf{A}(\mathbf{r}_a) = \mathbf{r}_a \times \mathbf{H}_0/2$. We then have

$$(\mathbf{r}_a \times \mathbf{H}_0)\cdot\hat{\mathbf{p}}_a = -\mathbf{H}_0\cdot\left(\mathbf{r}_a \times \hat{\mathbf{p}}_a\right) = -\hbar\mathbf{H}_0\cdot\hat{\mathbf{l}}_a \tag{24.33}$$

where $\hat{\mathbf{l}}_a$ is the dimensionless angular momentum operator of the individual electrons. We can then write (24.32) as

$$\hat{H} = \hat{H}_0 - \hat{\mu}\cdot\mathbf{H}_0 + \sum_{i=a}^{N}\frac{e^2}{8mc^2}\left(\mathbf{H}_0 \times \mathbf{r}_a\right)^2, \tag{24.34}$$

where we have introduced the magnetic moment operator

$$\begin{aligned}\hat{\mu} &= -|\mu_B|\sum_{a=1}^{N}\left(\hat{\mathbf{l}}_a + 2\hat{\mathbf{s}}_a\right) \\ &= -|\mu_B|\left(\hat{\mathbf{L}} + 2\hat{\mathbf{S}}\right).\end{aligned} \tag{24.35}$$

Here, $\mu_B = e\hbar/2mc$ is the usual Bohr magneton and $\hat{\mathbf{L}}$ and $\hat{\mathbf{S}}$ are the total orbital and spin angular momentum operators respectively.

Assuming the availability of the eigenfunctions $\psi_k^{(0)}(\{\mathbf{r}_a\})$ of $\hat{H}_0$, we may use first- and second-order perturbation theory to calculate the energy shift, $\Delta E_k = E_k - E_k^{(0)}$, in the eigenvalues, E_k of $\hat{H}$, from their zeroth-order values, $E_k^{(0)}$ through second order in the magnetic field $\mathbf{H}_0$; we write the results of such a calculation as

$$\Delta E_k = A_k H_0 + \frac{1}{2}H_0 B_k H_0, \tag{24.36}$$

[10] This statement does not hold for ferromagnets or strongly paramagnetic systems at low temperatures.

where

$$A_k = \left\langle \psi_k^{(0)} \middle| \hat{\mu}_z \middle| \psi_k^{(0)} \right\rangle, \tag{24.37a}$$

with $\hat{\mu}_z$ as the projection of the operator $\hat{\mu}$ onto $\mathbf{H}_0$, and

$$B_k = 2\sum_{k'}{}' \frac{\left|\left\langle \psi_k^{(0)} \middle| \hat{\mu}_z \middle| \psi_{k'}^{(0)} \right\rangle\right|^2}{E_{k'}^{(0)} - E_k^{(0)}} - \frac{e^2}{4mc^2}\left\langle \psi_k^{(0)} \middle| \sum_{i=1}^{N}\left(x_a^2 + y_a^2\right) \middle| \psi_k^{(0)} \right\rangle. \tag{24.37b}$$

To calculate the statistical properties we must evaluate the partition function, Eq. (4.3),

$$Z = \sum_k \exp\left[-E_k/k_BT\right] = \sum_k \exp\left[-\left(E_k^{(0)} + \Delta E_k\right)/k_BT\right]. \tag{24.38}$$

Assuming $\Delta E_k \ll k_BT$ and expanding through second order in the magnetic field, we obtain

$$Z = \sum_k \exp\left[-E_k^{(0)}/k_BT\right]\left[1 + \frac{A_kH_0}{k_BT} + \frac{A_k^2H_0^2}{2k_B^2T^2} + \frac{B_kH_0^2}{2k_BT}\right]. \tag{24.39}$$

The sum over k includes a sum over the quantum numbers M and S_z, associated with the projection of the operator $\hat{\mu}$ onto the external field $\mathbf{H}_0$; this results in a vanishing of the term linear in the magnetic field (24.39); i.e., $\bar{A} = 0$. The average of the remaining terms in (24.39) may be written

$$Z = \left[1 + \frac{H_0^2}{2k_BT}\left(\frac{\overline{A^2}}{k_BT} + \bar{B}\right)\right]\sum_k \exp\left[-\frac{E_k^{(0)}}{k_BT}\right]. \tag{24.40}$$

The partition function is related to the free energy as $Z = \exp(-F(H)/k_BT)$, or to second order in H:

$$\begin{aligned} F &= -k_BT\ln Z = -k_BT\ln Z\,(H_0 = 0) - \frac{H_0^2}{2}\left(\frac{\overline{A^2}}{k_BT} + \bar{B}\right) \\ &= F_0 - \frac{H_0^2}{2}\left(\frac{\overline{A^2}}{k_BT} + \bar{B}\right). \end{aligned} \tag{24.41}$$

If we restrict this sum to the states of a single molecule (assuming our gas consists of a single species), F will be the free energy per molecule. The magnetization is given by $M = -\partial F/\partial H$ and hence the susceptibility per molecule is given by $\partial M/\partial H = -\partial^2 F/\partial H^2$. If we have n molecules per unit volume then we can write the volume susceptibility as

$$\chi = n\left(\frac{\overline{A^2}}{k_BT} + \bar{B}\right); \tag{24.42}$$

this result is due to Van Vleck (1927). We recognize the first term in (24.42), which is positive definite and therefore corresponds to *paramagnetism*, as Curie's law discussed earlier in Sections 5.2 and 5.4. The second term in (24.42) may have either sign depending on the relative magnitude of the two contributions to B_k in (24.37b). If the associated matrix elements do not vanish, the first term in (24.37b) yields a *temperature-independent* paramagnetic (positive) contribution which is referred to as the Van Vleck temperature-independent paramagnetism. The second term, which is also temperature independent, corresponds to the Langevin diamagnetism, discussed in Section 5.1.

For an atom with no spin or orbital angular momentum, as one has with the rare gases, we would have spherical symmetry; in this case A vanishes and B involves only the second term in (24.37b) where, due to the spherical symmetry, we can write $x_a^2 = y_a^2 = r_a^2/3$. We then have

$$\chi = -\frac{ne^2}{6mc^2}\sum_{a=1}^{N}\left\langle r_a^2\right\rangle_{00}, \tag{24.43}$$

where $\left\langle r_a^2\right\rangle_{00}$is the matrix element of the ath electron with respect to the ground state wave function; note that this result is the same as we obtained earlier as Eq. (5.9). This result can also be applied to the contribution from the cores of atoms, even when they have non-zero spin and orbital angular momentum in their outer shells.

In the case of an atom with spin J, we would write $A = g_L \mu_B M_J$, where M_J is the total magnetic quantum number and g_L has the Lande form (given earlier as Eq. (5.20)):

$$g_L = 1 + \frac{J(1+1) - L(L+1) + S(S+1)}{2J(1+1)}. \tag{24.44}$$

Noting that

$$\overline{M_J^2} = \frac{1}{2J+1}\sum_{M_J=-J}^{+J} M_J^2 = \frac{1}{3}J(1+1),$$

we have

$$\chi = \frac{g_L^2 \mu_B^2}{3k_B T}J(J+1), \tag{24.45}$$

given earlier as Eq. (5.24).

When $|\mathbf{L}| = |\mathbf{S}| \neq 0$ and $\mathbf{L} + \mathbf{S} = \mathbf{J} = 0$, the diagonal matrix element of the magnetic moment operator vanishes (i.e., $A = 0$). However the matrix elements connecting the states L, S, J with L, S, J ± 1 within the multiplet do not vanish and involve energies only slightly higher than the ground state. Since the corresponding energy denominators in the first term in (24.37b) will then be small, this term will dominate B; i.e., we will have temperature-independent paramagnetism.

ADDITIONAL READING

The magnetic susceptibility of atoms and molecules is discussed in Section 52 of *Statistical Physics Part I*, L.D. Landau and E.M. Lifshitz, Pergamon Press, Oxford (1980).

Linear and nonlinear electric susceptibilities are discussed in Chapter 2 of *The Principles of Nonlinear Optics*, Y.R. Shen, John Wiley & Sons (1984).

Still valuable is the treatise by J.H. Van Vleck: *The Theory of Electric and Magnetic Susceptibilities*, Oxford University Press (1932).

Piezoelectricity, Pyroelectricity, and Ferroelectricity

25

For isotropic (amorphous) and many crystalline dielectrics the internal stresses are proportional to the square of an applied external electric field, a phenomenon called *electrostriction*, which is discussed in most texts on electricity and magnetism in connection with the Maxwell stress tensor. However for certain crystal symmetries (to be discussed below) strains develop which are *linear* in the applied electric field. Three such material classes can be distinguished: pyroelectrics, ferroelectrics, and piezoelectrics. The first two have the additional property that they involve the presence of a *spontaneous* electric polarization (analogous to the spontaneous magnetization that appears in a ferromagnet), and we begin with a discussion of these materials.

25.1 Ferroelectricity and pyroelectricity

As noted, ferroelectric and pyroelectric materials have a spontaneous electric polarization in the absence of an external electric field; i.e., $\mathbf{P}(\mathbf{E} = 0) \neq 0$. For it to have this property the crystal must possess *no more than a single symmetry axis with the possibility of a mirror plane passing through that axis*; i.e., they have the symmetry C_n (with $n = 1, 2, 3, 4, 6$) with the possible addition of the symmetry σ_v; hence crystals with a center of symmetry cannot be pyroelectric. Of the 32 crystal classes only the point groups, shown in Table 25.1, satisfy the required properties. If a symmetry allowing a spontaneous polarization arises as the result of a thermodynamic phase transition, where the crystal does not have a spontaneous polarization above some transition temperature, we refer to it as a *ferroelectric*; this name arises from the analogy with ferromagnetism which also results from a phase transition. Alternatively, the material may grow (say from a melt) with one of these structures "frozen in" and we then refer to it as a *pyroelectric*. In either case $\mathbf{P}(T)$ generally increases as the temperature is reduced.

A well-known example of a ferroelectric is $BaTiO_3$. Above a transition temperature it has the cubic perovskite structure, which is shown in Fig. 25.1. Below the transition the Ti and O atoms translate parallel to a unit cell edge relative to the Ba atoms; the crystal structure is then lowered to tetragonal.

Table 25.1 *Point groups compatible with pyroelectricity.*

Triclinic: C_1	Tetragonal: C_4, C_{4v}
Monoclinic: C_s, C_2	Rhombohedral: C_3, C_{3v}
Orthorhombic: C_{2v}	Hexagonal: C_6, C_{6v}

The Physics of Solids. First Edition. J. B. Ketterson.

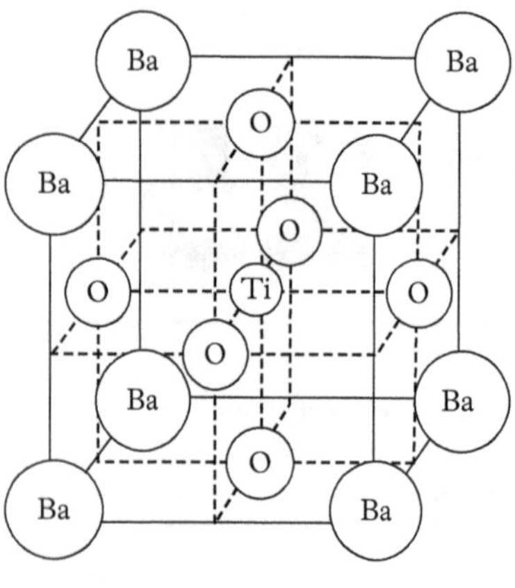

Figure 25.1 *The perovskite structure of* $BaTiO_3$.

25.2 Landau theory of second-order ferroelectric transitions

We now describe a general theory, due to Landau (1937), which can model the appearance of ferroelectricity below some specific temperature, T_c. A similar theory describes ferromagnets, superconductors, superfluids, and many other phase transitions.[1] When a polarization is present the material is said to be *ordered*; the polarization, which is a measure of the degree of ordering, is then interpreted as an *order parameter*, since it is zero above the transition and finite below.[2] If the order parameter grows continuously from zero at T_c the transition is said to be *second order*; if it jumps discontinuously we call it a *first-order* transition. The phase transition leading to ferroelectricity may be either first or second order, but we begin with the second-order case.

The basic idea is to expand the free energy in a power series in the order parameter. If the high temperature phase has inversion symmetry, only terms even in **P** can appear in the expansion: the free energy, being a scalar, cannot change sign on reversing **P**, irrespective of whether we are above or below the transition temperature. The expansion must then have the form

$$F(\mathbf{P},T) = F_0(T) + \alpha(T)P^2 + \beta(T)P^4 + \gamma(T)P^6 + \cdots. \tag{25.1}$$

We initially neglect the sixth-order term. Thermodynamic stability then requires that $\beta > 0$; otherwise the system could "run away" to arbitrarily large polarizations. We "drive" the transition by arranging for α to change sign at the transition temperature, $T_c : \alpha > 0$ for $T > T_c$, and $\alpha < 0$ for $T < T_c$; the simplest form is[3]

$$\alpha = a(T - T_c). \tag{25.2}$$

Inserting this form in Eq. (25.1), approximating β by its value at T_c, and minimizing F with respect to P, we have

$$a\,(T - T_c)\,P + 2\beta(T_c)P^3 = 0. \tag{25.3}$$

This equation has two solutions:

$$P = \begin{cases} 0 & ;T > T_c \\ \left[-\dfrac{a\,(T - T_c)}{2\beta(T_c)}\right]^{\frac{1}{2}} & ;T < T_c \end{cases}. \tag{25.4}$$

The first corresponds to $T > T_c$, where we expect and obtain $P = 0$, and the second to $T < T_c$, where $P \neq 0$. Note the singular square-root-like behavior of P in the low temperature phase. The dependence of the free energy on the order parameter for $T > T_c$, $T = T_c$, and $T < T_c$ is shown in Fig. 25.2.

[1] The associated theories will be discussed in Chapters 27, 33, 36, and 38.

[2] The term originates from ferromagnetism, where the electronic spins order so as to yield a net magnetic moment.

[3] When the effect of fluctuations is taken into account this assumed form breaks down near T_c.

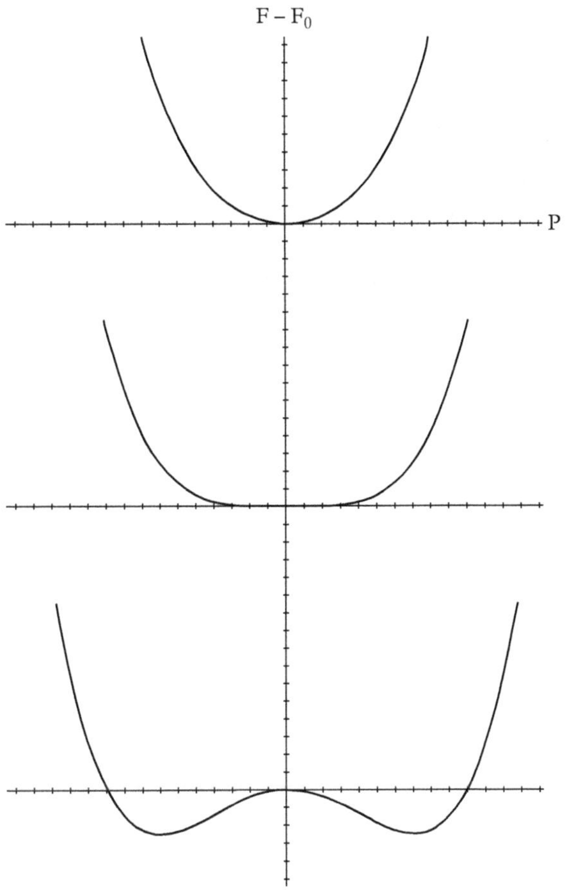

Figure 25.2 *Plots of* F *versus* P *for* $T > T_c$ *(upper),* $T = T_c$*, and* $T < T_c$ *(lower).*

Inserting (25.4) into (25.1) gives

$$F = \begin{cases} F_0(T) & ; T > T_c \\ F_0(T) - \dfrac{a^2(T - T_c)^2}{4\beta(T_c)} & ; T < T_c \end{cases} . \tag{25.5}$$

The heat capacity is given by the usual thermodynamic expression:

$$C = T\frac{\partial S}{\partial T} = -T\frac{\partial^2 F}{\partial T^2}, \tag{25.6}$$

which gives

$$C(T) = \begin{cases} C_0(T) & ; T > T_c \\ C_0(T) + T\dfrac{a^2}{2\beta} & ; T < T_c \end{cases} ; \tag{25.7}$$

in particular, the heat capacity undergoes a *discontinuity* at the transition temperature and is higher by an amount $Ta^2/2\beta$ in the ordered (low temperature) phase.

25.3 The discontinuous ferroelectric transition: the first-order transition

Transitions that otherwise satisfy the criteria for being second order do not always behave in that manner, and here we discuss a modification of the theory in which it occurs discontinuously. As a function of some second variable, which for F we can take to be the volume V, the parameter $\beta(T,V)$ may evolve so as to pass from positive to negative values, which would render the behavior unstable if the expansion (25.1) were terminated with the quartic term. It is then necessary to retain the term proportional to γ in the expansion and require it to be positive. Four distinct situations can occur, which are depicted in Fig. 25.3.

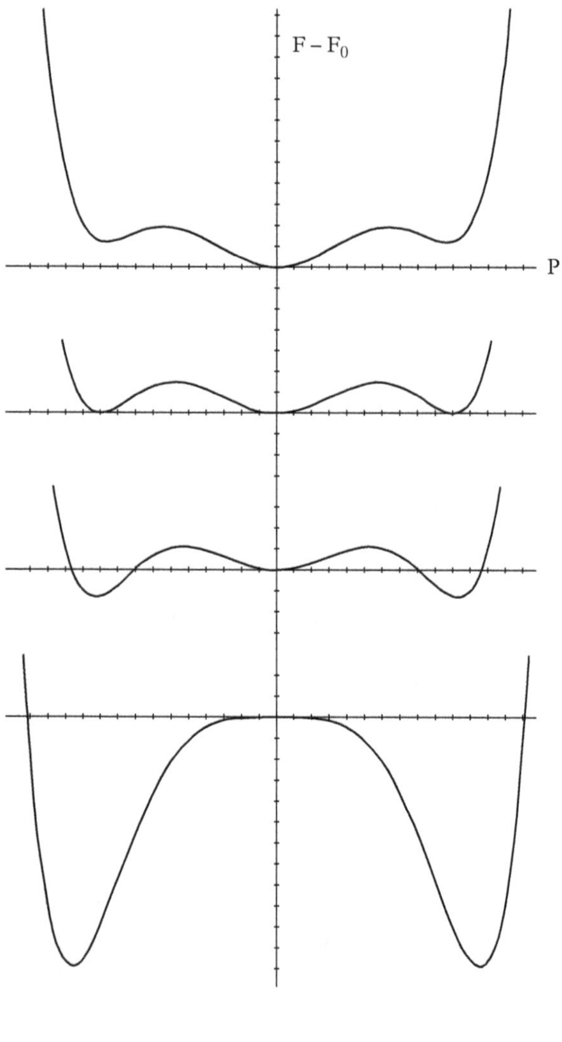

Figure 25.3 *The behavior of F(P,T,V) for a case where only first-order transitions can occur, in which $\beta < 0$ and $\gamma > 0$ and α decreases toward 0. The upper trace shows a temperature above the first-order transition where a metastable order exists. Below this is shown the temperature at which the energy of the ordered phase is degenerate with the unordered phase. Second from the bottom is a case where the ordered phase is more stable but the unordered phase may exist as a metastable phase. Finally at the bottom, we have a case where α passes through zero and only the ordered phase is stable.*

To analyze the behavior we again set the derivative of (25.1) with respect to P equal to zero, obtaining

$$\alpha P + 2\beta P^3 + 3\gamma P^5 = 0; \tag{25.8}$$

in place of (25.4) we now have the solutions

$$P^2 = \begin{cases} 0 \\ \dfrac{-\beta \pm \sqrt{\beta^2 - 3\alpha\gamma}}{3\gamma} \end{cases} . \tag{25.9}$$

As previously noted, we require $\gamma > 0$ for stability and note that the positive sign in front of the radical corresponds to the minimum in F. We continue to allow α to have either sign via (25.2). Real non-vanishing P requires both $\beta^2 > 3\alpha\gamma$ and $-\beta + \sqrt{\beta^2 - 3\alpha\gamma} > 0$. For positive β a non-vanishing P exists only for negative α, in agreement with our findings for the second-order transition (i.e., the transition temperature is not affected). For negative β, we have a non-vanishing P for all $\alpha < 0$, while for $\alpha > 0$ a minimum first appears at $\alpha = \beta^2/3\gamma$.

The free energy follows from inserting (25.9) into (25.1), and after some algebra we obtain

$$F = \begin{cases} F_0 ; \\ F_0 + \left[2\beta^3 - 9\alpha\beta\gamma + \sqrt{\beta^2 - 3\alpha\gamma}\left(6\alpha\gamma - 2\beta^2\right)\right]/27\gamma^2 \end{cases} . \tag{25.10}$$

When the quantity in square brackets vanishes in (25.10) we have a special case, where the phase with $P = 0$ has the same free energy as the ordered phase with $P \neq 0$. The system then has two phases that can coexist at the same temperature *in equilibrium*, which is the signature of a first-order transition (see Fig. 25.3). After some additional agebra this leads to the condition

$$\alpha = \frac{\beta^2}{4\gamma} \tag{25.11a}$$

or

$$T_{c1} = T_{c2} + \frac{\beta^2}{4a\gamma}; \tag{25.11b}$$

i.e., the system can undergo a first-order transition on cooling to T_{c1} *before* getting to the lower temperature T_{c2}, where the minimum at $P = 0$ disappears (and a second-order transition would occur were β to be positive). For temperatures satisfying

$$0 \leq \alpha \leq \beta^2/3\gamma, \tag{25.12}$$

either phase may exist, one as a "globally" stable phase and the other, having higher free energy, as a so-called *metastable phase.* For

$$0 \le \alpha \le \frac{\beta^2}{4\gamma} \tag{25.13}$$

we have $F(P \neq 0) < F(P = 0)$, while for

$$\frac{\beta^2}{4\gamma} \le \alpha \le \frac{\beta^2}{3\gamma}, \tag{25.14}$$

$F(P \neq 0) > F(P = 0)$. Finally, for $\alpha < 0$ the unpolarized state is absolutely unstable, as previously noted. Note that when the system is in a metastable phase and a small region of the stable phase appears, called a nucleus, the surrounding media will grow onto, and adopt the phase of, the nucleus, exchanging latent heat in the process.[4]

The entropy is given by

$$S(T) = -\left(\frac{\partial F(T,P)}{\partial T}\right)_P, \tag{25.15}$$

which, from (25.1) and (25.2), is

$$S = S_0 - aP^2, \tag{25.16a}$$

where

$$S_0(T) = -\left(\frac{\partial F_0(T,P)}{\partial T}\right)_P. \tag{25.16b}$$

At the first-order transition, T_{c1} is given by (25.11b), where P changes by an amount $P^2 = -\beta/2\gamma$ (recall β is negative), the entropy would therefore undergo a discontinuous change and the associated *latent heat* Q would be given by

$$Q = -aT_{c1}\beta/2\gamma. \tag{25.17}$$

Again, discontinuous transitions of the type we are discussing, where a latent heat is involved, are called first-order transitions.

[4] Note that the nucleus must be of sufficient size to overcome the effects of the surface tension that is always present between two phases in contact. As an example, when water is above the boiling temperature and a bubble appears, the hydrostatic pressure within the bubble must be high enough to overcome the surface pressure tending to collapse the bubble.

25.4 The effect of an applied electric field on a ferroelectric; the susceptibility

We next describe the effect of an electric field, and its response via the dielectric constant near the ferroelectric transition. To do this we must extend the functional dependence of F to include the electric and displacement fields. The basic thermodynamic identity is then given by

$$dF = TdS + \mathbf{E} \cdot d\mathbf{D}/4\pi. \tag{25.18}$$

In the same way, one performs a Legendre transformation to change the independent variable from volume V to the pressure P, in defining the Gibbs from the Helmholtz free energy of a P,V,T thermodynamic system, we define a free energy:

$$\tilde{F} = F - \mathbf{E} \cdot \mathbf{D}/4\pi, \tag{25.19}$$

for which the thermodynamic identity is

$$d\tilde{F} = TdS - \mathbf{D} \cdot d\mathbf{E}/4\pi; \tag{25.20a}$$

the displacement field then follows from

$$\frac{\partial \tilde{F}}{\partial \mathbf{E}} = -\mathbf{D}/4\pi = -\mathbf{P} - \mathbf{E}/4\pi. \tag{25.20b}$$

Integrating (25.20b) at *fixed polarization* and including the terms from (25.1) gives

$$\tilde{F} = F_0(T) + \alpha(T)P^2 + \beta(T)P^4 + \gamma(T)P^6 + \cdots - PE - E^2/8\pi, \tag{25.21}$$

where we confine ourselves to the case where **E** is parallel to **P**. Restricting to the case $\beta > 0$ (where we can set $\gamma = 0$) and again minimizing with respect to P, we obtain

$$2a(T - T_c)P + 4\beta P^3 = E. \tag{25.22}$$

This cubic form (for the case, $2\alpha = 4\beta = 1$) is plotted in Fig. 25.4.

The electric susceptibility is given by $\chi = P/E$ and for small fields and temperatures above T_c, we have

$$\chi = \frac{1}{2a(T - T_c)}, T > T_c. \tag{25.23}$$

Below T_c only the *differential susceptibility*, $\chi = \partial P/\partial E$, is meaningful; evaluating this derivative from Eq. (25.22) and then substituting (25.4), we obtain

$$\chi = \frac{1}{4a(T_c - T)}, T < T_c; \tag{25.24}$$

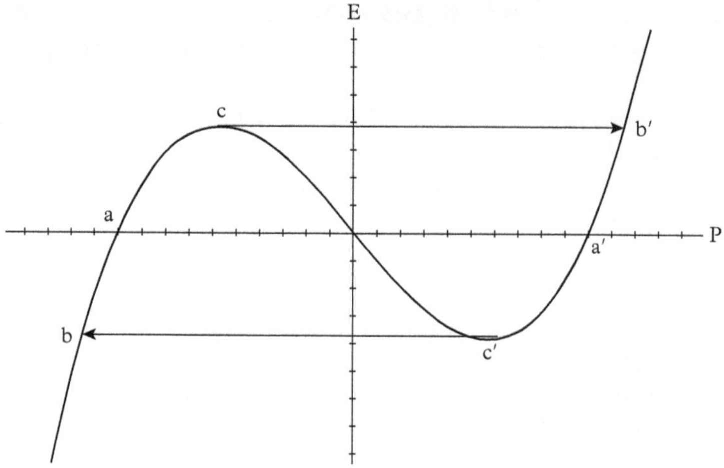

Figure 25.4 *A plot of the form,* $E = -P + P^3$. *The region between c and c′ is unstable.*

Note that this is half the value obtained for $T > T_C$. Importantly, we see that χ diverges as $(T_C - T)^{-1}$ on both sides of the transition.[5]

The large dielectric constants accompanying ferroelectricity can be exploited to make capacitors with a large capacitance. This same property allows a greater amount of charge to be stored in a smaller volume and hence arrays of ferroelectric elements have been utilized as information storage media. However, near a second-order phase transition the response slows down and an electrical phase shift develops between the applied oscillatory electric field, **E**, and the material's response, **P**. As a result dissipation occurs; hence ferroelectric-based capacitors are not generally used for high frequency applications.

25.5 Piezoelectricity

Ferroelectrics and pyroelectrics have the property that an external electric field induces a stress. This property is called piezoelectricity. There is a third class of materials that has this property, but does not have a polarization in the absence of a stress; these materials, strictly speaking, are the piezoelectrics. All three of these materials classes possess an accompanying *inverse* response: an applied stress produces an electric field. Note that the application of a stress can formally alter the equilibrium crystal structure of a piezoelectric so that it corresponds to that of a pyroelectric.

These properties are of great practical importance. Using the first property one can generate mechanical oscillations, sound waves, by applying an oscillatory electric field to a piezoelectric. The piezoelectric material may, in turn, be bonded to a non-piezoelectric material into which the sound can propagate; it is then referred to as a *transducer*. By measuring the propagation velocities of shear-like

[5] When the effects of fluctuations are included, rather than having $\chi \propto (T_C - T)^{-1}$, the behavior approaches the form $(T_C-T)^{-\gamma}$ as $T \to T_C$, where γ differs from one.

and longitudinal-like waves in various directions one can go on to determine the elements of the elastic tensor c_{ijkl}, as discussed in Chapter 1. The piezoelectric property persists, at some level, up to the highest sound frequencies that solids can support, which lie in the infrared. Using the inverse effect one can convert mechanical oscillations into an oscillating electric voltage that can then be amplified and rectified; i.e., we have a sound detector or receiver. Ferroelectrics and *pyroelectrics* usually have larger piezoelectric coefficients; however they generally have larger losses than the non-ferroelectric piezoelectrics and are therefore less useful for high frequency applications.

Of the 32 crystal classes only 20 allow piezoelectricity. Ten of these are also pyroelectric and were given earlier in Table 25.1. The ten additional classes making up the piezoelectrics are given in Table 25.2.

Table 25.2 *Additional point groups compatible with piezoelectricity.*

Rhombic: $\mathbf{D_2}$	Hexagonal: $\mathbf{D_6}$, $\mathbf{C_{3h}}$, $\mathbf{D_{3h}}$
Tetragonal: $\mathbf{D_4}$, $\mathbf{D_{2d}}$, $\mathbf{S_4}$	Cubic: $\mathbf{T}$, $\mathbf{T_d}$
Rhombohedral: $\mathbf{D_3}$	

We now discuss the thermodynamics associated with piezoelectricity. We write the combined first and second laws of thermodynamics as

$$dE = TdS - \sum_{ij} \sigma_{ij} du_{ij} - \frac{1}{4\pi}\mathbf{D}\cdot d\mathbf{E}, \tag{25.25}$$

where E is the energy per unit volume, S is the entropy per unit volume, σ_{ij} and u_{ij} are the stress and strain tensors (see Chapter 1),[6] and **D** and **E** are the displacement and electric fields. The corresponding free energy per unit volume, $F = E - TS$, is then

$$dF(T, u_{ij}, \mathbf{E}) = -SdT - \sum_{ij} \sigma_{ij} du_{ij} - \frac{1}{4\pi}\mathbf{D}\cdot d\mathbf{E}. \tag{25.26}$$

The strain and displacement fields are then given by

$$\sigma_{ij}(T, \mathbf{E}) = -\left(\frac{\partial F}{\partial u_{ij}}\right)_{T,E}; \quad \mathbf{D} = -\left(\frac{\partial F}{d\mathbf{E}}\right)_{T,u_{ij}}. \tag{25.27a,b}$$

We can also define a Gibbs free energy per unit volume G by making a Legendre transformation using the form $-u_{ij}\sigma_{ij}$, obtaining

$$dG(T, \sigma_{ij}, \mathbf{E}) = -SdT + \sum_{ij} u_{ij} d\sigma_{ij} - \frac{1}{4\pi}\mathbf{D}\cdot d\mathbf{E}, \tag{25.28}$$

which gives

$$u_{ij}(T, \mathbf{E}) = \left(\frac{\partial G}{\partial \sigma_{ij}}\right)_{T,E}; \quad \mathbf{D} = -4\pi\left(\frac{\partial G}{d\mathbf{E}}\right)_{T,\sigma_{ij}}. \tag{25.29a,b}$$

[6] Note that in this section we use u_{ij} rather than ε_{ij} for the strain tensor, since the latter will here denote the dielectric constant tensor.

For a deformable dielectric crystal that supports piezoelectricity the strain is given in leading order by

$$u_{ij}(T,\sigma_{ij},\mathbf{E}) = u_{ij}(T,0,0) + \sum_{kl}\left(\frac{\partial u_{ij}}{\partial \sigma_{kl}}\right)_{T,\mathbf{E}}\sigma_{kl} + \frac{1}{4\pi}\sum_{k}\left(\frac{\partial u_{ij}}{\partial E_k}\right)_{T,\sigma_{ij}} E_k$$
$$= \sum_{kl} s_{ijkl}\sigma_{kl} + \sum_{k} d_{k,ij}E_k, \tag{25.30}$$

where in the second step we have assumed no intrinsic strain, and introduce the fourth-rank elastic compliance tensor, $s_{ijkl} = (\partial u_{ij}/\partial\sigma_{kl})_{T,E}$(see Eq. (1.4a)), and the third-rank piezoelectric constant tensor, $d_{k,ij} \equiv 4\pi\,(\partial u_{ij}/\partial E_k)_{T,\sigma_{ij}}$, where the comma is inserted to separate indices referring to σ_{ij} and E_k, which are different thermodynamic variables. Note that since u_{ij} is a symmetric tensor we must have

$$d_{k,ij} = d_{k,ji}. \tag{25.31}$$

Similarly, to leading order the displacement field is given by

$$D_i(T,u_{ij},\mathbf{E}) = D_i(T,0,0) + \sum_k \varepsilon_{ik}E_k + 4\pi\sum_{jk} d_{i,jk}\sigma_{jk}, \tag{25.32}$$

where the first term, $D_i(T,0,0) = 4\pi P_i(T,0,0)$, is non-zero only for pyroelectric materials; in the remaining two terms we have introduced the second-rank dielectric constant, $\varepsilon_{ij} = (\partial D_i/\partial E_k)_{T,u_{ij}}$, and made use of the fact that $d_{k,ij}$ is a cross derivative of G with respect to E_k and u_{ij}, which is independent of the order of differentiation.

In static equilibrium one has the electrostatic conditions:

$$\nabla\cdot\mathbf{D} = 0, \quad \nabla\times\mathbf{E} = 0, \tag{25.33a,b}$$

while the condition for mechanical equilibrium is given by

$$\sum_j \frac{\partial\sigma_{ij}}{\partial x_j} = 0; \tag{25.34}$$

these equations would have to be solved simultaneously along with applicable electrostatic and mechanical boundary conditions, leading to much complexity in all but the simplest cases.

For completeness, we will list the symmetry classes for which a material can be piezoelectric and the associated non-zero components of the matrix $d_{k,ij}$. In the most general (monoclinic) case the symmetry condition (25.31) reduces the 27 elements to 18. The non-zero elements of the tensor $d_{k,ij}$ are those that remain unchanged under all symmetry operations of the crystal. We list those for the non-pyroelectric piezoelectrics in Table 25.3 and for the pyroelectrics in Table 25.4.[7]

[7] The z axis is chosen as the second-, third-, fourth-, or sixth-order symmetry axis; for $\mathbf{C_s}$ it is the axis perpendicular to the mirror plane. The x and y axes are chosen to make the maximum number of coefficients vanish. When σ_v is present the x axis lies in the mirror plane.

Table 25.3 *Non-vanishing components of the tensor* $\mathbf{d}_{i,jk}$ *for non-pyroelectric piezoelectrics.*

D_2	$d_{x,yz}, d_{z,xy}, d_{y,zx}$
D_{2d}	$d_{z,xy}, d_{x,yz} = d_{y,zx}$
T	$d_{x,yz} = d_{z,xy} = d_{y,xz}$
T_d	$d_{x,yz} = d_{z,xy} = d_{y,xz}$
D_4	$d_{x,yz} = -d_{y,xz}$
D_6	$d_{x,yz} = -d_{y,xz}$
S_4	$d_{z,xy}, d_{x,yz} = d_{y,zx}, d_{z,xx} = -d_{z,yy}, d_{x,zx} = -d_{y,zy}$
D_3	$d_{y,zx} = -d_{x,yz}, d_{y,xy} = -d_{x,xx} = d_{x,yy}$
D_{3h}	$d_{y,xy} = -d_{x,xx} = d_{x,yy}$
C_{3h}	single parameter with the right choice of coordinates

Table 25.4 *Non-vanishing components of the tensor* $\mathbf{d}_{i,jk}$ *for pyroelectrics.*

C_1	all $d_{i,jk} \neq 0$
C_s	all $d_{i,jk} \neq 0$ for which the suffix z occurs twice or is absent
C_2	$d_{z,xx}, d_{z,yy}, d_{z,zz}, d_{x,xz}, d_{y,yz}, d_{x,yz}, d_{y,xz}, d_{z,yx}$
C_{2v}	$d_{z,xx}, d_{z,yy}, d_{z,zz}, d_{x,xz}, d_{y,yz}$
C_3	$d_{z,zz}, d_{x,xz} = d_{y,yz}, d_{x,xx} = -d_{x,yy} = -d_{y,xy}, d_{z,xx} = d_{z,yy}, d_{x,yz} = -d_{y,xz}, d_{y,xx} = -d_{y,yy} = d_{x,xy}$
C_{3v}	$d_{z,zz}, d_{x,xz} = d_{y,yz}, d_{x,xx} = -d_{x,yy} = -d_{y,xy}, d_{z,xx} = d_{z,yy}$
C_4	$d_{z,xx} = d_{z,yy}, d_{z,zz}, d_{x,xz} = d_{y,yz}, d_{x,yz} = -d_{y,xz}$
C_{4v}	$d_{z,xx} = d_{z,yy}, d_{z,zz}, d_{x,xz} = d_{y,yz}$
C_6	$d_{z,zz}, d_{x,xz} = d_{y,yz}, d_{z,xx} = d_{z,yy}, d_{x,yz} = -d_{y,zx}$
C_{6v}	$d_{z,zz}, d_{x,xz} = d_{y,yz}, d_{z,xx} = d_{z,yy}$

ADDITIONAL READING

Pyroelectricity and ferroelectricity are elegantly discussed in Sections 17 and 19 of *Electrodynamics of Continuous Media*, L.D. Landau, E.M. Lifshitz, and L.P. Pitaevskii, Pergamon Press, Oxford (1984).

The mean field theory of second-order phase transitions, as formulated by its originator, is treated in detail in Chapter XIV of *Statistical Physics Part I*, L.D. Landau and E.M. Lifshitz, Pergamon Press, Oxford (1980).

The properties of tensors as they apply to piezoelectrics and other symmetry related phenomena in solids are discussed in *Physical Properties of Crystals: Their Representation by Tensors and Matrices*, J.F. Nye, Oxford Science Publishers (1985).

Piezoelectric Crystals and Their Application to Ultrasonics, W.P. Mason, Van Nostrand, NY (1950).

Ferroelectric Phenomena in Crystals: Physical Foundations, Boris A. Strukov and Arkadi P. Levanyuk, Springer (2011).

Problem

25.1. Verify Eq.'s (25.10) and (25.11a) as well as the stability regimes defined by Eq.'s (25.12), (25.13) and (25.14).

Appendix 25A Piezoelectric transducers

An important application of piezoelectricity is in the generation and detection of sound waves, as discussed earlier in Section 1.10. Applying an electrical voltage to a piezoelectric transducer generates a sound wave; in the inverse effect a sound wave produces an electrical voltage. Figure 25A.1 shows, schematically, a piezoelectric slab that is clamped (cemented) to a second material of infinite extent into which is launched a propagating plane wave. A uniform external sinusoidal voltage, $V_g(t) = V_g e^{-i\omega t}$, produced by an external generator, is applied to the two electrodes that are plated onto the slab.[8] We consider a situation where the electric

[8] By assuming a uniform electric field we are ignoring the effects of electromagnetic wave propagation, which is valid in the present case since the electromagnetic wavelength is typically 10^5 times larger.

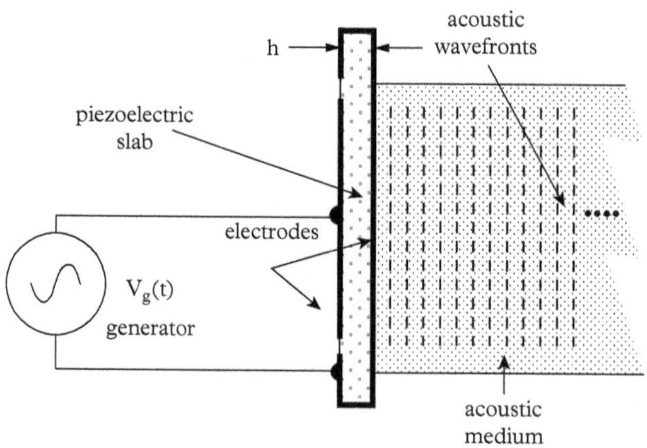

Figure 25A.1 *A piezoelectric resonator loaded by a semi-infinite acoustic medium in which a longitudinal plane wave is launched. Note the electrode at the interface wraps around the perimeter of the piezoelectric platelet so as to allow electrical contact from the free side, a scheme referred to as coaxial plating.*

field acts along a crystal axis such that we only generate a longitudinal wave propagating perpendicular to the slab, which we take as the z axis; the stress and strain tensors (see Chapter 1) will each involve a single element, $\sigma_{33}(z,t) \equiv \sigma(z,t)$ and $u_{33}(z,t) = \partial u(z,t)/\partial z$. We write the three relevant material constants as $s_{3333} \equiv s$, $\varepsilon_{33} \equiv \varepsilon$, and $d_{3,33} \equiv d$.

With these restrictions, the piezoelectric equations, (25.30) and (25.32), take the form

$$D(z,t) = \varepsilon E(z,t) + 4\pi d\sigma(z,t) \tag{25A.1}$$

and

$$\frac{\partial u(z,t)}{\partial z} = s\sigma(z,t) + dE(z,t). \tag{25A.2}$$

For later use, we combine these equations to express D(z,t) as a function of $\partial u(z,t)/\partial z$ and $\sigma(z,t)$,

$$D(z,t) = \varepsilon d^{-1}\frac{\partial u(z,t)}{\partial z} + \left(4\pi d - \varepsilon s d^{-1}\right)\sigma(z,t), \tag{25A.3}$$

and as a function of $\partial u(z,t)/\partial z$ and $E(z,t)$,

$$D(z,t) = \frac{4\pi d}{s}\frac{\partial u(z,t)}{\partial z} + \left(\varepsilon - \frac{4\pi d^2}{s}\right)E(z,t). \tag{25A.4}$$

We start our analysis by noting there is no free charge within the transducer (the only free charge being on the electrode plates). Therefore we can set $\nabla \cdot \mathbf{D} = 0$, implying $D = D(t)$. Equation (25A.3) then gives

$$\frac{\partial^2 u(z,t)}{\partial z^2} = \left(s - \frac{4\pi d^2}{\varepsilon}\right)\frac{\partial \sigma(z,t)}{\partial z}. \tag{25A.5}$$

Noting that $\partial\sigma(z,t)/\partial z$ is the volume force, which from Newton's law (see Eq. (1.27)) is equal to $\rho\partial^2 u(z,t)/\partial t^2$, we obtain the wave equation

$$\frac{\partial^2 u(z,t)}{\partial z^2} - \frac{1}{v_l^2}\frac{\partial^2 u(z,t)}{\partial t^2} = 0, \tag{25A.6}$$

where

$$v_l = \left[\rho\left(s - \frac{4\pi d^2}{\varepsilon}\right)\right]^{-\frac{1}{2}} \cong (\rho s)^{-1/2} \tag{25A.7}$$

is the effective velocity of longitudinal sound.

Since we are restricting to a single frequency associated with the external generator, we write $u(z,t) = e^{-i\omega t}u(z)$. In the piezoelectric material we write

$$u(z) = (u_+ e^{+ikz} + u_- e^{-ikz}), \tag{25A.8a}$$

where $k = \omega/v_l$. In the material loading the piezoelectric we have

$$u'(z) = u' e^{+ik'z}, \tag{25A.8b}$$

where all primed quantities henceforth apply to this media. The associated strains are

$$\frac{du(z)}{dz} = ik(u_+ e^{+ikz} - u_- e^{-ikz}) \tag{25A.9a}$$

and

$$\frac{du'(z)}{dz} = ik' u' e^{+ik'z}, \tag{25A.9b}$$

with $k' = \omega/v'_l$.

We locate the interface between the transducer and the load at $z = 0$ and the free surface at $z = -h$. Continuity of the displacements at $z = 0$ gives

$$u_+ + u_- = u'. \tag{25A.10}$$

The stress must vanish at the free surface, located at $z = -h$, and be continuous at $z = 0$. Because E is position dependent, whereas D is constant, we apply these constraints using Eq. (25A.3); in the approximation $4\pi d^2/s\varepsilon \ll 1$ we have used, we can rewrite this equation as

$$\sigma(z) = s^{-1}\frac{\partial u(z)}{\partial z} - ds^{-1}\frac{D}{\varepsilon}. \tag{25A.11}$$

We then have the following two conditions:

$$iks^{-1}(u_+ e^{-ikh} - u_- e^{+ikh}) - ds^{-1}\frac{D}{\varepsilon} = 0 \tag{25A.12}$$

and

$$iks^{-1}(u_+ - u_-) - \frac{ds^{-1}}{\varepsilon}D = ik's'^{-1}u'. \tag{25A.13}$$

Solving (25A.10), (25A.12), and (25A.13), we obtain

$$u_- = \frac{Z - Z' - Ze^{-ikh}}{i\omega Z\left[(Z+Z')e^{-ikh} - (Z-Z')e^{+ikh}\right]}\frac{ds^{-1}}{\varepsilon}D, \tag{25A.14a}$$

$$u_+ = \frac{Z + Z' - Ze^{+ikh}}{i\omega Z\left[(Z+Z')e^{-ikh} - (Z-Z')e^{+ikh}\right]}\frac{ds^{-1}}{\varepsilon}D, \tag{25A.14b}$$

and

$$u' = \frac{\left(2 - e^{-ikh} - e^{+ikh}\right)}{i\omega\left[(Z+Z')e^{-ikh} - (Z-Z')e^{+ikh}\right]} \frac{ds^{-1}}{\varepsilon} D, \qquad (25A.14c)$$

where we write $k\,s^{-1} = \omega\,\rho\,v_l = \omega\,Z$; Z is called the *acoustic impedance*.

Now the charge per unit area on the electrodes is $D/4\pi$, for a total electrode charge of $Q = AD/4\pi$, where A is the active area of the electrode; the current is then $I = -i\omega Q$. On the other hand, the voltage is given by $V = \int_{-h}^{0} E(z)dz$, which from Eq. (25A.4) takes the form

$$\begin{aligned} V &= \int_{-h}^{0} dz \left[\frac{D}{\varepsilon} - \frac{4\pi d}{\varepsilon s}\frac{\partial u(z)}{\partial z}\right] \\ &= \frac{h}{\varepsilon}D - \frac{4\pi d}{\varepsilon s}\left(u(0) - u(-h)\right). \end{aligned} \qquad (25A.15)$$

From Eq. (25A.8a) and Eqs. (25A.14a,b) we can write (25A.15) as

$$\begin{aligned} V &= \frac{D}{\varepsilon}h - \frac{4\pi d}{\varepsilon s}\left(u_+(1-e^{-ikh}) + u_-(1-e^{+ikh})\right) \\ &= \frac{h}{\varepsilon}D - \frac{4\pi d^2}{s^2\varepsilon^2}\frac{4Z - 2Z(e^{+ikh}+e^{-ikh}) + Z'(e^{+ikh}-e^{-ikh})}{i\omega Z\left[-(Z-Z')e^{+ikh} + (Z+Z')e^{-ikh}\right]}D. \end{aligned} \qquad (25A.16)$$

A special case is the unloaded transducer, corresponding to $Z' = 0$, for which we find

$$\begin{aligned} V &= \frac{h}{\varepsilon}D + \frac{4\pi d^2}{s^2\varepsilon^2}\frac{4 - 2(e^{+ikh}+e^{-ikh})}{i\omega Z\left[e^{+ikh} - e^{-ikh}\right]}D \\ &= \frac{h}{\varepsilon}D - \frac{8\pi d^2}{\omega\, s^2\varepsilon^2 Z}\frac{1-\cos kh}{\sin kh}D \\ &= \frac{h}{\varepsilon}D - \frac{8\pi d^2}{\omega\, s^2\varepsilon^2 Z}\tan\frac{kh}{2}D \end{aligned} \qquad (25A.17)$$

From this discussion the electrical impedance, $Z_E \equiv V/I$, is then given by

$$Z_E = \frac{V}{I} - \frac{V}{i\omega Q} = -\frac{4\pi}{i\omega A}\frac{V}{D}, \qquad (25A.18)$$

which on inserting (25A.16) gives

$$Z_E = \frac{4\pi}{i\omega A}\left\{\frac{h}{\varepsilon} - \frac{4\pi d^2}{s^2\varepsilon^2}\frac{4Z - 2Z(e^{+ikh}+e^{-ikh}) + Z'(e^{+ikh}-e^{-ikh})}{i\omega Z\left[-(Z-Z')e^{+ikh} + (Z+Z')e^{-ikh}\right]}\right\}. \qquad (25A.19)$$

For the unloaded case we would have

$$Z_E = -\frac{4\pi}{i\omega A}\left\{\frac{h}{\varepsilon} - \frac{8\pi d^2}{\omega\, s^2\varepsilon^2 Z}\tan\frac{kh}{2}\right\}. \qquad (25A.20)$$

Note that the impedance, which is purely reactive for the acoustically unloaded case, becomes infinite when $kh/2 = (2n + 1)(\pi/2)$, or equivalently when $h = (2n + 1)(\lambda/2)$, where $\lambda = v_l/f$, $f = \omega/(2\pi)$ and v_l is the longitudinal sound velocity. Acoustic losses within the piezoelectric material, or coupling to the environment, limit the response and result in a resistive component to the impedance. This resonant behavior is used to control the frequency of electrical oscillators. By choosing crystallographic orientations of a material like quartz such that the temperature dependence of the resonant frequency vanishes, excellent stability can be achieved. For many years such oscillators served as frequency standards.

Problems

25.1. Show that the electrical impedance of an unloaded transducer can vanish as well as diverge and evaluate the frequencies at which this occurs. Construct an equivalent circuit composed of capacitors and inductors that will represent the electrical impedance of a transducer in the vicinity of its fundamental resonance and assign values to the components.

25.2. Obtain the expression for the electrical impedance, Z_E, of a piezoelectric transducer when it is loaded on *both* sides with material having the same acoustic impedance Z'.

Part IX

Magnetism

Ferromagnetism and Antiferromagnetism

26

26.1 A survey of magnetic properties

The vast majority of materials are either paramagnetic or diamagnetic (with susceptibilities greater than or less than zero respectively; see Section 24.3), and have no magnetic moment in the absence of an external magnetic field. However, a small set of materials can have a spontaneous magnetic structure (i.e., in the absence of an external magnetic field). The two most well-known magnetic states pertain to *ferromagnets* and *antiferromagnets*.[1] The first has been known since antiquity and is the term loosely used for materials which display a macroscopic magnetic moment. It derives its name from the fact that the early-known materials contained iron (which in Latin is ferrum). The second term designates those materials which have no macroscopic magnetic moment, but which consist of two or more crystallographic *sub-lattices*, each of which has an average moment, but which when taken together cancel each other. These are but two of a number of magnetic states. The magnetic structure disappears above some well-defined *critical temperature* (for a pure compound) which is designated as either the *Curie temperature*, for a ferromagnet, or the *Néel temperature*, for an antiferromagnet; both temperatures will be designated as T_C. The transition is an example of a thermodynamic phase transition, with the lower temperature (magnetic) phase being more ordered, in that the lattice sites possess magnetic moments that either disappear or randomize for temperatures above T_C.

Spontaneous magnetic states arise from the spin (and accompanying magnetic moment) degrees of freedom of the atoms making up a solid. The closed inner shells of atoms contain equal numbers of spin up and spin down electrons and hence cannot result in a spontaneous magnetic moment. If the electrons of the outer (valence) shells fully participate in forming chemical bonds with neighboring atoms, we expect no magnetism from these electrons either, as we now argue. In the older empirical (Lewis) view, a chemical bond arises from the sharing of electrons between a pair of neighboring atoms, a single bond involving one electron from each atom. In the quantum mechanical Heitler–London theory of the covalent bond (as applied to, say, the hydrogen molecule), the spins of the two atoms in the bound (singlet) state are oppositely directed (the interaction in the triplet state is generally repulsive). The Pauli principle then allows the electrons

[1] The 1970 Nobel prize was awarded to Louis Eugène Félix Néel "for fundamental work and discoveries concerning antiferromagnetism and ferrimagnetism which have led to important applications in solid state physics."

to come more closely together in the region between the atoms, which in turn screens the repulsive nuclear charges, thereby allowing the nuclei to more closely approach each other. One then says that the exchange interaction plays a key role in chemical bond formation. The electrons participating in covalent bond formation, being paired (one with spin up the other with spin down) contribute no net magnetic moment. Nor would a moment arise with ionic bonding: the valence electrons leave the cations and enter the outer shell of the anions resulting in filled, rare-gas-like shells for both atoms, which have no net magnetic moment.

Magnetism generally occurs in materials containing atoms with unfilled d and f shells (primarily the 3d transition series and 4f rare earth atoms). These atoms may be thought of as being "magnetic" due to Hund's rule, which states that an atom will have the highest possible spin angular momentum, S, that still satisfies the Pauli principle.[2] Specifically, as the 3d shell fills with increasing atomic number, the spins of the electrons will all be aligned up to the point where the d shell is half full (5 electrons), where the total spin would be 5/2, and we would have a spin magnetic moment of $5\mu_B$;[3] since the spin up d states are now all full, any additional electrons must go into spin down states, which reduces the total spin; a full d shell (10 electrons) would then have no moment.[4]

To understand Hund's rule qualitatively, consider the case of two electrons in an orbital with a given l and n where the total spin can be either 0 or 1. For a symmetric S = 1 state it follows from the Pauli principle that the orbital part of the wave function would be antisymmetric; since the spatial wave function $\varphi(\mathbf{r}_1, \mathbf{r}_2)$ then vanishes for $\mathbf{r}_1 = \mathbf{r}_2$, and is small in the immediate vicinity, this implies that the electrostatic Coulomb energy,

$$U_{\text{Coulomb}} = e^2 \int d^3r_1 d^3r_2 \, |\varphi(\mathbf{r}_1, \mathbf{r}_2)|^2 \, / \, |\mathbf{r}_1 - \mathbf{r}_2| \, ,$$

will be minimal, thereby favoring the high spin state. By extension when there are more than two electrons one would expect the most antisymmetrical orbital state to have the lowest energy, and that is typically the one with the highest spin.

If, after forming the chemical bonds stabilizing the solid, typically involving s and p electrons (with an admixture of some of the d states in the transition metals), there are unfilled d or f shell orbitals lying largely inside the valence bond orbitals, the atoms can carry over a magnetic moment into the solid phase. If, in addition, there exists sufficient *coupling* between the moments on *neighboring* magnetic atoms through the valence electrons to overcome the thermal tendency to disorder them, then the solid can be magnetic. Such arguments serve only to rationalize the existence of magnetic crystals. Whether a given material will be magnetic, and what magnetic structure it will adopt, is in general difficult to predict, although with modern density functional methods great progress has been made in recent years. The case of metallic nickel will be briefly discussed in Section 26.2.

The remainder of our discussion in this section will primarily be phenomenological. In insulators,[5] where one can approximately assign a spin (and

[2] Subservient to this rule is the corollary that the atom will also have the highest total *orbital* angular momentum, L, consistent with the given total *spin* angular momentum, S. A third rule states that if the incomplete shell is not more than half filled, the lowest state has the minimum value of $J = |L - S|$; if the shell is more than half filled the lowest energy has $J = |L - S|$.

[3] We ignore any orbital contribution to the magnetic moment here.

[4] This does not mean that magnetism cannot occur in a compound containing e.g., Ni, where the atom can have a filled d shell (in a $3d^{10}4s^0$ configuration); for Ni^{++}, two electrons are "promoted" to the 4s or 4p shells to form bonds leaving two "holes" in the d shell, which would then have a magnetic moment.

[5] In metals the magnetic and electronic structure must be treated on an equal footing, and a spin Hamiltonian treatment is not in general valid.

accompanying moment) to each magnetic atom, it is common to model the interaction between the spins on neighboring magnetic sites (which leads to the magnetism) by the following *spin, or exchange, Hamiltonian*:

$$\hat{H} = -\frac{1}{2}\sum_{i,j} J_{ij}\hat{\mathbf{S}}_i \cdot \hat{\mathbf{S}}_j; \tag{26.1a}$$

here the sum extends over all atoms in the crystal (indicated by our use of vector or bold face indices) and the factor 1/2 avoids double counting. Our choice of the negative sign in (26.1) conforms with the standard convention.

The exchange constants J_{ij} measure the strength of the interaction between sites. Note that this form depends only on the angle between the spin operators. In particular, this form is independent of the orientation of the coordinates used to orient spin space and those involving the axes of the crystal in coordinate space. This reflects the physical fact that, in the absence of spin-orbit coupling, a relativistic effect, these spaces are physically *unconnected*. Hence, at least in materials involving lighter elements, the magnetization, e.g., in a ferromagnet, can point in an arbitrary direction relative to the crystal axes. If we limit the interaction to nearest spin-bearing neighbors only, Eq. (26.1a) becomes

$$\hat{H} = -\frac{1}{2}\sum_{i=1}^{N}\sum_{j=1}^{z} J\hat{\mathbf{S}}_i \cdot \hat{\mathbf{S}}_j, \tag{26.1b}$$

where $\hat{\mathbf{S}}_i$ and $\hat{\mathbf{S}}_j$ denote the spin operators on neighboring sites, N is the total number of spins, and z is the number of nearest neighbors.

In some very anisotropic materials (see Section 26.5) the spins are constrained to project along a single axis, which for simplicity can be chosen as the z axis, thereby involving the operator $\hat{\sigma}_z$. In this case Eq. (26.1a) would be replaced by

$$\hat{H} = -\frac{1}{2}\sum_{ij} J_{ij}\hat{\sigma}_{zi}\hat{\sigma}_{zj} \tag{26.2a}$$

and Eq. (26.1b) by

$$\hat{H} = -\frac{1}{2}\sum_{i}\sum_{j=1}^{z} J\,\hat{\sigma}_{zi}\hat{\sigma}_{zj}. \tag{26.2b}$$

This approximation is referred to as the *Ising model.*

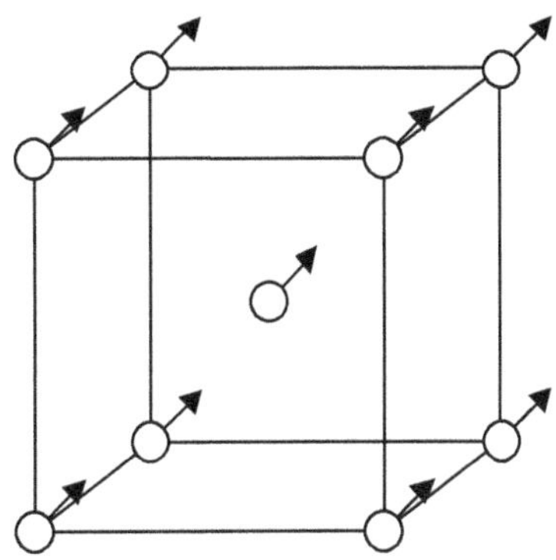

Figure 26.1 *A ferromagnetic lattice as occurs in bcc Fe. The individual spins are shown projected along the* [111] *direction.*

26.1.1 Ferromagnetism

If J is positive in Eq. (26.1b), the lowest energy state will involve parallel spins; i.e., we will have a ferromagnet. A simple ferromagnet is bcc iron, which is shown in Fig. 26.1. Below the Curie temperature (1043 K for Fe) the material acquires a

Table 26.1 *Selected ferromagnets.*

Material	Critical temperature (K)	Saturation magnetization (G)
Fe	1043	1752
Co	1388	1466
Ni	627	510
Gd	293	1980
Dy	85	3000
$CrBr_3$	37	270
Au_2MnAl	200	323
Cu_2MnAl	630	726
Cu_2MnIn	500	613
EuO	77	1910
EuS	16.5	1184
MnAs	318	870
MnBi	670	675
$GdCl_3$	2.2	550

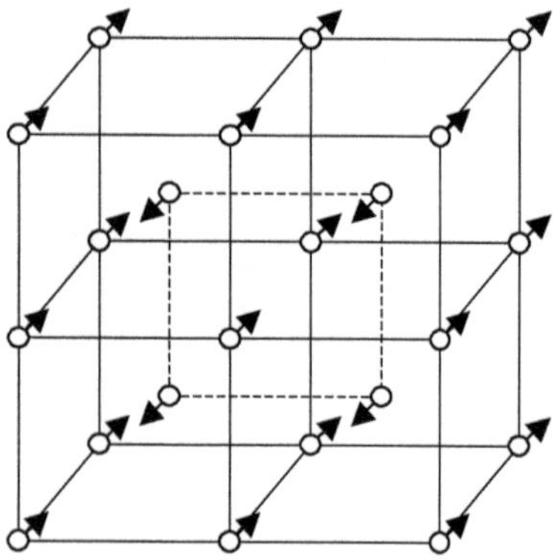

Figure 26.2 *Four bcc unit cells of antiferromagnetic* Cr. *The magnetic structure consists of two interpenetrating simple cubic lattices with oppositely directed spins.*

spontaneous magnetization. Magnetic phase transitions are generally second order, so the magnetization grows from zero at the Curie temperature, approaching a *saturation* value (e.g., $M(T = 0) = 1752$ Oe for Fe). Table 26.1 lists a number of ferromagnets along with their saturation magnetizations and Curie temperatures.

26.1.2 Antiferromagnetism

If J is negative in Eq. (26.1b), an antiferromagnetic state is favored.[6] Figure 26.2 shows four unit cells of an antiferromagnetic structure based on the bcc lattice, which approximates the magnetic structure of Cr;[7] the spins are shown directed along the [111] axis. Note that the spin up and spin down sublattices consist of two interpenetrating simple cubic lattices; the oppositely directed spins in the cube centers are connected by dotted lines. Table 26.2 lists some antiferromagnetic compounds along with their Neel temperatures.

26.1.3 Ferrimagnetism

A *ferrimagnet* is a structure that usually consists of two different magnetic species. When we have atoms with different spins (magnetic moments), occupying sites

[6] In the case of antiferromagnetic coupling, it is possible for some crystal structures to not permit two interpenetrating sub-lattices, a situation that is referred to as frustration.

[7] The magnetic structure of Cr is actually more complicated. At low temperatures there is an approximately 5% difference between the magnetic unit cell and the real-space unit cell due to the formation of a standing spin wave. The addition of a small amount of V results in a commensurate structure.

Table 26.2 *Some antiferromagnetic materials.*

Material	Critical Temperature(K)	Material	Critical Temperature(K)
Cr	311	$KMnF_3$	88.3
MnO	122	$KCoF_3$	125
FeO	198	MnF_2	67.4
CoO	291	FeF_2	78.4
NiO	600	CoF_2	37.2
$RbMnF_3$	54.5	$MnCl_2$	2
$KFeF_3$	115	VS	1040

with different local symmetry, they can, depending on how they are aligned, add (as in a ferromagnet), completely anti-align (as in an antiferromagnet), or some combination of align and anti-align, such that the total moment *does not vanish.* The ferrimagnet is an example of the latter. Perhaps the most important material of this kind is *magnetite,* a reddish-colored mineral, which has the chemical formula Fe_3O_4 and that was commonly used as the storage medium in tape recorders (and older disc drives). Written in this way, the formula disguises both the chemistry and the magnetic structure, so we rewrite it in the more transparent form, $Fe^{2+}O^{2-} \cdot Fe_2^{3+}O_3^{2-}$. The ferrous ($Fe^{2+}$) ions are in a d^6 configuration with spin 2 ($4\mu_B$) while the ferric ions (Fe_2^{3+}) are in a d^5 configuration with spin 5/2 ($5\mu_B$).

Magnetite is a cubic structure consisting of eight fcc cells involving a total of 32 nearly-close-packed O^{2-} ions; it is called the *inverse spinel structure* and is related to the spinel structure, $MgAl_2O_4$ (See Fig. 26.3). The iron atoms occupy the interstices between the O^{2-} ions. Of the 16 Fe_2^{3+} ions, eight occupy A-sites with tetrahedral symmetry, while the other eight Fe_2^{3+} along with the eight Fe^{2+}ions occupy B-sites, with octrahedral symmetry. We do not need to concern ourselves with the details of the actual structure, only with the alignment of the spins. This is shown schematically in Fig. 26.4.

26.1.4 Complex structures

Ferro-, antiferro-, and ferrimagnetism by no means exhaust the possible magnetic structures. In Fig. 26.5, we show the twisted magnetic structure of the metallic rare earth elements Dy and Eu. The arrangement on the left shows the "spiral" magnetic structure which exists for temperatures 90 K $<$ T $<$ 180 K in Dy, while the figure on the right shows the "conical" structure existing in the element Eu for temperatures below 50 K. The moment would average to zero in the first case while in the second case it would have a non-zero value along the twist axis (the c axis of this hexagonal crystal). Clearly, the magnetic unit cell is significantly

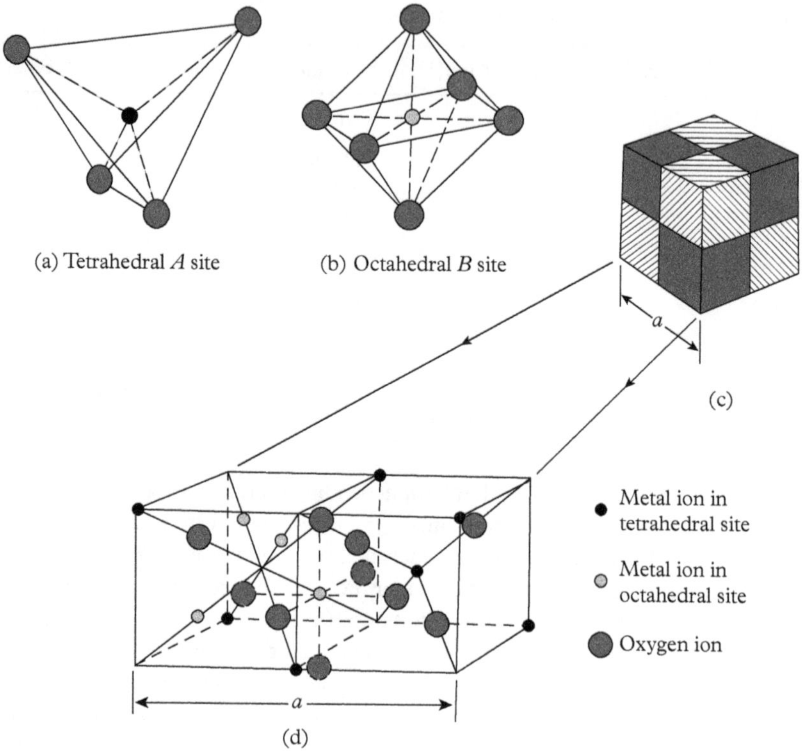

Figure 26.3 *The cubic unit cell of the mineral spinel with the chemical formula* $MgAl_2O_4$ *is shown in (c); to include other materials having the same structure involving metals with corresponding valences, we can write the more general formula* $A^{+2}B_2^{+3}O_4^{-2}$. *The full cubic unit cell involves 16 formula units. The structure can be regarded as eight alternating cubic sub-cells. The divalent A atoms lie at the center of a tetrahedron (and are said to have tetrahedral,* T_d *symmetry, see (a)) while the trivalent B atoms are at the center of an octahedron (octahedral symmetry,* O_h, *shown in (b)). (Two oxygen atoms lying outside of both of the unit cells in (d) are included to display the octahedron.) Typical A ions are* Mn, Fe, Co, Ni, Cu, Zn, Mg, Cd *while the B ions are typically* Fe, Al, Cr. *For the inverse spinel structure, associated with magnetite, the occupations are* $Fe^{+2}Fe_2^{+3}O_4^{-2}$. *Drawing by L.W. Leary.*

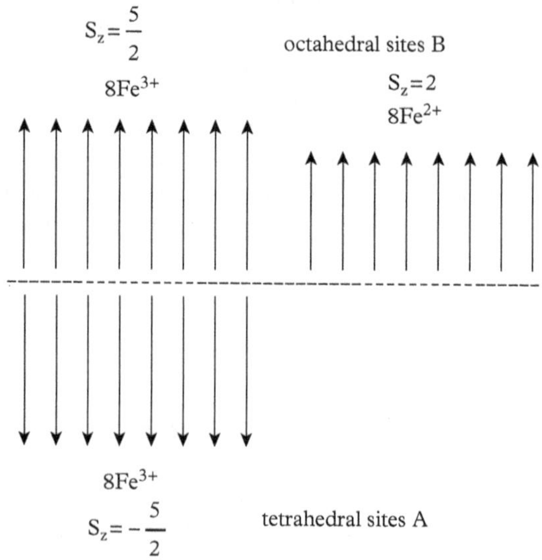

Figure 26.4 *A schematic drawing of the sites occupied and the accompanying spins in magnetite.*

larger than the crystallographic unit cell for this material. Such structures are determined by (elastic) neutron diffraction. The neutron, having a magnetic moment, is scattered by the internal magnetic moment distribution (f electrons in this case), and from the intensity of the various orders one can obtain Fourier components of the magnetic moment distribution.

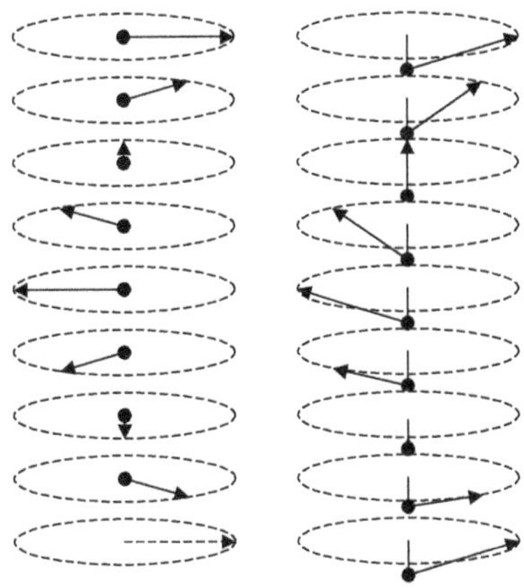

Figure 26.5 *The arrangement of the magnetic moments in the hexagonal rare earth metal, Dy; (a) shows the "spiral" magnetic structure, stable for 90 K < T < 180 K, while (b) shows the "conical" structure, stable in Eu below 50 K.*

26.2 Density functional theory of band magnetism

One of the more striking successes of the density functional theory discussed in Section 15.2 is its ability to calculate ground state magnetic properties of some materials. An early example was the calculations on Ni by Wang and Callaway.[8] These authors employed two different exchange/correlation potentials, both of which yielded very similar results.

As with any self-consistent band structure calculation, they constructed a starting charge density and an associated potential from which a set of wave functions was calculated; from these wave functions new charge densities and potentials were obtained, and the process was continued until it converged. However the process was generalized as follows: (i) the charge densities of spin up and spin down electrons were evaluated *separately*; (ii) these charge densities were then used to calculate corresponding exchange correlation potentials, *which will be spin dependent if the spin populations differ*;[9] (iii) to get the process started, a small difference in the spin populations was arbitrarily introduced; (iv) on carrying out successive iterations the net difference between the total spin up and spin down populations *evolved and ultimately stabilized* (as did the charge density and total energy). The resulting band structure along symmetry axes is shown in Fig. 26.6.

The difference between the spin up and spin down carrier densities when multiplied by the Bohr magneton is the *position-dependent magnetization*; i.e., $M_S(\mathbf{r}) = \mu_B\left(n_\uparrow(\mathbf{r}) - n_\downarrow(\mathbf{r})\right)$. The average of this quantity over a unit cell is the saturation magnetization that one would measure with a typical magnetometer, the experimental value of which corresponds to $0.56\mu_B$/atom. For the better of the two potentials used in the calculations the agreement is about 4%. In addition, with the availability of the position-dependent magnetization, one can calculate the *spin-dependent atomic form factors* (see Section 9.8) detailing how the moment is actually distributed in space; this dependence can be measured directly from spin polarized neutron diffraction measurements. Finally, the shapes of the spin up and spin down Fermi surfaces are obtained, data on which can be obtained by inverting dHvA data (see Section 20.3);[10] cuts through the (100) and (110) planes are shown in Fig. 26.7.

This spin polarized band structure technique is not limited to ferromagnets but can be applied equally well to antiferromagnetic and other magnetic structures. In such cases the size of the real-space unit cell must be expanded so as to coincide

[8] Energy bands in ferromagnetic Ni, C.S. Wang and J. Callaway, *Phys. Rev.* B15, 298 (1977).

[9] A local exchange-correlation potential for the spin polarized case: I, U. Von Barth and L. Hedin, *J. Phys. C: Solid State Physics* 5, 1629 (1972).

[10] De Haas-Van Alphen effect in ferromagnetic nickel, D.C. Tsui and R.W. Stark, *Phys. Rev. Lett.* 17, 871 (1966).

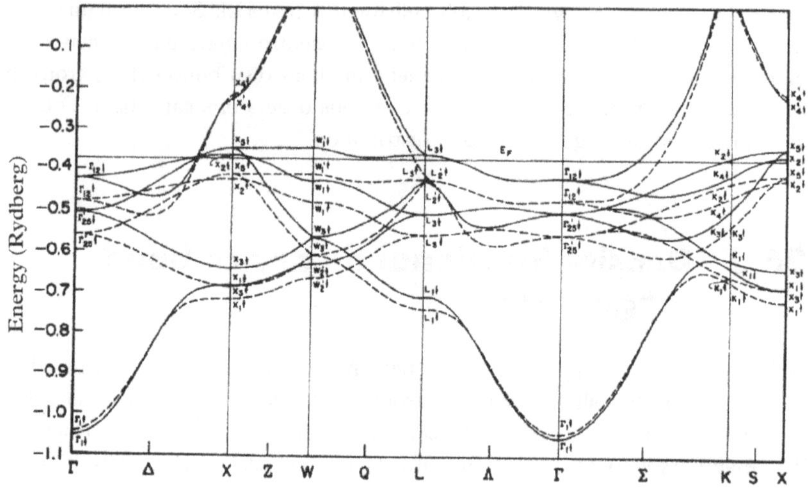

Figure 26.6 *The band structure of ferromagnetic nickel, as calculated by Wang and Callaway using the Kohn–Sham potential. Majority and minority states are shown by dashed and solid lines respectively.*

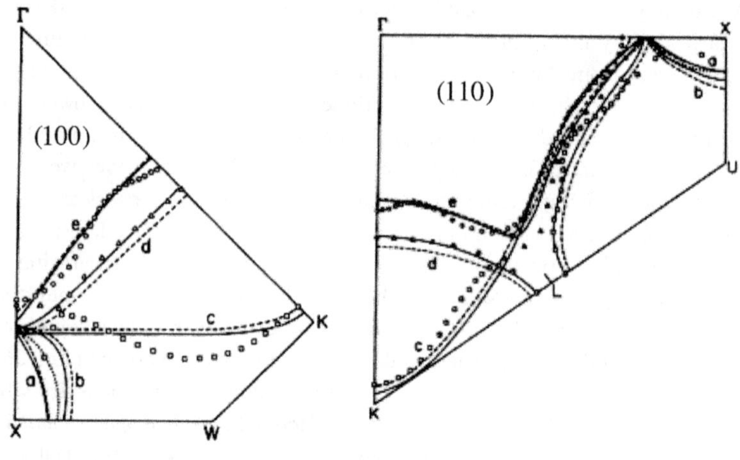

Figure 26.7 *Cross sections of the Fermi surface of* Ni *in the (100) and (110) planes for two different exchange correlation potentials. The squares and circles show dimensions obtained from the de Haas–van Alphen data of R. Stark and D. Tsui (unpublished).*

with that of the *magnetic structure*. For materials having a larger atomic number Z, spin–orbit coupling effects must be included from the outset.

26.3 The Landau model of ferromagnetism

As was discussed in Chapter 25 in connection with ferroelectricity, it is natural to expand the free energy near the critical temperature, T_C, in terms of a suitable *order parameter*. For a ferromagnet, the natural choice for this order parameter is the spontaneous magnetization, **M**. States in which the direction of magnetization is reversed must have the same free energy; i.e., $F(\mathbf{M}) = F(-\mathbf{M})$; hence a power series expansion of the free energy in **M** is expected to contain only even terms. As previously discussed, due the fact that magnetism arises from the exchange energy, which is independent of the spin quantization axis in the absence of spin–orbit coupling, it is a good approximation to ignore any small dependence of the energy on the vector direction of the magnetization and to regard the free energy as a function of the magnitude, $M = |\mathbf{M}|$, of the magnetization. (The phenomenological effects of the spin–orbit coupling are discussed in Section 26.5.) In the absence of magnetism the independent thermodynamic variables entering the (Helmholtz) free energy are (T,ρ). If there are no external magnetic fields, $\mathbf{H}$, present then we expect an expansion of F having the form

$$F(T, \rho, \mathbf{M}) = F_0(T, \rho, M = 0) + \alpha M^2 + \beta M^4. \tag{26.3}$$

As in the ferroelectric case we "drive" the transition by writing $\alpha = a(T-T_c)$, with a equal to a positive constant, such that the term quadratic in M becomes negative for $T < T_c$ in Eq. (26.3); to keep the energy bounded we must have $\beta > 0$. Plots of $F - F_0$ versus M are shown in Fig. 26.8 for $T > T_c$, $T = T_c$, and $T < T_c$. We see that for $T < T_c$, the solution at $M = 0$ is unstable and a new solution with $M \neq 0$ becomes the stable solution. We find these solutions by minimizing the free energy with respect to M: i.e.,

$$\frac{\partial F}{\partial M} = 0 = 2M\left(\alpha M + 2\beta M^2\right).$$

This has the solutions

$$M = \begin{cases} 0 & ; T > T_c \\ (-\alpha/2\beta)^{1/2} & ; T < T_c \end{cases}. \tag{26.4}$$

The solution with $M > 0$ is the only stable solution for $T < T_c$.

It is important to point out that the symmetry of the system was *spontaneously* lowered by the transition, in that it was isotropic above the transition temperature, but acquired a *vector direction* (associated with the direction of the magnetization) below the transition. Hence one says there is a *spontaneous symmetry breaking* at the transition. This is a general feature of second-order transitions; the symmetry of the low temperature phase is always lower than the high temperature phase.

We now extend our treatment to the case where a magnetic field $\mathbf{H}$ is present. Since we ultimately wish to treat *inhomogeneous* systems, the quantities of interest here must be *specific quantities*. Hence our free energy is to be interpreted as a *free energy per unit volume*. In place of the usual independent thermodynamic variables, T, V, characterizing this function, it is common to use the variables T, ρ, where ρ is the mass density, and for the dependent variables, the entropy per unit mass, s, and the chemical potential per unit mass, μ. The basic thermodynamic identity for the differential of free energy, $F(T, \rho)$, is then written

$$dF = -\rho s dT + \mu d\rho. \tag{26.5}$$

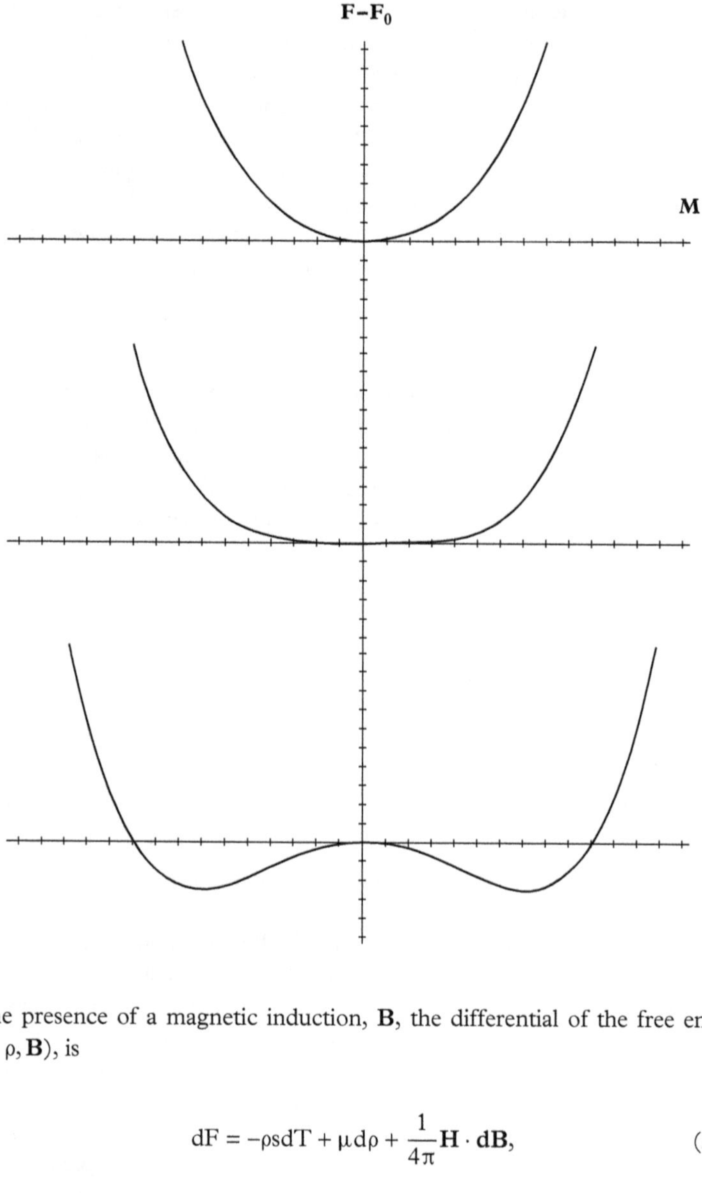

Figure 26.8 *Plots of F versus M for $T > T_c$ (upper), $T = T_c$, (middle), and $T < T_c$ (lower).*

In the presence of a magnetic induction, **B**, the differential of the free energy, $F(T, \rho, \mathbf{B})$, is

$$dF = -\rho s dT + \mu d\rho + \frac{1}{4\pi}\mathbf{H} \cdot \mathbf{dB}, \tag{26.6}$$

where **H** is the magnetic field. If we wish to have **H** as the independent variable we must perform a Legendre transformation and define a new free energy density through the definition[11]

$$\tilde{F}(T, \rho, \mathbf{H}) = F(T, \rho, \mathbf{B}) - \frac{1}{4\pi}\mathbf{H} \cdot \mathbf{B}. \tag{26.7}$$

[11] This transformation is analogous to defining the Gibbs free energy from the Helmoltz free energy through $G = F + PV$.

Differentiating this form yields

$$d\tilde{F} = -\rho s dT + \mu d\rho - \frac{1}{4\pi}\mathbf{B}\cdot d\mathbf{H}. \qquad (26.8)$$

Writing $\mathbf{B} = \mathbf{H} + 4\pi\mathbf{M}$ and integrating (26.8) with respect to $\mathbf{H}$ at constant $\mathbf{M}$ (which we take as being independently fixed by minimizing the free energy) we obtain

$$\tilde{F}(T, \rho, \mathbf{M}, \mathbf{H}) = F_0(T, \rho, \mathbf{M}, \mathbf{H} = 0) - \mathbf{M}\cdot\mathbf{H} - H^2/8\pi, \qquad (26.9)$$

where our writing, $F_0(T, \rho, \mathbf{M}, 0)$, allows for the presence of a spontaneous magnetization at $\mathbf{H} = 0$. Using (26.7) we can construct $F(T, \rho, \mathbf{M}, \mathbf{B})$:

$$\begin{aligned} F(T, \rho, \mathbf{M}, \mathbf{B}) &= \tilde{F}(T, \rho, \mathbf{M}, \mathbf{H}) + \frac{1}{4\pi}\mathbf{H}\cdot\mathbf{B} \\ &= F_0(T, \rho, \mathbf{M}) - \mathbf{M}\cdot\mathbf{H} - \frac{1}{8\pi}H^2 + \frac{1}{4\pi}\mathbf{H}\cdot(\mathbf{H} + 4\pi\mathbf{M}) \\ &= F_0(T, \rho, \mathbf{M}) + \frac{1}{8\pi}H^2 \\ &= F_0(T, \rho, \mathbf{M}) + \frac{1}{8\pi}(\mathbf{B} - 4\pi\mathbf{M})^2. \end{aligned} \qquad (26.10)$$

From Section 27.3 we have that $F_0(T, \rho, \mathbf{M}, \mathbf{H} = 0) = F_0(T, \rho, 0, 0) + \alpha M^2 + \beta M^4$ near T_c. We can then obtain the magnetization in the presence of a field $\mathbf{H}$ by minimizing $\tilde{F}(T, \rho, \mathbf{M}, \mathbf{H})$ with respect to $\mathbf{M}$ at constant $\mathbf{H}$, which yields

$$2\alpha M + 4\beta M^3 - H = 0. \qquad (26.11)$$

For $T > T_c$, where there is no zero-field spontaneous magnetization we may ignore the term in M^3. To measure the response to H we define the magnetic susceptibility, $\chi = M/H$; from (26.11),

$$\chi(T) = \frac{1}{2\alpha} = \frac{1}{2a(T - T_c)}. \qquad (26.12)$$

Note that this diverges as $T \to T_c$. However below the transition it is only meaningful to define a *differential* susceptibility, $\chi(T) = \partial M/\partial H$. Differentiating (26.11) yields

$$\frac{\partial M}{\partial H} = \frac{1}{(2\alpha + 12\beta M^2)} = \frac{1}{2\alpha + 12\beta(-\alpha/2\beta)}$$

or

$$\chi(T) = \frac{1}{4a(T_c - T)}. \qquad (26.13)$$

Note that Eqs. (26.12) and (26.13) differ by a factor of 2, as found earlier for the electric susceptibility of a ferroelectric.

Last we examine the specific heat. The heat capacity per unit volume at constant density is defined as

$$c_{\rho,\mathbf{H}} = T\left(\frac{\partial(\rho s)}{\partial T}\right)_{\rho,\mathbf{H}} = -T\left(\frac{\partial^2 \tilde{F}}{\partial T^2}\right)_{\rho,\mathbf{H}},$$

where we have used Eq. (26.8) in the second step. At zero field the heat capacity undergoes a discontinuity at the transition, which we obtain by twice differentiating Eq. (26.3) with (26.4) substituted for M and evaluating the resulting expression for $T = T_c + 0^+$ and $T = T_c - 0^+$:

$$c_{\rho,\mathbf{H}}(T = T_c + 0^+) = -T\left(\frac{\partial^2 F_0}{\partial T^2}\right)_{\rho,\mathbf{H}}; \qquad T = T_c + 0^+ \tag{26.14a}$$

and

$$c_{\rho,\mathbf{H}}(T = T_c - 0^+) = c_{\rho,\mathbf{H}}(T = T_c + 0^+) + \frac{a^2 T_c}{2\beta}; \qquad T = T_c - 0^+, \tag{26.14b}$$

where 0^+ is a positive infinitesimal. Hence the heat capacity of the low temperature phase is larger by the amount $a^2T_c/2\beta$.

26.4 The Weiss model of ferromagnetism

The classical expression for the magnetic susceptibility of a paramagnetic material, Curie's law, was obtained in Chapter 5 as

$$M(T) = \frac{C}{T}H. \tag{26.15}$$

As a model for ferromagnetism, P. Weiss (1907) assumed the presence of what he called a *molecular field* (a term still used), H_M, or, more commonly today, an *exchange field*. The basic assumption was that this molecular field is proportional to the magnetization; i.e.,

$$H_M = \lambda M. \tag{26.16a}$$

This field is added, on a purely phenomenological basis, to the magnetic field entering Eq. (5.12a); i.e.,

$$M(T) = \frac{C}{T}(H + \lambda M). \tag{26.16b}$$

Solving for the magnetic susceptibility, $\chi = M/H$, we obtain

$$\chi = \frac{C}{T - \lambda C}.$$

Defining a transition temperature, $T_c = \lambda C$, we may write this as

$$\chi = \frac{C}{T - T_c}. \tag{26.17}$$

If we set $C = (2a)^{-1}$, we obtain agreement with the Landau theory.

Equation (26.17) is known as the *Curie–Weiss law*; it gives a good representation of the magnetic susceptibility in the paramagnetic phase, as long as one does not approach the transition temperature too closely. By changing the sign of the molecular field (and with it the transition or Néel temperature) the Curie-Weiss law can also be used to describe the susceptibility in antiferromagnets.

This same theory can also be used to calculate the magnetization for $T < T_c$. To accomplish this we require a more accurate expression for the magnetic susceptibility of a "gas" of spins. As a model we may use the Langevin classical theory or the more accurate Brillouin theory, discussed in Chapter 5. Here we will limit the discussion to a spin 1/2 system where the magnetization is given by

$$M(T) = N\mu \tanh \frac{\mu H}{k_B T}, \tag{5.18b}$$

where N, μ, and k_B are the number of spins per unit volume, their magnetic moment, and Boltzmann's constant, respectively. To apply the Weiss theory all we do is replace H by an *effective field*, $H + H_M$. In the absence of an external magnetic field we would write

$$M(T) = N\mu \tanh \frac{\mu\lambda M(T)}{k_B T} \tag{26.18}$$

and solve this implicit equation for M. There is a maximum temperature for which this equation has a solution, which we identify as the transition temperature. Near this temperature M is small and we may expand the hyperbolic tangent as

$$M = N\mu \left[\frac{\mu\lambda M}{k_B T} - \frac{1}{3}\left(\frac{\mu\lambda M}{k_B T}\right)^3 \cdots \right].$$

Solving for M, we obtain

$$M = \sqrt{3} N\mu \sqrt{\frac{T_c - T}{T_c}} \tag{26.19}$$

where we write $T_c = N\mu^2\lambda/k_B$. Note that Eq. (26.19) has the same $M \propto (T_c - T)^{1/2}$ behavior as the Landau theory.

At very low temperatures we may write $\tanh x \cong 1 - 2e^{-2x}$, and the expression for the magnetization becomes

$$M = N\mu\left[1 - 2\exp\left(-\frac{2\lambda N\mu^2}{k_B T}\right)\right]$$
$$= N\mu\left[1 - 2\exp\left(-\frac{2T_c}{T}\right)\right]. \tag{26.20}$$

As we will see later, $M(T) - M(0) \propto AT^{3/2}$ due to the excitation of spin waves. However, overall, the Weiss model gives a somewhat better representation of M(T); the strength of the Landau theory lies in the generality of the approach.

26.5 Anisotropy energy

As discussed, magnetism arises from exchange interactions and in the absence of relativistic effects (spin–orbit coupling) the free energy is independent of the direction along which the magnetization aligns. However when these relativistic effects are included a dependence of the free energy on the alignment of the magnetization appears; this effect is referred to as the *anisotropy energy.*

Except in the immediate vicinity of the transition temperature, it is a very good approximation to regard the *magnitude* of the magnetization as constant and to assume that the anisotropy energy depends only on the direction cosines of the vector direction of **M**. Like all quantities entering the free energy, the form of the anisotropy energy must have the full symmetry of the high temperature (paramagnetic) phase. We write $\mathbf{M} = M\mathbf{m}$, where $\mathbf{m}$ is a unit vector:

$$\mathbf{m} = m_x\mathbf{x} + m_y\mathbf{y} + m_z\mathbf{z}, \tag{26.21a}$$

where **x**, **y**, and **z** are unit vectors along the three principal axis, and **m** is normalized as

$$m^2 = m_x^2 + m_y^2 + m_z^2 = 1. \tag{26.21b}$$

Measuring the direction in terms of the usual polar coordinates, we write $m_x = \sin\theta\cos\varphi$, $m_y = \sin\theta\sin\varphi$, and $m_z = \cos\theta$. Our task then is to construct expressions for the anisotropy energy in various orders as a function of the direction cosines, m_i.

In leading order this energy can only depend on quadratic forms involving the direction cosines, and hence we write

$$U_{aniso} = \sum_{i,j=1}^{3} K_{ij} m_i m_j. \tag{26.22}$$

The simplest case is that of a *uniaxial* material (e.g., a hexagonal or tetragonal crystal and essentially all thin films) where we may write this energy as

$$\begin{aligned} U_{aniso} &= K_1(m_x^2 + m_y^2) \\ &= K_1 \sin^2\theta. \end{aligned} \tag{26.23a}$$

We have adopted the usual convention here; we could just as well have written

$$\begin{aligned} U_{aniso} &= -K_1 m_z^2 \\ &= -K_1 \cos^2\theta = -K_1(1 - \sin^2\theta). \end{aligned} \tag{26.23b}$$

The constant can be absorbed into the remainder of F, in which case we have the same form as (26.23a). Clearly, for $K_1 > 0$ we will have $\mathbf{M} \,//\, \mathbf{z}$, which is referred to as an *easy axis* material; if $K_1 < 0$, we have an *easy plane* material. At this order in the expansion there will be no preferred direction within the plane.

Proceeding to the next order we would have the following forms for a hexagonal (having six-fold symmetry about **z**) and a tetragonal (having four-fold symmetry) material:

$$U_{aniso} = K_1(m_x^2 + m_y^2) + K_2(m_x^2 + m_y^2)^2 \qquad \text{(hexagonal)} \tag{26.23c}$$

and

$$\begin{aligned} U_{aniso} &= K_1(m_x^2 + m_y^2) + K_2(m_x^2 + m_y^2)^2 + K_2' m_x^2 m_y^2 \\ &= K_1 \sin^2\theta + K_2 \sin^4\theta + K_2' \sin^4\theta \sin^2\varphi \cos^2\varphi \qquad \text{(tetragonal)} \\ &= K_1 \sin^2\theta + K_2 \sin^4\theta + \frac{1}{4} K_2' \sin^4\theta \sin^2 2\varphi. \end{aligned} \tag{26.23d}$$

These forms follow from the requirement that the expressions be invariant under the allowed rotational symmetries of the crystals. Assuming $K_1 < 0$ (easy plane magnetization), in the tetragonal case, $K_2' < 0$ leads to $\sin^2 2\varphi = 1$; i.e., the magnetization aligns along the square diagonal. The case $K_2' > 0$ leads to $\sin^2 2\varphi = 0$; i.e., alignment along a square edge. For the hexagonal case there is still no favored in-plane direction at this (quartic) order of the expansion in m_i; to discriminate an in-plane direction in this case we must go to sixth order involving a term

$$\frac{1}{2}[(m_x + i m_y)^6 + (m_x - i m_y)^6] = \sin^6\theta \cos 6\varphi. \tag{26.24}$$

For the cubic case, which has the highest symmetry, the leading term in the anisotropy energy has the form

$$U_{aniso} = K\left(m_x^2 m_y^2 + m_y^2 m_z^2 + m_z^2 m_x^2\right); \tag{26.25a}$$

an equivalent form (differing again only by a constant) is

$$U_{aniso} = -\frac{1}{2}K\left(m_x^4 + m_y^4 + m_z^4\right). \tag{26.25b}$$

For $K > 0$ the energy will be a minimum for **M** // **x**, **y**, or **z** (i.e., along a cube axis); for $K < 0$, we would have $m_x^2 = m_y^2 = m_z^2 = \frac{1}{3}$ or $U_{aniso} = -\frac{1}{6}K$; i.e., **M** aligns along a cube diagonal. A good representation of the data for Fe and Ni involves carrying one more term in the expansion; in sixth order there is only one invariant and adding it to (26.25a) we have

$$U_{aniso} = K_1\left(m_x^2 m_y^2 + m_y^2 m_z^2 + m_z^2 m_x^2\right) + K_2 m_x^2 m_y^2 m_z^2. \tag{26.26}$$

Anisotropy energy is a required ingredient of permanent magnets; without it the material breaks up into small domains to minimize the overall magnetic energy, as we will discuss shortly. As previously noted, anisotropy energy arises from relativistic effects which increase with atomic number, Z. Although there are magnetic elements with larger Z, e.g., Gd, they are not magnetic at room temperature. Hence if we are to increase the anisotropy energy of 3d magnetic materials we should examine alloys or compounds containing higher Z constituents, preferably those having large magnetic moments themselves; examples of such materials are $SmCo_5$ and $Nd_2Fe_{14}B$, the latter having a tetragonal crystal structure.

26.6 The gradient energy

Shortly we will want to discuss situations where the direction of the magnetization varies in space. Up to now we have assumed our system is uniformly magnetized. If however the direction of magnetization were to vary as a function of position in the sample, it is natural to assign some increase in energy associated with this variation, in analogy to the Hook's law elastic energy associated with distorting a solid body. In constructing this energy we must exclude any terms that can be transformed by an integration by parts into a surface term, since these cannot contribute to a bulk thermodynamic energy (in which surface terms are excluded). The following form has the proper behavior:

$$U_{non-u} = \frac{1}{2}\sum_{i,j=1}^{3} \alpha_{ij}\frac{\partial \mathbf{M}}{\partial x_i}\cdot\frac{\partial \mathbf{M}}{\partial x_j}, \tag{26.27}$$

where α_{ij} is a symmetrical tensor. For ferromagnetic order to be stable the principal values of this tensor must be positive. For the case of a cubic crystal the tensor must be diagonal with all three principal values equal

$$U_{non-u} = \frac{1}{2}\alpha\left[\left(\frac{\partial \mathbf{M}}{\partial x}\right)^2 + \left(\frac{\partial \mathbf{M}}{\partial y}\right)^2 + \left(\frac{\partial \mathbf{M}}{\partial z}\right)^2\right]. \quad \text{(cubic)} \tag{26.28}$$

For the case of a uniaxial magnetic material we would have

$$U_{non\text{-}u} = \frac{1}{2}\alpha_1 \left[\left(\frac{\partial \mathbf{M}}{\partial x} \right)^2 + \left(\frac{\partial \mathbf{M}}{\partial y} \right)^2 \right] + \frac{1}{2}\alpha_2 \left(\frac{\partial \mathbf{M}}{\partial z} \right)^2 . \text{ (uniaxial)} \qquad (26.29)$$

Like the Landau expansion itself, the forms based on (26.27) can be regarded as the leading terms in a power expansion involving the derivatives of the magnetization. Near the transition both the magnitude of **M**, M, and its vector direction, **m**, become "soft" in some sense. However, far from the transition temperature, where the magnetization is treated as a constant, it can only *twist,* but its magnitude remains essentially fixed. In this limit we must regard forms based on (26.27) as the leading terms in a power series involving the spatial derivatives of the *vector direction* of **M**, **m** = **M**/M, and not the magnitude of the magnetization.

26.7 Surface tension of a domain wall

Suppose we have two regions in a ferromagnet where the magnetization has undergone a change in direction. The system is then said to consist of two *domains.* The region separating the two domains, which we will show involves a finite "transition width," is called the *domain wall or Bloch wall,* which is accompanied by a *wall energy,* which, in analogy with the energy of the interface between two thermodynamic phases, can be referred to as a *surface tension.*

There are a large number of ways domain walls can form, depending on crystal symmetry, sample shape, etc. We will treat the case where the energy associated with the domain wall arises from two competing contributions: the anisotropy energy, which we will assume has the uniaxial/easy axis form, and the gradient energy. Referring to Fig. 26.9, we take the z axis as the anisotropy axis and assume that the reversal takes place as we proceed along the x axis. We will assume that the magnitude of the magnetization remains fixed and thus it undergoes a rotation by π on passing through the boundary.

The structure of the domain wall is determined by a minimization of the free energy $\tilde{F}$, given by Eq. (26.9), with the added contributions of the gradient energy, (26.29), and the uniaxial anisotropy energy, (26.23a). We start by noting that we are considering a one-dimensional problem and hence **M** is a function of x only; i.e., **M** = **M**(x). The rotation will take place in the y–z plane as depicted in Fig. 26.9, as we now show. By Maxwell's second equation, **B**, which also depends only on x, obeys the equation $dB_x/dx = 0$, and hence we would have B_x = const. However, deep within the domains we have $M_x = 0$, and we are implicitly assuming that in these regions $H_x = 0$; hence we must have $B_x = 0$ everywhere. So if a non-zero $M_x(x)$ were to develop in the region of the domain wall it would have to be accompanied by a non-zero $H(x) = -4\pi M_x$; from Eq. (26.9) this would produce a contribution to $\tilde{F}$ of

$$-M_x H_x - H_x^2/8\pi = H_x^2/4\pi - H_x^2/8\pi = H_x^2/8\pi > 0.$$

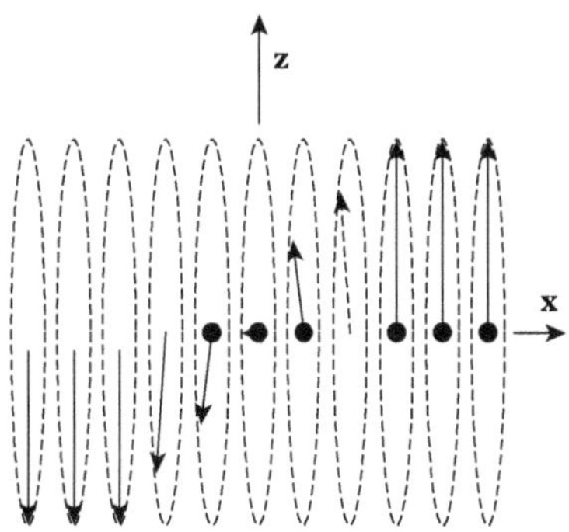

Figure 26.9 *The behavior of the magnetization at a domain wall in the z-y plane.*

Hence, $M_x = 0$ everywhere and the rotation occurs in a plane perpendicular to the x axis.

With these assumptions, and writing the total free energy as $\tilde{F}$, the added total *free energy per unit area*, $\Delta\tilde{F}/\mathcal{A}$, associated with the gradient and anisotropy terms can be written as

$$\Delta\tilde{F} = \mathcal{A}\int_{-\infty}^{+\infty} dx \left\{\frac{1}{2}\alpha_1\left[\left(\frac{\partial M_y}{\partial x}\right)^2 + \left(\frac{\partial M_z}{\partial x}\right)^2\right] + \frac{\beta}{2}M_y^2\right\} \tag{26.30a}$$

$$= \frac{1}{2}M^2\int_{-\infty}^{+\infty} dx \left[\alpha_1\left(\frac{d\theta}{dx}\right)^2 + \beta\sin^2\theta\right], \tag{26.30b}$$

where we wrote $K_1 = \beta\,|\mathbf{M}|^2$. The quantity $\Delta\tilde{\mathcal{F}}/\mathcal{A}$ is then interpreted as a *surface tension.* Assuming the two uniform regions are much thicker than the interface between them, we have the boundary conditions

$$\theta(+\infty) = 0; \theta(-\infty) = \pi; \frac{d\theta(x = \pm\infty)}{dx} = 0. \tag{26.31}$$

If we send $x \to t$, we see that (26.30) is the Lagrangian for a particle of mass α_1 moving in a potential $V(x) = -\beta\sin^2\theta$. Since the "velocity" and potential vanish at $\pm\infty$, the total energy is zero and we have the following first integral of (26.30) of

$$\left(\frac{d\theta}{dx}\right)^2 - \frac{\beta}{\alpha_1}\sin^2\theta = 0. \tag{26.32}$$

Integrating a second time yields

$$\cos\theta = \tanh[x/\delta], \tag{26.33}$$

where we have defined a *width parameter*, $\delta = \sqrt{\alpha_1/\beta}$. Inserting (26.32) into (26.30b), we have

$$\begin{aligned}\Delta\tilde{F}/\mathcal{A} &= M^2\alpha_1\int_{-\infty}^{+\infty} dx\left(\frac{d\theta}{dx}\right)^2\\ &= M^2\alpha_1\int_{-\infty}^{+\infty} dx\frac{d\theta}{dx}\cdot\sqrt{\frac{\beta}{\alpha_1}}\sin\theta\\ &= M^2\sqrt{\alpha_1\beta}\int_0^{\pi} d\theta\sin\theta\\ &= 2M^2\sqrt{\alpha_1\beta}.\end{aligned}$$

We may then write the domain wall surface tension as

$$\Delta\tilde{F}/\mathcal{A} = M^2\Delta, \tag{26.34}$$

where $\Delta \equiv 2\sqrt{\alpha_1\beta}$.[12]

[12] Some additional examples are treated in *Electrodynamics of Continuous Media*, L.D. Landau, E.M. Lifshitz, and L.P. Piteavskii, Pergamon Press, Oxford (1984).

26.8 Ferromagnetic domains

Above some critical size, a ferromagnetic specimen spontaneously breaks up into domains in which the direction of the magnetization differs. These domains are separated by domain walls, an example of which was considered in the previous section. There is a vast variety of such structures; we will restrict our discussion to a uniaxial ferromagnet of the easy axis type, having the geometry of a slab where the anisotropy axis lies parallel to the 1 axis (see Fig. 26.10) of a slab of thickness L_3; the dimensions L_1 and L_2, are assumed to be large compared to L_3. For the most part our discussion will be qualitative in character.

We consider two different limits: (a) strong anisotropy and (b) weak anisotropy; these cases are shown in Fig. 26.10. Note that, as drawn, the successive domains have thickness, **a**; not shown on this scale is the domain wall thickness, δ, which is assumed to be much smaller than **a**. The macroscopic dimensions, L_1, L_2, L_3, are assumed to be large compared with δ and a.

The thickness **a** follows from a minimization of the total free energy, $\tilde{F}$; this involves two contributions, (i) the energy associated with the domain walls themselves and (ii) the energy associated with the emergence of the field at the edges of the domain.

For the first of these, it was shown in the previous section that the formation of a single domain wall costs an energy $M^2 \Delta \times \mathcal{A}$, where $\mathcal{A} = L_2 L_3$ is the area of

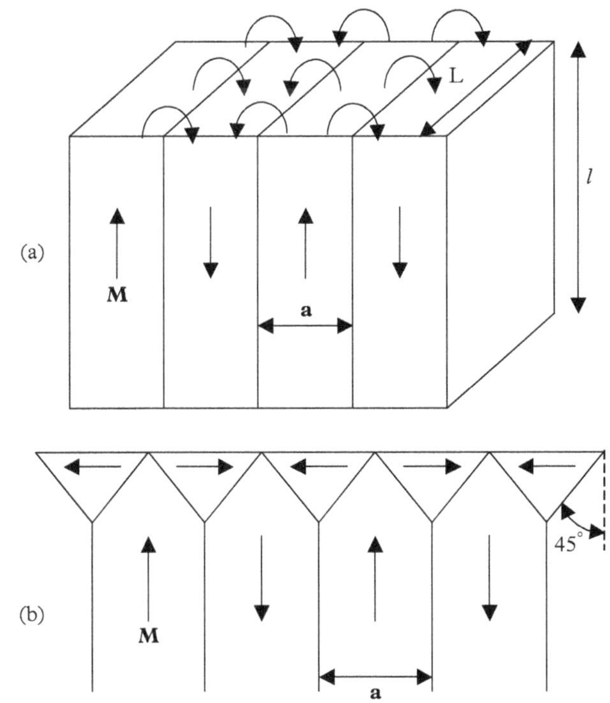

Figure 26.10 *The domain structure of an easy-axis, uniaxial ferromagnet (axis oriented vertically) for: (a) strong anisotropy (shown in 3d) and (b) weak anisotropy (shown as a cross section).*

the domain and Δ is defined by Eq. (26.34). Hence the energy of a single domain wall is of order $M^2L_2L_3\Delta$. However, the number of walls is $L_1/\mathbf{a}$ and therefore the total energy of all walls is $M^2L_1L_2L_3\Delta/\mathbf{a}$.

We now consider the second contribution arising from effects at the edges of the slab. In case (a) it is assumed that the anisotropy is sufficiently large that the magnetization is rigidly aligned parallel to the easy axis, right up to the edge of the sample. From the Maxwell boundary condition on the perpendicular component of the **B** field immediately above the slab, a periodic fringing field **H** of order **M** is produced in the region adjacent to the slab, reversing from domain to domain. This field persists for a distance of order **a** above the slab, after which it curls around to continuously merge into the magnetization of the adjacent domains. The increase in energy associated with this region is then proportional to the volume occupied by the fringing field; i.e., the product of **a** and the area of the upper surface: $\mathbf{a} \times L_1L_2$. We may then write the fringing field contribution to the total energy approximately as $\mathbf{a}\,L_1L_2M^2$.

The domain thickness now follows from minimizing the sum of the energies associated with the fringing field and the domain walls, as previously given,

$$\frac{d}{da}\left(\mathbf{a}\,L_1L_2M^2 + M^2L_1L_2L_3\Delta/\mathbf{a}\right) = 0,$$

from which it follows that $\mathbf{a} \propto (L_3\Delta)^{1/2}$. Thus *the domain size depends on a macroscopic dimension of the slab*: as the sample dimension L_3 increases, **a** scales as $(L_3)^{1/2}$ and the domain walls move farther apart.

In case (b) it is assumed that, rather than generating external fringing **H** fields, an array of successively reversing 45° domain walls forms. The anisotropy energy then produces a contribution from the upper surface of order $K_1 \times$ (edge volume) $\cong K_1\mathbf{a}\,L_1L_2$; this leads to a domain size such that $\mathbf{a} \propto \left(M^2\Delta L_3/K_1\right)^{1/2}$; i.e., **a** again scales as $(L_3)^{1/2}$.[13]

26.9 Total magnetic free energy of a body

If there are external currents producing an *external* magnetic field, $\mathbf{H}_0$, we must take account of the fact that this would produce an energy density $H_0^2/8\pi$ *in the absence of a magnetic body*. Noting that $\tilde{F}$ in (26.7) contains a term, $-H^2/8\pi$, it follows that if we are to remove the contribution of the external field to the energy density we must write the total free energy density in the presence of an external field as

$$\tilde{\mathcal{F}} = \int d^3r \left(\tilde{F} + \frac{H_0^2}{8\pi}\right). \tag{26.35}$$

The integral here would be taken *over all space*. It turns out that starting with this expression one can derive an alternative expression, where the integration is *only*

[13] Further analysis shows that for large thicknesses, **a**, additional domain structures encroach on the edges of the film; see: L.D. Landau, E.M. Lifshitz, and L.P. Pitaevskii, *Electrodynamics of Continuous Media*, Pergamon Press, NY (1984).

over the body. We do not reproduce this (somewhat subtle) derivation but only quote the final result:[14]

$$\tilde{F} = \int d^3r \left(\tilde{\mathrm{F}} + \frac{1}{8\pi}\mathbf{H}\cdot\mathbf{B} - \frac{1}{2}\mathbf{M}\cdot\mathbf{H}_0 \right), \tag{26.36a}$$

or equivalently,

$$F = \int d^3r \left(\mathrm{F} - \frac{1}{8\pi}\mathbf{H}\cdot\mathbf{B} - \frac{1}{2}\mathbf{M}\cdot\mathbf{H}_0 \right). \tag{26.36b}$$

26.10 Single domain particles

When a body becomes small enough it will consist of a single domain. Of course if a large enough field is applied, all bodies will be in a single domain state. Even in the absence of an external field, H_0, the magnetization of the body produces a field **H** proportional to **M**. This field interacts with the body's magnetization to produce an energy proportional to $MH \times V \propto M^2V$, where V is the sample volume. The gradient energy is of order $(\alpha M^2/d^2) \times V$, where d is an appropriate sample dimension. Assuming $V \propto d^3$, it is then clear that when $d \leq \alpha^{1/2}$, the particle will consist of a single domain. This does not mean that the particle will be uniformly magnetized, since it still needs to satisfy boundary conditions on its surface.

The only shape of a body that allows a uniform magnetization is the ellipsoid (or a degenerate form thereof). We will not solve the magneto-static problem of an ellipsoid in an external field, which is treated in most texts on electricity and magnetism, but will merely quote the result.[15] The field within the ellipsoid in the presence of an external field $\mathbf{H}_0$ is given by

$$\mathbf{H} = \mathbf{H}_0 - 4\pi\mathbf{n}\cdot\mathbf{M}, \tag{26.37}$$

where **n** is the so-called demagnetization tensor. We give some limiting cases for the demagnetization coefficients in the principal axis of the ellipsoid:

$$n_1 = n_2 = n_3 = \frac{1}{3} \quad \text{(sphere);} \tag{26.38a}$$

$$n_1 = n_2 = \frac{1}{2};\ n_3 = 0 \quad \text{(infinite cylinder);} \tag{26.38b}$$

$$n_1 = n_2 = 0;\ n_3 = 1 \quad \text{(infinite plate).} \tag{26.38c}$$

The last of these is equivalent to the Maxwell boundary conditions, in which the normal component of **B** and the tangential component of **H** are continuous.

[14] L.D. Landau, E.M. Lifshitz, and L.P. Pitaevskii, *Electrodynamics of Continuous Media*, Pergamon Press, NY (1984), Eq. (32.7).

[15] The problem is essentially identical to that of the dielectric ellipsoid, discussed in Section 2.2.

To proceed further we insert the **H**-dependent part of (26.9) into (26.36), obtaining

$$\tilde{F} = VU_{aniso} - \frac{1}{2}\mathbf{M}\cdot\int d^3r\,(\mathbf{H}+\mathbf{H}_0)\,, \tag{26.39}$$

where we again emphasize that the integration is only over the volume of the sample, V. Inserting (26.37) into (26.36a) then gives

$$\tilde{F} = 2\pi V\mathbf{M}\cdot\mathbf{n}\cdot\mathbf{M} - V\mathbf{M}\cdot\mathbf{H}_0 + VU_{aniso}. \tag{26.40}$$

The direction of the magnetization follows from minimizing (26.40) with respect to the direction of **M**. We rewrite the (quadratic) anisotropy energy in a slightly different form as

$$U_{aniso} = \frac{1}{2}\mathbf{M}\cdot\beta\cdot\mathbf{M}, \tag{26.41}$$

and, in component form, (26.40) then becomes

$$\tilde{F} = \frac{1}{2}V\sum_{i,j}(4\pi n_{ik} + \beta_{ik})\,M_iM_k - V\mathbf{H}_0\cdot\mathbf{M}. \tag{26.42}$$

26.11 The magnetization curve of an easy axis uniaxial ferromagnet

Tape recorders and older computer hard discs store data as a succession of discrete, in-plane magnetized regions. Higher storage densities can be achieved using so-called vertical magnetic recording, in which successive regions are magnetized perpendicular to the plane of the disc. We now examine the stability of vertically magnetized domains. We write the uniaxial anisotropy energy (26.23a) as $U_{aniso} = K_1\sin^2\theta = (1/2)\,\beta M^2\sin^2\theta$, where, as in Eq. (26.41), we have introduced a dimensionless constant, β. Adding this form to our free energy (26.9), we have

$$\tilde{F}(T,\rho,\mathbf{M},\mathbf{H}) = F_0(T,\rho,\mathbf{M}) + \frac{1}{2}\beta M^2\sin^2\theta - \mathbf{M}\cdot\mathbf{H} - H^2/8\pi. \tag{26.43}$$

We recall that well below T_c, the magnitude of **M** is assumed constant, independent of the direction of **H**, so we examine the energetics of (26.43) only with respect to the angle θ; we therefore write

$$\tilde{F}(T,\rho,\mathbf{M},\mathbf{H}) = F_0(T,\rho,\mathbf{M}) + \frac{1}{2}\beta M^2\sin^2\theta - M(H_x\sin\theta + H_z\cos\theta) - H^2/8\pi.$$

Minimizing this free energy by setting $\partial \tilde{F}/\partial\theta = 0$, we obtain

$$\beta M \cos\theta \sin\theta = H_x \cos\theta - H_z \sin\theta, \tag{26.44a}$$

which can be rewritten as

$$\frac{H_x}{\sin\theta} - \frac{H_z}{\cos\theta} = \beta M. \tag{26.44b}$$

Writing $s = \sin\theta$, this becomes the following fourth-order algebraic equation in s:

$$(\beta M s - H_x)^2(1 - s^2) = H_z^2 s^2. \tag{26.45}$$

For $H = 0$, (26.45) gives $s^2(1 - s^2) = 0$, which has the solutions $s = 0$ (or $\theta = 0, \pi$) and $s = \pm 1$ ($\theta = \pm\pi/2$). The first two solutions ($\theta = 0, \pi$) correspond to energy *minima*, with the magnetization pointing along one of the two easy axis directions, while the second pair ($\theta = \pm\pi/2$) correspond to energy *maxima*, with the magnetization perpendicular to the easy axis directions.

At a finite field that lies in an arbitrary direction, there continues to be a regime where there are two solutions, which is separated from a second one where there is only a single solution. At the boundary the energies in the two-solution region *merge*, implying that there is an inflexion point and with it an instability. Here, both $\partial\tilde{F}/\partial\theta$ and $\partial^2\tilde{F}/\partial\theta^2$ vanish and Fig. 26.11 shows the boundary generated by the solution of these two equations; outside this contour there is only a single solution.

We now consider in more detail the particular case where the field lies parallel or antiparallel to the z axis; here, (26.45) becomes

$$s^4 - s^2\left(1 - \frac{H_z^2}{(\beta M)^2}\right) = 0. \tag{26.46}$$

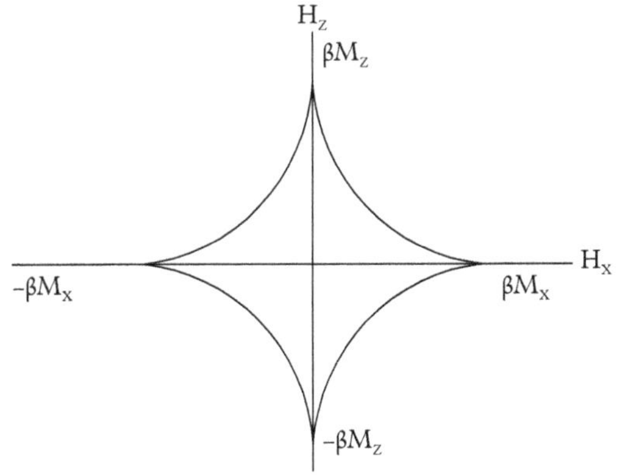

Figure 26.11 *The curve separating the regions where the magnetization can point in two directions (inside the curve) from that where it points in only a single direction (outside the curve).*

This equation has the solutions

$$s^2 = \left\{ \begin{matrix} 0 \\ 1 - \frac{H_z^2}{(\beta M)^2} \end{matrix} \right\}. \tag{26.47}$$

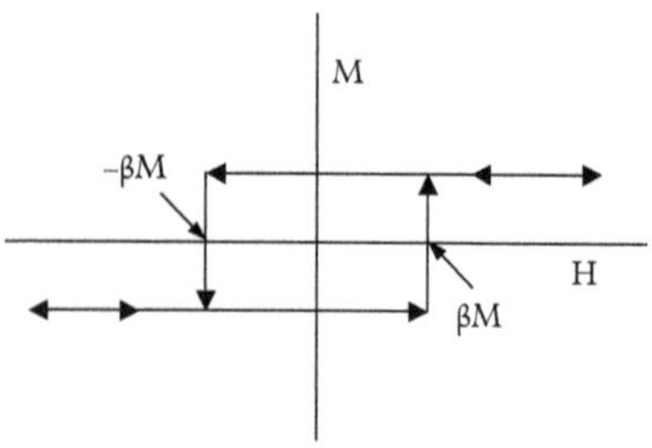

Figure 26.12 *Idealized behavior of the magnetization of a uniaxial ferromagnet showing the bistable behavior.*

The solution with s = 0, or equivalently θ = 0 or π, corresponds to the minima previously discussed for the H = 0 case; however they no longer have the same energy. The second solution corresponds to an energy maximum, where now $\theta \neq \pi/2$; i.e., it differs from the zero-field case. When $H_z^2/(\beta M)^2 = 1$, the second pair of solutions merges with the first marking the onset of an instability. If we start with the magnetization parallel to z and (a small) external field in the opposite direction, the magnetization will remain parallel up to a critical field βM, beyond which it will spontaneously reverse (switch). If the field is now reversed (to lie parallel to the z axis) the magnetization and field will again be antiparallel up to the critical field, above which the sample will return to its original orientation. This behavior is shown in Fig. 26.12; it forms a basis for a computer memory element.

26.12 Measuring the magnetization

Measurements of the static or quasistatic magnetic moment of materials are performed by a wide variety of techniques. The static methods involve the two basic laws of magnetostatics governing the energy, E, and torque, **N**, associated with a body of (total) magnetic moment **m**, placed in an external field **H**:

$$E = -\mathbf{m} \cdot \mathbf{H} \tag{26.48}$$

and

$$\mathbf{N} = \mathbf{m} \times \mathbf{H}. \tag{26.49}$$

If $\mathbf{H} = \mathbf{H}(\mathbf{r})$, a force, $\mathbf{F} = -\nabla E(\mathbf{r})$, will be exerted on the sample.[16] This is the basis for the oldest technique for measuring the magnetic moment in which a sample is placed in a field that is engineered to have a known gradient (it is also the reason why two magnets attract or repel each other). In practice, the sample is hung from a wire that extends outside the field region and is attached to a balance that measures the force. This arrangement constitutes the *Faraday magnetometer.*

If the sample has a magnetic moment, either spontaneous or induced, that points in a direction *different* from **H**, then a torque will be exerted that can be measured with a *torsion balance.* A situation where **M** and **H** are not collinear can arise from the fact that the sample is a crystal and the field does not lie along a symmetry axis, or, for a ferromagnet, where demagnetization effects (so-called shape anisotropy) or magneto-crystalline anisotropy can cause the magnetic moment to lie in a different direction. To measure the effect one can hang the sample

[16] In practice the magnetization can also be inhomogeneous, which complicates the analysis.

from a wire with a known torsion constant (torque versus twist angle) and record the angle of rotation by, e.g., deflecting a laser beam from a mirror attached to the sample. Both of the described methods can be miniaturized (e.g., by attaching the sample to a cantilever) to allow measurements in high magnetic fields or in other extreme environments.

Other, dynamic, approaches are now more commonly used that are based in one way or another on Faraday's law of induction, which states that the voltage induced in a coil is proportional to the rate of change of the total flux, Φ, it encloses according to

$$V = -\frac{1}{c}\frac{\partial \Phi}{\partial t}. \tag{26.50}$$

A change in flux can arise from a change in an external magnetic field, including the effects that the field has on any magnetic materials that couple flux into the coil; it can also be produced by moving a magnetic sample relative to a coil. In practice one may use one or two coils. Figure 26.13 shows an arrangement with two coaxial coils that are counter-wound (or serially opposed). In one realization, the sample is located midway between these two coils in such a way that the flux it produces passes equally through both coils; if the sample is now mechanically translated up and down in an oscillatory manner, an alternating voltage will be induced in the coils, the magnitude of which is proportional to the magnetic moment of the sample. This voltage can be amplified and detected (using a phase sensitive or lock-in detector); the resulting signal will be proportional to the magnetization. This is the basis of the *Foner magnetometer*, named after Simon Foner.

Alternatively the sample may be smoothly pulled from the center of one coil to the center of the adjacent one. If the time-dependent voltage induced across the coils, $V_{coil}(t)$, is applied to an analogue (operational amplifier) integrator, or

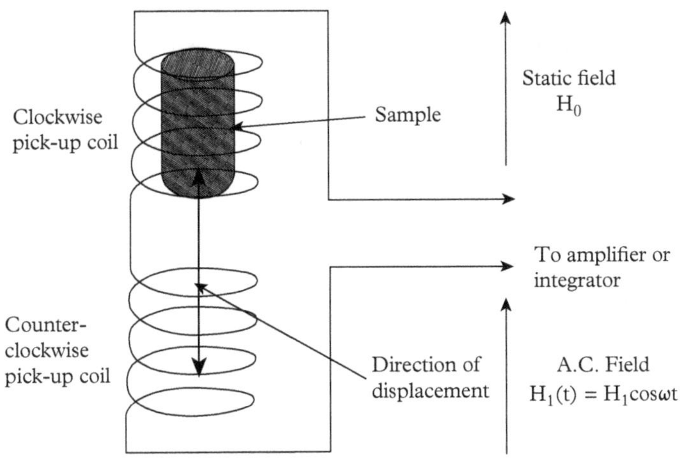

Figure 26.13 *The experimental arrangement used to measure the magnetization with either an integrating magnetometer or a superconducting SQUID.*

digitized and numerically integrated, the resulting voltage change, $\Delta V_{integrated}$, will be

$$\Delta V_{integrated} = \int_{t_{start}}^{t_{stop}} dt\, V_{coil}(t) = \frac{1}{c}\int_{t_{start}}^{t_{stop}} dt\, \frac{d\Phi}{dt} = \frac{1}{c}[\Phi_1 - \Phi_2] = \frac{1}{c}\Delta\Phi, \tag{26.51}$$

where $\Delta\Phi$ is the flux difference generated by moving the sample between the two coils, which is in turn proportional to the magnetic moment.[17] Either of these two techniques may be applied in the presence of a magnetic field, in which case one can obtain $\mathbf{m}(\mathbf{H})$. If the field translation direction and coil axes are all parallel one obtains the component of $\mathbf{m}$ parallel to $\mathbf{H}$; if the coils and translation direction are perpendicular to $\mathbf{H}$, perpendicular components of $\mathbf{m}$ can be obtained. In Section 33.5.2, the superconducting SQUID will be discussed, which can be used to measure flux changes directly with greatly enhanced sensitivity.

Finally we discuss an approach where a small field, $H_1 \cos \omega t$, is superimposed on a static field, H_0, in the two-coil arrangement. Since the number of turns is closely balanced (and any residual mismatch may be cancelled externally) the voltage induced in the coils will be proportional to the change in magnetization,

$$m(H + H_1 \cos \omega t) \cong m(H) + \chi(H)H_1 \cos \omega t, \tag{26.52}$$

where $\chi(\omega, H) = dm(\omega, H)/dH$ is the *differential* susceptibility. When measured at finite frequency, dissipative and other effects can enter and the susceptibility becomes complex,

$$\chi(\omega, H) = \chi'(\omega, H) + i\chi''(\omega, H),$$

where the imaginary part produces a signal that is $\pi/2$ out of phase with the real part; the two components are separated using a phase sensitive detector. Spurious out of phase components arise naturally when studying metallic samples, as a result of eddy currents induced by the driving field; various components in the system itself can also produce spurious out of phase components and establishing the true phase of the signal in the coils can be problematic. However, out of phase components arising from intrinsic causes in the sample associated with, e.g., magnetic hysteresis and slow relaxation phenomena in spin glass materials, are of interest.

[17] The proportionality constant involves a number of geometrical factors, including the number of turns in the coil, how close they are to each other, etc.

ADDITIONAL READING

A widely used book is *The Theory of Magnetism: An Introduction to the Study of Cooperative Phenomena* by D.C. Mattis, Harper's Publishing Co. (1965).

Various volumes and chapters in the series, *Magnetism*, Ed., G.T. Rado and H. Suhl, Academic Press, NY, contain a wealth of information on a wide range of topics.

Perhaps the best discussion of the thermodynamics of magnetism, together with a treatment of domains and numerous other topics, is given in *Electrodynamics of Continuous Media*, L.D. Landau, E.M. Lifshitz, and L.P. Pitaevskii, Chapters IV and V, Pergamon Press (1984).

A good treatment of several microscopic aspects of magnetism is given in *Statistical Physics Part II*, E.M. Lifshitz and L.P. Pitaevskii, Chapter VII, Pergamon Press (1980).

In applying density functional theory to predict magnetic structures, a widely used local density functional was developed by U. Von Barth and L. Hedin in the paper, A local exchange-correlation potential for the spin polarized case, *Journal of Physics C: Solid State Physics* 5, 1629 (1972).

Problems

26.1. Verify the forms (26.23c,d), (26.24), (26.25a,b), and (26.26).

26.2. Use the discussion of Section 26.10 to obtain an equation the solution of which will yield the direction in which the magnetization, **M** points in a thin platelet for an arbitrarily directed magnetic field $\mathbf{H}_0$. What is the solution for the case where $\mathbf{H}_0$ is perpendicular to the platelet?

26.3. Find the equations that generate the curve shown in Fig. 26.11.

26.4. Develop the theory of anisotropy energy to the point where it defines the in-plane orientation of an easy plane hexagonal material.

27

Dynamic Properties of Magnetic Materials

In Chapter 26 we focused on the static and thermodynamic properties of magnetic materials. Here we will extend the discussion to include dynamic, finite frequency, responses generated by external microwave fields. We will also treat the Bose spin wave excitations, which are excited thermally and contribute to the heat capacity.

27.1 The Landau–Lifshitz equation

At the macroscopic level the dynamics of spontaneously magnetized materials is governed by the so-called Landau–Lifshitz equation. In the next chapter we will study the dynamics of nuclear moments and (in the absence of a spontaneous moment) electron moments. Here we treat only spontaneously magnetized materials, since the presence of strong internal fields makes this topic significantly different from conventional magnetic resonance. We start by developing a semi-classical description.

The torque $\mathbf{N}$ on a magnetic dipole $\boldsymbol{\mu}$ is given by $\mathbf{N} = \boldsymbol{\mu} \times \mathbf{H}$. This must be equal to the rate of change of the angular momentum, which we write as $\hbar\dot{\mathbf{s}}$, where $\mathbf{s}$ is the intrinsic spin of the particle; i.e.,

$$\hbar\dot{\mathbf{s}} = \boldsymbol{\mu} \times \mathbf{H}. \tag{27.1}$$

Equation (27.1) is the *Larmor equation*. By writing $\boldsymbol{\mu} = g\mu_B\mathbf{s} = \hbar\gamma\,\mathbf{s}$, where we introduced the gyromagnetic ratio $\gamma \equiv g\mu_B/\hbar$, we can rewrite (27.1) as

$$\dot{\boldsymbol{\mu}} = \gamma\boldsymbol{\mu} \times \mathbf{H}. \tag{27.2}$$

If we sum over all (unpaired) moments in a unit volume of some material we obtain the magnetization (the dipole moment per unit volume) and (27.2) becomes

$$\dot{\mathbf{M}} = \gamma\mathbf{M} \times \mathbf{H}. \tag{27.3}$$

Equation (27.3) is called the *Landau–Lifshitz equation*. To generalize this equation further we would replace $\mathbf{H}$ by $\mathbf{H}_{\text{eff}}$, where the latter includes changes in an external field, $\mathbf{H}_0$, arising from shape demagnetization effects (see Section 26.10) as

The Physics of Solids. First Edition. J. B. Ketterson.

well as additional effective fields generated by U_{anis} and by taking the functional derivative of U_{non-u}.

27.2 Ferromagnetic resonance

In what follows we will derive the precession frequency, ω_0, for spins in the interior of a ferromagnet in the presence of an external magnetic field, $\mathbf{H}_0$, together with the demagnetization fields for several geometries. However, without a constant input of energy the magnetization would relax toward the equilibrium direction due to the influence of dissipation, which we discuss in Section 27.3. However, if an oscillatory field, $\mathbf{H}_1$ (typically much less than $\mathbf{H}_0$), is applied (typically perpendicular to $\mathbf{H}_0$) with a frequency, ω, in the vicinity of some natural precession frequency, ω_0, the magnetization will, under steady-state conditions, remain tipped with respect to $\mathbf{H}_0$ and precess at the applied frequency ω. The response (tipping angle) will of course be largest if $\omega = \omega_0$.

It is important to recognize that it is the fields *inside* the body that enter the Larmor equation of motion, which can differ from $\mathbf{H}_0$. In ordinary magnetic resonance experiments (see Chapter 28) the magnetization is usually assumed to be small enough that we can neglect the demagnetizing fields discussed in Section 26.10. However, as previously noted, this is not permissible for a ferromagnet. We start by examining the case of a ferromagnetic slab, as shown schematically in Fig. 27.1(a,b). For case (a) the external field, $\mathbf{H}_0$, and the internal magnetization, $\mathbf{M}_0$, are parallel to the slab and taken to lie along the x direction; for case (b) $\mathbf{H}_0$ and $\mathbf{M}_0$ are perpendicular to the slab[1] which we define as the z axis. We will solve the Landau–Lifshitz equation (27.3) for both of these cases.

Case (b) is the simplest to understand so we begin with this. The continuity of $\mathbf{B}_0$ at the interface tells us that the internal static field must be $H_z = H_0 - 4\pi M_z$. We will not write down the driving field $\mathbf{H}_1$, but instead assume the magnetization is tipped from the z axis and is freely precessing in the x–y plane, as previously noted. Our equation of motion is then

$$\dot{\mathbf{M}} = \gamma \begin{vmatrix} \mathbf{i} & \mathbf{j} & \mathbf{k} \\ M_x & M_y & M_z \\ 0 & 0 & H_0 - 4\pi M_z \end{vmatrix} . \tag{27.4}$$

The associated equations of motion are

$$\dot{M}_x = \gamma M_y (H_0 - 4\pi M_0) \tag{27.5a}$$

and

$$\dot{M}_y = -\gamma M_x (H_0 - 4\pi M_0) , \tag{27.5b}$$

[1] In the absence of a large perpendicular anisotropy, this requires $H_0 > 4\pi M$.

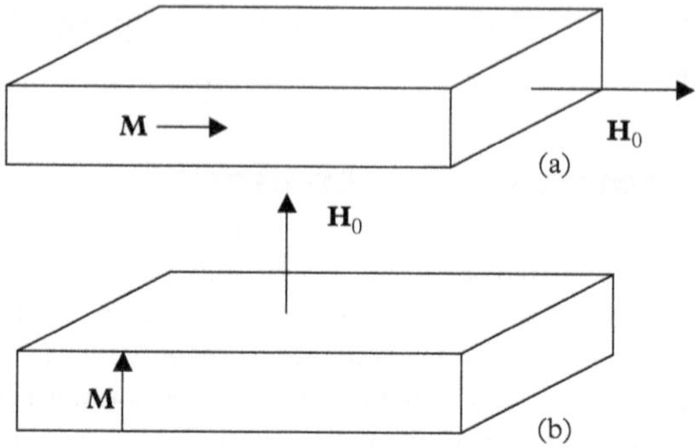

Figure 27.1 *A ferromagnetic slab for the case when* H_0 *is (a) parallel, and (b) perpendicular to the plane of the slab.*

where we have assumed weak excitation and set $M_z \cong M_0$; solving this set of coupled equations yields the precession frequency:

$$\omega = \gamma(H_0 - 4\pi M_0). \tag{27.6}$$

For larger tipping angles the precession frequency would become a function of the angle, unlike the conventional resonance case; however other phenomena occur for large tipping angles which we will not discuss here.

Case (a) is considerably different. The continuity of the tangential field requires that $H_x = H_0$. However, when the magnetization is tipped so that it has a component in the z direction, requiring that this produces no external field B_z (which again must be continuous) generates an H_z field:

$$B_z = 0 = H_z + 4\pi M_z \rightarrow H_z = -4\pi M_z.$$

In place of (27.4) we now have

$$\dot{\mathbf{M}} = \gamma \begin{vmatrix} \mathbf{i} & \mathbf{j} & \mathbf{k} \\ M_x & M_y & M_z \\ H_0 & 0 & -4\pi M_z \end{vmatrix}, \tag{27.7}$$

corresponding to the three equations

$$\dot{M}_x = -4\pi\gamma M_y M_z, \tag{27.8a}$$

$$\dot{M}_y = \gamma\,(M_z H_0 + 4\pi M_x M_z)\,, \tag{27.8b}$$

and

$$\dot{M}_z = -\gamma M_y H_0. \tag{27.8c}$$

Equation (27.8a) is second order in the small quantities, M_y and M_z, and may be neglected in a linear theory. Setting $M_x \cong M_0$, the equations become

$$\dot{M}_y = \gamma M_z\,(H_0 + 4\pi M_0) \tag{27.9a}$$

and

$$\dot{M}_z = -\gamma M_y H_0. \tag{27.9b}$$

Solving this pair of equations gives

$$\omega = \gamma\sqrt{H_0 (H_0 + 4\pi M_0)}. \tag{27.10}$$

Note that the resonance occurs at a *higher* field (than for a free dipole) when H_0 is perpendicular to the slab and at a *lower* field when H_0 is parallel to the slab.

Let us now consider the more general case of an ellipsoid; however we will limit the discussion to the case where the external field is parallel to a principal axis of the ellipsoid, which we take as the z axis. Our equation of motion is then

$$\dot{\mathbf{M}} = \gamma \begin{vmatrix} \mathbf{i} & \mathbf{j} & \mathbf{k} \\ M_x & M_y & M_z \\ -4\pi n_x M_x & -4\pi n_y M_y & H_0 - 4\pi n_z M_z \end{vmatrix}, \tag{27.11}$$

where n_i are the three demagnetization factors discussed in Section 26.10. Again approximating M_z by M_0 the equations of motion for the transverse components are

$$\dot{M}_x = \gamma \left[M_y (H_0 - 4\pi n_z M_0) + 4\pi n_y M_y M_0\right] \tag{27.12a}$$

and

$$\dot{M}_y = \gamma \left[-4\pi n_x M_x M_0 - M_x (H_0 - 4\pi n_z M_0)\right], \tag{27.12b}$$

which yields the resonance frequency

$$\omega = \gamma\sqrt{\left[H_0 + 4\pi \left(n_y - n_z\right) M_0\right] \left[H_0 + 4\pi \left(n_x - n_z\right) M_0\right]}. \tag{27.13}$$

We can extend our discussion to include the presence of uniaxial anisotropy. From Eq. (26.23a), $U_{aniso} = K_1 \sin^2 \theta$, where we will measure θ from the z axis. This produces a torque, $-\partial U_{aniso}/\partial\theta = -2K_1 \sin\theta \cos\theta$, in a direction perpendicular to $\mathbf{M}_\perp$ and $\mathbf{z}$ which we may write as $-(2K/M_0^2)(M_x\mathbf{i} + M_y\mathbf{j}) \times M_z\mathbf{k}$, leading to the equation of motion

$$\dot{\mathbf{M}} = \gamma \left\{ \begin{vmatrix} \mathbf{i} & \mathbf{j} & \mathbf{k} \\ M_x & M_y & M_z \\ -4\pi n_x M_x & -4\pi n_y M_y & H_0 - 4\pi n_z M_z \end{vmatrix} - \frac{2K}{M_0^2} \begin{vmatrix} \mathbf{i} & \mathbf{j} & \mathbf{k} \\ M_x & M_y & 0 \\ 0 & 0 & M_z \end{vmatrix} \right\}, \tag{27.14}$$

which for small tipping angles has the components

$$\dot{M}_x = \gamma \left[M_y \left(H_0 + 4\pi \left(n_y - n_z \right) M_0 \right) - \frac{2K}{M_0} M_y \right] \tag{27.15a}$$

and

$$\dot{M}_y = \gamma \left[-M_x \left(H_0 + 4\pi \left(n_x - n_z \right) M_0 \right) + \frac{2K}{M_0} M_x \right], \tag{27.15b}$$

leading to the resonance frequency

$$\omega = \gamma \sqrt{\left[H_0 + 4\pi \left(n_y - n_z \right) M_0 - \frac{2K}{M_0} \right] \left[H_0 + 4\pi \left(n_x - n_z \right) M_0 - \frac{2K}{M_0} \right]}. \tag{27.16}$$

Equations (27.6), (27.10) and (27.16) are commonmy referred to as the Kittel equations.

27.3 Magnetic relaxation

Up to this point we have neglected relaxation effects in ferromagnets. It turns out that this is a vast and complicated topic that we will avoid by only describing a phenomenological model due to Landau and Lifshitz which is still widely employed. We rewrite the equation of motion by initially assuming that terms are present in the equation of motion for each of the three orthogonal axes that can be defined with the vectors **H** and **M**:

$$\dot{\mathbf{M}} = \alpha \mathbf{M} + \gamma \mathbf{M} \times \mathbf{H} - \lambda' \mathbf{M} \times (\mathbf{M} \times \mathbf{H}). \tag{27.17}$$

If **M** is strictly conserved, as Landau and Lifshitz argued, based on the strong nature of the exchange interaction, then we must have $\alpha = 0$.[2] We may then write (27.17) as

$$\dot{\mathbf{M}} = \gamma \mathbf{M} \times \mathbf{H} - \frac{\lambda}{M_s} \mathbf{M} \times (\mathbf{M} \times \mathbf{H}), \tag{27.18a}$$

where the sign of the last term has been chosen so as to correspond to dissipation and we write $\lambda' \equiv \lambda/M_s$. In place of (27.18a), the equation of motion is also commonly written as

$$\dot{\mathbf{M}} = \gamma' \mathbf{M} \times \mathbf{H} + (\alpha'/M_s)\mathbf{M} \times \dot{\mathbf{M}}, \tag{27.18b}$$

which is known as the *Gilbert equation*.[3] When the second term on the right-hand side in either of these equations is included, it results in a width to the FMR resonance line. Equations (27.18a,b), which are phenomenological, hide the underlying complexity of magnetic relaxation. Microscopically, magnetic relaxation in a ferromagnet involves a number of processes including magnon–magnon, magnon–phonon, and in metals, magnon–electron and electron–electron interactions; magnons will be discussed shortly, in Section 27.7.

[2] Note that this differs from magnetic relaxation in non-ferromagnetic systems, where the magnetization can differ from its equilibrium value toward which it relaxes with a characteristic time, T_1, in the Bloch equations (see Section 28.2). Close to the transition temperature longitudinal relaxation also occurs in ferromagnets.

[3] T.L. Gilbert, Thesis, Illinois Institute for Technology (1956); also see: *IEEE Trans. Magn.* **40**, 3443 (2004); it is equivalent to the Landau–Lifshitz equation under the substations, $\gamma = \gamma'(1 + \alpha'^2)$ and $\lambda = \gamma'\alpha'(1 + \alpha'^2)$.

27.4 Spin wave resonance

Spin waves are propagating oscillations in the magnetization. At long wavelengths they arise from effects caused by the Maxwell boundary conditions at the surfaces of a sample and depend on its geometry; they are then called magnetostatic modes and will be treated in Appendix 16A, for the case of a slab. Here, we limit ourselves to short wavelengths where exchange effects dominate. Furthermore, we only discuss the case of an isotropic ferromagnet having a slab geometry with both the wave vector of the spin wave and the magnetic field *perpendicular* to the slab; such modes can be excited with uniform H_1 fields due to the presence of sample boundaries. Since anisotropy energy is not necessary for the modes to exist it will be ignored.

When the magnetization is inhomogeneous, as when a spin wave is present, the free energy density, $\tilde{F}$, includes a contribution, $\frac{1}{2}\alpha\sum_{i=1}^{3}(\partial\mathbf{M}/\partial x_i)\cdot(\partial\mathbf{M}/\partial x_i)$ (for the isotropic case, see Section 26.6). This term results in an effective field that is given by the *functional derivative* of the free energy (see Eq. (A.19b) in Appendix A):

$$\mathbf{H}_{grad} = \frac{\delta\tilde{F}}{\delta\mathbf{M}} = -\sum_{i=1}^{3}\frac{\partial}{\partial x_i}\frac{\partial\tilde{F}}{\partial\mathbf{M}/\partial x_i},$$

which yields $\mathbf{H}_{grad} = \alpha\nabla^2\mathbf{M}$. Our equation of motion is then

$$\dot{\mathbf{M}} = \gamma\,\mathbf{M}\times\left[\mathbf{H}+\alpha\nabla^2\mathbf{M}\right], \tag{27.19}$$

where $\mathbf{H}$ is the effective magnetic field. For a geometry with $\mathbf{H}_0$ perpendicular to a slab and waves propagating along z, this leads to the equations of motion:

$$\frac{\partial M_x}{\partial t} = \gamma M_y(H_0-4\pi M_0)-\gamma\alpha M_0\frac{\partial^2 M_y}{\partial z^2} \tag{27.20a}$$

$$\frac{\partial M_y}{\partial t} = -\gamma M_x(H_0-4\pi M_0)+\gamma\alpha M_0\frac{\partial^2 M_x}{\partial z^2}. \tag{27.20b}$$

We again form the combination, $M_+(z,t) = M_x(z,t)+iM_y(z,t)$, and assume the dependence, $e^{-i\omega t+ikz}$, that yields

$$\omega = \omega_0+\gamma\alpha M_0 k^2, \tag{27.21}$$

where $\omega_0 = \gamma(H_0-4\pi M_0)$. This form of the long wavelength spectrum for exchange dominated spin waves is quite general, although the frequency, ω_0, is a function of the geometry, as we have seen earlier. Because of the abrupt change in the environment at the boundaries of the slab we will assume that the spin wave amplitude vanishes there; this is referred to as *pinning*. We then have the boundary condition, $n\lambda = d/2$, where n is an integer and $k = 2\pi/\lambda$. However, if we try to excite the spin waves with a *uniform* external field it is necessary that we have an

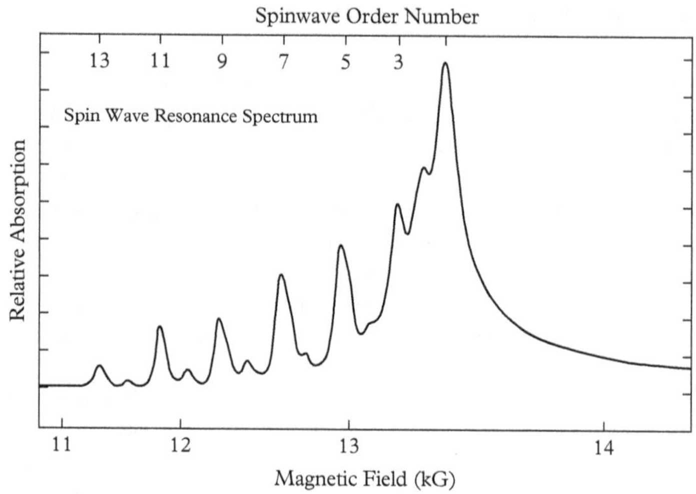

Figure 27.2 *Ferromagnetic resonance spectrum of a* 5600A *film of permalloy at* 8.89 GHz. *The mode number and magnetic field are shown on the top and bottom scales respectively;* n = 0 *corresponds to the uniform precession mode. (After M.H. Seavey and P.E. Tennenwald (1958).)*

odd number of half-wavelengths, since this results in an *asymmetry* in the instantaneous magnetization of the spin wave averaged over its path. Hence we require $(2n + 1)\lambda = d/2$, where n is again an integer. An example of spin wave satellites surrounding the main FMR line for a permalloy film is shown in Fig. 27.2.[4] Note that there is a small contribution from the even harmonics.

27.5 Antiferromagnetic resonance

One can also observe magnetic resonance in antiferromagnets.[5] We restrict ourselves to a two-sublattice model of an antiferromagnet where the two sublattices have magnetizations, $\mathbf{M}_1$ and $\mathbf{M}_2$ and $|\mathbf{M}_1| = |\mathbf{M}_2| = M$. Each sublattice is assumed to produce a Weiss *molecular field* or an *exchange* field (see Eq. (26.16a)) on the other sublattice. In addition, we must assume the presence of a uniaxial anisotropy field, $\mathbf{H}_A$, since the phenomenon of antiferromagnetic resonance does not exist without it, as we will see from the final result. Thus, in the absence of an external field, $\mathbf{H}_0$, the first sublattice experiences an effective field:

$$\mathbf{H}_1 = -\lambda\mathbf{M}_2 + H_A\mathbf{z}, \tag{27.22a}$$

where λ, which is positive for an antiferromagnet, is the Weiss parameter in the exchange energy $U_{ex} = \lambda\mathbf{M}_1 \cdot \mathbf{M}_2$, and **z** is a unit vector along the z axis, which we take as the anisotropy field axis. The corresponding expression for the field on the second sublattice is

$$\mathbf{H}_2 = -\lambda\mathbf{M}_1 - H_A\mathbf{z}. \tag{27.22b}$$

[4] Direct observation of spin-wave resonance, M.H. Seavey and P.E. Tennenwald, *Phys. Rev. Lett.* **1**, 168 (1958).

[5] Theory of antiferromagnetic resonance, C. Kittel, *Phys. Rev.* **82**, 565 (1951); Theory of antiferromagnetic resonance, F. Keffer and C. Kittel, *Phys. Rev.* **85**, 329 (1952).

The resulting Larmor equation of motion for $\mathbf{M}_1$ is

$$\dot{\mathbf{M}}_1 = \gamma \begin{vmatrix} x & y & z \\ M_{1x} & M_{1y} & M_{1z} \\ -\lambda M_{2x} & -\lambda M_{2y} & \lambda M + H_A \end{vmatrix}, \tag{27.23}$$

resulting in the pair of equations

$$\dot{M}_{1x} = \gamma \left[(\lambda M + H_A) M_{1y} - \lambda M (-M_{2y}) \right] \tag{27.24a}$$

and

$$\dot{M}_{1y} = \gamma \left[-(\lambda M + H_A) M_{1x} + M (-\lambda M_{2x}) \right]. \tag{27.24b}$$

A similar pair of equations holds for $\mathbf{M}_2$:

$$\dot{M}_{2x} = \gamma \left[-(-M)(-\lambda M_{1y}) + (-\lambda M - H_A) M_{2y} \right] \tag{27.25a}$$

and

$$\dot{M}_{2y} = \gamma \left[(-M)(-\lambda M_{1x}) - (-\lambda M - H_A) M_{2x} \right]. \tag{27.25b}$$

We define new variables, $M_{1+} = M_{1x} + iM_{1y}$ and $M_{2+} = M_{2x} + iM_{2y}$. Assuming a time dependence, $e^{-i\omega t}$, gives

$$-i\omega M_{1+} = -i\gamma \left[(\lambda M + H_A) M_{1+} + \lambda M M_{2+} \right] \tag{27.26a}$$

and

$$-i\omega M_{2+} = i\gamma \left[\lambda M M_{1+} + (\lambda M + H_A) M_{2+} \right]. \tag{27.26b}$$

Setting the determinant to zero yields

$$\begin{aligned} \omega^2 &= \gamma^2 H_A (H_A + 2\lambda M) \\ &= \gamma^2 H_A (H_A + H_E), \end{aligned} \tag{27.27}$$

where $H_E \equiv 2\lambda M$ is an exchange field.[6] Note the frequency vanishes in the absence of anisotropy, as mentioned earlier.

[6] Including the effect of a small external field, H_0 yields $\omega = \gamma \left[\pm H_0 + (H_A(H_A + H_E))^{1/2} \right]$.

27.6 Thermodynamic properties

We now examine the effect of spin waves in a ferromagnet on the thermodynamic properties. From Eq. (27.21), the spectrum in the absence of an external magnetic field is given by

$$\omega = \gamma\alpha M_0 k^2. \tag{27.28}$$

These excitations must obey the Planck relation, $\varepsilon(k) = \hbar\omega = \hbar\gamma\alpha k^2$. Since the associated quanta, referred to as *magnons*, can appear and disappear one at a time, they correspond to bosons. At low temperatures, where the density of excitations is sufficiently low that we can neglect their interaction, and where the wave vectors are small enough that our phenomenological gradient expansion is applicable, we can calculate the magnon contribution to the energy using the Planck distribution law:

$$E(T) = \frac{L^3}{(2\pi)^3}\int d^3k \frac{\hbar\omega(\mathbf{k})}{e^{\hbar\omega(\mathbf{k})/k_B T} - 1}. \tag{27.29}$$

In our isotropic model we have

$$d^3k \to 4\pi k^2 dk = \frac{4\pi}{2(\gamma\alpha M_0)^{3/2}}\omega^{1/2} d\omega$$

and introducing $x \equiv \hbar\omega/k_B T$, we can rewrite this integral as

$$\begin{aligned} E(T) &= \frac{L^3 (k_B T)^{5/2}}{(2\pi)^2 (\hbar\gamma\alpha M_0)^{3/2}}\int_0^\infty dx \frac{x^{3/2}}{e^x - 1} \\ &= \frac{L^3 (k_B T)^{5/2}}{(2\pi)^2 (\hbar\gamma\alpha M_0)^{3/2}}\Gamma(5/2)\zeta(5/2). \end{aligned} \tag{27.30}$$

The heat capacity is given by

$$C(T) = \frac{\partial E(T)}{\partial T} = \frac{5L^3 (k_B)^{5/2}}{8\pi^2 (\hbar\gamma\alpha M_0)^{3/2}}\Gamma(5/2)\zeta(5/2)T^{3/2}, \tag{27.31}$$

a result first obtained by Bloch and referred to as the Bloch $T^{3/2}$ law. This behavior is best studied in an insulating ferromagnet where the linear electronic specific heat, which would compete at low temperatures, is absent; the $T^{3/2}$ magnon heat capacity would then dominate over the T^3 contribution from the lattice at low temperatures.

27.7 The microscopic theory of magnons

We now discuss the spin excitations in a ferromagnetic material from a microscopic point of view. As a model Hamiltonian, we use the form (26.1) to describe the spin–spin interaction and add to it the interaction with an external magnetic field:

$$\hat{H} = -\frac{1}{2}\sum_{\substack{i,j=1\\ i\neq j}}^{N} J_{ij}\hat{\mathbf{S}}_i \cdot \hat{\mathbf{S}}_j - g\mu_B H_0 \sum_{i=1}^{N} \hat{S}_{iz}, \tag{27.32}$$

where we assume the external field, $\mathbf{H}_0$, is parallel to $\mathbf{z}$, and the bold subscripts imply a lattice with sites denoted by three independent integers. In the presence of ferromagnetic exchange interactions, the spin alignment will have its maximum value in the ground state.[7] Let the ground state eigenfunction of this Hamiltonian be denoted by $|\chi_0\rangle$. The total spin would then be NS. Since this is only achieved if each spin is maximally aligned, we must have

$$\hat{S}_{zi}\,|\chi_0\rangle = S\,|\chi_0\rangle\,. \tag{27.33}$$

We introduce the raising and lowering operators

$$\hat{S}_+ = \hat{S}_x + i\hat{S}_y;\;\; \hat{S}_- = \hat{S}_x - i\hat{S}_y, \tag{27.34}$$

which obey the usual commutation relations

$$\hat{S}_+\hat{S}_- - \hat{S}_-\hat{S}_+ = 2\hat{S}_z;\; \hat{S}_z\hat{S}_\pm - \hat{S}_\pm\hat{S}_z = \pm\hat{S}_\pm, \tag{27.35a,b}$$

and have the following matrix elements with respect to eigenfunctions quantized along $\mathbf{z}$:

$$\langle S_z|\,\hat{S}_+\,|S_z - 1\rangle = \langle S_z - 1|\,\hat{S}_-\,|S_z\rangle = \sqrt{(S + S_z)\,(S - S_z + 1)}. \tag{27.36}$$

In terms of these operators, the operator $\hat{\mathbf{S}}_i \cdot \hat{\mathbf{S}}_j$ in (27.32) can be rewritten in the form

$$\begin{aligned}\hat{\mathbf{S}}_i \cdot \hat{\mathbf{S}}_j &= \hat{S}_{iz}\hat{S}_{jz} + \hat{S}_{ix}\hat{S}_{jx} + \hat{S}_{iy}\hat{S}_{jy}\\ &= \hat{S}_{iz}\hat{S}_{jz} + \frac{1}{2}\left[\hat{S}_{i+}\hat{S}_{j-} + \hat{S}_{i-}\hat{S}_{j+}\right].\end{aligned} \tag{27.37}$$

Using (27.37) and the fact that $\hat{\mathbf{S}}_i$ and $\hat{\mathbf{S}}_j$ commute (since they involve different spins), the Hamiltonian becomes

$$\hat{H} = -\frac{1}{2}\sum_{\substack{i,j=1\\ i\neq j}}^{N} J_{ij}\left[\hat{S}_{iz}\hat{S}_{jz} + \hat{S}_{i+}\hat{S}_{j-}\right] - g\mu_B H_0 \sum_{i} \hat{S}_{iz}, \tag{27.38}$$

[7] We neglect the possible presence of domain walls, which form when the system minimizes the total energy arising from both the spin–spin interactions and the magnetic fields produced by the spins.

where we have used the symmetry $J_{ij} = J_{ji}$. Since S_z is maximal for $|\chi_0\rangle$ we must have

$$S_+ |\chi_0\rangle = 0; \tag{27.39}$$

therefore,

$$\hat{H} |\chi_0\rangle = \left[-\sum_{\substack{i,j=1 \\ i\neq j}}^{N} J_{ij} S^2 - g\mu_B H_0 NS \right] |\chi_0\rangle , \tag{27.40}$$

where the term in the square brackets is the ground state energy. We now change variables from **i**, **j**, to **j** and $l = \mathbf{i} - \mathbf{j}$; we may then write the ground state energy as

$$E_0 = -\frac{1}{2} NS^2 \sum_{l} J_l - g\mu_B H_0 NS, \tag{27.41}$$

with the total magnetic moment being $g\mu_B NS$.

We must now seek a description of the excited states. A state in which the ith spin is flipped is obtained using the lowering operator, $\hat{S}_{i-}$. The normalization of such a state follows from considering

$$\begin{aligned} (S_{i-} |\chi_0\rangle)^* (S_{i-} |\chi_0\rangle) &= \langle S| \hat{S}_{i+} \hat{S}_{i-} |S\rangle \\ &= \langle S| \hat{S}_{i+} |S-1\rangle \langle S-1| \hat{S}_{i-} |S\rangle \\ &= 2S, \end{aligned} \tag{27.42}$$

where we have used (27.36) in the last step. However such a state is not an eigenstate of our Hamiltonian: the interactions act so as to mix in states for which any of the other N spins is flipped. This suggests that we seek a solution which is a superposition of all such states. The situation is entirely analogous to our earlier discussion of the tight binding approximation for electronic wave functions. In the presence of translational symmetry of the spin lattice an appropriate variational form is clearly

$$|\chi_{\mathbf{k}}\rangle = \frac{1}{\sqrt{2SN}} \sum_{j} e^{i\mathbf{k}\cdot\mathbf{r}_j} \hat{S}_{j-} |\chi_0\rangle , \tag{27.43}$$

where the factor $N^{-1/2}$ is again for normalization. The magnon energy is the expectation value of $\hat{H}$ with respect to this trial wave function minus the ground state energy:

$$\begin{aligned} \varepsilon(\mathbf{k}) &= \langle \chi_{\mathbf{k}} | \hat{H} - E_0 | \chi_{\mathbf{k}} \rangle \\ &= \langle \chi_0 | \frac{1}{\sqrt{2SN}} \sum_{j'} e^{i\mathbf{k}\cdot\mathbf{r}_{j'}} \hat{S}_{j'+} \left[-\frac{1}{2} \sum_{\substack{i,j=1 \\ i\neq j}}^{N} J_{ij} [\hat{S}_{iz}\hat{S}_{jz} + \hat{S}_{i+}\hat{S}_{j-}] \right. \\ &\quad \left. - g\mu_B H_0 \sum_{i} \hat{S}_{iz} - E_0 \right] \frac{1}{\sqrt{2SN}} \sum_{j''} e^{i\mathbf{k}\cdot\mathbf{r}_{j''}} \hat{S}_{j''-} |\chi_0\rangle . \end{aligned} \tag{27.44}$$

To evaluate this expression we commute all $\hat{S}_+$ operators to the right (using Eqs. (27.35a,b)), where they annihilate the ground state and retain the remaining terms. A short calculation gives (F. Bloch (1930))

$$\begin{aligned}\varepsilon(\mathbf{k}) &= S\sum_{l} J_l\left(1 - e^{i\mathbf{k}\cdot\mathbf{r}_l}\right) + g\mu_B H_0 \\ &= S\sum_{l>0} J_l\left[1 - \cos\left(i\mathbf{k}\cdot\mathbf{r}_l\right)\right] + g\mu_B H_0. \end{aligned} \tag{27.45}$$

Note that as $\mathbf{k} \to 0$, $\varepsilon(\mathbf{k})$ has the same form as we obtained with our gradient expansion, Eq. (27.28).

27.8 Micromagnetics

Calculating the static and low frequency dynamic response of magnetic bodies having an arbitrary shape is non-trivial. Micromagnetics is a field that concerns itself with carrying out such calculations in terms of various model parameters designed to capture the effects of exchange, dipolar coupling, anisotropy, magnetic relaxation, etc. While many models of magnetic response are based on a continuum description, as we have for the most part used in our previous discussions, we here describe a powerful alternative approach based on the so-called *discrete-dipole approximation*,[8] with which one can *numerically* calculate the static and dynamic response of magnetic bodies of arbitrary shape.[9] We start with the calculation of equilibrium properties.

27.8.1 Equilibrium magnetization of an arbitrarily shaped particle

As discussed in Section 26.10, only ellipsoids and degenerate forms thereof can have a uniform magnetization. All other shapes are non-uniformly magnetized. The exchange interaction acts so as to keep the magnetization of all the spins parallel, but dipolar fields arising from shape effects compete to make this direction position dependent in a non-ellipsoidal body (or within any body for fields and sizes where domains are present). One usually assumes the magnitude of the magnetization to be constant and we will assume this, although it may not be the case for small (near-superparamagnetic) particles, or near the transition temperature.

The discrete-dipole model is based on simply summing the magnetic fields from N discrete point dipoles that uniformly fill the body in question on some grid. The magnetic moment of these particles is fixed so as to reproduce the magnetization, $\mathbf{M}$ (the magnetic moment per unit volume), of the body in question. Thus the field on site i is given by summing over the fields produced by the remaining dipoles j and any external fields. As we discuss, to account for the

[8] This method was first applied to study the radiation properties of arrays of electric dipoles in astrophysical applications: Scattering and absorption of light by non-spherical dielectric grains, E.M. Purcell and C.R. Pennypacker, *Astrophys. J.* **186**, 705 (1973).

[9] Micromagnetic simulation of absorption spectra, K. Rivkin and J.B. Ketterson, *J. Magn. Magn. Mater.* **306**, 204 (2006).

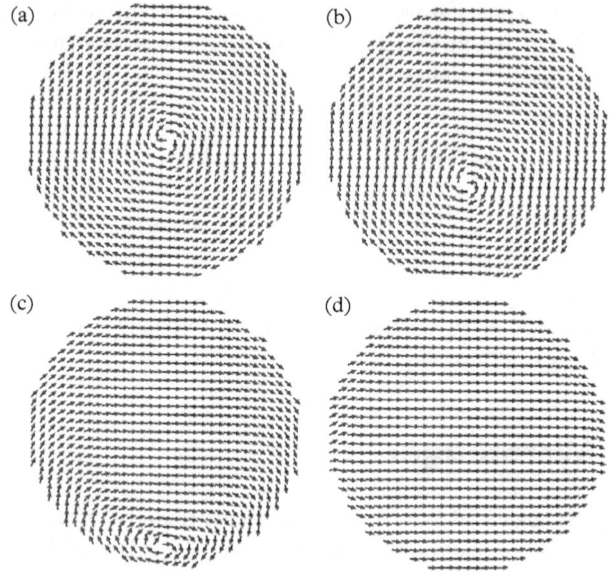

Figure 27.3 *Equilibrium configuration of a* 70 nm *permalloy disc for different values of the external d.c. field:* (I.) H_0 = 0Oe; *(II.)* H_0 = 109Oe; *(III.)* H_0 = 403 Oe; *(IV.)* H_0 = 1181 Oe. *(After Rivkin et al. (2005).)*

exchange interaction we add a nearest-neighbor exchange field with a strength measured by an exchange constant, J. Anisotropy fields can also be easily incorporated. Pinning of surface moments arises naturally since these moments have no neighbors; it can also be adjusted by altering J for surface or near-surface moments.

Since the equilibrium vector direction of the individual dipoles is not initially known, we might start by assuming that they are all aligned parallel to an external field, $\mathbf{H}_0$. We then calculate the field at positions i and reorient the dipoles so as to lie parallel to this local field; this procedure is repeated with the new orientations until the local fields stabilize. This yields a stable (sometimes metastable) magnetic configuration of the particles. As an alternative one can start with arbitrary spin directions (obtained from a random number generator) for each particle and then relax the system; this procedure is useful to determine which configurations are metastable (by examining the total magnetic energy), or when an external field is absent.

Figure 27.3 shows an example of the equilibrium configuration of discrete spins in a 700 nm diameter permalloy *disc* that was relaxed from a random starting configuration.[10] Note that the equilibrium configuration at zero field is a *vortex* positioned in the center of the disc; however, as the external field increases the vortex moves in a perpendicular direction, finally leaving the disc and giving way to an essentially uniform magnetization at high fields.

One can use the described methods to calculate the magnetic state of an arbitrary object for both increasing and decreasing fields. In general the magnetic states differ for these two histories, so-called hysteretic behavior.

[10] Analysis of ferromagnetic resonance response of square arrays of permalloy nanodots, K. Rivkin, Wentao Xu, L.E. De Long, V.V. Metlushkoc, B. Ilic, and J.B. Ketterson, *J. Magn. Magn. Mater.* **309**, 317 (2007).

27.8.2 The dynamic magnetic response in the linear regime

In this section we will assume the spin dynamics follows a modified Larmor or Landau–Lifshitz form (see Eq. (27.2))

$$\dot{\mathbf{m}}_i = -\gamma \mathbf{m}_i \times \mathbf{h}_i^{total} - \frac{\beta\gamma}{M_s}\mathbf{m}_i \times \left(\mathbf{m}_i \times \mathbf{h}_i^{total}\right), \qquad (27.46)$$

where γ is the gyromagnetic ratio and $\mathbf{m}_i$ is obtained from the saturation magnetization M_s together with the chosen grid for the macrospins. The second term on the right of (27.46) accounts for dissipation, with β a dimensionless parameter measuring the strength, where $\beta\gamma \equiv \lambda$ (see Eq. (27.18a)). We will write the total field as the sum of three contributions:

$$\mathbf{h}_i^{total} = \mathbf{h}_i^{exchange} + \mathbf{h}_i^{dipole} + \mathbf{H}_0, \qquad (27.47)$$

where

$$\mathbf{h}_i^{dipole} = \sum_{j \neq i}\left[\frac{3\mathbf{r}_{ij}(\mathbf{m}_j \cdot \mathbf{r}_{ij})}{\mathbf{r}_{ij}^5} - \frac{\mathbf{m}_j}{r_{ij}^3}\right] \qquad (27.48)$$

and

$$\mathbf{h}_i^{exchange} = J\sum_{NN}\mathbf{m}_i, \qquad (27.49a)$$

with

$$J = 2\frac{A}{M_s^2 a^2}, \qquad (27.49b)$$

where A is an exchange stiffness parameter and a is the chosen distance between nearest-neighbor dipoles. One can also add an anisotropy field; for the uniaxial case the crystalline anisotropy energy has the form $E_A = K\sin^2\theta$ (see Eq. (26.23a)). The resultant torque, $\mathbf{N}_A = \partial E_A/\partial\boldsymbol{\theta}$, may be written in a vector form as

$$\mathbf{N}_A = K\frac{(\mathbf{m}_i \cdot \hat{\mathbf{H}}_A)\mathbf{m}_i \times \hat{\mathbf{H}}_A}{M_s^2},$$

or as an effective magnetic field,

$$\mathbf{h}_i^{anisotropy} = \frac{K(\mathbf{m}_i \cdot \hat{\mathbf{H}}_A)}{M_s^2}\hat{\mathbf{H}}_A, \qquad (27.50)$$

where the unit vector, $\hat{\mathbf{H}}_A$, points along the anisotropy axis; for an easy axis system, K is positive while for an easy plane system it is negative. In our example we will limit ourselves to dipole–dipole and exchange interactions only.

As with our discussion of ferromagnetic resonance in slabs, in Section 27.2, we linearize the problem by writing both the applied fields and magnetic moments as the sum of a zeroth-order (static) part and a small first-order, time-dependent perturbation:

$$\mathbf{m}_i = \mathbf{m}_i^{(0)} + \mathbf{m}_i^{(1)}(t) \tag{27.51a}$$

and

$$\mathbf{h}_i = \mathbf{h}_i^{(0)} + \mathbf{h}_i^{(1)}(t). \tag{27.51b}$$

Substituting these forms into (27.46) and retaining only linear terms, we have

$$-\frac{d\mathbf{m}_i^{(1)}}{dt} = \gamma\left[\mathbf{m}_i^{(0)} \times \mathbf{h}_i^{(1)} + \mathbf{m}_i^{(1)} \times \mathbf{h}_i^{(0)}\right] + \frac{\gamma\beta}{M_s}\mathbf{m}_i^{(0)} \times \left[\mathbf{m}_i^{(0)} \times \mathbf{h}_i^{(1)} + \mathbf{m}_i^{(1)} \times \mathbf{h}_i^{(0)}\right]. \tag{27.52}$$

We assume a solution of the form

$$\mathbf{m}_i^{(1)}(t) = \mathbf{V}_i^{(k)} e^{-i\omega^{(k)}t}, \tag{27.53}$$

where

$$\omega^{(k)} = \omega'^{(k)} - i\beta\omega''^{(k)}, \tag{27.54}$$

and $\omega^{(k)}$ and $\mathbf{V}^{(k)}$ are, respectively, the eigenvalues and (right) eigenvectors, which can be obtained by numerically solving Eq. (27.52). The index k numbers the modes, the total number of which equals the number of spins N used in the simulation. After finding the eigenvectors and eigenvalues, one can apply the formalism used for solving a set of N-coupled, first-order, linear differential equations in the presence of an inhomogeneous driving term, represented here by an r.f. (or microwave) magnetic field, $\mathbf{h}^{(rf)}(t)$, with the result

$$\mathbf{m}^{(1)}(t) = \mathbf{V}^{(k)}\left[e^{-i\omega^{(k)}t}c_k + e^{-i\omega^{(k)}t}\int e^{i\omega^{(k)}t}\int \mathbf{V}_L^{(k)*}\cdot \mathbf{g}(t)dx^3dt\right], \tag{27.55}$$

where

$$\mathbf{g}(t) = -\gamma\mathbf{m}^{(0)} \times \mathbf{h}^{(rf)}(t); \tag{27.56}$$

note that it is the so-called left eigenvector, $\mathbf{V}_L^{(k)}$, that appears in this expression.[11] The first term is the transient response and the constants, c_k, depend on the initial conditions; the second term describes the steady-state behavior. Knowing

[11] Note that since the matrix associated with the set of equations (27.52) is non Hermitian, the eigenvectors when operating to the right and left differ.

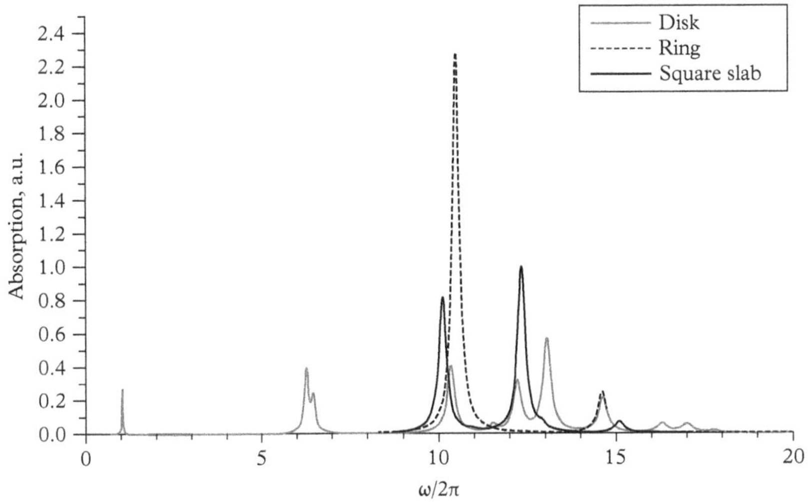

Figure 27.4 *Resonant modes of a thin ferromagnetic disc, ring, and square slab in the absence of an external magnetic field. (After Rivkin et al. (2005).)*

the spectrum of spin waves excited by the external RF field, one can calculate the average power absorption per unit time; this takes the form (which we simply state)

$$\dot{E} = \omega\gamma \mathrm{Re}\left(-\frac{\int \mathbf{V}_L^{(k)*}\cdot\left(\mathbf{m}^{(0)}\times\mathbf{h}^{(rf)}\right)dx^3}{\omega-\omega'^{(k)}+i\beta\omega''^{(k)}}\int \mathbf{V}^{(k)}\cdot\mathbf{H}^{(rf)*}dx^3\right), \qquad (27.57)$$

where here both right and left eigenvectors appear.

The method has been applied to calculate the absorption spectra of a variety of objects. One application was to calculate the *absorption spectra* in permalloy at zero field of a circular disc (175 nm in diameter), a circular ring (175 nm OD, 35nm ID), and a square slab (175 nm edge), all with a 25 nm thickness; the results are shown in Fig. 27.4.[12] The resonance at low fields arises from the *gyroscopic precession of the core* of the *single-vortex state* of the disc and the square.

27.8.3 The direct approach

An alternative to the eigenvalue approach, previously discussed, is to directly integrate the coupled Landau–Lifshitz equations with respect to time for an assembly of spins using, say, the Runge–Kutta (R–K) algorithm. This is a powerful approach and a computer program that performs this integration was developed some time ago by the National Institute for Standards and Technology (NIST).[13] A disadvantage of the method is that it does not yield eigenfrequencies directly; they must be inferred via a large response to an external drive at specific frequencies.[14] However, this approach is applicable to the nonlinear regime. The dynamic switching of memory elements is an important example of an intrinsically nonlinear phenomenon.

[12] Microscopic study of magnetostatic spin waves, K. Rivkin, L.E. DeLong, J.B. Ketterson, *J. App. Phys.* 97, 10E309 Part 2 (2005).

[13] This is the so-called OOMMF code, developed by the National Institute of Standards and Technology (NIST), that is in the public domain.

[14] Micromagnetic calculations of ferromagnetic resonance in submicron ferromagnetic particles, S. Jung, J.B. Ketterson, and V. Chandrasekhar, *Phys. Rev. B* **66**, 132405 (2002).

ADDITIONAL READING

The series *Magnetism*, cited in the previous chapter, also contains chapters on the dynamic response of magnetic materials.

Spin waves are discussed in the following two books: *Spin Waves*, A.I. Akhiezer, V.G. Bar'yakhtar, and S.V. Peletminskii, translated from the Russian by S. Chomet; North-Holland, Amsterdam. *Spin Waves: Theory and Applications*, Daniel D. Stancil and Anil Prabhakar, Springer, Berlin (2009).

Much of the current research on magnetization dynamics involves nanostructures, which are discussed in *Nonlinear Magnetization Dynamics in Nanosystems*, Isaak D. Mayergoyz, Giorgio Bertotti, and Claudio Serpico, Elsevier, Amsterdam, the Netherlands (2009).

An older but excellent book on micromagnetics is *Micromagnetics*, William Fuller Brown, Interscience Publishers (1963).

Problems

27.1. Demonstrate the equivalence of Eqs. (27.18a) and (27.18b).

27.2. Find the ferromagnetic resonance frequency for the case where the magnetic field lies in the plane of the slab, but the uniaxial symmetry axis is directed perpendicular to the slab.

Appendix 27A Magnetostatic modes

In Section 27.2 we discussed ferromagnetic resonance, where effects arising from the Maxwell boundary conditions governing the internal field alter the local Larmor frequency seen by the precessing spins, which results in the Kittel equations. In Section 27.4 we showed how the presence of exchange results in wave-like excitations, the frequencies of which lie adjacent to the spatially uniform precession frequency at long wavelengths and which are termed spin-wave excitations. However at still longer wavelengths, effects arising from a more careful treatment of the local field seen by the precessing spins enter; the resulting modes are referred to variously as Walker modes[15] (in ellipsoids) or Damon and Eshbach modes (in rectangular plates).[16]

[15] Magnetostatic modes in ferromagnetic resonance, L.R. Walker, *Phys. Rev.* **105**, 390 (1957).

[16] Magnetostatic modes in a ferromagnetic slab, R.W. Damon and J.R. Eshbach, *J. Chem. Phys. Sol.* **19**, 308 (1961). See also, Considerations on the propagation of magnetostatic waves and spin waves, P.C. Fletcher and C. Kittel, *Phys. Rev.* **120**, 2004 (1960). The topic is also discussed in the book, *Light Scattering in Magnetic Solids*, Ch. 8, M.G. Cottam and D.J. Lockwood, Wiley (1986).

27A.1 Damon and Eshbach theory of spin waves in a platelet

27A.1.1 Basic equations

The geometry utilized in many experiments is that of a platelet of thickness s and we will only consider the case of an in-plane external magnetic field, which in our chosen coordinate system (see Fig. 27A.1) we write as $\mathbf{H}_0 = H_0\mathbf{z}$, where $\mathbf{z}$ is a unit vector along the z axis. The modes must simultaneously satisfy the Landau–Lifshitz equation for $\mathbf{M}$ as well as Maxwell's equations. The first of these, which governs the precession dynamics of the magnetization $\mathbf{M}$, is given in the absence of dissipation by

$$\frac{d\mathbf{M}}{dt} = \gamma \mathbf{M} \times \mathbf{H}, \tag{27A.1}$$

where γ is the electron gyromagnetic ratio and $\mathbf{H}$ is the *internal* magnetic field (with the geometry-dependent demagnetization effects included). Since the phase velocities of spin waves turn out to be many orders of magnitude smaller than the velocity of light (and here we assume insulating samples where there are no ohmic currents), the Maxwell equations reduce to

$$\nabla \cdot \mathbf{B} = \nabla \cdot (\mathbf{H} + 4\pi\mathbf{M}) = 0 \tag{27A.2a}$$

and

$$\nabla \times \mathbf{H} = 0; \tag{27A.2b}$$

this is referred to as the *magnetostatic limit*.

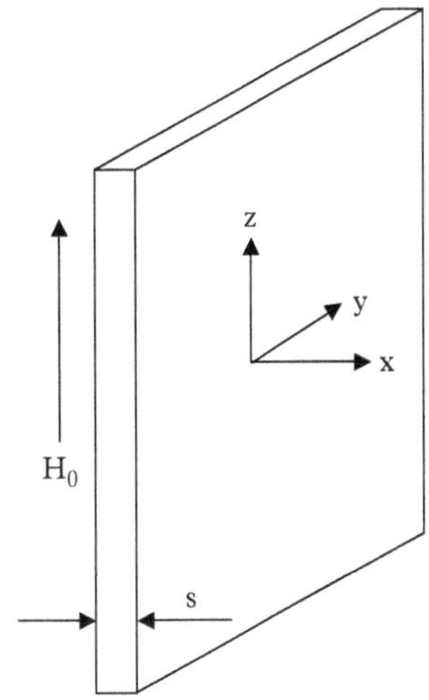

Figure 27A.1 *Coordinate system for spin waves propagating in a platelet of thickness s.*

We will assume that $\mathbf{H}_0$ is large enough that the internal static magnetization, $\mathbf{M}_0$, is uniform in our (infinite) platelet geometry. From the Maxwell boundary condition on the continuity of the tangential fields, the static internal field is also $\mathbf{H}_0$, along which $\mathbf{M}_0$ aligns in equilibrium. Allowing for dynamic contributions to these two fields, which we write as $\mathbf{h}(\mathbf{r}, t)$ and $\mathbf{m}(\mathbf{r}, t)$, and restricting ourselves to a single frequency, we can then write

$$\mathbf{H}(\mathbf{r}, t) = H_0\mathbf{z} + \mathbf{h}(\mathbf{r})e^{i\omega t} \tag{27A.3a}$$

and

$$\mathbf{M}(\mathbf{r}, t) = M_0\mathbf{z} + \mathbf{m}(\mathbf{r})\, e^{i\omega t}. \tag{27A.3b}$$

Here we ignore any external driving field, $\mathbf{h}^{ext}(\mathbf{r}, t)$, and hence $\mathbf{h}(\mathbf{r}, t)$ will arise solely from $\mathbf{m}(\mathbf{r}, t)$; i.e., we solve for the homogeneous problem for the collective modes of the system. Equation (27A.2b) allows us to write $\mathbf{h}(\mathbf{r}, t) = \nabla\psi(\mathbf{r}, t)$, where $\psi(\mathbf{r}, t) = \psi(\mathbf{r})e^{i\omega t}$ is a magnetic scalar potential; inserting this form for $\mathbf{h}(\mathbf{r}, t)$ into Eq. (27A.2a), Maxwell's equations reduce to

$$\nabla^2\psi + 4\pi\nabla \cdot \mathbf{m} = 0. \tag{27A.4}$$

Returning to Eq. (27A.1), substituting our forms for **H** and **M** from (27A.3a,b), with **h** written in terms of ψ, and retaining only terms linear in **m** and ψ yields, after some algebra,

$$4\pi m_x = \kappa\frac{\partial\psi}{\partial x} - i\nu\frac{\partial\psi}{\partial y} \tag{27A.5a}$$

and

$$4\pi m_y = i\nu\frac{\partial\psi}{\partial x} + \kappa\frac{\partial\psi}{\partial y}, \tag{27A.5b}$$

where we have introduced the dimensionless parameters (used in much of the literature)

$$\kappa = \frac{\Omega_H}{\Omega_H^2 - \Omega^2}; \quad \nu = \frac{\Omega}{\Omega_H^2 - \Omega^2}; \tag{27A.6a,b}$$

with

$$\Omega_H = \frac{H_0}{4\pi M_0}; \quad \Omega = \frac{\omega}{4\pi\gamma M_0}. \tag{27A.7a,b}$$

Substituting Eqs. (27A.5a,b) for **m** into Eq. (27A.4), we obtain the following equation for ψ:

$$(1+\kappa)\left(\frac{\partial^2\psi^i}{\partial x^2} + \frac{\partial^2\psi^i}{\partial y^2}\right) + \frac{\partial^2\psi^i}{\partial z^2} = 0; \tag{27A.8a}$$

outside the sample, where $M_0 = 0$ and hence $\kappa = 0$, we have

$$\nabla^2\psi^o = 0. \tag{27A.8b}$$

27A.1.2 Boundary conditions, solution forms, and dispersion relations

Matching the tangential component, H_y, requires (apart from an irrelevant additive constant)

$$\psi^i\Big|_{x=\pm s/2} = \psi^o\Big|_{x=\pm s/2}. \tag{27A.9a}$$

Matching the normal component B_x requires $h_x^i + 4\pi m_x = h_x^o$ or, on using (27A.5a),

$$\left[(1+\kappa)\frac{\partial\psi^i}{\partial x} - i\nu\frac{\partial\psi^i}{\partial y}\right]\Bigg|_{x=\pm s/2} = \frac{\partial\psi^o}{\partial x}\Bigg|_{x=\pm s/2}. \tag{27A.9b}$$

We seek a solution to these equations of the form

$$\psi(x, y, z) = X(x)Y(y)Z(z). \qquad (27A.10)$$

Given that Eqs. (27A.8a,b) are linear, second-order, partial differential equations with constant coefficients, the solutions X(x), Y(y), and Z(z) will either be oscillatory or exponential forms. Demanding that the boundary condition (27A.9a) be satisfied for all y and z requires

$$\psi^i(\mathbf{r}) = X^i(x)Y(y)Z(z) \qquad (27A.11a)$$

and

$$\psi^o(\mathbf{r}) = X^o(x)Y(y)Z(z), \qquad (27A.11b)$$

where X^i and X^o apply to the regions inside and outside the sample respectively. Our interest here is in modes propagating in the sample plane, where we can write

$$Y(y) \propto e^{-ik_y y} \qquad (27A.12a)$$

and

$$Z(z) \propto e^{-ik_z z}. \qquad (27A.12b)$$

Outside the sample, X(x) must fall off exponentially; hence,

$$X^o(x) = Ce^{-\kappa_x x}, \quad x > s/2, \qquad (27A.13a)$$

and

$$X^o(x) = De^{+\kappa_x x}, \quad x < -s/2. \qquad (27A.13b)$$

Inside the sample we anticipate oscillatory behavior and write

$$X^i(x) = A \sin k_x x + B \cos k_x x \qquad (27A.14)$$

Inserting the forms (27A.12a,b), (27A.13a,b), and (27A.14) into (27A.8a,b) then gives

$$(1 + \kappa)(k_x^2 + k_y^2) + k_z^2 = 0 \qquad (27A.15a)$$

and

$$\kappa_x^2 - k_y^2 - k_z^2 = 0. \qquad (27A.15b)$$

Applying our first boundary condition (27A.9a), we obtain

$$A = \frac{C - D \exp(-\kappa_x s/2)}{2 \quad \sin(k_x/2)} \tag{27A.16a}$$

and

$$B = \frac{C + D \exp(-\kappa_x s/2)}{2 \quad \cos(k_x/2)}. \tag{27A.16b}$$

Applying our second boundary condition (27A.9b) at x = +s/2 while also eliminating A and B using (27A.16a,b) gives

$$(1+\kappa)k_x\left[\frac{(C-D)}{2}\cot\left(\frac{k_x s}{2}\right) - \frac{(C+D)}{2}\tan\left(\frac{k_x s}{2}\right)\right] - C\nu k_y = -C\kappa_x; \tag{27A.17a}$$

in the same way, matching x = −s/2 gives

$$(1+\kappa)k_x\left[\frac{(C-D)}{2}\cot\left(\frac{k_x s}{2}\right) + \frac{(C+D)}{2}\tan\left(\frac{k_x s}{2}\right)\right] - D\nu k_y = D\kappa_x. \tag{27A.17b}$$

Setting the determinant of the coefficients of the homogeneous set of equations (27A.17a,b) for C and D to zero yields

$$\kappa_x^2 + 2\kappa_x k_x(1+\kappa)\cot(k_x s) - k_x^2(1+\kappa)^2 - \nu^2 k_y^2 = 0. \tag{27A.18}$$

Equation (27A.18), together with Eqs. (27A.15a,b), governs the dispersion of in-plane propagating magnons in an infinite platelet in terms of the wave vector components κ_x, k_x, k_y, k_z, and the parameter $\kappa = \kappa(\omega, H_0, M_0)$. For fixed H_0 and M_0, we can in principle solve (27A.18) together with (27A.15a,b) for $\omega = \omega(k_x, k_y)$, or, equivalently, $\omega = \omega(k, \phi)$ where $k = \sin\phi k_z + \cos\phi k_y$. We start our analysis of Eq. (27A.18) by discussing propagation perpendicular and parallel to the applied field.

27A.1.3 Propagation perpendicular to the field

Here we have $k_z = 0$ ($\phi = 0$) and Eqs. (27A.15a,b), leading to

$$k_x^2 = -k_y^2; \kappa_x^2 = k_y^2 \tag{27A.19a,b}$$

or

$$\kappa = -1; \kappa_x^2 = k_y^2. \tag{27A.20a,b}$$

From the definition (27A.6a), Eq. (27A.20a) is equivalent to $\Omega^2 = \Omega_H^2(\Omega_H + 1)$, or on using (27A.7b),

$$\omega = \gamma\sqrt{H_0(H_0 + 4\pi M_0)}. \tag{27A.21}$$

Note that (27A.20b), together with (27A.18), require $\kappa_x^2 = k_y^2 = 0$, and hence (27A.21) corresponds to a *uniform precession* of the magnetization; it is the well-known Kittel equation, obtained earlier as Eq. (27.10).

For the case when (27A.19a) applies, k_x must be imaginary if we are to have a propagating mode (real k_y). Together with (27A.19b), we conclude that only a *surface wave* is allowed, i.e., one that decays exponentially on going into or out of the film (and in the present case is governed by the same decay length, κ_x^{-1}). For these waves, (27A.18) becomes

$$1 + (1+\kappa)^2 - \nu^2 + 2i(1+\kappa)\cot(ik_y s) = 0. \tag{27A.22}$$

In the thick film limit ($s \to \infty, \cot(ik_y s) \to -i$), we obtain $\nu = \pm(\kappa + 2)$, with the lower sign corresponding to the physical root which, in turn, leads to $\Omega = \Omega_H + 1/2$ or

$$\omega = \gamma\,(H + 2\pi M). \tag{27A.23}$$

This mode is generally referred to as the Daman–Eshbach mode; its frequency lies *higher* than the Kittel mode (or other volume modes to be discussed).

27A.1.4 Propagation parallel to the field

Here we have $k_y = 0$ or $\phi = \pi/2$, and Eqs. (27A.15a,b) require that

$$k_x^2 = -\frac{k_z^2}{(1+\kappa)}, \quad \kappa_x^2 = k_z^2. \tag{27A.24a,b}$$

The first of these says we have propagating bulk waves (real k_x) provided that $1 + \kappa < 0$ or $\kappa < -1$. From (27A.6a), this requires $\Omega^2 < \Omega_H(\Omega_H + 1)$ or, equivalently, $\omega < \gamma\sqrt{H_0(H_0 + 4\pi M_0)}$; i.e., all such modes lie *below* the Kittel frequency.

Inserting (27A.20b) into (27A.18) gives

$$\cot(k_z s) = \frac{(1+\kappa)^2 - 1}{2(1+\kappa)}. \tag{27A.25}$$

The numerical solution of this equation yields, $\kappa = \kappa(k_z s, n)$, which is multi-valued with branches numbered by the index n. For a given value of $k_z s$ and branch n, we can use the associated value of κ to find $\Omega = \Omega(\Omega_H)$ for those parameters using the definition (27A.6a). The uniform mode ($k_z = 0$) corresponds to $\kappa = -1$; i.e., we recover the Kittel mode. For $k_z s \to \infty, \Omega \to \Omega_H$; i.e., $\omega \to \gamma H_0$, the free spin Larmor frequency. These various solutions are referred to as *magnetostatic modes*. They are also called *backward volume modes* since the group and phase velocities are in opposite directions.

27A.1.5 The general case

For propagation at an arbitrary angle, ϕ, we return to (27A.15a,b) and solve for κ_x and k_x in terms of k_y and k_z, and substitute these expressions into (27A.18); introducing the quantity, $\eta = k_z/k_y$, where $\phi = \tan^{-1}\eta$ and is measured from the z axis., one can then eliminate k_z to obtain

$$(1+\eta^2)^{1/2} + 2|(1+\eta^2)^{1/2}|\left(-\frac{1+\eta^2+\kappa}{1+\kappa}\right)^{1/2}(1+\kappa)\cot\left[|k_y|s\left(-\frac{1+\eta^2+\kappa}{1+\kappa}\right)^{1/2}\right] + (1+\kappa)^2\frac{1+\eta^2+\kappa}{1+\kappa} - \nu^2 = 0, \tag{27A.26}$$

which for arbitrary parameters must be solved numerically. We note in passing that the surface mode does not exist at all angles, but merges with the highest bulk mode at an angle given by $\phi_s = \tan^{-1}\left(\Omega_H^{1/2}\right)$, and beyond this point ceases to exist as a surface wave. A curious property of the surface waves is that in the limit of large thicknesses they propagate in only one direction on a given surface and in the opposite direction in the opposing surface: switching the magnetic field direction reverses the directions on the two surfaces.

27A.2 Adding exchange

In the presence of exchange, and for a cubic material, one must add a contribution to the free energy, of the form

$$U_{non-u} = \frac{1}{2}\alpha\left[\left(\frac{\partial \mathbf{M}}{\partial x}\right)^2 + \left(\frac{\partial \mathbf{M}}{\partial y}\right)^2 + \left(\frac{\partial \mathbf{M}}{\partial z}\right)^2\right], \tag{27A.27}$$

where the parameter α measures the strength of the effect. Defining an effective magnetic field as the functional derivative of this expression gives

$$\begin{aligned}\mathbf{H}_{grad} &= -\frac{\delta U_{non-u}}{\delta \mathbf{M}} = \sum_{i=1}^{3}\frac{\partial}{\partial x_i}\frac{\partial U_{non-u}}{\partial \mathbf{M}/\partial x_i}\\ &= \alpha\nabla^2\mathbf{M}.\end{aligned} \tag{27A.28}$$

Extending the dipolar theory of Daman and Eshbach to include exchange greatly increases the algebraic complexity of the calculations.[17,18,19] A perturbative treatment of spin waves in a thin platelet that combines magnetostatic and exchange effects has been given by Kalinikos and Slavin.[20] From a different perspective, Kreisel et al.[21] studied a microscopic model of spins on a cubic lattice coupled by a nearest-neighbor Heisenberg interaction, together with the long range dipole–dipole interaction; the system was infinite in y and z directions, but had a different number of layers along the x direction, to represent the sample

[17] Ferromagnetic resonance in thin films I: Normal mode frequencies, M. Sparks, *Phys. Rev. B* **1**, 3831 (1970).

[18] For a discussion of exchange effects when the field is perpendicular to the sample plane see, Magnetoexchange branches and spin-wave resonance in conducting and insulating films: Perpendicular resonance, T. Wolfram and R.E. De Wames, *Phys. Rev. B* **4**, 3125 (1971).

[19] *Spin Waves*, A.I. Akhiezer, V.G. Bar'Yakhtar, and S, Peletimnskii, North-Holland, Amsterdam (1968).

[20] Theory of dipole-exchange spin wave spectrum for ferromagnetic films with mixed exchange boundary conditions, B.A. Kalinikos and A.N. Slavin, *J. Phys. C: Solid State Phys.* **19** (1986).

[21] Microscopic spin-wave theory for yttrium-iron garnet films, A. Kreisel, F. Sauli, L. Bartosch, and P. Kopietz, *Eur. Phys. J.* **B71**, 59 (2009).

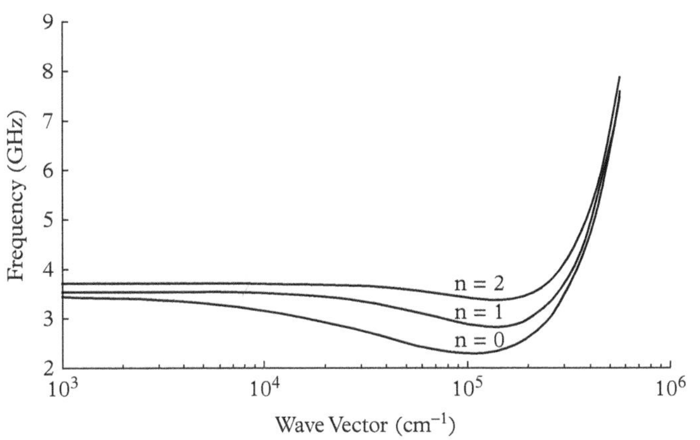

Figure 27A.2 *The three lowest magnon dispersion curves in yttrium iron garnet for propagation perpendicular to the magnetic field, as calculated for:* H = 700 G, s = 0.495 *microns,* $\mu = 2\mu_B$, $M_0 = 139.3G$, *and* $\gamma\alpha = 5.17 \times$ Gm2. *(J. Sklenar, unpublished.)*

thickness. The surface mode and the lowest lying (uniform in x) volume mode are in good agreement with the Damon–Eshbach theory for small in-plane k vectors. However, all of their higher-order volume modes, which involve progressively increasing numbers of nodes along x, are shifted upward in frequency. This arises because the wave vectors along x are quantized such that $k_x \cong n\pi/s$ and this generates an exchange contribution to the mode frequencies of $\gamma\alpha n^2\pi^2/s^2$. We find that the following form gives good overall agreement with exact calculations of Kreisel et al., *including the higher volume modes*,

$$\omega = \omega(n, k_y, k_z, H) + \gamma\alpha\frac{n^2\pi^2}{s}, \tag{27A.29}$$

where $k_{\parallel}$ is the in-plane component of the wave vector and

$$H = H_0 + \alpha k_{\parallel}^2. \tag{27A.30}$$

In Fig. 27A.2 we plot some of the low lying mode frequencies as a function of wave vector along z (// to $\mathbf{H}_0$), including the effects of exchange.

28

Magnetic Resonance

Magnetic resonance experiments involving both electrons and nuclei play a major role in solid-state physics. Electrons from uncompensated spins, particularly those involving d and f electrons, provide a powerful probe of local fields within a crystal, a subject we will not develop here.[1] Atomic nuclei play a similar role. In metals they probe the overlap of the conduction electron wave function on the nuclei and in superconductors they can be used as a probe of the energy gap.

28.1 Electron spin resonance (ESR) and nuclear magnetic resonance (NMR)

Classically, both atomic nuclei with non-zero spin (which always have a non-vanishing magnetic moment) and unpaired electrons will undergo Larmor precession if the spin axis is tipped relative to the local magnetic field. If in addition one applies an oscillating external magnetic field with a frequency that matches the Larmor precession frequency, we expect the system to resonantly absorb energy. This is the basic idea underlying magnetic resonance.[2] If nuclear spins are in resonance, we call it nuclear magnetic resonance (NMR);[3] if electron spins are in resonance, we call it electron spin resonance (ESR) or electron paramagnetic resonance (EPR). Many variants exist, some of which we discuss later in this chapter.

Quantum mechanically, the interaction of a magnetic moment, $\hat{\mu}$, with an external magnetic field, $\mathbf{H}$ (here assumed constant and uniform), can be described by a Hamiltonian, $\hat{\mathrm{H}}$, of the form

$$\hat{\mathrm{H}} = -\hat{\mu} \cdot \mathbf{H}. \tag{28.1}$$

Note that this expression is identical to the classical expression for the energy of a magnetic dipole in an external field, but where the magnetic moment is replaced by the associated quantum mechanical operator. We ignore for now any other perturbations coupling to spin. For the case of the electron, the magnetic moment operator is $\hat{\mu}_e = g\mu_B\hat{s}$, where g is the electron g-factor (2.0023), μ_B is the electron Bohr magneton, $-|e|\hbar/2m_e c$, and $\hat{s} = \hat{\sigma}/2$ is the electron spin operator, with $\hat{\sigma}$ being the vector formed from the three Pauli matrices. For nuclei, we write the magnetic moment operator as $\hat{\mu}_n = +\mu_n\hat{\mathrm{i}}$, where $\hat{\mathrm{i}}$ is the nuclear spin operator and μ_n is the empirically determined magnetic moment of the nucleus being studied,

[1] A classic text on this topic is *Electron Paramagnetic Resonance of Transition Ions*, A. Abragam and B. Bleaney, Clarendon Press, Oxford (1970).

[2] Some of the concepts discussed earlier, in Chapter 27, will be reintroduced here so that this chapter can be read independently.

[3] Four Nobel prizes have been awarded in the field of NMR: Physics 1952, Felix Bloch and Edward Mills Purcell "for their development of new methods for nuclear magnetic precision measurements and discoveries in connection therewith"; Chemistry 1991, Richard R. Ernst "for his contributions to the development of the methodology of high resolution nuclear magnetic resonance (NMR) spectroscopy"; Chemistry 2002, Kurt Wüthrich "for his development of nuclear magnetic resonance spectroscopy for determining the three-dimensional structure of biological macromolecules in solution"; Medicine 2003, Paul C. Lauterbu and Peter Mansfield "for their discoveries concerning magnetic resonance imaging."

The Physics of Solids. First Edition. J. B. Ketterson.

which is of order a so-called nuclear magneton, given by $\mu_N = +|e|\hbar/2M_pc$, where M_p is the mass of the proton. Hence nuclear moments are typically three to four orders of magnitude smaller than electron moments.

For a static magnetic field, $\mathbf{H}_0$, and in the absence of other perturbations coupling to spin (to be discussed later), the resulting spectrum consists of equally spaced levels, two for the electron (2s+1 = 2), and 2i+1 for the nuclei. These levels will be separated by an energy

$$\Delta E = \mu_{e,n} H_0, \tag{28.2}$$

which is also written in terms of an angular frequency:

$$\frac{\Delta E}{\hbar} = \omega = \gamma_{n,e} H_0, \tag{28.3}$$

where $\gamma_{n,e}$ is called the *gyromagnetic ratio.*

If a second, sinusoidally varying, magnetic field is applied, which we write as

$$\mathbf{H}_1(t) = \mathbf{H}_1(0)\cos\omega t, \tag{28.4}$$

it can induce transitions between these levels. The selection rule governing these transitions is the same as occurs in the Zeeman effect in atomic physics; i.e., $\Delta s = \pm 1, \Delta i = \pm 1$.

We start by treating the case of a single electron. We write the total Hamiltonian as[4]

$$\hat{H} = \hat{H}_0 + \hat{H}_1(t). \tag{28.5}$$

To simplify the calculations, rather than using the linearly polarized form (28.4), we will take the case of a *circularly polarized* $\mathbf{H}_1(t)$ field and write

$$\mathbf{H}_1(t) = H_1(0)(\mathbf{i}\cos\omega t - \mathbf{j}\sin\omega t), \tag{28.6}$$

where **i** and **j** are unit vectors in the x and y directions and the static field is presumed to be parallel to the z axis, $\mathbf{H}_0 = H_0\mathbf{k}$. Our total Hamiltonian is then

$$\hat{H} = \hat{H}_0 + \hat{H}_1 \tag{28.7a}$$

or, on substituting the Pauli matrices (and recalling that $\mu_e = -\mu_B$, as previously defined,

$$\hat{H} = -\mu_B \begin{pmatrix} H_0 & H_1(0)e^{i\omega t} \\ H_1(0)e^{-i\omega t} & -H_0 \end{pmatrix}. \tag{28.7b}$$

[4] Here we must be careful not confuse the Hamiltonian (an operator) with the magnetic field; similarly one must not confuse the unit vectors **i** and **j** (introduced below) with quantities involving angular momentum.

The solution to the well-known static problem is

$$\hat{H}_0 \begin{pmatrix} a_+ \\ a_- \end{pmatrix} = E \begin{pmatrix} a_+ \\ a_- \end{pmatrix}, \tag{28.8}$$

or

$$\mu_B \begin{pmatrix} H_0 & 0 \\ 0 & -H_0 \end{pmatrix} \begin{pmatrix} a_+ \\ a_- \end{pmatrix} = E \begin{pmatrix} a_+ \\ a_- \end{pmatrix}, \tag{28.9}$$

which has eigenvalues $E_\pm = \pm \mu_B H_0$ and eigenvectors

$$\psi_+ = \begin{pmatrix} 1 \\ 0 \end{pmatrix} \text{ and } \psi_- = \begin{pmatrix} 0 \\ 1 \end{pmatrix}. \tag{28.10}$$

The energy difference is therefore $\Delta E = E_+ - E_- = 2\mu_B H_0$.

To solve the time-dependent problem we must use the time-dependent Schrödinger equation:

$$i\hbar \begin{pmatrix} \dot{a}_+ \\ \dot{a}_- \end{pmatrix} = -\begin{pmatrix} \mu_B H_0 & \mu_B H_1(0) e^{i\omega t} \\ \mu_B H_1(0) e^{-i\omega t} & -\mu_B H_0 \end{pmatrix} \begin{pmatrix} a_+ \\ a_- \end{pmatrix}, \tag{28.11a}$$

or written out explicitly

$$\begin{aligned} -i\hbar \dot{a}_+ &= \mu_B H_0 a_+ + \mu_B H_1(0) e^{i\omega t} a_- \\ -i\hbar \dot{a}_- &= \mu_B H_1(0) e^{-i\omega t} a_+ - \mu_B H_0 a_-. \end{aligned} \tag{28.11b}$$

We make the change of variables, $a_+(t) = b_+(t) e^{i\mu_B H_0 t/\hbar}$ and $a_-(t) = b_-(t) e^{-i\mu_B H_0 t/\hbar}$, which we may call *going into the rotating frame*; substituting these forms into (28.11a,b), we have

$$\begin{aligned} i\hbar \dot{b}_+ &= -\mu_B H_1(0) e^{i(\omega-\omega_0)t} b_- \\ i\hbar \dot{b}_- &= -\mu_B H_1(0) e^{i(-\omega+\omega_0)t} b_+, \end{aligned} \tag{28.12}$$

where $\omega_0 \equiv 2\mu_B H_0/\hbar$. At the frequency, $\omega = \omega_0$, which we designate as the *resonance frequency*, these equations simplify to

$$\begin{aligned} i\dot{b}_+ &= -\frac{1}{2}\omega_1 b_- \\ i\dot{b}_- &= -\frac{1}{2}\omega_1 b_+, \end{aligned} \tag{28.13}$$

where we have defined the so-called *Rabi frequency*, $\omega_1 = 2\mu_B H_1/\hbar$, and we henceforth assume that H_1 is the field amplitude at $t = 0$. Eliminating b_-, we have

$$\ddot{b}_+ + (\omega_1/2)^2 b_+ = 0; \tag{28.14a}$$

thus,

$$b_+(t) = \cos\frac{1}{2}\omega_1 t$$
$$b_-(t) = i\sin\frac{1}{2}\omega_1 t. \qquad (28.14b)$$

These amplitudes satisfy the normalization $|b_+(t)|^2 + |b_-(t)|^2 = 1$ at all times. Note that the system periodically goes back and forth between the upper and lower levels with an angular frequency ω_1. In particular, one could "invert" the system by applying a pulsed H_1 field for a time corresponding to $\omega_1 t_\pi = \pi$, which is referred to as a *pi pulse*. Note that as the problem has been formulated no overall absorption occurs; energy absorbed from the field on making transitions to the upper level is returned by transitions to the lower level (the probability for stimulated absorption is equal to that for stimulated emission).

Let us develop a semi-classical description. We recall the classical Larmor equation of motion for the magnetization (Eq. (27.3)):

$$\dot{\mathbf{M}} = \gamma\,\mathbf{M} \times \mathbf{H}. \qquad (28.15)$$

In discussing (28.15), we suppose that initially we have a system of spins in equilibrium, where $\dot{\mathbf{M}} = 0$, in the presence of a static magnetic field $\mathbf{H}_0$ along the z direction. For a paramagnet in a weak field ($\mu H_0 \ll k_B T$), we expect an equilibrium magnetization, $M_0 = \chi H_0$, parallel to the z axis, where according to Curie's law, $\chi \propto T^{-1}$ (see Eq. (5.12b)). As before, we now assume that a small sinusoidal field, $\mathbf{H}_1(t)$, of the form (28.6) is applied in the x–y plane. Equation (28.15) then becomes

$$\dot{\mathbf{M}} = \gamma \begin{vmatrix} \mathbf{i} & \mathbf{j} & \mathbf{k} \\ M_x & M_y & M_z \\ H_1\cos\omega t & -H_1\sin\omega t & H_0 \end{vmatrix}, \qquad (28.16)$$

or in component form

$$\dot{M}_x = \gamma\left(M_y H_0 + M_z H_1 \sin\omega t\right) \qquad (28.17)$$
$$\dot{M}_y = \gamma\left(M_z H_1 \cos\omega t - M_x H_0\right) \qquad (28.18)$$

and

$$\dot{M}_z = \gamma\left(-M_x H_1 \sin\omega t - M_y H_1 \cos\omega t\right). \qquad (28.19)$$

We now proceed to directly solve these coupled equations of motion.[5] Suppose we initially assume that $M_z \cong M_0 \gg M_x, M_y$. Then in Eq. (28.19) we have terms of second order, $M_x H_1$, $M_y H_1$ on the right-hand side; if we neglect the time

[5] An alternative way to proceed is to go directly into a frame of reference rotating with angular velocity ω, using the operational equation $d/dt \to d/dt + \Omega\times$.

dependence of M_z, (28.17) and (28.18) become a set of coupled driven (inhomogeneous) equations. If we introduce a new variable, $M_+(t) = M_x(t) + iM_y(t) \propto e^{-i\omega t}$, and write $\omega_0 = \gamma H_0$, we can combine these two equations in the form

$$\dot{M}_+ + i\omega_0 M_+ = i\omega_1 M_0 e^{-i\omega t}, \tag{28.20}$$

where $\omega_1 = \gamma H_1 = 2\mu_B H_1/\hbar$ is again the Rabi frequency. Assuming a steady-state form, $M_+ \propto e^{-i\omega t}$, Eq. (28.20) becomes

$$M_+ = \frac{\omega_1 M_0}{\omega_0 - \omega} e^{-i\omega t}. \tag{28.21}$$

Equation (28.21) is divergent for $\omega - \omega_0 = 0$. Clearly our assumption that $M_z \cong M_0 \gg M_x, M_y$ has then broken down. (As we will see shortly, when relaxation effects are present, M_+ will remain finite at resonance.) Hence, we must remove our assumption that the time dependence of M_z can be neglected; to do this we write (28.20) more precisely as

$$\dot{M}_+ + i\omega_0 M_+ = i\omega_1 M_z(t) e^{-i\omega t}. \tag{28.22}$$

At this point we seek a particular solution to Eq. (28.22) of the form $M_+(t) = A(t)e^{-i\omega t}$. Inserting this form into (28.20), we have

$$\begin{aligned} \dot{M}_+(t) &= \dot{A}(t)e^{-i\omega t} - i\omega A(t)e^{-i\omega t} \\ &= -i\omega_0 A(t)e^{-i\omega t} + i\omega_1 M_z(t)e^{-i\omega t}. \end{aligned} \tag{28.23}$$

For the on-resonance case, $\omega = \omega_0$, this equation reduces to

$$\dot{A}(t) = i\gamma H_1 M_z. \tag{28.24}$$

Since M_z is real it follows that $A(t)$ is purely imaginary. From this fact, the definition $M_+(t) = M_x(t) + iM_y(t)$, and our form $M_+(t) = A(t)e^{-i\omega t}$, we obtain $M_x(t) = -iA(t)\sin\omega t$ and $M_y(t) = -iA(t)\cos\omega t$; inserting these in (28.19) gives

$$\begin{aligned} \dot{M}_z(t) &= i\omega_1 A(t)[\sin^2\omega t + \cos^2\omega t] \\ &= i\omega_1 A(t). \end{aligned} \tag{28.25}$$

Combining (28.24) and (28.25) gives

$$\ddot{A}(t) + \omega_1^2 A(t) = 0. \tag{28.26}$$

If we require that $M_z(t = 0) = M_0$, then from (28.25) we must have

$$A(t) = iM_0 \sin(\omega_1 t) \tag{28.27a}$$

or

$$M_z(t) = M_0 \cos(\omega_1 t). \tag{28.27b}$$

The corresponding solution for the perpendicular component of the magnetization is

$$\mathbf{M}_\perp(t) = M_0 \sin(\omega_1 t)\,(\mathbf{i} \sin \omega t + \mathbf{j} \cos \omega t)\,; \tag{28.28}$$

i.e., the perpendicular component of the magnetization vector precesses around the z axis at the frequency $\omega = \omega_0$, while both $|\mathbf{M}_\perp(t)|$ and $M_z(t)$ slowly oscillate 90° out of phase with each other, with an amplitude M_0 at a frequency $\omega_1 = \gamma H_1$. Note that $\mathbf{M}_\perp(t)$ and $\mathbf{H}_1(t)$ are 90° out of phase with each other, as is required since in the absence of relaxation effects no power is dissipated. The oscillation of M_x is analogous to what we found earlier for the spin 1/2 quantum mechanical case.

28.2 Relaxation effects

Let us next generalize the Larmor equation to include the effects of relaxation. The presence of an applied magnetic field lowers the symmetry of the system by inducing a magnetization along the z axis. Hence it is reasonable, on phenomenological grounds, to assume that the longitudinal and transverse components of **M** can relax (decay) with different characteristic times. Furthermore the transverse components must relax toward zero, while the longitudinal magnetization must relax toward the thermodynamic equilibrium value M_0 (typically given by Curie's law). We therefore generalize the z (longitudinal) component of the Larmor equation to the form

$$\frac{dM_z}{dt} = \gamma\,(\mathbf{M} \times \mathbf{H})_z + \frac{M_0 - M_z}{T_1}, \tag{28.29a}$$

where T_1 is called the *longitudinal relaxation time.* The two equations of motion for the transverse components of the magnetization are written as

$$\frac{dM_x}{dt} = \gamma\,(\mathbf{M} \times \mathbf{H})_x - \frac{M_x}{T_2} \tag{28.29b}$$

and

$$\frac{dM_y}{dt} = \gamma\,(\mathbf{M} \times \mathbf{H})_y - \frac{M_y}{T_2}, \tag{28.29c}$$

where T_2 is referred to as the *transverse relaxation time*; it is a general property that $T_2 \le T_1$. Equations (28.29a–c) are called the Bloch equations, after F. Bloch who proposed them.[6]

[6] Nuclear induction, F. Bloch, *Phys. Rev.* **70**, 460–473 (1946).

If we again assume that $M_z \cong M_0 >> M_x, M_y$, as we did earlier, then in place of Eq. (28.20), we have

$$\dot{M}_+ + i\omega_0 M_+ + \frac{M_+}{T_2} = i\omega_1 M_0 e^{-i\omega t} \tag{28.30}$$

and in place of (29.21), we have

$$M_+ = \frac{\omega_1 M_0}{\omega_0 - \omega - \frac{i}{T_2}} e^{-i\omega t}. \tag{28.31}$$

We may write the response in terms of a susceptibility as

$$\begin{aligned} M_+ &= \chi H_1 = (\chi' + i\chi'') H_1 \\ &= \left[\frac{\gamma M_0 (\omega_0 - \omega) T_2^2}{1 + (\omega_0 - \omega)^2 T_2^2} + i \frac{\gamma M_0 T_2}{1 + (\omega_0 - \omega)^2 T_2^2} \right] H_1 . \end{aligned} \tag{28.32}$$

The power absorbed involves the imaginary component of the susceptibility and is given by the time derivative of the average energy:

$$\overline{\frac{dU}{dt}} = \overline{\mathbf{H} \cdot \frac{d\mathbf{M}}{dt}} = \frac{\omega}{2} \chi''(\omega) H_1^2(t),$$

or

$$P(\omega) = \frac{\omega}{2} \frac{\gamma M_0 T_2}{1 + (\omega_0 - \omega)^2 T_2^2} H_{10}^2, \tag{28.33}$$

which forms the basis of continuous wave nmr.

28.3 Spin echoes

In solid-state research one seldom performs nuclear magnetic resonance experiments by simply observing the energy absorbed by a system from a continuously running oscillator, for reasons which will shortly become clear. Most experiments are performed using so-called *spin echo* techniques. The principles involved have much wider applicability, particularly in optics, so it is important to gain an understanding of them in at least one setting.

We start by examining the on-resonance case, where $\omega = \omega_0$ with $\omega_0 = \gamma H_0$, as given in Eqs. (28.27b) and (28.28), which we collect here for convenience:

$$M_z(t) = M_0 \cos(\omega_1 t) \tag{28.34a}$$

and

$$\mathbf{M}_\perp(t) = M_0 \sin(\omega_1 t) \left(\mathbf{i} \sin \omega_0 t + \mathbf{j} \cos \omega_0 t \right). \tag{28.34b}$$

First we make what may seem to be a curious comment: there are no "inertial effects" in magnetic resonance (the inertia, such as it is, is tied up with the quantum mechanical spinning of the particles, the nuclei, and electrons). The application of a magnetic field having a component perpendicular to the magnetic moment (spin axis) immediately generates precession at a rate governed by the strength of that field through the Larmor equation, and it stops immediately when that field is removed (i.e., there is no "coasting" afterward). The same is true when the system is precessing (at a rate ω_0) under the action of a static field, $\mathbf{H}_0$, and a second field, $\mathbf{H}_1$, which is itself precessing at the rate ω_0, is applied perpendicular to the first: the $\mathbf{H}_1$ field induces a precession at a rate $\omega_1 = \gamma H_1$ in the *frame of reference rotating at* ω_0, but now about an axis perpendicular to both $\mathbf{H}_1(t)$ and $\mathbf{M}_\perp$. This is the physical content of equations (28.34a,b).

The lifetimes, T_2, encountered in nuclear magnetic resonance experiments are often so long that the inhomogeneity in the applied H_0 field, ΔH_0, completely dominates the width of the resonance line (which is then of order ΔH_0 rather than $1/\gamma T_2$), such that it becomes impossible to determine T_2; i.e., it is difficult to determine T_2 when $\Delta H_0 > 1/\gamma T_2$. Since this line width does not correspond to a "real" line width, in the sense that there is dissipation associated with it, one speculates that there ought to be a way to "undo" the effects of this field inhomogeneity. This is but one of the many things that spin echoes allow one to do. We now address this specific problem.

Suppose we have a set of spins with a net spin which initially points along the static field, $\mathbf{H}_0$ (this corresponding to the state of thermal equilibrium), which we take as the z axis. At $t = 0$ we apply an $\mathbf{H}_1$ field at a frequency, ω_0, and leave it on for a time, $t_{\pi/2}$, corresponding to $\omega_1 t_{\pi/2} = \gamma H_1 t_{\pi/2} = \pi/2$, after which it is zero; we refer to this burst of r.f. magnetic field as a $\pi/2$ pulse. According to Eqs. (28.34a,b) we then have

$$\left.\begin{aligned} &M_z = 0 \\ &\mathbf{M}_\perp(t) = M_0\,(\mathbf{i}\sin\omega_0 t + \mathbf{j}\cos\omega_0 t) \end{aligned}\right\}\ \omega_1 t = \omega_1 t_{\pi/2} = \frac{\pi}{2}. \tag{28.35}$$

In writing this equation, we have tacitly assumed that $\gamma\Delta H_0 t_{\pi/2} \ll \pi/2$; i.e., even though the external $\mathbf{H}_0$ field is inhomogeneous, the precessing spins do not get "out of step" with each other during the time $t_{\pi/2}$ (mathematically, it assumes that the cancellation of the two terms that occurred in (28.23) is still valid during the time $t_{\pi/2}$). For subsequent times (much greater than $t_{\pi/2}$), the inhomogeneity in the magnetic field causes the spins in different parts of the sample to have different phases (or equivalently to point in different directions in the x–y plane). For simplicity, we might assume that the field varies linearly over the sample in the z direction only, i.e., $H_0(z) = H_0 + H'_0 z$, and that the sample extends symmetrically between $z = \pm L/2$. Then, in a time t in the rotating x–y frame, spins at different points in the sample having different values of z will have phase angles $\varphi(z,t) = \gamma H'_0 zt$ and are said to have "fanned out"; more generally, the precession frequency would be written $\varphi(z,t) = \gamma\Delta H(z)t$. After a sufficient time the

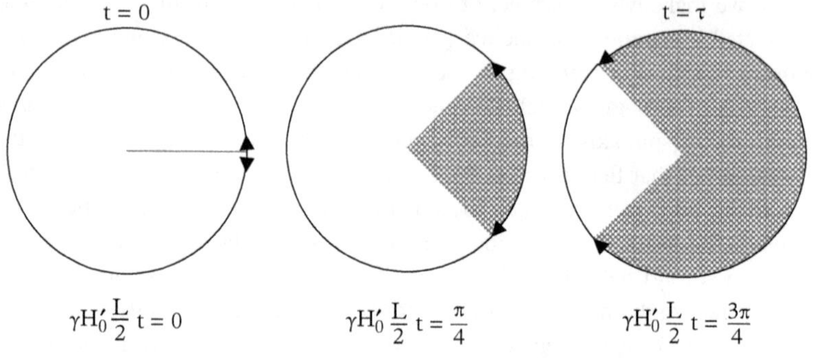

Figure 28.1 *Following the application of a* $\pi/2$ *pulse, the spins "fan out," as shown in the rotating* x-y *frame. The distribution after an arbitrary time* τ *is shown on the right. For much longer* τ*, the distribution would be nearly uniform.*

spins in different parts of the sample get spread out more or less uniformly over the entire x–y plane; i.e., $\varphi(z,t) \gg 2\pi$ throughout most of the sample. The initial stages of this fanning is shown schematically in Fig. 28.1. A signal that was earlier induced by the precessing magnetization following the $\pi/2$ pulse (often in the same coil that originally generated the $\mathbf{H}_1$ field and now being used to detect this magnetization) now disappears. The spins in different parts of the sample are still precessing in their local field; this precession continues until "real" irreversible T_2-relaxation-processes dissipate this locally macroscopic precession. So the magnetization is "there," and one just needs a trick to extract it.

As we will see, the way to recover the part of the precessing magnetization that has not been irreversibly lost[7] is to apply a second pulse, t_π, at a time τ after the first pulse. This second pulse is twice the length[8] of the first pulse, corresponding to a rotation of the spins by an angle π about an axis in the x–y plane, perpendicular to $\mathbf{H}_1$ in the rotating frame. This results in the spin system itself generating a pulse (in the coil that originally excited it) at a time 2τ.

To show this we start by rewriting (28.35), taking into account the position dependence of H_0:

$$\begin{aligned}\mathbf{M}_\perp(z,t) &= M_0\,[\sin(\omega_0 t + \varphi(z,t))\,\mathbf{i} + \cos(\omega_0 t + \varphi(z,t))\,\mathbf{j}] \\ &= M_0\,[\sin(\omega_0 t)\,\mathbf{i} + \cos(\omega_0 t)\,\mathbf{j}]\cos\varphi(z,t) \\ &\quad + M_0\,[\cos(\omega_0 t)\,\mathbf{i} - \sin(\omega_0 t)\,\mathbf{j}]\sin\varphi(z,t). \end{aligned} \tag{28.36}$$

The two square brackets in the second line correspond to components of the rotating magnetization that are at *right angles* and *parallel*, respectively, to the instantaneous direction of $\mathbf{H}_1$ (when it is activated). Thus when the t_π pulse is applied at the time τ it causes the first component to change sign, while it has no effect on the second term. Following the t_π pulse (and ignoring its width) we have

$$\begin{aligned}\mathbf{M}_\perp(z,\tau) &= -M_0\,[\sin(\omega_0\tau)\,\mathbf{i} + \cos(\omega_0\tau)\,\mathbf{j}]\cos\varphi(z,\tau) \\ &\quad + M_0\,[\cos(\omega_0\tau)\,\mathbf{i} - \sin(\omega_0\tau)\,\mathbf{j}]\sin\varphi(z,\tau) \\ &= -M_0\,[\sin(\omega_0\tau - \varphi(z,\tau))\,\mathbf{i} + \cos(\omega_0 t - \varphi(z,\tau))\,\mathbf{j}]\,; \end{aligned} \tag{28.37}$$

[7] As examples, irreversible changes in the magnetization can result from (i) the diffusion of a spin into a region where the external field differs, or (ii) fluctuations in the local field acting on one spin due to dipole–dipole coupling with neighboring spins that are themselves randomly moving and precessing.

[8] Alternatively it could be twice the amplitude of the first pulse.

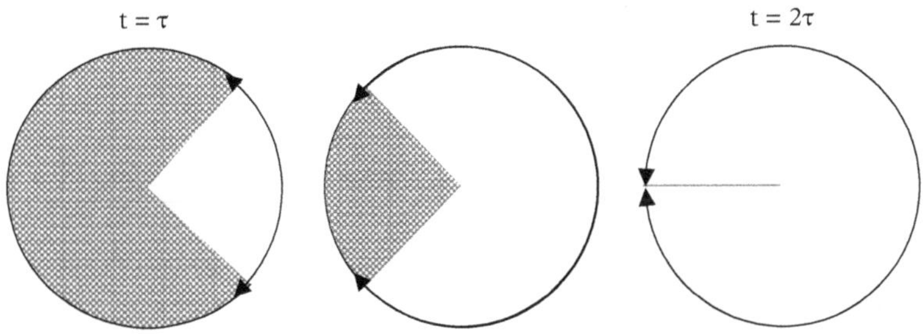

Figure 28.2 *Following the application of a π/2 pulse and waiting for a time τ, a π pulse is applied; the spins then reverse their precession and the fan collapses at a total time 2τ in the rotating x-y frame.*

after this time, the phase of the oscillatory terms continues to advance according to the form $\omega t + \varphi(z, t)$ (where t is now measured from τ) and therefore the signal is given by

$$\begin{aligned} \mathbf{M}_{\perp}(z, t) = &- M_0 \sin(\omega_0 (t + \tau) + \varphi(z, t) - \varphi(z, \tau))\, \mathbf{i} \\ &- M_0 \cos(\omega_0 (t + \tau) + \varphi(z, t) - \varphi(z, \tau))\, \mathbf{j}. \end{aligned} \tag{28.38}$$

As shown schematically in Fig. 28.2, at a time $t = \tau$ (corresponding to a total time 2τ since the original $t_{\pi/2}$ pulse), the two terms $\varphi(z, t) - \varphi(z, \tau)$ *cancel each other*, and all the dephasing that occurred due to the inhomogeneity in the $\mathbf{H}_0$ field likewise cancels; a voltage then develops across the coil (used to generate $\mathbf{H}_1$). This signal is referred to as an *echo* and hence the technique is called the *spin echo* method.[9] That part of the signal that is not recovered arises from irreversible effects associated with T_2. By measuring the echo amplitude as a function of τ one can determine T_2.

If our initial pulse corresponds to a time t_π, rather than $t_{\pi/2}$, this corresponds to *reversing* the direction of the initial magnetization; i.e., $M_z \to -M_z$. According to Eq. (28.29a) this component of the magnetization will then decay exponentially back toward positive values, with a characteristic time T_1. The progress of this decay, from which T_1 can be determined, can be monitored by applying a $t_{\pi/2}$ pulse so as to bring the remaining magnetization into the x–y plane where it is sensed by the pickup coils.

28.4 Nuclear quadrupole resonance

A nucleus with a spin $i \geq 1$ will possess a static electric quadrupole moment in addition to a magnetic dipole moment. We start by recalling the energy of a charge distribution, $\rho(\mathbf{r})$ (that of the nucleus), in the presence of an external potential, $V(\mathbf{r})$ (created by the surrounding electrons), which is given by

$$\Delta E = \int d^3r\, \rho(\mathbf{r}) V(\mathbf{r}). \tag{28.39}$$

[9] Spin echoes, E.L. Hahn, *Phys. Rev*, **80**, 580 (1950).

We expand the potential energy in the vicinity of the nucleus (located at $\mathbf{r} = 0$ for convenience) as

$$V(\mathbf{r}) = V(0) + \left.\frac{\partial V(\mathbf{r})}{\partial \mathbf{r}}\right|_{\mathbf{r}=0} \cdot \mathbf{r} + \frac{1}{2}\mathbf{r} \cdot \left.\frac{\partial^2 V(\mathbf{r})}{\partial \mathbf{r}\partial \mathbf{r}}\right|_{\mathbf{r}=0} \cdot \mathbf{r} + \cdots. \tag{28.40}$$

The zeroth-order term is absorbed into the ground state energy of the crystal and the linear (dipole) term vanishes if the center of charge (also the center of mass) of the nucleus is chosen as the origin. When the second-order term is substituted into (28.40), it leads to a contribution to the electrostatic energy of the form

$$\Delta E^{(2)} = \frac{1}{2}\sum_{i,j=1}^{3} V_{ij} \int d^3r\, \rho(\mathbf{r}) x_i x_j, \tag{28.41}$$

where

$$V_{ij} \equiv \left.\frac{\partial^2 V(\mathbf{r})}{\partial x_i \partial x_j}\right|_{\mathbf{r} = 0.} \tag{28.42}$$

We can diagonalize the symmetric second rank tensor V_{ij} via a principal axis transformation; applying Laplace's equation to (28.40) and evaluating at $\mathbf{r} = 0$ yields $\sum_{i=1}^{3}(\partial^2 V(\mathbf{r})/\partial x_i^2)|_{\mathbf{r}=0}$; i.e., the tensor V_{ij} is traceless. Since by symmetry all three components must be identical for a cubic crystal, we have immediately that $\Delta E^{(2)}$ vanishes in this case. We rewrite (28.41) in the form

$$\Delta E^{(2)} = \frac{1}{6}\left\{\sum_{i,j=1}^{3} V_{ij} \int d^3r\rho(\mathbf{r})\left[3x_i x_j - \delta_{ij} r^2\right] + \sum_{i=1}^{3} V_{ii} \int d^3r\, \rho(\mathbf{r}) r^2\right\}; \tag{28.43}$$

by the previous argument, the last term vanishes and we rewrite the first term as

$$\Delta E^{(2)} = \frac{1}{6}\sum_{i,j=1}^{3} V_{ij}\, Q_{ij}, \tag{28.44}$$

where we have defined a quadrupole tensor, Q_{ij}, as

$$Q_{ij} = \int d^3r\, \rho(\mathbf{r})\left[3x_i x_j - \delta_{ij} r^2\right]; \tag{28.45}$$

since $\bar{r}^2 = \sum_{i=1}^{3} \bar{x}_i^2$, we see that the trace of $\mathbf{Q}$ also vanishes.

The quantum mechanical operator corresponding to (28.45) follows from replacing the charge density by $\rho(\mathbf{r}) = |e| \sum_{p=1}^{Z} \delta(\mathbf{r} - \mathbf{r}_p)$, where $\mathbf{r}_p$ denotes the positions of the Z protons in the nucleus; hence,

$$Q_{ij} = |e| \sum_{p=1}^{Z} \left[3x_{ip}x_{jp} - \delta_{ij}r_p^2\right]. \tag{28.46}$$

The expectation value of the quadrupole operator is then

$$Q_{ij} = |e| \int d^3\{x_{ip}\}d^3\{x_{in}\} \sum_{p=1}^{Z} \psi^*(\{x_{in}\},\{x_{ip}\}) \left[3x_{ip}x_{jp} - \delta_{ij}r_p^2\right] \psi(\{x_{in}\},\{x_{ip}\}), \tag{28.47}$$

where $\{x_{in}\},\{x_{ip}\}$ denotes the three Cartesian coordinates of the N neutrons and Z protons. Now the quadrupole operator has the structure of an L = 2 spherical harmonic. From the theory of angular momentum, the product $[3x_ix_j - \delta_{ij}r^2]\psi_{i,i_z}$, where ψ_{i,i_z} denotes the angular momentum quantum numbers of the nuclear wave functions (with the coordinates suppressed), would be a superposition with states $\psi_{i'_z}$, where $i'_z = i+2, i+1, i, i-1$, and $i-2$. If originally $i_z = +1/2$, then the i'_z making up the superposition involve 7/2, 5/2, 3/2, −1/2, −3/2, all of which are orthogonal to the original state. On the other hand, if $i_z = +1$, then $i'_z = 3, 2, 1, 0$, and -1, which contains the needed state. Hence a nucleus must have $i = 1$ or greater in order to have a quadrupole moment (confirming the statement given at the beginning of this section).

The existence of a total angular momentum axis **i** implies rotational (cylindrical) symmetry about this axis,[10] which we take as the z axis and the principal element of the quadrupole tensor along that axis we write as Q_{zz}. The rotational symmetry requires that $Q_{xx} = Q_{yy}$. Combining this with the requirement that the trace of **Q** vanishes, we have $Q_{xx} = Q_{yy} = \frac{1}{2}Q_{zz}$; i.e., there is only one independent component of the quadrupole tensor. If we perform a principal axis transformation of our (classical) quadrupole tensor, the new elements would be given by

$$\mathbf{Q} = \mathbf{R}(\theta,\phi,\psi) \begin{pmatrix} -Q_{zz}/2 & 0 & 0 \\ 0 & -Q_{zz}/2 & 0 \\ 0 & 0 & Q_{zz} \end{pmatrix} \tilde{\mathbf{R}}(\theta,\phi,\psi), \tag{28.48a}$$

where $\mathbf{R}(\theta,\phi,\psi)$ is the Euler rotation matrix, familiar from rigid body mechanics and θ,ϕ,ψ denote the Euler angles. The cylindrical symmetry allows us to consider a single rotation about, say, the x axis, and write **R** as

$$\mathbf{R} = \begin{pmatrix} 1 & 0 & 0 \\ 0 & \cos\theta & \sin\theta \\ 0 & -\sin\theta & \cos\theta \end{pmatrix}, \tag{28.48b}$$

[10] Were nuclei like molecules, with the average nucleon positions fixed, this would not be the case. Nuclei, however, are liquid drops.

with the result that Q_{zz} along the new z axis is given by

$$Q_{z'z'} = \frac{1}{2}\left(3\cos^2\theta - 1\right)Q_{zz}, \tag{28.49a}$$

where

$$Q_{zz} \equiv \int d^3r\rho(\mathbf{r})[3z^2 - r^2]. \tag{28.49b}$$

We now make the transition to the quantum mechanical case. The angle Θ between the nuclear angular momentum axis, **i**, and the quantization axis (in this case the external magnetic field direction, $\mathbf{H}_0$) follows from the vector model as

$$\cos\Theta = \frac{i_z}{\sqrt{i(i+1)}}. \tag{28.50}$$

When this angle is substituted into (28.49a) it yields

$$Q_{z'z'} = \frac{1}{2}\left(\frac{3i_z^2 - i(i+1)}{i(i+1)}\right)Q_{zz}. \tag{28.51}$$

The largest value $Q_{z'z'}$ can have is for $i_z = i$, which gives

$$\begin{aligned} Q_{z'z'}^{max} &= \frac{1}{2}\left(\frac{i(2i-1)}{i(i+1)}\right)Q_{zz} \\ &\equiv Q. \end{aligned} \tag{28.52}$$

In terms of this quantity we may write

$$Q_{z'z'} = \frac{3i_z^2 - i(i+1)}{i(2i-1)}Q, \tag{28.53}$$

or as an operator,

$$\hat{Q}_{z'z'} = \frac{3\hat{i}_z^2 - \hat{i}^2}{i(2i-1)}Q. \tag{28.54}$$

Using (28.44) we can project $\hat{Q}_{zz}$ along the other two principal axes, obtaining the quadrupole tensor elements:

$$\hat{Q}_{x'x'} = \frac{3\hat{i}_x^2 - \hat{i}^2}{i(2i-1)}Q; \qquad \hat{Q}_{y'y'} = \frac{3\hat{i}_y^2 - \hat{i}^2}{i(2i-1)}Q, \tag{28.55a,b}$$

where Θ is now the angle that $\hat{\mathrm{i}}$ makes with respect to these two axes. The quadrupole Hamiltonian operator based on the classical energy (22.44), and in a coordinate system where V_{ij} is diagonal, can now be written as

$$\hat{H}^{(Q)} = \frac{Q}{6i(2i-1)}\left[V_{zz}\left(3\hat{i}_z^2 - \hat{i}^2\right) + V_{yy}\left(3\hat{i}_y^2 - 2\hat{i}^2\right) + V_{xx}\left(3\hat{i}_x^2 - 2\hat{i}^2\right)\right]. \quad (28.56)$$

Using $\sum_i V_{ii} = 0$ and $\sum_i \hat{i}_i^2 = \hat{i}^2$, Eq. (28.56) can be rewritten as

$$\hat{H}^{(Q)} = \frac{Q}{4i(2i-1)}\left[V_{zz}\left(3\hat{i}_z^2 - \hat{i}^2\right) + (V_{xx} - V_{yy})\left(\hat{i}_x^2 - \hat{i}_y^2\right)\right]. \quad (28.57)$$

We note that by introducing the raising and lowering operators,

$$\hat{i}_+ = \hat{i}_x + i\,\hat{i}_y; \; \hat{i}_- = \hat{i}_x - i\,\hat{i}_y \quad (28.58a,b)$$

and the parameters

$$V_0 = V_{zz}; \; V_{\pm 1} = V_{zx} \pm iV_{zy}; \; V_{\pm 2} = \frac{1}{2}\left(V_{xx} - V_{yy}\right) \pm iV_{xy}, \quad (28.59a–c)$$

one can write the quadrupole Hamiltonian for an *arbitrary coordinate system* as

$$\hat{H}^{(Q)} = \frac{Q}{4i(2i-1)}\left[V_0\left(3\hat{i}_z^2 - \hat{i}^2\right) + V_{+1}\left(\hat{i}_-\hat{i}_z + \hat{i}_z\hat{i}_-\right) + V_{-1}\left(\hat{i}_+\hat{i}_z + \hat{i}_z\hat{i}_+\right) + V_{+2}\hat{i}_-^2 + V_{-2}\hat{i}_+^2\right]. \quad (28.60)$$

The Hamiltonian (28.60) produces a splitting of the nuclear levels at zero magnetic field, which can be observed by sweeping the applied r.f. frequency.[11] It also shifts the positions of the lines at finite field and the energy levels follow from simultaneously diagonalizing the sum of the Zeeman and quadrupole Hamiltonians.

We will only treat the simplest case in which the system has cylindrical symmetry where, from (28.56), the total Hamiltonian is given by

$$\hat{H}^{(\text{total})} = -\gamma_n \hbar H_0 \hat{i}_{z'} + \frac{QV_o}{4i(2i-1)}\left(3\hat{i}_z^2 - \hat{i}^2\right); \quad (28.61)$$

in general $\mathbf{z}'$ is not parallel to $\mathbf{z}$. When $\mathbf{z}$ is parallel to $\mathbf{z}'$ the eigenvalues are given by

$$E_n = -\gamma_n \hbar H_0 i_z + \frac{QV_o}{4I(2i-1)}\left[3i_z^2 - i(i+1)\right]; \quad (28.62)$$

if these two axes are not parallel, the problem is more complex but an analytic result may be obtained using perturbation theory, either in a high or low field limit. We remark in passing that the computation of the V_{ii} is in general non-trivial.

[11] Nuclear electric quadrupole interactions in crystals, R.V. Pound, *Phys. Rev.* **79**, 685, 1950; Nuclear quadrupole resonance, H.G. Dehmelt, *Am. J. Phys.* **22**, 110 (1954).

28.5 The quantum mechanical expression for energy dissipation

From Eq. (28.33), the energy dissipated per unit time is given by $\dot{E} = \omega\chi''(\omega)H_1^2/2$ or $\chi''(\omega) = 2\dot{E}/\omega H_1^2$; here $\chi''(\omega)$ is the imaginary part of the magnetic susceptibility and H_1 is the amplitude of the applied radio frequency field. Microscopically, the net energy absorbed per unit time for a transition between two nuclear levels n and m is given by the photon energy associated with the transition multiplied by the difference between the rate that photons are absorbed and emitted by the nuclei:

$$\dot{E} = \hbar\omega\left[p_n w_{n\to m} - p_m w_{m\to n}\right], \tag{28.63}$$

where $w_{n\to m}$ is the transition rate from n to m and p_n is the probability that the state n is occupied. In terms of the imaginary part of the susceptibility (28.63) is then

$$\chi''(\omega) = \frac{2\hbar}{H_1^2}\sum_{n,m}\left[p_n w_{n\to m} - p_m w_{m\to n}\right]. \tag{28.64}$$

Assuming a linear polarization,

$$\mathbf{H}_1(t) = \mathbf{x}H_{10}\cos\omega t = \mathbf{x}\frac{H_{10}}{2}\left(e^{i\omega t} + e^{-i\omega t}\right) \tag{28.65}$$

where $\mathbf{x}$ is a Cartesian unit vector and only one of the exponentials results in resonant absorption for a given sign of ω; this yields the perturbing Hamiltonian

$$\hat{H}_1(t) = -H_{10}\frac{e^{i\omega t} + e^{-i\omega t}}{2}\hat{\mu}_x. \tag{28.66}$$

Nuclear wave functions have a negligible overlap between sites (allowing us to ignore their spin statistics) and hence p_n has the Boltzmann form (see Eq. (4.10))

$$p_n = Z^{-1}e^{-E_n/k_B T},$$

where

$$Z = \sum_n e^{-E_n/k_B T}$$

is the normalization factor (the partition function); combining this with the golden rule for the transition probability,

$$w_{n\to m} = \frac{2\pi}{\hbar}\left|\left(\hat{H}_1\right)_{mn}\right|^2 \delta\left(E_m - E_n\right), \tag{28.67}$$

we obtain

$$\chi''(\omega) \cong \frac{\pi}{Z}\sum_{n,m} \left|\langle m|\, \hat{\mu}_x \,|n\rangle\right|^2 \left[e^{-E_n/k_B T} - e^{-E_m/k_B T}\right] \delta\left(E_m - E_n - \hbar\omega\right). \quad (28.68)$$

At room temperature we generally have $\hbar\omega \ll E_n, E_m$, so we expand $e^{-E_m/k_B T}$ through first order which gives

$$\chi''(\omega) \cong \frac{\pi}{Z}\sum_{n,m} \left|\langle m|\, \hat{\mu}_x \,|n\rangle\right|^2 e^{-E_n/k_B T} \frac{E_m - E_n}{k_B T} \delta\left(E_m - E_n - \hbar\omega\right). \quad (28.69)$$

Finally, noting that the delta function will restrict $E_m - E_n = \hbar\omega$, we obtain our desired result:

$$\chi''(\omega) = \frac{\hbar\pi\omega}{k_B T Z}\sum_{n,m} \left|\langle m|\, \hat{\mu}_x \,|n\rangle\right|^2 e^{-E_n/k_B T} \delta\left(E_m - E_n - \hbar\omega\right). \quad (28.70)$$

Note that this is an odd function of ω, as required for it to represent dissipation.

28.6 The chemical and Knight shifts

The magnetic field experienced by the nucleus of an atom involves contributions from: (i) an external field, $\mathbf{H}_0$; (ii) fields produced by circulating currents associated with the surrounding electrons; and (iii) a direct dipole–dipole interaction from any unpaired spins in the atom. If the atom is embedded in some material we can have additional contributions, (iv), resulting from the other atoms. The second of these, which is present in atoms with $l > 0$, results in the orbital hyperfine interaction encountered in atomic physics, and for isolated atoms the associated field is generally larger than typical laboratory magnetic fields. However, any spontaneous current associated with orbital angular momentum is typically "quenched" in solids due to the breaking of spherical symmetry (from which the possibility of crystal field split states follows). However, as discussed in Sections 5.1 and 24.3, there will always be diamagnetic currents induced in the surrounding electrons in the presence of an external field which, in turn, produce an added field at the nucleus that shifts the Larmor precession frequency. This effect is called the *chemical shift* since the shifts so induced depend not only on the atom in which the particular nucleus is centered, but also on the chemical bonding of that atom with its neighbors; different configurations inducing different shifts that can yield information on the nature of the bonding. This has become a powerful analytical tool in deducing the structure of complex molecules, but being somewhat specialized, will not be discussed here.

Mechanism (iii) is important in metals, where it is known as the *Knight shift*,[12] and is typically an order of magnitude larger than the diamagnetic shift associated with the core electrons. Here the application of an external magnetic field produces an imbalance in the spin population of spin up and spin down conduction electrons through the Pauli paramagnetic spin susceptibility, which then acts on the nuclei. To treat this problem we start with the electron spin contribution to the hyperfine Hamiltonian:[13]

$$\hat{H}_{hf} = \frac{8\pi}{3}\mu_e\mu_n \sum_i \hat{\imath} \cdot \hat{s}\,\delta(\mathbf{r}_i - \mathbf{R}). \tag{28.71}$$

This contribution vanishes for all filled bands, since there are equal numbers of spin up and spin down electrons and hence the net effect vanishes. Hence we only need to deal with unfilled bands. The shift in energy of the nucleus produced by a particular Bloch state with wave vector $\mathbf{k}$, band index n, and spin σ, $\psi_{\mathbf{k}n\sigma} = e^{i\mathbf{k}\cdot\mathbf{r}}u_{\mathbf{k}n}(\mathbf{r})\chi_\sigma$, is $\Delta\varepsilon_{\mathbf{k}n\sigma} = \langle\psi_{\mathbf{k}n\sigma}|\hat{H}_{hf}|\psi_{\mathbf{k}n\sigma}\rangle$. Placing the nucleus at $\mathbf{R} = 0$ and inserting (28.71) into this expression, the total energy shift for a nucleus with spin i_z along the external field axis arising from all the conduction electrons is then

$$\Delta E_n = \frac{8\pi}{3}\mu_e\mu_n i_z \sum_{\mathbf{k}n\sigma} |u_{\mathbf{k}n}(0)|^2 \left(n(\varepsilon_{\mathbf{k}\uparrow}) - n(\varepsilon_{\mathbf{k}\uparrow})\right), \tag{28.72}$$

where $n(\varepsilon_{\mathbf{k}\sigma})$ is the Fermi occupation function. In order to be able to carry out the sum over $\mathbf{k}$ and n, we introduce the averaged form $\left\langle|u_{\mathbf{k}n}(0)|^2\right\rangle_{\varepsilon=\varepsilon_F}$, which allows us to factor it out, whereupon (28.72) becomes (see Section 6.7)

$$\begin{aligned}\Delta E_n &= \frac{8\pi}{3}\mu_e\mu_n i_z \left\langle|u_{\mathbf{k}n}(0)|^2\right\rangle_{\varepsilon=\varepsilon_F} N(\varepsilon_F)\mu_e H_0 \\ &= \frac{8\pi}{3}\mu_n i_z \left\langle|u_{\mathbf{k}n}(0)|^2\right\rangle_{\varepsilon=\varepsilon_F} \chi_P H_0,\end{aligned} \tag{28.73}$$

where $N(\varepsilon_F)$ is the density of states and χ_P is the Pauli spin susceptibility. Since both the precession frequency of the free nuclear spin and the Knight shift are proportional to H_0, the fraction $\Delta H/H_0$ is a constant and given by

$$\frac{\Delta H}{H_0} = \frac{8\pi}{3}\mu_n i_z \left\langle|u_{\mathbf{k}n}(0)|^2\right\rangle_{\varepsilon=\varepsilon_F} \chi_P. \tag{28.74}$$

Finally, in connection with mechanism (iii), we note that in a ferromagnet, there is an internal field associated with the magnetization and hence there would be a constant, essentially field-independent shift associated with this internal field.

As will be discussed in Section 44.5.2, the coupling of the nuclear spins to conduction electrons, which are free to move between atoms, produces a coupling between the nuclear spins that in turn contributes to the line width. This is the so-called RKKY interaction.

[12] Nuclear magnetic resonance shift in metals, W.D. Knight, *Phys. Rev.* **76**, 1259 (1949).

[13] This is treated in most texts on quantum mechanics.

FURTHER READING

The literature on NMR is huge and we will only suggest a few books. A quite readable treatment of NMR at a near undergraduate level is given in *Principles of Magnetic Resonance*, by C.P. Slichter, Harper and Row, NY (1963). Parts of our discussion are drawn from here. Other books include: *Nuclear Magnetic Resonance*, by E.R. Andrew, Cambridge University Press, Cambridge, UK (2009); *Nuclear Magnetic Resonance and Relaxation*, Brian Cowen, Cambridge University Press, Cambridge, UK (1997); and *Introduction to Solid-State NMR Spectroscopy*, Melinda J. Duer (Ed), Blackwell Publishers (2004). A classic and exhaustive treatment is *The Principles of Nuclear Magnetism*, A. Abragam, Oxford University Press (1961).

For a good discussion of the techniques of actually doing NMR experiments, see *Experimental Pulse NMR: A Nuts and Bolts Approach*, Eiichi Fukushima and Stephen B.W. Roeder, Westview Press (1981).

Appendix 28A Dipolar broadening

The previous discussion has neglected the magnetic field acting on the individual spins resulting from the remaining nuclei. This magnetic dipole–dipole contribution to the spin Hamiltonian is given by

$$\hat{H}^{(d-d)} = \frac{1}{2}\sum_{j,k}^{N}{}' \left[\frac{\hat{\mu}_j \cdot \hat{\mu}_k}{r_{jk}^3} - \frac{3\left(\hat{\mu}_j \cdot \mathbf{r}_{jk}\right)\left(\hat{\mu}_k \cdot \mathbf{r}_{jk}\right)}{r_{jk}^5}\right], \tag{28A.1}$$

where $\mathbf{r}_{jk}$ is the vector separation between sites j and k and the j = k term is omitted. The associated local magnetic fields, H_{loc}, are of order a gauss and under ordinary conditions do not affect the position of the resonance line, but do contribute to its width, which is the problem we will examine. We next write

$$\mathbf{r}_{jk} = r_{jk}\mathbf{n}_{jk}, \tag{28A.2a}$$

where

$$\mathbf{n}_{jk} = \left[\cos\theta_{jk}\mathbf{z} + \sin\theta_{jk}\cos\varphi_{jk}\mathbf{x} + \sin\theta_{jk}\sin\varphi_{jk}\mathbf{y}\right] \tag{28A.2b}$$

are unit vectors pointing from site j to k while **x**, **y**, **z**, are the Cartesian unit vectors. We introduce the magnet moment operator:

$$\hat{\mu}_j = \hbar\gamma_j\left[\hat{i}_{jx}\mathbf{x} + \hat{i}_{jy}\mathbf{y} + \hat{i}_{jz}\mathbf{z}\right],$$

and write

$$\hat{i}_{xj} = \frac{1}{2}\left(\hat{i}_{j+} + \hat{i}_{j-}\right); \qquad \hat{i}_{jy} = -\frac{i}{2}\left(\hat{i}_{j+} - \hat{i}_{j-}\right). \qquad (28A.3a,b)$$

Inserting this into (28A.1), and after some algebra, we arrive at

$$\hat{H}^{(d-d)} = \hbar^2 \sum_{j,k}^{N} \frac{\gamma_i\gamma_j}{r_{jk}^3}\left(1 - 3\cos^2\theta_{jk}\right)\left[\hat{i}_{jz}\hat{i}_{kz} - \frac{1}{4}\left(\hat{i}_{j+}\hat{i}_{k-} + \hat{i}_{k-}\hat{i}_{j+}\right)\right]$$
$$+ \left[\text{terms involving}: \hat{i}_{j+}\hat{i}_{kz}, \hat{i}_{j-}\hat{i}_{kz}, \hat{i}_{jz}\hat{i}_{k+}, \hat{i}_{jz}\hat{i}_{k-}, \hat{i}_{j+}\hat{i}_{k+}, \hat{i}_{j-}\hat{i}_{k-}\right]. \qquad (28A.4)$$

The first bracket in (28A.4) involves operator products of the form $\hat{i}_{jz}\hat{i}_{kz}$, $\hat{i}_{j+}\hat{i}_{k-}$ and $\hat{i}_{j-}\hat{i}_{k+}$, all of which have the property that they commute with the Zeeman Hamiltonian when **z** is chosen as the quantization axis; they leave the total spin unaffected, although the latter interchanges the spin projections on sites j and k. They contribute an energy of order $\hbar\gamma H_{loc}$. On the other hand, the first four terms in the second bracket of (28A.4) couple states differing by $i_z = \pm 1$, and the latter two couple states differing by $i_z = \pm 2$. In second-order perturbation theory and in an external field H_0 parallel to **z**, these terms would result in small admixtures of the $i_z \pm 1$ and $i_z \pm 2$ states into the i_z states, with amplitudes of order H_{loc}/H_0, and would produce energy shifts of order $\hbar\gamma H_{loc}^2/H_0$. In the presence of a time-dependent $H_1(t)$ field, the $i_z \pm 2$ admixtures would result in absorption at $\omega = 0$ and $\omega = 2\gamma H_0$ (with a transition rate reduced by $(H_{loc}/H_0)^2$), which is not of interest here. It is therefore a good approximation to neglect all the terms in the second bracket in Eq. (28A.4).

The remaining terms in Eq. (28A.4) constitute a "reduced" Hamiltonian, $\hat{H}_0^{(d-d)}$, which, on using the identity $\hat{\mathbf{i}}\cdot\hat{\mathbf{i}} = (1/2)(\hat{i}_+\hat{i}_- + \hat{i}_-\hat{i}_+) + \hat{i}_z\hat{i}_z$, can be written as

$$\hat{H}_0^{(d-d)} = \frac{\hbar^2}{4}\sum_{j,k}^{N}\gamma_i\gamma_j\left[\frac{\left(1 - 3\cos^2\theta_{jk}\right)}{r_{jk}^3}\left(3\hat{i}_{jz}\hat{i}_{kz} - \hat{\mathbf{i}}_j\cdot\hat{\mathbf{i}}_k\right)\right]. \qquad (28A.5)$$

We can now write the time-independent Schrödinger equation associated with the total Hamiltonian, $H^{(total)}$, involving the contributions $\hat{H}^{(Zeeman)}$ and $\hat{H}_0^{(d-d)}$, as

$$\left(\hat{H}^{(Zeeman)} + \hat{H}_0^{(d-d)}\right)\Phi(I_z, \nu) = (-\hbar\gamma H_0 I_z + \varepsilon_\nu)\,\Phi(I_z, \nu), \qquad (28A.6)$$

where $\hat{I}_z = \sum_{j=1}^{N}\hat{i}_{jz}$ is the total magnetic quantum number. Since the two contributions to $H^{(total)}$ commute, the corresponding energies must be additive; however the dipole eigenenergies, ε_ν, are associated with a set of "internal" quantum numbers ν (the totality of the quantum numbers running from 1 to $(2i+1)^N$), and can only be found by a numerical diagonalization of the $\hat{H}_0^{(d-d)}$ (no general solution can be obtained). For this reason we must use the trick, originally invented by Van Vleck, of expanding the line width in moments, if we are to obtain a satisfactory solution; we may regard this expansion as a kind of perturbation theory in the line width.

We start with Eq. (28.70), which we write in the form

$$\chi''(\omega) = \frac{\hbar\pi\omega}{k_B TZ} f(\omega), \tag{28A.7a}$$

where

$$f(\omega) \equiv \sum_{n,m} \left|\langle m|\,\hat{\mu}_x\,|n\rangle\right|^2 e^{-E_n/k_BT}\,\delta\left(E_m - E_n - \hbar\omega\right), \tag{28A.7b}$$

where $f(\omega)$ is an even function of frequency. For the remainder of this appendix we will assume $E_n \ll k_B T$, in which case we set $e^{-E_n/k_BT} = 1$ and our final result will be independent of temperature; i.e., it will represent the line broadening for a completely disordered spin system. We define the nth *moment* of $f(\omega)$ as

$$\left\langle \omega^n \right\rangle = \frac{\int_0^{\infty} d\omega\, \omega^n f(\omega)}{\int_0^{\infty} d\omega\, f(\omega)} \tag{28A.8a}$$

and the nth *deviation*,

$$\left\langle \Delta\omega^n \right\rangle = \frac{\int_0^{\infty} d\omega\, \left(\omega - \langle\omega\rangle\right)^n f(\omega)}{\int_0^{\infty} d\omega\, f(\omega)}; \tag{28A.8b}$$

$\left\langle \Delta\omega^2 \right\rangle$ is clearly a measure of the square of the width of the resonance line, where the latter is our goal, and we expect on physical grounds that

$$\left\langle \Delta\omega^2 \right\rangle \cong \left(\gamma H_{loc}\right)^2. \tag{28A.9}$$

Now, $\left(\omega - \langle\omega\rangle\right)^2 = \omega^2 - 2\omega\,\langle\omega\rangle + \langle\omega\rangle^2$; since $f(\omega)$ is an even function of ω the moment of the second term vanishes and we have

$$\left\langle \Delta\omega^2 \right\rangle = \left\langle \omega^2 \right\rangle - \langle\omega\rangle^2. \tag{28A.10}$$

We start with an evaluation of the zeroth moment; exploiting the property that $f(\omega)$ is an even function of ω, we write

$$\begin{aligned}\int_0^{\infty} d\omega\, f(\omega) &= \frac{1}{2}\int_{\infty}^{\infty} d\omega\, f(\omega)\\ &= \frac{1}{2}\int_{\infty}^{\infty} d\omega \sum_{n,m} \left|\langle m|\,\hat{\mu}_x\,|n\rangle\right|^2 \delta\left(E_m - E_n - \hbar\omega\right),\end{aligned}$$

or

$$\int_0^\infty d\omega\, f(\omega) = \frac{1}{2\hbar} \mathrm{Tr}\, \hat{\mu}_x^2. \tag{28A.11}$$

Now the trace may be evaluated with respect to an arbitrary set of complete states (spanning the same space), so we use the simplest possible form which is the product of the eigenfunctions, $|\phi_i\rangle = \begin{pmatrix} 1 \\ 0 \end{pmatrix}_i$ or $\begin{pmatrix} 0 \\ 1 \end{pmatrix}_i$, of a single-spin quantized along the external field axis: i.e.,

$$\Phi(i_{1z} \cdots i_{Nz}) = \prod_{i=1}^{N} |\phi_i\rangle. \tag{28A.12}$$

In evaluating the trace of

$$\hat{\mu}_x^2 = \left(\sum_{j=1}^{N} \hat{\mu}_{jx} \right)^2,$$

we encounter intermediate states involving the products $\hat{\mu}_{jx}\hat{\mu}_{kx}$, in which $j \neq k$ and $j = k$. The trace of the former vanishes and the trace of the latter is independent of the quantization axis, in which case, $\hat{\mu}_{jx}^2 = \hat{\mu}_{jy}^2 = \hat{\mu}_{jz}^2 = \hat{\mu}_j^2/3$. Since there are (2i+1) eigenfunctions for each j and $\hat{\mu}_j^2 = \hbar^2\gamma^2\hat{i}_j^2$, we have

$$\mathrm{Tr}\hat{\mu}_{jx}^2 = \frac{\hbar^2\gamma^2 i(i+1)}{3}(2i+1); \tag{28A.13}$$

since we obtain an identical result for all N spins, we have

$$\int_0^\infty d\omega\, f(\omega) = \frac{1}{2\hbar}\frac{N\hbar^2\gamma^2 i(i+1)}{3}(2i+1). \tag{28A.14}$$

Next, we proceed to calculate $\langle\omega\rangle$, which involves the integral

$$\begin{aligned}\int_0^\infty d\omega\, \omega\, f(\omega) &= \frac{1}{\hbar^2}\int_0^\infty d(\hbar\omega) \sum_{m,n} (\hbar\omega) \left|\langle m|\, \hat{\mu}_x\, |n\rangle\right|^2 \delta\,(E_m - E_n - \hbar\omega) \\ &= \frac{1}{\hbar^2}\int_0^\infty d(\hbar\omega) \sum_{m,n} (E_m - E_n) \left|\langle m|\, \hat{\mu}_x\, |n\rangle\right|^2 \delta\,(E_m - E_n - \hbar\omega)\end{aligned}$$

or

$$\int_0^\infty d\omega\, \omega f(\omega) = \frac{1}{\hbar^2} \sum_{I_z,\nu,\mu}^{N} \left(\hbar\omega + \varepsilon_\mu - \varepsilon_\nu\right) \left|\langle I_z + 1, \mu|\, \hat{\mu}_x\, |I_z, \nu\rangle\right|^2, \qquad (28A.15)$$

where in the last line, μ and ν denote the internal quantum numbers and we have utilized the fact that $E_m > E_n$, together with the property that the dipole moment operator only couples states with $\Delta I_z = \pm 1$. Introducing $\hat{\mu}_\pm = \hat{\mu}_x \pm i\hat{\mu}_y$, or $\hat{\mu}_x = \frac{1}{2}\left(\hat{\mu}_+ + \hat{\mu}_-\right)$, and using the matrix elements for $\hat{\mu}_\pm$ (see Eqs. (28A.3a,b)), we have

$$\int_0^\infty d\omega\, \omega\, f(\omega) = \frac{1}{4\hbar^2} \sum_{I_z,\nu,\mu}^{N} \left(\hbar\omega_0 + \varepsilon_\mu - \varepsilon_\nu\right) \langle I_z + 1, \mu|\, \hat{\mu}_+\, |I_z, \nu\rangle\, \langle I_z, \nu|\, \hat{\mu}_-\, |I_z + 1, \mu\rangle\,. \qquad (28A.16)$$

The term in (28A.16) involving ω_0 is given by

$$\begin{aligned}
&\frac{1}{4\hbar} \sum_{I_z,\nu,\mu}^{N} \omega_0 \left\langle I_z + 1, \mu \left|\hat{\mu}_+\right| I_z, \nu\right\rangle \left\langle I_z, \nu \left|\hat{\mu}_-\right| I_z + 1, \mu\right\rangle \\
&\quad = \frac{\omega_0}{4\hbar} \mathrm{Tr}\hat{\mu}_+\hat{\mu}_- \\
&\quad = \frac{1}{4\hbar^2} \mathrm{Tr}\left[\hat{\mu}_x^2 + \hat{\mu}_y^2 + i\left(\hat{\mu}_x\hat{\mu}_y - \hat{\mu}_y\hat{\mu}_x\right)\right] \\
&\quad = \frac{\omega_0}{2\hbar} \mathrm{Tr}\hat{\mu}_x^2, \qquad (28A.17)
\end{aligned}$$

where, in the bracket in the third line, the first two terms give an identical result and the last vanishes. $\mathrm{Tr}\hat{\mu}_x^2$ in the last line in (28A.17) can be evaluated using (28A.13). To calculate the term involving $\varepsilon_\mu - \varepsilon_\nu$ in (28A.16), we note that these energies are the eigenvalues of $\hat{H}_0^{(d\text{–}d)}$; thus,

$$\begin{aligned}
&\frac{1}{4\hbar^2} \sum_{I_z,\nu,\mu}^{N} \left(\varepsilon_\mu - \varepsilon_\nu\right) \langle I_z + 1, \mu|\, \hat{\mu}_+\, |I_z, \nu\rangle\, \langle I_z, \nu|\, \hat{\mu}_-\, |I_z + 1, \mu\rangle \\
&\quad = \frac{1}{4\hbar^2} \sum_{I_z,\nu,\mu}^{N} \langle I_z + 1, \mu| \left[\hat{H}_0^{(d\text{–}d)}, \hat{\mu}_+\right] |I_z, \nu\rangle\, \langle I_z, \nu|\, \hat{\mu}_-\, |I_z + 1, \mu\rangle \\
&\quad = \frac{1}{4\hbar^2} \mathrm{Tr}\left[\hat{H}_0^{(d\text{–}d)}, \hat{\mu}_+\right] \hat{\mu}_-. \qquad (28A.18)
\end{aligned}$$

A term by term evaluation of this expression using Eq. (28A.5) for $\hat{H}_0^{(d\text{–}d)}$ shows that it vanishes. Using (28A.11) and (28A.17), we then have

$$\langle\omega\rangle = \frac{\int_0^\infty \omega f(\omega) d\omega}{\int_0^\infty f(\omega) d\omega} = \omega_0. \qquad (28A.19)$$

To complete our task we need to calculate

$$\int_0^\infty d\omega\,\omega^2 f(\omega) = \frac{1}{2\hbar^3}\int_{-\infty}^{\infty} d(\hbar\omega)\sum_{m,n}(\hbar\omega)^2\left|\langle m|\,\hat{\mu}_x\,|n\rangle\right|^2\delta(E_m - E_n - \hbar\omega)$$
$$= \frac{1}{2\hbar^3}\sum_{n,m}^{N}(E_n - E_m)^2\left|\langle m|\,\hat{\mu}_x\,|n\rangle\right|^2. \tag{28A.20}$$

Introducing the total Hamiltonian operator, $\hat{H}^{(\mathrm{total})}$, which generates the energies E_n and E_m, we rewrite (28A.20) as

$$\int_0^\infty d\omega\,\omega^2 f(\omega) = -\frac{1}{2\hbar^3}\sum_{n,m}^{N}\left|\langle m|\left[\hat{\mu}_x,\hat{H}^{(\mathrm{total})}\right]|n\rangle\right|^2 = -\frac{1}{2\hbar^3}\mathrm{Tr}\left[\hat{\mu}_x,\hat{H}^{(\mathrm{total})}\right]^2$$
$$= -\frac{1}{2\hbar^3}\mathrm{Tr}\left\{\left[\hat{\mu}_x,\hat{H}^{(\mathrm{Zeeman})}\right]^2 + 2\left[\hat{\mu}_x,\hat{H}_0^{(d\text{–}d)}\right]\left[\hat{\mu}_x,\hat{H}^{(\mathrm{Zeeman})}\right]\right.$$
$$\left.+\left[\hat{\mu}_x,\hat{H}_0^{(d\text{–}d)}\right]^2\right\}. \tag{28A.21}$$

Since the first term remains in the absence of the dipole–dipole interaction, it must be equal to ω_0^2, as can be confirmed by explicitly evaluating it; this term will cancel the $\langle\omega\rangle^2$ term in (28A.10), where the latter is given by (28A.19). The cross term vanishes as it involves traces of the operator $\hat{\mu}_x$. The line width then results from the last term in (28A.21), which must again be evaluated explicitly using (28A.5). When the result is divided by the normalization factor, (28A.14), we have our final result for the square of the line width:

$$\left\langle\Delta\omega^2\right\rangle = \frac{3}{4}\gamma^4\hbar^2 i(i+1)\frac{1}{N}\sum_{j,k}\frac{\left(1 - 3\cos\theta_{jk}\right)^2}{r_{jk}^6}. \tag{28A.22}$$

If all sites are identical, resulting in N identical sums, we may write (28A.22) as

$$\left\langle\Delta\omega^2\right\rangle = \frac{3}{4}\gamma^4\hbar^2 i(i+1)\sum_{k}\frac{\left(1 - 3\cos\theta_{jk}\right)^2}{r_{jk}^6}. \tag{28A.23}$$

As noted earlier this result is independent of temperature.

We note in passing that at very low temperatures (≈ 10 nK) the nuclear dipole–dipole interaction, and any electronic coupling brought on by the interaction of the nuclei with the conduction electrons, results in nuclear ordering. In that limit, rather than producing random shifts of the nuclear energy levels (leading to the linewidth), the fields add coherently to shift the positions of the lines.[14]

[14] At ultra-low temperatures, where nuclei spontaneously align as a result of dipole–dipole and RKKY interactions, the line positions do shift. For a review of experiments in metals see: Nuclear magnetic ordering in simple metals at positive and negative nanokelvin temperatures, A.S. Oja and O.V. Lounasmaa, *Rev. Mod. Phys.* **69**, 1 (1997).

Appendix 28B Microscopic theory of spin-lattice relaxation; the Korringa relation

28B.1 Nuclear Zeeman level transitions induced by electrons

In Eq. (28.29a) we introduced the relaxation time, T_1, which phenomenologically describes the decay of the magnetic polarization along the external field direction. In this appendix we will calculate this rate for the case where it is dominated by the interaction of the nuclear spins with conduction electrons in a metal. We start by writing the transition probability per unit time from an initial state with quantum numbers $i_z, \mathbf{k}, \sigma$, to some final state $i'_z, \mathbf{k}', \sigma'$. Here i_z is the z axis projection of the nuclear spin, $\mathbf{k}$ is the wave vector of the electron, and σ is the electron spin quantum number; we will suppress the band index, n. The Hamiltonian governing this process is the same as that leading to the Knight shift and is given as Eq. (28.71). The transition rate follows from the standard expression from time-dependent perturbation theory:

$$w_{i_z\mathbf{k}\sigma\to i'_z\mathbf{k}'\sigma'} = \frac{2\pi}{\hbar}\left|\langle i'_z\mathbf{k}'\sigma'|\,\hat{H}_{hf}\,|i_z\mathbf{k}\sigma\rangle\right|^2 \delta(\varepsilon_{i_z} + \varepsilon_{\mathbf{k}\sigma} - \varepsilon_{i'_z} - \varepsilon_{\mathbf{k}'\sigma'}), \tag{28B.1}$$

where ε_{i_z} and $\varepsilon_{\mathbf{k}\sigma}$ are the energy levels of the nuclei and the electrons respectively. The full rate for $i_z \to i'_z$ involving all the conduction electrons will then be given by

$$W_{i_z\to i'_z} = \sum_{\mathbf{k}\sigma,\mathbf{k}'\sigma'} w_{i_z\mathbf{k}\sigma\to i'_z\mathbf{k}'\sigma'} n(\mathbf{k},\sigma)\left(1 - n(\mathbf{k}',\sigma')\right), \tag{28B.2}$$

where $n(\mathbf{k},\sigma)$ is the Fermi function and we have incorporated the requirement that the initial state must be occupied and the final state empty. The matrix element itself is

$$\langle i'_z\mathbf{k}'\sigma'|\,\hat{H}_{hf}\,|i_z\mathbf{k}\sigma\rangle = \frac{8\pi}{3}\mu_e\mu_n \langle i'_z|\,\hat{\mathbf{i}}\,|i_z\rangle\langle\sigma'|\,\hat{\mathbf{s}}\,|\sigma\rangle\, u^*_{\mathbf{k}'}(0)u_{\mathbf{k}}(0), \tag{28B.3}$$

and inserting it into (28B.1), we have

$$\begin{aligned} w_{i_z\mathbf{k}\sigma\to i'_z\mathbf{k}'\sigma'} = {}& \frac{2\pi}{\hbar}\frac{64\pi^2}{9}\mu_e^2\mu_n^2 \sum_{\alpha\alpha'} |u_{\mathbf{k}'}(0)|^2\,|u_{\mathbf{k}}(0)|^2 \langle i_z|\,\hat{i}_\alpha\,|i'_z\rangle\langle i'_z|\,\hat{i}_{\alpha'}\,|i_z\rangle \\ & \times \langle\sigma'|\,\hat{s}_\alpha\,|\sigma'\rangle\langle\sigma'|\,\hat{s}_{\alpha'}\,|\sigma\rangle\,\delta(\varepsilon_{i_z} + \varepsilon_{\mathbf{k}\sigma} - \varepsilon_{i'_z} - \varepsilon_{\mathbf{k}'\sigma'}). \end{aligned} \tag{28B.4}$$

This expression is now inserted into (28B.2). In carrying out the sums over $\mathbf{k}$ and $\mathbf{k}'$ we again introduce the averages $\langle|u_{\mathbf{k}}(0)|^2\rangle$, so that they can be pulled from the summations.

Now experiments are typically carried out in a regime where $k_B T \gg \mu_n H_0$, so we can neglect the energy difference, $\varepsilon_{i_z} - \varepsilon_{i'_z}$, in the delta function. In the limit $T \ll T_F$ we can also write

$$n(\varepsilon)\,(1 - n(\varepsilon)) = -k_B T \frac{\partial n(\varepsilon)}{\partial \varepsilon} \cong k_B T \delta(\varepsilon - \varepsilon_F). \tag{28B.5}$$

The sums over $\mathbf{k}$ and $\mathbf{k}'$ are now identical and can be performed by introducing the density of states, $N(\varepsilon)$. Carrying out the sums over the energies, the transition rate between the nuclear levels becomes

$$W_{i_z \to i'_z} = \frac{2\pi}{\hbar} \frac{64\pi^2}{9} \mu_e^2 \mu_n^2 \left\langle |u_{\mathbf{k}}(0)|^2 \right\rangle^2 N^2(\varepsilon_F)(k_B T) \times \sum_{\alpha\alpha' ss'} \langle i_z | \hat{i}_\alpha | i'_z \rangle \langle i'_z | \hat{i}_{\alpha'} | i_z \rangle \langle \sigma | \hat{s}_\alpha | \sigma' \rangle \langle \sigma' | \hat{s}_{\alpha'} | \sigma \rangle . \tag{28B.6}$$

The sum over σ and σ' for our spin 1/2 electrons can be performed directly as

$$\sum_{\sigma\sigma'} \langle \sigma | \hat{s}_\alpha | \sigma' \rangle \langle \sigma' | \hat{s}_{\alpha'} | \sigma \rangle = \mathrm{Tr}(\hat{s}_\alpha \hat{s}_{\alpha'}) = \frac{\delta_{\alpha\alpha'}}{2}, \tag{28B.7}$$

resulting in

$$W_{i_z \to i'_z} = \frac{64\pi^3}{9\hbar} \mu_e^2 \mu_n^2 \left\langle |u_{\mathbf{k}}(0)|^2 \right\rangle^2 N^2(\varepsilon_F) k_B T \sum_\alpha \left| \langle i_z | \hat{i}_\alpha | i'_z \rangle \right|^2 . \tag{28B.8}$$

28B.2 Formal expression for the spin–lattice relaxation time

We now develop an expression for the spin-lattice relaxation time, T_1. Our approach assumes that the spin system can be characterized by some *effective temperature*, T_S, even if it is out of equilibrium with the lattice temperature, T_L. This is usually a reasonable approximation since the time scale for spin–spin coupling, T_2, is typically much shorter than T_1, allowing the spin system to be in equilibrium with itself and simultaneously out of equilibrium with the lattice. When characterized by its own temperature, the ratio of the occupation fraction of any two nuclear energy levels n_{i_z} and $n_{i'_z}$, characterized by quantum numbers i_z and i'_z, is governed by Boltzmann occupation factors according to

$$\frac{n_{i_z}}{n_{i'_z}} = \frac{\exp(-\beta_S \varepsilon_{i_z})}{\exp(-\beta_S \varepsilon_{i'_z})}, \tag{28B.9}$$

where $\beta_S = 1/k_B T_S$. Summing this expression over i'_z yields

$$n_{i_z} = \frac{\exp(-\beta_S \varepsilon_{i_z})}{Z}, \tag{28B.10}$$

where we have used the property that the occupation fractions must satisfy

$$\sum_{i'_z} n_{i'_z} = 1, \tag{28B.11}$$

since, collectively, all spins are distributed over the $2i_z + 1$ nuclear Zeeman sublevels. We have also defined the quantity Z,

$$Z = \sum_{i'_z} \exp\left(-\beta_S \varepsilon_{i'_z}\right), \tag{28B.12}$$

which is the partition function (see Eq. (4.11)). Now the average energy of the system is given by

$$\bar{\varepsilon} = \sum_{i_z} n_{i_z} \varepsilon_{i_z}, \tag{28B.13}$$

and for convenience we will assume the zero of energy is fixed such that

$$\sum_{i_z} \varepsilon_{i_z} = 0. \tag{28B.14}$$

The rate of change of the average energy, which we take as a measure of the rate at which the system comes into equilibrium, is given by

$$\frac{d\bar{\varepsilon}}{dt} = \sum_{i_z} \varepsilon_{i_z} \frac{dn_{i_z}}{dt}. \tag{28B.15}$$

Now the rate of change of the occupation fractions, n_{i_z}, is given by the difference between the rates for transitions into and out of the state i_z:

$$\frac{dn_{i_z}}{dt} = \sum_{i'_z} \left(n_{i'_z} W_{i'_z \to i_z} - n_{i_z} W_{i_z \to i'_z}\right). \tag{28B.16}$$

Combining (28B.15) and (28B.16), we have

$$\begin{aligned} \frac{d\bar{\varepsilon}}{dt} &= \sum_{i_z i'_z} \varepsilon_{i_z} \left(n_{i'_z} W_{i'_z \to i_z} - n_{i_z} W_{i_z \to i'_z}\right) \\ &= \frac{1}{2} \sum_{i_z i'_z} \left(n_{i'_z} W_{i'_z \to i_z} - n_{i_z} W_{i_z \to i'_z}\right)\left(\varepsilon_{i_z} - \varepsilon_{i'_z}\right). \end{aligned} \tag{28B.17}$$

Our goal is to determine $\dot{\beta}$, which is a measure of the rate at which the spin temperature approaches the lattice temperature. For this purpose we write $\dot{\bar{\varepsilon}}$ in the form

$$\frac{d\bar{\varepsilon}}{dt} = \frac{d\bar{\varepsilon}}{d\beta} \frac{d\beta}{dT}. \tag{28B.18}$$

The first factor in (28B.18) follows from (28B.13) and (28B.10) as

$$\frac{d\bar{\varepsilon}}{d\beta} = \frac{d}{d\beta}\sum \varepsilon_{i_z}\frac{\exp(-\beta_S\varepsilon_{i_z})}{Z}. \tag{28B.19}$$

We first examine the denominator, Z. At temperatures such that $\beta\mu_n H_0 \ll 1$ we can expand Z as

$$Z = \sum_{i_z}\left(1 - \beta\varepsilon_{i_z} + \frac{1}{2}\beta^2\varepsilon_{i_z}^2 + \cdots\right). \tag{28B.20}$$

The second term on the right vanishes due to the condition (28B.14), while the first term gives the total degeneracy of the nuclear levels, (2i + 1); i.e.,

$$Z \cong 2i + 1. \tag{28B.21}$$

Hence to leading order we can neglect the temperature dependence of Z; differentiating the remaining factor in the numerator of (28B.19) we obtain

$$\begin{aligned}\frac{d\bar{\varepsilon}}{d\beta} &= -\frac{1}{2i+1}\sum_{i_z}\varepsilon_{i_z}^2\exp(-\beta_S\varepsilon_{i_z})\\ &\cong -\frac{1}{2i+1}\sum_{i_z}\varepsilon_{i_z}^2.\end{aligned} \tag{28B.22}$$

Proceeding with the calculation of $\dot{\varepsilon}$, we note that $\dot{n}_{i_z}$ must vanish at the equilibrium lattice temperature, T_L; since this is true for any of the levels, Eq. (28B.16) requires that

$$n_{i'_z}W_{i'_z\to i_z} - n_{i_z}W_{i_z\to i'_z} = 0. \tag{28B.23}$$

This property is called the principle of *detailed balancing*. Writing

$$W_{i'_z\to i_z} = W_{i_z\to i'_z}\frac{n_{i_z}}{n_{i'_z}} = W_{i_z\to i'_z}\exp\left(-\beta_L(\varepsilon_{i_z} - \varepsilon_{i'_z})\right) \tag{28B.24}$$

and substituting this expression into (28B.17) gives

$$\begin{aligned}\frac{d\bar{\varepsilon}}{dt} &= \frac{1}{2}\sum_{i_z i'_z}\left(n_{i'_z}W_{i'_z\to i_z} - n_{i_z}W_{i_z\to i'_z}\right)(\varepsilon_{i_z} - \varepsilon_{i'_z});\\ &= \frac{1}{2}\sum_{i_z i'_z}W_{i_z\to i'_z}\left[n_{i'_z}\exp\left(-\beta_L(\varepsilon_{i_z} - \varepsilon_{i'_z})\right) - n_{i_z}\right](\varepsilon_{i_z} - \varepsilon_{i'_z})\\ &= \frac{1}{2}\sum_{i_z i'_z}W_{i_z\to i'_z}\left[n_{i'_z}\exp\left(-\beta_L(\varepsilon_{i_z} - \varepsilon_{i'_z})\right) - n_{i_z}\right](\varepsilon_{i_z} - \varepsilon_{i'_z}).\end{aligned} \tag{28B.25}$$

Using (28B.10), with the understanding that the spin populations have the temperature T_S, (28B.25) becomes

$$\begin{aligned}\frac{d\bar{\varepsilon}}{dt} &= \frac{1}{2}\sum_{i_z i'_z} n_{i'_z} W_{i_z \to i'_z}\left[\exp\left(-\beta_L(\varepsilon_{i_z}-\varepsilon_{i'_z})\right)-\exp\left(-\beta_S(\varepsilon_{i_z}-\varepsilon_{i'_z})\right)\right](\varepsilon_{i_z}-\varepsilon_{i'_z}) \\ &\cong -\frac{1}{2}(\beta_L-\beta_S)\sum_{i_z i'_z} n_{i'_z} W_{i_z\to i'_z}(\varepsilon_{i_z}-\varepsilon_{i'_z})^2 \\ &= \frac{-(\beta_L-\beta_S)}{2(2i+1)}\sum_{i_z i'_z}\exp(-\beta_S\varepsilon_{i'_z})W_{i_z\to i'_z}(\varepsilon_{i_z}-\varepsilon_{i'_z})^2 \\ &\cong -\frac{(\beta_L-\beta_S)}{2(2i+1)}\sum_{i_z i'_z} W_{i_z\to i'_z}(\varepsilon_{i_z}-\varepsilon_{i'_z})^2. \end{aligned} \tag{28B.26}$$

Combining (28B.22) and (28B.26), we obtain $\dot{\beta}_S$ from Eq. (28B.18) as

$$\begin{aligned}\frac{d\beta_S}{dt} &= \frac{1}{2}(\beta_L-\beta_S)\frac{\sum_{i_z i'_z} W_{i_z\to i'_z}(\varepsilon_{i_z}-\varepsilon_{i'_z})^2}{\sum_{i_z}\varepsilon_{i_z}^2} \\ &= \frac{(\beta_L-\beta_S)}{T_1},\end{aligned} \tag{28B.27}$$

where we have identified the relaxation time governing the rate of change of β_S with time as

$$\frac{1}{T_1} = \frac{1}{2}\frac{\sum_{i_z i'_z} W_{i_z\to i'_z}(\varepsilon_{i_z}-\varepsilon_{i'_z})^2}{\sum_{i_z}\varepsilon_{i_z}^2}; \tag{28B.28}$$

this expression is quite general in that it is independent of the mechanism inducing the transitions.

28B.3 Evaluating T_1 for conduction electrons; the Korringa relation

The relaxation time, T_1, follows from inserting (28B.8) into (28B.28):

$$\frac{1}{T_1} = \frac{32\pi^3}{9\hbar}\mu_e^2\mu_n^2\left\langle|u_k(0)|^2\right\rangle^2 N^2(\varepsilon_F)k_B T\frac{\sum_{i_z i'_z \alpha}\left|\langle i_z|\hat{i}_\alpha|i'_z\rangle\right|^2(\varepsilon_{i_z}-\varepsilon_{i'_z})^2}{\sum_{i_z}\varepsilon_{i_z}^2} \tag{28B.29}$$

The factor involving the two sums can be rewritten as

$$\frac{\sum\limits_{i_z i_z' \alpha} \left|\langle i_z|\,\hat{i}_\alpha\,|i_z'\rangle\right|^2 (\varepsilon_{i_z} - \varepsilon_{i_z'})^2}{\sum\limits_{i_z} \varepsilon_{i_z}^2} = -\frac{\sum\limits_{i_z i_z' \alpha} \langle i_z|\left[\hat{H}_Z, \hat{i}_\alpha\right]|i_z'\rangle\langle i_z'|\left[\hat{H}_Z, \hat{i}_\alpha\right]|i_z\rangle}{\sum\limits_{i_z} \varepsilon_{i_z}^2}$$

$$= -\frac{\sum\limits_\alpha \mathrm{Tr}\left[\hat{H}_Z, \hat{i}_\alpha\right]^2}{\sum\limits_\alpha \mathrm{Tr}\,\hat{H}_Z^2},$$

where $\hat{H}_Z$ is the nuclear Zeeman Hamiltonian, $\hat{H}_Z = \mu_n H_0 \hat{i}_z$. Using the commutation relations, $\left[\hat{i}_x, \hat{i}_y\right] = i\,\hat{i}_z$, and permutations thereof, we have

$$\sum_\alpha \mathrm{Tr}\left[\hat{H}_Z, \hat{i}_\alpha\right]^2 = -\mu_n^2 H_0 \mathrm{Tr}\left[\hat{i}_x^2 + \hat{i}_y^2\right]; \sum_\alpha \mathrm{Tr}\,\hat{H}_Z^2 = \mu_n^2 H_0 \mathrm{Tr}\,\hat{i}_z^2;$$

together with the property, $\mathrm{Tr}\,\hat{i}_x^2 = \mathrm{Tr}\,\hat{i}_y^2 = \mathrm{Tr}\,\hat{i}_z^2$, we then have

$$-\frac{\sum\limits_\alpha \mathrm{Tr}\left[\hat{H}_Z, \hat{i}_\alpha\right]^2}{\sum\limits_\alpha \mathrm{Tr}\,\hat{H}_Z^2} = -2.$$

Inserting this result in Eq. (28B.29) then gives[15]

$$\frac{1}{T_1} = \frac{64\pi^3}{9\hbar}\mu_e^2\mu_n^2 \left\langle |u_k(0)|^2 \right\rangle^2 N^2(\varepsilon_F) k_B T. \tag{28B.30}$$

Hence, $T_1 T = \text{const}$. This result can be combined with Eq. (28.74) for the Knight shift, ΔH, with the result that the product $T_1\,(\Delta H/H_0)^2$ is also a constant.

[15] Nuclear magnetic relaxation and resonance line shift in metals, *J. Korringa, Phys.* **16**, 601 (1950).

Part X
Optical Properties

Optical Responses

29

29.1 The matrix element for optical absorption spectra

Important information about the excitation spectra of semiconductors and insulators can be obtained by studying the frequency dependence of the absorption of light passing through the material. At low temperatures, for which room temperature is a good approximation when $\varepsilon_g >> k_B T$, transitions take place only between the occupied valence band states and the unoccupied conduction band states. According to the so-called *golden rule* of time-dependent perturbation theory, the transition probability per unit time, w, between two such states is given by

$$w = \frac{2\pi}{\hbar} |M_{if}|^2 \, \delta(\varepsilon_i + \hbar\omega - \varepsilon_f), \tag{29.1}$$

where ε_i is the initial (valance band) state, ε_f is the final (conduction band) state, $\hbar\omega$ is the energy of the light quanta absorbed, and M_{if} is the matrix element connecting the two states. The matrix element involves the perturbation Hamiltonian derived in Section 24.2 as (see Eq. (24.9))

$$\hat{H}_1 = \frac{i\hbar e}{mc}\mathbf{A}(\mathbf{r}, t) \cdot \nabla = -\frac{e}{mc}\mathbf{A}(\mathbf{r}, t) \cdot \hat{\mathbf{p}}, \tag{29.2a}$$

leading to the absorption matrix element

$$M_{if} = \left\langle \psi_{\mathbf{k}_f n_f} \left| -\frac{e}{mc}\mathbf{A}(\mathbf{r}, t) \cdot \hat{\mathbf{p}} \right| \psi_{\mathbf{k}_i n_i} \right\rangle, \tag{29.2b}$$

where $\psi_{\mathbf{k}_i, n_i}(\mathbf{r}) = e^{i\mathbf{k}_i \cdot \mathbf{r}} u_{\mathbf{k}_i, n_i}(\mathbf{r})$, and similarly for $\psi_{\mathbf{k}_f, n_f}(\mathbf{r})$. We write the vector potential as we did in Eq. (24.11a),

$$\mathbf{A}(\mathbf{r}, t) = \frac{1}{2}\left[\mathbf{A}(\mathbf{r})e^{-i\omega t} + \mathbf{A}(\mathbf{r})^* e^{i\omega t}\right], \tag{29.3}$$

which for a plane wave takes the form

$$\mathbf{A}(\mathbf{r}, t) = \frac{1}{2}\left[\mathbf{A}_0 e^{-i\omega t + i\mathbf{q}\cdot\mathbf{r}} + \mathbf{A}_0^* e^{i\omega t - i\mathbf{q}\cdot\mathbf{r}}\right]. \tag{29.4}$$

The Physics of Solids. First Edition. J. B. Ketterson.

Suppressing the time-dependent factor (which is already included in (29.1)), the corresponding matrix element involving the first term in (29.4) (which is responsible for absorption) is given by

$$M_{if} = -\frac{\hbar e}{2mc}\left\langle u_{k_f n_f}\left| e^{i(k_i-k_f+q)\cdot r}\mathbf{A}_0\cdot(\mathbf{k}_i - i\nabla)\right| u_{k_i n_i}\right\rangle. \tag{29.5}$$

The integral extends over all space and we may exploit the periodicity of the u_{kn} to rewrite it as (where we include the implicit normalization factor $L^{-3/2} = (NV_{cell})^{-1/2}$ in the wave functions)

$$\begin{aligned}&\frac{1}{NV_{cell}}\int_{\text{all space}} d^3r\, e^{i(k_i-k_f+q).r}u^*_{k_f n_f}(\mathbf{k}_i - i\nabla)\,u_{k_i n_i}\\ &= \sum_{\mathbf{R}} e^{i(k_i-k_f+q)\cdot\mathbf{R}}\int_{\text{unit cell}} d^3r\, e^{i(k_i-k_f+q)\cdot r}u^*_{k_f n_f}(\mathbf{k}_i - i\nabla)\,u_{k_i n_i}.\end{aligned} \tag{29.6}$$

The sum over **R** is oscillatory but diverges when $\mathbf{k}_i - \mathbf{k}_f + \mathbf{q} = 0$, giving, formally, the number of unit cells, N, which cancels the corresponding factor in the denominator. The net result is that the integral vanishes unless momentum is conserved. Since the **q**-vectors associated with light are of order 10^3cm^{-1}, whereas wave vectors in the Brillouin zone are to be compared with a number of order a zone dimension, $k_{BZ} = \pi/a \cong 10^8\text{cm}^{-1}$, it is usually a good approximation to ignore **q** and write $\mathbf{k}_i = \mathbf{k}_f = \mathbf{k}$. We then speak of **k**-conserving or *vertical* transitions. We are left with the integral over the unit cell, $\int_{cell} d^3r\, u^*_{kn_f}(\mathbf{k} - i\nabla)\,u_{k,n_i}$; note that due to the orthogonality of the $u_{kn}(\mathbf{r})$ for different n, the factor proportional to **k** vanishes. Assuming the matrix element does not vanish on symmetry grounds, we may neglect higher-order terms.[1] This corresponds to the electric dipole approximation. Using $\mathbf{E} = -\partial\mathbf{A}/c\partial t$ and again limiting ourselves to the first term in (29.4), we have

$$M_{if} = -i\frac{E_0}{2\omega}\left\langle u_{k,n_f}(\mathbf{r})\left|-\frac{e}{m}\varepsilon\cdot\hat{\mathbf{p}}\right| u_{k,n_i}(\mathbf{r})\right\rangle, \tag{29.7}$$

where ε is the polarization vector and E_0 is the amplitude of the electric field. To obtain the optical absorption we must sum (29.7) over all initial and final states that satisfy the energy conservation constraint, $\varepsilon_f = \varepsilon_i + \hbar\omega$; i.e., this sum involves the factor

$$\begin{aligned}\mathcal{N}_J(\varepsilon) &= \sum_{k,\sigma}\delta(\varepsilon_v(\mathbf{k}) + \varepsilon - \varepsilon_c(\mathbf{k}))\\ &= \frac{2}{(2\pi)^3}\int d^3k\,\delta(\varepsilon_v(\mathbf{k}) + \varepsilon - \varepsilon_c(\mathbf{k})),\end{aligned} \tag{29.8}$$

where $\varepsilon = \hbar\omega$; this quantity is called the *joint density of states*.

[1] If the matrix element does vanish, we may expand $u_{k_f,n_f}(\mathbf{r})$ as

$$\begin{aligned}u_{k_f,n_f}(\mathbf{r}) &= u_{k_i+q,n_f}(\mathbf{r})\\ &= u_{k_i,n_f}(\mathbf{r}) + \mathbf{q}\cdot\nabla_k u_{k_i,n_f}(\mathbf{r}) + \cdots,\end{aligned}$$

the leading term of which corresponds to an electric quadrupole or magnetic dipole transition.

29.2 Critical points in the optical absorption

The joint density of states exhibits certain singularities, first noted by Van Hove and hence called *Van Hove singularities*.[2] They are associated with points in **k**-space where the quantity $\varepsilon_c(\mathbf{k}) - \varepsilon_v(\mathbf{k})$ is a local *maximum, minimum, or saddle-point*; here the number of states satisfying the energy conservation criteria is an extremum. Figure 29.1(a) shows schematically an electron and hole band centered at the Γ point in **k**-space. Near this point we may write

$$\begin{aligned}\varepsilon_c(\mathbf{k}) - \varepsilon_v(\mathbf{k}) &= \varepsilon_g + \frac{\hbar^2 k^2}{2m_e} + \frac{\hbar^2 k^2}{2m_h} \\ &= \varepsilon_g + \frac{\hbar^2 k^2}{2\mu},\end{aligned} \tag{29.9}$$

where $1/\mu = 1/m_e + 1/m_h$. The joint density of states thus has the same form as one would have for a single carrier, but with m replaced by μ:

$$\mathcal{N}_J(\varepsilon) = \frac{2^{1/2}\mu^{3/2}}{\pi^2\hbar^2}\sqrt{\varepsilon - \varepsilon_g}. \tag{29.10}$$

With increasing photon energy a square-root singularity in the optical absorption occurs at $\varepsilon = \varepsilon_g$.

More generally, there will be points in the Brillouin zone involving bands that can have extrema at *different* points, as in Fig. 29.1(b), but where the difference $\varepsilon_c(\mathbf{k}) - \varepsilon_v(\mathbf{k})$ has an extremum at yet another point; i.e., where $\nabla_k\left(\varepsilon_c(\mathbf{k}) - \varepsilon_v(\mathbf{k})\right) = 0$; near these points, designated $\mathbf{k}_0$, we can expand $\varepsilon_c(\mathbf{k}) - \varepsilon_v(\mathbf{k})$ in a power series which has the form

$$\varepsilon_c(\mathbf{k}) - \varepsilon_v(\mathbf{k}) = \varepsilon_0 + \frac{1}{2}(\mathbf{k} - \mathbf{k}_0)\cdot\mathbf{A}\cdot(\mathbf{k} - \mathbf{k}_0), \tag{29.11}$$

where $A_{ij} = \partial^2/\partial k_i \partial k_j\big|_{\mathbf{k}=\mathbf{k}_0}$. Defining $\boldsymbol{\kappa} \equiv (\mathbf{k} - \mathbf{k}_0)$ and performing a principal axis transformation, we may write the expansion as

$$\varepsilon_c(\mathbf{k}) - \varepsilon_v(\mathbf{k}) = \varepsilon_0 + \frac{\hbar^2}{2m}\left[\alpha_1\kappa_1^2 + \alpha_2\kappa_2^2 + \alpha_3\kappa_3^2\right]. \tag{29.12}$$

There are four different cases one must consider, depending on whether there are 0, 1, 2, or 3 negative eigenvalues, α_i, of the matrix **A**; these are designated, M_0, M_1, M_2, and M_3. The first and last of these correspond to an ellipsoidal region of **k** space where the possibility of optical transitions is "opening up" or "closing down" with increasing ω. The curve to the right in Fig. 29.2 corresponds to an M_0 case, while that at the left corresponds to M_3. The form of the density of states is given by

$$\mathcal{N}_J(\varepsilon) = \frac{2^{1/2}m^{3/2}}{\pi^2\hbar^2\left(|\alpha_1\alpha_2\alpha_3|\right)^{1/2}}\sqrt{\pm(\varepsilon - \varepsilon_0)}, \tag{29.13}$$

where the + and − signs correspond to M_0 and M_3.

[2] The occurrence of singularities in the elastic frequency distribution of a crystal, L. Van Hove, *Phys. Rev.* **89**, 1189 (1953).

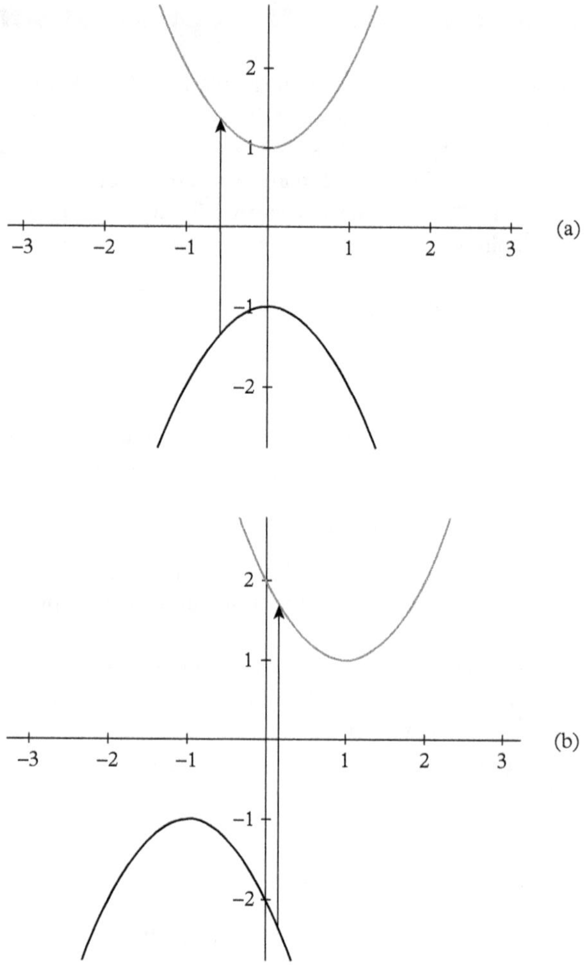

Figure 29.1 *A model band structure involving a filled valance band and an empty electron band. In (a) the maxima and minima are located at the same point in the Brillouin zone. In (b) the extrema are located at different points in k space. The arrows show optical transitions.*

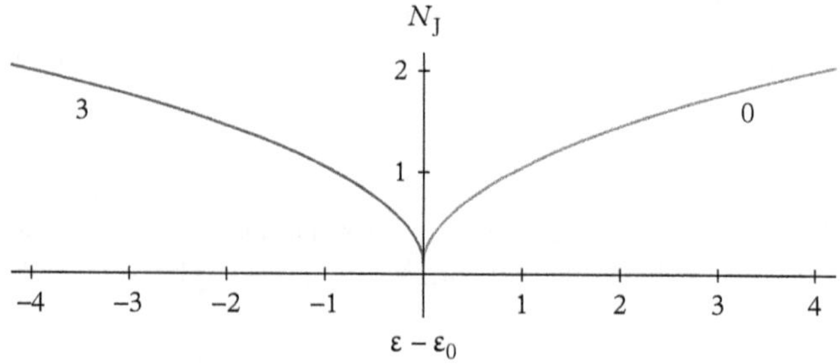

Figure 29.2 *The curves at the right and left correspond to a critical region of k space where the possibility of optical transitions is "opening up" or "closing down", respectively, with increasing ω; the numbers correspond to the number of negative eigenvalues of the A matrix.*

Next we consider the cases when one or two of the eigenvalues α_i are negative. The function $\varepsilon_c(\mathbf{k}) - \varepsilon_v(\mathbf{k})$ is then, locally, a hyperboloid that, as a function of energy, changes from one to two sheets or vice versa. We scale the coordinates and use cylindrical coordinates ρ and ζ with $\rho^2 - \zeta^2 = 2m(\varepsilon - \varepsilon_0)$; the joint density of states may then be written as

$$\begin{aligned} N_J(\varepsilon) &= \frac{2}{(2\pi)^3}\frac{\partial}{\partial\varepsilon}\int_0^{\varepsilon} d^3\kappa = \frac{2}{(2\pi)^3}\frac{\partial}{\partial\varepsilon}\int 2\pi\kappa_\perp d\kappa_\perp d\kappa_{//} \\ &= \frac{2}{(2\pi)^2\hbar^2\left(|\alpha_1\alpha_2\alpha_3|\right)^{1/2}}\frac{\partial}{\partial\varepsilon}\int_{\zeta_l}^{\zeta_u} d\zeta\int\rho d\rho \\ &= \frac{2}{(2\pi)^2\hbar^2\left(|\alpha_1\alpha_2\alpha_3|\right)^{1/2}}\frac{\partial}{\partial\varepsilon}\frac{1}{2}\int_{\zeta_l}^{\zeta_u} d\zeta\left[\zeta^2 + 2m(\varepsilon-\varepsilon_0)\right]. \end{aligned} \tag{29.14}$$

The upper limit of the integral, ζ_u, approaches infinity; the lower limit, ζ_l, depends on whether we have one or two sheets of the hyperboloid, for which cases we have $\sqrt{2m(\varepsilon-\varepsilon_0)}$, and 0 respectively. The integrals for these two cases are given by

$$\frac{1}{2}\int_0^{\zeta_u} d\zeta\left(\zeta^2 - 2m(\varepsilon-\varepsilon_0)\right) = \frac{\zeta_u^3}{6} - m(\varepsilon-\varepsilon_0)\zeta_u \tag{29.15a}$$

and

$$\begin{aligned} \frac{1}{2}\int_{\sqrt{2m(\varepsilon-\varepsilon_0)}}^{\zeta_u} d\zeta\left(\zeta^2 - m(\varepsilon-\varepsilon_0)\right) &= \frac{\zeta_u^3}{6} - \frac{1}{6}\left(2m(\varepsilon-\varepsilon_0)\right)^{3/2} \\ &\quad - m(\varepsilon-\varepsilon_0)\zeta_u + \frac{1}{2}\left(2m(\varepsilon-\varepsilon_0)\right)^{3/2}. \end{aligned} \tag{29.15b}$$

Taking the derivative of the difference between (29.15a,b), including a factor of 2 to account for integration over negative values of ζ_l, and inserting the result in (29.14), we have

$$N_J(\varepsilon) = \text{const.} - \frac{2^{1/2}m^{3/2}}{\pi^2\hbar^2\left(|\alpha_1\alpha_2\alpha_3|\right)^{1/2}}\sqrt{\varepsilon-\varepsilon_0}. \tag{29.16}$$

Including the case where the square root has the form, $\sqrt{\varepsilon_0-\varepsilon}$, these forms correspond to the M_1 and M_2 cases and are shown in Fig. 29.3. The optical transitions associated with the four different Van Hove singularities can be used to test the accuracy of band structure calculations. By "differentiating" the observed optical absorption spectra the strength of the singularities can be enhanced.[3]

[3] Two such techniques are to measure the reflectivity as the temperature is caused to oscillate or an oscillating electric field is applied across the sample. The latter is called the Franz–Keldysh effect: W. Franz, Einfluß eines elektrischen Feldes auf eine optische Absorptionskante, *Z. Naturforschung* **13a**, 484 (1958) and L.V. Keldysh, Behaviour of non-metallic crystals in strong electric fields, *J. Exptl. Theoret. Phys. (USSR)* **33**, 994 (1957), translation: *Sov. Phys. JETP* **6**, 763 (1958).

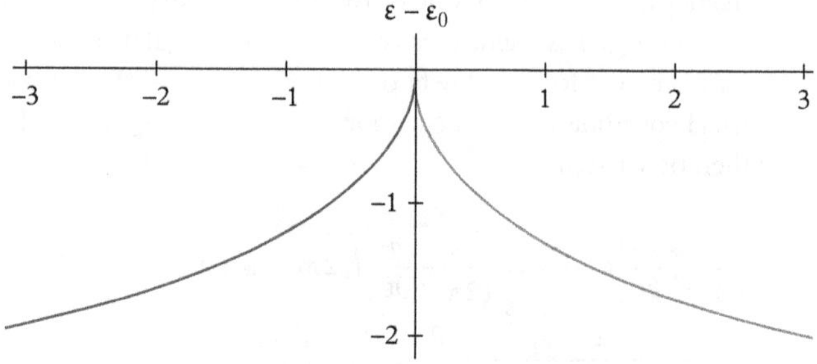

Figure 29.3 *The case where there is a saddle-point in the joint density of states for two different cases. The right side shows the decrease of the density of states on going from a hyperboloid of one to two sheets, while the left side shows the converse.*

29.3 The frequency-dependent susceptibility

The dielectric response of a material is characterized by the polarization induced by an external electromagnetic field according to the usual expression relating the Maxwell **D** and **E** fields through the relation, $\mathbf{D} = \mathbf{E} + 4\pi\mathbf{P}$. In general **P** will be a function of **E** and if the fields are sufficiently small we may expand **P** as a power series in **E** and retain only the leading term. However, we cannot simply write $P_i = \chi_{ij}E_j$,[4] since it must be borne in mind that the polarization itself arises from displacements in the positions of the electrons and ions making up the material, and that these displacements are necessarily delayed in time with respect to the instantaneous value of the applied electric field. We express this physical requirement by writing the polarization in a more general linear form as an integration over previous times, t':

$$\begin{aligned} P_i(t) &= \int_t^{-\infty} dt' \chi_{ij}(t-t')E_j(t') \\ &= \int_0^{\infty} d\tau \chi_{ij}(\tau)E_j(t-\tau), \end{aligned} \tag{29.17}$$

where we write $\tau = t - t'$ in the second line. This form expresses the fact that the polarization existing at the present time arose from the time dependence of the electric field at past times; it also builds in the constraint that **P** cannot depend on the electric field at future times, which is referred to as causality. An instantaneous interaction would correspond to $\chi_{ij}(t-t') = \chi_{ij}\delta(t-t')$.

We now Fourier transform Eq. (29.17):

$$\begin{aligned} \int_{-\infty}^{+\infty} dt\, e^{i\omega t}P_i(t) &= \int_0^{\infty} d\tau \chi_{ij}(\tau) \int_{-\infty}^{+\infty} dt\, e^{i\omega t}E_j(t-\tau) \\ &= \int_0^{\infty} d\tau e^{i\omega\tau}\chi_{ij}(\tau) \int_{-\infty}^{+\infty} d(t-\tau)e^{i\omega(t-\tau)}E_j(t-\tau). \end{aligned}$$

Defining $\chi_{ij}(\tau) = 0$ for negative τ (thereby insuring causality), the first integral can be extended over all time and written as $\chi_{ij}(\omega)$; we then have

$$P_i(\omega) = \chi_{ij}(\omega)E_j(\omega). \tag{29.18}$$

[4] In all expressions involving tensor operations we will here employ the summation convention.

We henceforth assume an isotopic or cubic system where $\chi_{ij}(\omega) = \delta_{ij}\chi(\omega)$, and then (29.18) becomes

$$\mathbf{P}(\omega) = \chi(\omega)\mathbf{E}(\omega). \tag{29.19}$$

Since both $\mathbf{P}(t)$ and $\mathbf{E}(t)$ are real we must have $\mathbf{P}(\omega) = \mathbf{P}^*(-\omega)$ and $\mathbf{E}(\omega) = \mathbf{E}^*(-\omega)$, which leads to

$$\chi^*(-\omega) = \chi(\omega). \tag{29.20}$$

We separate $\chi(\omega)$ into real and imaginary parts as

$$\chi(\omega) = \chi'(\omega) + i\chi''(\omega). \tag{29.21}$$

Applying (29.20) to (29.21), we have

$$\chi'(\omega) = \chi'(-\omega) \tag{29.22a}$$

and

$$\chi''(\omega) = -\chi''(-\omega); \tag{29.22b}$$

i.e., $\chi'(\omega)$ and $\chi''(\omega)$ are *even* and *odd* functions of the frequency respectively.

We will assume that $\chi(\omega) \to 0$ as $\omega \to \infty$; simple models satisfy this criterion and it is reasonable on physical grounds in that *particle inertia* suppresses displacement at high frequencies.

We now derive an expression for the energy dissipated. From thermodynamics the work done on the dielectric is[5]

$$dE = -\mathbf{P} \cdot d\mathbf{E} \tag{29.23}$$

or

$$\frac{dE}{dt} = -\mathbf{P} \cdot \frac{d\mathbf{E}}{dt}. \tag{29.24}$$

We write $\mathbf{E}(t) = \frac{1}{2}\left(\mathbf{E}_0 e^{-i\omega t} + \mathbf{E}_0^* e^{i\omega t}\right)$, in which case (29.24) can be written

$$\frac{dE}{dt} = -\left[\frac{1}{2}\left(\chi(\omega)\mathbf{E}_0 e^{-i\omega t} + \chi^*(\omega)\mathbf{E}_0^* e^{i\omega t}\right)\right]\frac{1}{2}\left(-i\omega\mathbf{E}_0 e^{-i\omega t} + i\omega\mathbf{E}_0^* e^{i\omega t}\right),$$

which we write as

$$\frac{dE}{dt} = \frac{1}{4}i\omega\big(\chi^*(\omega) - \chi(\omega)\big)\mathbf{E}_0\mathbf{E}_0^* + \frac{1}{4}i\omega\left(\chi(\omega)\mathbf{E}_0^2 e^{-2i\omega t} - \chi^*(\omega)\mathbf{E}_0^{*2} e^{2i\omega t}\right). \tag{29.25}$$

The term in the first bracket represents a steady-state energy loss, while the oscillatory terms average zero; neglecting the latter and using (29.22b) we have

$$\frac{dE}{dt} = \frac{1}{2}\omega\,\chi''(\omega)\,|\mathbf{E}_0|^2, \tag{29.26}$$

which represents the energy, $\dot{Q}$, dissipated (absorbed) by the system. Clearly an increase in the energy of the body with time is associated with a positive $\chi''(\omega)$.

[5] At some risk of confusion but for continuity of notation we use E for energy and **E** for electric field.

29.4 Some analytic properties of the susceptibility

We now examine the properties of $\chi(\omega)$ in the complex ω plane, where we write, $\omega = \omega' + i\omega''$, from which an important connection between the real and imaginary parts of $\chi(\omega)$ can be established. The Fourier transform of $\chi(\tau)$, when considered in the complex ω plane, has the form

$$\chi(\omega) = \int_0^\infty d\tau\, \chi(\tau)\, e^{(i\omega' - \omega'')\tau}. \tag{29.27}$$

Now all physical systems show some dissipation and hence there are no divergent resonant responses; therefore $\chi(t)$ remains finite at all times. We can therefore assert that the integral (29.27) converges for all positive ω''; i.e., it converges in the upper half of the complex ω plane.[6] The function $\chi(\omega)$ has no singularities on the real axis,[7] except possibly at $\omega = 0$; this follows from the causality principle. It further follows from (29.27) that

$$\chi^*(\omega) = \int_0^\infty d\tau\, \chi(\tau) \left(e^{i(\omega' + i\omega'')\tau}\right)^*$$
$$= \int_0^\infty d\tau\, \chi(\tau)\, e^{-i(\omega' - i\omega'')\tau} = \int_0^\infty dt\, \chi(t) e^{-i\omega^*\tau};$$

i.e.,

$$\chi^*(\omega) = \chi(-\omega^*), \tag{29.28}$$

which generalizes (29.22a,b) to the complex plane. An immediate result of (29.28) is that $\chi(\omega)$ is *real* on the imaginary ω axis.

A stronger statement can be made: $\chi(\omega)$ is real *only* on the imaginary axis and furthermore decreases monotonically for increasing $i\omega''$. To prove these statements we first write Cauchy's integral theorem in the form

$$\oint \frac{dz}{z - c} = 2\pi i n, \tag{29.29}$$

where n is the number of times the contour encircles the point c in the complex plane. We let $z = \chi(\omega)$ and restrict ourselves to the upper half plane where $\chi(\omega)$ is analytic in the variable ω. We may then rewrite (29.29) as

$$\oint \frac{d\chi}{d\omega} \frac{d\omega}{\chi(\omega) - c} = 2\pi i n. \tag{29.30}$$

Since $\chi(\omega)$ is analytic in the upper half plane there are no poles there, and we may assume that c is real; i.e., any poles lie on the real axis. Since $\chi(\omega) \to 0$ at least as fast as ω^{-2} (see Eq. (3.19)), $d\chi/d\omega \to 0$ at least as fast as ω^{-3}. We may

[6] As an example, consider the classical form for the Drude–Lorentz dielectric constant, given in Eq. (3.21), for which the associated susceptibility may be written as

$$\chi(\omega) = \frac{Ne^2/m}{(\omega_0^2 - \omega^2) - i\omega\gamma},$$

which has a pole in the lower half plane at $\omega'' = -\gamma/2$. Also note that for large ω, $\chi \propto -\omega^{-2}$; i.e., the response goes to zero at high frequencies, which is the inertial effect mentioned earlier.

[7] The function $\chi(\omega)$ is not defined in the lower half of the complex ω plane since the integral (29.27) would diverge there for positive times. $\chi(\omega)$ could be analytically continued from the upper half plane, but would then contain singularities in general.

therefore extend the semicircle in the upper half of the ω plane to infinity where it's contribution will vanish. The contour in the ω plane is plotted in Fig. 29.4.

We now examine what the contour looks like in the complex χ plane; the semicircle in the complex ω plane maps onto zero in the complex χ plane (since $\chi(\omega) \to 0$ as $\omega \to \infty$). On the other hand, the origin in the complex ω plane maps to the point $\chi_0 \equiv \chi(\omega = 0)$. Along the real axis, ω, the function $\chi(\omega)$ is real only at $\omega = 0$, where as just stated it takes the value χ_0, and at $\omega = \infty$, where it vanishes. Because of this property the contour can intersect the real axis only at *these two points*. We also note that the curve is symmetrical with respect to the real axis since, from (29.28), $\chi^*(\omega) = \chi(-\omega^*)$. It then follows that the point $\chi(\omega) = c$ can only be encircled *once*. A hypothetical contour in the complex χ plane having these properties is shown in Fig. 29.5.

Thus we have

$$\frac{1}{2\pi i}\oint \frac{d\chi}{\chi - c} = \begin{cases} 0; c > \chi_0 \\ 1; c < \chi_0 \end{cases} \tag{29.31}$$

Since the function χ takes on the value c only once, it follows that on the imaginary axis χ cannot have a maximum of a minimum. Therefore, $\chi(\omega)$ can only decrease monotonically on the imaginary axis.

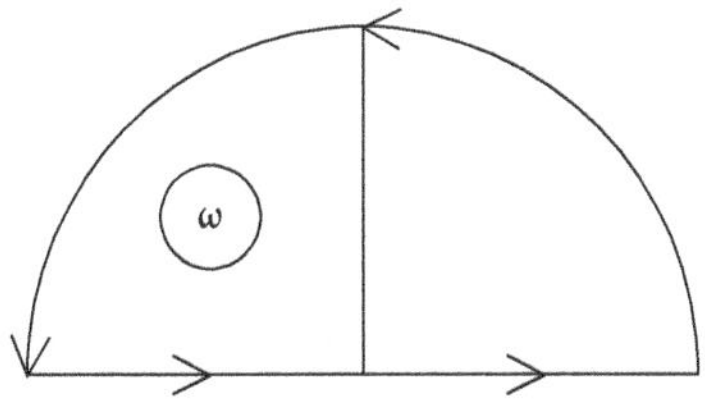

Figure 29.4 *Integration contour in the complex*

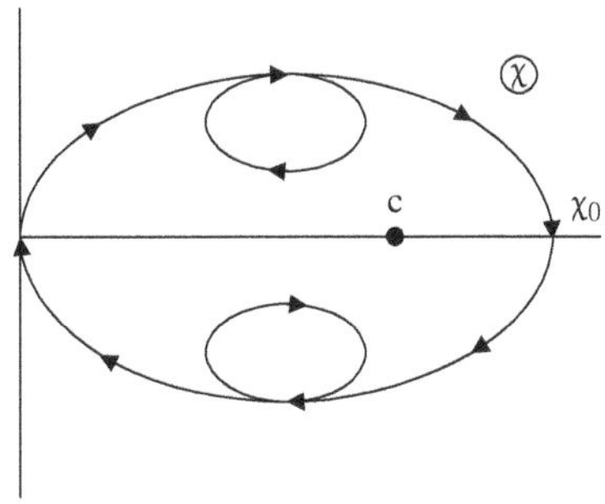

Figure 29.5 *A hypothetical contour in the complex χ plane.*

29.5 The Kramers-Kronig relations

Suppose we choose some frequency, $\omega = \omega_0$, and integrate the quantity $\chi(\omega)/(\omega - \omega_0)$ around a contour in the upper half plane where $\chi(\omega)$ is analytic. This contour (see Fig. 29.6) includes all the real ω axis, except the point $\omega = \omega_0$, and an infinite semicircle. From the previous discussion, the integral along the semicircle vanishes. We go around the point at ω_0 by using an infinitesimal semicircle, which also extends into the upper half plane. Since the contour contains no singularities the integral along the real axis likewise vanishes. Using the standard prescription for treating such integrals (see Eq. (14.59)), we have

$$P\int_{-\infty}^{+\infty} d\omega \frac{\chi(\omega)}{\omega - \omega_0} - i\pi\chi(\omega = \omega_0) = 0. \tag{29.32}$$

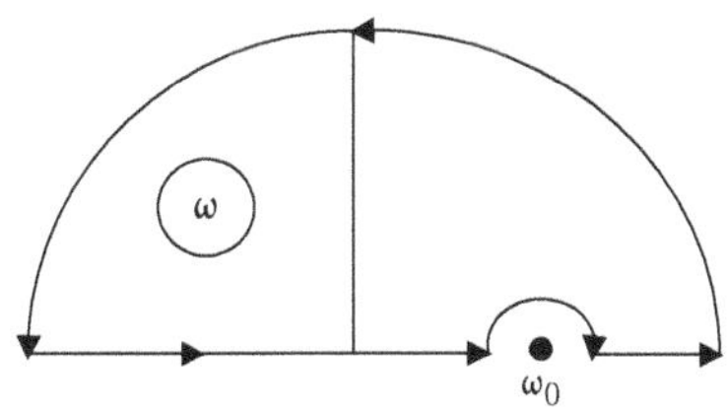

Figure 29.6 *The contour in the complex plane used to obtain the Kramers-Kronig relations.*

Again, writing $\omega = \omega' + i\omega''$, introducing a variable ξ in place of ω, writing ω in place of ω_0, and setting the real and imaginary parts of (29.32) separately to zero, we obtain

$$\chi'(\omega) = \frac{1}{\pi} P\int_{-\infty}^{+\infty} d\xi \frac{\chi''(\xi)}{\xi - \omega} \tag{29.33}$$

and

$$\chi''(\omega) = -\frac{1}{\pi} P\int_{-\infty}^{+\infty} d\xi \frac{\chi'(\xi)}{\xi - \omega}. \tag{29.34a}$$

Using the fact that $\chi''(\omega)$ is an odd function of ω, we can rewrite (29.33) as

$$\chi'(\omega) = \frac{1}{\pi} P \int_0^{+\infty} d\xi \frac{\chi''(\xi)}{\xi - \omega} + \frac{1}{\pi} P \int_0^{+\infty} d\xi \frac{\chi''(\xi)}{\xi + \omega}$$

or

$$\chi'(\omega) = \frac{2}{\pi} P \int_0^{+\infty} d\xi \frac{\xi \chi''(\xi)}{\xi^2 - \omega^2}. \tag{29.34b}$$

These are the celebrated Kramers–Kronig relations. What they tell us is that if we have one component of the susceptibility, either χ'' or χ', at *all* frequencies, then we may obtain the other component by performing one of the integrations, using (29.34a,b). Even when data is available only over a more limited range, one can sometimes use known asymptotic forms for χ'' or χ' to complete the integration.

In Section 29.1 we discussed the transition rate for the absorption of photons from an electromagnetic field. As an application of the formalism we now show how, by using (29.34b), we can obtain the real part of the dielectric constant from the frequency dependence of the absorption. Combining (29.1), (29.3), and (29.4), we obtain for the total transition rate $w(\omega)$:

$$w(\omega) = \frac{2\pi}{\hbar} \left(\frac{e}{m\omega}\right)^2 \left(\frac{E_0(\omega)}{2}\right)^2 \sum_{\mathbf{k}} \left|\langle u_{\mathbf{k},n_f}(\mathbf{r}) \left|\varepsilon \cdot \hat{\mathbf{p}}\right| u_{\mathbf{k},n_i}(\mathbf{r})\rangle\right|^2 \delta(\varepsilon_i(\mathbf{k}) + \hbar\omega - \varepsilon_f(\mathbf{k})). \tag{29.35a}$$

If we neglect the $\mathbf{k}$ dependence of the matrix elements,[8] we can write this expression as

$$w(\omega) = \frac{2\pi}{\hbar} \left(\frac{e}{m\omega}\right)^2 \left(\frac{E_0(\omega)}{2}\right)^2 \left|\langle u_{\mathbf{k},n_f}(\mathbf{r}) \left|\varepsilon \cdot \hat{\mathbf{p}}\right| u_{\mathbf{k},n_i}(\mathbf{r})\rangle\right|^2 \mathcal{N}_J(\omega). \tag{29.35b}$$

The energy loss per unit time, dE/dt, is $\hbar\omega$ times the number of photons absorbed per unit time, $w(\omega)\hbar\omega$, or

$$\frac{dE}{dt} = 2\pi\omega \left(\frac{e}{m\omega}\right)^2 \left(\frac{E_0(\omega)}{2}\right)^2 \sum_{\mathbf{k}} \left|\langle u_{\mathbf{k},n_f}(\mathbf{r}) \left|\varepsilon \cdot \hat{\mathbf{p}}\right| u_{\mathbf{k},n_i}(\mathbf{r})\rangle\right|^2 \delta(\varepsilon_i(\mathbf{k}) + \hbar\omega - \varepsilon_f(\mathbf{k})). \tag{29.36}$$

Rewriting (29.26) as $dE/dt = \omega\, \varepsilon''(\omega) \left|\mathbf{E}_0\right|^2 / 8\pi$, where $\varepsilon''(\omega)$ is the imaginary part of the dielectric constant, and equating it to (29.36), we have

$$\varepsilon''(\omega) = \left(\frac{2\pi e}{m\omega}\right)^2 \sum_{\mathbf{k}} \left|\langle u_{\mathbf{k},n_f}(\mathbf{r}) \left|\varepsilon \cdot \hat{\mathbf{p}}\right| u_{\mathbf{k},n_i}(\mathbf{r})\rangle\right|^2 \delta(\varepsilon_i(\mathbf{k}) + \hbar\omega - \varepsilon_f(\mathbf{k})). \tag{29.37}$$

Inserting this expression in (29.34b) and carrying out the integration, we obtain the real part of the dielectric constant, $\varepsilon'(\omega) = 1 + 4\pi\chi'(\omega)$, as

$$\varepsilon'(\omega) = 1 + \frac{4\pi e^2}{m} \sum_{\mathbf{k}} \left(\frac{2}{m\hbar\omega_{if}}\right) \frac{\left|\langle u_{\mathbf{k},n_f}(\mathbf{r}) \left|\boldsymbol{\varepsilon} \cdot \hat{\mathbf{p}}\right| u_{\mathbf{k},n_i}(\mathbf{r})\rangle\right|^2}{\omega_{if}^2 - \omega^2}. \tag{29.38}$$

Note the resemblance of this form to the Drude form, discussed in Chapter 3

[8] Provided the matrix elements vary smoothly, the singular behavior associated with the Van Hove singularities will still be present.

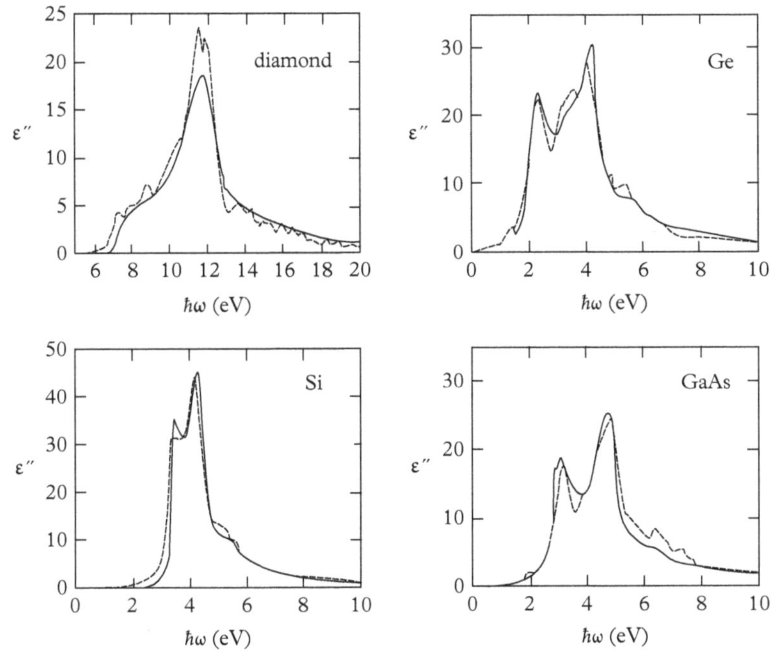

Figure 29.7 *The measured (solid line) and calculated (dashed line) imaginary part of the dielectric constant, ε″, for C (diamond) Si, Ge, and GaAs (after Benedict, Shirley, and Bohn (1998)).*

29.6 Some examples of optical absorption spectra

Figure 29.7 shows the calculated and measured imaginary part of the dielectric constant, $\varepsilon''(\omega)$, which measures the optical absorption, for diamond, silicon, germanium, and gallium arsenide as given by Benedict, Shirley, and Bohn (1998).[9] The peaks arise from van Hove singularities. The calculations include the effects of a screened (see Section 15.3.2) electron–hole interaction, which considerably improves the agreement between theory and experiment.[10]

[9] Theory of optical absorption in diamond, Si, Ge, and GaAs, L.X. Benedict, E.L. Shirley and R.B. Bohn, *Phys. Rev.* 57, R9385 (1998).

[10] In direct band gap semiconductors like GaAs, the Coulomb interaction between electrons and holes has a strong effect on the nature of the absorption edge, altering it from a square-root singularity to a discontinuity when the excitation energy rises above the band gap. This interaction also leads to the formation of bound states called excitons, which are treated in Chapter 30. For a discussion, see A. Stahl and I. Balslev, *Electrodynamics of the Semiconductor Band Edge*, Springer Tracts in Modern Physics, Vol. 110 Springer, Berlin (1987).

ADDITIONAL READING

Some books treating the optical properties of semiconductors are: *Fundamentals of Semiconductors*, P.Y. Yu and M. Cardona, Springer (2001); *Semiconductor Optics*, C.F. Klingshirn, Springer (1997 and 2004), and H. Haug and S.W. Koch, *Quantum Theory of the Optical and Electronic Properties of Semiconductors*, World Scientific (1994).

The analytic properties of the susceptibility are discussed in *Statistical Physics Part I*, L.D. Landau and E.M. Lifshitz, Section 123, Pergamon Press, Oxford (1980).

Problems

29.1. You are given an electron and a hole band with the behaviors,

$$\varepsilon_e = \varepsilon_c + \frac{\hbar^2\left[\alpha_{xx}(k_x - k_0)^2 + \alpha_{yy}k_y^2 + \alpha_{zz}k_z^2\right]}{2m},$$

$$\varepsilon_h = \varepsilon_v - \frac{\hbar^2 k^2}{2m_h}$$

where $\varepsilon_c > \varepsilon_v$ and $\varepsilon_c - \varepsilon_v \equiv \varepsilon_g$. Calculate the joint density of states, N_J, at the critical point and the location of this point in **k** space. To which of the four critical points does this correspond?

29.2. Same as Problem 29.1, only now,

$$\varepsilon_e = \varepsilon_c - \frac{\hbar^2\left[\alpha_{xx}(k_x - k_0)^2 + \alpha_{yy}k_y^2 + \alpha_{zz}k_z^2\right]}{2m}$$

$$\varepsilon_h = \varepsilon_v - \frac{\hbar^2 k^2}{2m_h}.$$

Polaritons, Excitons, and Plasmons

30

30.1 The optical phonon–polariton

In Section 24.1 we described a simple model for the contribution to the dielectric constant arising from the polarization of the ions in a material like NaCl. Such materials, which contain two different atoms having *opposite* charges, interact strongly with light when the frequencies of the light wave and the optical vibrational mode are comparable; the essentially uniform light field, on the scale of the atom separation, pulls the positively and negatively charged atoms in opposite directions.

We recall that from the four Maxwell equations one obtains the electromagnetic wave equation as

$$\nabla^2 \mathbf{E} - \frac{1}{c^2}\frac{\partial^2 \mathbf{D}}{\partial t^2} = 0. \tag{30.1}$$

We write $\mathbf{D} = \mathbf{E} + 4\pi\mathbf{P}$, but note that the response of $\mathbf{P}$ to $\mathbf{E}$ is governed by the Newtonian equations of the lattice. We assume that waves are propagating parallel to the z axis and polarized parallel to the x axis. We denote the characteristic frequency of the optical mode as ω_T. The equation of motion for the normal coordinate, U_x, representing the relative displacement of the two atoms of our diatomic lattice, is then

$$\ddot{U}_x + \omega_T^2 U_x = \frac{q}{M} E_x, \tag{30.2}$$

where q and M are the effective charge and reduced mass appropriate to the optical mode in question. The frequency of the optical mode may be taken as a constant for the long light wavelengths involved. Defining the polarization in the usual way as $P_x = nqU_x$, where n is the number of atom pairs per unit volume, and assuming a light wave of the form $\mathbf{E}(z,t) = \mathrm{Re}E_0\hat{\mathbf{x}}e^{ikz-i\omega t}$, we can write the pair of equations (30.1) and (30.2) as

$$\left(-k^2 + \frac{\omega^2}{c^2}\right)E_0 + \frac{4\pi\omega^2}{c^2}P_0 = 0, \tag{30.3a}$$

The Physics of Solids. First Edition. J. B. Ketterson.

and

$$-\frac{nq^2}{M}E_0 + \left(-\omega^2 + \omega_T^2\right)P_0 = 0. \tag{30.3b}$$

Setting the determinant of the coefficients of E_0 and P_0 to zero, we obtain the dispersion relation as

$$\left(\frac{\omega^2}{c^2} - k^2\right)\left(\omega_T^2 - \omega^2\right) + \frac{4\pi nq^2}{M}\frac{\omega^2}{c^2} = 0. \tag{30.4}$$

The two branches associated with this equation are plotted in Fig. 30.1. At $k = 0$ the solution is

$$\omega^2 = \omega_T^2 + \frac{4\pi nq^2}{M} \equiv \omega_L^2, \tag{30.5}$$

where we have defined another frequency, ω_L; the frequencies ω_L and ω_T are referred to as the longitudinal and transverse optical frequencies, respectively, and are identified in the figure.

The dielectric constant is given by

$$\varepsilon(\omega) = \varepsilon_\infty + \frac{4\pi P}{E} = \varepsilon_\infty + \frac{4\pi nq^2/M}{\omega_T^2 - \omega^2}, \tag{30.6}$$

where we have added a background dielectric constant, ε_∞, arising from the electronic polarization (which is generally frequency independent in the range under consideration). If we define the dielectric constant at $\omega = 0$ as ε_0, then from (30.6) we have

$$\varepsilon_0 = \varepsilon_\infty + 4\pi nq^2/\omega_T^2 M. \tag{30.7}$$

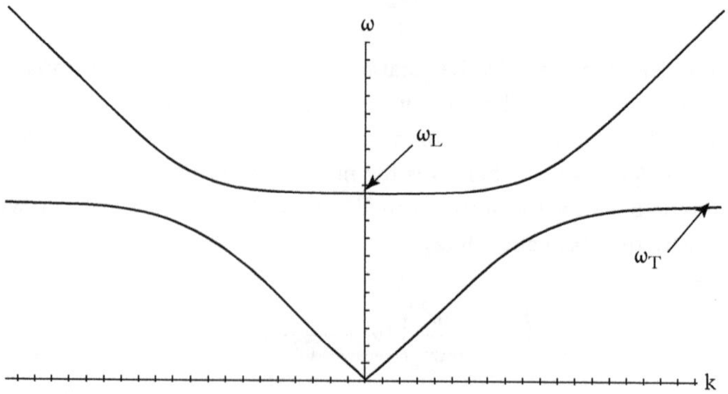

Figure 30.1 *The frequency versus wave vector for a coupled system involving an optical phonon and a light wave. Note the splitting between the lower and upper branches.*

In terms of these quantities we can write $\varepsilon(\omega)$ in the form

$$\varepsilon(\omega) = \varepsilon_\infty + (\varepsilon_0 - \varepsilon_\infty)\frac{\omega_T^2}{\omega_T^2 - \omega^2} = \varepsilon_\infty \frac{\omega_L^2 - \omega^2}{\omega_T^2 - \omega^2}; \tag{30.8}$$

note that this has a zero at $\omega = \omega_L$, and a pole at ω_T (if we add a damping term of the form $\dot{U}_x/\tau$ to the left-hand side of (30.2) the unphysical divergence in (30.8) will be removed). Waves do not propagate between these two frequencies where $\varepsilon(\omega)$ is negative,[1] since this corresponds to an imaginary k vector, implying that the waves are exponentially damped; this is referred to as the *reststrahl* effect. The frequency dependence of the dielectric constant is plotted schematically in Fig. 30.2. Equation (30.8) gives us the following relation between ω_L and ω_T:

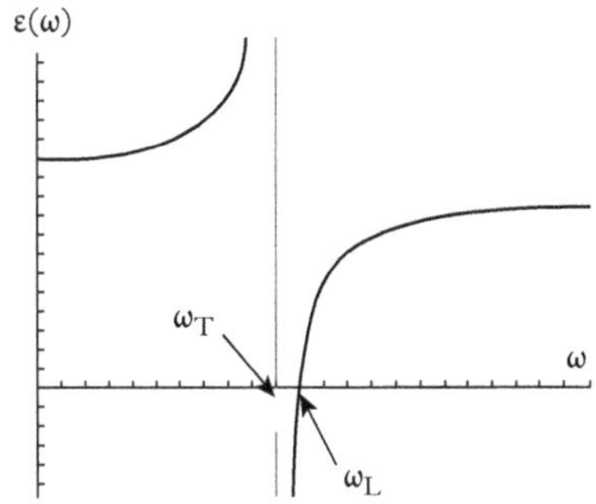

Figure 30.2 *The frequency dependence of the dielectric constant, $\varepsilon(\omega)$; note that it goes through zero at ω_L and, in the absence of damping, goes through infinity at ω_T.*

$$\varepsilon_\infty + (\varepsilon_0 - \varepsilon_\infty)\frac{\omega_T^2}{\omega_T^2 - \omega_L^2} = 0. \tag{30.9}$$

Eq. (30.9) may also be written in the form

$$\frac{\omega_L^2}{\omega_T^2} = \frac{\varepsilon_0}{\varepsilon_\infty}; \tag{30.10}$$

this expression is referred to as the *Lyddane–Sachs–Teller relation.*[2]

30.2 Excitons

In Chapter 10 we studied the electron states of a periodic crystal which, in the single-electron approximation, have the Bloch form. Excitations from the ground state of such a system can occur in various ways. One way is to bring an electron from infinitely far away into the crystal and place it in an unoccupied Bloch state, in which case the charge changes by $-|e|$ (this of course requires the presence of a free surface); conversely, one can remove an electron from an occupied Bloch state and carry it off to infinity, thereby changing the charge by $|e|$. The excitations created in this way can formally be termed *electron-like* and *hole-like* excitations respectively; they are also called single-particle excitations.

Another way to excite the system is to take an electron from an occupied state and promote it to an unoccupied state, in which case we say we make an *electron–hole pair.*[3] These are referred to as two-particle excitations. The question then arises as to what are the effects of the (attractive) Coulomb interaction between the pair. In a semiconductor we may regard the band gap as that energy required to excite an electron from the top of the highest hole band to the bottom of the lowest conduction band, in a manner such that the final electron and hole states are uncorrelated, requiring that they be widely separated. But it is possible to imagine another kind of excitation in which an electron and hole are created, but they do not escape their mutual Coulomb attraction and end up in

[1] Surface waves, so-called surface polaritons, can propagate when the dielectric constant is negative, an example being the surface plasmon, which we discuss in Section 30.5.

[2] On the polar vibrations of alkali halides, R.H. Lyddane, T.G, Sachs, and E. Teller, *Phys. Rev.* **59**, 673 (1941).

[3] The optical excitations discussed in Chapter 29 are formally electron–hole pair excitations.

a bound electron–hole state. Such states are termed *excitons*. The binding energy and characteristic size of excitons varies widely among different materials. One may construct simple models for the limiting cases where the size of the excitons is (i) much larger than, or (ii) of the order of, the spacing between the atoms. These limiting cases are referred to as Wannier and Frenkel excitons respectively. We begin with a discussion of the Wannier exciton.

30.2.1 Wannier excitons

For the Wannier exciton one assumes that the electron and hole interact via the usual classical Coulomb potential, but modified by the dielectric constant of the host crystal, ε, i.e., $V(r) = -e^2/\varepsilon r$. To keep the treatment as simple as possible we assume that the electron and hole are, respectively, associated with nondegenerate band minima and maxima located at the Γ point of a cubic crystal, in which case we can assume locally parabolic bands with isotropic effective masses m_e and m_h. Replacing e^2 by e^2/ε and introducing the reduced mass, $\mu = m_e m_h/(m_e + m_h)$, we can then write the energy levels immediately in the hydrogenic form:

$$\varepsilon_n = \varepsilon_g - \frac{\mu e^4}{2\hbar^2\varepsilon^2 n^2} + \frac{\hbar^2 k^2}{2(m_e + m_h)}, \tag{30.11a}$$

leading to various Rydberg-like spectral series; the last term in (30.11a) accounts for the translational kinetic energy of the exciton. The corresponding Bohr radius is given by

$$a_0 = \frac{\hbar^2 \varepsilon}{e^2 \mu}. \tag{30.11b}$$

Since the excitons may form in a singlet ($S = 0$ or paraexciton) or triplet ($S = 1$ or orthoexciton), spin splitting (usually referred to as exchange splitting) of the excitonic levels occurs. In addition to the excited states of the exciton, the electron and hole can recombine through various processes, leading to exciton annihilation. When the coupling to light via stimulated emission and absorption of excitons is included, one obtains the analogue of a polariton, which is usually called an exciton–polariton. We will discuss exciton–polaritons shortly.

A well-known exciton spectral series is the so-called yellow series in cuprous oxide (Cu_2O). The exciton spectrum is shown in Fig. 30.3;[4] a set of parameters describing these excitations is given in Table 30.1. In Cu_2O there are other low lying bands, leading to additional exciton spectral series (the so-called green, blue, and violet series). Cuprous oxide is further distinguished by the fact that the symmetry of the electron and hole levels leading to yellow exciton formation is such that ground state annihilation of the $S = 1$ spin triplet state proceeds only by *quadruple* emission, while the $S = 0$ ground state annihilation is electromagnetically forbidden in all orders. For this reason, Cu_2O has been the center of much attention in the quest to achieve excitonic Bose condensation. The weaker coupling associated with the quadrupole coupling to light of the $n = 1$ orthoexciton makes it an ideal material to study exciton–polariton formation.

[4] Optical absorption of cuprous oxide, P. Baumeister, *Phys. Rev.*, 121, 359 (1961).

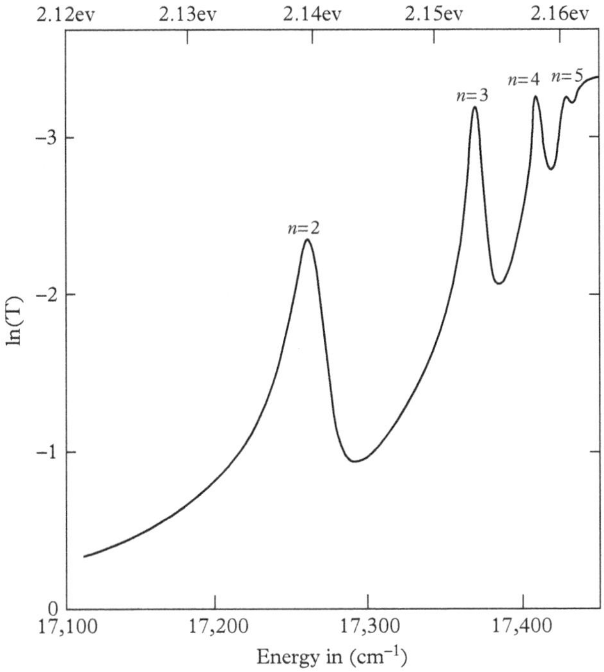

Figure 30.3 *The optical absorption of cuprous oxide, showing the Rydberg levels n = 2 through n = 5 in Eq. (30.11). (After P.W. Baumeister (1961).)*

Table 30.1 *Parameters characterizing the yellow exciton series in* Cu_2O.

Quantity	Symbol	Value
Dielectric constant	ε_∞	7
Electron mass	m_e	1.0 m_o
Hole mass	m_h	0.7 m_o
n = 1 Exciton mass	m_{ex}	2.7 m_o
Rydberg (n>1)	Ry_{ex}	97 meV
Bohr radius (n > 1)	a_{ex}	11.1 A
n = 1 Binding energy	Ry_{ex}	150 meV
n = 1 Bohr radius	a_{ex}	5.0 A
Band gap	E_{gap}	2.17 eV(4 K)
Ortho-para exchange energy	Δ	2 meV

30.2.2 The Frenkel exciton

We now consider the limit in which the electron–hole separation of the exciton bound state approaches atomic or molecular dimensions. Here, we might picture the electron and hole as actually residing on the same molecule or atom; i.e., we have an excited state orbital (the electron) and the vacant orbital which it previously occupied (the hole). This picture is applicable to rare gas crystals (e.g., solid krypton) and various molecular crystals (e.g., anthracene) where the coupling between neighboring atoms/molecules is small. Nonetheless there is some overlap of orbitals on adjacent sites and hence this combined electron–hole excitation can hop (make transitions) between neighboring atoms/molecules.

For simplicity we assume our material is made up of rare-gas-like atoms rather than molecules. We construct a tight-binding-like model for the Frenkel exciton as follows. Neglecting overlap entirely, we would write the ground state wave function as a product of the wave functions of all the atoms making up the solid; the *atomic* wave functions might themselves be represented as a Hartree product of all the occupied orbitals of the atom (see Section 15.1.1). Hence the atomic wave function of the atom at the origin would have the form

$$\begin{aligned}\psi(\mathbf{r}_1,\mathbf{r}_2,\ldots\mathbf{r}_N) &= u_1(\mathbf{r}_1)u_2(\mathbf{r}_2)u_3(\mathbf{r}_3)\ldots u_N(\mathbf{r}_N)\\ &= \prod_{\mathbf{r}_i} u_i(\mathbf{r}_i),\end{aligned} \tag{30.12}$$

where $\mathbf{r}_1,\mathbf{r}_2,\ldots\mathbf{r}_N$ denote the coordinates on an N-electron atom. Orbitals on other lattice sites have the same form but with all coordinates displaced by a real-space lattice vector, $\mathbf{R}$; hence they are written as $\psi(\mathbf{r}_1-\mathbf{R},\mathbf{r}_2-\mathbf{R},\ldots\mathbf{r}_N-\mathbf{R})$. We would then write the ground state crystal wave function as

$$\begin{aligned}\Psi(\mathbf{r}_1,\mathbf{r}_2,\ldots\mathbf{r}_N) &= \prod_{\mathbf{R}} \psi(\mathbf{r}_1-\mathbf{R},\mathbf{r}_2-\mathbf{R},\ldots\mathbf{r}_N-\mathbf{R})\\ &= \prod_{\mathbf{R}}\prod_{\mathbf{r}_i} u_i(\mathbf{r}_i).\end{aligned} \tag{30.13}$$

The excited state of the atom at the origin in this approximation is written as

$$\phi(\mathbf{r}_1,\mathbf{r}_2,\ldots\mathbf{r}_N) = u_1(\mathbf{r}_1)u_2(\mathbf{r}_2)\ldots v_j(\mathbf{r}_j)\ldots u_N(\mathbf{r}_N); \tag{30.14}$$

what we have here is a state where the jth occupied orbital, $u_j(\mathbf{r}_j)$, is now empty and that electron now occupies an excited state orbital, $v_j(\mathbf{r}_j)$. For convenience we suppress writing all the remaining occupied orbitals[5] (and assume that they are

[5] In practice, the remaining orbitals are altered by exciting a given electron to a higher orbital; this is referred to as relaxation, since, with one less unit of charge in the inner orbitals, the screening is reduced and the remaining orbitals move closer to the nucleus.

not altered by exciting the jth electron to the excited state, $v_j(\mathbf{r}_j)$). An excited state of the crystal which is fully localized at the site $\mathbf{R}'$ is then

$$\phi\left(\mathbf{r},\mathbf{R}'\right)=v\left(\mathbf{r}-\mathbf{R}'\right)\prod_{\mathbf{R}\neq\mathbf{R}'}u(\mathbf{r}-\mathbf{R}). \tag{30.15}$$

At this point we turn on an interaction between neighboring sites. For simplicity we will assume that neither the ground state nor the excited state of the electron in question is degenerate.[6] We further restrict the discussion to a one-dimensional lattice with spacing a. With these restrictions we may write the Hamiltonian as a matrix acting in lattice site space, with the sites designated by n, which in turn denotes a lattice position, $X_n = na$; thus,

$$\hat{H}\phi_n(x)=\varepsilon\phi_n(x)+T\left[\phi_{n+1}(x)+\phi_{n-1}(x)\right], \tag{30.16}$$

where $\phi_n(x) = \phi(x-X_n)$, ε is the energy of the excited (v) state relative to the ground (u) state and T measures the strength of the hopping, which is in turn a measure of the overlap of the orbitals on neighboring sites. We seek a solution as a superposition of states $\phi_n(x)$ on all atoms of the form used earlier in connection with the tight-binding model (Section 11.4) and the vibrational modes of a one-dimensional lattice (Section 16.1),

$$\Phi_k(x)=\sum_n e^{ikX_n}\phi_n(x). \tag{30.17}$$

Multiplying (30.16) by e^{ikX_n} and summing over n, we have

$$\hat{H}\Phi_k(x)=\sum_n e^{ikX_n}\left[\varepsilon\phi_n(x)+T\left(\phi_{n+1}(x)+\phi_{n-1}(x)\right)\right]. \tag{30.18}$$

This may be rearranged in the form

$$\begin{aligned}\hat{H}\Phi_k(x)&=\sum_n e^{ikX_n}\left[\varepsilon+T\left(e^{ika}+e^{-ika}\right)\right]\phi_n(x)\\&=\sum_n e^{ikX_n}\left[\varepsilon+2T\cos ka\right]\phi_n(x)\\&=\left[\varepsilon+2T\cos ka\right]\Phi_k.\end{aligned} \tag{30.19}$$

At this point we can invoke periodic boundary conditions, as in earlier discussions. Hence the exciton eigenfunctions are waves with wave vector, k, and energies, $\varepsilon(k) = \varepsilon + 2T\cos ka$; for small k we have $\varepsilon(k) \cong \varepsilon + 2T + Ta^2k^2$; the second term on the right represents a correction to the atomic excitation energy, ε, while the third is a k-dependent translational kinetic energy which we may write in term of an effective mass, $m^* = \hbar^2/2Ta^2$.

[6] If the states are degenerate we must diagonalize the matrices in the subspaces involved, as in Section 11.4.2.

30.3 The exciton-polariton

The dispersion relation given by Eq. (30.11a) shares the property with the optical phonon, in that it will intersect that of the photon dispersion. We will not treat the interaction of light with the exciton microscopically, but use a so-called mechanical or classical Drude-like model in which we picture the electron and hole making up the exciton as bound with a spring and having a resonant frequency, ω_T, as assumed earlier for the phonon–polariton case. The kinetic energy can be added by replacing $\omega_T \rightarrow \omega_T + \hbar k^2/2\,(m_e + m_e)$. The justification for such a crude treatment lies in the quantum mechanical result for the polarizibility governed by (24.31a), which has the same form as the Drude model, albeit with an altered oscillator strength (involving the dipole matrix element). The treatment is then identical except that the frequencies are now much higher, by three to four orders of magnitude, due to the lighter mass of the electron relative to the ions.

Cu_2O represents a special case since the orthoexcitons couple to light only via a quadrupole interaction,[7] while the paraexcitons do not couple at all. In this case there is no splitting between $\omega = 0$ and $\omega = \infty$.

30.4 Plasmons

A limiting case of the polariton formalism is free electrons in a metal where ω_T vanishes (i.e., the spring constant binding the electrons to the atoms vanishes). The dielectric constant is then given by

$$\varepsilon(\omega) = \varepsilon_\infty + \frac{4\pi P}{E} = \varepsilon_\infty - \frac{4\pi ne^2/m}{\omega^2} \qquad (30.20a)$$

or, defining the square of the plasma frequency by $\omega_p^2 = 4\pi ne^2/m$,

$$\varepsilon(\omega) = \varepsilon_\infty - \frac{\omega_p^2}{\omega^2}. \qquad (30.20b)$$

Note that the dielectric constant is now negative for all frequencies below the plasma frequency, corresponding to an imaginary wave vector, as discussed previously. The plasma frequency lies in the ultraviolet for metals; i.e., metals are totally reflecting at visible light frequencies and below ω_p and become transparent above ω_p. We can include the effects of electron scattering via a phenomenological scattering time by including a damping term in the Drude-like equation of motion, in which case (30.20b) becomes

$$\varepsilon(\omega) = \varepsilon_\infty - \frac{\omega_p^2}{\omega^2 + i\omega/\tau}; \qquad (30.21)$$

this results in a purely imaginary dielectric constant as $\omega \rightarrow 0$, corresponding to the electrical conductivity, $\sigma = ne^2\tau/m$.

[7] Coherent propagation and quantum beats of quadrupole polaritons in Cu_2O, D. Frohlich, K. Kulik, B. Uebbing, A. Mysyrowicz, V. Langer, H. Stolz, and W. von der Osten, *Phys. Rev. Lett.* 67, 2343 (1961). Here the behavior is modeled by replacing the coupling constant in the expression for the dielectric function with the prescription, $\omega_p^2 \rightarrow fc^2k^2$.

Small particles, particularly those made from Ag and Au, can exhibit a plasmon resonance at a specific frequency. This property can be derived from the expression for the polarization of a sphere, which is given in texts on electromagnetism as

$$\mathbf{P} = R^3 \frac{\varepsilon_2 - \varepsilon_1}{\varepsilon_1 + 2\varepsilon_2} \mathbf{E}, \tag{30.22}$$

where a is the sphere radius, ε_1 is the dielectric constant of the background medium, and ε_2 is that of the sphere itself. For the special case where $\varepsilon_1 = 1$ (a surrounding vacuum), and taking $\varepsilon_\infty = 1$ in Eq. (30.21), the real part of the denominator of (30.22) will vanish for $\omega = \omega_p/\sqrt{3}$, leading to a resonance at that frequency with the height limited only by τ or ultimately dipole radiation. The plasmon resonance gives glass containing numerous small gold particles its distinctive red color. Equation (30.22) is valid only in the limit $a \ll \lambda$, where λ is the wavelength of light; when a becomes comparable to λ one must use the so-called Mie theory for scattering from a dielectric sphere.[8]

30.5 Surface polaritons

Under suitable conditions polaritons can exist at the interface between two media. Consider a half space of a material with a frequency-dependent dielectric constant, $\varepsilon_1(\omega)$, for $z < 0$ on top of which is a second material with a frequency-independent dielectric constant, ε_2. When $\varepsilon_1(\omega) < 0$, the wave vector *becomes imaginary* for normal incidence. In addition to the case of metals, discussed in Section 30.4, this can also happen in some semiconductors for a narrow interval of frequencies. In both cases the incoming wave would be totally reflected. However, under appropriate conditions a propagating transverse magnetic (TM) *surface polariton mode* can exist in this regime, which for $z > 0$ and for propagation parallel to x has the form

$$H_y = H_0 \exp(ik_{//}x - \kappa_> z - i\omega t), \tag{30.23}$$

where $k_{//}$ is the in-plane wave vector and κ measures the rate of decay perpendicular to the surface. (For $z < 0$ we have a similar form with the same $k_{//}$ but with $\kappa_>$ replaced by $-\kappa_<$.) The fields in both media must satisfy the wave equation; for the first medium we will have

$$k_{//}^2 - \kappa_>^2 = \frac{\varepsilon(\omega)\omega^2}{c^2} \tag{30.24}$$

and similarly for the second. By demanding that the fields satisfy the Maxwell boundary conditions, we can relate $\kappa_>$ and $\kappa_<$ and obtain the following dispersion relation (given as Problem 30.1):

$$k_{//} = \frac{\omega}{c}\sqrt{\frac{\varepsilon_2\varepsilon_1(\omega)}{\varepsilon_2 + \varepsilon_1(\omega)}}, \tag{30.25}$$

with the accompanying condition, $\varepsilon_1(\omega) < -\varepsilon_2$.

[8] This theory is described in Section 9.25 of Chapter IX in *Electromagnetic Theory*, J.A. Stratton, McGraw-Hill, NY (1941).

In the Lorentz model the dielectric constant is given by (see Eq. (3.21))

$$\varepsilon(\omega) = \varepsilon_\infty + \frac{\omega_p^2}{\omega_0^2 - \omega^2 - i\omega\gamma}, \tag{30.26}$$

where $\gamma = 1/\tau$ measures the damping, and for a metal we take $\omega_0 = 0$. For the surface plasmon polaritons we have the limiting behaviors, $\omega = ck/\sqrt{\varepsilon_1(\omega)}$ as $k \to 0$ and $\omega = \omega_p\sqrt{\varepsilon_\infty\varepsilon_2/(\varepsilon_\infty + \varepsilon_2)}$ as $k_{//} \to \infty$. For the special case, $\varepsilon_2 = 1, \varepsilon_\infty = 1$, we have $\omega = \omega_p/\sqrt{2}$; this is to be compared to the case of a small spherical metallic particle in a similar limit, where we found $\omega = \omega_p/\sqrt{3}$. For the case of exciton–polaritons, surface modes exist in the range

$$\omega_0^2 < \omega^2 \le \omega_0^2 + \omega_p^2 \frac{\varepsilon_\infty\varepsilon_2}{\varepsilon_\infty + \varepsilon_2}. \tag{30.27}$$

A striking confirmation of plasmon polaritons arises in the so-called *attenuated total reflection* (ATR) phenomenon observed in thin Ag and Au films deposited on a hemispherical prism, when the light enters from the prism side. The experimental geometry is shown in Fig 30.4.[9] The effect occurs at a specific angle, where the component $k_{//}$ of the wave vector of the incoming electromagnetic wave projected onto the plane of the film matches the wave vector of a propagating surface plasmon polariton; this occurs at an angle *above* the angle of total internal reflection and when the condition is satisfied, the energy in the incoming beam is partially absorbed by the surface plasmon and hence the amount of light reflected is greatly attenuated; by properly adjusting the thickness of the metal film one can achieve a condition where the reflected amplitude vanishes (i.e., the incoming wave is totally absorbed by the film). Some typical data are shown in Fig. 30.5.[10]

[9] Determination of optical constants of metals by excitation of surface plasmons, E. Kretschmann, *Zeitschrift fur Physik* **241**, 313 (1971).

[10] A similar condition can be achieved when exciting surface exciton–polaritons.

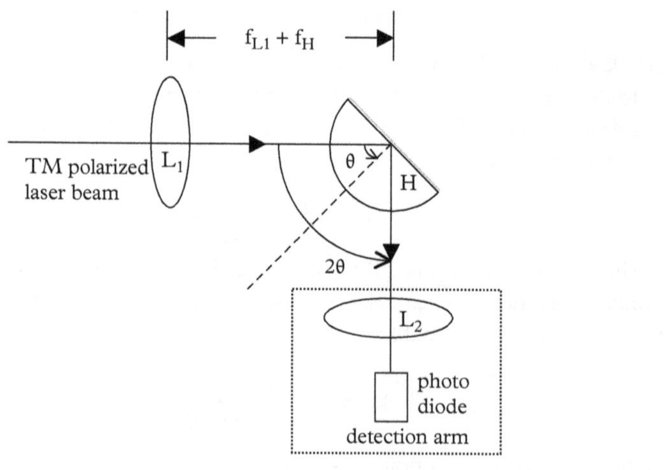

Figure 30.4 *The experimental set-up used to study attenuated total reflection.* L_1*: focusing lens;* H*: hemispherical lens;* L_2*: focusing lens; PD: photodiode.* L_1 *and* H *are placed so that their spacing equals the sum of their focal lengths. The hemispherical lens rotates by an angle,* θ*, while the detection arm rotates by* 2θ*. (after L. Luan (unpublished).)*

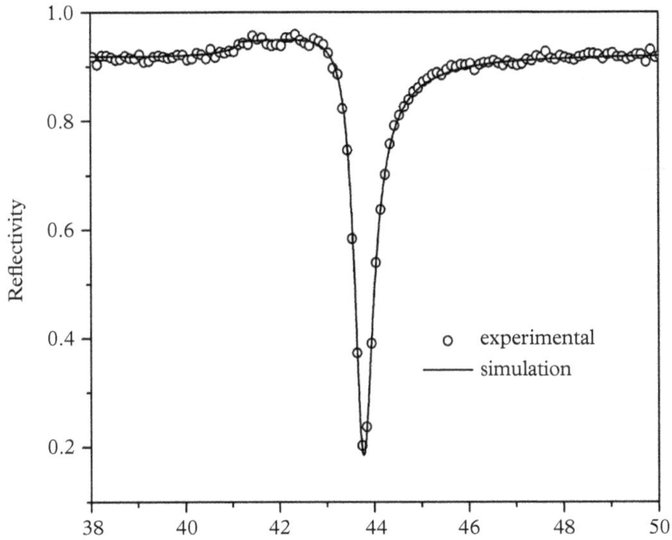

Figure 30.5 *The reflectivity versus angle of 532 nm laser light reflecting from BK7 glass ($\varepsilon = 2.309$) covered by a 61.1 nm* Ag *film. The sharp minimum occurs when the projected component of the light wave vector matches that of the surface plasmon polariton. The curve shows a simulation to the theory based on the Fresnel coefficients of the media involved. (After L. Luan, unpublished).*

ADDITIONAL READING

Plasmons are discussed at length in *Surface Plasmons on Smooth and Rough Surfaces and on Gratings*, Heinz Raether, Springer Tracts in Modern Physics (1988).

Excitons are discussed, largely from the mechanical model, in *Crystal Optics with Spatial Dispersion, and Excitons,* Vladimir M. Agranovich and V. Ginzburg, Springer, Berlin. For a discussion of BEC as it relates to Cu_2O and CuCl, see *Bose-Einstein Condensation of Excitons and Bi-excitons*, S.A. Moskalenko and D.W. Snoke, Cambridge University Press (2000).

Problems

30.1. Derive Eq. (30.25).

30.2. Polariton modes can persist in thin films and they can be treated in a manner analogous to that described in Section 30.5 for surface polaritons. Assume that we have a thin film with thickness d and dielectric constant $\varepsilon_1(\omega)$, that is surrounded by semi-infinite media with dielectric constant ε_2. By symmetry, there will be two independent modes: symmetric and antisymmetric. Show that evanescent decay lengths for these modes (inside and outside the film) are given by the equations

$$\kappa_2\varepsilon_1 + \kappa_1\varepsilon_2 \coth(\kappa_1 d/2) = 0; \; \kappa_1\varepsilon_2 + \kappa_2\varepsilon_1 \tanh(\kappa_2 d/2) = 0.$$

The dispersion relation then follows from requiring the solutions to satisfy the wave equation

$$k_{//}^2 - \kappa_i^2 = \varepsilon_i(\omega/c)^2.$$

30.3. Obtain the equations describing the propagation of surface optical phonon polaritons.

Behavior Under Intense Illumination: NLO, the e–h Liquid and Excitonic BEC

31

In Chapter 29, we examined the linear response of insulators and semiconductors to light, particularly with respect to optical absorption. With the availability of high intensity pulsed lasers it is now possible to study responses in the nonlinear regime where phenomena such as harmonic generation and frequency mixing occur, as well as collective behaviors of the high densities of excitations so produced. The latter is of interest in connection with the electron–hole liquid and Bose–Einstein condensation (BEC) of excitons, both of which will be discussed here.

31.1 The nonlinear susceptibility

The linear relation between the polarization, **P**, and the electric field, **E**, given by Eqs. (2.28) and (2.29) becomes less accurate as the magnitude of the field increases and then it becomes necessary to include higher-order terms in a power series expansion of **P**(**E**). We will limit ourselves to the quadratic and cubic corrections; the polarization is then written as

$$\mathbf{P} = \mathbf{P}^{(1)} + \mathbf{P}^{(2)} + \mathbf{P}^{(3)}. \tag{31.1}$$

As previously noted, the large fields required to generate nonlinear responses are commonly encountered in experiments with pulsed lasers. Building in the physical requirement that the response of a material only *follows* the excitation that causes it (termed causality), we can write these terms in the expansion of the polarization in component form as

$$P_i^{(1)}(t) = \int_0^\infty d\tau \chi_{ij}^{(1)}(\tau) E_j(t-\tau), \tag{31.2}$$

$$P_i^{(2)}(t) = \int_0^\infty d\tau_1 \int_0^\infty d\tau_2 \chi_{i,jk}^{(2)}(\tau_1,\tau_2) E_j(t-\tau_1) E_k(t-\tau_2), \tag{31.3}$$

and

$$P_i^{(3)}(t) = \int_0^\infty d\tau_1 \int_0^\infty d\tau_2 \int_0^\infty d\tau_3 \chi_{i,jkl}^{(3)}(\tau_1,\tau_2,\tau_3) E_j(t-\tau_1) E_k(t-\tau_2) E_l(t-\tau_3). \tag{31.4}$$

The Physics of Solids. First Edition. J. B. Ketterson.

Here, $\chi^{(1)}$, $\chi^{(2)}$, and $\chi^{(3)}$ are referred to as the linear (Chi 1), quadratic (second order or Chi 2), and cubic (third order or Chi 3) susceptibilities, respectively. Equation (31.1), which describes the linear response, is identical to (29.17). $\chi^{(2)}$ vanishes in a medium having inversion symmetry: the operation $\mathbf{E} \to -\mathbf{E}$ must then cause $\mathbf{P} \to -\mathbf{P}$, which requires that $\chi^{(2)}$ vanishes identically. Hence the pyroelectric and piezoelectric materials discussed in Chapter 26 can have a non-vanishing $\chi^{(2)}$. An exception is when the susceptibility contains dynamic quadrupole (as is the case with cuprous oxide) and magnetic dipole contributions. For all other media the leading correction is $\chi^{(3)}$.

Since $\chi^{(2)}(\tau_1, \tau_2)$ depends on two times, we introduce the double Fourier transform of (31.3) according to

$$\chi^{(2)}_{i,jk}(\omega_1, \omega_2) = \int_0^\infty d\tau_1 \int_0^\infty d\tau_2\, e^{i(\omega_1\tau_1+\omega_2\tau_2)} \chi^{(2)}_{i,jk}(\tau_1, \tau_2). \tag{31.5}$$

Similarly, we will have a triple Fourier transform for $\chi^{(3)}$,

$$\chi^{(3)}_{i,jkl}(\omega_1, \omega_2, \omega_3) = \int_0^\infty d\tau_1 \int_0^\infty d\tau_2 \int_0^\infty d\tau_3\, e^{i(\omega_1\tau_1+\omega_2\tau_2+\omega_3\tau_3)} \chi^{(3)}_{i,jkl}(\tau_1, \tau_2, \tau_3). \tag{31.6}$$

The susceptibilities $\chi^{(2)}_{i,jk}(\omega_1, \omega_2)$ and $\chi^{(3)}_{i,jkl}(\omega_1, \omega_2, \omega_3)$ may be calculated using perturbation theory in a manner similar to that used to obtain the linear susceptibility given by Eq. (24.28).[1]

The higher-order polarizations must be invariant with respect to interchange of the fields; for $\chi^{(2)}$, the interchange involves the operations $j \to k$, $k \to j$, $\tau_1 \to \tau_2$, $\tau_2 \to \tau_1$; hence,

$$\chi^{(2)}_{i,jk}(\omega_1, \omega_2) = \chi^{(2)}_{i,kj}(\omega_2, \omega_1), \tag{31.7}$$

with similar requirements for $\chi^{(3)}$. Since the susceptibilities are real, we have the further requirement

$$\chi^{(2)}_{i,jk}(\omega_1, \omega_2)^* = \int_0^\infty d\tau_1 \int_0^\infty d\tau_2\, e^{-i(\omega_1\tau_1+\omega_2\tau_2)} \chi^{(2)}_{i,jk}(\tau_1, \tau_2) = \chi^{(2)}_{i,jk}(-\omega_1, -\omega_2). \tag{31.8}$$

31.2 Effects of $\chi^{(2)}$ and $\chi^{(3)}$ on harmonic fields

We first examine the effect of $\chi^{(2)}$ for the cases where we have a single field of frequency ω; this will be followed by fields having two frequencies, ω_1 and ω_2. In the presence of nonlinearity the fields must always be expressed as real quantities.

[1] The expressions are given in *Nonlinear Optics* by N. Bloembergen, Addison-Wesley (1992).

To begin we limit ourselves to a monochromatic field. We may write such a field as $E_i(t) = \mathrm{Re}E_{0i}e^{-i\omega t}$ or

$$E_i(t) = \frac{1}{2}\left[E_{0i}e^{-i\omega t} + E^*_{0i}e^{+i\omega t}\right]. \tag{31.9}$$

Inserting this latter form in Eq. (31.3), we have

$$P^{(2)}_i(t) = \frac{1}{4}\int_0^\infty d\tau_1 \int_0^\infty d\tau_2 \chi^{(2)}_{i,jk}(\tau_1,\tau_2)\left[E_{0j}E_{0k}\,e^{-2i\omega t+i\omega(\tau_1+\tau_2)}\right.$$
$$\left. + E_{0j}E^*_{0k}\,e^{+i\omega(\tau_1-\tau_2)} + E^*_{0j}E_{0k}e^{-i\omega(\tau_1-\tau_2)} + E^*_{0j}E^*_{0k}e^{2i\omega t-i\omega(\tau_1+\tau_2)}\right].$$

From Eq. (31.5), we have

$$P^{(2)}_i(t) = \frac{1}{4}\left[e^{-2i\omega t}\chi^{(2)}_{i,jk}(\omega,\omega)E_{0j}E_{0k} + \chi^{(2)}_{i,jk}(\omega,-\omega)E_{0j}E^*_{0k}\right.$$
$$\left. +\chi^{(2)}_{i,jk}(-\omega,\omega)E^*_{0j}E_{0k} + e^{2i\omega t}\chi^{(2)}_{i,jk}(-\omega,-\omega)E^*_{0j}E^*_{0k}\right].$$

Using Eq. (5.9) we can write this as

$$P^{(2)}_i(t) = \frac{1}{2}\sum_{jk}\mathrm{Re}\left[e^{-2i\omega t}\chi^{(2)}_{i,jk}(\omega,\omega)E_{0i}E_{0j} + \chi^{(2)}_{i,jk}(\omega,-\omega)E_{0i}E^*_{0j}\right]. \tag{31.10}$$

Thus the second-order polarization contains terms at *twice the frequency* (2ω or second harmonic) and *zero frequency* (a d.c. polarization); the latter is the familiar "rectification" associated with a quadratic (square-law) nonlinearity.

Consider now the case where there are two electric fields with components, E_{1i} and E_{2i}, having incommensurate frequencies, ω_1 and ω_2. The total electric field, $E_i(t)$, may then be written as

$$E_i(t) = \frac{1}{2}\left[E_{01i}e^{-i\omega_1 t} + E^*_{01i}e^{i\omega_1 t} + E_{02i}e^{-i\omega_2 t} + E^*_{02i}e^{i\omega_2 t}\right]. \tag{31.11}$$

The total polarization, $P^{(2)}_i(t)$, resulting from these two fields is

$$P^{(2)}_i(t) = \frac{1}{4}\int_0^\infty d\tau_1 \int_0^\infty d\tau_2 \chi^{(2)}_{i,jk}(\tau_1,\tau_2)\left[E_{01j}\,e^{-i\omega_1(t-\tau_1)} + E^*_{01j}\,e^{+i\omega_1(t-\tau_1)}\right.$$
$$\left. + E_{02j}\,e^{-i\omega_2(t-\tau_1)} + E^*_{02j}\,e^{+i\omega_2(t-\tau_1)}\right] \times \left[E_{01k}\,e^{-i\omega_1(t-\tau_2)} + E^*_{01k}\,e^{+i\omega_1(t-\tau_2)}\right.$$
$$\left. + E_{02k}\,e^{-i\omega_2(t-\tau_2)} + E^*_{02k}\,e^{+i\omega_2(t-\tau_2)}\right].$$

In forming the products of the quantities in the two square brackets, we generate five classes of terms involving the exponentials, $e^{\pm 2i\omega_1 t}$, $e^{\pm 2i\omega_2 t}$, e^0, $e^{\pm i(\omega_1+\omega_2)t}$, and $e^{\pm i(\omega_1-\omega_2)t}$. We have already encountered the terms associated with the first three forms in the previous example. The last two forms represent an additional effect called *frequency mixing*: they involve the *sum and difference frequencies*, $\omega_1 \pm \omega_2$. We collect all the terms involving the mixed frequencies and denote the resulting polarization as $P_i^{(2)(\text{mix})}$, which is given by (where we exploit the symmetry properties of $\chi^{(2)}$)

$$P_i^{(2)(\text{mix})}(t) = \frac{1}{4}\int_0^\infty d\tau_1 \int_0^\infty d\tau_2 \chi_{i,jk}^{(2)}(\tau_1, \tau_2)\Big[2E_{01j}E_{02k}e^{i[\omega_1\tau_1+\omega_2\tau_2-(\omega_1+\omega_2)t]} + 2E_{01j}E_{02k}^* e^{i[\omega_1\tau_1-\omega_2\tau_2-(\omega_1-\omega_2)t]} + \text{c.c.}\Big].$$

On carrying out the integrals, we obtain

$$P_i^{(2)(\text{mix})}(\omega_1, \omega_2) = \frac{1}{2}\Big[\chi_{i,jk}^{(2)}(\omega_1, \omega_2)E_{01j}E_{02k}e^{-i(\omega_1+\omega_2)t} + \chi_{i,jk}^{(2)}(\omega_1, -\omega_2)E_{01j}E_{02k}^* e^{-i(\omega_1-\omega_2)t} + \chi_{i,jk}^{(2)}(-\omega_1, \omega_2)E_{01j}^*E_{02k}e^{i(\omega_1-\omega_2)t} + \chi_{i,jk}^{(2)}(-\omega_1, -\omega_2)E_{01j}^*E_{02k}^*e^{i(\omega_1+\omega_2)t}\Big]$$

which we can rewrite as

$$P_i^{(2)(\text{mix})}(\omega_1, \omega_2) = \text{Re}\Big[\chi_{i,jk}^{(2)}(\omega_1, \omega_2)E_{01j}E_{02k}e^{-i(\omega_1+\omega_2)t} + \chi_{i,jk}^{(2)}(\omega_1, -\omega_2)E_{01j}E_{02k}^*e^{-i(\omega_1-\omega_2)t}\Big], \quad (31.12)$$

where the two terms are referred to as the sum frequency and difference frequency terms respectively.

Finally, we discuss third harmonic generation by $\chi^{(3)}$ from a single harmonic field of the form (31.9),

$$\begin{aligned}
P_i^{(3)}(t) &= \frac{1}{8}\int_0^\infty d\tau_1 \int_0^\infty d\tau_2 \int_0^\infty d\tau_3\, \chi_{i,jkl}^{(3)}(\tau_1, \tau_2, \tau_3)\Big[E_{0j}\, e^{-i\omega(t-\tau_1)} + E_{0j}^*\, e^{i\omega(t-\tau_1)}\Big] \\
&\quad \times \Big[E_{0k}e^{-i\omega(t-\tau_2)} + E_{0k}^*e^{i\omega(t-\tau_2)}\Big]\Big[E_{0l}e^{-i\omega(t-\tau_3)} + E_{0l}^*e^{i\omega(t-\tau_3)}\Big] \\
&= \frac{1}{8}\Big\{\Big[\chi_{i,jkl}^{(3)}(\omega, \omega, \omega)E_{0j}E_{0k}E_{0l}e^{-3i\omega t} + \chi_{i,jkl}^{(3)}(-\omega, -\omega, -\omega)E_{0j}^*E_{0k}^*E_{0l}^*e^{3i\omega t}\Big] \\
&\quad + \Big[\chi_{i,jkl}^{(3)}(\omega, \omega, -\omega)E_{0j}E_{0k}E_{0l}^* + \chi_{i,jkl}^{(3)}(\omega, -\omega, \omega)E_{0j}E_{0k}^*E_{0l} \\
&\quad + \chi_{i,jkl}^{(3)}(-\omega, \omega, \omega)E_{0j}^*E_{0k}E_{0l}\Big]e^{-i\omega t} + \Big[\chi_{i,jkl}^{(3)}(\omega, -\omega, -\omega)E_{0j}E_{0k}^*E_{0l}^* \\
&\quad + \chi_{i,jkl}^{(3)}(-\omega, -\omega, \omega)E_{0j}^*E_{0k}^*E_{0l} + \chi_{i,jkl}^{(3)}(-\omega, \omega, -\omega)E_{0j}^*E_{0k}E_{0l}^*\Big]e^{i\omega t}\Big\}.
\end{aligned}$$

Using the relation

$$\chi^{(3)*}_{i,jkl}(\omega_1,\omega_2,\omega_3) = \chi^{(3)}_{i,jkl}(-\omega_1,-\omega_2,-\omega_3), \tag{31.13}$$

and permuting the indices, jk$l$, in the 3ω terms (which is permissible since all the arguments of χ have the same frequency), we may rewrite $P^{(3)}(t)$ as

$$P^{(3)}_i(t) = \frac{1}{4}\sum_{jkl} \mathrm{Re}\Big\{\chi^{(3)}_{i,jkl}(\omega,\omega,\omega)E_{0j}E_{0k}E_{0l}e^{-3i\omega t} + \Big[\chi^{(3)}_{i,jkl}(-\omega,\omega,\omega)E^*_{0j}E_{0k}E_{0l}$$
$$+ \chi^{(3)}_{i,jkl}(\omega,-\omega,\omega)E_{0j}E^*_{0k}E_{0l} + \chi^{(3)}_{i,jkl}(\omega,\omega,-\omega)E_{0j}E_{0k}E^*_{0l}\Big]e^{-i\omega t}\Big\}. \tag{31.14}$$

Hence, $\chi^{(3)}$ generates fields at the third harmonic (3ω) and modifies the field at the applied frequency ω; the latter is known as the *Kerr effect*.[2] The discussion is easily generalized to treat the case where multiple frequencies are present.

The second- and third-order nonlinearities are utilized to make various optical devices such as sum and difference frequency generators, switches, and parametric amplifiers.

31.3 Collective effects in semiconductors under intense illumination

Under very intense over-the-energy-gap laser illumination of a semiconductor, electrons and holes are produced at a high rate. These production rates are balanced by a variety of decay mechanisms available to the electron–hole pairs. Depending on the light intensity, and the semiconductor together with its doping, the system may evolve into various states of dynamic equilibrium. If, on the other hand, we have some high initial concentration of electron–hole pairs following the application of a strong laser pulse, we may then study the time dependence of various populations following the pulse. In a pure material such decay processes include direct electron–hole recombination, indirect electron–hole recombination,[3] or various other processes; the latter include phonon-assisted transitions of carriers to different bands, or exciton formation from which subsequent decay processes ultimately result in recombination. When impurities/dopants are present, electrons and holes may be trapped on the associated sites.

In the case of continuous, high-intensity, illumination of a pure semiconductor, one has a steady-state population of electrons and holes along with excitons, which often achieve temperatures approximately that of the lattice. An interesting case is germanium. Here, electrons at Γ resulting from electron–hole pairs created by direct optical transitions, quickly decay via phonon assisted processes to the electron pockets at the L points of the Brillouin zone, where they can bind with the Γ-point holes to form excitons. As the intensity is increased, and with it

[2] The contribution form $\chi^{(3)}$ produces a change in the index of refraction, Δn, which is proportional to E^2. If Δn is positive it can cause a propagating laser beam to come to a focus, thereby greatly increasing the local field intensity and lead to a breakdown of the medium; this phenomenon is called self-focusing.

[3] If the carriers are located at different points in the Brillouin zone, recombination requires the participation of a lattice phonon to conserve crystal momentum.

the populations in the resulting electron–hole/exciton gas, a collective transition from a gas to *liquid electron–hole droplets* occurs within the host crystal, a remarkable phenomenon, that was predicted by Keldysh.[4] We will discuss this state in Section 31.4.

With a tunable laser one may resonantly create a steady-state exciton population; alternatively one may study a weakly decaying exciton population that evolves from an intense, over-the-gap laser pulse after the high energy electron–hole pair population has decayed. Here cuprous oxide, Cu_2O, is a material which has received much attention over the years since direct recombination of the n = 0 orthoexcitons is only quadrupole allowed, and direct paraexciton decay is forbidden in all orders.[5] One then has the exciting possibility of observing coherent exciton phenomena, the most challenging being *Bose condensation* (BEC). However this possibility remains controversial in bulk systems.

An alternative route to BEC was provided by the development of a method to engineer two-dimensional cavities in which the optical decay of excitons is suppressed, and BEC has been observed in such systems. As necessary background we will review BEC in an ideal gas in Section 31.5 and the extension to a weakly interacting Bose gas in Sections 31.6 and 31.7, which is of interest as a model for superfluidity in general.[6] In Section 31.8 we take the opportunity to discuss the Gross–Pitaevskii equation, which is widely used to treat inhomogeneous superfluids. In Section 31.9 we discuss excitons in a 2d photonic cavity. Finally, in Section 31.10 we describe the experiments which led to excitonic BEC.

31.4 The electron–hole liquid

As the density of an exciton gas increases, interactions become important, e.g. excitonic molecules (biexcitons) can appear, a state which has been most studied in CuCl. We can loosely think of a high density exciton or biexciton gas as corresponding to a gas of Li atoms or H_2 molecules above their liquefaction temperatures; at higher densities these latter materials undergo a *first-order transition* to a liquid state, the first being a metal and the second an insulator. So the question arises as to whether a high density exciton gas can also *phase separate* into a liquid–gas mixture (prior to reaching a density where BEC might occur) and what the nature of that liquid phase is. As opposed to bulk excitonic BEC, which remains controversial, the formation of liquid electron–hole "droplets" in some semiconductors has been overwhelmingly confirmed experimentally, with the largest body of work involving Ge and Si; furthermore, the condensed state turns out to be a *metallic electron–hole liquid*, as first proposed by L.V. Keldysh.

The band structure of Ge was shown in Fig. 11.14(b). The heavy and light holes for both Si and Ge are located at the Γ point and the nature of the constant energy surfaces in the presence of spin–orbit coupling was discussed in Section 12.1.7 and given as Eqs. (12.39) and (22.16); the associated parameters were given in Table 22.3. The electrons in Si lie along the Γ – X line while

[4] For a review, see: The electron hole liquid in semiconductors, L.V. Keldysh, *Contemp. Phys.*, **27**, 395 (1986).

[5] Phonon assisted decays are dipole allowed for both species.

[6] This model is also important in achieving a microscopic understanding superfluid ^{4}He, which we take up in Chapter 36.

those in Ge are located at the L point, resulting in a total of six and four ellipsoids of revolution respectively. The longitudinal and transverse effective masses were given in Table 22.2.

At temperatures provided by liquid helium (typically 1.2 K to 4.2 K) and at laser pump intensities high enough that electron–hole droplets form, the electrons and holes form interpenetrating degenerate Fermi seas, as shown schematically in Fig. 31.1. Because of the large spacing of the electron and hole valleys in the Brillouin zone in Si and Ge, electron–hole recombination cannot proceed by a process in which only a photon is emitted; energy and momentum conservation require both a phonon (having a large momentum but small energy, $\hbar\Omega$) and a photon (which has a large energy, $\hbar\omega$, but small momentum). For this reason the excitons have a long lifetime and would be a good candidate for BEC if e–h droplet formation did not occur.

Figure 31.2 shows the luminescence spectrum of Ge at 4.2 K in a two-phase region where a gas of free electrons and holes together with excitons coexists with e–h liquid droplets. Optical luminescence peaks arising from exciton and e–h recombination will be shifted to lower energies relative to the band gap, since they involve the energy carried away by the emitted phonons, as well as the binding energies of the excitons or the e–h liquid. The observed processes in Fig. 31.2 involve transverse acoustic (TA), longitudinal acoustic (LA), and transverse optical (TO) phonons; if we neglect the wave vector, $\mathbf{q}$, of the light as well as the spread in wave vectors, $\mathbf{k}_e$ and $\mathbf{k}_h$, of the electrons and holes about their local band extrema, these phonons will have wave vectors, $\mathbf{K}$, that extend from the zone center to the location of the electron extrema (the point L in Ge). (The phonon spectrum of

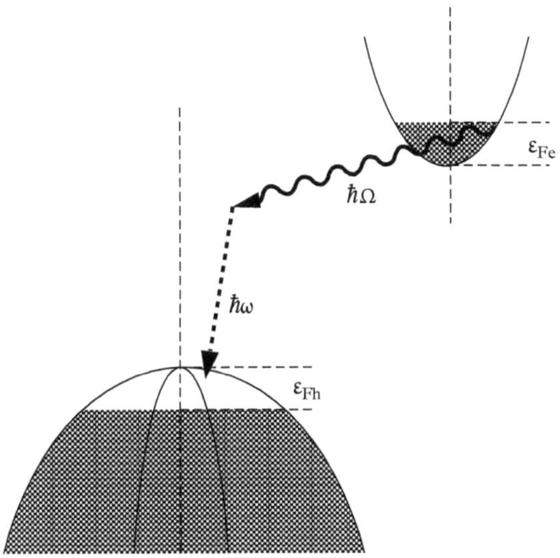

Figure 31.1 *Schematic drawing showing the separate Fermi seas associated with the electrons and holes. Because of the large displacement of the electron and hole valleys in the Brillouin zone, recombination requires a two-particle process involving a phonon (with low energy and a large wave vector) and a photon (with large energy and a small wave vector) to conserve energy and momentum.*

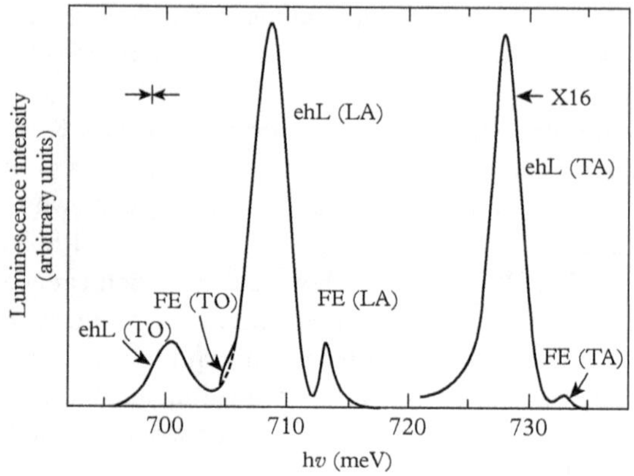

Figure 31.2 *A luminescence spectrum from Ge at 4.2 K. The peaks labeled ehL arise from the electron-hole liquid; the larger one involves a longitudinal acoustic (LA), while the smaller involves a transverse optical (TO) phonon. Also seen are the free exciton (FE) decays involving the same phonons. The spectra to the right involve transverse acoustic (TA) phonons. (After Thomas et. al..)*

Si was shown in Fig. 16.7 and is similar to that of Ge.) Since the band gap in Ge is smaller than Si the phonon induced shifts in the free exciton lines will be proportionally larger. Note that in all cases the e–h liquid peaks lie at *lower* energies relative to a corresponding free exciton peak. In addition, due to the range of the energies associated with the individual electron and hole states in the two Fermi seas, the associated lines have an added broadening relative to the associated exciton transition. This added width contains important information on the carrier concentrations in the respective Fermi seas.

The line shape follows from the electron and hole recombination rate that is given by the usual golden rule expression

$$w = \frac{2\pi}{\hbar}|M|^2 \int_0^\infty d\varepsilon_e \int_0^\infty d\varepsilon_h N_e(\varepsilon_e)N_h(\varepsilon_h)f(\varepsilon_e)f(\varepsilon_h) \times\delta(\varepsilon_e + \varepsilon_h + \varepsilon_G - \hbar\omega - \hbar\Omega), \tag{31.15}$$

where $\hbar\omega(\mathbf{q})$ is the energy of the light quantum, $\hbar\Omega(\mathbf{K})$ is the energy of the phonon for the wave vector connecting the electron extrema to the zone center, and $\varepsilon_G(n)$ is the band gap *in the liquid* at local droplet carrier density, n (which differs from $\varepsilon_G(n \to 0)$). The density of states for the electrons and holes are given by

$$N_e(\varepsilon) = \nu_e\left(1/2\pi^2\right)\left(2m_e/\hbar^2\right)^{3/2}\varepsilon^{1/2} \tag{31.16a}$$

and

$$N_h(\varepsilon) = (1/2\pi^2)(2m_h/\hbar^2)^{3/2}\varepsilon^{1/2}, \tag{31.16b}$$

where ν_e is the number of electron ellipsoids, and m_e and m_h are the density of states effective masses for an electron ellipsoid and the holes.[7]

If we ignore the energy dependence arising from the matrix element M that governs the processes and treat it as a constant, we can directly fit the data of

[7] Here, $m_h \equiv m_{hH}[1 + (m_{hL}/m_{hH})^{3/2}]^{2/3}$ and the required parameters are given in Tables 22.2 and 22.3.

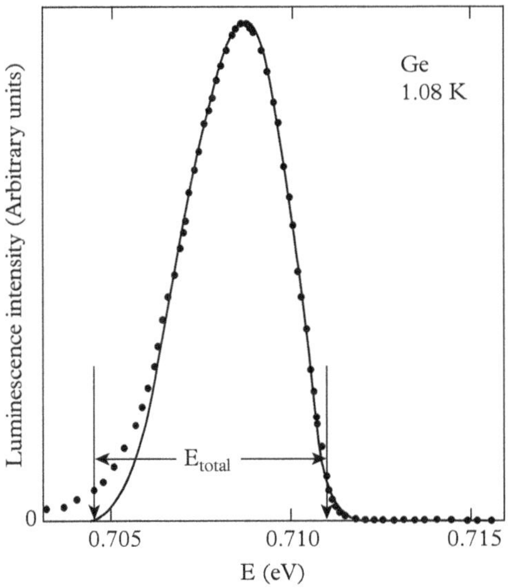

Figure 31.3 *The points show the e-h luminescence arising from the LA phonon assisted decay process at* T = 1.08 K. *The curve shows a fit to Eq. (31.15) under the assumption of an energy-independent matrix element. (Thomas et al. (1973).)*

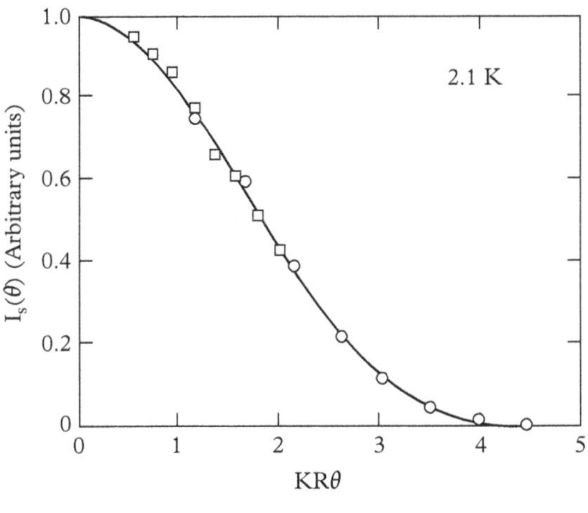

Figure 31.4 *Angular dependence of the light scattering at a wavelength of* 3.39 μm *from an ensemble of electron-hole droplets taken at two different depths within a Ge crystal. (After Pokrovskii and Svistunova (1971).)*

Fig. 31.3 to Eq. (31.5).[8] At low temperatures, where the Fermi surfaces are relatively sharp, the width of the peak will then be a measure of a combined Fermi energy, $\varepsilon_{total} = \varepsilon_e + \varepsilon_h$, while the shift relative to $\varepsilon_G(n \to 0)$ gives a measure of the total binding energy of the droplets (ignoring any surface energy).

In addition to the line width of the luminescence many other properties of the electron–hole liquid have been studied. By monitoring the extinction coefficient of transmitted infrared light, the onset of droplet condensation can be observed. This extinction is accompanied by an onset of scattering from these droplets.[9] Figure 31.4 shows the results of light scattering from Ge collected from two different depths within the crystal where the droplet sizes differ. The line is

[8] Temperature-dependent luminescence from electron-hole liquid in Ge, G.A. Thomas, T.G. Phillips, T.M. Rice, and J.C. Hensel, *Phys Rev. Lett.* **31**, 386 (1973).

[9] Infrared probing of a large electron-hole drop in germanium, Y.E. Pokrovskii and K.I. Svistunova, *JETP Lett.* **13**, 212 (1971).

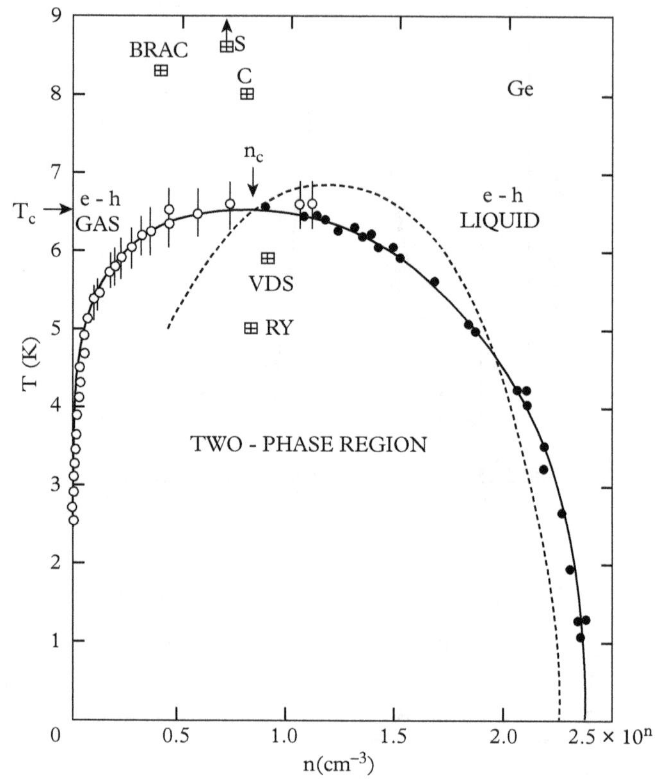

Figure 31.5 *The electron liquid-gas phase diagram in Ge. (After Thomas, et al. (1974).)*

universal and is based on a modified Rayleigh scattering theory. The circles and squares correspond to fits for radii of 3.4 μm and 7.6 μm respectively. If a stress is induced near a surface by pressing on it with a suitable object, it shifts the position of the band edges. Small droplets can then coalesce into a large droplet that can be visualized via a video camera. Another sensitive probe is a shift in the intensity of the cyclotron resonance of free carriers at the onset of droplet condensation. By carefully interpreting the results of numerous experiments a liquid–gas *coexistence curve* has been constructed, that is shown in Fig. 31.5.[10]

By applying a uniform stress to Si and Ge one can remove the degeneracy between the various ellipsoids, and furthermore, split the degenerate hole complex into isotropic light and heavy hole sheets. We will not discuss this fascinating topic and refer the reader to the review by Wong.[11]

There has been much interest in the electron–hole liquids as model systems to test the accuracy of various theories of an interacting electron gas, since the effects of band structure enter only through model parameters such as the dielectric constant and the various effective masses.

[10] Liquid-gas phase-diagram of an electron-hole fluid, G.A. Thomas, T.M. Rice, and J.C. Hensel, *Phys. Rev. Lett.* 33, 219 (1974).

[11] See "Additional Reading" section.

31.5 Bose condensation in an ideal gas

Bose–Einstein condensation (BEC), or Bose condensation for short, is the property of an ideal gas with a fixed number of Bose particles (those having integral spin) that some fraction of the particles must collect (condense) in a zero momentum state below some temperature. Excitons can be regarded as bosons at densities where the spacing between them is large compared to the size of the electron–hole bound state. Under the additional assumptions that the interaction among them is neglected and that (on realizable experimental time scales) they can achieve a dynamic equilibrium that is characterized by some temperature, T, and particle number, N (both of which can be externally controlled), the possibility of BEC arises. For simplicity we assume that the excitons are located at the Γ point and that the bound electrons and holes arise from non-degenerate bands; otherwise we would have to include the associated band or valley degeneracy. The energy levels will be written as $\varepsilon(\mathbf{k}) = \hbar^2 k^2/2m$, where their energy is measured relative to an exciton at rest. The total number of excitons, N, with mass, m, degeneracy, g (three for a spin one ortho-exciton), in a volume, L^3, is then given by

$$N = \frac{gL^3}{(2\pi)^3} \int \frac{d^3k}{e^{[\varepsilon(\mathbf{k})-\mu]/k_B T} - 1}; \tag{31.17}$$

crucial here is our inclusion of the chemical potential, μ, in the Bose distribution, unlike our earlier discussion of phonons (see Eq. (16.79)) where there is no constraint on the number of phonons present. Rewriting this in terms of the energy ε alone, we have

$$N = \frac{gL^3 m^{3/2}}{2^{1/2}\pi^2\hbar^3} \int_0^\infty \frac{\varepsilon^{1/2} d\varepsilon}{e^{[\varepsilon-\mu]/k_B T} - 1}. \tag{31.18}$$

The corresponding expression for the total energy, E, is

$$E = \frac{gL^3}{(2\pi)^3} \int_0^\infty \frac{\varepsilon(\mathbf{k}) d^3k}{e^{[\varepsilon(\mathbf{k})-\mu]/k_B T} - 1} \tag{31.19}$$

or in terms of the energy alone,

$$E = \frac{gL^3 m^{3/2}}{2^{1/2}\pi^2\hbar^3} \int_0^\infty \frac{\varepsilon^{3/2} d\varepsilon}{e^{[\varepsilon-\mu]/k_B T} - 1}. \tag{31.20}$$

Introducing the dimensionless variable $z = \varepsilon/k_B T$, we may rewrite (31.18) as

$$N = \frac{g\,(mk_B T)^{3/2}\,L^3}{2^{1/2}\pi^2\hbar^3} \int_0^\infty \frac{z^{1/2} dz}{e^{z-(\mu/k_B T)} - 1}. \tag{31.21}$$

and (31.20) as

$$E = \frac{g\,(mk_BT)^{5/2}L^3}{2^{1/2}\pi^2\hbar^3}\int_0^\infty \frac{z^{3/2}dz}{e^{z-(\mu/k_BT)}-1}. \tag{31.22}$$

In the absence of a condensate, the chemical potential must be *less than zero,*[12] since otherwise there will be a value of ε for which $\varepsilon - \mu = 0$ or $\exp[(\varepsilon-\mu)/k_BT] = 1$, resulting in a vanishing denominator and hence a diverging integral. The case $\mu = 0$ is special (we then have a zero in the numerator *and* denominator of (31.18), the ratio of which, by itself, is indeterminate); as we will see shortly, this situation corresponds to the presence of what is called a Bose condensate. When the total number of particles is fixed, the chemical potential is that value which satisfies (31.17). As the temperature falls, the chemical potential approaches zero from below. However, at some critical temperature, $T = T_c$, μ reaches zero and for $T < T_c$ Eq. (31.17) can no longer be satisfied unless a macroscopic number, N_0, of the particles have the energy zero, corresponding to a wave vector $k = 0$; these excitons are said to be in the *condensate.* As the temperature continues to fall, the number of excitons in the condensate grows. At $T = 0$ we have $N = N_0$; i.e., all particles are in the condensate. The number of particles in the condensate below T_c follows from the solution of the equation

$$N = \frac{gL^3m^{3/2}}{2^{1/2}\pi^2\hbar^3}\int_0^\infty \frac{\varepsilon^{1/2}d\varepsilon}{e^{\varepsilon/k_BT}-1} + N_0 \tag{31.23a}$$

or

$$N = \frac{g\,(mk_BT)^{3/2}L^3}{2^{1/2}\pi^2\hbar^3}\int_0^\infty \frac{z^{1/2}dz}{e^z-1} + N_0. \tag{31.23b}$$

The *critical* temperature, T_c, is the solution of the equation

$$N = \frac{gL^3m^{3/2}}{2^{1/2}\pi^2\hbar^3}\int_0^\infty \frac{\varepsilon^{1/2}d\varepsilon}{e^{\varepsilon/k_BT_c}-1} \tag{31.24a}$$

or

$$\frac{N}{L^3} = \frac{g\,(mk_BT_c)^{3/2}}{2^{1/2}\pi^2\hbar^3}\int_0^\infty \frac{z^{1/2}dz}{e^z-1}, \tag{31.24b}$$

which corresponds to a temperature where $\mu = 0$, but where the number of particles in the condensate is zero. Equation (31.24b) can be solved numerically with the result

$$k_BT_c = \frac{3.31}{g^{2/3}}\frac{\hbar^2}{m}\left(\frac{N}{L^3}\right)^{2/3}. \tag{31.25}$$

[12] For the case of a Fermi gas, the denominator in Eq. (31.2) is replaced by $e^{[\varepsilon(\mathbf{k})-\mu]/k_BT} + 1$; the denominator cannot vanish in this case and hence the chemical potential may have either sign. At $T = 0$, μ corresponds to the Fermi energy, ε_F.

Note that, in order of magnitude, the onset of a condensate occurs when the thermal de Broglie wavelength, $\lambdabar = \hbar/(2mk_BT)^{1/2}$, is of order the mean separation of the excitons, $\left(L^3/N\right)^{1/3}$.[13] From (31.23b) and (31.24b) it follows that

$$N_0 = N\left[1 - \left(\frac{T}{T_c}\right)^{3/2}\right]. \tag{31.26}$$

Further analysis shows that there is a discontinuity in the *derivative* of the heat capacity, rather than the heat capacity itself, as at a second-order phase transition; we might refer to this as a *third-order* phase transition.

31.6 The weakly interacting Bose gas

The previous discussion neglects interactions, the effects of which were first treated consistently by Bogoliubov.[14] The method he used was profound in character and later, after being adapted to fermions, provided a route to reformulate the BCS theory of superconductivity. Since we will use the latter method in our discussion of superconductivity, it is appropriate to discuss the boson version of the theory here. We will limit ourselves to the S = 0 case.

Bogoliubov used a model Hamiltonian of the form (see Eq. (B.29))

$$\hat{H} = \sum_{ik} H^{(1)}_{ik}\hat{a}^\dagger_i\hat{a}_k + \frac{1}{2}\sum_{ikl m} U^{(2)}_{ikl m}\hat{a}^\dagger_i\hat{a}^\dagger_k\hat{a}_m\hat{a}_l + \ldots, \tag{31.27}$$

where (see Eq. (B.26))

$$\begin{aligned} U^{(2)}_{ikl m} &= \langle ik|U^{(2)}|lm\rangle \\ &= \iint d^3r_1 d^3r_2 \psi^*_i(\mathbf{r}_1)\psi^*_k(\mathbf{r}_2)U^{(2)}(\mathbf{r}_1-\mathbf{r}_2)\,\psi_l(\mathbf{r}_1)\psi_m(\mathbf{r}_2) \end{aligned} \tag{31.28}$$

and the $\psi_i(\mathbf{r})$ denote the wave functions of the single particle states with quantum number(s) i, associated with the (destruction) operators $\hat{a}_i$. Since we are here dealing with a gas, plane waves of the form $\psi_{\mathbf{k}}(\mathbf{r}) = L^{-3/2}e^{i\mathbf{k}\cdot\mathbf{r}}$ are appropriate basis functions; hence we can write

$$\begin{aligned} U^{(2)}_{\mathbf{k}'_1\mathbf{k}'_2\mathbf{k}_2\mathbf{k}_1} &= \frac{1}{L^6}\iint e^{-i\mathbf{k}'_1\cdot\mathbf{r}_1}e^{-i\mathbf{k}'_2\cdot\mathbf{r}_2}U(\mathbf{r}_1-\mathbf{r}_2)e^{i\mathbf{k}_2\cdot\mathbf{r}_2}e^{i\mathbf{k}_1\cdot\mathbf{r}_1}d^3r_1d^3r_2 \\ &= \frac{1}{L^6}\iint e^{-i\mathbf{k}'_1\cdot(\mathbf{r}_1-\mathbf{r}_2)}e^{-i(\mathbf{k}'_2+\mathbf{k}'_1)\cdot\mathbf{r}_2}U(\mathbf{r}_1-\mathbf{r}_2)e^{i(\mathbf{k}_2+\mathbf{k}_1)\cdot\mathbf{r}_2}e^{i\mathbf{k}_1\cdot(\mathbf{r}_1-\mathbf{r}_2)}d(\mathbf{r}_1-\mathbf{r}_2)d^3r_2 \\ &= \frac{1}{L^6}\iint U(\mathbf{k}_1-\mathbf{k}'_1)e^{i(\mathbf{k}_2+\mathbf{k}_1-\mathbf{k}'_2-\mathbf{k}'_1)\cdot\mathbf{r}_2}d^3r_2, \end{aligned}$$

[13] This criterion also applies to the onset of degeneracy in a Fermi system.

[14] On the theory of superfluidity, N.N. Bogoliubov, *J. Phys.* 11, 23 (1947).

where in the third step we have introduced the Fourier transform of the two-body potential; the remaining integral vanishes unless $\mathbf{k}_2 + \mathbf{k}_1 - \mathbf{k}'_2 - \mathbf{k}'_1 = 0$ (corresponding to momentum conservation in our translationally invariant system in the presence of two-body interactions). When this condition is satisfied, the integral results in a factor L^3 or $U^{(2)}_{k'_1 k'_2 k_2 k_1} = L^{-3} U(\mathbf{k}_1 - \mathbf{k}'_1)\delta_{k_2+k_1-k'_2-k'_1}$. Inserting this matrix element into (31.27), and writing our single particle exciton energies as $H^{(1)}_{kk'} = \left(\hbar^2 k^2/2m\right)\delta_{kk'}$, our Hamiltonian becomes

$$\hat{H} = \sum_{k,k'} \left(\frac{\hbar^2 k^2}{2m}\right) \delta_{kk'} \hat{a}^\dagger_{k'} \hat{a}_k$$
$$+ \frac{1}{2L^3} \sum_{k_1,k_2,k'_1,k'_2} U(\mathbf{k}_1 - \mathbf{k}'_1)\delta_{k_2+k_1-k'_2-k'_1} \hat{a}^\dagger_{k'_1} \hat{a}^\dagger_{k'_2} \hat{a}_{k_2} \hat{a}_{k_1};$$

defining $\mathbf{k}'_1 = \mathbf{k}_1 - \mathbf{q}$ and $\mathbf{k}'_2 = \mathbf{k}_2 + \mathbf{q}$, this expression can then be written as

$$\hat{H} = \sum_k \left(\frac{\hbar^2 k^2}{2m}\right) \hat{a}^\dagger_k \hat{a}_k + \frac{1}{2L^3} \sum_{k_1,k_2,q} U(\mathbf{q}) \hat{a}^\dagger_{k_1-q} \hat{a}^\dagger_{k_2+q} \hat{a}_{k_2} \hat{a}_{k_1}. \tag{31.29}$$

We have gone through this derivation in some detail since this model Hamiltonian is widely used and is valid for fermions as well as bosons. For the case at hand we have the boson commutation relations

$$\hat{a}_k \hat{a}^\dagger_{k'} - \hat{a}^\dagger_{k'} \hat{a}_k = \delta_{kk'}; \; \hat{a}_k \hat{a}_{k'} - \hat{a}_{k'} \hat{a}_k = 0. \tag{31.30a,b}$$

We initially confine ourselves to $T = 0$ and in any case we limit our discussion to weak interactions; we will further set $U(\mathbf{q}) = U(\mathbf{q} = 0) \equiv U_0$. With these changes our Hamiltonian becomes

$$\hat{H} = \sum_k \frac{\hbar^2 k^2}{2m} \hat{a}^\dagger_k \hat{a}_k + \frac{U_0}{2L^3} \sum_{k_1,k_2,q} \hat{a}^\dagger_{k_1-q} \hat{a}^\dagger_{k_2+q} \hat{a}_{k_2} \hat{a}_{k_1}. \tag{31.31}$$

Bogoliubov (1947) made the following important observation: if there is a macroscopic number of particles N_0 in the $\mathbf{k} = 0$ state, then (recalling that the number operator is $\hat{N}_k = \hat{a}^\dagger_k \hat{a}_k$) we will have $\hat{a}^\dagger_0 \hat{a}_0 = N_0 \cong N$. Furthermore, $\hat{a}^\dagger_0 \cong \hat{a}_0 = N_0^{1/2}$ and, since $N_0 \gg 1$, the order of the operators in the commutation relation, $\hat{a}_k \hat{a}^\dagger_k - \hat{a}^\dagger_k \hat{a}_k = 1$, becomes *irrelevant*; i.e., we can treat the $\mathbf{k} = 0$ occupation numbers as *macroscopic* or *thermodynamic* quantities. On the other hand, for $\mathbf{k} \neq 0$, the order of $\hat{a}_k$ and $\hat{a}^\dagger_k$ remains important and both are much smaller than $\hat{a}^\dagger_0 \cong \hat{a}_0 = N_0^{1/2}$. The leading contribution to the interaction term in (31.31) involves the form

$$\hat{a}^\dagger_0 \hat{a}^\dagger_0 \hat{a}_0 \hat{a}_0 = a_0^4. \tag{31.32}$$

The next largest terms would be of order in a_0^3; however, these vanish since they do not conserve momentum. The second-order terms, which is as far as we will carry the calculation, have the form

$$a_0^2 \sum_{k\neq 0} \left(\hat{a}_k\hat{a}_{-k} + \hat{a}_k^\dagger\hat{a}_{-k}^\dagger + 4\hat{a}_k^\dagger\hat{a}_k\right); \tag{31.33}$$

the terms in this expression arise from the following six combinations in (31.31): $\mathbf{k}_1 - \mathbf{q} = \mathbf{k}_2 + \mathbf{q} = 0$; $\mathbf{k}_1 = -\mathbf{k}_2 = 0, \mathbf{q} = \mathbf{k}_1 = 0$; $\mathbf{q} = \mathbf{k}_2 = 0$; $\mathbf{k}_1 - \mathbf{q} = \mathbf{k}_2 = 0$; and $\mathbf{k}_2 + \mathbf{q} = \mathbf{k}_1 = 0$. We now examine the relation between N, N_0, a_0^2, and a_0^4. We start by writing

$$N = \sum_k \hat{a}_k^\dagger\hat{a}_k = N_0 + \sum_{k\neq 0} \hat{a}_k^\dagger\hat{a}_k = a_0^2 + \sum_{k\neq 0} \hat{a}_k^\dagger\hat{a}_k.$$

Therefore,

$$a_0^2 = N - \sum_{k\neq 0} \hat{a}_k^\dagger\hat{a}_k \tag{31.34a}$$

and

$$a_0^4 = N^2 - 2N \sum_{k\neq 0} \hat{a}_k^\dagger\hat{a}_k. \tag{31.34b}$$

The forms (31.32) and (31.33) may then be combined as

$$N^2 + N \sum_{k\neq 0} \left(\hat{a}_k\hat{a}_{-k} + \hat{a}_k^\dagger\hat{a}_{-k}^\dagger + 2\hat{a}_k^\dagger\hat{a}_k\right), \tag{31.35}$$

where, to the same accuracy, we have replaced N_0 by N in (31.33). Our Hamiltonian then becomes

$$\hat{H} = \frac{N^2}{2L^3}U_0 + \sum_k \frac{\hbar^2 k^2}{2m}\hat{a}_k^\dagger\hat{a}_k + \frac{N}{2L^3}U_0 \sum_{k\neq 0} \left(\hat{a}_k\hat{a}_{-k} + \hat{a}_k^\dagger\hat{a}_{-k}^\dagger + 2\hat{a}_k^\dagger\hat{a}_k\right). \tag{31.36}$$

To lowest order, the first term gives the ground state energy at $T = 0$, and its derivative with respect to N is the chemical potential:

$$E_0 = \frac{N^2}{2L^3}U_0; \mu = \frac{N}{L^3}U_0. \tag{31.37a,b}$$

The Hamiltonian (31.36) is *quadratic* in the three forms, $\hat{a}_k\hat{a}_{-k}$, $\hat{a}_k^\dagger\hat{a}_{-k}^\dagger$, and $\hat{a}_k^\dagger\hat{a}_k$; however, the first two of these products do not have the form of a number operator. We now show that by an appropriate *canonical transformation* the

Hamiltonian can be brought into a free-particle-like form in which a new number operator appears. The transformation that will accomplish this has the form

$$\hat{a}_{\mathbf{k}} = u_{\mathbf{k}}\hat{b}_{\mathbf{k}} + v_{\mathbf{k}}\hat{b}^{\dagger}_{-\mathbf{k}} \tag{31.38a}$$

and

$$\hat{a}^{\dagger}_{\mathbf{k}} = u^*_{\mathbf{k}}\hat{b}^{\dagger}_{\mathbf{k}} + v^*_{\mathbf{k}}\hat{b}_{-\mathbf{k}}. \tag{31.38b}$$

This procedure is referred to as a *Bogoliubov transformation*; a similar trick (although now adapted to Fermi operators) plays an important role in the theory of superconductivity (see Eqs. (35.22) and (35.23)). When Eqs. (31.38a,b) are substituted into Eqs. (31.30a,b) it is found that the operators $\hat{b}^{\dagger}_{\mathbf{k}}$ and $\hat{b}_{\mathbf{k}}$ will satisfy the *same* commutation relations (31.30a,b)

$$\hat{b}_{\mathbf{k}}\hat{b}^{\dagger}_{\mathbf{k}'} - \hat{b}^{\dagger}_{\mathbf{k}'}\hat{b}_{\mathbf{k}} = \delta_{\mathbf{k}\mathbf{k}'};\ \hat{b}_{\mathbf{k}}\hat{b}_{\mathbf{k}'} - \hat{b}_{\mathbf{k}'}\hat{b}_{\mathbf{k}} = 0 \tag{31.39a,b}$$

provided that we require $u_{\mathbf{k}}$ and $v_{\mathbf{k}}$ to satisfy the relation[15]

$$|u_{\mathbf{k}}|^2 - |v_{\mathbf{k}}|^2 = 1 \tag{31.40}$$

The phases of $u_{\mathbf{k}}$ and $v_{\mathbf{k}}$ may be chosen to be real here, but we will return to this point later. We will not go through the algebra in detail but simply remark that the transformation (31.38a,b) that diagonalizes the Hamiltonian involves the forms

$$u_{\mathbf{k}} = \frac{1}{\sqrt{1 - L_{\mathbf{k}}^2}}, \tag{31.41a}$$

and

$$v_{\mathbf{k}} = \frac{L_{\mathbf{k}}}{\sqrt{1 - L_{\mathbf{k}}^2}}, \tag{31.41b}$$

where

$$L_{\mathbf{k}} = \frac{1}{mu^2}\left[\varepsilon(\mathbf{k}) - \frac{p^2}{2m} - mu^2\right], \tag{31.42}$$

with

$$\xi(\mathbf{k}) = \frac{\hbar^2 k^2}{2m}, \tag{31.43a}$$

$$\varepsilon(\mathbf{k}) = \sqrt{u^2\hbar^2k^2 + \left(\frac{\hbar^2k^2}{2m}\right)^2}, \tag{31.43b}$$

[15] If a transformation of coordinates is applied to the generalized momenta and coordinates of a mechanical system which preserves the form of Hamilton's equations, it is said to be canonical; hence a transformation of the operators of a quantum system that preserves the commutation relations is said to be canonical.

and

$$u = \sqrt{\frac{U_0 N}{mL^3}}; \tag{31.43c}$$

this can be verified by substituting (31.41a,b) and (31.38a,b) into (31.36). The quantities $u_{\mathbf{k}}$ and $v_{\mathbf{k}}$ can also be rewritten as

$$u_{\mathbf{k}}^2 = \frac{1}{2}\left[\frac{\xi(k) + mu^2}{\varepsilon(k)}\right] \tag{31.44a}$$

and

$$v_{\mathbf{k}}^2 = \frac{1}{2}\left[\frac{\xi(\mathbf{k}) + mu^2}{\varepsilon(\mathbf{k})} - 1\right], \tag{31.44b}$$

which in this form bear some similarity to corresponding expressions that arise in the theory of superconductivity (see Eqs. (35.33) and (35.34)). The Hamiltonian then takes the form

$$\hat{H} = E_0 + \sum_{k \neq 0} \varepsilon(\mathbf{k})\, \hat{b}_{\mathbf{k}}^{\dagger} \hat{b}_{\mathbf{k}}, \tag{31.45}$$

where

$$E_0 = \frac{1}{2} Nmu^2 + \frac{1}{2}\sum_{k \neq 0}\left[\varepsilon(\mathbf{k}) - \frac{\hbar^2 k^2}{2m} - mu^2 + \frac{m^3 u^4}{\hbar^2 k^2}\right]. \tag{31.46}$$

This Hamiltonian has the same *structure* as non-interacting particles; the only difference is that the single-particle energies, $\hbar^2 k^2/2m$, have been replaced by $\varepsilon(\mathbf{k})$ as given by (31.43b), and we have a shift in the zero of energy. The quantity u, which has the dimensions of velocity, will be discussed shortly.

Qualitatively, the spectrum $\varepsilon(\mathbf{k})$ given by Eq. (31.43b) has the following properties. At long wavelengths the effects of the interparticle potential dominate and the spectrum has a linear frequency versus wavevector behavior; this is characteristic of sound waves with a velocity u. We expect on physical grounds that any interacting system will possess low frequency sound waves. At the other extreme the high-energy excitations behave as free particles with energy $\hbar^2 k^2/2m$, which is also expected on physical grounds. At intermediate wave vectors there is a gradual transition between these two limiting behaviors.

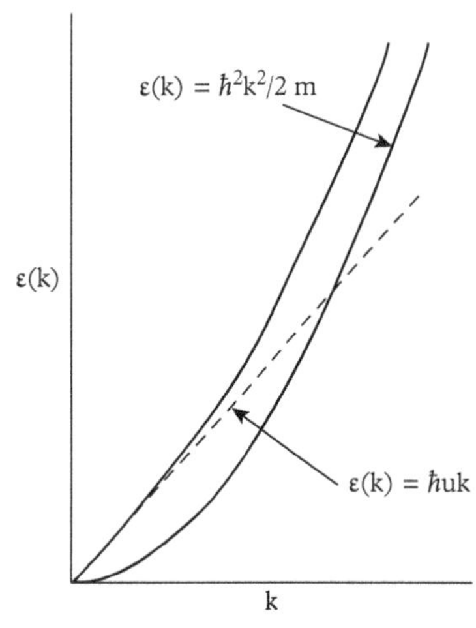

Figure 31.6 *The energy spectrum of a weakly interacting Bose gas. The spectrum for a non-interacting gas is also shown for comparison.*

31.7 The weakly interacting Bose gas for non-zero temperatures

At absolute zero there are no excitations relative to the interacting ground state; i.e.,

$$n_k = \left\langle \hat{b}_k^\dagger \hat{b}_k \right\rangle = 0 \text{ (ground state)}. \tag{31.47}$$

The operator $\hat{b}_k^\dagger \hat{b}_k$ is the number operator for what are called *elementary excitations*; they are also called *quasiparticle* or *dressed particle* excitations. Since, to the order of the approximation given by (31.45), the interactions have been removed, the mean number of excitations at finite temperature would be the same as the expression for an ideal gas

$$n(\mathbf{k}) = \frac{1}{e^{\varepsilon(\mathbf{k})/k_B T} - 1}; \tag{31.48}$$

however the energy spectrum $\varepsilon(\mathbf{k})$ is now given by (31.43b) rather than the free particle form $\hbar^2 k^2/2m$. Note that the chemical potential is now equal to zero, since there is no constraint on the number of excitations, as there was earlier on the number of atoms in Section 31.1; the situation is entirely analogous to the case of phonons in a crystal or, more relevant to the present context, phonon-like excitations in superfluid ^{4}He (phonons in ^{4}He will be discussed in Section 36.1).

We can also calculate the occupation numbers $\langle \hat{a}_k^\dagger \hat{a}_k \rangle$ of the original or "bare" particles, which we write as $\tilde{n}_k$, and how they are affected by temperature. From (31.38) we have

$$\tilde{n}_k = \left\langle (u_k \hat{b}_k^\dagger + v_k \hat{b}_{-k})(u_k \hat{b}_k + v_k \hat{b}_{-k}^\dagger) \right\rangle$$

Since the operators $\hat{b}_k^\dagger \hat{b}_{-k}^\dagger$ and $\hat{b}_{-k}\hat{b}_k$ have no diagonal matrix elements, we have from (31.41a,b)

$$\begin{aligned} \tilde{n}_k &= u_k^2 \left\langle \hat{b}_k^\dagger \hat{b}_k \right\rangle + v_k^2 \left\langle \hat{b}_{-k} \hat{b}_{-k}^\dagger \right\rangle \\ &= [n_k + L_k^2 (n_k + 1)]/(1 - L_k^2), \end{aligned} \tag{31.49}$$

where we have used $\langle \hat{b}_k^\dagger \hat{b}_k \rangle = n_k$, and $\langle \hat{b}_{-k} \hat{b}_{-k}^\dagger \rangle = n_k + 1$, and the expression is of course valid only for $\mathbf{k} \neq 0$. For the T = 0 case where $n_k = 0$, and using (31.42) and (31.49), one finds

$$\tilde{n}_k = \frac{m^2 u^4}{2\varepsilon(\mathbf{k}) \left[\varepsilon(\mathbf{k}) + \hbar^2 k^2/2m + mu^2 \right]}. \tag{31.50}$$

The number of particles in the $\mathbf{k} = 0$ condensate is

$$N_0 = N - \sum_{k \neq 0} \tilde{n}_k = N - \frac{L^3}{(2\pi)^3} \int \tilde{n}_k d^3k. \quad (31.51)$$

Carrying out the integration (31.51) gives

$$N_0 = N\left[1 - \frac{8}{3}\sqrt{\frac{Na^3}{\pi L^3}}\right], \quad (31.52)$$

where we introduced the scattering length, a, which is given in lowest order by $U_0 = 4\pi\hbar^2 a/m$; to second order the relation is

$$U_0 = \frac{4\pi\hbar^2 a}{m}\left(1 + \frac{4\pi a}{L^3}\sum_{k \neq 0}\frac{1}{k^2}\right). \quad (31.53)$$

At non-zero temperature, $\tilde{n}_k$ increases; at some temperature we would reach a point where N_0 in Eq. (31.51) would vanish. Although the validity of the theory would break down prior to this point, this would qualitatively mark the disappearance of the condensate (and the superfluidity that is associated with it).

31.8 The Gross–Pitaevskii equation

The ground state wave function of a weakly interacting gas of Bose condensed particles was studied independently by Gross and Pitaevskii. The Hamiltonian for a system of interacting bosons in an external potential, V_{trap}, in second quantized form is given in (B.35),

$$\hat{H} = \int d^3r \hat{\psi}(\mathbf{r})\left[-\frac{\hbar^2}{2m}\nabla^2 + V_{trap}(\mathbf{r})\right]\hat{\psi}(\mathbf{r}) + \frac{1}{2}\int d^3r \int d^3r' \hat{\psi}^\dagger(\mathbf{r})\hat{\psi}^\dagger(\mathbf{r}')U(\mathbf{r}-\mathbf{r}')\hat{\psi}'(\mathbf{r}')\hat{\psi}'(\mathbf{r}), \quad (31.54)$$

where $\hat{\psi}(\mathbf{r})$ and $\hat{\psi}^\dagger(\mathbf{r})$ are the boson field operators that annihilate and create a particle at the position $\mathbf{r}$ and $V_{trap}(r)$ is some external potential that confines the particles. Considering s-wave scattering only, the interaction potential is

$$U(\mathbf{r}-\mathbf{r}') = U_0\delta(\mathbf{r}-\mathbf{r}'), \quad (31.55)$$

where $U_0 = 4\pi\hbar^2 a/m$ is the interaction strength (see Eq. (31.53)) and a is the scattering length.

The Heisenberg equation of motion is

$$\frac{\partial \hat{\psi}(\mathbf{r},t)}{\partial t} = \frac{i}{\hbar}\left[\hat{H}, \hat{\psi}(\mathbf{r},t)\right]. \tag{31.56}$$

Plugging in the Hamiltonian and using the commutation relations (B39a,b) for boson particles, we obtain

$$i\hbar\frac{\partial \hat{\psi}(\mathbf{r},t)}{\partial t} = \left[-\frac{\hbar^2}{2m}\nabla^2 + V_{trap}(\mathbf{r})\right]\psi'(\mathbf{r},t) + U_0\hat{\psi}^{\dagger}(\mathbf{r},t)\hat{\psi}(\mathbf{r},t)\hat{\psi}(\mathbf{r},t). \tag{31.57}$$

At this point one assumes the presence of a condensate and that the operator $\hat{\psi}$ can be written as

$$\hat{\psi}(\mathbf{r},t) = \psi(\mathbf{r},t) + \hat{\psi}'(\mathbf{r},t). \tag{31.58a}$$

Here ψ is a classical or "c-number" quantity, i.e., an ordinary (but in general complex) function, which is related to the condensed fraction through $|\psi|^2 = N_0/V$, and $\hat{\psi}'(\mathbf{r})$ is an operator describing the remaining, or "fluctuating," part of the system. Neglecting the fluctuation part, we obtain the nonlinear Schrödinger equation or Gross–Pitaevskii equation (GP)

$$i\hbar\frac{\partial \psi(\mathbf{r},t)}{\partial t} = \left[-\frac{\hbar^2}{2m}\nabla^2 + V_{trap}(\mathbf{r}) + NU_0\left|\psi(\mathbf{r},t)^2\right|\right]\psi(\mathbf{r},t). \tag{31.58b}$$

To find a stationary solution for the condensate wave function in this mean field theory, one can substitute the form $\psi(\mathbf{r},t) = \exp(-i\mu t/\hbar)\psi(\mathbf{r})$, to give the time-independent equation

$$\left[-\frac{\hbar^2}{2m}\nabla^2 + V_{trap}(\mathbf{r}) + NU_0\left|\psi(\mathbf{r},t)^2\right|\right]\psi(\mathbf{r},t) = \mu\psi(\mathbf{r},t). \tag{31.59}$$

31.9 Polaritons in a 2d photonic cavity

As discussed in Section 30.3, excitons strongly couple with light in the region where their dispersion relations, ω versus $\mathbf{k}$, would intersect in the absence of coupling, the resulting quantum particles being the exciton–polaritons (or polaritons for short). The curvatures of the resulting branches, $\partial^2\omega(\mathbf{k})/\partial k^2$, are generally large in the vicinity of the crossing (much larger than for the excitons by themselves), implying small effective masses, which would ordinarily favor BEC. However, in the absence of a local minimum in $\omega(\mathbf{k})$ at which excitations can accumulate, BEC of polaritons will not occur. An additional problem is that in most semiconductors the dipole coupling with the light results in very short exciton

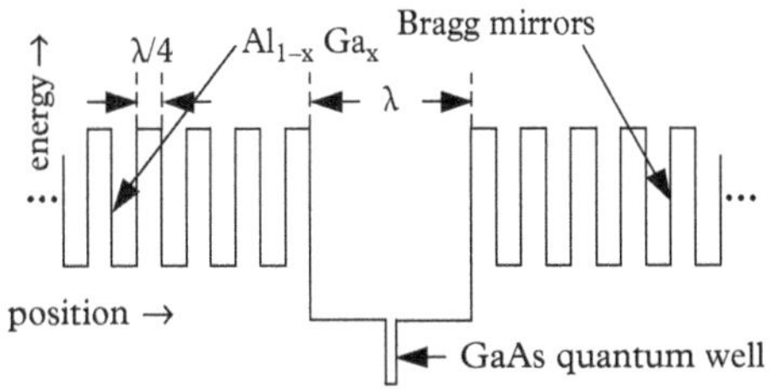

Figure 31.7 *A photonic structure consisting of a GaAs quantum well confined between* $Al_{1-x}Ga_xAs/AlAs$ *multilayer Bragg mirrors which suppress spontaneous exciton-polariton decay.*

lifetimes. Both of these problems can be addressed by confining the excitons to two dimensions and engineering an appropriate photonic structure to keep the light confined, as we now discuss.[16,17]

Figure 31.7 shows a schematic diagram of a two-dimensional GaAs quantum well sandwiched between two multilayer dielectric Bragg mirrors.[18] The inclusion of several closely spaced quantum wells can increase exciton lifetimes while minimizing their direct interaction for a given overall density. For GaAs quantum wells, the alternating layers of the mirrors involve AlAs and an $Al_{1-x}Ga_xAs$ alloy. Since the materials making up the mirrors have a larger band gap than the GaAs, within which the excitons lie slightly lower in energy, they will not absorb energy. The layer spacing is chosen to reflect light at the polariton frequencies themselves, and is approximately 1/4 the average wavelength of light in the media. In the absence of coupling to excitons we can write the dispersion relation of the light confined between the mirrors as

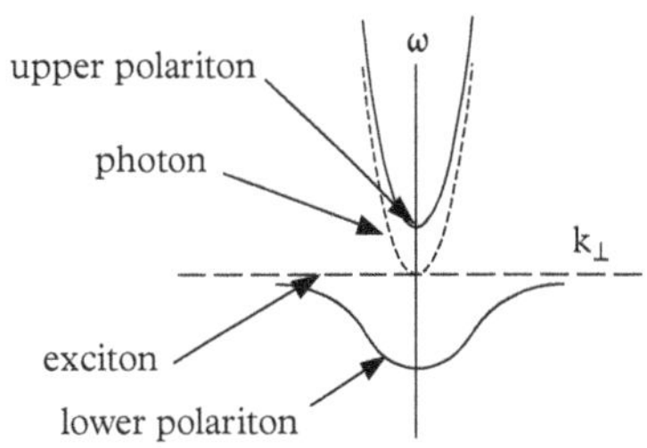

Figure 31.8 *The band structure in the vicinity of the upper branch of the gap produced by the Bragg mirrors, together with the additional splittings introduced by photon-exciton coupling, to produce polaritons possessing a local extremum.*

$$\omega(k_z, k_\perp) = c\sqrt{k_z^2 + k_\perp^2}, \tag{31.60}$$

where c is the velocity of light in GaAs, while k_z and $k_\perp$ are the components of the wavevector parallel and perpendicular to the layers. The presence of the mirrors quantizes k_z in terms of the mirror separation, d, yielding the frequencies

$$\begin{aligned} \omega(n, d, k_\perp) &= c\sqrt{\frac{n^2\pi^2}{d^2} + k_\perp^2} \\ &= \omega_0 + \frac{1}{2}\frac{c^2}{\omega_0}k^2 - \cdots \end{aligned}, \tag{31.61}$$

where we have ignored a correction accounting for the fact that the nodes will not be precisely at the leading edges of the mirrors, and we have also defined $\omega_0 = cn\pi/d$.

We now focus our attention on Fig. 31.8. The flat and parabolic dashed curves show, schematically, the dispersion curves for the 2d quantum-well excitons and the mirror-confined photons respectively. Note the curvature of the excitons is negligible relative to the photons, and that the mirror spacing has been chosen so that at $k_\perp = 0$ the excitons and photons are degenerate, or nearly so.

[16] Observation of the coupled exciton-photon mode splitting in a semiconductor quantum microcavity, C. Weisbuch, M. Nishioka, A. Ishikawa, and Y. Arakawa, *Phys. Rev. Lett.* **69**, 3314 (1992).

[17] Exciton-polaritons in microcavities: present and future, A. Kavokin, *Appl. Phys. A* **89**, 241 (2007).

[18] Quantum wells are discussed in Section 47.4, while Bragg mirrors are treated in Section 47.5.

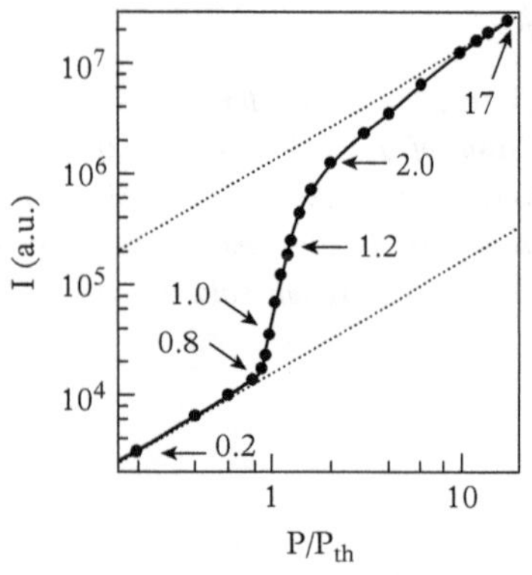

Figure 31.9 *Lower polariton emission intensity versus the normalized pump intensity,* P/P_{th}*, for light leaving perpendicular to the sample (corresponding to* $k = 0$*); note the superluminal rise at the critical pumping level. Various specific pumping levels are shown with the arrows. (After Deng et al. (2002).)*

When the coupling of the light to the excitons is turned on, and taking account of the fact that k_z is quantized through Eq. (31.61), in place of Eq. (30.3a) we now have

$$\left(-k_\perp^2 - \frac{n^2\pi^2}{d^2} + \frac{\omega^2}{c^2}\right) E_0 + \frac{4\pi\omega^2}{c^2} P_0 = 0 \tag{31.62}$$

and in place of Eq. (30.3b),

$$\left(\frac{\omega^2}{c^2} - k_\perp^2 - \frac{n^2\pi^2}{d^2}\right)\left(\omega_T^2 - \omega^2\right) + \frac{\omega^2\omega_X^2}{c^2} = 0. \tag{31.63}$$

For our assumption of degeneracy at $k_\perp = 0$, i.e., $\omega_0 = \omega_T$, we would have

$$\left(\omega_0^2 - \omega^2 + c^2k_\perp^2\right)\left(\omega_0^2 - \omega^2\right) + \omega^2\omega_X^2 = 0. \tag{31.64}$$

The mode repulsion resulting from the coupling manifests itself in the two solid curves in Fig. 31.8, now referred to as the upper and lower polariton branches. Excitons can now accumulate in the minimum of the lower polariton branch (which can be the global minimum) and BEC can occur. In addition, the mixing with the light component *greatly decreases* the effective mass of the excitations thereby allowing BEC at higher temperatures and lower densities.

31.10 BEC of exciton–polaritons in two dimensions

BEC of exciton–polaritons in cavities of the kind just described has been reported in GaAs based devices by the Yamamoto group[19] and by the Deng group[20] for CdTe structures. Condensation in a spatial trap created by the application of a localized external strain has also been studied;[21] here we picture a pin pressed against the surface of a quantum well device. As a result of these and other experiments work in this field has greatly expanded.[22]

The first step in creating a BEC is to flood the quantum well(s) with excitons, followed by monitoring their momentum distribution. This initial pumping is best achieved by bringing in laser light at a large angle such that it is not reflected by the Bragg mirrors, but has an energy sufficient to create excitons in the quantum well for the corresponding value of $k_\perp$ (which is fixed by the incoming angle). These excitons will lie above the minimum energy at $k_\perp = 0$ and must therefore lose momentum and energy through inelastic processes if they are to condense; these processes involve acoustic and optical phonon interactions as well as exciton–exciton scattering and will not be discussed here. The distribution in momentum space can be monitored from the angle of the emerging recombination emission of the excitons in the well, vertical emission corresponding to $k_\perp = 0$. Although the mirrors transmit only a fraction of this light there is still sufficient emission to record the angular dependence of the emitted light which, in turn, corresponds to the in-plane momentum distribution.

Figure 31.9 shows the intensity of light emitted perpendicular to a GaAs based device similar to that described earlier (however with a total of 12 wells), as a

[19] Condensation of semiconductor microcavity exciton polaritons, H. Deng, G. Weihs, C. Santori, J. Bloch, and Y. Yamamoto, *Science* **298**, 199 (2002).

[20] Bose–Einstein condensation of exciton polaritons, J. Kasprzak, M. Richard, S. Kundermann, A. Baas, P. Jeambrun, J.M.J. Keeling, F.M. Marchetti, M.H. Szymanska, R. Andre, J.L. Staehli, V. Savona, P.B. Littlewood, B. Deveaud, and Le Si Dang, *Nature* **443**, 409 (2006).

[21] Bose-Einstein condensation of microcavity polaritons in a trap, R. Balili, V. Hartwell, D. Snoke, L. Pfeiffer, and K. West, *Science* **316**, 1007 (2007).

[22] Exciton-polariton Bose-Einstein condensation, H. Deng, H. Haug, and Y. Yamamoto, *Rev. Mod. Phys.* **82**, (2010).

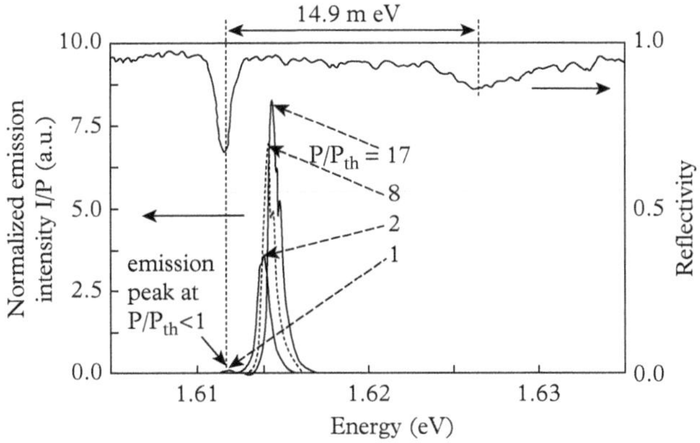

Figure 31.10 *The upper curve shows a reflection spectrum arising from the lower (narrow) and upper (broad) polariton bands. The lower curves show the emission spectrum for four different pumping levels. (After Deng et al, (2002).)*

function of the pump power, as observed by the Yamamoto group. We note that at a critical intensity an abrupt change of slope is observed in the emission versus pump power behavior, *signaling the formation of a condensate*. It is important that the densities involved in these experiments are much smaller than the so-called Mott limit where the exciton separation approaches the exciton radius, at which point the excitons would lose their identity.

The $\omega(\mathbf{k}_\perp)$ relation for both the upper and lower polariton branches can be measured through the angular dependence of the reflected light; when conservation of energy between the incoming light and the resulting cavity excitations can be satisfied, a dip occurs in the reflectivity. This energy dependence of the reflectivity for a selected angle is shown as the upper trace in Fig. 31.10. Both the lower and upper branches are seen. The spectrum of the lower branch, where the excitons reside following pumping, can be observed from the emission spectrum; this is also shown in Fig. 31.10, for four different pumping levels. Note that at higher levels it is blue shifted due to a phase space filling effect (which we do not discuss). By assembling such data the authors construct the dispersion curve for the lower polariton branch shown in Fig. 31.11. Shown as the dashed lines are the predictions of the uncoupled photon and exciton modes, although shifted vertically for easy comparison with the predictions for the lower polariton branch, shown by the solid line. In addition to the static properties we have described, Deng et al. also obtained evidence for the formation of macroscopic quantum coherence by measuring the second-order coherence function of the light emitted from the polaritons, in the time domain, which is beyond the scope of our discussion here.

Figure 31.12 shows a rendition of the angular distribution of the recombination radiation from a CdTe quantum well cavity for three different pumping powers by the Deng group. Note that at the highest pump intensity, on the far right, a narrow peak emerges which is taken as a signature of the formation of a condensate.

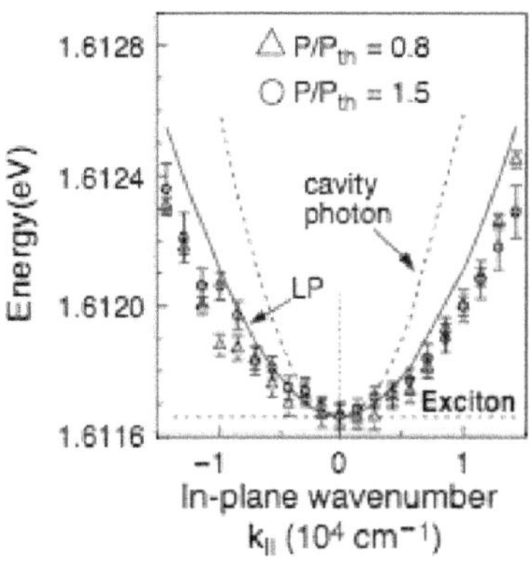

Figure 31.11 *The solid curve shows the calculated lower polariton curve in the presence of coupling, together with the data above and below the threshold pump power. The dotted lines represent the uncoupled cavity and exciton dispersion curves, which have been shifted for easy comparison with the coupled case. (After Deng et al. (2002).)*

Figure 31.12 *Far-field emission measured at 5 K for three excitation intensities which are increasing to the right. Shown as a pseudo-3d image of the far-field emission within an angular cone of ±23° with the emission intensity displayed on the vertical axis (in arbitrary units). With increasing excitation power, a sharp and intense peak is formed in the centre of the emission distribution, corresponding to emission angles* $\theta_x = \theta_y = 0$, *implying that they are associated with* $k_\perp = 0$. *(After Kasprzak et al. (2006).)*

ADDITIONAL READING

Principles of Non-linear Optics, Y.R. Shen, John Wiley and Sons, NY (1984).

Electrodynamics of Continuous Media, L.D. Landau, E.M. Lifshitz, and L.P. Pitaevskii, Chapter XIII, Pergamon Press (1984).

The electron–hole liquid is discussed in the following articles: T.M. Rice, *Solid State Physics*, Vol. 32, p. 1, Ed. H. Ehrenreich, F. Seitz, and D. Turnbull, Academic Press (1977); J.C. Hensel, T.G. Phillips, and G.A. Thomas, *Solid State Physics*, Vol. 32, p. 88, Ed. H. Ehrenreich, F. Seitz, and D. Turnbull, Academic Press (1977); G.K. Wong, in *Novel Materials and Techniques in Condensed Matter*, p. 227, Ed. G.W. Crabtree and P. Vashista, North Holland (1982); L.V. Keldysh in *Bose-Einstein Condensation*, p. 246, Ed. A. Griffin, D.W. Snoke, and S. Stringari, Cambridge University Press (1995).

The Bogoliubov theory and excitonic BEC in photonic cavities are both discussed in Chapter 2 of *Novel Superfluids*, Vol. 1, Ed. K.H. Bennemann and J.B. Ketterson, Oxford University Press (2013); Chapter 9 by Y. Yamamoto is devoted to 2d polariton condensates.

Part XI

Superconductivity and Superfluidity

A Phenomenological Theory of Superconductivity: The London Equations

32

32.1 The superconducting state

Many metals, alloys, and intermetallic compounds[1] undergo a phase transition at some (generally low) temperature, T_c, to a state having zero electrical resistance, a phenomenon which is called superconductivity. Superconductivity was discovered by Kamerlingh Onnes in 1911; it has been sensitively probed by observing the magnetic field produced by a circulating supercurrent, using a sensitive technique such as nuclear magnetic resonance. Favorable materials exhibit no detectable decay of this current for periods limited by the patience of the observer (~years). However, superconductivity is better defined by the nature of the associated phase transition, and it then includes materials or measurement conditions where the resistance may not vanish; here we postpone the discussion of these exceptions and initially restrict ourselves to perfect superconductors.

A second fundamental characteristic associated with a superconductor is the exclusion of magnetic flux, discovered by W. Meissner and R. Ochsenfeld[2] (which we will often refer to simply as the Meissner effect in what follows). The degree of flux exclusion can depend on the material or measurement conditions; we will also assume that this property is nearly perfectly displayed for our initial discussion. Materials displaying this property are called type I superconductors.

The combination of zero resistance and perfect diamagnetism results in a clear distinction between a superconductor (which as we will see is in a thermodynamic state) and a hypothetical *perfect conductor* (which would have the unique *transport property* of zero resistance); this is illustrated in Figure 32.1 and involves the different responses that each would have for different histories of cooling below the transition temperature and applying a magnetic field. Here we only discuss the case of an infinite cylinder.[3] If we start by cooling through T_c and then apply a magnetic field, both the superconductor and the perfect conductor would exclude the field. For the perfect conductor, induced currents arising from Faraday's law would screen (exclude) the flux. Flux exclusion in the superconductor could be assigned to the same mechanism or (more fundamentally) to the Meissner effect itself. If we reverse the order by first applying the field and then cooling through

[1] An alloy is a solid solution of two or more elements at least one of which is a metal. An intermetallic compound involves a metal and one or more other elements which form a chemical compound having nearly precise ratios of its constituents.

[2] Short initial announcements, W. Meissner, and R. Ochsenfeld, *Naturwissen* **21**, 787 (1933).

[3] This avoids demagnetization effects, field distortions caused by currents flowing in the sample itself.

The Physics of Solids. First Edition. J. B. Ketterson.

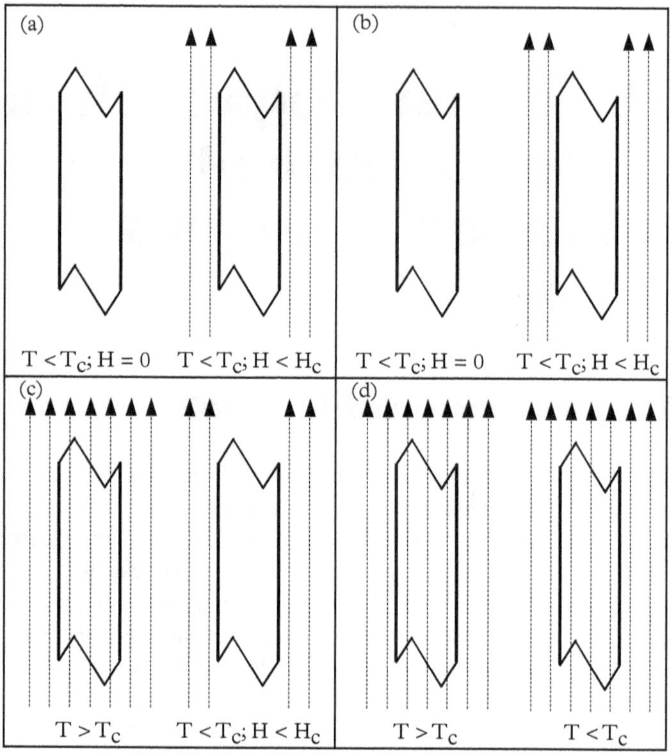

Figure 32.1 *(a) A field* H < H_c *is applied to a superconductor when* T < T_c. *(b) A field* H < H_c *is applied to a perfect conductor when* T < T_c. *(c) A superconductor is cooled below* T_c *in a fixed field* H < H_c . *(d) A perfect conductor is cooled below* T_c *in a fixed field* H < H_c.

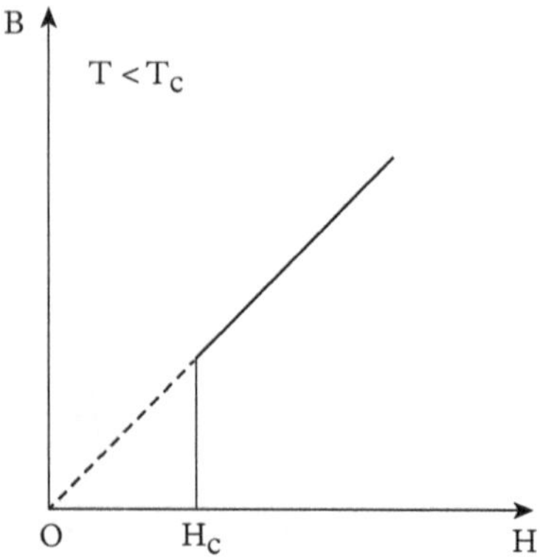

Figure 32.2 *The* B *versus* H *curve for a type* I *superconductor.*

T_c, the superconductor and the perfect conductor behave differently: the superconductor excludes the flux (the Meissner effect); the perfect conductor would remain fully permeated by the field.

These experimental observations argue that the transition associated with superconductivity is indeed a phase transition, since an equilibrium thermodynamic state is defined by its independent thermodynamic variables (in this case T and H), and is *independent of its history* (which as we see is not true for the perfect conductor).

Superconductivity, and with it the Meissner effect, does not persist to arbitrarily high magnetic fields. For each temperature there is a well-defined critical field, $H_c(T)$, at which superconductivity disappears.[4] Figure 32.2 shows a universal curve of the behavior of B versus H. Since $B = H + 4\pi M$, having $B = 0$ below H_c implies $M = -H/4\pi$, or equivalently a susceptibility, $\chi = M/H = -1/4\pi$.

A superconductor also exhibits a discontinuous increase in its specific heat at T_c (there is no latent heat at zero field), below which it drops rapidly (approaching an exponential dependence at low temperatures). As we encountered in our discussions of the ferroelectric and the ferromagnetic transitions in Chapters 25 and 26, a discontinuity in the specific heat is a signature of a second-order phase

[4] This statement is restricted to so-called type I materials, having a shape for which there is no demagnetization effect, as will be discussed in later sections.

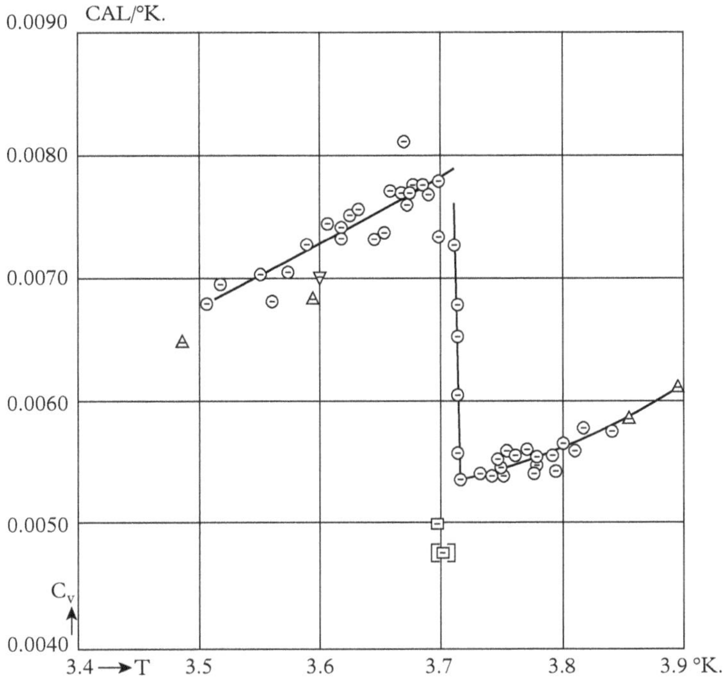

Figure 32.3 *Specific heat of tin as function of temperature. (After Keesom and van Kok (1932).)*

transition, providing added evidence that superconductivity is associated with a distinct thermodynamic phase. Figure 32.3 shows the heat capacity of tin, a typical superconductor.[5]

There is a related low temperature phenomenon known as superfluidity which occurs in liquid helium. There are actually two such liquids involving the isotopes ^{4}He (see Chapter 36) and ^{3}He (see Chapter 38) having superfluid transitions near 2 K and 2×10^{-3} K respectively, depending on the pressure. For some purposes a superconductor may be regarded as a superfluid having an electric charge. ^{4}He superfluidity involves a Bose condensation, discussed in Chapter 31, a phenomenon which is related to superconductivity in a somewhat subtle or indirect way. The superfluidity of ^{3}He, which we will take up in Chapter 38, is intimately related to superconductivity. Another related superfluid is the neutron liquid in the interior of neutron stars.

32.2 The London equations

Here we present two derivations of equations which are useful in describing many of the magnetic properties of superconductors. The older approach, used by F. London and H. London,[6] starts with the Drude–Lorentz equation of motion for electrons in a metal which, as discussed in Chapter 3, is Newton's law for the

[5] On the change of the specific heat of tin when becoming supra conductive, W.H. Keesom, and J.A. Kok, *Proc. Koninklijke Nederlandse akademie van wetenschappen* **35**, 743 (1932).

[6] The electromagnetic equations of the supraconductor, F. London and H. London, *Proc. R. Soc. Lond.* **A149**, 71 (1935).

velocity, v, of an electron with mass, m, and charge, e, in an electric field, **E**, with a phenomenological viscous drag force proportional to **v**/τ:

$$m\left(\dot{\mathbf{v}} + \frac{1}{\tau}\mathbf{v}\right) = e\mathbf{E}. \qquad (32.1)$$

For a perfect conductor, $\tau \to \infty$. Introducing the current density $\mathbf{j} = ne\mathbf{v}$, where n is the conduction electron number density, Eq. (32.1) can be written as

$$\frac{d\mathbf{j}}{dt} = \frac{ne^2}{m}\mathbf{E}, \qquad (32.2)$$

which is referred to as the first London equation. We next examine the *time derivative* of Maxwell's fourth equation which is (in c.g.s. units)

$$\nabla \times \frac{\partial \mathbf{H}}{\partial t} = \frac{4\pi}{c}\frac{\partial \mathbf{j}}{\partial t} + \frac{\varepsilon}{c}\frac{\partial^2 \mathbf{E}}{\partial t^2}, \qquad (32.3)$$

where ε is the background dielectric constant. Taking the curl of (32.3) and using (32.2) we have

$$\nabla \times \left(\nabla \times \frac{\partial \mathbf{H}}{\partial t}\right) = \left(\frac{4\pi ne^2}{mc} + \frac{\varepsilon}{c}\frac{\partial^2}{\partial t^2}\right)\nabla \times \mathbf{E}; \qquad (32.4)$$

using Maxwell's third equation, $\nabla \times \mathbf{E} = -\frac{1}{c}\frac{\partial \mathbf{H}}{\partial t}$, we have

$$\nabla \times \left(\nabla \times \frac{\partial}{\partial t}\mathbf{H}\right) + \left(\frac{1}{\lambda_L^2} + \frac{\varepsilon}{c^2}\frac{\partial^2}{\partial t^2}\right)\frac{\partial}{\partial t}\mathbf{H} = 0, \qquad (32.5)$$

where we have introduced a unit of length called the *London depth*, λ_L, defined by

$$\frac{1}{\lambda_L^2} = \frac{4\pi ne^2}{mc^2}. \qquad (32.6)$$

Equation (32.5) has been obtained for a perfect conductor model. In order to conform with the experimentally observed Meissner effect, we must exclude time-independent field solutions arising from integrating (32.5) once with respect to time and we therefore write

$$\nabla \times (\nabla \times \mathbf{H}) + \left(\frac{1}{\lambda_L^2} + \frac{\varepsilon}{c^2}\frac{\partial^2}{\partial t^2}\right)\mathbf{H} = 0; \qquad (32.7)$$

this is referred to as the second London equation. In what follows we will refer to Eq. (32.7) simply as the London equation.

An alternative derivation of (32.7) is motivated by the idea that some of the moving electrons behave collectively as a superfluid, a liquid possessing no viscosity.[7] This concept is borrowed from the physics of liquid ^{4}He; below 2.19 K

[7] *Superconductivity of Metals and Alloys*, P.G. deGennes, Benjamin, NY (1966).

(at atmospheric pressure) this liquid behaves as if it were composed of a mixture of two liquids: a superfluid, having no viscosity, and a normal liquid, having a finite viscosity (see Section 36.2). We assume that the *total* free energy of a superfluid consists of three parts

$$F = F_N + E_{kin} + E_{mag}, \tag{32.8}$$

where F_N is the free energy associated with the normal liquid, E_{kin} is the kinetic energy of the moving superfluid, and E_{mag} is the magnetic field energy. We may write these latter two terms as

$$E_{mag} = \frac{1}{8\pi}\int H^2(\mathbf{r})d^3r \tag{32.9}$$

and

$$E_{kin} = \frac{1}{2}\int \rho v^2(\mathbf{r})d^3\mathbf{r}, \tag{32.10}$$

where $\rho(\mathbf{r})$ is the mass density associated with the superfluid. Writing $\rho = nm$ and $\mathbf{v}(\mathbf{r}) = (1/ne)\mathbf{j}(\mathbf{r})$, and using the fourth Maxwell equation, $\nabla \times \mathbf{H} = (4\pi/c)\mathbf{j}$, Eq. (32.10) becomes

$$E_{kin} = \frac{1}{8\pi}\int \frac{mc^2}{4\pi ne^2}(\nabla \times \mathbf{H})^2\, d^3r; \tag{32.11}$$

n is now interpreted as the density of *superconducting* electrons. We will assume that the superconducting electrons adjust their motion so as to *minimize the total free energy*; this requires, $\delta(E_{mag} + E_{kin}) = 0$ or

$$\int \left[\mathbf{H}(\mathbf{r})\cdot\delta\,\mathbf{H}(\mathbf{r}) + \frac{mc^2}{4\pi ne^2}(\nabla \times \mathbf{H}(\mathbf{r}))\cdot(\nabla \times \delta\,\mathbf{H}(\mathbf{r}))\right] d^3r = 0, \tag{32.12}$$

where $\delta\mathbf{H}(\mathbf{r})$ is a variation of the (initially unknown) function $\mathbf{H}(\mathbf{r})$. Integrating the second term by parts (and placing the resulting surface outside the superconductor) and introducing the London depth, we obtain

$$\int d^3r \left[\mathbf{H}(\mathbf{r}) + \lambda_L^2 \nabla \times (\nabla \times \mathbf{H})\right]\delta\,\mathbf{H}(\mathbf{r}) = 0. \tag{32.13}$$

Since the variation $\delta\mathbf{H}(\mathbf{r})$ is arbitrary, the term in the square brackets must vanish; therefore,

$$\nabla \times (\nabla \times \mathbf{H}) + \frac{1}{\lambda_L^2}\mathbf{H} = 0, \tag{32.14a}$$

which is equivalent to (32.7) (including the displacement term in Maxwell's equation yields the last term in (32.7), which is negligible for most applications). Using the vector identity

$$\nabla \times (\nabla \times \mathbf{H}) = \nabla(\nabla \cdot \mathbf{H}) - \nabla^2\mathbf{H},$$

and Maxwell's second equation, we can write (32.14a) as

$$\nabla^2\mathbf{H} - \frac{1}{\lambda_L^2}\mathbf{H} = 0. \tag{32.14b}$$

Another way to present (32.14a) is to write $\nabla \times \mathbf{H} = (4\pi/c)\mathbf{j}$ in the first term, and $\mathbf{H} = \nabla \times \mathbf{A}$ in the second term and work in a gauge where $\nabla \cdot \mathbf{A} = 0$, which yields the equation,

$$\mathbf{j} = -\frac{ne^2}{mc}\mathbf{A}. \tag{32.15}$$

As a simple application of Eq. (32.14a), we now discuss the behavior of a superconductor in a magnetic field near a plane boundary. Consider first the case of a uniform field perpendicular to a superconductor surface lying in the x–y plane. From the second Maxwell equation, $\nabla \cdot \mathbf{H} = 0$, we obtain $\partial H_z/\partial z = 0$ or $\mathbf{H} = \text{const}$. For constant $\mathbf{H}$ the first term in (32.14b) vanishes and hence $\mathbf{H} = 0$ is the only solution. Thus a superconductor exhibiting the Meissner effect cannot have a field component perpendicular to its surface.

As the second example, consider a field lying parallel to the superconductor surface, e.g., $\mathbf{H}\|\hat{\mathbf{x}}$ which we may write as $\mathbf{H} = H(z)\hat{\mathbf{x}}$ (which satisfies $\nabla \cdot \mathbf{H} = 0$). For this case Eq. (32.14b) becomes

$$\frac{\partial^2 H_x}{\partial z^2} - \frac{1}{\lambda_L^2}H_x = 0 \tag{32.16}$$

or

$$\mathbf{H}_x(z) = \hat{\mathbf{x}}H_x(0)e^{-z/\lambda_L}. \tag{32.17}$$

A field parallel to the surface is therefore allowed; however it decays exponentially, with a characteristic length, λ_L, in the interior. The length, $\lambda_L(T = 0)$ ranges from 500–10,000 Å, depending on the material.[8] Accompanying this parallel field is a current density concentrated in the vicinity of the surface, which, from Maxwell's fourth equation, is

$$\mathbf{j}(z) = \frac{-\lambda_L c}{4\pi}H_x(0)e^{-z/\lambda_L}\hat{\mathbf{y}}. \tag{32.18}$$

This current density shields or screens the magnetic field from the interior of the superconductor; alternatively, this current density is required by the Meissner effect to expel this field from the sample interior.

[8] In high temperature superconductors, even larger London depths are encountered.

32.3 Thermodynamics of a type I superconductor

To simplify our discussion of the thermodynamics of type I superconductors,[9] we restrict the geometry to one for which the external field, **H**, is not distorted by the shape of the superconductor (examples being an infinitely long cylinder with **H** parallel to the axis, or a plane slab of infinite extent with **H** parallel to its surface). Far inside the superconductor (i.e., several London depths from the surface), the magnetic field essentially vanishes in the superconducting state and is equal to **H** in the normal state. In the thermodynamic identities that follow we identify this interior field as **B**, the flux density; the **H** field will be taken as the applied external field. The relation between **B** and **H** was shown earlier as Fig. 32.2.

We recall the thermodynamic identity for the response of a system in a magnetic field

$$dE = TdS + \frac{1}{4\pi}\mathbf{H}\cdot d\mathbf{B}, \tag{32.19}$$

where E is the energy density and S is the entropy density. When T and **B** are the independent variables, we use the Helmholtz free energy density, $F = E - TS$, and when T and **H** are the independent variables, we use the Gibbs free energy density,

$$G = F - (1/4\pi)\,\mathbf{B}\cdot\mathbf{H}; \tag{32.20}$$

taking the differential of these two quantities and using (32.19) yields

$$dF = -SdT + \frac{1}{4\pi}\mathbf{H}\cdot d\mathbf{B} \tag{32.21a}$$

and

$$dG = -SdT - \frac{1}{4\pi}\mathbf{B}\cdot d\mathbf{H}. \tag{32.21b}$$

A type I superconductor displays the Meissner–Ochsenfeld properties,

$$B = 0; \quad H < H_c \tag{32.22a}$$

$$B = H; \quad H > H_c. \tag{32.22b}$$

H_c is called the *thermodynamic critical field*. Since **H** and T will be our independent variables, we integrate (32.21b) at constant T to obtain

$$G(T,H) = G(T,0) - \frac{1}{4\pi}\int \mathbf{B}\cdot d\mathbf{H}; \tag{32.23}$$

[9] On supraconductivity I, C.J. Gorter and H.B.G. Casimir, *Physica* 1, 306 (1934).

integrating this expression for a superconductor obeying Eqs. (32.22a,b), we have

$$G(T, H) = G(T, 0) - \frac{1}{8\pi}\left(H^2 - H_c^2\right) \quad (H > H_c) \tag{32.24}$$

or

$$G(T, H) = G(T, 0). \quad (H < H_c) \tag{32.25}$$

(Note G is continuous at the transition.) We define a function

$$G_0 \equiv G(T, 0) + \frac{1}{8\pi}H_c^2. \tag{32.26}$$

We may then write

$$\begin{aligned} G(T, H) &= G_0 - \frac{1}{8\pi}H^2 \quad (H > H_c) \\ &\equiv G_N \end{aligned} \tag{32.27}$$

and

$$\begin{aligned} G(T, H) &= G_0 - \frac{1}{8\pi}H_c^2 \quad (H < H_c), \\ &\equiv G_S \end{aligned} \tag{32.28}$$

where G_N and G_S denote the normal and superconducting states respectively. We may interpret G_0 as the Gibbs free energy that the normal metal would have at zero field (were it stable); hence the Gibbs free energy of the superconducting state is lower than that of the normal state by $(1/8\pi)H_c^2(T)$; this energy is referred to as the *condensation energy*. Since $G \equiv F - (1/4\pi)\,\mathbf{H} \cdot \mathbf{B}$, we have $F_S(T, 0) = G_0 - (1/8\pi)H_c^2$. The normal state Helmholtz free energy F is then

$$F_N(T, 0) = F_S(T, 0) + \frac{H_c^2}{8\pi}. \tag{32.29}$$

From Eq. (32.21b),

$$S = -\left(\frac{\partial G}{\partial T}\right)_H; \tag{32.30}$$

thus,

$$S_N = -\frac{\partial G_0}{\partial T} \tag{32.31}$$

and

$$S_S = -\frac{\partial G_0}{\partial T} + \frac{1}{4\pi}H_c\left(\frac{\partial H_c}{\partial T}\right)_H. \tag{32.32}$$

Note that the entropy is discontinuous across the transition, and hence we have a *first-order transition* (when $H \neq 0$):

$$S_N - S_S = -\frac{1}{4\pi} H_c(T) \left(\frac{\partial H_c(T)}{\partial T} \right)_H. \tag{32.33}$$

The heat of the transition is

$$Q = T(S_N - S_S) = -\frac{T}{4\pi} H_c(T) \left(\frac{\partial H_c(T)}{\partial T} \right)_H; \tag{32.34}$$

this equation corresponds to the Clausius–Clapeyron equation of a (P, V, T) system. The specific heat per unit volume (at constant H) is defined as

$$C_H = T \left(\frac{\partial S}{\partial T} \right)_H, \tag{32.35}$$

or

$$\begin{aligned} \Delta C_H &\equiv C_N - C_S \\ &= -\frac{T}{4\pi} \left[H_c \frac{\partial^2 H_c}{\partial T^2} + \left(\frac{\partial H_c}{\partial T} \right)^2 \right]. \end{aligned} \tag{32.36}$$

At $H = 0$, where the transition is second order,

$$\Delta C_H|_{T=T_c} = -\frac{T_c}{4\pi} \left(\frac{\partial H_c}{\partial T} \right)^2_{T=T_c}; \tag{32.37}$$

this is sometimes called the Rutgers formula (Rutgers (1933)).

32.4 The intermediate state

If a superconducting body of arbitrary shape is placed in a magnetic field, the flux exclusion associated with the Meissner effect will in general distort the magnetic field. Exceptions are an infinite cylinder with the field parallel to the axis, or a sheet or half space with H_0 parallel to the plane of symmetry. For situations involving different shapes the local magnetic field will vary over the surface. As an example consider the case of a spherical superconductor, shown in Fig. 32.4. From magnetostatics the field will be highest at the equator (on the circle C in Fig. 32.4), where it is $H = (3/2)\ H_0$. Hence flux enters the sample, not at the thermodynamic critical field, H_c, but at a value $H_0 = (2/3)\ H_c$. For magnetic fields $H_c > H_0 > (2/3)H_c$, the sample consists of alternating domains of normal metal and superconductor. A superconductor in such a regime is said to be in the *intermediate state* (Landau, 1937).

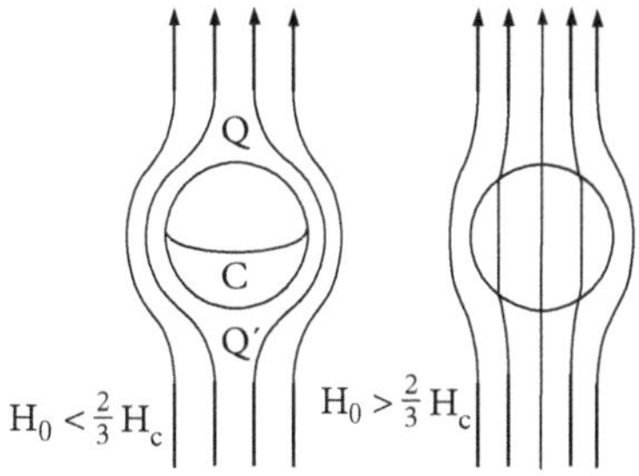

Figure 32.4 *The magnetic field distribution about a superconducting sphere of radius* a. *For an applied field,* $H_0 < 2H_c/3$, *there is a complete Meissner effect and the field at the equator (at any point on circle C) is* $3H_0/2$; *the field at the poles (Q,Q′) is zero. For* $2H_c/3 < H_0 < H_c$, *the sphere is in the intermediate state.*

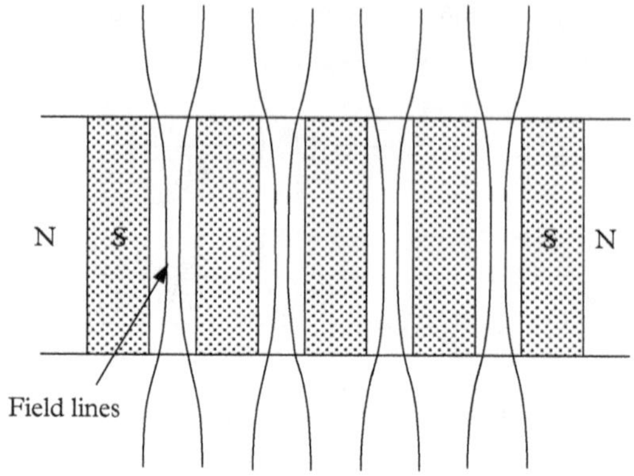

Figure 32.5 *Flux distribution in a slab in a perpendicular magnetic field.*

We will limit our discussion to the simple case of a plane superconducting sheet with H_0 perpendicular to the surface. From our earlier discussion we know that a superconductor cannot sustain a field component perpendicular to its surface. The resulting field behavior is shown qualitatively in Fig. 32.5. It has the following features:

(i) For a magnetic field, $0 < H_0 < H_c$, the sample consists of adjacent domains which are wholly superconducting (with no internal flux) or normal (with $H_{local} = H_c$).

(ii) In the interior of the superconductor and far from the surface, the domain walls are parallel to the applied field direction; the fraction of the cross section that is superconducting is fixed by H_0 and H_c such that the *total* flux through the sample is conserved,

$$\frac{\text{normal cross section}}{\text{total cross section}} = \frac{H_0}{H_c}. \tag{32.38}$$

Hence the superconducting fraction is $1 - H_0/H_c$ (which vanishes, as it should, for $H_0 = H_c$).

(iii) Near the surfaces the flux sheets "flare out" (which reduces the field curvature, which would otherwise raise the local fields at the interface). Were the local field to remain fixed at H_c, to sustain the average flux, needle-like superconducting domains would have to encroach on the normal domains in the vicinity of the surface. Such needles do not occur since the field profile is controlled by a minimization of the surface energy (between superconducting and free surface or the normal regions, see Section 32.5), the superfluid kinetic energy, and the total magnetic field energy.

A detailed treatment of the domain structure is mathematically complex and will not be dealt with here (see: Landau, Lifshitz, and Pitaevskii (1984); deGennes (1966); London (1950)).

32.5 Surface energy between a normal and a superconducting metal

Consider a slab parallel to the x–y plane in a perpendicular magnetic field parallel to z. Assume we have a normal-metal/superconductor phase boundary perpendicular to the x axis with the superconductor occupying the region $x > 0$, as in one of the domains of Fig. 32.5. The *total* free energy, F, in the London model is given by

$$F = \mathcal{A}\int_0^\infty dx\left[F_S + \frac{H^2(\mathbf{r})}{8\pi} + \frac{\lambda_L^2}{8\pi}(\nabla \times \mathbf{H}(\mathbf{r}))^2\right], \tag{32.39}$$

where $\mathcal{A}$ is the interface area; here F_S is the condensation energy density, $H_c^2(T)/8\pi$, and the second and third terms are respectively the magnet field energy and the kinetic energy densities, as given by Eqs. (32.9) and (32.11), both of which only contribute in the surface region and hence contribute to the surface energy. At our phase boundary in the intermediate state we have

$$G_N = G_S. \tag{32.40}$$

From the definition of the Gibbs free energy density, Eq. (32.20), and the fact that the independent variable $\mathbf{H}$ is here constant and given by $\mathbf{H}_c$, we can write the total Gibbs free energy as

$$G = F - \frac{\mathcal{A}}{4\pi}\int_0^\infty \mathbf{H}_c \cdot \mathbf{B}(x)dx; \tag{32.41}$$

in the present context we can use Eq. (32.17) and write

$$\mathbf{B}(x) = \mathbf{H}(x) = H_c\hat{\mathbf{z}}\exp[-x/\lambda_L].$$

Inserting this form in Eq. (32.41) and combining the latter with Eq. (32.39), we obtain

$$\begin{aligned} G &= \mathcal{A}\int_0^\infty \left[F_S + H_c^2\left(\frac{1}{8\pi}e^{-2x/\lambda_L} + \frac{1}{8\pi}e^{-2x/\lambda_L} - \frac{1}{4\pi}e^{-x/\lambda_L}\right)\right]dx \\ &\equiv \mathcal{A}\int_0^\infty F_S dx + \gamma\mathcal{A} \end{aligned}, \tag{32.42}$$

where γ is the surface energy per unit area, i.e., the "surface tension," given by

$$\begin{aligned} \gamma &= \frac{H_c^2}{4\pi}\left(-\frac{\lambda_L}{2}e^{-2x/\lambda_L} + \lambda_L e^{-x/\lambda_L}\right)\Bigg|_0^\infty \\ &= -\lambda_L\frac{H_c^2}{8\pi}. \end{aligned} \tag{32.43}$$

Note this surface energy is *negative*. This suggests that the system can lower its energy by *maximizing the interfacial area* (i.e., the system is unstable to the formation of multiple domains with associated interfaces).

Type I superconductors (for geometries not possessing an intermediate state, as in Figure 32.1) display a single domain for $H \le H_c$; hence they must have a *positive* domain wall energy. Physically, this positive surface energy arises because superconductivity is suppressed over a region of order ξ in the vicinity of a normal-metal/superconductor interface; i.e., we lose the condensation energy over a volume of order $\mathcal{A}\xi$, where ξ is a *coherence length*. This is equivalent to a positive contribution to the surface energy of order

$$\gamma = \xi \frac{H_c^2}{8\pi}. \tag{32.44}$$

In a type I material we have $\xi > \lambda_L$, and hence the positive contribution (32.44) *outweighs* the negative contribution (32.43) and the interface is stable. For so-called type II materials, to be discussed in Section 32.7, the balance is shifted and the system does, in some sense, try to *maximize* the amount of internal interfacial area above some field (referred to as the lower critical field); however it is subject to a constraint imposed by quantum mechanics, as we discuss in Sections 32.6 and 32.7.

32.6 Quantized vorticity

The previous discussion of the surface energy of a normal-metal/ superconductor interface suggests that type II materials, where $\xi \lesssim \lambda$, are unstable to the formation of domain structures which in some way maximize the amount of interface area. Two possible domain geometries are (i) an array of nested sheets (closed or open, depending on the geometry)[10] and (ii) a two-dimensional lattice of flux filaments. Calculations show the latter domain structure to be more stable.

Since the filaments (are presumed to) admit flux into the interior of the superconductor, we envision them as having a normal core with a diameter of order ξ, outside of which supercurrents flow within a region of order λ, which produce the internal field via Ampere's law.

As a primitive model of a single flux filament we consider the extreme limit $\xi \to 0$, for which the London approach should provide a good description. We recall Eq. (32.5) associated with our first derivation of the London equation,

$$\frac{\partial}{\partial t}\left[\nabla \times (\nabla \times \mathbf{H}) + \frac{4\pi n e^2}{mc^2}\mathbf{H}\right] = 0; \tag{32.45}$$

we next integrate (32.45) over an area A intersecting the filament (for convenience we choose a plane perpendicular to its axis) and use Ampere's law (the static form of Maxwell's fourth equation) obtaining

[10] For a superconducting slab we envision an array of interfaces parallel to the surface and for a cylindrical sample an array of coaxial cylinders. Other shapes would have more complex structures.

$$\frac{\partial}{\partial t}\int\left[\frac{mc}{ne^2}\nabla\times\mathbf{j}+\mathbf{H}\right]\cdot d^2\mathbf{r}=0. \tag{32.46}$$

In integrating (32.46) with respect to time we now allow the possibility of a *non-zero constant of integration*, Φ, with units of magnetic flux (since the flux filament phenomenon violates the Meissner behavior); thus

$$\int\left[\frac{mc}{ne^2}\nabla\times\mathbf{j}+\mathbf{H}\right]\cdot d^2\mathbf{r}=\Phi. \tag{32.47}$$

Applying Stokes' law to the first term in (32.47) yields

$$\frac{mc}{ne^2}\oint\mathbf{j}\cdot d\mathbf{l}+\int d^2\mathbf{r}\cdot\mathbf{H}=\Phi. \tag{32.48}$$

If we choose the contour to enclose a large area, we may expect the first term to be exponentially small (since the currents fall off exponentially with a characteristic length λ_L); we can then identify the constant of integration, Φ, as the total flux contained within the filament (most of which also falls inside a radius of order λ_L).

To gain further insight we substitute $\mathbf{H}=\nabla\times\mathbf{A}$ into (32.48) and again apply Stokes' law to obtain

$$\oint\left[\frac{mc}{ne^2}\mathbf{j}+\mathbf{A}\right]\cdot d\mathbf{l}=\Phi; \tag{32.49a}$$

Writing $\mathbf{j}=ne\mathbf{v}=ne\frac{\mathbf{p}}{m}$, (32.49a) becomes

$$\oint\left[\mathbf{p}+\frac{e}{c}\mathbf{A}\right]\cdot d\mathbf{l}=\frac{e}{c}\Phi. \tag{32.49b}$$

We identify the integrand as the *canonical momentum* associated with the motion of a charged particle in the Hamiltonian formulation of mechanics. F. London correctly concluded that superconductivity was a macroscopic quantum phenomenon, and guided by this insight he suggested that (32.49b) must conform with the Bohr–Sommerfeld quantization rule for the (quasiclassical) motion of an electron; i.e., $(|e|/c)\Phi=nh$ or $\Phi=nhc/|e|$, where n is an integer. However this assumes that the orbiting entities are single electrons; Ginzburg and Landau allowed for a more general case where $e\to e^*$; we then have

$$\Phi=n\Phi_0, \tag{32.50a}$$

where n is an integer and

$$\Phi_0=\frac{hc}{|e^*|}. \tag{32.50b}$$

From the BCS theory it is now known that $e^* = 2e$; i.e.,

$$\Phi_0 = \frac{hc}{2|e|} = 2.07 \times 10^{-7}\,\text{Gauss cm}^2, \qquad (32.50c)$$

which is referred to as the *flux quantum*. Hence flux could enter a type II superconductor as an array of quantized flux filaments; the lowest energy situation corresponds to singly quantized ($n = 1$) filaments each carrying a flux quantum, Φ_0.

We next examine Eq. (32.49) for a contour of radius $\lambda \gg r \gg \xi$; the amount of flux contained is then vanishing small and the first term in (32.49b) dominates, yielding the condition $2\pi pr = h$ or $p = \hbar/r$; BCS theory also dictates that the mass of the orbiting entity is $m^* = 2m$; hence,

$$\mathbf{v}(\mathbf{r}) = \frac{\hbar}{2mr}\hat{\boldsymbol{\theta}}, \qquad (32.51)$$

where $\hat{\boldsymbol{\theta}}$ is an azimuthally unit vector. This velocity profile corresponds to the large r behavior of a *vortex in a fluid*, although with the vorticity quantized. One then refers to the filaments as *quantized vortex lines* or *vortex lines* for short.

Vorticity involves a non-vanishing curl of the velocity; i.e., $\nabla \times \mathbf{v} \neq 0$; but the curl of $\mathbf{v}$ in (32.51) vanishes for $r \neq 0$. Hence the circulation, $\Gamma \equiv \oint \mathbf{v} \cdot d\boldsymbol{l} \neq 0$, and thus all the vorticity must be located in an *infinitesimal region near the origin*. Physically it would be spread out over a coherence length, ξ. In our $\xi \to 0$ model we can obtain an approximate description by adding a singular source term to the London equation in the form

$$\mathbf{H} + \lambda_L^2 \nabla \times (\nabla \times \mathbf{H}) = \Phi_0 \hat{\mathbf{z}} \delta^{(2)}(\mathbf{r}), \qquad (32.52)$$

where $\delta^{(2)}(\mathbf{r})$ is a two-dimensional δ function and $\hat{\mathbf{z}}$ is a unit vector (parallel to the vortex axis). Equation (32.52) may be written in cylindrical coordinates as

$$H_z - \frac{\lambda_L^2}{r}\frac{d}{dr}\left(r\frac{dH_z}{dr}\right) = \Phi_0 \delta^{(2)}(\mathbf{r}). \qquad (32.53)$$

The left side of (32.53) is a special case of Bessel's equation[11] and the solution having the required singular behavior near $r = 0$ is

$$H_z = \frac{\Phi_0}{2\pi\lambda_L^2} K_0\left(\frac{r}{\lambda_L}\right), \qquad (32.54)$$

where K_0 is the zeroth-order modified Bessel function with an imaginary argument. From Ampere's law, $\mathbf{j} = (c/4\pi)\nabla \times \mathbf{H} = -(c/4\pi)\cdot(dH_z/dr)\hat{\boldsymbol{\theta}}$, and for small x $K_0(x) \cong \ln(1/x)$; hence, $j_\theta = (c/4\pi)\cdot(\Phi_0/2\pi\lambda_L^2 r) = neh/4\pi mr$ or $v = h/2mr$, in

[11] Here we adopt the definitions of Abramowitz and Stegun (1970), p. 374 ff.

agreement with (32.51). For large x we use the form $K_0(x) \cong (\pi/2)^{1/2}\exp(-x)$ or $H_z \cong (\Phi_0/2\pi\lambda_L^2)\cdot(\pi\lambda_L/2r)^{1/2}\cdot\exp(-r/\lambda_L)$ and the field drops off exponentially, as previously argued.

The energy of a vortex line follows from Eqs. (32.8)–(32.11),

$$E = \frac{L}{8\pi}\int\left[\mathbf{H}^2(\mathbf{r}) + \lambda_L^2(\nabla\times\mathbf{H})^2\right]d^2r, \tag{32.55}$$

where L is the length of the vortex line. We integrate (32.55) by parts to obtain

$$\frac{E}{L} = \frac{\lambda_L^2}{8\pi}\oint \mathbf{H}\times(\nabla\times\mathbf{H})\cdot dl + \frac{1}{8\pi}\int\mathbf{H}\cdot\left[\mathbf{H} + \lambda_L^2\nabla\times(\nabla\times\mathbf{H})\right]d^2r. \tag{32.56}$$

The quantity in square brackets in the second term of (32.56) is equal to the left side of (32.52). Were we to simply replace it by $\delta^{(2)}(\mathbf{r})$ and integrate, we would obtain $H(r\to 0)$, which is logarithmically divergent. This indicates that our simple model for the vortex core is not sufficiently accurate to evaluate the energy. To avoid the divergence problem we assume that H is finite everywhere, thus eliminating $\delta^{(2)}(\mathbf{r})$ from the right side of (32.52), which results in the second term in (32.56) vanishing. We must still account for the energy in the core of the vortex, however, which we do by separating the line integral in (32.55) into three parts: a circle at a very large radius (which makes a vanishingly small contribution due to the exponential fall-off of $\mathbf{H}(\mathbf{r})$ at large r), two counter traversing radial paths from the outer circle to an inner circle of a very small radius (which cancel each other), and finally a path around the inner circle (which makes the only non-vanishing contribution). For small r the first term in (32.56) is

$$\frac{E}{L} = \frac{\lambda_L^2}{8\pi}\frac{\Phi_0^2}{(2\pi\lambda_L^2)^2}\oint dl\frac{1}{r}ln\left(\frac{\lambda_L}{r}\right) \tag{32.57}$$

or

$$\frac{E}{L} = \frac{\Phi_0^2}{(4\pi\lambda_L)^2}ln\frac{\lambda_L}{\xi}, \tag{32.58}$$

where we choose a radius, $r=\xi$, for the (inner) line integral; this corresponds to the physically reasonable assumption that the field divergence is removed at the coherence length scale (as a more complete theory confirms). We expect that the 1/r divergence of the superfluid velocity ultimately destroys superconductivity in the vortex core; we may model this effect by assuming that the density of superconducting electrons, n(r), approaches zero (sufficiently rapidly) as $r\to 0$. Figure 32.6 shows the qualitative behavior of H(r) near the center (core) of a vortex filament; also shown is the *superconducting order parameter*, $|\psi(\mathbf{r})|$, which behaves like a quantum mechanical wave function, and will be introduced in Chapter 33; it is then natural to associate $n(\mathbf{r}) = |\psi(\mathbf{r})|^2$.

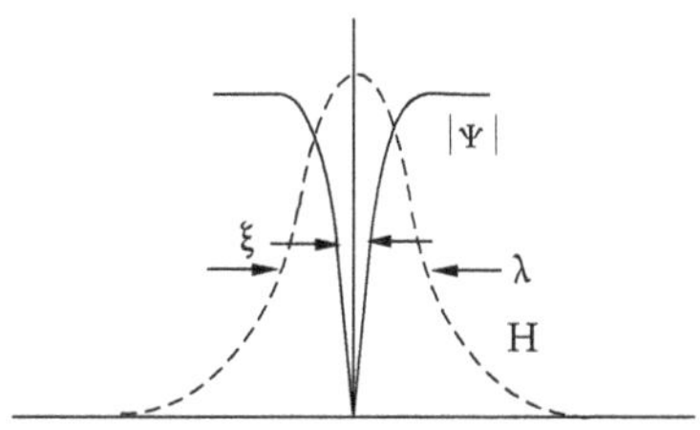

Figure 32.6 *The structure of an isolated Abrikosov vortex line in a type II superconductor showing, qualitatively, the position dependence of the order parameter, ψ, and the magnetic field,* H.

In Section 32.7, where we discuss the mixed state of a type II superconductor, we will require an expression for the interaction energy of two vortices. Returning to our London-like model, we generalize Eq. (32.52) to the case of two (parallel) vortices

$$\mathbf{H} + \lambda_L^2 \nabla \times (\nabla \times \mathbf{H}) = \Phi_0 \hat{\mathbf{z}} \left[\delta^{(2)}(\mathbf{r}-\mathbf{r}_1) + \delta^{(2)}(\mathbf{r}-\mathbf{r}_2) \right]. \tag{32.59}$$

Since this equation is linear the resulting magnetic field will be the sum of two terms having the same form as (32.54),

$$\begin{aligned} H_z &= \frac{\Phi_0}{2\pi\lambda_L^2} \left[K_0\left(\frac{|\mathbf{r}-\mathbf{r}_1|}{\lambda_L}\right) + K_0\left(\frac{|\mathbf{r}-\mathbf{r}_2|}{\lambda_L}\right) \right] \\ &= H_z^{(1)} + H_z^{(2)}. \end{aligned} \tag{32.60}$$

The total energy (which is quadratic in H) will involve three terms: two of these correspond to the “self energies” of the individual vortices (as given by (32.58)) and the third results from their interaction. The interaction energy in the extreme London limit can be evaluated from the second term in (32.56), along with Eq. (32.52) (we assume the path of the first integral involves a single circle of very large radius encircling both vortices); the interaction energy is then given by

$$\begin{aligned} \frac{E^{(12)}}{L} &= \frac{1}{8\pi} \int d^2\mathbf{r} \cdot \left\{ \mathbf{H}^{(1)} \cdot \left[\mathbf{H}^{(2)} + \lambda_L^2 \nabla \times (\nabla \times \mathbf{H}^{(2)}) \right] \right. \\ &\quad \left. + \mathbf{H}^{(2)} \cdot \left[\mathbf{H}^{(1)} + \lambda_L^2 \nabla \times \left(\nabla \times \mathbf{H}^{(1)} \right) \right] \right\} \\ &= \frac{\Phi_0}{8\pi} \int \left[H^{(1)}(|\mathbf{r}-\mathbf{r}_1|)\delta^{(2)}(\mathbf{r}-\mathbf{r}_2) + \mathbf{H}^{(2)}(|\mathbf{r}-\mathbf{r}_2|)\delta^{(2)}(\mathbf{r}-\mathbf{r}_1) \right] \cdot d^2\mathbf{r}; \end{aligned}$$

on integrating and using Eq. (32.54), we obtain

$$\frac{E^{(12)}}{L} = 2\left(\frac{\Phi_0}{4\pi\lambda_L}\right)^2 K_0\left(\frac{|\mathbf{r}_1-\mathbf{r}_2|}{\lambda_L}\right), \tag{32.61}$$

where, due to our assumptions, we require $|\mathbf{r}_1 - \mathbf{r}_2| \geq \xi$. The sign of (32.61) is positive and hence the force (per unit length) between two vortices is *repulsive*. We may rewrite Eq. (32.61) in the form

$$\frac{E^{(12)}}{L} = \frac{\Phi_0}{4\pi} H^{(12)}, \tag{32.62}$$

where $H^{(12)} \equiv H^{(1)}(\mathbf{r}_2) = H^{(2)}(\mathbf{r}_1)$, with the latter given by (32.54); i.e., $H^{(12)}$ is the contribution to the field at one vortex resulting from the presence of another. Regarding $-(\Phi_0 L/4\pi)\hat{\mathbf{z}}$ as a magnetic moment, μ, we could write $E^{(12)} = -\mu \cdot \mathbf{H}$.

32.7 Type II superconductivity

Superconductors with $\xi < \lambda$, termed type II materials, behave differently in a magnetic field from those with $\xi > \lambda$ (type I materials). Figure 32.7 shows the field dependence of the magnetization of an ideal type II superconductor (the sample geometry is assumed to be a plane slab or a cylinder to avoid geometrically induced field inhomogeneities, as noted earlier in our discussion of the intermediate state). For low fields we have $-4\pi M \propto H$; i.e., the sample displays a Meissner-like behavior. However at a field $H = H_{c1}$, called the *lower critical field*, flux abruptly enters the sample; for an ideal sample the susceptibility, $\chi = (dM/dH) \to -\infty$ at this field, as we will comment on further, shortly. The magnetization decreases continuously above H_{c1} and reaches zero at a field $H = H_{c2}$, called the *upper critical field*. The regime $H_{c1} < H < H_{c2}$ represents a new thermodynamic superconducting state called the mixed state or *Shubnikov phase*.[12]

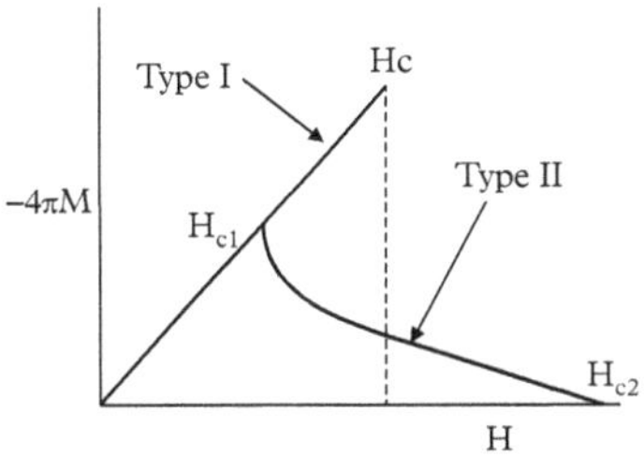

Figure 32.7 *The reversible magnetization curves for a type I and a type II cylindrical superconductor. The magnetic field is applied parallel to the cylinder axis.*

We now show that the Meissner state of a type II superconductor becomes unstable to the entry of vortex filaments at a field which we identify with H_{c1}. For magnetic fields which are only slightly above H_{c1} we may write the Gibbs free energy per unit volume as

$$G(H) = G(H = 0) + n_L \frac{E}{L} + \frac{1}{L} \sum_{i<j} E^{(i,j)} - \frac{\mathbf{B} \cdot \mathbf{H}}{4\pi}. \tag{32.63}$$

The second term represents the self energy of the individual lines, where n_L is the number of lines per cm^2 with energy E/L per cm, the third term is the (repulsive) interaction energy between the lines (and is summed over a unit of area), and the last term is the usual field term relating the Gibbs and Helmholtz free energies (see Eq. (32.20)). In the presence of a uniform array of flux lines the magnetic induction is

$$B = n_L \Phi_0. \tag{32.64}$$

Near H_{c1} (above the value at which flux first enters), we can ask what happens if we neglect the interaction term; using (32.64), the Gibbs energy density will then be

$$G(H) = G(H = 0) + n_L \left(\frac{E}{L} - \frac{1}{4\pi} H \Phi_0 \right). \tag{32.65}$$

For $H < 4\pi E/(\Phi_0 L)$, (32.65) is minimized by setting $n_L = 0$. However for $H > 4\pi E/(\Phi_0 L)$, Eq. (32.65) would suggest that $G \propto -n_L$; i.e., the system can lower its free energy indefinitely simply by creating more flux lines. It is clear that we can identify H_{c1} as

$$H_{c_1} = \frac{4\pi E}{\Phi_0 L} \tag{32.66a}$$

[12] L.V. Shubnikov, V.I. Khotkevich, Y.D. Shepelev, and Y.N. Riabinin, *Zh. Eksp. Teor. Fiz.* 7, 221 (1937).

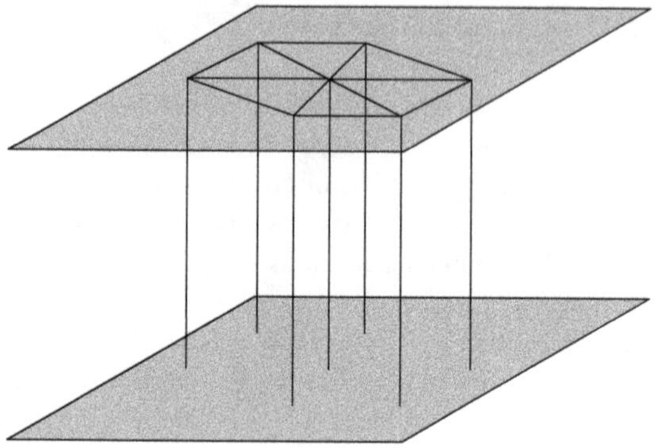

Figure 32.8 *Schematic representation of the flux lattice in a type II superconductor.*

or

$$H_{c_1} = \frac{\Phi_0}{4\pi\lambda_L^2} ln\left(\frac{\lambda_L}{\xi}\right), \tag{32.66b}$$

where we have used (32.58) for the vortex self energy in (32.66b).[13] The negative divergence of G for $H > H_{c1}$ is eliminated if we include the vortex–vortex interaction (repulsion) effects; i.e., we must seek the minimum with respect to B of the quantity

$$\Delta G = \frac{B}{4\pi}\left[H_{c1} - H + \frac{1}{2}z\frac{\Phi_0}{2\pi\lambda_L^2}K_0\left(\frac{d}{\lambda_L}\right)\right]. \tag{32.67}$$

The first two terms are the same as in (32.65). The last term is the result of the vortex–vortex repulsion; the factor 1/2 assigns half of the interaction energy to each vortex, z is the number of nearest neighbors in the vortex lattice (the exponential fall-off of K_0 justifies including only nearest-neighbor interactions at low fields where n_L is small), and d (appearing in the argument of K_0) is the *vortex–vortex spacing*, which depends on the symmetry of the vortex lattice.

Calculations show that a triangular (centered hexagonal) arrangement of lines has the lowest energy, as shown in Fig. 32.8. For this lattice, $n_L = 2/(\sqrt{3}d^2) = B/\Phi_0$ or $d^2 = 2\Phi_0/(\sqrt{3}B)$, and z = 6. On substituting these values of d and z in Eq. (32.67) we can obtain $\Delta G = \Delta G(B)$. By minimizing ΔG one can obtain the magnetic induction, B, for a given external field, H. Carrying out the minimization process for each H we can develop the function B = B(H) (or M = M(H) through $B = H + 4\pi M$). Further analysis of this model shows that $\chi = (\partial M/\partial H)_T$ does diverge as $H \to H_{c1}$. The overall behavior is shown in Fig. 32.7. However, in practice the transition at H_{c1} is not singular, but is rounded to some extent, a result of the *pinning of flux lines on crystalline defects*.

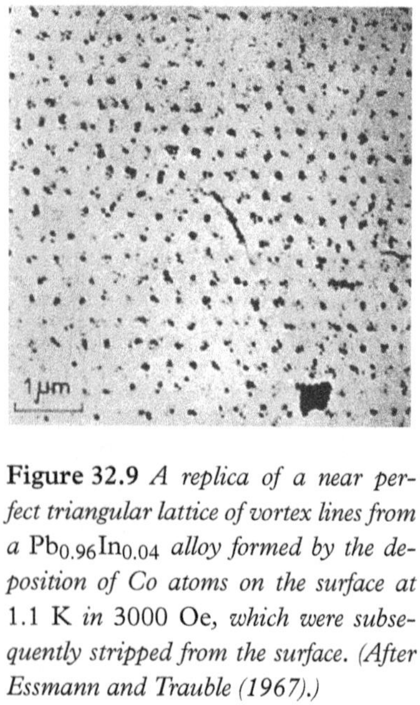

Figure 32.9 *A replica of a near perfect triangular lattice of vortex lines from a* $Pb_{0.96}In_{0.04}$ *alloy formed by the deposition of Co atoms on the surface at* 1.1 K *in* 3000 Oe, *which were subsequently stripped from the surface. (After Essmann and Trauble (1967).)*

[13] We can now make contact with the comment made at the end of Section 31.5. Taking the classical limit as $h \to 0$, we see that the lower critical field would approach zero. At any finite field we would then have a divergent number of flux lines; i.e., the system would have a divergent internal interface area. Thus quantum mechanics imposes the constraint on the maximal amount of internal interfacial area.

Figure 32.9 shows an electron micrograph of a flux lattice. This image was formed by the deposition of magnetic cobalt atoms on the surface of a $Pb_{0.96}In_{0.04}$ alloy held in the superconducting state at 1.1 K and in a magnetic field of 3 kOe; note that the Co atoms are drawn to the vortex cores where the field is concentrated. The Co atoms were subsequently stripped from the surface via the later deposition of a carbon layer to form a replica.[14]

32.8 Superconducting magnets

In the presence of an externally generated electric current density, **j**, the flux lattice of a type II superconductor will experience a net Lorentz volume force, $\mathbf{f}_V = (1/c)\mathbf{j} \times \mathbf{B}$. In the absence of some counterbalancing force the flux lattice would accelerate in a direction perpendicular to the current, and to maintain the net magnetic flux vortex lines would enter (nucleate) on one side of a sample (typically a wire in a magnet) and leave (annihilate) on the opposite side. In practice, the lattice acquires a steady-state velocity due to the presence of a viscous drag force which can be modeled by a term $\eta\mathbf{v}$, where **v** is an average drift velocity of the lattice (J. Bardeen and M.J. Stephen).[15] Since this is a dissipative processes the material is, strictly speaking, no longer a superconductor.

In practice, flux lines can "pin" on inhomogeneities (pinning sites) in the sample that are there naturally or, more commonly, are introduced through processing to enhance the effect. The material is then superconducting. All high field superconducting magnets exploit this mechanism. However at higher current densities the flux lines can "break loose" from the pinning sites. This motion typically involves a "bundle" of lines that collectively depin, producing a local burst of heat which can drive that region of the sample normal. Unless this heat is rapidly dissipated the region can grow via ohmic heating and spread throughout the material driving the entire sample normal, a process referred to as a *quench*. On a much longer time scale one can encounter the phenomenon of *flux creep* in the presence of a current; thermal activation over local potential barriers to a lower energy configuration can always occur and with it dissipation; it is just a matter of the associated time scales. Magnets utilized for precision nuclear magnetic resonance applications are extraordinarily stable. They characteristically contain a superconducting switch which, after energizing the magnet, connects the input and output current leads so that current flows in the absence of an external current source. This trick avoids heat input to the cryogenic environment from dissipation in the current leads connecting the magnet to room temperature.

ADDITIONAL READING

Superfluids, Vol. I, Fritz London, Dover, NY (1964).
Superconductivity, J.B. Ketterson and S.N. Song, Cambridge University Press (2000).

[14] The direct observation of individual flux lines in Type II superconductors, U. Essmann and H. Tråuble, *Phys. Lett.* **24A**, 526 (1967).

[15] Theory of the motion of vortices in superconductors, J. Bardeen and M.J. Stephen, *Phys. Rev.* **140**, 1197 (1965).

Superconductivity of Metals and Alloys, P.G. deGennes, W.A. Benjamin, NY (1966).

Introduction to Superconductivity, M. Tinkham, McGraw-Hill, NY (1975).

Superfluidity and Superconductivity, D.R. Tilley and J. Tilley, Adam Hilger, Bristol (1986).

Superconductivity: Conventional and Unconventional Superconductors, Vol. 1, K.H. Bennemann and J.B. Ketterson, Eds. Springer (2008).

The intermediate state is treated in *The Electrodynamics of Continuous Media*, L.D. Landau, E.M. Lifshitz, and L.P. Pitaevskii, Pergamon Press, Oxford (1984).

Magnetic Flux Structures in Superconductors, Rudolf Huebener, Springer, Berlin (2001).

A Phenomenological Theory of Superconductivity: The Ginzburg–Landau Theory and the Josephson Effects

33

33.1 Ginzburg–Landau theory

33.1.1 Basic equations

In the previous chapter we introduced the idea that superconductivity is some kind of macroscopic quantum state. Ginzburg and Landau[1,2] built this idea into the Landau second-order phase transition theory by assuming the existence of a macroscopic "wave function," ψ, which they took as the order parameter associated with superconductivity.[3] Since wave functions can be complex, only the form $\psi\psi^*$ may enter the expansion of the free energy per unit volume; we therefore write

$$\mathrm{F} = \mathrm{F(T)} + \alpha\,|\psi|^2 + \frac{1}{2}\beta\,|\psi|^4. \tag{33.1}$$

The minimization proceeds exactly as discussed earlier in Sections 25.2 and 26.3; i.e.,

$$|\psi| = 0 \qquad \mathrm{T} > \mathrm{T_c} \tag{33.2a}$$

$$|\psi| = \left[a\frac{(\mathrm{T_c}-\mathrm{T})}{\beta}\right]^{1/2}. \qquad \mathrm{T} < \mathrm{T_c} \tag{33.2b}$$

To describe situations where the superconducting state is *inhomogeneous* we must generalize (33.1). The total free energy, F, then involves an integration of F over the sample and hence we write

$$\begin{aligned} F &= F_0 + \int \mathrm{F}(\mathbf{r})d^3r \\ &= F_0 + \int d^3r\left[\alpha\,|\psi(\mathbf{r})|^2 + \frac{1}{2}\beta\,|\psi(\mathbf{r})|^4\right]. \end{aligned} \tag{33.3}$$

[1] V.L. Ginzburg and L.D. Landau, *Zh. Eksp. Teor. Fiz.* **20**, 106 (1950).

[2] The 2003 Nobel Prize in Physics was awarded to Vitaly Lazarevich Ginzburg and Anthony James Leggett "for pioneering contributions to the theory of superconductors and superfluids."

[3] A microscopic definition of the order parameter will be given in Section 35.3.

The Physics of Solids. First Edition. J. B. Ketterson.

Equation (33.3) in its present form does not model the increase in energy associated with a spatial distortion of the order parameter, i.e., effects associated with what is called a coherence length, ξ, as introduced in Section 32.5. To account for such effects Ginzburg and Landau added a "gradient energy" term to (33.3) of the form

$$F_G = \int \frac{\hbar^2}{2m^*} |\nabla \psi(\mathbf{r})|^2 \, d^3r, \tag{33.4}$$

with m* as a parameter; the choice of the coefficient, $\hbar^2/2m^*$, makes (33.4) mimic the quantum mechanical kinetic energy (introduced earlier in Eq. (32.10)). Ginzburg and Landau assumed that if (33.4) was to be regarded as the kinetic energy contribution to the Hamiltonian density of the superconducting electrons, then (as in Hamiltonian mechanics) the interaction of the electrons with an electromagnetic field would be accomplished by the Hamiltonian prescription (see Section 18.1),

$$\nabla \to \nabla - \frac{ie^*}{\hbar c}\mathbf{A}; \tag{33.5}$$

the use of e* allows the superconducting entities to carry a different charge ($e^* = 2e$ in BCS theory). Combining these we have

$$F_G = \int d^3r \frac{\hbar^2}{2m^*} \left| \left(\nabla - \frac{ie^*}{\hbar c}\mathbf{A}(\mathbf{r}) \right) \psi(\mathbf{r}) \right|^2 . \tag{33.6}$$

Finally we must add the contribution of the magnetic field to the energy density,

$$F_H = \int d^3r \frac{1}{8\pi} H^2(\mathbf{r}).$$

Combining these we have

$$\begin{aligned} F = F_0 + \int d^3r \Big[& \alpha |\psi(\mathbf{r})|^2 + \frac{1}{2}\beta |\psi(\mathbf{r})|^4 \\ & + \frac{\hbar^2}{2m^*} \left| (\nabla - \frac{ie^*}{\hbar c} \mathbf{A}(\mathbf{r}))\, \psi(\mathbf{r}) \right|^2 + \frac{1}{8\pi} H^2(\mathbf{r}) \Big] \end{aligned}. \tag{33.7}$$

The minimization of (33.7) must be carried out using the methods of the calculus of variations (see Appendix A, Eq. (A.18), since F is a *functional* involving the free energy density,

$$F = F\big[\psi(\mathbf{r}), \psi^*(\mathbf{r}), \mathbf{A}(\mathbf{r}), \nabla\psi(\mathbf{r}), \nabla\psi^*(\mathbf{r}), \nabla \cdot \mathbf{A}(\mathbf{r}), \mathbf{H}(\mathbf{r}) = \nabla \times \mathbf{A}(\mathbf{r})\big],$$

which in turn involves the unknown functions, $\psi(\mathbf{r})$, $\psi^*(\mathbf{r})$, $\mathbf{A}(\mathbf{r})$, and associated derivatives.

Taking a variation of (33.7) with respect to $\psi^*(\mathbf{r})$ yields

$$\delta F = \int d^3r \left[-\frac{\hbar^2}{2m^*}\left(\nabla - \frac{ie^*}{\hbar c}\mathbf{A}(\mathbf{r})\right)^2 \psi(\mathbf{r}) + \alpha\psi(\mathbf{r}) + \beta\,|\psi(\mathbf{r})|^2\,\psi(\mathbf{r})\right]\delta\psi^*(\mathbf{r})$$

$$+ \int d^2\mathbf{r}\cdot\frac{\hbar^2}{2m^*}\left(\nabla - \frac{ie^*}{\hbar c}\mathbf{A}(\mathbf{r})\right)\psi(\mathbf{r})\delta\psi^*(\mathbf{r}); \tag{33.8}$$

(a variation with respect to ψ, which is an independent variable, yields the complex conjugate of (33.8)). To minimize F we set the integrand of the first part of (33.8) to zero; this yields the first Ginzburg–Landau equation,

$$-\frac{\hbar^2}{2m^*}\left(\nabla - \frac{ie^*}{\hbar c}\mathbf{A}(\mathbf{r})\right)^2 \psi(\mathbf{r}) + \alpha\psi(\mathbf{r}) + \beta\,|\psi(\mathbf{r})|^2\,\psi(\mathbf{r}) = 0. \tag{33.9}$$

The surface term (which was generated through an integration by parts) can be used (with caution) to establish certain boundary conditions and will be discussed later.

Variation of (33.7) with respect to $\mathbf{A}$ (with $\mathbf{H} = \nabla\times\mathbf{A}(\mathbf{r})$) yields Ampere's law,

$$\nabla\times\mathbf{H}(\mathbf{r}) = \frac{4\pi}{c}\mathbf{j}(\mathbf{r}) \tag{33.10}$$

provided that we identify $\mathbf{j}(\mathbf{r})$ as

$$\mathbf{j}(\mathbf{r}) = \frac{e^*}{2m^*}\left[\psi^*(\mathbf{r})\left(-i\hbar\nabla - \frac{e^*}{c}\mathbf{A}(\mathbf{r})\right)\psi(\mathbf{r}) + \psi(\mathbf{r})\left(+i\hbar\nabla - \frac{e^*}{c}\mathbf{A}(\mathbf{r})\right)\psi^*(\mathbf{r})\right],$$

or equivalently

$$\mathbf{j}(\mathbf{r}) = \frac{-ie^*\hbar}{2m^*}\left(\psi^*(\mathbf{r})\nabla\psi(\mathbf{r}) - \psi(\mathbf{r})\nabla\psi^*(\mathbf{r})\right) - \frac{e^{*2}}{m^*c}\,|\psi(\mathbf{r})|^2\,\mathbf{A}(\mathbf{r}). \tag{33.11}$$

This is the second G–L equation; it corresponds to the expression for the current density in quantum mechanics. Also note that the current density satisfies the equation $j(\mathbf{r}) = c\,\delta F/\delta\mathbf{A}(\mathbf{r})$; i.e., it is the variable conjugate to $\mathbf{A}(\mathbf{r})$.

33.1.2 Gauge invariance

The simplest solution of (33.9) is the case of a uniform superconductor where ψ = const. and $\mathbf{A} = 0$. However, (33.9) possesses a continuum of other solutions having the same free energy, which we now show. As we can with any complex function, we write $\psi(\mathbf{r}) = \mathcal{A}(\mathbf{r})\exp(i\phi(\mathbf{r}))$, where $\mathcal{A}(\mathbf{r})$ and $\phi(\mathbf{r})$ are a position-dependent amplitude and phase respectively. Let us examine a class of solutions which satisfy the complex equation

$$\left(\nabla - \frac{ie^*}{\hbar c}\mathbf{A}(\mathbf{r})\right)\psi(\mathbf{r}) = 0, \tag{33.12}$$

which from Eq. (33.7) have the same free energy as a uniform superconductor when $\mathbf{H} = 0$, and from (33.11) correspond to a vanishing current density. Equation (33.12) is equivalent to the two real equations:

$$\nabla \mathcal{A}(\mathbf{r}) = 0 \tag{33.13a}$$

and

$$\left[\nabla\phi - \frac{e^*}{\hbar c}\mathbf{A}(\mathbf{r})\right] = 0. \tag{33.13b}$$

From (33.13a) we see that the only allowed solutions of (33.12) involve a constant amplitude, $\mathcal{A}$; Eq. (33.13b), on the other hand, has infinitely many solutions involving a vector potential and a position-dependent phase (which does not affect the free energy) related by

$$\mathbf{A} = \frac{\hbar c}{e^*}\nabla\phi. \tag{33.14}$$

Any vector potential satisfying (33.14) results in a uniform free energy and (on substituting (33.12) into (33.11)) a vanishing current density (note that $\mathbf{H} = \nabla \times \mathbf{A} = 0$ for all $\mathbf{A}$ of the form (33.14)).

This exercise shows that the symmetry broken in superconductivity is *gauge symmetry*, or equivalently, *phase symmetry*. Superconductors having different phase functions, $\phi(\mathbf{r})$, are in a real sense physically distinct;[4] this arbitrariness of the phase is the analogue for a superconductor of the property that the magnetization may point in any direction in an isotropic (liquid) ferromagnet.

33.2 Boundaries and boundary conditions

33.2.1 The pinned boundary and the coherence length

We first examine a simple case involving an inhomogeneous order parameter generated by the presence of a boundary, in the absence of a magnetic field. Assume that we have a superconducting *half space* occupying the region $x > 0$. We further assume that the order parameter is driven to (pinned at) zero at this interface. Experimentally this can be accomplished by coating the surface of the superconductor with a film of ferromagnetic material.[5] We then seek a solution to the one-dimensional Ginzburg–Landau equation,

$$-\frac{\hbar^2}{2m^*}\frac{d^2\psi}{dx^2} + \alpha\psi + \beta\psi^3 = 0. \tag{33.15}$$

[4] Strictly speaking we cannot determine the absolute phase of a superconductor, but in our later discussion of the Josephson effects we will show that phase differences can be measured.

[5] Paramagnetic impurities (those bearing a spin in a host material) or interfaces with a ferromagnetic metal strongly depress superconductivity. A normal metal interface has a much smaller effect and an insulator or vacuum has a negligible effect for most purposes.

Noting that α is negative in the superconducting state, ($\alpha = -|\alpha|$), defining a quantity called the Ginzburg–Landau coherence length as

$$\xi^2 \equiv \frac{\hbar^2}{2m^* |\alpha|}, \tag{33.16}$$

and writing $(\beta/|\alpha|)\psi^2 = f^2$, we may rewrite (33.15) as

$$-\xi^2 f'' - f + f^3 = 0. \tag{33.17}$$

Multiplying by f', we may rewrite (33.17) as

$$\frac{d}{dx}\left[-\frac{\xi^2 f'^2}{2} - \frac{1}{2}f^2 + \frac{1}{4}f^4\right] = 0; \tag{33.18}$$

hence the quantity in square brackets must be a constant. Far from the boundary, $f' = 0$ and $f^2 = 1$ (equivalent to $\psi^2 = |\alpha|/\beta$) and hence the constant of integration is 1/4; the first integral of Eq. (33.18) then becomes

$$\xi^2 f'^2 = \frac{1}{2}(1 - f^2)^2, \tag{33.19}$$

which has the solution, $f = \tanh\left[x/(\sqrt{2}\xi)\right]$ or

$$\psi = \left(\frac{|\alpha|}{\beta}\right)^{1/2} \tanh\left(\frac{x}{\sqrt{2}\xi}\right). \tag{33.20}$$

From (33.20) it follows that ξ is a measure of the distance over which the order parameter responds to a perturbation. Since $\alpha = a(T - T_c)$, we have

$$\xi(T) = \left(\frac{\hbar^2}{2m^* a T_c}\right)^{1/2} \left(1 - \frac{T}{T_c}\right)^{-1/2}. \tag{33.21}$$

We see that the G–L coherence length diverges as $(1 - T/T_c)^{-1/2}$; this divergence is a general property of the coherence length at all second-order phase transitions (although the exponent differs in general from this "mean field" value of 1/2 close to T_c).

33.2.2 The London equation and the penetration depth

We now examine Eq. (33.11) for $\mathbf{j}(\mathbf{r})$. In a limit where the first two terms are negligible, we recognize the remainder as the London equation, as written in the form, (32.15) provided that we identify $n = |\psi|^2$; this supports the identification of ψ

as a (condensate) wave function associated with the superconducting electrons. We can immediately reinterpret the London penetration depth as

$$\lambda_L^2 = \frac{m^* c^2}{4\pi e^{*2} |\psi|^2}; \tag{33.22a}$$

its temperature dependence now follows from that for ψ, given in (33.2b),

$$\lambda_L = \left(\frac{m^* c^2 \beta}{4\pi e^{*2} a T_c}\right)^{1/2} \left(1 - \frac{T}{T_c}\right)^{-1/2}. \tag{33.22b}$$

Comparing (33.21) and (33.22), we see that in the G–L theory λ_L and ξ both diverge as $(1-T/T_c)^{-1/2}$. Their ratio, called the *Ginzburg–Landau parameter*, is therefore a constant, which we write as

$$\kappa \equiv \frac{\lambda_L}{\xi} = \frac{m^* c}{e^* \hbar} \left(\frac{\beta}{2\pi}\right)^{1/2}. \tag{33.23}$$

33.2.3 The normal-metal/superconductor phase boundary; the G–L surface tension

Let us now return to the discussion of the surface tension of a normal–superconductor phase boundary, begun in Section 32.5. From Eq. (32.41) the total Gibbs free energy, which is a constant across the phase boundary, is given by

$$G = \mathcal{A} \int_{-\infty}^{+\infty} dx \left[F(x) - \frac{1}{4\pi} \mathbf{H}_c \cdot \mathbf{B}(x)\right]; \tag{33.24}$$

Here, $\mathcal{A}$ is the interface area. Unlike our discussion with the London model, we now have the G–L expression for the free energy density and due to the gradient term the superconducting properties do not turn on abruptly at the interface (which we still locate, nominally, at $x = 0$). Furthermore, the integral should also extend into the region $x < 0$ to allow for the possibility that superconductivity may penetrate there.[6] Inserting Eq. (33.7) into (33.24), employing a gauge, where $\mathbf{A} = A(x)\hat{\mathbf{z}}$, and using the definition of γ given by (32.43), we obtain the one-dimensional equation,

$$\gamma = \int_{-\infty}^{+\infty} dx \left[\frac{\alpha^2}{2\beta} + \frac{\hbar^2}{2m^*} \left|\frac{d\psi(x)}{dx}\right|^2 + \frac{e^{*2}}{2m^* c^2} A^2(x) |\psi(x)|^2 \right.$$

$$\left. + \alpha |\psi(x)|^2 + \frac{1}{2}\beta |\psi(x)|^4 + \frac{B^2(x)}{8\pi} - \frac{H_c B(x)}{4\pi}\right]. \tag{33.25}$$

The vanishing of the cross term, $(ie^*/\hbar c)\mathbf{A} \cdot \nabla$, in both (33.25) and the first G–L equation (33.9), allows us to choose ψ real; it then follows from (33.11) that

$$j_z(x) = -\frac{e^{*2}}{m^* c} \psi^2(x) A(x) \tag{33.26a}$$

[6] This happens in practice and is called the proximity effect.

and

$$j_x = j_y = 0. \tag{33.26b}$$

To compute γ we must simultaneously solve the first and second G–L equations, (33.9) and (33.10), for $\psi(x)$ and $B(x)$ $(= -dA(x)/dx)$, subject to boundary conditions which will be specified shortly, and insert the results in (33.25). To eliminate various parameters in the subsequent calculations we rewrite all equations in terms of the scaled variables,

$$\bar{x} = x/\lambda_L,\ \bar{\psi} = \left(\frac{\beta}{|\alpha|}\right)^{1/2} \psi,\ \bar{A} = \frac{A}{H_c\lambda_L},\ \text{and}\ \bar{B} = \frac{B}{H_c}. \tag{33.27}$$

In terms of these variables the first and second G–L equations become (where we now drop the bars over the variables)

$$\psi'' = \kappa^2\left[\left(\frac{1}{2}A^2 - 1\right)\psi + \psi^3\right] \tag{33.28}$$

and

$$A'' = A\psi^2. \tag{33.29}$$

Using $\alpha^2/2\beta = H_c^2/8\pi$ (the bulk condensation energy) and the scaled variables of Eq. (33.27), we may rewrite Eq. (33.25) for the surface energy as

$$\gamma = \frac{\lambda_L H_c^2}{8\pi}\int_{-\infty}^{+\infty}\left[\frac{2}{\kappa^2}\psi'^2 + \left(A^2 - 2\right)\psi^2 + \psi^4 + (A' - 1)^2\right]dx. \tag{33.30}$$

Appropriate boundary conditions are

$$\psi = 0, \quad B = A' = 1 \quad \text{at} \quad x = -\infty \tag{33.31a}$$

and

$$\psi = 1, \quad A' = 0 \quad \text{at} \quad x = +\infty \tag{33.31b}$$

(where $x > 0$ is nominally the superconducting side).[7]

The solutions of (33.28) and (33.29) for arbitrary κ must be obtained numerically. The resulting behaviors of $B(x)$ and $\psi(x)$ are shown qualitatively in Fig. 33.1. The behavior for $\kappa \ll 1$ is shown to the left; here the field varies more rapidly than the order parameter. The behavior for $\kappa \gg 1$ is shown to the right and here we have the opposite behavior.

[7] From the structure of Eqs. (33.28) and (33.29), it follows from the boundary conditions (33.30a,b) that $\psi' = 0$ at $x = \pm\infty$. Our restriction to real ψ requires that the constant, A (which is equivalent to a phase ϕ), vanishes at $x = +\infty$.

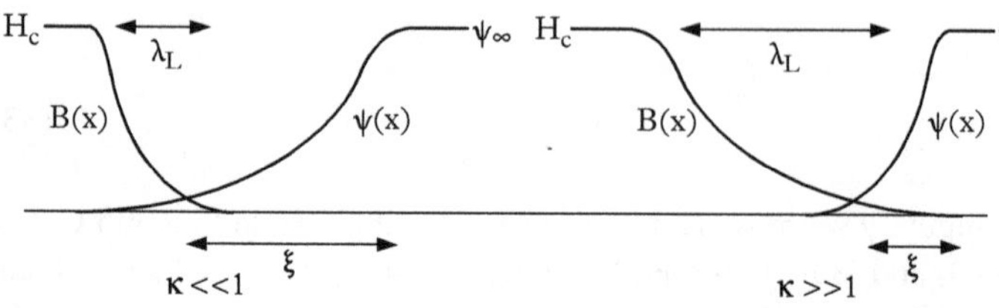

Figure 33.1 *Schematic diagram of the variation of B and ψ in a domain wall. The case κ ≪ 1 refers to a type I superconductor with a positive surface energy; the case κ ≫ 1 refers to a type II superconductor with negative surface energy.*

Equations (33.28) and (33.29) can be manipulated to rewrite the integrand entering Eq. (33.30). First we multiply (33.28) by ψ', which immediately leads to

$$\frac{1}{\kappa^2}\frac{d}{dx}(\psi'^2) = \frac{d}{dx}\left[\left(\frac{1}{2}A^2 - 1\right)\psi^2 + \frac{1}{2}\psi^4\right] - \psi^2\frac{d}{dx}\left(\frac{A^2}{2}\right).$$

Similarly, we multiply (33.29) by A', which yields

$$\frac{d}{dx}\left(\frac{A'^2}{2}\right) = \psi^2\frac{d}{dx}\left(\frac{A^2}{2}\right).$$

Combining these two equations we obtain a first integral as

$$\frac{2}{\kappa^2}\psi'^2 + (2 - A^2)\psi^2 - \psi^4 + A'^2 = \text{const.} = 1; \qquad (33.32)$$

the value of the constant was fixed by the boundary conditions at either $+\infty$ or $-\infty$. Combining this expression with Eq. (33.32) gives

$$\gamma = \frac{\lambda_L H_c^2}{4\pi}\int_{-\infty}^{+\infty}\left[\frac{2}{\kappa^2}\psi'^2 + A'(A' - 1)\right]dx; \qquad (33.33)$$

we can compute γ, once ψ' and $A' = -B(x)$ are available.

Returning to Eq. (33.32) we see that in the limit $A' = A = 0$ we recover the dimensionless form of Eq. (33.19);

$$\psi' = \frac{\kappa}{\sqrt{2}}\left(1 - \psi^2\right). \qquad (33.34)$$

The solution of this equation is

$$\psi'(x) = \tanh\frac{\kappa(x - x_0)}{\sqrt{2}} \qquad (33.35)$$

(which is the analogue of Eq. (33.20)); here x_0 is the nominal position of the boundary. The characteristic length scale of the order parameter variation associated with this normal/superconductor phase boundary is ξ (in unscaled units), as discussed earlier in connection with Eq. (33.20). This form does not satisfy the boundary conditions (33.30a,b) (as expected since a stable phase boundary in an unbounded superconductor can exist only in the presence of a field), but instead satisfies the boundary conditions $\psi = +1$ or -1 as $x \to +\infty$. However, assuming the presence of an order parameter quenching mechanism at $x_0 = 0$ (e.g., a thin ferromagnetic plane embedded in an otherwise homogeneous superconductor), we may evaluate the associated surface tension for this case by substituting (33.35) into (33.33); carrying out the integral (with $A = 0$), we obtain

$$\gamma = \frac{H_c^2 \lambda_L}{3\sqrt{2}\pi\kappa} = \frac{1.9\lambda_L}{\kappa}\frac{H_c^2}{8\pi}. \tag{33.36}$$

From the definition of κ given in (33.23) we see that $\gamma \sim \xi H_c^2/8\pi$, as anticipated earlier in Eq. (32.44).

When A is non-zero the characteristic length scale of a field variation may be estimated from Eq. (33.29). In a type I material at a vacuum/superconductor interface, ψ may be regarded as constant over distances where the field varies. Since the length scale of Eq. (33.30) is unity this corresponds to the field varying over a distance λ_L (the London depth) in unscaled units. However, near a normal-metal/superconductor phase boundary the field variation occurs in a region where ψ is small. If we seek an approximate solution to (33.29) as an exponentially decaying form, $A \sim e^{-x/\delta}$, where δ is a characteristic length, then $1/\delta^2 \cong |\psi|^2$. From an expansion of Eq. (33.35), $A \sim \exp(-x/\delta)$ about $x = x_0$, $\psi \sim \kappa(x - x_0)$. The average value of ψ^2 in a region of width δ would then be of order $(\kappa\delta)^2$. Combining these forms we see that the field decay would be governed by $(\kappa\delta)^2 \cong 1/\delta^2$ or $\delta \sim \kappa^{-1/2}$. This would result in a (negative) surface tension contribution of order $-\kappa^{1/2}\gamma$ which, although less than the expression (33.36) in a type I material, decreases slowly with κ and hence limits the accuracy of this expression. With increasing κ the surface tension continues to decrease and further analysis (see Lifschitz and Pitaevskii (1980), Section 46) shows that it passes through zero for $\kappa = 1/\sqrt{2}$.

An important application of the theory concerns the boundary between two dissimilar materials, at least one of which is superconducting. A boundary condition which is appropriate for the case when no current flows parallel to the surface normal, $\hat{\mathbf{n}}$, is[8]

$$\left[-i\hbar\nabla - \frac{e^*}{c}\mathbf{A}\right]\cdot\hat{\mathbf{n}}\psi = -\frac{1}{b}\psi, \tag{33.37}$$

where b is a parameter having the units of length. The superconductor/insulator case corresponds to the limit $b \to \infty$. Equation (33.37) can be seen as a generalization of the boundary condition that would follow from blindly requiring the surface term in Eq. (33.8) to vanish.

[8] We emphasize that Eq. (33.37) is a *macroscopic* boundary condition. The behavior of the microscopic order parameter may differ substantially near the interface.

33.2.4 Superconductors in contact when a current is present

Lastly we consider contact between two superconductors in the presence of a current normal to the boundary. In the most general case the boundary, which we will locate at $x = 0$, may have properties different from both superconductors, an example being a thin layer of a third (normal) metal, which is referred to as an SNS (superconductor–normal–superconductor) junction. The most general (Cauchy) boundary conditions connecting the two sides of the junction, denoted 1 ($x < 0$) and 2 ($x > 0$), have the form[9]

$$\psi_1(0) = M_{11}\psi_2(0) + M_{12}\left(\frac{d}{dx} - \frac{ie^*}{\hbar c}A_x\right)\psi_2(0) \tag{33.38a}$$

and

$$\left(\frac{d}{dx} - \frac{ie^*}{\hbar c}A_x\right)\psi_1(0) = M_{21}\psi_2(0) + M_{22}\left(\frac{d}{dx} - \frac{ie^*}{\hbar c}A_x\right)\psi_2(0). \tag{33.38b}$$

If we choose a single gauge for regions 1 and 2 the vector potential will also be continuous. The coefficients, M_{ij}, are not independent but are constrained by the requirement of current conservation through the boundary

$$j_1(0) = j_2(0); \tag{33.39}$$

as a result we may choose the M_{ij} to be real and they must satisfy the condition

$$M_{11}M_{22} - M_{12}M_{21} = 1. \tag{33.40}$$

The resulting expression for the current through the boundary, j_n, is

$$\begin{aligned} j_n &= -\frac{ie^*\hbar}{2m}\left[\psi_2^*(0)\left(\frac{1}{M_{12}}\psi_1(0) - \frac{M_{11}}{M_{12}}\psi_2(0)\right) - \text{c.c.}\right] \\ &= -\frac{ie^*\hbar}{2m^*M_{12}}\left[\psi_1^*(0)\psi_2(0) - \psi_1(0)\psi_2^*(0)\right] \\ &= \frac{e^*\hbar}{m^*M_{12}}|\psi_1(0)||\psi_2(0)|\sin\left(\phi_2(0) - \phi_1(0)\right) \end{aligned} \tag{33.41}$$

where ϕ_1 and ϕ_2 denote the order parameter phases. The maximum value of this current is

$$j_{max} = \frac{e^*\hbar}{m^*M_{12}}|\psi_1(0)||\psi_2(0)|. \tag{33.42}$$

If j_{max} is much less than the bulk critical currents in the two superconductors then we may approximate the $\psi_i(x)$ by the $\mathbf{A} = 0$ forms analogous to (33.20),

$$\psi_1(x) = \psi_1(-\infty)\tanh\left(\frac{x + x_1}{\sqrt{2}\xi_1}\right) \qquad x \leq 0 \tag{33.43a}$$

9 Boundary effects in superconductors, P.G. de Gennes, *Rev. Mod. Phys.* **36**, 225 (1964).

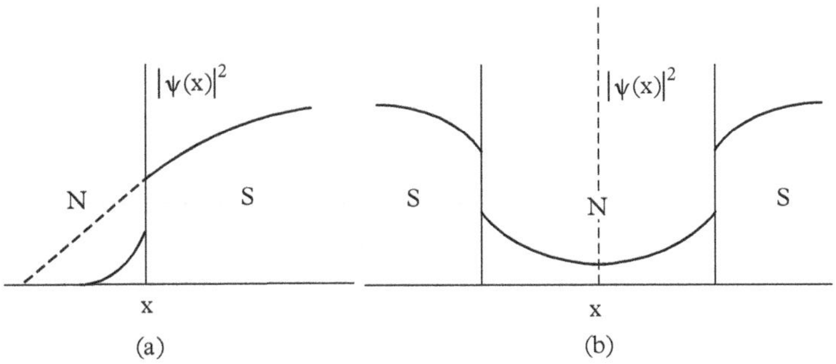

Figure 33.2 *Schematic of the order parameter behavior: (a) near an NS boundary; (b) in an SNS junction.*

and

$$\psi_2(x) = \psi_2(+\infty)\tanh\left(\frac{-x + x_2}{\sqrt{2}\xi_2}\right); \qquad x \geq 0. \qquad (33.43b)$$

Here, x_1 and x_2 are free parameters and $\xi_{1,2}$ are the G–L coherence lengths in the two media. From (33.43) we have immediately

$$\psi_1(0) = \psi_1(-\infty)\tanh\left(\frac{x_1}{\sqrt{2}\xi_1}\right);\ \psi_2(0) = \psi_2(+\infty)\tanh\left(\frac{x_2}{\sqrt{2}\xi_2}\right) \qquad (33.44a)$$

and

$$\psi_1'(0) = \frac{\psi_1(-\infty)}{\sqrt{2}\xi_1}\left[1 - \tanh^2\left(\frac{x_1}{\sqrt{2}\xi_1}\right)\right];\ \psi_2'(0) = -\frac{\psi_2(+\infty)}{\sqrt{2}\xi_2}\left[1 - \tanh^2\left(\frac{x_2}{\sqrt{2}\xi_2}\right)\right]. \qquad (33.44b)$$

By substituting these expressions into the $\mathbf{A} = 0$ limit of Eqs. (33.38a,b) we may numerically solve for the parameters x_1 and x_2, assuming that the coupling coefficients M_{ij} are known. The behavior of $|\psi_i(x)|$ is shown schematically for an SNS junction in Fig. 33.2. Note that with these more general boundary conditions $\psi(x)$ can be discontinuous, reflecting the effects of an interface layer.

We will encounter Eq. (33.41) in Section 33.4, when we discuss the Josephson effects at an SIS junction (where I designates a thin insulating barrier). If the coupling is weak, as in the case of an SIS junction, we set $M_{11} = M_{22} \cong 0$ and $M_{21} = -1/M_{12}$. In this case the values of $|\psi_1(0)|$ and $|\psi_2(0)|$ are only slightly shifted from their equilibrium bulk values, $|\psi_1(-\infty)|$ and $|\psi_2(+\infty)|$.

33.3 The upper critical field and the phase diagram of a type II superconductor

As the magnetic field is raised above H_{c1}, and the density of flux lines n_L increases, a point is eventually reached where the distance between flux lines becomes of the order of the vortex core diameter; i.e., $n_L^{-1/2} \sim \xi$. One would then expect a transition to the normal state, and the field at which this occurs, called the upper critical field, is designated H_{c2}. In the region just below H_{c2} the superconducting order parameter must be small and this is the regime where the Ginzburg–Landau approach should provide a good description.

We may calculate H_{c2} by linearizing the first G–L equation, since the order parameter is vanishingly small just at H_{c2}. Equation (33.9) then becomes

$$-\frac{\hbar^2}{2m^*}\left(\nabla - \frac{ie^*}{\hbar c}\mathbf{A}\right)^2 \psi + \alpha\psi = 0; \tag{33.45}$$

this equation is the same as the Schrödinger equation for a particle with energy, $-\alpha$, mass, m^*, and charge e^*, in a magnetic field, $\mathbf{H}$, with an associated vector potential, $\mathbf{A}$. We will assume a uniform field, $\mathbf{H}_0 \| \hat{\mathbf{z}}$. The solution of (33.45) is easiest in the so-called Landau gauge where we write the vector potential as

$$\mathbf{A} = H_0 \hat{\mathbf{y}}\, x; \tag{33.46}$$

the resulting G–L equation is then[10]

$$-\frac{\hbar^2}{2m^*}\left[\frac{\partial^2}{\partial x^2} + \left(\frac{\partial}{\partial y} - \frac{ie^* H_0}{\hbar c}x\right)^2 + \frac{\partial^2}{\partial z^2}\right]\psi = -\alpha\psi. \tag{33.47}$$

We may separate the variables in (33.47) by writing $\psi_{n,k_z,k_y}(x,y,z) = e^{ik_z z + ik_y y}\, u_n(x)$ (we do not normalize ψ at this point as this property follows only from solving the nonlinear G–L equations). Inserting this wave function in (33.47) we obtain

$$\left[-\frac{\hbar^2}{2m^*}\frac{d^2}{dx^2} + \frac{1}{2}m^*\omega_c^2 (x - x_0)^2\right] u_n(x) = \varepsilon_n u_n(x), \tag{33.48}$$

where

$$x_0 \equiv \frac{\hbar c}{e^* H_0} k_y \left(= \frac{\Phi_0}{2\pi H_0} k_y\right), \qquad \varepsilon_n = -\alpha + \frac{\hbar^2 k_z^2}{2m^*}.$$

Equation (33.48) is the Schrödinger equation of a harmonic oscillator with frequency $\omega_c = (|e^*| H_0)/(m^* c)$ (the cyclotron frequency of the particle), with energies

$$\varepsilon_n = \left(n + \frac{1}{2}\right)\hbar\omega_c, \tag{33.49}$$

having its origin at the point x_0. The full (unnormalized) wave functions are

$$\psi_{n,k_z,k_y}(x,y,z) = e^{i(k_y y + k_z z)} e^{-(x-x_0)^2/2a_H^2} H_n\left(\frac{x - x_0}{a_H}\right), \tag{33.50}$$

where $a_H^2 \equiv \hbar/m^*\omega_c = \hbar c/e^* H_0 = \Phi_0/2\pi H_0$ and H_n are the Hermit polynomials.

[10] Apart from including the z-dependence our discussion here is essentially identical to that in Section 48.1.3.

Only the smallest eigenvalue, $n = 0$, $k_z = 0$ (corresponding to the highest allowed transition temperature) is meaningful, since our linearized theory is valid only as a description of the *onset* of superconductivity. Hence,

$$-\alpha = \frac{1}{2}\hbar\omega_c \tag{33.51a}$$

or

$$\begin{aligned} H_{c2} &= -\frac{2m^* c\alpha}{\hbar e^*} \\ &= \frac{2m^* c}{\hbar e^*} a(T_c - T) \end{aligned}, \tag{33.51b}$$

or, as it is more commonly written as

$$H_{c2} = \frac{\Phi_0}{2\pi\xi^2}. \tag{33.51c}$$

where ξ is the coherence length given by Eq. (33.16). From (33.51b) we see that H_{c2} goes to zero at T_c and increases linearly below that temperature. Recalling, $n_L = B/\Phi_0$, we see that (33.51c) is in accord with the order-of-magnitude estimate of H_{c2} made at the beginning of this section. However, we are then struck by the fact that in our particle analogy, we have considered only a single quantum state ($n = 0$, the "ground state"), and yet the material here contains the largest possible number of flux lines. The resolution to this apparent contradiction is that our $n = 0$ quantum state is highly degenerate (as evidenced by the fact that the quantum number, k_y, does not affect the energy). This high degeneracy is utilized in the solution to the nonlinear problem of constructing the high density of flux lines associated with the Abrikosov flux lattice, which we will not discuss here.

Having identified two of the characteristic fields, H_{c_1} and H_{c_2}, we are in a position to discuss the overall thermodynamic behavior of a type II superconductor in the H–T plane. If we restrict ourselves to the case of a long cylinder (with the field parallel to the axis to avoid the complications of the intermediate state), then for $0 < H < H_{c_1}$ (see Eq. (32.66b)) we have complete flux exclusion. For $H_{c_1} < H < H_{c_2}$ we are in the mixed state (also called the vortex or Shubnikov phase) where the superconductor is permeated with an array of vortex lines; as noted earlier, in most materials the lines form a triangular (centered hexagonal) lattice, the spacing of which decreases with increasing magnetic field in order to accommodate the associated flux. Finally, for $H > H_{c_2}$ superconductivity disappears. This behavior is shown schematically in Fig. 33.3. Additional phases that occur in high temperature superconductors will be discussed in Section 47.8.

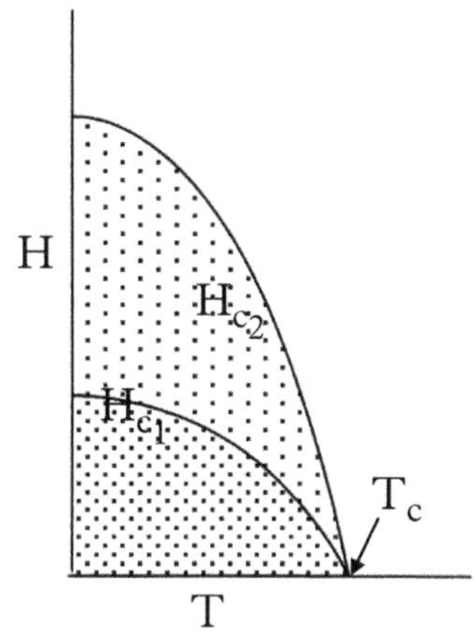

Figure 33.3 *The phase diagram of a type II superconductor.*

Interestingly, superconductivity persists to a higher field,

$$H_{c_3} \cong 1.7 H_{c_2} \tag{33.52}$$

at the surface, a phenomenon we will not discuss.

33.4 The Josephson effects

Suppose we have a "junction" which weakly couples two superconductors. Such junctions can be formed from a thin insulating layer that separates two superconductors, through which electrons can tunnel.[11] A common example is the Nb/AlO$_x$/Nb junction. These can be prepared by first depositing a thick layer of Nb followed by a thin layer of Al; a small amount of oxygen is then admitted into the deposition chamber thereby oxidizing the Al layer, after which a final layer of Nb is deposited.[12] Such junctions can have remarkable properties, as first predicted by B. Josephson.[13]

33.4.1 The Josephson equations

The G–L boundary conditions between two superconductors were given earlier as Eqs. (33.38a,b). At the end of that section it was pointed out that at a superconductor/insulator/superconductor junction, where the magnitudes of the order parameter on each side are minimally perturbed, the parameters in that expression satisfy the conditions, $M_{11} = M_{22} \cong 0$ and $M_{21} = -1/M_{12}$. Equations (33.38a,b) then take the form[14]

$$\frac{\partial \psi_1}{\partial x} - \frac{ie^*}{\hbar c} A_x \psi_1 = \frac{\psi_2}{\lambda} \tag{33.53a}$$

and

$$\frac{\partial \psi_2}{\partial x} - \frac{ie^*}{\hbar c} A_x \psi_2 = -\frac{\psi_1}{\lambda}, \tag{33.53b}$$

where we write $\lambda = M_{21}^{-1} = -M_{12}$. We insert (33.53a) into Eq. (33.11), yielding

$$\begin{aligned} j_x &= -\frac{ie^*\hbar}{2m^*}\left[\psi_1^*\frac{\partial \psi_1}{\partial x} - \psi_1 \frac{\partial \psi_1^*}{\partial x}\right] - \frac{e^{*2}}{m^* c}|\psi_1|^2 A_x \\ &= -\frac{ie^*\hbar}{2m^*}\left[\psi_1^*\left(\frac{\psi_2}{\lambda} + \frac{ie^*}{\hbar c}A_x\psi_1\right) - \psi_1\left(\frac{\psi_2^*}{\lambda^*} - \frac{ie^*}{\hbar c}A_x\psi_1^*\right)\right] - \frac{e^{*2}}{m^* c}|\psi_1|^2 A_x. \end{aligned} \tag{33.54}$$

In the absence of magnetic atoms, the superconducting properties are invariant under time reversal, which results in $\psi \to \psi^*$, $\mathbf{j} \to -\mathbf{j}$, and $\mathbf{A} \to -\mathbf{A}$. Under these operations both sides of Eqs. (33.53a,b) turn into their complex conjugates and hence λ must be real. We then obtain from (33.54)

$$j_x = \frac{-ie^*\hbar}{2m^*\lambda}\left(\psi_1^*\psi_2 - \psi_1\psi_2^*\right). \tag{33.55}$$

[11] The 1973 Nobel Prize in Physics was divided, one half, jointly, to Leo Esaki and Ivar Giaever "for their experimental discoveries regarding tunneling phenomena in semiconductors and superconductors, respectively" and the other half to Brian David Josephson "for his theoretical predictions of the properties of a supercurrent through a tunnel barrier, in particular those phenomena which are generally known as the Josephson effects."

[12] Preparation of Josephson junctions with thin Al layers, M. Gurvitch, M.A. Washington, H.A. Huggins, and J.M. Rowell, *IEEE Trans. Magn.* **MAG-19**, 791 (1983).

[13] Possible new effects in superconductive tunneling, B.D. Josephson, *Phys. Lett.* 1, 251 (1962).

[14] The form of Eqs. (33.53a,b) can be directly argued from the zero current condition, Eq. (33.12), by adding a phenomenological coupling to the adjacent superconducting layer through a parameter, λ.

Writing $\psi_i = |\psi_i| e^{i\phi_i}$, and assuming that both sides are prepared from the same kind of superconducting material, $|\psi_1| = |\psi_2|$, we have

$$j = j_m \sin \phi_{21}, \tag{33.56}$$

where

$$j_m = \frac{e^* \hbar}{m^* \lambda} |\psi|^2 \tag{33.57a}$$

and

$$\phi_{21} = \phi_2 - \phi_1. \tag{33.57b}$$

Note that j_m is the maximum current density that may be carried by the junction.

In deriving Eq. (33.56) we have assumed that no electric field and magnetic flux density are present in the junction. When no electric field is present in the junction, the phase is time independent. We generalize to the case when a field is present by using a gauge invariance argument. Under a gauge transformation (recall $\mathbf{H} = \nabla \times \mathbf{A}$ and $\mathbf{E} = -\nabla V - \partial \mathbf{A}/c\partial t$),

$$\mathbf{A} \rightarrow \mathbf{A} + \nabla \chi \tag{33.58}$$

and

$$V \rightarrow V - \frac{1}{c}\frac{\partial \chi}{\partial t}, \tag{33.59}$$

where $\chi(\mathbf{r}, t)$ is an arbitrary single-valued function and V is the potential. Since the G–L equation contains the form $\nabla - (ie^*/\hbar c)\mathbf{A}$, (see Eq. (33.5), we must change the phase of the G–L wave function by

$$\phi \rightarrow \phi + \frac{e^*}{\hbar c}\chi(t). \tag{33.60}$$

Comparing Eqs. (33.59) and (33.60) we see that the form

$$\frac{\partial \phi}{\partial t} - \frac{e^*}{\hbar} V = 0 \tag{33.61}$$

is *gauge invariant*. If V is (initially) assumed to be independent of time and denoted as V_{21}, then integration of (33.61) yields

$$\phi_{21} = \phi_{21}^{(0)} - \frac{e^*}{\hbar} V_{21} t \tag{33.62}$$

or

$$j = j_m \sin\left[\phi_{21}^{(0)} - \frac{e^*}{\hbar} V_{21} t\right]. \tag{33.63}$$

Introducing the frequency, $\omega_J = \partial\phi_{21}/\partial t$, we see that (33.62) leads to

$$\omega_J = \frac{|e^*|}{\hbar}V_{21} = \frac{2|e|}{\hbar}V_{21}. \tag{33.64}$$

The oscillating current, $j(t)$, caused by the static voltage, V_{21}, given by (33.63), will be associated with an oscillating voltage which will be superimposed on the static voltage, V_{21}. As a result, electromagnetic radiation is emitted from a Josephson junction in the presence of a potential. Equation (33.64) now forms the basis for defining the standard volt in terms of a measured frequency and the fundamental constants, e and h.[15]

We next examine the Josephson effects in the presence of a magnetic field. We will restrict ourselves to the case of a relatively weak field where a quasiclassical description is adequate; i.e., the dominant effect of a field, which is described by a vector potential **A**, is to make the phase position dependent. From the discussion surrounding Eqs. (33.13a,b) we know that for a "pure" gauge field (one not involving a field $\mathbf{H}(\mathbf{r})$) the *only* effect of the vector potential is to produce a position-dependent phase; this suggests that in the presence of a vector potential associated with a weak field, $\mathbf{H}(\mathbf{r})$, the effect may be approximately incorporated in the phase of the wave function. Comparing Eqs. (33.58) and (33.60) we have the gauge-invariant form analogous to (33.61) as

$$\nabla\phi - \frac{e^*}{\hbar c}\mathbf{A} = 0. \tag{33.65}$$

Therefore, the gauge-invariant *phase difference* is given by

$$\phi_{21} = \phi_{21}^{(0)} + \frac{2\pi}{\Phi_0}\int_1^2 \mathbf{A}\cdot dl, \tag{33.66}$$

where $\Phi_0 \equiv (hc/2e)$ is the flux quantum.

The fundamental equations governing the behavior of Josephson junctions are the current phase relation (33.56), the voltage phase relation (33.62), and the gauge-invariant phase relation (33.66). They are believed to be exact. In the subsequent sections these equations are applied to some simple junction structures and circuits.

33.4.2 Magnetic field effects: the two-junction SQUID

Consider the superconducting circuit shown in Fig. 33.4 involving two Josephson junctions connected by superconducting leads. The loop formed by this circuit is assumed to contain a magnetic flux, Φ, arising from some field distribution,

[15] Determination of e/h, using macroscopic quantum phase coherence in superconductors – implications for quantum electrodynamics and fundamental physical constants, B.N. Taylor, W.H. Parker, and D.N. Langenberg, *Rev. Mod. Phys.* **41**, 375 (1969).

$\mathbf{H}(\mathbf{r})$, with an associated vector potential $\mathbf{A}(\mathbf{r})$. On entering the loop on the left in Fig. 33.4 (point 1), the current splits into two components i_1 and i_2, with associated current densities, j_u and j_l, where the subscripts refer to the upper and lower paths respectively. From Eq. (33.56) the total current arriving at point 2 is

$$\begin{aligned} j &= j_u + j_l \\ &= j_{mu} \sin \Delta\phi_u + j_{ml} \sin \Delta\phi_l, \end{aligned} \qquad (33.67)$$

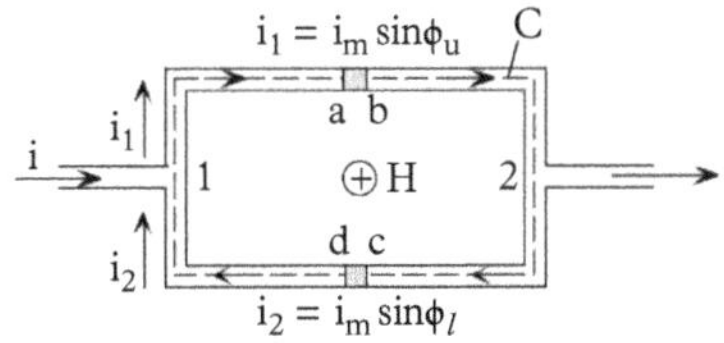

Figure 33.4 *Schematic of a dc SQUID consisting of two Josephson junctions connected in by superconducting links. The path of integration, C, is shown by the dashed line.*

where j_{mu} and j_{ml} correspond to the maximum current densities associated with the upper and lower junctions, and $\Delta\phi_u$ and $\Delta\phi_l$ are the corresponding gauge-invariant phase shifts. Assuming a matched pair of junctions for simplicity ($j_{mu} = j_{ml} = j_m$), we may rewrite (33.67) as

$$j = 2j_m \sin\left(\frac{\Delta\phi_u + \Delta\phi_l}{2}\right) \cos\left(\frac{\Delta\phi_u - \Delta\phi_l}{2}\right). \qquad (33.68)$$

The gauge-invariant phase shifts may be obtained by integrating $\nabla\phi$ around the closed path C, shown in Fig. 33.4. Noting that ϕ is a multivalued function that can change by $2\pi n$ upon completing the path, we have

$$\oint_C \nabla\phi \cdot dl = (\phi_b - \phi_a) + (\phi_c - \phi_b) + (\phi_d - \phi_c) + (\phi_a - \phi_d) = 2\pi n, \qquad (33.69)$$

where n is an integer. The phase differences across the upper and lower Josephson junctions are given by Eq. (33.66) as

$$\phi_b - \phi_a = \Delta\phi_u + \frac{2\pi}{\Phi_0} \int_a^b \mathbf{A} \cdot dl$$

and

$$\phi_d - \phi_c = -\Delta\phi_l + \frac{2\pi}{\Phi_0} \int_c^d \mathbf{A} \cdot dl.$$

The second and fourth terms in Eq. (33.69) are phase differences in the superconducting leads themselves, and are found by using the supercurrent equation (33.52) and the expression for the London penetration depth, Eq. (32.6),

$$\phi_c - \phi_b = \int_b^c \nabla\phi \cdot dl = \frac{2\pi}{\Phi_0} \int_b^c \left(\mathbf{A} + \frac{4\pi\lambda_L^2}{c}\mathbf{j}\right) \cdot dl$$

and

$$\phi_a - \phi_d = \int_d^a \nabla\phi \cdot dl = \frac{2\pi}{\Phi_0} \int_d^a \left(\mathbf{A} + \frac{4\pi\lambda_L^2}{c}\mathbf{j}\right) \cdot dl.$$

Substituting these four equations into Eq. (33.69) gives

$$\Delta\phi_u - \Delta\phi_l = 2\pi n + \frac{2\pi}{\Phi_0}\oint_C \mathbf{A}\cdot d\mathbf{l} + \frac{2\pi}{\Phi_0}\frac{4\pi\lambda_L^2}{c}\int_{C'} \mathbf{j}\cdot d\mathbf{l}. \tag{33.70a}$$

The integration of **A** is around a complete closed path C, and is equal to the total flux, Φ, inside the area enclosed by the contour. The integration of **j** follows the path C′ which excludes the integration over the insulators. If the superconducting leads are thicker than the London penetration depth, the integration path can be taken deep inside the superconductors where the integral involving the supercurrent density is negligible. The phase difference is then simply related to the total flux by

$$\Delta\phi_u - \Delta\phi_l = 2\pi n + \frac{2\pi\Phi}{\Phi_0}. \tag{33.70b}$$

Using this equation to eliminate $\Delta\phi_u$ from Eq. (33.68), the total current is

$$j = 2j_m \sin\left(\Delta\phi_l + \frac{\pi\Phi}{\Phi_0}\right)\cos\left(\frac{\pi\Phi}{\Phi_0}\right). \tag{33.71}$$

When the inductance, L, of the loop is taken into account, the total flux in Eq. (33.70b) consists of the externally applied flux, Φ_{ext}, and the flux generated by the screening circulating current density, j_{cir}; i.e.,

$$\Phi = \Phi_{ext} + Lj_{cir}. \tag{33.72}$$

For the identical junction case,

$$2j_{cir} = j_m(\sin\Delta\phi_u - \sin\Delta\phi_l) \tag{33.73}$$

In general, Eqs. (33.71), (33.72), and (33.73) must be solved self-consistently to describe the behavior of the two-Josephson-junction loop. For simplicity, we assume the loop inductance is negligible and consider only the effect of the externally applied flux on the characteristics of the loop. The maximum supercurrent density which can be carried by the loop is found by maximizing Eq. (33.71) with respect to $\Delta\phi_l$; that is,

$$\Delta\phi_l + \frac{\pi\Phi_{ext}}{\Phi_0} = (n + 1/2)\pi. \tag{33.74}$$

Hence the maximum supercurrent density, j_{max}, is given by

$$j_{max} = 2j_m\left|\cos\left(\frac{\pi\Phi_{ext}}{\Phi_0}\right)\right|, \tag{33.75}$$

which is periodic in the external flux. Since $\Phi_0 \sim 2.07\times10^{-7}$ gauss $\cdot$ cm^2, it is clear that the device pictured in Fig. 33.4 can be used to measure very small changes in magnetic field. It is sometimes referred to as a *two-junction SQUID*, where the latter is an acronym constructed from the words superconducting quantum interference device.

33.4.3 The extended Josephson junction

We next discuss the behavior of a single planar Josephson junction in a magnetic field. We refer to the junction cross section, depicted in Fig. 33.5. The magnetic field is directed into the page along $\hat{y}$. The middle of the junction is taken as the origin and the vector potential in the three regions is

$$A_z = \begin{cases} -Hxe^{-(z-\frac{a}{2})/\lambda_L} & \left(z > \frac{a}{2}\right) \\ -Hx & \left(\frac{a}{2} > z > -\frac{a}{2}\right), \\ -Hxe^{(z+\frac{a}{2})/\lambda_L} & \left(z < -\frac{a}{2}\right) \end{cases}$$

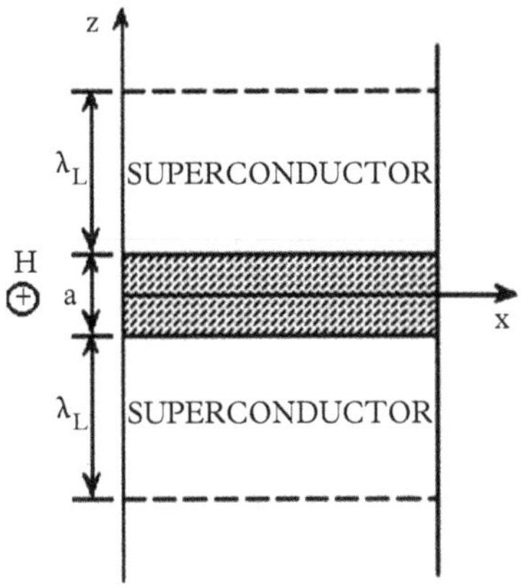

Figure 33.5 *Schematic diagram of an SIS Josephson tunnel junction.*

where the exponential dependencies arise from the Meissner effect in the two superconductors. The phase difference encountered in crossing the junction from $z = -\infty$ to $z = +\infty$ at a given horizontal coordinate x is

$$\Delta\phi(x) = \phi_{21}^{(0)} - \frac{e^*}{\hbar c}\int_{-\infty}^{\infty} A_z(x,z)dz$$
$$= \phi_{21}^{(0)} + \frac{e^*}{\hbar c}H_0(a + 2\lambda_L)x. \quad (33.76)$$

Assuming a rectangular junction of dimensions L_x and L_y, the total current, I, through the junction is

$$I = j_m L_y \int_{-L_x/2}^{L_x/2} \sin\Delta\phi(x)dx$$
$$= j_m L_y \int_{-L_x/2}^{L_x/2} \sin\left[2\pi H_0(a + 2\lambda_L)x/\Phi_0 + \phi_{21}^{(0)}\right]dx$$

or

$$I = I_m \left|\frac{\sin(\pi\Phi/\Phi_0)}{(\pi\Phi/\Phi_0)} \sin\phi_{21}^{(0)}\right|, \quad (33.77)$$

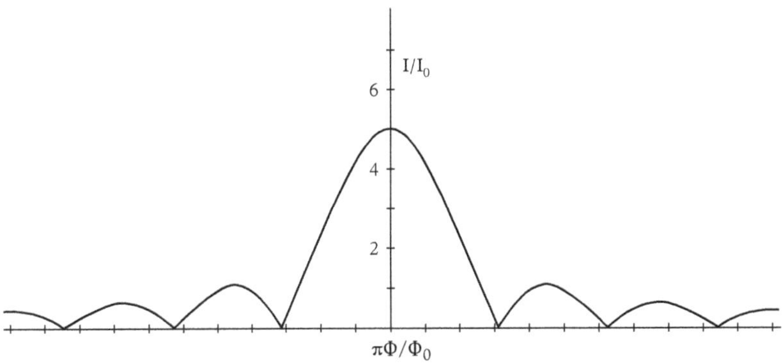

Figure 33.6 *The behavior of the critical current of a Josephson interferometer as a function of the flux.*

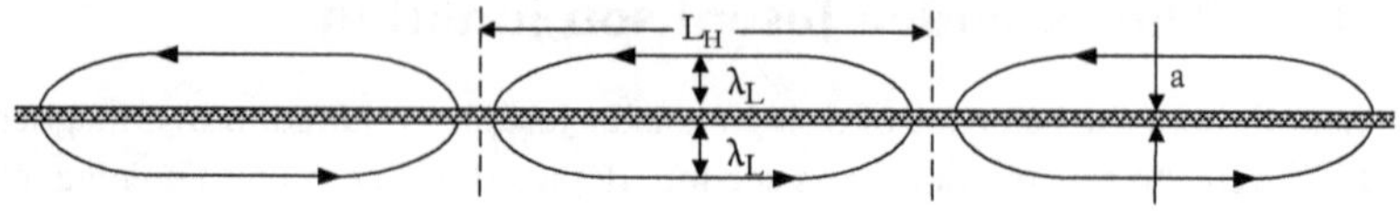

Figure 33.7 *The one-dimensional Josephson flux lattice that is present in a long superconducting junction at high fields. The closed curves show, schematically, one of the constant current density contours that sustain the internal quantized flux.*

where $\Phi = H_0(a + 2\lambda_L) L_x$ and $I_m = j_m L_x L_y$. We note that we obtain a "diffraction-like" (sinx/x) pattern (see Fig. 33.6) involving the variable, Φ/Φ_0, where Φ is the flux contained within the junction.

The junction area entering the definition of ϕ involves an effective thickness, made up of the insulator thickness plus a contribution of one London depth in each superconductor, times the junction width, L_x. For a long junction and large magnetic fields (although well below any critical fields associated with the superconductors themselves) we can interpret ϕ/ϕ_0 as the number of flux quanta residing in the junction, each of which will have a characteristic length such that $\Phi_0 = H_0(a + 2\lambda_L)L_H$, as shown schematically in Fig. 33.7. We will encounter a related structure when we discuss vortices in high temperature superconductors.

ADDITIONAL READING

Superconductivity, J.B. Ketterson and S.N. Song, Cambridge University Press (2000).

Superconductivity of Metals and Alloys, P.G. deGennes, W.A. Benjamin, NY (1966).

Introduction to Superconductivity, M. Tinkham, McGraw-Hill, NY (1975).

Superfluidity and Superconductivity, D.R. Tilley and J. Tilley, Adam Hilger, Bristol (1986).

Superconductivity: Conventional and Unconventional Superconductors, Vol. 1, K.H. Bennemann and J.B. Ketterson, Eds., Springer (2008).

Physics and Applications of the Josephson Effect, A. Barone and G. Paterno, John Wiley & Sons (1982).

Problem

33.1. You are given an infinite superconducting slab of thickness d. Assuming a solution of the form $\psi = u(z)\exp(ik_x x + ik_y y)$, solve the first Ginzburg–Landau equation under the assumption $d \ll \xi$ to show that the upper critical field for H parallel to the slab is given by $H_{c2\parallel} = \sqrt{12}\Phi_0/2\pi\xi d$; the resulting temperature dependence is then $H_{c2\parallel} \propto \sqrt{T_c - T}$, which is referred to as a *two-dimensional behavior*.

The Microscopic Theory of Superconductivity: Cooper Pairing and the Bardeen–Cooper–Schrieffer Theory

34

34.1 Pairing of two electrons above a filled Fermi sea

Historically, one of the important steps leading to the successful formulation of the microscopic theory of superconductivity[1] was the model calculation performed by Cooper (1956), which showed that two electrons, interacting above a filled Fermi sea, had a bound state in the presence of an arbitrarily weak attractive interaction.[2] It was assumed that the only effect of the remaining electrons was to exclude Fourier components of the pair wave function with wave vectors $k \leq k_F$, as shown in Fig. 34.1. We now discuss Cooper's calculation.

We write the wave equation of the two electrons selected, having coordinates $\mathbf{r}_1$ and $\mathbf{r}_2$, in the form

$$\psi(\mathbf{r}_1, \mathbf{r}_2) = \varphi_q(\rho) e^{i\mathbf{q}\cdot\mathbf{R}} \chi(\sigma_1, \sigma_2), \tag{34.1}$$

where $\varphi_q(\rho)$ is the wave function in the center of mass, $\mathbf{R}$ is the center of mass coordinate, $\mathbf{R} \equiv (\mathbf{r}_1 + \mathbf{r}_2)/2$, $\mathbf{q}$ is the wave vector of the center of mass, $\boldsymbol{\rho} \equiv \mathbf{r}_1 - \mathbf{r}_2$, σ_1 and σ_2 denote the spins of the two electrons ($\uparrow$ or $\downarrow$), and χ denotes the spin wave function (any spin–orbit coupling is neglected). In what follows we will assume that pairing takes place in a singlet state for which

$$\chi = \frac{1}{\sqrt{2}} \left[\begin{pmatrix} 1 \\ 0 \end{pmatrix} \begin{pmatrix} 0 \\ 1 \end{pmatrix} - \begin{pmatrix} 0 \\ 1 \end{pmatrix} \begin{pmatrix} 1 \\ 0 \end{pmatrix} \right]. \tag{34.2}$$

We initially restrict ourselves to the case $\mathbf{q} = 0$ (i.e., the center of mass of the pair is at rest with respect to the Fermi sea). We first write φ as a Fourier expansion,[3]

[1] The 1972 Nobel Prize in Physics was awarded jointly to John Bardeen, Leon Neil Cooper, and John Robert Schrieffer "for their jointly developed theory of superconductivity, usually called the BCS-theory."

[2] Bound electron pairs in a degenerate Fermi gas, L.N. Cooper, *Phys. Rev.* **104** 1189 (1956).

[3] Since the singlet spin wave function is antisymmetric, $\varphi_q(\rho)$ must be symmetric in order to satisfy the Pauli principle; i.e., $\varphi_q(\rho) = \varphi_q(-\rho)$, which in turn requires that $g(\mathbf{k}) = g(-\mathbf{k})$. If there is an orbital angular momentum associated with the state we could have multiple amplitudes, $g_i(\mathbf{k})$, where i numbers the amplitudes of the various orbital angular momenta (or in a crystal, the appropriate irreducible representations) associated with the state. Finally, the electrons could pair in a triplet state, in which case $\chi_{\sigma_1\sigma_2} = \chi_{\sigma_2\sigma_1}$; we would then have amplitudes, $g_i(\mathbf{k})$, satisfying $g_i(\mathbf{k}) = -g_i(-\mathbf{k})$. This latter case occurs in liquid ^{3}He.

The Physics of Solids. First Edition. J. B. Ketterson.

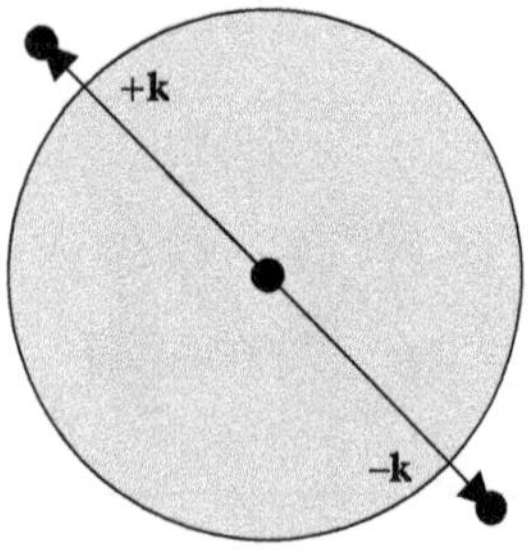

Figure 34.1 *Two electrons with momenta +k and −k interacting above a filled Fermi sea (shaded region). Fourier amplitudes of the pair wave function with* $k < k_F$ *must be excluded.*

$$\begin{aligned}\varphi(\rho) &= \sum_{\mathbf{k}} g(\mathbf{k})e^{i\mathbf{k}\cdot\boldsymbol{\rho}} \\ &= \sum_{\mathbf{k}} g(\mathbf{k})e^{i\mathbf{k}\cdot\mathbf{r}_1}e^{-i\mathbf{k}\cdot\mathbf{r}_2};\end{aligned} \tag{34.3}$$

written in this way, we see that the pair wave function can be regarded as being made up of plane wave states with *equal and opposite momentum* (and opposite spins for our assumed singlet pairing). The Schrödinger equation for the pair wave function is

$$\begin{aligned}&\left[-\frac{\hbar^2}{2m}\left(\nabla_1^2 + \nabla_2^2\right) + V(\mathbf{r}_1, \mathbf{r}_2)\right]\psi(\mathbf{r}_1, \mathbf{r}_2) \\ &= \left(\varepsilon + 2\frac{\hbar^2 k_F^2}{2m}\right)\psi(\mathbf{r}_1, \mathbf{r}_2)\end{aligned}, \tag{34.4}$$

where we have defined the energy, ε, relative to the combined energy of two electrons at the Fermi energy. Inserting (34.3) into (34.4) yields

$$\begin{aligned}&\sum_{\mathbf{k}'}\left[\frac{\hbar^2}{2m}2k'^2 + V(\mathbf{r}_1, \mathbf{r}_2)\right]g(\mathbf{k}')e^{i\mathbf{k}'\cdot(\mathbf{r}_1-\mathbf{r}_2)} \\ &= \left(\varepsilon + 2\frac{\hbar^2 k_F^2}{2m}\right)\sum_{\mathbf{k}'} g(\mathbf{k}')e^{i\mathbf{k}'\cdot(\mathbf{r}_1-\mathbf{r}_2)}.\end{aligned} \tag{34.5}$$

We assume translational invariance and write $V(\mathbf{r}_1, \mathbf{r}_2) = V(\rho)$. We then multiply by $e^{-i\mathbf{k}\cdot\boldsymbol{\rho}}$ and integrate over $d^3\rho$,

$$\begin{aligned}&\frac{\hbar^2}{m}\int\sum_{\mathbf{k}'} e^{i(\mathbf{k}'-\mathbf{k})\cdot\boldsymbol{\rho}}g(\mathbf{k}')k'^2 d^3\rho + \int\sum_{\mathbf{k}'} V(\rho)e^{i(\mathbf{k}'-\mathbf{k})\cdot\boldsymbol{\rho}}g(\mathbf{k}')d^3\rho \\ &= \left(\varepsilon + \frac{\hbar^2 k_F^2}{m}\right)\int\sum_{\mathbf{k}'} e^{i(\mathbf{k}'-\mathbf{k})\cdot\boldsymbol{\rho}}g(\mathbf{k}')d^3\rho.\end{aligned} \tag{34.6}$$

Now,

$$\begin{aligned}\int e^{i(\mathbf{k}-\mathbf{k}')\cdot\mathbf{r}}d^3\rho &= \begin{cases}0; & \mathbf{k} \neq \mathbf{k}' \\ L^3; & \mathbf{k} = \mathbf{k}'\end{cases} \\ &= L^3\,\delta_{\mathbf{k}\mathbf{k}'},\end{aligned} \tag{34.7}$$

where L^3 is the volume of the system. Equation (34.6) then becomes

$$\frac{\hbar^2}{m}k^2 g(\mathbf{k}) + \sum_{\mathbf{k}'} g(\mathbf{k}')V_{\mathbf{k}\mathbf{k}'} = (\varepsilon + 2\varepsilon_F)g(\mathbf{k}), \tag{34.8}$$

where we have defined

$$V_{\mathbf{kk}'} = \frac{1}{L^3}\int V(\rho)\, e^{i(\mathbf{k}'-\mathbf{k})\cdot\rho} d^3\rho. \tag{34.9}$$

Equation (34.8) is referred to as the Bethe–Goldstone equation for the two-electron problem. The Pauli principle forbids our inclusion of any plane wave states with $k < k_F$ since, according to our assumption of a filled Fermi sea, these states are already occupied; i.e., $g(\mathbf{k}) = 0$ for $k < k_F$. For our model calculation we adopt the simple potential,

$$V_{\mathbf{kk}'} = \begin{cases} -V, \left|\hbar^2k^2/2m - \varepsilon_F\right| \text{ and } \left|\hbar^2k'^2/2m - \varepsilon_F\right| < \hbar\omega_D \\ 0, \left|\hbar^2k^2/2m - \varepsilon_F\right| \text{ or } \left|\hbar^2k'^2/2m - \varepsilon_F\right| > \hbar\omega_D \end{cases}, \tag{34.10}$$

where $\hbar\omega_D$ is a "cut-off frequency" and we have assumed that the pairing is attractive. In conventional superconductors this attraction arises from a phonon-induced electron–electron interaction (see Section 21.3), which acts for energies such that $\hbar\omega \leq \hbar\omega_D \equiv k_B\theta_D$, where k_B is the Boltzmann constant and θ_D is the Debye temperature.[4] Substituting (34.10) into Eq. (34.8) and carrying out the integration, we obtain

$$\left(-\frac{\hbar^2k^2}{m} + \varepsilon + 2\varepsilon_F\right) g(\mathbf{k}) = -V\sum_{\mathbf{k}'} g(\mathbf{k}'). \tag{34.11}$$

The right side of (34.11), being summed over all $\mathbf{k}'$, must be a constant. Dividing (34.11) by the factor $\left(-\hbar^2k^2/m + \varepsilon + 2\varepsilon_F\right)$ leaves $g(\mathbf{k})$ on the left side of (34.11). If we sum both sides over $\mathbf{k}$, the constant $\sum_{\mathbf{k}} g(\mathbf{k})$ appears on both sides and may be canceled leaving the "self-consistency" condition

$$V\sum_{\mathbf{k}} \frac{1}{\dfrac{\hbar^2k^2}{m} - \varepsilon - 2\varepsilon_F} = 1. \tag{34.12}$$

Writing $\hbar^2k^2/2m - \varepsilon_F = \xi$, converting the sum to an integration and introducing the density of states, $N(\xi)$ yields

$$V\int_0^{\hbar\omega_D} \frac{N(\xi)d\xi}{2\xi - \varepsilon} = 1.$$

$N(\xi)$ may be taken as a constant over the range of integration and taken from under the integral sign; carrying out the integration we obtain

$$\frac{1}{2}N(0)V\ln\left(\frac{\varepsilon - 2\hbar\omega_D}{\varepsilon}\right) = 1; \tag{34.13}$$

assuming $|\varepsilon| \ll \hbar\omega_D$ (transition temperatures are typically much less than θ_D), we obtain

$$\varepsilon = -2\hbar\omega_D e^{-\dfrac{2}{N(0)V}}. \tag{34.14}$$

[4] That superconductivity might arise from the electron–phonon interaction was first pointed out by H. Fröhlich (1950).

We note the following properties of Eq. (34.14):

(i) The sign of ε is negative and hence (34.14) corresponds to a bound state.

(ii) A bound state suggests a gap in the energy spectrum, which is consistent with an exponentially falling heat capacity observed experimentally.

(iii) We obtain a bound state regardless of how small V is; this is in contrast to the two-body problem in the absence of a Fermi sea where $N(\xi) \propto \xi^{1/2}$ and we obtain a bound state only if the attraction is greater than some minimum value.

(iv) The result (34.14) cannot be expanded in a power series in V and hence could not be obtained by a perturbation theory argument.

It is instructive to calculate the mean square radius of the pair wave function,

$$\overline{\rho^2} = \frac{\int |\varphi(\rho)|^2 \rho^2 d^3\rho}{\int |\varphi(\rho)|^2 d^3\rho}. \tag{34.15}$$

Now,

$$\varphi(\rho) = \sum_{\mathbf{k}} g(\mathbf{k}) e^{i\mathbf{k}\cdot\boldsymbol{\rho}}, \quad |\varphi(\rho)|^2 = \sum_{\mathbf{k},\mathbf{k}'} g(\mathbf{k}) g^*(\mathbf{k}') e^{i(\mathbf{k}-\mathbf{k}')\cdot\boldsymbol{\rho}};$$

we then have $\int |\varphi(\rho)|^2 d^3\rho = L^3 \sum_{\mathbf{k}} |g(\mathbf{k})|^2$, which is the denominator of (34.15). The numerator is evaluated by noting that ρ can be generated by operating on the exponential factor with $-i\nabla_{\mathbf{k}}$: $\rho e^{i\mathbf{k}\cdot\boldsymbol{\rho}} = -i\nabla_{\mathbf{k}} e^{i\mathbf{k}\cdot\boldsymbol{\rho}}$; we generate ρ^2 by incorporating two such terms acting with respect to $\mathbf{k}$ and $\mathbf{k}'$. Performing two integrations by parts in the variables $\mathbf{k}$ and $\mathbf{k}'$ yields

$$\begin{aligned}\int |\varphi(\rho)|^2 \rho^2 d^3\rho &= \int \sum_{\mathbf{k},\mathbf{k}'} (-i\nabla_{\mathbf{k}'} g^*(\mathbf{k}'))\,(i\nabla_{\mathbf{k}} g(\mathbf{k}))\, e^{i(\mathbf{k}-\mathbf{k}')\cdot\boldsymbol{\rho}} d^3\rho \\ &= L^3 \sum_{\mathbf{k}} |\nabla_{\mathbf{k}} g(\mathbf{k})|^2,\end{aligned}$$

where we have again used Eq. (34.7). Thus,

$$\overline{\rho^2} = \frac{\sum_{\mathbf{k}} |\nabla_{\mathbf{k}} g(\mathbf{k})|^2}{\sum_{\mathbf{k}} |g(\mathbf{k})|^2}. \tag{34.16}$$

From Eq. (34.11), $g(\mathbf{k}) \propto \text{const}/(2\xi-\varepsilon)$; noting $\partial/\partial k = (\partial\xi/\partial k)\cdot\partial/\partial\xi = \hbar v_F\cdot\partial/\partial\xi$ (where v_F is the Fermi velocity); Eq. (34.16) then takes the form

$$\overline{\rho^2} = \frac{\hbar^2 v_F^2 \int \frac{4d\xi}{(2\xi-\varepsilon)^4}}{\int \frac{d\xi}{(2\xi-\varepsilon)^2}} = \frac{-\frac{2}{3}(\hbar v_F)^2 \frac{1}{(2\xi-\varepsilon)^3}\Big|_0^\infty}{-\frac{1}{2}\frac{1}{2\xi-\varepsilon}\Big|_0^\infty} = \frac{4}{3}\frac{\hbar^2 v_F^2}{\varepsilon^2}. \tag{34.17}$$

Assuming $\varepsilon \sim k_B T_c$, $T_c \sim 10$ K and $v_F \sim 10^8$ cm/sec, we obtain $\left(\overline{\rho^2}\right)^{1/2} \sim 10^{-4}$cm or 10^4Å; this value is of the same order as a coherence length introduced by Pippard, ξ_0, which lends an interpretation to that quantity in terms of the size of the Cooper pair wave function. However, since an electron typically occupies a volume $\sim(2\text{Å})^3$ there would be *of order* 10^{11} *other electrons within a "coherence volume"* and hence it is certainly not reasonable to ignore these electrons in constructing a pair wave function. This defect is corrected by the BCS theory which treats all electrons on an equal footing.

If we were to compute the binding energy for the case when $\mathbf{q} \neq 0$ we would obtain

$$\varepsilon(q) = \varepsilon(0) + \frac{\hbar v_F q}{2}. \tag{34.18}$$

Recalling that $\varepsilon(q)$ is negative, we see that $\varepsilon(q) = 0$ for $q = 2|\varepsilon(0)|/\hbar v_F$; hence if q^{-1} has a value of order the Pippard coherence length, the bound state (and presumably superconductivity) will *disappear*. More importantly, (34.18) shows that the maximum binding comes from pairing two electrons with *equal and opposite* momenta; this strongly suggests that in constructing a many body ground state, the feature of pairing electrons with no net momenta (equal and opposite $\mathbf{k}$) should be retained.

If we convert the wave vector q into a velocity we obtain $v_c = 2\varepsilon(0)/p_F$. This suggests that there is a critical velocity for the destruction of superfluidity. A similar criterion was obtained by Landau (1941) for superfluid ^{4}He (see Section 36.2). However in a superconductor, the magnetic field created by a current density complicates a simple application of this idea to estimate the critical current.

34.2 The Bardeen–Cooper–Schrieffer superconducting ground state

The success of the Cooper model calculation in producing a bound state, with its implication that a proper many-body theory would contain a gap in the excitation spectrum (as dictated by various experiments), strongly suggests that the ground state wave function should be constructed from pairs of electrons[5]; i.e., we might examine an N-particle form like

$$\Psi_N = \psi(1,2)\psi(3,4)\psi(5,6)\ldots\psi(N-1,N), \tag{34.19a}$$

where we define the two-particle pair wave function as

$$\psi(1,2) \equiv \varphi(\mathbf{r}_1,\mathbf{r}_2)\chi(\sigma_1,\sigma_2). \tag{34.19b}$$

[5] Theory of superconductivity, J. Bardeen, L.N. Cooper, and J.R. Schrieffer, *Phys. Rev.* **108**, 1175 (1957).

To ensure that (34.19a) satisfies the Pauli principle we must make it antisymmetric. To do this we use the antisymmetrization operator, $\hat{A} = \Sigma_P(-1)^P\hat{P}$, where $\hat{P}$ is the permutation operator, $\hat{P}\,\psi(1,2) = \psi(2,1)$, and sum over all permutations, the number of which increases rapidly[6]. Including the normalization factor our wave function then takes the form

$$\Psi_N = \frac{1}{\sqrt{N}}\sum_P (-1)^P\,\hat{P}\,\psi(1,2)\psi(3,4)\psi(5,6)\ldots\psi(N-1,N). \tag{34.20}$$

From our examination of the Cooper problem, we take $\varphi(\mathbf{r}_1, \mathbf{r}_2) = \varphi(\mathbf{r}_1 - \mathbf{r}_2)$ and assume the center of mass of each pair is at rest; we also assume translational invariance. We again perform a Fourier expansion

$$\varphi(\mathbf{r}) = \sum_{\mathbf{k}} g_{\mathbf{k}} e^{i\mathbf{k}\cdot\mathbf{r}}. \tag{34.21}$$

On substituting (34.21) into (34.20), we then obtain

$$\Psi_N = \frac{1}{\sqrt{N}}\sum_{\mathbf{k}_1}\cdots\sum_{\mathbf{k}_{N/2}} g_{\mathbf{k}_1}\cdots\cdots g_{\mathbf{k}_{N/2}}\sum_P(-1)^P\hat{P}\left\{e^{i\mathbf{k}_1\cdot(\mathbf{r}_1-\mathbf{r}_2)}\cdots\cdots e^{i\mathbf{k}_{N/2}\cdot(\mathbf{r}_{N-1}-\mathbf{r}_N)}\right\}$$
$$\times\,(1\uparrow)(2\downarrow)\cdots(N-1\uparrow)(N\downarrow)\} \tag{34.22}$$

where we again restrict ourselves to the pairing of spin up and spin down electrons (singlet pairing).[7]

Rather than writing (34.22) in the coordinate representation, as we have, it is usually easier to perform calculations in the occupation number (Wigner–Jordan) representation; i.e., to use the language of second quantization. (This technique is reviewed in Appendix B.) Instead of writing $\hat{A}\,e^{i\mathbf{k}_1\cdot(\mathbf{r}_1-\mathbf{r}_2)}(1\uparrow)(2\downarrow)$, we write $\hat{c}^\dagger_{\mathbf{k}_1\uparrow}\hat{c}^\dagger_{-\mathbf{k}_1\downarrow}|\phi_0\rangle$, where $|\phi_0\rangle$ is the vacuum state and $\hat{c}^\dagger_{\mathbf{k}_1\sigma_1}$ is the Fermi creation operator for an electron i having momentum $\mathbf{k}_i$ and spin projection $\sigma_i(=\uparrow$ or $\downarrow)$. We recall the commutation relations for these operators,

$$\hat{c}^\dagger_{\mathbf{k}\alpha}\hat{c}^\dagger_{l\beta} + \hat{c}^\dagger_{l\beta}\hat{c}^\dagger_{\mathbf{k}\alpha} = 0, \tag{34.23a}$$

$$\hat{c}_{\mathbf{k}\alpha}\hat{c}_{l\beta} + \hat{c}_{l\beta}\hat{c}_{\mathbf{k}\alpha} = 0, \tag{34.23b}$$

and

$$\hat{c}^\dagger_{\mathbf{k}\alpha}\hat{c}_{l\beta} + \hat{c}_{l\beta}\hat{c}^\dagger_{\mathbf{k}\alpha} = \delta_{\mathbf{k}l}\delta_{\alpha\beta}, \tag{34.23c}$$

where $\mathbf{k}$ and l denote wave vectors and α and β, spin projections. In the second quantized notation the many-body pair wave function (34.22) can then be written as

$$\Psi_N = \sum_{\mathbf{k}_1}\cdots\sum_{\mathbf{k}_{N/2}} g_{\mathbf{k}_1}\cdots g_{\mathbf{k}_{N/2}}\hat{c}^\dagger_{\mathbf{k}_1\uparrow}\hat{c}^\dagger_{-\mathbf{k}_1\downarrow}\cdots\hat{c}^\dagger_{\mathbf{k}_{N/2}\uparrow}\hat{c}^\dagger_{-\mathbf{k}_{N/2}\downarrow}|\phi_0\rangle. \tag{34.24}$$

[6] For four electrons, the wave function $A\{\psi(1,2)\psi(3,4)\}$ would already contain twelve terms.

[7] The antisymmetrization operation may be written more explicitly using the Slater determinant notation given in Eq. (C12). The quantum numbers, p_1, p_2, p_3, p_4, ... correspond in the present case to $\mathbf{k}_1\uparrow$, $-\mathbf{k}_1\downarrow$, $\mathbf{k}_2\uparrow$, $-\mathbf{k}_2\downarrow$, ... We again have $g(\mathbf{k}_i) = g(-\mathbf{k}_i)$ for the singlet pairs being considered here.

It is quite difficult to perform calculations with (34.24) and Bardeen, Cooper, and Schrieffer (BCS) proposed an alternative wave function,

$$\Psi_{BCS} = \prod_{\mathbf{k}} (u_{\mathbf{k}} + v_{\mathbf{k}} \hat{c}^{\dagger}_{\mathbf{k}\uparrow} \hat{c}^{\dagger}_{-\mathbf{k}\downarrow}) \,|\, \phi_0 \rangle, \tag{34.25}$$

where the product extends over all plane wave states.[8] It is much easier to perform calculations with the BCS wave function. (We have collected the required calculations involving the BCS wave function in Appendix 34A.) The (in general complex) quantities $u_{\mathbf{k}}$ and $v_{\mathbf{k}}$ are not independent, but are fixed by the normalization condition, $\langle \Psi_{BCS} \,|\, \Psi_{BCS} \rangle = 1$, which yields

$$|u_{\mathbf{k}}|^2 + |v_{\mathbf{k}}|^2 = 1. \tag{34.26}$$

The wave function (34.24) for the special case,

$$g(\mathbf{k}) = 1, \qquad |\mathbf{k}| < k_F,$$

and

$$g(\mathbf{k}) = 0, \qquad |\mathbf{k}| > k_F, \tag{34.27}$$

describes the "Fermi vacuum" or Fermi sphere: all states with $|\mathbf{k}| < k_F$ are filled and those with $|\mathbf{k}| > k_F$ are empty. The corresponding form for Eq. (34.25) is

$$u_{\mathbf{k}} = 0, \quad v_{\mathbf{k}} = 1, \qquad |\mathbf{k}| < k_F;$$

and

$$u_{\mathbf{k}} = 1, \quad v_{\mathbf{k}} = 0 \qquad |\mathbf{k}| > k_F. \tag{34.28}$$

The difference between (34.24) and (34.25) for the general case is that the first defines a state with precisely N/2 pairs (N electrons), while the second is a superposition of pair states containing 2, 4, 6, ..., N, ..., ∞ electrons; i.e., it does not describe a state with a fixed number of particles. However by superposition, we may relate the two wave functions as

$$\Psi_{BCS} = \sum_{N} \lambda_N \Psi_N, \tag{34.29}$$

with the normalization condition, $\sum_N |\lambda_N|^2 = 1$. The average number of particles, $\overline{N}$, associated with the BCS wave function is

$$\overline{N} = \left\langle \Psi_{BCS} \,|\, \hat{N} \,|\, \Psi_{BCS} \right\rangle, \tag{34.30}$$

[8] Writing the factors entering Eq. (34.25) explicitly as

$$\prod_{\mathbf{k}} \left(u_{\mathbf{k}} + v_{\mathbf{k}} \hat{c}^{\dagger}_{\mathbf{k}\uparrow} \hat{c}^{\dagger}_{-\mathbf{k}\downarrow} \right) |\, \phi_0 \rangle$$
$$= \left(u_{\mathbf{k}_1} + v_{\mathbf{k}_1} \hat{c}^{\dagger}_{\mathbf{k}_1\uparrow} \hat{c}^{\dagger}_{-\mathbf{k}_1\downarrow} \right) \left(u_{\mathbf{k}_2} + v_{\mathbf{k}_2} \hat{c}^{\dagger}_{\mathbf{k}_2\uparrow} \hat{c}^{\dagger}_{-\mathbf{k}_2\downarrow} \right) \left(u_{\mathbf{k}_2} + v_{\mathbf{k}_2} \hat{c}^{\dagger}_{\mathbf{k}_2\uparrow} \hat{c}^{\dagger}_{-\mathbf{k}_2\downarrow} \right) \cdots$$

and forming the resulting product, we see that it contains terms with one, two, three, ⋯ pairs; i.e., it does not have a unique number of particles; an initially strange conclusion that we address later.

where we have defined the number operator,

$$\hat{N} \equiv \sum_{k,\sigma} \hat{c}^\dagger_{k\sigma}\hat{c}_{k\sigma}. \tag{34.31}$$

Evaluating (34.30) using the state function (34.25), gives

$$\bar{N} = \sum_{k} 2|v_k|^2. \tag{34.32}$$

The mean square fluctuation (or the variance) of the number of particles is defined as

$$\begin{aligned}\overline{\delta N^2} \equiv \overline{N^2} - \bar{N}^2 &= \left\langle \Psi_{BCS} \left| \hat{N}\hat{N} \right| \Psi_{BCS} \right\rangle - \left\langle \Psi_{BCS} \left| \hat{N} \right| \Psi_{BCS} \right\rangle^2 \\ &= \sum_{k} 4|v_k|^2 |u_k|^2. \end{aligned} \tag{34.33}$$

Now, $\sum_k \to (V/(2\pi)^3)\int d^3k$; therefore (34.33) is proportional to the volume or, equivalently $\bar{N}$. Hence,

$$\frac{\overline{\delta N^2}}{\overline{N}} \sim \overline{N}^{-1/2}.$$

For a typical 1 cm^3 metallic sample, $\overline{N} \sim 10^{22}$ and therefore $\overline{N}^{-1/2}$ is of order 10^{-11}. This says that the coefficients in the expansion (34.27) will be *highly peaked around* $\overline{N}$, as shown schematically in Fig. 34.2. We therefore reach the very important conclusion that calculations performed using (34.24) or (34.25) for macroscopic samples should differ by negligible amounts. Since it is far easier to calculate using (34.25), this form was adopted by BCS.

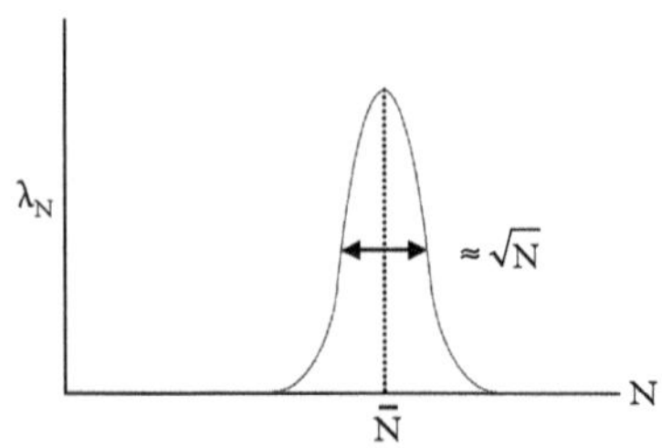

Figure 34.2 *The probability amplitude,* λ_N*, of finding a configuration with* N *particles in the BCS wave function. Note that* λ_N *is strongly peaked around* $\overline{N}$.

The approximate ground state energy is obtained by using (34.25) as a trial wave function and evaluating the expectation value of the Hamiltonian of the system of electrons; this is followed by minimization of the resultant energy with respect to the free parameter, v_k, subject to the condition that the average number of particles is $\overline{N}$. We may write the Hamiltonian of our (translationally invariant) system of electrons as (see Eq. (B.51), generalized to include spin)

$$\hat{H} = \sum_{k,\sigma} \frac{\hbar^2 k^2}{2m} \hat{c}^\dagger_{k\sigma}\hat{c}_{k\sigma} + \sum_{\substack{k,k',q \\ \sigma,\sigma'}} V_{k,k',q;\sigma,\sigma'} \hat{c}^\dagger_{k+q\sigma}\hat{c}^\dagger_{k'-q\sigma'}\hat{c}_{k'\sigma'}\hat{c}_{k\sigma}, \tag{34.34}$$

where the second term accounts for the electron–electron interactions and momentum conservation (and transfer by an amount **q**) has been built into the scattering potential, V. In many-body theory it is conventional to measure all excitation energies from the Fermi surface. Since our wave function does not fix the number of electrons we must introduce a constraint that the *average* number of particles is $\overline{N}$ i.e., $\left\langle \Psi_{BCS} \middle| \hat{N} \middle| \Psi_{BCS} \right\rangle = \overline{N}$ (see Eq. (34.32)). This constraint is introduced through a Lagrange multiplier, μ, which we know from statistical physics is the chemical potential. Incorporating these considerations, we write the Hamiltonian as

$$\hat{H}' = \hat{H} - \mu\hat{N} = \sum_{\mathbf{k},\sigma} \xi_{\mathbf{k}} \hat{c}^{\dagger}_{\mathbf{k}\sigma} \hat{c}_{\mathbf{k}\sigma} + \sum_{\substack{\mathbf{k},\mathbf{k}',\mathbf{q};\\ \sigma,\sigma'}} V_{\mathbf{k},\mathbf{k}',\mathbf{q};\sigma,\sigma'} \hat{c}^{\dagger}_{\mathbf{k}+\mathbf{q}\sigma} \hat{c}^{\dagger}_{\mathbf{k}'-\mathbf{q}\sigma'} \hat{c}_{\mathbf{k}'\sigma'} \hat{c}_{\mathbf{k}\sigma} \tag{34.35a}$$

which we separate as

$$\hat{H}' = \hat{H}_0 + \hat{H}_I \tag{34.35b}$$

where

$$\xi_{\mathbf{k}} = \frac{\hbar^2 k^2}{2m} - \mu \tag{34.36a}$$

This calculation is still too complex to be carried out with the "full" Hamiltonian (34.34) and it is necessary to introduce further approximations (beyond the variational approximation). The first approximation assumes that the Landau–Fermi liquid picture applies to the electrons. In this model, the majority of the effects of normal state electron–electron interactions may be incorporated by assuming that the $\hat{c}^{\dagger}$ operators create "quasiparticles" rather than electrons. The energy, $\varepsilon(\mathbf{k})$, which for free electrons is $\hbar^2\mathbf{k}^2/2m$, is then altered to $\hbar^2\mathbf{k}^2/2m^*$, where m^* is a *many-body corrected* effective mass (not to be confused with the effective mass in semiconductors, which arises from band structure effects); i.e.,

$$\begin{aligned} \xi_{\mathbf{k}} &= \frac{\hbar^2 k^2}{2m^*} - \mu \\ &= \frac{\hbar^2}{2m^*}\left(k^2 - k_F^2\right), \\ &\cong \hbar \mathbf{v}_F \cdot \delta\mathbf{k} \end{aligned} \tag{34.36b}$$

where k_F is the Fermi wave vector, v_F is the Fermi velocity, and $\delta\mathbf{k} \equiv \mathbf{k} - \mathbf{k}_F$. A quasiparticle is pictured as an electron together with an accompanying polarization/correlation "cloud", which accounts for the coordinated motion of all the remaining electrons as they "get out of the way" of the electron considered (because of either electron–electron repulsion or the Pauli principle). This idea allows us to regard the electron–electron scattering potential as weak (in some sense) since it then contains only the "left over" component, which cannot be incorporated into an effective mass.

In addition to the electron–electron contribution to the effective mass m^*, there is a contribution from the electron–phonon interaction.[9] Physically this contribution results from a polarization of the lattice of ions by the electron. In most metals this contribution is actually larger than the electron–electron contribution.[10]

[9] This contribution is limited to excitation energies, $\xi \leq \hbar\omega_D$, where ω_D is the Debye energy. Above this energy the lattice can no longer follow the electron's motion and this mechanism is suppressed.

[10] Exceptions are the highly correlated so-called heavy Fermion materials, where very large electron–electron mass enhancements are encountered (> 10); these materials will be discussed in Section 45.3

The "left over" or residual electron–electron interaction, in addition to being weaker, also has a much more complicated structure. (The bare Coulomb potential has the form $4\pi e^2/q^2$, whereas the residual potential depends on $\mathbf{k}, \mathbf{k}'$, and $\mathbf{q}$, as indicated in Eq. (34.35).) When the effect of the indirect electron–phonon interaction is included to form the overall electron–electron interaction, the interaction becomes attractive for low excitation energies, as was discussed in Section 21.3.[11]

We now restrict ourselves to that part of the interaction Hamiltonian which contributes to superconductivity; i.e., we retain only the attractive part leading to the formation of pairs with opposite momenta and opposite spins. To accomplish this we replace $\mathbf{k}'$ by $-\mathbf{k}$, followed by replacing $\mathbf{k}+\mathbf{q}$ by $\mathbf{k}'$ in Eq. (34.35a); writing $\alpha = \uparrow$ and $\beta = \downarrow$, our "reduced" Hamiltonian then becomes[12]

$$\begin{aligned}\hat{H}_R &= \sum_{\mathbf{k},\sigma} \xi_{\mathbf{k}} \hat{c}^\dagger_{\mathbf{k}\sigma} \hat{c}_{\mathbf{k}\sigma} + \sum_{\mathbf{k},\mathbf{k}'} V_{\mathbf{k}\mathbf{k}'} \hat{c}^\dagger_{\mathbf{k}'\uparrow} \hat{c}^\dagger_{-\mathbf{k}'\downarrow} \hat{c}_{-\mathbf{k}\downarrow} \hat{c}_{\mathbf{k}\uparrow} \\ &= \hat{H}_0 + \hat{H}_{IR}. \end{aligned} \tag{34.37}$$

The ground state energy is the expectation value of this quantity calculated with Ψ_{BCS},

$$E' = E - \mu \overline{N} = \left\langle \Psi_{BCS} \left| \hat{H}'_R \right| \Psi_{BCS} \right\rangle. \tag{34.38}$$

The matrix elements of $\hat{H}'_0$ and $\hat{H}_{IR}$ are evaluated in Appendix 34A as

$$\left\langle \Psi_{BCS} \left| \hat{H}'_0 \right| \Psi_{BCS} \right\rangle = 2 \sum_{\mathbf{k}} |v_{\mathbf{k}}|^2 \xi_{\mathbf{k}} \tag{34.39a}$$

and

$$\left\langle \overline{\Psi}_{BCS} \left| \hat{H}_{IR} \right| \overline{\Psi}_{BCS} \right\rangle = \sum_{\mathbf{k},\mathbf{k}'} V_{\mathbf{k}\mathbf{k}'} v^*_{\mathbf{k}} u_{\mathbf{k}} v_{\mathbf{k}'} u^*_{\mathbf{k}'}. \tag{34.39b}$$

To complete our evaluation of the ground state energy using this BCS variational wave function, we must minimize the sum of (34.39a) and (34.39b) subject to the condition, $|u^2_{\mathbf{k}}| + |v^2_{\mathbf{k}}| = 1$. In what follows we choose the phases so $u^2_{\mathbf{k}}$ and $v^2_{\mathbf{k}}$ are real; the normalization condition is conveniently handled by introducing the definitions

$$|u_{\mathbf{k}}| = \cos\theta_{\mathbf{k}} \tag{34.40a}$$

and

$$|v_{\mathbf{k}}| = \sin\theta_{\mathbf{k}}. \tag{34.40b}$$

[11] Strictly speaking, the form of the quasiparticle interaction employed in Eq. (34.34) is not sufficiently general to account for the indirect effects of the electron–phonon interaction, and, to a lesser degree, the "direct" electron–electron interaction. To properly account for such effects one must introduce into the total Hamiltonian all the dynamical variables associated with the motion of both the phonons and the electrons (including the interaction terms). Both the electron system and the phonon system require a finite time to polarize and hence all effective interactions are "retarded" or delayed in time; this translates into a frequency dependence of this interaction. The electrons, being light, respond rapidly and it is a good approximation to regard their response as instantaneous. The ion motion (associated with the phonons) on the other hand is slow and cannot be ignored. We can qualitatively capture this effect by cutting off the effective electron–electron interaction in Eq. (34.34) for electron excitation energies, $\xi > \hbar\omega_D$, where ω_D is a characteristic phonon frequency. For the electron system the cut-off would be of order $\hbar\omega_p$ or μ, where ω_p is the plasmon frequency. Both of these energies are of order 10 eV. Further analysis (by Bogoliubov) shows that the direct effects are suppressed in the presence of a low frequency attraction. A useful picture is to think of one electron moving into some region of space and locally polarizing the lattice. Slightly later, after the first electron has left, but before the lattice can relax, a second electron enters and feels the polarization. This interaction, which is second order in perturbation theory, is necessarily attractive.

[12] Written out fully, the potential $V_{\mathbf{k},\mathbf{k}',\mathbf{q}}$, coupling initial states $\mathbf{k}, \mathbf{k}'$ to final states $\mathbf{k}+\mathbf{q}, \mathbf{k}'-\mathbf{q}$ would be (suppressing spin indices) $V_{\mathbf{k},\mathbf{k}';\mathbf{k}+\mathbf{q},\mathbf{k}'-\mathbf{q}}$; after the substitutions we have $V_{\mathbf{k},-\mathbf{k};\mathbf{k}',-\mathbf{k}'}$, which we denote simply as $V_{\mathbf{k}\mathbf{k}'}$.

Thus,

$$\left\langle \Psi_{\mathrm{BCS}} \left| \hat{\mathrm{H}}_{\mathrm{R}}' \right| \Psi_{\mathrm{BCS}} \right\rangle = 2 \sum_{\mathbf{k}} \xi_{\mathbf{k}} \sin^2 \theta_{\mathbf{k}} + \frac{1}{4} \sum_{\mathbf{kk}'} \mathrm{V}_{\mathbf{kk}'} \sin 2\theta_{\mathbf{k}} \sin 2\theta_{\mathbf{k}'}. \tag{34.41}$$

Minimizing with respect to the parameters $\theta_{\mathbf{k}}$ yields

$$2\xi_{\mathbf{k}} \sin 2\theta_{\mathbf{k}} + \cos 2\theta_{\mathbf{k}} \sum_{\mathbf{k}'} \mathrm{V}_{\mathbf{k},\mathbf{k}'} \sin 2\theta_{\mathbf{k}'} = 0$$

or

$$\xi_{\mathbf{k}} \tan 2\theta_{\mathbf{k}} = -\frac{1}{2} \sum_{\mathbf{k}'} \mathrm{V}_{\mathbf{kk}'} \sin 2\theta_{\mathbf{k}'}. \tag{34.42a}$$

Anticipating the final outcome we define the function

$$\Delta_{\mathbf{k}} = -\sum_{\mathbf{k}'} \mathrm{V}_{\mathbf{kk}'} u_{\mathbf{k}'} v_{\mathbf{k}'} = -\frac{1}{2} \sum_{\mathbf{k}'} \mathrm{V}_{\mathbf{kk}'} \sin 2\theta_{\mathbf{k}'}, \tag{34.43}$$

which is called the *gap function*, and (34.33) becomes

$$\tan 2\theta_{\mathbf{k}} = \frac{\Delta_{\mathbf{k}}}{\xi_{\mathbf{k}}}. \tag{34.42b}$$

In addition to the gap function (34.43) we define a second quantity

$$\varepsilon_{\mathbf{k}} = \sqrt{\xi_{\mathbf{k}}^2 + \Delta_{\mathbf{k}}^2}; \tag{34.44}$$

in Chapter 35 we will show that $\varepsilon_{\mathbf{k}}$ is the energy required to add an electron to the system in a state **k**. With these definitions, Eq. (34.42b) becomes

$$\sin 2\theta_{\mathbf{k}} = 2u_{\mathbf{k}} v_{\mathbf{k}} = \frac{\Delta_{\mathbf{k}}}{\varepsilon_{\mathbf{k}}}. \tag{34.45}$$

We may also calculate

$$\cos 2\theta_{\mathbf{k}} = u_{\mathbf{k}}^2 - v_{\mathbf{k}}^2 = \frac{\xi_{\mathbf{k}}}{\varepsilon_{\mathbf{k}}}. \tag{34.46}$$

The combinations (34.45) and (34.46) occur regularly in evaluating various properties of superconductors; they are often referred to as coherence factors. From these expressions we obtain $u_{\mathbf{k}}^2$ and $v_{\mathbf{k}}^2$ individually as

$$u_{\mathbf{k}}^2 = \frac{1}{2} \left(1 + \frac{\xi_{\mathbf{k}}}{\varepsilon_{\mathbf{k}}} \right) \tag{34.47}$$

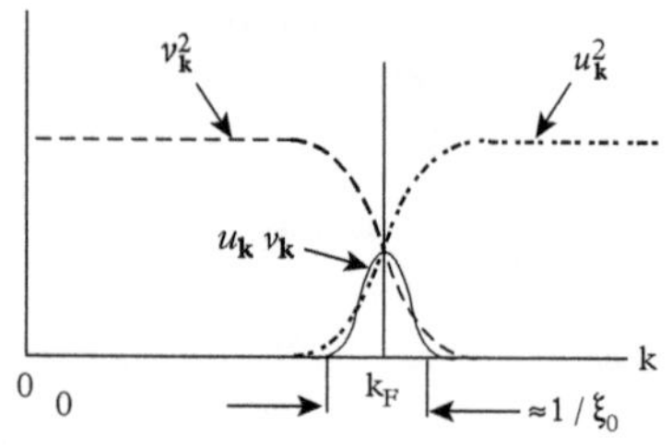

Figure 34.3 *The coherence factors, v_k^2 and u_k^2; also shown is the condensation amplitude, $u_k v_k$, in the BCS ground state.*

and

$$v_k^2 = \frac{1}{2}\left(1 - \frac{\xi_k}{\varepsilon_k}\right). \tag{34.48}$$

The quantities u_k^2, v_k^2 and the condensation amplitude, $u_k v_k$ (also called the pair amplitude) are sketched in Fig. 34.3. Note that rather than being step functions, as was the case for the Fermi vacuum, u_k and v_k are "smeared out" in the vicinity of k_F. We associate the width of this smearing with ξ_0^{-1}, the Pippard or BCS coherence length. The smearing is a consequence of the *correlation* resulting from the pairing interaction; qualitatively, correlating the positions of the electrons introduces uncertainty in their momenta.

Substituting (34.45) into (34.43) yields

$$\Delta_k = -\sum_{k'} V_{kk'} \frac{\Delta_{k'}}{2\varepsilon_{k'}} = -\sum_{k'} V_{kk'} \frac{\Delta_{k'}}{2\sqrt{\xi_{k'}^2 + \Delta_{k'}^2}}. \tag{34.49}$$

Equation (34.49) is a nonlinear integral equation for the gap function, Δ_k; it is called the *gap equation*.

Equation (34.49) has the trivial solution, $\Delta_k = 0$ and using (34.47) and (34.48) we have

$$v_k = 1, \quad u_k = 0, \quad \text{for} \quad \xi_k < 0 \tag{34.50}$$

and

$$v_k = 0, \quad u_k = 1, \quad \text{for} \quad \xi_k > 0. \tag{34.51}$$

The resulting BCS wave function is then

$$\psi_{BCS} = \prod_{k<k_F} \hat{c}_{k\uparrow}^{\dagger} \hat{c}_{-k\downarrow}^{\dagger} \,|\phi_0\rangle; \tag{34.52}$$

Eq. (34.52) is just the usual ground state wave function of a filled Fermi sea (also called the Fermi vacuum). We will discuss this case again in Chapter 35.

The simplest example of a non-trivial solution to Eq. (34.49) is to use the Cooper model potential,

$$V_{kk'} = \begin{cases} -V, & |\xi_k| \text{ and } |\xi_{k'}| \le \hbar\omega_D; \\ 0, & |\xi_k| \ge \hbar\omega_D \text{ or } |\xi_{k'}| \ge \hbar\omega_D. \end{cases} \tag{34.53}$$

The function Δ_k will then have the form

$$\Delta_k = \Delta, \quad \xi_k < \hbar\omega_D; \tag{34.54a}$$

$$\Delta_k = 0, \quad \xi_k > \hbar\omega_D. \tag{34.54b}$$

Provided that we cut the integral off at $\pm\hbar\omega_D$, we may then take Δ_k from under the integral sign in (22.29) and cancel it from both sides, yielding the condition

$$1 = N(0)V \int_{-\hbar\omega_D}^{+\hbar\omega_D} \frac{d\xi}{2\left(\xi^2+\Delta^2\right)^{1/2}}$$
$$= N(0)V \sinh^{-1}\left(\frac{\hbar\omega_D}{\Delta}\right) \qquad (34.55)$$

or

$$\frac{\Delta}{\hbar\omega_D} = \left[\sinh\left(\frac{1}{N(0)V}\right)\right]^{-1}, \qquad (34.56)$$

where $N(0)$ is the density of states at the Fermi energy, which is assumed to be constant over energies of order $\hbar\omega_D$. Since $\theta_D = \hbar\omega_D/k_B \sim 300\,K$ and $\Delta \sim T_c \sim 10\,K$, we may approximate (34.56) as

$$\Delta \cong 2\hbar\omega_D e^{-\frac{1}{N(0)V}}. \qquad (34.57)$$

Note the close similarity between the "gap" Δ and the bound state energy found by Cooper. Having the expressions (34.47) and (34.48) for u_k and v_k, we can evaluate the ground state energy using (34.39a) and (34.39b),

$$E - \mu\bar{N} = 2\sum_k \frac{\xi_k}{2}\left(1-\frac{\xi_k}{\varepsilon_k}\right) + \frac{1}{4}\sum_{k,k'} V_{kk'}\frac{\Delta_k\Delta_{k'}}{\varepsilon_k\varepsilon_{k'}}. \qquad (34.58)$$

The first (single-particle) term in (34.58) may be rewritten as[13]

$$-\Delta^2 N(0)\int_{(-\hbar\omega_D/\Delta)}^{(\hbar\omega_D/\Delta)} \frac{x^2dx}{(1+x^2)^{1/2}} + N(0)\int_{-\hbar\omega_D}^{\hbar\omega_D} |\xi|\,d\xi + \int_{-\mu}^{\infty} N(\xi)(\xi-|\xi|)d\xi$$
$$= -\Delta^2 N(0)\frac{\hbar\omega_D}{\Delta}\left(\frac{\hbar^2\omega_D^2}{\Delta^2}+1\right)^{1/2} + \Delta^2 N(0)\sinh^{-1}\left(\frac{\hbar\omega_D}{\Delta}\right) + N(0)\hbar^2\omega_D^2,$$
$$+2\int_{-\mu}^{0} N(\xi)\xi d\xi$$

where we have again assumed the density of states to be constant in the interval $\hbar\omega_D > \xi > -\hbar\omega_D$ Expanding the first term to first order in $(\Delta/\hbar\omega_D)^2$, using (34.55) for the second term, and noting that the last term is the total normal ground state energy, E_0, we obtain the form

$$E = E_0 + \frac{\Delta^2}{V} - \frac{1}{2}N(0)\Delta^2. \qquad (34.59)$$

13 $\int \frac{x^2}{(x^2+1)^{1/2}}dx = \frac{1}{2}\left[x(1+x^2)^{1/2} - \sinh^{-1}x\right].$

Using (34.49) twice to evaluate the second (interaction) term in (34.58), we obtain $-\Delta^2/V$. The total "condensation energy" is then

$$(E - \mu\overline{N}) - (E - \mu\overline{N})_{\Delta=0} = -\frac{1}{2}N(0)\Delta^2. \tag{34.60}$$

Referring to Eq. (32.29), we obtain the condensation energy in the BCS model as

$$\frac{1}{8\pi}H_c^2(T = 0) = \frac{1}{2}N(0)\Delta^2. \tag{34.61}$$

ADDITIONAL READING

Superconductivity, J.B. Ketterson and S.N. Song, Cambridge University Press (2000).

Theory of Superconductivity J. Robert Schrieffer, Westview Press (Advanced Books Classics), Boulder, CO (1999).

Superconductivity of Metals and Alloys, P.G. deGennes, W.A. Benjamin, NY (1966).

Introduction to Superconductivity, M. Tinkham, McGraw-Hill, NY (1975).

Superfluidity and Superconductivity, D.R. Tilley and J. Tilley, Adam Hilger, Bristol (1986).

Superconductivity: Conventional and Unconventional Superconductors, *Vol. 1*, K.H. Bennemann and J.B. Ketterson, Eds. Springer (2008).

Much of the ground work leading up to the BCS theory was summarized in Theory of superconductivity: theoretical part, J. Bardeen, *Encyclopedia of Physics*, *Vol. 15*, p. 274. Springer-Verlag, Berlin (1956).

For some interesting commentary on the BCS theory from the point of view of one of its main elaborators, see: BCS: The scientific "love of my life", Philip W. Anderson, *Int. J. Mod. Phys. B* 24, 3983 (2010).

Problem

34.1. Calculate $\overline{\delta N^2}$ as given by Eq. (34.33).

Appendix 34A Some calculations involving the BCS wave function

BCS introduced *pairing operators* defined as

$$\hat{b}_k^\dagger = \hat{c}_{k\uparrow}^\dagger \hat{c}_{-k\downarrow}^\dagger \tag{34A.1a}$$

and

$$\hat{b}_k = \hat{c}_{-k\downarrow}\hat{c}_{k\uparrow}, \tag{34A.1b}$$

where $\hat{c}^{\dagger}_{k\sigma}$ and $\hat{c}_{k\sigma}$ are the usual fermion creation and destruction operators obeying the anti-commutation relations (34.23a–c). The pairing operators, on the other hand, obey commutation relations (which are verified by direct calculation) given by

$$\hat{b}_k\hat{b}^{\dagger}_{k'} - \hat{b}^{\dagger}_{k'}\hat{b}_k = 0, \mathbf{k} \neq \mathbf{k}' \tag{34A.2a}$$

$$\hat{b}_k\hat{b}^{\dagger}_{k} - \hat{b}^{\dagger}_{k}\hat{b}_k = 1 - \hat{n}_{k\uparrow} - \hat{n}_{-k\downarrow}, \tag{34A.2b}$$

and

$$\hat{b}_k\hat{b}_{k'} - \hat{b}_{k'}\hat{b}_k = 0, \tag{34A.2c}$$

where $\hat{n}_{k\uparrow}$ and $\hat{n}_{k\downarrow}$ are the number operators for spin up and spin down electrons respectively. We may write the total number operator (for both spins) as

$$\hat{n}_{k\uparrow} + \hat{n}_{k\downarrow} = 2\hat{b}^{\dagger}_{k}\hat{b}_k. \tag{34A.3}$$

In terms of the pairing operators, the reduced Hamiltonian, $\hat{H}_R$, can be written as

$$\hat{H}_R = 2\sum_{k} \xi_k\hat{b}^{\dagger}_{k}\hat{b}_k - \sum_{k,k'} V_{kk'}\hat{b}^{\dagger}_{k}\hat{b}_{k'}. \tag{34A.4}$$

To obtain the ground state energy we must calculate the expectation value of $\hat{H}_R$ with the BCS wave function

$$\Psi = \prod_{k}\left(1 + g_k b^{\dagger}_{k}\right)|0\rangle, \tag{34A.5}$$

subject to the constraint that the expectation value of the total particle number operator is fixed at the average particle number, N. Let us first normalize Ψ:

$$\langle\Psi\,|\,\Psi\rangle = \prod_{k,k'}\left\langle 0\left|\left(1 + g^{*}_{k}\hat{b}_k\right)\left(1 + g_{k'}\hat{b}^{\dagger}_{k'}\right)\right|0\right\rangle$$

$$= \prod_{k,k'}\left\langle 0\left|1 + g^{*}_{k}\hat{b}_k + g_{k'}\hat{b}^{\dagger}_{k'} + g^{*}_{k}g_{k'}\hat{b}_k\hat{b}^{\dagger}_{k'}\right|0\right\rangle.$$

The terms with an odd number of operators vanish; using (34A.2b) to commute the operators in the last term gives

$$\langle\Psi\,|\,\Psi\rangle = \prod_{k,k'}\left\langle 0\left|1 + g^{*}_{k}g_{k'}(1 - \hat{n}_{k\uparrow} - \hat{n}_{k\downarrow})\delta_{kk'} + \hat{b}^{\dagger}_{k}\hat{b}_{k'}\right|0\right\rangle.$$

The operator $\hat{b}_{\mathbf{k}'}$ annihilates the vacuum; furthermore $\hat{n}_{\mathbf{k}}|0\rangle$ vanishes since the vacuum state has no particles. Therefore,

$$\langle\Psi|\Psi\rangle = \prod_{\mathbf{k}}(1 + g_{\mathbf{k}}^* g_{\mathbf{k}})$$

and the normalized ground state wave function is

$$|\Psi\rangle = \prod_{\mathbf{k}} \frac{\left(1 + g_{\mathbf{k}} b_{\mathbf{k}}^{\dagger}\right)|0\rangle}{\left(1 + g_{\mathbf{k}}^* g_{\mathbf{k}}\right)^{1/2}}. \tag{34A.6}$$

It is useful to introduce the quantities

$$u_{\mathbf{k}} \equiv \frac{1}{\left(1 + |g_{\mathbf{k}}|^2\right)^{1/2}} \tag{34A.7a}$$

and

$$v_{\mathbf{k}} \equiv \frac{g_{\mathbf{k}}}{\left(1 + |g_{\mathbf{k}}|^2\right)^{1/2}}, \tag{34A.7b}$$

whereupon (34A.6) becomes

$$|\psi\rangle = \prod_{\mathbf{k}}\left(u_{\mathbf{k}} + v_{\mathbf{k}} b_{\mathbf{k}}^{\dagger}\right)|0\rangle, \tag{34A.8}$$

where $u_{\mathbf{k}}$ and $v_{\mathbf{k}}$ satisfy

$$|u_{\mathbf{k}}|^2 + |v_{\mathbf{k}}|^2 = 1. \tag{34A.9}$$

To evaluate the ground state energy we require the expectation value of

$$\langle\Psi|\hat{H}_R|\Psi\rangle = \langle\Psi|\hat{H}_0|\Psi\rangle + \langle\Psi|\hat{H}_{IR}|\Psi\rangle, \tag{34A.10}$$

where

$$\hat{H}_0 = 2\sum_{\mathbf{k}} \xi_{\mathbf{k}} \hat{b}_{\mathbf{k}}^{\dagger} \hat{b}_{\mathbf{k}} \tag{34A.11a}$$

and

$$H_{IR} = -\sum_{\mathbf{k},\mathbf{k}'} V_{\mathbf{k}\mathbf{k}'} \hat{b}_{\mathbf{k}}^{\dagger} \hat{b}_{\mathbf{k}'}. \tag{34A.11b}$$

Let us write the product of the terms contributing to the wave function (34A.8) explicitly:

$$|\psi\rangle = \left(u_{\mathbf{k}_1} + v_{\mathbf{k}_1} b_{\mathbf{k}_1}^{\dagger}\right)\left(u_{\mathbf{k}_2} + v_{\mathbf{k}_2} b_{\mathbf{k}_2}^{\dagger}\right)\ldots\left(u_{\mathbf{k}_i} + v_{\mathbf{k}_i} b_{\mathbf{k}_i}^{\dagger}\right)\ldots|0\rangle,$$

where $\mathbf{k}_1, \mathbf{k}_2, \ldots$ correspond to the set of normal state eigenfunctions of $\hat{H}_0$. In evaluating the expectation values in Eq. (34A.10), those states which are not affected by $\hat{H}_0$ or $\hat{H}_{IR}$ will result in successive factors of unity due to the normalization condition (34A.9). Only terms involving the pairing of a $\hat{b}_k$ and $\hat{b}_k^\dagger$ (associated with the *same* configuration space eigenfunction) give a non-zero contribution to the expectation value. The expectation value $\langle \Psi | \hat{H}_{IR} | \Psi \rangle$ involves terms of the form

$$\left[\cdots\left(u_{k_i}^* + v_{k_i}^* \hat{b}_{k_i}\right)\cdots\left(u_{k_j}^* + v_{k_j}^* \hat{b}_{k_j}\right)\right] V_{qq'} \hat{b}_q^\dagger \hat{b}_{q'} \left[\cdots\left(u_{k'_i} + v_{k'_i} \hat{b}_{k'_i}^\dagger\right)\cdots \left(u_{k'_j} + v_{k'_j} \hat{b}_{k'_j}^\dagger\right)\cdots\right] |0\rangle .$$

The terms involving $\hat{b}_k^\dagger \hat{b}_k$ or $\hat{b}_k \hat{b}_k^\dagger$ pairs are

$$\langle 0 | u_q^* u_q V_{qq} \hat{b}_q^\dagger \hat{b}_q + v_q^* v_{q'} V_{qq'} \hat{b}_q \hat{b}_q^\dagger \hat{b}_{q'} \hat{b}_{q'}^\dagger + u_q^* v_{q'}^* u_{q'} v_q V_{qq'} \hat{b}_{q'} \hat{b}_q^\dagger \hat{b}_{q'} \hat{b}_q^\dagger (1 - \delta_{qq'}) | 0 \rangle . \quad (34A.12)$$

The first term vanishes, since $\hat{b}|0\rangle = 0$. Using Eq. (34A.2b) the second term may be written as $(1 - \hat{n}_q)(1 - \hat{n}_{q'}) v_q^* v_{q'} V_{qq'}$; it corresponds to the Hartree–Fock contribution to the total energy and may be discarded if we assume that the ξ_k are quasiparticle energies (which already include electron–electron interaction effects). The last term contributes to the pairing energy (and vanishes in the normal state where $u_k^* v_k = 0$); again using (34A.2b) and $\hat{n}_k | 0 \rangle = 0$, we have

$$\langle \Psi | \hat{H}_{IR} | \Psi \rangle = \sum_{k,k'} u_k^* v_k u_{k'} v_{k'}^* V_{kk'} . \quad (34A.13)$$

We may use (34A.12) to evaluate the contribution of $\hat{H}_0$ to (34A.10) by replacing $V_{kk'}$ by $2\xi_k \delta_{kk'}$, to obtain

$$\langle \Psi | \hat{H}_0 | \Psi \rangle = 2 \sum_k \xi_k \, |v_k|^2 . \quad (34A.14)$$

35 Elementary Excitations and the Thermodynamic Properties of Superconductors

To treat the properties of a superconductor at finite temperatures we need to develop techniques to describe excitations from the BCS ground state. The most powerful approach is through a transformation of the creation and destruction operators, $\hat{c}^{\dagger}_{\mathbf{k}\sigma}$ and $\hat{c}_{\mathbf{k}\sigma}$, that was introduced independently by Bogoliubov and Valatin.

35.1 The Bogoliubov–Valatin transformation

We recall expressions (34.39a,b) which together give the expectation value of the reduced Hamiltonian:

$$\left\langle \Psi_{BCS} \middle| \hat{H}_R \middle| \Psi_{BCS} \right\rangle = 2\sum_{\mathbf{k}} v_{\mathbf{k}}^2 \xi_{\mathbf{k}} + \sum_{\mathbf{k},\mathbf{k}'} u_{\mathbf{k}} v_{\mathbf{k}} u_{\mathbf{k}'} v_{\mathbf{k}'} V_{\mathbf{k}\mathbf{k}'}. \tag{35.1}$$

Let us calculate the energy, call it $\tilde{\varepsilon}_{\mathbf{k}}$, required to add an electron to the system in a state $|\mathbf{k}\uparrow\rangle$, assuming that its companion state, $|-\mathbf{k}\uparrow\rangle$, is empty. In order to do this we must: (i) account for the energy change due to removal of the amplitude of the (bound) pair[1] associated with this wave vector, **k** and (ii) add the energy of the lone electron introduced into the state $|\mathbf{k}\uparrow\rangle$. When we remove the bound pair from the ground state, according to (35.1) the energy of the system changes by an amount

$$-2v_{\mathbf{k}}^2\xi_{\mathbf{k}} - 2u_{\mathbf{k}}v_{\mathbf{k}}\sum_{\mathbf{k}'} u_{\mathbf{k}'}v_{\mathbf{k}'}V_{\mathbf{k}\mathbf{k}'} \tag{35.2}$$

(the factor 2 in the second term arises because the chosen pair state (**k**, –**k**) occurs twice, since we have a double sum). Using Eq. (34.43) we may write the form (35.2) as

$$-2v_{\mathbf{k}}^2\xi_{\mathbf{k}} + 2u_{\mathbf{k}}v_{\mathbf{k}}\Delta_{\mathbf{k}}. \tag{35.3}$$

[1] Note that we cannot simply subtract the contribution from the interaction term alone, because the pairing also affects the single particle energies.

Adding to (35.3) the energy, $\xi_{\mathbf{k}}$, of one (unbound) electron then yields the quasiparticle excitation energy, $\tilde{\varepsilon}_{\mathbf{k}}$,

$$\begin{aligned}\tilde{\varepsilon}_{\mathbf{k}} &= \xi_{\mathbf{k}}\left(1-2v_{\mathbf{k}}^2\right)+2u_{\mathbf{k}}v_{\mathbf{k}}\Delta_{\mathbf{k}}\\ &= \xi_{\mathbf{k}}\left[1-\left(1-\frac{\xi_{\mathbf{k}}}{\varepsilon_{\mathbf{k}}}\right)\right]+\frac{\Delta_{\mathbf{k}}^2}{\varepsilon_{\mathbf{k}}}\\ &= \varepsilon_{\mathbf{k}}\end{aligned} \tag{35.4}$$

where we have used Eqs. (34.45) and (34.48) together with definition (34.44). Thus the energy needed to add an electron in state $|\mathbf{k}\uparrow\rangle$ is $\varepsilon_{\mathbf{k}}$. If we calculate the energy required to *remove* an electron in a state $|-\mathbf{k}\downarrow\rangle$, we also obtain $\varepsilon_{\mathbf{k}}$. Note that the minimum excitation energy, $\varepsilon_{\mathbf{k}}$, is $\Delta_{\mathbf{k}}$; i.e., the excitation spectrum has an *energy gap*. Note also that since the pairing interaction smears out the single particle occupation factors ($u_{\mathbf{k}}$ and $v_{\mathbf{k}}$) it is possible to add electrons to the system for states ξ both above and below the chemical potential, μ. (In a normal ground state the states with $\mathbf{k} < k_F$ would be filled at $T = 0$ and hence no electrons could be added by the Pauli principle.)

An important case involves the energy required to excite the superconducting ground state by absorbing a Bose-like particle, such as a microwave photon or thermal phonon from the lattice. This process combines the two processes discussed: we must (i) *remove* an electron from the state $\mathbf{k}$, which requires a minimum energy Δ, and (ii) *add* this electron to some state $\mathbf{k}'$ which also requires a minimum energy Δ; *the total minimum energy required is then* 2Δ (and not Δ).[2]

We next examine the wave functions associated with various excitations. We may write the wave function of a particle-like state (an electron) with spin $\uparrow$ and wave vector $\mathbf{k}$ as

$$\begin{aligned}\hat{c}^{\dagger}_{\mathbf{k}\uparrow}\,|\Psi_{BCS}\rangle &= \left(u_{\mathbf{k}}c^{\dagger}_{\mathbf{k}\uparrow}+v_{\mathbf{k}}\hat{c}^{\dagger}_{\mathbf{k}\uparrow}\hat{c}^{\dagger}_{\mathbf{k}\uparrow}\hat{c}^{\dagger}_{-\mathbf{k}\downarrow}\right)\prod_{\mathbf{k}'\neq\mathbf{k}}\left(u_{\mathbf{k}'}+v_{\mathbf{k}'}\hat{c}^{\dagger}_{\mathbf{k}'\uparrow}\hat{c}^{\dagger}_{-\mathbf{k}'\downarrow}\right)|\phi_0\rangle\,.\\ &= u_{\mathbf{k}}\hat{c}^{\dagger}_{\mathbf{k}\uparrow}\prod_{\mathbf{k}'\neq\mathbf{k}}\left(u_{\mathbf{k}'}+v_{\mathbf{k}'}\hat{c}^{\dagger}_{\mathbf{k}'\uparrow}\hat{c}^{\dagger}_{-\mathbf{k}'\downarrow}\right)|\phi_0\rangle\,.\\ &= u_{\mathbf{k}}\,|\mathbf{k}\uparrow\rangle\end{aligned} \tag{35.5}$$

where we write

$$|\mathbf{k}\sigma\rangle \equiv \hat{c}^{\dagger}_{\mathbf{k}\sigma}\prod_{\mathbf{k}'\neq\mathbf{k}}\left(u_{\mathbf{k}'}+v_{\mathbf{k}'}\hat{c}^{\dagger}_{\mathbf{k}'\uparrow}\hat{c}^{\dagger}_{-\mathbf{k}'\uparrow}\right)|\phi_0\rangle. \tag{35.6}$$

Similarly, we may write a hole state with spin $\downarrow$ and wave vector $-\mathbf{k}$ as

$$\begin{aligned}\hat{c}_{-\mathbf{k}\downarrow}\,|\Psi_{BCS}\rangle &= \left(u_{\mathbf{k}}\hat{c}_{-\mathbf{k}\downarrow}+v_{\mathbf{k}}\hat{c}_{-\mathbf{k}\downarrow}\hat{c}^{\dagger}_{\mathbf{k}\uparrow}\hat{c}^{\dagger}_{-\mathbf{k}\downarrow}\right)\prod_{\mathbf{k}'\neq\mathbf{k}}\left(u_{\mathbf{k}'}+v_{\mathbf{k}'}\hat{c}^{\dagger}_{\mathbf{k}'\uparrow}\hat{c}^{\dagger}_{-\mathbf{k}'\downarrow}\right)|\phi_0\rangle\\ &= -v_{\mathbf{k}}\,|\mathbf{k}\uparrow\rangle\,.\end{aligned} \tag{35.7}$$

[2] There is some analogy here with the creation of an electron–hole pair by a gamma ray in the earlier Dirac theory involving filled negative energy states where the excitation energy is $2mc^2$. Formally such excitations are two-particle excitations, sometimes referred to as excitons.

Note that we have generated the same state (apart from normalization) by either adding an electron into $+\,\mathbf{k}\uparrow$ or removing it from $-\,\mathbf{k}\downarrow$. Similarly we have

$$\hat{c}_{\mathbf{k}\uparrow}\,|\Psi_{\mathrm{BCS}}\rangle = v_{\mathbf{k}}\,|-\mathbf{k}\downarrow\rangle \tag{35.8}$$

and

$$\hat{c}^{\dagger}_{-\mathbf{k}\downarrow}\,|\Psi_{\mathrm{BCS}}\rangle = u_{\mathbf{k}}\,|-\mathbf{k}\downarrow\rangle. \tag{35.9}$$

If we multiply (35.5) by $v_{\mathbf{k}}$ and (35.7) by $u_{\mathbf{k}}$ and add, we obtain

$$\left(u_{\mathbf{k}}\hat{c}_{-\mathbf{k}\downarrow} + v_{\mathbf{k}}\hat{c}^{\dagger}_{\mathbf{k}\uparrow}\right)|\Psi_{\mathrm{BCS}}\rangle = 0 \tag{35.10}$$

or

$$\hat{\gamma}_{-\mathbf{k}\downarrow}\,|\Psi_{\mathrm{BCS}}\rangle = 0\,, \tag{35.11}$$

where we have introduced a new operator

$$\hat{\gamma}_{-\mathbf{k}\downarrow} \equiv u_{\mathbf{k}}\hat{c}_{-\mathbf{k}\downarrow} + v_{\mathbf{k}}\hat{c}^{\dagger}_{\mathbf{k}\uparrow}. \tag{35.12}$$

On the other hand, multiplying (35.8) by $v_{\mathbf{k}}$ and (35.9) by $u_{\mathbf{k}}$, adding and applying the normalization condition yields

$$\hat{\gamma}^{\dagger}_{-\mathbf{k}\downarrow}\,|\Psi_{\mathrm{BCS}}\rangle = |-\mathbf{k}\downarrow\rangle. \tag{35.13}$$

Similarly, using (35.5) and (35.7) yields

$$\left(u_{\mathbf{k}}\hat{c}^{\dagger}_{\mathbf{k}\uparrow} - v_{\mathbf{k}}\hat{c}_{-\mathbf{k}\downarrow}\right)|\Psi_{\mathrm{BCS}}\rangle = |\mathbf{k}\uparrow\rangle \tag{35.14}$$

or

$$\hat{\gamma}^{\dagger}_{\mathbf{k}\uparrow}\,|\Psi_{\mathrm{BCS}}\rangle = |\mathbf{k}\uparrow\rangle, \tag{35.15}$$

where

$$\hat{\gamma}^{\dagger}_{\mathbf{k}\uparrow} \equiv u_{\mathbf{k}}\hat{c}^{\dagger}_{\mathbf{k}\uparrow} - v_{\mathbf{k}}\hat{c}_{-\mathbf{k}\downarrow}. \tag{35.16}$$

Using (35.8) and (35.9), we obtain

$$\hat{\gamma}_{\mathbf{k}\uparrow}\,|\Psi_{\mathrm{BCS}}\rangle = 0. \tag{35.17}$$

In a superconductor, the two pairs of operators, $\hat{\gamma}_{\mathbf{k}\uparrow}$, $\hat{\gamma}_{-\mathbf{k}\downarrow}$ and $\hat{\gamma}^{\dagger}_{\mathbf{k}\uparrow}$, $\hat{\gamma}^{\dagger}_{-\mathbf{k}\downarrow}$, play the role of *quasiparticle destruction and creation operators*. Since the ground state contains no quasiparticle excitations, we must have $\hat{\gamma}\,|\Psi_{\mathrm{BCS}}\rangle = 0$; on the other hand $\hat{\gamma}^{\dagger}\,|\Psi_{\mathrm{BCS}}\rangle = |\mathbf{k}\rangle$ is an excited state with one quasiparticle. The operators $\hat{\gamma}^{\dagger}\,|\Psi_{\mathrm{BCS}}\rangle = |\mathbf{k}\rangle$ are called Bogoliubov–Valatin (BV) operators, after Bogoliubov[3] and Valatin[4] who first introduced them; it has become common to simply call them Bogoliubov operators.[5] It is easy to verify that these operators satisfy the Fermi anticommutation rules,

$$\left[\hat{\gamma}_{\mathbf{k}\sigma},\, \hat{\gamma}_{\mathbf{k}'\sigma'}\right]_{+} = 0 \tag{35.18a}$$

and

$$\left[\hat{\gamma}^{\dagger}_{\mathbf{k}\sigma},\, \hat{\gamma}^{\dagger}_{\mathbf{k}'\sigma'}\right]_{+} = 0 \tag{35.18b}$$

and

$$\left[\hat{\gamma}^{\dagger}_{\mathbf{k}\sigma},\, \hat{\gamma}_{\mathbf{k}'\sigma'}\right]_{+} = \delta_{\mathbf{k}\mathbf{k}'}\delta_{\sigma\sigma'}. \tag{35.18c}$$

Orthonormal states involving more than one quasiparticle excitation may be generated by applying successive Bogoliubov quasiparticle creation operators to the ground state:

$$\left|\mathbf{k}_1\sigma_1,\, \mathbf{k}_2\sigma_2, \mathbf{k}_3\sigma_3, \cdots\right\rangle = \left(\hat{\gamma}^{\dagger}_{\mathbf{k}_1\sigma_1}\hat{\gamma}^{\dagger}_{\mathbf{k}_2\sigma_2}\hat{\gamma}^{\dagger}_{\mathbf{k}_3\sigma_3}\cdots\right)\left|\Psi_{\mathrm{BCS}}\right\rangle. \tag{35.19}$$

It is important to note that the Bogoliubov quasiparticle operators incorporate two features. The first feature is apparent in the absence of a pairing correlation ($\Delta = 0$), where the operators become

$$\hat{\gamma}^{\dagger}_{\mathbf{k}\uparrow} = \begin{cases} \hat{c}^{\dagger}_{\mathbf{k}\uparrow}; & k > k_F \\ -\hat{c}_{-\mathbf{k}\downarrow}; & k < k_F \end{cases} \tag{35.20a,b}$$

and

$$\hat{\gamma}^{\dagger}_{-\mathbf{k}\downarrow} = \begin{cases} \hat{c}^{\dagger}_{-\mathbf{k}\downarrow} & k > k_F \\ \hat{c}_{\mathbf{k}\uparrow} & k < k_F. \end{cases} \tag{35.21a,b}$$

Consider (35.20a), which says that to create an electron excitation $|\mathbf{k}\uparrow\rangle$ for $k > k_F$ we simply add an electron with quantum numbers, **k**, ↑. However, (35.20b) says that to add an excitation with quantum numbers **k**, ↑ when $k < k_F$, we must *destroy* the *existing* electron state with quantum numbers –**k**, ↓, to *create* a *hole* state with quantum numbers **k**, ↑. The negative sign with (35.20b) reflects both an arbitrary phase convention and the fact that our pairing has been restricted to singlet states. Equations (35.21a,b) are interpreted in a similar manner.

The second feature of the Bogoliubov quasiparticle operators is more subtle. In the superconductor ($\Delta \neq 0$), the operators $\hat{c}^{\dagger}_{\mathbf{k}\uparrow}$ (electron-like) and $\hat{c}_{-\mathbf{k}\downarrow}$ (hole-like) become *mixed*, there being some amplitude for each to be non-zero for a

[3] A new method in the theory of superconductivity. 1, N.N. Bogoliubov, *Sov. Phys. JETP* 7, 41 (1958).

[4] Comments on the theory of superconductivity, J.G. Valatin, *Nuovo Cimento* 7, 843 (1958).

[5] Analogous operators were introduced earlier by Bogoliubov in his treatment of the weakly interacting Bose gas, as discussed in Chapter 31.

region $\Delta k \sim 1/\xi_0$ about k_F. This property reflects correlations introduced by the pairing interactions.

The commutation relations (35.18) resulted from our definitions of the $\hat{\gamma}$ operators and the normalization property obeyed by u_k and v_k. We could also proceed in the opposite direction by demanding that the $\hat{\gamma}$ operators satisfy the Fermi commutation rules, in which case the normalization condition would be a resultant property. Since the transformations $\hat{\gamma} \Leftrightarrow \hat{c}$ preserve the Fermi commutation rules they are said to be *canonical* (a word commonly invoked in physics when a transformation retains the overall structure of the resulting equations).

There is an alternative way (Bogoliubov's method) to obtain the ground state energy of a superconductor in which the BCS wave function is not explicitly involved. To do this we: (i) rewrite the Hamiltonian, $\hat{H}_R$, in terms of the Bogoliubov operators, $\hat{\gamma}, \hat{\gamma}^\dagger$ and (ii) commute, using (35.18), all destruction operators, $\hat{\gamma}$ to the right where we assume they annihilate the ground state, which by definition contains no excited states. The resulting c-number expression is then minimized (with respect to u_k, v_k) to obtain the ground state energy.

Implementing this strategy requires the inverse of the transformations (35.12) and (35.16), which are easily found to be

$$\hat{c}_{-k\downarrow} = u_k \hat{\gamma}_{-k\downarrow} - v_k \hat{\gamma}^\dagger_{k\uparrow} \tag{35.22}$$

and

$$\hat{c}^\dagger_{k\uparrow} = u_k \hat{\gamma}^\dagger_{k\uparrow} + v_k \hat{\gamma}_{-k\downarrow}. \tag{35.23}$$

We will not carry out the calculation of the ground state energy using the Bogoliubov method because, with just a little more effort, we can obtain the free energy for any temperature, which we do in Section 35.2. One could, in fact, dispense with the BCS wave function entirely and introduce the Bogoliubov transformation from the outset. However some physical insight is lost in this approach, in that a rationale for introducing the transformation is then lacking.

Equations (35.22) and (35.23) may be written in matrix form; introducing the notation of Eq. (34.40a,b), we have

$$\begin{pmatrix} \hat{c}^\dagger_{k\uparrow} \\ \hat{c}_{-k\downarrow} \end{pmatrix} = \begin{pmatrix} \cos\theta_k & \sin\theta_k \\ -\sin\theta_k & \cos\theta_k \end{pmatrix} \begin{pmatrix} \hat{\gamma}^\dagger_{k\uparrow} \\ \hat{\gamma}_{-k\downarrow} \end{pmatrix}. \tag{35.24}$$

The inverse transformations, (35.12) and (35.16), are

$$\begin{pmatrix} \hat{\gamma}^\dagger_{k\uparrow} \\ \hat{\gamma}_{-k\downarrow} \end{pmatrix} = \begin{pmatrix} \cos\theta_k & -\sin\theta_k \\ \sin\theta_k & \cos\theta_k \end{pmatrix} \begin{pmatrix} \hat{c}^\dagger_{k\uparrow} \\ \hat{c}_{-k\downarrow} \end{pmatrix}. \tag{35.25}$$

Equations (35.24) and (35.25) differ only in the sign of the angle of rotation in "operator space."

35.2 Thermodynamic properties

As discussed in the previous section, excited states are easily expressed using the Bogoliubov–Valatin operators. We can also rewrite the reduced Hamiltonian of the system in terms of the Bogoliubov quasiparticle operators. In doing so we generate two types of terms: those containing products of an equal number of creation $\left(\hat{\gamma}^{\dagger}_{\mathrm{k}\sigma}\right)$ and destruction $(\hat{\gamma}_{\mathrm{k}\sigma})$ operators and those containing an unequal number. When taking thermally averaged quantum-expectation values of such products, only the first forms yield non-zero averages. Furthermore, since the fermion–quasiparticle excitations do not interact (in our mean-field-like theory), the averages are given by

$$\left\langle \hat{\gamma}^{\dagger}_{\mathrm{k}\sigma}\hat{\gamma}_{\mathrm{k}\sigma}\right\rangle = \mathrm{f}_{\mathrm{k}\sigma} \tag{35.26a}$$

and

$$\left\langle \hat{\gamma}_{\mathrm{k}\sigma}\hat{\gamma}^{\dagger}_{\mathrm{k}\sigma}\right\rangle = 1 - \mathrm{f}_{\mathrm{k}\sigma}, \tag{35.26b}$$

where $\mathrm{f}_{\mathrm{k}\sigma}$ will turn out to be the usual ideal Fermi gas occupation factor,

$$\mathrm{f}_{\mathrm{k}\sigma} = \frac{1}{e^{\varepsilon_{\mathrm{k}}/k_B T} + 1}. \tag{35.27}$$

Inserting Eqs. (35.22) and (35.23) into our reduced Hamiltonian, (34.37) yields

$$\hat{H}_R = \hat{H}_0 + \hat{H}_{IR},$$

where

$$\hat{H}_0 = \sum_{\mathrm{k}} \xi_{\mathrm{k}}\left[2v_{\mathrm{k}}^2 + \left(u_{\mathrm{k}}^2 - v_{\mathrm{k}}^2\right)\left(\hat{\gamma}^{\dagger}_{\mathrm{k}\uparrow}\hat{\gamma}_{\mathrm{k}\uparrow} + \hat{\gamma}^{\dagger}_{-\mathrm{k}\downarrow}\hat{\gamma}_{-\mathrm{k}\downarrow}\right)\right] + \left[\text{terms with unequal numbers of } \hat{\gamma}^{\dagger}_{\mathrm{k}\sigma} \text{ and } \hat{\gamma}_{\mathrm{k}\sigma}\right] \tag{35.28}$$

and

$$\hat{H}_{IR} = \sum_{\mathrm{k},\mathrm{k}'} V_{\mathrm{k}\mathrm{k}'}\left\{u_{\mathrm{k}}v_{\mathrm{k}}u_{\mathrm{k}'}v_{\mathrm{k}'}\left[1 - \hat{\gamma}^{\dagger}_{\mathrm{k}\uparrow}\hat{\gamma}_{\mathrm{k}\uparrow} - \hat{\gamma}^{\dagger}_{-\mathrm{k}\downarrow}\hat{\gamma}_{-\mathrm{k}\downarrow}\right]\left[1 - \hat{\gamma}^{\dagger}_{\mathrm{k}'\uparrow}\hat{\gamma}_{\mathrm{k}'\uparrow} - \hat{\gamma}^{\dagger}_{-\mathrm{k}'\downarrow}\hat{\gamma}_{-\mathrm{k}'\downarrow}\right] + \left[\text{terms with unequal numbers of } \hat{\gamma}^{\dagger}_{\mathrm{k}\sigma} \text{ and } \hat{\gamma}_{\mathrm{k}\sigma}\right]\right\}. \tag{35.29}$$

Using Eq. (35.26a,b), the expectation value of our reduced Hamiltonian is[6]

$$\left\langle \hat{H}_R \right\rangle = 2\sum_{k} \xi_k \left[v_k^2 + \left(u_k^2 - v_k^2 \right) f_k \right] + \sum_{k,k'} V_{kk'} u_k v_k u_{k'} v_{k'} \left[1 - 2f_k \right] \left[1 - 2f_{k'} \right]. \tag{35.30}$$

The Helmholtz free energy in the pairing approximation is given by

$$F = \left\langle \hat{H}_R \right\rangle - TS, \tag{35.31}$$

where the entropy for fermions is given by the usual expression from statistical mechanics:

$$\begin{aligned} S &= -k_B \sum_{k,\sigma} \left[f_k \ln f_k + (1 - f_k) \ln (1 - f_k) \right] \\ &= -2k_B \sum_{k} \left[f_k \ln f_k + (1 - f_k) \ln (1 - f_k) \right]. \end{aligned} \tag{35.32}$$

Again, introducing $u_k = \cos\theta_k$ and $v_k = \sin\theta_k$ and minimizing F with respect to the parameters θ_k yields

$$u_k^2 = \frac{1}{2}\left(1 + \frac{\xi_k}{\varepsilon_k} \right) \tag{35.33}$$

and

$$v_k^2 = \frac{1}{2}\left(1 - \frac{\xi_k}{\varepsilon_k} \right), \tag{35.34}$$

where we have defined

$$\varepsilon_k = \sqrt{\xi_k^2 + \Delta_k^2}; \tag{35.35}$$

these expressions are the same as those obtained before as Eqs. (34.47), (34.48), and (34.44), but where in place of Eq. (34.43), Δ_k is now defined by

$$\Delta_k = -\sum_{k'} V_{kk'} u_{k'} v_{k'} (1 - 2f_{k'}). \tag{35.36a}$$

This equation may also be written as

$$\Delta_k = -\sum_{k'} V_{kk'} u_{k'} v_{k'} \tanh\left(\frac{\varepsilon_{k'}}{2k_B T} \right). \tag{35.36b}$$

We note again that all quantities have the same form as for the T = 0 (ground state) case, except (35.36a), which contains an extra factor, $(1 - 2f_{k'})$.[7] If we

[6] In the Landau theory of a Fermi liquid, the quasiparticle energies are defined as the functional derivative of the total energy with respect to the distribution function; i.e., $\varepsilon_k \equiv \delta E/\delta f_{k\sigma}$. Differentiating (35.39) with respect to δf_k, using Eqs. (35.33), (35.34), and (35.37), for u_k^2, v_k^2, and Δ_k, and dividing by 2 (so as to include only a single spin state), we obtain $\varepsilon_k = (\xi^2 + \Delta^2)^{1/2}$, in agreement with Eq. (35.4).

[7] It should be noted that since ε_k is positive definite, $f_k = 0$, for all **k** at T = 0.

minimize F with respect to f_k (i.e., $\partial F/\partial f_k = 0$) we obtain (35.27): the derivative of (35.30) yields ε_k and the Fermi occupation factor then follows from expression (35.32) for the entropy, S, which is based on the exclusion principle.

Using the expressions for u_k and v_k, we may rewrite the gap equation, (35.36a), as

$$\Delta_k = -\sum_{k'} V_{kk'} \frac{(1-2f_{k'})}{2\varepsilon'_k} \Delta_{k'}. \tag{35.37}$$

Let us first discuss the behavior of the gap function in the Cooper model,

$$V_{kk'} = \begin{cases} -\dfrac{V}{L^3} & \text{for} \quad |\xi_k| \text{ and } |\xi_{k'}| \le \hbar\omega_D \\ 0 & \text{for} \quad |\xi_k| \text{ or } |\xi_{k'}| > \hbar\omega_D. \end{cases} \tag{35.38}$$

Equation (35.37) may then be written as the condition

$$1 = N(0)V \int_{-\hbar\omega_D}^{+\hbar\omega_D} d\xi \frac{1 - 2f\left(\sqrt{\xi^2 + \Delta^2(T)}\right)}{2\sqrt{\xi^2 + \Delta^2(T)}}. \tag{35.39}$$

The solution of Eq. (35.39), which must be obtained numerically, gives $\Delta(T)$, which is shown in Fig. 35.1.

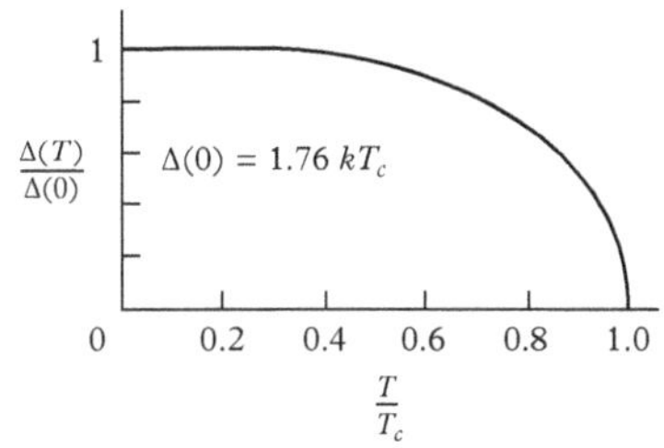

Figure 35.1 *Temperature dependence of the energy gap in the BCS theory*

The transition temperature, T_c, corresponds to $\Delta(T_c) = 0$, which from (35.39) is given by

$$1 = N(0)V \int_{-\hbar\omega_D}^{+\hbar\omega_D} d\xi \frac{1 - 2f(\xi, T_c)}{2\xi}$$

$$= N(0)V \int_0^{+\hbar\omega_D} \frac{d\xi}{\xi} \tanh\left(\frac{\xi}{2k_B T_c}\right). \tag{35.40}$$

For large ξ, $\tanh(\xi/2k_BT_c) \to 1$, and the integral has the asymptotic form, $ln(\hbar\omega_D/k_BT_c) + C$; a numerical calculation yields $C = ln(1.13)$. Therefore,

$$1 = N(0)V ln \frac{1.13\hbar\omega_D}{k_B T_c}$$

or

$$k_B T_c = 1.13\hbar\omega_D e^{-1/N_0 V}. \tag{35.41}$$

If we combine Eq. (34.57) for $\Delta(0)$ with (35.41), we obtain

$$\Delta(0) = 1.76 k_B T_c \tag{35.42}$$

We remark in passing that when the potential causing the Cooper pairing involves phonons, we expect the cut-off frequency to scale as $\omega_D \propto M^{-1/2}$; this is called the isotope effect, and the fact that T_c in many metals scales in this way was an early clue that pairing arises through the exchange of phonons.

We now evaluate the free energy. The $-TS$ contribution to F (Eq. (35.32)) can be rewritten as

$$-TS = \sum_k \left[\varepsilon_k (1 - 2f_k) - 2k_B T\, ln \left(2 \cosh \frac{\varepsilon_k}{2k_B T} \right) \right]. \tag{35.43}$$

The contribution of $\langle \hat{H}_{IR} \rangle$ to the energy, E', follows from Eqs. (35.30) and (35.36):

$$\begin{aligned} \langle \hat{H}_{IR} \rangle &= -\sum_k \frac{\Delta_k^2}{2\varepsilon_k}(1 - 2f_k), \\ &= -\frac{\Delta^2(T)}{V} \end{aligned} \tag{35.44}$$

From Eq. (35.28) we have

$$\langle \hat{H}'_0 \rangle = 2 \sum_k \left[\xi_k v_k^2 + f_k \varepsilon_k - \frac{\Delta^2}{\varepsilon_k} f_k \right]. \tag{35.45}$$

From (35.44) and (35.45) we obtain the energy $E' = \langle H'_R \rangle$ as

$$\langle \hat{H}'_R \rangle = 2 \sum_k \left[\xi_k v_k^2 + f_k \varepsilon_k - \frac{\Delta^2}{\varepsilon_k} f_k \right] + \frac{\Delta^2}{V}. \tag{35.46a}$$

This expression can be rewritten in a more convenient form as

$$\langle \hat{H}'_R \rangle = 2 \sum_k \frac{1}{2\varepsilon_k} \left[(\varepsilon_k + \xi_k)^2 f_k - (\varepsilon_k - \xi_k)^2 (1 - f_k) \right]. \tag{35.46b}$$

Combining (35.46) with (35.43) yields the free energy as

$$F' = \sum_k \left[2\varepsilon_k v_k + \varepsilon_k - \frac{\Delta^2(T)}{\varepsilon_k} - 2k_B T\, ln \left(2 \cosh \frac{\varepsilon_k}{2k_B T} \right) \right] - \frac{\Delta^2(T)}{V}. \tag{35.47}$$

The condensation free energy is related to the thermodynamic critical field by $F_N - F_S = (1/8\pi) H_c^2(T)$, as given by Eq. (32.29).

From thermodynamics, the heat capacity is $C = TdS/dT$ or

$$\begin{aligned} C(T) &= -2k_B T \sum_k [ln f_k - ln(1 - f_k)] \frac{\partial f_k}{\partial T} \\ &= 2k_B T \sum_k \frac{\varepsilon_k}{k_B T} \frac{d}{dT} \left(\frac{1}{e^{\varepsilon_k/k_B T} + 1} \right). \end{aligned} \tag{35.48}$$

Noting that $\varepsilon_k = \varepsilon_k(T)$ and writing $f'(x)$ in terms of $f(x)$, we obtain

$$C = \frac{-2}{k_B T^2} \sum_k f_k(1-f_k)\left(-\varepsilon_k^2 + T\Delta \frac{d\Delta}{dT}\right)$$

$$= \frac{2}{k_B T^2} N(0) \int_0^\infty d\xi f(\varepsilon)(1-f(\varepsilon))\left(\varepsilon^2 - T\Delta\frac{d\Delta}{dT}\right). \tag{35.49}$$

At T_c the heat capacity is discontinuous and from (35.48) the increase, $\Delta C(T_c) = C_s - C_n$, is given by

$$\Delta C(T_c) = -\frac{2}{k_B T_c} N(0)\Delta\frac{d\Delta}{dT}) \int_0^\infty d\xi f(|\xi|)(1-f(|\xi|))$$

$$= -N(0)\left.\frac{d\Delta^2}{dT}\right|_{T=T_c}. \tag{35.50}$$

To evaluate this quantity we need the behavior of $\Delta(T)$ in the vicinity of T_c. This can be obtained by expanding (35.39) and using Eq. (34.57), with the result:

$$\Delta(T) = 1.74\Delta(0)\sqrt{1-\frac{T}{T_c}}, \tag{35.51}$$

in agreement in form with the Landau theory of a second-order phase transition. Using this expression together with the expression for the normal state heat capacity, $C_n = (2\pi^2/3)N(0)k_B^2 T$, then yields

$$\frac{\Delta C(T_c)}{C_n} = 1.43. \tag{35.52}$$

At very low temperature, $f_k \ll 1$ and $\Delta\, d\Delta/dT \cong 0$, the heat capacity then approaches the limiting form

$$C \cong \frac{2N(0)\Delta^2(0)e^{-\frac{\Delta(0)}{k_B T}}}{k_B T^2} \int_0^\infty e^{-\frac{\xi^2}{2k_B T\Delta(0)}} d\xi$$

$$\cong 2N(0)\Delta(0)\sqrt{2\pi}k_B\left(\frac{\Delta(0)}{k_B T}\right)^{3/2} e^{-\frac{\Delta(0)}{k_B T}}; \tag{35.53}$$

we see that the heat capacity approaches zero asymptotically as $\exp(-\Delta(0)/k_B T)$ due to the presence of the gap in the energy spectrum, as discussed in Chapter 32.

35.3 The pair amplitude and the order parameter

We now define a quantity called the *anomalous pair amplitude*, as the expectation value of the operator, $\hat{c}_{k\uparrow}\hat{c}_{-k\downarrow}$, which we write as $\langle \hat{c}_{k\uparrow}\hat{c}_{-k\downarrow}\rangle$. Ordinarily, any expectation value involving an unequal number of creation and destruction operators will vanish as it cannot preserve the number of particles. However, in treating a superconductor we used a wave function (more generally a statistical ensemble) which does not describe a system with a fixed particle number. To evaluate the pair amplitude we insert Eq. (35.22) and the Hermitian conjugate of (35.23) in our definition and retain only combinations involving the Bogoliubov number operator:

$$\begin{aligned}\langle \hat{c}_{\mathbf{k}\uparrow}\hat{c}_{-\mathbf{k}\downarrow}\rangle &= \left\langle \left(u_{\mathbf{k}}\hat{\gamma}_{\mathbf{k}\uparrow} + v_{\mathbf{k}}^{*}\hat{\gamma}_{-\mathbf{k}\downarrow}^{\dagger}\right)\left(u_{\mathbf{k}}\hat{\gamma}_{-\mathbf{k}\downarrow} - v_{\mathbf{k}}^{*}\hat{\gamma}_{\mathbf{k}\uparrow}^{\dagger}\right)\right\rangle \\ &= \left\langle u_{\mathbf{k}}v_{\mathbf{k}}^{*}\left(-\hat{\gamma}_{\mathbf{k}\uparrow}\hat{\gamma}_{\mathbf{k}\uparrow}^{\dagger} + \hat{\gamma}_{-\mathbf{k}\downarrow}^{\dagger}\hat{\gamma}_{-\mathbf{k}\downarrow}\right)\right\rangle \\ &= \left\langle u_{\mathbf{k}}v_{\mathbf{k}}^{*}\left(\mathrm{f}_{-\mathbf{k}\downarrow}(\mathrm{T}) + \mathrm{f}_{\mathbf{k}\uparrow}(\mathrm{T}) - 1\right)\right\rangle. \end{aligned} \tag{35.54}$$

Note that since the product $u_{\mathbf{k}}v_{\mathbf{k}}^{*}$ vanishes above the transition temperature it is common to identify (35.54) with the *order parameter* introduced earlier in Chapter 33, in our discussion of the Ginzburg–Landau theory. If we assign global phases according to $u_{\mathbf{k}} = e^{i\phi/2}\,|u_{\mathbf{k}}|$ and $v_{\mathbf{k}} = e^{-i\phi/2}\,|v_{\mathbf{k}}|$ we can write

$$\begin{aligned}\psi(\mathrm{T}) &= \sum_{\mathbf{k}}\langle \hat{c}_{\mathbf{k}\uparrow}\hat{c}_{-\mathbf{k}\downarrow}\rangle \\ &= e^{i\phi}\,|\psi(\mathrm{T})|, \end{aligned} \tag{35.55}$$

where

$$|\psi(\mathrm{T})| = \sum_{\mathbf{k}}|u_{\mathbf{k}}|\,|v_{\mathbf{k}}|\left(\mathrm{f}_{-\mathbf{k}\downarrow}(\mathrm{T}) + \mathrm{f}_{\mathbf{k}\uparrow}(\mathrm{T}) - 1\right). \tag{35.56}$$

Taking $\mathrm{f}_{-\mathbf{k}\downarrow} = \mathrm{f}_{\mathbf{k}\uparrow}$, we can write (35.52) as

$$|\psi(\mathrm{T})| = \sum_{\mathbf{k}}|u_{\mathbf{k}}|\,|v_{\mathbf{k}}|\tanh\left(\frac{\varepsilon_{\mathbf{k}}}{2k_B\mathrm{T}}\right). \tag{35.57}$$

Note the similarity of this result with the equation for the gap, Eqs. (35.36a,b); in fact the gap function is also commonly taken as the order parameter.

ADDITIONAL READING

Superconductivity, J.B. Ketterson and S.N. Song, Cambridge University Press (2000).

Superconductivity of Metals and Alloys, P.G. deGennes, W.A. Benjamin, NY (1966).

Introduction to Superconductivity, M. Tinkham, McGraw-Hill, NY (1975).

Superconductivity: Conventional and Unconventional Superconductors, Vol. 1, K.H. Bennemann and J.B. Ketterson, Eds., Springer (2008).

Superfluid ^{4}He

36

Liquid He is the only substance which, below some temperature-dependent pressure, does not solidify on cooling to absolute zero.[1] The fact that it does not freeze is due to the large quantum zero-point energy and the weak, isotropic, interatomic force. Two isotopes of helium exist, ^{4}He and ^{3}He. Both nuclei have two protons but ^{4}He and ^{3}He have, respectively, two neutrons and one neutron, corresponding to nuclear angular momenta of $I = 0$ and $I = 1/2$. The $S = 0$ electronic ground state has no angular momentum and hence the total quantum number, F, is identical to I. The very low temperature properties of the liquid phase of these two isotopes are governed by the associated quantum statistics: Bose–Einstein for ^{4}He and Fermi–Dirac for ^{3}He. Since quantum effects play such a dominant role the helium liquids are often called *quantum liquids*.

On cooling the gas at atmospheric pressure, liquefaction of ^{4}He occurs at 4.2 K. On further cooling, a second-order liquid–liquid phase transition is observed at 2.19 K. This is a unique phase transition not occurring elsewhere in nature and the properties of the new (low temperature) liquid are quite unusual. Most important is the property of superfluidity, where, depending on the experimental conditions, a flow can take place with no associated viscosity. However, under a different set of conditions the liquid may behave as an ordinary viscous liquid. In fact it turns out that, in a mathematically consistent way, the properties of the liquid may be described in terms of two inter-penetrating and non-interacting fluids: a normal viscous fluid, and a *superfluid* which has no viscosity. To rationalize this picture requires a microscopic model. The one we will use here is the phenomenological model of Landau.[2,3,4]

36.1 Landau's phenomenological model for ^{4}He

We can extend some of the reasoning used in Sections 31.5 and 31.6 for the weakly interacting Bose gas to the case of the "Bose liquid," ^{4}He. A basic notion in low temperature physics is the idea that quantum many-body systems may be described in terms of *elementary excitations*. At absolute zero there are no excitations, and the system is said to be in its ground state. On heating the system, excitations will appear. The translational invariance of liquid helium suggests that we associate a momentum and energy with each of these excitations and, provided they are few in number, they will behave as a weakly interacting gas. Being

[1] The 1913 Nobel Prize in Physics was awarded to Heike Kamerlingh Onnes "for his investigations on the properties of matter at low temperatures which led, inter alia, to the production of liquid helium."

[2] L. Landau, The theory of superfluidity in helium II, *J. Phys. USSR* **V**, 71 (1941).

[3] L.D. Landau, On the theory of superfluidity in helium II, *J. Phys. USSR* **XI**, 91, (1947).

[4] The 1962 Nobel Prize in Physics was awarded to Lev Landau "for his pioneering theories for condensed matter, especially liquid helium."

The Physics of Solids. First Edition. J. B. Ketterson.

made up of atoms obeying Bose statistics, our excitations will also be bosons. If we have the energy–momentum dispersion relation, $\varepsilon(\mathbf{p})$, for the excitations, which we expect to differ in some ways from that of the weakly interacting Bose gas, we can use the Bose distribution for an ideal gas to determine the thermal properties. The total energy will then be given by

$$E = \frac{V}{(2\pi\hbar)^3} \int d^3p\, n(\varepsilon, T)\, \varepsilon, \tag{36.1}$$

where $n(\varepsilon)$ is the Bose occupation probability for the state, $\varepsilon(p)$, given by

$$n(\varepsilon, T) = \frac{1}{e^{\varepsilon(p)/k_B T} - 1}. \tag{36.2}$$

Unlike the Bose gas, there is no chemical potential since there is no constraint on the excitations.

At long wavelengths (small p) the excitations must correspond to longitudinal sound waves.[5] Here, our excitations are the *phonons* that we encountered for the weakly interacting Bose gas, which are familiar from the Debye theory of specific heats. Since the frequency–wave vector relation of a sound wave is (where we now replace u by c)

[5] A liquid, which will not support a shear, can have no transverse waves.

$$\omega = ck, \tag{36.3a}$$

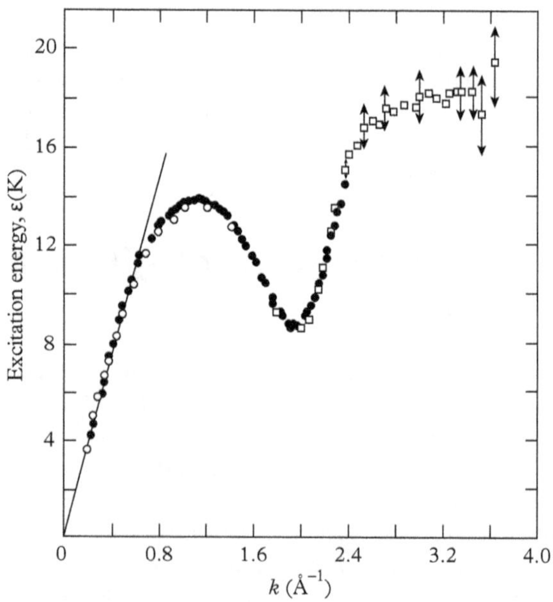

Figure 36.1 *The dispersion curve for He II for the elementary excitations at* T = 1.1 K. *(From Cowley and Woods (1971).)*

where $c = 2.38 \times 10^4$ cm/sec is the sound velocity (for $T = P = 0$), our phonons obey the dispersion relation

$$\varepsilon = cp. \tag{36.3b}$$

Now, at zero temperature and pressure the average interatomic spacing in ^4He is 3.8 A, and when the wavelength begins to approach this value significant deviations can be expected to occur.[6] Fortunately, the method of inelastic neutron scattering permits a complete determination of $\varepsilon(p)$, the results of which are shown in Fig. 36.1.[7,8] For historical reasons the excitations near the minimum (at $p = p_0$) are called *rotons*. For temperatures above 1 K they dominate the thermal properties.[9] In the vicinity of p_0 we may write the roton energy as

$$\varepsilon(p) = \Delta + \frac{(p - p_0)^2}{2\mu}, \tag{36.4}$$

where $\Delta = 8.9$ K, $p_0 = 2.1 \times 10^{-19}$ gcm/sec; $\mu = 1.72 \times 10^{-24}$ g.

Let us first calculate the thermal properties of the phonon gas, where ε is given by Eq. (36.3b). Doing the angular integration and defining the dimensionless variable, $x = cp/k_B T$,

$$E_p = \frac{V(k_B T)^4}{2\pi^2 \hbar^3 c^3} \int dx \frac{x^3}{e^x - 1};$$

at low temperature the upper limit of the integration, which is ordinarily cut off by the number of degrees of freedom in the Debye theory, may be extended to ∞. Using $\int dx\, x^3/(e^x - 1) = \pi^4/15$, we obtain

$$E_p = \frac{V\pi^2 (k_B T)^4}{30\hbar^3 c^3}. \tag{36.5}$$

The phonon heat capacity, $c_p = \partial E/\partial T$, is given by

$$c_p = \frac{2}{15} \frac{V\pi^2 k_B^4 T^3}{\hbar^3 c^3}. \tag{36.6}$$

We next calculate the roton thermal properties. Since $\Delta \gg k_B T$, we may use the Boltzmann occupation factor. Using Eq. (36.4) for $\varepsilon(p)$, Eq. (36.1) becomes

$$E_r = \frac{V}{(2\pi\hbar)^3} \int d^3p e^{-\left[\Delta + \frac{1}{2\mu}(p - p_0)^2\right]/k_B T} \left[\Delta + \frac{1}{2\mu}(p - p_0)^2\right]. \tag{36.7}$$

To sufficient accuracy we may write $p^2 dp = p_0^2\, dp$, and using

$$\int_0^\infty dx e^{-ax^2} = \left(\frac{\pi}{a}\right)^{1/2} \quad \text{and} \quad \int_0^\infty dx e^{-ax^2} x^2 = \frac{\pi^{1/2}}{2a^{3/2}},$$

[6] Although first principles many-body quantum theory permits a reasonably accurate calculation of $\varepsilon(p)$, the overall shape may be rationalized by considering two effects. Since the liquid possesses a structure on an interatomic distance scale, one might expect a diffraction-like effect analogous to the turning down of a phonon dispersion curve as it approaches the first Brillouin zone; however for larger momenta the curve should ultimately take up a free particle behavior, where $\varepsilon(p) = p^2/2m$.

[7] R.A. Cowley and A.D.B. Woods, *Phys. Rev. Lett.* **21**, 787 (1968); *Can. J. Phys.* **49**, 177 (1971).

[8] A.D.B. Woods and R.A. Cowley, Structure and excitations of liquid helium, *Rep. Prog. Phys.* **36**, 1135 (1973).

[9] This is a "phase space effect"; since all statistical integrations involve a factor, $d^3p = 4\pi p^2 dp$, excitations with large p can be important even though separated from the ground state by a large energy.

we obtain

$$E_r = \frac{2Vp_0^2\,(\mu k_B T)^{1/2}\,e^{-\Delta/k_B T}}{(2\pi)^{3/2}\quad \hbar^3}\left[\Delta + \frac{1}{2}k_B T\right]. \tag{36.8}$$

The roton heat capacity is given by

$$C_r = \frac{2k_B\,Vp_0^2\,(\mu k_B T)^{1/2}}{(2\pi)^{3/2}\quad \hbar^3}\,e^{-\Delta/k_B T}\left[\frac{3}{4} + \frac{\Delta}{k_B T} + \left(\frac{\Delta}{k_B T}\right)^2\right]. \tag{36.9}$$

36.2 Superfluidity

We now examine the property of superfluidity in more detail.[10] Consider the situation shown in Fig. 36.2(a); liquid helium in its ground state (i.e., at $T = 0$) is flowing uniformly[11] through a pipe with velocity **v**. Let us now go into a frame of reference in which the liquid is at rest; in this frame (Fig. 36.2(b)) the pipe moves with velocity $-\mathbf{v}$. Suppose we now introduce an excitation of energy, $\varepsilon(p)$, into the liquid in the rest frame. From the law of transformation of energy to a moving frame (Galilean invariance), the energy E in the lab frame is given by

$$E = \varepsilon + \mathbf{p}\cdot\mathbf{v} + \frac{Mv^2}{2}, \tag{36.10}$$

where $Mv^2/2$ is just the kinetic energy of the liquid as a whole. On the other hand, $\varepsilon + \mathbf{p}\cdot\mathbf{v}$ is the change in energy due to the appearance of the excitation, which must be less than zero if the liquid is to decrease its energy, i.e., if drag is to occur. Thus,

$$\varepsilon + \mathbf{p}\cdot\mathbf{v} < 0 \tag{36.11}$$

this condition is most easy to satisfy if **p** and **v** are antiparallel; i.e., if

$$\varepsilon - pv < 0 \tag{36.12a}$$

[10] Part of the 1978 Nobel Prize in Physics was awarded to Pyotr Leonidovich Kapitsa "for his basic inventions and discoveries in the area of low-temperature physics" which included superfluidity in ^4He. This phenomenon was independently discovered by J.F. Allen and A.D. Misener, *Nature* **141**, 75 (1937).

[11] Such a flow is possible only if the viscosity vanishes, which we will shortly rationalize.

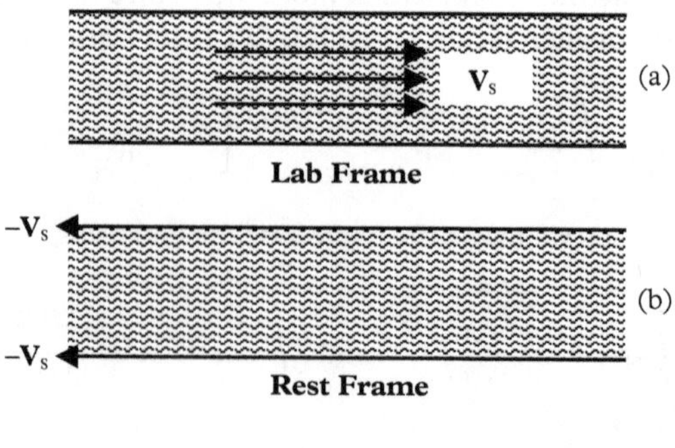

Figure 36.2 *(a) A superfluid moving uniformly through a pipe with velocity $\mathbf{v}_s$ in the laboratory frame; (b) The same fluid as viewed from a frame where it is at rest but the pipe moves with a velocity $-\mathbf{v}_s$.*

or

$$v > \frac{\varepsilon}{p} . \qquad (36.12b)$$

This inequality is first satisfied for the excitations that lie at the point of tangency of a line drawn from the origin to the curve $\varepsilon(p)$, which occurs near the roton minimum. Since the roton minimum is relatively sharp, we can approximate this so-called critical velocity as

$$v_{crit} \cong \frac{\Delta}{p_0}; \qquad (36.13)$$

v_{crit} is about 5×10^3 cm/sec. For fluid velocities less than this value the energy of the fluid cannot be lowered by the introduction of excitations; the fluid will therefore flow with no viscous loss, i.e., it will be a *superfluid*.[12]

36.3 The two-fluid model

At finite temperature some distribution of excitations exists. Let us calculate the effect of moving this gas of excitations with some velocity, $-\mathbf{v}$, relative to the fixed ground state (Fig. 36.2(b)). The total momentum associated with the moving excitation gas is

$$\mathbf{P} = \frac{V}{(2\pi\hbar)^3} \int d^3p\, \mathbf{p}\, n(\varepsilon') , \qquad (36.14)$$

where ε' is the energy of the excitation in a frame moving with the gas, which we take to be the equilibrium distribution. We use our Galilean transformation to evaluate the energies in terms of quantities measured in the frame where the ground state is at rest, i.e., $\varepsilon' = \varepsilon - \mathbf{p} \cdot \mathbf{v}$. Thus,

$$\mathbf{P} = \frac{V}{(2\pi\hbar)^3} \int d^3p\, \mathbf{p}\, n(\varepsilon - \mathbf{p} \cdot \mathbf{v}), \qquad (36.15)$$

or on expanding to leading order in $\mathbf{v}$,

$$\mathbf{P} \cong \frac{V}{(2\pi\hbar)^3} \int d^3p\, \mathbf{p} \left[n(\varepsilon) - \frac{\partial n}{\partial \varepsilon}\, \mathbf{p} \cdot \mathbf{v} + \cdots \right]. \qquad (36.16)$$

Noting that the average of $\mathbf{p}$ for an equilibrium distribution vanishes, and performing the angular integration on the remaining term, we have

$$\mathbf{P} = -\mathbf{v}\, \frac{4\pi}{3} \frac{V}{(2\pi\hbar)^3} \int dp\, p^4 \frac{\partial n}{\partial \varepsilon} . \qquad (36.17)$$

[12] Some superconductors, along with the Fermi superfluid ^{3}He, can have a BCS-like energy gap with nodes; i.e., the slope of the line implied by (36.13) is formally zero. The criterion for superfluidity then lies at a deeper level and is associated with the existence of a condensate and an associated macroscopic phase function.

We now evaluate Eq. (36.17) for the phonons. Integrating by parts and using $\varepsilon = cp$, we obtain

$$\mathbf{P} = \mathrm{v}\,\frac{16\pi}{3c}\,\frac{V}{(2\pi\hbar)^3}\int p^3 n(p)\,dp = \frac{4}{3c^2}\,E_p\,\mathrm{v}, \tag{36.18}$$

where E_p is again the total energy of the phonon gas, given by Eq. (36.5). If we associate $|\mathbf{P}|/|\mathbf{v}|$ with a *mass*, then the corresponding mass per unit volume will be given by

$$\rho_{np} = \frac{4}{3c^2}\,\frac{E_p}{V} = \frac{2\pi^2}{45\hbar^3 c^5}\,(k_B T)^4. \tag{36.19}$$

This quantity we identify as the phonon contribution to a *normal mass density* of the liquid. To evaluate the roton contributions we again employ Boltzmann statistics and note that $\partial n/\partial\varepsilon = -n/k_B T$; using Eq. (36.17) we have

$$\rho_{nr} = \frac{4\pi}{3k_B T(2\pi\hbar)^3}\int dp\,p^4\,n(\varepsilon)\,. \tag{36.20}$$

We approximate p^4 by $p_0{}^4$ and the resulting integral is

$$\rho_{nr} = \frac{2\mu^{1/2}\,p_0^4}{3(2\pi)^{3/2}(k_B T)^{1/2}\hbar^3}\,e^{-\Delta/k_B T}. \tag{36.21}$$

The total normal density is then

$$\rho_n = \rho_{np} + \rho_{nr}. \tag{36.22}$$

That fraction of the liquid which is not normal we call the *superfluid*; i.e.,

$$\rho_s = \rho - \rho_n, \tag{36.23}$$

where ρ is the density of the fluid itself and, for velocities below the critical velocity, it moves with *no friction.* At $T = 0$ all of the liquid is superfluid.

The transition temperature, T_λ, to the normal state (He I), or λ point as it is called,[13] is identified as the temperature at which $\rho_n(T_\lambda) = \rho$. For temperatures below the phase transition, $\rho_n < \rho$. As was the case for the dilute Bose gas, interactions of the elementary excitations would make such an estimate of T_c only qualitative.

36.4 The thermo-mechanical effect

Consider two vessels which are connected by a super leak,[14] as shown in Fig. 36.3. It is observed that if a temperature differential is established between the two

[13] The name arises from the shape of the heat capacity curve, which looks like the Greek letter λ near the phase transition.

[14] A super leak is a connection constructed so that the resistance to the flow of normal fluid is exceedingly high and hence only superfluid can be transported.

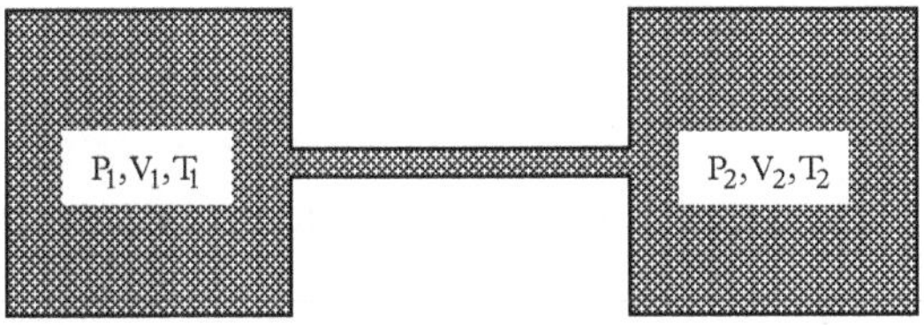

Figure 36.3 *Two vessels maintained at different temperatures, connected by a super leak.*

vessels, an exchange of superfluid occurs, resulting in a pressure differential ΔP. We will analyze this situation thermodynamically. The combined first and second laws may be written as

$$dE_i = T_i\, dS_i - P_i dV_i + \mu_i\, dN_i, \tag{36.24}$$

where the index i distinguishes the two vessels and E_i, S_i, V_i, and N_i are the energy, entropy, volume, and particle number for each vessel. Since the superfluid cannot transport entropy (it is in the ground state), the entropies of the two vessels must remain constant, i.e., $dS_1 = dS_2 = 0$; also the volumes are here assumed to be constant, requiring $dV_1 = dV_2 = 0$. Since the total energy remains constant, we have

$$\mu_1 dN_1 + \mu_2 dN_2 = 0. \tag{36.25}$$

But the total number of particles, $N_1 + N_2$, is also constant, which requires $dN_1 = -dN_2$; thus,

$$\mu_1(T_1, P_1) = \mu_2(T_2, P_2)\,. \tag{36.26}$$

Since the pressure dependence of μ is small, we can expand it, obtaining

$$\begin{aligned} \mu(T_2, P_2) &= \mu(T_2, P_1) + \left(\frac{\partial \mu}{\partial P}\right)_T (P_2 - P_1) \\ &= \mu(T_1, P_1) + \int_{T_1}^{T_2} dT \left(\frac{\partial \mu}{\partial T}\right)_P + \left(\frac{\partial \mu}{\partial P}\right)_T (P_2 - P_1). \end{aligned} \tag{36.27}$$

Using the standard expression for the differential of the Gibbs free energy, we have

$$\left(\frac{\partial \mu}{\partial T}\right)_P = -\frac{S}{N}; \quad \left(\frac{\partial \mu}{\partial P}\right)_T = \frac{V}{N}. \tag{36.28}$$

Defining the mass density, $\rho = M/V$, introducing an entropy per unit mass,[15] $s = S/M$, and using Eq. (36.26), Eq. (36.27) becomes

$$P_2 - P_1 = \int_{T_1}^{T_2} dT\, \rho s. \tag{36.29}$$

The thermo-mechanical effect, or fountain effect as it is also called, is analogous to the *osmotic pressure* which occurs in solutions where the solvent (here the superfluid) can be exchanged between two vessels containing different quantities of solute (the normal fluid) through a semi-permeable membrane (the super leak).

[15] Note that in hydrodynamics it is common to introduce an *entropy per unit mass*, which in some ways is more natural than an entropy per unit volume; the former is constant under an adiabatic volume change (as when an adiabatic sound wave propagates), whereas the latter would not be constant.

36.5 Hydrodynamics of superfluids and sound propagation

Associated with the densities, $\rho_n(\mathbf{r}, t)$ and $\rho_s(\mathbf{r}, t)$, characterizing the normal and superfluid components, we will in general have two velocity fields, $\mathbf{v}_n(\mathbf{r}, t)$ and $\mathbf{v}_s(\mathbf{r}, t)$, and we now give a simplified discussion of the accompanying two-fluid hydrodynamics.[16] Since the superfluid component represents the fraction of the liquid in its ground state, we associate the transport of entropy with the normal fluid moving at $\mathbf{v}_n$. Continuing to write the entropy per unit volume as ρs, where ρ is the mass per unit volume and s is the entropy per unit mass, and ignoring dissipative (entropy generating) effects, the rate of change of entropy within some volume, V, arises from a flux, $\rho s\mathbf{v}_n$, of entropy through the surrounding surface, S, and is therefore given by

$$\frac{\partial}{\partial t}\int_V \rho s\, d^3r = -\int_S \rho s\mathbf{v}_n \cdot d^2\mathbf{r}; \tag{36.30}$$

recognizing that this must hold independent of the chosen volume and applying Green's theorem, we obtain a law for conservation of entropy as

$$\frac{\partial}{\partial t}(\rho s) + \nabla \cdot (\rho s\mathbf{v}_n) = 0. \tag{36.31}$$

Our next task is to obtain equations of motion that describe the time evolution of our two velocity fields. In what follows we will continue to ignore dissipation and limit our discussion to the linear approximation using the intuitive approach given by London, which in turn utilized earlier ideas of Tisza[17] (rather than the more formal derivation of Landau). For a general state of motion, the kinetic energy density, e_K, and momentum density, $\mathbf{g}$, are given by

$$e_K = \frac{1}{2}\rho_n v_n^2 + \frac{1}{2}\rho_s v_s^2 \tag{36.32}$$

and

$$\mathbf{g} = \rho_n\mathbf{v}_n + \rho_s\mathbf{v}_s. \tag{36.33}$$

We can rewrite Eq. (36.32) in the form

$$e_K = \frac{1}{2\rho}(\rho_s\mathbf{v}_s + \rho_n\mathbf{v}_n)^2 + \frac{\rho_s\rho_n}{2\rho}(\mathbf{v}_s - \mathbf{v}_n)^2 \tag{36.34}$$

which we rewrite as

$$e_K = e_K^+ + e_K^-; \tag{36.35}$$

comparing (36.33) and (36.34), we have $e_K^+ = \mathbf{g}^2/2\rho$. We therefore associate e_K^+ with the kinetic energy density of the liquid as a whole, whereas e_K^- can be viewed as an "internal" kinetic energy density.

[16] For a brief history, see, The two-fluid theory and second sound in liquid helium, R.J. Donnelly, *Phys. Today* October 2009, p. 34.

[17] L. Tisza, *Nature*, **141**, 913 (1938); *Compt. Rend.* **207**, 1035 (1938); *Compt. Rend.* **207**, 1186 (1938); *J. Phys. Radium* **1**, 164 (1940); *J. Phys. Radium* **1**, 350 (1940). It seems that Tisza was the first to propose a two-fluid model of He II, which he based on a Bose–Einstein condensation. Although this model does not give the correct excitation spectrum, which is better captured by Landau's model, it does introduce the idea that He II possesses a condensate, which is now generally accepted as the defining characteristic of a superfluid. Curiously, Landau appears to have not placed much emphasis on this feature. For additional discussion of these points, see, Allan Griffin, Laszlo Tisza (1907–2009): An Appreciation, *J. Low. Temp. Phys.* **157**, 1 (2009).

36.5.1 Second sound

He II supports two kinds of propagating collective modes. The first is the usual sound waves, where the momentum and density oscillate out of phase. The other mode is called *second sound* and corresponds to a counter motion of the normal and superfluid velocities, but with no change in the overall momentum density or mass density of the fluid; since this is a new feature, unique to the superfluid state, we will treat it first. A consistent treatment of this mode emerges by assuming that $\mathbf{g} = 0$, from which it follows that $\mathbf{v}_s = -(\rho_n/\rho_s)\mathbf{v}_n$, $e_K^+ = 0$, and $e_K^- = \rho(\rho_n/2\rho_s)\mathbf{v}_n^2$. Next we recall the thermodynamic identity which, for a fixed volume of fluid, takes the form

$$d\varepsilon = Td(\rho s) - \mu'\, d\rho, \tag{36.36}$$

where ε is the energy per unit volume and μ' is the chemical potential per unit mass. Carrying out a Legendre transformation to define the Gibbs free energy per unit volume, $\rho\mu' = \varepsilon + P - T\rho s$, where P is the pressure, we obtain

$$\rho d\mu' = dP - \rho s dT. \tag{36.37}$$

Now the change in free energy density, $d(\rho\mu')$, at constant pressure and density must arise from the change in the internal kinetic energy density, $de_K^- = (\rho\rho_n/\rho_s)\,\mathbf{v}_n \cdot d\mathbf{v}_n$. Suppose we consider a slab of area A and thickness dx. The change in free energy of this slab is then given by $A de_K^- dx$. This change must result from an influx of entropy density taken as moving with velocity, v_n, normal to the slab, which is driven by a temperature differential, dT, that acts for a time, dt, yielding $Av_n\, \rho s\, dTdt$. Equating these two forms, we obtain $(\rho_n/\rho_s)\,(\partial v_n/\partial t) = -s\partial T/\partial x$ or, extending to an arbitrary volume element,

$$\frac{\rho_n}{\rho_s}\frac{\partial \mathbf{v}_n}{\partial t} + s\nabla T = 0. \tag{36.38}$$

Taking the divergence of (36.38), using (36.31), and ignoring terms of second order gives

$$\frac{\partial^2 s}{\partial t^2} - \frac{\rho_s}{\rho_n} s^2 \nabla^2 T = 0;$$

introducing the heat capacity per unit mass at constant pressure, $c_P = T\partial s/\partial T$, this equation becomes

$$\frac{\partial^2 T}{\partial t^2} - \frac{T\rho_s}{c_P\rho_n} s^2 \nabla^2 T = 0. \tag{36.39}$$

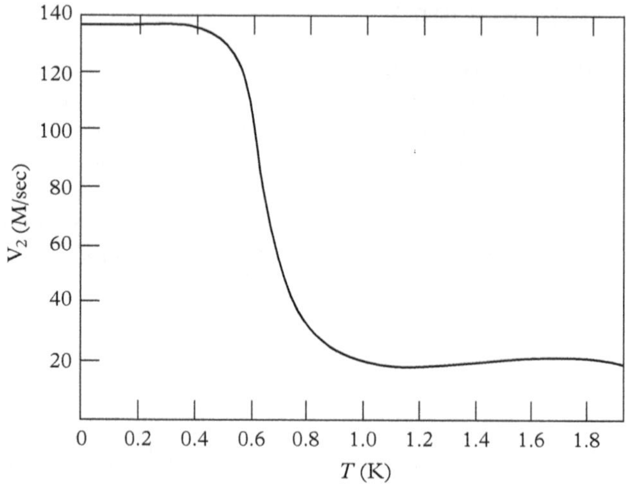

Figure 36.4 *The temperature dependence of the velocity of second sound in superfluid* ^{4}He. *(Adapted from V. Peshkov (1960).)*

Equation (36.39) is a wave equation having solutions with a mode velocity

$$c_2^2 = \frac{\rho_s}{\rho_n}\frac{s^2 T}{c_P} \tag{36.40}$$

in which the temperature oscillates in space and time. Since the density and pressure associated with the propagation of this mode remain constant it cannot be excited by a conventional piezoelectric transducer. However, passing an oscillating electrical current through a thin resistive film will cause its temperature to oscillate and will launch a second sound wave (propagating at twice the frequency); the mode can be detected using a second film having a high differential change in resistance with temperature (such as a superconductor biased close to its transition temperature with a magnetic field). Figure 36.4 shows the temperature dependence of the second sound velocity; the phenomenon was first observed by V. Peshkov in 1944.[18]

36.5.2 First sound and the equations of motion

To complete our discussion of the linearized hydrodynamics of He II we must include motions in which $\mathbf{g} \neq 0$. When $\mathbf{g} = 0$ we have $\tilde{\mathbf{v}} \equiv \mathbf{v}_n - \mathbf{v}_s = (\rho/\rho_s)\mathbf{v}_n$, and from Eq. (36.38) it then follows that

$$\begin{aligned} \dot{\mathbf{v}} &= \dot{\mathbf{v}}_n - \dot{\mathbf{v}}_s \\ &= -(\rho/\rho_n)\, s\, \nabla T. \end{aligned} \tag{36.41}$$

This equation involving the relative motion of the two fluids should also hold in a frame where $\mathbf{g} \neq 0$, moving with a velocity, $\mathbf{v}$, defined by $\rho\mathbf{v} = \rho_s\mathbf{v}_s + \rho_n\mathbf{v}_n$.

[18] Second sound in helium II, V. Peshkov, *J. Phys. USSR*, 8, 381 (1944); *Sov. Phys. JETP* 11, 580 (1960).

Now the Euler equation of conventional hydrodynamics should continue to hold for the motion of the fluid as a whole; hence,

$$\rho \dot{\mathbf{v}} = -\nabla P. \tag{36.42}$$

Using Eqs. (36.41) and (36.42) to separately solve for $\mathbf{v}_n$ and $\mathbf{v}_s$, we obtain

$$\dot{\mathbf{v}}_s = -\frac{1}{\rho}\nabla P + s\,\nabla T \tag{36.43a}$$

and

$$\dot{\mathbf{v}}_n = -\frac{1}{\rho}\nabla P - \frac{s\rho_s}{\rho_n}\nabla T. \tag{36.43b}$$

To complete the description we require the equations for conservation of mass and entropy:

$$\frac{\partial \rho}{\partial t} + \nabla \cdot (\rho \mathbf{v}) = 0 \tag{36.44a}$$

and

$$\frac{\partial}{\partial t}(\rho s) + \nabla \cdot (\rho s \mathbf{v}_n) = 0. \tag{36.44b}$$

By combining Eqs. (36.42) and (36.44a) and writing $c_1^2 = \partial P/\partial \rho$, we obtain the wave equation for first sound,

$$\nabla^2 \rho - \frac{1}{c_1^2}\frac{\partial^2 \rho}{\partial t^2} = 0, \tag{36.45}$$

where c_1 is the velocity of ordinary density waves in a fluid, which in the present case is designated as *first sound.*

36.5.3 Fourth sound

Analogous to our discussion of the fountain effect, we can identify a propagating mode, *called fourth sound,* in which the superfluid oscillates but the normal fluid is "locked"; the latter is achieved by filling the environment with a finely divided powder or by propagating through a capillary, which via the normal fluid viscosity suppresses its motion. We start by writing (36.44a) in the form $\partial\rho/\partial t + \nabla \cdot (\rho_s \mathbf{v}_s + \rho_n \mathbf{v}_n) = 0$ and setting $\mathbf{v}_n = 0$, which gives $\partial\rho/\partial t = -\rho_s \nabla \cdot \mathbf{v}_s$. Taking the divergence of (36.43a) and inserting this last form gives

$$\frac{\partial^2 \rho}{\partial t^2} = \frac{\rho_s}{\rho}\nabla^2 P - \rho_s s\,\nabla^2 T. \tag{36.46}$$

On the other hand, setting $\dot{\mathbf{v}}_n = 0$ in (36.43b) gives

$$\frac{\partial}{\partial t}(\rho s) = 0, \tag{36.47}$$

or $\delta s = -(s/\rho)\delta\rho$. Writing $P = P(\rho, s)$, expanding the first term in (36.46), and then using (36.47), we have

$$\begin{aligned}\nabla^2 P &= \left(\frac{\partial P}{\partial \rho}\right)_s \nabla^2 \rho + \left(\frac{\partial P}{\partial s}\right)_\rho \nabla^2 s \\ &= \left[\left(\frac{\partial P}{\partial \rho}\right)_s - \left(\frac{\partial P}{\partial s}\right)_\rho \frac{s}{\rho}\right] \nabla^2 \rho;\end{aligned} \tag{36.47a}$$

similarly,

$$\nabla^2 T = \left[\left(\frac{\partial T}{\partial \rho}\right)_s - \left(\frac{\partial T}{\partial s}\right)_\rho \frac{s}{\rho}\right] \nabla^2 \rho. \tag{36.47b}$$

Equation (36.46) now becomes

$$\frac{\partial^2 \rho}{\partial t^2} = \left[\frac{\rho_s}{\rho}\left(\frac{\partial P}{\partial \rho}\right)_s - \frac{\rho_s s}{\rho^2}\left(\frac{\partial P}{\partial s}\right)_\rho - \rho_s s\left(\frac{\partial T}{\partial \rho}\right)_s + \frac{\rho_s s^2}{\rho}\left(\frac{\partial T}{\partial s}\right)_\rho\right] \nabla^2 \rho. \tag{36.48}$$

Assuming propagating waves and using the definitions of c_1 and c_2 we may rewrite this as

$$\omega^2 - k^2\left\{\frac{\rho_s}{\rho} c_1^2 + \frac{\rho_n}{\rho} c_2^2\left[1 - \frac{1}{\rho s}\left(\frac{\partial P}{\partial T}\right)_\rho - \frac{\rho}{s}\left(\frac{\partial T}{\partial \rho}\right)_s\left(\frac{\partial s}{\partial T}\right)_\rho\right]\right\} = 0. \tag{36.49}$$

The last term may be rewritten as

$$-\frac{\rho}{s}\left(\frac{\partial T}{\partial \rho}\right)_s\left(\frac{\partial s}{\partial T}\right)_\rho = \frac{\rho}{s}\left(\frac{\partial T}{\partial \rho}\right)_s\left(\frac{\partial s}{\partial \rho}\right)_T\left(\frac{\partial \rho}{\partial T}\right)_s = \frac{\rho}{s}\left(\frac{\partial s}{\partial \rho}\right)_T = -\frac{1}{\rho s}\left(\frac{\partial p}{\partial T}\right)_\rho, \tag{36.50}$$

where in the last step we have used a Maxwell relation. Defining $\beta = (\partial P/\partial T)_\rho$, we may write

$$c_4^2 = \frac{\omega^2}{k^2} = \frac{\rho_s}{\rho} c_1^2 + \frac{\rho_n}{\rho} c_2^2\left(1 - \frac{2\beta}{\rho s}\right). \tag{36.51}$$

The second term is usually small and to a good approximation we have $c_4 = (\rho_s/\rho)\, c_1$. Fourth sound has been a powerful probe in studying the superfluid density.

36.6 The Landau–Ginzburg theory of ^{4}He

As in earlier applications of Landau's theory of a second-order phase transition, we assume that the free energy may be expanded as a power series in an *order parameter*. For ^{4}He it is taken to be the complex scalar function, $\psi(\mathbf{r}, t)$, the condensate fraction, that was introduced in Section 31.8, in our discussion of the Gross–Pitaevskii equation (see Eq. (31.58)).[19] We again assume that the expansion of F involves only even powers of $|\psi|$, and for a uniform system we write it as

$$F = F_0(T) + \int d^3r \left[\alpha(T)\,|\psi|^2 + \beta(T)\,|\psi|^4\right] \qquad (36.52)$$

or

$$F = F_0(T) + \int d^3r\,[F_2(T) + F_4(T)], \qquad (36.53)$$

where $F_0(T)$ is the free energy of the normal phase and F_2 and F_4 are quadratic and quartic contributions to the free energy density of the superfluid phase.

To assure overall system stability we take $\beta > 0$. When $\alpha(T) > 0$, the minimum energy obviously occurs for $\psi = 0$; however for $\alpha < 0$, ψ will be finite. Thus, we can model a second-order phase transition by writing $\alpha = a(T - T_c)$ and $\beta(T) = \beta(T_c)$. The magnitude of ψ is determined by minimizing Eq. (36.52) with respect to ψ^*:

$$\alpha(T)\,\psi + 2\beta\psi\,|\psi|^2 = 0, \qquad (36.54)$$

which has two solutions:

$$\psi = 0; \qquad T > T_c \qquad (36.55)$$

$$|\psi| = \left[\frac{a(T_c - T)}{2\beta}\right]^{1/2}; \quad T < T_c. \qquad (36.56)$$

Thus the system orders at T_c with a strength proportional to $(T_c - T)^{1/2}$.

In the presence of spatial inhomogeneities we must include an additional term in F to account for the extra free energy that arises when the order parameter is position dependent; such a contribution involves $\nabla\psi$. From symmetry, the new contribution cannot change sign under $\mathbf{r} \to -\mathbf{r}$ and to the lowest order we may then write

$$F_{g2} = \frac{1}{2}K\,|\nabla\psi|^2, \qquad (36.57)$$

[19] The analogous theory for the case of a superconductor is discussed in Chapter 33 and for superfluid ^{3}He in Chapter 38.

where now,

$$F = F_0(T) + \int d^3r \left[F_2 + F_4 + F_{g2}\right]. \tag{36.58}$$

In their treatment of superconductivity, Landau and Ginzburg (see Chapter 33) treated ψ phenomenologically as a *macroscopic wave function* in the spirit of F. London's idea. The effect of a magnetic field is then incorporated by noting that in quantum mechanics one uses the prescription $\nabla \to \nabla - (ie/\hbar c)\mathbf{A}$, where $\mathbf{A}$ is the vector potential. One may derive the particle current using the following trick. Suppose we assign an infinitesimal *fictitious charge* e^* *to each He atom.* Classically, the change in free energy produced by a change in $\delta\mathbf{A}$, in the presence of an electric current $\mathbf{j}$, is given by

$$\delta F = \frac{1}{c}\int d^3r\, \mathbf{j}\cdot\delta\mathbf{A}. \tag{36.59}$$

Thus, we define the superfluid mass current, $\mathbf{g}$, by the functional derivative

$$\mathbf{g} = \frac{M_4 c}{e^*}\left(\frac{\delta F}{\delta \mathbf{A}}\right)_{e^*=0}, \tag{36.60}$$

where M_4 is the mass of the ^{4}He atom. In the presence of a vector potential we have

$$F_{g2} = \frac{1}{2}K\left|\left(\nabla - \frac{ie^*}{\hbar c}\mathbf{A}\right)\psi\right|^2, \tag{36.61}$$

which gives

$$\mathbf{g} = \frac{i}{2}\left(\frac{M_4}{\hbar}\right)K\left[-\psi^*\,\nabla\,\psi + \psi\,\nabla\,\psi^*\right]. \tag{36.62}$$

Writing $\psi(\mathbf{r}) = |\psi(\mathbf{r})|\, e^{i\phi(\mathbf{r})}$, we obtain

$$\mathbf{g} = K\left(\frac{M_4}{\hbar}\right)|\psi|^2\,\nabla\phi. \tag{36.63}$$

To make contact with the Landau theory where the superfluid momentum is $\rho_s \mathbf{v}_s$, we write

$$\mathbf{v}_s = \frac{\hbar}{M_4}\nabla\phi \tag{36.64}$$

and

$$\rho_s = K\left(\frac{M_4}{\hbar}\right)^2|\psi|^2. \tag{36.65}$$

Note that if we literally interpret ψ as a wave function and use the quantum mechanical expressions for particle flux, we are led to the expression

$$\mathbf{g} = \frac{\hbar}{2i}\left(\psi^*\nabla\psi - \psi\nabla\psi^*\right), \tag{36.66}$$

where $\rho_s = M_4\,|\psi|^2$ and $\mathbf{v}_s = (\hbar/M_4)\,\nabla\varphi$, implying that $K = \hbar^2/M_4$.

If the theory is generalized to account for time-dependent phenomena, the Gross–Pitaevskii equation, discussed in Chapter 31, can serve as a starting point.

36.7 Quantized vorticity

A basic assumption of the Landau hydrodynamics was that $\nabla \times \mathbf{v}_s = 0$ (so-called irrotational flow) and we now look at this assumption in more detail. We start by examining the integral

$$\int d^2\mathbf{r} \cdot \nabla \times \mathbf{v}_s = \oint d\boldsymbol{l} \cdot \mathbf{v}_s, \tag{36.67}$$

which is sometimes called the *circulation*. If we consider individual superfluid (condensate) atoms moving with velocity, $\mathbf{v}_s$, each having a momentum, $\mathbf{p} = M_4\mathbf{v}_s$, we can rewrite the integral on the right-hand side of (36.67) in the form $(1/M_4)\oint d\boldsymbol{l}\cdot\mathbf{p}$. We now assume that our condensate atoms are executing circular "orbits" around some imaginary central line, as one has for a vortex line in conventional hydrodynamics; if we now interpret the line integral in the light of the Bohr–Sommerfeld quantization rule, $\oint d\boldsymbol{l}\cdot\mathbf{p} = 2\pi n\hbar$, and carry out this integral on a circle of radius r around the central line, we obtain the condition

$$\oint d\boldsymbol{l}\cdot\mathbf{p} = M_4\oint d\boldsymbol{l}\cdot\mathbf{v}_s = 2\pi r M_4 v_s = 2\pi n\hbar \tag{36.68}$$

or

$$v_s = \frac{nh}{M_4}\frac{1}{r}, \tag{36.69}$$

where $h = 2\pi\hbar$ is the usual Planck constant. We have obtained the somewhat remarkable result that the flow is *quantized* for a superfluid vortex. Note that for all $r \neq 0$ the curl of (36.69) vanishes; in this sense the flow is still irrotational. But the velocity predicted by this expression diverges for small r and hence it must break down; we expect this breakdown to occur at a radius, r_c, where the superfluid velocity, v_s, is of the order of the Landau critical velocity (50 m/sec):

$$r_c = \frac{nh}{M_4}\frac{1}{v_{crit}}. \tag{36.70}$$

We can interpret r_c as a *core radius* for our vortex and it is of the order of an angstrom. For values of r of the order of or smaller than r_c, we must have $\nabla \times \mathbf{v}_s \neq 0$ when $n \neq 0$. In practice, vortices with $n > 1$ are unstable to a break-up into vortices with $n = 1$.

We can equally well consider a vortex from the perspective of the Ginzburg–Landau theory. According to Eq. (36.64) the superfluid velocity is related to the gradient of the phase of the superfluid (condensate) wave function. Requiring that the change in phase on circumventing the vortex core be restricted to a multiple of 2π also leads to Eq. (36.69). For the case of a superconductor one has quantized flux lines, as was discussed in Section 32.6.

Problem

36.1. Suppose we write the critical current of a superconductor in the form $\mathbf{j}_{crit} = ne\mathbf{v}_{crit}$, where n is the carrier density and v_{crit} is given by Eq. (36.13), with Δ and p_0 interpreted as the BCS energy gap, Δ, and Fermi momentum, p_F, respectively. Using Eq. (34.61), relating the thermodynamic condensation energy (and the associated critical field, H_c) to that predicted by the BCS theory, show that the critical current density can be written in the form

$$j_{crit} = \text{const.} \times cH_c/\lambda_L,$$

where λ_L is the London penetration depth, as given by Eq. (32.6). Obtain the proportionality constant.

ADDITIONAL READING

The Physics of Liquid and Solid Helium. Parts I and II. K.H. Bennemann and J.B. Ketterson, Eds., Wiley Interscience, NY (1975, 1978).

Introduction to the Theory of Superfluidity, I.M. Khalatnikov, W.A. Benjamin and Sons, NY (1965).

Superfluidity and Superconductivity, D.R. Tilley and J. Tilley, Adam Hilger, Bristol (1986).

The Properties of Liquid and Solid Helium, J.W. Wilks, Oxford University Press, Oxford (1967).

The Theory of Quantum Liquids, Vols. I and II, P. Nozieres and D. Pines, Addison-Wesley, Reading, MA (1966, 1990).

Superfluids, Vol. II, Fritz London, Dover, NY (1964).

Landau's Theory of a Fermi Liquid

37

37.1 Equilibrium properties

37.1.1 The challenge

When examined more carefully, it is quite surprising that the Pauli–Sommerfeld free electron model, discussed in Chapter 6, was able to explain the experimentally observed linear behavior of the specific heat and the temperature-independent paramagnetic susceptibility of a metal. This is because *the energy associated with the Coulomb repulsion between electrons is of the same order as the kinetic energy*[1] and it is therefore not at all obvious that such a model should work in practice.

37.1.2 Basic ideas

Liquid ^{3}He (at temperatures below about 0.1 K) is another system which has some features in common with a Fermi gas. The long range nature of the Coulomb potential adds an initially unnecessary feature to understanding the success of the independent (or free) electron model, so we will use (neutral) liquid ^{3}He as our model for what Landau termed a Fermi "liquid."[2] We recall that ^{3}He atoms have no electronic spin and their statistics is therefore governed by the $I = 1/2$ nuclear spin. Atoms in this liquid may be thought of as having an even larger interaction than electrons in metals since the "empty volume" for them to move around in is small; most of the liquid volume is occupied by the essentially impenetrable "hard core" associated with the filled $1s^2$ shell of the He atoms.[3]

Figure 37.1 shows plots of the ratio of the constant volume heat capacity divided by the temperature as a function of the temperature for various pressures. We note that this ratio approaches a constant at low temperatures, just as one would predict if the Pauli–Sommerfeld model were applicable.[4]

The success of the Pauli–Sommerfeld model rests on there being a sharp discontinuity separating the occupied from the unoccupied states at $T = 0$, which is associated with the Fermi occupation factor, $n(\varepsilon)$.[5] A sharp discontinuity, in turn, requires that the energy of the "one-particle" states at ε_F be precisely defined; i.e., there can be no uncertainty in the energy, $\delta\varepsilon$, of these states caused by scattering. From the uncertainty principle, $\delta\varepsilon \sim \hbar/\tau$, where τ is some ^{3}He–^{3}He scattering time; therefore we must have $\tau \to \infty$ at $T = 0$.

[1] This follows from the viral theorem of classical mechanics; alternatively one can compare the Fermi energy with the inter-electron Coulomb repulsion for a distance corresponding to the average separation between the electrons.

[2] One sometimes refers to the conduction electrons in a metal as the electron liquid, since they are bound to the metal yet free to move within it.

[3] The potential acting between two ^{3}He atoms is well represented by the form given by Eq. (7.2) that is strongly repulsive at short distances.

[4] Specific heat of normal liquid He^3, D. Greywall, *Phys. Rev. B* 27, 2747 (1983).

[5] To avoid confusion we use the notation $n(\mathbf{p})$ for the distribution function so as to distinguish it from the Landau quasiparticle interaction function, $f(\mathbf{p}, \mathbf{p}')$, which will be introduced shortly; we also use the momentum, $\mathbf{p}$, rather than the wave vector, $\mathbf{k}$, to denote the states of our Fermi system. These notations conform with those conventionally used in describing ^{3}He.

The Physics of Solids. First Edition. J. B. Ketterson.

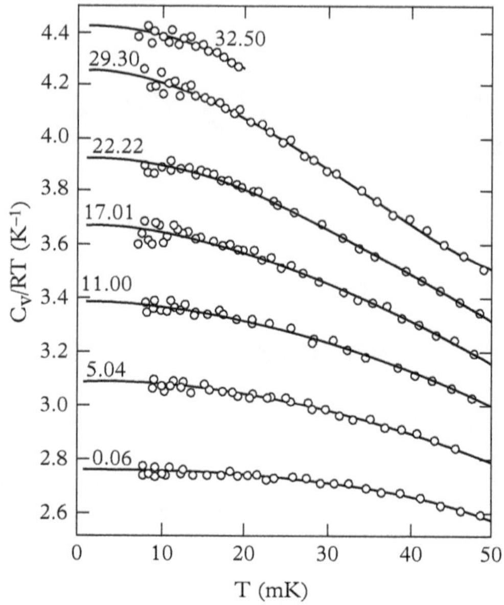

Figure 37.1 *The ratio of the heat capacity of ^{3}He at constant volume divided by temperature as a function of the temperature for various pressures. (After Greywall (1983).)*

Following Landau's argument,[6] let us consider a gas of ^{3}He atoms at absolute zero with the interatomic potential "switched off." All states with an energy less than the Fermi energy (or chemical potential), $\mu = p_F^2/2m$, where $p_F = \hbar(3\pi^2 N/V)^{1/3}$ (see Eqs. (6.14a,b)), will be occupied and all states of higher energy will be empty; here N and V are the number of particles and the volume of the system respectively and N/V is therefore the particle number density. At a temperature $T \ll T_F$ (where $k_B T_F \equiv \mu$), the excitations will consist of a few atoms in states with energies above ε_F (called *particles*), and some unoccupied or empty states with energies less than ε_F (called *holes*). These particle and hole states have energies in an interval of order $k_B T$ about μ. If we now slowly (adiabatically) turn on the interatomic potential, two things happen: (i) the nature of the excited states "deforms," and rather than involving a single particle an excitation can be pictured as a particle and a "coherent or adiabatic" motion of the surrounding particles (called *backflow*) as they "get out of the way" of the excited particle.[7] The total entity (particle + backflow) is called a *quasiparticle*.[8] The second thing that happens is that the quasiparticles may scatter from each other; i.e., transitions occur between quasiparticle states. As noted earlier, this results in a broadening of the quasiparticle energies, $\delta\varepsilon$, by an amount of the order of $\hbar/\tau$ where τ is the collision time. By considering the transition rate for the scattering of particles in the background of a Fermi sea one can show that[9]

$$\tau^{-1} \propto \left\{ \begin{array}{ll} (\varepsilon - \varepsilon_F)^2; & k_B T \ll |\varepsilon - \varepsilon_F| \\ (k_B T)^2; & k_B T \gg |\varepsilon - \varepsilon_F| \end{array} \right\}. \qquad (37.1)$$

[6] The theory of a Fermi liquid, L.D. Landau, *Sov. Phys. JETP* 3, 920 (1957).

[7] We emphasize that this is only a classical picture carried over from the hydrodynamics of a body moving in a fluid.

[8] From the Latin, quasi means "as if."

[9] See: *The Theory of Quantum Liquids*, D. Pines and P. Nozières, Benjamin, NY (1966).

This fundamental behavior has its origin in the Pauli exclusion principle: at low temperatures the majority of the particles in the Fermi sea cannot scatter, since the final states (which energetically they might scatter into) are *filled*. Crudely put, one says that the number of excited particles is proportional to T and the probability of finding an empty state in the vicinity of ε_F for the particles to scatter into also scales as T, giving an overall T^2 behavior of the scattering rate.[10]

Based on this one can make the following very important observation: as $T \to 0$ the ratio of the energy level uncertainty to the excitation energy itself (which must be considered *relative* to the Fermi energy itself, since the latter represents the ground state) approaches zero; i.e., $\delta\varepsilon/|\varepsilon - \varepsilon_F| \to 0$. In this sense the few excited states of the system behave as *independent* "particles." However, as previously noted, the motion of the excited particles really involves a coordinated motion in the background of the remaining (unexcited) particles.

The quasiparticle energies must be associated with various quantum numbers. A fundamental postulate of Landau's Fermi liquid theory is that energies are a function of the momentum, p, and that the momentum, p_F, corresponding to the Fermi energy continues to be given by the *free particle* expression,[11]

$$p_F = \hbar(3\pi^2 N/V)^{1/3}. \qquad (37.2)$$

Thus the quasiparticle states are in one-to-one correspondence with those of an ideal Fermi gas. In addition, each quasiparticle has a spin quantum number, $\sigma = \pm 1/2$.[12]

37.1.3 The quasiparticle interaction function

Since the quasiparticle states are well defined only in the immediate vicinity of the Fermi energy, μ (due to the broadening associated with the collisions), it is not possible to write the total energy of the liquid as a sum over the quasiparticle energies;[13] this is to be contrasted with the free particle case where, from Eq. (6.35), $E_{total} = (3/5)N\mu$. However it is possible to express the *change* in the total energy density of the system, $\delta(E_{total}/V)$, in terms of a *change* in the occupation (i.e., a change in the distribution function, $\delta n(\mathbf{p})$) of the states near μ, which we write in the form

$$\delta\left(\frac{E_{total}}{V}\right) = \frac{2}{(2\pi\hbar)^3}\int d^3p\,\varepsilon(\mathbf{p})\delta n(\mathbf{p}); \qquad (37.3)$$

here, V is the volume of the system and $\varepsilon(\mathbf{p})$ is defined as the *quasiparticle energy*.[14]

In a many-body system we expect the excitation energies themselves to depend on the occupation of the remaining states (e.g., since they generate the previously mentioned backflow). We will write this dependence as

$$\delta\varepsilon(\mathbf{p}) = \frac{2}{(2\pi\hbar)^3}\int d^3p\, f(\mathbf{p},\mathbf{p}')\delta n(\mathbf{p}'), \qquad (37.4)$$

where $f(\mathbf{p},\mathbf{p}')$ is a function characteristic of the system.

[10] Our argument here is qualitative; a convincing argument can only be made by examining the structure of the collision integral.

[11] This postulate was later proved formally using Green's function methods: On the theory of the Fermi liquid, L.D. Landau, *Sov. Phys. JETP-USSR* 8, 70 (1959).

[12] The arguments concerning the lifetime of quasiparticles near the Fermi surface and the preservation of the number of states within it also apply to the electron gas, as shown by Luttinger. The ideas are commonly carried over to metals in general, and referred to as Landau–Fermi liquid theory; the circumstances under which this is legitimate are widely debated.

[13] This behavior also occurs in the Hartree and Hartree–Fock approximations.

[14] Consider a change in the distribution function, $\delta n(\mathbf{p}) \sim (1/2)(2\pi)^3\delta(\mathbf{p}-\mathbf{p}_0)$, representing the excitation of a single quasiparticle with momentum, $\mathbf{p}_0$; with (37.3) we then have immediately that the change in energy of the system is $\varepsilon(\mathbf{p})$, as it must be.

The corrections to $\varepsilon(\mathbf{p})$ represented by (37.4) are essential in considering collective modes (e.g., sound waves) which result in a position- and time-dependent distribution function, $n(\mathbf{p},\mathbf{r},t)$, as will be discussed in Section 37.2; in general both the shape and volume of the Fermi surface change. The change in the quasiparticle energies resulting from the change in the distribution function is often referred to as a *molecular field effect.* The quantities, E_{total} and $\varepsilon(\mathbf{p})$, are termed *functionals* since they involve integrals over the unspecified function, $\delta n(\mathbf{p})$. Formally we write

$$\varepsilon(\mathbf{p}) = \frac{\delta(E_{total}/V)}{\delta n(\mathbf{p})} \tag{37.5}$$

and

$$f(\mathbf{p},\mathbf{p}') = \frac{\delta\varepsilon(\mathbf{p})}{\delta n(\mathbf{p}')} = \frac{\delta^2(E_{total}/V)}{\delta n(\mathbf{p})\delta n(\mathbf{p}')}; \tag{37.6}$$

the latter property shows $f(\mathbf{p},\mathbf{p}')$ is symmetrical in the variables $\mathbf{p}$ and $\mathbf{p}'$.

The quasiparticle velocity is taken to be given by the usual quantum mechanical expression for the group velocity,

$$\mathbf{v} = \frac{\partial\varepsilon(\mathbf{p})}{\partial\mathbf{p}}. \tag{37.7}$$

We *define* an effective mass[15] as the ratio of the momentum to the velocity,

$$m^* \equiv \frac{p}{v}. \tag{37.8}$$

Depending on the pressure, this quantity is three to six times greater than the mass of a bare ^{3}He atom and this large enhancement is another manifestation of the strongly interacting character of ^{3}He.

We now consider the form of the equilibrium distribution function, $n(\mathbf{p})$, which follows in a manner similar to that for the free particle case by maximizing the entropy, $S = k_B\, ln\, W$, according to (see Eq. (6.20) with $g_l = 1$)

$$\delta\left(\frac{S}{k_B V}\right) = \delta\sum_{\mathbf{k}}(-n(\mathbf{p})\, ln\, n(\mathbf{p}) - (1-n(\mathbf{p}))\, ln\,(1-n(\mathbf{p}))); \tag{37.9}$$

this must be done subject to the constraints that the total energy, E_{total}, and particle number, N, of the system are constant. Introducing the usual Lagrange multipliers μ and β, as in Eq. (6.21), we then have

$$\delta\left(\frac{S}{k_B V}\right) - \beta\mu\delta\left(\frac{N}{V}\right) + \beta\delta\left(\frac{E_{total}}{V}\right) = 0, \tag{37.10}$$

which leads to the distribution function[16]

$$n(\mathbf{p}) = \frac{1}{e^{\beta(\varepsilon(\mathbf{p})-\mu)}+1}, \tag{37.11}$$

where $\beta = 1/k_B T$. Note that since $\varepsilon(\mathbf{p})$ is a functional of $n(\mathbf{p})$, Eq. (37.11) is more complicated than it initially appears.

[15] This is a "many-body" effective mass and should not be confused with that arising from the effects of an underlying band structure, which is absent in ^{3}He.

[16] The preceding derivation of a distribution function, $n(\mathbf{p})$, that is defined for all $\mathbf{p}$ can be criticized on the grounds that the states are not well defined far off the Fermi surface. For a discussion of this and other subtle aspects of Fermi liquid theory, see: *The Theory of Quantum Liquids* by D. Pines and P. Noziers, W.A. Benjamin, NY (1961).

37.1.4 Sum rules

Information concerning the function, $f(\mathbf{p},\mathbf{p}')$, must, at present, be obtained experimentally. Landau derived two "sum rules" which fixed two "moments" of this function, which are referred to as the compressibility and effective mass sum rules, which we now derive.

The first of these is derived from the expression for the sound velocity, $c^2 = \partial P/\partial \rho$, where P and ρ are the pressure and density respectively. From the differential relation for the Gibbs free energy, $d(N\mu) = VdP - SdT + \mu dN$, we obtain (at constant T), $N\partial\mu/\partial N = V\partial P/\partial N$; noting that $\rho = Nm/V$, we have $c^2 = (N/m)\partial\mu/\partial N$.[17] In evaluating the derivative of μ with respect to N we must include: (i) the explicit dependence resulting from the density dependence of p_F, Eq. (37.2), and (ii) an implicit dependence arising from Eq. (37.4). The change in p_F associated with the first effect causes the second effect, in that a shift in the occupation of states in the vicinity of p_F alters the average molecular field associated with a quasiparticle excitation. Combining the two effects we have

$$\delta\mu = \frac{\partial \varepsilon_F}{\partial p_F}\frac{\partial p_F}{\partial N}\delta N + \frac{2}{(2\pi\hbar)^3}\int d^3pp'\, f(\mathbf{p}_F,\mathbf{p}')\delta n\,(\mathbf{p}'); \qquad (37.12)$$

Writing[18] $\delta n(\mathbf{p}') = (\partial\theta(p_F - p')/\partial p_F)(\partial p_F/\partial N)\delta N = \delta(p_F - p')(\partial p_F/\partial N)\delta N$ and defining

$$\bar{f} = \frac{1}{4\pi}\int d\Omega'\, f(\hat{\mathbf{p}}_F,\hat{\mathbf{p}}_F'),$$

we have

$$\delta\mu = \frac{p_F}{3N}\left(v_F + \frac{p_F^2}{\hbar^3\pi^2}\bar{f}\right)\delta N$$

or

$$\bar{f} = \frac{3m\pi^2\hbar^3}{p_F^3}\left(c^2 - \frac{p_F^2}{3mm^*}\right), \qquad (37.13)$$

which is our first sum rule.

The second sum rule follows from the fact that the laws of mechanics must be invariant under a transformation to a uniformly moving reference frame, which is called *Galilean invariance* in non-relativistic mechanics. As a result, the relation between the (total) momentum, $\mathbf{P}_0$, in a frame which is at rest and the momentum, $\mathbf{P}$, in a system moving with velocity, $\mathbf{V}$, is given by

$$\mathbf{P} = \mathbf{P}_0 + M\mathbf{V}, \qquad (37.14)$$

where M is the total mass of the fluid. For reference, the corresponding relation between the energy, $E_0(\mathbf{P}_0)$, in the rest frame and the energy, E, in the moving frame is

$$E = E_0(\mathbf{P}_0) + \mathbf{P}_0 \cdot \mathbf{V} + \frac{1}{2}MV^2. \qquad (37.15)$$

[17] Strictly speaking sound propagates adiabatically; however the distinction between isothermal and adiabatic sound is negligible at low temperatures.

[18] $\theta(x)$ is the unit step function having the value 1 for $x > 0$ and 0 for $x < 0$.

Assume that our system, taken as an infinite sample of liquid ^{3}He, is at rest in the lab frame (i.e., $\mathbf{P}_0 = 0$). We then have $\mathbf{P} = M\mathbf{V}$ in the moving frame; the left-hand side of Eq. (37.14) is the sum of the individual particle momenta,

$$\mathbf{P} = \frac{2}{(2\pi\hbar)^3}\int d^3p\,\mathbf{p}\,n(\mathbf{p}),$$

and the right-hand side is a sum involving their velocities,

$$M\mathbf{V} = \frac{2}{(2\pi\hbar)^3}\int d^3p\;m\mathbf{v}(\mathbf{p})n(\mathbf{p}).$$

Writing $\mathbf{v}(\mathbf{p}) = \partial\varepsilon(\mathbf{p})/\partial\mathbf{p}$ from Eq. (37.7) and equating these two forms we have

$$\frac{2}{(2\pi\hbar)^3}\int d^3p\,\mathbf{p}n(\mathbf{p}) = \frac{2m}{(2\pi\hbar)^3}\int d^3p\frac{\partial\varepsilon\,(\mathbf{p})}{\partial\mathbf{p}}n(\mathbf{p}). \tag{37.16}$$

The trick is to take a variation of both sides, and recall that ε is a functional of $n(\mathbf{p})$:

$$\frac{2}{(2\pi\hbar)^3}\int d^3p\,\mathbf{p}\delta n(\mathbf{p}) = m\frac{2}{(2\pi\hbar)^3}\int d^3p\frac{\partial\delta\varepsilon(\mathbf{p})}{\partial\mathbf{p}}n(\mathbf{p}) + m\frac{2}{(2\pi\hbar)^3}\int d^3p\frac{\partial\varepsilon(\mathbf{p})}{\partial\mathbf{p}}\delta n(\mathbf{p}). \tag{37.17}$$

Using (37.4) we obtain

$$\frac{1}{m}\frac{2}{(2\pi\hbar)^3}\int d^3p\,\mathbf{p}\delta n(\mathbf{p}) = \left[\frac{2}{(2\pi\hbar)^3}\right]^2\int d^3p\int d^3p'\frac{\partial}{\partial\mathbf{p}}f(\mathbf{p},\mathbf{p}')\delta n(\mathbf{p}')n(\mathbf{p}) + \frac{2}{(2\pi\hbar)^3}\int d^3p\frac{\partial\varepsilon(\mathbf{p})}{\partial\mathbf{p}}\delta n(\mathbf{p}). \tag{37.18}$$

Integrating the first term on the right hand side by parts, placing the surface outside the Fermi sphere, relabeling $\mathbf{p}$ and $\mathbf{p}'$, and requiring (37.18) to hold for an arbitrary variation, we have

$$\frac{\mathbf{p}}{m} = -\frac{2}{(2\pi\hbar)^3}\int d^3p'\frac{\partial n(\mathbf{p}')}{\partial\mathbf{p}}f(\mathbf{p}',\mathbf{p}) + \frac{\mathbf{p}_F}{m^*}, \tag{37.19}$$

where we have used Eqs. (37.7) and (37.8) to rewrite the second term on the right. We again write $n(p') = \theta(p_F - p')$, from which we obtain $\partial n(\mathbf{p}')/\partial\mathbf{p}' = -\delta(p_F - p')\mathbf{p}'/p'$. Inserting this form into Eq. (37.19), carrying out the integration with respect to dp', and canceling a common factor of p_F, we obtain

$$\frac{1}{m} = \frac{1}{m^*} + \frac{2p_F}{(2\pi\hbar)^3}\int d\Omega' f(\mathbf{p}_F,\mathbf{p}'_F)\cos\theta', \tag{37.20}$$

where have used the fact that the δ-function, and limiting ourselves to infinitesimal variations δf about p_F, restrict the values of $\mathbf{p}$ and $\mathbf{p}'$ to the Fermi surface.[19] Furthermore rotational invariance, and the fact that $f(\mathbf{p}, \mathbf{p}')$ is symmetric, require that $f(\mathbf{p}_F, \mathbf{p}'_F) = f(\cos\chi)$, where χ is the angle between two quasiparticles on the Fermi surface. Eq, (37.21) is our second sum rule.

The physical idea implicit in the above derivation is that a moving reference frame results in a displacement of the entire Fermi sphere by a momentum $m\mathbf{V}$. However, the displaced sphere may be regarded as the original undisplaced sphere with the addition of a "crescent" of particle-like excitations on the half sphere in the direction of motion, and a crescent of holes on the opposite half sphere, as shown in Fig. 37.2. These crescents clearly represent a macroscopic change in the distribution function, $\delta n(\mathbf{p})$, which, in turn, causes a corresponding shift in the quasiparticle energies, $\delta\varepsilon(\mathbf{p})$. This energy shift results in $m^* \neq m$.

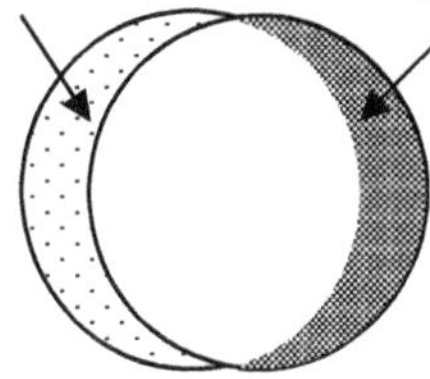

Figure 37.2 *The Fermi sphere in a frame at rest and a frame moving to the right with velocity* V.

In order to parameterize $f(\cos\chi)$ we first introduce a dimensionless function, $F(\cos\chi)$, defined as

$$F(\cos\chi) = N(\varepsilon)f(\cos\chi), \tag{37.21}$$

where $N(\varepsilon)$ is the density of states (including both spins): $N(\varepsilon) = m^* p_F/\pi^2\hbar^3$. Second, we expand this function in Legendre polynomials,

$$\begin{aligned} F(\cos\chi) &= \sum_n F_n P_n(\cos\chi) \\ &= F_0 + F_1 \cos\chi + \dots. \end{aligned} \tag{37.22}$$

In terms of this expansion our two sum rules, Eqs. (37.13) and (37.20) become

$$F_0 = \frac{3mm^*c^2}{p_F^2} - 1 \tag{37.23a}$$

and

$$F_1 = 3\left(\frac{m^*}{m} - 1\right). \tag{37.23b}$$

The Sommerfeld heat capacity (see (6.37a)) in the presence of Fermi liquid effects is then modified to the form

$$C_V(T) = \frac{\pi^2}{3} k_B^2 \frac{N[(\mu(0))]}{1 + \frac{F_1}{3}} T, \tag{37.24a}$$

or, equivalently, we replace the *bare mass* by a *many-body enhanced mass*,

$$m^* = \frac{1}{1 + \frac{F_1}{3}} m. \tag{37.24b}$$

[19] The derivation presented here can be criticized in that (37.20) involves an integration over $n(\mathbf{p})$ which is not strictly valid; for a more rigorous derivation, see: Noziéres and Pines (1964).

This enhancement is generally small in ordinary metals; in ^{3}He it is approximately six near the melting curve. However in so-called *heavy Fermion* intermetallic compounds (to be discussed in Chapter 45), mass enhancements of 1000 or more have been reported (although part of this is due to band structure effects).

The two sum rules, (37.23a,b), exhaust the information that may be obtained from purely thermodynamic measurements (at zero field). Additional information concerning F may be obtained from zero sound measurements, to be discussed in Section 37.2.

The remaining thermodynamic property of ^{3}He which we have not discussed is the magnetic susceptibility. In the presence of a magnetic field, which we will take along the **z** axis, each quasiparticle will have its energy shifted according to

$$\delta\varepsilon(\mathbf{p}) = -\frac{1}{2}\hbar\gamma_0 H + \frac{1}{(2\pi\hbar)^3}\int d^3p f^{(a)}(\mathbf{p},\mathbf{p}')\left[\delta n_\uparrow(\mathbf{p}') - \delta n_\downarrow(\mathbf{p}')\right], \qquad (37.25)$$

where $f^{(a)}(\mathbf{p},\mathbf{p}')$ is a *different* (the antisymmetric[20]) Fermi liquid interaction function which accounts for the additional energy associated with splitting the spin up and spin down Fermi surfaces. The first term on the right is a *direct* shift of the quasiparticle energy levels due to the *bare* magnetic moments, $(\hbar/2)\,\gamma_0 = \mu_3$, where γ_0 is the gyromagnetic ratio of the ^{3}He nuclei. The second term, involving the independent Fermi liquid interaction function is a "molecular field" resulting from a spin polarized distribution function.[21] We write the quasiparticle energy shift in the form $\delta\varepsilon = -(\hbar/2)\gamma H$, which is valid in a linear (low-field) approximation; $(\hbar/2)\gamma$ is an *effective* quasiparticle magnetic moment that includes the effects of the molecular field, $f^{(a)}$. We may write the change in the distribution function as

$$\left[\delta n_\uparrow(\mathbf{p}') - \delta n_\downarrow(\mathbf{p}')\right] \cong \frac{\partial n}{\partial \varepsilon}\delta\varepsilon, \qquad (37.26)$$

and Eq. (37.25) becomes

$$\gamma_0 = \gamma\left(1 + \frac{m^* p_F}{\pi^2\hbar^3}\frac{1}{4\pi}\int d\Omega' f^{(a)}(\mathbf{p},\mathbf{p}')\right). \qquad (37.27)$$

We expand the function $f^{(a)}(\mathbf{p},\mathbf{p}')$ in Legendre polynomials analogous to Eq. (37.22),

$$F^{(a)} = \sum_n F_n^{(a)} P_n(\cos\chi) \qquad (37.28)$$

and we may then write (37.32) in the form

$$\gamma = \frac{\gamma_0}{1 + F_0^{(a)}}. \qquad (37.29)$$

[20] What we wrote earlier as $f(\mathbf{p},\mathbf{p}')$ is often written as $f^{(s)}(\mathbf{p},\mathbf{p}')$ and referred to as the *symmetric* part of the Landau interaction function.

[21] It is related to the molecular field encountered in the Bloch–Stoner theory of the exchange enhancement of the magnetic susceptibility of electrons in metals.

This immediately leads to a generalization of the usual expression for the Pauli paramagnetic susceptibility, $\chi_0 = (\hbar/2)^2\gamma_0^2 N(\varepsilon_F)$, given by Eq. (6.31), to the form

$$\chi = \frac{1}{4}\frac{\hbar^2\gamma_0^2 N(\varepsilon_F)}{1 + F_0^{(a)}}. \tag{37.30}$$

Equation (37.30), which we may call the magnetic susceptibility sum rule, can be used to fix $F_0^{(a)}$ from the measured magnetic susceptibility. $F_0^{(a)}$ is negative in He^3 so the magnetic susceptibility is enhanced.[22]

37.2 Collisionless collective modes

We may loosely define a collective mode as a time- (or frequency-) and space- (or wave-vector-) dependent oscillation of the fluid involving a coordinated motion of its constituent particles. We can distinguish collective modes according to whether they are hydrodynamic or collisionless.

Hydrodynamic collective modes owe their existence to various conservation laws (momentum, energy, and mass). Ordinary sound waves are the best-known hydrodynamic mode, but the damped viscous shear waves and the thermal conduction wave are also classed as hydrodynamic modes. If the liquid has a continuously broken symmetry, as in superfluid ^{3}He and ^{4}He or liquid crystals, additional modes appear (e.g., the second sound mode of He II and superfluid ^{3}He). Hydrodynamic modes are restricted to frequencies, ω, satisfying the condition $\omega\tau \ll 1$, where τ is a characteristic relaxation time.[23]

A necessary (but not sufficient) condition for the existence of collisionless collective modes is $\omega\tau \gg 1$. According to the argument given in Section 37.2, the collisions between excited quasiparticles become more infrequent as the temperature is lowered. Hence at sufficiently low temperatures, the fluid will enter the region $\omega\tau \gg 1$. As shown by Landau,[24] collisionless collective modes in ^{3}He arise from the restoring force associated with the molecular field, which occurs when the distribution function of the quasiparticles is distorted from its equilibrium form. The time evolution of the distribution function, $n = n(\mathbf{p}, \mathbf{r}, t)$, is governed by the Boltzmann equation. Although we have argued that collisions between quasiparticles become unimportant at low temperatures, this does not mean that the particles do not interact. This interaction manifests itself through a *coherent* molecular field rather than through *incoherent* quasiparticle scattering events. In fact, one may neglect collisions altogether in deriving the frequency–wave vector relation of the collective modes; collisions only lead to damping.

37.2.1 The kinetic equation

We obtain the Boltzmann or kinetic equation (see Section 19.2) by equating the total time derivative of the distribution function to a collision integral, $I[n]$,

$$\frac{\partial n}{\partial t} + \frac{\partial n}{\partial \mathbf{r}}\cdot\dot{\mathbf{r}} + \frac{\partial n}{\partial \mathbf{p}}\cdot\dot{\mathbf{p}} = I[n]; \tag{37.31}$$

[22] A large enhancement of the *electronic* susceptibility occurs in some metals, such as Pd, and is referred to as the *Stoner* or *exchange enhancement*.

[23] One may understand this condition by calculating the time needed to equilibrate the temperature oscillations associated with successive nodes and antinodes of a sound wave in a gas (from kinetic theory) and comparing it with the period, $t_0 = 2\pi/\omega$. One finds that adiabatic (hydrodynamic) propagation requires $t_0 \gg \tau$.

[24] Oscillations in a Fermi liquid, L.D. Landau, *Sov. Phys. JETP* 5, 101 (1957).

as previously discussed we will set I[n] = 0. From Hamilton's equations, $\dot{\mathbf{r}} = \partial\varepsilon/\partial\mathbf{p}$ and $\dot{\mathbf{p}} = -\partial\varepsilon/\partial\mathbf{r}$, we obtain

$$\frac{\partial n}{\partial t} + \frac{\partial n}{\partial \mathbf{r}} \cdot \frac{\partial \varepsilon}{\partial \mathbf{p}} - \frac{\partial n}{\partial \mathbf{p}} \cdot \frac{\partial \varepsilon}{\partial \mathbf{r}} = 0. \tag{37.32}$$

Writing the distribution function as the sum of an equilibrium part and a small deviation, $n(\mathbf{p}, \mathbf{r}, t) = n_0(p) + n'(\mathbf{p}, \mathbf{r}, t)$, and noting that $\varepsilon = \varepsilon[n]$, through Eq. (37.4), we obtain, on retaining only the linear terms in n′,

$$\frac{\partial n'(\mathbf{p},\mathbf{r},t)}{\partial t} + \frac{\partial n'(\mathbf{p},\mathbf{r},t)}{\partial \mathbf{r}} \cdot \mathbf{v} - \frac{\partial n_0}{\partial \varepsilon}\mathbf{v} \cdot \frac{2}{(2\pi\hbar)^3}\int f(\mathbf{p},\mathbf{p}')\frac{\partial n'}{\partial \mathbf{r}} d^3p' = 0. \tag{37.33}$$

Note that in a non-interacting Fermi gas the last term, which (as we will show) is the restoring force required for the existence of a collective mode, would be absent.[25] The collective modes are the *eigensolutions* of Eq. (37.33), which is an *integro-differential* equation. We seek solutions to the form $n'(\mathbf{p}, \mathbf{r}, t) \sim n'(\mathbf{p})e^{-i(\omega t - \mathbf{k}\cdot\mathbf{r})}$, which on substituting into (37.33) yields

$$(\omega - \mathbf{k}\cdot\mathbf{v})n'(\mathbf{p}) - \frac{\partial n_0}{\partial \varepsilon}\mathbf{k}\cdot\mathbf{v}\frac{2}{(2\pi\hbar)^3}\int d^3p f(\mathbf{p},\mathbf{p}')n'(\mathbf{p}') = 0. \tag{37.34}$$

The equilibrium distribution function is a step or theta function, $n_0(\mathbf{p}) = \theta(\mathbf{p}_F - \mathbf{p})$; we expect the deviation, $n'(\mathbf{p})$, to be proportional to a delta function which we write in the form

$$n'(\mathbf{p}) = \nu(\hat{\mathbf{p}})\delta(\mathbf{p} - \mathbf{p}_F)/v_F = \nu(\hat{\mathbf{p}})\frac{\partial n_0}{\partial \varepsilon}. \tag{37.35}$$

Operationally, $n'(\mathbf{p})$ is equivalent to an angular-dependent shift of the Fermi radius by an amount, $\delta\mathbf{p} = \nu(\hat{\mathbf{p}})\hat{\mathbf{p}}/v_F$:

$$\begin{aligned} n(\mathbf{p}_F - \mathbf{p} - \nu(\hat{\mathbf{p}})/v_F) &\cong n(\mathbf{p}) + \frac{\partial n_0}{\partial \mathbf{p}} \cdot \delta\mathbf{p} + \ldots \\ &= \theta(\mathbf{p}_F - \mathbf{p}) + \delta(\mathbf{p}_F - \mathbf{p})\frac{\nu(\hat{\mathbf{p}})}{v_F}\ldots \end{aligned}$$

Specifically, the eigensolutions we seek will correspond to a time-, space-, and *angular*-dependent (shape) change of the Fermi surface. Inserting (37.35) into (37.33) and using (37.21), we have

$$(\omega - \mathbf{k}\cdot\mathbf{v})\nu(\hat{\mathbf{p}}) - \mathbf{k}\cdot\mathbf{v}\frac{1}{4\pi}\int d\Omega' F(\hat{\mathbf{p}}, \hat{\mathbf{p}})\nu(\hat{\mathbf{p}}') = 0. \tag{37.36}$$

All possible collective modes, which involve density–velocity oscillations, follow from the solution of (37.36). We rewrite the linear homogeneous integral equation (37.36) in the standard form by defining a new variable

$$\psi(\hat{\mathbf{p}}) = \frac{s - \cos\theta}{\cos\theta}\nu(\hat{\mathbf{p}}), \tag{37.37}$$

[25] Equation (37.33) is valid only for a neutral Fermi liquid, such as ^3He, where the self-consistent field arises from short range forces. In a charged system one must explicitly include the added contribution of the long range Coulomb potential which is obtained by simultaneously solving the kinetic equation and Maxwell's equations (or Poisson's equation in the quasistatic limit); the oscillations then occur at a characteristic $\omega_p = (4\pi ne^2/m)^{1/2}$; see Section 19.13.

which leads to the equation

$$\psi(\hat{\mathbf{p}}) = \frac{1}{4\pi}\int d\Omega' F(\hat{\mathbf{p}},\hat{\mathbf{p}}')\frac{\cos\theta'}{s-\cos\theta'}\psi(\hat{\mathbf{p}}'), \tag{37.38}$$

where θ is the angle between **k** and **v** and we have defined a dimensionless velocity, $s = \omega/kv_F$; the quantity $(1/4\pi)F(\hat{\mathbf{p}},\hat{\mathbf{p}}')\cos\theta'/(s-\cos\theta')$ is the kernel of our integral equation.

37.2.2 Collisionless longitudinal zero sound

The simplest possible situation leading to the existence of longitudinal zero sound[26] is where the expansion of the Landau interaction function consists of a single term, F_0. Integrating the right side of (37.38) then yields a constant; this implies that $\psi(\hat{\mathbf{p}})$ is a constant or

$$\nu(\theta) = \text{const}\cdot\frac{\cos\theta}{s-\cos\theta}. \tag{37.39}$$

The condition for a solution to exist is then

$$\frac{F_0}{2}\int_{-1}^{+1}\frac{x dx}{s-x} = 1. \tag{37.40}$$

Carrying out the integration we obtain

$$\frac{1}{F_0} = \frac{s}{2}ln\left(\frac{s+1}{s-1}\right) - 1. \tag{37.41}$$

Equation (37.41) has solutions only for $F_0 > 0$, which in this simplified model, is the criterion for the existence of zero sound; i.e., in the absence of a repulsive interaction, F_0, there is no zero sound. We examine the limiting cases of $F_0 \to 0$ and $F_0 \to \infty$. For the first of these we have

$$s = 1 + 2e^{-2/F_0}; \tag{37.42}$$

in this limit, the phase velocity of the longitudinal zero sound mode, $\omega/k \equiv c_0^\ell$, approaches the Fermi velocity, v_F. In the other limit, $F_0 \to \infty$, $s \gg 1$, and we expand the logarithm (to the third order in s^{-1}) to obtain

$$s^2 = \frac{F_0}{3} \tag{37.43}$$

or $c_0^\ell = v_F(F_0/3)^{1/2}$.

Figure 37.3 shows the distortion of the Fermi surface on alternate half cycles associated with the passage of a longitudinal zero sound wave for the case $s = 4$. The largest distortion occurs at the leading edge of the Fermi surface. Note that in addition to a shape change there is a motion of the Fermi sphere as a whole. In the limit $s \to 1$, the distortion approaches a dimple on the leading edge of

[26] Here we will repeat parts of the discussion given earlier, in Section 14.2.3, where the phenomenon of zero sound emerged in a different context.

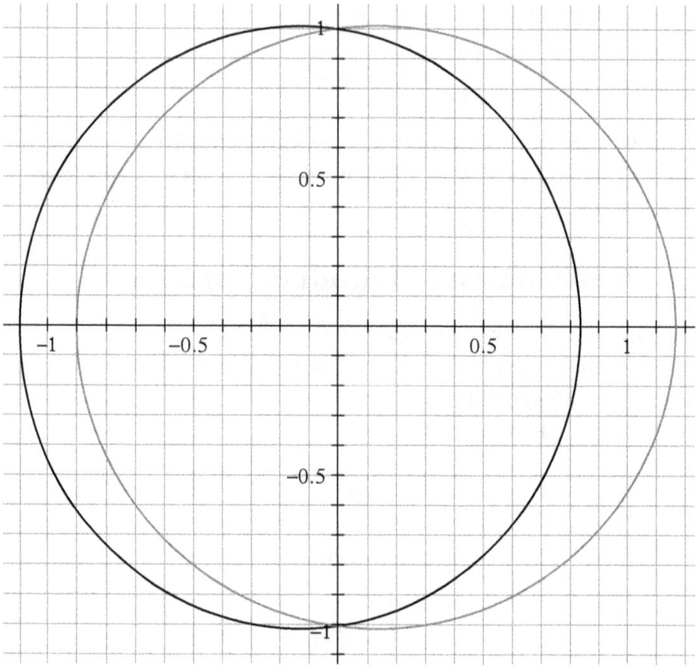

Figure 37.3 *The distortion of the Fermi surface associated with longitudinal zero sound on alternate half cycles for the case,* s = 4.

the Fermi surface. From Fig. 37.3 it is clear that the distortion corresponds to a longitudinal excitation of the system.

Figure 37.4 shows the temperature dependence of the change in the velocity and the attenuation of sound in liquid ^{3}He.[27,28] Note that at temperatures below approximately 40 mK the attenuation increases in proportion to T^{-2}. To understand this behavior we recall Eq. (1.58), where we found that the attenuation of sound in a material is proportional to the product of the square of the frequency and the imaginary part of the compliance, where we may interpret the latter as the viscosity. From the kinetic theory of gases we know that the viscosity is proportional to the collision time, τ; hence in a Fermi liquid the attenuation should scale as T^{-2}, in agreement with the experiment. As the temperature continues to fall a point is reached where $\omega\tau = 1$ and we enter the zero sound regime. Here it turns out the attenuation scales as τ or equivalently T^2, again in agreement with experiment. The increase in the velocity in the zero sound regime follows from the fact that when $\omega\tau > 1$, the system does not fully relax during sound propagation and effectively becomes "stiffer."

[27] Zero sound was first reported in: Propagation of zero sound in liquid He^3 at low temperatures, W.R. Abel, A.C. Anderson, and J.C. Wheatley, *Phys. Rev. Lett.* 17, 74 (1966).

[28] J.B. Ketterson, P.R. Roach, B.M. Abraham, and P.D. Roach, in *Quantum Statistics and the Many Body Problem*, p. 35, S.B. Trickey, W.P. Kirk, and J.W. Dufty, Eds., Plenum Press, NY (1975); the theoretical fit to the data is due to P. Wolfle.

37.2.3 Collisionless transverse zero sound

We now consider the case where $F_1 \neq 0$ in (37.23). We may write the function $\cos\chi = \hat{\mathbf{p}} \cdot \hat{\mathbf{p}}'$ as

$$\cos\chi = \cos\theta\cos\theta' + \sin\theta\sin\theta'(\cos\phi\cos\phi' + \sin\phi\sin\phi'). \tag{37.44}$$

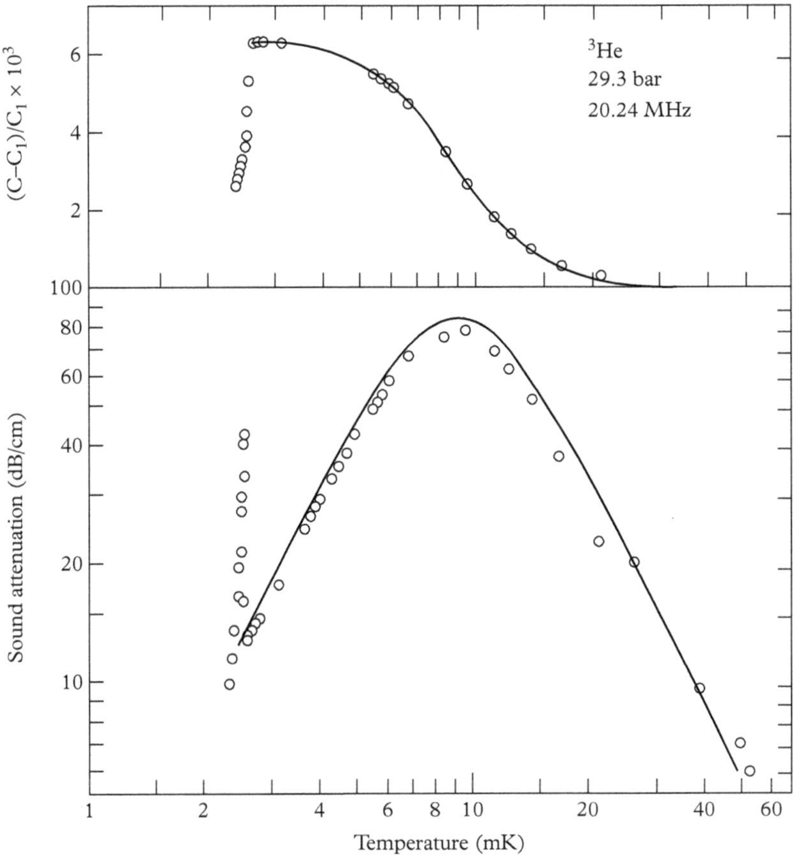

Figure 37.4 *The velocity shift and the attenuation of sound in* ^{3}He. *The crossover from first sound to zero sound occurs near 10 mK. The sharp features near 2.5 mK mark the onset of superfluidity. (Data from the Argonne group).*

We seek solutions to Eq. (37.38) of the form

$$\psi(\theta, \phi) = \text{const} \cdot \sin\theta \begin{pmatrix} \cos\phi \\ \sin\phi \end{pmatrix}, \tag{37.45}$$

where the two forms correspond to the two orthogonal polarizations associated with a transverse mode; the corresponding form for $\nu(\hat{\mathbf{p}})$ is

$$\nu(\theta, \phi) = \text{const} \cdot \frac{\sin\theta\cos\theta}{s - \cos\theta} \begin{pmatrix} \cos\phi \\ \sin\phi \end{pmatrix}. \tag{37.46}$$

Inserting (37.46) into (37.38) and using (37.44), we obtain, on integrating over $d\phi$,

$$1 = \frac{F_1}{4} \int_{-1}^{+1} \frac{x(1 - x^2)}{s - x} dx \tag{37.47}$$

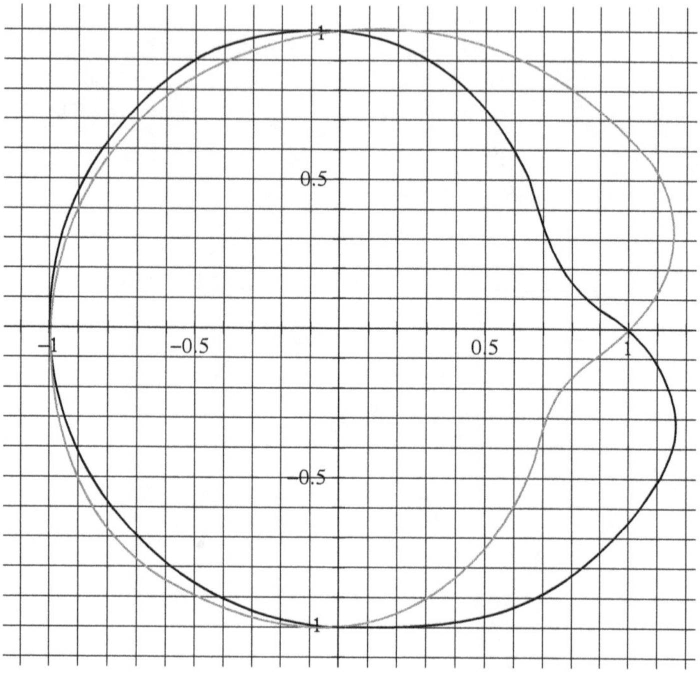

Figure 37.5 *The distortion of the Fermi surface in the plane of the polarization associated with transverse zero sound on alternate half cycles for the case,* s = 1.1.

where x = cos θ (note that for a mode having the structure of (37.46) the integral involving the F_0 term vanishes by symmetry). Carrying out the integration, we obtain

$$\frac{F_1}{4}s(1-s^2)\,ln\left(\frac{s+1}{s-1}\right)+2(s^2-1)+\frac{2}{3}=1. \tag{37.48}$$

The threshold for transverse zero sound propagation corresponding to the limit $s \to 1$ is $F_1 > 6$. Since this criterion is only barely satisfied in ^{3}He, we do not explore the large s limit. Figure 37.5 shows the (maximal) distortion of the Fermi surface associated with transverse zero sound for the case s = 1.2.

ADDITIONAL READING

Fermi liquid theory is discussed by G. Baym and C. Pethick in Chapter 1 of *The Physics of Liquid and Solid Helium, Part II.* K.H. Bennemann and J.B. Ketterson, Eds., Wiley Interscience, NY (1975, 1978).

The Superfluid Phases of Helium 3, D. Vollhardt and P. Wolfle, Taylor Francis, London (1990).

Problems

37.1. Using the Euler equation of motion from hydrodynamics together with the equation of mass continuity, show that the velocity of sound, c, is given by, $c^2 = \partial P/\partial \rho$.

37.2. Carry out the expansions required to obtain Eqs. (37.41) and (37.43).

Appendix 37A Collisionless spin waves in ^{3}He

Having considered collective modes involving the distribution function, one can ask if there are modes associated with its spin degrees of freedom; i.e., spin waves. Spin waves in liquid helium three were first considered by Silin (1958).[29] It is appropriate to discuss the propagation of spin waves since, (i) they prepare us for more general types of collective modes occurring in superfluid ^{3}He, involving the many components of the full distribution function including its anomalous components, and (ii) they occur (along with other modes) as eigenfunctions of the kinetic equations of the superfluid.

Our discussion of the density/velocity collective modes made use of the Boltzmann equation, wherein one treats the time evolution of a distribution function, $n(\mathbf{p}, \mathbf{r}, t)$. Such a treatment is semi-classical in that we simultaneously specify the values of $\mathbf{p}$ and $\mathbf{r}$ of the quasiparticles which, since they are associated with non-commuting operators, is not strictly valid in quantum mechanics. A formal analysis, due to Wigner, makes clear the sense in which a distribution function treatment is valid. The following arguments are not intended to be rigorous, but rather to physically motivate the resulting equations of motion for the spin generalized distribution function, Eq. (37A.3).

When including the spin degrees of freedom the functions, $n(\mathbf{p}, \mathbf{r}, t)$ and $\varepsilon(\mathbf{p}, \mathbf{r}, t)$, become matrices, $\underline{n}(\mathbf{p}, \mathbf{r}, t)$ and $\underline{\varepsilon}(\mathbf{p}, \mathbf{r}, t)$. In general these matrices do not commute. Therefore terms in the spin generalized kinetic equation like $(\partial \underline{n}/\partial \mathbf{r}) \cdot (\partial \underline{\varepsilon}/\partial \mathbf{p})$ are ambiguous; the ambiguity is removed by the usual quantum mechanical prescription of replacing classical expressions involving non-commuting operators by symmetrized expressions. Thus for the time dependence of $\underline{n}$ arising from the *quasiclassical* $(\mathbf{p}, \mathbf{r})$ degrees of freedom, $\dot{\underline{n}}^c$, we write

$$\begin{aligned}\frac{d\underline{n}^c}{dt} &= \frac{\partial \underline{n}}{\partial t} + \frac{1}{2}\left(\frac{\partial \underline{n}}{\partial \mathbf{r}} \cdot \frac{\partial \underline{\varepsilon}}{\partial \mathbf{p}} + \frac{\partial \underline{\varepsilon}}{\partial \mathbf{p}} \cdot \frac{\partial \underline{n}}{\partial \mathbf{r}}\right) - \frac{1}{2}\left(\frac{\partial \underline{n}}{\partial \mathbf{p}} \cdot \frac{\partial \underline{\varepsilon}}{\partial \mathbf{r}} + \frac{\partial \underline{\varepsilon}}{\partial \mathbf{r}} \cdot \frac{\partial \underline{n}}{\partial \mathbf{p}}\right)\\ &= \frac{\partial \underline{n}}{\partial t} + \frac{1}{2}\left\{\underline{n}, \underline{\varepsilon}\right\}_{\mathbf{r},\mathbf{p}} - \frac{1}{2}\left\{\underline{\varepsilon}\,\underline{n}\right\}_{\mathbf{r},\mathbf{p}},\end{aligned} \tag{37A.1a}$$

29 Oscillations of a Fermi liquid in a magnetic field, V.P. Silin, *Sov. Phys. JETP* 6, 945 (1958).

where in writing the latter we have employed the Poisson bracket notation,

$$\left\{\underset{\sim}{n}, \underset{\sim}{\varepsilon}\right\}_{\mathbf{r},\mathbf{p}} = \frac{\partial \underset{\sim}{n}}{\partial \mathbf{r}} \cdot \frac{\partial \underset{\sim}{\varepsilon}}{\partial \mathbf{p}} - \frac{\partial \underset{\sim}{n}}{\partial \mathbf{p}} \cdot \frac{\partial \underset{\sim}{\varepsilon}}{\partial \mathbf{r}}. \tag{37A.1b}$$

When treating the time evolution of the spin degrees of freedom, which have no semi-classical limit, we must be more careful. Thus the time dependence arising from the strictly quantum mechanical character of the spin degrees of freedom, $\dot{\underset{\sim}{n}}^{q}$, must be obtained from the density matrix equation of motion. Here the energy matrix, $\underset{\sim}{\varepsilon}$, plays the role of a Hamiltonian and we write

$$\frac{d\underset{\sim}{n}^{q}}{dt} = \frac{i}{\hbar}\left[\underset{\sim}{\varepsilon}, \underset{\sim}{n}\right], \tag{37A.2}$$

where $[\underset{\sim}{\varepsilon}, \underset{\sim}{n}]$ denotes the commutator of $\underset{\sim}{\varepsilon}$ and $\underset{\sim}{n}$. The total time derivative of the distribution function is given by the sum of the terms on the right in (37A.1a) and (37A.2) and must be equated to a spin-generalized collision integral, $\underset{\sim}{I}(\underset{\sim}{n})$:

$$\frac{d\underset{\sim}{n}}{dt} = \frac{d\underset{\sim}{n}^{c}}{dt} + \frac{d\underset{\sim}{n}^{q}}{dt} = \underset{\sim}{I}(\underset{\sim}{n}). \tag{37A.3}$$

We now separate the spin (traceless) and density components of the matrices $\underset{\sim}{n}$ and $\underset{\sim}{\varepsilon}$ by writing them in the form

$$\underset{\sim}{n} = n\underset{\sim}{1} + \mathbf{m} \cdot \underset{\sim}{\sigma} \tag{37A.4a}$$

and

$$\underset{\sim}{\varepsilon} = \varepsilon\underset{\sim}{1} + \mathbf{e} \cdot \underset{\sim}{\sigma}. \tag{37A.4b}$$

To evaluate (37A.3), we exploit the identity

$$(\mathbf{a} \cdot \underset{\sim}{\sigma})(\mathbf{b} \cdot \underset{\sim}{\sigma}) = (\mathbf{a} \cdot \mathbf{b})\underset{\sim}{1} + i(\mathbf{a} \times \mathbf{b}) \cdot \underset{\sim}{\sigma}, \tag{37A.5}$$

where **a** and **b** are any two vectors; we then have

$$\underset{\sim}{\varepsilon}\underset{\sim}{n} - \underset{\sim}{n}\underset{\sim}{\varepsilon} = 2i(\mathbf{e} \times \mathbf{m}) \cdot \underset{\sim}{\sigma} \tag{37A.6a}$$

and

$$\underset{\sim}{\varepsilon}\underset{\sim}{n} + \underset{\sim}{n}\underset{\sim}{\varepsilon} = 2(n\varepsilon + \mathbf{m} \cdot \mathbf{e})\underset{\sim}{1} + 2(n\mathbf{e} + \varepsilon\mathbf{m}) \cdot \underset{\sim}{\sigma} \tag{37A.6b}$$

The equation of motion separates into a scalar and a vector equation, first derived by Silin (1958):

$$\frac{\partial n}{\partial t} + \frac{\partial n}{\partial \mathbf{r}} \cdot \frac{\partial \varepsilon}{\partial \mathbf{p}} + \frac{\partial m_i}{\partial r_j}\frac{\partial e_i}{\partial p_j} - \frac{\partial n}{\partial \mathbf{p}} \cdot \frac{\partial \varepsilon}{\partial \mathbf{r}} - \frac{\partial m_i}{\partial p_j}\frac{\partial e_i}{\partial r_j} = I(\mathbf{p}) \tag{37A.7}$$

and

$$\frac{\partial \mathbf{m}}{\partial t} + \frac{\partial n}{\partial r_i}\frac{\partial \mathbf{e}}{\partial p_i} + \frac{\partial \mathbf{m}}{\partial r_i}\frac{\partial \varepsilon}{\partial p_i} - \frac{\partial n}{\partial p_i}\frac{\partial \mathbf{e}}{\partial r_i} - \frac{\partial \mathbf{m}}{\partial p_i}\frac{\partial \varepsilon}{\partial r_i} - \frac{2}{\hbar}\mathbf{e} \times \mathbf{m} = \mathbf{J}(\mathbf{p}), \tag{37A.8}$$

where the repeated indices are summed. $I(\mathbf{p})$ and $\mathbf{J}(\mathbf{p})$ are collision integrals, which will be neglected in the following discussion. Using Eq. (37.25), we have[30]

$$\delta\varepsilon(\mathbf{p}) = \frac{2}{(2\pi\hbar)^3}\int d^3p'\, f^{(s)}(\mathbf{p},\mathbf{p}')\delta n(\mathbf{p}') \tag{37A.9}$$

and

$$\mathbf{e} = -\frac{1}{2}\hbar\gamma_0\mathbf{H} + \frac{2}{(2\pi\hbar)^3}\int d^3p' f^{(a)}(\mathbf{p},\mathbf{p}')\mathbf{m}(\mathbf{p}'). \tag{37A.10}$$

We now examine small amplitude spin waves. We write

$$\mathbf{m} = \mathbf{m}_0 + \mathbf{m}' \tag{37A.11a}$$

and

$$\mathbf{e} = \mathbf{e}^{ext} + \mathbf{e}_0 + \mathbf{e}', \tag{37A.11b}$$

where $\mathbf{m}_0$ is the static component of the vector distribution function,[31] $-(\hbar/2)(\partial n_0/\partial\varepsilon)\gamma\mathbf{H}$, $\mathbf{m}'$ is the dynamic change associated with the spin wave, $\mathbf{e}^{ext} = -(\hbar/2)\gamma_0\mathbf{H}$, $\mathbf{e}_0 = (2/(2\pi\hbar)^3)\int d^3p' f^{(a)}(\mathbf{p},\mathbf{p}')\mathbf{m}_0(\mathbf{p}')$, and $\mathbf{e}' = (2/(2\pi\hbar)^3)\int d^3p' f^{(a)}(\mathbf{p},\mathbf{p}')\mathbf{m}'(\mathbf{p}')$.

The linearized form of Eq. (37A.8) is

$$\frac{\partial \mathbf{m}'}{\partial t} + v_i\frac{\partial \mathbf{m}'}{\partial x_i} - \frac{\partial n_0}{\partial p_i}\frac{\partial \mathbf{e}'}{\partial x_i} - \frac{2}{\hbar}\mathbf{e}_0 \times \mathbf{m}' - \frac{2}{\hbar}\mathbf{e}' \times \mathbf{m}_0 + \gamma_0\mathbf{H} \times \mathbf{m}' = 0, \tag{37A.12}$$

where the repeated indices are again summed. The fourth and sixth terms may be combined as $\gamma\mathbf{H} \times \mathbf{m}'$. Defining $m'_\pm = m'_x \pm i m'_y$, our vector equation of motion becomes

$$\frac{\partial m'_z}{\partial t} + \mathbf{v}\cdot\nabla m'_z - \frac{\partial n_0}{\partial \varepsilon}\frac{2}{(2\pi\hbar)^3}\mathbf{v}\cdot\nabla\int d^3p f^{(a)}(\mathbf{p},\mathbf{p}')m'_z(\mathbf{p}') = 0 \tag{37A.13}$$

[30] Use of the form (37.25) is restricted to a weakly polarized liquid; i.e., for magnetic fields such that $\hbar\gamma_0 H \ll \mu$.

[31] More generally, it may be a uniformly processing distribution resulting from the application of a "tipping" pulse, as is commonly used in nuclear magnetic resonance.

and

$$\frac{\partial m_\pm}{\partial t}+\left(\mathbf{v}\cdot\nabla\pm i\Omega_0^a\right)m'_\pm-\left(\mathbf{v}\cdot\nabla\pm i\Omega_0^a\right)\frac{\partial n_0}{\partial\varepsilon}\frac{2}{(2\pi\hbar)^3}\int d^3p' f^{(a)}(\mathbf{p},\mathbf{p}')m'_\pm(\mathbf{p}')=0. \tag{37A.14}$$

Here, $\Omega_0^a = \Omega_0/(1+F_0^a)$ and $\Omega_0 = \gamma_0 H$, the Larmor frequency.

The first of these equations is identical in structure to that for zero sound; however, since $F_0^a < 0$ it has no propagating solutions.

The second equation has propagating solutions only for long wavelengths and for $H \neq 0$. Assuming a form

$$m_\pm(\mathbf{p},\mathbf{r},t)=\frac{\partial n_0}{\partial\varepsilon}\mu_\pm(\hat{\mathbf{p}})e^{-i\omega t+i\mathbf{k}\cdot\mathbf{r}},$$

Eq. (37A.14) becomes

$$\left(-\omega+\mathbf{k}\cdot\mathbf{v}\pm\Omega_0^a\right)\mu_\pm(\hat{\mathbf{p}})=-\left(\mathbf{k}\cdot\mathbf{v}\pm\Omega_0^a\right)\frac{1}{4\pi}\int d\hat{\mathbf{p}}' F^a\left(\hat{\mathbf{p}},\hat{\mathbf{p}'}\right)\mu_\pm(\hat{\mathbf{p}}). \tag{37A.15}$$

We define the variables

$$\psi_\pm(\hat{\mathbf{p}})=\frac{-\omega+\mathbf{k}\cdot\mathbf{v}\pm\Omega_0^a}{\mathbf{k}\cdot\mathbf{v}\pm\Omega_0^a}\mu_\pm(\hat{\mathbf{p}})$$

and obtain

$$\psi_\pm(\hat{\mathbf{p}})=-\frac{1}{4\pi}\int d\hat{\mathbf{p}}' F^a\left(\hat{\mathbf{p}},\hat{\mathbf{p}'}\right)\frac{kv\cos\theta'\pm\Omega_0^a}{-\omega+kv\cos\theta'\pm\Omega_0^a}\psi_\pm(\hat{\mathbf{p}}). \tag{37A.16}$$

We first examine the limit $k=0$. One of the many solutions to (37A.16) is $\psi(\hat{\mathbf{p}})=$ const., yielding

$$-\frac{1}{4\pi}\frac{\pm\Omega_0^a}{\pm\Omega_0^a-\omega}\int d\hat{\mathbf{p}}' F^a\left(\hat{\mathbf{p}},\hat{\mathbf{p}'}\right)=1, \tag{37A.17}$$

which leads to $\omega = \pm\Omega_0$; the spins process at the *free particle* Larmor frequency. For $k \neq 0$ we limit ourselves to a single Fermi liquid parameter, F_0^a; the form $\psi(\hat{\mathbf{p}})=$ const. is still a solution and the dispersion is governed by

$$+\frac{F_0^a}{2}\int_0^\pi \frac{kv\cos\theta'\pm\Omega_0^a}{-\omega+kv\cos\theta'\pm\Omega_0^a}d\cos\theta'=1 \tag{37A.18}$$

or

$$-\frac{F_0^a}{2}\left[2+\frac{\omega}{kv}\ln\left(\frac{-\omega\pm\Omega_0^a+kv}{-\omega\pm\Omega_0^a-kv}\right)\right]=1. \tag{37A.19}$$

Carrying out the expansion of the logarithm (again to third order) yields the long wavelength spin wave dispersion relation,

$$\omega(k) = \pm\Omega_0\left[1 + \frac{(1+F_0^a)^2}{3F_0^a}\left(\frac{kv}{\Omega_0}\right)^2\right]; \qquad (37A.20)$$

note that since $F_0^a < 0$, $\omega(k)$ *decreases* with increasing wave number, unlike spin waves in ferromagnets.

Spin waves in liquid ^3He in an external magnetic field have been observed experimentally, particularly in the presence of a magnetic field gradient.[32]

[32] Collisionless spin waves in liquid ^3He, N. Masuhara, D. Candela, D.O. Edwards, R.F. Hoyt, H.N. Scholz, D.S. Sherrill, and R. Combescot, *Phys. Rev. Lett.* 53, 1168 (1984).

38

Superfluid ^{3}He

38.1 Basic ideas

Since liquid ^{3}He is a Fermi liquid, as discussed in Chapter 37, it is natural to expect that at some temperature it might become a superfluid, in analogy with BCS superconductors.[1,2] As first observed by Osheroff, Richardson, and Lee[3], this indeed happens at temperatures between approximately 1×10^{-3}K and 3×10^{-3}K, depending on the pressure. The phase diagram in the P–T plane is shown in Fig. 38.1. At zero magnetic field there are two different superfluid phases, referred to as the A phase and B phase (for historical reasons). The first of these occurs in a narrow wedge in P–T space at higher temperatures and near the melting curve; the latter exists over a broader region.

However, the BCS pairing in ^{3}He differs from the commonly encountered electron case in that the two fermions are bound in a state with an orbital angular momentum of $l = 1$. There are potentially three such states associated with the magnetic quantum number, m $(0, \pm 1)$. Since the parity of such states is odd on interchanging the two ^{3}He atoms (the parity of spherical harmonics is $(-1)^l$), the spin part of the quantum mechanical state function must be even in order for the overall state function to be antisymmetric and thereby satisfy the Pauli principle; i.e., it must involve the s = 1 spin states, of which there are also potentially three. The most general state would then be a superposition of all of the possible nine products that can be formed from the three orbital and three spin states with complex amplitudes (18 parameters in all). The two equilibrium states actually occurring in nature at zero field correspond to very specific combinations of subsets of these nine states, which will be treated shortly in this chapter. The validity of the states identified with the A and B phases has been unambiguously established as a result of numerous experiments, those involving nuclear magnetic resonance and ultrasonic propagation yielding the most conclusive evidence.

Before discussing the nature of the two BCS superfluid ground states, we seek a qualitative explanation of the underlying factors leading to the pairing in a higher angular momentum state. This result has its origin in the combination of the attractive Van der Waals-like He–He interaction at large distances, together with the strongly repulsive potential at short separations (the so-called hard core effect, see Fig. 7.1). One can consider the scattering of two ^{3}He atoms with momenta $+k_F$ and $-k_F$ by such a potential (from which a Cooper pair might form) and

[1] The 1996 Nobel Prize in Physics was awarded jointly to David M. Lee, Douglas D. Osheroff, and Robert C. Richardson "for their discovery of superfluidity in helium-3."

[2] For a discussion of the history preceding the discovery of superfluidity in ^{3}He, see the chapter by Anderson and Brinkman listed in the Additional Reading section at the end of this chapter.

[3] New magnetic phenomena in liquid He-3 below 3 mK , D.D. Osheroff, R.C. Richardson, and D.M. Lee, *Phys. Rev. Lett.* **29**, 920 (1972).

The Physics of Solids. First Edition. J. B. Ketterson.

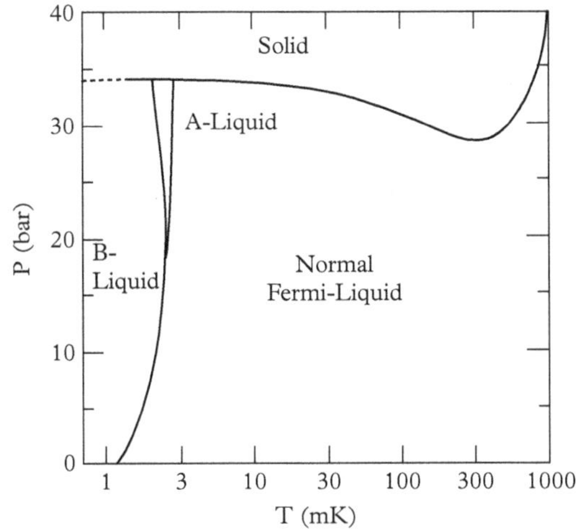

Figure 38.1 *The phase diagram of liquid and solid ^{3}He in the P-T plane. Only temperatures below the liquid-gas line are shown. In this regime we have three liquid phases: normal liquid and two superfluid phases, denoted as A and B. At pressures above about 33 bar we encounter the solid phase, which is also divided into several phases.*

examine the various partial wave phase shifts, $\delta_l(2k_F)$, governing the scattering.[4] The net result of such a calculation is that the s-wave phase shift, δ_0, is negative corresponding to repulsion, while for the next highest angular momentum state, δ_1, associated with $l = 1$ or p-wave scattering, it is positive, corresponding to attraction. Physically, this occurs because the particles would "see" the repulsive barrier for s-wave scattering; however the centrifugal barrier suppresses the overlap of the $l = 1$ and higher states on the repulsive hard core region of the potential,[5] and thereby allows the attractive van der Waals tail to win out. As noted above, particles in an $l = 1$ orbital must be in a triplet spin state.[6]

There is an alternative way to think about the attraction needed to form a Cooper pair. ^{3}He atoms in the liquid ground state can locally avoid each other by aligning their spins—the Pauli principle will then keep them further apart, thereby avoiding an increase in energy arising from the hard core repulsion. This property favors local spin alignment (on a fluctuating basis), thereby enhancing the magnetic susceptibility, but not to the point where the liquid acquires a spontaneous moment (which would onset at $F_0^a = -1$, see Eq. (37.30)). Nevertheless, another ^{3}He atom encountering a polarized region would have its energy lowered by temporarily aligning; this idea forms the basis of the spin fluctuation model, or paramagnon model as it is sometimes called, of the attractive interaction leading to superfluidity in ^{3}He. This qualitative idea is supported by a model, but realistic, many-body calculation.

38.2 The generalized order parameter of a superfluid

As discussed in Section 35.3, a superconductor has the unusual property that the expectation value of the destruction operator of a pair of electrons of opposite

[4] Possible phase transition in liquid He-3, V.J. Emery and A.M. Sessler, *Phys. Rev.* **119**, 43 (1960).

[5] At momenta significantly higher than the Fermi momentum the particles can of course approach each other more closely and thereby sample the hard core.

[6] Exploiting the centrifugal barrier to avoid a short range repulsion and open up a higher l attractive interaction represents an alternative strategy to achieve pairing, that is also exploited in high temperature and heavy fermion superconductors, to be discussed in Chapters 45 and 46. ^{3}He was the first system discovered that exploits this strategy.

momenta and spin, $\hat{c}_{\mathbf{k}\uparrow}\hat{c}_{-\mathbf{k}\downarrow}$, called the pair amplitude, has a non-zero value. This occurs because of the unique form of the superconducting wave function in that it does not have a fixed number of particles. Here we alter the definition of the order parameter introduced in Eq. (35.55) and write it as

$$\psi(\mathbf{p}, T) \equiv \sum_{|\mathbf{k}|} \langle \hat{c}_{\mathbf{k}\uparrow}\hat{c}_{-\mathbf{k}\downarrow} \rangle, \tag{38.1}$$

where $\mathbf{p}$ is a unit vector parallel to $\mathbf{k}$ at the Fermi surface; the temperature dependence is suppressed in what follows.[7] The form (38.1), where we only sum over the *magnitude* of $\mathbf{k}$, has the advantage of explicitly allowing for an *anisotropy* of the order parameter over the Fermi surface. Note that ψ has the property that it vanishes in the normal state, as required.

The superfluid ^{3}He order parameter is more complicated since the ^{3}He quasiparticles are paired in an $l = 1$, $s = 1$ state. A natural generalization of Eq. (38.1) is

$$\psi_{\alpha\beta}(\mathbf{p}, T) \equiv \sum_{|\mathbf{k}|} \langle \hat{c}_{\mathbf{k}\alpha}\hat{c}_{-\mathbf{k}\beta} \rangle, \tag{38.2}$$

where α and β refer to the two possible spin states of each ^{3}He quasiparticle. From the commutation relations obeyed by $\hat{c}_{\mathbf{k}\alpha}$, $\psi_{\alpha\beta}$ has the property,

$$\begin{aligned} \psi_{\alpha\beta}(\mathbf{p}) &= \sum_{|\mathbf{k}|} \langle \hat{c}_{\mathbf{k}\alpha}\hat{c}_{-\mathbf{k}\beta} \rangle \\ &= -\sum_{|\mathbf{k}|} \langle \hat{c}_{-\mathbf{k}\beta}\hat{c}_{\mathbf{k}\alpha} \rangle. \\ &= -\psi_{\beta\alpha}(-\mathbf{p}). \end{aligned} \tag{38.3}$$

The behavior of $\psi_{\alpha\beta}(\mathbf{p})$ under inversion of $\mathbf{k}$ involves two cases:

Singlet Pairing (space symmetric, spin antisymmetric)

$$\psi_{\alpha\beta}(\mathbf{p}) = -\psi_{\beta\alpha}(\mathbf{p}) = \psi_{\alpha\beta}(-\mathbf{p}). \tag{38.4}$$

Triplet Pairing (space antisymmetric, spin symmetric)

$$\psi_{\alpha\beta}(\mathbf{p}) = \psi_{\beta\alpha}(\mathbf{p}) = -\psi_{\alpha\beta}(-\mathbf{p}). \tag{38.5}$$

For the spin part of $\psi_{\alpha\beta}$ we will use a two by two matrix representation that can be written in the form $i\hat{\sigma}_i\hat{\sigma}_2$, where the subscript i runs from zero to three with $i = 0$ defined as the unit matrix and $i = 1$ to 3 as the three Pauli matrices:

$$i\hat{\sigma}_0\hat{\sigma}_2 = i\begin{pmatrix} 1 & 0 \\ 0 & 1 \end{pmatrix}\begin{pmatrix} 0 & -i \\ i & 0 \end{pmatrix} = \begin{pmatrix} 0 & 1 \\ -1 & 0 \end{pmatrix} \equiv \chi_0 \tag{38.6}$$

[7] The material in this section is based, in part, on Ginzburg-Landau Approach to L ≠ 0 Pairing, N. D. Mermin and C. Stare, *Phys. Rev. Letters 30, 1135 (1973)*, the thesis of G. B. Stare, P-Wave Models for the A Phase of Superfluid Helium-3, Cornell University, 1974, and some informal notes prepared by V. Ambegaokar.

and

$$
\begin{aligned}
i\hat{\sigma}_1\hat{\sigma}_2 &= i\begin{pmatrix} 0 & 1 \\ 1 & 0 \end{pmatrix}\begin{pmatrix} 0 & -i \\ i & 0 \end{pmatrix} = \begin{pmatrix} -1 & 0 \\ 0 & 1 \end{pmatrix} \equiv \chi_1, \\
i\hat{\sigma}_2\hat{\sigma}_2 &= i\begin{pmatrix} 0 & -i \\ i & 0 \end{pmatrix}\begin{pmatrix} 0 & -i \\ i & 0 \end{pmatrix} = \begin{pmatrix} i & 0 \\ 0 & i \end{pmatrix} \equiv \chi_2, \\
i\hat{\sigma}_3\hat{\sigma}_2 &= i\begin{pmatrix} 1 & 0 \\ 0 & -1 \end{pmatrix}\begin{pmatrix} 0 & -i \\ i & 0 \end{pmatrix} = \begin{pmatrix} 0 & 1 \\ 1 & 0 \end{pmatrix} \equiv \chi_3.
\end{aligned} \tag{38.7}
$$

We will show shortly that this representation is equivalent to a more conventional representation. The form given in Eq. (38.6) is antisymmetric on the interchange of α and β and thus represents the singlet state; the states (38.7) are symmetric and together form a representation of the three triplet states.

In matrix form our order parameter may be written as

$$
\underset{\sim}{\psi} \equiv \begin{pmatrix} \sum\limits_{|\mathbf{k}|}\langle \hat{c}_{\mathbf{k}\uparrow}\hat{c}_{-\mathbf{k}\uparrow}\rangle & \sum\limits_{|\mathbf{k}|}\langle \hat{c}_{\mathbf{k}\uparrow}\hat{c}_{-\mathbf{k}\downarrow}\rangle \\ \sum\limits_{|\mathbf{k}|}\langle \hat{c}_{\mathbf{k}\downarrow}\hat{c}_{-\mathbf{k}\uparrow}\rangle & \sum\limits_{|\mathbf{k}|}\langle \hat{c}_{\mathbf{k}\downarrow}\hat{c}_{-\mathbf{k}\downarrow}\rangle \end{pmatrix} \tag{38.8}
$$

$$
\equiv \begin{pmatrix} \psi_{\uparrow\uparrow} & \psi_{\uparrow\downarrow} \\ \psi_{\downarrow\uparrow} & \psi_{\downarrow\downarrow} \end{pmatrix}. \tag{38.9}
$$

In the absence of spin orbit coupling we can write our order parameter as a product of a vector containing the spatial variation, $d_i(\mathbf{p})$, and spin parts taken from Eq. (38.6) and (38.7). For our two cases we have

$$
\underset{\sim}{\psi}^{\text{singlet}} = d_0(\mathbf{p})i\hat{\sigma}_2 = \begin{pmatrix} 0 & d_0(\mathbf{p}) \\ -d_0(\mathbf{p}) & 0 \end{pmatrix} \tag{38.10}
$$

and

$$
\underset{\sim}{\psi}^{\text{triplet}} = \mathbf{d}(\mathbf{p}) \cdot i\hat{\sigma}_2\hat{\sigma} = \begin{pmatrix} -d_x + id_y & d_z \\ d_z & d_x + id_y \end{pmatrix}. \tag{38.11}
$$

$\underset{\sim}{\psi}^{\text{singlet}}$ and $\underset{\sim}{\psi}^{\text{triplet}}$ are, respectively, spin scalar (singlet) and spin vector (triplet) *order parameters*; the d_i are, in general, complex. For the triplet case, our writing **d** as a vector is a convenient way of denoting the fact that the triplet order parameter is in general a linear combination of the three spin states.

In the more conventional notation, our three spin states may be written (including a normalization factor) as

$$\chi_1 = \frac{1}{\sqrt{2}}\left[-\begin{pmatrix}1\\0\end{pmatrix}\begin{pmatrix}1\\0\end{pmatrix}+\begin{pmatrix}0\\1\end{pmatrix}\begin{pmatrix}0\\1\end{pmatrix}\right]$$

$$\chi_2 = \frac{i}{\sqrt{2}}\left[\begin{pmatrix}1\\0\end{pmatrix}\begin{pmatrix}1\\0\end{pmatrix}+\begin{pmatrix}0\\1\end{pmatrix}\begin{pmatrix}0\\1\end{pmatrix}\right]$$

$$\chi_3 = \frac{1}{\sqrt{2}}\left[\begin{pmatrix}1\\0\end{pmatrix}\begin{pmatrix}0\\1\end{pmatrix}+\begin{pmatrix}0\\1\end{pmatrix}\begin{pmatrix}1\\0\end{pmatrix}\right]. \tag{38.12}$$

In addition, we have the antisymmetric scalar spin state:

$$\chi_0 = \frac{1}{\sqrt{2}}\left[\begin{pmatrix}1\\0\end{pmatrix}\begin{pmatrix}0\\1\end{pmatrix}-\begin{pmatrix}0\\1\end{pmatrix}\begin{pmatrix}1\\0\end{pmatrix}\right]. \tag{38.13}$$

Note that χ_0 is the usual $S = 0$ singlet state while χ_3 is the $S = 1; S_z = 0$ state, the conventional $S = 1, S_z = 1$ states may be formed by taking linear combinations of χ_1 and χ_2. It is easy to verify that the three states (38.12) have the property

$$\hat{S}_1\chi_1 = \hat{S}_2\chi_2 = \hat{S}_3\chi_3 = 0, \tag{38.14}$$

i.e., $\hat{\mathbf{S}} \cdot \chi = 0$. For our two-particle system the total spin operator is

$$\hat{\mathbf{S}} = \frac{1}{2}\left(\hat{\sigma}^{(1)} + \hat{\sigma}^{(2)}\right), \tag{38.15}$$

where (1) and (2) refer to the two quasiparticles of a pair. It then follows that **w** *transforms as a vector*, which is convenient when considering how the order parameter transforms under rotations which we take up next.

38.3 Rotation properties of the order parameter

In what follows we will want to know the behavior of the order parameter under rotations. Since the orbital state can in principle correspond to any l value, whereas the spin of our pair can only be zero or one, we will restrict ourselves to rotations of the spin system relative to a fixed orbital system. In quantum mechanics the rotation operator in spin space is $\hat{R}(\boldsymbol{\theta}) = e^{i\hat{\mathbf{S}}\cdot\boldsymbol{\theta}}$. By direct calculation one can verify that the effect of operating with $\hat{\boldsymbol{\sigma}}^{(1)}$ on the state functions given by Eqs. (38.12) and (38.13) or with $\hat{\sigma}$ on their matrix representation in Eqs. (38.10) and (38.11) is identical. Similarly the effect of operating with $\hat{\boldsymbol{\sigma}}^{(2)}$ on Eqs. (38.13) and (38.14), and with $\hat{\sigma}^{tr}$ on the right on Eqs. (38.10) and (38.11) is the same. Thus we can write

$$\hat{R}\underset{\sim}{\psi} \equiv \underset{\sim}{\psi}' = e^{\frac{i}{2}\hat{\sigma}\cdot\boldsymbol{\theta}}\underset{\sim}{\psi}e^{\frac{i}{2}\hat{\sigma}^{tr}\cdot\boldsymbol{\theta}} \tag{38.16}$$

or, in component form,

$$\psi'_{\alpha\beta} = \left(e^{\frac{i}{2}\hat{\sigma}\cdot\theta}\right)_{\alpha\delta}\left(e^{\frac{i}{2}\hat{\sigma}\cdot\boldsymbol{\theta}}\right)_{\beta\gamma}\psi_{\delta\gamma}. \tag{38.17}$$

For an infinitesimal rotation we have

$$\underset{\sim}{\psi}' = \underset{\sim}{\psi}' + \frac{1}{2}\hat{\sigma}\cdot\boldsymbol{\theta}\underset{\sim}{\psi} + \frac{1}{2}\underset{\sim}{\psi}(\hat{\sigma}\cdot\boldsymbol{\theta})^{tr}. \tag{38.18}$$

Let us examine such infinitesimal rotations for our two cases. The following identity, which is easily verified, is useful,

$$\hat{\sigma}_2\hat{\sigma}^{tr} = -\hat{\sigma}\hat{\sigma}_2. \tag{38.19}$$

Singlet case:

$$\begin{aligned}\underset{\sim}{\psi}^{singlet} &= d_0 i\hat{\sigma}_2\\ \underset{\sim}{\psi}' &= \underset{\sim}{\psi} + \frac{1}{2}(\hat{\sigma}\cdot\boldsymbol{\theta})\,(d_0 i\hat{\sigma}_2) + \frac{1}{2}(d_0 i\hat{\sigma}_2)(\hat{\sigma}\cdot\boldsymbol{\theta})^{tr}.\end{aligned}$$

Using Eq. (38.19) we have immediately

$$\underset{\sim}{\psi} = \underset{\sim}{\psi}' \tag{38.20}$$

or

$$\hat{\sigma}\cdot\boldsymbol{\theta}\underset{\sim}{\psi}^{singlet} = 0 \tag{38.21}$$

Triplet case:

$$\begin{aligned}\underset{\sim}{\psi}^{triplet} &= \mathbf{d}\cdot\hat{\sigma}\, i\hat{\sigma}_2\\ \underset{\sim}{\psi}' &= \underset{\sim}{\psi} + \frac{1}{2}\,(\hat{\sigma}\cdot\boldsymbol{\theta})(\mathbf{d}\cdot\hat{\sigma})\, i\hat{\sigma}_2 + \frac{1}{2}(\mathbf{d}\cdot\hat{\sigma})i\hat{\sigma}_2(\hat{\sigma}^{tr}\cdot\boldsymbol{\theta})\\ &= \underset{\sim}{\psi} + \frac{1}{2}\,(\hat{\sigma}\cdot\boldsymbol{\theta})\,(\mathbf{d}\cdot\hat{\sigma})\, i\hat{\sigma}_2 - \frac{1}{2}\,(\mathbf{d}\cdot\hat{\sigma})\,(\hat{\sigma}\cdot\boldsymbol{\theta})\, i\hat{\sigma}_2.\end{aligned}$$

Using the identity[8]

$$(\hat{\sigma}\cdot\mathbf{a})(\hat{\sigma}\cdot\mathbf{b}) = \mathbf{a}\cdot\mathbf{b} + i(\mathbf{a}\times\mathbf{b})\cdot\hat{\sigma} \tag{38.22}$$

[8] Landau and Litshitz, *Quantum Mechanics*, p. 191.

we have

$$\underset{\sim}{\psi}' = \underset{\sim}{\psi} - (\boldsymbol{\theta} \times \mathbf{d}) \cdot \hat{\sigma}\, i\hat{\sigma}_2 \tag{38.23}$$

or

$$\underset{\sim}{\psi}' = (\mathbf{1} - \boldsymbol{\theta}\times)\, \mathbf{d} \cdot \hat{\sigma}\, i\hat{\sigma}_2. \tag{38.24}$$

The operator $(\mathbf{1} - \boldsymbol{\theta}\times)$ is an infinitesimal three-dimensional rotation matrix, $\mathbf{R}(-\boldsymbol{\theta})$; i.e., we can define a vector,

$$\begin{aligned} \mathbf{d}' &= \mathbf{R}(-\boldsymbol{\theta})\mathbf{d} \\ &= \mathbf{R}^{-1}(\boldsymbol{\theta})\mathbf{d} \end{aligned} \tag{38.25}$$

Rotation of the space coordinates by an angle $\boldsymbol{\theta}$ is equivalent to rotating the spin by an angle $-\boldsymbol{\theta}$; thus $\mathbf{d}$ behaves like a spin vector.

38.4 Ginzburg–Landau theory of the superfluid phases of ^{3}He

To proceed further we must assume an explicit form for $\mathbf{d}(\mathbf{p})$. Triplet state superfluidity could, in principle, occur for any odd orbital angular momentum; since the parity is $(-1)^l$, and the spin state is symmetric, the overall wave function will be antisymmetric. However, in ^{3}He the evidence is overwhelming that the pairs are in a p state. The $l = 1$ spherical harmonics of the orbital wave function (written as a vector) are

$$\begin{aligned} p_z &= \cos\theta \\ p_x &= \sin\theta\cos\phi \\ p_y &= \sin\theta\sin\phi \end{aligned} \tag{38.26}$$

and can be taken ascomponents of a vector $\mathbf{p}$.

The spin vector, $\mathbf{d}$, must be made up of linear combinations of the $l = 1$ spherical harmonics. The most general form would then be

$$\mathbf{d} = \mathbf{A} \cdot \mathbf{p}, \tag{38.27}$$

where $\mathbf{A}$ is a 3×3 complex matrix; written in component form, Eq. (38.27) is

$$d_\mu = \sum_i A_{i\mu} p_i, \tag{38.28}$$

where we use Greek and Roman indices for spin and real space respectively.

Since **d** is complex, the matrix **A** contains a total of 18 real parameters, which could in principle result in 18 different states (superfluids). However, only a much smaller number of states actually occur in ^{3}He.

To parameterize the free energy in a Landau–Ginzburg theory for ^{3}He we must form all of the quadratic and quartic invariants associated with the order parameter[9]. In the absence of spin orbit coupling or magnetic dipole forces (which couple spin and real space) the free energy must be invariant to separate rotations of spin and real space. The free energy must be a real scalar: thus we must always couple a **d** with a $\mathbf{d}^*$. For the quadratic terms we must examine the form $\mathbf{d}\cdot\mathbf{d}^* = \sum_{\mu,i,j} A_{\mu i}A^*_{\mu j}p_i p_j$. If we average over the angles, θ and ϕ, occurring in the definition of the p_i, we obtain $\overline{\mathbf{d}\cdot\mathbf{d}^*} = \sum_{\mu,i,j} A_{\mu i}A^*_{\mu j}\delta_{ij}$. Hence, the only allowed combination is $\mathrm{Tr}(\mathbf{A}\cdot\mathbf{A}^\dagger) = \sum_{i\mu} A_{i\mu}A^*_{i\mu}$.

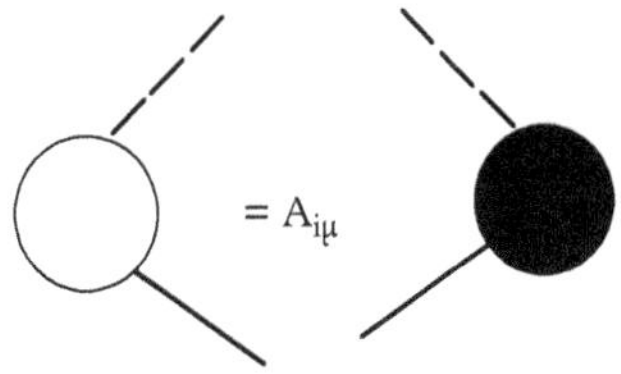

Figure 38.2 *Diagrams representing* $A_{i\mu}$ *and* $A^*_{i\mu}$.

In fourth order there are five invariants. They are most easily enumerated by introducing the diagrams shown in Fig. 38.2. Here the dashed and solid lines refer to the spin and orbit indices respectively. The five fourth order invariants correspond to the diagrams shown in Fig. 38.3(a) through (e). Connecting a line between two circles corresponds to summing over an index. Since we exclude terms mixing the spin and orbital degrees of freedom we connect a solid line only to a solid line and a dashed line only to a dashed line. Note that we could also draw diagram (d) in the "twisted" form shown in Fig. 38.4. The diagrams (a) through (e) correspond to the following five quartic invariants:

$$|\mathrm{Tr}(\mathbf{A}\cdot\mathbf{A}^{tr})|^2 = \left(\sum_{i\mu} A_{i\mu}A_{i\mu}\right)\left(\sum_{i\mu} A^*_{i\mu}A^*_{i\mu}\right) \tag{38.29}$$

$$\left[\mathrm{Tr}\left(\mathbf{A}\cdot\mathbf{A}^\dagger\right)\right]^2 = \left(\sum_{i\mu} A_{i\mu}A^*_{i\mu}\right)^2 \tag{38.30}$$

$$\mathrm{Tr}\left[(\mathbf{A}\cdot\mathbf{A}^{tr})\cdot(\mathbf{A}\cdot\mathbf{A}^{tr})^*\right] = \sum_{ij\mu\nu} A_{i\mu}A_{i\nu}A^*_{j\nu}A^*_{j\mu} \tag{38.31}$$

$$\mathrm{Tr}\left[\left(\mathbf{A}\cdot\mathbf{A}^\dagger\right)\cdot\left(\mathbf{A}\cdot\mathbf{A}^\dagger\right)\right] = \sum_{ij\mu\nu} A_{i\mu}A^*_{i\nu}A_{j\nu}A^*_{j\mu} \tag{38.32}$$

$$\mathrm{Tr}\left[\left(\mathbf{A}\cdot\mathbf{A}^\dagger\right)\left(\mathbf{A}\cdot\mathbf{A}^\dagger\right)^*\right] = \sum_{ij\mu\nu} A_{i\mu}A^*_{i\nu}A^*_{j\nu}A_{j\mu} \tag{38.33}$$

where Tr and tr stand for trace and transpose respectively. Thus the free energy density may be written as

$$\begin{aligned} F(\psi) &= F(0) - \alpha\mathrm{Tr}\left(\mathbf{A}\cdot\mathbf{A}^\dagger\right) + \beta_1\left|\mathrm{Tr}(\mathbf{A}\cdot\mathbf{A}^{tr})\right|^2 \\ &+ \beta_2\left[\mathrm{Tr}\left(\mathbf{A}\cdot\mathbf{A}^\dagger\right)\right]^2 + \beta_3\,\mathrm{Tr}\left[(\mathbf{A}\cdot\mathbf{A}^{tr})\cdot(\mathbf{A}\cdot\mathbf{A}^{tr})^*\right]. \\ &+ \beta_4\mathrm{Tr}\left[\left(\mathbf{A}\cdot\mathbf{A}^\dagger\right)\cdot\left(\mathbf{A}\cdot\mathbf{A}^\dagger\right)\right] + \beta_5\mathrm{Tr}\left[\left(\mathbf{A}\cdot\mathbf{A}^\dagger\right)\cdot\left(\mathbf{A}\cdot\mathbf{A}^\dagger\right)^*\right] \end{aligned} \tag{38.34}$$

[9] The G–L like theory of ^{3}He described here follows that given in: Ginzburg–Landau approach to *l* not equal 0 pairing, N.D. Mermin and C. Stare, *Phys. Rev. Lett.* 30, 1135 (1973).

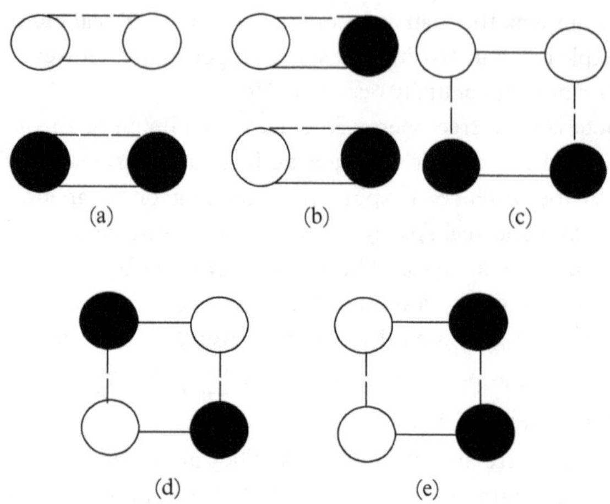

Figure 38.3 *The five quadratic invariants involving $A^*_{i\mu}$ and $A_{i\mu}$.*

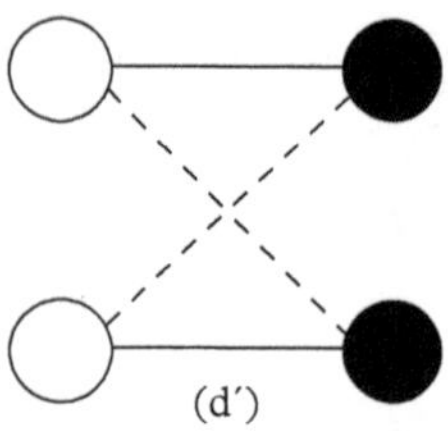

Figure 38.4 *An alternative form for diagram 38.2d.*

For the general case we would minimize Eq. (38.34) with respect to the nine complex constants characterizing **A**. Here we will restrict ourselves to so-called *unitary states* which we now discuss. The unitarity condition is

$$\underset{\sim}{\psi}\underset{\sim}{\psi}^\dagger = 1; \tag{38.35}$$

for the case of a triplet order parameter given by Eq. (38.11),

$$\begin{aligned}\underset{\sim}{\psi}\underset{\sim}{\psi}^\dagger &= (\mathbf{d}\cdot\boldsymbol{\sigma}\sigma_2)\left(\sigma_2\,\mathbf{d}^*\cdot\sigma\right)\\ &= \begin{pmatrix} d_z & d_x + id_y \\ d_x - id_y & -d_z \end{pmatrix}\begin{pmatrix} d_z^* & d_x^* + id_y^* \\ d_x^* - id_y^* & -d_z^* \end{pmatrix}\\ &= |\mathbf{d}|^2\,\sigma_0 - i\left(\mathbf{d}\times\mathbf{d}^*\right)\cdot\sigma.\end{aligned} \tag{38.36}$$

Thus unitarity requires that $\mathbf{d}\times\mathbf{d} = 0$.

We now show that a unitary state has no net spin. To do this we examine the expectation value of $\hat{\mathbf{S}}\cdot\boldsymbol{\theta}$; using Eq. (38.24) we have

$$\begin{aligned}\mathrm{Tr}\left[\underset{\sim}{\psi}^\dagger\hat{\mathbf{S}}\cdot\boldsymbol{\theta}\,\underset{\sim}{\psi}\right] &= \mathrm{Tr}\left[(\mathbf{d}\cdot\boldsymbol{\sigma}i\sigma_2)^\dagger\;((\mathbf{d}\times\boldsymbol{\theta})\cdot\boldsymbol{\sigma}\,i\sigma_2)\right]\\ &= \mathrm{Tr}\left[i\sigma_2\left(\mathbf{d}^*\cdot\boldsymbol{\sigma}\right)\;((\mathbf{d}\times\boldsymbol{\theta})\cdot\boldsymbol{\sigma}\,i\sigma_2)\right]\end{aligned} \tag{38.37}$$

Using the cyclic invariance of the trace and the properties $\sigma_i^2 = 1$, $\sigma_1\sigma_2 = i\sigma_3$, and $\mathrm{Tr}\,\sigma_i = 0$, we obtain

$$\begin{aligned}\mathrm{Tr}\left[\underset{\sim}{\psi}^\dagger\hat{\mathbf{S}}\cdot\boldsymbol{\theta}\underset{\sim}{\psi}\right] &= 2\mathbf{d}^*\cdot(\mathbf{d}\times\boldsymbol{\theta}) = 2\left(\mathbf{d}^*\times\mathbf{d}\right)\cdot\theta\\ &= 0,\end{aligned} \tag{38.38}$$

where in the last step we have used the unitarity condition, $\mathbf{d} \times \mathbf{d}^* = 0$. Thus a unitary state has no net spin polarization at any point on the Fermi surface and consequently could not have an overall spin polarization (which would correspond to a "nuclear ferromagnet").

The property, $\mathbf{d} \times \mathbf{d}^* = 0$, will be satisfied if $\mathbf{d}$ has the form

$$\mathbf{d} = a\,\mathbf{n}, \tag{38.39}$$

where $\mathbf{n}$ is a real vector and a is a complex scalar. We also know that for p-wave pairing, $\mathbf{d}$ is linear in $\mathbf{p}$. There are then two possible cases:

$$d_\mu = a n_{i\mu} p_i, \qquad \text{(case I)} \tag{38.40}$$

where $n_{i\mu}$ are components of a real matrix and a is a complex constant; in this case the vector $\mathbf{n}$ in Eq. (38.39) depends on the angles defining the p_i and a is independent of the angles. Alternatively we could have

$$d_\mu = a_i p_i n_\mu \qquad \text{(case II)} \tag{38.41}$$

where a_i is a complex vector and n_μ are components of a real vector. In this latter case the parameter a in (38.39) depends on the angles defining p_i, and n_μ is independent of the angles.

38.5 Minimization of the free energy for an $l = 1$, $s = 1$ superfluid

To obtain the order parameter for the observed phases we must carry out a minimization of the free energy given by Eq. (38.42) for the cases I and II. The details are somewhat tedious and some readers may wish to go to the final results given in Section 38.6.

Case I

For this case we have that $\mathbf{A}$ is real except for an irrelevant complex constant. It follows that $\mathbf{A} \cdot \mathbf{A}^\dagger = \mathbf{A} \cdot \mathbf{A}^{tr}$ is a *real symmetric matrix*, which can always be diagonalized by a proper rotation of the coordinate system. Since the trace is invariant under such a rotation, the free energy may be expressed in terms of the three real, non-negative, eigenvalues of $\mathbf{A} \cdot \mathbf{A}^{tr}$: λ_1, λ_2 , and λ_3. In evaluating the free energy given in Eq. (38.34) we need

$$\mathrm{Tr}\left[\mathbf{A} \cdot \mathbf{A}^{tr}\right] = \sum_i \lambda_i \tag{38.42}$$

$$\mathrm{Tr}\left[(\mathbf{A} \cdot \mathbf{A}^{tr}) \cdot (\mathbf{A} \cdot \mathbf{A}^{tr})\right] = \sum_i \lambda_i^2. \tag{38.43}$$

Using Eq. (38.42) and (38.43), we may write Eq. (38.34) as

$$\Delta F \equiv F - F_0$$

$$\Delta F = -\alpha \sum_i \lambda_i + (\beta_1 + \beta_2)\left(\sum_i \lambda_i\right)^2 + (\beta_3 + \beta_4 + \beta_5)\sum_i \lambda_i^2. \tag{38.44}$$

There are three different cases, depending on whether no eigenvalues vanish (three-dimensional case), one eigenvalue vanishes (two-dimensional case), or two eigenvalues vanish (one-dimensional case). Three non-vanishing λ_i corresponds to a superfluid involving all three orbital states, p_i; analogously two non-vanishing λ_i corresponds to two orbital states, and a single λ_i to one state. We first minimize Eq. (38.44) for the case of no vanishing eigenvalues:

$$\frac{\partial \Delta F}{\partial \lambda_i} = -\alpha + 2(\beta_1 + \beta_2)\sum_{j=1}^{3} \lambda_j + 2(\beta_3 + \beta_4 + \beta_5)\,\lambda_i = 0 \tag{38.45}$$

$$\lambda_i = \frac{-\alpha + 2(\beta_1 + \beta_2)\sum_{j=1}^{3} \lambda_j}{2(\beta_3 + \beta_4 + \beta_5)}. \tag{38.46}$$

Summing Eq. (38.45) over i and solving for $\sum_{i=1}^{3} \lambda_i$, we obtain

$$\sum_{i=1}^{3} \lambda_i = \frac{3\alpha}{6\,(\beta_1 + \beta_2) + 2(\beta_3 + \beta_4 + \beta_5)}; \tag{38.47}$$

inserting Eq. (38.47) into Eq. (38.46), we obtain

$$\lambda_i = \frac{\alpha}{6(\beta_1 + \beta_2) + 2(\beta_3 + \beta_4 + \beta_5)}. \tag{38.48}$$

Using Eq. (38.48) we evaluate the free energy given by Eq. (38.44), obtaining

$$\Delta F = -\frac{3}{4}\alpha^2 \frac{1}{3(\beta_1 + \beta_2) + (\beta_3 + \beta_4 + \beta_5)}. \tag{38.49}$$

The fact that all three eigenvalues are equal implies that $\mathbf{A} \propto \mathbf{R}$, since $\mathbf{R} \cdot \mathbf{R}^{tr} = \mathbf{1}$. We may express $\mathbf{d}$ in the form

$$\mathbf{d} = \Delta \mathbf{R}(\theta) \cdot \mathbf{p}, \tag{38.50}$$

where Δ is a complex constant (gap parameter) and $\mathbf{R}(\theta)$ is a rotation matrix for an arbitrary angle about an arbitrary axis; this state is referred to as the *three dimensional, Balian–Werthamer (BW)*, or *isotropic state*. We can also write (with $\mathbf{m}_i$

forming an orthogonal basis in one coordinate system and l_i a basis in a rotated system)

$$\mathbf{d} = \Delta[l_1(\mathbf{m}_1 \cdot \mathbf{p}) + l_2(\mathbf{m}_2 \cdot \mathbf{p}) + l_3(\mathbf{m}_3 \cdot \mathbf{p})] \tag{38.51}$$

or

$$d_\mu = \sum_{ij} l^j_\mu m^i_j p_i. \tag{38.52}$$

We next minimize Eq. (38.44) in two dimensions; in this case the analogues of Eqs. (38.47) and (38.48) are

$$\sum_{j=1}^{2} \lambda_j = \frac{\alpha}{2(\beta_1 + \beta_2) + (\beta_3 + \beta_4 + \beta_5)}, \tag{38.53}$$

$$\lambda_i = \frac{\alpha}{2} \frac{1}{2(\beta_1 + \beta_2) + (\beta_3 + \beta_4 + \beta_5)}. \tag{38.54}$$

In place of (38.49) we have

$$\Delta F = \frac{\alpha^2}{2} \frac{1}{2(\beta_1 + \beta_2) + (\beta_3 + \beta_4 + \beta_5)}. \tag{38.55}$$

For the two-dimensional case we have

$$\mathbf{A} \cdot \mathbf{A}^{\mathrm{tr}} = P_2, \tag{38.56}$$

where P_2 is a two-dimensional projection operator. The **d** vector may be written in the form

$$\mathbf{d} = \Delta[l_1(\mathbf{m}_1 \cdot \mathbf{p}) + l_2(\mathbf{m}_2 \cdot \mathbf{p})] \tag{38.57}$$

or

$$d_\mu = \Delta \sum_{j=1}^{2} \sum_{i=1}^{3} l^j_\mu m^j_i p_i. \tag{38.58}$$

This state is also called the *planar state.*

In one dimension, minimization of Eq. (38.44) gives

$$\lambda = \frac{\alpha}{2(\beta_1 + \beta_2 + \beta_3 + \beta_4 + \beta_5)}, \tag{38.59}$$

$$\Delta F = \frac{\alpha^2}{4(\beta_1 + \beta_2 + \beta_3 + \beta_4 + \beta_5)} \tag{38.60}$$

and

$$\mathbf{d} = \Delta l a(\mathbf{m} \cdot \mathbf{p}). \tag{38.61}$$

This state is also called the *one-dimensional* or *polar state.*

Case II

In this case we have $\mathbf{d} = (\mathbf{a}\cdot\mathbf{p})\mathbf{n} = \mathbf{A}\cdot\mathbf{p}$, where $\mathbf{A}$ has the form

$$\mathbf{A} = \begin{pmatrix} a_x n_x & a_y n_x & a_z n_x \\ a_x n_y & a_y n_y & a_z n_y \\ a_x n_z & a_y n_z & a_z n_z \end{pmatrix}, \tag{38.62}$$

for which we calculate

$$\begin{aligned} \mathrm{Tr}\left[\mathbf{A}\cdot\mathbf{A}^{\dagger}\right] &= \left(a_x a_x^* + a_y a_y^* + a_z a_z^*\right)\left(n_x^2 + n_y^2 + n_z^2\right) \\ &= \mathbf{a}\cdot\mathbf{a}^*; \end{aligned}, \tag{38.63}$$

similarly,

$$\mathrm{Tr}\ \left|\mathbf{A}\cdot\mathbf{A}^{\mathrm{tr}}\right|^2 = |\mathbf{a}\cdot\mathbf{a}|^2 \tag{38.64}$$

$$\mathrm{Tr}\left(\mathbf{A}\cdot\mathbf{A}^{\dagger}\right)^2 = \left(\mathbf{a}\cdot\mathbf{a}^*\right)^2 \tag{38.65}$$

$$\mathrm{Tr}\left[\left(\mathbf{A}\cdot\mathbf{A}^{\mathrm{tr}}\right)\cdot\left(\mathbf{A}\cdot\mathbf{A}^{\mathrm{tr}}\right)^*\right] = |\mathbf{a}\cdot\mathbf{a}|^2 \tag{38.66}$$

$$\mathrm{Tr}\left[\left(\mathbf{A}\cdot\mathbf{A}^{\dagger}\right)\cdot\left(\mathbf{A}\cdot\mathbf{A}^{\dagger}\right)\right] = \left(\mathbf{a}\cdot\mathbf{a}^*\right)^2 \tag{38.67}$$

$$\mathrm{Tr}\left[\left(\mathbf{A}\cdot\mathbf{A}^{\dagger}\right)\left(\mathbf{A}\cdot\mathbf{A}^{\dagger}\right)^*\right] = \left(\mathbf{a}\cdot\mathbf{a}^*\right)^2. \tag{38.68}$$

Substituting Eqs. (38.63) to (38.68) into Eq. (38.34), we obtain

$$\Delta F = -\alpha\left(\mathbf{a}\cdot\mathbf{a}^*\right) + (\beta_1 + \beta_3)\,|\mathbf{a}\cdot\mathbf{a}|^2 + (\beta_2 + \beta_4 + \beta_5)\left(\mathbf{a}\cdot\mathbf{a}^*\right)^2. \tag{38.69}$$

Note that $0 \leq |\mathbf{a}\cdot\mathbf{a}| \leq (\mathbf{a}\cdot\mathbf{a}^*)$; in particular $|\mathbf{a}\cdot\mathbf{a}|$ can be zero. Therefore if $(\beta_1 + \beta_3) > 0$, the free energy is minimized by having $|\mathbf{a}\cdot\mathbf{a}| = 0$. We write $\mathbf{a} = \mathbf{a}_1 + i\mathbf{a}_2$; then $\mathbf{a}\cdot\mathbf{a} = a_1^2 - a_2^2 + 2\,i\mathbf{a}_1\cdot\mathbf{a}_2$. Thus if $|\mathbf{a}_1| = |\mathbf{a}_2|$ and $\mathbf{a}_1 \perp \mathbf{a}_2$, we have $\mathbf{a}\cdot\mathbf{a} = 0$ and $\mathbf{a}\cdot\mathbf{a}^* = a_1^2 + a_2^2 = +2a_1^2$. In this case,

$$\Delta F = 2\alpha a_1^2 + 4(\beta_2 + \beta_4 + \beta_5)a_1^4. \tag{38.70}$$

Minimizing Eq. (38.70) with respect to a_1, we obtain

$$a_1^2 = \frac{\alpha}{4(\beta_2 + \beta_4 + \beta_5)}, \tag{38.71}$$

from which it follows that

$$\Delta F = -\frac{\alpha^2}{4}\frac{1}{\beta_2 + \beta_4 + \beta_5}. \tag{38.72}$$

This state is called the *axial* or *Anderson–Brinkman–Morel (ABM)* state and the vector **d** has the form (where we now write $\hat{\mathbf{d}}$ in place of **n**)

$$\mathbf{d} = \hat{\mathbf{d}}(\mathbf{a}\cdot\mathbf{p}), \tag{38.73}$$

where we write

$$\mathbf{a} = (\mathbf{m} + i\,\mathbf{n})\Delta, \tag{38.74}$$

where **m** and **n** are two orthogonal unit vectors which project to independent combinations of orbital states; we define a third direction, $\boldsymbol{l}$, the gap axis (which microscopically is also the angular momentum axis), by

$$\boldsymbol{l} = \mathbf{m}\times\mathbf{n}. \tag{38.75}$$

For the case, $\beta_1 + \beta_3 < 0$, the minimal occurs when $|\mathbf{a}\cdot\mathbf{a}|$ is maximum, i.e., when it is equal to $\mathbf{a}\cdot\mathbf{a}^*$;[10] in this case we have

$$\Delta F = -\alpha(\mathbf{a}\cdot\mathbf{a}^*) + (\beta_1+\beta_2+\beta_3+\beta_4+\beta_5)(\mathbf{a}\cdot\mathbf{a}^*)^2. \tag{38.76}$$

Minimization of Eq. (38.76) results in the same free energy as encountered in the one-dimensional case.

Let us collect the free energies of the various states as contained in Eqs. (38.49), (38.55), (38.60), and (38.72):

$$\text{3D: } \Delta F = -\frac{\alpha^2}{2}\frac{1}{\frac{2}{3}(\beta_3+\beta_4+\beta_5)+2(\beta_1+\beta_2)} \tag{38.77}$$

$$\text{2D: } \Delta F = -\frac{\alpha^2}{2}\frac{1}{(\beta_3+\beta_4+\beta_5)+2(\beta_1+\beta_2)} \tag{38.78}$$

$$\text{1D: } \Delta F = -\frac{\alpha^2}{2}\frac{1}{2(\beta_3+\beta_4+\beta_5)+2(\beta_1+\beta_2)} \tag{38.79}$$

$$\text{Axial: } \Delta F = -\frac{\alpha^2}{2}\frac{1}{2(\beta_2+\beta_4+\beta_5)}. \tag{38.80}$$

It follows that, regardless of the signs of $(\beta_3+\beta_4+\beta_5)$ and $(\beta_1+\beta_2)$, the 2d state is never stable. For $\beta_1+\beta_2>0$, and $\beta_3+\beta_4+\beta_5>0$ the 3d state will have a lower energy than the Id state.

From the weak coupling theory of superconductivity (which we do not discuss) one can show that $\beta_1 = -s$, $\beta_2 = \beta_3 = \beta_4 = 2s$, and $\beta_5 = -2s$; here, s is a parameter. Inserting these results in Eqs. (38.77), (38.79), and (38.80), we obtain

$$\left.\begin{aligned} \text{3D: } \quad \Delta F &= -\frac{3\alpha^2}{20s} \\ \text{1D: } \quad \Delta F &= -\frac{\alpha^2}{12s} \\ \text{Axial: } \quad \Delta F &= -\frac{\alpha^2}{8s} \end{aligned}\right\}\text{ weak coupling.} \tag{38.81}$$

[10] For this case the vector **a** may be written as $\mathbf{a} = \boldsymbol{\Delta}e^{i\phi}$, where $\boldsymbol{\Delta}$ is a real vector.

It follows that in weak coupling theory, the 3d or Balian–Werthamer (BW) state has the lowest energy and ^{3}He B is identified with this state. However, in the strong coupling theories of Anderson and Brinkman the possibility that the axial phase may be more stable in some situations is rationalized, and indeed ^{3}He A is identified as the Anderson–Brinkman–Morel phase.

38.6 Anisotropy of the energy gap in an $l = 1$, $s = 1$ superfluid; order parameter of ^{3}He

For the 3d phase, $\mathbf{d} = \Delta\mathbf{R}\cdot\mathbf{p}$, by Eq. (38.50); since the axes are arbitrary we take

$$\mathbf{R} = \begin{pmatrix} 1 & 0 & 0 \\ 0 & 1 & 0 \\ 0 & 0 & 1 \end{pmatrix}, \mathbf{A} = \Delta \begin{pmatrix} 1 & 0 & 0 \\ 0 & 1 & 0 \\ 0 & 0 & 1 \end{pmatrix}. \tag{38.82}$$

Using $\underset{\sim}{\psi} = \mathbf{d}\cdot\boldsymbol{\sigma} i\sigma_2$, we obtain

$$\underset{\sim}{\psi} = \Delta \begin{pmatrix} p_x + ip_y & -p_z \\ -p_z & -p_x + ip_y \end{pmatrix}. \tag{38.83}$$

The square of the gap energy is proportional to $\mathrm{Tr}[\underset{\sim}{\psi}\,\underset{\sim}{\psi}^\dagger]$, for which we have

$$\begin{aligned} \mathrm{Tr}[\underset{\sim}{\psi}\,\underset{\sim}{\psi}^\dagger] &= 2\Delta^2\hat{k}^2 \\ &= 2\Delta^2; \end{aligned} \tag{38.84}$$

thus we see that the gap is *isotropic* in the 3d phase.

For the 1d case, $\mathbf{d} = \boldsymbol{l}(\mathbf{m}\cdot\mathbf{p})\Delta$, by Eq. (38.61), again the axes are arbitrary so we choose $\boldsymbol{l} \parallel \mathbf{m} \parallel \mathbf{z}$, obtaining

$$\mathbf{A} = \Delta \begin{pmatrix} 0 & 0 & 0 \\ 0 & 0 & 0 \\ 0 & 0 & 1 \end{pmatrix} \tag{38.85}$$

and

$$\underset{\sim}{\psi} = \Delta \begin{pmatrix} 0 & -p_z \\ -p_z & 0 \end{pmatrix}, \tag{38.86}$$

from which it follows, using $k_z = \cos\theta$,

$$\begin{aligned} \mathrm{Tr}[\underset{\sim}{\psi}\,\underset{\sim}{\psi}^\dagger] &= 2\Delta^2 k_z^2 \\ &= 2\Delta^2\cos^2\theta; \end{aligned} \tag{38.87}$$

thus the gap has a maximum at the pole and vanishes along the equator and it is for this reason that the state is referred to as the polar state.

Combining Eqs. (38.73) and (38.74), we can write the axial state as $\mathbf{d} = \hat{\mathbf{d}}\Delta(\mathbf{m} + i\mathbf{n}) \cdot \mathbf{p}$. If we take $\mathbf{d} \parallel \mathbf{y}$, then $\mathbf{A}$, $\underset{\sim}{\psi}$, and $\mathrm{Tr}[\underset{\sim}{\psi}\,\underset{\sim}{\psi}^\dagger]$ take the form

$$\mathbf{A} = i\Delta \begin{pmatrix} 0 & 0 & 0 \\ 1 & i & 0 \\ 0 & 0 & 0 \end{pmatrix}, \tag{38.88}$$

$$\underset{\sim}{\psi} = i\,\Delta \begin{pmatrix} p_x + ip_y & 0 \\ 0 & p_x + ip_y \end{pmatrix}, \tag{38.89}$$

and

$$\begin{aligned} \mathrm{Tr}[\underset{\sim}{\psi}\,\underset{\sim}{\psi}^\dagger] &= 2\left(p_x^2 + p_y^2\right) \\ &= 2\Delta^2 \sin^2\theta; \end{aligned} \tag{38.90}$$

thus the gap vanishes on the pole in the axial state and has a maximum on the equator. The gap structure for the BW and ABM phases are shown schematically in Fig. 38.5.

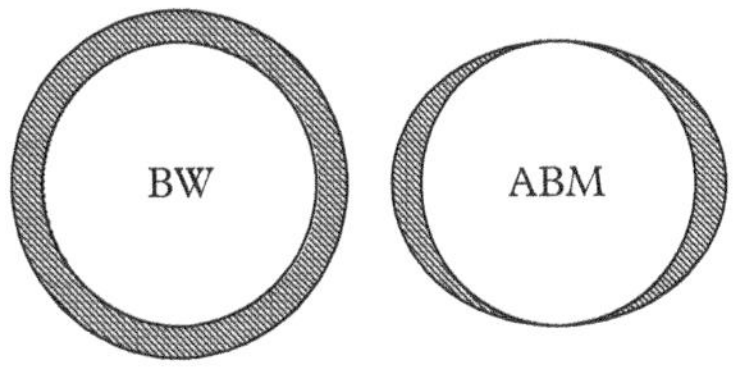

Figure 38.5 *The energy gap of the 3d (or Balian-Werthamer) and axial (or Anderson-Brinkman-Morel) phases of superfluid ³He.*

Among the more striking properties of superfluid ^{3}He are various collective modes that are associated with dynamic distortions of the energy gaps.[11] These include the longitudinal zero sound and spin waves modes of the normal fluid that were described in Chapter 37; however, in the superfluid there is a total of 18 such modes associated with oscillations of various linear combinations of the nine complex parameters that make up the **A** matrix of Eq. (38.91). For a detailed microscopic discussion of these modes the reader is referred to the book by Vollhardt and Wolfle. For the case of the B phase, they have also been described by a phenomenological time-dependent Ginzburg–Landau theory.[12]

ADDITIONAL READING

The literature on superfluid helium three is vast and we have limited ourselves in this chapter to the nature of the paired state. More complete coverage is contained in the following:

The Superfluid Phases of Helium 3, D. Vollhardt and P. Wolfle, Taylor and Francis, London (1990). Experimental methods as well as many results are covered in R. Dobbs, *Solid Helium Three*, Oxford University Press, Oxford (1994).

The Physics of Liquid and Solid Helium, Part II. K.H. Bennemann and J.B. Ketterson, Eds., Wiley Interscience, NY (1978). Chapter 3 by Philip W. Anderson

[11] Sound propagation in heavy fermion superconductors, B.K. Sarma, M. Levy, S. Adenwalla, and J.B. Ketterson, *Physical Acoustics*, Vol. 20, p. 1601, R.N. Thurston, Ed., Academic Press, NY (1992).

[12] Sound propagation in superfluid ^{3}He, Z. Zhao, S. Adenwalla, B.K. Sarma, and J.B. Ketterson, *Adv. Phys.* **41**, 147 (1992).

and William F. Brinkman reviews the theory of superfluidity in ^{3}He together with the history leading up to the discovery, while Chapter 4 by David M. Lee and Robert C. Richardson covers the major experimental results together with the methods used to obtain them.

For an early review see: A theoretical description of the new phases of liquid ^{3}He, Anthony J. Leggett, *Rev. Mod. Phys.* 47, 331 (1975).

Problem

38.1. Verify Eq. (38.16).

Part XII

Disordered Materials

Alloys

39

Many advantageous properties of solids can be enhanced by forming mixtures of two or more elements. Such a mixture is termed an alloy (also called a solution, although this latter term is sometimes reserved for fluids). Metallurgy is basically about mixtures involving at least one metal and, through composition and processing, strategies to achieve better properties for various applications. Such properties include enhanced hardness or ductility, which involve suppressing or enhancing permanent deformations under the influence of external stresses (and must be distinguished from a reversible elastic deformation). The origins of metallurgy stretch back into antiquity, even defining various developmental stages (the bronze and iron ages) through which civilizations evolved. One can also include the alteration of the electrical properties, which in semiconductors led to the creation of important devices by inhomogeneous doping with impurities as discussed in Chapters 22 and 23.

39.1 Phase equilibrium

39.1.1 Phases in contact: the Gibbs phase rule

There is generally a limit on the solubility of one species (a molecule or atom) in another, and above this limit a second thermodynamic phase must be present (with an associated surface of separation having specific properties of its own). Even for a pure substance (termed a one-component system), we are familiar with the idea that more than one phase can be present under appropriate conditions (e.g., a liquid in equilibrium with its own vapor), as we now review.

We start by writing the basic differential thermodynamic identity involving the total energy, E, for a single-species system in a single phase,[1]

$$dE = TdS - PdV + \mu dN. \tag{39.1a}$$

or

$$dS = \frac{dE}{T} + \frac{P}{T}dV - \frac{\mu}{T}dN. \tag{39.1b}$$

Suppose we now have two phases in equilibrium and in contact; therefore they will have the same temperature. To further specify the problem we put the material (consisting of the two phases) in a box with a fixed *total* volume and

[1] There is a subtle interdependence of the second and third terms in Eq. (39.1a); see Landau and Lifshitz, *Stat. Phys.*, Section 24.

The Physics of Solids. First Edition. J. B. Ketterson.

furthermore isolate this box from the environment so that no energy is exchanged; in such a case, $dE = 0$ and $dV = 0$ in Eq. (39.2b). In equilibrium the total entropy, $S = S^{(1)} + S^{(2)}$, will be a maximum, where we use a superscript to denote the two different phases. We write the last term in (39.1b) as $(\mu^{(1)}/T)dN^{(1)} + (\mu^{(2)}/T)dN^{(2)}$; if we assume our box does not exchange particles with the environment (say through permeable walls), the total number of particles, $N = N^{(1)} + N^{(2)}$, will be constant, leading to $dN^{(1)} = -dN^{(2)}$. Maximizing the total entropy with respect to the only remaining variable, which we can choose as $N^{(1)}$, we have

$$\begin{aligned} \frac{dS}{dN^{(1)}} &= -\left(\frac{\mu^{(1)}}{T} + \frac{\mu^{(2)}}{T}\frac{dN^{(2)}}{dN^{(1)}}\right) \\ &= -\frac{\mu^{(1)}}{T} + \frac{\mu^{(2)}}{T} = 0 \end{aligned} \tag{39.2}$$

We therefore have

$$\mu^{(1)} = \mu^{(2)}. \tag{39.3}$$

We now ask how many phases can exist simultaneously in our one-species (one-component) system. Geometrically, the condition (39.3) describes the intersection of the two surfaces, $\mu^{(1)}(P,T)$ and $\mu^{(2)}(P,T)$, in μ, P, and T space; this intersection can be projected onto the P–T plane as a line, which we can write as $P = P(T)$, and is referred to as a *line of phase transitions or a coexistence line*. If *three* phases are in contact, then in place of equation (39.2) we would have a pair of equations,

$$\mu^{(1)}(P,T) = \mu^{(2)}(P,T) = \mu^{(3)}(P,T). \tag{39.4}$$

This can be viewed as the intersection of three surfaces in μ, P, and T space, which can only occur at isolated points; these points are referred to as *triple points*. Note the number of allowed phases is by no means limited to two or three; a material may have many different phases as a function of P and T (liquid, vapor, and solids with various crystal structures), each with its own surface, $\mu^{(i)} = \mu^{(i)}(P,T)$. The earlier discussion simply tells us how many phases may be *simultaneously* present.

If additional species (also called components) are present, the last term in Eq. (39.1a) takes the form $\sum_{j=1}^{n} \mu_j dN_j$, where n is the *number of different species.* More phases can then exist simultaneously and Eq. (35.3) then takes the more general form

$$\begin{aligned} \mu_1^{(1)} &= \mu_1^{(2)} = \mu_1^{(3)} = \cdots \mu_1^{(r)} \\ \mu_2^{(1)} &= \mu_2^{(2)} = \mu_2^{(3)} = \cdots \mu_2^{(r)} \\ \mu_3^{(1)} &= \mu_3^{(2)} = \mu_3^{(3)} = \cdots \mu_3^{(r)} \\ &\text{etc.} \end{aligned} \tag{39.5}$$

where r is the number of different phases *simultaneously* present. Equations (39.5) constitute a set of *independent equations*, or constraints, the number of which is

$$\#\,\text{constraints} = n(r-1). \tag{39.6}$$

But the $\mu_j^{(i)}$ themselves can only be functions of *intensive* variables; we may take these to be the *concentrations*, defined as

$$c_j^{(i)} = \frac{N_j^{(i)}}{N^{(i)}}, \tag{39.7a}$$

where $N_j^{(i)}$ is the number of atoms of species j in phase i and

$$N^{(i)} \equiv \sum_{j=1}^{n} N_j^{(i)} \tag{39.7b}$$

is the total number of atoms of all species in phase i. Clearly,

$$\sum_{j=1}^{n} c_j^{(i)} = 1, \tag{39.8}$$

and with this relation the total number of independent concentrations (or compositions) is then

$$\#\,\text{compositions} = r(n-1). \tag{39.9}$$

We also define

$$c_j \equiv \sum_{i=1}^{r} c_j^{(i)}; \tag{39.10a}$$

$$N_j \equiv \sum_{i=1}^{r} N_j^{(i)}; \tag{39.10b}$$

$$N \equiv \sum_{j=1}^{n} \sum_{i=1}^{r} N_j^{(i)}; \tag{39.10c}$$

Here, c_j and N_j are respectively the total concentration and number of atoms of species j in all phases and N is the total number of atoms (of all species) in the system. Including the temperature and pressure with Eq. (39.9) the total number of *unknowns* is then

$$\#\,\text{unknowns} = 2 + r(n-1). \tag{39.11}$$

Combining (39.11) and (39.6), the number of *free* variables, f (those which may be varied arbitrarily *without* destroying thermodynamic equilibrium) is $2 + r(n-1) - n(r-1)$, or

$$f = 2 - r + n. \tag{39.12}$$

Since the number of free variables must be greater than or equal to zero, we arrive at the inequality

$$r \leq n + 2. \tag{39.13}$$

This is the famous Gibbs phase rule.

When we have a single-phase system ($r = 1$) the maximum value of f is $n + 1$. These $n + 1$ variables may take on arbitrary values within some region (corresponding to the range over which the phase exists); typically we would write $\mu = \mu\,(P, T, c_1 \cdots c_{n-1})$. The presence of additional phases reduces the number of free variables and when r takes on the value $n + 2$ we will have $f = 0$; this corresponds to the maximum number of phases that may be present simultaneously, and represents a *point* in this $(n + 1)$-dimensional space.

The *complete* thermodynamic state of a system can be specified by the Gibbs potential,

$$\Phi = \sum_{i=1}^{r}\sum_{j=1}^{n} \mu_j^{(i)} N_j^{(i)}, \tag{39.14}$$

which, with temperature and pressure, involves $2 + rn$ parameters; however the constraints implied by (39.6) leave us with $[2 + rn] - [n(r-1)] = 2 + n = [2 - r + n] + r$ independent variables, which exceeds the requirements of the Gibbs phase rule by r; i.e., *we have* r *undefined variables over and above those needed to specify thermodynamic equilibrium.* However a choice of $r(n - 1)$ independent *concentrations*, $c_j^{(i)}$, leaves the r quantities, $N^{(i)}$ (see (39.7b)), undefined; these are just the total numbers of atoms in (the amount of) the phase i and they correspond to the r undefined (extensive) variables. For a one-component single-phase system this is simply the total number of atoms.

39.1.2 Binary mixtures

For the case of a single-phase binary mixture ($r = 1; n = 2$), Eq. (39.13) involves four variables: T, P, N_1, and N_2; the concentrations defined by (39.7) are given by $c_1 = N_1/N$, and $c_2 = N_2/N$, where $N = N_1 + N_2$, but with (39.8) only one of these is independent, which we take as c_1. We may then write the chemical potentials of the two species, which depend on three intensive variables, as

$$\mu_1 = \mu_1\,(P, T, c_1);\ \mu_2 = \mu_2\,(P, T, c_1). \tag{39.15}$$

In a two-phase binary alloy the equilibrium conditions (39.5) take the form

$$\mu_1^{(1)}\left(P, T, c_1^{(1)}\right) = \mu_1^{(2)}\left(P, T, c_1^{(2)}\right) \tag{39.16a}$$

and

$$\mu_2^{(1)}\left(P, T, c_1^{(1)}\right) = \mu_2^{(2)}\left(P, T, c_1^{(2)}\right), \tag{39.16b}$$

where we have again used condition (39.8). As these equations stand we have two independent concentrations. Using (39.16b) we can eliminate one of these concentrations, say $c_1^{(2)}$; this leaves us with $c_1^{(1)} \equiv c$ as the only free concentration. Equation (39.16a) now has the form

$$\mu_1^{(1)}(P, T, c) = \mu_1^{(2)}(P, T, c); \tag{39.17}$$

this dependence on three variables is consistent with our general result (39.12). We emphasize again that in writing Eq. (39.17) in the form we have, it does not mean that c is the same in both phases, but only that once it is specified in one phase, here taken as $\mu_1^{(1)}(P, T, c)$ the value in the second phase, $\mu_1^{(2)}(P, T, c)$, is fixed. This is shown graphically in Fig 39.1. The shaded region[2] corresponds to forbidden concentrations and we have a mixture of two phases with compositions given by those at the phase boundary lines in this region.

39.1.3 The chemical potential of a dilute alloy

In order to discuss the properties of alloys in the dilute limit we will need to develop an expression for the change in the Gibbs free energy brought on by dissolving N' solute molecules in N host solvent molecules, where $N' \ll N$; the interaction between impurities can then be neglected. Our treatment will involve intuitive arguments.[3]

We write the total free energy in the form

$$\Phi(P, T, N, N') = \Phi_0(P, T, N) + \Phi_1(P, T, N, N'), \tag{39.18}$$

where Φ_0 is that for the pure solvent and Φ_1 is the correction due to the solute; i.e., the interactions between the impurities can be ignored. One contribution to Φ_1 arises from the effect of adding a single impurity to the solute, which we write as $\alpha(P, T, N)$; adding N' such impurities would then contribute a term $N'\alpha(P, T, N)$ to Φ_1 (note the N-dependence of α will turn out to be non-trivial). There is also a contribution to Φ_1 arising from the entropy (which enters the free energy through

[2] The shading is included here for increased clarity; in published compilations of phase diagrams this shading is generally omitted.

[3] It is based on that given by Landau and Lifshitz in *Stat. Phys. Part I*, Section 88.

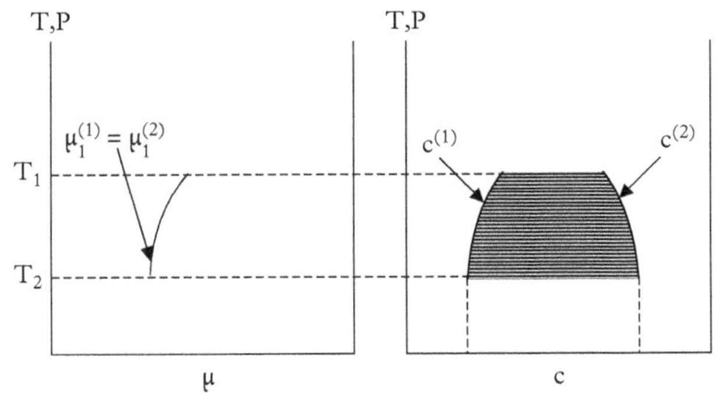

Figure 39.1 *Discontinuous behavior associated with the concentration of different phases in chemical equilibrium.*

the $-TS$ term in its definition), sometimes called the "mixing entropy," given by $S' = k_B\, ln\bar{W} = k_B\, ln\left(1/N'!\right)$, which is associated with the different configurations of the N' identical impurities[4] (here $\bar{W}$ is the probability function entering the Boltzmann definition of the entropy). However, N' is still a macroscopic quantity so we may apply Sterling's approximation, where $ln\left(N'!\right) = N' ln(N') - N' = N' ln(N'/e)$. Hence,

$$-TS' = k_B T N' ln(N'/e). \tag{39.19}$$

Combining the two contributions, we have

$$\begin{aligned}\Phi\left(P, T, N, N'\right) &= \Phi_0\left(P, T, N\right) + N'\alpha(P, T, N) + k_B T N'\, ln(N'/e) \\ &= N\mu_0\left(P, T\right) + N'\, k_B T\, ln\left[\frac{N'}{e} e^{\alpha\,(P,T,N)/k_B T)}\right]\end{aligned} \tag{39.20}$$

where we write $\Phi_0 = N\mu_0\left(P, T\right)$ in the second step, with $\mu_0\left(P, T\right)$ the chemical potential of the pure solvent. Now, on general grounds (39.23) must have the form

$$\Phi\left(P, T, N, N'\right) = N\mu\left(P, T, c\right) + N'\mu'\left(P, T, c\right), \tag{39.21}$$

where we have introduced the concentration by $c \equiv N'/N$. To make (39.20) and (39.21) consistent we must argue that the N-dependence of α(P,T,N) is such that $e^{\alpha(P,T,N)/k_B T)} = f\left(P, T\right)/N$ (implying that $\alpha\left(P, T, N\right) = k_B T\, ln\left(f\left(P, T\right)/N\right)$, where $f\left(P, T\right)$ is some function of P and T only); assuming this form we have

$$\begin{aligned}\Phi\left(P, T, N, N'\right) &= N\mu_0(P, T) + N' k_B T\, ln\left[\frac{N'}{eN} f(P, T)\right] \\ &= N\mu_0(P, T) + N' k_B T ln\left[\frac{N'}{eN}\right] + N'\psi\left(P, T\right),\end{aligned} \tag{39.22}$$

where we have defined yet another function, $\psi\left(P, T\right) \equiv k_B T\, ln\left[f(P, T)\right]$. The chemical potentials follow from the definition $\mu_i = \left(\partial\Phi/\partial N_i\right)_{P,T}$ and operating on Eq. (39.19) we obtain

$$\begin{aligned}\mu\left(P, T, c\right) &= \frac{\partial\Phi}{\partial N} = \mu_0\left(P, T\right) - k_B T\frac{N'}{N} \\ &= \mu_0\left(P, T\right) - k_B T c\end{aligned} \tag{39.23a}$$

and

$$\begin{aligned}\mu'\left(P, T, c\right) &= \frac{\partial\Phi}{\partial N'} = k_B T\, ln\left(\frac{N'}{eN}\right) + k_B T + \psi\left(P, T\right) \\ &= k_B T\, ln\, c + \psi\left(P, T\right).\end{aligned} \tag{39.23b}$$

If our solution phase separates at a small value of c we may utilize Eqs. (39.23a,b) to analyze this situation. If the solute has limited solubility in the

[4] As an example, this follows from Eq. (6.16) when $n \gg g$, where the statistics become irrelevant.

host phase, the phase separated solute phase will generally have limited solubility for the original solvent phase.

39.1.4 Characteristic binary phase diagrams

Equation (39.17) can be viewed, geometrically, as the intersection of the two surfaces, $\mu_1^{(1)}(P,T,c)$ and $\mu_1^{(2)}(P,T,c)$, in the four-dimensional space of μ, P, T, and c. This intersection, which is a *surface* in three-dimensional space, may be represented by any of the following functions:

$$T = T(P,c);\ P = P(T,c);\ c = c(P,T), \tag{39.24a,b,c}$$

depending on our choice of independent variables, (P, c), (T, c), or (P, T); thus we have two free parameters, again consistent with (39.12). We will assume our system is held at a constant pressure and henceforth suppress this variable (which in practice is harder to significantly alter). The phase boundaries are then *lines* in temperature–composition space.

As shown in Fig. 39.1, phase diagrams in T–c space will in general contain regions where only a mixture of two phases having different compositions occurs. These regions commence at, or undergo a singular behavior at, discrete temperatures where three phases meet. There are four different behaviors that can occur at such temperatures: (i) *pure compound points*; (ii) *triple points*; (iii) *critical points*; and (iv) *equal concentration points.*

We begin with the pure compound point. Suppose we have two phases, (1) and (2) of a pure compound (the solvent) that are in contact, requiring $\mu_1^{(1)}(T_P) = \mu_1^{(2)}(T_P)$, where T_P denotes the transition temperature of the pure compound. If now a small amount of a second material is introduced, then by Eq. (39.23a), the chemical potentials of the solvent host in each phase shifts, and with it the temperature of the equilibrium point; expanding around T_P we have

$$\frac{\partial \mu_1^{(1)}(T_P)}{\partial T}\Delta T - k_B T_P c^{(1)} = \frac{\partial \mu_1^{(2)}(T_P)}{\partial T}\Delta T - k_B T_P c^{(2)}. \tag{39.25}$$

Now, $\left(\partial \mu^{(1,2)}/\partial T\right)_{P,c} = -s^{(1,2)}(T)$, where s(T) is the entropy per solvent molecule. We can then rewrite (39.25) as

$$-\left(s^{(1)} - s^{(2)}\right)\Delta T = k_B T_P\left(c^{(1)} - c^{(2)}\right); \tag{39.26}$$

introducing the latent heat via the definition $q = T_P\left(s^{(2)} - s^{(1)}\right)$, we obtain the concentration dependence of the splitting of the equilibrium line,

$$\Delta T = \frac{k_B T_P^2}{q}\Delta c, \tag{39.27}$$

where $\Delta c = \left(c^{(1)} - c^{(2)}\right)$. This situation is shown schematically in Fig. 39.2; note that the excluded concentration regime may open either upward or downward, depending on the sign of q in (39.27).

Figure 39.2 *Temperature versus concentration behavior of the phase boundaries near a pure compound point.*

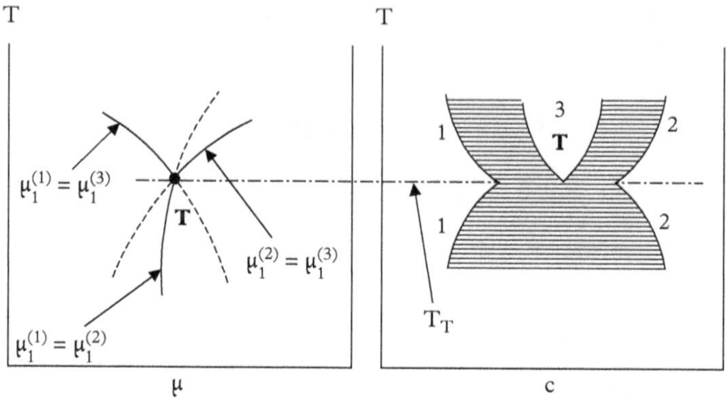

Figure 39.3 *Behavior in the* T-μ *and* T-c *planes in the vicinity of a triple point.*

We next consider the triple point, shown schematically in Fig. 39.3, which arises from the intersection, at a temperature, $T = T_T$, of the three curves, $\mu_1^{(1)}(T) = \mu_1^{(2)}(T)$, $\mu_1^{(1)}(T) = \mu_1^{(3)}(T)$, and $\mu_1^{(2)}(T) = \mu_1^{(3)}(T)$ (only two of which are independent), in the μ–T plane. As shown, for temperatures below T_T, the $\mu_1^{(1)}(T) = \mu_1^{(2)}(T)$ phase boundary is assumed to be the equilibrium curve (the remaining two being shown as dashed lines, which may support metastable phases for some range of parameters), and in the c–T plane behaves as shown earlier, in Fig. 39.1. For temperatures above the triple point, $T = T_T$, we have two equilibrium curves governing the behavior of the concentrations, depending on whether c is greater or less than a critical value, $c = c_T$, at which the third phase appears and either the first or the second disappears. Note that for the situation being considered, the equilibrium concentration of the third phase must evolve from the value, $c = c_T$; the concentrations of the first or third phase remain continuous but their derivative with respect to temperature must change discontinuously. Note that some systems show a behavior which is obtained from Fig. 39.3, by reflecting through the line $T = T_T$.

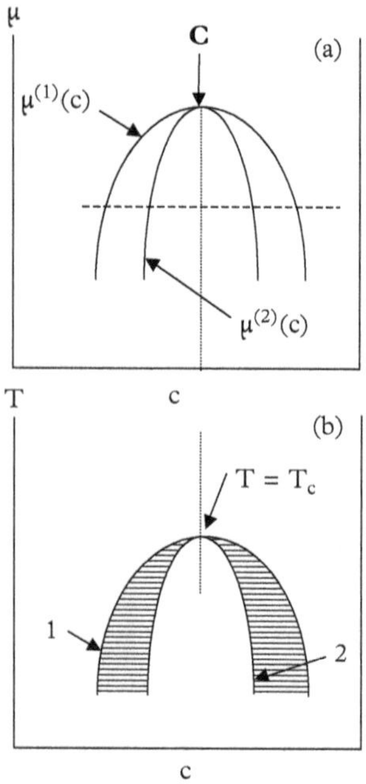

Figure 39.4 *(a): The two curves,* $\mu^{(1)}(c)$ *and* $\mu^{(2)}(c)$ *resulting from eliminating* T *using the relation,* $\mu^{(1)}(T,c) = \mu^{(2)}(T,c)$*; here,* T *is a parameter and the dashed line shows the intersection for some arbitrarily chosen value below* T_c*. (b) The phase separation curve associated with* a point of equal concentration, *which can occur for two phases having different symmetry.*

The equal concentration point arises from the behavior shown in Fig. 39.4 and may be viewed as a special case of a two-phase equilibrium. As true generally, for a two-phase equilibrium, we may eliminate the variable, T, using the condition, $\mu_1^{(1)}(T,c) = \mu_1^{(2)}(T,c)$, which results in two curves, $\mu_1^{(1)}(c)$ and $\mu_1^{(2)}(c)$. If either of these curves has a maximum at some temperature, $T = T_c$, the equilibrium condition requires that the other also has a maximum, since $\mu_1^{(1)}(c+dc) = \mu_1^{(2)}(c+dc)$, leading to

$$\frac{d\mu_1^{(1)}(c)}{dc} = \frac{d\mu_1^{(2)}(c)}{dc};$$

therefore the two curves must touch as shown in Fig. 39.4(a). Since $\partial\mu/\partial c = (\partial\mu/\partial T)\cdot(\partial T/\partial c) = -s\partial T/\partial c$, where s is the entropy per particle (which is always greater than zero), it follows that if $d\mu/dc$ is zero, dT/dc must also vanish. In the

T–c plane, shown in Fig. 39.4(b), the equilibrium phases exist outside the outer curve and inside the inner curve.

Figure 39.5 shows an alternative way in which a forbidden region can emerge within an alloy phase diagram, called a *critical point* (C), which defines a temperature below which phase separation occurs. From the behavior shown in Fig. 39.5(a), we may think of the phase boundary as arising from a *single function*, $\mu(T, c)$ (rather than from the equality of *two independent functions*, $\mu^{(1)}(T, c)$ and $\mu^{(2)}(T, c)$). For this type of behavior to occur the *symmetry* of the two phases must be *identical*,[5] since there is a continuous path around the critical point connecting the separated phases. However for temperatures below the critical point, the function $\mu(T, c)$ has two solutions resulting in two phases which are *distinguishable*.[6] (The familiar liquid–vapor phase diagram has this behavior in the T–V plane.) Because of this symmetry requirement, critical points in solids are uncommon; however they commonly occur in fluid systems.

Various limiting cases of the four behaviors discussed can occur in which the composition of one of the phases is for all practical purposes fixed (typically because of strong chemical binding considerations). These are shown in Fig. 39.6 for the: (a) pure compound point; (b) triple point; and (c) equal concentration point. In all of these cases the range of allowable concentrations collapses to a line. As noted all of these cases may occur as reflections through T_c.

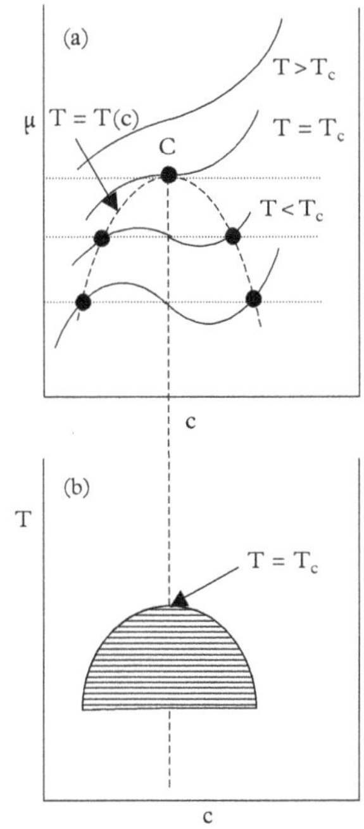

Figure 39.5 *(a) the behavior of* $\mu(T, c)$ *in the* μ–c *plane for several temperatures (shown as dotted lines) in the vicinity of* T_c. *The dashed lines show the concentrations in the two phases below* T_c. *(b) The region of excluded concentrations in the* T–c.

39.1.5 Examples of phase diagrams

The phase diagrams of real materials can all be decomposed into parts involving the four singular features discussed earlier. We will now discuss some specific examples.[7] We start with the Pb–Sn system which is ordinary solder and is shown in Fig. 39.7. This phase diagram is easily seen to consist of a triple-point central region, where the 1, 2, and 3 phases of Fig. 39.3 are, respectively, tetragonal (A6) Sn, fcc (A1) Pb, and a melt which is a solution of both. The upper two forbidden regions evolve into pure compound behaviors emanating from the melting points of Sn on the left and Pb on the right. The minimum in the melting temperature (the 3 phase) is referred to as the *eutectic*; it is this relatively low melting temperature (and the ability of the Pb and Sn to wet many other metals) that makes solder useful for joining metals.

[5] The phase boundary of a fluid as a function of temperature and volume has this form since the symmetries of the liquid and vapor are the same.

[6] As the liquid and vapor are for the liquid–vapor transition.

[7] For a collection of phase diagrams, see *Constitution of Binary Alloys*, M. Hansen, McGraw-Hill, NY (1958). Those shown are actually computer model diagrams. For a description of the methods used, see *CALPHAD (Calculation of Phase Diagrams): A Comprehensive Guide*, N. Saunders and A.P. Miodownik, Pergamon (1998).

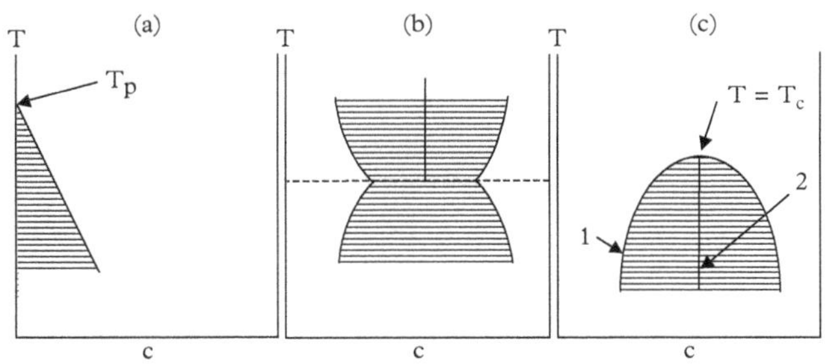

Figure 39.6 *Limiting phase separation behaviors for the cases of the (a) pure compound point; (b) triple point; and (c) equal concentration point.*

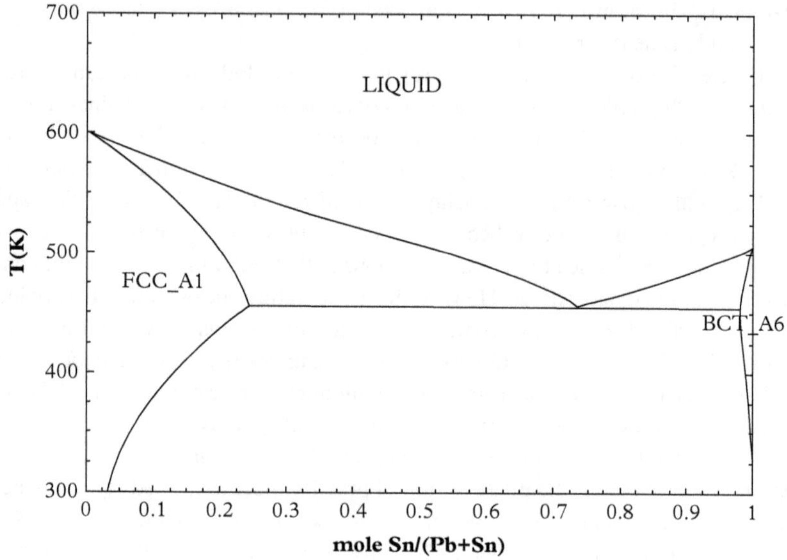

Figure 39.7 *The* Pb–Sn *phase diagram.*

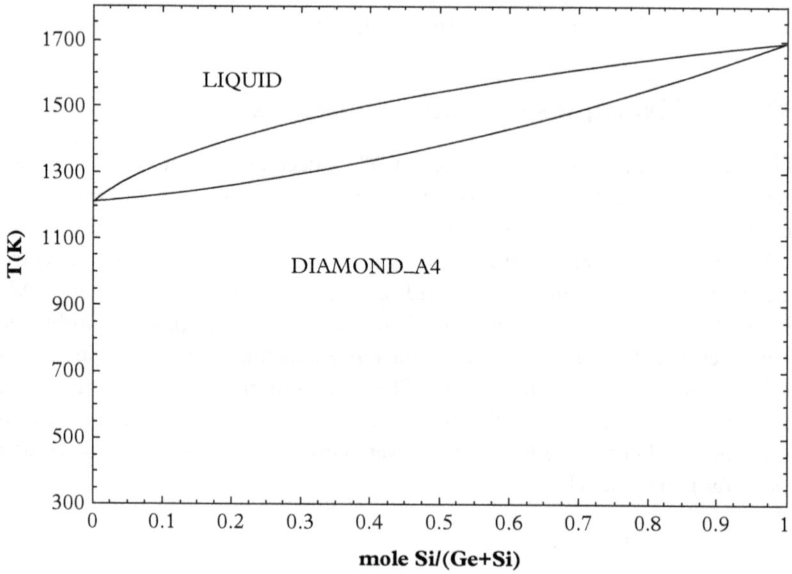

Figure 39.8 *The phase diagram of* Si-Ge.

Figure 39.8 shows the phase diagram of Si–Ge (A4 or diamond structure). It is formed from two pure compound regions, which merge to form a forbidden region between the melt and a continuous solid solution of the two constituents; such mixtures are said to be miscible. Si–Ge alloys are of interest since they can combine the higher mobilities associated with Ge while not reducing the band gap excessively, which would degrade thermal stability.

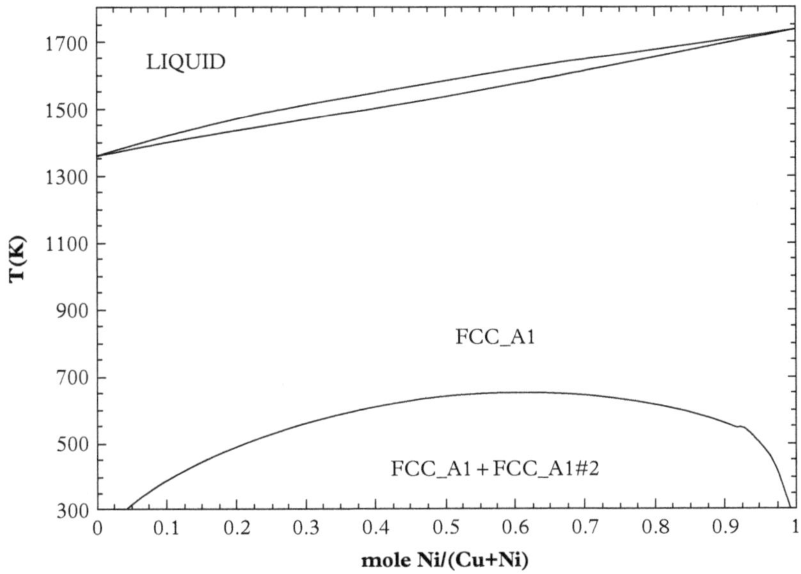

Figure 39.9 *The phase diagram of* Cu-Ni.

Figure 39.9 shows the phase diagram of Cu–Ni. At higher temperatures it shows a miscibility behavior similar to that of Si–Ge. However, at lower temperatures it displays a critical point and phase separates into two phases with the same face centered cubic (A1) symmetry, in the manner shown in Fig. 39.5.

Finally, Fig. 39.10 shows the phase diagram of the In–Sb system. The composition $In_{0.5}Sb_{0.5}$ corresponds to the III–V semiconducting compound InSb; here indium has the A6 structure while antimony is A7. Since the stoichiometry of

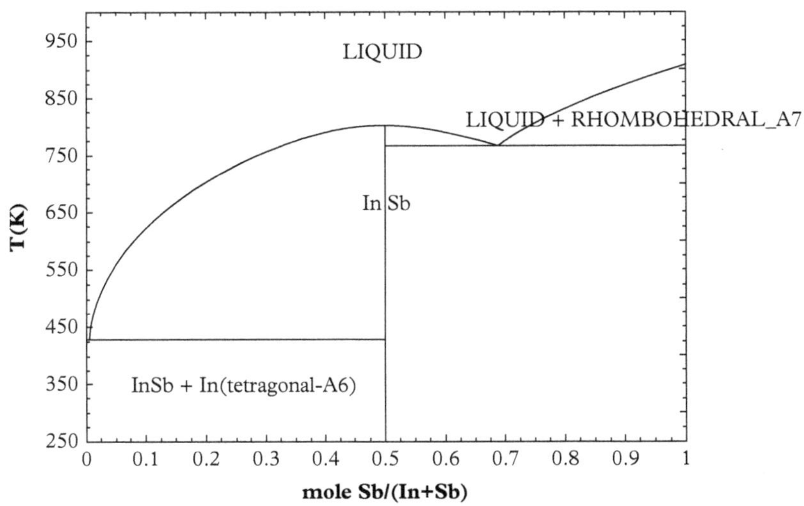

Figure 39.10 *The phase diagram of* In-Sb.

this compound is quite precise, this system displays the limiting form of the equal composition feature, shown in Fig. 39.6(b). On each side of the equal composition feature we have triple points (at 500°C and 155°C) followed by the limiting form of pure compound behavior shown in Fig. 39.6(a) at the two extremes.

39.2 Order–disorder transitions

The atoms making up the lattice of a particular alloy can be ordered or disordered depending on circumstances. Classic examples are the intermetallic compounds Cu_3Au and CuZn; we will focus on the latter and restrict ourselves to the case where we have exactly equal numbers of copper and zinc atoms. At higher temperatures the Cu and Zn atoms randomly occupy sites on a body centered cubic (bcc) Bravais lattice. But when the temperature is lowered, the material undergoes an ordering in which two sublattices form onto which the Cu and Zn atoms segregate (as proposed by Gustav Tammann in 1919); the Bravais lattice is then cubic and has a basis with two atoms per unit cell, as shown in Fig. 39.11. When this happens new lines appear in X-ray diffraction (C.H. Johansson and J.O. Linde, 1925) corresponding to the forbidden lines of the bcc lattice (see Problem 9.3). In what follows we describe a simple mean field theory due to W.L. Bragg and E.J. Williams.[8] An alternative theory is that of Bethe,[9] and more recently the problem has been treated with renormalization group methods applied to the Ising model of magnetism, to which the order–disorder transition is related.

Assume that we have two sublattices, α and β. If all the A atoms are on α sites and the B atoms are on β sites, we have perfect order. We will call an A atom on an α site, or a B atom on a β site, a R (right) atom; correspondingly an A atom on β site, or a B atom on α a site, will be called a W (wrong) atom. At any given temperature some fraction, r, of the atoms will be R atoms, and some fraction, w, will be W atoms. As a measure of the *long range order* in the system we define a quantity s according to

$$s = \frac{r - w}{r + w}. \tag{39.28}$$

[8] The effect of thermal agitation on atomic arrangement in alloys I & II, W.L. Bragg and E.J. Williams, *Proc. R. Soc. A* **145**, 699 (1934); *Proc. R. Soc. A* **151**, 540 (1935).

[9] Statistical theory of superlattices, H.A. Bethe, *Proc. R. Soc.* **150**, 552 (1935).

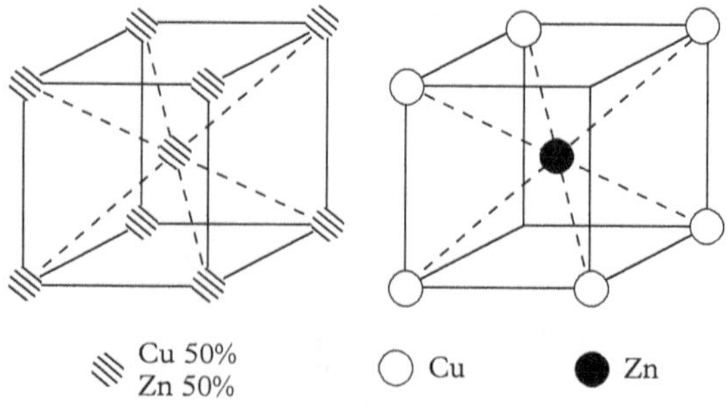

Figure 39.11 *The site occupancy of the binary alloy,* CuZn, *in the fully disordered and fully ordered limits. (Courtesy of C. Cayron.)*

Clearly, r + w = 1. If all atoms are r atoms then s = 1 and we have perfect order; obviously, if all atoms are w atoms, yielding s = −1, we also have perfect order, but with the designation of sublattices reversed.

Suppose we have two nearest-neighbor atoms and that each is on a right site; i.e., we have an A atom on an α site and a B atom on a β site. If we interchange these two atoms we would expect the (free) energy of the system to increase, more so at lower temperatures where the thermal energy activating such interchanges is smaller; to model this effect we introduce a temperature-dependent potential energy, V = V(T). From Boltzmann statistics we might expect the ratio of the w and r populations to be given by

$$\frac{w}{r} = \exp\left(-\frac{V}{2k_B T}\right), \tag{39.29}$$

where the factor 2 corrects for the fact that V corresponds to a pair of atoms. Combining (39.28) and (39.29) yields

$$s(T) = \tanh\left(\frac{V(T)}{4k_B T}\right). \tag{39.30}$$

Note the similarity of this expression with the magnetization of a spin 1/2 system.

The Bragg–Williams model makes the assumption

$$V(T) = V_0 s(T). \tag{39.31}$$

Solving (39.30) and (39.31) yields s(T).

The Bragg–Williams theory for the long range order, s(T), is essentially identical to the Weiss theory of ferromagnetism, discussed in Section 26.4. As in that theory, there is a critical temperature above which s(T) vanishes. The critical temperature is given by

$$T_c = \frac{4V_0}{k_B}. \tag{39.32}$$

39.3 The isolated charged impurity: the Friedel sum rule

Consider an isolated impurity having a charge z′ (in units of |e|) embedded in a free electron-like metal formed from atoms having a charge z; here z and z′ denote the number of valence electrons outside a filled shell.[10] If z differs from z′ we expect the charge difference to be screened by the conduction electrons; in the Thomas–Fermi model this screening would occur within a characteristic distance governed by $\mathbf{r}_{TF}$, as discussed in Section 14.2.1. However this model does not predict some subtleties of the behavior, as we will see. Physically, the screening

[10] J. Friedel, *Phil. Mag. Suppl.* **3**, 446 (1954); *Nuovo Cimento Supplemento* 7, 287 (1954).

involves a "pile-up" of conduction electron charge (for the case $z' > z$) in the vicinity of the impurity atom.

For a spherically symmetric impurity potential we may write the solution to the Schrödinger equation as $\psi_{klm}(r, \theta, \varphi) = Y_{lm}(\theta, \varphi) R_{kl}(r)$ where the functions $Y_{lm}(\theta, \varphi)$ are spherical harmonics normalized such that $\int d\Omega \, |Y_{lm}(\theta, \varphi)|^2 = 1$. If we rewrite the radial part of the wave function as $R_{kl}(r) = \chi_{kl}(r)/r$, the differential equation for $\chi_{kl}(r)$ has the form

$$\frac{d^2\chi_{kl}(r)}{dr^2} + \left[k^2 - \frac{l(l+1)}{r^2} - \frac{2m}{\hbar^2}V(r)\right]\chi_{kl}(r) = 0, \tag{39.33}$$

where $k^2 = 2mE/\hbar^2$, and to ensure that the wave function is everywhere finite we must require that $\chi_{kl}(0) = 0$. *Outside* the range of the electrostatic potential, $V(\mathbf{r})$, and the centrifugal barrier potential, $l(l+1)/r^2$, the *asymptotic* form of $\chi_{kl}(r)$ for a positive energy *stationary* state with wave vector k has the form[11]

$$\chi_{kl}(r \to \infty) \cong A_{kl} \sin\left(kr - \frac{1}{2}l\pi + \delta_l(k)\right). \tag{39.34}$$

Here, A_{kl} is a normalization factor and the parameters $\delta_l(k)$, which arise from the presence of the scattering potential and are referred to as phase shifts, are wave vector (or energy) dependent. Note that these asymptotic forms do not vanish at the origin. In the absence of the potential when all the phase shifts vanish, (39.34) is clearly

$$\chi_{kl}^{(0)}(r \to \infty) \cong A_{kl}^{(0)} \sin\left(kr - \frac{1}{2}l\pi\right). \tag{39.35}$$

We now fix the normalization. Strictly speaking this would involve an integration of $|\chi_{kl}(r)|^2$ over all r, which is divergent; for this reason we normalize the total wave function over a sphere of radius r_0,

$$\int_0^{r_0} dr \left|\chi_{kl}^{(0)}(r)\right|^2 = 1; \tag{39.36}$$

choosing the normalization constant to be real gives $A_{kl}^{(0)} = \sqrt{2/r_0}$ (apart from an oscillatory term; see Eq. (39.42)). Note that with this normalization, states with differing k are not orthogonal.

Quantitatively, if the quantity

$$\Delta P_{kl}(r) = |\chi_{kl}(r)|^2 - \left|\chi_{kl}^{(0)}(r)\right|^2 \tag{39.37}$$

(where $\chi_{kl}(r)$ and $\chi_{kl}^{(0)}(r)$ are the forms valid for all r) is positive (negative), it represents an increase (decrease) in the probability density, $\Delta P_{kl}(r)$, of finding a

[11] See Landau and Lifshitz, *Quantum Mechanics*, Sections 32 and 33.

scattered particle with particular k, l values in the vicinity of the scattering center. For an attractive scattering center, if the strength of the potential increases, such that an additional *node* in the wave function of a *state* with wave vector, k, and angular momentum, l, occurs, the phase shift, δ_l, associated with the lth partial passes through an odd multiple of $\pi/2$, a phenomenon referred to as a *resonance*.[12] The additional nodes so created are referred to as *resonant states*. So if r_0 is chosen to be much larger than the range of the excess particle density near the scattering center,[13] that fraction of the extra probability of finding a particle in state k, l associated with the scattering center can then be calculated from (39.47). This probability is given by

$$\Delta N_{kl} = \int_0^{r_0} dr \Delta P(r) = \int_0^{r_0} dr \left[|\chi_{kl}(r)|^2 - \left|\chi_{kl}^{(0)}(r)\right|^2 \right]. \qquad (39.38)$$

If we attempt to evaluate Eq. (39.38) using the asymptotic form of the wave functions, a problem arises in that $\chi_{kl}(r)$ is then finite at $r = 0$ and hence $R_{kl}(0)$ diverges. If, however, we appropriately force $\chi_{kl}(r = 0)$ to vanish, the error made using the asymptotic forms in (39.38) is negligible since the integrals of both $|\chi_{kl}(r)|^2$ and $\left|\chi_{kl}^{(0)}(r)\right|^2$ are proportional to r_0 and are dominated by their large r behavior.

We now pursue this behavior systematically.[14] We start by considering two states, χ_{kl} and $\chi_{k'l}$; we multiply (39.33) for the state k by $\chi_{k'l}(r)$ and the complex conjugate of (39.33) for k' by $\chi_{kl}^*(r)$, subtract, and integrate out to our radius, r_0, obtaining

$$\int_0^{r_0} dr \left[\chi_{k'l}^*(r) \frac{d^2\chi_{kl}(r)}{dr^2} - \chi_{kl}(r) \frac{d^2\chi_{k'l}^*(r)}{dr^2} \right] + \left(k^2 - k'^2\right) \int_0^{r_0} dr \chi_{k'l}^*(r)\chi_{kl}(r) = 0. \qquad (39.39)$$

On integrating the left-hand side by parts, we obtain (where we henceforth assume real wave functions)

$$\left[\chi_{k'l}(r) \frac{d\chi_{kl}(r)}{dr} - \chi_{kl}(r) \frac{d\chi_{k'l}(r)}{dr} \right]\Bigg|_0^{r_0}. \qquad (39.40)$$

Next we assume that k and k′ do not differ significantly and expand $\chi_{k'l}(r)$ and $d\chi_{k'l}(r)/dk$ to first order in $(k - k')$:

$$\chi_{k'l}(r) = \left[\chi_{kl}(r) + \frac{d\chi_{kl}(r)}{dk}\left(k - k'\right) \right];$$

and

$$\frac{d\chi_{k'l}(r)}{dk} = \left[\frac{d\chi_{kl}(r)}{dk} + \frac{d^2\chi_{kl}(r)}{dkdr}\left(k - k'\right) \right].$$

[12] This is sometimes called Levinson's theorem. The scattering cross section associated l^{th} partial wave is a maximum at resonance. Qualitatively, over a wider range of energy the total change in δ_l on passing through the resonance is π. The properties that the response of a system is largest for a phase shift of $\pi/2$ and that the total change on passing fully through the resonance is π, are common general characteristics of a resonance.

[13] The range of a bound or resonant state wave function may be considerably larger than the range of the potential. The deuteron is a good example in which the two particles largely reside outside the range of the potential, which is sometimes approximated as a delta function.

[14] Part of the approach we will use here is employed in treating the scattering of slow nucleons (see references).

Substituting these forms into Eq. (39.39), we get

$$\left[\frac{d\chi_{kl}(r)}{dk}\frac{d\chi_{kl}(r)}{dr} - \chi_{kl}(r)\frac{d^2\chi_{kl}(r)}{dkdr}\right]\Bigg|_0^{r_0} = 2k\int_0^{r_0} dr\,(\chi_{kl}(r))^2. \tag{39.41}$$

If we evaluate the left-hand side of (39.41) using the asymptotic form (39.34) for $\chi_{kl}(r)$, we obtain

$$\int_0^{r_0} dr\,(\chi_{kl}(r))^2 \cong \frac{1}{r_0}\left[r_0 + \frac{d\delta_l(k)}{dk} - \frac{1}{2k}\sin 2\left(kr - \frac{1}{2}l\pi + \delta_l(k)\right)\right], \tag{39.42}$$

where, as discussed, we assume the integral evaluated at r = 0 vanishes. On the other hand, in the absence of the scattering center we would have

$$\int_0^{r_0} dr\left(\chi_{kl}^{(0)}(r)\right)^2 \cong \frac{1}{r_0}\left[r_0 - \frac{1}{2k}\sin 2\left(kr - \frac{1}{2}l\pi\right)\right]. \tag{39.43}$$

As previously argued, the change in the probability of finding a particle, ΔN_{kl}, within our sphere of radius, r_0, is given by the *difference* between (39.39) and (39.35),

$$\begin{aligned}\Delta N_{kl} &\cong \frac{1}{r_0}\left[\frac{d\delta_l(k)}{dk} - \frac{1}{2k}\sin 2\left(kr_0 - \frac{1}{2}l\pi + \delta_l(k)\right) + \frac{1}{2k}\sin 2\left(kr_0 - \frac{1}{2}l\pi\right)\right] \\ &= \frac{1}{r_0}\left[\frac{d\delta_l(k)}{dk} - \frac{1}{k}\sin(\delta_l(k))\cos(2kr_0 - l\pi + \delta_l(k))\right],\end{aligned} \tag{39.44}$$

where, as noted, we require the value of the integral at r=0 to vanish.

We now carry over this concept to the case where the region outside of our scattering center is occupied by a free electron gas, along with the associated atoms producing it (which then constitute the remainder of the lattice). We take V(r) to be the *difference* between the host and the impurity potentials. To find the total excess number of electrons associated with this difference potential we must integrate ΔN_{kl} over all states, $k \le k_F$, weighted by the appropriate density of states which is given by $2(2l+1)r_0 dk/\pi$; here the first factor accounts for spin, the second for the degeneracy associated with the m quantum number, and the last is the density of standing wave states in one-dimensional k space. We then have

$$\begin{aligned}\Delta N &= \frac{2}{\pi}\sum_l (2l+1)\int_0^{k_F} dk\Delta N_{kl} \\ &= \frac{2}{\pi}\sum_l (2l+1)\int_0^{k_F} dk\left[\frac{d\delta_l(k)}{dk} - \frac{1}{k}\sin(\delta_l(k))\cos(2kr_0 - l\pi + \delta_l(k))\right]\end{aligned} \tag{39.45}$$

or

$$\Delta N(r_0) = \frac{2}{\pi}\sum_l (2l+1)\delta_l(k_F) + \Delta N^{(osc)}(r_0). \qquad (39.46)$$

The last term can be associated with an oscillating number density; to obtain this quantity at the point, r_0, we differentiate $\Delta N^{(osc)}(r_0)$ with respect to r_0 and divide by the area of a shell, $4\pi r_0^2$, with the result,

$$\begin{aligned}\delta n(r_0) &\equiv \frac{1}{4\pi r_0^2}\frac{\partial \Delta N^{(osc)}(r_0)}{\partial r_0} \\ &= -\frac{1}{2\pi^2 r_0^3}\sum_l (2l+1)\left[\sin(\delta_l(k_F))(\cos(2k_F r_0 - l\pi + \delta_l(k)) \right. \\ &\quad \left. -\cos(2k_F r_0 - l\pi))\right] \end{aligned} \qquad (39.47)$$

and would represent an oscillatory behavior of the number density. We argue that the tendency of the electron gas to screen out any fluctuations in charge density would strongly suppress this latter term and we would then be left with

$$z' - z \equiv \Delta z = \frac{2}{\pi}\sum_l (2l+1)\delta_l(k_F). \qquad (39.48)$$

This is the Friedel sum rule. In practice, this sum rule is well satisfied by restricting the l-values to those associated with the valence electrons; thus in a simple metal host we would retain the s, p, and d ($l = 0$ to 2) or the s, p, d, and f (l = 0 to 3) phase shifts for a transition or a rare earth impurity respectively.

Under similar assumptions (a free-electron-like host) we can also represent the electrical conductivity in terms of scattering phase shifts. We recall that the differential scattering cross section is given by the relation[15]

$$\sigma(\theta) = \frac{1}{k^2}\left|\sum_{l=0}^{\infty}(2l+1)\,e^{i\delta_l}\sin\delta_l P_l(\theta)\right|^2. \qquad (39.49)$$

If we are given the phase shifts, this expression can be used to estimate the electrical conductivity resulting from impurity scattering by evaluating the *transport cross section*, introduced in Section 19.10,

$$\sigma_{tr} = \int d\Omega\,\sigma(\theta)\,(1-\cos(\theta)). \qquad (39.50)$$

We write the electrical conductivity as $\sigma = ne^2\tau/m$ with $\tau = l v_F$ and $l = 1/\sigma_{tr} n$, where τ, l, v_F, and n, are the electron relaxation time, mean free path, Fermi velocity, and number density respectively. We leave it as a problem to show that σ_{tr} can be obtained from $\sigma(\theta)$ using the expression

$$\sigma_{tr} = \frac{2}{k_F^2}\sum_l (l+1)\sin^2(\delta_l(k_F) - \delta_{l+1}(k_F)). \qquad (39.51)$$

15 See Landau and Lifshitz, *Quantum Mechanics*, Section 115.

39.4 The electronic structure of alloys

Translational symmetry is absent in an alloy and hence Bloch's theorem does not apply; i.e., the electronic states, even in the single particle picture, can no longer be numbered by the reduced wave vector (in the first Brillouin zone) and the band index. However, techniques to describe the electronic properties of alloys are clearly important, since many of the most beneficial material properties are a direct result of alloying. A chemical potential still divides the occupied from the unoccupied states at absolute zero and the concept of a density of states survives (whatever the nature of the electronic states), from which the electronic heat capacity and the magnetic susceptibility can be estimated. The density of states is also useful for studying the systematics of the occurrence of superconductivity and magnetism. However the Fermi surface, as a sharp structure in momentum space, quickly loses its meaning: we expect an uncertainty in the Fermi wave vectors, $\Delta k \approx l^{-1}$; this does not mean that the electronic states themselves are ill-defined, but only that they are no longer classified by their wave vectors and any attempt to do so requires a superposition of Bloch-like states. Various schemes have been proposed to deal with this problem, some of which we will discuss in this section. We begin with the simplest.

39.4.1 The virtual crystal approximation

For simplicity we will restrict our discussion to binary alloys and suppose our crystal is composed of atoms, which we designate A and B, having concentrations x and $1 - x$ respectively. The simplest approximation is to assume the electrons move in some sort of average potential and that we may calculate an associated band structure (using Bloch states) and from it a density of states (an observable quantity). Such a potential might be written as

$$V(\mathbf{r}) = xV_A(\mathbf{r}) + (1 - x)\, V_B(\mathbf{r}), \tag{39.52}$$

although some prescription to construct it must be given. Calculations based on this approach are said to be in the *virtual crystal approximation.*

Earlier, we discussed the nearly free electron approximation, applicable to metals having only s–p valence electrons. In lowest order the crystal potential is ignored and one draws a free electron sphere, having a radius calculated from the number of valence electrons, within a Brillouin zone, the symmetry and dimensions of which follow from the lattice constants. The Fermi surface is constructed by mapping the various sheets back into the first Brillouin zone, as discussed in Chapter 11. The metals for which this scheme is a useful approximation were termed the *simple metals*. For an alloy involving two simple metals it would be natural to calculate a Fermi radius for the alloy from the average electron concentration of its constituents: $n = xn_A + (1 - x)\, n_B$. Ignoring the structure resulting from Brillouin zone contact, the density of states is that for a free electron gas at the conduction electron density resulting from this prescription.

The free electron model is essentially useless for the 3d, 4d, and 5d transition metals. With few exceptions (e.g., Mn) these elements have the bcc (e.g., Cr, Fe, Mo, Nb, Ta, Ti, W, V, Zr), hcp (e.g., Co, Ti), and fcc (e.g., Co, Fe, Ni, Pd, Pt) structures (those listed twice have differing low and high temperature structures, but in some cases the high temperature phase may exist as a metastable phase at low temperatures). As a simple model we might think of a system like Cu_xNi_{1-x}, which we will take as miscible. As a first approximation (and ignoring the magnetic response) we might try to *interpolate* between the band structures of Cu and Ni (both being fcc) and place the Fermi level according to the average electron concentration. We could proceed similarly to treat changes occurring in alloys involving Ti, V, Cr and Fe. This is sometimes called the *rigid band approximation.* The density of states calculated in this way can be useful as a guide in interpreting experimental data (and was all that was available in the era before more refined techniques, to be discussed further).

But the electrons themselves do not move in an idealized potential of the form (39.47). Resorting to a semi-classical, tight-binding, picture an electron "hopping" from one transition metal ion to a neighboring one might spend significantly more time (orbit the nucleus more times) on one ion (for which it might be more tightly bound) than on its neighbor. So the real motion of an electron may differ significantly from that resulting from an averaged potential. Better methods now exist and the virtual crystal approximation (or the related rigid band approximation) is now, for the most part, only of historical interest.

The next two subsections discuss some approaches that have been used to address the above issues. They are however rather formal and some readers may prefer to pass on to section 39.4.4 which is more approachable.

39.4.2 The average T-matrix approximation (ATA)

Equation (39.52) suggests a possibly improved model for an alloy, beyond the virtual crystal approximation (VCA) introduced in Section 39.4.1. Rather than forming a concentration weighted superposition of the individual potentials, V_A and V_B, in a binary alloy, perhaps we can form a weighted average of the scattering amplitudes, $t_l^{(A)}$ and $t_l^{(B)}$,

$$\bar{t}_l = xt_l^{(A)} + (1-x)t_l^{(B)} \tag{39.53}$$

or on using Eq. (15A.11),

$$\bar{t}_l = \frac{1}{\kappa}\left(xe^{i\delta_l^{(A)}}\sin\delta_l^{(A)} + (1-x)e^{i\delta_l^{(B)}}\sin\delta_l^{(B)}\right), \tag{39.54}$$

where the phase shifts, $\delta_l^{(A)}$ and $\delta_l^{(B)}$, are obtained from integrating the radial Schrödinger equation for the potentials, $V_A(r)$ and $V_B(r)$. This is the averaged

T-matrix approximation (ATA) proposed by Korringa[16]. Note, however, that if one writes

$$\bar{t}_l = \frac{1}{\kappa} e^{i\bar{\delta}_l} \sin \bar{\delta}_l \tag{39.55}$$

and solves for the "average" phase shifts, $\bar{\delta}_l$, they turn out to be *complex* for positive energies (with a positive imaginary part), implying that the associated potential which produces them is also complex.This is physically unacceptable, implying that the ATA is not suitable as a theory for alloys. In any case the quantities, $\cot \bar{\delta}_l$ (which will also be complex), enter the diagonal terms of the KKR matrix, and the real and imaginary parts of the associated determinant do not in general vanish simultaneously as a function of the (real) energy; therefore we cannot calculate energies of our system with this approach. This might have been anticipated since **k** is not a good quantum number, due to the loss of translational invariance and, strictly speaking, a function, $\varepsilon(\mathbf{k})$, does not exist.

Associated with the KKR technique is a wave function and were the technique to work it would imply the existence of some kind of average wave function. However, in the real system the true wave functions of various states do not have the same amplitudes on the A and B atoms; in particular in a material like Cu_xNi_{1-x} we would expect the d orbitals to be largely filled on the Cu sites, but only partially so on the Ni sites. Such behavior is expected on physical grounds, since the atomic potentials dominate locally, and we conclude that the concept of an "average wave function" has little or no meaning. Hence we require a more powerful approach.

The exact solution of the multiple scattering equations for some configuration of atoms in an alloy would be very complicated; furthermore the information to specify all the positions in a given material is generally not available. But more important it would yield too much information. More practical would be information for *ensembles* of atoms arising from the many different configurations the alloy might have. Hence we will seek a kind of *average description* wherein the system is in no particular quantum state. This requires a description in terms of a *density matrix*, or more generally a Green's function, that is averaged over configurations in some way, and from which expectation values of certain observables can be calculated. This is beyond the scope of our presentation.

39.4.3 The coherent potential approximation (CPA)

Given the unsatisfactory characteristics of the ATA, one needs an alternative way to construct a configuration averaged description of an alloy. From the earlier discussion it is clear that it should be based on a Green's function approach. We begin by postulating the existence of a Green's function, $G^{(W)}$, which describes a system in which electrons propagate coherently, i.e., without scattering, in an effective medium, and from which configuration averaged properties of an alloy, such as the density of states can be accurately calculated.[17] Associated with this Green's function is a potential energy function,

$$W(\mathbf{r}, \varepsilon) = \sum_n w(\mathbf{r}_n, \varepsilon), \tag{39.56}$$

[16] J. Korringa, *J. Phys. Chem. Solids* 7, 252 (1958).

[17] Coherent potential model of substitutional disordered alloys, P. Soven, *Phys. Rev.* **156**, 809 (1967).

where the local functions, $w(\mathbf{r}, \varepsilon)$, which are in general energy dependent and complex, are the same on all sites.

To do this one demands that an electron propagating in the coherent system according to $G^{(W)}$ sees only the *difference* between coherent potential, $w(\mathbf{r}, \varepsilon)$, and the true potential when it encounters a site where the potential is either $v_A(\mathbf{r})$ or $v_B(\mathbf{r})$. We denote the effective scattering amplitudes for these two cases as $\tilde{t}_A(\mathbf{r}, \mathbf{r}'; \varepsilon)$ and $\tilde{t}_B(\mathbf{r}, \mathbf{r}'; \varepsilon)$. For the case of an A atom on the central site we have an equation for $\tilde{t}_A$ of the form

$$\tilde{t}_A(\mathbf{r}, \mathbf{r}'; \varepsilon) = (v_A(\mathbf{r}_n) - w(\mathbf{r}_n, \varepsilon))\delta(\mathbf{r} - \mathbf{r}') + (v_A(\mathbf{r}_n) - w(\mathbf{r}_n, \varepsilon)) \int d^3 r_1 G^{(W)}(\mathbf{r}, \mathbf{r}_1; \varepsilon)\tilde{t}_A(\mathbf{r}_1, \mathbf{r}'; \varepsilon). \quad (39.57)$$

A similar one applies for the case when a B atom is on the central site. To complete the specification of the problem we impose a third equation requiring that the *concentration weighted average* of these scattering amplitudes *vanishes*; i.e.,

$$\overline{t^{(w)}} = x\tilde{t}_n^{(A)} + (1 - x)\tilde{t}_n^{(B)} \quad (39.58)$$

Equations (39.57) and (39.58) formally define the coherent potential model for a binary alloy.[18] However they are deceptively simple in appearance and their solution is rather complex; we refer the reader to the review by Faulkner for the details. The density of states for copper–nickel alloys of three different compositions calculated with the CPA KKR method is shown in Fig. 39.12.

[18] It turns out that one can carry through such a program without actually constructing the potential, $w(\mathbf{r}, \varepsilon)$.

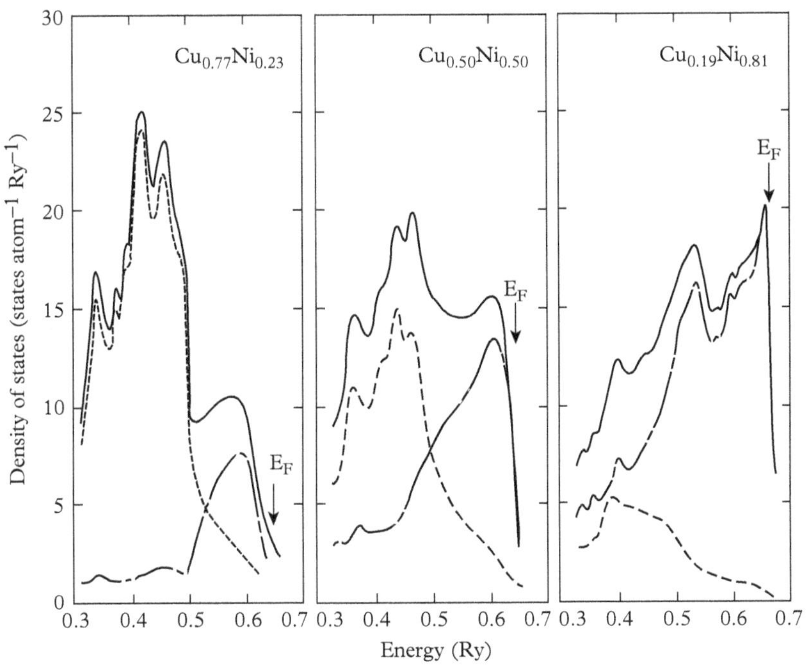

Figure 39.12 *The solid line shows the density of states for three different copper-nickel alloys as calculated with the KKR CPA method. The dashed line shows the projected density of states on the copper sites while the long/short dashed line shows the contribution from the nickel sites. (After J.S. Faulkner,* Prog. in Mat. Sci. *27, 1 (1982).)*

39.4.4 The supercell approach

With the availability of ever increasing computational power one can adopt a "brute force" approach to calculating the properties of alloys that is based on the supercell technique, discussed earlier in Section 16.7 in connection with phonons. In the present application, one expands the nominal unit cell of the material under consideration to a larger unit cell containing integral multiples of the original cell. As an example, if the alloy nominally has a bcc crystal structure with lattice constant a and two atoms per cell (as in the CuZn case), one could consider a cubic supercell containing copies of the basic bcc cell with a lattice constant 4a which would then contain 128 atoms. While maintaining the average concentration of the constituents, one can then distribute the constituent atoms within this larger cell in some manner and calculate electronic properties, such as the total energy and the density of states. By examining various configurations and performing

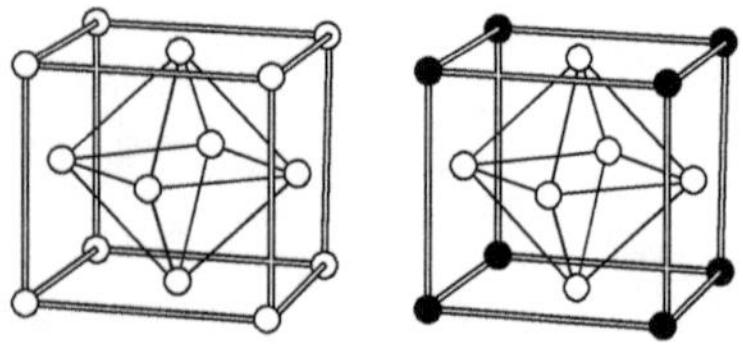

Figure 39.13 *(above) Crystal structures of the face centered cubic (fcc) and (b) L1$_2$ structures. (After Lu et al. (1992).)*

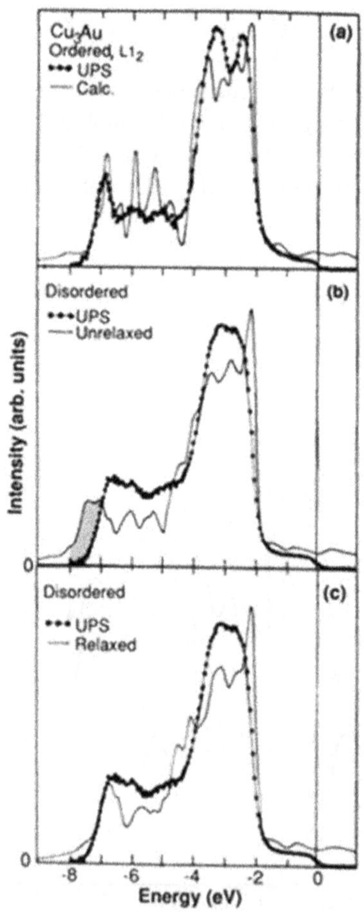

Figure 39.14 *(right) Comparison of the relativistic DOS and the measured valence band photoemission for Cu$_3$Au. Thin lines are the calculated results, while the connected dotted lines are the experimental results. (a) The ordered L1$_2$ Cu$_3$Au calculated at a = 3.743 A. (b) The unrelaxed Cu$_3$Au DOS calculated at a = 3.755 A. (c) The relaxed Cu$_3$Au DOS calculated at a = 3.755 A. The shaded area in (b) highlights the 1 eV discrepancy with experiment, which is rectified by including positional relaxation (c). (After Lu et al. (1992) and Krummacher et al. (1989).)*

some average over the results obtained (including some smoothing of the density of states, which generally contains sharp features) one can calculate alloy properties. One attractive feature is that the atoms will in general want to reside away from the nominal extensions of the basic unit cell sites since an alloy is never perfectly periodic but rather each atom responds to its local environment. One can then perform total energy calculations for various displacements and seek out the minimum energy for a given, fully relaxed, configuration. Another attractive feature is that by using a sufficiently large unit cell we can examine different concentrations, e.g., $Cu_{1-x}Zn_x$, as long as the values of x are compatible with the size of the supercell.

Two problems immediately arise: (1) the number of configurations increases rapidly with the size of the supercell and (2) atoms will in general have a preference as to their neighbors;[19] this is especially true of systems that form stoichiometric compounds, as one has, for example, in compound semiconductor alloys where chemical preferences come into play. Although the second of these can, in principle, be addressed through total energy calculations, computational constraints place limits on how many configurations can be examined. In addition, alloys form at finite temperatures where entropic disorder competes with energetics. For this reason much effort has gone into choosing some small set of "special quasirandom structures" to examine.[20]

As an example of the supercell approach we show results from the alloy Cu_3Au.[21] Figure 39.13(a) shows the crystal structure of the disordered fcc lattice, while Fig. 39.13(b) shows the $L1_2$ structure of the ordered phase. Figure 39.14 shows density of states along with the photoemission spectrum.[22] Of particular interest is that significantly better agreement is obtained for the relaxed configuration.

ADDITIONAL READING

A standard source of alloy phase diagrams is *Constitution of Binary Alloys*, F.A. Shunk, M. Hansen, and K. Anderko, McGraw-Hill (1969).

Still valuable is the discussion of alloys in Section 5 of Chapter I of *Properties of Metals and Alloys*, N.F. Mott and H. Jones, Oxford University Press (1936).

The Friedel sum rule is discussed quantitatively in *Quantum Theory of Solids*, p. 341, C. Kittel, J. Wiley and Sons, NY (1963), and qualitatively in *Principles of the Theory of Solids*, p. 135, J.M. Ziman, Cambridge University Press, Cambridge (1964).

The scattering of slowly moving nucleons is discussed in *Elementary Nuclear Theory*, p. 55, H. A. Bethe and P. Morrison, John Wiley & Sons, NY (1956), or *Theoretical Nuclear Physics*, p. 56, J.M. Blatt and V.F. Weisskopf, John Wiley & Sons, NY (1952).

[19] At absolute zero one's intuition is that all systems either phase separate or form ordered structures, although the equilibrium times can be exceedingly long. More random configurations present at higher temperatures can sometimes be "quenched in," so-called quenched disorder, by rapid cooling.

[20] Electronic properties of random alloys: special quasirandom structures, S.-H. Wei, L.G. Ferreira, J.E. Bernard, and A. Zunger, *Phys. Rev. B* **42**, 9622 (1990).

[21] Electronic structure of ordered and disordered Cu_3Au and Cu_3Pd, Z.W. Lu, S.-H. Wei, and A. Zunger, *Phys. Rev. B* **45**, 10314 (1992).

[22] Comparative-study of the electronic-structure of the ordered and disordered $Cu_3Au(100)$ and $Cu_3Au(110)$ surfaces, S. Krummacher, N. Sen, W. Gudat, R. Johnson, F. Grey, and J. Ghijsen, *Z. Phys. B* **75**, 235 (1989).

For a description of important historical milestones in the development of multiple scattering theory that led up to the KKR formalism, see: Early history of multiple scattering theory for ordered systems, J. Korringa, *Phys. Rep.* **238**, 341 (1994).

The CPA theory has been extended to go beyond the muffin tin approximation: see Green's functions in full-potential multiple-scattering theory, A. Rusanu, G.M. Stocks, Y. Wang, and J.S. Faulkner, *Phys. Rev. B* **84**, 035102 (2011).

For a review of computational methods to study the electronic structure of alloys, see: The modern theory of alloys, J.S. Faulkner, *Prog. Mater. Sci.*, Vol. 27, pp. 1–187 (1982).

Problem

39.1. Obtain Eq. (39.51) for the transport scattering cross section from (39.49) and (39.50).

Defects and Diffusion in Crystalline Solids

Color Centers

40

40.1 Point defects

Much of our earlier discussion of crystal lattices assumed that they are infinite in extent possessing complete translational symmetry; i.e., they were assumed to be free from internal defects, broadly defined as departures from perfect periodicity. Even when treating the thermodynamics of alloys, where perfect periodicity is lost, symmetry entered in an average sense when distinguishing the different phases in contact. But many important properties, including electrical transport and material strength, depend implicitly on the presence of defects.

In this chapter we limit ourselves to so-called point defects; larger scale atomic rearrangements will be discussed in Chapter 41. A dilute concentration of impurities can be regarded as a kind of point defect with respect to a host. But other kinds of point defect can be present, even in a pure material, examples of which are given. Although the word defect is generally used for various imperfections in solids, they always occur in real materials and their presence cannot be ignored in achieving a realistic understanding of the properties of solids. There is a variety of point defects of which we will only discuss some of the simplest, together with some material properties they impart.

Figure 40.1 shows a 2d square monatomic array in which one atom has been removed from a lattice site and placed inbetween atoms elsewhere in the lattice. The empty site is referred to as a *vacancy* and the site it moves to as an *interstitial*. Such defects are classed as point defects. Either of these two defects may occur in the absence of the other, depending on the overall energetics of the material. Not shown in the figure is the small rearrangement in the local positions of the atoms surrounding the vacancy and interstitial sites that is caused by the imbalance of the local forces resulting from their creation, an effect referred to as relaxation.

One must regard point defects as always being present. Although it generally raises the energy of the crystal to create such defects, the increase in entropy, which makes a negative contribution, $-TS$, to the free energy, guarantees their presence at some concentration for finite temperatures. Since most crystals are prepared from a higher temperature liquid state, they start life with some

The Physics of Solids. First Edition. J. B. Ketterson.

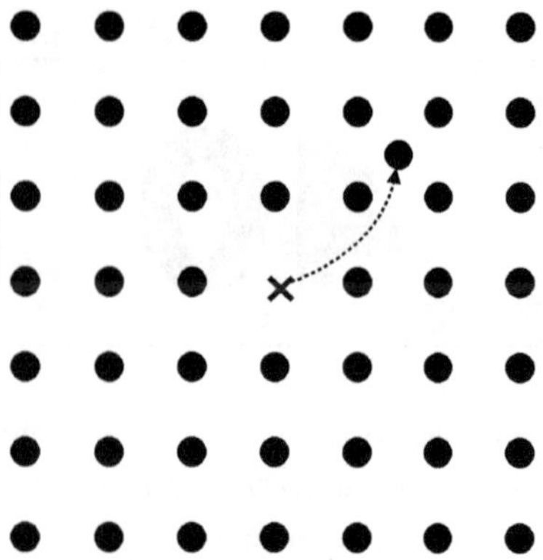

Figure 40.1 *A vacancy and an interstitial in a square crystal lattice.*

concentration of defects. As the temperature is lowered the number of vacancies and interstitials will tend to decrease, either by recombining (whereupon a vacancy and interstitial annihilate), or by either species wandering to a free surface (or grain boundary, to be discussed in Chapter 41) where it disappears. However, they do not disappear altogether in the bulk of the material; the kinetic processes that allow them to move slow down as the temperature falls, and the motion ultimately becomes so slow that the defects are *immobilized.* The remaining defects are in effect "frozen in"; i.e., the crystal never achieves equilibrium.[1]

40.2 Concentration of point defects

We now calculate the number of defects *of a specific type* that will be present in a crystal in thermodynamic equilibrium. We do this by minimizing the appropriate free energy with respect to the number of such defects; for constant pressure this is the Gibbs free energy: $G = E - TS + PV$. We start by calculating the entropy from statistical considerations. In Chapter 6 we determined the number of ways in which n objects could be arranged in g cells when only one object can be in each cell, with the result

$$W = \frac{g!}{n!(g-n)!}. \tag{6.16}$$

Here the lattice sites play the role of g cells and the presence of two defects on the same site is energetically highly suppressed. If N is the number of atoms and n the number of defects, then the number of lattice sites is $g = N + n$ and Eq. (6.16) becomes

$$w = \frac{(N+n)!}{N!n!}. \tag{40.1}$$

Hence we have a configurational entropy

$$\begin{aligned} S^{\text{config.}} &= k_B \, ln\left(\frac{(N+n)!}{N!n!}\right) \\ &\cong k_B \left[(N+n)\, ln\,(N+n) - n\, ln\,(n) - N\, ln\,(N)\right] \end{aligned}, \tag{40.2}$$

where in the second step we have used Stirling's approximation, $ln(n!) = n\, ln(n) - n$, and similarly for N!. We define the formation energy of a single defect as $\varepsilon(T)$ (we will neglect the generally weak temperature dependence in what follows); then the energy of n such defects (assuming that they are dilute enough that we can neglect any interactions) is $n\varepsilon$. We define the volume occupied by an atom in the perfect crystal as $v(T)$ (where we will again neglect any temperature dependence arising from thermal expansion); this will also be

[1] Solid helium at $T = 0$ is an exception, where the motion of defects involves quantum zero-point motion. This fact has dramatic consequences; e.g., defects can be present and move via quantum fluctuations. Perhaps most surprising is that *collective motions* (surface waves) involving the coherent rearrangement of the atoms at a free surface can occur.

the total volume increase on introducing this defect. The Gibbs free energy associated with the n defects in a crystal exposed to a pressure, P, is then

$$G = E - TS + PV.$$
$$= n\varepsilon - k_B T\,[(N+n)\,ln(N+n) - n\,ln(n) - N\,ln(N)] + Pnv. \qquad (40.3)$$

We find the equilibrium number of defects by minimizing this expression with respect to n, with the result (assuming $n \ll N$)[2],

$$\frac{\partial G}{\partial n} = 0 = \varepsilon - k_B T\,ln\left(\frac{N}{n}\right) + Pv.$$

Solving for n we obtain

$$n = Ne^{-(\varepsilon+Pv)/k_B T}$$
$$\cong Ne^{-\varepsilon/k_B T}; \qquad (40.4)$$

where we have neglected the term, Pv, in the exponent in the second step since it is usually small at atmospheric pressure.

40.3 Point defects in lattices with more than one species

Historically, much effort has been focused on defects in insulating ionic compounds, in part to achieve a better understanding of the photographic process, where the absorption of light results in the chemical dissociation of compounds such as AgBr. But other ionic materials are also of interest because the defects often have associated optical resonances, called color centers, which alter the color of the material.

For ionic compounds involving two species we may have four defects involving vacancies and interstitials for both ions. Commonly occurring combinations are equal numbers of positive and negative *ion vacancies* (Schottky case) and equal numbers of positive ion vacancies together with positive ion interstitials.[3] Both of these combinations preserve the necessary charge neutrality. In particular, the empty lattice site (left behind by a positively charged interstitial) in the Frankel case acquires a net negative charge when viewed from some distance away, due to the excess of negative charge in the vicinity of the vacancy. Figure 40.2 shows a 2d NaCl-like lattice with two pairs of Schottky (a) and Frankel (b) defects.

The argument leading to Eq. (40.4) may be repeated for the case when several species are present with the result

$$n_j = N_j\, e^{-\varepsilon_j/k_B T}, \qquad (40.5)$$

[2] Formally, the derivative of the Gibbs free energy with respect to the number n_i of some species i is the chemical potential; i.e., $\mu_i = \partial G/\partial n_i$; if there is no constraint on the number n_i the system will minimize itself with respect to that number.

[3] Negative ions generally have much larger ionic radii and hence an interstitial would have a high formation energy.

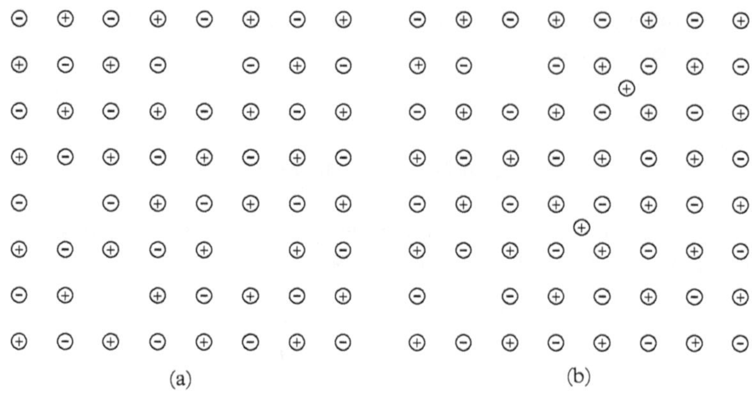

Figure 40.2 *A 2d* NaCl *like lattice: (a) a pair of Schottky defects (positive and negative vacancies); (b) a pair of Frankel defects (a positive vacancy and a positive interstitial).*

where n_j is the number of defects of species j and N_j is the number of sites they may occupy. This expression is a good model for metals with uncharged defects. However when the defects are charged, as in the NaCl lattice, overall electrical neutrality must be maintained which requires

$$\sum_j n_j q_j = 0, \tag{40.6}$$

where q_j are the charges of the impurities; note the magnitude of the effective charges will generally be multiples of the electronic charge, e. The condition (40.6) must now be imposed as a *constraint* when minimizing the total Gibbs free energy,

$$G = \sum_j n_j \varepsilon_j - \sum_j k_B T \left[(N_j + n_j)\, ln(N_j + n_j) - n_j\, ln\, n_j - N_j\, ln\, N_j \right] + \sum_j P\, n_j v_j. \tag{40.7}$$

The constraint is incorporated using the Lagrange multiplier method leading to the condition

$$\frac{\partial}{\partial n_j}\left[G + \lambda \sum_j n_j q_j \right] = 0, \tag{40.8}$$

which yields (in the approximations discussed earlier) the defect concentrations

$$n_j = N_j\, e^{-(\varepsilon_j + \lambda q_j)/k_B T}. \tag{40.9}$$

The parameter λ follows from substituting this expression into Eq. (40.6),

$$\sum_j N_j\, e^{-(\varepsilon_j + \lambda q_j)/k_B T} q_j = 0. \tag{40.10}$$

Equations (40.9) and (40.10) are then solved iteratively for n_j and λ.

The energies, ε_j, are usually sufficiently different (on the scale of k_BT) that the defect with the smallest energy controls the overall population; defect pairs involving this defect and the defect with the opposite charge, having the lowest energy for that charge, then dominate. We will assume monovalent defects; i.e., $q_+ = -q_- = |e|$. We designate the associated two energies as $\varepsilon_\pm$ and Eq. (40.10) becomes $n_+ = n_-$, with

$$n_+ = N_+ e^{-(\varepsilon_+ + \lambda)/k_BT} \tag{40.11a}$$

and

$$n_- = N_- e^{-(\varepsilon_- - \lambda)/k_BT}. \tag{40.11b}$$

Taking the product of these two expressions yields[4]

$$n_+ n_- = N_+ N_- e^{-(\varepsilon_+ + \varepsilon_-)/k_BT} \tag{40.12}$$

or

$$n_+ = n_- = \sqrt{N_+ N_-} e^{-(\varepsilon_+ + \varepsilon_-)/2k_BT}. \tag{40.13}$$

Equilibrium defect pairs make an experimentally observable contribution to the heat capacity. On the other hand, defects can be "quenched-in" by rapidly reducing a sample from high temperature (where the defects are present in equilibrium) to a much lower temperature (where their annihilation is greatly slowed down). On slowly reheating the sample this quenched-in energy is released and the amount of energy released can be measured experimentally by comparing with a subsequent warming curve. Non-equilibrium defect concentrations are also produced by neutron irradiation (and subsequent reactions leading to recoil), resulting in "swelling" (a problem with materials used in nuclear reactors). Foreign atoms may be *implanted* by exposing a sample to a high energy (kilovolt) flux of the associated ions (a process called ion implantation); this is an important strategy in doping semiconductors.

40.4 Diffusion: microscopic theory

The locations of point defects do not remain fixed in time; since they are subject to random thermal motions they vibrate around but also regularly jump (or hop) to neighboring lattice sites. Their behavior is a classic example of what is called a *random walk*. To describe this phenomenon, picture a one-dimensional lattice, the lattice sites of which are numbered by the collection of all positive and negative integers, n, on which a "particle" (in the present case the defect) is initially placed at $n = 0$. After some time, $t \gg \tau$, where τ is some characteristic time to be defined, the particle will have "walked" or "diffused" away to some remote site, n. We need to determine the probability, P, for finding the particle on a given site, n, as a function of time.

[4] Very similar expressions were obtained in our discussion of the number of holes and electrons in a semiconductor (with λ playing the role of the Fermi energy).

We describe the behavior in terms of a series of steps. Our model will be that following each step, the probabilities are 50% that the particle remains on its lattice site (where it was) and 50% that the particle moves to a neighboring site (of which there are two).[5,6] At the start the particle is assumed to be on the lattice site n = 0. After the first step it is on lattice sites n = −1, 0, 1, with normalized probabilities 1/4, 2/4, 1/4; after the second step it is on lattice sites n = −2, −1, 0, 1, 2, with probabilities 1/16, 4/16, 6/16, 4/16, 1/16;[7] and so on. Defining the step number as g, this sequence of probabilities can be represented by the expression[8]

$$P(n,g) = \frac{1}{2^{2g}}\frac{(2g)!}{(g+n)!\,(g-n)!}; \quad g \neq 0; |n| \leq g, \tag{40.14}$$

where n numbers the lattice sites and therefore takes on both positive and negative values. Since we will be interested in the continuum (macroscopic) behavior we examine the limit $g \gg n \gg 1$ of (40.14). To do this we use a more accurate version of Stirling's approximation, given by $n! \cong n^n e^{-n}\sqrt{2\pi n}$; inserting this form in (40.14) yields

$$\begin{aligned} P(n,g) &= \frac{1}{2^{2g}}\frac{(2g)^{2g}\,e^{-2g}}{(g+n)^{(g+n)}\,e^{-(g+n)}\,(g-n)^{(g-n)}\,e^{-(g-n)}}\sqrt{\frac{g}{\pi\,(g^2-n^2)}} \\ &= \frac{1}{2^{2g}}\frac{(2g)^{2g}}{g^{g+n}\left(1+\frac{n}{g}\right)^{(g+n)} g^{g-n}\left(1-\frac{n}{g}\right)^{(g-n)}}\sqrt{\frac{1}{\pi g\left(1-\frac{n^2}{g^2}\right)}} \\ &= \sqrt{\frac{1}{\pi g\left(1-\frac{n^2}{g^2}\right)}}\frac{1}{\left(1-\frac{n^2}{g^2}\right)^g\left(1+\frac{n}{g}\right)^n\left(1-\frac{n}{g}\right)^{-n}}. \end{aligned}$$

Using the expression $\lim_{\eta\to\infty}(1 \pm \alpha/\eta)^{\pm\eta} = e^{\pm\alpha}$, and examining the limit where $n^2/g = \text{const.}$ as $g \to \infty$, we may rewrite this expression as

$$P(n,g) = \sqrt{\frac{1}{\pi g}}e^{-\frac{n^2}{g}}. \tag{40.15}$$

Introducing time by writing[9] $g = 2t/\tau$ and $n = x/a$, we have

$$P(x,t) = \sqrt{\frac{\tau}{2\pi t}}e^{-\frac{\tau}{2a^2}\frac{x^2}{t}}\theta(t), \tag{40.16}$$

where we have introduced the theta function, implying that we set the probability for negative times to zero. Assuming the hopping to be independent of spatial

[5] The problem reduces to that of a double coin toss: one toss for movements to the left (heads move left, tails stay put) and a second for movements to the right (heads move right, tails stay put).

[6] One might also assume that the particle jumps to neighboring lattice sites with 100% probability in successive steps; this model has the awkward feature that the particle is on strictly even or strictly odd lattice sites on successive steps, which is non-physical.

[7] Of course the particles do not have to move at times τ; it is only that on the average they jump in this time.

[8] This expression follows from the binomial form given in Eq. (6.16a), ${}_gC_n = g!/n!(g-n)!$, with the substitutions, $g \to 2g$ (to account for tossing of two coins), followed by an origin shift (to account for negative lattice site designations), $n \to n+g$; the factor $1/2^{2g}$ normalizes the total probability of finding the particle to unity.

[9] The extra factor of 2 here is to conform with the commonly used definition of D, to be introduced shortly, and is related to our choice of the jump probabilities associated with successive steps, g.

direction we can generate a theory in three dimensions simply by multiplying three forms like (40.16) together,

$$P(\mathbf{r},t) = \left(\frac{\tau}{2\pi t}\right)^{3/2} e^{-\frac{\tau}{2a^2}\frac{x^2+y^2+z^2}{t}}\theta(t). \tag{40.17}$$

The Gaussian forms, (40.15) and (40.16), tell us that the probability for finding a particle, initially located at some origin, spreads out into the surrounding lattice symmetrically such that the average distance traveled is proportional to the square root of the time.

40.5 Diffusion: macroscopic theory

In a continuum approximation one introduces a concentration variable, $c(\mathbf{r},t)$, defined as the number of defects per unit volume with $\int dVc(\mathbf{r},t) = N$, where N is the total number of defects. One further assumes that the flux of these defects, $\mathbf{f}$ (defined as the number of defects crossing unit area perpendicular to $\mathbf{f}$ in unit time), is proportional to the gradient of the concentration; i.e., $\mathbf{f} = -D\nabla c$, where D is called the *diffusion constant* (this relation is sometimes called *Fick's law*). Balancing the rate at which defects leave some volume, V, enclosed by the surface, S, we have

$$\int_S d\mathbf{S}\cdot \mathbf{f}(\mathbf{r},t) = -\frac{\partial}{\partial t}\int_V dVc(\mathbf{r},t). \tag{40.18}$$

We use Green's theorem to convert the surface integral to a volume integral; noting that this relation must hold for any volume, it must then hold for the integrands themselves. Assembling these ideas we obtain the *diffusion equation*:

$$\frac{\partial c(\mathbf{r},t)}{\partial t} - D\nabla^2 c(\mathbf{r},t) = 0. \tag{40.19}$$

If there are several kinds of defects we must introduce a concentration variable, $c_i(\mathbf{r},t)$, for each species. If defects appear or disappear then we must introduce so-called source or sink terms, $s_i(\mathbf{r},t)$, describing the rate of change of the concentrations associated with these processes, and Eq. (40.19) then becomes

$$\frac{\partial c_i(\mathbf{r},t)}{\partial t} - D_i\nabla^2 c_i(\mathbf{r},t) = s_i(\mathbf{r},t), \tag{40.20}$$

where $s_i > 0$ corresponds to the production of defects. If these defects are generated spontaneously, in order to maintain their thermal equilibrium concentrations, $c_i^{(0)}(T)$, an appropriate form (similar to that encountered in the spontaneous generation of holes and electrons encountered in our discussion of inhomogeneous semiconductors) would be

$$s_i(\mathbf{r},t) = -\frac{c_i(\mathbf{r},t) - c_i^{(0)}}{\tau_{gi}}, \tag{40.21}$$

where τ_{gi} is a spontaneous *generation rate* (generally much longer than a hopping time). The concentrations may be interconnected; e.g., a vacancy could annihilate with an interstitial in a monatomic system; the probability of such an encounter would be proportional to the product of the concentrations of the two species (rendering the resulting equations nonlinear). Finally, if the defects are charged we must introduce a charge neutrality constraint. Note that a net flux of vacancies in one direction corresponds to a net mass flow in the opposite direction. Consistent with our definition, Eq. (40.19) refers to some arbitrary distribution of defects within the solid, whereas Eq. (40.17) refers to a single defect located at some origin at $t = 0$. If we consider the special case where $c(\mathbf{r},t)$ involves only a single impurity, where $\int dV\, c(\mathbf{r},t) = 1$, we can relate the *probability*, $P(\mathbf{r},t)$, given by (40.17), to a *probability density* defined as $c(\mathbf{r},t) \equiv P(\mathbf{r},t)/a^3$; i.e.,

$$c(\mathbf{r},t) = \left(\frac{\tau}{2\pi t a^2}\right)^{3/2} e^{-\frac{r^2}{4Dt}}\,\theta(t) = \left(\frac{1}{4\pi Dt}\right)^{3/2} e^{-\frac{r^2}{4Dt}}\,\theta(t), \tag{40.22}$$

where we have *defined* the diffusion constant as $D = a^2/2\tau$, thus making a connection between the macroscopic and microscopic theories.[10] Equation (40.22) is actually the Green's function for the diffusion equation in the absence of boundaries; rewriting it in terms of an arbitrary origin, $(\mathbf{r}',t')$, we have

$$G(\mathbf{r},t;\ \mathbf{r}',t') = \left(\frac{1}{4\pi D(t-t')}\right)^{3/2} e^{-\frac{|\mathbf{r}-\mathbf{r}'|^2}{4D(t-t')}}\,\theta(t-t'), \tag{40.23}$$

which satisfies the *inhomogeneous* equation

$$\frac{\partial G(\mathbf{r},t;\mathbf{r}',t')}{\partial t} - D\nabla^2 G(\mathbf{r},t;\ \mathbf{r}',t') = \delta^{(3)}(\mathbf{r}-\mathbf{r}')\delta(t-t'). \tag{40.24}$$

We verify this by substituting (40.23) into (40.24), carrying out the implied differentiations, and noting that it satisfies the homogeneous part of the equation, with the exception of a term generated by the time differentiation of $\theta(t-t')$:

$$\left(\frac{1}{4\pi D(t-t')}\right)^{3/2} e^{-\frac{|\mathbf{r}-\mathbf{r}'|^2}{4D(t-t')}}\,\delta(t-t').$$

Clearly his term vanishes unless $t = t'$. But

$$\left(\frac{1}{4\pi D(t-t')}\right)^{3/2}\int d^3R\, e^{-\frac{R^2}{4D(t-t')}} = 1$$

and, furthermore, for small $t-t'$ the integrand is non-zero only in the immediate vicinity of $R = 0$ and hence has the expected behavior of $\delta^{(3)}(\mathbf{r}-\mathbf{r}')$; i.e.,

[10] Other definitions of the diffusion constant are found in the literature.

$G(\mathbf{r},t;\ \mathbf{r}',t')$ satisfies (40.24). Note the mean square probability of finding a defect, which is at the origin at t = 0, is clearly given by

$$\overline{r^2} \equiv \int dVr^2c(\mathbf{r},t) =$$

$$4\pi\int r^2dr\left(r^2\left(\frac{\tau}{2\pi ta^2}\right)^{3/2} e^{-\frac{r^2}{4Dt}}\right)$$

or

$$\overline{r^2} = 6Dt. \qquad (40.25)$$

40.6 Temperature dependence of the diffusion constant

In Sections 40.2 and 40.3 we discussed the temperature dependence of the equilibrium concentration of defects in solids. However, their motion is a transport process and requires overcoming an energy barrier associated with hopping of the defect from one site to another; the same holds for the transport of impurity atoms. Such processes are sometimes called *activated.* If one were simply to force a defect or impurity atom to go from one lattice site to a neighboring one without relaxing, *in a coordinated manner* the positions of the atoms near to which it passes, then the associated energy barrier would be forbiddingly high. Qualitatively, what has to happen during diffusion is that a configuration of the atoms in the vicinity of the defect occurs, where the hopping barrier is lowered, while simultaneously the atoms have sufficient kinetic energy to reconfigure. Hence if a vacancy is involved an atom from a neighboring site (of the appropriate charge in an ionic lattice) can hop into the vacancy site thus transporting it in the opposite direction, as noted earlier. A commonly used model for the hopping time entering the diffusion constant, D, is

$$\tau^{-1} = \nu e^{-\varepsilon_b/k_BT}, \qquad (40.26)$$

where ν is an *attempt frequency* (typically an Einstein or Debye frequency) which measures the rate at which the defect *attempts* to climb over the barrier, and the factor, $e^{-\varepsilon_b/k_BT}$, measures the Boltzmann probability that the amplitude of the impurity vibration will be large enough to climb the barrier (associated with a favorable local configuration), which is assigned a height, ε_b.

40.7 Ionic conduction

If our defects are charged, as in our example of an alkali halide, then in the presence of an external electric field these defects will move. This results in a charge transport mechanism that acts in parallel with any electronic conduction and is generally referred to as *ionic conduction*. It is extremely important in the operation of batteries where atomic charge transport is essential (see later).

We recall that in our discussion of inhomogeneous semiconductors we obtained a relation between the diffusion constant of a charged carrier and its mobility in an electric field; in the present context we may have multiple species, i, with different charges, and we write this relation as

$$D_i = \frac{k_B T \mu_i}{|e|}. \tag{40.27}$$

Typically these mobilities are several orders of magnitude smaller than those associated with electrons or holes; however the *concentration* of electrons and holes is negligible in a typical wide band gap insulator and hence the ionic conductivity then dominates.

Let us first consider a sample of pure NaCl at a temperature such that Frankel pairs (positively charged interstitial Na^+ ions and negatively charged vacancies) are present. In the presence of an electric field the positive Na^+ ions are drawn to the negatively charged electrode (the cathode) while the vacancies are drawn to the positively charged electrode (the anode). But the interstitial positive ions are not nearly as mobile as the vacancies. So rather than traversing the sample, positive ions in the vicinity of the cathode can be chemically reduced whereby they deposit as metallic sodium on the electrode. The amount of metal deposited is related to the total charge passing through the sample via Faraday's law: $N_{Na} = i \times t/|e|$, where N_{Na} is the number of Na atoms deposited, i is the current (assumed constant), and t is the total time the current passes; e is the electronic charge. To preserve charge neutrality a negatively charged vacancy is generated for each Na atom deposited. These vacancies then drift toward the anode; two vacancies arriving at the cathode are filled by two adjacent Na atoms, releasing a Cl_2 molecule in the process.

Since the electrical transport is dominated by the concentration and mobility of the vacancy, we may write the electrical conductivity as

$$\begin{aligned} \sigma(T) &= n_v(T)|e|\mu_v(T) \\ &= \frac{n_v(T)e^2}{k_B T} D_v(T), \end{aligned} \tag{40.28}$$

where we have used (40.27) in the second step and n_v, μ_v, and D_v are, respectively, the number density, mobility, and diffusion constant of the vacancies, all of which are strongly temperature dependent. If we combine (40.13) with (40.28), we obtain

$$\sigma(T) \propto e^{-(\bar{\varepsilon}+\varepsilon_b)/k_B T},$$

where $\bar{\varepsilon} \equiv (\varepsilon_+ + \varepsilon_-)/2$; if $\bar{\varepsilon}$ can be obtained independently we can determine ε_b from the temperature-dependent ionic conductivity.

Suppose a NaCl crystal is doped with $CaCl_2$; now the Ca^{++} ions occupy Na^+ sites and each of them will be associated with one Cl^- ion (in its usual lattice position) and one negatively charged vacancy (on an Na^+ lattice site). Hence we create one vacancy for each $CaCl_2$ molecule dissolved. The increased number of vacancies greatly increases the electrical conductivity (by providing hopping sites for Na^+ ions) and, provided the temperature is not too high, the number of vacancies may be taken as equal to the number of Ca atoms. By measuring the temperature dependence of the conductivity we then obtain the barrier height directly from a semi-logarithmic plot of σ versus $1/T$.

Of special interest is the case where foreign ions move through a host material (called the electrolyte), which itself remains *unaltered*; this process is fundamental to the operation of batteries and such materials are called *super-ionic conductors*. These materials are usually covalently bonded (so as to not participate directly in ion transport) while simultaneously having relatively open crystal structures that allow positive or negative ions to pass; the most useful materials are those which allow only one sign of foreign carrier to pass. The Li^+ ion, being small and hence more mobile, plays a major role in modern battery technology.

40.8 Color centers

When transition metal and rare earth ions are incorporated in an otherwise optically transparent dielectric host, new energy levels often result involving d- or f-like levels localized on the impurity atoms themselves. Optical transitions between such levels can then result in absorption and emission within the band gap of the host. Examples are Ti and Cr atoms in sapphire (Al_2O_3). The study of such levels is important; it forms the basis for various schemes to make lasers (especially when transitions involving phonons are also present). Just how the d and f levels split in the presence of the perturbing potential of the host crystal is the subject of crystal field theory, which we do not address here.

A pure alkali halide is optically transparent in the visible region of the spectrum, even at higher temperatures where an equilibrium concentration of point defects is present. However if certain impurities are introduced the crystal may become colored and the associated microscopic defects are called *color centers*. In addition to impurities, various radiations (X-rays, γ-rays, neutrons, energetic particles) also create color centers. We consider some of the simpler color centers.

40.8.1 The F center: a vacancy bound electron

Our first example is the F center.[11] It is observed that if an NaCl crystal is heated in the presence of Na vapor and then rapidly cooled, a metastable incorporation of excess Na atoms results (as verified by direct chemical analysis) which imparts a yellow color to the host. Similarly the incorporation of excess K in KCl imparts

[11] The word for color is farbe in German.

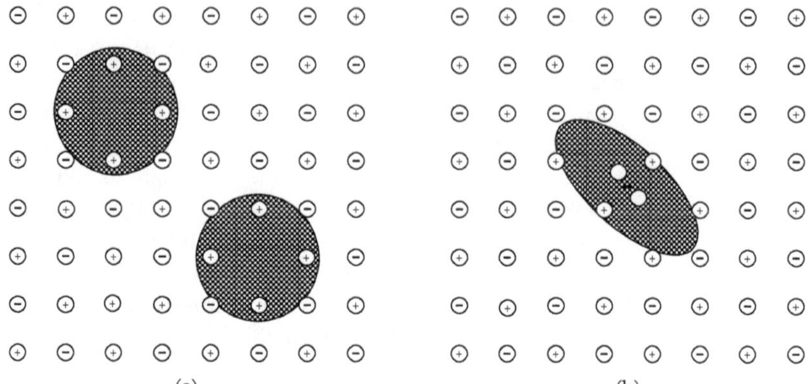

Figure 40.3 *A 2d NaCl-like lattice showing the following cases: (a) The F center. Here, two extra Na atoms are present in the lattice which in turn generate two* $\mathbf{Cl^-}$ *vacancies; the negative electrons from the two excess Na atoms bind in the vicinity of the vacancies (with their charge clouds shown as shaded circles). (b) The* $\mathbf{V_K}$ *center. Here, a* $\mathbf{Cl_2^-}$ *ion is formed from two nearest-neighbor* $\mathbf{Cl^-}$ *ions and a positive hole (shown as the shaded region).*

a magenta color. It is believed that the neutral alkali atoms so introduced occupy the alkali sites of the host crystal. An accompanying volume increase of the host is also observed, which is interpreted as arising from the spontaneous co-generation of an equal number of Cl^- vacancies (see Fig. 40.3).

To understand these responses we picture the following idealized sequence of steps: first we replace a positively charged alkali atom with a neutral alkali atom thereby reducing the amount of positive charge in the lattice as a whole by one unit; at large distances this neutral alkali atom then behaves as though it has a net negative charge. To compensate this charge a halogen vacancy forms (at what we will assume is a distant site). This vacancy behaves as though it has a net positive charge at large distances. The halogen vacancy and the neutral alkali atom together maintain global charge neutrality of the lattice as a whole, but at the expense of the Coulomb energy associated with their separation. At this point the outermost s electron associated with the added (neutral) alkali atom is still attached to that atom. If, however, this electron can bind more tightly to the positively charged halogen vacancy then it will transfer to that site. (Note that this electron transfer then destroys the identity of the imagined neutral alkali site, thereby putting all alkali sites on a more equal footing.) This *electron-bound to-a-halogen-vacancy* is the accepted model for the F center ground state, and is shown schematically in Fig 40.3(a). Some experimental absorption energies are given in Table 40.1.

Light absorption (which imparts the apparent color) results from the excitation of the vacancy-trapped electrons into higher bound or "free" (conduction band) states.[12] In some ways these excited states mimic those of an isolated atom; however because the states are formed in a crystal, a splitting of levels that would otherwise be degenerate in the atom occurs. For instance, in a cubic environment the three-fold degenerate p-states do not split, but the five-fold degenerate d-states split into a three-fold degenerate (designated as either T_{2g} or Γ_5^+) and a two-fold degenerate (designated E_g or Γ_3^+) set of levels. Application of electric or magnetic fields, and more importantly strain, to the crystal introduces further splitting;

[12] Re-radiation (called luminescence) from previously excited electrons (as they fall back into lower states) also occurs, but the frequencies may be shifted due to the simultaneous emission of phonons.

Table 40.1 *Experimental* F*-center absorption energies (eV).*

LiF	5.0	RbCl	2.0
NaF	3.6	CsCl	2.0
KF	2.7	LiBr	2.7
LiCl	3.1	NaBr	2.3
NaCl	2.7	KBr	2.0
KCl	2.2	RbBr	1.8

these can be useful tools for investigating site symmetry and other characteristics of the defects.

40.8.2 Additional color centers

One might think that by heating an alkali halide in a halogen atmosphere we could form the hole-analogue of the F center: a *hole-bound-to-an-alkali-vacancy* complex, historically dubbed the V center. We would then envision an excess neutral halogen atom substituting on a host halogen site along with a spontaneous co-generation of a negatively charged alkali site vacancy. This neutral halogen site would exhibit a negative charge at large distances. By generating a hole and transferring that hole to the alkali vacancy, the neutral halogen would acquire a negative charge and be indistinguishable from other halogen sites. Historically, certain features in the optical spectra resulting from halogen exposure were interpreted using this V-center scenario; however it is now believed that such centers do not exist.

A center that is thought to form in the presence of excess chlorine is the H center. Here a Cl_2^- ion replaces a single Cl^- on its site. Another proposed defect involving a Cl_2^--like ion is the V_K center. However this center can form even in a stoichiometric alkali halide crystal (one with equal numbers of cations and anions). Here it is assumed that two neighboring Cl^- sites (with a total charge of −2) move closer to each other and chemically bond to form a Cl_2^- ion;[13] this occurs by spontaneously generating a hole in the surrounding medium which is simultaneously trapped and which cancels one unit of charge (the remaining unit of charge being equally distributed between the two Cl atoms making up the ion). Both the H and the V_K centers have similar spectral features which, in turn, bare some similarity to the spectrum of the free Cl_2^- molecular ion.

Two final color centers we will mention are the M center and the R Center. These centers are thought to involve, respectively, a complex of two vacancies with two bound electrons and three vacancies with three bound electrons. For the M center the nearest-neighbor vacancies lie in a (100) plane while for the R center they lie in a (111) plane.

[13] Note that Cl_2 is a strongly bound molecule, as compared to, say, the Na_2 molecule and hence there is some tendency for excess Cl atoms to bind within the lattice.

FURTHER READING

P.G. Shewmon, *Diffusion in Solids*, McGraw-Hill, NY (1963).

F. Seitz, Color centers in alkali halide crystals, *Rev. Mod. Phys.* **18**, 384 (1946); **26**, 7 (1954).

F. Seitz, Speculation on the properties of silver halide crystals, *Rev. Mod. Phys.* **23**, 328 (1951).

S. Chandrasekhar, Stochastic problems in physics and astronomy, *Rev. Mod. Phys.* **43**, 1 (1943).

Some books discussing color centers are: JH Schulman and WD Compton, Color centers in solids, Pergamon Press, 1962; WB Fowler, Ed., Physics of color centers, Academic Press, 1968; JH Crawford, Jr. and LM Slifkin, Eds., Point defects in solids, Plenum Press, 1972. Vol. 1; A.M. Stoneham, Defect and defect processes in nonmetallic solids, W. Hayes, Wiley, 1985.

Dislocations and Grain Boundaries

41

In addition to the point defects described in the previous chapter, crystal lattices may (and generally do) contain other classes of defects; among them are *dislocations and grain boundaries*. Two fundamental dislocations: the *edge* dislocation and the *screw* dislocation will be treated in some detail. The subject of grain boundaries is exceedingly complex and we will limit ourselves to a few remarks at the end of this chapter.

Unlike some of the point defects discussed in the previous chapter, which may be produced by thermal equilibrium processes, the presence of dislocations and grain boundaries generally involves the past history of the crystal. Specifically, exposing the crystal to various external stresses (above some threshold value) will introduce dislocations. In addition, since a crystal grows faster in the presence of dislocations these will present unless the crystal is grown from a dislocation-free seed (a small sample used to initiate growth in some specific direction), and at a very slow rate. On the other hand, rapid cooling from the melt usually leads to the nucleation of multiple crystallites (grains), resulting in a so-called polycrystalline material.

41.1 The edge dislocation

Figure 41.1 shows a two-dimensional cross section through a simple cubic lattice containing a single edge dislocation. A three-dimensional representation is shown in Fig. 41.2, where the line intersections may be interpreted as atom positions. Note that an extra crystallographic plane has been inserted in the crystal; far from the leading edge of the inserted plane this will cause the planes on each side of it to spread by 1/2 a plane spacing, for a total displacement of one inter-plane spacing.

In static equilibrium the net force on each atom vanishes. The positions of the atoms in the immediate vicinity of the leading edge, which we may call the *core* of the dislocation, will depend on the details of the interatomic forces. Since these forces differ widely among materials, and are often not known in detail, evaluation of the increase in the energy density of the crystal in this region cannot be carried out in a general way. However, very far from this leading edge the interatomic

The Physics of Solids. First Edition. J. B. Ketterson.

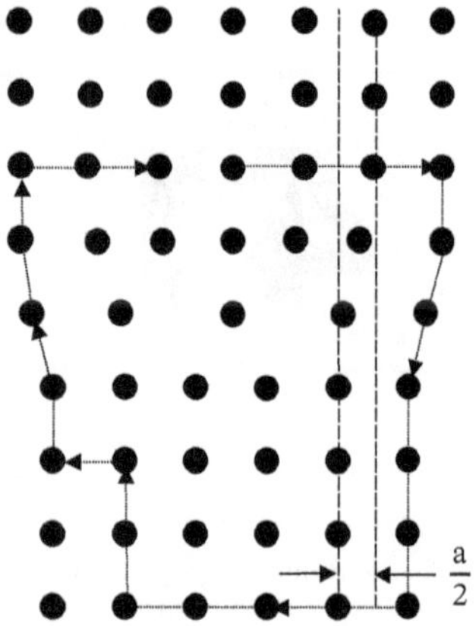

Figure 41.1 *An edge dislocation in a cubic crystal lattice resulting in a lateral expansion of one lattice spacing, a. Shown also is a contour (dotted line) encircling the defect resulting from equal numbers of forward/backward, up/down lattice displacements, which results in a net deficit of one primitive lattice displacement.*

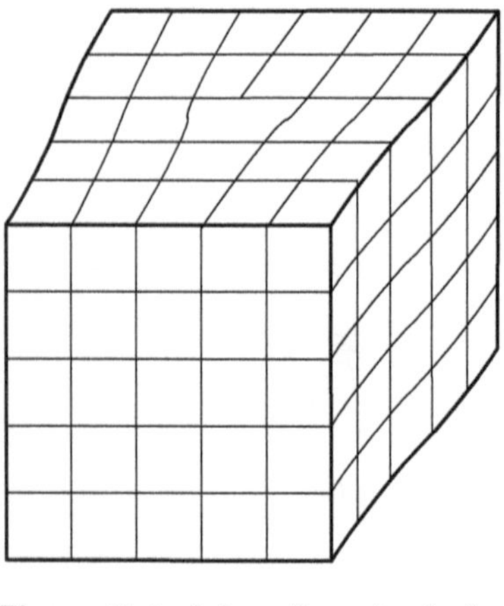

Figure 41.2 *A three-dimensional schematic of an edge dislocation; here we imagine that the atoms are located at the intersection of the grid lines.*

spacings will differ very little from the equilibrium lattice spacings and in this "far field" region we may evaluate the increase in energy density from elasticity theory, as we will see in Section 41.3.

It is useful to introduce a systematic way of determining whether a dislocation is present in the lattice from information which is, in principle, available far from the core. Imagine that we identify some distant atom and follow a contour which passes around the core while everywhere remaining far away from it. Now the position vectors of the lattice sites in the undistorted crystal are given in Chapter 8 as (see Eq. 8.2)

$$\mathbf{R} = n_1\mathbf{a} + n_2\mathbf{b} + n_3\mathbf{c}.$$

If we now constrain ourselves to positions on the *distorted* lattice and follow a contour involving equal numbers of displacements, $\pm n_1$, $\pm n_2$, $\pm n_3$, in such a way that the core is circumscribed, then we find there will be a net *deficit* of one lattice spacing in a direction perpendicular to the inserted plane (if we start at an origin above the added plane). Such a contour is depicted as the light dotted line in Fig. 41.1. In the present case this would be a vector of length a, perpendicular to the dislocation plane. If two lattice planes are inserted from the same side and the contour encloses both cores, this vector will have a length 2a. If, on the other hand, the two planes are inserted from opposite sides and the contour again encloses both cores, the vector will be zero. (If the cores of two such dislocations encounter each other they can disappear as a pair.) As a third example, two planes inserted at right angles will result in a vector of length $\sqrt{2}a$ along a face diagonal of the cube. The vector quantity associated with a general contour (which may enclose many dislocations of different types) is called the *Burgers vector*[1] and it is denoted by **b**. Clearly **b** will be a vector of the form **R** in the primitive lattice.

It is common to idealize the core of an edge dislocation as a line and refer to it in this way. We now consider the special case where we have an array of such lines which lie in a plane and are equally spaced by a distance, d. Now each line displaces the lattice on each side of this plane by a distance, a/2; the cumulative effect from the presence of a succession of such lines is therefore to incline the crystals on each side of the plane by an angle, $\theta = a/d$, with respect to each other. The plane of separation is then referred to as a *low angle grain boundary.*[2]

A little reflection shows that dislocation lines cannot initiate or terminate within a crystal. There are only two possibilities: either the lines terminate at a crystal boundary, or they form closed loops (called dislocation loops). Such loops may have arbitrary shapes and different loops may be entangled.

Dislocations can be introduced and displaced by the action of an external stress field. In particular, when a material is permanently deformed, dislocations are intimately involved. Figure 41.3 shows, schematically, how a material may be irreversibly "bent" by the introduction of successive edge dislocations. Another phenomenon involves the nucleation of an edge dislocation at one surface, and its motion through the crystal to an opposing surface, where it disappears; which

[1] Named after the Dutch physicist Johannes Burgers.

[2] See Section 41.4.

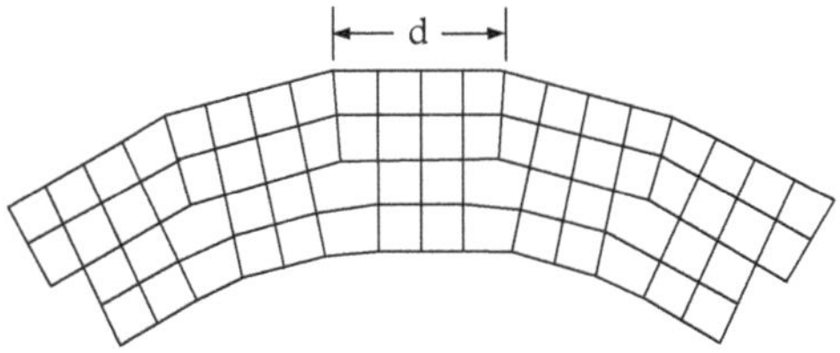

Figure 41.3 *An array of four edge dislocations which facilitate permanent bending in a crystal.*

permits one section of a crystal to displace with respect to a plane, advancing it from another by one lattice constant. A succession of such displacements can lead to a macroscopic displacement of the two parts relative to each other. The phenomenon is referred to as a *slip* and is depicted schematically in Fig. 41.4.

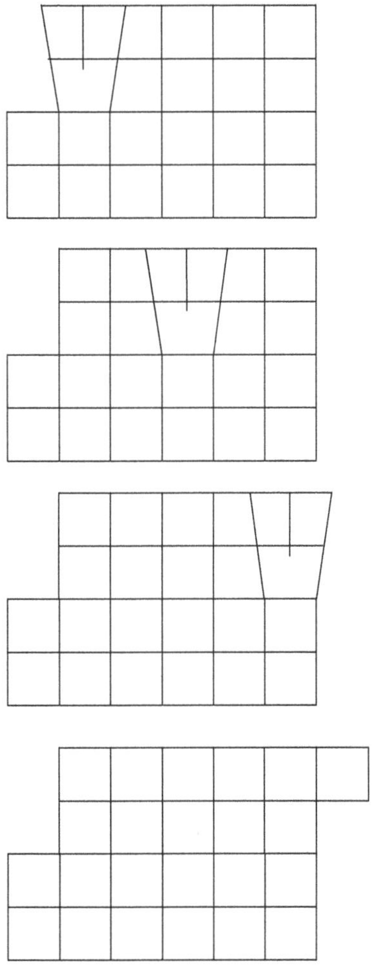

Figure 41.4 *The progression of an edge dislocation leading to a slip by one lattice spacing.*

41.2 The screw dislocation

The second important dislocation we discuss is the screw dislocation; an example of a defect of this kind is shown in Fig. 41.5. To visualize it we imagine that the crystal is cut part way through along some principal lattice plane terminating on a line of atoms and that the pieces of the crystal on each side of this cut are displaced in opposite directions tangent to this line by a total of one lattice spacing. In this case the Burgers vector will be directed along the line of atoms. These defects are often present during the growth of a crystal from the melt. Since atoms from the melt tend to deposit on the crystal along the exposed edge, this causes this edge to spiral around the core during subsequent growth, somewhat like a screw.

41.3 The continuum theory of the dislocation displacement field

For a few particularly simple geometries one can obtain analytic solutions for properties such as the displacement field and total energy of isolated dislocations. Such solutions are interesting as models, but cannot represent the behavior near the core where the displacements are comparable with interatomic spacings and an atomistic treatment becomes necessary. Here we obtain a general equation for the local displacement in an elastically isotropic medium.

In the continuum limit the core becomes a line singularity. Let us construct a closed contour, L, that encircles this singularity. From our previous discussion it is clear that we must impose the constraint that the vector line integral of the displacement around the singularity satisfies

$$\oint_L du_i = \sum_j \oint_L dx_j \frac{\partial u_i}{\partial x_j} = -b_i, \tag{41.1}$$

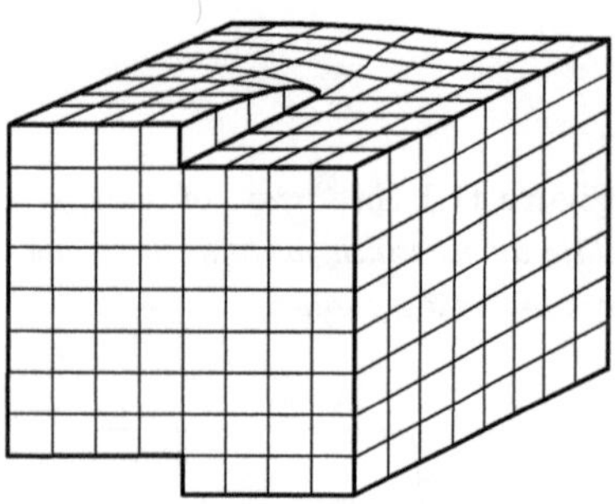

Figure 41.5 *A screw dislocation in a cubic crystal with its Burgess vector parallel to (001). The primitive defect results from cutting part way through the crystal and then displacing the crystal on each side of the cut by one lattice spacing.*

where du_i is the ith component of the differential of the displacement vector, introduced in Chapter 1 (see Eq. 1.2) and b_i is the ith component of the Burgers vector.[3] To simplify the notation we introduce the definition

$$w_{ij} = \frac{\partial u_i}{\partial x_j}, \tag{41.2}$$

which we call the *deformation tensor.* From the definitions given in Eqs. (1.3a) and (1.3b), we may write

$$w_{ij} = \varepsilon_{ij} + \omega_{ij}, \tag{41.3}$$

where

$$\varepsilon_{ij} = \frac{1}{2}\left(\frac{\partial u_i}{\partial x_j} + \frac{\partial u_j}{\partial x_i}\right) = \frac{1}{2}\left(w_{ij} + w_{ji}\right) \tag{41.4a}$$

is the symmetric strain tensor and

$$\omega_{ij} = \frac{1}{2}\left(\frac{\partial u_i}{\partial x_j} - \frac{\partial u_j}{\partial x_i}\right) = \frac{1}{2}\left(w_{ij} - w_{ji}\right) \tag{41.4b}$$

is the antisymmetric rotation tensor. Converting the line integral (41.1) to a surface integral using Stokes' theorem, we have

$$\sum_j \int_L dx_j w_{ji} = \sum_{j,k,l} \int_S ds_j\, e_{jkl} \frac{\partial w_{li}}{\partial x_k} = -b_i; \tag{41.5a}$$

here S is a surface enclosed by the contour, L, ds_i is the ith component of the vector differential element of surface area, and e_{jkl} is the completely antisymmetric third rank tensor, where $e_{123} = -e_{213} = 1$, etc. Now, $\partial w_{li}/\partial x_k = \partial^2 u_i/\partial x_k \partial x_l$, and assuming $u_i(\mathbf{r})$ is twice differentiable outside the core, the quantity $\partial^2 u_i/\partial x_k \partial x_l$ is symmetric in k, l; when contracted with the antisymmetric tensor, e_{jkl}, the result must then vanish. Hence the integrand of the surface integral in (41.5a) vanishes everywhere with the exception of the core, where it is presumed to be singular. In order that the constraint (41.5a) be satisfied we must therefore have

$$\sum_{k,l} e_{jkl} \frac{\partial w_{li}}{\partial x_k} = -n_j b_i \delta^{(2)}(\boldsymbol{\rho}), \tag{41.5b}$$

[3] There is a strong mathematical similarity between the condition (43.1) and the flux quantization condition introduced in connection with the superconducting vortex (see Eq. 32.49).

where n_j denote the components of a unit vector directed along the core of the dislocation and $\delta^{(2)}(\boldsymbol{\rho})$ is a two-dimensional delta function (with $\boldsymbol{\rho}$ denoting two appropriate coordinates lying in the surface, S). Equation (41.5b) is the differential form corresponding to the integral constraint (41.1).

We now derive the differential equation for the displacement vector in the presence of a dislocation. In equilibrium the volume force (introduced in connection with Eq. (1.24)) must vanish; this quantity is given by the divergence of the stress tensor, σ_{ij}, and therefore we require

$$\sum_j \frac{\partial \sigma_{ij}(\mathbf{r}, t)}{\partial x_j} = 0. \tag{41.6}$$

For an elastically isotropic system, σ_{ij} is given as (see Problem 1.3)

$$\sigma_{ij} = \frac{Y}{1+\sigma}\left(\varepsilon_{ij} + \frac{\sigma}{1-2\sigma}\delta_{ij}\sum_k \varepsilon_{kk}\right), \tag{41.7}$$

where Y is Young's modulus and σ is Poisson's ratio. Setting the divergence of (41.7) to zero, we obtain

$$\sum_j \frac{\partial \varepsilon_{ij}}{\partial x_j} + \frac{\sigma}{1-2\sigma}\sum_k \frac{\partial \varepsilon_{kk}}{\partial x_i} = 0. \tag{41.8}$$

Using Eq. (41.4a) we can rewrite (41.8) as

$$\sum_j \frac{1}{2}\left(\frac{\partial w_{ij}}{\partial x_j} + \frac{\partial w_{ji}}{\partial x_j}\right) + \frac{\sigma}{1-2\sigma}\sum_k \frac{\partial w_{kk}}{\partial x_i} = 0. \tag{41.9}$$

Next we rewrite our differential condition (41.5) by multiplying by e_{jim}, summing over j and i, and using the identity, $\sum_j e_{jkl}e_{jim} = \delta_{ki}\delta_{lm} - \delta_{km}\delta_{li}$, to obtain

$$\sum_i \frac{\partial w_{mi}}{\partial x_i} - \sum_i \frac{\partial w_{ii}}{\partial x_m} = -(\mathbf{n}\times\mathbf{b})_m\,\delta^{(2)}(\boldsymbol{\rho}). \tag{41.10}$$

Solving (41.10) for $\sum_i \partial w_{mi}/\partial x_i$, relabeling subscripts (i→j; m→i), and inserting into Eq. (41.9), we obtain

$$\sum_j \frac{\partial w_{ji}}{\partial x_j} + \frac{1}{1-2\sigma}\sum_k \frac{\partial w_{kk}}{\partial x_i} = (\mathbf{n}\times\mathbf{b})_i\,\delta^{(2)}(\boldsymbol{\rho}). \tag{41.11}$$

Inserting the definition (41.2) for w_{ij}, we obtain the differential equation for the displacement u_i as

$$\sum_j \frac{\partial^2 u_i}{\partial x_j^2} + \frac{1}{1-2\sigma}\frac{\partial}{\partial x_i}\sum_k \frac{\partial u_k}{\partial x_k} = (\mathbf{n}\times\mathbf{b})_i\,\delta^{(2)}(\boldsymbol{\rho}) \tag{41.12a}$$

or in vector form

$$\nabla^2 \mathbf{u} + \frac{1}{1-2\sigma} \nabla (\nabla \cdot \mathbf{u}) = \mathbf{n} \times \mathbf{b}\, \delta^{(2)}(\boldsymbol{\rho}). \tag{41.12b}$$

Note that in seeking a solution to Eq. (41.12b) we must still require condition (41.5b) to be satisfied. In Sections 41.3.1 and 41.3.2 we apply this theory to the case of screw and edge dislocations.

41.3.1 Continuum elasticity theory of the isolated screw dislocation

The simplest example of a dislocation is the screw dislocation. To parameterize the strain field we introduce the cylindrical coordinates, ρ, φ, and z, where the z axis is parallel to the dislocation axis and ρ is measured from this axis; we measure the azimuthal angle, φ, from the x axis. By definition, the Burgers vector for this kind of dislocation is given by $b_x = b_y = 0$; $b_z = b$. Choosing the vector, **n**, in (41.5b) to lie in the negative z direction, it follows by symmetry that $u_x = u_y = 0$ and $u_z = b\varphi/2\pi$. The only non-zero component of u_{ij} is

$$u_{z\varphi} = \frac{1}{2}\left(\frac{1}{\rho}\frac{\partial u_z}{\partial \varphi} + \frac{\partial u_\varphi}{\partial z}\right) = b/4\pi\rho; \tag{41.13a}$$

from Eq. (1.43) the corresponding component of the stress tensor is

$$\sigma_{z\varphi} = 2\mu u_{z\varphi} = b/2\pi\rho. \tag{41.13b}$$

Note that the stress is a pure shear.

The total energy is given by

$$\begin{aligned} E &= \frac{1}{2}\int d^3r\, 2\sigma_{z\varphi} u_{z\varphi} \\ &= \frac{\mu b^2 L}{4\pi\rho}\int \frac{d\rho}{\rho} = \frac{\mu b^2 L}{4\pi} ln\left(\frac{L}{b}\right), \end{aligned} \tag{41.14}$$

where we cut the logarithmically divergent integral off at the lower limit at the Burgers length, b, which is of order a lattice spacing, and at the dislocation length, L, at the upper limit; both of these length scales being somewhat arbitrary. We can assign an energy of order $\approx \mu b^2 L$ to the region that has been ignored in the vicinity of the dislocation core; when $ln\,(L/b) \gg 1$ this core energy may be neglected. The logarithmic dependence of the energy on these large and small length scales is characteristic of all line dislocations.[4]

[4] As an example, we note the similarity between the energy per unit length associated with the dislocation and that for the vortex line in a superconductor in Section 32.6, given by Eq. (32.58). Both arise from a line singularity in the medium.

41.3.2 Continuum elasticity theory of the isolated edge dislocation

The treatment of the edge dislocation is somewhat more complex. In this case a 2π circuit in the x–y plane is deficient by a Burgers vector, b, which we assume lies along x. Equation (41.12b) now has the form

$$\nabla^2 \mathbf{u} + \frac{1}{1-2\sigma}\nabla\nabla \cdot \mathbf{u} = -b\mathbf{j}\delta^{(2)}(\boldsymbol{\rho}), \tag{41.15}$$

where $\rho \equiv \sqrt{x^2 + y^2}$. We begin by dividing the vector displacement field into two contributions:

$$\nabla^2 \mathbf{u} + \frac{1}{1-2\sigma}\nabla\nabla \cdot \mathbf{u} = -b\mathbf{j}\delta^{(2)}(\boldsymbol{\rho})\mathbf{u} = \mathbf{u}_1 + \tilde{\mathbf{u}}, \tag{41.16}$$

where $\mathbf{u}_1$ is constructed to satisfy (41.5) and $\tilde{\mathbf{u}}$ is an additional contribution needed to satisfy (41.15). We leave it as an exercise to show that (41.5) is satisfied by a displacement field having the form $(b/2\pi)\varphi\,\mathbf{i}$, and that $\nabla^2\varphi = 0$, where φ is the angle measured from the y axis. However, if instead we choose $\mathbf{u}_1$ to have the form

$$\mathbf{u}_1 = \frac{b}{2\pi}\left[\varphi\,\mathbf{i} + ln(\rho)\,\mathbf{j}\right], \tag{41.17}$$

which has a vanishing curl, and substitute it into (41.15), the resulting equation for $\tilde{\mathbf{u}}$ has the same form as (41.15), with the exception of the strength of the source term; thus

$$\nabla^2 \tilde{\mathbf{u}} + \frac{1}{1-2\sigma}\nabla\nabla \cdot \tilde{\mathbf{u}} = -2b\mathbf{j}\;\delta^{(2)}(\boldsymbol{\rho}), \tag{41.18}$$

where we have used the property, $\nabla^2 ln(\rho) = 2\pi\delta^{(2)}(\rho)$. We now write $\tilde{\mathbf{u}} = \mathbf{u}_2 + \mathbf{u}_3$, where $\mathbf{u}_2$ is chosen to satisfy the equation

$$\nabla^2 \mathbf{u}_2 = -2b\mathbf{j}\;\delta^{(2)}(\boldsymbol{\rho}). \tag{41.19}$$

Equation (41.18) then becomes

$$\nabla^2 \mathbf{u}_3 + \frac{1}{1-2\sigma}\nabla\nabla \cdot \mathbf{u}_3 = -\frac{1}{1-2\sigma}\nabla\nabla \cdot \mathbf{u}_2. \tag{41.20}$$

By introducing a potential, ϕ, according to $\mathbf{u}_3 = \nabla\phi$, we can write (41.20) as

$$\begin{aligned} \nabla^2\nabla\phi + \frac{1}{1-2\sigma}\nabla\left(\nabla \cdot \nabla\phi\right) &= \frac{2(1-\sigma)}{1-2\sigma}\nabla\left(\nabla^2\phi\right) \\ &= -\frac{1}{1-2\sigma}\nabla\left(\nabla \cdot \mathbf{u}_2\right). \end{aligned}$$

Equating the terms under the gradient sign, we have

$$\nabla^2\phi = -\frac{1}{2(1-\sigma)}\nabla\cdot\mathbf{u}_2 = -\frac{2b}{2\pi}\frac{1}{2(1-\sigma)}\mathbf{j}\cdot\nabla ln(\rho). \tag{41.21}$$

We introduce a function, $\psi(\rho)$, through the equation

$$\nabla^2\psi = ln(\rho), \tag{41.22}$$

which on substituting into (41.21) and equating the terms under the Laplacian yields

$$\phi = -\frac{2b}{2\pi}\frac{1}{2(1-\sigma)}\mathbf{j}\cdot\nabla\psi. \tag{41.23}$$

A suitable solution of (41.22) is

$$\psi = \frac{1}{4}\left(\rho^2\, ln(\rho) - \rho^2\right), \tag{41.24}$$

which gives

$$\mathbf{u}_3 = \frac{2b}{2\pi}\frac{1}{2(1-\sigma)}\frac{1}{4}\left[\frac{2xy}{\rho^2}\mathbf{i} - \left(\frac{y^2-x^2}{\rho^2} + 2ln(\rho)\right)\mathbf{j}\right]. \tag{41.25}$$

Including the earlier contributions obtained for $\mathbf{u}_2$ and $\mathbf{u}_3$, our complete expression for the displacement field, $\mathbf{u}$, is

$$\mathbf{u} = \frac{b}{2\pi}\left[\left(\varphi - \frac{1}{4(1-\sigma)}\frac{2xy}{\rho^2}\right)\mathbf{i} - \left(\frac{1-2\sigma}{2(1-\sigma)}ln(\rho) + \frac{1}{4(1-\sigma)}\frac{y^2-x^2}{\rho^2}\right)\mathbf{j}\right]. \tag{41.26}$$

The corresponding stress tensor elements are

$$\sigma_{xx} = \frac{b}{2\pi}\frac{\mu}{1-\sigma}\frac{y\left(3x^2+y^2\right)}{\rho^4}, \tag{41.27a}$$

$$\sigma_{yy} = \frac{b}{2\pi}\frac{\mu}{1-\sigma}\frac{y\left(x^2-y^2\right)}{\rho^4}, \tag{41.27b}$$

and

$$\sigma_{xy} = \frac{b}{2\pi}\frac{\mu}{1-\sigma}\frac{x\left(x^2-y^2\right)}{\rho^4}. \tag{41.27c}$$

One could go on to calculate the total energy which we do not do here; however it will have a logarithmic structure similar in form to (41.14).

41.4 Grain boundaries

The study of interfaces between different materials is a subject of great generality. To set sensible limits, one can restrict enquiry to those interfaces that separate different regions of materials that are in thermodynamic equilibrium with each other, where the chemical potentials of the individual constituents are equal.[5]

If stable interfaces are to exist, the constituents must be able to chemically or physically bond to each other. In a solid this requires interactions at the atomic level extending throughout the interface. Everyday experience tells us that these interactions must generally be established when the material is originally prepared; broken (cracked or fractured) materials undergo local atomic rearrangements and are seldom able to realign at the level required for them to again join with each other.[6] The actual processes involved in fracture form a discipline in themselves.

The individual crystallites making up a macroscopic solid are referred to as grains, and the interfaces which separate them are called *grain boundaries*. Grain boundaries have thermodynamic properties of their own. As an example, it is usually possible to assign an interface energy or surface tension to them. However the description of solid–solid interfaces is much more complex than that associated with fluid surfaces, since the specification of how such crystallites meet involves many more parameters.

At a simpler level, one can consider different phases of a pure element in contact with each other. The simplest case is that of a solid in equilibrium with its own liquid phase. But even for this case the interface energy (surface tension) will depend on the orientation of the particular surface involved, which can be specified by its Miller indices (hkl). If the material has two solid phases, then the interface energy for a locally flat region would involve the crystal structures and the associated Miller indices of both sides; i.e., one must specify the plane orientations, (hk$l$) and (h′k′l'), together with an azimuthal angle measuring their relative in-plane orientation (additional parameters specifying in-plane atomic registry can also enter).[7] But the complete surface of an individual grain embedded in a bulk host must present a minimum of four interfaces (a tetrahedron) to the host (all of which can be different) in order to be fully enclosed. In practice much more complex shapes occur.

Interfaces tend to be detrimental to electrical transport since they will scatter the electrons; however they can add to the mechanical strength by blocking the macroscopic displacement of one part of a crystal with respect to another, caused by crystallographic planes slipping with respect to each other (e.g., by the motion and nucleation of dislocations). Much of the field of metallurgy is concerned with grain boundaries and how their growth (e.g., by annealing) and motion affect the mechanical properties, particularly in alloys.

Grain boundaries can also provide paths for host or foreign atoms to diffuse along in a polycrystalline material. In addition, concentrations of the various atomic or molecular species making up the solid may differ from their bulk values

[5] Keep in mind that the interfaces involved with most useful devices, such as those in semiconductor electronics, are not in thermal equilibrium but are simply stable for long times due to slow interdiffusion of the constituents.

[6] Chemical interactions with environmental gases also inhibit the rejoining of materials.

[7] As will be discussed in Section 47.10, reconstruction can occur at surfaces that further lowers their symmetry.

in the immediate vicinity of an interface. As a consequence, interface adsorbed species sometimes diffuse at significantly higher rates along grain boundaries than they do in the bulk, a phenomenon called *grain boundary diffusion.* This phenomenon complicates the interpretation of measurements aimed at determining a bulk diffusion constant.

A well-known example of grain boundary segregation involves hydrogen in metals; hydrogen can concentrate at the boundaries where it can weaken the bonding, a process called hydrogen embrittlement. Under the action of strain, hydrogen is thought to concentrate at the leading edge of a crack, thereby aiding its propagation and leading to overall fracture.

ADDITIONAL READING

Theory of Dislocations in Solids, F.R.N. Nabarro, Oxford University Press (1967).
Dislocations, J. Friedel, Addison-Wesley (1964).
Elementary Theory of Dislocations, J. Weertman and J.R. Weertman, Macmillan (1964).
Theory of Dislocations, J.P. Hirth and J. Lothe, McGraw-Hill (1968).
Theory of Elasticity, Chapter IV, L.D. Landau and E.M. Lifshitz (1986).
Interfaces in Crystalline Materials, AP Sutton and RW Balluffi, Oxford (1995).
The Nature and Behavior of Grain Boundaries, EW Hart, Plenum, (1972).

Problems

41.1. Using Hooke's law for an isotropic medium, as given by Eq. (1.38), verify the elements of the stress tensor for an edge dislocation given in Eqs. (41.27a–c) from the displacement vector given by Eq. (41.25).

41.2. Show that in cylindrical coordinates, $\nabla^2 \ln \rho = 2\pi\delta^{(2)}(\boldsymbol{\rho})$, where ρ is the radius vector and $\delta^{(2)}(\boldsymbol{\rho})$ is a two-dimensional delta function.

41.3. Show that for an edge dislocation, Eq. (41.5) is satisfied by a displacement field of the form $\mathbf{u} = (b/2\pi)\varphi\mathbf{i}$ and that $\nabla^2\varphi = 0$ where φ is the angle measured from the y-axis.

41.4. There is a simpler treatment of the screw dislocation than the formal one given in Section 41.5. We picture the dislocation as a series of nested cylinders each formed by rolling up a rectangular platelet; in addition, each platelet is sheared along the screw axis by the Burgers vector, b, on going around the circumference of the cylinders so formed. First calculate the shear energy of an individual platelet of length, L, width, $2\pi\rho$ (where ρ is its radius), and thickness, $d\rho$. Then sum (integrate) the energies of successive cylinders from an inner to an outer radius to obtain the total energy.

Quantum Theory of Electrical Transport in Dilute Alloys

42

The approach used to treat electron transport phenomena in Chapter 19 was based on the semi-classical Boltzmann transport equation. Other than implicitly assuming plane-wave-like electron states, as numbered via their wave vectors, **k**, the primary place where quantum mechanics entered the model was in the use of the Fermi–Dirac distribution function for filling these states, so as to systematically incorporate the Pauli exclusion principle.

Here we will develop a more fully quantum mechanical treatment of electrical transport, utilizing the relatively straightforward approach of Mattis and Bardeen.[1,2,3] As we will see, it is based on averaging a quantity involving a position-dependent correlation of the single particle wave functions, that enters the quantum mechanical expression for the current density.[4] After introducing a simple assumption for the mathematical form of this correlation we will recover some results obtained in Chapter 19 that were based on the Boltzmann equation. In particular we will re-derive the Chambers expression for the conductivity.

The Mattis–Bardeen approach is limited to a single particle picture and will ultimately be restricted to plane wave like states. In spite of this limitation the approach sets the stage for examining impurity scattering at a deeper level, specifically phenomena that arise in highly disordered systems, which we take up in Chapter 43. There we will find that an explicitly quantum mechanical property, that wave functions have a phase, will play an important role.

42.1 The vector potential as a time-dependent perturbation

As discussed in Section 18.1, the vector, **A**, and scalar, ϕ, potentials are the fundamental quantities entering any quantum mechanical description of the electromagnetic response of matter. In the models to be discussed, we will take the charge density as being uniform, in which case the scalar potential can be set to zero, which is consistent with the choice of the gauge, $\nabla \cdot \mathbf{A} = 0$, that we will utilize. The electric and magnetic fields are then given by

$$\mathbf{E}(\mathbf{r}, t) = -\frac{1}{c}\frac{\partial \mathbf{A}(\mathbf{r}, t)}{\partial t}; \tag{42.1a}$$

$$\mathbf{H}(\mathbf{r}, t) = \nabla \times \mathbf{A}(\mathbf{r}, t). \tag{42.1b}$$

[1] Theory of the anomalous skin effect in normal and superconducting metals, D.C. Mattis and J. Bardeen, *Phys. Rev.* **111**, 412 (1958). Here, our discussion will be limited to normal metals; however it can straightforwardly be extended to superconductors to treat the Meissner effect and microwave absorption.

[2] *Conductivity Problems Quantum Theory*, Daniel Charles Mattis, Thesis, University of Illinois (1957) (available through ProQuest).

[3] For an alternative approach based on the density matrix which gives the same result, see: Anomalous Skin Effect in Metals, D.C. Mattis and G. Dresselhaus, *Phys. Rev.* **111**, 403 (1958).

[4] A more rigorous description of quantum transport is provided by the Green's function approach, which we do not take up in this book.

The Physics of Solids. First Edition. J. B. Ketterson.

If the external fields are sufficiently weak, we may regard the vector potential as a perturbation in the time-dependent Schrödinger equation,[5]

$$\hat{H}\Psi(\mathbf{r},t) = i\hbar\frac{\partial\Psi(\mathbf{r},t)}{\partial t},$$

where

$$\hat{H} = \frac{1}{2m}\left(\frac{\hbar}{i}\nabla - \frac{e}{c}\mathbf{A}(\mathbf{r},t)\right)^2 + U(\mathbf{r})$$

and $U(\mathbf{r})$ is the crystal potential, including the effects of impurities, as we will elaborate on further. We expand the one-electron time-dependent wave functions to first order as $\Psi_n(\mathbf{r},t) \cong \Psi_n^{(0)}(\mathbf{r},t) + \Psi_n^{(1)}(\mathbf{r},t)$, where $\Psi_n^{(0)}(\mathbf{r},t) = \psi_n^{(0)}(\mathbf{r})\,e^{-(i/\hbar)\varepsilon_n^{(0)}t}$ are the states in the absence of the external perturbation, $\mathbf{A}(\mathbf{r},t)$, but in the presence of impurities; i.e.,

$$\left[\frac{\hbar^2}{2m}\nabla^2 + U(\mathbf{r})\right]\Psi_n^{(0)}(\mathbf{r}) = \varepsilon_n\Psi_n^{(0)}(\mathbf{r}).$$

The first-order perturbation in the wave function, $\Psi_n^{(1)}(\mathbf{r},t)$, arising from $\mathbf{A}(\mathbf{r},t)$ then follows from the solution of the inhomogeneous Schrödinger-like equation,

$$\begin{aligned}\left[i\hbar\frac{\partial}{\partial t} + \frac{\hbar^2}{2m}\nabla^2 - U(\mathbf{r})\right]\Psi_n^{(1)}(\mathbf{r},t) &= \frac{ie\hbar}{2mc}\left[\mathbf{A}(\mathbf{r},t)\cdot\nabla\Psi_n^{(0)}(\mathbf{r},t) + \nabla\cdot\left(\mathbf{A}(\mathbf{r},t)\Psi_n^{(0)}(\mathbf{r},t)\right)\right]\\ &= \frac{ie\hbar}{mc}\mathbf{A}(\mathbf{r},t)\cdot\nabla\Psi_n^{(0)}(\mathbf{r},t),\end{aligned} \tag{42.2}$$

where we have adopted the Lorentz–London gauge, $\nabla\cdot\mathbf{A} = 0$, in the second step. We now expand $\Psi_n^{(1)}(\mathbf{r},t)$ in terms of the complete set of time-dependent eigenstates, $\psi_l^{(0)}(\mathbf{r})e^{-(i/\hbar)\varepsilon_l t}$, that are formally zeroth order in the vector potential, $\mathbf{A}(\mathbf{r},t)$, where again they include the effects of impurities:

$$\Psi_n^{(1)}(\mathbf{r},t) = \sum_l a_{ln}(t)\,\psi_l^{(0)}(\mathbf{r})e^{-\frac{i}{\hbar}\varepsilon_l^{(0)}t}. \tag{42.3}$$

Inserting (42.3) into (42.2), multiplying by $\psi_m^{(0)*}(\mathbf{r})$, integrating over all $\mathbf{r}$, and using the orthonormality of the wave functions yields

$$i\hbar\dot{a}_{mn}(t) = \frac{ie\hbar}{mc}e^{-\frac{i}{\hbar}\left(\varepsilon_n^{(0)}-\varepsilon_m^{(0)}\right)t}\int d^3r\,\psi_m^{(0)}(\mathbf{r})\mathbf{A}(\mathbf{r},t)\cdot\nabla\psi_n^{(0)}(\mathbf{r}). \tag{42.4}$$

[5] The fact that Ohm's law is a linear relation between the perturbation (the electric field) and the response (the electric current) provides assurance that first-order perturbation will be sufficient to obtain this behavior.

In order to proceed with the discussion we will assume that $\mathbf{A}(\mathbf{r}, t)$ involves only a single frequency,[6] ω and write it in the strictly real form

$$\mathbf{A}(\mathbf{r}, t) = \lim_{\delta \to 0} \frac{1}{2} \left[\mathbf{A}(\mathbf{r}) e^{-i\omega t + \delta t} + \mathbf{A}^*(\mathbf{r}) e^{+i\omega t + \delta t} \right]; \qquad (42.5)$$

the factor, $e^{\delta t}$, assures that the perturbation vanishes for $t = -\infty$ and will be adiabatic in the limit $\delta \to +0$ Carrying out the integration of (42.4) we obtain

$$a_{mn} = \frac{ie\hbar}{2mc} e^{-\frac{i}{\hbar}\left(\varepsilon_n^{(0)} - \varepsilon_m^{(0)}\right)t} \left[\frac{e^{-i\omega t}}{\varepsilon_n^{(0)} - \varepsilon_m^{(0)} + \hbar\omega + i0} \int d^3r\, \psi_m^{(0)*}(\mathbf{r}) \mathbf{A}(\mathbf{r}) \cdot \nabla \psi_n^{(0)}(\mathbf{r}) \right.$$

$$\left. + \frac{e^{+i\omega t}}{\varepsilon_n^{(0)} - \varepsilon_m^{(0)} - \hbar\omega + i0} \int d^3r\, \psi_m^{(0)*}(\mathbf{r}) \mathbf{A}^*(\mathbf{r}) \cdot \nabla \psi_n^{(0)}(\mathbf{r}) \right], \qquad (42.6)$$

where 0 denotes the positive infinitesimal, which we have set to zero in the exponentials.

42.2 Relation between the current density and the vector potential

The quantum mechanical expression for the electrical current in the one-electron approximation is given by

$$j(\mathbf{r}, t) = -\frac{e\hbar}{2im} \sum_n f_n \left[\Psi_n^*(\mathbf{r}, t) \nabla \Psi_n(\mathbf{r}, t) - \Psi_n(\mathbf{r}, t) \nabla \Psi_n^*(\mathbf{r}, t) \right]$$

$$- \frac{e^2}{mc} \mathbf{A}(\mathbf{r}, t) \sum_n f_n \left| \Psi_n(\mathbf{r}, t) \right|^2,$$

where $f_n = f(\varepsilon_n)$ is the Fermi function for the occupation probability[7] of the states. To first order in $\mathbf{A}(\mathbf{r}, t)$, the current is

$$j(\mathbf{r}, t) = -\frac{e\hbar}{2im} \sum_n f_n \left[\Psi_n^{(0)*}(\mathbf{r}, t) \nabla \Psi_n^{(1)}(\mathbf{r}, t) + \Psi_n^{(1)*}(\mathbf{r}, t) \nabla \Psi_n^{(0)}(\mathbf{r}, t) \right.$$

$$\left. - \Psi_n^{(0)}(\mathbf{r}, t) \nabla \Psi_n^{(1)*}(\mathbf{r}, t) - \Psi_n^{(1)}(\mathbf{r}, t) \nabla \Psi_n^{(0)*}(\mathbf{r}, t) \right].$$

$$- \frac{e^2}{mc} \mathbf{A}(\mathbf{r}, t) \sum_n f_n \Psi_n^{(0)*}(\mathbf{r}, t) \Psi_n^{(0)*}(\mathbf{r}, t). \qquad (42.7)$$

On substituting (42.3) we obtain $j(\mathbf{r}, t)$ to first order as

$$j(\mathbf{r}, t) = -\frac{e\hbar}{2im} \sum_{n,m} f_n \left\{ \left[\psi_n^*(\mathbf{r}) \nabla \psi_m(\mathbf{r}) - \psi_m(\mathbf{r}) \nabla \psi_n^*(\mathbf{r}) \right] e^{+\frac{i}{\hbar}(\varepsilon_n - \varepsilon_m)t} a_{mn}(t) \right.$$

$$\left. - \left[\psi_n(\mathbf{r}) \nabla \psi_m^*(\mathbf{r}) - \psi_m^*(\mathbf{r}) \nabla \psi_n(\mathbf{r}) \right] e^{-\frac{i}{\hbar}(\varepsilon_n - \varepsilon_m)t} a_{mn}^*(t) \right\} - \frac{ne^2}{mc} \mathbf{A}(\mathbf{r}, t), \qquad (42.8)$$

[6] An arbitrary time variation may then be formed by a Fourier superposition of various frequencies.

[7] Note we have taken f_n as time independent here, which is valid only in lowest order.

where in the last term, $n = \sum_n f_n \psi_n^* \psi_n = N/L^3$ is the number density of electrons and we have dropped the designations of zeroth order quantities.[8] Note that a_{mn} is a *functional* of $\mathbf{A}(\mathbf{r}, t)$ through (42.6); hence $\mathbf{j}(\mathbf{r},t)$ is a functional of $\mathbf{A}(\mathbf{r}, t)$.[9] Writing $\mathbf{j}(\mathbf{r}, t)$ in a form analogous to (42.5) as

$$\mathbf{j}(\mathbf{r}, t) = \frac{1}{2}\left[\mathbf{j}(\mathbf{r})e^{-i\omega t} + \mathbf{j}^*(\mathbf{r})e^{+i\omega t}\right], \tag{42.9}$$

we can rewrite (42.8) and (42.9) in the non-local form[10]

$$\mathbf{j}(\mathbf{r}, \omega) = \int d^3\mathbf{r}'\, \mathbf{s}\left(\mathbf{r}, \mathbf{r}'; \omega\right) \cdot \mathbf{A}\left(\mathbf{r}', \omega\right), \tag{42.10}$$

where

$$s_{ij}(\mathbf{r}, \mathbf{r}', \omega)$$

$$= -\frac{e^2\hbar^2}{2m^2c}\sum_{n,m} f_n \left\{ \left[\psi_n^*(\mathbf{r})\frac{\partial}{\partial x_i}\psi_m(\mathbf{r}) - \psi_m(\mathbf{r})\frac{\partial}{\partial x_i}\psi_n^*(\mathbf{r})\right] \frac{\psi_m^*(\mathbf{r}')\frac{\partial}{\partial x'_j}\psi_n(\mathbf{r}')}{\varepsilon_n - \varepsilon_m + \hbar\omega + i0} \right.$$

$$\left. + \left[\psi_n(\mathbf{r})\frac{\partial}{\partial x_i}\psi_m^*(\mathbf{r}) - \psi_m^*(\mathbf{r})\frac{\partial}{\partial x_i}\psi_n(\mathbf{r})\right] \frac{\psi_m(\mathbf{r}')\frac{\partial}{\partial x'_j}\psi_n^*(\mathbf{r}')}{\varepsilon_n - \varepsilon_m - \hbar\omega - i0} \right\} - \frac{ne^2}{mc}\delta_{ij}. \tag{42.11}$$

42.3 Case of a uniform electric field: the local conductivity

As an initial application of (42.11) let us assume that the vector potential is spatially uniform, i.e., $\mathbf{A}\left(\mathbf{r}', \omega\right) \rightarrow \mathbf{A}(\omega)$, in which case it factors out of the integral in (42.10) and we can carry out the integration directly, over $d^3\mathbf{r}'$. Furthermore, in considering the conductivity of a system involving impurities, particular local configurations are not of interest so we will take an average over some volume, V, that is large compared with typical inter-impurity spacings; i.e., we are interested in the form

$$\bar{\mathbf{j}}(\omega) = \bar{\mathbf{s}}(\omega) \cdot \mathbf{A}(\omega), \tag{42.12a}$$

where this average is denoted as

$$\bar{\mathbf{s}}(\omega) = \frac{1}{V}\int d^3r d^3\mathbf{r}'\, \mathbf{s}(\mathbf{r}, \mathbf{r}', \omega). \tag{42.12b}$$

Recalling that $\hat{\mathbf{p}} = (\hbar/i)\nabla$ is a Hermitian operator with matrix elements satisfying $(\hat{\mathbf{p}})_{nm} = (\hat{\mathbf{p}})^*_{mn}$, (42.11) becomes[11]

$$\bar{s}_{ij}(\omega) = \frac{e^2}{m^2cV}\sum_{n,m} f_n \left\{ \frac{(\hat{p}_i)_{nm}(\hat{p}_j)_{mn}}{\varepsilon_n - \varepsilon_m + \hbar\omega + i0} + \frac{(\hat{p}_i)_{mn}(\hat{p}_j)_{nm}}{\varepsilon_n - \varepsilon_m - \hbar\omega - i0} \right\} - \frac{ne^2}{mc}\delta_{ij}.$$

8 Formally, we should include a factor, $1-f_m$, to ensure that the intermediate states are unoccupied. It turns out that the additional terms that are generated cancel, so for convenience we will not include this factor.

9 One also says that the relation between **A** and **j** is *non-local* in that the current at some point depends on the vector potential at neighboring points.

10 This equation is related to the so-called Green–Kubo relations: Markoff random processes and the statistical mechanics of time-dependent phenomena. II. Irreversible processes in fluids, M.S. Green, *J. Chem. Phys.* **22**, 98 (1954); Statistical-mechanical theory of irreversible processes. I. General theory and simple applications to magnetic and conduction problems, R. Kubo, *J. Phys. Soc. Jpn.* **12**, 570 (1957).

11 To establish this, one must argue that the surface integral associated with the integration by parts vanishes; here we invoke periodic boundary conditions or growing incoherence associated with scattering.

Writing $\bar{\mathbf{j}}(\omega) = \bar{\sigma}(\omega) \cdot \mathbf{E}(\omega)$, using (42.1a) (with ω taken as $\omega + i0$), and defining a velocity operator, $\hat{\mathbf{v}} = \hat{\mathbf{p}}/m$, we can introduce a local conductivity, $\bar{\sigma}_{ij}(\omega) = (c/i(\omega + i0))\ \bar{s}_{ij}(\omega)$, as

$$\bar{\sigma}_{ij}(\omega) = \frac{e^2}{i(\omega + i0)V} \sum_{n,m} f_n \left\{ \frac{\langle n|\,\hat{v}_i\,|m\rangle\,\langle m|\,\hat{v}_j\,|n\rangle}{\varepsilon_n - \varepsilon_m + \hbar\omega + i0} + \frac{\langle m|\,\hat{v}_i\,|n\rangle\,\langle n|\,\hat{v}_j\,|m\rangle}{\varepsilon_n - \varepsilon_m - \hbar\omega - i0} \right\} - \frac{ne^2}{i(\omega + i0)m}\delta_{ij}. \tag{42.13}$$

To discuss this equation, we start by recalling our expression for the electrical conductivity, obtained earlier using the Boltzmann equation for single-band Bloch-like states,

$$\sigma_{ij} = -\frac{2e^2\tau}{(2\pi)^3} \int d^3k v_i v_j \frac{\partial f(\varepsilon)}{\partial \varepsilon}. \tag{19.22}$$

Writing $(\partial f(\varepsilon)/\partial\varepsilon)\, v_j = \partial f(\varepsilon)/\partial(\hbar k_j)$, this expression becomes

$$\sigma_{ij} = -\frac{2e^2\tau}{(2\pi)^3} \int d^3k \frac{\partial \varepsilon}{\partial(\hbar k_i)} \frac{\partial f_0(\varepsilon)}{\partial(\hbar k_j)} = \frac{2e^2\tau}{(2\pi)^3\hbar^2} \int d^3k \frac{\partial^2 \varepsilon}{\partial k_i \partial k_j} f(\varepsilon), \tag{42.14}$$

where we have performed an integration by parts in the second step. Using Eq. (10.42) we have the energy through second order in $\mathbf{k}$ as

$$\varepsilon_k \cong \frac{\hbar^2}{2m}k^2 + \hbar^2 \sum_{m \neq n} \frac{\left(\mathbf{k} \cdot (\hat{\mathbf{v}})_{nm}\right)\left((\hat{\mathbf{v}})_{mn} \cdot \mathbf{k}\right)}{\varepsilon_n - \varepsilon_m}. \tag{42.15}$$

Differentiating (42.15) twice with respect to $\mathbf{k}$ and combining it with (42.14) gives

$$\sigma_{ij} = \frac{ne^2\tau}{m}\delta_{ij} + \frac{e^2\tau}{V} \sum_{n,m} f_n \left\{ \frac{\langle n|\,\hat{v}_i\,|m\rangle\langle m|\,\hat{v}_j\,|n\rangle}{\varepsilon_n - \varepsilon_m} + \frac{\langle n|\,\hat{v}_j\,|m\rangle\langle m|\,\hat{v}_i\,|n\rangle}{\varepsilon_n - \varepsilon_m} \right\}. \tag{42.16}$$

The first term in (42.15) is the free electron Drude form, while the second may be viewed as a correction arising from replacing 1/m by the reciprocal effective mass tensor (see Eq. (10.37b)).

We can make contact between Eq. (43.16) and the $\omega \to 0$ limit of (42.13) by replacing i0 with i/τ; this is not unreasonable on phenomenological grounds since the "past is forgotten" in the integration leading to Eq. (42.6) in a time of order, τ, due to scattering (we will introduce a better model later). Note that at finite frequencies such that $\omega \gg 1/\tau$, but less than the band gap, the conductivity becomes purely imaginary, implying that the electric field and current move out of phase with each other, which is non dissipative. To understand this we note that it is common to incorporate the effects of a medium in Maxwell's fourth equation, through either $\mathbf{j}$ or $\partial\mathbf{P}/\partial t$; it is to some extent arbitrary whether we do it through one or the other. An imaginary current can then be regarded as arising from a time-dependent field-induced electrical polarization that is in turn characterized by a dielectric constant.

An alternative way to take the zero frequency limit of the conductivity is to expand the denominators to first order in ω (where here we suppress the infinitesimal),

$$\frac{1}{\varepsilon_n - \varepsilon_m \pm \hbar\omega} \cong \frac{1}{\varepsilon_n - \varepsilon_m}\left(1 \mp \frac{\hbar\omega}{\varepsilon_n - \varepsilon_m}\right),$$

which on inserting in (42.13) gives

$$\bar{\sigma}_{ij}(\omega) = \frac{e^2}{i\omega V}\sum_{n,m} f_n \left[\frac{\langle n|\hat{v}_i|m\rangle\langle m|\hat{v}_j|n\rangle + \langle n|\hat{v}_i|m\rangle\langle m|\hat{v}_j|n\rangle}{\varepsilon_n - \varepsilon_m}\right.$$
$$\left. + \hbar\omega\frac{\langle n|\hat{v}_i|m\rangle\langle m|\hat{v}_j|n\rangle - \langle n|\hat{v}_i|m\rangle\langle m|\hat{v}_j|n\rangle}{(\varepsilon_n - \varepsilon_m)^2}\right] - \frac{ne^2}{i\omega m}\delta_{ij}. \quad (42.17)$$

From the Heisenberg equation of motion we can write $\langle n|\hat{v}_i|m\rangle = (i/\hbar)(\varepsilon_n - \varepsilon_m)\langle n|\hat{x}_i|m\rangle$. Combining this with the commutation relation, $[\hat{p}_i, \hat{x}_j] = i\hbar\delta_{ij}$, and the completeness relation, $\sum_m |m\rangle\langle m| = 1$, we see that the first term in (42.17) cancels the last. We are then left with[12]

$$\bar{\sigma}_{ij}(\omega) = \frac{ie^2\hbar}{V}\sum_{n,m} f_n \frac{\langle n|\hat{v}_i|m\rangle\langle m|\hat{v}_j|n\rangle - \langle n|\hat{v}_i|m\rangle\langle m|\hat{v}_j|n\rangle}{(\varepsilon_n - \varepsilon_m)^2}$$
$$= \frac{ie^2\hbar}{V}\sum_{n,m}(f_n - f_m)\frac{\langle n|\hat{v}_i|m\rangle\langle m|\hat{v}_j|n\rangle}{(\varepsilon_n - \varepsilon_m)^2}, \quad (42.18)$$

where we have symmetrized with respect to n and m in the second step. Note that the diagonal elements of (42.18) vanish; hence it includes no effects of scattering and only enters into discussions of the Hall effect, which we do not pursue here.

42.4 A non-local model

Our introduction of a relaxation time into Eq. (42.13), by replacing i0 with i/τ, was rather artificial and here we will develop the alternative model used by Mattis and Bardeen, mentioned previously. We start by noting that we can integrate the second and fourth terms in Eq. (42.11) by parts in the variable, r', since it extends over all space;[13] if in addition we symmetrize with respect to the dummy indices, m, n, it becomes

$$s_{ij}(\mathbf{r},\mathbf{r}',\omega) = -\frac{e^2\hbar^2}{4m^2c}\sum_{n,m}(f_n - f_m)\left\{\frac{\psi_n^*(\mathbf{r})\frac{\partial}{\partial x_i}\psi_m(\mathbf{r})\,\psi_m^*(\mathbf{r}')\frac{\partial}{\partial x'_j}\psi_n(\mathbf{r}')}{\varepsilon_n - \varepsilon_m + \hbar\omega + i0}\right.$$
$$\left. + \frac{\psi_m^*(\mathbf{r})\frac{\partial}{\partial x_i}\psi_n(\mathbf{r})\,\psi_n^*(\mathbf{r}')\frac{\partial}{\partial x'_j}\psi_m(\mathbf{r}')}{\varepsilon_n - \varepsilon_m - \hbar\omega - i0}\right\} - \frac{ne^2}{mc}\delta_{ij}\delta(\mathbf{r}-\mathbf{r}'). \quad (42.19)$$

12 Quantized Hall conductance in a two-dimensional periodic potential, D.J. Thouless, M. Kohmoto, M.P. Nightingale, and M. den Nijs, *Phys. Rev. Lett.* **49**, 406 (1982).

13 Here we ignore the effect of boundaries, which can greatly complicate the discussion.

Our goal here is to develop an approximate theory of the effects of impurity scattering on the electrical conductivity. The true one-electron wave functions will involve the positions of *all the impurities*; i.e., $\psi_n = \psi_n(\mathbf{r}; \{\mathbf{R}_\alpha\})$, where $\{\mathbf{R}_\alpha\}$ denotes the set of all impurity coordinates as parameters. These would be different for every material prepared (or different macroscopic volume elements of the same material), even though the concentrations and preparation conditions are nominally identical. Under most experimental conditions the positioning of the coordinates, $\{\mathbf{R}_\alpha\}$, would be unobservable[14] and therefore one seeks a description where these positions are *averaged over*. Now it is not meaningful to average the one-electron wave functions of the individual electrons, since they oscillate rapidly on a scale dictated by their de Broglie wavelengths; such an average wave function would not extend throughout the crystal and hence the associated current flux would vanish, which is unphysical. More meaningful is to seek an average of the wave functions *with themselves*, i.e., *their amplitude correlation function*. Such a measure is provided by the density matrix in the position representation associated with the state, n, which is defined as

$$\rho_n\left(\mathbf{r},\mathbf{r}';\{\mathbf{R}_\alpha\}\right) = \psi_n^*\left(\mathbf{r};\{\mathbf{R}_\alpha\}\right)\psi_n\left(\mathbf{r}';\{\mathbf{R}_\alpha\}\right), \tag{42.20}$$

in terms of which we can rewrite (42.19) as (before averaging)

$$s_{ij}\left(\mathbf{r},\mathbf{r}',\omega\right) = -\frac{e^2\hbar^2}{2m^2c}\sum_{n,m}\left(f_n - f_m\right)\left\{\frac{\left(\frac{\partial}{\partial x_i}\rho_m\left(\mathbf{r}',\mathbf{r};\{\mathbf{R}_\alpha\}\right)\right)\left(\frac{\partial}{\partial x'_j}\rho_n\left(\mathbf{r},\mathbf{r}';\{\mathbf{R}_\alpha\}\right)\right)}{\varepsilon_n - \varepsilon_m + \hbar\omega + i0}\right.$$
$$\left. + \frac{\left(\frac{\partial}{\partial x'_j}\rho_n\left(\mathbf{r}',\mathbf{r};\{\mathbf{R}_\alpha\}\right)\right)\left(\frac{\partial}{\partial x_i}\rho_m\left(\mathbf{r},\mathbf{r}';\{\mathbf{R}_\alpha\}\right)\right)}{\varepsilon_n - \varepsilon_m - \hbar\omega - i0}\right\} - \frac{ne^2}{mc}\delta_{ij}\delta\left(\mathbf{r}-\mathbf{r}'\right). \tag{42.21}$$

In performing a configuration average of (42.21), we will assume that the average of the forms containing density matrix products can be approximated as a product of their averages; for the product entering the first term we have

$$\overline{\left(\frac{\partial}{\partial x_i}\rho_m(\mathbf{r}',\mathbf{r};\{\mathbf{R}_\alpha\})\right)\left(\frac{\partial}{\partial x'_j}\rho_n\left(\mathbf{r},\mathbf{r}';\{\mathbf{R}_\alpha\}\right)\right)}$$
$$= \left(\frac{\partial}{\partial x_i}\overline{\rho_m\left(\mathbf{r}',\mathbf{r};\{\mathbf{R}_\alpha\}\right)}\right)\left(\frac{\partial}{\partial x'_j}\overline{\left(\rho_n\left(\mathbf{r},\mathbf{r}';\{\mathbf{R}_\alpha\}\right)\right)}\right), \tag{42.22}$$

and similarly for the product in the second term.

[14] We note that under special circumstances and at very low temperatures, quantum interference effects associated with the change in the coordinates of a single impurity can be observed.

42.5 A free electron density matrix with scattering: the Bardeen ansatz

Formally, point impurities destroy the microscopic invariance of a system under the primitive lattice operations involving translations and rotations, although they are still preserved in some average sense. Even when assuming that impurities are located randomly on lattice sites, calculating the configuration averages entering (42.22) in real systems is a very complicated problem, involving multiple scattering theory for band electrons.[15] Our goal here will be limited to extracting the general features of electrical transport in the presence of impurities at low temperatures, without reference to any specific system. Toward this end it will be sufficient to assume that, in the absence of impurity scattering, we have free-electron-like states with quantum numbers, **k**. When impurities are present, we lose the full translational and rotational symmetry of the electron gas. Now, at low temperatures where electron–phonon scattering is suppressed the scattering is largely elastic. Electrons then sample (scatter among) states lying on an energy shell, $\varepsilon(\mathbf{k})$, on a time scale measured by the relaxation time, τ. The configuration averaged quantity, $\overline{\rho_n(\mathbf{r},\mathbf{r}';\{\mathbf{R}_\alpha\})}$, then involves an average over all states, n, with the same energy and, furthermore, adopts the translationally and rotationally invariant form $\rho(R,\varepsilon)$, where $R = |\mathbf{r}-\mathbf{r}'|$.

Our task is now to construct a reasonable form for $\rho(R,\varepsilon)$. If we write our plane wave states as $\varphi_{\mathbf{k}}(\mathbf{r}) = e^{i\mathbf{k}\cdot\mathbf{r}}$, the associated density matrix in the absence of scattering is $\rho_{\mathbf{k}}(\mathbf{r}',\mathbf{r}) = \varphi^*_{\mathbf{k}}(\mathbf{r})\varphi_{\mathbf{k}}(\mathbf{r}') = e^{-i\mathbf{k}\cdot\mathbf{r}}e^{i\mathbf{k}\cdot\mathbf{r}'}$, which we write as $\rho_{\mathbf{k}}(\mathbf{R}) = e^{-i\mathbf{k}\cdot\mathbf{R}}$, where $\mathbf{R} \equiv \mathbf{r}-\mathbf{r}'$. Note that for free electrons we have the property, $\partial\rho_{\mathbf{k}}(\mathbf{R})/\partial\mathbf{r} = \partial\rho_{\mathbf{k}}(\mathbf{R})/\partial\mathbf{R} = -\partial\rho_{\mathbf{k}}(\mathbf{R})/\partial\mathbf{r}'$, which we can use to combine the numerators in (42.21). Since elastic scattering redistributes electrons with energies, $\varepsilon(\mathbf{k})$, over the associated energy shell, $\varepsilon(k)$, as a *first step* in representing the effects of configuration averaging, we integrate over all angles,

$$\rho(R,\varepsilon) = \frac{1}{4\pi}\int d\Omega\,\rho_{\mathbf{k}}(\mathbf{R}) = \frac{1}{4\pi}\int d\Omega\, e^{-i\mathbf{k}\cdot\mathbf{R}} = \frac{\sin kR}{kR}. \tag{42.23}$$

Formally, Eq. (42.23) does not contain the effects of real scattering, since it is only the angular average of that part of the sum over states involved with the free electron density matrix. To explicitly include the effects of scattering we will draw upon a result from the more formal multiple scattering methods which replace (42.23) with the following expression which we have taken from Bardeen:[16,17]

$$\rho(R,\varepsilon) = \frac{\sin kR}{kR}e^{-R/2l}, \tag{42.24}$$

[15] Accompanying this problem are all the issues identified in Section 19.11.

[16] This result was obtained by Bardeen by using a simple argument concerning the transmission of a plane wave through a slab of material containing impurities: J. Bardeen, *Encyclopedia of Physics* (Springer Verlag, Berlin, 1956), Vol. 15, p. 274. It was obtained formally by P.R. Weiss and E. Abrahams in, Correlation of electron amplitudes in impure metals, *Phys. Rev.* **111**, 722 (1958). It is also contained in Multiple scattering of waves, M. Lax, *Rev. Mod. Phys.* **23**, 287 (1951).

[17] By analogy, we can picture light propagating through an inhomogeneous medium: a cloudy crystal ball or a fog; in the absence of true absorption, a forward propagating flux is eventually scattered in all directions resulting in a locally bright background illumination.

where l is a mean free path.[18] With (42.24) we can rewrite Eq. (42.21) for our free electrons as

$$s_{ij}(\mathbf{R},\omega) = \frac{e^2\hbar^2}{2m^2c}\frac{(4\pi)^2}{(2\pi)^6}\int_0^\infty k^2 dk \int_0^\infty k'^2 dk' \frac{\partial\rho(R,\varepsilon)}{\partial R}\frac{\partial\rho(R,\varepsilon')}{\partial R}$$

$$\times\left[\frac{f(\varepsilon')-f(\varepsilon)}{\varepsilon-\varepsilon'+\hbar\omega+i0}+\frac{f(\varepsilon')-f(\varepsilon)}{\varepsilon-\varepsilon'-\hbar\omega-i0}\right]n_i n_j - \frac{ne^2}{mc}\delta_{ij}\delta(\mathbf{r}-\mathbf{r}'), \tag{42.25}$$

where $\mathbf{n}$ is a unit vector along $\mathbf{A}$.

We must now examine the behavior of the form involving the derivative of the two density matrices (42.24) contained in (42.25); carrying out the differentiations, we have

$$\frac{\partial\rho(R,\varepsilon)}{\partial R}\frac{\partial\rho(R,\varepsilon')}{\partial R}$$

$$= \frac{e^{-R/l}}{kk'R^4}\left[kR\cos kR - \left(1+\frac{R}{2l}\right)\sin kR\right]\left[k'R\cos k'R - \left(1+\frac{R}{2l}\right)\sin k'R\right].$$

In a regime where k and k′ are near k_F, and both $k_F R$ and $k_F l >> 1$, the leading term in this expression is $(e^{-R/l}/R^2)\cos kR\cos k'R$. Writing $k' = k + \delta k$ where $\delta k = k' - k$, we have

$$\cos kR\cos k'R = \cos kR\cos((k+\delta k)R)$$
$$= \cos^2 kR\cos\delta kR - \cos kR\sin kR\sin\delta kR.$$

For the large kR limit considered here the second term is rapidly oscillating through zero; on the other hand, the factor, $\cos^2 kR$, in the first term can be approximated by its average value of 1/2 while the factor, $\cos\delta kR$, slowly oscillates. We then have

$$\frac{\partial\rho(R,\varepsilon)}{\partial R}\frac{\partial\rho(R,\varepsilon')}{\partial R} \cong \frac{e^{-R/l}}{2R^2}\cos\left[(k-k')R\right]$$

$$= \frac{e^{-R/l}}{2R^2}\cos\left[\frac{dk}{d\varepsilon}(\varepsilon-\varepsilon')R\right]. \tag{42.26}$$

42.6 The Chambers path integral

We are now in a position to calculate $s_{ij}(\mathbf{R};\omega)$ from (42.25) using (42.26). It will turn out that we can ignore the gauge current if we subtract from the term in the square brackets of Eq. (42.23) its value at $\omega = 0$. This factor is then written as

$$(f(\varepsilon')-f(\varepsilon))\left[\frac{1}{\varepsilon-\varepsilon'+\hbar\omega+i0}+\frac{1}{\varepsilon-\varepsilon'-\hbar\omega-i0}-\frac{2}{\varepsilon-\varepsilon'}\right]. \tag{42.27a}$$

[18] This behavior of our correlation function emerges more rigorously using the Green's function approach, as described in *Methods of Quantum Field Theory in Statistical Physics*, A.A. Abrikosov, L.P. Gorkov, and I.E. Dzyaloshinski, Pergamon Press, Oxford (1965).

Noting that here the factor in square brackets is antisymmetric in the variables ε and ε′, and by requiring that we treat the subsequent integrations symmetrically, we may rewrite this expression as

$$-2f(\varepsilon)\left[\frac{1}{\varepsilon-\varepsilon'+\hbar\omega+i0}+\frac{1}{\varepsilon-\varepsilon'-\hbar\omega-i0}-\frac{2}{\varepsilon-\varepsilon'}\right]. \tag{42.27b}$$

The expression (42.27b) is appreciable only for energies of order $\hbar\omega$ or k_BT of the Fermi energy and in what follows we will measure energies from the Fermi energy. To facilitate convergence we introduce the factor

$$\lim_{a\to\infty}\frac{a^2}{\varepsilon^2+\varepsilon'^2+a^2} \tag{42.28}$$

and take the limit $a \to \infty$ at the end of the calculation.

Using (42.26), (42.27b), and (42.28), we can rewrite Eq. (42.25) as

$$s_{ij}(\mathbf{R};\omega) = -\frac{e^2\hbar^2}{2m^2c}\frac{(4\pi)^2}{(2\pi)^6}\frac{e^{-R/l}}{2R^2}\left(2\pi^2N(0)\right)^2 n_in_j\lim_{a\to\infty}\int_{-\infty}^{\infty}d\varepsilon\int_{-\infty}^{\infty}d\varepsilon' 2f(\varepsilon)$$
$$\times\frac{a^2\cos\left[\frac{dk}{d\varepsilon}(\varepsilon-\varepsilon')R\right]}{\varepsilon^2+\varepsilon'^2+a^2}\left[\frac{1}{\varepsilon-\varepsilon'+\hbar\omega+i0}+\frac{1}{\varepsilon-\varepsilon'-\hbar\omega-i0}-\frac{2}{\varepsilon-\varepsilon'}\right], \tag{42.29}$$

where we write $k^2dk \cong k_F^2dk = k_F^2(dk/d\varepsilon)d\varepsilon = 2\pi^2N(0)d\varepsilon$, where $N(0)$ is the density of states for a single spin at the Fermi energy. The first factor in the second line has poles at $\varepsilon' = \pm i(\varepsilon^2 + a^2)^{1/2}$; however their contribution vanishes in the limit $\mathbf{a} \to \infty$. The second factor (in the square brackets) involves three terms. The third of these results is a delta function in **R**, as discussed previously. The poles of the first two terms are at $\varepsilon' = \varepsilon \pm \hbar(\omega + i0)$, and on writing the cosine function in exponential form, and carrying out the integration over $d\varepsilon'$ via contour integration, yield

$$s_{ij}(\mathbf{R},\omega) = -\frac{e^2\hbar^2}{2m^2c}\frac{e^{-R/l}}{R^2}(N(0))^2 n_in_j\lim_{a\to\infty}2\pi i\int_{-\infty}^{\infty}d\varepsilon\, f(\varepsilon)\exp\left(i\frac{dk}{d\varepsilon}\hbar\omega R\right)$$
$$\times\left[\frac{a^2}{\varepsilon^2+a^2+(\varepsilon-\hbar\omega)^2}-\frac{a^2}{\varepsilon^2+a^2+(\varepsilon+\hbar\omega)^2}\right]$$
$$= -\frac{e^2\hbar^2}{2m^2c}\frac{e^{-R/l}}{R^2}(N(0))^2 n_in_j\lim_{a\to\infty}2\pi i\exp\left(i\frac{dk}{d\varepsilon}\hbar\omega R\right)$$
$$\times\int_{-\infty}^{\infty}d\varepsilon\frac{a^2\left[f(\varepsilon)-f(\varepsilon-\hbar\omega)\right]}{\varepsilon^2+a^2+(\varepsilon-\hbar\omega)^2},$$

where in the second step we have moved the origin of ε in the second term by $\hbar\omega$. Approximating $f(\varepsilon)-f(\varepsilon-\hbar\omega) \cong (\partial f/\partial\varepsilon)\hbar\omega$ and taking the limit $a \to \infty$, we obtain

$$s_{ij}(\mathbf{R},\omega) = -\frac{e^2\hbar^2}{2m^2c}(N(0))^2(2\pi i\hbar\omega)\frac{e^{-R/l}e^{i\omega R/V_F}}{R^2}n_in_j \tag{42.30}$$

where we have introduced the Fermi velocity, $V_F = d\varepsilon/\hbar dk$. From Eq. (42.1a), we have $\mathbf{A}(\mathbf{r},\omega) = (ic/\omega)\mathbf{E}(\mathbf{r},\omega)$ and in place of $s_{ij}(\mathbf{R},\omega)$ we can introduce the *non-local electrical conductivity* kernel, $\sigma_{ij}(\mathbf{R},\omega) = (ic/\omega)\, s_{ij}(\mathbf{R},\omega)$, as

$$\sigma_{ij}(\mathbf{R},\omega) = \frac{e^2\hbar^2}{m^2}(N(0))^2\pi\hbar\frac{e^{-R/l}e^{i\omega R/V_F}}{R^2}n_in_j; \tag{42.31a}$$

the relation between the current and the electric field then becomes

$$\mathbf{j}(\mathbf{r},\omega) = \int d^3\mathbf{r}'\sigma(\mathbf{r}-\mathbf{r}',\omega)\cdot\mathbf{E}(\mathbf{r}',\omega). \tag{42.31b}$$

This is Chambers' formula[19] in the frequency domain.[20] Carrying out the integral in Eq. (42.31b) for the case of a homogeneous field at zero frequency yields $\sigma = (2/3)e^2N(0)V_Fl$, which for free electrons gives the Drude form $\sigma = ne^2\tau/m$ where $\tau = l/V_F$.

This quantum treatment of electrical transport, while having the virtue of being relatively straightforward, has a number of shortcomings. To begin with the procedure used to obtain $\rho(R,\varepsilon)$ was not given, nor was the factorization, implicit in Eq. (42.22), justified. Furthermore we have no recipe to calculate the mean free path, l, or equivalently the relaxation time, τ; in particular it is not clear if this relaxation time corresponds to the "total relaxation time," τ, as calculated from Eq. (19.68), or the "transport relaxation time," τ_{tr}, given by Eq. (19.67). In addition, the theory does not yield so-called *quantum corrections* to the conductivity. To address such problems one must use the more powerful Green's function methods developed by Abrikosov and Gorkov[21] and independently by Edwards[22], which go beyond the scope of this book. Application of this formalism shows that the time, τ, entering the conductivity is indeed the transport time, τ_{tr}, introduced in Section 19.10.

[19] The kinetic formulation of conduction problems, R.G. Chambers, *Proc. Phys. Soc. Lond.* A**65**, 458 (1952); Magnetoresistance effects in the group-i metals at high fields, R.G. Chambers, **A238**, 344 (1958).

[20] Our earlier discussion, in Section 19.13, was carried out in the time domain on the basis of the Boltzmann equation.

[21] See Section 39 in *Methods of Quantum Field Theory in Statistical Physics*, A.A. Abrikosov, L.P. Gorkov, and I.E. Dzyaloshinski, Pergamon Press, Oxford (1965). These methods are also discussed in *Green's Functions for Solid State Physics*, S. Doniach and E.H. Sondheimer, World Scientific Publishing, Singapore (1998).

[22] A new method for the evaluation of electric conductivity in metals, S.F. Edwards, *Philos. Mag.* **3**, 1020 (1958).

ADDITIONAL READING

Ashcroft and Mermin briefly discuss the electrical conductivity at finite frequencies utilizing perturbation theory in Chapter 13 of their book.

Problem

42.1. Obtain the electrical conductivity, $\sigma(\omega)$, at a finite frequency for the case of a uniform electric field using Eq. (42.31a).

Appendix 42A Zero-frequency response; Landau diamagnetism

Suppose that rather than having an applied *electric* field we have a constant *magnetic* field, **H**. Formally this is just a special case of the previous formalism in the limit $\omega = 0$. Here there can be no dissipation and hence the only response will turn out to be the diamagnetism associated with the electrons, which we will assume to be free. We can then take $\rho(R, \varepsilon)$ from (42.23) and Eq. (42.25) takes the form

$$s_{ij}(\mathbf{R}) = \frac{e^2\hbar^2}{2m^2c}\frac{(4\pi)^2}{(2\pi)^6}\int_0^\infty k^2 dk \int_0^\infty k'^2 dk' \left(\frac{\partial}{\partial R}\frac{\sin kR}{kR}\right)\left(\frac{\partial}{\partial R}\frac{\sin k'R}{k'R}\right)$$
$$\times\left[\frac{f(\varepsilon')-f(\varepsilon)}{\varepsilon-\varepsilon'+i0}+\frac{f(\varepsilon')-f(\varepsilon)}{\varepsilon-\varepsilon'-i0}\right]n_i n_j - \frac{ne^2}{mc}\delta_{ij}\delta(\mathbf{r}-\mathbf{r}'). \qquad (42A.1)$$

For those terms in the integrand which involve $f(\varepsilon')$ we first integrate over k and similarly for the factors involving $f(\varepsilon)$. Writing $\sin k'R$ in exponential form and integrating over dk' using Cauchy's theorem, closing the contours in the convergent halves of the complex plane, and exploiting the rapidly oscillating character of the integrand for large k', yields

$$s_{ij}(\mathbf{R}) = -\frac{ne^2}{2mc}\delta_{ij}\delta(\mathbf{r}-\mathbf{r}') + \frac{e^2\hbar^2}{2m^2c}\frac{(4\pi)^2}{(2\pi)^6}\int_0^\infty k^2 dk$$
$$\times f(\varepsilon)\left(\frac{\partial}{\partial R}\frac{\sin kR}{kR}\right)\left[-\pi k\left(\frac{2m}{\hbar^2}\right)\left(\frac{\partial}{\partial R}\frac{\cos kR}{kR}\right)\right]n_i n_j$$

or

$$s_{ij}(\mathbf{R}) = -\frac{ne^2}{2mc}\delta_{ij}\delta(\mathbf{r}-\mathbf{r}') + \frac{e^2\hbar^2}{2m^2c}\frac{(4\pi)^2}{(2\pi)^6}\left(\frac{2m\pi}{\hbar^2R^2}\right)\int_0^{k_F} k^2 dk$$
$$\times\left[\frac{\cos 2kR}{R}+\frac{k}{2}\left(1-\frac{1}{k^2R^2}\right)\sin 2kR\right]n_i n_j. \qquad (42A.2)$$

Integrating over dk, we obtain

$$s_{ij}(\mathbf{R}) = -\frac{ne^2}{2mc}\delta_{ij}\delta(\mathbf{r}-\mathbf{r}') + \frac{e^2\hbar^2}{2m^2c}\frac{(4\pi)^2}{(2\pi)^6}\left(\frac{2m\pi}{\hbar^2}\right)\left[\left(\frac{n_i n_j}{R^2}\right)\left(-\frac{k_F^3}{4R}\cos 2k_FR\right.\right.$$
$$\left.\left.+\frac{7k_F^2}{8R^2}\sin 2k_FR+\frac{9k_F}{8R^3}\cos 2k_FR-\frac{9}{16R^4}\sin 2k_FR\right)\right]. \qquad (42A.3)$$

From Ampere's law it follows that the diamagnetic current density and the magnetization are related by $\nabla \times \mathbf{M} = (1/c)\mathbf{j}$; substituting this into the zero frequency form of Eq. (42.10) and taking the curl of both sides of the resulting equation, we have

$$\nabla \times (\nabla \times \mathbf{M}) = \frac{1}{c}\nabla \times \int d^3\mathbf{r}'\, \mathbf{s}(\mathbf{R}) \cdot \mathbf{A}(\mathbf{r}'). \tag{42A.4}$$

On performing an integration by parts, using Eq. (42.1b), and applying a standard vector identity, this expression can be written as

$$-\nabla^2 \mathbf{M}(\mathbf{r}) = \frac{1}{c}\int d^3\mathbf{r}'\, \mathbf{s}(\mathbf{R}) \cdot \mathbf{H}(\mathbf{r}'). \tag{42A.5}$$

At this point we take the Fourier transform of (42A.5) yielding

$$\mathbf{M}(\mathbf{q}) = \frac{1}{cq^2}\, \mathbf{s}(\mathbf{q}) \cdot \mathbf{H}(\mathbf{q}), \tag{42A.6}$$

where

$$\mathbf{s}(\mathbf{q}) = \int d^3R\, e^{-i\mathbf{q}\cdot\mathbf{R}}\, \mathbf{s}(\mathbf{R}). \tag{42A.7}$$

We introduce the Fourier transform of the *magnetic susceptibility* as

$$\chi(\mathbf{q}) = M(\mathbf{q})/H(\mathbf{q}) = \frac{1}{cq^2} s(\mathbf{q}), \tag{42A.8}$$

and expand $\chi(\mathbf{q})$ in a power series in $\mathbf{q}$ as

$$\chi(\mathbf{q}) = \frac{1}{cq^2}\left[s_0 + s_1 + s_2 + \cdots\right], \tag{42A.9}$$

where the coefficients, s_n, are given by

$$s_n = \frac{1}{n!}\int d^3R\, (-i\mathbf{q}\cdot\mathbf{R})^n\, \mathbf{s}(\mathbf{R}). \tag{42A.10}$$

Since we are interested in the macroscopic susceptibility (corresponding to small $\mathbf{q}$) we will retain only the first non-vanishing term in (42A.9).

The term involving s_0 in (42A.9) is simply an integration of $\mathbf{s}(\mathbf{R})$ over d^3R, which involves two contributions. One of them, associated with the first term in (42A.3), is the so-called *gauge current*, $-(ne^2/2mc)\mathbf{A}(\mathbf{r},t)$. Integration over d^3R of the factor in square brackets in the second term of (42A.3) involves an angular part, $\int d\Omega\, n_i n_j = (4\pi/3)\,\delta_{ij}$, and a radial part given by $k_F^3/4$; collecting all the factors, we have $(e^2/2mc)\cdot(2/(2\pi)^3)\cdot\left((4\pi/3)k_F^3\right)\mathbf{A}(\mathbf{r},t)$ or equivalently $(ne^2/2mc)\mathbf{A}(\mathbf{r},t)$; hence this term *exactly cancels* the gauge current. The next term in (42A.9), involving the integral s_1, vanishes directly since it is odd. We are then left with the term arising from s_2, for which the corresponding angular and radial integrals are 2/5 and $-5k_F/8$.[23] Collecting factors we obtain the magnetic susceptibility in leading order as

$$\chi = -\frac{e^2 k_F}{12\pi^2 mc^2} \tag{42A.11}$$

(L.D. Landau, 1930). Note that the sign is negative: the orbital response of an electron gas is *diamagnetic.* Originally (42A.11) was obtained by quantizing the energy levels of the electrons resulting from the applied magnetic field (the Landau levels discussed in Sections 18.5 and 20.1.1). However at low fields, where $\omega_c\tau \ll 1$, the electrons do not complete their orbits before scattering, and it is then more appropriate to use a perturbation treatment starting from the free electron gas, as we have done here.[24]

As a model to handle the case of a superconductor, the previous formalism can be generalized by artificially introducing a gap in the excitation spectrum of the electrons. Rather than obtaining the weak Landau diamagnetism, one then obtains the Meissner effect.

[23] The associated integral does not converge for large R and it is necessary to cut it off at a distance of order the mean free path, l. The integral is well behaved if one uses the more accurate form for $\rho(R,\varepsilon)$, introduced in Eq. (42.25). The resulting corrections are of order $(k_F l)^{-1}$ which we here assume to be much less than one.

[24] Landau diamagnetism was first derived along the lines discussed here in: Theory of superconductivity, J. Bardeen, *Encyclopedia of Physics*, Vol. 15, p. 274, Springer Verlag, Berlin (1956).

Electrical Transport in Highly Disordered Media: Localization/Interaction Effects

43

The methods we have developed to describe electrical transport, either semi-classical via the Boltzmann transport equation, as in Chapter 19, or more quantum mechanical, as in Chapter 42, generally fail as the effects of scattering increase and as the temperature approaches absolute zero. This is especially true in one or two dimensions, as one approximates experimentally with a long wire or a thin film. In particular, experiments show that the resistance of highly disordered materials,[1] that appear metallic at high temperatures, increases as the temperature is lowered, as opposed to settling down to a constant characteristic of the residual resistance regime. This increase may occur slowly, as in metals with a high concentration of impurities, so-called dirty metals; here it is called *weak localization*. Alternatively it may rise much more dramatically, as in a highly doped semiconductor (that at higher temperatures might mimic a metal), where it is called *strong localization*. Before entering into a discussion of some theoretical techniques that have been developed to model these behaviors, we give some experimental examples, beginning with a doped semiconductor example of strong localization.

43.1 Temperature dependence of the electrical resistivity of a highly doped semiconductor: strong localization

Figure 43.1 shows a plot of log(R(T) versus $T^{-1/4}$ and $T^{-1/2}$ of the resistance of a series of heavily In-doped CdSe samples.[2] As discussed in Section 22.8, when the doping increases one expects conduction within an impurity band; this first occurs via a temperature activated hopping of electrons or holes between surrounding sites, and ultimately by metallic-like conduction at higher concentrations due to impurity wave function overlap. However, the temperature dependence

[1] Philip W. Anderson, Sir Nevill F. Mott, and John H. van Vleck shared the 1977 Nobel Prize in Physics for "their fundamental theoretical investigations of the electronic structure of magnetic and disordered systems."

[2] Probing the Coulomb gap in insulating n-type CdSe, Youzhu Zhang, Oei-hua Dai, Miguel Levy, and M.P. Sarachik, *Phys. Rev. Lett.* **64**, 2687 (1990).

The Physics of Solids. First Edition. J. B. Ketterson.

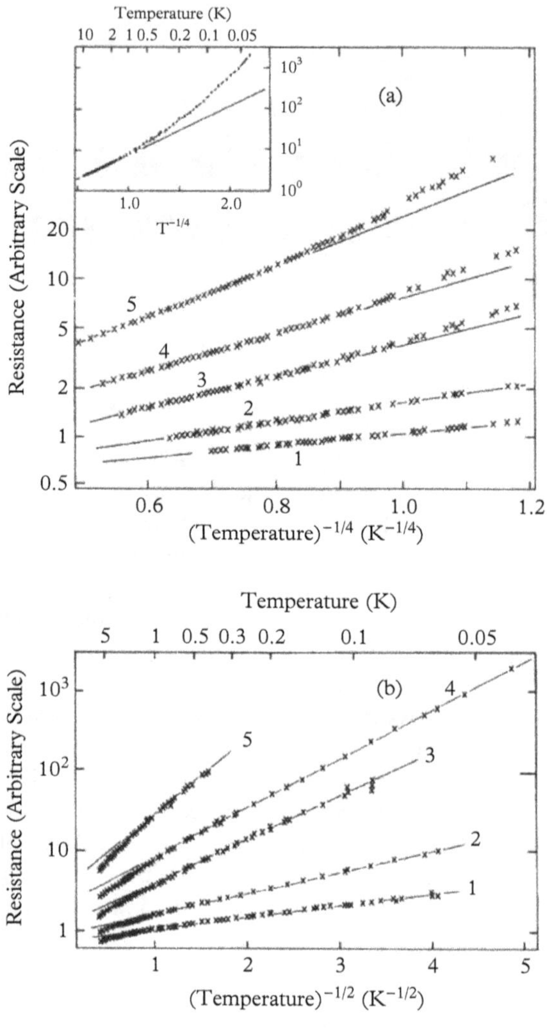

Figure 43.1 *The log of the resistance of five* n-*type* CdSe *samples doped with* In *at concentrations of* 2.8×10^{18} *(1),* 2.4×10^{18} *(2),* 2.2×10^{18} *(3),* 2.2×10^{18} *(4) and* 2.18×10^{18} *(5) plotted against* $T^{1/4}$ *(a) and* $T^{-1/2}$ *(b) to reflect two different models for the temperature dependence. (After Zhang et al. (1990)).*

shown in Fig. 43.1 is clearly not a simple Arrhenius-like activated behavior in which we would have $\rho \propto \exp(\varepsilon_b/k_B T)$, where ε_b is some typical barrier height; rather than being characterized by a T^{-1} dependence of the exponent, the data are more consistent with a $T^{-1/2}$ or $T^{-1/4}$ behavior. This still relatively strong "stretched exponential" increase in the resistance as the temperature falls can be interpreted as the electrons become progressively more localized (on the impurity sites) as the temperature falls; such exponential behaviors are a defining characteristic of strong localization. Many models have been developed to explain the observed behaviors, some of which we will take up.

43.2 Temperature dependence of the electrical resistivity of a disordered metal and an alloy: weak localization

Figure 43.2 shows the temperature dependence of the resistance for the case of thin copper films with various thicknesses as a function of temperature.[3] As plotted it is clear that the data are consistent with a *logarithmic* rise at low temperatures. This kind of behavior is widely observed in many metallic thin film systems;[4] it is essentially universal. For historical reasons and because of the smaller (when compared to the strong localization, previously described) size of this effect, it is referred to as weak localization. We recall that in metals the resistance of a metal typically drops as the temperature falls and one enters what is called the residual resistance regime, where elastic impurity scattering dominates over inelastic

[3] Localization and negative magnetoresistance in thin copper-films. C. Van Haesendonck, L, Van den Dries, Y. Bruynseraede, and G. Deutscher, *Phys. Rev.* 25, 5090 (1982).

[4] Films approximate two-dimensional systems at low temperatures; see Eq. (43.16) and the discussion at the end of Section 43.4.

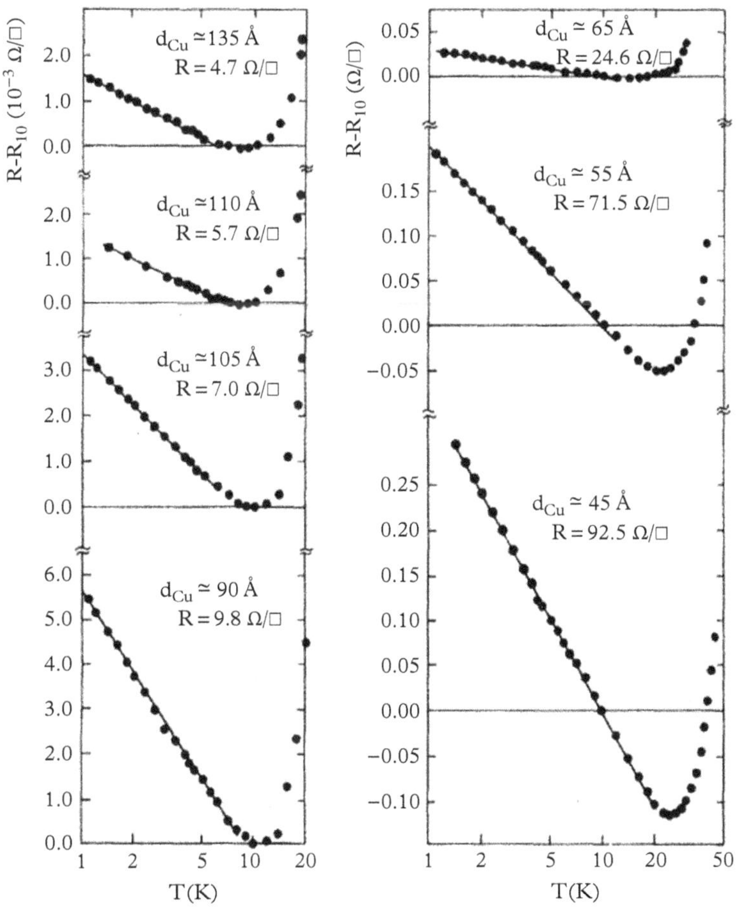

Figure 43.2 *Resistance* R(T) – R(T = 10 K) *of thin copper films having different thicknesses versus the log of the temperature (after Van Haesendonck et al. (1982)).*

electron–phonon scattering (where we encounter the Bloch $\rho \propto T^5$ behavior, see Section 21.5). However, in highly impure metals it is often seen experimentally that as the temperature is further reduced the resistance passes through a minimum and then slowly rises. As previously noted, the effects are more pronounced in films and wires, where the dimensionality is effectively reduced to 2d and 1d respectively and each has its own characteristic temperature dependence (e.g., logarithmic in 2d).

These observed low temperature behaviors cannot be described by the Boltzmann transport equation, where *inelastic* collisions (involving phonons) are infrequent but where elastic collisions are frequent. To systematically describe these effects a more quantum mechanical description of the system is needed than the Boltzmann approach can provide. Since the quantum mechanical treatment of Chapter 42, to the level it was taken, also gave the same Boltzmann result, we anticipate that the treatment has to be extended,[5] as will be discussed *qualitatively* in Section 43.4. We will find that the observed effects arise from a subtle *self-interference effect* associated with the repeated elastic scattering of diffusing electron waves in dirty metals. This effect was anticipated in Section 19.11 where we critiqued the Boltzmann approach.

43.3 Electrical transport via hopping; strong localization

In our discussion in Section 22.6 of localized impurity states in crystalline group IV semiconductors such as Si, we constructed a model in which an isolated group III or group V impurity would form four covalent bonds with its neighbors, together with a loosely bound hole or electron state involving the dielectrically screened Coulomb interaction. As discussed in Section 22.8, at high impurity concentrations, overlap of neighboring orbitals can produce sufficient level mixing that the hole or electron states spread out into a near continuum; what is loosely called an impurity band (lying near the valence or conduction band edge); note this is not a band in the usual Bloch sense, but rather a high density of strongly coupled impurity states.

Now consider the case of an amorphous semiconductor. Based on chemical considerations each atom would prefer to adopt locally tetrahedral bonds with the adjacent atoms. But the random nature of the amorphous structure results in some of the atoms being *frustrated*, in the sense that there are either too many or too few nearest neighbors; this results in so-called *dangling bonds* which can, in turn, result in localized states, some, depending on the local environment, lying far from where the valence or conduction band edges would lie in the periodic system.

The relation between the electrical transport properties and the continuum of impurity levels formed at high concentrations in a periodic system, or that formed by dangling bonds in an amorphous structure, has attracted much attention over

[5] A key needed ingredient was lost in making the approximation given in Eq. (42.22).

the years. Because of the lack of translational symmetry, simple considerations, like those which immediately emerge from the nature of Bloch states and the presence or absence of partially filled bands, are not available to dictate (for cases where electron–electron interactions can be ignored) what the transport properties will be in such disordered systems.

In 1958, Anderson considered the motion of electrons moving in a random potential and came to the rather broad conclusion that, whatever the number of dimensions, there are quantum states which *do not exhibit a diffusion*, i.e., that show zero mobility.[6] Specifically, he constructed a model for tight-binding, but otherwise independent, electrons moving in a lattice of random potentials. This calculation launched lines of inquiry that continue to the present day. In what immediately follows we confine ourselves to a metal. Anderson posed the following query: Suppose we are given (see Eq. (11.55a)) on-site tight-binding parameters, $h_{0i} \equiv \varepsilon_i$, varying randomly in some interval, $\pm V_0$, about an average value, $\bar{\varepsilon}_i$, and nearest-neighbor matrix elements, h_1 (also called hopping integrals), which are assumed to be site independent. In the absence of disorder the latter lead to a band width, $B = 4h_1$, in one dimension (see Eq. (11.55b)). In a three-dimensional simple cubic lattice we would have $B = 2zh_1$, where $z = 6$ is the number of nearest neighbors. What Anderson found (and we will not reproduce his calculation) was that the states near the center of the band became *localized* at a value, $V_0/B = 1.7$. The general conclusion that the electronic states become localized in the presence of sufficient disorder is now supported by a vast body of theoretical and experimental research. Because such states are qualitatively different from extended states, Bloch states being an example of the latter, Anderson's conclusion represented a radical departure.

In the presence of weak disorder or low impurity concentrations, electron scattering in metals is usually characterized by some matrix element, $|V_{\mathbf{kk}'}|^2$, connecting states **k** and **k**′, leading to a concentration-dependent mean free path, l; through the position–momentum uncertainty principle this leads to an indeterminacy in the electron wave vectors, $\Delta k \cong l$; note that if the impurities have no internal levels lying near the Fermi energy (that could be excited) the energy will remain fixed. With increasing concentration a point may be reached where $l \cong a$, where a is a lattice spacing. Since it makes little sense to speak of a mean free path shorter than an atomic spacing, it can be argued that this limiting value must define some limiting or *minimum metallic conductivity (maximum resistivity)*; the criterion, $l \cong a$, is now called the Ioffe–Regel limit.[7] But in Anderson's model nothing limits the size of the parameter V_0 (characterizing the disorder) from continuing to increase; hence it is natural to expect a *qualitative* change in the *character* of the conduction process at some point.

Empirical studies and various models suggest that regions of extended and localized states can be simultaneously present in the same system (albeit at separate energies). If the Fermi level falls in a range where the states are extended, then the conduction process is more or less conventional, and the conductivity tends toward a constant at low temperatures. If, on the other hand, the Fermi energy

[6] Absence of diffusion in certain random lattices, P.W. Anderson, *Phys Rev.* **109**, 1492, (1958).

[7] In a typical metal the lattice spacing, a, is of order the de Broglie wavelength of electrons at the Fermi surface, $\bar{\lambda}_F = \hbar/p_F$, so in a more modern context, $a \approx D_F$ rather than $a \approx l$, provides a more universal demarcation between the extended and localized states in 3d.

falls in a region dominated by localized states, electrical conduction will occur via an *activated process* (in which the energy is not fixed); characteristically, activated processes fall off with temperature, although the process need not be dominated by a simple Boltzmann factor, as we will see.

In the presence of both extended and localized states, the energy, ε_c, separating the two regimes is called the *mobility edge*,[8] since states with energies above ε_c have much higher mobilities. If the Fermi level lies in the region of localized states, but at the same time ε_c is not too much larger than ε_F, we might expect the conductivity to vary as

$$\sigma = \sigma_0 e^{-(\varepsilon_c - \varepsilon_F)/k_B T}, \tag{43.1}$$

the extended states being much more effective in transporting charge. Equation (43.1) has the same form as in a conventional clean semiconductor.

Another activated process involves transitions (or hopping as it is called) from one localized state to another caused by the absorption/emission of a phonon. As the temperature falls the number of electrons excited to the mobility edge falls rapidly and conduction ultimately gives way to transport by hopping. To discuss hopping we will assume, as previously anticipated, that the energies, ε_i, of the localized states densely occupy some range of energies in the vicinity of the Fermi energy, and that the wave functions associated with the states decay as $e^{-\alpha r}$(we will assume the decay parameter, α, is independent of energy for simplicity). Surrounding some site with energy, ε_i, are other sites with energies, ε_j, located distances, r_{ij}, away.

The simplest approximation is to assume that conduction occurs only via *nearest-neighbor hopping*;[9] let such neighbors be separated by an average energy, $\Delta\varepsilon$, and an average distance, r_0. Now the electrical current density may be written,

$$j = e\, n_{\text{eff}} \times (v_+ - v_-). \tag{43.2}$$

The first factor in (43.2) can be written as $eN(\varepsilon_F)k_B T$, where $N(\varepsilon_F)$ is the density of states. The effective forward/backward hopping velocities, $v_\pm$, can be written as the product of the distance hopped, r_0, and the forward/backward hopping rates; the hopping rate can, in turn, be written as the product of a phonon-assisted attempt frequency, ν_{ph}, and the probability for a successful hop in the presence of an applied external field, **E**; these factors combine to give the velocities as

$$v_\pm = r_0\, \nu_{ph} \exp\left(-2\alpha r_0 - \frac{\Delta\varepsilon \pm e\mathbf{E}r_0}{k_B T}\right). \tag{43.3}$$

The first term in the exponent accounts for the wave function overlap between the two states (as it would enter the matrix element coupling the two states), the second term accounts for the thermal activation needed to cause a transition between

[8] Simple band model for amorphous semiconducting alloys, M.H. Cohen, H. Fritzsch, and S.R. Ovshinsky, *Phys. Rev. Lett.* **22**, 1065 (1969).

[9] Impurity conduction at low concentrations, A. Miller and E. Abrahams, *Phys. Rev.* **120**, 745 (1960).

the states whose energies differ by $\Delta\varepsilon$, together with the shift induced by the applied electric field. Assembling these factors, we obtain the current density, j, as

$$j = eN(\varepsilon_F)k_B T r_0 \nu_{ph} e^{-2\alpha r_0 - \frac{\Delta\varepsilon}{k_B T}} \sinh\left(\frac{e E r_0}{k_B T}\right), \tag{43.4}$$

which on expanding to first order in the applied electric field yields a conductivity,

$$\sigma = 2e^2 N(\varepsilon_F)\nu_{ph} r_0^2 \, e^{-2\alpha r_0 - \frac{\Delta\varepsilon}{k_B T}}. \tag{43.5}$$

At low temperatures the second term in the exponential becomes very large and the conductivity would go to zero exponentially with T^{-1}.

If, on the other hand, we consider hopping to more remote sites (i.e., $r_{ij} > r_0$), we can find a state with a smaller $\Delta\varepsilon$; however the wave function overlap will then be smaller, tending to offset the gain from the smaller $\Delta\varepsilon$. This leads to what is called *variable-range hopping*.[10] Clearly what we want to do is optimize the competing effects of a smaller overlap and a smaller energy difference. This will occur when the exponent in (43.5) satisfies

$$2\alpha r + \frac{\Delta\varepsilon}{k_B T} = \text{minimum}, \tag{43.6}$$

where r is some effective hopping distance, to be determined. We now need a strategy to calculate the likelihood of finding a smaller $\Delta\varepsilon$. For a system in d dimensions, the number of states in an interval, $\Delta\varepsilon$, in a volume, V, surrounding the impurity[11] follows from the density of states, $N(\varepsilon)$, as

$$\#\text{states} = N(\varepsilon) V \Delta\varepsilon = N(\varepsilon)\gamma_d r^d \Delta\varepsilon, \tag{43.7}$$

(where γ_d is given by $4\pi/3$, π, and 2 for the cases of 3, 2, and 1 dimension respectively). If we set $\Delta\varepsilon = k_B T$ (the available energy to drive the transition) and seek the one state with the smallest energy difference within this volume (which we then assume dominates the conductivity due to its smallness), we obtain

$$\Delta\varepsilon = \frac{1}{N(\varepsilon)\gamma_d r^d}. \tag{43.8}$$

Inserting this expression into Eq. (43.8) and minimizing with respect to r gives

$$r = \left[\frac{d}{2\alpha N(\varepsilon)\gamma_d k_B T}\right]^{\frac{1}{d+1}}. \tag{43.9}$$

Substituting this value into Eq. (43.6), we obtain the exponent as

$$-2\alpha\left(1+\frac{1}{d}\right)\left(\frac{d}{2\alpha N(\varepsilon)\gamma_d k_B T}\right)^{\frac{1}{d+1}} \equiv -\left(\frac{T_0}{T}\right)^{\frac{1}{d+1}}, \tag{43.10}$$

[10] Conduction in non-crystalline systems 1: localized electronic states in disordered systems, N.F. Mott, *Philos. Mag.* 17, 1259 (1968); Conduction in non-crystalline systems 2: metal-insulator transition in a random array of centres, N.F. Mott and E.A. Davis, *Philos. Mag.* 17, 1269 (1968); Metal-insulator transition, N.F. Mott, *Rev. Mod. Phys.* 40, 677 (1968).

[11] One could just as well have asked for the number states lying within a d-dimensional spherical shell of radius r and thickness δr (to be defined), rather than an enclosed volume, as we do here; this would alter the temperature dependence that we obtain and only goes to show the qualitative nature of the model presented here.

where we regard T_0 as a parameter. Our expression for the temperature dependence of the conductivity in the variable-range hopping regime is then

$$\sigma(T) = 2e^2 N(\varepsilon_F) \nu_{ph} r^2 e^{-\left(\frac{T_0}{T}\right)^{\frac{1}{d+1}}}. \tag{43.11}$$

The most important feature of (43.11) is that it varies significantly more slowly with temperature than a simple Boltzmann-like form. For the case of $d = 3$, the exponent is proportional to $T^{-1/4}$ which, as seen from Fig. 43.1, gives a reasonably good representation of the data. It is clear that the result involves numerous assumptions and other models lead to an exponent that differs from the value, $(d+1)^{-1}$, obtained here. The key feature is that much weaker temperature dependencies relative to the simple Arrhenius behavior are to be expected at very low temperatures.

43.4 Weak localization

Here we give a qualitative discussion of the physical origin of several phenomena that are observed in the electrical transport of disordered metals at very low temperatures, which cannot be explained on the basis of the Boltzmann equation.[12] They are particularly apparent when the impurity concentration is high or other disorder is present; the phenomena are further enhanced when the effective dimensionality is lower, as in thin films or wires, which under appropriate conditions can be approximated as two- and one-dimensional systems respectively. The effects are collectively referred to as *weak localization* since they are viewed as being associated with a gradual transition from a metal to an insulator, arising from a change in the character of the electron wave functions (from extended to localized) as the disorder increases. However this possible feature will not be entering our discussion.

One phenomenon is an unusual logarithmic temperature-dependent correction to the low temperature behavior of the resistivity in thin films, an example of which was shown in Fig. 43.2. Based on the semi-classical Boltzmann equation, discussed in Chapter 19, the resistivity should approach a temperature-independent value (referred to as Matthiessen's rule). Another effect is a *negative* magnetoresistance; this is the opposite of the classical behavior associated with curved orbits in a magnetic field, which leads to an increase of the resistance with field (positive magnetoresistance, see Eqs. (19.65b) and (19.66a)). Puzzles such as these extended over a period of about 40 years starting in the 1930s with semi-classical Boltzmann equation treatments. Contributing to the resolution of these problems was the continued evolution of the many-body Green's function based methods; however at the level they were initially treated, they still gave the Drude form.[13]

Clearly something is missing and the key ingredient that is intrinsically left out in the Boltzmann approach, and not accounted for in the earlier quantum

[12] The arguments we present were derived in part from some notes provided by D.E. Khmelnitskii.

[13] For a derivation see: *Methods of Quantum Field Theory in Statistical Physics*, A.A. Abrikosov, L.P. Gorkov, and I.E. Dzialoshinskii, Dover Publications Inc., Dover (1975).

theories, is that *wave functions have phases*. Effects arising from quantum interference cannot enter a description based on the semi-classical Boltzmann equation since the distribution function only measures the *probability* of occupancy of the underlying quantum states; their phases do not enter.

As a way to introduce phases we will utilize the concept of Feynman path amplitudes (with which we assume some familiarity), which collectively describe the quantum mechanical amplitude of an electron at some space–time point, $(\mathbf{r}, t)$, that was "released" at some other space–time point, $(\mathbf{r}', t')$.[14] Figure 43.3 shows, schematically, two of the many paths (which are actually lines in a d+1-dimensional space of paths) which leave (r', t') and arrive at $(\mathbf{r}, t)$. If each path has an amplitude, A_i, the total probability, W, of our electron arriving at $(\mathbf{r}, t)$ is then a sum over such paths, which we write as

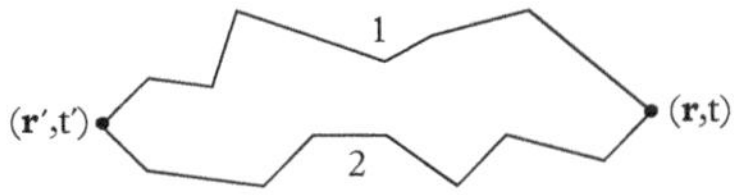

Figure 43.3 *Two different paths connecting the space-time points* $(\mathbf{r}', t')$ *and* $(\mathbf{r}, t)$.

$$W = \left|\sum_i A_i\right|^2 = \sum_i |A_i|^2 + \sum_{i \neq j} A_i A_j^*. \tag{43.12}$$

In the limit of large quantum numbers, we will treat the sum in the first term on the right-hand side of (43.12) as being the classical probability of the electron diffusing from $(\mathbf{r}', t')$ to $(\mathbf{r}, t)$. This probability is simply related to the Green's function for the diffusion equation;[15]

$$\frac{\partial G(\mathbf{r}, t; \mathbf{r}', t')}{\partial t} - D\nabla^2 G(\mathbf{r}, t; \mathbf{r}', t') = \delta(\mathbf{r} - \mathbf{r}')\delta(t - t')$$

in d space dimensions, and with t' set to 0 it is given by

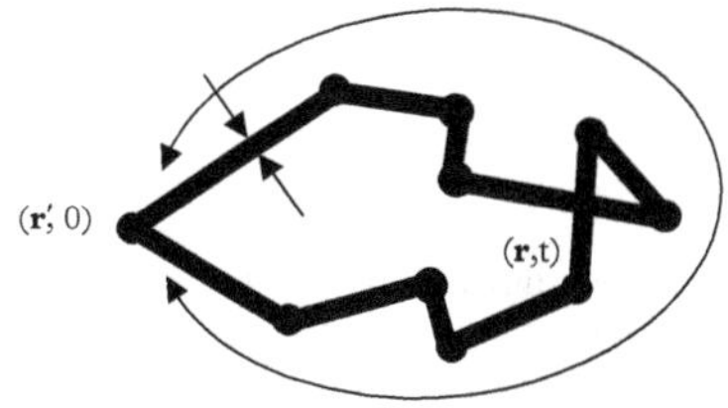

Figure 43.4 *An example of a path, and its accompanying tube of diameter,* λ_F, *that returns to the origin. The arrows indicate that the path can be traversed in either direction. In the presence of time reversal symmetry, the two paths will have the same phase and thus interfere constructively, irrespective of the path taken. In the presence of a magnetic field, which destroys the time reversal symmetry, the phases will differ (due to the Aharonov–Bohm effect), by a different amount for each possible path taken and, collectively, rapidly average to zero.*

$$G(\mathbf{r}, t; \mathbf{r}', 0) = \left(\frac{1}{4\pi D t}\right)^{d/2} e^{-\frac{|\mathbf{r}-\mathbf{r}'|^2}{4Dt}}\,\theta(t), \tag{43.13}$$

where D is the diffusion constant, which for a degenerate Fermi gas is given by $D = V_F^2\tau/d$ (a variant of the Einstein equation).

The second term in (43.12) represents an *interference* between the different paths. Now, in a single particle picture electrons arriving at $(\mathbf{r}, t)$ along different paths would have well-defined phases even though they experienced multiple elastic scattering events along the way; however, each path would involve a different phase which fluctuates wildly between 0 and 2π for large $\mathbf{r}$, and the net amplitude at the point $(\mathbf{r}, t)$ would average to zero.

An exception to this argument occurs when the observation and source points are *identical*, i.e., when $\mathbf{r} = \mathbf{r}'$, although we consider all times $t \geq 0$. In principle, electrons associated with a pair of paths, A_i and A_j, that are *strictly time reversed* and return to $\mathbf{r}'$, will arrive *in phase*, as shown schematically in Fig. 43.4. This would be the case *independent of the path*; i.e., *all* returning paths would contribute in the absence of phase destroying processes. We then have a kind of multi-path "coherent back scattering" which lowers the transmission into the sample and thereby increases the resistance.

[14] Formally, this is the Green's function, $G(\mathbf{r}', t'; \mathbf{r}, t)$.

[15] Classical diffusion of atoms was treated in Section 40.5, see Eq. (40.23).

In what follows we give only a *semi-quantitative* discussion of how the coherent backscattering affects the conductivity of a metal.[16] Now the classical differential probability, dW, of finding an electron in a volume element, d^dx, at some time, $t > 0$, in the limit of many scattering events follows from the Green's function for the diffusion equation, $G(\mathbf{r}, t;\ \mathbf{r}', 0)$. It turns out that we can relax the criteria of strictly time reversed paths somewhat. Qualitatively and semi-classically, electrons traversing neighboring paths which lie within "tubes" having a radius of the order the Fermi wavelength, λ_F, around a classical trajectory, will largely arrive *in phase* back at the source point, $\mathbf{r}'$. The differential volume of these tubes is of order, $d^dx \approx \lambda_F^{d-1} V_F dt$. We now assert that the total fraction of electrons that return to the starting point, which involves an integral of dW over such tubes, decreases the outward diffusion, thereby diminishing the charge transported. The net effect will be to reduce the conductivity by the fraction,[17]

$$\frac{\delta\sigma}{\sigma} \approx -\int dW(\mathbf{r} = \mathbf{r}', t) \approx -\int_{t_1}^{t_2} dt\, \lambda_F^{d-1} V_F G(0, t;\ 0, 0)$$

or on using Eq. (43.13),

$$\frac{\delta\sigma}{\sigma} = -\int_{t_1}^{t_2} dt \frac{\lambda_F^{d-1} V_F}{(4\pi Dt)^{d/2}}.$$

We must now choose reasonable values for the limits entering this integral. Since our treatment is based on the classical diffusion equation it certainly does not apply for times shorter than a typical impurity scattering time (and physically there are few returning electrons for $t < \tau$, since they then propagate ballistically); hence we take $t_1 = \tau$. The upper limit is more subtle since in the presence of inelastic electron–phonon[18] or electron–electron scattering processes, phase information is irreversibly lost, and the time average of the interference term is diminished. The loss of information on the path-dependent phases, ϕ_i, is characterized by a different relaxation time which we designate as τ_ϕ, a *phase-breaking time*. At low temperatures, τ_ϕ grows due to the gradual disappearance of phonons and the Fermi liquid suppression of electron–electron scattering; this leads to the inequality, $\tau_\phi \gg \tau, \tau_{tr}$, where the latter two times are those arising from impurity scattering in our discussion of electrical conductivity in Chapters 19 and 42. With the limits now defined we can write the correction to the conductivity as

$$\frac{\delta\sigma}{\sigma} \approx -\int_{\tau}^{\tau_\phi} \frac{\lambda_F^{d-1} V_F}{(4\pi Dt)^{d/2}} dt. \tag{43.14}$$

In evaluating this expression we will drop numerical factors of order unity. Using the Einstein relation,[19] $\sigma = e^2 N(\varepsilon_F) D$, $\lambda\!\!\!^-_F = \hbar/p_F$, and writing $N(\varepsilon_F) \approx \hbar^{-d} p_F^{d-1}/V_F$, Eq. (43.15a) takes the form

$$\delta\sigma \approx \frac{e^2}{\hbar} \int_{\tau}^{\tau_\phi} \frac{dt}{D^{(d-2)/2} t^{d/2}}; \tag{43.15}$$

[16] A rigorous discussion requires many-body Green's function methods; e.g., see: Particle conductivity in a two-dimensional random potential, L.P. Gorkov, A.I. Larkin, and D.E. Khmelnitskii, *JETP Letters* **30**, 228 (1979) and, Magnetoresistance and Hall-effect in a disordered 2-dimensional electron-gas, B.L. Altshuler, D. Khmelnitzkii, A.I. Larkin, and P.A. Lee, *Phys. Rev. B* **22**, 5142 (1980).

[17] Anderson localization and anomalous magnetoresistance at low temperatures, A.I. Larkin and D.E. Khmelnitskii, *Sov. Phys. Uspehi.* **136**, 536 (1982) and, Localization and coherent scattering of electrons, D.E. Khmelnitskii, *Physica B+C*, **126B**, 235 (1984).

[18] At low temperatures, where phonons have long wavelengths, neighboring paths can suffer nearly identical phase shifts and the inelastic scattering is suppressed.

[19] Earlier in a non-degenerate system obeying Boltzmann statistics, the Einstein relation, Eq. (23.27), was given as $D = k_B T\mu/e$; inserting the definition of the mobility, $\mu = e\tau/m$, we have $D = k_B T\tau/m$. However for a degenerate system, the characteristic energy per particle is $n/N(\varepsilon)$ rather than $k_B T$, yielding, $D = \tau n/N(\varepsilon)m$. We can then write the Drude conductivity as $\sigma = ne^2\tau/m = e^2 D N(\varepsilon)$.

carrying out the integration, we obtain[20]

$$\delta\sigma \approx -\frac{e^2}{\hbar}\begin{cases} L_\phi; & d = 1 \\ \ln\left(\tau_\phi/\tau\right); & d = 2 \\ \text{const.} - L_\phi^{-1}; & d = 3 \end{cases}, \tag{43.16}$$

where we have defined a *phase-breaking length*, $L_\phi = \sqrt{D\tau_\phi}$, also called the Thouless length.

Note that in all three cases, the conductivity decreases as τ_ϕ increases, implying that the resistivity, ρ, increases; in the case of one and two dimensions this increase in ρ appears to be unbounded (the lower the dimensionality the more likely the particle finds its way back to the origin), while in three dimensions the resistivity increases but ultimately saturates. The effect is quite large in this sense, given that it arises from the small fraction of returning paths. A diverging resistivity in the 1d- and 2d-systems indicates a (gradual) transition to an insulating state, the initial behavior of which is then labeled weak localization. It is common to approximate $\tau_\phi \propto T^{-p}$. In two dimensions this results in a linear behavior with a slope of $-pe^2/\hbar$ at low temperatures when plotting the resistance against the log of the temperature;[21] the data on Cu films shown in Fig. 43.1 display this behavior, as do many other systems.

The question naturally arises as to what dimensionality actually applies to a conductor having a given geometry. Suppose we have a slab parallel to the x–y plane with thickness, z_0, in the z direction. For times, τ_ϕ, satisfying the condition, $(D\tau_\phi)^{1/2} = L_\phi \ll z_0$, the system will effectively be three dimensional. On the other hand, if $L_\phi \gg z_0$, the system behaves two dimensionally. Since $\tau_\phi = \tau_\phi(T)$, one can have a crossover from three- to two-dimensional behavior as the temperature is lowered and L_ϕ becomes greater than z_0. Similarly, if there is a second lateral dimension, y_0, along the y direction, as with a sheet of finite width, a second crossover from two to one dimension would occur at a still lower temperature when $L_\phi \geq y_0$. The implication here is that a 3d system with one large dimension would be an insulator at absolute zero; such structures are then, effectively, one dimensional.

43.5 Magnetoresistance and the Aharonov–Bohm effect

Our earlier discussion of longitudinal magnetoresistance in Sections 19.8 and 19.9 was based on the Boltzmann equation, together with the Lorentz force. This effect vanishes for the special case of free electrons and an isotropic relaxation time, but for more complicated systems (an example being given in Section (19.9)) a *positive* shift, $\delta\rho(H) \approx (\omega_c\tau)^2\rho(H = 0)$, is obtained at low fields; here $\omega_c\tau$ can be viewed as the average angle through which the electron trajectories

[20] In addition to corrections to the conductivity arising from weak localization, electron–electron interactions can lead to effects having similar temperature dependencies: see, Interaction effects in disordered Fermi systems in 2 dimensions, B.L. Altshuler, A.G. Aronov, and P.A. Lee, *Phys. Rev. Lett.* **44**, 1288 (1980).

[21] In other connections one introduces a *quantum of resistance* as $\pi\hbar/e^2 = h/2e^2 = 12.9\,k\Omega$.

are bent by the magnetic field between collisions, with ω_c and τ being a characteristic cyclotron frequency and collision time respectively. The fact that the shift is positive turns out to be a model-independent feature of the Boltzmann approach.

It is widely observed that metals in the weak localization regime show a *negative magnetoresistance*. As seen in the previous section, wave function phase plays a dominant role in the low temperature behavior of dirty metals with long inelastic scattering times, so it is natural to examine the effect of a magnetic field on the phase of the trajectories considered. A magnetic field-dependent wave function phase (in that case of the order parameter) was encountered earlier when we discussed the two-junction SQUID, in Section 33.4.2, but we will reconstruct the needed arguments here for completeness.

In the presence of a magnetic field, as represented by a vector potential, **A**, the wave vector of an electron acquires an additional position-dependent component (see Eq. (18.7)), given by $(e/\hbar c)\mathbf{A}(\mathbf{r})$. Pairs of amplitudes, A_i and A_j, contributing to the second term on the right of Eq. (43.12), that are associated with time-reversed paths, traversing closed paths which start and return to the origin (clockwise and counterclockwise in Fig. (43.4)), will suffer net phase shifts, ϕ,

$$A_{i,j} \rightarrow A_{i,j}e^{\pm i\phi} = A_{i,j}e^{\pm ie/\hbar c)\int \mathbf{A}(\mathbf{r})\cdot d\mathbf{r}} \tag{43.17}$$

Using Stokes' theorem for our case of closed loops, we have

$$\begin{aligned}\phi &= \frac{e}{\hbar c}\oint \mathbf{A}(\mathbf{r})\cdot d\mathbf{r} = \frac{e}{\hbar c}\int \nabla \times \mathbf{A}\cdot d\mathbf{S}\\ &= \frac{e}{\hbar c}\int \mathbf{H}\cdot d\mathbf{S} = \frac{e}{\hbar c}\Phi,\end{aligned} \tag{43.18}$$

where Φ denotes the magnetic flux contained within a given closed trajectory starting and ending at the origin; for the mesoscopic devices of interest here, where the external field, **H**, is essentially constant, the flux can be written, $\Phi = \mathbf{H}\cdot\mathbf{S}$. For our diffusing electrons the enclosed area, S, depends on the time, t, and is approximately given by Dt (recall $r^2 \cong Dt$), so the net phase shift is $(e/\hbar c)HDt$. In place of Eq. (43.14) we now have

$$\frac{\delta\sigma}{\sigma} \approx -\int_{\tau}^{\tau_\phi} dt \frac{\lambda_F^{d-1} v_F}{(4\pi Dt)^{d/2}} \cos\left(\frac{eHDt}{c\hbar}\right). \tag{43.19}$$

Proceeding as we did after Eq. (43.14) and dropping numerical factors of order unity, we can express the shift in the conductivity as

$$\delta\sigma \approx \frac{e^2}{\hbar}\int_{\tau}^{\tau_\phi} \frac{dt}{D^{(d-2)/2}t^{d/2}}\left[1 - \cos\left(\frac{eHDt}{c\hbar}\right)\right]. \tag{43.20}$$

Noting that the integrand now vanishes at the lower limit and rewriting in a dimensionless form, we have

$$\Delta\sigma_{2d} \approx \frac{e^2}{\hbar} f_{2d}\left(\frac{eHD\tau_\phi}{c\hbar}\right), \quad f_{2d}(x) = \int_0^x \frac{dz}{z}\,[1-\cos z]; \tag{43.21a}$$

and

$$\Delta\sigma_{3d} \approx \frac{e^2}{\hbar}\sqrt{\frac{eH}{c\hbar}} f_{3d}\left(\frac{eHD\tau_\phi}{c\hbar}\right), \quad f_{3d}(x) = \int_0^x \frac{dz}{z^{3/2}}\,[1-\cos z], \tag{43.21b}$$

with $x = eHD\tau_\phi/c\hbar = L_\phi^2/l_H^2$, with $\lambda_H \equiv \sqrt{c\hbar/eH}$ having units of length.[22] The limiting forms of the integrals are

$$f_{2d}(x) \cong \begin{cases} x^2; & x \ll 1, \\ \log x; & x \gg 1, \end{cases} \tag{43.22a}$$

and

$$f_{3d}(x) \cong \begin{cases} x^{3/2}; & x \ll 1, \\ x^{1/2}; & x \gg 1. \end{cases} \tag{43.22b}$$

For a clean 3d system the orbits lie in planes perpendicular to the field and the magnetoresistance depends on the field direction. However, diffusing electrons in a 3d disordered system traverse paths that wander in *three dimensions*; hence the phase cancellation effects will occur *independent of the magnetic field direction.*

For the 2d case, where the electrons are constrained to move within a film of thickness, $z_0 \ll L_\phi$, the dominant role is played by the component of the field *perpendicular* to the film. A smaller effect occurs when the field lies parallel to the film; electrons diffusing for a time, τ_ϕ, will tend to return more often to the origin since they are reflected from opposing film surfaces (see Fig. 43.5); i.e., there will be approximately $N_\phi = L_\phi^2/z_0^2$ "sub-loops" in this time, each containing a smaller flux of order $\Phi \cong Hz_0^2$. The cosine factor entering Eq. (43.19) then has the form

$$\begin{aligned} \cos\left(ez_0^2H/c\hbar\right) &\cong 1 - \frac{1}{2}\left(ez_0^2H/c\hbar\right)^2 \\ &= 1 - \frac{1}{2}(z_0/l_H)^2 \end{aligned} \tag{43.23}$$

However the fluxes within these loops will have a *random sign* and hence tend to cancel, and it turns out that this reduces the amplitude by a factor,

$$e^{-t/\tau_H}; \quad \frac{1}{\tau_H} \approx D\left(\frac{eHz_0}{c\hbar}\right)^2 = \frac{Dz_0^2}{\lambda_H^4}. \tag{43.24}$$

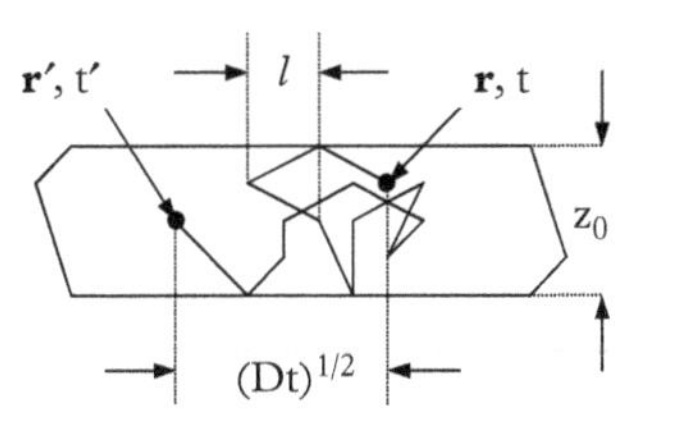

Figure 43.5 *The projection perpendicular to the magnetic field of a diffusive path of total length, $L_\phi = V_F\tau_\phi$, traced out by a particle moving in a slab of thickness, z_0. Shown also is the mean free path, l, and a typical distance covered in a time, τ_ϕ.*

[22] This same length is introduced in the treatment of Landau levels; see the discussion following Eq. (48.5a). It is related to the flux quantum by $\Phi_0 = \pi\lambda_H^2 H$.

We combine the time, τ_H, with τ_ϕ to make a time, $\tau_\phi(H)$, in a manner such that the shortest time dominates the response,

$$\frac{1}{\tau_\phi(H)} = \frac{1}{\tau_\phi} + \frac{1}{\tau_H}. \tag{43.25}$$

The resulting form for the field-dependent conductivity shift is then

$$\Delta\sigma(H) \approx \frac{e^2}{\hbar}\ln\left(\tau_\phi/\tau_\phi(H)\right); H \parallel \text{film}, \tag{43.26}$$

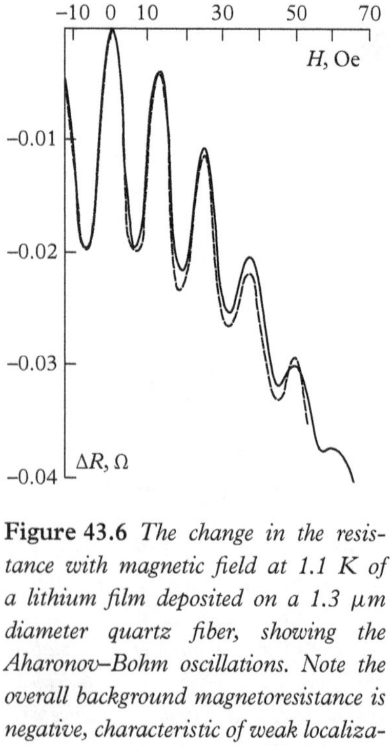

Figure 43.6 *The change in the resistance with magnetic field at 1.1 K of a lithium film deposited on a 1.3 μm diameter quartz fiber, showing the Aharonov–Bohm oscillations. Note the overall background magnetoresistance is negative, characteristic of weak localization. The dashed line shows a fit to the theory. (After Altshuler, Aronov, Spivak, Sharvin, and Sharvin (1988).)*

where $\tau_\phi(H) \leq \tau_\phi$ and hence we continue to have a negative magnetoresistance. Equation (43.26) behaves quadratically and logarithmically for small and large H respectively.

In the presence of spin–orbit coupling (which is enhanced in metals with a higher atomic number) the diffusing electrons can be "scattered" between the two spin projections at a rate, τ_{so}^{-1}.[23] We will not go through the associated analysis but only remark that when $\tau_{so} \gg c\hbar/DeH$, the *sign* of the change in conductivity changes from positive to negative.

43.6 The experiment of Yu. Sharvin and D. Sharvin

A special case is that of a thin conducting film with thickness, $a \ll \lambda_H$, deposited on the surface of an insulating cylinder with both the current and the magnetic field directed along the cylinder axis. In addition to returning directly to the origin, electrons can diffuse around the perimeter of the cylinder and return to where they started modulo an azimuthal angle of 2π. The Green's function for the diffusion equation for motion confined to the surface of cylinder of radius R has the form

$$G(\varphi, z, t; 0, 0, 0) = \frac{1}{4\pi Dt} e^{-\frac{z^2+\varphi^2R^2}{4Dt}} \sum_{n=0}^{\infty} e^{-\frac{(2\pi nR)^2}{4Dt}} \theta(t);\ 0 \leq \varphi < 2\pi. \tag{43.27}$$

In the limit where the film thickness is small compared with the radius, the Aharonov–Bohm phase factor becomes $\phi = e\Phi/\hbar c = 4\pi\Phi/\Phi_0$, where $\Phi = \pi R^2$ and $\Phi_0 = hc/2e$. Analogous to our argument in obtaining Eq. (43.19), the probability of returning to the origin involves $G(0, 0, t; 0, 0, 0)$, along with the Aharonov–Bohm factor. The equation corresponding to (43.20) is then

$$\Delta\sigma(H) \approx \frac{e^2}{\hbar}\int_0^{\tau_\phi}\frac{dt}{t}\sum_{n=0}^{\infty} e^{-\frac{(2\pi nR)^2}{Dt}}\left[1 - \cos\left(4\pi\frac{\Phi}{\Phi_0}\right)\right]. \tag{43.28}$$

[23] Scattering between different bands in polyvalent metals can also affect the magnetoresistance.

In the limit $D\tau_\phi \ll (2\pi R)^2$, where the electrons dephase before diffusing around the perimeter of the cylinder, Eq. (43.28) must reduce to the 2d case (corresponding to the $z_0 = 0$ in (43.26)). In the opposite limit, where $D\tau_\phi \gg (2\pi R)^2$, corresponding to the electron passing multiple times around the cylinder before undergoing a phase destroying process, $\Delta\sigma(H)$ *will oscillate with the magnetic field* with a period given by $2\pi R^2 H/\Phi_0$. At sufficiently high temperatures only a single harmonic is present, but the harmonic content grows rapidly as the temperature is lowered. The effect was predicted by Altshuler, Aronov, and Spivak;[24] it was first observed in magnesium by Yu. Sharvin and D. Sharvin.[25] It turns out that in spite of its relatively low atomic number the spin–orbit coupling is large enough to contribute a positive background to the magnetoresistance (see discussion at the end of the previous section). The experiment was repeated for lithium,[26] as shown in Fig. (43.6).

Aharonov–Bohm oscillations with magnetic field are also observed in the conductivity of small rings patterned between an input and an output lead.[27] Such oscillations, rather than arising from coherent backscattering, are analogous to those observed in the two-junction SQUID, discussed in Section 33.4.2. But rather than having a period involving $h/e^* = h/2e$, the period is given by h/e, since the interfering entities are electrons traversing both halves of the ring, as opposed to Cooper pairs.

43.7 Quantization effects in mesoscopic systems

The discussions of electrical transport in this chapter were limited to bulk materials and thin films. However, studies of systems, both clean and disordered, in one dimension, as wires or narrow strips,[28] and in "zero" dimensions, as point contacts,[29] can also involve quantization effects. Here we refer the reader to the books by Imry and Datta, in the Additional Reading section

43.8 Localization meets superconductivity

Superconductivity and electron localization represent two dramatic extremes of electrical transport. In disordered two-dimensional thin film materials, the experimental evidence is consistent with there being an abrupt phase transition between these two states at absolute zero as a function of the sheet resistance,[30] the latter being varied via the film thickness.[31] Figure 43.7(a) shows the temperature dependence of the electrical resistance of the material molybdenum carbide for various film thicknesses, which supports this conclusion.[32] Note that as the temperature is reduced there is a threshold thickness (13 A in this material) which divides thicker films, where the resistance rapidly falls, from thinner films, where it rapidly rises. Figure 43.7(b) shows the behavior extended to lower temperatures

[24] The Aharonov–Bohm effect in disordered conductors, B.L. Altshuler, A.G. Aronov, and B.Z. Spivak, *JETP Lett.* 33, 94 (1981).

[25] Magnetic-flux quantization in a cylindrical film of a normal metal, Yu. Sharvin and D. Sharvin, *JETP Lett.* **34**, 272 (1981).

[26] Observation of the Aharonov–Bohm effect in hollow metal cylinders, B.L. Altshuler, A.G. Aronov, B.Z. Spivak, D.Y. Sharvin, and Y.V. Sharvin, *JETP Lett.* 35, 588 (1982).

[27] h/e Aharonov–Bohm effect in gold rings, R.A. Webb, S. Washburn, C.P. Umbach, and R.B. Laibowitz, *J. Magn. Magn. Mater.* **54–57**, 1423 (1986).

[28] One-dimensional transport and the quantization of the ballistic resistance, D.A. Wharam, T.J. Thornton, R. Newbury, M. Pepper, H. Ahmed, J.E.F. Frost, D.C. Peacock, D.A. Ritchie, and G.A. Jones, *J. Phys. C Solid State Phys.* **21**, L209 (1988).

[29] Quantized conductance of point contacts in a two-dimensional electron-gas, B.J. Van Wees, H. van Houten, C.W.J. Beenakker, J.G. Williamson, L.P. Kouwenhoven, D. van der Maral, and C.T. Foxon, *Phys. Rev. Lett.* **60**, 848 (1988).

[30] The sheet resistance is that resistance presented by a square film. It is easy to verify that it is independent of the size of the square.

[31] For a popular discussion, see: Superconductor – insulator transitions in the two-dimensional limit, A.M. Goldman and N. Markovi, *Phys. Tod.* **51**, 39 (2008), and references therein.

[32] Critical sheet resistance for the suppression of superconductivity in thin Mo-C films, S.J. Lee and J.B. Ketterson, *Phys. Rev. Lett.* **64**, 3078 (1990); Superconductor-insulator transition in Mo-C films, Proceedings of the international conference on high temperature superconductivity and localization phenomena; book series: *Progress in High Temperature Superconductivity*, **32**, 36 (1992).

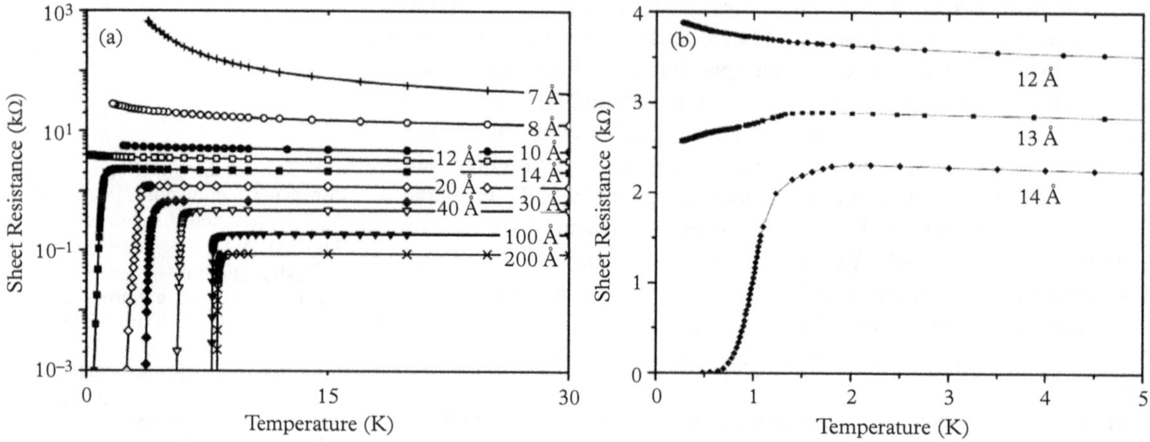

Figure 43.7 *(a) Temperature dependence of the sheet resistance of molybdenum carbide films for a range of thicknesses. (b) The behavior near the critical thickness extended to 0.25 K. Note the appearance of a relatively abrupt transition between the superconducting and insulating states for films in the vicinity of 13 A. (After Lee et al. (1990).)*

in the vicinity of the critical thickness. Note the shift between superconductivity and localization occurs near a sheet resistance of 3 kΩ, which in order of magnitude is similar to a quantity with the same dimensions assembled from fundamental constants; $h/2e^2 = 12.9$ kΩ. We will not pursue this behavior further, although it is expected to remain a center of much theoretical and experimental activity.

ADDITIONAL READING

Fundamentals of the Theory of Metals, A.A. Abrikosov, North Holland (1988).

Electronic Processes in Non-Crystalline Solids, N.F. Mott and E.A. Davis, Oxford University Press, Oxford (1979).

Introduction to Mesoscopic Physics, Yoseph Imry, Oxford University Press, Oxford (1997).

Electronic Transport in Mesoscopic Systems, Supriyo Datta, Cambridge University Press, Cambridge (1995).

Semiconductors, David K. Ferry, Macmillan, NY (1991).

We also note the following review articles:

Weak localization in thin films: a time-of-flight experiment with conduction electrons, G. Bergmann, *Phys. Rep.* **107**, 1 (1984). Here many important results are obtained via physical arguments without invoking Green's functions or other many-body formalisms.

Electron-electron interaction in disordered conductors, B.L. Altshuler and A.G. Aronov in *Electron-Electron Interactions in Disordered Systems*, Ed. A.L. Efros and M. Pollak, Elsevier Science Publishers B.V. (1985).
The Anderson-Mott transition, D. Belitz and T.R. Kirkpatrick, *Rev. Mod. Phys.* **66**, 261 (1994).
Disordered Electronic Systems, P.A. Lee and T.V. Ramakrishnan, *Rev. Mod. Phys.* **57**, 287 (1985).
Scattering and Interference effects in variable range hopping conduction, in *Hopping Transport in Solids* Ed. M. Pollak and B. Shklovskii, Elsevier Science Publishers B.V. (1991).
Universalities: From Anderson localization to quantum chaos, by Boris L. Altshuler, B.D. Simons, in Course 1, *Mesoscopic Quantum Physics*, Ed. E. Akkermans, G. Montambaux, J.L. Prichard and J. Zinn-Justin, Les Houches, Session LXI, Elsevier Science Publishers B.V. (1995).

Problem

43.1. The Green's function for the diffusion equation for a film parallel to the x–y plane with thickness, z_0, is given by

$$G(x, y, z, t; 0, 0, z', 0) = \frac{1}{4\pi Dtz_0} e^{-\frac{x^2+y^2}{4Dt}} \sum_{n=0}^{\infty} \cos\left(\frac{\pi nz}{z_0}\right) \cos\left(\frac{\pi nz'}{z_0}\right) e^{-\frac{\pi^2 n^2 Dt}{z_0^2}}.$$

Show that for $Dt \gg z_0^2$ and $Dt \ll z_0^2$ we obtain 2d and 3d behaviors respectively.

44

Magnetic Impurities and their Interactions: The Anderson Model, the Kondo Effect, and the RKKY Interaction

It is known experimentally that transition and rare earth atoms (those with unfilled d and f shells) may have a *net magnetic moment* when dissolved in various host metals.[1] However we note that such moments do not always form, and the conditions for their existence are somewhat subtle.[2] Of course isolated atoms can have a moment (and for the atoms of interest a rather large moment as described by Hund's rules), and the issue here is under what circumstances this impurity moment can survive in the host metal when the interaction with the conduction electrons is taken into account, with the latter thought of as host atom s–p shells broadened out into bands in forming the metallic state. The simplest case is where the metal contains only a single impurity atom which interacts with a sea of free electrons. This problem was first treated satisfactorily using a model Hamiltonian proposed by P.W. Anderson.[3] Since it forms the basis for other phenomena and is a component of subsequent theories for strongly correlated systems, this chapter will begin with an extended discussion of this model. Solutions in the mean field approximation follow in Section 44.2.

Once present, such moments can scatter the host electrons thereby contributing to the residual electrical resistivity at low temperatures in proportion to their concentration. Connecting the model Hamiltonian used to describe this scattering, the so-called s–d Hamiltonian, with the Anderson Hamiltonian is discussed in Section 44.3. However along with a temperature-independent contribution, an additional contribution is observed experimentally, which is proportional to the log of the temperature. It is observed in many alloys containing magnetic impurities and was partially explained, theoretically, by J. Kondo and is now referred to as the Kondo effect, which will be discussed in Section 44.4. A complete theoretical treatment, which removes the divergence at very low temperatures, is more demanding and will not be discussed.

At higher concentrations spin–spin interaction effects acting through the host conduction electrons become important. The strength and sign of this interaction

[1] Insulators involving d and f shell atoms (including adjacent ions such as Cu^{+2}) commonly have a moment if, after bond formation, these shells are unfilled. Such ions are referred to as paramagnetic ions.

[2] With the spin-polarized supercell approach, discussed in Chapter 40, one can now predict, with reasonable accuracy, whether a moment will form; of course the size of the unit cell must be large enough to keep the impurities in neighboring unit cells sufficiently separated.

[3] Localized magnetic states in metals, P.W. Anderson, *Phys. Rev.* **124**, 41 (1961).

varies as a function of the separation between the spins and we will consider a simple model for this effect, the so-called RKKY interaction, in Section 44.5. This same interaction can also couple nuclear spins.

At low temperatures the interaction between spins leads to the "freezing-in" of a random arrangement of their orientations, a system known as a *spin glass*,[4] which is discussed briefly in Section 44.6. Higher concentrations can result in a ferromagnetic alloy.

44.1 The Anderson Hamiltonian

In what follows we construct the simplest model that is general enough to capture the observed behavior qualitatively. Our first assumption is that the impurity atoms have only a single localized "d" or "f" orbital, or two states when spin is included; this assumption vastly simplifies the calculation.[5] However the energy of this state differs, depending on whether it is occupied by one or two electrons. If a single electron (say a d electron) is present we denote the energy as ε_{0d}. However if two electrons are present (with opposite spins as required by the Pauli principle and hence there will be no net magnetic moment), the additional Coulomb repulsion energy associated with double occupancy raises the energy by an amount we designate as U. If we denote the single orbital of our model d electron as $\varphi(\mathbf{r})$, then the constant, U (usually called the Hubbard interaction[6]), may be written as

$$U = \int d^3r_1 \int d^3r_2 \, |\varphi(\mathbf{r}_1)|^2 \, |\varphi(\mathbf{r}_2)|^2 \frac{e^2}{|\mathbf{r}_1 - \mathbf{r}_2|}, \tag{44.1}$$

which is positive definite. Using the second quantized notation we can write this term in the Hamiltonian in the form

$$U\hat{n}_{d\uparrow}\hat{n}_{d\downarrow}, \tag{44.2}$$

where $\hat{n}_{d\sigma} \equiv \hat{c}^{\dagger}_{d\sigma}\hat{c}_{d\sigma}$ and $\hat{c}^{\dagger}_{d\sigma}$ and $\hat{c}_{d\sigma}$ create and destroy a d electron with spin σ respectively. Including the one-electron energy, ε_{0d}, that part of the total Hamiltonian which involves only the d electron is written as

$$\hat{H}_d = \varepsilon_{0d}\left(\hat{n}_{d\uparrow} + \hat{n}_{d\downarrow}\right) + U\hat{n}_{d\uparrow}\hat{n}_{d\downarrow}. \tag{44.3}$$

To complete our model Hamiltonian we need two more ingredients. The first, $\hat{H}_k$, is associated with the one-electron kinetic energies of the conduction electrons (which for simplicity we take as being derived from the s valence electrons of the host, as in a monovalent metal). The second is an (off-diagonal) interaction, $\hat{H}_{kd}$, between our single d state and the host conduction electrons (called the *s–d interaction*). We write these two contributions as

$$\hat{H}_k + \hat{H}_{kd} = \sum_{\mathbf{k},\sigma} \varepsilon_k \hat{c}^{\dagger}_{k\sigma}\hat{c}_{k\sigma} + \sum_{\mathbf{k},\sigma}\left(V_{kd}\hat{c}^{\dagger}_{k\sigma}\hat{c}_{d,\sigma} + V^{*}_{kd}\hat{c}^{\dagger}_{d\sigma}\hat{c}_{k\sigma}\right). \tag{44.4}$$

[4] The name derives from the state of order present in an ordinary glass, where only short-range correlations are present. Hence for the spin glass we expect no macroscopic moment. The number of independent random arrangements of such spins is vast and many of these may have energies close to the globally lowest state; hence no unique ground state exists in practice.

[5] Inclusion of additional states is sometimes called the multi-state Anderson model. But even in a cubic environment, the five d levels split into a three-fold t_{2g} and a two-fold e_g state (see Figure 45.4). On lowering the symmetry to tetragonal, single states occur. The copper d states in high temperature superconductors (see Chapter 46) can for many purposes be limited to the $d_{x^2-y^2}$ orbital.

[6] Electron correlations in narrow energy bands, J. Hubbard, *Proc. R. Soc. A* 276, 238 (1963).

Our complete model Hamiltonian is therefore

$$\hat{H} = \hat{H}_k + \hat{H}_{k\,d} + \hat{H}_d$$
$$= \sum_{k,\sigma} \varepsilon_k \hat{c}^\dagger_{k\sigma} \hat{c}_{k\sigma} + \sum_{k,\sigma} \left(V_{kd} \hat{c}^\dagger_{k\sigma} \hat{c}_{d\sigma} + V^*_{kd} \hat{c}^\dagger_{d\sigma} \hat{c}_{k\sigma} \right)$$
$$+ \varepsilon_{0d} \left(\hat{n}_{d\uparrow} + \hat{n}_{d\downarrow} \right) + U \hat{n}_{d\uparrow} \hat{n}_{d\downarrow}; \tag{44.5}$$

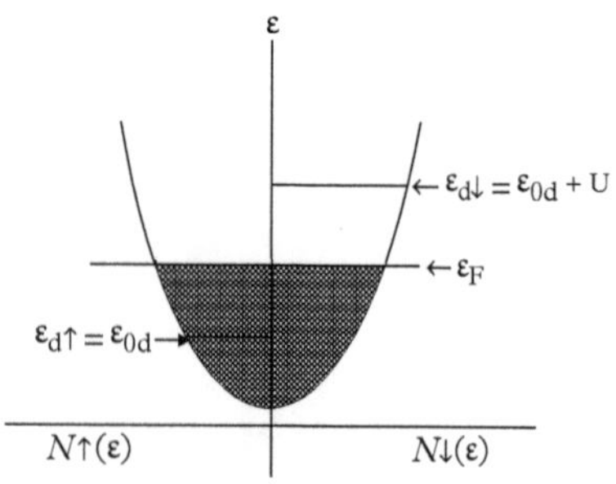

Figure 44.1 *Unperturbed energy levels in the absence of* s–d *admixture.*

for simplicity we will locate the zero of energy at the Fermi level (i.e., $\varepsilon_F = 0$). The relevant energy scales are usually taken such that

$$U \gg \varepsilon_{0d} \gg N(0)\, \overline{|V_{kd}|^2},$$

where $N(0)$ is the density of states at the Fermi level and $\overline{|V_{kd}|^2}$ is an average of $|V_{kd}|^2$ over occupied states in the conduction band; as we will see shortly, $N(0)\,\overline{|V_{kd}|^2}$ is a measure of the mixing between our two impurity states and all the conduction band states. We will also assume that V_{kd} becomes small for $|\varepsilon_k|$ greater than a cut-off energy of order ε_{0d}.

To set the stage for what follows, let us assume that $V_{kd} = 0$. If $\varepsilon_{0d} > 0$ (recall $\varepsilon_F = 0$) neither d state is occupied at absolute zero and hence there will be no moment. If $\varepsilon_{0d} < 0$ but $\varepsilon_{0d} + U > 0$ the occupied levels would appear as in Fig. 44.1. Then one of the d states (say spin up) would be *occupied* with energy ε_{0d} and the second would be *empty*. Clearly the impurity site involving the d electron would then possess a net magnetic moment. Ignoring any interaction between a concentration of such impurities in a non-magnetic metallic host, we would expect a paramagnetic contribution to the magnetic susceptibility having a Curie-like temperature dependence, in addition to the temperature-independent Pauli form (from the host electrons). Such a contribution is sometimes (but not always) observed experimentally for transition and rare earth impurities dissolved in simple (s–p) metals. Finally, if $\varepsilon_{0d} < 0$ and $\varepsilon_{0d} + U < 0$ both states would be occupied, each with energy, $\varepsilon_{0d} + U/2$, and we would again have no moment. These possibilities partially account for the fact that not all d or f electron impurities result in a net magnetic moment. However when $V_{kd} \neq 0$ these conclusions are strongly affected, as we will see.

44.2 The mean-field theory

We now construct a mean-field theory to treat the case when there is coupling between our single d electron and the conduction band. Our first task will be to calculate the ground state energy. We will do this using a variational approach involving a state function of the form

$$|\Phi_0\rangle = \prod_{\sigma, n < n_{max}} \hat{c}^\dagger_{n\sigma} |0\rangle, \tag{44.6}$$

where the $\hat{c}^{\dagger}_{n\sigma}$ operators will create one-electron states which are yet to be determined and n_{max} denotes the highest occupied state, which is fixed by the Fermi energy. In this approximation the ground state energy is given by

$$E = \frac{\langle \Phi_0 | \hat{H} | \Phi_0 \rangle}{\langle \Phi_0 \, | \Phi_0 \rangle}. \tag{44.7}$$

We will assume that the operators, $\hat{c}^{\dagger}_{n\sigma}$, for the states in the presence of the interactions can be written as a linear combination of the operators, $\hat{c}_{k\sigma}$ and $\hat{c}_{d\sigma}$, which appear in the Hamiltonian; i.e.,

$$\hat{c}_{n\sigma} = \sum_{\mathbf{k}} a_{nk\sigma} \hat{c}_{k\sigma} + a_{nd\sigma} \hat{c}_{d\sigma} \tag{44.8a}$$

and

$$\hat{c}^{\dagger}_{n\sigma} = \sum_{\mathbf{k}} a^{*}_{nk\sigma} \hat{c}^{\dagger}_{k\sigma} + a^{*}_{nd\sigma} \hat{c}^{\dagger}_{d\sigma}. \tag{44.8b}$$

For an infinite system the quantum numbers, n and **k**, are of course continuous, whereas d denotes our single impurity state; the index σ takes on two values corresponding to up and down spin states.

The best ground state energy, E, and wave function, $|\Phi_0\rangle$, are obtained by minimizing (44.7) with $a_{nk\sigma}$ and $a_{nd\sigma}$, and their complex conjugates, $a^{*}_{nk\sigma}$ and $a^{*}_{nd\sigma}$, treated as variational parameters. However, it is sufficient to minimize with respect to one of these parameter sets, since they yield complex conjugate conditions. Rather than minimizing Eq. (44.7) directly, we can introduce a Lagrange multiplier, $\varepsilon_{n\sigma}$, associated with the constraint, $\langle \Phi_0 | \Phi_0 \rangle = 1$, and our minimization then involves the equations[7]

$$\frac{\partial}{\partial a^{*}_{nk\sigma}} \left(\langle \Phi_0 | \hat{H} | \Phi_0 \rangle + \varepsilon_{n\sigma} \langle \Phi_0 | \, \Phi_0 \rangle \right) = 0 \tag{44.9a}$$

and

$$\frac{\partial}{\partial a^{*}_{nd\sigma}} \left(\langle \Phi_0 | \hat{H} | \Phi_0 \rangle + \varepsilon_{n\sigma} \langle \Phi_0 | \, \Phi_0 \rangle \right) = 0; \tag{44.9b}$$

using (44.6) and (44.8) leads to the set of equations

$$\langle \hat{c}_{k\sigma} | \, \hat{H} \, | \Phi_0 \rangle - \varepsilon_{n\sigma} \langle \hat{c}_{k\sigma} | \, \Phi_0 \rangle = 0 \tag{44.10a}$$

and

$$\langle \hat{c}_{d\sigma} | \, \hat{H} \, | \Phi_0 \rangle - \varepsilon_{n\sigma} \langle \hat{c}_{d\sigma} | \, \Phi_0 \rangle = 0. \tag{44.10b}$$

Evaluating (44.10a,b) using the Hamiltonian (44.5) and the trial ground state (44.6) results in the equations[8]

$$\varepsilon_{n\sigma} a_{nk\sigma} = \varepsilon_{k} a_{nk\sigma} + V_{kd} a_{nd\sigma} \tag{44.11}$$

[7] A similar parameter was introduced in our earlier discussion of the Hartree and Hartree–Fock theory (see Section 15.1.1).

[8] This is accomplished by commuting the destruction operators, $\hat{c}_{k\sigma}$ and $\hat{c}_{d\sigma}$, to the far right where when acting on the vacuum state they give zero.

and

$$\varepsilon_{n\sigma} a_{nd\sigma} = \left(\varepsilon_{0d} + U\langle n_{d,-\sigma}\rangle\right) a_{nd\sigma} + \sum_k V_{kd} a_{nk\sigma}. \qquad (44.12)$$

Equations (44.11) and (44.12) form a set of nonlinear (when $U \neq 0$) homogeneous equations that must be solved for the coefficients, $a_{nd\sigma}$ and $a_{nk\sigma}$, and the parameters (eigenvalues), $\varepsilon_{n\sigma}$.

44.2.1 Case $V_{kd} = 0$, $U \neq 0$

When $V_{kd} = 0$, the solution of Eq. (44.11) is trivial and corresponds to the continuum of solutions,

$$\varepsilon_{n\sigma} = \varepsilon_{k\sigma}, \qquad (44.13a)$$

associated with all the conduction electron states. When $V_{kd} = 0$ and we also have $U = 0$, Eq. (44.12) has the two solutions,

$$\varepsilon_{d\uparrow,\downarrow} = \varepsilon_{0d} \qquad (44.13b)$$

associated with either the spin up or spin down state of our d level. When $U \neq 0$ the solutions are

$$\varepsilon_{n\uparrow} = \varepsilon_{0d\uparrow} + U\langle n_{d\downarrow}\rangle,\ \varepsilon_{n\downarrow} = \varepsilon_{0d\downarrow} + U\langle n_{d\uparrow}\rangle. \qquad (44.13c)$$

As noted earlier, if $\varepsilon_{0d} < \varepsilon_F$ and $\varepsilon_{0d} + U > \varepsilon_F$, only one of the spin projections will be occupied and that state will have an energy, ε_{0d}. If $\varepsilon_{0d} + U < \varepsilon_F$, both spin states will be occupied; from (44.3) the total energy will be

$$E_d = \varepsilon_{0d}\left[\langle n_{d\downarrow}\rangle + \langle n_{d\downarrow}\rangle\right] + U\langle n_{d\downarrow}\rangle\langle n_{d\uparrow}\rangle = 2\varepsilon_{0d} + U;$$

however, the single particle excitation energies[9] will be $\varepsilon_{d\uparrow,\downarrow} = \varepsilon_{0d} + U$. If $\varepsilon_{0d} > \varepsilon_F$, both states are empty.

44.2.2 Case $V_{kd} \neq 0$, $U = 0$

We start by solving (44.11) for $a_{nk\sigma}$, yielding

$$a_{nk\sigma} = -\frac{V_{kd}\, a_{nd\sigma}}{(\varepsilon_k - \varepsilon_{n\sigma})}, \qquad (44.14)$$

which on substituting into (44.12) gives coupled nonlinear equations for the two spin states,

$$\left[-\left(\sum_k \frac{|V_{kd}|^2}{(\varepsilon_k - \varepsilon_{n\sigma})}\right) + \left(\varepsilon_{0d} - \varepsilon_{n\sigma} + U\left|a_{nd,-\sigma}\right|^2\right)\right] a_{nd\sigma} = 0. \qquad (44.15)$$

[9] As in Fermi liquid theory, we define the excitation energies as $\varepsilon_{d\uparrow,\downarrow} = \partial E_d / \partial \langle n_{d\uparrow,\downarrow}\rangle$.

Let us now study the solution of (44.15) for the case U = 0 where the two spin states are uncoupled, in which case we have

$$-\int_0^{\varepsilon_F} d\varepsilon' \frac{N(\varepsilon')\,\overline{|V_{\mathbf{k}d}|^2}}{(\varepsilon' - \varepsilon_{n\sigma})} + (\varepsilon_{0d} - \varepsilon_{n\sigma}) = 0, \qquad (44.16)$$

where $\overline{|V_{\mathbf{k}d}|^2}$ implies an average over a *constant energy* surface in **k** space and $N(\varepsilon)$ is the density of states of that surface. We must now seek the eigenvalues, $\varepsilon_{n\sigma}$, of this integral equation.

As previously noted, our conduction electron states have a continuous energy spectrum; some uncertainty then arises as to how we interpret the vanishing denominator, which occurs in performing the integration in Eq. (44.16). If we write $\varepsilon_{nd\sigma} \to \varepsilon_{nd\sigma} + i0$, where $i0$ is a positive infinitesimal, the singularity is avoided;[10] we may then perform the integration using the prescription,

$$\frac{1}{x - a - i0} = P\frac{1}{x - a} + i\pi\delta(x - a), \qquad (44.17)$$

where P denotes the *principal value* of the associated integral. This principal value contribution to the integral depends on the details of the host band structure, which are not of interest to us here, since we seek only a qualitative description of the effects of a magnetic impurity. We denote it by $\delta\varepsilon_{n\sigma}$, which is interpreted as a *level shift* from the impurity energy, ε_{0d}; formally this contribution is given by

$$\delta\varepsilon_{n\sigma} = -P\int_0^{\varepsilon_F} d\varepsilon' \frac{N(\varepsilon')\,\overline{|V^2_{\mathbf{k}d}|}}{(\varepsilon' - \varepsilon_{n\sigma})}. \qquad (44.18)$$

Including the contribution of the delta function in (44.17), Eq. (44.16) has the solution

$$\varepsilon_{n\sigma} = \varepsilon_{0d} + \delta\varepsilon_{0d} + i\Gamma_{nd}, \qquad (44.19)$$

where we have introduced the *level width*,

$$\Gamma_{nd} = \pi N(\varepsilon_n)\,\overline{|V_{\mathbf{k}d}|^2}, \qquad (44.20)$$

which we henceforth take to be a constant, Γ.

In the absence of the coupling, $V_{\mathbf{k}d}$, between the conduction electrons and our impurity, the density of states *associated with our d level* would be $\rho_d(\varepsilon) = \delta(\varepsilon - \varepsilon_{d\sigma})$.[11] When $V_{\mathbf{k}d} \neq 0$ we will define the density of states as

$$\rho_d(\varepsilon) = \sum_n |a_{nd\sigma}|^2\,\delta(\varepsilon - \varepsilon_{n\sigma}). \qquad (44.21)$$

In leading order, the amplitudes of the conduction states are altered only infinitesimally and we may take $a_{n\mathbf{k}\sigma} = \delta_{n\mathbf{k}}\theta(\varepsilon_F - \varepsilon_{\mathbf{k}})$, where the theta function ensures

[10] Formally, this problem is resolved by regarding the perturbation, $V_{\mathbf{k}d}$, as being turned on infinitely slowly, starting at $-\infty$, according to

$$V_{\mathbf{k}d}(t) = \lim_{\alpha\to 0} V_{\mathbf{k}d}e^{\alpha t}$$

and calculating the wave functions using *time-dependent* perturbation theory.

[11] We use the symbol, ρ_d, to note that it is the density of states associated with a single d state, as opposed to the density of states, $N(\varepsilon)$, associated with the conduction electrons that formally diverges if we take it as applying to all N electrons of an infinite system.

that only states below the Fermi energy are occupied; inserting this form into Eq. (44.12), we get

$$a_{nd\sigma} = -\frac{1}{(\varepsilon_{0d} - \varepsilon_{n\sigma})} \sum_k V_{kd} \delta_{nk} \theta(\varepsilon_F - \varepsilon_k)$$

$$= -\frac{V_{kd}}{(\varepsilon_{0d} - \varepsilon_{n\sigma})}; \qquad n \leq n_{max}. \tag{44.22}$$

Substituting (44.22) into (44.21), and using (44.19) and the definition (44.20) gives

$$\rho_d(\varepsilon) = \sum_n |a_{nd\sigma}|^2 \delta(\varepsilon - \varepsilon_{n\sigma})$$

$$= \int_0^{\varepsilon_F} N(\varepsilon) d\varepsilon \left| \frac{V_{kd}}{(\varepsilon_{0d} - \varepsilon_{n\sigma})} \right|^2 \delta(\varepsilon - \varepsilon_{n\sigma})$$

$$= \frac{1}{\pi} \frac{\Gamma}{(\varepsilon_{0d} + \delta\varepsilon_d - \varepsilon)^2 + \Gamma^2}; \tag{44.23}$$

note that (44.23) has the familiar Lorentzian form,[12] which is encountered in the theory resonant atomic energy levels, and it satisfies the normalization condition, $\int_{-\infty}^{+\infty} d\varepsilon\, \rho_d(\varepsilon) = 1$. The physical content of this equation is central to our remaining discussion.

Similarly, we would define the density of states of the conduction electrons as $\rho_k(\varepsilon) = \sum_n |a_{nk\sigma}|^2 \delta(\varepsilon - \varepsilon_n)$. Substituting $a_{nk\sigma}$ from (44.14), we have

$$\rho_k(\varepsilon) = \sum_n \left| \frac{V_{kd}}{(\varepsilon_k - \varepsilon_{n\sigma})} \right|^2 |a_{nd\sigma}|^2 \delta(\varepsilon - \varepsilon_{n\sigma}),$$

which on inserting $a_{nd\sigma}$ from (44.22) gives

$$\rho_k(\varepsilon) = \left| \frac{V_{kd}}{(\varepsilon_k - \varepsilon_{n\sigma})} \right|^2 \rho_d(\varepsilon). \tag{44.24}$$

In the limit, $V_{kd} \to 0$, $\rho_k(\varepsilon) \to N(\varepsilon)$, the "bare" density of states, as must happen; in the presence of the interaction it represents the effect of a small admixture of the d state. However we will not require this equation in what follows.

44.2.3 Case $V_{kd} \neq 0, U \neq 0$

In the presence of the "mixing" potential, V_{kd}, and the resulting broadening of the energy level, ε_{0d}, this state will in general have a *fractional occupation.* The occupancy for the state having spin σ is given by integrating the associated density

[12] Sur la structure electronique des metaux et alliages de transition et des metaux lourds, J. Friedel, *J. Phys. Radium* **19**, 573 (1958).

of states, $\rho_d(\varepsilon)$, as given by (44.23), from the bottom of the band (which to the necessary accuracy we may place at $-\infty$) to the Fermi energy, ε_F; i.e.,

$$\langle n_{d\sigma}\rangle = \frac{1}{\pi}\int_{-\infty}^{\varepsilon_F} d\varepsilon \frac{\Gamma}{(\varepsilon_{d\sigma} - \varepsilon)^2 + \Gamma^2},$$

which gives

$$\langle n_{d\sigma}\rangle = \frac{1}{\pi}\cot^{-1}\left[\frac{\varepsilon_{d\sigma} - \varepsilon_F}{\Gamma}\right], \tag{44.25}$$

where we henceforth write $\varepsilon_d \equiv \varepsilon_{d0} + \delta\varepsilon_d$. To complete the model we must now include the effects of the Coulomb repulsion energy, U; within a *self-consistent* framework this is accomplished by substituting

$$\varepsilon_{d+} \equiv \varepsilon_d + U\langle n_{d-}\rangle \tag{44.26a}$$

$$\varepsilon_{d-} \equiv \varepsilon_d + U\langle n_{d+}\rangle \tag{44.26b}$$

into Eq. (44.25) and solving the resulting pair of equations

$$\langle n_{d+}\rangle = \frac{1}{\pi}\cot^{-1}\left[\frac{\varepsilon_d - \varepsilon_F + U\langle n_{d-}\rangle}{\Gamma}\right] \tag{44.27a}$$

and

$$\langle n_{d-}\rangle = \frac{1}{\pi}\cot^{-1}\left[\frac{\varepsilon_d - \varepsilon_F + U\langle n_{d+}\rangle}{\Gamma}\right], \tag{44.27b}$$

where $\langle n_{d+}\rangle$ and $\langle n_{d-}\rangle$ denote the expectation values of the spin up and spin down population of our d state, which we henceforth write as $n_{d\pm}$. We note that on physical grounds the solutions are restricted to the range, $0 \le n_{d\pm} \le 1$.

We start by recalling, $\lim_{x\to\pm\infty}\cot^{-1} x = 0, \pi$, and consider the limit where $|V_{\mathbf{k}d}| \to 0$ (i.e., $\Gamma \to 0$) where the argument of the cotangent in (44.27a,b) diverges. For the case $\varepsilon_d < \varepsilon_F$ but $\varepsilon_d - \varepsilon_F + U > 0$, we have the solution, $n_{d+} = 1$, $n_{d-} = 0$, or an essentially equivalent solution, $n_{d+} = 0$, $n_{d-} = 1$. On the other hand when $|V_{\mathbf{k}d}| \to 0$ with $\varepsilon_d > \varepsilon_F$, we have $n_{d+} = n_{d-} = 0$. These are of course the two cases discussed earlier in Section 44.3.

In examining the behavior of the resulting solutions for intermediate cases where $|V_{\mathbf{k}d}| \neq 0$, it is convenient to introduce the dimensionless ratios

$$\alpha \equiv \frac{\varepsilon_F - \varepsilon_d}{U} \tag{44.28a}$$

and

$$\beta \equiv \frac{U}{\Gamma}; \tag{44.28b}$$

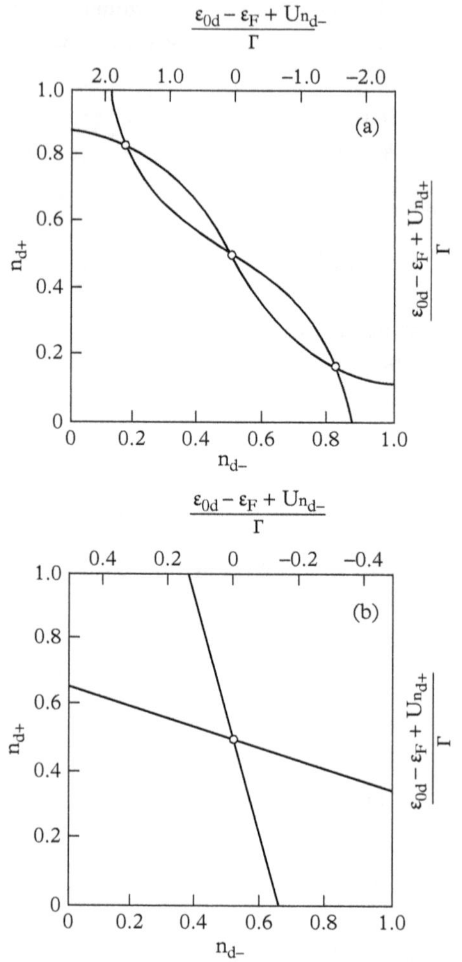

Figure 44.2 *(a)* $\alpha = 1/2, \quad \beta = 5$; *(b)* $\alpha = 12, \quad \beta = 1$. *(After Anderson (1961).)*

the limit, $\alpha = 0$, means the impurity level is positioned right at the Fermi level. In terms of the parameters α and β, Eqs. (44.27a,b) become

$$\pi n_{d\pm} = \cot^{-1}\left(\beta(n_{d\mp} - \alpha)\right). \qquad (44.29)$$

These equations must be solved either numerically or graphically. Figure 44.2 shows a plot of n_{d+} (vertical axis) versus $(\varepsilon_{0d} - \varepsilon_F + Un_{d-})/\Gamma$ (horizontal axis), and simultaneously, n_{d-} (horizontal axis) versus $(\varepsilon_{0d} - \varepsilon_F + Un_{d+})/\Gamma$ (vertical axis). The upper pair show a magnetic case ($\alpha = 1/2, \ \beta = 5$) and the lower pair a non-magnetic case ($\alpha = 1/2, \ \beta = 1$); the intersection points represent solutions to the combined equations. Note that for the magnetic case we actually have three solutions; however the non-magnetic, $n_{d+} = n_{d-} = 1/2$, solution is *unstable* (it is an energy maximum); the two asymmetric solutions represent the essentially equivalent spin up dominant and spin down dominant magnetic states. We now develop some approximate solutions for these two regimes which are valid in certain limits:

(a) The magnetic limit: $n_{d+} \cong 1$, $n_{d-} \cong 0$
Equation (44.29) may now be approximated as

$$\pi n_{d+} = \pi - \frac{1}{\beta(\alpha - n_{d-})}; \quad \pi n_{d-} = -\frac{1}{\beta(\alpha - n_{d+})}, \tag{44.30a.b}$$

from which we obtain the approximate solution,

$$\alpha(1 - n_{d+}) = (1 - \alpha)n_{d-} \cong \frac{1}{\pi\beta\left(1 - \frac{1}{\pi\alpha\beta(1-\alpha)}\right)}. \tag{44.31}$$

If we define a quantity, $m \equiv n_{d+} - n_{d-}$, that corresponds to the *effective magnetic moment* in Bohr magnetons (we take $g = 2$), these equations give

$$\begin{aligned} m &\cong 1 - \frac{1}{\pi\alpha\beta(1-\alpha) - 1} \\ &\cong 1 - \frac{1}{\pi\frac{\varepsilon_F - \varepsilon_d}{\Gamma}\left(1 - \frac{\varepsilon_F - \varepsilon_d}{U}\right) - 1}. \end{aligned} \tag{44.32}$$

Note that this differs from one when $\Gamma = 0$; i.e., the moment can differ from the free-impurity value. A schematic diagram of the split and broadened levels for the magnetic case is shown in Fig. 44.3.

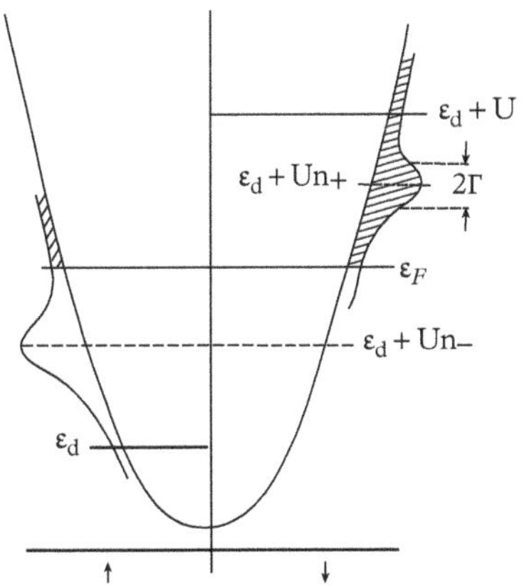

Figure 44.3 *A schematic plot of the broadened levels in the presence of the Coulomb repulsion, U; here the shaded regions are unoccupied portions of the* d *state resonance. (After Anderson, (1961).)*

(b) Non-magnetic cases: $n_{d+} = n_{d-} = n$
The possibility of having a non-magnetic state when $\varepsilon_d < \varepsilon_F$ but $\varepsilon_d + U > \varepsilon_F$ is a new feature produced when $V_{kd} \neq 0$. Equation (44.29) then takes the form

$$\cot(\pi n_d) = \beta(n_d - \alpha). \tag{44.33}$$

For the case where ε_d lies close to ε_F, but is spread out both above and below, we expect $n_d \cong 1/2$ and in this case we may expand the cotangent in (44.33) to leading order as

$$\cot(\pi n_d) \cong \pi\left(\frac{1}{2} - n_d\right), \tag{44.34}$$

which on equating to the right-hand side of (44.33) and solving for n_d gives

$$\begin{aligned} n_d &= \frac{1}{2}\left[\frac{1 + 2\alpha\beta/\pi}{1 + \beta/\pi}\right] \\ &= \frac{1}{2}\left[\frac{\pi\Gamma + 2(\varepsilon_F - \varepsilon_d)}{\pi\Gamma + U}\right]. \end{aligned} \tag{44.35}$$

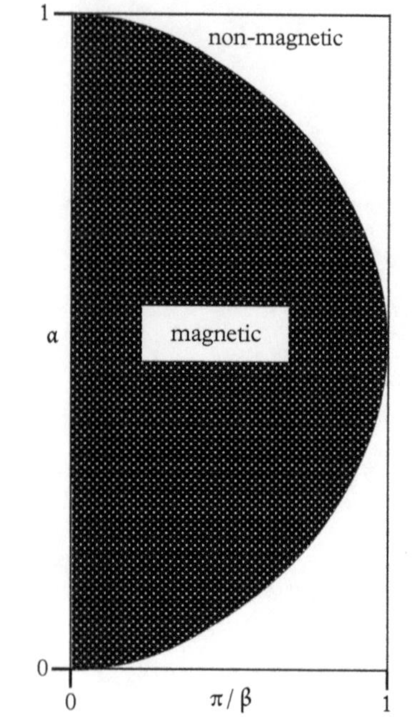

Figure 44.4 *A plot of α versus π/β showing the boundary between a polarized (magnetic) and unpolarized (non-magnetic) impurity spin. (After Anderson, (1961).)*

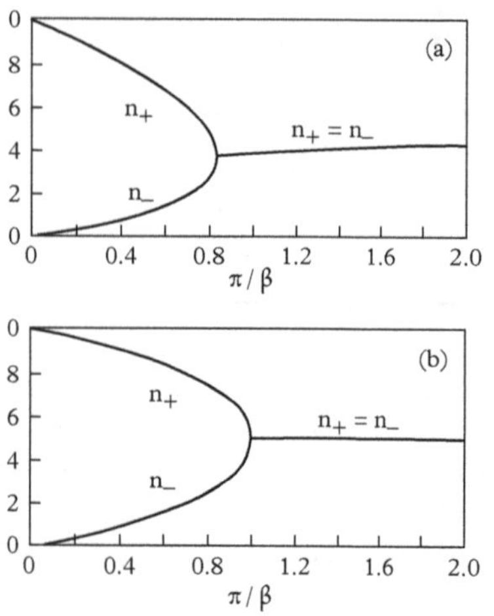

Figure 44.5 *The ratio of n_+/n_- as a function of π/β for the cases (a) α = 1/2; (b) α = 1/4. (After Anderson, (1961).)*

For $n_d \cong 1/2$ we want large s–d mixing (large Γ) and our d level close to the Fermi level, which were our initial assumptions for this case.

Another interesting non-magnetic limit is one leading to small n_d; we then have the approximation, $\cot(\pi n_d) \cong 1/\pi n_d$; inserting this into Eq. (44.33) we get $1/\pi n_d \cong \beta(n_d - \alpha)$. Solving the resulting quadratic equation for n_d and expanding in the limit $\beta\alpha^2 \ll 1$ gives

$$\begin{aligned} n_d &= \frac{1}{\pi^{1/2}\beta^{1/2}} + \frac{\alpha}{2} + \cdots \\ &= \frac{1}{\pi^{1/2}}\left(\frac{\Gamma}{U}\right)^{1/2} + \frac{\varepsilon_F - \varepsilon_d}{2U} + \cdots . \end{aligned} \tag{44.36}$$

(c) The transition curve

We next solve for the curve that represents the boundary separating the polarized and unpolarized impurity states. On this curve we still have the condition, $n_{d+} = n_{d-} = n$, that led to Eq. (44.33). To obtain a necessary second condition we note that the two equations (44.27a,b) have the same slope at the transition; differentiating and equating these equations we obtain

$$\frac{\pi}{\sin^2(\pi n_c)} = \beta_c, \tag{44.37}$$

where the subscript, c, denotes the critical value of these parameters. Solving Eq. (44.33) for β_c and equating it to (44.37) gives

$$\frac{\pi}{\sin^2(\pi n_c)} = \frac{\cot(\pi n_c)}{n_c - \alpha},$$

which is equivalent to

$$\sin(2\pi n_c) = 2\pi(n_c - \alpha). \tag{44.38}$$

Solution of this equation gives n_c as a function of α; inserting this value of n_c into (44.37) gives the corresponding β and together they generate a transition curve in α, β space. Figure 44.4 shows this curve. Figure 44.5 shows a plot of n_+ and n_- as a function of π/β for the cases α = 1/2 (upper) and α = 1/4 (lower).

In the absence of crystal field splitting we cannot ignore the effects of orbital degeneracy, and exchange effects enter the model Hamiltonian; we refer the reader to Anderson's original paper for a discussion of this rather complex topic. One qualitative effect is that in the presence of exchange, involving a coupling constant J, the condition for the existence of a magnetic state is *less stringent*. In place of Eq. (44.37) one has

$$\beta_c \geq \frac{\pi}{\sin^2(\pi n_c)} - \frac{J}{2\Gamma}. \tag{44.39}$$

It is noted that exchange alone would generally not be strong enough to produce a polarized state; i.e., U is the dominant term.

44.3 The s–d exchange model

In addition to the model Hamiltonian of Anderson, Zener's *s–d exchange Hamiltonian*[13] is also commonly used to describe effects of magnetic impurities in metals, especially with respect to the electrical transport properties. For the case of a single d impurity the s–d Hamiltonian describing the impurity-induced scattering between conduction electrons is given by

$$\hat{H}_{sd} = \sum_{k',k} J_{k,k'} \left[\hat{S}_+ \hat{c}^\dagger_{k\downarrow} \hat{c}_{k'\uparrow} + \hat{S}_- \hat{c}^\dagger_{k\uparrow} \hat{c}_{k'\downarrow} + \hat{S}_z \left(\hat{c}^\dagger_{k\uparrow} \hat{c}_{k'\uparrow} - \hat{c}^\dagger_{k\downarrow} \hat{c}_{k'\downarrow} \right) \right]; \qquad (44.40)$$

here S_+, S_-, and S_z are the spin-1/2 raising, lowering, and z-projection operators. This Hamiltonian will be used in Section 44.4 to discuss the temperature-dependent minimum in the resistance observed in metals containing magnetic impurities (called the Kondo effect). However, to provide continuity we now seek a connection between the Anderson and s–d Hamiltonians, as was first made by Schrieffer and Wolff (1966),[14] the connection being referred to as the Schrieffer–Wolff transformation.[15] Here we will establish this connection using an approach based on d state projection operators.

To effect this transformation we start by writing the total ground state wave function, $|\Phi_0\rangle$, given by (44.6) as the sum of terms with amplitudes involving, zero, one, and two electrons on the d impurity site:

$$|\Phi_0\rangle = |\Phi_{00}\rangle + |\Phi_{01}\rangle + |\Phi_{02}\rangle\,. \qquad (44.41a)$$

To carry out this decomposition we introduce a three-component "projection operator," $\hat{P}$, which when operating on $|\Phi_0\rangle$ creates these amplitudes; i.e.,

$$|\Phi_0\rangle = \sum_{j=1}^{3} \hat{P}_j\, |\Phi_0\rangle\,. \qquad (44.41b)$$

The components, $\hat{P}_j$, can be written by inspection as

$$\hat{P}_0 = (1 - \hat{n}_{d\uparrow})(1 - \hat{n}_{d\downarrow})$$

$$\hat{P}_1 = \hat{n}_{d\uparrow} + \hat{n}_{d\downarrow} - 2\hat{n}_{d\uparrow}\hat{n}_{d\downarrow}$$

$$\hat{P}_2 = \hat{n}_{d\uparrow}\hat{n}_{d\downarrow} \qquad (44.42a–c)$$

and have the properties

$$\hat{P}_i \hat{P}_j = \hat{P}_i \delta_{ij}, \qquad (44.42d)$$

$$\sum_{i=1}^{3} \hat{P}_i = \hat{1}. \qquad (44.42e)$$

[13] Interaction between the d-shells in the transition metals, C. Zener, *Phys. Rev.* **81**, 440 (1950).

[14] Relation between Anderson and Kondo Hamiltonians, J.R. Schrieffer and P.A. Wolff, *Phys. Rev.* **149**, 491 (1966).

[15] Here we will follow the approach given in *The Kondo Problem to Heavy Fermions*, A. C. Hewson, Cambridge University Press, Cambridge (1993), rather than that originally used by Schrieffer and Wolff.

Operating on the Schrödinger equation, $\hat{H}\,|\Phi_0\rangle = E\,|\Phi_0\rangle$, associated with the Anderson Hamiltonian (44.5) with any of the three operators, $\hat{P}_i$, gives $\hat{P}_i\hat{H}\,|\Phi_0\rangle = E\,|\Phi_{0i}\rangle$; using (44.42d) we can then rearrange the left-hand side as

$$\hat{P}_i\hat{H}\sum_{j=1}^{3}\hat{P}_j\,|\Phi_0\rangle = \hat{P}_i\hat{H}\sum_{j=1}^{3}\sum_{k=1}^{3}\delta_{jk}\hat{P}_j\hat{P}_k\,|\Phi_0\rangle = \sum_{j=1}^{3}\left(\hat{P}_i\hat{H}\hat{P}_j\right)\left(\hat{P}_j\,|\Phi_0\rangle\right),$$

or equivalently

$$\sum_{j=1}^{3}\left(\hat{P}_i\hat{H}\hat{P}_j\right)|\Phi_{0j}\rangle = E\hat{P}_i\,|\Phi_0\rangle\,, \tag{44.43}$$

which when written in matrix form is

$$\begin{pmatrix} \hat{H}_{00} & \hat{H}_{01} & \hat{H}_{02} \\ \hat{H}_{10} & \hat{H}_{11} & \hat{H}_{12} \\ \hat{H}_{20} & \hat{H}_{21} & \hat{H}_{22} \end{pmatrix}\begin{pmatrix} |\Phi_{00}\rangle \\ |\Phi_{01}\rangle \\ |\Phi_{02}\rangle \end{pmatrix} = E\begin{pmatrix} |\Phi_{00}\rangle \\ |\Phi_{01}\rangle \\ |\Phi_{02}\rangle \end{pmatrix}, \tag{44.44}$$

where $\hat{H}_{ij} = (\hat{P}_i\hat{H}\hat{P}_j)$. Since there are no terms in the Hamiltonian creating or destroying two electrons in the d state, $\hat{H}_{02} = \hat{H}_{20} = 0$. The Hamiltonian matrix then takes the form

$$\overset{\leftrightarrow}{\hat{H}} = \begin{pmatrix} \hat{H}_{\mathbf{k}}(1-\hat{n}_{d\uparrow})(1-\hat{n}_{d\downarrow}) & 0 & 0 \\ 0 & (\hat{H}_{\mathbf{k}}+\varepsilon_{0d})(\hat{n}_{d\uparrow}+\hat{n}_{d\downarrow}-2\hat{n}_{d\uparrow}\hat{n}_{d\downarrow}) & 0 \\ 0 & 0 & (\hat{H}_{\mathbf{k}}+2\varepsilon_{0d}+U)\hat{n}_{d\uparrow}\hat{n}_{d\downarrow} \end{pmatrix}$$
$$+\begin{pmatrix} 0 & \sum\limits_{\mathbf{k},\sigma}V^*_{\mathbf{k}d}\hat{c}^\dagger_{\mathbf{k}\sigma}(1-\hat{n}_{d,-\sigma})\hat{c}_{d\sigma} & 0 \\ \sum\limits_{\mathbf{k},\sigma}V_{\mathbf{k}d}\hat{c}^\dagger_{d\sigma}(1-\hat{n}_{d,-\sigma})\hat{c}_{\mathbf{k}\sigma} & 0 & \sum\limits_{\mathbf{k},\sigma}V_{\mathbf{k}d}\hat{c}^\dagger_{\mathbf{k}\sigma}\hat{n}_{d,-\sigma}\hat{c}_{d\sigma} \\ 0 & \sum\limits_{\mathbf{k},\sigma}V^*_{\mathbf{k}d}\hat{c}^\dagger_{d\sigma}\hat{n}_{d,-\sigma}\hat{c}_{\mathbf{k}\sigma} & 0 \end{pmatrix}, \tag{44.45}$$

where $\hat{H}_{\mathbf{k}} = \sum_{\mathbf{k},\sigma}\varepsilon_{\mathbf{k}}\hat{c}^\dagger_{\mathbf{k}\sigma}\hat{c}_{\mathbf{k}\sigma}$ is the band Hamiltonian and we used the property $\hat{c}_{d\sigma}\hat{n}_{d\sigma'} = \left(\delta_{\sigma\sigma'}\hat{c}_{d\sigma'} + \hat{n}_{d\sigma'}\hat{c}_{d\sigma}\right)$ in obtaining the off-diagonal elements.

We will be interested in the limit where the effective Hamiltonian involves a singly occupied d level. By introducing appropriate inverse operators we can eliminate $|\Phi_{00}\rangle$ and $|\Phi_{02}\rangle$, obtaining the following equation for $|\Phi_{01}\rangle$:

$$\left[\hat{H}_{11} + \hat{H}_{12}(E-\hat{H}_{22})^{-1}\hat{H}_{21} + \hat{H}_{10}(E-\hat{H}_{00})^{-1}\hat{H}_{01}\right]|\Phi_{01}\rangle = E\,|\Phi_{01}\rangle\,; \tag{44.46}$$

up to this point no approximations have been made.

A short calculation on the second term in (44.46) yields the form

$$\sum_{k,k',\sigma,\sigma'} \frac{V_{kd}^{*} V_{k'd}}{E-\left(\sum_{k''} \varepsilon_{k''}-\varepsilon_{k'}\right)-2\varepsilon_d-U} \hat{c}_{k\sigma}^{\dagger} \hat{c}_{k'\sigma'} \hat{c}_{d\sigma} \hat{c}_{d\sigma'}^{\dagger} \hat{n}_{d,-\sigma'}. \tag{44.47}$$

Note that the intermediate state energy, $\varepsilon_{k'} = 2\varepsilon_d + U$, is absent since it would correspond to a doubly occupied d-state state. The factor preceding the operators in (44.47) can be rearranged as

$$\left(E-\left(\sum_{k''} \varepsilon_{k''}-\varepsilon_{k'}\right)-2\varepsilon_d-U\right)^{-1} = -\frac{1}{U+\varepsilon_d-\varepsilon_{k'}}\left(1-\frac{E-\varepsilon_d-\sum_{k''} \varepsilon_{k''}}{U+\varepsilon_d-\varepsilon_{k'}}\right)^{-1}. \tag{44.48}$$

Now $U + \varepsilon_d - \varepsilon_{k'}$ in (44.48) is the energy needed to raise a single conduction electron from the filled Fermi sea to the unoccupied second d state, which lies above ε_F at an energy, $U + \varepsilon_d$. On the other hand, $\varepsilon_d + \sum_{k''} \varepsilon_{k''}$ *would* be the total ground state energy, E, were it not for the s–d mixing caused by V_{kd}, which from (44.47) is second order in this quantity. Hence the second term occurring in the brackets in (44.48) can be neglected and our perturbation (44.47) becomes

$$\hat{H}_{12}\frac{1}{E-\hat{H}_{22}}\hat{H}_{21} \cong \sum_{k,k',\sigma,\sigma'} \frac{-V_{kd}\, V_{k'd}}{U+\varepsilon_d-\varepsilon_{k'}} \hat{c}_{k\sigma}^{\dagger} \hat{c}_{k'\sigma'} \hat{c}_{d\sigma} \hat{c}_{d\sigma'}^{\dagger} \hat{n}_{d,-\sigma'}. \tag{44.49}$$

In our singly occupied d state space, the term associated with the operator $\hat{c}_{d\downarrow}^{\dagger}\hat{c}_{d\uparrow}$ occurring in (44.49) flips the spin of the d electron from up to down and may be identified with the operator $\hat{S}_+$ in Eq. (44.40); similarly, $\hat{c}_{d\uparrow}^{\dagger}\hat{c}_{d\downarrow}$, which flips from down to up, is identified as $\hat{S}_-$. Treating the third term in Eq. (44.46) in a similar manner and combining it with (44.49), we can then obtain the constants $J_{k,k'}$ for the first two terms entering our s–d Hamiltonian (44.40) as

$$J_{k,k'} = V_{kd} V_{k'd}\left(\frac{1}{U+\varepsilon_d-\varepsilon_{k'}}+\frac{1}{\varepsilon_k-\varepsilon_d}\right). \tag{44.50}$$

Finally, we form the following combinations:

$$\hat{S}_z = \hat{c}_{d\uparrow}^{\dagger}\hat{c}_{d\uparrow} - \hat{c}_{d\downarrow}^{\dagger}\hat{c}_{d\downarrow} \tag{44.51a}$$

and

$$1 = \hat{c}_{d\uparrow}^{\dagger}\hat{c}_{d\uparrow} + \hat{c}_{d\downarrow}^{\dagger}\hat{c}_{d\downarrow}. \tag{44.51b}$$

Equation (44.51a) yields the third term in (44.40), while (44.51b) results in a (spin-independent) pure potential scattering contribution,

$$\hat{H}_{sd}^{(pot)} = \sum_{k',k} K_{k,k'} \left[\hat{c}_{k\uparrow}^{\dagger} \hat{c}_{k'\uparrow} + \hat{c}_{k\downarrow}^{\dagger} \hat{c}_{k'\downarrow} \right], \tag{44.52}$$

where

$$K_{k,k'} = \frac{1}{2} V_{kd} V_{k'd} \left(\frac{1}{\varepsilon_k - \varepsilon_d} - \frac{1}{U + \varepsilon_d - \varepsilon_{k'}} \right). \tag{44.53}$$

The expressions obtained are valid in the range $|\varepsilon_k| \ll |\varepsilon_d - \varepsilon_F|$ and $|\varepsilon_k| \ll |U + \varepsilon_d - \varepsilon_F|$. In the local moment regime where $U + \varepsilon_d > \varepsilon_F$ and $\varepsilon_d < \varepsilon_F$, and only states ε_k close to ε_F are thermally excited, the effective coupling constants, $J_{k,k'}$, between the localized spins and the conduction electrons are *positive*, corresponding to an *antiferromagnetic* interaction. This coupling leads to an unusual temperature dependence of the resistivity, as first explained by J. Kondo.

44.4 The Kondo effect: electrical resistance versus temperature from paramagnetic impurities

In Chapter 19 we discussed the origin of electrical resistivity in metals as it arises from *static* scattering centers, i.e., those having no internal dynamical degrees of freedom. A key feature of the theory presented there is that the electrical resistivity approaches a constant as the temperature is lowered toward absolute zero. However, experimentally, it is observed that the resistivity of metals containing magnetic impurities has the unusual property of passing through a *minimum*, after which it increases, ultimately approaching a constant at much lower temperatures.[16] If a sufficiently large magnetic field is applied, so as to fully align the moments, the minimum disappears from which we can conclude that the effect arises from the internal spin degrees of freedom of the impurities as they interact with the host conduction electrons.

Figure 44.6 shows the resistivity as a function of temperature for a series of Mo_xNb_{1-x} alloys containing 1% Fe.[17] Note that for larger Mo concentrations a resistivity minimum develops. For comparison the onset and magnitude of the Fe moment in the Mo-Nb-Re system is shown in Fig. 44.7.[18] What is significant here is that a resistance minimum is only observed in those alloys containing impurities possessing a magnetic moment. As we will show, the presence of such impurities results in a contribution to the resistivity proportional to the logarithm of the temperature such that it rises as the temperature falls. In addition, there is the

[16] It is important to recognize that the Kondo effect, which is strictly connected with paramagnetic impurities, is distinct from the localization effect covered in Section 43.4 which occurs universally for nonmagnetic impurities.

[17] Resistivity of Mo-Nb and Mo-Re alloys containing Fe, M.P. Sarachik, E. Corenzwit, and L.D. Longinotti, *Phys. Rev.* 135, A1041 (1964).

[18] Local magnetic moment associated with an iron atom dissolved in various transition metal alloys, A.M. Clogston, H.J. Williams, R.C. Sherwood, E. Corenzwit, M. Peter, and B.T. Matthias, *Phys. Rev.* 125, 541 (1962).

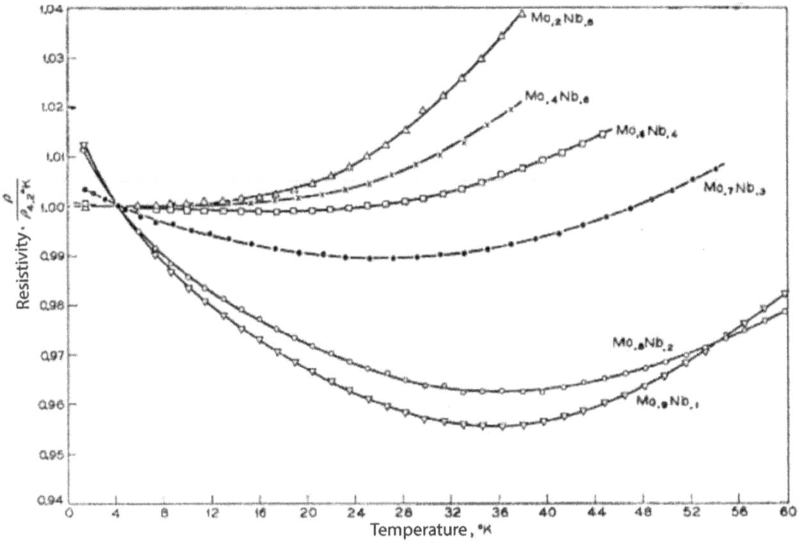

Figure 44.6 *Resistance versus temperature showing the phenomenon of a minimum for 1%* Fe *in various alloys of* Mo *and* Nb. *(After Sarachik, Corenzwit, and Longinotti (1964).)*

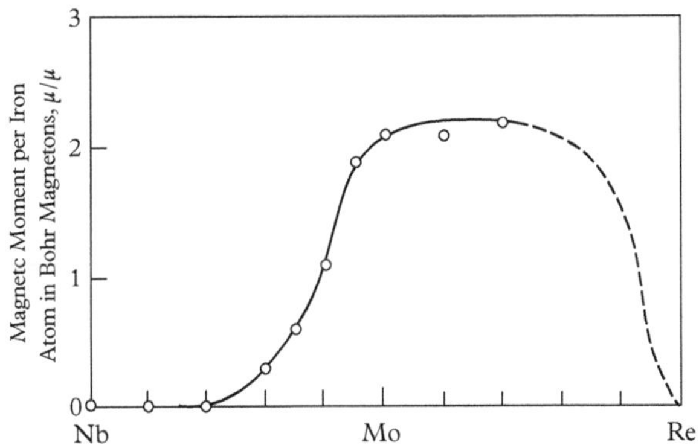

Figure 44.7 *The magnetic moment per iron atom versus concentration for* MoNb *and* MoRe *alloys. (After Clogston et al. (1962).)*

Bloch contribution arising from electron–phonon scattering which is proportional to T^5 (see Section 21.5). Hence we may write

$$\rho(T) = \rho_0 + aT^5 - cb\ \ln(T/T_0), \tag{44.54}$$

where ρ_0 is the residual resistance arising from host impurities, a, b, and T_0 are constants, and c is the magnetic impurity concentration. Setting the derivative of this form to zero we see that the resistivity has a minimum at a temperature, T_{min}, given by

$$T_{min} = (cb/5a)^{1/5}. \tag{44.55}$$

A theoretical explanation of the logarithmic rise in resistance was first given by J. Kondo and involves going beyond the leading contribution to conduction electron/paramagnetic impurity scattering. Because the calculations are somewhat involved they have been relegated to Appendix 44A.

44.5 Polarization of conduction electrons by an embedded spin; the RKKY interaction

If a particle with spin, S, is embedded in a metal host it will polarize the surrounding conduction electrons, as we will show shortly. One example is the nuclei of host or impurity atoms which have a non-vanishing nuclear spin; in this case S would correspond to the nuclear spin, I. From another point of view, atoms with unfilled d or f shells, which retain a spin within the solid (as discussed at the beginning of this section), can induce a (much stronger) polarization of the surrounding conduction electrons.

44.5.1 Conduction electron polarization

As a model for the coupling we assume a so-called contact interaction which acts only when the conduction electron spin, s, is at the lattice site, **R**, of the "impurity" spin, S:

$$\hat{H}' = \frac{J}{\hbar^2}\delta^{(3)}(\mathbf{R}-\mathbf{r})\,\hat{\mathbf{S}}\cdot\hat{\mathbf{s}}, \tag{44.56}$$

where the coupling strength, J, has units of energy times volume, and in nuclei arises from the Fermi contact interaction. Our plan here is to calculate the spin polarization in the conduction electrons arising from (44.56) using first-order time-independent perturbation theory; in Section 44.5.2 we discuss how this polarization results in an interaction with neighboring spins. To simplify the treatment we will treat the impurity spin operator, $\hat{\mathbf{S}}$, as a classical (external) vector, **S**; in addition we will limit ourselves to otherwise free electrons having wave functions in the absence of **S** of the form

$$\psi^{(0)}_{\mathbf{k},s_z}(\mathbf{r},\alpha) = \frac{1}{L^{3/2}}e^{i\mathbf{k}\cdot\mathbf{r}}\chi_{s_z}(\alpha), \tag{44.57}$$

where s_z denotes the projections, $\uparrow, \downarrow$, of electron spin states, $\chi_{s_z}(\alpha)$, with components α, of the operator, $\hat{s}_z$, quantized along the direction of **S**, where the latter is taken as parallel to z. The wave function through first order in the presence of $\hat{H}'$ is then given by

$$\begin{aligned}\psi_{\mathbf{k}s_z}(\mathbf{r},\alpha) &= \psi^{(0)}_{\mathbf{k}s_z}(\mathbf{r},\alpha) + \psi^{(1)}_{\mathbf{k}s_z}(\mathbf{r},\alpha)\\ &= \psi^{(0)}_{\mathbf{k}s_z}(\mathbf{r},\alpha) + \sum_{\mathbf{k}'s'_z}\frac{\left(\sum_{\alpha'}\int d^3r'\,\psi^{(0)*}_{\mathbf{k}'s'_z}(\mathbf{r}',\alpha')\hat{H}'(\mathbf{S})\psi^{(0)}_{\mathbf{k}s_z}(\mathbf{r}',\alpha')\right)}{\varepsilon_{\mathbf{k}}-\varepsilon_{\mathbf{k}'}}\psi^{(0)}_{\mathbf{k}s'_z}(\mathbf{r},\alpha).\end{aligned} \tag{44.58}$$

The spin up and spin down electron states are then

$$\psi_{k\uparrow}(\mathbf{r},\alpha) = \psi^{(0)}_{k\uparrow}(\mathbf{r},\alpha) + \frac{JSe^{ik\cdot R}}{2\hbar L^{3/2}} \sum_{k'_z} \frac{e^{ik'\cdot(r-R)}\chi_\uparrow(\alpha)}{\varepsilon_k - \varepsilon_{k'}} \tag{44.59a}$$

and

$$\psi_{k\downarrow}(\mathbf{r},\alpha) = \psi^{(0)}_{k\downarrow}(\mathbf{r},\alpha) - \frac{JSe^{ik\cdot R}}{2\hbar L^{3/2}} \sum_{k'_z} \frac{e^{ik'\cdot(r-R)}\chi_\downarrow(\alpha)}{\varepsilon_k - \varepsilon_{k'}}. \tag{44.59b}$$

The electron number density associated with all electrons in a given spin state, s_z, is

$$n_{s_z}(\mathbf{r}) = \sum_{k\alpha} f(\varepsilon_k)\, \psi^*_{ks_z}(\mathbf{r},\alpha)\psi_{ks_z}(\mathbf{r},\alpha), \tag{44.60}$$

where $f(\varepsilon_k)$ is the Fermi occupation factor for a single spin. Retaining terms to order J, the total number density and spin density are

$$n(\mathbf{r}) = n_\uparrow(\mathbf{r}) + n_\downarrow(\mathbf{r}) = n_0 + \mathbf{O}(J^2) \tag{44.61a}$$

and

$$\begin{aligned} m(\mathbf{r}) &= n_\uparrow(\mathbf{r}) - n_\downarrow(\mathbf{r}) \\ &= \frac{2JS}{L^6} \sum_k f(\varepsilon_k)\, e^{-ik\cdot(r-R)} \sum_{k'} \frac{e^{ik'\cdot(r-R)}}{\varepsilon_k - \varepsilon_{k'}} \end{aligned} \tag{44.61b}$$

where n_0 is the density in the absence of the spin, S. If in the sum over $\mathbf{k}'$ we regard $L^{-3/2}e^{ik'\cdot r}$ and $L^{-3/2}e^{-ik'\cdot R}$ as the plane wave eigenstates, $\psi_k(\mathbf{r})$ and $\psi^*_k(\mathbf{R})$, then the associated Green's function is[19]

$$\begin{aligned} G(\mathbf{r},\mathbf{R}) &= \frac{1}{L^3} \sum_{k'} \frac{e^{ik'\cdot(r-R)}}{\varepsilon_k - \varepsilon_{k'}} \\ &= \frac{2m}{\hbar^2} \frac{\cos[k\,|\mathbf{r}-\mathbf{R}|]}{4\pi\,|\mathbf{r}-\mathbf{R}|}. \end{aligned} \tag{44.62}$$

Writing $\rho = |\mathbf{r}-\mathbf{R}|$, the remaining sum over $\mathbf{k}$ in spherical coordinates involves the integral

$$\begin{aligned} 2\pi \int_0^{2k_F} k^2 dk \int_0^{\pi} \sin\theta d\theta e^{-ik\rho\cos\theta} &= \frac{4\pi}{\rho} \int_0^{2k_F} kdk \cos k\rho \sin k\rho \\ &= \frac{\pi k_F}{\rho^2}\left[\frac{\sin(2k_F\rho)}{2k_F\rho} - \cos(2k_F\rho)\right], \end{aligned}$$

[19] The resulting integral is interpreted as the principal value.

which gives

$$m(\mathbf{r}) = \frac{JSk_F}{(2\pi)^3}\frac{m}{\hbar^2}\frac{1}{\rho^3}\left[\frac{\sin(2k_F\rho)}{2k_F\rho} - \cos(2k_F\rho)\right]. \tag{44.63}$$

It is common to rewrite this expression as

$$m(\mathbf{r}) = n_0\frac{3JSmk_F}{\pi\hbar^2}F(2k_F\,|\mathbf{r}-\mathbf{R}|), \tag{44.64}$$

where $n_0 = k_F^3/3\pi^2$ and we have introduced the function

$$F(x) = \frac{1}{x^4}(\sin x - x\cos x)\,; \tag{44.65}$$

Note that the behavior is oscillatory. Physically this oscillatory behavior is produced by the sharp "cut-off" of the Fermi occupation function, as evidenced by the fact that the associated wave vector is $2k_F$. This is reminiscent of the Kohn effect, discussed in Section 14.2.4.

44.5.2 The RKKY interaction

The spin polarization of the conduction electrons resulting from the coupling to an embedded spin will act on neighboring spins, thereby creating an indirect interaction between them. In addition, we will have the magnetic dipole–dipole interaction which directly couples the spins. Spin–spin coupling via conduction electron polarization, referred to as the RKKY interaction,[20] can be calculated by evaluating the first-order shift in energy of a second spin, $\mathbf{S}^{(2)}$, located at the point, $\mathbf{R}$, arising from the first-order shift in the wave functions of the conduction electrons, as given by (44.57), resulting from a spin, $\mathbf{S}^{(1)}$, located at $\mathbf{R} = 0$; an overall second-order process:

$$\sum_\alpha \int d^3r\,\psi^*_{\mathbf{k}s_z}(\mathbf{r},\alpha)\hat{H}'(\mathbf{S}^{(2)})\psi_{\mathbf{k}s_z}(\mathbf{r},\alpha)\quad .$$

$$\cong \sum_\alpha \frac{J}{\hbar^2}\int d^3r\,\psi^{(0)*}_{\mathbf{k}s_z}(\mathbf{r},\alpha)\left(\mathbf{S}^{(2)}\cdot\hat{\mathbf{s}}\right)\delta^{(3)}(\mathbf{r}-\mathbf{R})\psi_{\mathbf{k}s_z}(\mathbf{r},\alpha) + \text{c.c.} \tag{44.66}$$

Summing this energy shift over all the occupied states involves operations similar to those previously detailed; noting that the classical vector, $\mathbf{S}^{(2)}$, can lie at an angle, θ, relative to our quantization axes, $\mathbf{S}^{(1)}$, the resulting energy shift is then

$$\Delta E = \mathbf{S}^{(1)}\cdot\mathbf{S}^{(2)}\frac{2J^2}{(2\pi)^6}\sum_{\mathbf{k}} f(\varepsilon_{\mathbf{k}})\sum_{\mathbf{k}'}\frac{e^{-i(\mathbf{k}-\mathbf{k}')\cdot\mathbf{R})}}{\varepsilon_{\mathbf{k}}-\varepsilon_{\mathbf{k}'}}$$

$$= \mathbf{S}^{(1)}\cdot\mathbf{S}^{(2)}\frac{4J^2mk_F^4}{(2\pi)^3}F(2k_FR). \tag{44.67}$$

[20] M.A. Ruderman and C. Kittel, *Phys. Rev.* **96**, 99 (1954); K. Yoshida. *Phys. Rev.* **106**, 893 (1957).

Interpreting $\mathbf{S}^{(1)} \cdot \mathbf{S}^{(2)}$ as an operator, the energy shift becomes a spin–spin interaction Hamiltonian,

$$\hat{H} = \hat{\mathbf{S}}^{(1)} \cdot \hat{\mathbf{S}}^{(2)} \frac{4J^2 m k_F^4}{(2\pi)^3} F(2k_F R). \tag{44.68}$$

The behavior of $F(2k_F R)$ results in an interaction that oscillates between ferromagnetic and antiferromagnetic; it is ferromagnetic at short range and the first zero is at $2k_F R = 4.49$.

When applying this theory to embedded d and f electron spins, where the associated wave function is more spread out, the delta function contact interaction (44.56) is a poor approximation and should be replaced by a more spread out function which we can write as $J(\mathbf{R} - \mathbf{r})$. Equation (44.59) would then have the form

$$\psi_{\mathbf{k}s_z}(\mathbf{r}, \alpha) = \psi_{\mathbf{k}\uparrow}^{(0)}(\mathbf{r}, \alpha) + \frac{S e^{i\mathbf{k}\cdot\mathbf{R}}}{2\hbar L^{3/2}} \sum_{\mathbf{k}' s'_z} J(\mathbf{k}' - \mathbf{k}) \frac{e^{i\mathbf{k}'\cdot(\mathbf{r}-\mathbf{R})} \chi_{s'_z}(\alpha)}{\varepsilon_{\mathbf{k}} - \varepsilon_{\mathbf{k}'}}, \tag{44.69}$$

where $J(\mathbf{k}' - \mathbf{k})$ is the Fourier transform of $J(\mathbf{R}-\mathbf{r})$. As important at this level would be to replace plane wave conduction electron states by the real Bloch states. We will not pursue either of these refinements here. However we do note that if the Fermi surface has flat portions these can lead to much stronger oscillations. The RKKY mechanism plays an important role, along with the Kondo effect, in the physics of heavy fermion systems, to be discussed in Section 45.3.

44.6 Spin glasses

In this section we will briefly consider the behavior of an alloy containing a dilute concentration of spins, typically at the 1% level, that are embedded in an otherwise non-magnetic host and that couple to each other via an interaction which, qualitatively, has the form of (44.68). Typical examples are Mn or Fe atoms in a Cu or Au host. The key point is that the oscillations in the spin–spin coupling, and the presence of many neighbors produce competing interactions with random sign and in random directions. This leads to an enormous number of spin configurations of the system as a whole having *essentially the same free energy*. In order to find the lowest free energy configuration on being cooled to some temperature, the system would need to explore a very large number of configurations, referred to as an energy landscape; and in practice this never happens, even when the cooling is carried out very slowly. To explore other configurations the system must surmount barriers involving multiple paths in the multi-dimensional space of spin configurations. A similar behavior occurs in conventional glasses, hence the term spin glasses.

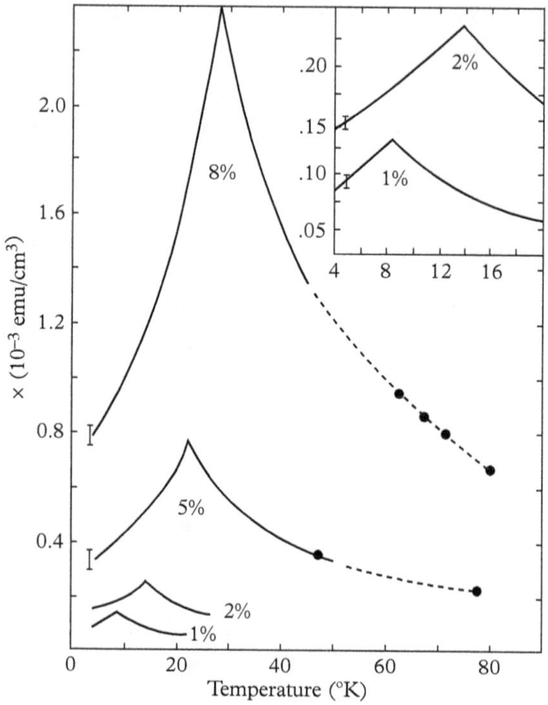

Figure 44.8 *The temperature dependence of the magnetic susceptibility of dilute alloys of* Fe *in* Au. *(After Cannella et al. (1971).)*

At sufficiently high temperatures the spins are randomly directed and a natural question is whether on cooling a phase transition occurs. Suggesting that possibility is observation of a sharp cusp in the magnetic susceptibility, an example of which is shown in Fig. 44.8.[21] The sharp drop can be viewed as a freezing out of the spin–flip degrees of freedom as the system is cooled. However thermal measurements show no accompanying heat capacity or latent heat anomaly, arguing against a phase transition; the subject remains controversial and has attracted great attention (see Additional Reading section); the associated phenomena carry over to other branches of science.

[21] Magnetic susceptibility of Au-Fe Alloys, V. Cannella, J.A. Mydosh, and J.I. Budnick, *J. Appl. Phys.* 42, 1689 (1971).

ADDITIONAL READING

For an extended discussion of the Anderson model and the Kondo effect, see *The Kondo Problem to Heavy Fermions*, A.C. Hewson, Cambridge University Press, Cambridge (1993).

Local moments and the Kondo effect are treated in Chapters 6 and 7 of *Advanced Solid State Physics*, Philip Phillips, Westview Press, Bolder, CO (2003).

Localized magnetic states in metals, P.W. Anderson, *Phys. Rev.* **124**, 41 (1961). This early paper is quite readable with respect to the underlying physics; unlike our treatment here, it employs a combined equation of motion and Green's function approach.

Resistance minimum in dilute magnetic alloys, J. Kondo, *Prog. Theor. Phys.* **32**, 37 (1964).

For an accessible discussion of spin glasses and associated phenomena, see: *Spin Glasses and Complexity*, D.L. Stein and C.M. Newman, Princeton University Press (2013). For more technical discussions, see: Spin glasses: Experimental facts, theoretical concepts and open questions, K. Binder and A.P. Young, *Rev. Mod. Phys.* **58**, 801 (1986); *Spin Glasses and other Frustrated Systems*, D. Chowdhury, J. Wiley, NY (1986); *Spin Glasses*, K.H. Fischer and J.A, Hertz, Cambridge University Press (1991).

Problems

44.1. Carry out the calculations leading to Eqs. (44.12) and (44.13).

44.2. Assuming the density of states is constant in the range, $\varepsilon_0 < \varepsilon_k < \varepsilon_F$, and given by, $N(\varepsilon)$, calculate the level shift, $\delta\varepsilon_{n\sigma}$, using Eq. (44.18).

44.3. Verify Eq. (44.10).

44.4. Verify that Eq. (44.24) does approach the non-interacting density of states in the limit, $V_{kd} \to \infty$.

44.5. In the presence of a magnetic field, H, the positions of the up and down levels are shifted from ε_{0d} by amounts,

$$\delta\varepsilon_{0d+} = -\mu H + U\delta n_{d+} \text{ and } \delta\varepsilon_{0d-} = \mu H + U\delta n_{d-},$$

where μ is the magnetic moment of the electrons. Using Eqs. (44.27a,b) and defining the magnetic susceptibility as $\chi = \mu(\delta n_{d+} - \delta n_{d-})/H$ show that

$$\chi = \frac{2\mu^2}{(\pi\Gamma/\sin^2(\pi n)) - U}.$$

44.6. Carry out the calculations of the off-diagonal elements in the matrix given in Eq. (44.45).

44.7. Obtain the form given in Eq. (44.47).

Appendix 44A Kondo's second-order perturbation theory calculation

A partial explanation of the effects of spin-dependent scattering was first given by J. Kondo (1964).[22] This work launched extensive theoretical activity and a

[22] Resistance minimum in dilute magnetic alloys, J. Kondo, *Prog. Theor. Phys.* 32, 37 (1964).

vast number of experimental studies covering numerous alloy systems. A description valid for all temperatures involves renormalization group methods which are beyond the scope of this book.[23,24,25] Here we reconstruct Kondo's original calculations, which are based on perturbation theory.

To begin with we simplify the s–d Hamiltonian, $\hat{H}_{\mathbf{k}\,d}$, given in Eq. (44.40) by assuming a **k**-independent exchange constant, J; we must also generalize this expression to an array of impurities. The resulting Hamiltonian, as used by Kondo, is then

$$\hat{H}_{Kondo} = -\frac{J}{N}\sum_{n,k',k} e^{i(k-k')\cdot \mathbf{R}_n}\Big[\hat{S}_+\hat{c}^\dagger_{k\downarrow}\hat{c}_{k'\uparrow} + \hat{S}_-\hat{c}^\dagger_{k\uparrow}\hat{c}_{k'\downarrow} + \hat{S}_z\left(\hat{c}^\dagger_{k\uparrow}\hat{c}_{k'\uparrow} - \hat{c}^\dagger_{k\downarrow}\hat{c}_{k'\downarrow}\right)\Big], \tag{44A.1}$$

where the factor, $e^{i(k-k')\cdot \mathbf{R}_n}$, accounts for the shift in phase of the conduction states in translating from the origin to the impurity positioned at lattice vector, $\mathbf{R}_n$.

44A.1 Calculating the transition rate

It will turn out that the temperature-dependent effect we seek does not arise in leading order and we must therefore generalize the usual expression for the transition rate between initial and final states, a and b, to read[26]

$$w_{a\to b} = \frac{2\pi}{\hbar}\left|\left(\hat{H}'\right)_{ba} + \sum_c \frac{\left(\hat{H}'\right)_{bc}\left(\hat{H}'\right)_{ca}}{E_a - E_b}\right|^2 \delta(E_a - E_b) \tag{44A.2}$$

$$\cong \frac{2\pi}{\hbar}\left[\left|(H')_{ba}\right|^2 + (H')_{ba}\sum_c \frac{\left(\hat{H}'\right)_{bc}\left(\hat{H}'\right)_{ca}}{E_a - E_b} + (H')_{ab}\sum_c \frac{\left(\hat{H}'\right)_{cb}\left(\hat{H}'\right)_{ac}}{E_a - E_b}\right]\delta(E_a - E_b)$$

$$= w^{(2)}_{a\to b} + w^{(3)}_{a\to b} + \cdots;$$

Formally, E_a is the total energy of the state with quantum numbers, a; however in our model all transitions will only involve the single particle energies, ε_k.

The effect we seek involves $w^{(3)}_{a\to b}$ which is third order in $\hat{H}'$ and involves a product of first- and second-order contributions. In order to understand the structure of the second-order contributions (the terms involving a summation over intermediate states) we initially suppress the spin indices and restrict ourselves to pure potential scattering, as represented by the Hamiltonian (44A.52);

[23] Renormalization group – critical phenomena and Kondo problem, K.G. Wilson, *Rev. Mod. Phys.* 47, 773 (1975).

[24] Temperature-dependent susceptibility of symmetric Anderson model – connection to Kondo model, H.R. Krishnamurthy, K.G. Wilson, and J.W. Wilkins, *Phys. Rev. Lett.* 35, 1101 (1975).

[25] The 1982 Nobel Prize in Physics was awarded to Kenneth G. Wilson "for his theory for critical phenomena in connection with phase transitions."

[26] Landau and Lifshitz, *Quantum Mechanics* (1977), p. 156.

as a simplification we assume the $K_{kk'}$ are constant. The second-order contribution involves two classes of intermediate states that we denote schematically as

$$|\cdots\mathbf{q}\cdots\mathbf{k}\cdots\rangle \rightarrow |\cdots\mathbf{q}\cdots\mathbf{q}'\cdots\rangle \rightarrow |\cdots\mathbf{q}\cdots\mathbf{k}'\cdots\rangle, \text{ (a)} \qquad (44A.3)$$

and

$$|\cdots\mathbf{q}\cdots\mathbf{k}\cdots\rangle \rightarrow |\cdots\mathbf{k}'\cdots\mathbf{k}\cdots\rangle \rightarrow |\cdots\mathbf{k}'\cdots\mathbf{q}\cdots\rangle. \text{ (b)} \qquad (44A.4)$$

In the class (a), the initial state, **k**, is emptied while the intermediate state, $\mathbf{q}'$, is filled, followed by emptying the state, $\mathbf{q}'$, and filling the final state, $\mathbf{k}'$; for this to occur the state, **k** must initially be full and the states, $\mathbf{q}'$ and $\mathbf{k}'$, empty (here the state labeled **q** is simply a spectator). The intermediate states, $\mathbf{q}'$, and the final state, $\mathbf{k}'$, are generated by the forms, $K\hat{c}^\dagger_{q'}\hat{c}_k$, followed by, $K\hat{c}^\dagger_{k'}\hat{c}_{q'}$, occurring in the double sum over the (dummy) wave vectors in $\hat{H}'$. In class (b) an existing state, **q**, is emptied and the final state, $\mathbf{k}'$, filled, followed by emptying the initial state, **k**, and filling that intermediate state, **q**; these processes arises from the forms, $K\hat{c}^\dagger_{k'}\hat{c}_q$, followed by $K\hat{c}^\dagger_q\hat{c}_k$ in $\hat{H}'$. In addition we have the operator, $K\hat{c}^\dagger_{k'}\hat{c}_k$, entering the first-order matrix elements (which multiply the second-order terms in (44A.2)). The resulting matrix elements are unity and the contributions from the second and third terms in (44A.2) then become

$$w^{(3)}_{k'\to k} \cong +2\frac{2\pi}{\hbar}\left(\frac{K}{N}\right)^3 \sum_n \left[\sum_{q'} \frac{(1-f(\mathbf{q}'))}{\varepsilon_{k'}-\varepsilon_{q'}} + \sum_q \frac{f(\mathbf{q})}{\varepsilon_{k'}-\varepsilon_q}\right]\delta(\varepsilon_{k'}-\varepsilon_k), \quad (44A.5)$$

where the factor 2 arises from the two spin states. The first term in (44A.5) corresponds to class (a) transitions and the second to class (b) transitions; the Fermi factors, $f(\mathbf{q})$ and $(1-f(\mathbf{q}'))$, ensure the presence or absence respectively of the initial states, **q**, or intermediate states, $\mathbf{q}'$, in accordance with the Pauli principle. Noting that $\sum_n = cN$, where c is the concentration and N is the number of atoms, Eq. (44A.5) takes the form

$$w^{(3)}_{k\to k'} = +2\frac{2\pi}{\hbar}\left(\frac{K}{N}\right)^3 cN \sum_q \frac{1}{\varepsilon_k-\varepsilon_q}\delta(\varepsilon_{k'}-\varepsilon_k); \qquad (44A.6)$$

note that the $f(\mathbf{q})$ factors in the sums *cancel* and hence *no anomalous temperature dependence is encountered* for pure potential scattering.[27] We also note that the summation involves a principal value contribution and, in addition, is logarithmically divergent for large ε_q; the latter would be cut off by an energy dependence of K, that has been suppressed.

We now consider the effect of spin-dependent scattering induced by the Hamiltonian (44A.1). In addition to the wave vectors associated with the various conduction band states, the quantum numbers now include their spins, ↑ or ↓, as well as the spins associated with n^{th} impurity, $n_\uparrow$ or $n_\downarrow$ (we will restrict

[27] The sum over **q**, in addition to involving a principal value integral, is also logarithmically divergent; the latter will be cut off by the **k**-dependence of the scattering potential K, which has been suppressed.

ourselves to spin-1/2 impurities in what follows). There are two (overall) classes of transitions (labeled A and B): those in which the spin of the initial and final states is the same and those in which it is opposite; within each of these classes there are four sub-classes (labeled (a) through (d)). In class A, the initial and final spins are parallel. For up–up scattering the second-order processes involve the following four transitions:

$$\left|\cdots q_\sigma \cdots k_\uparrow \cdots;\ \cdots n_{\sigma'} \cdots\right\rangle \to \left|\cdots q_\sigma \cdots q'_\uparrow \cdots;\ \cdots n_{\sigma'} \cdots\right\rangle \to \left|\cdots q_\sigma \cdots k'_\uparrow \cdots;\ \cdots n_{\sigma'} \cdots\right\rangle, \tag{a}$$

$$\left|\cdots q_\uparrow \cdots k_\uparrow \cdots;\ \cdots n_{\sigma'} \cdots\right\rangle \to \left|\cdots k'_\uparrow \cdots k_\uparrow \cdots;\ \cdots n_{\sigma'} \cdots\right\rangle \to \left|\cdots q_\uparrow \cdots k'_\uparrow \cdots;\ \cdots n_{\sigma'} \cdots\right\rangle, \tag{b}$$

$$\left|\cdots q_\sigma \cdots k_\uparrow \cdots;\ \cdots n_\downarrow \cdots\right\rangle \to \left|\cdots q_\sigma \cdots q'_\downarrow \cdots;\ \cdots n_\uparrow \cdots\right\rangle \to \left|\cdots q_\sigma \cdots k'_\uparrow \cdots;\ \cdots n_\downarrow \cdots\right\rangle, \tag{c}$$

$$\left|\cdots q_\downarrow \cdots k_\uparrow \cdots;\ \cdots n_\uparrow \cdots\right\rangle \to \left|\cdots k'_\uparrow \cdots k_\uparrow \cdots;\ \cdots n_\downarrow \cdots\right\rangle \to \left|\cdots q_\downarrow \cdots k'_\uparrow \cdots;\ \cdots n_\uparrow \cdots\right\rangle, \tag{d}$$

(44A.7)

with a corresponding set for down–down scattering. Note that processes (a) and (b) involve no impurity spin flips, while processes (c) and (d) involve two such spin flips. For all four of these processes the first-order matrix elements involves no spin flips. As previously, the transitions (a) and (c) require an empty intermediate state and hence a factor $(1-f(\mathbf{q}'))$ enters while processes (b) and (d) require an occupied initial state, thereby requiring a factor $f(\mathbf{q})$. The Fermi factors for (a) and (b) (that to not involve spin flips) will cancel; as was the case for pure potential scattering, discussed previously; they will not contribute any temperature dependence and can be neglected. The two processes involving two successive spin flips, (c) and (d), are generated by the operators, $\hat{S}_{n-}\hat{S}_{n+}$ and $\hat{S}_{n+}\hat{S}_{n-}$ respectively; here the associated Fermi factors *do not cancel* because these operators do not commute (we recall $[\hat{S}_{n+}, \hat{S}_{n-}] = 2\hat{S}_z$, which is ± 1 in the present case); hence these processes are central to our discussion. Carrying out the implied calculations for the transition rate and retaining only the terms proportional to $f(\mathbf{q})$, we obtain

$$\begin{aligned} w_{k\uparrow\to k'\uparrow} &= -\frac{2\pi}{\hbar}\left(\frac{J}{N}\right)^3 4\sum_n s_{nz}^2 \sum_q \frac{f(\mathbf{q})}{\varepsilon_k - \varepsilon_q}\delta(\varepsilon_k - \varepsilon_{k'}) \\ &= -\frac{2\pi}{\hbar}\left(\frac{J}{N}\right)^3 cN \sum_q \frac{f(\mathbf{q})}{\varepsilon_k - \varepsilon_q}\delta(\varepsilon_k - \varepsilon_{k'}), \end{aligned} \tag{44A.8}$$

where we have taken $J_{kk'}$ to be a constant and we write

$$\overline{\sum_n s_{nz}^2} = \frac{1}{3}\cdot\frac{1}{2}\left(\frac{1}{2}+1\right)cN,$$

which is valid for randomly oriented spins. The same expression applies to the rate, $w_{k\downarrow\to k'\downarrow}$. These results, which now involve the temperature through the occupation factors, $f(\mathbf{q}, T)$, are a manifestation of the dynamics associated with the spin–flip processes.

The spin–flip scattering of the conduction electrons falls into class B; those for up→down scattering involve the following four transitions:

$$\left|\cdots q_\sigma \cdots k_\uparrow \cdots ; \cdots n_\downarrow \cdots\right\rangle \to \left|\cdots q_\sigma \cdots q'_\uparrow \cdots ; \cdots n_\downarrow \cdots\right\rangle \to \left|\cdots q_\sigma \cdots k'_\downarrow \cdots ; \cdots n_\uparrow \cdots\right\rangle, \quad \text{(a)}$$

$$\left|\cdots q_\uparrow \cdots k_\uparrow \cdots ; \cdots n_\downarrow \cdots\right\rangle \to \left|\cdots k'_\downarrow \cdots k_\uparrow \cdots ; \cdots n_\uparrow \cdots\right\rangle \to \left|\cdots q_\uparrow \cdots k'_\uparrow \cdots ; \cdots n_\uparrow \cdots\right\rangle, \quad \text{(b)}$$

$$\left|\cdots q_\sigma \cdots k_\uparrow \cdots ; \cdots n_\downarrow \cdots\right\rangle \to \left|\cdots q_\sigma \cdots q'_\downarrow \cdots ; \cdots n_\uparrow \cdots\right\rangle \to \left|\cdots q_\sigma \cdots k'_\downarrow \cdots ; \cdots n_\uparrow \cdots\right\rangle, \quad \text{(c)}$$

$$\left|\cdots q_\downarrow \cdots k_\uparrow \cdots ; \cdots n_\uparrow \cdots\right\rangle \to \left|\cdots k'_\downarrow \cdots k_\uparrow \cdots ; \cdots n_\uparrow \cdots\right\rangle \to \left|\cdots q_\uparrow \cdots k'_\downarrow \cdots ; \cdots n_\uparrow \cdots\right\rangle, \quad \text{(d)}$$

$$\text{(44A.9)}$$

with a corresponding set for down→ up scattering. All of these transitions involve a single impurity spin–flip occurring in the second-order terms. However the first-order matrix elements involve the same spin–flip, both being facilitated by the operator, $\hat{S}_+$, where we recall, $\hat{S}_+\left|n_\downarrow\right\rangle = \left|n_\uparrow\right\rangle$ (with the $\mathbf{k}_\downarrow \to \mathbf{k}'_\uparrow$ transitions involving the operator $\hat{S}_-$); these operators therefore enter twice. Carrying out the calculations and again retaining only the term proportional to $f(\mathbf{q})$, we obtain

$$w^{(3)}_{k\uparrow\to k'\downarrow} = w^{(3)}_{k\downarrow\to k'\uparrow} = \frac{2\pi}{\hbar}\left(\frac{J}{N}\right)^3 cN \sum_{q} \frac{f(q)}{\varepsilon_k - \varepsilon_q}\delta(\varepsilon_k - \varepsilon_{k'}). \tag{44A.10}$$

Combining (44A.8) and (44A.10), the total scattering rate involving the Fermi occupation factor, $f(\mathbf{q})$, is then

$$\begin{aligned} w^{(3)}_{k\to k'} &= w^{(3)}_{k\uparrow\to k'\uparrow} + w^{(3)}_{k\downarrow\to k'\downarrow} + 2w^{(3)}_{k\uparrow\to k'\downarrow} \\ &= \frac{2\pi}{\hbar}\left(\frac{J}{N}\right)^3 4cN \sum_{q} \frac{f(q)}{\varepsilon_k - \varepsilon_q}\delta(\varepsilon_k - \varepsilon_{k'}, \\ &= \frac{2\pi}{\hbar}\left(\frac{J}{N}\right)^3 4cN^2 g(\varepsilon) \end{aligned} \tag{44A.11}$$

where we have introduced the quantity

$$g(\varepsilon) \equiv -\frac{1}{N}\sum_{q} \frac{f(q)}{\varepsilon - \varepsilon_q}. \tag{44A.12}$$

44A.2 Calculating the conductivity

We combine the second and third order processes by writing $w = w^{(2)} + w^{(3)}$, leading to $\tau_{total}^{-1} = \tau^{-1} + \tau_K^{-1}$, where τ is the usual relaxation time that contains all the potential scattering mechanisms, including the non-singular contributions from the magnetic impurities. On the other hand, τ_K arises solely from the singular contribution, as given by (44A.11), and for our case where $w^{(3)}(\varepsilon)$ is not a function of angle, $\tau_K(\varepsilon)$, is given by (see Section 19.10)

$$\frac{1}{\tau_K(\varepsilon)} = \frac{1}{(2\pi)^3}\int d^3k' w^{(3)}(\varepsilon') = \frac{mk_F}{2\pi^2\hbar^2}\int d\varepsilon' w^{(3)}(\varepsilon'). \tag{44A.13}$$

Inserting Eq. (44A.11) for $w^{(3)}(\varepsilon')$, we obtain

$$\frac{1}{\tau_K} = \frac{9\pi J^3 c}{2\varepsilon_F\hbar} g(\varepsilon). \tag{44A.14}$$

When $\tau = \tau(\varepsilon)$, the electrical conductivity must be written as (contrast this with Eq. (19.22))

$$\sigma = -\frac{e^2}{12\pi^2}\int d^3k\, \tau(\varepsilon) v^2 \frac{\partial f^0(\varepsilon)}{\partial \varepsilon}. \tag{44A.15}$$

Writing $\tau_{total} = \tau\tau_K/(\tau + \tau_K) \cong \tau(1 - \tau/\tau_K)$ and $\sigma = \sigma_0 + \sigma_K$, we have

$$\begin{aligned}
\sigma_K &= -\frac{e^2}{12\pi^2}\tau^2 \int d^3k \frac{v^2}{\tau_K}\frac{\partial f^0(\varepsilon)}{\partial \varepsilon} \\
&= -\frac{3e^2 J^3 c\tau^2}{8\pi\varepsilon_F\hbar}\int d^3k\, g(\varepsilon) v^2 \frac{\partial f^0(\varepsilon)}{\partial \varepsilon} \\
&= \frac{3e^2 J^3 \tau^2}{8\pi\varepsilon_F\hbar}\frac{cV}{(2\pi)^3 N}\int d^3k \int d^3k' \frac{f(\varepsilon')}{\varepsilon' - \varepsilon}\frac{\partial f^0(\varepsilon)}{\partial \varepsilon}.
\end{aligned} \tag{44A.16}$$

Rewriting the integrals in this expression in terms of the energies and integrating by parts according to the principal value prescription gives

$$\begin{aligned}
\int d^3k \int d^3k' \frac{f(\varepsilon')}{\varepsilon' - \varepsilon}\frac{\partial f^0(\varepsilon)}{\partial \varepsilon} &= \frac{32\pi^2 m^3}{\hbar^6}\int d\varepsilon \int d\varepsilon' f_0(\varepsilon)\frac{\partial f_0(\varepsilon')}{\partial \varepsilon'}\frac{1}{\varepsilon - \varepsilon'} \\
&= \frac{32\pi^2 m^3}{\hbar^6}\left(2 + \int d\varepsilon \int d\varepsilon' ln\left(\frac{\sqrt{\varepsilon} - \sqrt{\varepsilon'}}{\sqrt{\varepsilon} + \sqrt{\varepsilon'}}\right)\frac{\partial f_0(\varepsilon)}{\partial \varepsilon}\frac{\partial f_0(\varepsilon')}{\partial \varepsilon'}\right).
\end{aligned} \tag{44A.17}$$

The remaining double integral may accurately be approximated as

$$\int d\varepsilon \int d\varepsilon' ln\left(\frac{\varepsilon - \varepsilon'}{4\varepsilon_F}\right)\frac{\partial f_0(\varepsilon)}{\partial \varepsilon}\frac{\partial f_0(\varepsilon')}{\partial \varepsilon'} = ln(T/T_0),$$

where T_0 is a constant. Retaining only the singular contribution, we have

$$\sigma_K \cong \frac{3e^2J^3\tau^2}{8\pi\varepsilon_F\hbar}\frac{cV}{(2\pi)^3N}\frac{32\pi^2m^3}{\hbar^6}ln(T/T_0) \tag{44A.18a}$$

or

$$\sigma_K \cong \sigma_{K0}ln(T/T_0), \tag{44A.18b}$$

where we have absorbed all the constants into a parameter, σ_{K0}.

From Eqs. (44A.18a,b), which is negative for $T < T_0$, we see that there will be a slow rise in the electrical resistivity at low temperatures. This behavior is in agreement with that seen experimentally in Fig. 44.6 and a host of other metallic systems with magnetic impurities. Qualitatively it is also a temperature at which the magnet susceptibility changes from a Curie-like behavior to the temperature-independent Pauli behavior.

However we are immediately presented with a problem: we would not expect the resistivity to increase without limit as $T \rightarrow 0$. Experimentally, this increase ultimately flattens out and approaches a constant value. However, the associated theoretical discussion is complex and as previously mentioned, is best treated with renormalization group methods, which we will not take up here (see Additional Reading). Nevertheless, what happens qualitatively is that the sea of conduction electrons surrounding the impurity spin becomes locally polarized such that at very low temperatures it collectively *cancels* the spin of the impurity; the impurity is then magnetically "screened," resulting in an overall spin singlet state.

Part XIII

Special Topics

Strongly Correlated Systems

45

Up to this point, the quantum mechanics of electrons in solids has largely involved properties that can be treated within the framework of self-consistent, one-electron, band theory. This approach provides a good description of many metals, including those with d electrons, and various semiconductors with constituents having only s and p valence electrons. But this approach usually fails for the f electrons in rare earth metals (which tend to be immobile and magnetically polarized), and is problematic for the so-called heavy-fermion intermetallic compounds, formed with certain 4f and 5f elements (where very large effective masses are observed), as well as a host of chemical compounds containing atoms with d and f electrons, particularly oxides. This list must also include the high temperature cuprate (and other) superconductors. Qualitatively, as we discuss, we expect new effects to occur when the added Hubbard electrostatic energy, U, between two electrons on the same atom in the same otherwise unfilled band becomes larger than the overall energy width, W, of that band. The energies, U, are expected to be largest when the electrons are tightly bound to the atoms, resulting in strong overlap of the orbitals within a given atomic shell; a property of 3d and 5f electrons.

The basis for the success of one-electron physics near the Fermi surface lies in Landau's argument, based on the Pauli principle, that the states near the Fermi surface, the quasiparticles, have long lifetimes (see Section 37.1). As we move off the Fermi surface such one-electron-like states decay; i.e., they develop a width which can quickly become comparable to their energy relative to the Fermi energy. In addition, local atomic and molecular considerations start to dominate. High energy photons, say those used in ARPES to study off-Fermi surface behavior, are for all practical purposes absorbed locally, and hence probe the local chemical and ultimately the atomic environment and its subsequent response. Curiously, even a ground state property like superconductivity intimately involves off-Fermi surface properties since it is based on Cooper pairing, the strength of which involves virtual fluctuations in either the lattice (phonon pairing mechanisms) or the collective electron spin/charge density (electronic pairing mechanisms).

To understand strongly correlated systems one must directly confront the physics that is left out of the one-electron models. As argued above we must start with the atoms themselves. From Hund's rules (see Section 5.5) we know that the mutual repulsion of electrons in the same angular momentum shell, l, favors each electron having a different magnetic quantum number, m (leading to the highest spin consistent with the Pauli principle), thereby minimizing the orbital overlap

The Physics of Solids. First Edition. J. B. Ketterson.

and with it the Coulomb repulsion. So when filling a d shell, where $l = 2$ and the total shell capacity is 10, the maximum spin would be 5/2. In considering a periodic solid composed of atoms with tightly localized orbitals, such as those associated with atoms having a partly filled 3d shell[1] lying within the more extended 4s and 4p orbitals (that can form bonds to hold the solid together), it is natural to ask what happens to the spin associated with such atoms.

In Bloch's tight-binding model, the electronic states for the crystal are constructed as linear combinations of atomic orbitals and the electrons "coherently hop" from atom to atom as a result of overlap of these wave functions with their neighbors, thereby forming the bands. If these bands are full we expect insulating behavior; if they are partially full we expect to have a metal. This picture ignores the effect of the Coulomb repulsion present when more than one electron is in an orbital from which an associated tight-binding band is derived. The Anderson impurity model, discussed in Section 44.1, contains the physical ideas needed to make a start on dealing with this problem. There we considered the simplest case of an impurity atom with a *single* orbital level lying at an energy, ε_{0d}, near but below ε_F, so that it would be occupied. The intra-orbital electron–electron Coulomb repulsion was incorporated through the energy, U, which acts when two electrons occupy that orbital state. If there is no coupling to the itinerant electron sea, and if $U > \varepsilon_F - \varepsilon_{0d}$, the level will be singly occupied: the atom is magnetic. If we now add coupling to the electron sea through the term containing V_{kd} (see Eq. (44.5)) such that, qualitatively, $N(0)\left\langle V_{kd}^2 \right\rangle > U > \varepsilon_F - \varepsilon_d$, the level is "mixed into" the background electron sea[2] and two electrons with net spin zero can occupy the level: the atom is non-magnetic.

[1] The fourth row of the periodic table starts with $K(4s^1)$ followed by $Ca(4s^2)$; but because the 3d levels lie lower than the 4p levels, the 3d shell then fills, the order being: $Sc(4s^2 3d^1)$, $Ti(4s^2 3d^2)$, $V(4s^2 d^3)$, $Cr(4s^1 3d^5)$ $Mn(4s^2 3d^5)$, $Fe(4s^2 3d^6)$, $Co(4s^2 3d^7)$, $Ni(4s^2 3d^8)$, $Cu(4s^1 3d^{10})$, $Zn(4s^2 3d^{10})$. Note, the levels do not fill precisely in the expected order since there can be more than one electronic configuration in the vicinity of the ground state. Now the d electrons have smaller orbitals relative to s–p electrons and for this reason play a lesser role in chemical bonding. In simplified terms, however, varying numbers of electrons from the 3d shell may be "promoted" to the s–p shell, resulting in *multiple* (and sometimes mixed) valence states; Sc^{+3} (at the left) and Zn^{+2} (at the right) are exceptions. The most complex atoms of this group are V and Mn, which display four different valences, their oxides being especially complicated. The Hund's rule coupling among the d electrons can result in a net spin, often causing the solids to be magnetically ordered.

[2] Qualitatively, $N(0)\left\langle V_{kd}^2 \right\rangle$ is a level width, and would correspond to the band width of a lattice formed from such impurities that couple through a background sea of free (s–p) electrons.

45.1 The Mott insulator and the Mott transition

In the band theory of solids we distinguish metals from insulators according to whether the Fermi level falls within one or more partially filled bands (a metal), or lies between the filled bands and higher unoccupied bands (an insulator). This picture works well for a wide range of materials but fails for some materials, particularly 3d transition metal oxides (de Boer and Verway (1937)). An oft-cited example, first discussed by Mott (1937), is nickel oxide, which we will use as an example in much of the discussion below. This material has the cubic NaCl structure in which the Ni sites are octahedrally coordinated with respect to each other, and similarly for the O sites (as defined in Section 8.13). The atomic $4s^2\ 3d^8$ and $4s^1\ 3d^9$ states of Ni lie very close to each other. From the chemical point of view, 3d or 4s electrons from the Ni can bond with oxygen p orbitals and form the compound, $Ni^{+2}O^{-2}$, leaving holes in the 3d shell; from a tight-binding perspective we then expect either d–d overlap or coupling through the O p states to result in *unfilled* d-like bands in the solid and hence *metallic* behavior. This expectation is consistent with band structure calculations that will be discussed below. However experimentally, NiO is an *insulator*. This "contradiction" extends to many transition metal oxide and other (e.g. sulfide) systems; in addition transitions between metallic and insulating phases can occur as a function of the temperature or electron concentration (doping). This class of phenomenon has been termed *the metal–insulator transition* (MIT). What happens physically is that, in the absence of sufficient screening, electrostatic repulsion between the electrons causes them to localize on the transition metal ions, *even for the case of a half-filled band*:[3] in order to transport charge an electron on one ion must transit across a neighboring ion, and we anticipate that if the Coulomb repulsion of the extra electron on that neighbor is sufficiently large it cannot do this; i.e., collectively, the electrons will be localized. Since this phenomenon involves the coordinates of at least two (and ultimately many) electrons, it is said to arise from electron correlations, which are incorporated in only a mean field sense in the usual one-electron picture of solids.

A way to treat this problem is to generalize the Anderson impurity model to a lattice. When a finite U value is assigned to, say, all the Ni atoms in Mott's NiO example (since these are the atoms contributing holes for possible band formation), we might anticipate that these ions can become magnetic. If the ions adopt an *antiferromagnetic* structure, the unit cell doubles and an originally half-filled band would then be fully occupied.[4] As a result of the *superexchange* mechanism,[5] to be discussed shortly, NiO is indeed antiferromagnetic. At this point one might be tempted to ignore or suppress the effects of U in electronic structure calculations and be satisfied with using, say, spin polarized local density functional theory (DFT), as discussed in Section 26.2. However there are cases where insulating behavior occurs, whereas band theory predicts metallic behavior and long range antiferromagnetic behavior is *not* observed, or, alternatively, insulating behavior persists above the Néel temperature, T_N. One way to think about this behavior is to picture the electrons as being "bound" to the empty hole states within the band, and that this binding is stronger than interatomic interactions leading to antiferromagnetism, an idea that is supported by optical absorption experiments in which such bound states are broken. Alternatively, one can view the unpaired spins on the atoms as being present, and to some extent correlated, but their directions fluctuate resulting in no long range order. Clearly such behaviors go beyond any independent electron picture. More generally, in many materials involving d and f shell valence electrons, where magnetism may or may not arise, ignoring Coulomb correlation effects arising from these orbitals causes density functional theory, as it is conventionally practiced, to fail. In particular the density of states, as observed in photoemission experiments, can strongly deviate from the DFT predictions.

45.1.1 A simple model

In an attempt to capture some of the above effects we now introduce a simple model, the so-called t–U model. Consider a system consisting of atoms having

[3] Mott credits Peierls with the following observation: "It is quite possible that the electrostatic interaction between the electrons prevents them from moving at all. At low temperatures the majority of the electrons are in their proper places in the ions. The minority which have happened to cross the potential barrier find therefore all the other atoms occupied, and in order to get through the lattice have to spend a long time in ions already occupied by other electrons. This needs a considerable addition of energy and so is extremely improbable at low temperatures." This statement captures the basic physics.

[4] Transition-metal monoxides – band or Mott insulators, K. Terakura, A.R. Williams, T. Oguchi, and J.K. Kubler, *Phys. Rev. Lett.* **52**, 1830 (1984); Band theory of insulating transition-metal monoxides – band-structure calculations, K.Terakura, T. Oguchi, A.R. Williams, and J. Kubler, *Phys. Rev.* **8**, 4734 (1984).

[5] Apart from the ferrimagnets, the vast majority of insulating magnetic oxides are antiferromagnetic; this is in accord with Anderson's superexchange model: New approach to the theory of superexchange interactions, P.W. Anderson, *Phys. Rev.* **115**, 2 (1959).

electrons occupying a single tight-binding atomic "d" state. We then picture that electron as hopping between sites on a simple cubic lattice. From the discussion in Section 11.4 (and a simple extension of Eq. (11.55b) to three dimensions) it follows that for the case of nearest-neighbor interactions we can write the energies of the states in the associated band in the form

$$\varepsilon(\mathbf{k}) = \varepsilon_0 - 2t(\cos k_x a + \cos k_y a + \cos k_z a), \tag{45.1}$$

where t is the overlap integral (denoted h_1 in Eq. (11.55b)) and a is the lattice constant; we will fix the energy scale such that $\varepsilon_0 = 0$. In the second quantized form the resulting Hamiltonian is then written as

$$\hat{H}_t = \sum_{k,\sigma} \varepsilon(\mathbf{k}) \hat{c}^\dagger_{k\sigma} \hat{c}_{k\sigma}. \tag{45.2}$$

Because the repulsion associated with the Hubbard U parameter acts only when two electrons are on the same site, i, it is convenient to transform our band Hamiltonian (45.2) into a "site Hamiltonian" by Fourier transforming the **k**-space operators, $\hat{c}_{k\sigma}$, into site operators, $\hat{c}_{j\sigma}$, as

$$\hat{c}_{k\sigma} = \sum_j e^{i\mathbf{k}\cdot\mathbf{r}} \hat{c}_{j\sigma}; \tag{45.3}$$

in this representation, $\hat{H}_t$ becomes

$$\hat{H}_t = -t \sum_{i,\sigma} \sum_{j,\sigma}^{nn} \hat{c}^\dagger_{i\sigma} \hat{c}_{j\sigma} + \text{c.c.} \tag{45.4}$$

To account for the Coulomb repulsion between two electrons on the same site (having opposite spin by the Pauli principle), we add a Hubbard U term, yielding a contribution,[6]

$$\hat{H}_U = U \sum_i \hat{n}_{i\uparrow} \hat{n}_{i\downarrow}, \tag{45.5}$$

where $\hat{n}_{i\sigma} = \hat{c}^\dagger_{i\sigma} \hat{c}_{i\sigma}$ is the number operator for an electron on site, i, with spin, σ, and U is the on-site Coulomb integral, which is a generalization of Eq. (44.1) to all sites,

$$U = \iint d^3r d^3r' \, \varphi^*_{i,\sigma}(\mathbf{r}) \varphi_{i,\sigma}(\mathbf{r}) \frac{e^2}{|\mathbf{r} - \mathbf{r}'|} \varphi^*_{i,-\sigma}(\mathbf{r}')_{i,-\sigma} \varphi(\mathbf{r}'). \tag{45.6}$$

The total Hamiltonian is then

$$\hat{H} = \hat{H}_t + \hat{H}_U - \mu\hat{N} = -t \sum_{i,\sigma} \sum_{j,\sigma}^{nn} \hat{c}^\dagger_{i\sigma} \hat{c}_{j\sigma} + \text{c.c} + U \sum_i \hat{n}_{i\uparrow} \hat{n}_{i\downarrow} - \mu N, \tag{45.7}$$

[6] Electron correlations in narrow energy bands, J. Hubbard, *Proc. R. Soc. Lond. A*, **276**, 238 (1963).

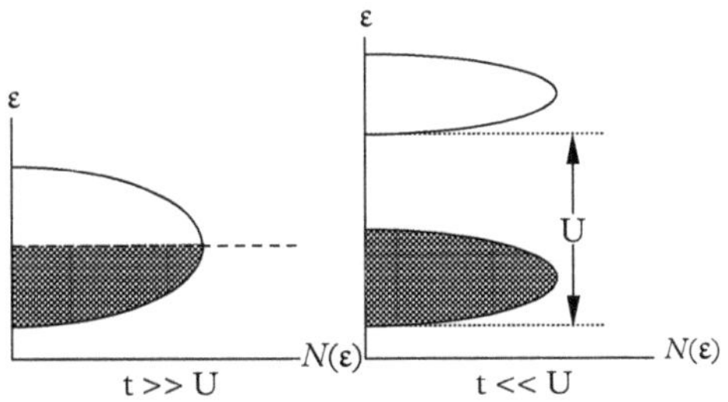

Figure 45.1 *Plots of the density of states, $N(\varepsilon)$, versus the single particle energy, ε, for the single-band Mott–Hubbard model. The shaded areas show the filled states for the case of one electron per s-state atom. When $t \gg U$, the un-split doubly degenerate s band is half filled; however when $t \ll U$, the sub-bands associated with the two spin polarizations are split by an energy U, with one state filled and the other empty.*

where our inclusion of the last term allows for an arbitrary filling of the lattice through the chemical potential, μ, with $\hat{N}$ being the total number operator which is given by

$$\hat{N} = \sum_{i,\sigma} \hat{n}_{i\sigma}. \tag{45.8}$$

Exact solutions for the ground and excited states of a system governed by the Hamiltonian (45.7) cannot be obtained analytically. Great effort has gone into developing approximate treatments as well as obtaining numerical solutions.[7] However in the limit that $U \gg W = 4t$, where W is the band width, the band structure consists of two sub-bands *split by the energy*, U. For the case of one electron per s-state atom, where the material would ordinarily be a metal, the sub-band associated with one spin polarization would be full while the other would be empty. Excitations would then involve transfer of charge between these bands. This situation is shown schematically in Fig. 45.1. If t vanishes the occupied band has zero width. Depending on the residual exchange energy acting between spins, as governed by a parameter, J (see Eq. (26.1a)), the system could be ferromagnetic ($J > 0$) or antiferromagnetic ($J < 0$). In the latter case the unit cell would be enlarged, thereby opening a band gap on its own.

Figure 45.2 shows, schematically, the line separating the Mott insulating phase from the metallic phase as a function of the chemical potential, μ (which controls the filling), and the parameter, t/U; the larger this latter parameter, the greater the tendency toward metallic behavior. Note that the vertical trajectory designated by A is primarily driven by the carrier concentration, n, and would involve a first-order transition; horizontal trajectories are driven by the band width, W, and for the special case shown in B would involve a second-order transition. This latter behavior is referred to as a *quantum critical point.* In the neighborhood of such points large quantum fluctuations in the charge and/or spin degrees of freedom occur *along* with accompanying enhancements in the specific heat and/or spin susceptibility. In the presence of orbital degeneracy (discussed below) bond orientation fluctuations can occur.

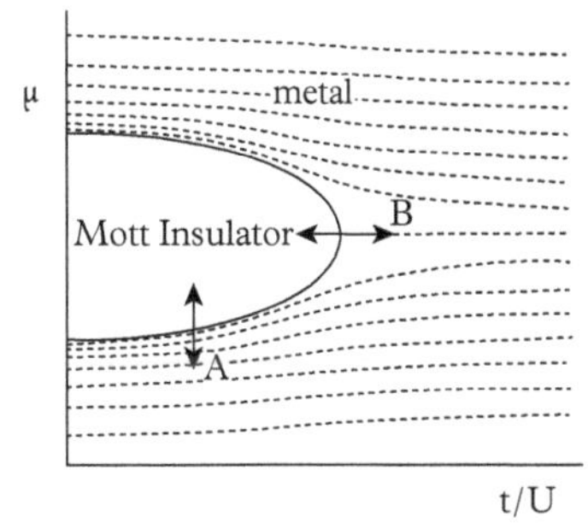

Figure 45.2 *The phase diagram of a single band Mott system showing the boundary between the metallic and insulating phases as a function of the band filling, controlled by μ, and t/U.*

The behavior at intermediate values of t/U at half filling is shown schematically in Fig. 45.3.[8] Starting from a non-interacting system, increasing U has two effects. Initially it leads to band narrowing, because hopping is suppressed due

[7] See the review articles cited at the end of this chapter.

[8] Strongly correlated materials: insights from dynamical mean-field theory, G. Kotliar and D. Vollhardt, *Phys. Today* 3, 53 (2004).

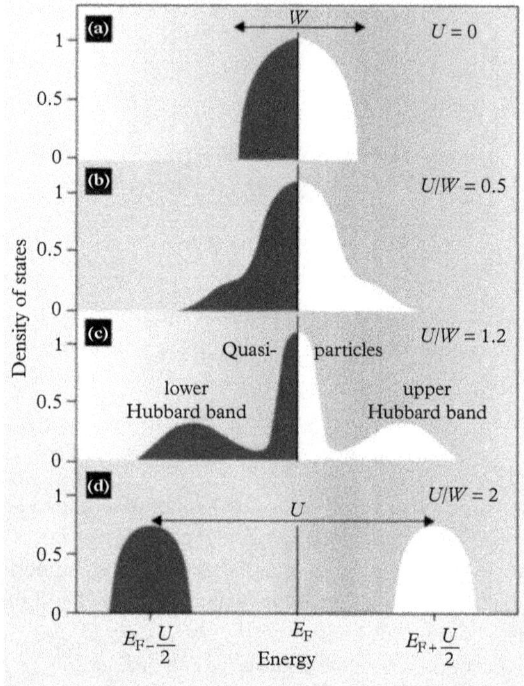

Figure 45.3 *The redistribution of the density of states under a Mott transition (after Kotliar and Vollhardt (2004)).*

to the energy cost of double occupancy arising from U.[9] Secondly, high energy excitations gradually emerge in the metallic state, due to the states lost in the single-particle band (the total number of states being preserved). This "spectral-weight" redistribution is a fundamental feature of strongly correlated electron systems. The single-particle density of states disappears at the MI transition.

We now return to the question of whether neighboring spins in a material like NiO tend to align or anti align. If they were parallel a spin from one site could not temporarily hop onto its neighbor, as that spin state would already be occupied, the process being forbidden by the Pauli principle. However if the spin on the neighboring site were opposite, the neighboring electron could temporarily occupy the state, allowing it to delocalize and lower its energy. By second-order perturbation theory, the energy is lowered for the antiparallel alignment by an amount of order $-t^2/U$, thereby favoring antiferromagnetism; this is the physics underlying the superexchange mechanism, noted above in Section 45.1.

45.1.2 A more realistic model

Real examples of MITs involve chemical compounds, commonly 3d transition metal oxides, such as antiferromagnetic NiO, ferromagnetic V_2O_3, and non-magnetic VO_2 and Ti_2O_3. In order to focus the discussion we will limit ourselves to those transition metal monoxides which have the NaCl structure: TiO,

[9] Application of Gutzwiller's variational method to the metal insulator transition, W.F. Brinkman and T.M. Rice, *Phys. Rev. B* 2, 4302 (1970).

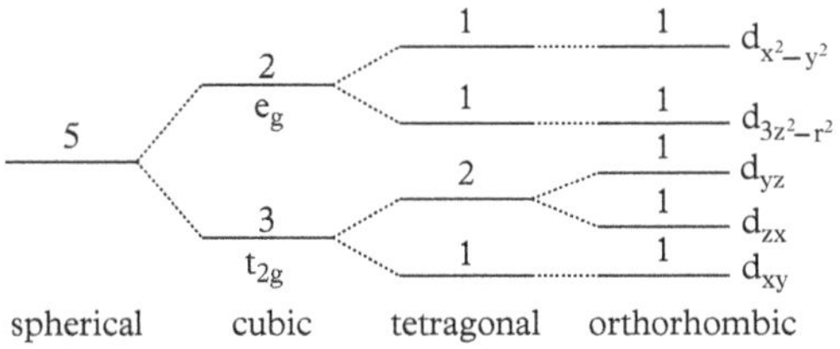

Figure 45.4 *The splitting of the* d *levels for progressively lower symmetries of the atomic site.*

VO, MnO, FeO, CoO, and NiO. Of these the first two are metallic while the last four are classified as Mott insulators. In a tight-binding language, the d electron states, which are degenerate in the spherical symmetry of the atom, will be split by the underlying crystal potential (referred to as a crystal field effect). Lower symmetry leads to lower degeneracy and with it an increasing likelihood that a gap can be opened when narrow bands and strong on-site Coulomb repulsion are present. Figure 45.4 shows, schematically, how the splitting proceeds as the symmetry of the atomic site is lowered; note that each level is doubly degenerate and spin–orbit coupling has been neglected. The energies of the various levels typically occur in the order shown. Hence for the octahedral site symmetry associated with cubic NaCl structures, the Fermi level lies near or in the bands formed from the lower lying t_{2g} orbitals, d_{xy}, d_{yz}, and d_{zx}, for the lighter 3d atoms such as Ti, V, Cr, . . . on the left-hand side of the periodic table, and near or in those formed from the e_g orbitals, $d_{3z^2-r^2}$ and $d_{x^2-y^2}$, for the heavier 3d atoms such as Cu and Ni on the right-hand side, the lower levels now being filled.[10] Partially filled, degenerate t_{2g} or e_g levels can lead to a structural (called a Jahn–Teller) distortion that partly lifts the degeneracy.[11]

If we ignore the effects of the anions (e.g., O atoms) in the lattice (implying that the associated orbitals are filled and energetically far from the d levels), generalizing the single band Hamiltonian given in Eq. (45.7) would require inclusion of the following effects: (i) overlap integrals, $t_{ij}^{mm'}$, connecting all the d states (numbered by the magnetic quantum number, m) on one atom to those on the nearest neighbors; (ii) matrix elements of the screened intra-site Coulomb, $U^{mm'}$, and exchange, $J^{mm'}$ electron–electron interaction, which enter whenever more than one of the d orbitals is occupied on a given atom (and which together generate the Hund's rule couplings acting within the d shell, as noted above); and (iii) intersite Coulomb integrals, $V_{ij}^{mm'}$.[12] It is common to replace the orbital-dependent parameters, $U_{mm'}$ and $J_{mm'}$, by spherically averaged matrix elements, $\bar{U}$ and $\bar{J}$, and sometimes treat them as adjustable parameters to fit experimental data (we drop the bar over these quantities in what follows). If we neglect the interatomic Coulomb integrals the resulting Hamiltonian has the form

$$\hat{H} = \sum_{i,j,\sigma,m,m'} t_{ij}^{mm'} \hat{c}_{i\sigma m}^{\dagger} \hat{c}_{j\sigma m'} + \text{h.c.} + \frac{\bar{U}}{2} \sum_{i,m,m',\sigma} \hat{n}_{im\sigma} \hat{n}_{im-\sigma} + \frac{(U-J)}{2} \sum_{i,m\neq m',\sigma} \hat{n}_{im\sigma} \hat{n}_{im'\sigma}. \tag{45.9}$$

Even with this simplification the mathematical complexities are clearly enormous.

[10] When presenting non-relativistic band structure, calculations involving d electrons (see Chapter 13), it is more common to use the notations Γ'_{25} and Γ_{12}, rather than t_{2g} and e_g.

[11] The Jahn–Teller distortion sets in when the reduction of the electronic energy due to removing a degeneracy is linear in the distortion; it will then exceed the quadric increase in the elastic energy of the lattice.

[12] For an example and discussion of a Hamiltonian containing such parameters, see C. Castellani, C.R. Natoli, and J. Ranninger, Magnetic structure of V_2O_3 in the insulating phase, *Phys. Rev. B* 18, 4945 (1978).

45.1.3 Incorporating the p states

We now expand the discussion to include, qualitatively, the effects of the anions, called *ligands*, which bridge the metallic ions in the compounds. For cases where these are oxygen, which has the atomic configuration, $1s^2 2s^2 2p^4$, the 2s states in the crystal are filled, as they are in the atom, and lie deep (approximately 14 eV below the p states); they play no role near the Fermi level. Only the 2p states are involved in bonding. In the NaCl structure the O atoms also have octahedral symmetry and the three-fold degenerate atomic 2p states (six-fold with spin) form bonding orbitals directed along the orthogonal x, y, and z axes and do not split. For the case of NiO, and from an ionic chemical perspective, the two 4s electrons from the $4s^2 3d^8$ ground state configuration of Ni fill the two holes of the oxygen 2p shell (and the associated bands), leaving two holes in the higher lying nickel e_g levels arising from the crystal field split nickel 3d levels and the resulting bands associated with their overlap; the t_{2g} bands are full. For the monatomic 3d oxides, more generally, the midpoint of the (filled) oxygen p band, ε_p, relative to the midpoint of the (partially filled) d band, ε_d, as measured by the parameter, $\Delta = |\varepsilon_d - \varepsilon_p|$, plays an important role. If $U < \Delta$ then d–d coupling occurs by direct overlap of these orbitals and we say that we have a Mott–Hubbard system. On the other hand, if $U > \Delta$ then d–d coupling occurs in second-order perturbation theory *through* the p orbitals and we say we have a *charge transfer insulator*; the associated spin–spin coupling is called *super exchange* and leads to antiferromagnetism (Anderson (1959)). These two cases are shown schematically in Fig. 45.5.[13] As we will see below, Δ decreases as we move to the right in the 3d series and there is a tendency to pass between these two regimes. To include the effects of ligands in a tight-binding model would require additional overlap

[13] Metal-insulator transitions, M. Imada, A. Fujimori, and Y. Tokura, *Rev. Mod. Phys.* 70, 1039 (1998).

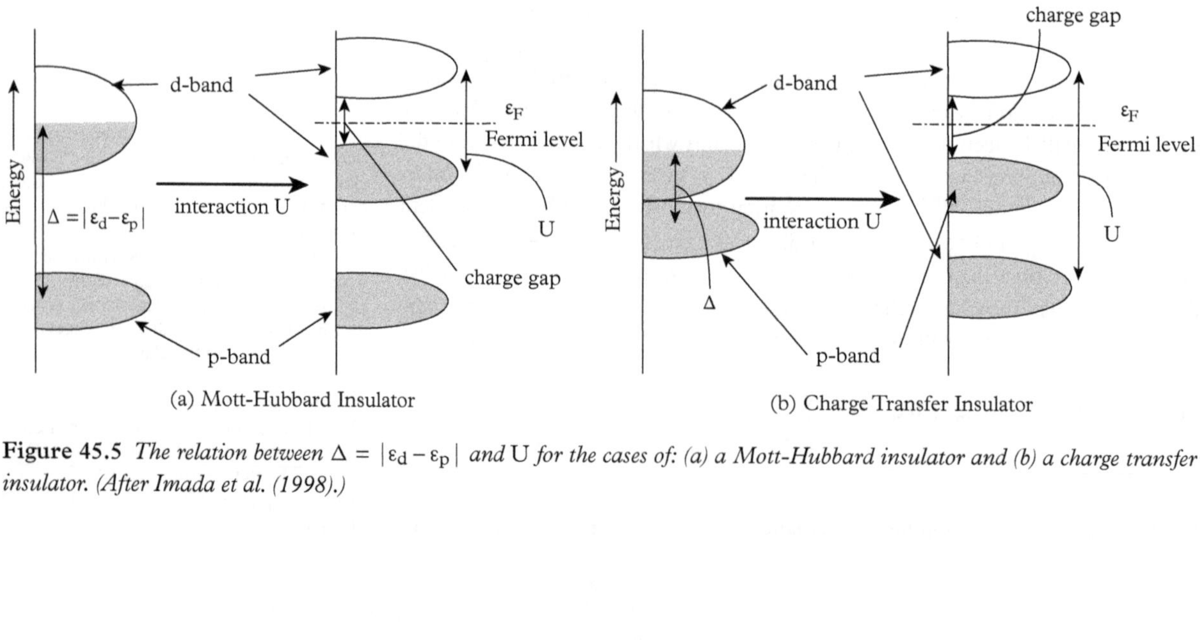

Figure 45.5 *The relation between* $\Delta = |\varepsilon_d - \varepsilon_p|$ *and* U *for the cases of: (a) a Mott-Hubbard insulator and (b) a charge transfer insulator. (After Imada et al. (1998).)*

integrals connecting all of the d and p orbitals and other parameters, rendering calculations even more difficult. Clearly something more systematic is desirable.

45.2 The LDA implementation of DFT

The local density approximation to density functional theory (DFT) has been wildly successful in generating many ground state properties, such as total energies, spin polarizations, and (in metals) Fermi surfaces. When applied to the transition metal oxides, the LDA generally predicts the right crystal structures and yields good lattice constants; this also holds for other strongly correlated systems. Here we start by summarizing the overall behavior of the transition metal monoxides emerging from conventional LDA band structure calculations.

Figure 45.6 shows the general behavior of the local density augmented plane wave (see Section 13.1) energy bands along the [100] $\Gamma - X$ direction (the

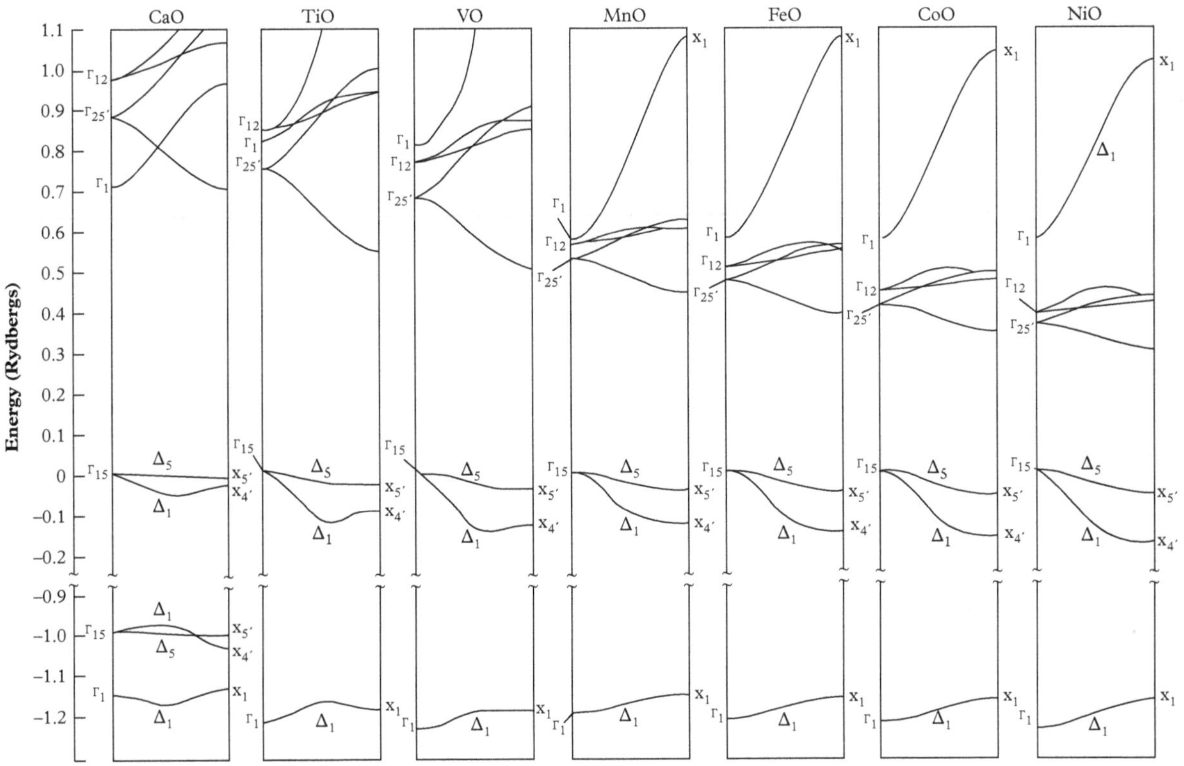

Figure 45.6 *The augmented plane wave (APW) energy bands of the* 3d *transition metal oxides in the* NaCl *structure along the* $\Gamma - X$ *direction (after Mattheiss (1972)).*

Δ line) in the 3d transition metal oxides with the NaCl structure, as calculated by Matthias (1957).[14] Starting from the lowest energy at the Γ point for Ca, we have the levels $\Gamma_1, \Gamma_{15}, \Gamma_{15}, \Gamma_1, \Gamma_{25'}$, and Γ_{12}; these correspond to the atomic-derived O 2s, Ca 3p, O 2p, Ca 4s, Ca 3d (t_{2g}), and Ca 3d (e_g) levels, the latter two split by the octahedral crystal field. As we move to the right, what was the Ca 3p levels move to lower energies (out of the range plotted), while the original Ca 4s level moves to higher energies, into and ultimately above the 3d-band complex; in addition the O 2p levels move closer to the 3d levels. For CaO, the Fermi energy lies between the highest O (Γ_{15}) band and the lowest Ca (Γ_1) band; for all of the remaining compounds, ε_F lies within the s–d complex, rising as the d shell fills.

As we pointed out, the last four compounds are experimentally found to be insulators, in contradiction to these (and most other) band structure predictions. Also, as noted above, DFT predicts metallic behavior for FeO and CoO, whereas experimentally they are insulators, and although NiO in the antiferromagnetic state is predicted to be an insulator, the calculated band gap, $\approx$0.3 eV, is far smaller than the observed value, $\approx$4.3eV.[15,16] Clearly what we are missing is the effects of the Coulomb repulsion parameter, U, that was incorporated into the above tight-binding based formulations to model this problem.

45.2.1 An LDA + U implementation of DFT without orbital polarization

We now ask whether the LDA theory, if altered in some way, is capable of describing transition metal oxides and other strongly correlated systems. To simplify the discussion we will initially treat all d orbitals on the same footing; i.e., they will interact with each other via a single, orbital-independent, Coulomb interaction, U, for which we adopt the definition[17]

$$U = E(d^{n+1}) + E(d^{n-1}) - 2E(d^n), \tag{45.10}$$

where $E(d^n)$ is the total energy of a system with n_d electrons in the 3d shell of some particular site. Note that the values of these energies are somewhat loosely defined and can depend on the "time scale." When s, p conduction electrons are present and moving in wide bands they can quickly redistribute their motions so as to screen, and hence lower, the "bare" Coulomb energies that are present in the isolated 3d atoms; this effect enters naturally when performing self-consistent LDA calculations. Experimental values for $E(d^{n-1})$ and $E(d^{n+1})$ relative to $E(d^n)$ can be taken from the thresholds in photo-emission and inverse photo-emission spectra (PES and IPS, see Section 20.2), where one removes or adds an electron to the system; it is natural to question if the screening is complete in such experiments.

A relatively simple model for obtaining a value for U is by constructing a supercell (see Section 39.4.4) and removing or adding one electron in a judicious way within this larger unit; i.e. the system is treated as if it contains an "impurity atom" associated with one less or one more electron in the 3d shell. Such calculations

[14] Electronic structure of the 3d transition-metal monoxides. I. Energy-band results, L.F. Mattheiss, *Phys. Rev. B* 5, 290 (1972).

[15] Magnitude and origin of the band-gap in NiO, G.A. Sawatzky and J.W. Allen, *Phys. Rev. Lett.* **53**, 2339 (1984); Band gaps and electronic structure of transition-metal compounds, J. Zaanen, G.A. Sawatzky, and J.W. Allen, *Phys. Rev. Lett.* 55, 419 (1985).

[16] Structure of valence and conduction levels in NiO, J.M. McKay and V.E. Henrich, *Phys. Rev. Lett.* **53**, 2343 (1984).

[17] C. Herring, in *Magnetism, Vol. IV*, Ed. G.T. Rado and H. Suhl, Academic Press, NY(1966).

have been carried out for monatomic metals such as Fe and Ce, from which U values were extracted.[18] Of interest to us here are Mott insulators, and our example case of NiO has been examined with a supercell consisting of eight atoms per cell (four formula units) with encouraging results.[19] For the PES simulation, one 3d electron is removed from the central unit cell, resulting in a d^7 configuration, while adding one unit of charge to the supercell as a whole to maintain charge neutrality. The excitation energy was calculated using the prescription,[20]

$$\Delta\varepsilon = \frac{|\varepsilon_{GS} - \varepsilon_{ES}|}{2}, \tag{45.11}$$

where the difference is taken between the 3d LDA eigenvalues in the ground and excited states. The resulting excitation energy is 3.8 eV. To simulate the IPS spectrum an electron is added to the central atom 3d shell, creating a d^9 configuration, while removing an electron from the oxygen 2p band; this results in an excitation energy of 4.1 eV (versus the experimental value of 4.3eV). The sum of the two calculated energies, 3.8eV + 4.1eV = 7.9eV, can be interpreted as the Mott gap, U, a value consistent with other estimates. This result suggests that DFT, if appropriately modified, can be used to describe aspects of strongly correlated systems.

Although the preceding calculation is suggestive, we clearly need to formulate the theory for incorporating U on a more systematic basis. We recall that the LDA utilizes an exchange–correlation potential, $V^{XC}(\mathbf{r})$, that is based on the density of the uniform electron gas, the exchange contribution having the Dirac–Slater form,[21] which varies as $n(\mathbf{r})^{1/3}$. On the other hand, the exact *operator* form of $\hat{V}^{XC}$ is expected to differ for each Kohn–Sham orbital (as does the exchange term in the Hartree–Fock approximation), and not be simply a function of the local charge (and spin) density of the electrons. In addition this charge density is usually spherically averaged, whereas the actual density is anisotropic. We should also expect that the more anisotropic and localized a given K–S orbital is, the more likely it will experience a pronounced, orbital-dependent, exchange–correlation potential.

Several approaches have been proposed to improve the LDA for d electrons in solids that *directly* incorporate orbital-dependent intra-shell Coulomb and exchange interactions.[22,23] Schemes implementing this idea are called LDA + U. One might think that these intra-orbital contributions could be calculated from the properties of an isolated atom with which the localized orbitals are associated, say using a Hartree–Fock approach. But the resulting contributions turn out to be much too large, 15–20 eV, whereas a typical U parameter used to fit experimental spectra is closer to 8 eV or less. The reason, noted above, is that in the solid, loosely bound s and p electrons from a specific and neighboring sites overlap the localized d or f electrons on that site and *screen* the intra-atom Coulomb interactions. Hence any formalism that attempts to simultaneously account for effects arising from localized and extended orbitals must be carried out self-consistently.

[18] Density-functional calculation of effective Coulomb interactions in metals, V.I. Anisirnov' and O. Gunnarsson, *Phys. Rev. B* **43**, 7570 (1981).

[19] Model supercell local-density calculations of the 3d excitation spectra in NiO, M.R. Norman and A.J. Freeman, *Phys. Rev. B* **33**, 8896 (1986).

[20] The correct prescription for calculating excitation energies is to subtract the total energies for the two configurations as defined in Eq. (45.10); in practice this approach requires a very high level of numerical precision.

[21] See Section 15.2.3.

[22] Band theory and Mott insulators: Hubbard U instead of Stoner I, V.I. Anisimov, J. Zaanen, and O.K. Andersen, *Phys. Rev. B* **44**, 943 (1991).

[23] First-principles calculations of the electronic structure and spectra of strongly correlated systems: the LDA + U method, V.I. Anisimov, F. Aryasetiawan, and A.I. Lichtenstein, *J. Phys.: Condens. Matter* **9**, 767 (1997).

One strategy is to "subtract off" an "average" contribution of the d shell to the total energy and then "add back" an orbital-dependent form. If we initially neglect exchange and m-dependent orbital polarization effects, we would then write the total energy arising from a Hubbard-like contribution as

$$E = E_{LDA} - \frac{1}{2}Un_d(n_d - 1) + \frac{1}{2}U\sum_{i \neq j} n_i n_j; \tag{45.12}$$

here, n_d is the total d shell occupancy,

$$n_d = \sum_i n_i, \tag{45.13}$$

n_i is the occupancy of the ith orbital (a number), and $n_d(n_d - 1)/2$ is the total number of d–d interactions. Differentiation with respect to the state occupancy gives the orbital energies,

$$\varepsilon = \frac{\delta E}{\delta n_i} = \varepsilon^{LDA} + U\left(\frac{1}{2} - n_i\right). \tag{45.14}$$

Note that this expression gives corrections to ε^{LDA} of +U/2 and –U/2 for empty ($n_i = 0$) and fully occupied ($n_i = 1$) orbitals respectively; hence the expected feature of having the energy shift by U for an integer change in occupancy is now built in.[24] The LDA potential associated with the ith d orbital in the K–S equations is

$$V_i(\mathbf{r}) = \frac{\delta E}{\delta n_i(\mathbf{r})} = V^{LDA}(\mathbf{r}) + U\left(\frac{1}{2} - n_i(\mathbf{r})\right); \tag{45.15}$$

here the variation is taken with respect to $n_i(\mathbf{r})$, the density of the ith orbital (not to be confused with the *total* occupancy of that orbital, n_i), and not the full electron density, $n(\mathbf{r})$.

To complete an LDA + U calculation we require a prescription to calculate U (and J if intra-orbital exchange is included). Here one uses the supercell technique described previously. An approach is to use a band structure formalism based on atomic spheres (in which the d electron wave functions are largely confined) that also allows the coupling between the d states in the central unit cell and the rest of the supercell to be "switched off"; for convenience we can picture a tight-binding calculation in which the d state hopping integrals are set equal to zero.[25] One then performs self-consistent calculations in which the central d shell occupation numbers differ by one, from which U values are extracted as total energy differences. Note that this procedure preserves the overall charge neutrality, as the total charge within the supercell is maintained: those unit cells surrounding the central cell relax to accommodate the charge shift in the central cell.[26]

[24] In a solid the d-state occupancies are not integers. Although the total energies are a continuous function of n_d, the slope, which relates to the eigenvalues and the effective potential, changes discontinuously as the occupation passes through integer values; for a proof see: Density-functional theory for fractional particle number: derivative discontinuities of the energy, J.P. Perdew, R.G. Parr, M. Levy, and J.L. Balduz, Jr., *Phys. Rev. Lett.* **49**, 1691 (1982).

[25] In practice, one can use a tight-binding-like formalism based on so-called muffin tin orbitals, a formalism we do not treat here; for a discussion, see: Explicit, first-principles tight-binding theory, O.K. Andersen and O. Jepsen, *Phys. Rev. Lett.* 53, 2571 (1984).

[26] Clearly the larger the unit cell the better the approximation to an infinite system.

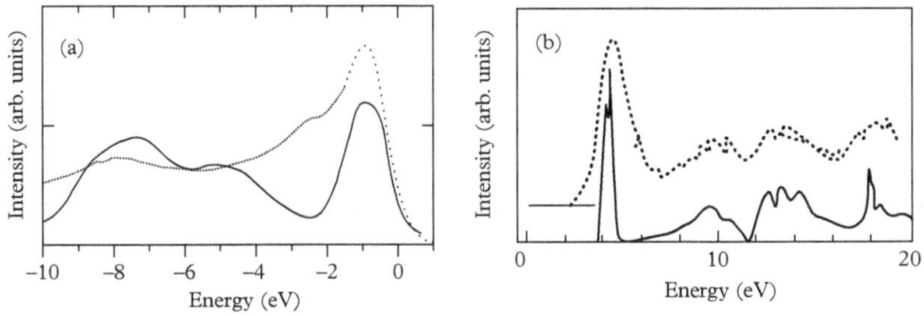

Figure 45.7 *The measured (points) and calculated spectrum for (a) photo-emission and (b) inverse photo-emission in* NiO. *(After Anisimov et al. (1993) and Swansky and Allen (1984).)*

Figure 45.7 shows the LDA + U calculations of the photo-emission spectrum (a) and the inverse photo-emission spectrum (b) of NiO, as calculated by Anisimov et al.;[27] also shown are the experimental data of Swansky and Allen.[28] Note the presence of a U-induced gap in the inverse photo-emission spectrum; i.e. we have an insulator whereas the LDA calculations predict metallic behavior. Depending on the approximations made, the predicted band gaps are 3.4 and 3.7 eV; this is to be compared to an experimental value of 4.3 eV. The magnetic moment per Ni atom falls in the range 1.6–1.7 μ_B. Another interesting property of the LDA + U calculation is that the p-like levels are pushed to higher energies than they have in the LDA.

45.2.2 Generalized m-dependent Coulomb interactions; orbital polarization

In the presence of a non-spherical potential, the parameters, U and J, become matricies, $U^{mm'}$ and $J^{mm'}$, and the techniques discussed above have been generalized to cover this case, although we will not discuss them here. At the next level of generalization,[29] the direct term in the energy is rewritten as

$$\sum_{m,m'\sigma} U_{mm'} n_{m\sigma} n_{m'-\sigma} \rightarrow \sum_{m,m',m''m'''\sigma} U_{mm'';m'm'''} n^{\sigma}_{mm'} n^{-\sigma}_{m''m'''}, \tag{45.16}$$

where a *density matrix*, $n^{\sigma}_{mm'}$, replaces $n_{m\sigma}$; a similar generalization applies to the exchange terms. Note the level occupations have now been replaced by a density matrix; this, in principle, allows the incorporation of intra-d-shell correlation effects, although the calculations are challenging. The inclusion of such effects can lead to an electronically driven *spontaneous orbital polarization*, the analogue of a spontaneous magnetic polarization. A specific and much studied case is $KCuF_3$ which has the perovskite structure. In agreement with the experiments, the generalized theory predicts a shift in the positions of the Cu atoms within the unit cell, unlike the LDA prediction which has them at their conventional positions; this is an example of a case where the LDA fails at the level of giving the right crystal structure.

[27] Density-functional theory and NiO photoemission spectra, V.I. Anisimov, I.V. Solovyev, M.A. Korotin, M.T. Czyżyk, and G.A. Sawatzky, *Phys. Rev. B* 48, 16929 (1993).

[28] Magnitude and origin of the band gap in NiO, G.A. Sawatzky and J.W. Allen, *Phys. Rev. Lett.* **53**, 2239 (1984).

[29] Density-functional theory and strong interactions: Orbital ordering in Mott-Hubbard insulators, A.I. Liechtenstein, V.I. Anisimov, and J. Zaanen, *Phys. Rev. B* **52**, R5467 (1995). This article also contains references to related work.

45.3 Strongly correlated electrons in metals

In our treatment of the Mott transition we focused on NiO. It is convenient to think of the material as being held together by the oxygen p levels bonding with the Ni d or s electrons and associated bands. This would result in an unfilled Ni d shell with only a small coupling to neighboring Ni atoms, narrow unfilled d bands, and metallic behavior. However, with this narrow band width the on-site d–d repulsion effects dominate and localize the electrons within the unfilled Ni 3d shell. Here the U-induced correlation effects turned a metal into an insulator (and simultaneously an antiferromagnet).

In the rare earth metals the 4f electrons are almost universally localized, irrespective of the predictions of LDA calculations. Intra-f-shell correlations and associated spin alignments play a dominant role, but direct inter-atom f–f coupling or indirect coupling via RKKY-like interactions involving the unfilled s–p electron bands results in a wide range of interesting magnetic behaviors.

In what follows we will focus on an interesting class of metallic compounds, the heavy fermion intermetallics, containing atoms with highly localized 4f or 5f electrons that lie close to the Fermi energy, and one or more other constituents which render the overall compound metallic. Although many behaviors are found, some of which will be described in Section 45.3.1, a unique property is the emergence of narrow bands with extraordinarily large electronic specific heats, corresponding to effective masses that can be as high as a few thousand electron masses. At high temperatures the f electrons tend to be localized, yielding Curie-like paramagnetic behavior; but at low temperatures they coherently propagate through the lattice and form narrow bands (albeit with hybridization effects) and display Pauli paramagnetism. Many display no magnetic order. Weak antiferromagnetism is seen in some compounds (e.g., URu_2Si_2 and UPt_3) while weak ferromagnetism is found in others (e.g., UGe_2, URhGe, UCoGe). Finally, a transition between antiferromagnetism and itinerant ferromagnetism has been observed in the alloys, $URu_{2-x}M_xSi_2$ with M = Mn, Tc, and Re.[30]

For some systems one can continuously pass between non-magnetic and magnetic phases at very low temperatures by applying hydrostatic pressure, a high magnetic field, or through alloying (which can act as a "chemical pressure"). In the vicinity of the transition, large quantum mechanical fluctuations are present, as opposed to temperature driven fluctuations characteristic of ordinary second-order phase transitions. The transition point is called a *quantum critical point*, the behavior of which is sketched in Section 45.3. We will start our discussion of heavy fermions by briefly reviewing some observed thermal, magnetic, and transport properties of three of the first compounds discovered to have such unusual properties: $CeCu_2Si_2$, UBe_{13}, and UPt_3.

[30] Ferromagnetic instability in the heavy-electron compound URu_2Si_2 doped with Re or Tc, Y. Dalichaouch, M.B. Maple, M.S. Torikachvili, and A.L. Giorgi, *Phys. Rev. B* 39, 2423 (1989).

45.3.1 A brief survey of some heavy fermion properties

One of the first compounds to attract wide attention was $CeCu_2Si_2$, due to the rapid increase in its specific heat below about 15 K.[31] Figure 45.8 shows a plot of C(T)/T versus T^2 which anticipates a Debye-Sommerfeld $C(T)/T = \gamma+\beta T^2$ behavior (see Section 6.6).[32] The sudden departure from this form arises from the 4f electrons of the Ce atoms, which in the atom have the configuration $6s^2 5d^1 4f^1$. These electrons are taken as being localized, or at least uncorrelated at high temperature. However, as the temperature falls below some characteristic value they begin to form a partly filled Bloch conduction band, thereby becoming mobile and contributing to the overall itinerant electronic density of states. However the associated effective mass turns out to be surprisingly large, $m^* \approx 100\, m_e$; this material and other compounds having high effective masses are now termed *heavy fermion materials*. That these electrons are indeed mobile was confirmed by the fact that at lower temperatures this compound becomes superconducting, and the ratio, $\Delta C(T_c)/\gamma T_c = 1.6$, is large, implying that the electrons involved in the transition include the heavier ones. We recall that in the BCS theory, $\Delta C(T_c)/\gamma T_c = 1.43$ (see Eq. (35.52)); larger values are a signature of what is termed *strong coupling superconductivity*.

Additional evidence for itinerant f electron behavior comes from the magnetic susceptibility: at high temperature the f electrons behave as a collection of independent localized spins. This interpretation follows from an observed Curie-like behavior of the magnetic susceptibility, from which one obtains an effective magnetic moment of $\mu_{\text{eff}} = 2.68\, \mu_B$.[33] However at low temperatures the susceptibility approaches a more temperature-independent form characteristic of the Pauli behavior for conduction electrons.

Our second compound showing heavy fermion characteristics is UBe_{13}. However, here it is the uranium 5f electrons that are involved: U has the ground state configuration, $7s^2 6d^1 5f^3$. The low temperature specific heat is shown in

[31] Superconductivity in the presence of strong Pauli paramagnetism: $CeCu_2Si_2$. F. Steglich, J. Aarts, C.D. Bredl, W. Lieke, D. Meschede, W. Franz, and J. Schafer, *Phys. Rev. Lett.* 43, 1892 (1979).

[32] Characterization of single crystals of $CeCu_2Si_2$. A source of new perspectives, G.R. Stewart, Z. Fisk, and J.O. Willis, *Phys. Rev. B* 28, 172 (1983).

[33] This can be compared with the value 2.4 for the Ce^{3+} ion (see Table 5.2).

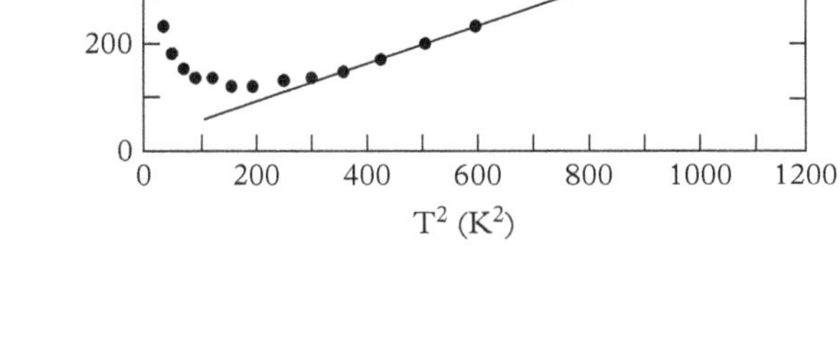

Figure 45.8 *The specific heat of the intermetallic compound,* $CeCu_2Si_2$*, fitted to the form,* $C(T) = \gamma + \beta T^2$*. Note the rapid rise at temperatures below about 15 K. (After Stewart et al. (1983).)*

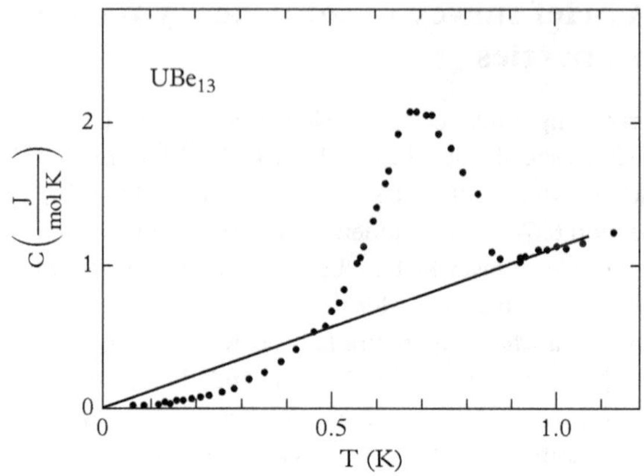

Figure 45.9 *The specific heat of the intermetallic compound,* UBe_{13}*; the peak is associated with a broad superconducting transition. (After Ott et al. (1983).)*

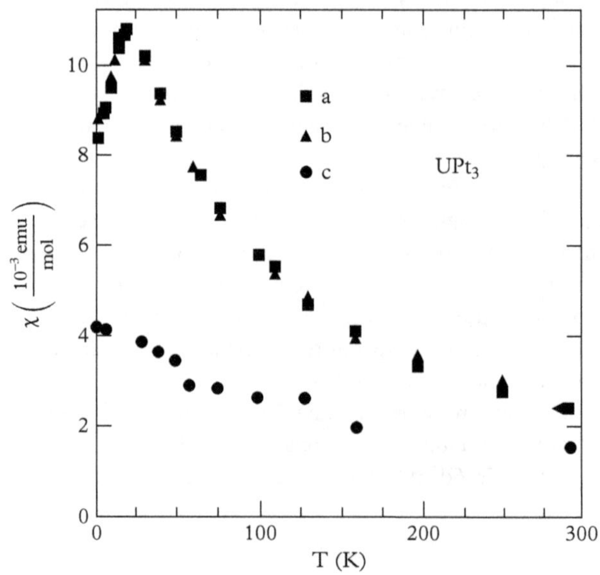

Figure 45.10 *A plot of the magnetic susceptibility of* UPt_3 *as a function of temperature along its three principal axes. Note the departure at low temperature from the* T^{-1} *Curie-like high temperature behavior due to an onset of the temperature-independent Pauli behavior. (After Frings et al. (1983).)*

Fig. 45.9.[34] This material also becomes superconducting, at a temperature, $T_c \approx 0.7\,K$, which accounts for the large peak seen in Fig. 45.9. From the linear portion just above T_c, one can assign an effective mass, $m^* \approx 2000m_e$, among the largest observed for the heavy fermion materials.

Our third intermetallic compound displaying large specific heat masses is the hexagonal material, UPt_3; the measurements yield an effective mass, $m^* \approx 100m_e$. A superconducting transition is observed at $T_c \cong 0.5\,K$ (see below). Figure 45.10 shows the temperature dependence of the magnetic susceptibility, which displays Curie-like behavior at high temperatures with $\mu_{eff} = 2.9\mu_B$, changing over to Pauli behavior at low temperatures.[35]

[34] UBe_{13}: an unconventional actinide superconductor, H.R. Ott, H. Rudigier, Z. Fisk, and J.L. Smith, *Phys. Rev. Lett.* **50**, 1595 (1983).

[35] Magnetic properties of U_xPt_y compounds, P.H. Frings, J.J.M. Franse, F.R. de Boer, and A. Menovsky, *J. Magn. Magn. Mater.* **31**, 240 (1983).

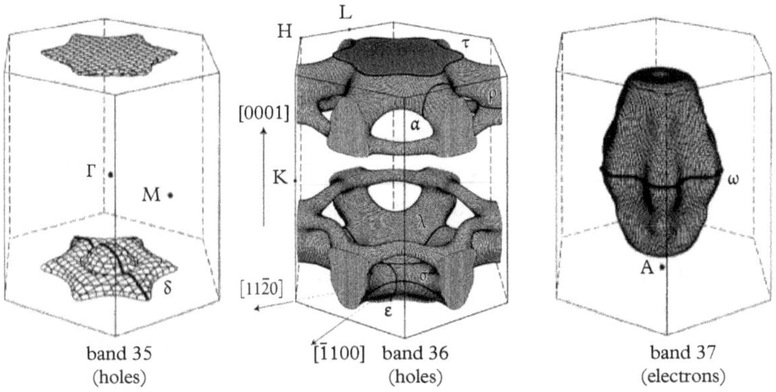

Figure 45.11 *The Fermi surface of* UPt_3. *Orbits observed in the de Haas-van Alphen experiments are identified by Greek letters. A particularly large area orbit is that designated as* ω.

UPt_3 can be prepared in a relatively pure form with good stoichiometry, leading to long electron scattering times thereby permitting detailed studies of the Fermi surface and related properties using the de Hass–van Alphen effect (see Section 20.1).[36,37,38] The surface consists of three sheets, which are shown in Fig. 45.11, along with the principal orbits measured. The extremal cross-sectional areas and cyclotron effective masses have been measured for many orbits, in particular some involving the large sheets of the surface, which can then be compared with the corresponding quantities obtained from band structure calculations.[39,40] Table 45.1 lists some of the dHvA frequencies (proportional to the extremal cross-sectional areas, see Eq. (18.36b)) and effective masses measured in the experiments. Note that the agreement between the frequencies is quite good. However, the experimentally measured effective masses are far larger than the calculated values. This is a direct manifestation of a large enhancement from many-body effects, which are thought to arise from the strong f state electron–electron interactions.[41] For the ω orbit the enhancement factor is

Table 45.1 *Cross-sectional areas and effective masses in* UPt_3.

	F (MG)		m^*/m_e	
Branch:*FS* orbit	Expt.	Calc.	Expt.	Calc.
		a axis (ΓK)		
α:*ML*	5.4(3)	10.4	25(3)	2.2
β:*L*4	6.0(4)	5.2	. . .	1.0
γ:Γ1	7.3(3)	8.2	40(7)	2.0
δ:*A*5	14.0(3)	9.1	50(8)	1.9
ϵ:Γ2	21.0(3)	24.0	60(8)	4.6
ω:Γ3	58.5(5)	52.8	90(15)	5.3

36 Heavy-fermion quasiparticles in UPt_3, L. Taillefer and G.G. Lonzarich, *Phys. Rev. Lett.* **60**, 1570 (1988).

37 Observation of a main Fermi surface in UPt_3, N. Kimura, T. Komatsubara, D. Aoki1, Y. Onuki, Y. Haga, E. Yamamoto, H. Aoki, and H. Harima , *J. Phys. Soc. Jpn.* **67**, 2185 (1998).

38 For a brief review of de Hass–van Alphen measurements in other heavy fermion compounds, see: de Haas–van Alphen effect in strongly correlated electron systems, Y. Onuki and R. Settai, *Physica B* **300**, 61 (2001).

39 Local density band approach to f-electron systems – heavy fermion superconductor UPt_3, T. Oguchi and A.J. Freeman, *J. Mag. Mag. Matl.* **52**, 174 (1985).

40 Fermi surface of UPt_3 within the local-density approximation, C.S. Wang, M.R. Norman, R.C. Albers, A.M. Boring, W.E. Pickett, H. Krakauer, and N.E. Christensen, *Phys. Rev. B* **35**, 7260 (1987).

41 The electron–phonon mass enhancement discussed in Section 21.4 is expected to be far too small to explain the observed values.

approximately 17. We must keep in mind that the band density of states already partially includes effects from narrow bands associated with the d and f electrons.

The superconducting properties of UPt_3 are unusual in that there is strong evidence for the existence of a multiplicity of different phases. This was convincingly demonstrated through anomalies observed in the attenuation of ultrasound,[42,43] as shown in Fig. 45.12, where the measured points lie on the lines that separate the different phases. There is now a consensus that the underlying pairing mechanism is electronic in origin and that the pairing occurs in a higher orbital angular momentum state, although its exact nature remains controversial. In these respects UPt_3 is similar to superfluid ^3He, which also has multiple phases and BCS-like pairing, there occurring in an $L = 1, S = 1$ state (see Section 38).

[42] Phase diagram of UPt_3 from ultrasonic velocity measurements, S. Adenwalla, S.W. Lin, Q.Z. Ran, Z. Zhao, J.B. Ketterson, J.A. Sauls, L. Taillefer, D.G. Hinks, M. Levy, and B.K. Sarma, *Phys. Rev. Lett.* **65**, 2298 (1990).

[43] Strain-order-parameter coupling and phase-diagrams in superconducting UPt_3, G. Bruls, D. Weber, B. Wolf, P. Thalmeier, B. Lüthi, A. de Visser, and A. Menovsky, *Phys. Rev. Lett.* **65**, 2294 (1990).

As a final topic in this brief review of heavy fermion properties, Fig. 45.13 shows the temperature dependence of the electrical resistance for our three chosen compounds: $CeCu_2Si_2$, UBe_{13}, and UPt_3. We see that for the first two of

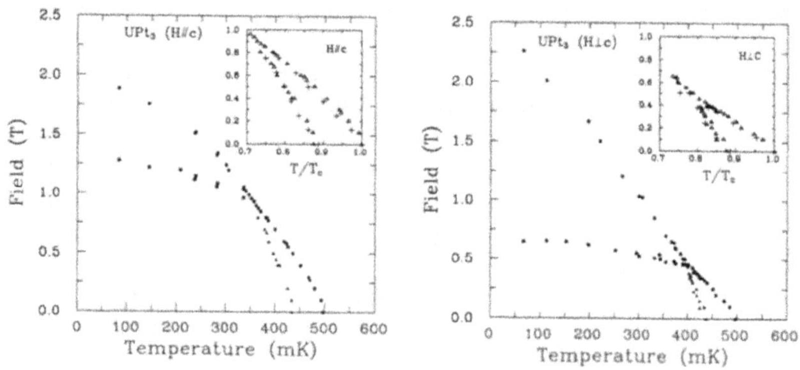

Figure 45.12 *The superconducting phase diagram of* UPt_3 *obtained from features in the attenuation of sound. Note the presence of multiple phases. (After Adenwalla et al. (1990).)*

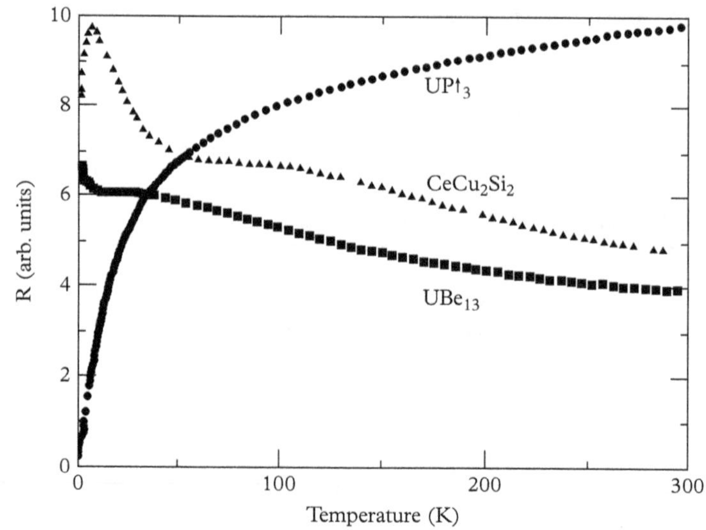

Figure 45.13 *The resistivity (in arbitrary units) of* $CeCu_2Si_2$ *(Stewart, Fisk, and Willis, 1983);* UBe_{13} *(Ott et al., 1983); and* UPt_3 *(after Stewart et al. (1984)).*

these compounds the resistivity rises as the temperature falls. Now on the basis of electron–phonon scattering the resistivity is expected to fall, but here the high temperature scattering is dominated by fluctuating localized spins associated with the U and Ce f electrons, the effect of which *increases* as the temperature falls. As the temperature continues to fall the resistivity goes through a maximum at some characteristic temperature, below which it decreases as the f levels become "coherent" and form into bands with high effective mass. UPt_3 behaves like a more conventional metal, decreasing monotonically as the temperature falls.

45.3.2 Physical effects underlying heavy fermion behavior

In starting to think about heavy fermion compounds it is useful to generalize the Anderson model to a lattice with a single f level having energy ε_f at sites numbered by j that are coupled to conduction electron states, **k**, through a potential, V_{kf}, the Hamiltonian then having the form (see Eq. (44.5) for comparison)

$$\hat{H} = \sum_{k\sigma} \varepsilon_k \hat{c}^{\dagger}_{k\sigma} \hat{c}_{k\sigma} + \sum_{kj\sigma} \left[V_{kf} \hat{c}^{\dagger}_{k\sigma} \hat{f}_{j\sigma} + V^{*}_{kf} \hat{f}^{\dagger}_{j\sigma} \hat{c}_{k\sigma} \right] + \sum_{j\sigma} \varepsilon_f \hat{f}^{\dagger}_{j\sigma} \hat{f}_{j\sigma} + U \sum_{j} \hat{f}^{\dagger}_{j\uparrow} \hat{f}_{j\uparrow} \hat{f}^{\dagger}_{j\downarrow} \hat{f}_{j\downarrow}, \tag{45.17}$$

where σ labels the two spin states and U accounts for on-site f-state Coulomb repulsion. If $\varepsilon_f < 0 \ll V_{kf}$ (with ε_f measured relative to the Fermi energy) and $U = 0$, the hybridization will broaden the f state on each atom into a resonance with a width of order $\Gamma = \pi N(\varepsilon_F)\,\overline{|V_{kf}|^2}$ (see Eq. (44.20)). For the case when a single electron occupies our single f level these resonances will then be half occupied by both spin up and spin down electrons and the system will be non-magnetic.[44] If now the contribution from V_{kf} is evaluated in second-order perturbation theory, the local f resonances will couple to each other through the conduction electrons resulting in narrow hybridized bands. If, in addition, we now slowly turn on U these already narrow bands would narrow still further (see the discussion surrounding Fig. 45.3), giving us a heavy fermion like metallic state. But if U becomes too large, the f level will be split with one spin state lying above the Fermi level and the other occupied, rendering the system paramagnetic. In both of these cases the f levels mix with the conduction electrons but do not directly overlap each other, which would add complexity (as would multiple f states). Even with a simplified model like this the system behavior depends in a complex way on the various parameters.

Of course the real heavy fermion materials involve multiple bands derived from the occupied s, p, d, and f atomic electrons on all atoms in the unit cell. As noted above, the Fermi surface is often well represented by DFT calculations. The bands are on the whole very narrow and this partly accounts for the large observed heat capacities and the high effective masses observed in de Hass–van Alphen experiments. However, a large U can again result in a Mott localization of the spins on the f sites and ferromagnetic or antiferromagnetic ground states

[44] If two electrons occupy the f level the system is trivially non-magnetic; in addition, the associated band which forms when inter-site f-state coupling is included would be full and hence an insulator.

of these spins. At temperatures above the ordering temperature the heat capacity then involves the entropy of the spin system as well as that from the non-f conduction electrons.

A model used to describe the formation of a heavy fermion ground state in the presence of U-driven f electron spin is to assume that at low temperatures the non-f conduction electrons *screen* the spins of the f sites, resulting in a near spin-less singlet state, similar to the Kondo ground state, where the conduction electrons screen the spin of an isolated f or d state. As the temperature is reduced the resistivity passes through a maximum, designated as the *coherence temperature*, T^*, marking the transition into a coherent state where screened f electrons *delocalize* and form hybridized bands with the remaining electrons. This ground state is termed a *Kondo lattice*.[45] However, a residual spin polarization and associated antiferromagnetism often remain.[46]

In our discussion of normal ^{3}He (see Section 37.1.2) we introduced the idea of a "back flow" of the surrounding fluid (when an excited atom moves through the fluid) as a classical mechanism for many-body mass enhancement;[47] this results from the hard core repulsion that forces surrounding atoms out of the path of the excited atom. At pressures just below the solidification pressure the mass is approximately six times the atomic mass. In a similar way the Kondo screening cloud involves the surrounding environment and provides a mechanism for mass enhancement in heavy fermion systems that can yield values which greatly exceed those calculated with conventional DFT; this Kondo screening associated with an interacting (dressed) electron system involves a polarization of the surrounding electrons which, qualitatively, we may think of as adding extra mass to the band electrons. However it is quickly "shaken off" as the temperature rises.

The low temperature properties of metallic electrons are modeled after Landau's theory of liquid ^{3}He, the classic Fermi liquid, from which the following temperature dependencies are expected for the heat capacity, magnetic susceptibility and electrical resistivity: $C = \gamma T$, $\chi = \chi_0$, and $\rho = \rho_0 + aT^2$. However in the presence of very narrow bands, where we have a high density of final states, the scattering rates of the elementary excitations lying off the Fermi surface rapidly become large and the lifetimes correspondingly short, quickly leading to large energy uncertainties, $\delta\varepsilon(\mathbf{k})$, that are comparable to the excitation energies, $|\varepsilon_F - \varepsilon(\mathbf{k})|$, themselves. At temperatures where this happens we have a breakdown of the Landau–Fermi liquid theory, where the latter forms the basis for the quasi-particle picture. Qualitatively, as the temperature increases the f electrons can rapidly "shed" the correlated conduction electron cloud that screens their spins and disrupt the indirect f–f coupling that leads to band formation.

45.4 Quantum phase transitions and marginal Fermi liquids

Weak antiferromagnetism is often present in heavy fermion metals. For some compounds the application of a magnetic field or a hydrostatic pressure can

[45] Kondo lattice and weak antiferromagnetism, S. Doniach, *Physica B & C* **91**, 231 (1977).

[46] The required inter-site f–f coupling is sometimes viewed as an RKKY interaction.

[47] Exchange also plays an important role by keeping atoms with like spins from the same point in space.

continuously depress the associated transition temperature, T_N, to absolute zero; alloying (chemical pressure) can produce the same behavior, depending on the system. In the vicinity of this transition fluctuations of a purely quantum mechanical nature are expected and the Landau Fermi liquid theory breaks down. A way to think about the transition is to picture the critical temperature of a second-order continuous phase transition being depressed to absolute zero by varying some parameter; the critical fluctuations then become quantum mechanical in character. One possibility is that at low but finite temperatures we have two regions where Fermi liquid (FL) behavior is expected, that are separated by a "disordered" non-Fermi-liquid (NFL) region, as shown in Fig. 45.14; the boundaries are not sharp. In the disordered region we picture quantum mechanically fluctuating bubbles of one phase forming in the background of the other (at multiple length scales), similar to the thermally driven fluctuating bubbles that form within the critical region of a classical second-order phase transition. At absolute zero the NFL region shrinks to zero at a critical value, δ_c, which is referred to as a *quantum critical point* (QCP).[48]

If we designate the parameter tuning the system between the ordered (say antiferromagnetic) and normal states (both here presumed to be Fermi liquids) and a non-Fermi-liquid regime as δ, the behavior shown in Fig. 45.14 is observed in some systems. Signatures of NFL behavior are taken as

$$\rho(T) = \rho_0 + A(T/T_0)^n, \quad n \approx 1 - 1.5, \tag{45.18a}$$

in the electrical resistivity, $\rho(T)$, and as

$$C(T)/T \approx (-1/T_0)ln(T_0) \text{ or } T^{-1+\lambda}, \quad \lambda \cong 0.7 - 0.8, \tag{45.18b}$$

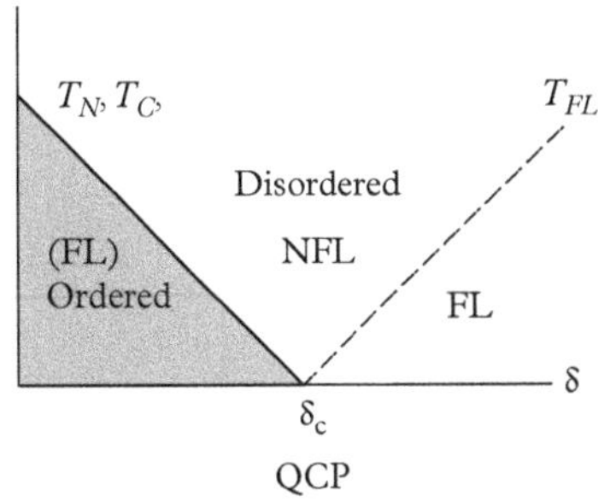

Figure 45.14 *Schematic phase diagram associated with the quantum critical region in the δ − T plane, showing ordered and non-ordered Fermi liquid (FL) regions separated by a non-Fermi-liquid (NFL) region that shrinks to a point, the quantum critical point (QCP), at absolute zero. (Courtesy of B. Maple)*

[48] For reviews, see: Non-Fermi liquid regimes and superconductivity in the low temperature phase diagrams of strongly correlated d- and f-electron materials, M.B. Maple, R.E. Baumbach, N.P. Butch, J.J. Hamlin, and M. Janoschek, *J. Low Temp. Phys.* **161**, 4 (2010) and Quantum phase transitions, Hilbert V. Lohneysen and Peter Wolfle, *AIP Conf Proc* **1014**, 107 (2008). This line of study was initiated in Quantum critical phenomena, J.A. Hertz, *Phys. Rev.* B **14**, 1165 (1976).

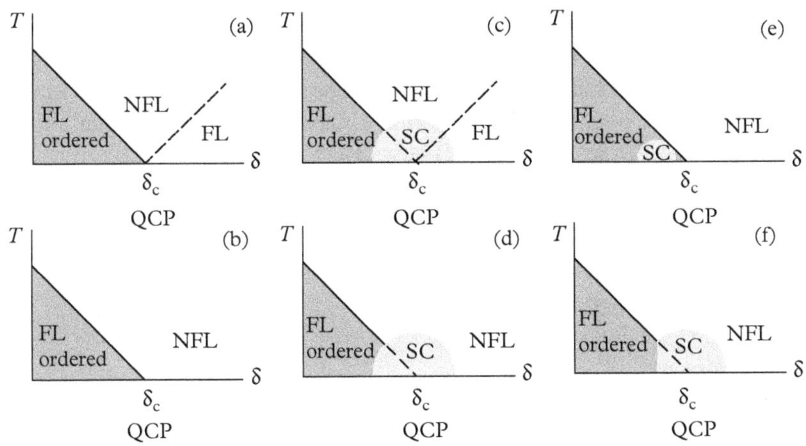

Figure 45.15 *Possible phase diagrams in the vicinity of a QCP with and without the appearance of superconductivity. In (c), (d), and (f) the superconducting "dome" obscures the QCP. The appearance of superconductivity in the vicinity of the QCP, as in (c), (d), and (f) is sometimes said to "protect" the transition by suppressing the associated fluctuations. Form (d) is characteristic of several high temperature superconductor systems (see Section 46). (After Maple et al. (2010).)*

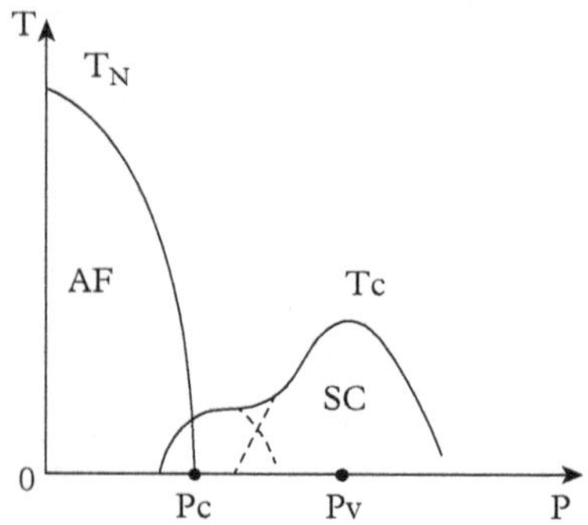

Figure 45.16 *Schematic* P-T *phase diagram for* $CeCu_2(Si/Ge)_2$ *showing the two critical pressures* P_c *and* P_v. *At* P_c, *where the Néel temperature* $T_N \to 0$, *superconductivity in region* SC *I is mediated by* AF *spin fluctuations; around* P_v, *in the region* SC *II, critical valence fluctuations provide the pairing mechanism. (After Miyake (2007).)*

in the heat capacity, C(T); here A is a constant and T_0 is some characteristic, scale-setting, temperature. A variety of forms have been suggested for the magnetic susceptibility. As noted above, the behaviors expected for a Fermi liquid are $\chi = \chi_0$, $C(T) = \gamma T$ while the resistivity would have an $n = 2$ behavior.

Superconductivity is often observed in the neighborhood of the transition, lending support to the idea that exchange of these bosonic fluctuations provides an electronic mechanism for Cooper pairing, and with it the opportunity for unconventional (non s-wave) superconducting states, as one has in ^{3}He. Possible phase diagrams in the presence of the QCP are shown in Fig. 45.15; panel (a) repeats the situation shown in Fig. 45.14.

45.5 Possible mechanisms underlying heavy fermion superconductivity

Although density functional calculations accurately reproduce the Fermi surface of UPt_3 they strongly underestimate the effective masses. In more conventional metals the electron–phonon interaction leads to a mass enhancement (see Section 21.3), but it is generally less than a factor two (see Section 20.1.4 for a discussion of Nb). However, any collective fluctuations a system can undergo will lead to a mass enhancement. One important example is spin fluctuations in a material near a magnetic instability, as proposed long ago for Pd.[49] Such fluctuations can also produce superconductivity through a purely magnetic pairing mechanism;[50] a dramatic example of the latter is superfluid ^{3}He.[51]

The mass enhancements in heavy fermion materials cover a wide range, extending up to values of a few thousand. However evidence is accumulating that the underlying physics may involve *two* mechanisms:[52] one in which the valence or average occupation of the f shell is non-integral and one in which it is integral. In the former a valence fluctuation approach is used, while in the latter one uses a Kondo-based approach.[53] Both approaches lead to heavy fermion behavior and, in principle, the occurrence of superconductivity near a quantum critical point. In fact, there is evidence for two domes of superconductivity in some cerium-based f-electron systems, one associated with an antiferromagnetic quantum critical point and the other with a valence transition quantum critical point; this is shown schematically in Fig. 45.16.

[49] Low-temperature properties of nearly ferromagnetic Fermi liquids, S. Doniach and S. Engelsberg, *Phys. Rev. Lett.* 17, 750 (1966); Effect of ferromagnetic spin correlations on superconductivity. F. Berk and J.R. Schrieffer, *Phys. Rev. Lett.* 17, 733 (1966).

[50] Possibility of triplet pairing in palladium, D. Fay and J. Appel, *Phys. Rev. B* 16, 2325 (1977).

[51] Anisotropic superfluidity in ^{3}He – possible interpretation of its stability as a spin-fluctuation effect, P.W. Anderson and W.F. Brinkman, *Phys. Rev. Lett.* 30, 1108 (1973).

[52] This point of view was emphasized to the author by B. Maple.

[53] For a review, see: New trend of superconductivity in strongly correlated electron systems, K. Miyake, *J. Phys.: Condens. Matter* 19, 125201 (2007).

SUPPLEMENTARY READING

Mott transitions:

The literature on the Mott transition is vast and the level of theoretical sophistication varies. We cite the following review articles: The Anderson-Mott transition,

D. Belitz and T.R. Kirkpatrick, *Rev. Mod. Phys.* **66**, 261 (1994); Metal-insulator transitions, M. Imada, A. Fujimori, and Y. Tokura, *Rev. Mod. Phys.* **70**, 1039 (1998); Doping a Mott insulator: Physics of high-temperature superconductivity, P.A. Lee, N. Nagaosa, and X.-G. Wen, *Rev. Mod. Phys.* **78**, 17 (2006); Mott, N.F., 1937, *Proc. Phys. Soc. Lond. A* **49**, 72; N.F. Mott, *Metal-Insulator Transitions*, Taylor & Francis, London/Philadelphia (1974). Two important papers, historically, are: J.H. de Boer and E.J.W. Verway, *Proc. Phys. Soc. Lond. A* **49**, 59 (1937) and L. Néel, Propriétées magnétiques des ferrites; Férrimagnétisme et anti-ferromagnétisme, Annales de Physique (Paris) **3**, 137 (1948).

Electronic structure of oxides:
For a discussion of augmented plane wave (APW) calculations of the transition metal monoxides and the resulting trends, see: Electronic structure of the 3d transition-metal monoxides: I energy-band results, L.F. Matthiass, *Phys. Rev.* **B 5**, 290 (1972).

Heavy fermions:
For a still valuable early review, see: Heavy-fermion systems, G.R. Stewart, *Rev. Mod. Phys.* **56**, 755 (1984). Concerning superconductivity in these systems, see: Heavy-fermion superconductivity, by P. Riseborough, G.M. Schmiedeshoff, and J.L. Smith in *Superconductivity: Novel Superconductors*, Ed. K.H. Bennemann and J.B. Ketterson, Springer (2008). For a discussion of spin–orbit effects in strongly correlated systems see: Correlated quantum phenomena in the strong spin-orbit regime, William Witczak-Krempa, Gang Chen, Yong Baek Kim, and Leon Balents, *Annu. Rev. Condens. Matter Phys.* **5**, 57 (2014) and Strong correlations from Hund's coupling, Antoine Georges, Luca de Medici, and Jernej Mravlje, *Annu. Rev. Condens. Matter Phys.* **4**, 13778 (2013).

Non Fermi liquid behavior:
For a review of non-Fermi-liquid behavior and quantum phase transitions see: Non-Fermi liquid regimes and superconductivity in the low temperature phase diagrams of strongly correlated d- and f-electron materials, M.B. Maple, R.E. Baumbach, N.P. Butch, J.J. Hamlin, and M. Janoschek, *Journal of Low Temp. Phys.* **161**, 4 (2010); Heavy fermions and quantum phase transitions, Qimiao Si and Frank Steglich, *Science* **329**, 1161 (2010); Non-Fermi-liquid behavior in d- and f-electron metals, G.R. Stewart, *Rev. Mod. Phys.* **73**, 797 (2001).

Quantum phase transitions:
The theory of quantum phase transitions is reviewed in Heavy fermions: electrons at the edge of magnetism, Piers Coleman, in *Handbook of Magnetism and Advanced Magnetic Materials*, Ed. Helmut Kronmüller and Stuart Parkin, John Wiley and Sons (2007). A book focusing on the topic is *Quantum Phase Transitions*, Subir Sachdev, Cambridge University Press (2011).

46 High Temperature Superconductors

Prior to 1986, high superconducting transition temperatures were largely limited to the intermetallic compounds having the A-15 structure, which is shown in Fig. 46.1. The material having the highest transition temperature is Nb_3Ge with $T_c = 23$ K. A related compound, Nb_3Sn, with a transition temperature of 18.3 K, has a high critical current and is widely used to make superconducting magnets, which can produce magnetic fields in the neighborhood of 225 kG. Experiments strongly suggest that the pairing in these materials involves the exchange of phonons and it is widely believed that they are conventional BCS superconductors.

46.1 The discovery of cuprate superconductors

In 1986, Bednorz and Müller discovered a material with a composition that we will write as $La^{+3}_{2-x}Ba^{+2}_{x}Cu^{+2}O_4$ (referred to as 214, for short) which had a transition temperature of approximately 35 K;[1,2] their original data for the temperature dependence of the resistivity is shown in Fig. 46.2. In addition to the superconducting behavior, we note that above the transition temperature the resistivity is approximately linear with temperature, a property shared by many of the high temperature superconductors. The 214 stoichiometric parent compound, $La^{+3}_{2}Cu^{+2}O^{-2}_{4}$ (LSCO), is an antiferromagnetic insulator, but when barium is added the valences no longer sum to zero and the material can be metallic. The sense of this imbalance is to make the material *electron deficient*: the carriers are holes. A closely related compound in which isovalent strontium replaces the barium is $La^{+3}_{2-x}Sr^{+2}_{x}Cu^{+2}O_4$. The crystal structure of the host compound, $La^{+3}_{2}Cu^{+2}O^{-2}_{4}$, is tetragonal[3] and is shown in Fig. 46.3(a).

The 214 discovery was quickly followed by the observation of a superconducting transition above 90 K in a compound with a composition which is now known to be $YBa_2Cu_3O_{7-\delta}$, referred to as 123 or YBCO. The parameter, δ, often called the *doping*, strongly affects the superconducting properties and is controlled through the oxygen partial pressure protocol utilized in preparing the material.[4] Similar to the 214 compound, when δ differs from zero, the valences do not sum to zero and the system can be a conductor. The structure is orthorhombic and

[1] The 1987 Nobel Prize in Physics was awarded jointly to J. Georg Bednorz and K. Alexander Müller "for their important break-through in the discovery of superconductivity in ceramic materials."

[2] Possible high T_c superconductivity in the Ba-La-Cu-O system, J.G. Bednorz and K.A. Müller, *Z. Phys* **B64**, 189 (1986).

[3] When the temperature is lowered below about 500 K, La_2CuO_4 transforms to a face centered orthorhombic structure.

[4] Superconductivity at 93 K in a new mixed-phase Y-Ba-Cu-O compound system at ambient pressure, M.K. Wu, J.R. Ashburn, C.J. Torng, P.H. Hor, R.L. Meng, L. Gao, Z.J. Huang, Y.Q. Wang, and C.W. Chu, *Phys. Rev. Lett.* 58, 908 (1987).

The Physics of Solids. First Edition. J. B. Ketterson.

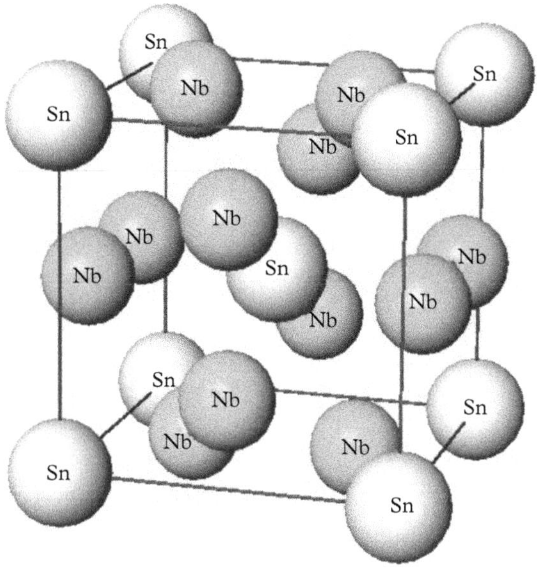

Figure 46.1 *The cubic A–15 structure of the superconducting compound* Nb_3Sn.

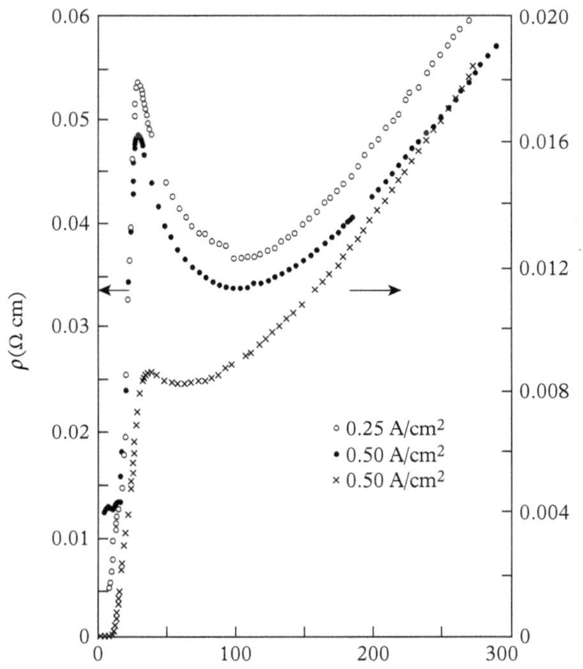

Figure 46.2 *The original data of Bednorz and Müller on the temperature dependence of resistivity in their compound with the stated composition* $Ba_xLa_{5-x}Cu_5)O_{5(3-Y)}$ *for ceramic samples with* x(Ba) = 1 *(upper curves, left scale) and* x(Ba) = 0.75 *(lower curve, right scale). The first two cases also show the influence of the current density.*

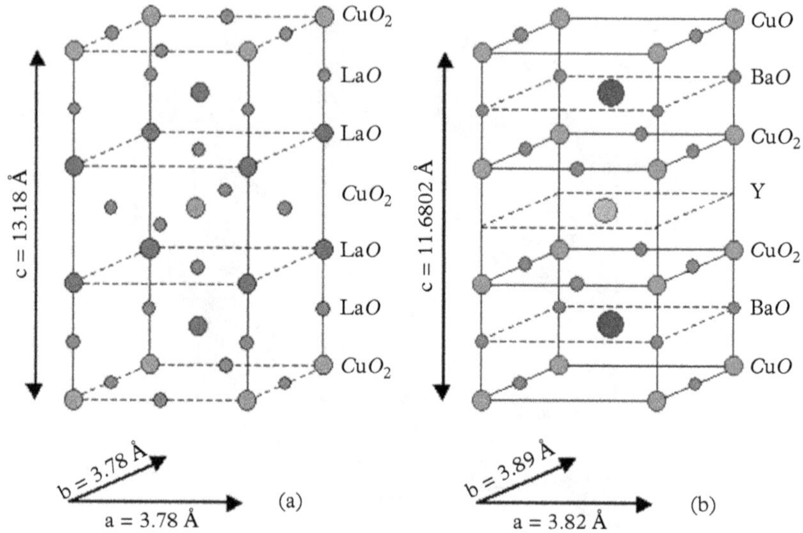

Figure 46.3 *The crystal structure of: (a)* La_2CuO_4 *the parent compound of the superconductors,* $La_{2-x}Ba_xCuO_4$ *and* $La_{2-x}Sr_xCuO_4$*; and (b)* $YBa_{2-x}Cu_3Cu_3O_{7-\delta}$*; here, note the presence of both planes and lines of oxygen atoms. (Courtesy of J. Hoffman.)*

Table 46.1 *Critical temperature and crystal structure of some high-*T_c *superconductors.*

Formula	Notation	T_c (K)	Cu-O planes per unit cell	Crystal structure
$YBa_2Cu_3O_7$	123	92	2	Orthorhombic
$Bi_2Sr_2CuO_6$	Bi-2201	20	1	Tetragonal
$Bi_2Sr_2CaCu_2O_8$	Bi-2212	85	2	Tetragonal
$Bi_2Sr_2Ca_2Cu_3O_{10}$	Bi-2223	110	3	Tetragonal
$Tl_2Ba_2CuO_6$	Tl-2201	80	1	Tetragonal
$Tl_2Ba_2CaCu_2O_8$	Tl-2212	108	2	Tetragonal
$Tl_2Ba_2Ca_2Cu_3O_{10}$	Tl-2223	125	3	Tetragonal
$TlBa_2Ca_3Cu_4O_{11}$	Tl-1234	122	4	Tetragonal
$HgBa_2CuO_4$	Hg-1201	94	1	Tetragonal
$HgBa_2CaCu_2O_6$	Hg-1212	128	2	Tetragonal
$HgBa_2Ca_2Cu_3O_8$	Hg-1223	134	3	Tetragonal

is shown in Fig. 46.3(b). Note the similarity of both the 214 and 123 with the cubic perovskite structure, shown earlier in Fig. 8.18, although here it consists of three such cells stacked on top of each other, with the important difference that there are some missing oxygen atoms. Subsequent work showed that the 214 and 123 compounds are members of a large family of copper-based, cuprate superconductors;[5] they are similar in that they all contain CuO_2 planes. Table 46.1 contains a list of some of the cuprate superconductors along with their transition temperatures.[6]

46.2 Some phenomenology of cuprate superconductors

Figure 46.4 shows the phase diagram of $La_{2-x}Sr_xCuO_4$; it is generic to most high T_c materials. We note the following features, some of which are unusual. (i) We have already pointed out that the parent compound, La_2CuO_4, is antiferromagnetic,[7] but this phase terminates at $x \cong 2\%$. (ii) The region within the semicircular "dome" corresponds to the superconducting phase. Hole doping at levels smaller than or larger than that corresponding to the maximum T_c are referred to as underdoping or overdoping respectively. (iii) Various experiments (particularly NMR) indicate that a gap forms in the electronic excitation spectrum to the left of the dashed line, termed the *pseudo gap*; strictly speaking it is not a phase transition since it is not accompanied by a thermal anomaly. The extrapolation of this line into the superconducting dome down to absolute zero, in the spirit of Fig. 45.15(c), has been speculated to be where a quantum critical point would lie, were it not blocked by the onset of superconductivity. Various experiments suggest that above the dome the regions to the right and left of the pseudo gap line may correspond to a Fermi liquid and a non-Fermi liquid respectively.

Rather than plotting the behavior as a function of the composition, x, the phase diagram shows more commonality with other cuprates if we use p, the number of holes, as the independent variable. Superconductivity then occurs in the range, $0.05 \leq p \leq 0.27$, while the dome lies at $p = 0.16$. However the temperature at the dome, T_{max}, does not vary in a systematic way. For the case of LSCO, antiferromagnetism disappears at $p = 0.04$.

Nuclear magnetic resonance has proved to be a particularly powerful probe of both the normal and superconducting properties. Figure 46.5 shows the Knight shift, designated ^{63}K, of the ^{63}Cu nucleus in the material, $YBa_2Cu_3O_{7-y}$.[8] In a conventional metal this quantity is temperature independent and is a measure of the conduction electron wave function at the nucleus which is, in turn, proportional to the magnetic susceptibility (the Knight shift, see Section 28.6.). For $y = 0$, corresponding to the stoichiometric compound, this behavior is observed above the transition temperature. However for $y = 0.63$, this quantity falls off as the temperature is lowered, a phenomenon interpreted as the formation of a gap in the excitation spectrum for some population of the electrons, the pseudo gap

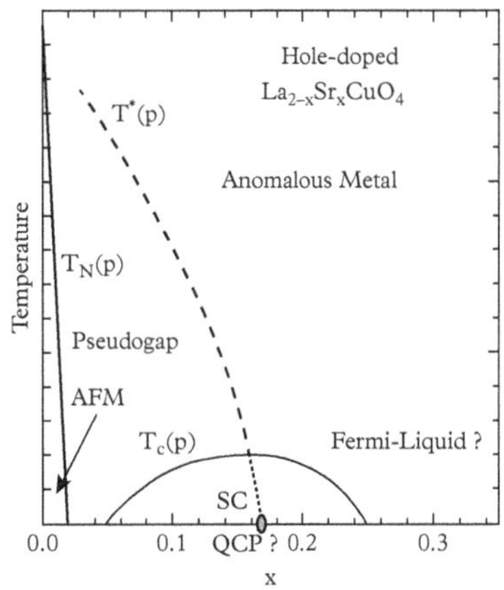

Figure 46.4 *A proposed phase diagram of the compound* $La_{2-x}Sr_xCuO_4$ *as a function of the composition,* x.

[5] For a general review of experiments on high temperature superconductors, see: High T_c superconductivity, H.R. Ott, Ch. 14 in *Superconductivity: Novel Superconductors*, Ed. K.H. Bennemann and J.B. Ketterson, Springer (2008).

[6] Under pressure, Hg1223 has a T_c of approximately 160 K, more than halfway to room temperature.

[7] It is also a Mott insulator (see Section 45.1), implying the importance of electron correlations.

[8] Cu and O NMR studies of the magnetic properties of $YBa_2Cu_3O_{6.63}$ ($T_c = 62$ K), M. Takigawa, A.P. Reyes, P.C. Hammel, J.D. Thompson, R.H. Heffner, Z. Fisk, and K.C. Ott, *Phys. Rev. B* **43**, 247 (1991).

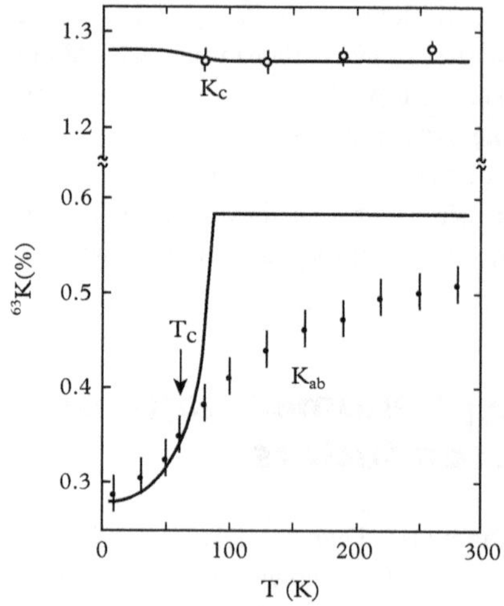

Figure 46.5 *The temperature dependence of the Knight shift of the* ^{63}Cu *nucleus in* $YBa_2Cu_3O_{7-y}$ *for the field along both the c axis and in the ab plane for* $y = 0.63$. *Also shown by the solid line is the behavior for* $y = 0$. *(After Takigawa et al. (1991).*

mentioned above. Many other experiments, including NMR relaxation rates (via the Korringa relation, see Section 28B.3) and anomalies in sound propagation,[9] also support this conclusion. It has been speculated that bosonic Cooper pair molecules form at temperatures below this line, which however do not then "condense" into a superconducting state. An alternative explanation is that there is a major reconstruction of the Fermi surface on the underdoped side that is driven by the appearance of a spin/charge density wave, as will be discussed shortly.

The approximately linear temperature dependence of the resistivity of the LBCO material above the transition temperature was pointed out in Fig. 46.2.[10] The resistivity of $La_{2-x}Sr_xCuO_4$ for various values of the strontium concentration, x, is shown in Fig. 46.6.[11] Over a wide range of temperatures the resistivity can be described by a form, $\rho \propto T^n$, with $n \cong 1.5$ (see Eq. (45.22a)).

[9] Bounding the pseudogap with a line of phase transitions in *$YBa_2Cu_3O_{6+\delta}$*, A. Shekhter, B.J. Ramshaw, R. Liang, W.N. Hardy, D.A. Bonn, F.F. Balakirev, R.D. McDonald, J.B. Betts, S.C. Riggs, and A. Migliori, *Nature* **498**, 75 (2013).

[10] We hasten to point out that a linear dependence of the resistivity over a wide temperature range is not uncommon, and is even observed in elemental transition metals; for a discussion, see: DC transport in metals, P.B. Allen, T.P. Beaulac, F.S. Khan, W.H. Butler, F.J. Pinski, and J.C. Swihart, *Phys. Rev. B* **34**, 4331 (1986).

[11] Systematic evolution of temperature-dependent resistivity in $La_{2-x}Sr_xCuO_4$, H. Takagi, B. Batlogg, H.L. Kao, J. Kwo, R.J. Cava, J.J. Krajewski, and W.F. Peck, Jr., *Phys. Rev. Lett.* **69**, 2975 (1992).

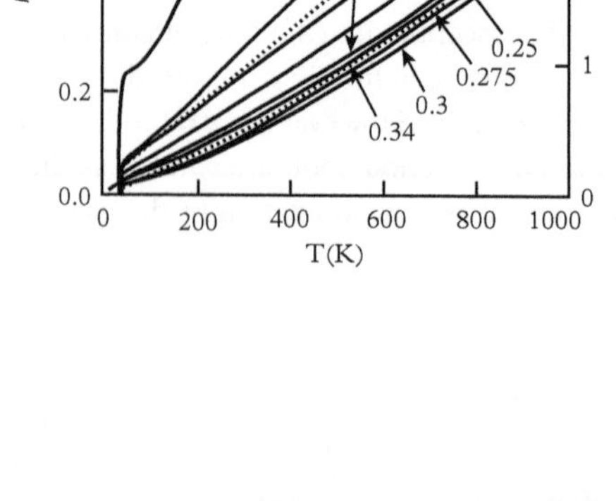

Figure 46.6 *The resistivity of bulk polycrystalline samples (scale to right) and (100) plane single crystal films (scale to left) of LSCO for various values of* x. *(After Takagi et al. (1992).)*

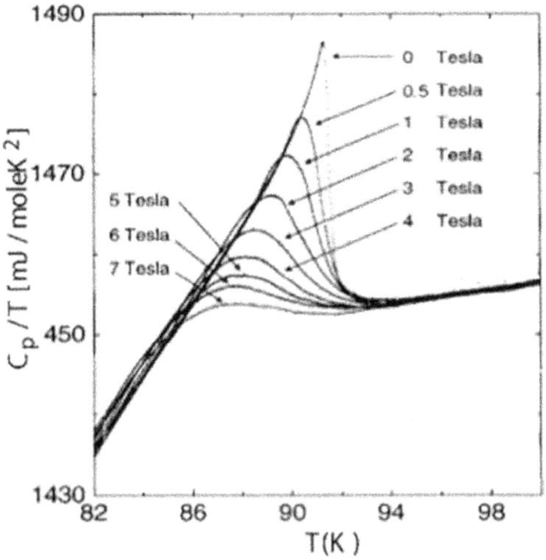

Figure 46.7 *The heat capacity of a single crystal of YBCO at various magnetic fields near the superconducting transition. (After Jeandupeux et al. (1996).)*

This may imply a non-Fermi-liquid behavior and, furthermore, the possibility of a hidden quantum phase transition; the latter would presumably have been under the superconducting dome. This interpretation is supported by the presence of a neighboring antiferromagnetic transition.

Figure 46.7 shows the heat capacity of a single crystal, $YBa_2Cu_3O_7$, for various magnetic fields in the vicinity of the superconducting transition.[12] It shows the expected discontinuity associated with the second-order superconducting transition. Later measurements by Schilling et al.[13] showed an additional sharp feature associated with a transition in the vortex lattice.

46.3 Electronic structure of cuprates

We will again restrict ourselves to the $La_{2-x}Sr_xCuO_4$ system, starting with a discussion of the crystal chemistry.[14] For $x = 0$ we will picture the system as consisting of $5s^25p^6La^{3+}$, $3s^23p^63d^9$ Cu^{2+}, and $2s^22p^6$ O^{2-}ions. The La core orbitals lie low in energy, as do those associated with the oxygen s orbitals; both have small overlap with neighboring atoms and result in narrow filled bands which play a minimal role in bonding and no role in electrical transport. We emphasize that La_2CuO_4 is a strongly layered structure with large ionic interactions and hence we expect a large crystal field splitting between oxygen p_z and the p_x, p_y orbitals.[15] More importantly, the copper d orbitals will split, here in the order: d_{yz}, d_{zx}; d_{xy}; $d_{3z^2-r^2}$; and $d_{x^2-y^2}(e_g)$. The antibonding state of the latter lies highest in energy. The first three correspond to the three-fold degenerate t_{2g} and the latter two to the two-fold degenerate e_g levels, that are encountered in a cubic environment; this level splitting scheme was shown earlier in Fig. 45.4.[16]

In addition to the crystal field effects we will have the shifts resulting from the overlap of the copper d states and the oxygen p states, with the crystal potential

[12] Scaling of the specific heat and magnetization of $YBa_2Cu_3O_7$ in magnetic fields up to 7T, O. Jeandupeux, A. Schilling, and H.R. Ott, *Phys. Rev. B* **53**, 12475 (1996).

[13] Anisotropic latent heat of vortex-lattice melting in untwinned $YBa_2Cu_3O_7$ A. Schilling, R.A. Fisher, N.E. Phillips, U. Welp, W.K. Kwok, and G.W. Crabtree, *Phys. Rev. Lett.* 78, 4833 (1997).

[14] Electronic structure of the high-temperature oxide superconductors, W.E. Pickett, *Rev. Mod. Phys.* **61**, 433 (1989).

[15] There has been some discussion in the high temperature superconductivity literature as to the role of the so-called Jahn–Teller effect, which can split multiply degenerate states in certain otherwise high symmetry structures; in the present case, the degeneracy is largely lifted by the overall crystal structure.

[16] Note that this order differs with respect to the two lowest levels from that shown in Fig. 45.4. For a more detailed discussion on level splitting see: Theory of lanthanum copper oxide (LSCO) superconductors, H. Kamimura, H. Ushio, and S. Matsuno, *Physica C* **460**, 991 (2007).

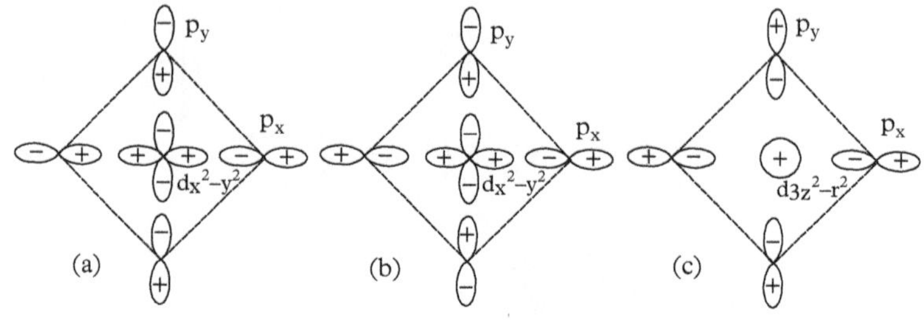

Figure 46.8 *La_2CuO_4 dp-σ orbitals in the copper-oxygen planes. (a) The $d_{x^2}-y^2$ case when the wave function has the full translational symmetry. (b) The fully antibonding case. (c) The $d_{3z^2-r^2}$ case.*

resulting in bonding and antibonding orbitals. The largest overlap, and hence the widest overall band width, is associated with the copper $d_{x^2-y^2}$ orbital and the neighboring oxygen p_x and p_y orbitals, as shown in Fig. 46.8; (a) shows the case of full translational symmetry whereas in (b) the signs of the wave function overlap correspond to the highest energy (antibonding) state. Bonding arises when the signs of the overlapping orbitals are the same (recall the potential is negative); the bonding configuration accounts for much of the cohesive energy. Also shown in Fig. 46.8(c) is the $d_{3z^2-r^2}$ orbital, which is also incorporated in some models.

Since the unit cell contains an odd number of electrons, paramagnetic band structure calculations must yield metallic behavior. Figure 46.9 shows the Brillouin zone (a) and the band structure (b) of the tetragonal phase of La_2CuO_4, as calculated in the local density approximation.[17] The Fermi surface, shown in Fig. 46.10, has a half-filled band; when viewed as holes it is centered at the X point of the Brillouin zone. As previously argued, this band is primarily associated with the copper antibonding $d_{x^2-y^2}$-like level. Calculations in the orthorhombic phase

[17] Tight-binding Hamiltonians for high-temperature superconductors and applications to coherent-potential-approximation calculations of the electronic-properties of $La_{2-x}Ba_xCuO_{4-y}$, D.A. Papaconstantopoulos, W.E. Pickett, and M.J. DeWeert, *Phys. Rev. B* **39**, 4235–4248 (1989).

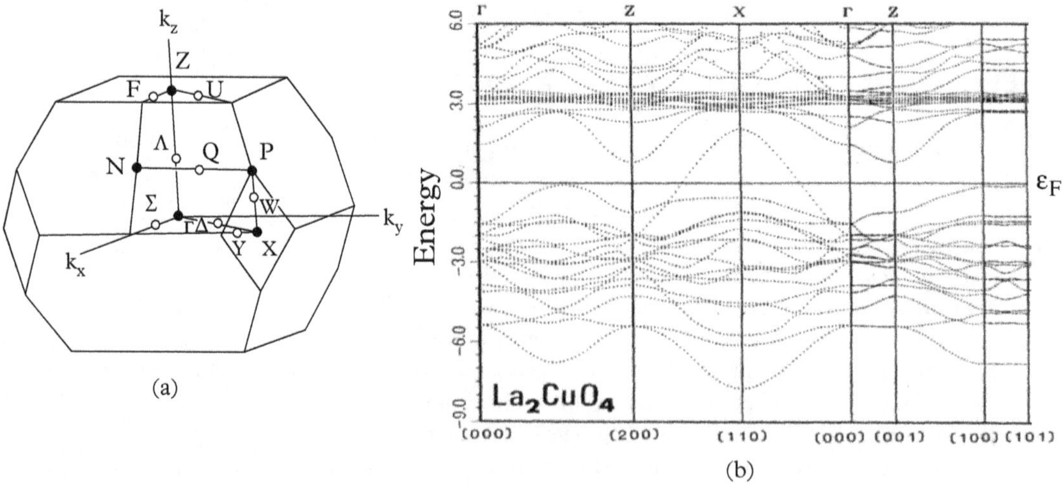

Figure 46.9 *(a) The tetragonal Brillouin zone showing the major symmetry lines and points. (b) The band structure of lanthanum copper oxide as calculated by the APW method, along with a line indicating the position of the Fermi level. (After Papaconstantopoulos, Pickett, and DeWeert (1988).)*

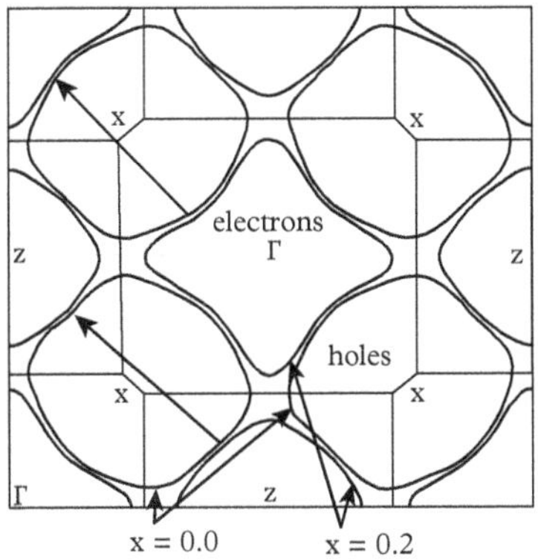

Figure 46.10 *The Fermi surface of paramagnetic* $La_{2-x}Sr_xCuO_4$ *for* x = 0.0 *and* 0.2. *For* x = 0 *the electron surface is open, whereas the corresponding hole surface is approximately circular and centered at* X. *For* x = 0.2 *the Fermi energy is lower and the electron surface is then closed and centered at* Γ. *The arrows show wave vectors connecting flat-like regions of the Fermi surface. (From Pickett, Krakauer, and Cohen (unpublished).)*

also predict metallic behavior. More interesting is that spin polarized calculations predict that La_2CuO_4 remains metallic in the antiferromagnetic phase where the size of the unit cell is increased.

46.4 Model Hamiltonians

That LSCO is an insulator can be taken as clear evidence that on-site Coulomb repulsion is introducing a correlation gap. For this reason much attention has been given to tight-binding models which incorporate a Hubbard interaction. At an intermediate level, the starting Hamiltonian near the Fermi level could include terms arising from: (i) the bare energies of the copper $d_{x^2-y^2}$ states, ε_d, and the oxygen p_x, p_y states, ε_p; (ii) the nearest-neighbor overlap integrals, V_{ij}, connecting these states; and (iii) the on-site copper Coulomb repulsion, U. Combining these contributions and noting that there is one Cu and two O atoms per square planar unit cell, we can write[18]

$$\hat{H} = \sum_{i\sigma} \varepsilon_d \hat{d}_i^\dagger \hat{d}_i + \sum_{j\sigma} \varepsilon_p \hat{p}_{j\sigma}^\dagger \hat{p}_{j\sigma} + \sum_{i\sigma} \sum_{j\sigma}^{nn} V_{ij} \left(\hat{d}_{i\sigma}^\dagger \hat{p}_{j\sigma} + \hat{p}_{j\sigma}^\dagger \hat{d}_{i\sigma} \right) + U \sum_i \hat{d}_{i\uparrow}^\dagger \hat{d}_{i\uparrow} \hat{d}_{i\downarrow}^\dagger \hat{d}_{i\downarrow}, \tag{46.1}$$

where $U > 0$, only nearest-neighbor $dp - \sigma$ terms are included, and $V_{ij} = \pm t_0$, where the differing signs arise from the relative phases of the d and p orbitals on opposite sides of the central Cu atom (see Fig. 46.8(a)). Band structure and other information show that $\varepsilon_p < \varepsilon_d$, so when $t_0 \ll \varepsilon_p, U$ and in the absence of doping, we expect to have an insulator with one $d_{x^2-y^2}$ electron (or equivalently one hole) per Cu atom and filled O levels (no holes).

One can go on to treat the overlap, t_0, as a perturbation. Zhang and Rice show that this leads to a term in the Hamiltonian that couples the Cu spins, **S**, through the neighboring oxygen p states having the Heisenberg-like form[19]

$$\hat{H} = J \sum_i \sum_j^{nn} \hat{S}_i \cdot \hat{S}_j. \tag{46.2a}$$

[18] Effective Hamiltonian for the superconducting Cu oxides, F.C. Zhang and T.M. Rice, *Phys. Rev. B* 37, 3759 (1988).

[19] This coupling is called *super-exchange* and plays a major role in magnetic oxides; for an extended discussion, see: New approach to the theory of superexchange interactions, P.W. Anderson, *Phys. Rev.* 115, 2 (1959).

Here, the sum is restricted to nearest-neighbor planar Cu atoms and to lowest order they find

$$J = \frac{4t_0^4}{\varepsilon_p^2 U} + \frac{4t_0^4}{2\varepsilon_p^3}. \tag{46.2b}$$

Note that J is positive and hence the interaction is *antiferromagnetic*; here we recall that, experimentally, La_2CuO_4 is antiferromagnetic.

As can be seen from Fig. 46.9, one hybridized level cuts through the Fermi energy and hence the low energy dynamics (superconductivity and electrical transport) may be describable by a "one-band" model. Hence as an alternative to Eq. (46.1) we might include direct coupling between the copper ions via second nearest-neighbor interactions along with a second-order coupling through the oxygen p states and write

$$\hat{H} = t\sum_{i\sigma}\sum_{j\sigma}^{nn} \hat{d}_{i\sigma}^{\dagger}\hat{d}_{j\sigma} + U\sum_{i} \hat{d}_{i\uparrow}^{\dagger}\hat{d}_{i\uparrow}\hat{d}_{i\downarrow}^{\dagger}\hat{d}_{i\downarrow}, \tag{46.3}$$

where the $\hat{d}$ operators now create the hybridized orbitals and t is an inter-copper hopping parameter which can be obtained from the band structure calculations in the vicinity of ε_F;[20] note the similarity to Eq. (45.7). It is widely believed that this model contains the necessary physics to explain the essential properties of both the normal and superconducting states of high T_c materials. In the limit of large t/U one can again do perturbation theory, obtaining the Hamiltonian,

$$\hat{H} = t\sum_{i\sigma}\sum_{j\sigma}^{nn} \hat{d}_{i\sigma}^{\dagger}\hat{d}_{j\sigma} + J\sum_{i}\sum_{j}^{nn} \hat{S}_i \cdot \hat{S}_j, \tag{46.4}$$

where the effect of the large U is incorporated through a constraint that requires no double occupancy of the Cu sites; this is referred to as the t–j model, which is also widely used to treat the high T_c materials.

46.5 Doping

As previously noted, metallic behavior in the cuprate materials (and when it occurs, superconductivity) is generally achieved by doping. For the case of La_2CuO_4 this is achieved by alloying with Sr^{2+} or Ba^{2+}, where the latter, having a valence one less than La^{3+}, lowers the Fermi level. Ba and Sr, which we take as fully ionized, then create holes in the copper–oxygen planes and one refers to the La-doped plane as the "charge reservoir." Although the levels are strongly hybridized, the holes tend to be localized on the copper atoms. For small doping levels the holes are initially trapped and hence the system remains insulting,

[20] For an example of a multi-parameter fit see: One-band tight-binding model parametrization of the high-Tc cuprates including the effect of k_z dispersion, R.S. Markiewicz, S. Sahrakorpi, M. Lindroos, Hsin Lin, and A. Bansil, *Phys. Rev. B* 72, 054519 (2005).

although the antiferromagnetism is progressively suppressed. On further doping the holes become mobile.

For compounds like $Bi_2^{+3}Sr_2^{+2}Ca^{+2}Cu_2^{+2}O_{8-y}^{-2}$ (Bi2212) and many other cuprates (including YBCO), doping is achieved by introducing oxygen vacancies rather than by alloying. Because of the strong in-plane copper–oxygen bonds, these vacancies tend to be situated out of the planes; they then play the role of the charge reservoir. For y = 0, Bi2212 has one d hole per Cu atom and the material is again an insulator. Doping in these materials can be controlled by adjusting the oxygen partial pressure during the preparation. Bi2212 is an ideal material for studies involving a free surface, such as photo-emission and scanning tunneling microscopy, since it cleaves cleanly.

46.6 Determining the Fermi surface of cuprates

An important property to be established in the metallic cuprates is whether a Fermi surface exists, and if so what is the associated shape and what are the accompanying effective masses. These are key questions, given the possibility of non-Fermi-liquid behavior. The first measurements that addressed this question involved angle-resolved photo-electron spectroscopy (ARPES), a technique[21] which was discussed in Section 20.2. Figure 46.11 shows the data of the Stanford group[22] on underdoped and overdoped $Bi_2Sr_2CaCu_2O_{8-\delta}$; this material is especially suitable for such measurements since it cleaves well (a process carried out in vacuum) to expose a clean flat surface. The data on the underdoped material suggest that the Fermi surface consists of a series of "arcs."[23] However these

[21] For a discussion of the ARPES technique and its interpretation see: Photo emission in high T_c superconductors, J.C. Campuzano, M.R. Norman, and M. Randeria, in *Superconductivity*, Ed. K.H. Bennemann and J.B. Ketterson, Springer (2008).

[22] Unconventional electronic structure evolution with hole doping in $Bi_2Sr_2CaCu_2O_{8-\delta}$; angle-resolved photoemission results, D.S. Marshall, D.S. Dessau, A.G. Loeser, C-H. Park, A.Y. Matsuura, J.N. Eckstein, I. Bozovic, P. Fournier, A. Kapitulnik, W.E. Spicer and Z.-X. Shen, *Phys. Rev. Lett.* **76**, 4841 (1996).

[23] Destruction of the Fermi surface in underdoped high-T_c superconductors, M.R. Norman, H. Ding, M. Randeria, J.C. Campuzano, T. Yokoya, T. Takeuchik, T. Takahashi, T. Mochiku, K. Kadowaki, P. Guptasarma, and D.G. Hinks, *Nature* **392**, 157 (1998).

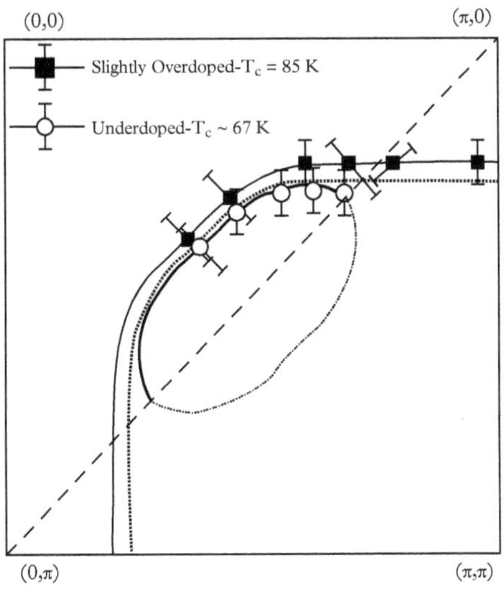

Figure 46.11 *The Fermi surface of* $Bi_2Sr_2CaCu_2O_{8-\delta}$. *Note that the underdoped material is consistent with the presence of small pockets, while the over doped sample closely resembles the band structure predictions. (After Marshall et al. (1996).)*

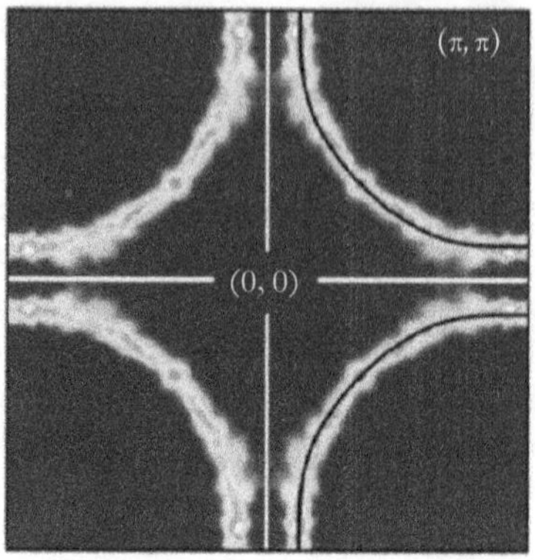

Figure 46.12 *The Fermi surface of* $Tl_2Ba_2CuO_{6+\delta}$ *from angle resolved photo-emission, as extracted from the position of the peak of the k–space spectral function. (After Plate et al. (2005).)*

can also be interpreted as small sheets of a much smaller closed surface, as completed by the dotted line in the figure. For the overdoped material the surface has the shape predicted by the band structure calculations (see Fig. 46.10). Another overdoped case, which we will return to, is the compound, $Tl_2Ba_2CuO_{6+\delta}$, the ARPES data from which is shown in Fig. 46.12, here drawn with respect to the zone center.[24]

As emphasized in Section 20.1, the de Haas–van Alphen (dHvA) effect is generally the most powerful probe of Fermi surface properties. However, relative to pure metals, the scattering times tend to be short in the cuprates, which suppresses the amplitude of the dHvA oscillations. With the availability of better samples, higher magnetic fields, and lower temperatures, dHvA oscillations have now been seen in several systems. Here materials with lower transition temperatures minimize complications in interpreting data (which is then in the vortex phase).

Figure 46.13(a) shows dHvA oscillations, obtained via torque measurements, in underdoped $YBa_2Cu_3O_{6.5}$ and overdoped $Tl_2Ba_2CuO_{6+\delta}$,[25] where the Fermi surface of the latter was pictured above. Note that the dHvA frequency, F, of the overdoped thallium material, seen in the upper trace in Fig. 46.13(a), is large.

[24] Fermi surface and quasiparticle excitations of overdoped $Tl_2Ba_2CuO_{6+\delta}$, M. Platé, J.D.F. Mottershead, I.S. Elfimov, D.C. Peets, Ruixing Liang, D.A. Bonn, W.N. Hardy, S. Chiuzbaian, M. Falub, M. Shi, L. Patthey, and A. Damascelli, *Phys. Rev. Lett.* **95**, 077001 (2005).

[25] Quantum oscillations and the Fermi surface of high-temperature cuprate superconductors, B. Vignolle, D. Vignolles, D. LeBoeuf, S. Lepault, B. Ramshaw, R. Liang, D. A. Bonn, W. N. Hardy, N. Doiron-Leyraud, A. Carrington, N. E. Hussey, L. Taillefer, and C. Proust, *Comptes Rendus Phys.* **12**, 446 (2011).

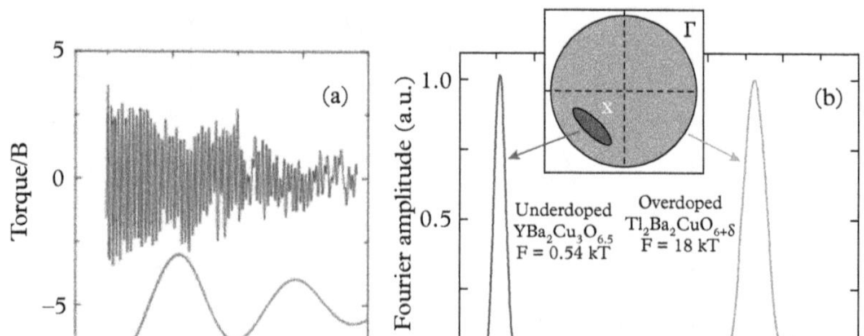

Figure 46.13 *(a) de Hass-van Alphen oscillations (dHvA) of the torque from underdoped* $YBa_2Cu_3O_{6.5}$ *(slow oscillations) and overdoped* $Tl_2Ba_2CuO_{6+\delta}$ *(fast oscillations). (b) The Fourier transform of the oscillations is shown to the left, from which the frequency is obtained and with it the cross-sectional area of the Fermi surface. Shown also are the Fermi surface cross sections, where the large surface has been centered at* X *to agree with band structure calculations, while the shape and center of the small surface are positioned to cover the arcs observed in ARPES measurements. (After Vignolle et al. (2011).)*

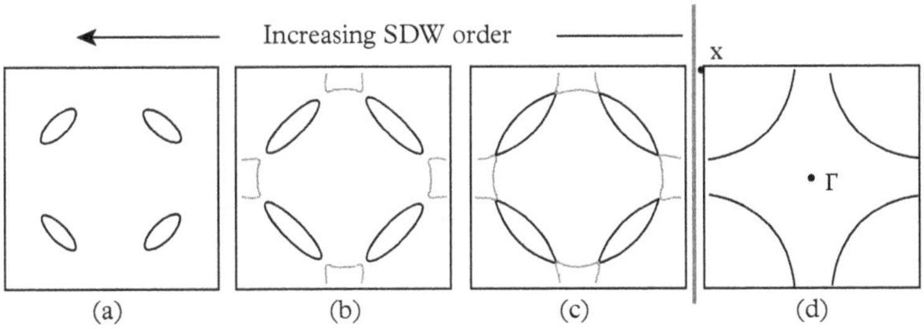

Figure 46.14 *A schematic depiction of the possible evolution of the Fermi surface in cuprates with increasing strength of a spin density wave (SDW) with a commensurate wave vector,* $Q = (\pi/a, \pi/a)$*, which has the effect of increasing the size of the unit cell in real space and reducing the size of the Brillouin zone. Panel (d) to the right shows the unperturbed Fermi surface which is pictured, for simplicity, as circles centered at* X *(see Fig. 46.10 for a more realistic shape). The presence of the wave couple states at* k *and* k + Q*, which can be represented by a second hole surface centered at* Γ*, is shown in (c). The resulting mixing reconstructs the larger unperturbed Fermi surface into hole and electron pockets of much smaller size, as shown in (b). Doping can alter the electron-hole balance, as in panel (a). (After Moon and Sachdev (2009).)*

Through the Onsager–Lifshitz, $F = c\hbar A_k/2\pi e$ (see Eq. (18.36b)), this implies that the extremal cross-sectional area of the Fermi surface, A_k, is large, consistent with what is observed using ARPES on the same compound and predicted by the band structure. On the other hand, the underdoped $YBa_2Cu_3O_{6.5}$ material, shown in the lower trace in Fig. 46.13(a), shows a much smaller frequency and with it a smaller area, which can be compatible with what was observed via ARPES on the underdoped bismuth compound.

The differences between the underdoped and overdoped materials are far too large to be accounted for through the marginally differing hole concentrations, p, on opposing sides of the superconducting dome. An explanation that has wide acceptance, and for which there is supporting experimental evidence, is that the system spontaneously develops a static *spin or charge density wave*. When this happens the basic unit cell is altered and the Fermi surface will *reconstruct*; this can be accompanied by a reduction in the density of states which would partially explain the pseudo gap phenomenon.

A possible reconstruction is shown schematically in Fig. 46.14. What has been proposed[26] is that a charge or spin density wave (or some combination of the two) develops along the square diagonal, which has a wave vector, $Q = (\pi/a, \pi/a)$. This would be most natural on the underdoped side due to the proximity of the antiferromagnetic phase. Such a wave would introduce gaps in the Fermi surface, analogous to those introduced by the lattice potential in the nearly free electron approximation (see Section 11.1). Taking the hole surface as a circle for simplicity, and recognizing that translation by the wave vector, Q, is equivalent to placing another circle at the X point, as shown in Fig. 46.14(c), the two surfaces then intersect and the perturbation accompanying the wave causes the Fermi surface to reconstruct, as shown in Figure 46.14(b). A shift in the Fermi level could then "squeeze out" the electron surface.

46.7 Josephson coupled layered systems

46.7.1 The Lawrence–Doniach model

Here we generalize the Ginzburg–Landau theory, discussed in Chapter 33, so that it can be applied to layered materials like the cuprate superconductors; related

[26] Competition between spin density wave order and superconductivity in the underdoped cuprates, E.G. Moon and S. Sachdev, *Phys. Rev. B* **80**, 035117 (2009).

materials include transition metal dichalcogenides[27] (e.g., $NbSe_2$) and artificially prepared superconductor/insulator superlattices. As a simple model, we assume that the superconductivity resides within infinitesimally thin planes (the CuO_2 layers) and that these planes are weakly coupled via order parameter tunneling (Josephson coupling) through the "insulating charge reservoir" layers to which we assign the total unit cell z-axis dimension, denoted as s.[28] In place of Eq. (33.7), and following Lawrence and Doniach (1971),[29,30] we introduce a free-energy functional of the form

$$F = F_n + F_s, \tag{46.5}$$

where

$$F_s = \sum_n s \int d^2r \left[\frac{\hbar^2}{2m^*} \left| \left(\nabla_\perp - \frac{ie^*}{\hbar c} \mathbf{A}_\perp(\mathbf{r}) \right) \psi_n(\mathbf{r}) \right|^2 + \alpha |\psi_n(\mathbf{r})|^2 + \frac{1}{2}\beta |\psi_n(\mathbf{r})|^4 \right.$$
$$\left. + \frac{\hbar^2}{2m_z^* s^2} \left| \psi_{n+1} \exp\left(-\frac{ie^*}{\hbar c} \bar{A}_z s \right) - \psi_n \right|^2 \right] + \frac{1}{8\pi} \int d^3r\, H^2(\mathbf{r}). \tag{46.6}$$

Here, the order parameter, $\psi_n(\mathbf{r})$, has a discrete dependence on the layer index, n, and a continuous dependence on the in-plane position coordinate, $\mathbf{r} = x\hat{\mathbf{x}} + y\hat{\mathbf{y}}$; the total vector potential, $\mathbf{A} = \mathbf{A}_\perp + A_z\hat{\mathbf{z}}$, is, however, defined at all points and we have assumed A_z to be slowly varying and written $\bar{A}_z = (1/s) \int_{ns}^{(n+1)s} A_z dz$. The structure of Eq. (44.6) is similar to Eq. (33.7) with respect to the in-plane components, while the interplane coupling is seen to be a finite-differences form of the operator

$$\frac{\partial}{\partial z} - \frac{ie^*}{\hbar c} A_z.$$

Variation with respect to ψ^* yields

$$-\frac{\hbar^2}{2m^*} \left(\nabla_\perp - \frac{ie^*}{\hbar c} \mathbf{A}_\perp \right)^2 \psi_n$$
$$- \frac{\hbar^2}{2m_z^* s^2} \left[\psi_{n+1} \exp\left(-\frac{ie^*}{\hbar c} \bar{A}_z s \right) - 2\psi_n + \psi_{n-1} \exp\left(+\frac{ie^*}{\hbar c} \bar{A}_z s \right) \right]$$
$$+ \alpha\psi_n + \beta\psi_n |\psi_n|^2 = 0. \tag{46.7}$$

Variation with respect to $\mathbf{A}_\perp$ and A_z yields Eq. (33.10) with

$$\mathbf{j}_\perp = -\frac{ie^*\hbar}{2m^*} (\psi_n^* \nabla_\perp \psi_n - \psi_n \nabla_\perp \psi_n^*) - \frac{e^{*2}}{m^* c} \mathbf{A}_\perp |\psi_n|^2 \tag{46.8}$$

and

$$j_{\|,n+1,n} = -\frac{ie^*\hbar}{2m_z^* s} \left[\psi_{n+1} \exp\left(-\frac{ie^*}{\hbar c} \bar{A}_z s \right) \psi_n^* - \psi_{n+1}^* \exp\left(+\frac{ie^*}{\hbar c} \bar{A}_z s \right) \psi_n \right]. \tag{46.9}$$

Equation (46.9) for the tunneling current flowing between planes n and n + 1 is equivalent to Eq. (33.55) (when rewritten in a gauge-invariant form).

[27] The chalcogenides are the group VI elements S, Te, and Se.

[28] Such a model, which is appropriate for a single-plane material like LSCO, can be generalized to the case where there are multiple Cu_2O planes per unit cell.

[29] W.E. Lawrence and S. Doniach, *Proceedings of the 16th International Conference on Low Temperature Physics*, p.361, Ed. E. Kanda, Academic Press of Japan.

[30] For a more extended discussion of this model, see Chapter 16 in *Superconductivity*, J.B. Ketterson and S.N. Song, Cambridge University Press, Cambridge (2000).

46.7.2 The upper critical fields

Solutions of Eq. (46.7) were first examined in detail by Klemm et al.[31] In this subsection we confine our interest to the evaluation of H_{c2}, in which case we may neglect the last (nonlinear) term. The external magnetic field will be assumed to lie in the x–z plane. In order to take advantage of the layer periodicity, we choose a gauge for which the vector potential has no z dependence:

$$\mathbf{A}_{\perp} + A_z\hat{\mathbf{z}} = Hx(\cos\theta\hat{\mathbf{y}} - \sin\theta\hat{\mathbf{z}}). \tag{46.10}$$

We seek a solution of the form

$$\psi_n = \exp(+ik_z ns + ik_y y)u_n(x);$$

noting that $\bar{A}_{\perp} = -Hx\sin\theta$, we find that

$$\left\{-\frac{\hbar^2}{2m^*}\frac{d^2}{dx^2} + \frac{1}{2}m^*\omega_c^2(x - x_0)^2\cos^2\theta \right.$$
$$\left. -\frac{\hbar^2}{m_z^* s^2}\left[\cos\left(k_z s + \frac{e^*Hs}{\hbar c}x\sin\theta\right) - 1\right] + \alpha\right\} u_n(x) = 0, \tag{46.11}$$

where $x_0 = \hbar c k_y/(e^*H\cos\theta$ is an "orbit center."

For H perpendicular to the layers, we again have a harmonic oscillator problem and we recover our earlier result (33.51b), obtained in Section 33, where $-\alpha = \hbar\omega_c(0)/2$. The associated perpendicular upper critical field is then

$$H_{c_2\perp} = -\frac{2m^*c\alpha}{\hbar e^*}, \tag{46.12a}$$

or in terms of the coherence length,

$$H_{c_2\perp} = \frac{\Phi_0}{2\pi\xi^2}, \tag{46.12b}$$

where $\xi \equiv (\hbar^2/2m^*|\alpha|)^{1/2}$.

For H parallel to the planes, Eq. (46.11) becomes

$$\left\{-\frac{\hbar^2}{2m^*}\frac{d^2}{dx^2} - \frac{\hbar^2}{m_z^* s^2}\left[\cos\left(k_z s + \frac{e^*Hs}{\hbar c}x\right) - 1\right] + \alpha\right\} u_n = 0. \tag{46.13}$$

Changing k_z has the effect of shifting the origin on the x axis. We shall initially fix the origin such that the argument of the cosine vanishes when x = 0. Near the zero-field transition temperature, where α is small, we expect the critical magnetic field, $H_{c2\parallel}$, to also be small and further that the lowest eigenfunction, u, which by

[31] Theory of the upper critical field in a layered superconductor, R.A. Klemm, A. Luher, and M.R. Beasley, Phys. Rev. B 12, 877 (1975).

Floquet's theorem is periodic (since the potential is periodic), is concentrated near the minimum of the *potential*,

$$V(x) = \frac{\hbar^2}{m_z^* s^2}\left[1 - \cos\left(\frac{e^* H s}{\hbar c}x\right)\right]. \tag{46.14}$$

Expanding Eq. (46.14) to second order, we obtain a harmonic oscillator potential which, on substitution into Eq. (46.12), yields

$$-\frac{\hbar^2}{2m^*}u'' + \left(\frac{1}{2}m^*\omega_c^2\left(\frac{\pi}{2}\right)x^2 + \alpha\right)u = 0, \tag{46.15}$$

where

$$\omega_c\left(\frac{\pi}{2}\right) \equiv \frac{|e^*|H}{(m^* m_z^*)^{1/2}c}.$$

The lowest eigenvalue is $-\alpha = (1/2)\hbar\omega_c(\pi/2)$, yielding an upper critical field,

$$H_{c2\|} = \Phi_0/2\pi\xi\xi_z, \tag{46.16}$$

where $\xi_z \equiv (\hbar^2/2m_z^*|\alpha|)^{1/2}$. Note that the upper critical field has the behavior, $H_{c2\|} \propto (T_c - T)$, characteristic of a thin film for an in-plane magnetic field.

For temperatures further from T_c, where the eigenvalue, $-\alpha$, and the associated parallel critical field are both large, u becomes less concentrated at the potential minimum and our expansion of V(x) breaks down; it then turns out that $H_{c2\|}$ becomes infinite at a temperature, T^*, such that $\xi_z(T^*) = s/\sqrt{2}$.[32] Hence the model must be expanded to fix this problem.

46.7.3 Finite layer thicknesses

Deutcher and Entin-Wohlman[33] generalized the Lawrence–Doniach model to the case of a superlattice of thin slabs of thickness, d, separated by Josephson-coupled insulating layers of thickness, s. Again taking the z axis normal to the layers, and placing z = 0 at the interface between the lower surface of a metallic layer, the free energy is then given by

$$F = F_{\text{in-layer}} + F_{\text{inter-layer}} + F_{\text{field}}, \tag{46.17}$$

where

$$F_{\text{in-layer}} = \sum_n \int dx\, dy \left\{\int_{nD}^{nD+d} dz \left[\frac{\hbar^2}{2m^*}\left|\left(\nabla - \frac{ie^*}{\hbar c}\mathbf{A}\right)\psi_n(x,y,z)\right|^2 + \alpha|\psi_n(x,y,z)|^2 + \frac{\beta}{2}|\psi_n(x,y,z)|^4,\right.\right. \tag{46.18a}$$

[32] In a real system this divergence would be removed by the effect of paramagnetic limiting, where the energy associated with Pauli paramagnetism is equal to the superconducting condensation energy.

[33] Critical fields of weakly coupled superconductors, G. Deutscher and O. Entin-Wohlman, *Phys. Rev. B* 17, 1249 (1978).

$$F_{\text{inter-layer}} = \sum_n \int dx\,dy \int_{nD}^{nD+d} dz \frac{\hbar^2}{2m_z^* s^2} \times$$
$$\left[\left| \psi_{n+1}(x, y, (n+1)D) \exp\left(-(ie^* s/\hbar c)\bar{A}_{n+1}\right) - \psi_n(x, y, nD + d) \right|^2 \right], \tag{46.18b}$$

and

$$F_{\text{field}} = \frac{1}{8\pi} \int dx\,dy\,dz\,H^2(x, y, z). \tag{46.18c}$$

Here, $\bar{A}_{n+1} \equiv (1/s) \int_{nD+d}^{(n+1)D} A_z(x,y,z)\,dz$ and $D = d + s$. We will not discuss the subsequent analysis, but simply quote two important results. In the limit $H \to 0 (T \to T_c)$, they obtain

$$H_{c2\|} = \frac{\Phi_0}{2\pi\xi\xi_z(D/s)}, \tag{46.19}$$

which is to be compared with (46.16). At lower temperatures, rather than encountering the divergence associated with the Lawrence–Doniach model, they find that $H_{c2\|}$ undergoes a *crossover* from a $(T_c - T)$ behavior to one that is proportional, $(T_c - T)^{1/2}$. This is referred to as a 3d to 2d crossover, the latter temperature dependence being characteristic of a thin (2d) superconducting slab, as previously noted (also see Problem 33.1).

46.7.4 An isolated vortex for the field parallel to the layers

We will start with Eqs. (46.7) and (46.8), which were derived from the Lawrence–Doniach free-energy functional (46.6). For high-κ superconductors in a weak magnetic field, the influence of the field on the value of $|\psi_n|$ can be neglected, and we can regard $|\psi_n|$ as a constant; i.e., we use the London approximation. We take the field along the $\hat{x}$ direction and choose a gauge such that $A_z = 0$. Writing $\psi_n = |\psi_n| e^{i\phi_n}$, Eqs. (46.7) and (46.8) give

$$\mathbf{A}_\perp = -\frac{4\pi\lambda_L^2}{c} \mathbf{j}_\perp + \frac{\Phi_0}{2\pi} \nabla\phi_n, \tag{46.20}$$

and

$$j_{z,n+1,n} = j_m \sin(\phi_{n+1} - \phi_n), \tag{46.21}$$

where $\lambda_L^2 = mc^2/4\pi|\psi|^2 e^2$ is the London penetration depth of the superconducting layers, ϕ_n is the phase in the nth superconducting layer, and $j_m \equiv e^*\hbar|\psi|^2/m_z^* s$ is the maximum Josephson supercurrent density. Note that in this model it is assumed, for simplicity, that each superconducting layer is isotropic with an intrinsic bulk penetration depth, λ_L. We use the rectangular contour, C,

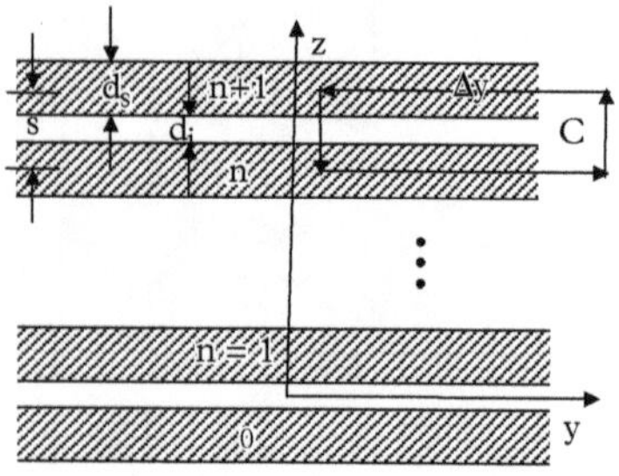

Figure 46.15 *Schematic of a Josephson-coupled superconductor/insulator superlattice. The insulating layers of thickness,* d_i, *alternate with superconducting layers (crosshatched) of thickness,* d_s. *The modulation wavelength is* $s = d_i + d_s$ *(note the designation of these thicknesses differ from those of Deutscher and Entin-Wohlman). The rectangular contour, C, is used to compute the magnetic flux in Eq. (46.22).*

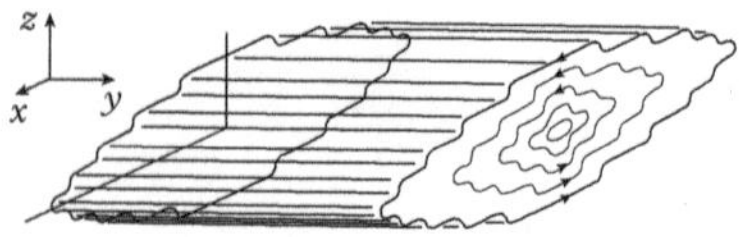

Figure 46.16 *A schematic representation of the contours of constant supercurrent in an isolated vortex in a layered superconductor arising from an in-plane magnetic field. The segments parallel to x lie in the superconducting layers, while those lying approximately parallel to z pass through the insulating layers.*

shown in Fig. 46.15, to compute the phase difference across one unit cell: $\Delta\phi \equiv \phi_{n+1} - \phi_n$ From Eq. (46.20), we have

$$\oint_C \mathbf{A}_\perp \cdot dl = \int_y^{y+\Delta y} dy' \int_{ns}^{(n+1)s} dz' H(y', z')$$

$$= -\frac{4\pi\lambda_L^2}{c}\left(\int_y^{y+\Delta y} j_{y,n} dy' + \int_{y+\Delta y}^{y} j_{y,n+1} dy'\right) + \frac{\Phi_0}{2\pi}(\phi(y) - \phi(y+\Delta y)).$$

Approximating the net flux as $s\Delta y H(y,z)$, we obtain

$$s\Delta y H(y,z) \approx \frac{4\pi\lambda_L^2}{c}(j_{y,n+1} - j_{y,n})\Delta y + \frac{\Phi_0}{2\pi}(\phi(y) - \phi(y+\Delta y)).$$

Hence,

$$\frac{4\pi\lambda_L^2}{c}\frac{\partial j_y}{\partial z} - \frac{\Phi_0}{2\pi s}\frac{\partial \phi}{\partial y} \approx H(y,z), \tag{46.22}$$

where we have used the approximations, $\partial j_y/\partial z \approx (j_{y,n+1} - j_{y,n})/s$ and $\partial\phi(y)/\partial y = (\phi(y+\Delta y) - \phi(y))/\Delta y\ (y))/\Delta y$. For small ϕ the Josephson current relation (46.21) may be approximated as $j_z = j_m \Delta\phi$, and substituting this form into Eq. (46.22), we obtain

$$\frac{4\pi\lambda_L^2}{c}\frac{\partial j_y}{\partial z} - \frac{4\pi\lambda_z^2}{c}\frac{\partial j_z}{\partial y} = H(y,z), \tag{46.23}$$

where we have defined

$$\lambda_z^2 = \lambda_J^2 \equiv \frac{c\Phi_0}{8\pi^2 s j_m}; \tag{46.24}$$

λ_J is called the *Josephson penetration depth.* Using the fourth Maxwell equation we obtain from (46.23) an anisotropic London equation,

$$\lambda_J^2 \frac{\partial^2 H}{\partial y^2} + \lambda_L^2 \frac{\partial^2 H}{\partial z^2} = H(y,z). \tag{46.25}$$

The single vortex solution of Eq. (46.25) and the lower critical field can be computed by procedures similar to those outlined in Section 32.7. Note that in the Lawrence–Doniach model, the vortex core *resides in an insulating layer.* There is no normal core since there is no variation in the amplitude of the order parameter associated with the vortex. Figure 46.16 shows the contours of constant supercurrent for an isolated vortex arising from an in-plane magnetic field. To calculate the flux line energy and H_{c1}, we would assume that the natural cut-off scale is the periodicity of the superlattice, s. The resulting parallel lower critical field is then

$$H_{c1\parallel} = \frac{\Phi_0}{4\pi\lambda_L\lambda_J} ln\left(\frac{\lambda_L}{s}\right). \tag{46.26}$$

The above model may be extended to take into account the effect of a finite superconducting layer thickness (Clem and Coffey (1990)).[34]

[34] Viscous flux motion in a Josephson-coupled layer model of high-Tc superconductors, John R. Clem and Mark W. Coffey, *Phys. Rev. B* 42, 6209 (1990).

46.7.5 An isolated vortex for the field perpendicular to a thin film: the pancake vortex

In order to treat the layered system with H perpendicular to the layers we need to first obtain the magnetic field generated by the so-called 2d pancake vortex[35] in a thin superconducting film, as first discussed by Pearl (1964). We again employ the London approximation, where we regard $|\psi(\mathbf{r})|$ as a constant except in the small normal-core region. Writing $\psi(\mathbf{r}) = |\psi|e^{i\phi(\mathbf{r})}$, where ϕ is the phase of the order parameter, the second GL equation can then be written as

$$j = -\frac{c}{4\pi\lambda_L^2}\left(\mathbf{A} - \frac{\phi_0}{2\pi}\nabla\phi\right), \tag{46.27}$$

and

$$\lambda_L^2 = \frac{m^* c^2}{4\pi e^{*2}|\psi|^2} \tag{46.28}$$

is again the London depth. Consider a superconducting film of thickness, d, and infinite extent, located in the x–y plane (see Fig. 46.17). When $d \to 0$, we can define a sheet current density, **K**, as $\mathbf{K} = jd$ and replace Eq. (46.27) by

$$\mathbf{K} = -\frac{c}{2\pi\Lambda}\left(\mathbf{A} - \frac{\Phi_0}{2\pi}\nabla\phi\right); \tag{46.29}$$

here we have introduced a 2d screening length, defined as

$$\Lambda = \frac{2\lambda_L^2}{d} = \frac{m^* c^2}{2\pi e^{*2} n_{2D}^*} \tag{46.30}$$

and the sheet density of Cooper pairs as $n_{2D}^* = d|\psi(\mathbf{r})|^2$. We consider the case of a film containing a single vortex with one fluxoid, $(\hat{\mathbf{z}}\,\Phi_0)$, at the origin. The Maxwell equation is

$$\nabla \times \mathbf{H} = 0, \qquad z \neq 0. \tag{46.31}$$

Using $\mathbf{H} = \nabla \times \mathbf{A}$, we obtain

$$\nabla \times (\nabla \times \mathbf{A}) = 0, \qquad z \neq 0. \tag{46.32}$$

In cylindrical coordinates (ρ, θ, z) with the unit vectors, $\hat{\boldsymbol{\rho}} = \hat{\mathbf{x}}\cos\theta + \hat{\mathbf{y}}\sin\theta$, $\hat{\boldsymbol{\theta}} = \hat{\mathbf{y}}\cos\theta - \hat{\mathbf{x}}\sin\theta$, and $\hat{\mathbf{z}}$, the vector potential $\mathbf{A} = \hat{\boldsymbol{\theta}}A_\theta(\rho, z)$, and Eq. (46.32) take the form

$$\left(\frac{\partial^2}{\partial\rho^2} + \frac{1}{\rho}\frac{\partial}{\partial\rho} + \frac{\partial^2}{\partial z^2} - \frac{1}{\rho^2}\right)A_\theta(\rho, z) = 0; \qquad (z \neq 0). \tag{46.33}$$

[35] Pancake Vortices, J. R. Clem, *Journal of Supercond.*, **17**, 613 (2004).

This equation can be solved by separation of variables. We will not carry out the calculations (see Pearl), which involve matching the components of the magnetic field associated with the solution to (46.33) to the surface current densities, **K**, using the Maxwell boundary condition. Limiting expressions for the sheet current density, $K_\theta(\rho)$, are

$$K_\theta(\rho) \approx c\Phi_0/4\pi^2\Lambda\rho, \qquad \rho \ll \Lambda, \tag{46.34a}$$

and

$$K_\theta(\rho) \approx c\Phi_0/4\pi^2\rho^2, \qquad \rho \gg \Lambda. \tag{46.34b}$$

From Eq. (46.35a), the magnetic flux, $\Phi_z(\rho,0)$, through a circle of radius, ρ is found to be

$$\Phi_z(\rho) \approx \Phi_0(\rho/\Lambda), \qquad \rho \ll \Lambda \tag{46.35a}$$

and

$$\Phi_z(\rho) \approx \Phi_0(1 - \Lambda/\rho), \qquad \rho \gg \Lambda. \tag{46.35b}$$

The associated magnetic field,

$$\mathbf{H} = \hat{\boldsymbol{\rho}}H_\rho(\rho, z) + \hat{\mathbf{z}}H_z(\rho, z),$$

is sketched in Fig. 46.17.

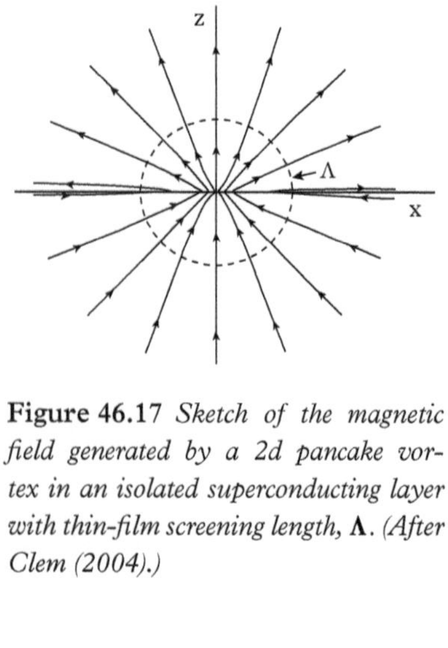

Figure 46.17 *Sketch of the magnetic field generated by a 2d pancake vortex in an isolated superconducting layer with thin-film screening length,* **Λ**. *(After Clem (2004).)*

From Eq. (46.34) we see that for small values of ρ the current is identical with the Abrikosov solution for bulk superconductors. However, the behavior at large distances is drastically different: instead of an exponential cut-off, we now find a slow $1/\rho^2$ decay of currents. The reason for this is that in films distant regions are electromagnetically coupled through free space, while in bulk superconductors the fields are diamagnetically screened by circulating superconducting electrons. Knowing the current distribution around a single vortex, one can calculated the free energy associated with a single vortex, E_1d. It turns out that E_1d has almost the same form as Eq. (32.58) and is given by $E_1d \approx (\Phi_0/4\pi\lambda_L)^2 \, ln(\lambda_L/\xi)$. However, the magnetization, M, associated with a single vortex in thin films is larger than that in a bulk superconductor by a factor R/d; in particular $M \approx \Phi_0 R/4\pi d$, where R is the film radius and d is the film thickness. Hence, the lower critical field, $H_{c1\perp}$ or a thin film superconductor is

$$H_{c1\perp} \approx \frac{\Phi_0}{4\pi\lambda_L^2} \cdot \frac{d}{R} ln\frac{\lambda_L}{\xi}, \tag{46.36}$$

which is a factor of d/R smaller than the bulk value.

46.7.6 An isolated vortex for the field perpendicular to a stack of layers

If we now go over to a lattice of superconducting planes and neglect the interlayer Josephson coupling, then we can obtain the field arising from a stack of pancake vortices simply by adding the fields from the Pearl solution. The resulting field lines arising from this construction are shown in Fig. 46.18.[36]

Flux lines in high temperature superconductors can be thought of as being built from stacks of pancake vortices, as shown in Fig. 46.19. Interestingly, if the magnetic field is tipped away from the z axis the circulating currents must stay within the individual layers in the absence of tunneling, and this tendency is retained in the presence of weak inter-layer coupling. Only the pancake origins shift to define the angle of the vortex axis. This situation persists to large angles before the pancake model breaks down and the currents penetrate the layers. Other arrangements are also predicted, as described by Clem.

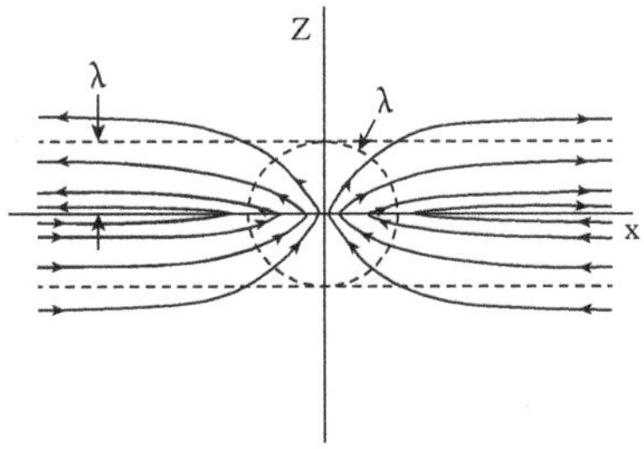

Figure 46.18 *The flux lines from a single pancake vortex together with that arising from an infinite stack of identical vortices positioned in layers above and below, spaced by distance, $s \ll \lambda$. (After Clem (1991).)*

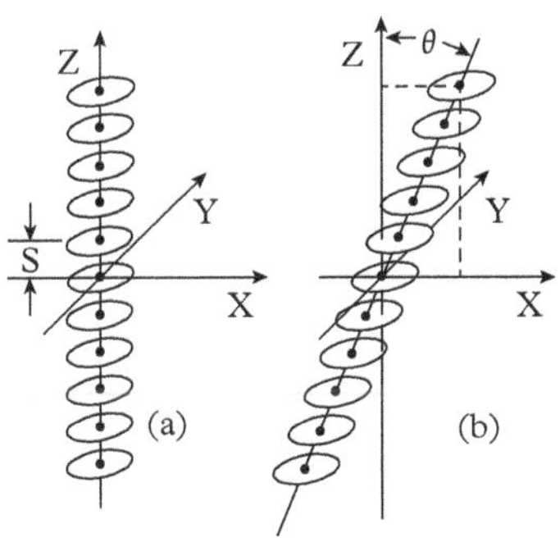

Figure 46.19 *(a) The array of stacked pancake vortices that constitute a single flux line in a layered superconductor for the case where the magnetic field lies along the z axis. (b) The array of stacked vortices that constitute a single flux line when the magnetic field is tilted by an angle, θ, relative to the z axis. Note that the plane normal associated with each pancake tends to remain aligned with the z axis. Only the origins shift to define the tilt angle. (After Clem (1991).)*

46.8 Vortex matter and phase diagrams

In this section we will discuss, qualitatively, the kinds of vortex matter states that are encountered in high temperature superconductors.[37] The associated phase diagrams tend to be complex and are still somewhat controversial. Two features combine to produce new features. Firstly, the much shorter coherence lengths, which can have atomic dimensions, lead to greatly weakened pinning to the larger-scale defects such as crystal grain boundaries, strain fields, and precipitates, which effectively pin vortices in conventional superconductors. Secondly, since vortices can be present at much higher temperatures in these materials, the lines tend to "vibrate" and can "shake loose" from their pinning sites, which as mentioned are already weaker.

Without pinning, and at higher temperatures even with pinning, the thermal agitation can cause the vortex lattice to melt for some range of fields and temperatures; we then have a *vortex liquid*, a new thermodynamic state. There will be

[36] Two-dimensional vortices in a stack of thin superconducting films: A model for high-temperature superconducting multilayers, J.R. Clem, *Phys. Rev. B* **43**, 7837 (1991).

[37] For a detailed discussion, see Vortex matter, by G. Blatter and V.B. Geshkenbein, in *Superconductivity, Volume 1*, K.H. Bennemann and J.B. Ketterson, Eds., Springer (2008).

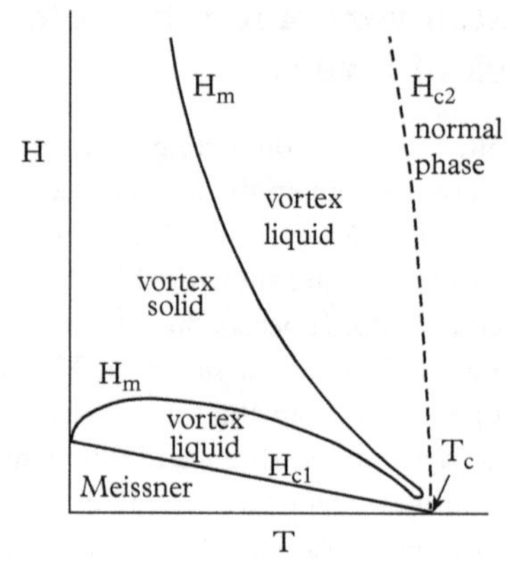

Figure 46.20 *Phase diagram of a high temperature superconductor having minimal pinning.*

some average number of lines per unit area but their cores will quickly become disordered at long range. In particular, this will be true just above the Meissner phase where the vortices are relatively far apart (and hence exponentially weakly interacting, see Section 32.6), as shown in Fig. 46.20; note the difference between this phase diagram and that for a conventional superconductor, shown earlier in Fig. 33.3. As the field increases, and the lines move closer to each other, their stronger mutual repulsion results in the formation of a lattice (a kind of 2d solid). On further increasing the field and temperature, the combination of a large London depth (that accompanies the short coherence length) results in a weakly modulated magnetic interaction, and the lattice melts. Finally, at still higher temperatures and fields we enter the normal phase; however the fluctuations lead to a broad transition (formally a crossover) which we have represented as a dashed line in Fig. 46.20. Resistivity experiments on high temperature superconductors at high fields always show a broadened transition.

When sufficient pinning is present the vortex solid phase is replaced by vortex glass phases. At lower fields, where the vortex–vortex interactions are stronger, a phase that has been termed a Bragg glass is deduced from the presence of 2d Bragg peaks in neutron diffraction experiments.[38] However, as is typical of 2d systems in general, the peaks are not sharp due to the (algebraic) behavior of the positional correlation function of the vortices at long range, as well as disorder along the vortex axes due to random local pinning. At higher fields where the vortex interactions are weaker but pinning remains sufficiently strong, lateral positional order gives way to a vortex glass phase. At higher temperatures both of the glass-like phases melt into a vortex liquid. A schematic phase diagram incorporating these elements is shown in Fig. 46.21.

The phenomenon of vortex creep in the presence of an applied current (see Section 32.8) is especially enhanced in high temperature superconductors due to the weak pinning. For this reason pinning sites, e.g., those introduced by radiation damage from energetic charged particles, are sometimes artificially introduced.

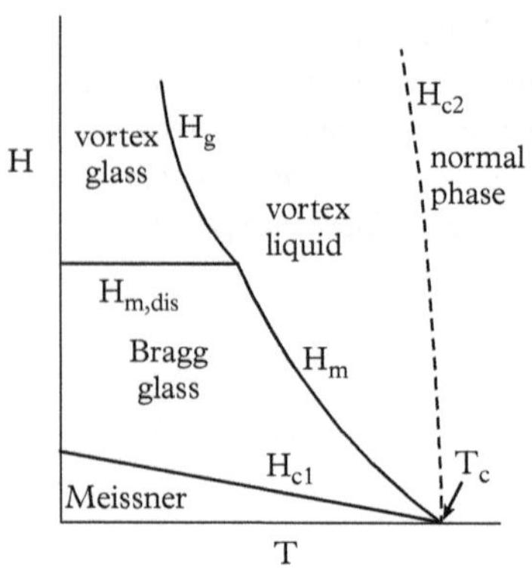

Figure 46.21 *Phase diagram of a high temperature superconductor having strong pinning.*

[38] A Bragg glass phase in the vortex lattice of a type II superconductor, T. Klein, I. Joumard, S. Blanchard, J. Marcus, R. Cubitt, T. Giamarchi, and P. Le Doussal, *Nature* 413, 404 (2001).

46.9 Non-cuprate high temperature superconductors

In addition to the cuprates, new classes of high temperature superconductors have emerged. Considerable excitement was generated when it was discovered that C_{60} becomes superconducting when intercalated with alkali metal atoms (see Chapter 49). The resulting compounds, A_3C_{60}, have transition temperatures of 18 K (K_3C_{60}),[39] 28 K (Rb_3C_{60}),[40] and 40 K (Cs_3C_{60}).[41,42] It is now generally accepted that these materials involve an extreme case of s-wave BCS phonon-mediated pairing.[43]

Somewhat later, the iron based pnictides (with transition temperatures as high as 56 K)[44,45,46] were discovered. The compounds have the chemical formula, REFeXO, where RE refers to a rare earth atom and X to As, P, or a chalcogen.

Two such iron based materials are LaFeAsO and $BaFe_2As_2$. Both materials have an atomic structure consisting of layers of Fe atoms sandwiched between As layers, as shown in Fig. 46.22. The pnictides are especially interesting since magnetic ions are generally regarded as being detrimental to superconductivity, suggesting the possibility of an unconventional (non-phonon-mediated) Cooper pairing. In fact the materials exhibit a strong interplay between magnetism and superconductivity, as demonstrated by phase diagrams, most easily generated by alloying. Figure 46.23 shows the phase diagram of $BaFe_2As_2$ generated by alloying with cobalt. Similar phase diagrams result from replacing As with F in LaFeAsO.

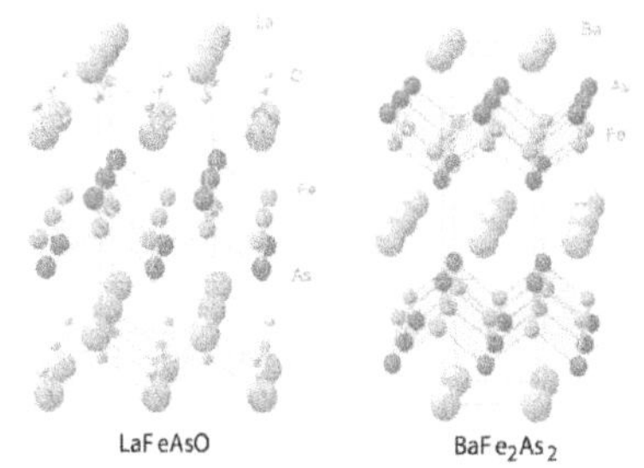

Figure 46.22 *The crystal structure of two pnictide superconductors*

[39] Superconductivity at 18 K in potassium doped C_{60}, A.F. Hebard, M.J. Rosseinsky, R.C. Haddon, D.W. Murphy, S.H. Glarum, T.T.M. Palstra, A.P. Ramirez, and A.R. Kortan, *Nature* **350**, 600 (1991).

[40] Superconductivity at 28 K in Rb_xC_{60}, M.J. Rosseinsky, A.P. Ramirez, S.H. Glarum, D.W. Murphy, R.C. Haddon, A.F. Hebard, T.T.M. Palstra, A.R. Kortan, S.M. Zahurak, and A.V. Makhija, *Phys. Rev. Lett.* **66**, 2830 (1991).

[41] Bulk superconductivity at 38 K in a molecular system, A.Y. Ganin, Y. Takabayashi, Y.Z. Khimyak, S. Margadonna, A. Tamai, M.J. Rosseinsky and. K. Prassides, *Nat Mater* 7 367 (2008).

[42] Ganin et al. report that the C_{60} molecules form a bcc lattice; together with the A atoms, the overall structure is tentatively identified as A-15. The technologically important compound Nb_3Sn, with $T_c = 18$ K also has this structure, as noted at the beginning of this chapter.

[43] Superconductivity in fullerides, O. Gunnarsson, *Rev. Mod. Phys.* **69**, 575 (1997).

[44] A.A. Kordyuk, *Low Temp Phys* **38**, 888 (2012).

46.10 Historical evolution of superconducting transition temperatures

We conclude this chapter with an overview of advances in the transition temperatures of superconducting materials. The highest superconducting transition temperature among the elements is for niobium: $T_c = 9.5$ K. At a much lower temperature we have tungsten, with $T_c = 0.012$ K.[47] In both cases the pairing mechanism is taken to be the electron–phonon interaction. Assuming no competition from magnetic ordering one cannot theoretically set a lower limit for a

[45] Iron based superconductors, Y. Kamihara, *J. Am. Chem. Soc* **128** 1012 (2006); Y. Kamihara, T. Watanabe, M. Hirano, and H. Hosono, *J. Am. Chem. Soc* **130** 3296 (2008).

[46] Superconductivity at 43 K in $LaO_{1-x}F_xFeAs$, H. Takahashi, et al., *Nature* **453**, 376 (2008); K. Ishida et al., Progress report on Fe– pnictides superconductivity, *J. Phys. Soc. Jpn.* 78, 062001 (2009).

[47] Critical-field measurement of superconducting tungsten as related to cerium-magnesium-nitrate temperature scale, W.C. Black, *Phys. Rev. Lett.* **21**, 28 (1968).

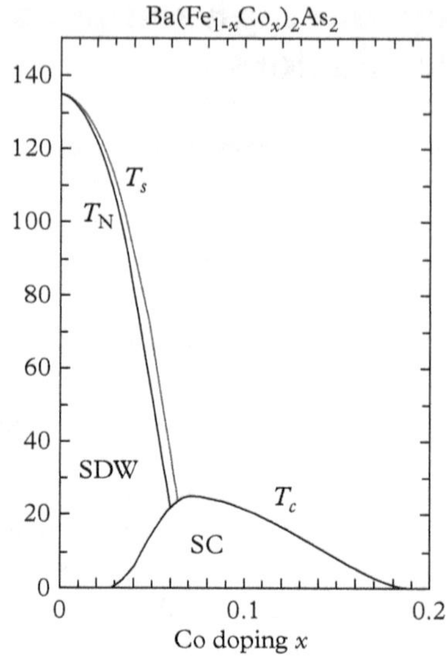

Figure 46.23 *The phase diagram of* $BaFe_2As_2$*, obtained by alloying with cobalt;* T_s*,* T_c*, and* T_N *denote structural, superconducting, and antiferromagnetic transitions.*

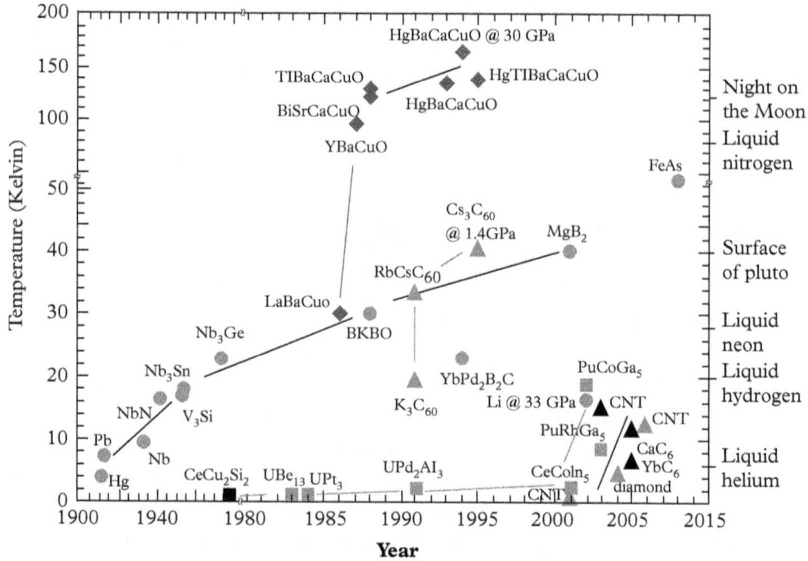

Figure 46.24 *A plot of materials with higher transition temperatures versus time. Important landmarks include the discoveries of the A15, cuprate, fullerene,* MgB_2*, and pnictide compounds. Shown also are heavy fermion and other materials which are of interest from the perspective of alternative pairing mechanisms, or are otherwise unique. (From the 2007 DOE report:* Basic Research Needs for Superconductivity.*)*

material like copper. The possibility that spin fluctuations might lead to unconventional superconductivity has been examined theoretically[48] and searched for experimentally[49] in materials like palladium, without success.

Figure 46.24 collects the transition temperature of trendsetting materials plotted according to the year in which they were discovered. Of course the most abrupt departure from a gradual overall increase was the discovery of the

[48] Possibility of triplet pairing in palladium, D. Fay and J. Appel, *Phys. Rev. B* **16**, 2325 (1977).

[49] Very low-temperature search for superconductivity in Pd, Pt, and Rh, R.A. Webb, J.B. Ketterson, W.P, Halperin, J.J. Vuillemin, and N.B. Sandesara, *J. Low Temp. Phys.* **32**, 659 (1978).

cuprates. But this should not deflect attention from other unusual classes of compounds that may set the stage for future advances. Historically, theory has not reliably pointed the way toward higher transition temperatures; such advances remain in the hands of experimentalists.

SUPPLEMENTARY READING

Experimental and theoretical aspects of high temperature superconductivity are extensively discussed in the various chapters of *Superconductivity, Volumes 1 and 2*, edited by K.H. Bennemann and J.B. Ketterson, Springer (2008). A standard reference is *Physical Properties of High Temperature Superconductors Vol. I –V*, D.M. Ginsburg, World Scientific, New York. P.W. Anderson has summarized a wide variety of experimental facts, with arguments as to their relative importance, along with his perspective on how the theory should be formulated, in *The Theory of Superconductivity in High T_c Cuprates*, Princeton University Press (1997). The electronic structure of cuprates is extensively discussed in: *Theory of Copper Oxide Superconductors*, H. Kamimura, H. Ushio, S. Matsuno, and T. Hamada, Springer (2005). Finally, the experimental and theoretical aspects of the subject are objectively reviewed in, *Quantum Liquids: Bose Condensation and Cooper Pairing in Condensed Matter Systems*, A.J. Leggett, Oxford University Press (2006).

47 Artificially Structured and Patterned Materials; Surfaces and Interfaces

47.1 Artificially layered materials

Elements, compounds, and mixtures (alloys) formed under conditions of thermodynamic equilibrium have occupied the bulk of our attention in this book. However techniques now exist, and others are rapidly evolving, for growing materials under controlled but *non-equilibrium* conditions. One category is artificially layered materials, which can be formed in various ways by exposing some substrate to alternating incoming mass fluxes from two (or more) sources which sequentially deposit on that substrate. Such structures may be prepared as an individual layer buried in an otherwise uniform second constituent, or as systems consisting of multiple layers both periodic and non-periodic. Furthermore the constituents may be chosen from the same or different materials categories including: metals, superconductors, semiconductors, insulators, ferromagnets and antiferromagnets, and ferroelectrics and piezoelectrics. The structures may have varying degrees of crystallographic order, starting from a perfect artificial superlattice (see Section 44.2), through various degrees of disorder including textured (where one crystallographic axis remains oriented with respect to the growth direction), through polycrystalline, to amorphous.

47.2 Artificial superlattices

Consider an idealized structure composed of m atomic planes (or molecular planes for compounds) of a material, A, alternated with n planes of a second material, B; it can be designated as A_mB_n. The resulting materials have been referred to by a variety of names including composition modulated alloys and artificial superlattices (or superlattices for short).[1] Figure 47.1 shows, schematically, a structure A_3B_4, composed of three and four atomic planes of elements A and B (filled and unfilled circles), which repeats itself through N periods.

As depicted, the materials in Fig. 47.1 have the same crystal structure (in this case square) with the same atomic spacings. However in practice one never has

[1] Technically, the term superlattice refers to a structure that forms under thermodynamic equilibrium conditions and has an oscillating composition profile with a repeat distance larger than what would otherwise be the basic unit cell of the material.

The Physics of Solids. First Edition. J. B. Ketterson.

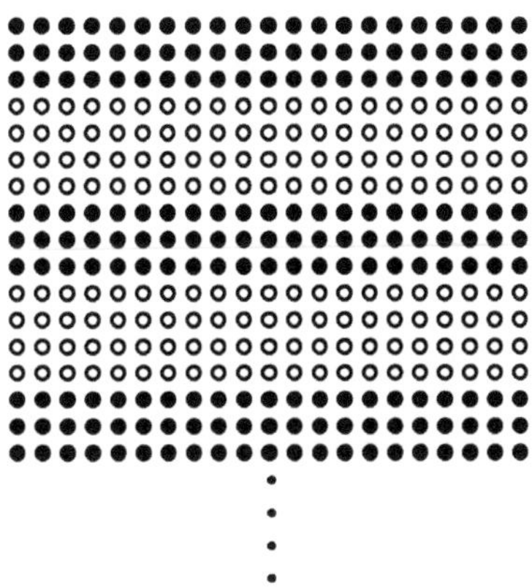

Figure 47.1 *An artificial superlattice, designated,* A_3B_4, *with alternating layers of materials A and B (filled and solid circles) having 3 and 4 atomic planes each and repeated through N periods.*

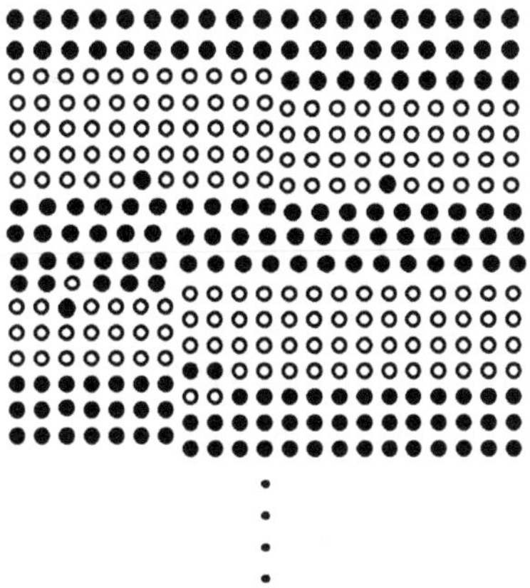

Figure 47.2 *A more realistic depiction of an artificial superlattice showing: incomplete planes, local distortions, and inter-diffusion.*

such idealized structures. For example, the atomic planes may not be complete,[2] and the materials may inter-diffuse or chemically react. A more representative situation involving different atomic sizes, incomplete planes, and interdiffusion is shown in Fig. 47.2. Even if the planes are complete, misfit dislocations can be present. Such dislocations involve a *miss-registry* between the two species at the interface due to the accumulated strain energy associated with a lattice mismatch and will always be present when the sublayers become thicker. The interfaces can be in complete registry, with no misfit dislocations present (so-called coherent structures) *if* the layers are thin enough and the lattice constants well enough matched.[3]

[2] Complete atomic planes would require an essentially unachievable level of control over the incoming fluxes and the shutters that interrupt them, which together control the number of atomic planes deposited. A special case is so-called atomic layer epitaxy in which, for reasons of chemical stoichiometry, only a single atomic plane of a chemical compound will deposit, say Cd, which is then followed by a second layer having the same property, say S, to form the compound, CdS in the present example. One can then form some specified number of molecular planes of this compound and then follow it with a specified number of molecular planes of a second chemical compound having similar deposition characteristics. A condition is that the separate atomic species have high vapor pressures relative to the compound(s) they form at the temperature the deposition is carried out; therefore the technique is of limited utility in practice.

47.3 Multilayer semiconductors

The most widely studied superlattices involve alternating semiconductor layers. Particular emphasis has been given to the III–V system, involving the semiconductors GaAs and AlAs or the various alloys, $Ga_{1-x}Al_xAs$. What makes this latter system unique is that the lattice constants of the constituents are exceedingly well matched, their room temperature zinc blende values being,

[3] The accumulated strain energy due to a lattice mismatch is proportional to the layer volume, L^2a, where a is a layer thickness and L is an in-plane dimension (L^2 being the interface area). The energy of a misfit dislocation is proportional to its length, L and the number of them is also proportional to L for a total contribution proportional to L^2. Hence if the thickness is small enough misfit dislocations will not nucleate.

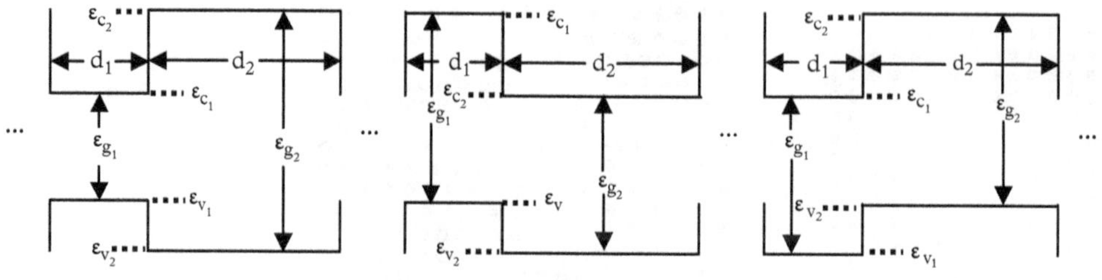

Figure 47.3 *Three classes of band alignments for left (1) and right (2) layers of a multilayer semiconductor system, as determined by the parameters* ε_{v1}, ε_{v2}, ε_{c1} *and* ε_{c2}. *They are classified as: Type I,* $\varepsilon_{v1} > \varepsilon_{v2}$, $\varepsilon_{c1} < \varepsilon_{c2}$; *Type II,* $\varepsilon_{v1} > \varepsilon_{v2}$, $\varepsilon_{c1} > \varepsilon_{c2}$; *and Type III,* $\varepsilon_{v1} < \varepsilon_{v2}$, $\varepsilon_{c1} < \varepsilon_{c2}$.

$a_{GaAs} = 5.6532$ and $a_{AlAs} = 5.6622$[4,5]. However their direct band gaps differ substantially: $\varepsilon_g^{GaAs} = 1.35$ eV and $\varepsilon_g^{AlAs} = 2.16$ eV. Both the lattice constants and band gaps for the alloys approximately obey a linear interpolation. The ability to engineer the band gap of the constituents of a layered structure through alloying is the basis of a wide array of device applications; particularly important are lasers, an example of which is discussed later in this chapter.

Other semiconductor combinations are also important and they can be sorted into three classes, as we now discuss. Figure 47.3 shows, schematically, the position-dependent band gap in two adjacent semiconductor layers for the three classes, designated as type I, type II, and type III. What differentiates them is the positioning of the valence and conduction band edges in the two materials: ε_{v_1}, ε_{v_2}, ε_{c_1}, and ε_{c_2}. AlAs/GaAs is a type I system. As we will see, the sub-layer thicknesses, d_1 and d_2, affect the electron and hole energy levels within the two layers.

47.4 Quantum wells; modulation doping

Here we will restrict our discussion to a type I quantum well system. Suppose we have a thin narrow gap layer, e.g., GaAs, sandwiched between two thick wider gap layers, $Al_xGa_{1-x}As$. For states lying between $\varepsilon_{c_{GaAs}}$ and $\varepsilon_{c_{Al_xGa_{1-x}As}}$, the electrons cannot propagate into the $Al_xGa_{1-x}As$ layer (the same is true for hole states lying between $\varepsilon_{v_{GaAs}}$ and $\varepsilon_{v_{Al_xGa_{1-x}As}}$, but that will not interest us here). We initially take the limit, $\varepsilon_{c_{Al_xGa_{1-x}As}} \to \infty$, in which case the conduction band wave functions are completely confined to the GaAs layers. Using an infinite "square well" model, we can then write the wave functions and energy levels as

$$\psi_{k,n} = \sqrt{\frac{2}{d_{GaAs}}} \sin \frac{n\pi z}{d_{GaAs}} e^{i(k_x x + ik_y y)} \tag{47.1a}$$

and

$$\varepsilon_{k,n} = \varepsilon_{c_{GaAs}} + \frac{\hbar^2\pi^2}{2d_{GaAs}^2 m_e} n^2 + \frac{\hbar^2}{2m_e}\left(k_x^2 + k_y^2\right), \tag{47.1b}$$

[4] The more stable structure for AlAs is wurtzite.

[5] This property is consistent with Pauling's covalent radii, listed in Table 7.5, where the values for Ga and Al are both given as 1.26 A. Although not as closely matched, the lattice constants for the other group V elements, N, P, and Sb when combined with group III, Ga and Al, are also well matched. Hg and Cd in the II–VI system also have similar covalent radii.

where d_{GaAs} is the GaAs layer thickness and n the number of half-wavelengths in the layer. An electron confined to a layer is said to be in a *quantum well* and the different n values are referred to as *minibands*. Note the wave function (47.1a) is in reality an "envelope wave function" in that full wave function is still modulated by the potentials associated with individual atoms. When $\varepsilon_{c_{Al_xGa_{1-x}As}}$ is finite, the electron wave function penetrates into the layers adjacent to the well. This is the analogue of the one-dimensional finite square well.

Now the band structure of bulk GaAs was discussed in Section 11.4 (see Fig. 11.15) and the band gap is given by the energy separating the doubly degenerate Γ_8 holes and the singly degenerate Γ_6 electrons. Under confinement, the Γ_8 level splits into a light and heavy level with the light level lying highest. In what follows we limit ourselves to treating the electron states, which for k vectors in the vicinity of the Γ point in the bulk can be written in the form

$$\varepsilon(\mathbf{k}) = \varepsilon_c + \frac{\hbar^2 k^2}{2m_e}; \tag{47.2}$$

this suggests a model Hamiltonian for the quantum well of the form

$$\varepsilon(\mathbf{k}) = \varepsilon_c + \frac{\hbar^2}{2m_e}\left(-\frac{\partial^2}{\partial z^2} + k_x^2 + k_y^2\right). \tag{47.3}$$

Here, ε_c is the energy of the conduction band edge and m_e is an (isotropic) effective mass. However, to treat this problem one needs to use different effective masses in the two layers together with the conduction band offset of the $Al_xGa_{1-x}As$ relative to GaAs layer (which represents the barrier height), along with some interface boundary conditions.[6] Figure 47.4 shows the results of such

[6] In addition to the effective masses, boundary conditions are required to determine the wave functions in successive layers. This topic is discussed in Superlattice band structure in the envelope-function approximation, G. Bastard, *Phys. Rev. B* **24**, 5693 (1981). One commonly used set is to match the wave function amplitude and the probability current at each interface. However, interface states (see Section 47.9) can form under some circumstances, which complicates matters, as discussed in Effects of quasi-interface states in HgTe-CdTe superlattices, Yia-Chung Chang, J.N. Schulman, G. Bastard, Y. Guldner, and M. Voos, *Phys. Rev. B* **31**, 2257 (1985) and Interface states and subbands in HgTe-CdTe Heterostructures, Y.R. Lin-Liu and L.J. Sham, *Phys. Rev. B* **32**, 5561 (1985).

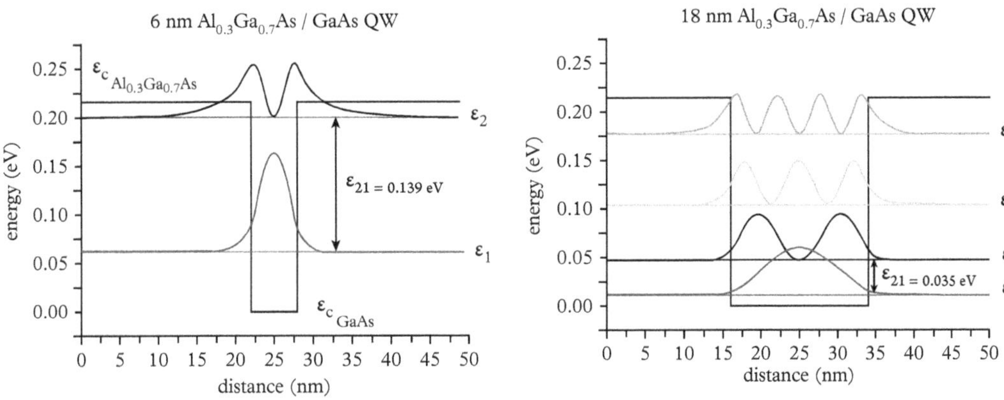

Figure 47.4 *The bound state probability density* $|\psi_{k=0,n}|^2$ *and energy levels for a GaAs quantum well sandwiched between* $Al_{0.3}Ga_{0.7}As$ *layers for two different well thicknesses, 6nm and 18nm, along with the energy of the first excited state. (Courtesy of Stefan Birner, nextnano GmbH.)*

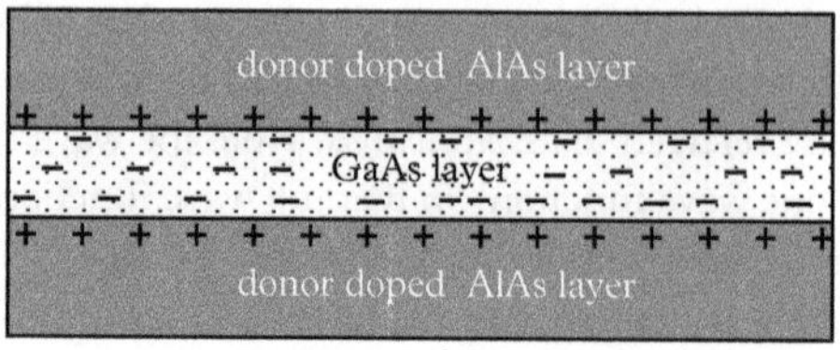

Figure 47.5 *Modulation doping: electrons from ionized donors in an $Al_{1-x}Ga_xAs$ layer fall into an adjacent GaAs layer where they create mobil electrons near the interface.*

a treatment for a GaAs layer between two $Al_{0.3}Ga_{0.7}As$ layers for two different thicknesses of the former. If the electron has an energy greater than $\varepsilon_{c_{AlAs}}$ then its motion is no longer confined to the GaAs well; i.e., the electron propagates throughout the structure.

Quantum wells allow a form of "remote doping" not available in a homogeneous semiconductor, as shown schematically in Fig. 47.5. By adding an n-type dopant to $Al_xGa_{1-x}As$ layers in the region immediately adjacent to a GaAs layer (so-called δ-function doping), Si being commonly used, electrons will ionize directly (non-thermally) and fall into miniband states (those quantized perpendicular to the layers, see Fig. 47.4) in the GaAs layers. Since the resulting positive donor ions are now remotely located from the electron gas in the GaAs layer, Coulomb scattering is thereby greatly reduced; the resulting electron mobilities can then greatly exceed those obtainable in bulk; values of more than $\approx 10^7 \mathrm{cm}^2/\mathrm{volt\,sec}$ having been achieved at low temperatures. Electrons are still confined to the interfaces by the donor potential with a density falling off with a characteristic screening length, λ_s (see Problem 47.1.); provided that $d_{GaAs} \ll \lambda_s$, this will minimally affect the position dependence of the electron density in the well. Figure 47.6 shows how the limiting mobilities have increased with time as material purity (largely governed by background pressure in the MBE deposition chamber during deposition[7]) has improved.[8] The high mobilities achieved have led to new physical phenomena, the most dramatic of which are the integer and fractional quantum Hall effects, to be discussed in Chapter 48.

47.5 Multiple wells and superlattices

One can go on to fabricate multiple quantum wells. Figure 47.7 shows the case of a double-well $Al_{0.2}Ga_{0.8}As/GaAs/Al_{0.2}Ga_{0.8}As/GaAs/Al_{0.2}Ga_{0.8}As$ structure. Note that the ground state wave function now involves symmetric and antisymmetric combinations relative to the point midway between the wells.

The limiting case of a multi-well structure is a superlattice; the levels, n, in Eq. (47.1b) then spread out into so-called minibands. As an approximate description for the motion perpendicular to the layers one can then use a Kronig–Penney-like model (see Section 10.8) by suitably matching the probability current at the interfaces and keeping in mind that the effective masses differ in the two layers.

[7] MBE, or molecular beam epitaxy, is simply the deposition from a heated source through a small orifice so that the emerging flux of atoms is approximately in thermal equilibrium with the source. This results in softer impacts of the atoms deposited on the substrate and, in many cases, higher crystalline quality. To minimize the contamination from background gases in the vacuum chamber, deposition is generally carried out in ultra high vacuum, in the 10^{-10} Torr range. The combination of these effects, together with pure constituents, then leads to the extraordinary mobilities achieved.

[8] The role of MBE in recent quantum Hall effect physics discoveries, L. Pfeiffer and K.W. West, *Physica E* **20**, 57 (2003).

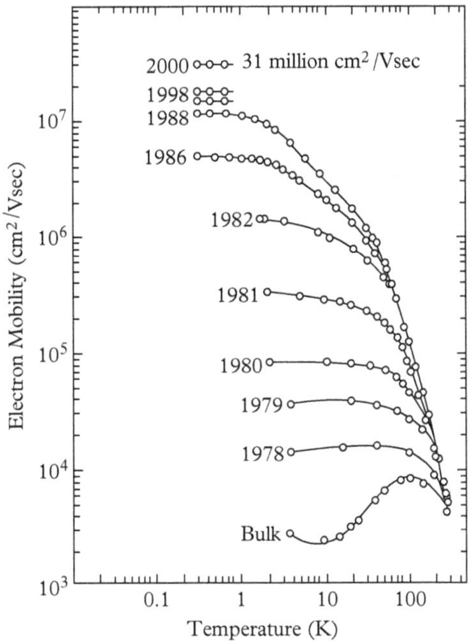

Figure 47.6 *The temperature dependence of the mobility of GaAs quantum wells as various improvements were introduced. (After L. Pfeiffer and K.W. West) (2003).*

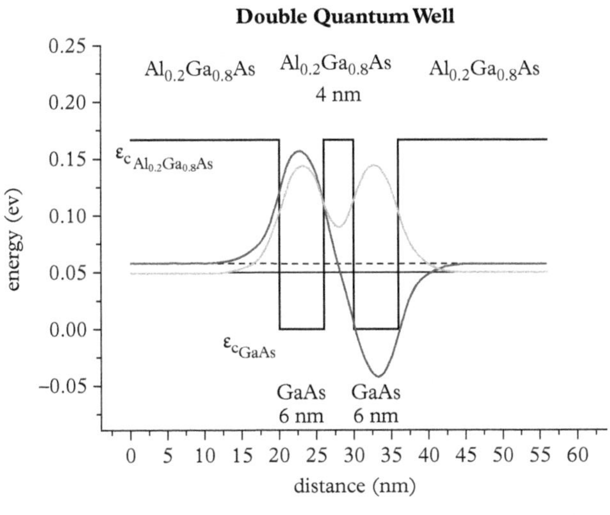

Figure 47.7 *A five-layer structure with a* $Al_{0.2}Ga_{0.8}$ *barrier and two* GaAs *wells. Note that what was the lowest state in the single well now splits into a symmetric and antisymmetric combination, due to inter-layer coupling. (Courtesy of Stefan Birner, nextnano GmbH.)*

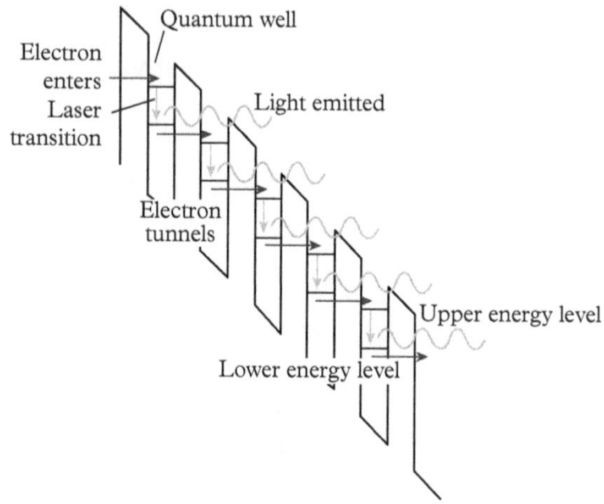

Figure 47.8 *The quantum cascade laser, in which electrons resonantly tunnel among successive quantum wells while simultaneously stimulating the growth of an accompanying coherent light wave.*

Multi-well structures have important applications. One interesting example is the so-called *quantum cascade laser* (QCL).[9] Figure 47.8 shows schematically how such a device operates. We first note that in the presence of a perpendicular electric field the bottom of successive wells is Stark shifted (tilted) by the applied field, the overall slope being fixed by the magnitude of the applied voltage. We now assume that an electron is injected into the first well. It is then free to tunnel into the adjacent well. However, the tunneling rate will be greatly enhanced if there is a miniband level in the second layer that is *degenerate* with one in the first layer, which we here assume to be the ground and first excited state respectively. This corresponds to *resonant tunneling*. To conserve energy the electron emits a light quantum, $\hbar\omega = \varepsilon_1^n - \varepsilon_2^{n+1}$, where n denotes successive layers. The electron is then free to tunnel to successive layers, and a coherent propagating light field builds up, which has the additional effect of enhancing the tunneling via stimulated emission.[10]

Another application of (insulating) multilayer structures is to make photonic cavities, where the layers serve as so-called Bragg mirrors; an example of which was discussed earlier, in Section 31.9. For normal incidence one can use the Fresnel equations of optics to match the electric and magnetic fields at the interfaces of successive layers. For periodic structures, one again has a Kronig–Penney-like model which has "stop bands" when the wavelength of the light lies in the optical band gap of this one-dimensional lattice. This effect is now widely used to make optical filters and high reflectivity mirrors.

47.6 Metallic multilayers

Modern thin film deposition methods, particularly sputtering,[11] allow the sequential deposition under clean conditions of successive layers of different classes of materials. The thickness of the resulting layers can range from subatomic

[9] Quantum cascade laser, J. Faist, F. Capasso, D.L. Sivco, C. Sirtori, A.L. Hutchinson, and A.Y. Cho, *Science* **264**, 553 (1994).

[10] Tunneling is still possible when the levels in adjacent wells are not degenerate; however, it will require the emission of phonons in addition to the light photon in order to allow energy conservation. Since multiple quanta are then involved the tunneling rate is greatly reduced.

[11] Sputtering is a technique whereby a plasma, typically formed from argon, is maintained above some target material, usually in the presence of a confining magnetic field. Impacts from ionized atoms on the substrate, resulting from an applied voltage, eject target atoms that then fall on a substrate. If the process is carried out in the presence of a reactive gas, compounds can be formed; e.g., Nb sputtered in a N environment results in the deposition of NbN. If sputtering is carried out with an r.f. field, an insulating target can be used.

(corresponding to a partial layer) to many microns, the latter limited only by the deposition time or film stability. As noted at the beginning of this chapter, this can involve a variety of material classes, depending on various physical and chemical compatibility factors. Either single or repeated interfaces between such materials can be formed.

Here we will briefly describe a few of the many different materials combinations that have been formed and studied. Examples include combinations consisting of normal metals (N), superconductors (S), and ferromagnets (F), or antiferromagnets (A). From the perspective of strongly affecting electronic properties, the combinations involving competing order parameters (or order versus no order) are most interesting; among the binary systems we then have N/S, N/F, N/A, F/S, A/S, and F/A. The broken symmetry associated with an S, F, or A symmetry has a characteristic coherence length and when this length becomes comparable with a layer thickness, the order present in one layer competes with that in the adjacent layer.

47.6.1 Superconducting multilayers

As an example of competing phases, consider the N/S system, Nb/Cu. Although these materials have different lattice constants and different crystal structures (Nb is bcc and Cu is fcc) it turns out that they form nicely segregated, but textured, layers ((110) for Nb and (111) for Cu). Figure 47.9 shows the superconducting transition temperature of equal thickness Nb/Cu multilayers as a function of the layer thickness.[12] Note that the transition temperature drops as the layer thickness approaches a characteristic coherence length in the superconductor, as is expected on physical grounds. Shown also is the result of applying[13] superconducting proximity effect theory[14] to this system, taking into account the natural suppression of the transition temperature in thin Nb films.[15]

[12] Superconductivity of Nb/Cu superlattices, I. Banerjee, Q.S. Yang, C.M. Falco, and I.K. Schuller, *Solid State Commun.* **41**, 805 (1982).

[13] Calculation of the transition temperature for artificial metallic superlattices in the dirty limit: application to Nb/Cu, P.R. Auvil and J.B. Ketterson, *Solid State Commun.* **67**, 1003 (1988).

[14] Theory of the upper critical-field of superconducting superlattices, S. Takahashi and M. Tachiki, *Phys. Rev. B* **33**, 4620 (1986).

[15] Superconducting tunneling in ultrathin Nb films, S.I. Park and T.H. Geballe, *Phys. Rev. Lett.* **57**, 901 (1986).

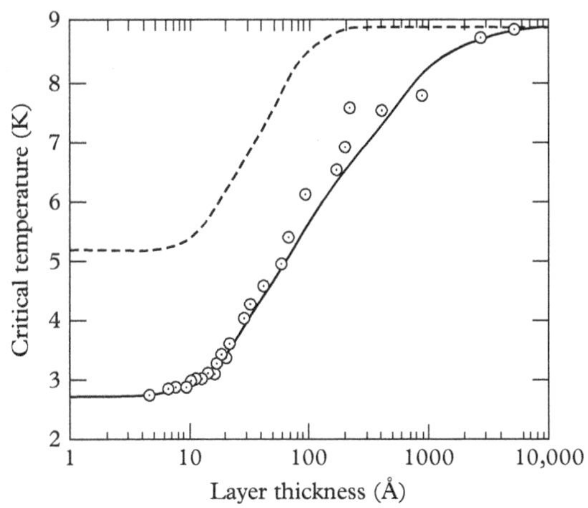

Figure 47.9 *The points show the data of Banerjee et al. (1982) for the superconducting transition temperature of artificial superlattices, as a function of the layer thickness, t. The solid line shows a fit to proximity effect theory by Auvil and Ketterson (1988). The dashed line shows the observed suppression of the transition temperature in thin Nb layers measured by Park and Geballe (1986) which is incorporated into the fit.*

47.6.2 Magnetic multilayers

The study of magnetic multilayers has turned out to be very important, particularly the phenomenon that came to be called giant magnetoresistance (GMR).[16] Although there were hints that something interesting could happen in earlier Cu/Ni data[17], observation of an unusually large magnetoresistance at 4.2 K in Fe/Cr multilayers by Fert and co-workers[18] focused interest. Figure 47.10 shows their measured in-plane magnetoresistance for three Fe/Cr multilayers. Note the significant change in resistance with magnetic field. This was followed by measurements on Fe/Cr/Fe trilayers by Gruenberg and co-workers,[19] which showed large effects even at room temperature, as shown in Fig. 47.11. With continuing experimentation on materials combinations, even larger changes in the room temperature resistance were discovered. The effect has been exploited to sense magnetized regions (bits) in the read heads of magnetic disc recorders.

These experiments involved in-plane electrical transport, as shown schematically in Fig. 47.12(a). One can also consider perpendicular transport where the layers are in series, as shown in Fig. 47.12(b).

Several effects contribute to the observed behaviors. In the absence of a magnetic field the lowest energy, magnetically, is for the magnetizations in the two layers (or alternate layers in a superlattice) to point in opposite directions, as is the tendency of a pair of adjacent bar magnets. This would be strongly reinforced in the presence of antiferromagnetic coupling through the thin Cr layer. For simplicity we will assume a bar geometry where, due to shape anisotropy, the

[16] The 2007 Nobel Prize in Physics was awarded to Albert Fert and Peter Grünberg "for the discovery of giant magnetoresistance." For a discussion of the history assembled see: *Scientific Background on the Nobel Prize in Physics 2007: The Discovery of Giant Magnetoresistance*, compiled by the Class for Physics of the Royal Swedish Academy of Sciences.

[17] Transport properties of the compositionally modulated alloy Cu/Ni, I.K. Schuller, C.M. Falco, J. Williard, J. Ketterson, B. Thaler, R. Lacos, and R. Dee, *AIP Conf. Proc.* **53**, 417 (1979).

[18] Enhanced magnetoresistance in layered magnetic structures with antiferromagnetic interlayer exchange, G. Binasch, P. Grünberg, F. Saurenbach, and W. Zinn, *Phys. Rev. B* **39**, 4828 (1989).

[19] Giant magnetoresistance of (001)Fe/(001)Cr magnetic superlattices, M.N. Baibich, J.M. Broto, A. Fert, F. Nguyen van Dau, F. Petroff, P. Eitenne, G. Creuzet, A. Friederich, and J. Chazelas, *Phys. Rev. Lett.* **61**, 2472 (1988).

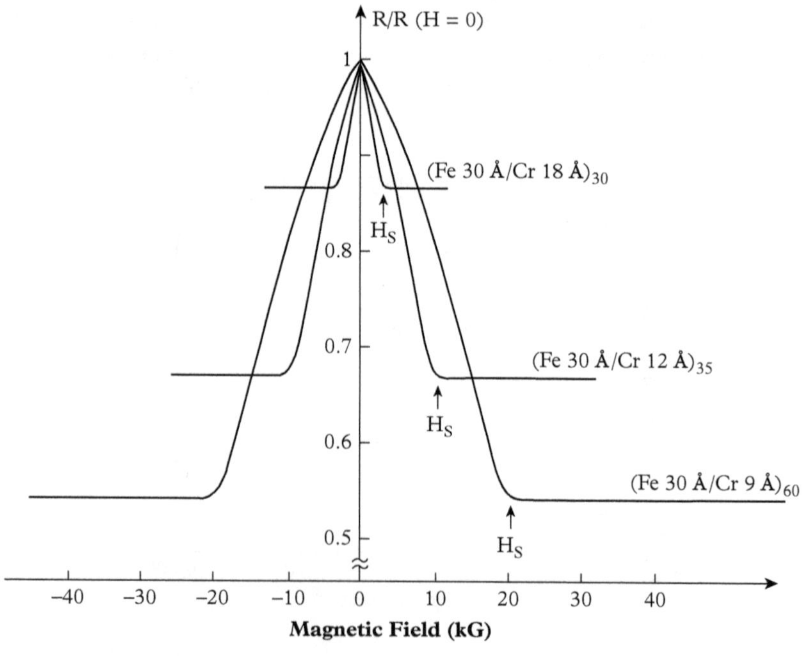

Figure 47.10 *The magnetoresistance of three Fe/Cr superlattices at 4.2 K for layer thicknesses and layer numbers indicated. The current and the applied field are along the same [110] axis in the plane of the layers. (After Baibich et al. (1988).)*

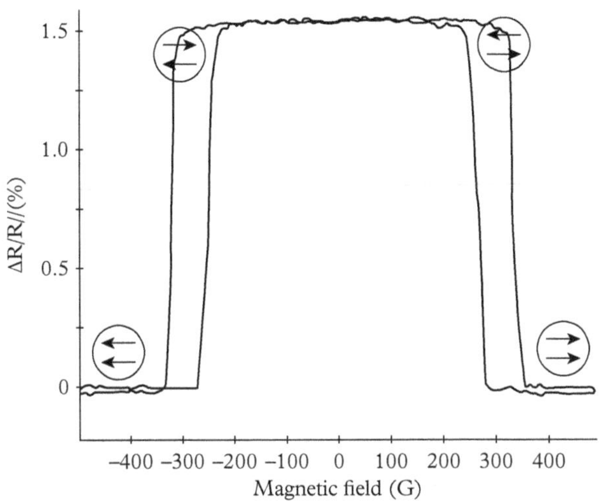

Figure 47.11 *The fractional change in magnetoresistance, $\Delta R/R_{//}$, of two 12 nm antiferromagnetically coupled Fe layers via 1 nm Cr. In the absence of a magnetic field the magnetization vectors in the layers are anti-aligned, while at high fields they are aligned. (After Binasch et al. (1989).)*

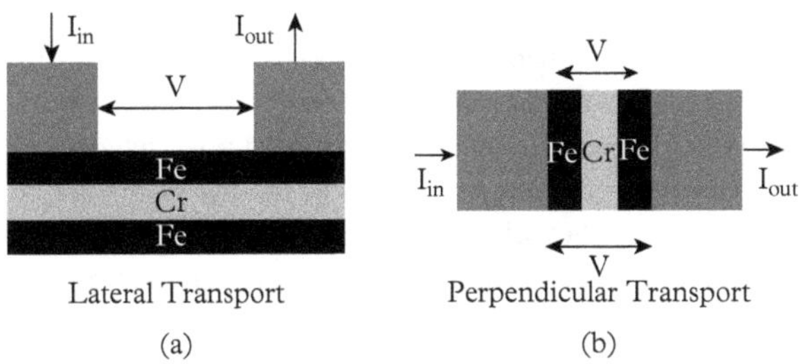

Figure 47.12 *A schematic representation of the two device geometries where: (a) the current flows in-plane, and (b) the current flows perpendicular to the layers. The overall resistance is larger for lateral transport but the fractional change in resistance can be larger for perpendicular transport. The contact layers are typically made from copper.*

magnetizations prefer to align parallel to the long axis.[20] If a magnetic field is applied along the bar axis there will be a field at which the anti-aligned layer reorients to lie parallel to the field. How this will affect the bar resistance is somewhat subtle and depends on various parameters, including the reflection and transmission characteristics through the layers of the spin up and spin down bands. On the basis of wave function symmetry alone, and in the absence of a spin flip on crossing the Cr, the matrix element associated with transmission of spin up and spin down electrons across the barrier to states with the *opposite* spin on the other side of the barrier (the configuration at low fields) is expected to be small. Electrons would then tend to be localized in the individual layers, with each layer having a resistance limited by interface scattering. On the other hand, at high fields transmission is more likely since it involves coupling of the same spin sub-bands. The effective mean free path then tends to be limited by the *combined* thickness of the layers. Based on this we expect a higher resistance for the (low-field) anti-aligned configuration, as observed experimentally. Devices based on the second geometry shown in Fig. 47.12(b) tend to be dominated by differing overall effective resistances of the middle layer when coupling either aligned or anti-aligned magnetic layers. Since such effects are greatly enhanced in insulating layers, we move directly to a discussion of those.

[20] Here we assume that the bar is short enough for axial domains not to form.

47.7 Spin tunneling

Rather than coupling two magnetic metal layers by conduction through a metallic layer, higher fractional changes in the resistance, $\Delta R/R$, as well as higher overall resistances, can be achieved by *quantum mechanical tunneling* through an insulating barrier.[21] Schematically, the metallic Cr layer is replaced by an insulator (Fig. 47.12(b)). Such structures are of interest as read heads in hard disk drives and as elements in magnetoresistive random access memory devices. Much research has gone into identifying optimal materials and conditions for maximizing the response.[22] Large responses at cryogenic temperatures have been obtained for various material combinations, but for most applications it is crucial that devices exploiting tunneling operate at room temperature. In this regard large responses have been obtained with 3d metals, particularly Fe and alloys thereof, combined with refractory oxides, such as Al_2O_3[23] and MgO;[24,25] the latter is particularly attractive since for growth along [100] there is a reasonably good lattice match between the MgO, which has the NaCl structure and bcc Fe, if the latter is rotated by 45° about [100].

Figure 47.13 shows the majority and minority spin sub-bands in a one-band model of the electronic structure of the ferromagnetic layers on each side of a barrier, across which there is an externally imposed potential drop, V. The case, $\mathbf{M}_L = \mathbf{M}_R$, applies when the magnetizations in the two layers are parallel, while the case, $\mathbf{M}_L = -\mathbf{M}_R$, applies when they are anti-parallel. It is important to emphasize that when the magnetizations are aligned, majority (minority) spins in one layer are aligned with majority (minority) spins in the other; however when

[21] Tunneling between ferromagnetic films, M. Jullière, *Phys. Lett. A* 54, 225–226 (1975).

[22] For a non-technical review, see Magnetic tunnel junctions, Jian-Gang Zhu and Chando Park, *Mater. Today* 9(11), 36 (2006).

[23] Large magnetoresistance at room temperature in ferromagnetic thin film tunnel junctions, J.S. Moodera, Lisa R. Kinder, Terrilyn M. Wong, and R. Meservey, *Phys. Rev. Lett.* 74, 3273 (1995).

[24] Giant room-temperature magnetoresistance in single-crystal Fe/MgO/Fe magnetic tunnel junctions, Shinji Yuasa, Taro Nagahama, Akio Fukushima, Yoshishige Suzuki, and Koji Ando, *Nat. Mater.* 3, 868 (2004).

[25] Giant tunnelling magnetoresistance at room temperature with MgO (100) tunnel barriers, S.P. Stuart Parkin, Christian Kaiser, Alex Panchula, Philip M. Rice, Brian Hughes, Mahesh Samant, and See-hun Yang, *Nat. Mater.* 3, 862 (2004).

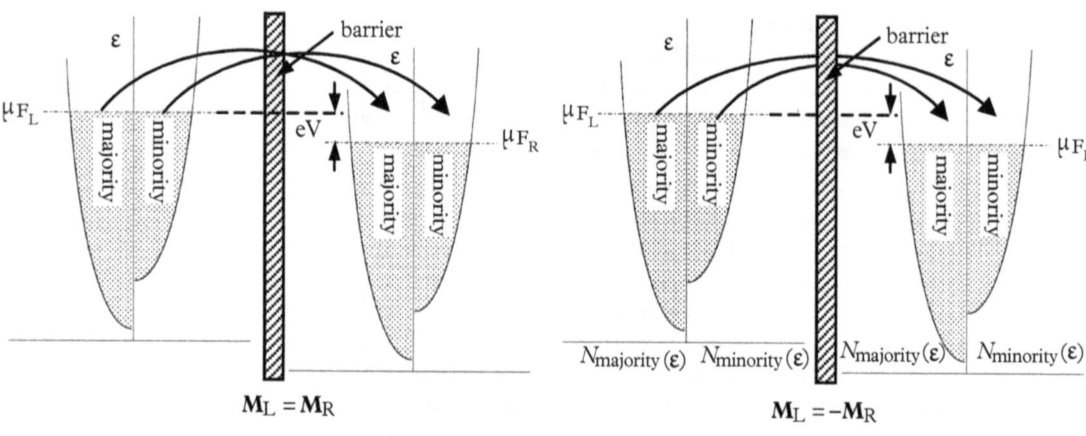

Figure 47.13 *The density of states for the majority and minority spins on the left and right sides of an insulating barrier in a simplified, one-band model of a magnetic tunnel junction. The curved arrows indicate a net inter-layer tunneling under the influence of an external potential,* V. *The case* $M_L = M_R$ *applies when the magnetizations in the two layers are parallel while the case* $M_L = -M_R$ *applies when they are anti-parallel.*

the magnetizations are anti-aligned the majority spins in one layer are aligned with minority spins in the other. The curved arrows indicate the states that can connect on tunneling for the aligned and anti-aligned cases, assuming the spins do not flip during tunneling. Intuitively, we expect larger matrix elements, and hence lower resistances, for the transitions majority → majority and minority → minority, which apply when the magnetizations are parallel, since the wave functions on each side of the barrier are then essentially the same. Likewise we expect higher resistances for transitions connecting majority → minority and minority → majority, which apply when the magnetizations are anti-parallel. Microscopic calculations[26] assuming perfectly ordered (fully crystalline) materials qualitatively support this idea. However they also show that the real situation is rather complex; among other things, the tunneling depends strongly on the angle of incidence of the electrons approaching the barrier. At this time the predicted changes in the magnetoresistance greatly exceed the experimental values, although realistic metallic interfaces contain a large amount of disorder.

Figure 47.14 shows the magnetic field dependence obtained by the Ando group for an Fe/MgO/Fe device.[27] Plotted is the product of the area and the magnetoresistance ratio, defined as $(R_{ap}-R_p) / R_p$, where R_{ap} and R_p denote the junction resistance for anti-parallel and parallel alignments of the Fe layer magnetizations. In addition, "pinning" of the orientation of the layer at the top of the device (referred to as the fixed layer) is achieved for a range of fields by depositing an antiferromagnetic Ir–Mn overlayer. In this way a reproducible anti-aligned state can be maintained over a range of applied fields. Shown also in Fig. 47.14 is a transmission electron microscope cross-section image of the junction region, demonstrating the high degree of alignment of the atoms in the three layers shown. Results on similar devices were obtained by an IBM group.

[26] Spin-dependent tunneling conductance of Fe MgO Fe sandwiches, W.H. Butler, X.-G. Zhang, T.C. Schulthess, and J.M. MacLaren, *Phys. Rev. B* **63**, 054416 (2001).

[27] Giant room-temperature magnetoresistance in single-crystal Fe/MgO/Fe magnetic tunnel junctions, Shinji Yuasa, Taro Nagahama, Akio Fukushima, Yoshishige Suzuki, and Koji Ando, *Nat. Mater.* **3**, 868 (2004).

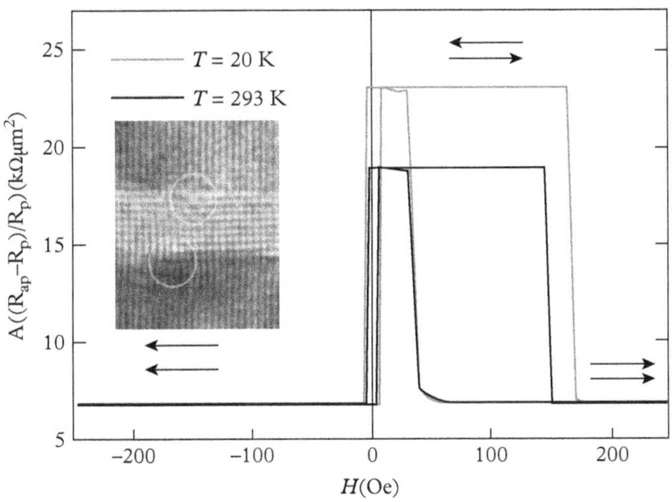

Figure 47.14 *The product of the magnetoresistance ratio for an Fe/MgO/Fe junction with the area versus an external applied magnetic field, applied parallel to the junction for parallel and anti-parallel alignments of the layer magnetizations. The inset shows a transmission electron microscope cross-section image of the junction region; the circles identify misfit dislocations arising from lattice mismatch. (After Yuasa et al. (2004).)*

The Ando group also performed a detailed study of the resistance ratio as a function of the MgO layer thickness, obtaining an overall exponential dependence, consistent with the expected behavior for quantum mechanical tunneling; in addition, small oscillations in the tunneling probability were observed, the origin of which is not well understood.

47.8 Spin torque oscillators

The dynamic response of coupled ferromagnetic layers in the presence of a current flow perpendicular to the layers is particularly interesting, since under the right conditions it can lead to a spontaneous precessional oscillation of the magnetization.[28] The geometry of two classes of devices is shown schematically in Fig. 47.15, the idea being that an applied d.c. current is concentrated into a very small region, thereby generating a high current density; this region is in contact with a larger region which facilitates electrical contact and also carries away locally generated heat. Here we picture electrons passing through a thick magnetic layer (serving as a fixed layer), which results in a spin polarized current in the adjacent copper layer[29] which, in turn, enters the second thinner magnetic layer (termed the free layer). In general we expect the current injected into the copper by the majority spin carriers, designated as spin up, to exceed that of the minority carriers due to their greater number; for convenience we here assume that the latter is absent.

In addition to the applied current an external magnetic field is also applied that is inclined by some angle, θ, with respect to the layer planes. It is assumed that this field does not alter the magnetization direction in the fixed layer, the latter being held in-plane by the film geometry itself (shape anisotropy), or by the addition of an engineered anisotropy energy. However the injected magnetization in the copper layer from the majority spin current will involve states quantized along the inclined field direction and hence will be a mixture of spin up and spin down states. Further analysis shows that the spin current arising from this mixed state will, on entering the free magnetic layer, induce a torque which has the form[30]

$$\left(\frac{d\hat{\mathbf{M}}}{dt}\right) = \eta(\theta)\frac{\mu\,I}{eV}\hat{\mathbf{M}}(t) \times \left(\hat{\mathbf{M}}(t) \times \hat{\mathbf{M}}_{\text{fixed}}\right), \tag{47.4}$$

[28] For a review, see: Spin transfer torques, D.C. Ralph and M.D. Stiles, *J. Magn. Magn. Mater.* **320**, 1190 (2008).

[29] Even in the absence of a current, a normal metal adjacent to a ferromagnetic metal will have a net spin polarization.

[30] Current-driven excitation of magnetic multilayers, J.C. Slonczewski, *J. Magn. Magn. Mater.* **159**, L1 (1996); Currents and torques in metallic magnetic multilayers, J.C. Slonczewski, *J. Magn. Magn. Mater.* **247**, 324 (2002).

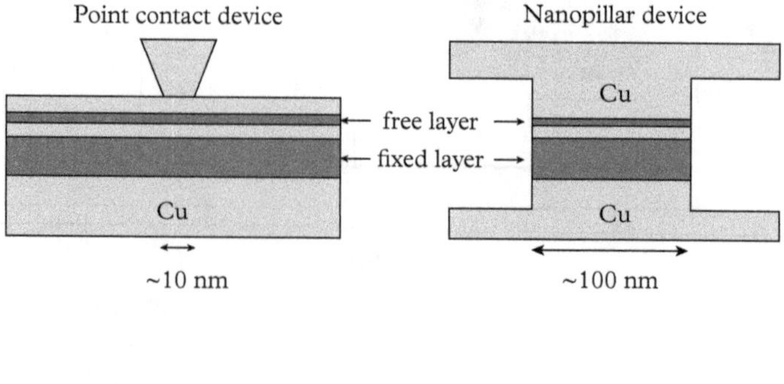

Figure 47.15 *Two geometries permitting a sustained high current density that flows between two magnetic layers separated by a thin, non-magnetic metal layer. (After Ralph and Stiles (2008).)*

where μ is the magnetic moment of the spins, I is the current through the nanopillar having volume V, e is the charge of the electron, and $\hat{\mathbf{M}}$ and $\hat{\mathbf{M}}_{\text{fixed}}$ are unit vectors in the direction of the magnetization in the free and fixed layers respectively; $\eta(\theta)$ is an angular-dependent factor that accounts for the effects of interface barrier transmission/reflection and transport between and within the layers. If the contribution from (47.4) exceeds the damping term, in the Landau–Lifshitz–Gilbert equation (27.18), the magnetization of the free layer will precess in the internal magnetic field, growing until nonlinear effects set in; this precession results in an emitted microwave signal, a process which is the inverse of ferromagnetic resonance (see Section 27.2). Figure 47.16 shows the spectrum of a signal from such a device, containing the fundamental FMR frequency and two harmonics, the latter arising from the nonlinear effects just mentioned.

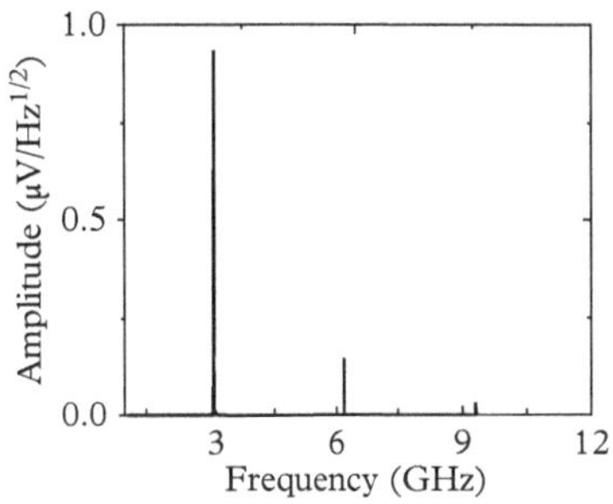

Figure 47.16 *Spectrum of voltage oscillations produced by a spin-torque-driven steady-state magnetic oscillator. Note the presence of several harmonics. (Courtesy of Ilya Krivorotov.)*

47.9 Surface states

Our description of one-electron quantum states in solids focused on Bloch waves, which are traveling waves and carry a current or equivalently a probability flux. In the absence of sources and sinks for electrons on two or more surfaces of a solid (e.g., wires connected to a battery), the probability flux must vanish to keep the overall electron charge or probability density constant in time. Hence, each Bloch state deep in the interior of the solid having quasi-momentum, **k**, must be balanced by another state with quasi-momentum, –**k**, that differs at most by a phase; all crystal states are then standing waves.

But near a surface the electrons no longer see the same potential energy function. Beyond some surface adjacent to where the strong positive nuclear charge abruptly ceases, or at a weaker level beyond the extent of the atomic core orbitals, the overall negative potential confining the outer electrons (referred to as valence electrons, those ultimately responsible for binding the solid) starts to fall off.[31] Since the electrons have kinetic energy, and they are bound by a finite potential, their wave functions penetrate into the surrounding vacuum and the associated charge redistribution results in an effective dipole moment per unit area. In addition, redistributions of the charge from the valence electrons changes the binding of surface atoms; crudely, surface atoms lose the binding arising from their neighboring atoms that were removed in forming the surface. As a result, they will reposition, a property we will examine in more detail.

In treating crystal binding in periodic solids using self-consistent one-electron theories, we almost exclusively use a basis of Bloch states. So a natural question is whether there are electron states that are *localized at a surface*, yet have a two-dimensional Bloch-like character that allows them to propagate parallel to the surface with an associated two-dimensional wave vector, $\mathbf{k}_{\parallel}$. As with bulk solids it is natural to examine the two limiting cases: nearly free electrons,[32] and tight-binding electrons.[33]

We will treat in some detail the free electron case involving a single Fourier component, V_1, of the lattice potential along with the shift, V_0, occurring in the

[31] In the free electron model (discussed in Chapter 6) this is the uniform positive background creating the potential that holds the electrons inside the imaginary box in which they move.

[32] On the surface states associated with a periodic potential, W. Shockley, *Phys. Rev.* **56**, 317 (1939).

[33] On the possible bound states of electrons on a crystal surface, I. Tamm, *Phys. Z. Soviet Union* **1**, 733 (1932).

Figure 47.17 *The potential associated with a single Fourier component, V_1, of the lattice potential, together with a discontinuity of magnitude, V_0, at the interface between a one-dimensional solid and the vacuum.*

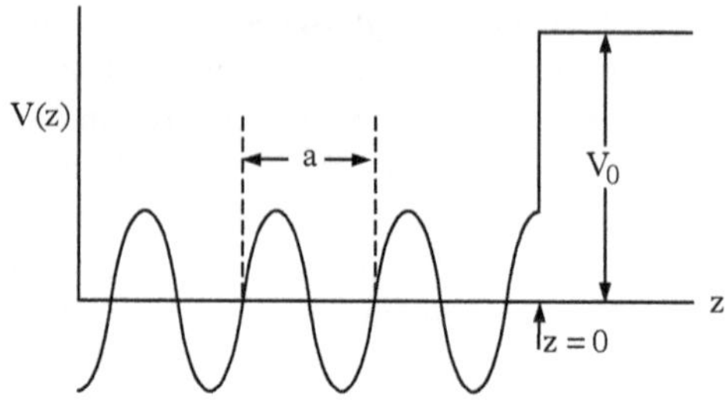

vacuum, as shown schematically in Fig. 47.17; we here limit ourselves to one dimension, where we expect to encounter at most zero-dimensional states. We recall the expression for the energy levels near the Brillouin zone points, Eq. (11.17), which for $K_1 = 2\pi/a$ is

$$E(k) = \frac{1}{2}\left\{\frac{\hbar^2}{2m}\left[k^2 + (k + 2\pi/a)^2\right] \pm \sqrt{\left[\frac{\hbar^2}{2m}(k^2 - (k + 2\pi/a)^2)\right]^2 + 4|V_1|^2}\right\}.$$

Writing $k = -\pi/a + q$, where q measures the departure from the first Brillouin zone point, this expression becomes

$$E(k) = \frac{\hbar^2}{2m}\left(\frac{\pi^2}{a^2} + q^2\right) \pm \sqrt{\left(\frac{\hbar^2 \pi q}{ma}\right)^2 + |V_1|^2}. \tag{47.5}$$

The eigenfunctions inside the lattice (where $z < 0$) that are associated with (47.5) (corresponding to the solution of Eqs. (11.14)) are given by

$$\psi_i = Ce^{iqz}\left[e^{i\pi z/a} + \left(-\frac{\hbar^2 \pi q}{ma|V_1|} \pm \sqrt{\left(\frac{\hbar^2 \pi q}{ma|V_1|}\right)^2 + 1}\right) e^{-i\pi z/a}\right], \tag{47.6}$$

where C is a normalization constant. For $q = 0$, this corresponds to symmetric and antisymmetric standing waves,

$$\psi_i = C\left(e^{i\pi z/a} \pm e^{-i\pi z/a}\right) \tag{47.7}$$

and for $q \neq 0$, the right or left propagating wave increasingly dominates, depending on the sign in front of the radical.

We now examine what happens in the region near the surface for energies lying *within the band gap* where there are no propagating states in the lattice. Suppose we write $q = i\kappa$, where κ is real, and ask if we can have real energies. From Eq. (47.5) it follows that for

$$-\frac{ma|V_1|}{\hbar^2\pi} < \kappa < \frac{ma|V_1|}{\hbar^2\pi}, \tag{47.8}$$

real solutions for $E(\kappa)$ exist for the appropriate choice of the sign of κ and the $\pm$ sign in the expression; in addition, the exponential in the prefactor, $e^{\kappa z}$, is real and for κ positive the amplitude will decay with negative z (the other sign covers the case where the lattice is located in the region $z > 0$). Defining

$$\sin 2\delta = \frac{\hbar^2 \pi \kappa}{ma\,|V_1|}, \tag{47.9}$$

Eq. (47.6) can be written as

$$\psi_i = Ce^{iqz}\left[e^{i\pi z/a} + \left(-i\sin 2\delta \pm \sqrt{1-\cos^2 2\delta}\right)e^{-i\pi z/a}\right]; \tag{47.10}$$

note that the modulus of the quantity $-i\sin 2\delta \pm \sqrt{1-\cos^2 2\delta}$ is unity. With a little algebra, Eq. (47.10) can be rewritten as

$$\psi_i = C'e^{\kappa z}\left[e^{i(\pi z/a \pm \delta)} + e^{-i(\pi z/a \pm \delta)}\right], \tag{47.11}$$

which allows us to identify δ as a phase shift. We see that, in addition to the attenuation associated with κ, the states oscillate with a period given by the lattice constant, a;[34] furthermore, apart from an overall phase factor, the wave functions are real, which is a characteristic of *one-dimensional bound states.*

To fix δ we would have to match (47.11) to the wave function outside the lattice (where $z > 0$), which is governed by V_0 and has the form

$$\psi_o = D\exp\left[-\sqrt{\frac{2m}{\hbar^2}}\,(V_0 - E(\kappa))z\right], \tag{47.12}$$

where D is again a normalization constant and $E(\kappa)$ follows from (47.5). By matching the logarithmic derivatives of Eqs. (47.11) and (46.12), the phase shift, δ, can be related to the surface bound state energies, $E(\kappa)$, and the well depth, V_0. Figure 47.18 shows the associated band structure along with a schematic representation of the surface states with an imaginary wave vector characterized by κ.

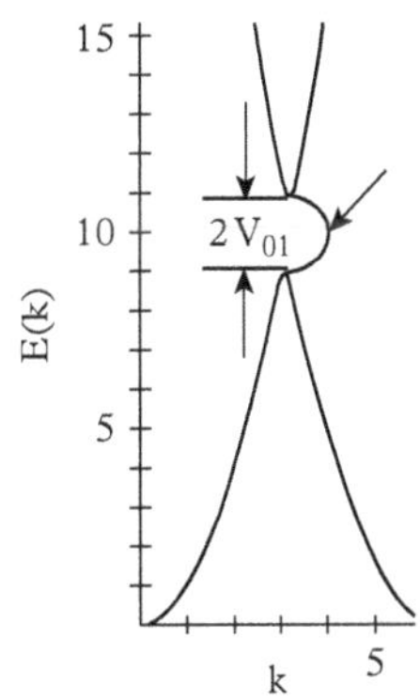

Figure 47.18 *The first and second energy bands in the repeated zone scheme. States with imaginary* k *in the gap are represented by the arc (indicated by the arrow).*

Note that no states strictly localized to the surface arise within the allowed range of the energies of the *propagating* states. However in a doped semiconductor, the presence of donors or acceptors can (at low temperatures) pin the Fermi level between the impurity levels and the conduction or valence band edge. In this case the Fermi level will also lie in the range of the surface states (lying between the bulk band edges), and the *surface will be metallic.* Interface states can be present between two bulk semiconductors provided there remains an energy gap when the materials are in contact; a single quantum well being an example.

We note in passing that at absolute zero, as long as $V_0 - E(k) > 0$, electrons will be bound to the lattice. At finite temperatures there will be electrons in the surrounding vacuum, although they will be bound by their electrostatic image

[34] These oscillations are analogous to the Friedel oscillations discussed in Section 39.3, although being one dimensional in character, they are stronger and longer range.

charges to the surface. In the presence of an applied potential drop, electrons can be drawn to an external electrode setting up a current in a closed circuit, as governed by the well-known Richardson–Duschman law of thermonic emission.

We will not discuss the case of tight-binding electrons, first treated by Tamm, but it can be shown that under suitable conditions surface states can also exist within the gap. The theory can be further generalized to apply to multiple orbitals, with the result that multiple surface states can be present. True surface states are still confined to regions of the electron spectrum where a gap is present. However, so-called resonant surface states can be present in energy ranges where there are bulk propagating states which have a finite width, analogous to the resonant states encountered in our discussion of d-like impurity levels in a free-electron-like metal (see Chapter 44). If there are surface atoms present that are chemically different from those of the host, additional surface states can exist.

47.10 Atoms at a free surface; reconstruction

Up to this point we have considered the surface atoms as being fixed at the positions they would have had before removing the neighboring atoms that result in some surface. This is shown for the case of a simple square lattice in Fig. 47.19(a). However as a consequence of the minimization of the energy of the solid as a whole (the Gibbs free energy at finite temperatures), surface atoms can relax (alter) their positions. In the simple case of a high symmetry surface this could simply involve a displacement of the surface atoms normal to the surface, as shown in Fig. 47.19(b), and in such cases the translational symmetry parallel to the surface would be retained. Alternatively atoms could translate parallel to the surface, as shown in Fig. 47.19(c). But in many materials, particularly those with

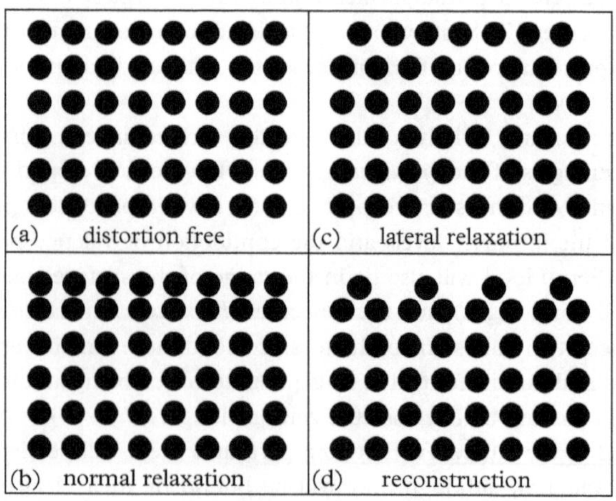

Figure 47.19 *Atom positions in a simple square two dimensional lattice having a surface. (a) Atoms retain their positions up to the surface. (b) Atoms relax by moving normal to the surface. (c) Atoms relax by moving parallel to the surface. (d) The system releases surface atoms and reconstructs, thereby breaking the translational symmetry parallel to the surface.*

directed bonds (such as in the group IV compounds, Si and Ge), the presence of the surface would leave "unsatisfied" or "dangling" bonds from the chemical perspective. In order to better satisfy local bonding considerations, the individual surface atoms may then choose to "rearrange" or "reconstruct" themselves, and in the process alter the translational symmetry parallel to the surface and even *shed atoms*; this process can create a larger in-plane unit cell, as shown in Fig. 47.19(d), where the cell size doubled.

In specifying a surface reconstruction we will first assume that the two-dimensional surface structure remains *commensurate* with the bulk structure. For convenience we will further assume that two of the three bulk primitive translation vectors, defined in Section 8.1, which we take as **a** and **b**, lie in the plane of the surface. The basis of the two-dimensional surface structure can then be specified by writing

$$\mathbf{a}_s = g_{11}\mathbf{a} + g_{12}\mathbf{b}$$
$$\mathbf{b}_s = g_{21}\mathbf{a} + g_{22}\mathbf{b}, \qquad (47.13)$$

where $\mathbf{a}_s$ and $\mathbf{b}_s$ are the *surface* primitive translation vectors, which in general will be larger than **a** and **b** since the surface unit cell is necessarily larger than that of the underlying crystal. As written, (47.13) also allows for a rotation by some angle, ϕ, through the off-diagonal elements. One then has three parameters: the principal components of the **g** matrix, g'_{11} and g'_{22}, and the rotation angle, ϕ, all of which are constrained registry with the underlying substrate.

Rather than specifying the parameters, g_{ij} (along with a plane in question), a commonly used notation is X(*hkl*) m × n Rϕ, where: X denotes the material (e.g., Si); *h, k, l*, the Miller indices of the surface plane; m and n, the surface unit cell dimensions in units of the primitive vectors **a** and **b**; and ϕ allows for a possible rotation through an angle. One often finds only the product m × n given. Note that this specification is minimal in that it does not include: (i) any information on distances normal to the surface, (ii) the behavior of planes adjacent to the outermost layer, and (iii) a basis (see Section 8.2) specifying atom positions *within* a surface unit cell.

Surface structures can be studied by low energy electron diffraction (LEED) and scanning tunneling microscopy (STM), the latter being easiest to interpret since it yields the structure in real space directly.[35] To avoid contamination from environmental atoms, such studies are generally carried out in ultra high vacuum on freshly prepared (usually in situ cleaved) surfaces. The principal surfaces of Si and Ge are known to reconstruct; as examples, the Si [111] surface is reported to reconstruct into 2×2, 5×5, 7×7, and 9×9 structures, while [111] Ge reconstructs as 2 × 8. Figure 47.20 shows a scanning tunneling microscope image of the 7 × 7 reconstruction of the silicon [111] surface.[36] The highest points are associated with so-called *ad-atoms*, those which protrude from more closely packed layers (and may be bound in an s state, as opposed to the more typical sp^3 bonds of group IV elements, see Section 7.4).

[35] An STM tip is not the "ideal point" and one usually has to account for the finite curvature and other effects arising from its structure.

[36] Atomic resolution of the silicon (111)-(7×7) surface by atomic force microscope, Franz J. Giessibl, *Science* **267**, 68 (1995). Also see: Scanning tunnelling microscopy study of Si(111)7×7 in the presence of multiple-step edges, R. Wiesendanger, G. Tarrach, D. Burgler, and H.J. Guntherodt, *Europhys. Lett.* **12**, 571 (1990).

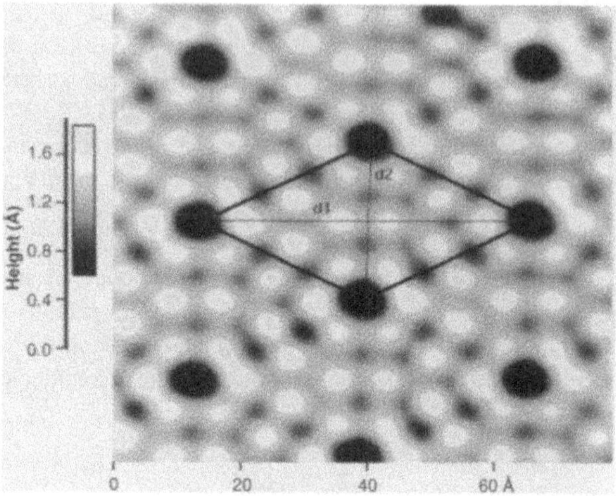

Figure 47.20 *An STM image of the Si(111)-(7 7) reconstruction. The unit cell is outlined by the black diamond and the lengths of the diagonals are 46.6 A and 26.9 A. The bright yellow spots are interpreted as ad-atoms, of which there are 12 per unit cell; the large black spots are voids having a depth ~2 A. (After Giessibl (1995).)*

The various surface crystal phases observed can depend on temperature and thermal history. In addition to studies of the surfaces of pure elements and compounds, there is much interest in where foreign atoms adsorb and how they order, a topic which is a part of the vast field of surface chemistry.

47.11 Calculating surface and interface electronic properties

The supercell technique, discussed earlier in Section 39.4.4, is ideally adapted to calculating the electronic properties of infinite *multilayers*, provided they have the necessary translational symmetry so that Bloch's theorem applies; i.e., the crystal structures of the different layers in the supercell can, by introducing some level of strain, accommodate each other in a fully periodic manner. All that is required is to make the "vertical" dimension of the supercell equal to the repeat distance of the superlattice and, on shorter distance scales, match the in-plane periods. For the (square-lattice based) superlattice shown in Fig. 47.1a, one would have a rectangular unit cell, $3 + 4 = 7$ atoms in height and one atom in width. Through self-consistent total energy calculations one can relax the positions of the atoms, particularly near the interfaces, to obtain the global equilibrium state. By incorporating a larger in-plane unit cell, some degree of reconstruction of the interface can be explored, again assuming the entire system remains periodic.

Multilayer semiconductor systems raise some interesting questions. Strictly speaking, and in a one-electron picture, we do not expect to have interface states in a multilayer system: interface states have exponential tails and would overlap between layers and therefore evolve into highly interface-localized *but still fully periodic* (and therefore propagating) Bloch states in an infinite multilayer.[37]

[37] The presence of a small amount of disorder would lead to long range, but not individual interface, localization.

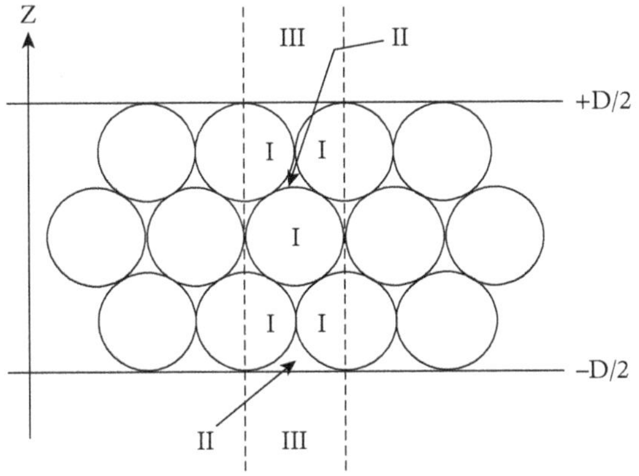

Figure 47.21 *Schematic representation of a two-dimensional threelayer slab showing how the potential is partitioned into three regions: (I) an assumed spherically symmetric "muffin tin" surrounding the individual atoms; (II) the space between the spheres where the potential is constant; and (III) the vacuum above and below the slab. (After Krakauer et. al. (1979)).*

However, one may be able to identify states that, due to their high degree of interface localization, would evolve into true interface states in the limit of large supercell periods.

The slab geometry shown in Fig. 47.21 offers an alternative to the supercell approach. Here one has strictly non-propagating states in the vertical direction. This scheme can be implemented through a modification of the augmented plane wave approach, discussed in Chapter 13, where the potential is now divided into *three* regions as shown in the figure; here will we only sketch a simplified version of the technique. The basis functions in regions I and II are similar to those for the bulk case given by Eq. (13.1), but with $k_\perp$ fixed according to

$$k_\perp^{(+)} = n\frac{2\pi}{D}, \quad k_\perp^{(-)} = \left(n + \frac{1}{2}\right)\frac{2\pi}{D}, \tag{47.14}$$

corresponding to even and odd standing waves in that direction. Region III represents the vacuum where one has evanescent states that are periodic in the plane of the slab but decay exponentially in the perpendicular directions. Such states can be constructed from superpositions of the form

$$\phi_{\mathbf{K}_\parallel}^{(III)}(\mathbf{r}_\parallel, z) = e^{i(\mathbf{k}_\parallel + \mathbf{K}_\parallel)\cdot\mathbf{r}_\parallel \mp \gamma_{\mathbf{K}_\parallel} z}\theta\left(z \mp \frac{D}{2}\right), \tag{47.15}$$

where $\mathbf{K}_\parallel$ denotes the in-plane reciprocal lattice vectors, $\mathbf{r}_\parallel$ is the in-plane coordinate, and z is the direction perpendicular to the slab; the wave vector components in this region are related to the energy as

$$E = \frac{\hbar^2}{2m}\left(k_\parallel^2 - \gamma^2\right) + V_{vac} \tag{47.16}$$

and V_{vac} is a vacuum reference energy. In addition to matching the forms in regions I and II at the muffin tin sphere radius, as in the bulk APW method, one also matches the forms in regions II and III at $z = \pm D/2$. The method is very powerful and has been applied to a number of systems. In particular it has been used to calculate the energetics of surface reconstruction. For the details of the subsequent analysis we refer the reader to the original work.[38]

47.12 The chemical potential; the work function

Earlier we introduced the electron chemical potential, μ, or equivalently the Fermi energy, ε_F, without asking how it might be measured on an absolute basis. Formally, we will define it as the energy required to *thermodynamically reversibly* move a single electron from a point deep inside some given system to a point infinitely far away, as measured relative to all dimensions characterizing the system. Alternatively we might picture the reverse process of moving an electron from infinitely far away to the interior of the system. Designing processes that reversibly add and remove electrons presents difficulties.

For a small system the energies required to reversibly add or remove an electron will be different. A striking example is the atom. The first process defines the minimum ionization energy, while the second requires the atom to be able to bind an additional electron; not all atoms can, and for those that do, the energy differs markedly from the ionization energy. However for a very large system, in the so-called thermodynamic limit, we expect these two energies to approach each other to order $1/N$, where N is the number of electrons in the system.

Let us examine more closely what is involved in removing an electron from a *metal*. To begin we restrict ourselves to absolute zero. Here no electrons have energies above the Fermi energy and removing an electron from below the Fermi energy would leave the system in an excited state that would subsequently decay, which is an irreversible process; hence the only electrons that can reversibly be removed are those at the Fermi energy. Now adiabatically transporting the electron within the bulk involves no energy: all unit cells are essentially identical. This remains true until the electron approaches some surface. As previously discussed, the presence of the surface causes: (i) the electronic charge to redistribute and (ii) the atoms to alter their positions (even reconstructing); the associated charge non-uniformity results in the formation of an electric dipole layer in the vicinity of the surface. From electrostatics we know that on passing through such a layer the potential shifts by some value, which we designate as W_S. This shift will be *different for each crystallographic plane*; i.e., the magnitude of the dipole layer and the associated potential energy shift, W_S, differs between surfaces. The minimum energy, W, required to remove an electron at the Fermi energy from the solid by transporting it from deep inside up to and through a given surface into the vacuum is then given by

$$W = W_S - \varepsilon_F. \tag{47.17}$$

[38] Band-structure of thin-films by linear augmented-plane-wave method, O. Jepsen, J. Madsen, and O.K. Andersen, *Phys. Rev. B* **18**, 605 (1978); Linearized augmented plane-wave method for the electronic band structure of thin films, H. Krakauer, M. Posternak, and A.J. Freeman, *Phys. Rev. B* **19**, 1706 (1979).

In the presence of an external field a metal acquires a surface charge density. More subtle is that to maintain overall equilibrium a surface can spontaneously acquire a local charge density in the absence of an external potential. We know from electrostatics that a metal object will be an equipotential but that the local charge density can vary depending on the curvature. In the present case a metallic sample can be overall charge neutral, but crystallographically, different surfaces can acquire different charge densities. Furthermore, the charge acquired by a given surface depends on a combination of which other surfaces are present and the accompanying areas of all surfaces, since they are in competition; i.e., the surface charging effect depends not only on the surface studied but the overall shape of the body itself. Given that the potential shift associated with a uniformly charged planar surface increases without limit perpendicular to that surface, we expect this increase to be offset by contributions from other surfaces such that the total potential shift arising from redistributed charge approaches a constant at distances large compared to all dimensions of the sample (and falling off faster than r in the process).[39] Since electrons emitted thermionically or via the photoelectric effect are typically collected by an external electrode close to the sample, the effects associated with surface charging are often ignored. Finally we note that if the surface is polycrystalline, or there are foreign atoms/molecules inhomogeneously adsorbed on the surface, the charge density, even on a flat surface, will be inhomogeneous. This phenomenon is referred to as the *patch effect*.

Calculating the contribution of the dipole layer in real materials can be done using density functional theory, for example using the modified APW method described in Section 47.11. It turns out that for the simple metals very good agreement can be obtained with a model incorporating a uniform positive background

[39] At this point we should also remark that the attractive force an electron experiences through its image charge should not explicitly appear in the energetics of the work function, since it is already a part of the binding energy of the electrons in the solid.

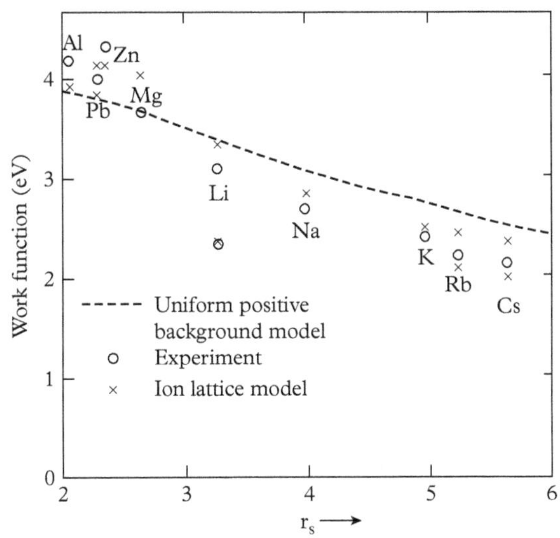

Figure 47.22 *The work function of the simple metals vs. an effective electron separation* r_s *(see section 15B.6) obtained from a jellium model. Also shown are the experimental values for the simple metals (open circles) together with the a correction produced by introducing a pseudopotential (the X's) to account for the discrete nature of the ions. (After Lang and Kohn (1971).)*

(jellium) that ends abruptly in a plane surface, and an accompanying interacting electron gas that in response adjusts its charge density near this surface, thereby generating a dipole layer, as shown by Lang and Kohn.[40] The predicted work functions are shown in Fig. 47.22, along with experimental values. The points marked as an X show the effect of a pseudo potential correction to account for an inhomogeneity in the positive background arising from the ions.

47.13 Methods to measure the work function

47.13.1 The photoelectric effect

In the photoelectric effect an incoming photon with energy, $\hbar\omega$, ejects an electron. Electrons ejected from states at the Fermi energy would have kinetic energy, K, which from conservation of energy would be given by

$$K = \hbar\omega - W, \tag{47.18a}$$

which on using (47.17) becomes

$$K = \hbar\omega - W_S + \varepsilon_F, \tag{47.18b}$$

where ε_F is here interpreted as a *negative* number since it is being measured *relative to the vacuum*. Electrons ejected below the Fermi level would have lower kinetic energy and would be emitted down to some cut-off energy $\varepsilon_c < \varepsilon_F$,

$$\hbar\omega = -W_S - \varepsilon_c. \tag{47.19a}$$

If the photon energy is fixed, an external retarding potential can be applied to cancel K and thereby determine W through a cut-off in the electron current. Alternatively, the photon energy can be reduced to a cut-off or threshold energy, $\hbar\omega_c$, below which no electrons are emitted,

$$\hbar\omega_c = W_S - \varepsilon_F. \tag{47.19b}$$

At finite temperatures the two cut-offs would be broadened, but this is usually minimal since we generally have $T \ll T_F$.

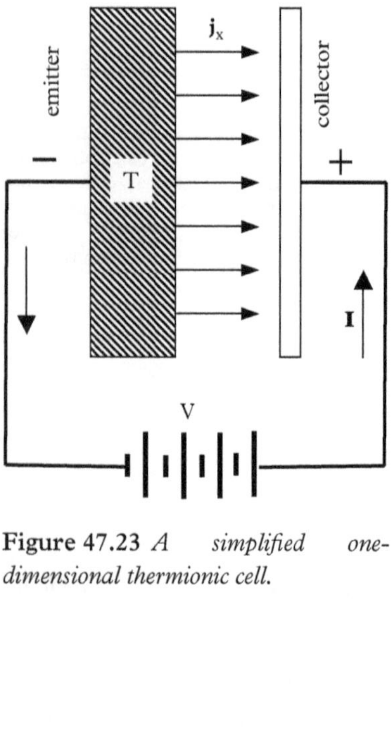

Figure 47.23 *A simplified one-dimensional thermionic cell.*

47.13.2 Thermionic emission

Referring to Fig. 47.23, assume we have a flat metal surface heated to some temperature, T, such that the electron distribution has a measurable tail that extends to energies greater than W. These electrons can escape the metal into the surrounding vacuum. Assume further that there is a collector electrode biased at a

[40] Theory of metal surfaces: work function, N.D. Lang and W. Kohn, *Phys. Rev. B* 3, 1215 (1971).

positive potential, V, such that the electrons are swept to that electrode and do not build up a charge density in the surrounding vacuum, which would otherwise alter the potential in the vicinity of the emitting surface.

In the limit where we can neglect V, the emitted current is given in terms of the Fermi distribution by

$$j_x = -\frac{2e}{(2\pi)^3}\int_{k_x>0} d^3k\, v_x f(\varepsilon_k)$$

or

$$j_x = -\frac{2e}{(2\pi)^3}\int_{k_x>0} d^3k\, v_x \frac{1}{e^{(\varepsilon_k+W_s-\mu(T))/k_BT}+1}, \qquad (47.20)$$

where $\mu(T)$ is evaluated at the temperature of the emitter. For the case of free electrons we leave it as a problem to show that

$$j_x = -\frac{em}{2\pi^2\hbar^3}(k_BT)^2 e^{-W/k_BT}. \qquad (47.21)$$

Incorporating the effects of the density of states of a real metal alters the coefficient but the behavior remains dominated by the factor, W, in the exponent. A plot of $ln\,(j_x/T)$ versus $1/k_BT$ yields a straight line with slope –W. This general behavior is referred to as the Richardson–Dushman law.

The work functions of the elements are listed in Table 47.1.[41] As previously emphasized, they can differ depending on the surface, resulting in a range of values. As an example, the work function of polycrystalline silver is 4.26 eV; however the measured values for some individual surfaces are: (100) face, 4.64 eV; (110) face, 4.52 eV; and (111) face, 4.74 eV.[42]

We mention in passing that the work function also enters the process of *field emission*, otherwise called *Fowler–Nordheim* tunneling.[43] Here, electrons quantum mechanically tunnel from a sharp metallic tip into the vacuum under the influence of an external field. However, uncertainties in the magnitude of the local electric field and the nature of the surface in regions of high curvature, where the tunneling occurs, limit the accuracy for determining W. However, the phenomenon is of interest in its own right as it lies at the heart of scanning tunneling microscopy (STM). In addition, field emission tips are commonly used in electron microscopes in order to have a more point-like source, which allows for a sharper focus. There has also been interest in making vacuum electronic devices where a field emission tip replaces a thermionic emitter.

47.13.3 The Kelvin method

Consider two different metals, 1 and 2, that are initially unconnected. If they are now connected via an ammeter (a device that formally has zero resistance),

[41] *CRC Handbook of Chemistry and Physics*, 2008.

[42] Work function measurements on (100) and (110) surfaces of silver, A.W. Dweydari and C.H.B. Mee, *Phys. Stat. Sol.* (a) 27, 223 (1975).

[43] Electron emission in intense electric fields, R.H. Fowler and L. Nordheim, *Proc. R. Soc. A* **119**, 173 (1928).

Table 47.1 *Work functions of the elements (eV)*

Ag	4.26–4.74	Al	4.06–4.26	As	3.75
Au	5.1–5.47	B	~4.45	Ba	2.52–2.7
Be	4.98	Bi	4.31	C	~5
Ca	2.87	Cd	4.08	Ce	2.9
Co	5	Cr	4.5	Cs	2.14
Cu	4.53–5.10	Eu	2.5	Fe	4.67–4.81
Ga	4.32	Gd	2.90	Hf	3.9
Hg	4.475	In	4.09	Ir	5.00–5.67
K	2.29	La	3.5	Li	2.9
Lu	~3.3	Mg	3.66	Mn	4.1
Mo	4.36–4.95	Na	2.36	Nb	3.95–4.87
Nd	3.2	Ni	5.04–5.35	Os	5.93
Pb	4.25	Pd	5.22–5.6	Pt	5.12–5.93
Rb	2.261	Re	4.72	Rh	4.98
Ru	4.71	Sb	4.55–4.7	Sc	3.5
Se	5.9	Si	4.60–4.85	Sm	2.7
Sn	4.42	Sr	~2.59	Ta	4.00–4.80
Tb	3.00	Te	4.95	Th	3.4
Ti	4.33	Tl	~3.84	U	3.63–3.90
V	4.3	W	4.32–5.22	Y	3.1
Yb	2.60	Zn	3.63–4.9	Zr	4.05

a quantity of charge will initially flow that will bring the Fermi energies of the two metals to the same value. The total charge that flows would involve the quantity of the materials contacted and is not of interest here. Let us assume that the two metals have been prepared such that, ideally, each is single crystals and has at least one flat surface aligned parallel to some crystallographic plane, the work function of which we want to measure. Suppose these surfaces are now aligned parallel to each other, separated by some distance, d, as shown in Fig. 47.24. To begin we will assume that the battery is replaced by a conducting wire. A (local) potential difference, $eV_{12} = W_1 - W_2$, will then be present between the surfaces, which we take to have the same area, A. Since the adjacent surfaces form a parallel plate capacitor there will be charges, $Q = CV$, on the adjacent surfaces where $C = A/4\pi d$. If now the separation between the plates is *varied* by an amount, δd, a charge, δQ, will flow through the circuit,

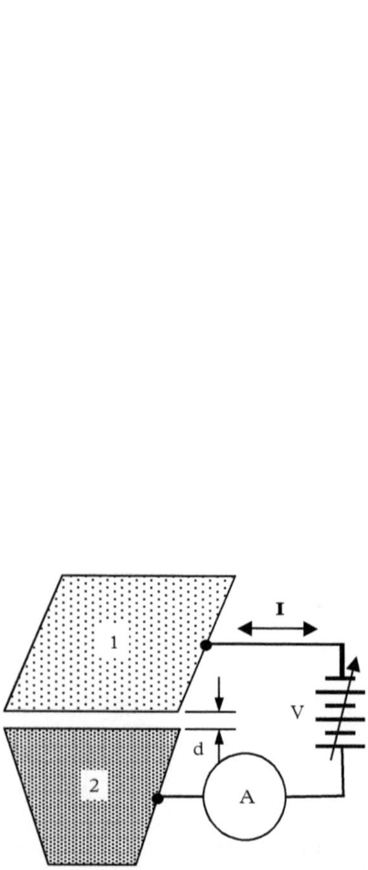

Figure 47.24 *Two metals, 1 and 2, having flat surfaces that are aligned parallel to each other and separated by a distance d. As d changes the current flowing in the ammeter A is monitored.*

$$\delta Q = V_{12}\left(\frac{\partial C}{\partial d}\right)\delta d = -V_{12}\frac{A}{4\pi d^2}\delta d, \tag{47.22}$$

since V_{12} is assumed to be constant. By integrating the output of the ammeter in time, we obtain $\delta Q = \int I(t)dt$. Knowing A, d, and δd, one can obtain V_{12} and hence the contact potential *difference*.[44] If one of the metals is a reference, the work function of which has been determined by, say, the photoelectric effect, the work function of the other metal can be determined. In practice, the voltage, V, in Fig. 47.24, which we set to zero, can be adjusted so as to *cancel* V_{12}, thereby making the current vanish, yielding a *null* measurement which can be carried out with higher precision.

This technique was suggested by Lord Kelvin. It forms the basis of a scanning probe microscopy technique (Kelvin probe force microscopy) that measures the position dependence of the contact potential of a surface, e.g., when studying surface topography.

47.14 Electron affinity; work function of a semiconductor

Up to this point our discussion of the work function has been restricted to metals at temperatures $T \ll T_F$. We now consider the case of a semiconductor. In the absence of surface effects the Fermi level lies within the band gap, where there are no (extended) states, so it is natural to ask what corresponds to the critical or threshold energy for photo-emission. In particular, is it still the Fermi energy?

In a metal (at $T = 0$), ε_F defines the energy at which a density of states for occupied electrons begins; therefore (neglecting the surface contribution), it fixes a threshold photon energy above which electrons can be ejected by incoming light. In a semiconductor, ε_F is fixed by a combination of doping and temperature and is *not related* to the onset of a high density of occupied electron states and, experimentally, the photoelectric threshold is observed to be *insensitive* to the position of the Fermi level. If we assume the valence band energy is measured relative to the vacuum and we are at absolute zero, we expect to see a threshold for photo-emission at a photon energy, $\hbar\omega_V = \varepsilon_V$, which we can interpret as a $T = 0$ work function. However, if we are at finite temperatures (or the material is doped strongly n-type) there will be an occupation of electron states in the conduction band and ejected electrons will be present for photon energies, $\hbar\omega_C \cong \varepsilon_C$, over a range involving the spread in occupation in the vicinity of the band edge. At higher photon energies the emission would drop off; with increasing energies we will again encounter the threshold associated with the valence band edge.[45]

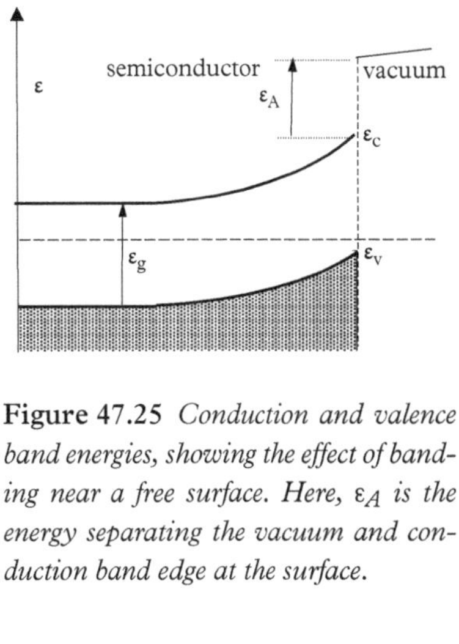

Figure 47.25 *Conduction and valence band energies, showing the effect of banding near a free surface. Here, ε_A is the energy separating the vacuum and conduction band edge at the surface.*

As with a metal, bringing an electron through the surface can involve surface contributions. In particular, ε_C and ε_V can shift near the surface, so-called band bending. Figure 47.25 shows a case where the band edges rise near the surface but the opposite can also occur. When this happens, their difference, $\varepsilon_C - \varepsilon_f$, tends to stay fixed at the band gap, ε_g. Band bending can be caused by various things such as surface states and reconstruction.

In examining the behavior near the surface one introduces the so-called *electron affinity*, ε_A, which separates the conduction band edge from the vacuum in the immediately vicinity of the surface. When the Fermi level lies near the conduction

[44] In practice the ammeter is replaced by an operational amplifier configured as a current integrator.

[45] From this we conclude that photo-emission in a semiconductor is not an "adiabatic" process, which would be governed by the chemical potential, but rather a "sudden" process. This is true at higher energies in all materials, where the electron is rapidly ejected leaving behind a hole which subsequently irreversibly collapses.

band edge (so we can ignore hole states), ε_A corresponds to the lowest energy at which an electron can be introduced into the semiconductor (in spite of the fact that ε_c may lie lower in the bulk).

A rise in the conduction band edge can also occur at a metal/semiconductor interface as a result of both interface states and Fermi energy alignment with the metal. The latter results in a charge transfer and an associated electrostatic potential barrier. The potential barrier associated with all these effects is called the *Schottkey barrier.* The barrier height can be raised or lowered by applying an external voltage of the appropriate sign across the junction which leads to an asymmetry in the barrier resistance; this is the basis of the *Schottkey diode.*

ADDITIONAL READING

For further discussion of quantum wells, including devices and computational methods, see: *Quantum Wells, Wires and Dots*, Paul Harrison, John Wiley and Sons (2009); *Theoretical and Computational Physics of Semiconductor Nanostructures*, Paul Harrison, John Wiley and Sons (2001). For one particular application, see *Quantum Cascade Lasers*, Jerome Faist, Oxford University Press (2013).

For a general review of layered metallic systems, see: Artificial metallic superlattices, B.Y. Jin and J.B. Ketterson, *Adv. Phys.* **38**, 191 (1989). A book covering many topics is: *Metallic Superlattices: Artificially Structured Materials*, Ed. T. Shinjo and T. Takada, Elsevier, Amsterdam (1987); there is also an extensive bibliography.

Applications of spin transfer effects are reviewed in: Spin-transfer effect and its use in spintronic components, B. Dieny, R.C. Sousa, J. Hérault, C. Papusoi, G. Prenat, U. Ebels, D. Houssameddine, B. Rodmacq, S. Auffret, and L.D. Buda-Prejbeanu, *Int. J. Nanotechnol.* 7, 591 (2010).

Exchange coupling between magnetic multilayers is a complex and strongly debated topic. For a review, see: Exchange bias, J. Nogués and I.K. Schuller, *J. Magn. Magn. Mater.* **192**, 203 (1999).

Various aspects of nanomagnetic structures are reviewed in: Colloquium: Opportunities in nanomagnetism, *S.D. Bader Rev. Mod. Phys.* 78, 1 (2006).

Surface states are discussed in the following books: *Basic Theory of Surface States*, Sidney G. Davison and Maria Steslicka, Clarendon Press, Oxford (1992); *Surface Science: An Introduction*, K. Oura, V.G. Lifshifts, A.A. Saranin, A.V. Zotov, and M. Katayama, Springer-Verlag (2003).

For a review of surface reconstruction and surface states in semiconductors, see the following: Semiconductor surface reconstruction: the structural chemistry of two-dimensional surface compounds, Charles B. Duke, *Chem. Rev.* **96**, 1237 (1996); and Theory of semiconductor surface reconstruction, G.P. Srivastava, *Rep. Prog. Phys.* **60**, 561 (1997).

The subtleties associated with the definition and measurement of the work function are well presented in *Solid State Physics*, N.W. Ashcroft and N.D. Mermin, Holt, Rinehaert and Winston, PA (1976).

Problems

47.1. Assume the presence of N_S single electron donor atoms per unit area, located infinitesimally inside an AlAs layer adjacent to a semi-infinite GaAs layer. Calculate the z-dependence of the electron concentration in the GaAs layer at absolute zero. Define a screening length, λ_S, that characterizes the position dependence of the resulting electron density.

47.2. Assume you are given a one-dimensional chain of alternating atoms with masses M_1 and M_2, connected by springs characterized by a spring constant, κ, similar to that discussed in Section 16.2. Can there be a *single* longitudinal vibrational mode residing at the end of such a chain and if so under what circumstances will it be present?

47.3. Obtain Eq. (47.21) from Eq. (47.20).

48 The Quantum Hall Effects

48.1 The integer quantum Hall effect

48.1.1 The phenomenon

The Hall effect was discussed earlier in Section 3.2, using the classical Drude model, and in Section 19.8, using a semi-classical free electron model. Under the assumption of a single carrier with an isotropic relaxation time, τ, either approach leads to the expression

$$\mathbf{E} = \rho_0 \mathbf{j} + R_H \mathbf{H} \times \mathbf{j}. \tag{48.1a}$$

If we take **j** // **x** and **H** // **z**, then **E** has non-zero x and y components, given by (see Eq. (19.56))

$$\begin{pmatrix} E_x \\ E_y \end{pmatrix} = \begin{pmatrix} \rho_{xx} & \rho_{xy} \\ \rho_{yx} & \rho_{yy} \end{pmatrix} \begin{pmatrix} j_x \\ 0 \end{pmatrix} = \begin{pmatrix} \rho_0 & R_H H \\ -R_H H & \rho_0 \end{pmatrix} \begin{pmatrix} j_x \\ 0 \end{pmatrix}, \tag{48.1b}$$

where $\rho_0 = 1/\sigma_0 = m_e/ne^2\tau$ is the resistivity, with σ_0, n, and m_e being the conductivity, carrier density, and effective mass (assumed isotropic) respectively; $R_H = -1/n|e|c$ is called the Hall constant. Note that the resistivity is independent of magnetic field in the Drude model. The conductivity matrix, σ, is the inverse of the resistivity matrix, ρ, and is given by

$$\sigma = \frac{\sigma_0}{1 + \omega_c^2\tau^2} \begin{pmatrix} 1 & -\omega_c\tau \\ \omega_c\tau & 1 \end{pmatrix}, \tag{48.2}$$

where $\omega_c = eH/m_e c$ is the cyclotron resonance frequency.

Suppose now that the motion is restricted along the z direction. Experimentally this can be realized in a MOSFET inversion layer, as discussed in Section 23.4.7, or in a GaAs quantum well sandwiched between $Ga_{1-x}Al_xAs$ layers, as discussed in Section 47.4. In either of these cases the motion perpendicular to the layers (making up the device) is quantized, as depicted schematically in Figs. 23.18 and 47.4[1]. One can adjust the external bias voltage of a MOSFET or the depth of a quantum well so that only a single state associated with the motion perpendicular to the film lies below the Fermi level. The only remaining degrees of freedom of the electrons in this state then involve the in-plane (x, y) translational motion and the spin projections ($s_z = \pm 1/2$).

[1] The 1985 Nobel Prize in Physics was awarded to Klaus von Klitzing "for the discovery of the quantized Hall effect."

The Physics of Solids. First Edition. J. B. Ketterson.

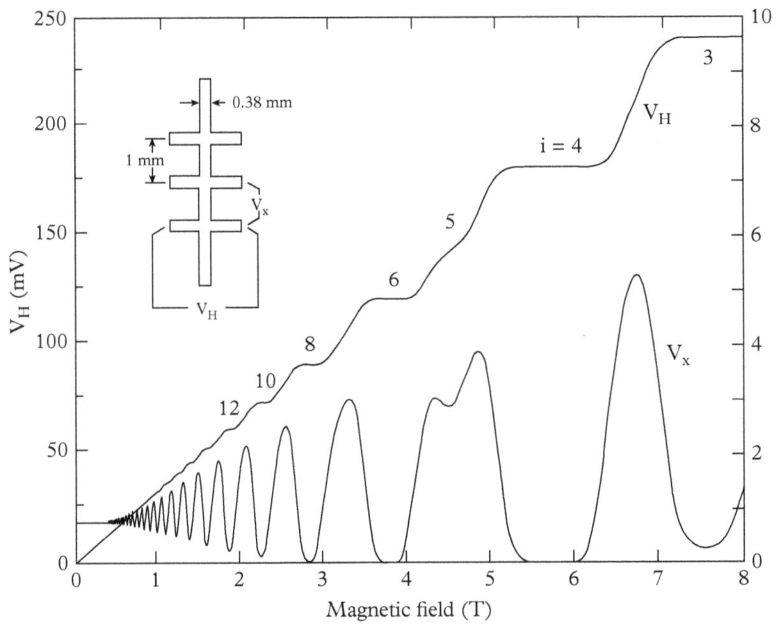

Figure 48.1 *The transverse Hall voltage,* V_H*, and the longitudinal voltage,* V_x*, at constant current for a* GaAs-$Al_xGa_{1-x}As$ *heterostructure at a temperature of 1.2 K. The carrier concentration is* $n = 5.6 \times 10^{11}$ *and the current is* $I = 5.5\ \mu A$*. (Cage et al. (1985).)*

Measurements on such two-dimensional systems at low temperatures and high magnetic fields display rather unusual properties, provided that the specimen is sufficiently pure (i.e., has a long elastic scattering time). Figure 48.1 shows measurements on a GaAs - $Al_xGa_{1-x}As$ structure.[2] The data have the following unique features (which are interconnected). First, the Hall voltage (see the scale to the left) shows discrete-like steps at higher fields; empirically it turns out that these voltages correspond to values of Hall resistance, $R_H = V_H/I$, given by

$$R_H = \frac{2\pi\hbar}{e^2 n} = \frac{h}{e^2 n}, \tag{48.3}$$

where n *is an integer*, within the accuracy to which these fundamental constants are known. Second, we note that at higher magnetic fields, the longitudinal resistance, $R = V_x/I$, drops by many orders of magnitude (depending on temperature, field, and sample purity). These effects were discovered by von Klitzing, Dorda, and Pepper (1980).[3] When supported by theoretical arguments it is now thought that Eq. (48.3) is exact; it can therefore be used as a way to determine the ratio of the fundamental constants involved, h/e^2. It then joins the Josephson effects as a solid-state phenomenon that can be used to refine the fundamental constants.

48.1.2 An electron in two dimensions in a magnetic field: rectangular coordinates

We begin by developing some results that will be basic to understanding, at least qualitatively, the observed behaviors, previously described. The presence of a

[2] A test of the quantum Hall effect as a resistance standard, M.E. Cage, R.F. Dziuba, and B.F. Field, *IEEE Trans. Instrum. Meas.* **IM-3**, 301 (1985).

[3] New method for high-accuracy determination of the fine-structure constant based on quantized Hall resistance, K. Klitzing, G Dorda, and M. Pepper, *Phys. Rev. Lett.* **45**, 494 (1980).

magnetic field, together with confinement in the direction perpendicular to the field, constrains the electrons to closed orbits in the classical limit, with the result that the energy levels become quantized, as was discussed earlier in Sections 18.5 and 20.1. Although there we worked in a quasi-classical approximation, we will here treat the single-electron problem fully quantum mechanically.[4] We will work in the effective mass approximation; furthermore we will assume that the effective mass tensor associated with the in-plane motion is diagonal with equal components, and denote them as m_e.[5] If we choose the gauge, $A_x = -Hy, A_y = 0$, the Schrödinger equation for such an electron in two dimensions is then

$$\frac{\hbar^2}{2m_e}\left[\left(\frac{1}{i}\frac{\partial}{\partial x}+\frac{eH}{\hbar c}y\right)^2-\frac{\partial^2}{\partial y^2}\right]\psi(x,y)=\varepsilon\psi(x,y). \tag{48.4}$$

We can separate variables in (48.4) by substituting the form $\psi(x,y) = e^{ikx}u(y)$, obtaining

$$\left[-\frac{\hbar^2}{2m_e}\frac{d^2}{dy^2}+\frac{1}{2}m_e\omega_c^2(y-y_0)^2\right]u(y)=\varepsilon u(y), \tag{48.5a}$$

where $y_0 \equiv -k\hbar c/eH = k\hbar c/|e|H$. We can also write $y_0 = kl_H^2$, where we have identified a *characteristic magnetic length*, $l_H \equiv \left(\hbar c/|e|H\right)^{1/2}$, which is independent of material parameters and for typical magnetic fields used in quantum Hall experiments is in the range 50–100 A. We can also write l_H in terms of the flux quantum, $\phi_0 \equiv hc/|e|$, introduced in Section 32.6: $l_H \equiv (\phi_0/2\pi H)^{1/2}$ (see Eq. (32.50b)).[6] In terms of l_H, Eq. (48.5a) can be written in the form

$$\frac{\hbar\omega_c}{2}\left[-l_H^2\frac{d^2}{dy^2}+\left(\frac{y}{l_H}-l_Hk\right)^2\right]u(y)=\varepsilon u(y), \tag{48.5b}$$

where we write $\omega_c = |e|H/m_ec$.

Equation (48.5a) has the form encountered in the one-dimensional harmonic oscillator problem, although here the origin is displaced to the point $y_0 = l_H^2k$. We may then immediately write the energies as

$$\varepsilon_n=\left(n+\frac{1}{2}\right)\hbar\omega_c, \tag{48.6}$$

which are referred to as *Landau levels* (L.D. Landau 1930); note they are independent of the quantum number, k. The corresponding (unnormalized) wave functions are

$$u(y)\propto H_n(y/l_H-l_Hk)e^{-(y-l_H^2k)^2/2l_H^2}, \tag{48.7}$$

where H_n are the Hermit polynomials. These wave functions are sketched for the first four levels in Fig. 48.2. As previously noted, the wave functions are centered

[4] Although using a different gauge, an analogous problem was treated in our discussion of the upper critical field of a type II superconductor in Section 33.3, leading to Eq. (33.47).

[5] This will be the case for electrons in GaAs which lie at the Γ-point. For Si, the electrons lie along the Γ–X directions. Only the ellipsoid associated with that Γ–X direction which is perpendicular to the layers is isotropic. Motion associated with the other ellipsoids is shifted in energy due to quantum confinement.

[6] The length, l_H, is identical to a_H of the G–L theory when $e^* = e$. Here we will write $\phi_0 = hc/e$ so that it can be distinguished from the superconducting flux quantum $\Phi_0 = hc/2e$.

on the points, $y_k = l_H^2 k$. The ground state Gaussian falls off with the characteristic length, l_H (along the y direction); the spatial extent of states with higher n approaches the classical amplitude. In what seems like a paradox, the states are not confined in the x direction, although classically our electron would execute a circular orbit; i.e., it would be confined in both the x and y directions; we will return to this point.

Since the energies do not involve k, the states must be degenerate with respect to this quantum number. We may evaluate the total degeneracy of a Landau level as follows. Since the orbit centers must be *inside the sample*, which we assume has a width, W in the *y direction*, we must clearly have $0 < y_k < W$. Substituting our expression for y_k, we have $0 < l_H^2 k < W$ or, equivalently, $0 < k < W/l_H^2$. The largest value of k allowed by this criterion is then

$$k_{max} = W/l_H^2. \tag{48.8a}$$

We now quantize the plane wave states involving motion along the *x direction* in terms of a quantum number, $n_k = 1, 2, \cdots n_{k\,max}$; $n_{k\,max}$ is then the total number of states involving the x degree of freedom. Applying periodic boundary conditions, the allowed values of k are then $k = (2\pi/L)n_k$, where L is the length of the sample.[7] Therefore,

$$k_{max} = \frac{2\pi}{L} n_{k\,max}. \tag{48.8b}$$

Combining (7a,b), we have

$$n_{k\,max} = \frac{LW}{2\pi l_H^2}. \tag{48.9}$$

Since all values of k have the same energy, $n_{k\,max}$ must correspond to the *total degeneracy* of a given Landau level, ε_n. The number of states *per unit area*, G, associated with a given Landau level, n, is given by $G = n_{k\,max}/LW$, and hence,[8]

$$G = \frac{1}{2\pi l_H^2} \tag{48.10a}$$

or

$$G = \frac{eH}{2\pi\hbar c} = \frac{H}{\phi_0}; \tag{48.10b}$$

note that this is independent of the quantum number, n, but *increases* with H. The total flux passing through the sample is $H \times LW$ and hence *each state is associated with one flux quantum*. Figure 48.3 shows, schematically, the succession of harmonic wells, along with the accompanying ground state wave functions around the origins, $y_{n_k} = (2\pi/L)\, l_H^2\, n_k$, resulting from the quantization of k.

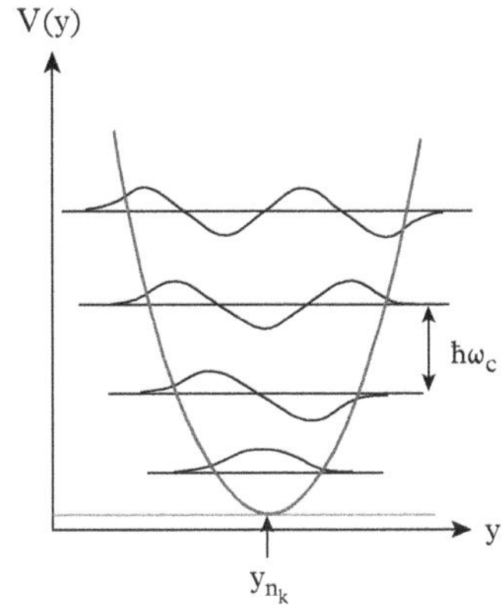

Figure 48.2 *The functions, $u_n(y)$, along with the harmonic potential provided by the magnetic field in the Landau gauge. (Courtesy M. Grayson.)*

[7] This choice of allowed k values also renders the states orthogonal:

$$\frac{1}{L}\int_0^L dx\, \psi_{k'}^*(x)\psi_k(x) = \frac{1}{L}\int_0^L dx\, e^{-ik'x}e^{ikx}$$
$$= \frac{1}{L}\int_0^L dx\, e^{-2\pi i n'_k x/L} e^{2\pi i n_k x/L}$$
$$= \delta_{n'_k n_k}.$$

[8] Note that the argument used here to obtain the level of degeneracy, which is somewhat more rigorous, differs from that used to obtain Eq. (20.5).

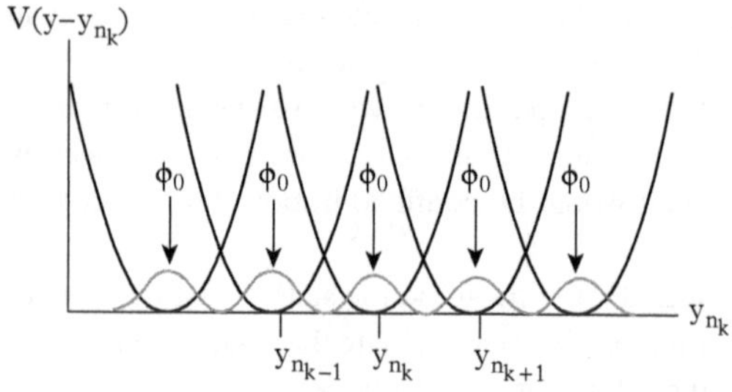

Figure 48.3 *The harmonic wells resulting from the vector potential in the Landau gauge with orbit centers* $y_k = l_H^2 k$ *when* k *quantized by periodic boundary conditions as* $k = (2\pi/L)n_k$. *Shown also are the Gaussian ground state wave functions. Each state is associated with one flux quantum. (Courtesy M. Grayson)*

We now return to the apparent paradox, noted previously. Since in quantum mechanics it is the vector potential that directly enters Schrödinger's equation, the form of the wave function will depend on the gauge chosen. In particular, had we chosen the gauge, $A_x = 0, A_y = Hx$ (as we did in Section 33.3 and which of course leads to the same magnetic field), we would have obtained plane wave like behavior along y. If we work in cylindrical coordinates, (r, θ), with a gauge involving only A_θ, which we take up in Section 48.1.6, we obtain wave functions which are localized in r but are oscillatory in θ. But the wave functions in all these gauges are highly degenerate. Since the states are degenerate, one might think of forming a linear superposition so as to form wave packets which are better localized. However, from Eq. (48.7) we see that if we superimpose a different k in an attempt to localize in x, we simultaneously move the origins along y, thereby smearing the wave function in this direction. Hence our ability to localize the orbits is intrinsically limited.

In thinking about this problem, we should not lose sight of the fact that semi-classically, the electron executes a *circular* cyclotron orbit around some origin. If at some time we were to localize the orbit center, as well as the position within the orbit, we should expect, physically, that both will spread out in time. To describe such behavior we would need to use a time-dependent formalism.

48.1.3 A two-dimensional electron gas in a magnetic field

We now move the discussion from that of a single electron moving in two dimensions to the case of a two-dimensional electron gas. We will assume, for the moment, that we are at absolute zero; also, we will ignore the electron–electron interaction. We further assume that the chemical potential, μ, is fixed by some *external reservoir.*[9] If μ lies between two Landau levels, all states with $\varepsilon_n < \mu$ will be filled and all those above will be empty. It is then common to refer to this state as *incompressible*, since a change in Fermi level does not change the electron concentration in a given Landau level; were the system a Fermi liquid and one increased the chemical potential (say by increasing the pressure or electron density), as with a compressible system, the radius of the Fermi sphere would increase.

[9] In two dimensions and in the absence of an external reservoir the Fermi level, $\varepsilon_F = (\hbar^2/2m_e)k_F^2$, at zero magnetic field is fixed by the two-dimensional electron density, $n^{(2d)}$, which is related to the Fermi wave vector as $n^{(2d)} = [2/(2\pi)^2]\pi k_F^2$, yielding $k_F = (2\pi n^{(2d)})^{1/2}$ or $\varepsilon_F = (\pi\hbar^2/m_e)n^{(2d)}$.

If we now increase the magnetic field, each Landau level will increase in energy and the levels become more widely separated. A level which earlier lay just below μ will therefore move up and, after it has passed through μ, must then empty. However, in an infinitesimally small field interval about the field at which this level passes through μ, the level occupancy will pass from the maximum degeneracy allowed by Eq. (48.9) to zero, and the system can transport charge.

48.1.4 Motion under orthogonal electric and magnetic fields

It follows from Newton's law and the Lorentz force that, in the absence of scattering, the motion of an electron in the x–y plane under the influence of an electric field, E_y, and a magnetic field, H (parallel to z), involves two components (see Fig. 48.4). First, there is the circular cyclotron orbit, with the azimuthal phase angle determined by the initial conditions, but second, the orbit center *simultaneously* drifts in the x direction with an average velocity given by[10]

$$\bar{v}_x = c\frac{E_y}{H}. \tag{48.11}$$

Although initially curious, it is E_y that will play the role of the Hall voltage, which we will clarify.

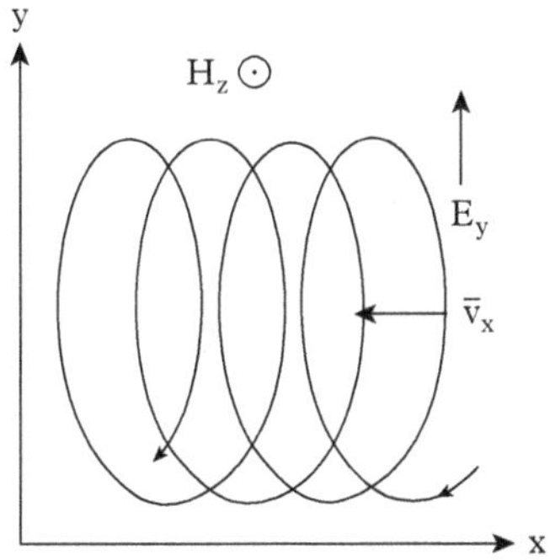

Figure 48.4 *Motion of an electron in two dimensions in the presence of an in plane electric and a perpendicular magnetic field magnetic field.*

Our previous quantum mechanical discussion can be generalized to the case where, simultaneously, we have a magnetic field in the z direction and an electric field in the y direction. The field, E_y, produces a potential energy, $V = -eyE_y$, and the Schrödinger equation (48.4) then becomes

$$\hat{H}_0(x, y)\psi(x, y) = \varepsilon\psi(x, y), \tag{48.12}$$

where

$$\hat{H}_0(x, y) = -\frac{\hbar^2}{2m}\left(\frac{\partial}{\partial x} + \frac{ieH}{\hbar c}y\right)^2 - \frac{\hbar^2}{2m}\frac{\partial^2}{\partial y^2} - eyE_y. \tag{48.13}$$

We again write the solution as a product

$$\psi(x, y) = e^{ikx}u(y),$$

obtaining

$$\frac{\hbar\omega_c}{2}\left[-l_H^2\frac{d^2}{dy^2} + \left(\frac{y}{l_H} - l_H k\right)^2 + \frac{2m_e cE_y}{\hbar H}y\right]u(y) = \varepsilon u(y).$$

By completing the square in the variable y we can rewrite this equation as

$$\frac{\hbar\omega_c}{2}\left[-l_H^2\frac{d^2}{dy^2} + \left(\frac{y}{l_H} - l_H k + \frac{m_e v_x l_H}{\hbar}\right)^2 + \frac{2l_H^2 k m_e v_x}{\hbar} - \frac{m_e^2 v_x^2 l_H^2}{\hbar^2}\right]u(y) = \varepsilon u(y),$$

[10] In the absence of scattering we might regard our system as having translational invariance. From the relativistic transformation properties of the Maxwell electromagnetic field it follows that, to lowest order, an electron moving with a velocity, v_x, in the presence of a field, H_z, will experience an electric field, $E_y = (v_x/c)H_z$, which is identical with the result (48.11).

where we have introduced the drift velocity, $v_x = cE_y/H$; for small currents we can neglect the term in v_x^2 and this equation becomes

$$\frac{\hbar\omega_c}{2}\left[-l_H^2\frac{d^2}{dy^2}+\left(\frac{y}{l_H}-l_Hk+\frac{m_e v_x l_H}{\hbar}\right)^2\right]u(y)=\left[\varepsilon-\omega_c l_H^2 k m_e v_x\right]u(y). \tag{48.14}$$

The solution to this equation can be written immediately on comparing (48.14) with Eqs. (48.5b) and (48.7),

$$\psi_{kn}(x,y;v)\propto e^{ikx}u_n\left(\frac{y}{l_H}-l_Hk+\frac{m_e v_x l_H}{\hbar}\right) \tag{48.15}$$

and

$$\varepsilon_{kn}=\hbar\omega_c\left(n+\frac{1}{2}+\frac{m_e v_x l_H^2 k}{\hbar}\right). \tag{48.16}$$

Note that the wave functions and energy levels now depend on the product of the drift velocity, v_x, and the quantum number, k; however the spacing between the levels numbered by n_k, given by $(m_e v_x l_H^2/\hbar)(2\pi/L)$, is *much smaller* (forming a quasi-continuum), than the spacing between the Landau levels, $\hbar\omega_c$, for the small v_x assumed here; i.e., each of the Landau levels can be treated as having the degeneracy, G.

48.1.5 Quantization of the Hall voltage: qualitative considerations

The presence of a drift velocity, v_x, will result in a two-dimensional current density (current per unit length),

$$j_x^{(2d)}=n^{(2d)}e\bar{v}_x=n^{(2d)}ec\frac{E_y}{H}, \tag{48.17}$$

where $n^{(2d)}$ is the two-dimensional electron density (number of electrons per unit area). It is common to introduce a filling fraction, ν, defined as

$$\nu=\frac{n^{(2d)}}{G}. \tag{48.18a}$$

If a specific level is full, together with those below it, while all the levels above are empty, ν will be an *integer*. From the expression for G, we have

$$\nu=\frac{n^{(2d)}}{H/\phi_0}. \tag{48.18b}$$

Combining (48.17) and (48.18b) and identifying E_y with the Hall voltage, E_H, we obtain the two-dimensional Hall resistivity as

$$\rho_{xy}^{(2d)} = \frac{1}{\sigma_{xy}^{(2d)}} = \frac{E_y}{j_x^{(2d)}} = \frac{1}{\nu}\frac{2\pi\hbar c}{e^2} = \frac{R_Q}{\nu}, \tag{48.19}$$

where we have introduced the *resistance quantum*, $R_Q = 2\pi\hbar c/e^2 = hc/e^2$;[11] note that it involves *no material parameters*, only fundamental constants. The resistance quantum enters other electrical transport phenomena, including localization theories (discussed in Chapter 43) and the so-called maximum metallic resistance in two dimensions. The same result follows if we simply insert Eq. (48.17) directly into Eq. (48.2) and take the limit $\omega_c\tau \to \infty$.

We are now in a position to give a *partial* explanation of the quantum Hall effect. Suppose there is an electric field in the y direction (i.e., the direction perpendicular to the direction of current transport). In the experiments shown in Fig. 48.1 the current is held fixed; from these results we see that on the flat portions of the quantum Hall steps the dissipation is very small; i.e., only a very small electric field, $E_x(\ll E_y)$, is required to sustain $j_x^{(2d)}$ (in the presence of a small amount of scattering) and the effective resistance is then very small.[12] One component of the explanation is that the dissipation is small because electrons in a filled Landau level *cannot scatter* to states of different energies at low temperature; this results in *low* resistances on the Hall plateaus. When a Landau level passes through the Fermi energy, scattering *between orbits with different centers* becomes possible and dissipation sets in; the electric field, E_x (and with it the voltage V_x), will then be much greater, as seen in the data.

For this scenario to work we must identify a mechanism for fixing the chemical potential at energies *between* Landau levels. There is strong evidence that this results from the presence of a continuum of *localized impurity states* (which are presumed to carry no current), in addition to the extended but discrete Landau levels. We will develop this idea more completely. In addition, there is *another channel* which is essentially free of dissipation that can carry current when the Fermi level lies between Landau levels; this will turn out to be *quantized edge states*, which we will also examine.

48.1.6 An electron in two dimensions in a magnetic field: cylindrical coordinates

When we come to discuss the fractional quantum Hall effect we will need the expression for the ground state wave function in cylindrical coordinates, and for completeness we will also develop a limiting case for all of the wave functions. As is easily verified a suitable form for the vector potential is

$$A_r = 0;\ A_\theta = \frac{1}{2}rH. \tag{48.20}$$

[11] Note that resistivity in two dimensions has the same units as resistance.

[12] Although it is E_x that "creates" E_y when minimum scattering is present, as when the Fermi level lies between Landau levels, the latter can be much smaller: the tail then wags the dog.

This form is commonly referred to as the symmetric gauge; it is equivalent to the form $\mathbf{A}(\mathbf{r}) = (1/2)\,\mathbf{r} \times \mathbf{H}$. Inserting (48.20) into the Schrödinger equation in cylindrical coordinates in the presence of a vector potential, we have

$$-\frac{\hbar^2}{2m_e}\left[\frac{1}{r}\frac{\partial}{\partial r}\left(r\frac{\partial}{\partial r}\right) + \left(\frac{1}{r}\frac{\partial}{\partial\theta} - \frac{ie}{2c}rH\right)^2\right]\psi(r,\theta) = \varepsilon\psi(r,\theta). \tag{48.21}$$

We will seek only an approximate solution to this equation. We start by writing $\psi(r,\theta) = \exp(i\,l\,\theta)R_l(r)$, where l is a positive integer (to ensure that the angular wave function is single valued) and substituting this form into (48.21) we obtain the equation for $R_l(r)$ as

$$-\frac{\hbar^2}{2m_e}\left[\frac{1}{r}\frac{d}{dr}\left(r\frac{d}{dr}\right) - \frac{1}{r^2}\left(l - \frac{r^2}{2l_H^2}\right)^2\right]R_{ln}(r) = \varepsilon\,R_{ln}(r), \tag{48.22}$$

where $l_H^2 = \hbar c/|e|H$, as previously. The second term acts as an attractive potential when $r^2 > 2ll_H^2$, while for small r it evolves into the centripetal barrier, thereby suppressing the amplitude of the wave function in that region; it vanishes at a value, $r_l = (2l\,l_H^2)^{1/2}$. In this region and to lowest order in $\Delta r = r - r_l$ we can write

$$\begin{aligned}\frac{1}{r^2}\left(l - \frac{r^2}{2l_H^2}\right)^2 &= \frac{1}{4l_H^4 r^2}\left(r_l^2 - r^2\right)^2 \\ &\cong \frac{1}{l_H^4}(r - r_l)^2.\end{aligned} \tag{48.23a}$$

For large l, accompanied by large r_l implying large r, we can write

$$\frac{1}{r}\frac{d}{dr}\left(r\frac{d}{dr}\right) \cong \frac{d^2}{dr^2}. \tag{48.23b}$$

Combining these approximations, Eq. (48.22) becomes

$$-\frac{\hbar^2}{2m_e}\frac{d^2R_{ln}(r)}{dr^2} + \frac{1}{2}m_e\omega_c^2\,(r - r_l)^2\,R_{ln}(r) = \varepsilon R_{ln}(r). \tag{48.24}$$

This equation is similar to (48.5a), except that rather than the electron being at the origin, $l_H^2 k$, it is now at $r_l = \sqrt{2l}\,l_H$. The unnormalized wave function will then have the approximate form

$$\psi_{ln}(r,\theta) \cong \exp(i\,l\,\theta)\,H_n\left((r - r_l)/l_H\right)\exp\left(-(r - r_l)^2/2l_H^2\right). \tag{48.25}$$

The wave functions of the electrons are now centered on *rings* of radius, r_l, but are azimuthally extended around the rings.

The exact wave functions are more complicated.[13] However, the (unnormalized) single electron ground state wave function is particularly simple and can be written in terms of the complex variable, $z = x - iy$, as

$$\psi_{l0} \propto z^l \exp\left[-|z|^2/4l_H^2\right]; \tag{48.26}$$

this can be easily verified by writing $z = re^{il\theta}$, substituting it into (48.22), and carrying out the differentiations. We will utilize this form extensively in discussing the fractional Hall effect.

48.1.7 Edge states

Landau levels that are centered close to the edges of a sample will be altered due to the potential that confines the electrons to the interior of the slab. Figure 48.5 shows some classical "skipping" orbits for the ideal case where the electron trajectory perfectly (referred to as specularly) reflects from a flat impenetrable boundary; three cases are shown for which the orbit center lies outside, inside, and at the boundary. Note that, as shown, all three orbits will experience an average drift velocity parallel to the wall; also note that the velocity is in the *opposite* direction at the opposing wall of the slab so there will be no net current transport through the slab.

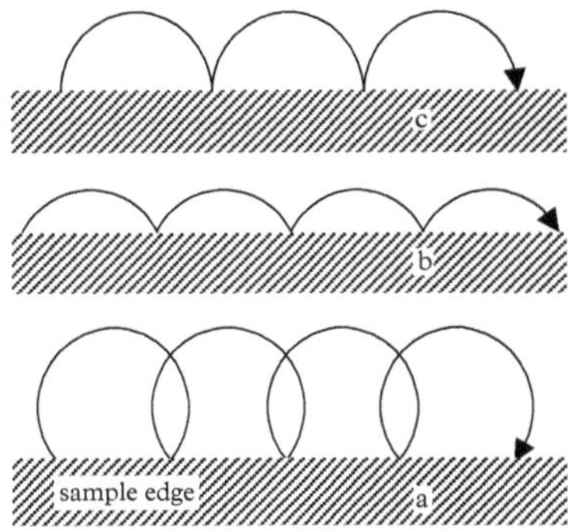

Figure 48.5 *Skipping trajectories when the orbit center is (a) outside, (b) inside, and (c) at a perfectly reflecting wall.*

We now examine the associated quantum mechanical problem. We begin by assuming that the boundary potential can be represented by the form

$$V(x,y) = \begin{cases} 0; & y < 0 \\ \infty; & y > 0 \end{cases}. \tag{48.27a}$$

For $y < 0$ the motion parallel to the boundary can still be represented by the plane wave form, e^{ikx}; simultaneously, the accompanying motion in the y direction will be governed by the one-dimensional Schrödinger equation (48.5a), but where now the potential has the modified form,

$$V(y) = \frac{1}{2}m_e\omega_c^2 \begin{cases} (y - y_k)^2, & y < 0; \\ \infty, & y > 0. \end{cases} \tag{48.27b}$$

A case where the origin, $y_k = l_H^2 k$, lies within the slab is shown in Fig. 48.6. The solutions are fixed by the boundary conditions

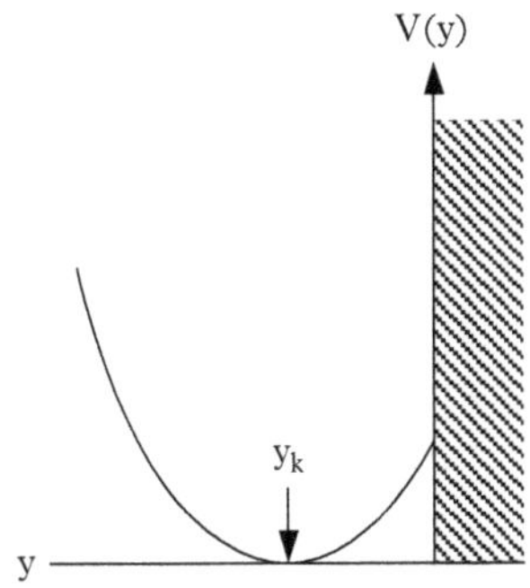

Figure 48.6 *The effective potential of an electron in a magnetic field near a boundary.*

$$u(y) = \begin{cases} 0, & y = 0; \\ \to 0, & y \to \infty. \end{cases} \tag{48.28}$$

The eigenvalues under these conditions no longer have the simple form, $\varepsilon_n = (n + 1/2)\hbar\omega_c$. In particular they depend on the origin, y_k, or, equivalently, they depend on k; i.e., $\varepsilon = \varepsilon_n(k)$ and for a given n are larger in energy due to the closer

[13] For the complete solution, see J.K. Jain, *Composite Fermions*, p. 29, Cambridge University Press, NY (2007).

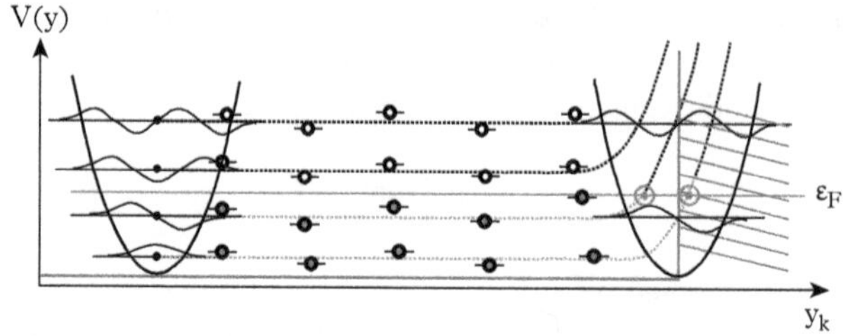

Figure 48.7 *Bulk and edge parabolic potentials that confine the motion of a particle in a magnetic field. The form at the left shows the potential along with the first four Landau levels associated with one of the many origins situated deep in the interior. To the right is shown the potential for the case when the origin is centered at the boundary. Here the states are all antisymmetric so as to create a* node *at the boundary; the portion within the interior is continued to illustrate this behavior but is in reality absent, as the wave function vanishes there. The solid line shows a chosen position for the Fermi level falling between two Landau levels. Shown also are localized states which pin the Fermi level between Landau levels. (Courtesy of M. Grayson.)*

confinement of the particle by the boundary. For the special case where $k = 0$, implying that $y_k = 0$, the solutions correspond to those far from the boundary but with *odd* n, since these have a *node* at the boundary; i.e., $\varepsilon_n(k = 0) = (2n + 1/2)\hbar\omega_c$. When $y_k \neq 0$, the eigenfunctions do not involve the Hermit polynomials and belong to a more general class. However, as the origin of a state in the interior of the slab is displaced toward an edge its energy will rise, connecting continuously to an energy corresponding to twice its bulk value. This situation is shown schematically in Fig. 48.7. The drift velocity of the nth edge state is given by the usual expression

$$v_n^{(edge)} = \frac{1}{\hbar}\frac{d\varepsilon_n(k)}{dk}. \tag{48.29}$$

Note that as the states lying below the Fermi level in the interior approach the edge their energy increases and they can intersect the Fermi level, thereby creating a current at the edge, since their velocity is finite according to Eq. (48.29); in equilibrium this current is canceled by one flowing on the opposing edge.

The edge currents provide the required additional channel, noted previously, which carries the current when the chemical potential lies between Landau levels.[14] Although electrons can scatter between edge states propagating in the *same* direction, scattering between states on *opposite* edges would require tunneling through an enormous potential barrier, which does not happen. The *net* current will then be the difference between the currents transported by opposing sides of the film.

48.1.8 Localized states

If there are irregularities in the potential that confines our two-dimensional gas they will produce a position-dependent electric field, $\mathbf{E}(x, y) = -e\nabla V(x, y)$. Depending on whether these irregularities produce a valley or hill in the local potential there will be an accompanying electric field that points inward or outward. We have just discussed how an electrical field will be associated with an in-plane velocity perpendicular to the electric field; provided this field varies

[14] This is the first example of a so-called "topological insulator," which is defined as a system that is insulating in the bulk but has current carrying states at the boundary.

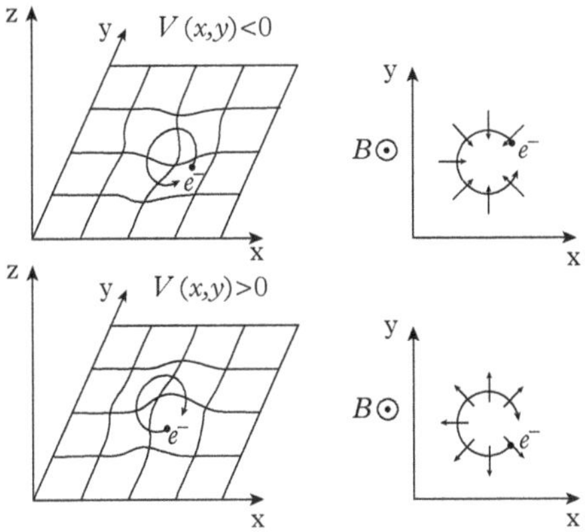

Figure 48.8 *To the left we show a valley, V(x,y) < 0, or hill, V(x,y) > 0, in the potential to form a localized state that can trap of an electron. The associated electric fields and orbits are shown to the right. (Courtesy of M. Grayson.)*

sufficiently slowly in space it can result in counter-clockwise or clockwise "orbits" about the extrema of the potential as shown schematically in Fig. 48.8. These can act as traps which in turn can pin the Fermi level between the Landau levels to produce quantized Hall plateaus.

48.2 The fractional quantum Hall effect

In very clean samples and at lower temperatures a new phenomenon sets in at fields above where the last Landau level passes through the Fermi level. As shown in Fig. 48.9 a new series of plateaus occur in the Hall voltage at non-integer, *rational fraction*, values of the filling fraction, ν, that are accompanied by a combination of spike-like features separated by low resistance regions in the magneto resistance.[15,16] The first such feature to be discovered was for $\nu = 1/3$, implying an effective charge of the same value.[17] This was followed by the observation of whole families of such fractions. The observation of this effect unleashed a wide theoretical and experimental effort to understand the origin of the observed fractions.

48.2.1 A semi-empirical accounting for observed filling fractions

The most comprehensive accounting for the observed fractions is the composite fermion model of J.K. Jain.[18] We will begin by developing a postulate that accounts for the vast majority of these fractions which we will then rationalize at various levels. We start by rewriting (48.19) for the Hall coefficient as

$$R_H \equiv R_Q/\nu, \tag{48.30}$$

[15] The Nobel Prize in Physics 1998 was awarded jointly to Robert B. Laughlin, Horst L. Störmer, and Daniel C. Tsui "for their discovery of a new form of quantum fluid with fractionally charged excitations."

[16] Composite Fermions in the Quantum Hall Regime, J.K. Jain, *Science* **266**, 1199 (1994).

[17] Two-dimensional magnetotransport in the extreme quantum limit, D.C. Tsui, H.L. Stormer, and A.C. Gossard, *Phys. Rev. Lett.* **48**, 1559 (1982).

[18] Composite-fermion approach for the fractional quantum Hall effect, J.K. Jain, *Phys. Rev. Lett.* **63**, 199 (1989).

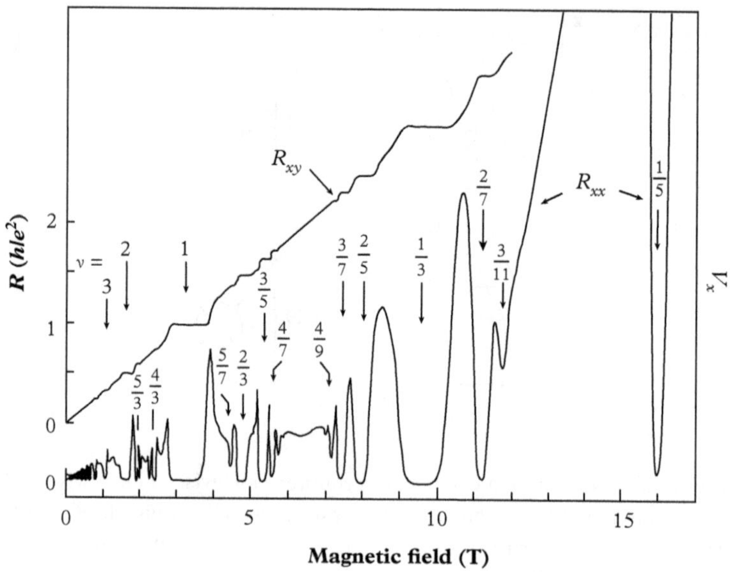

Figure 48.9 *The upper trace shows the magnetic field dependence of the Hall coefficient in a very pure GaAs quantum well. (After Goldman as quoted by Jain.)*

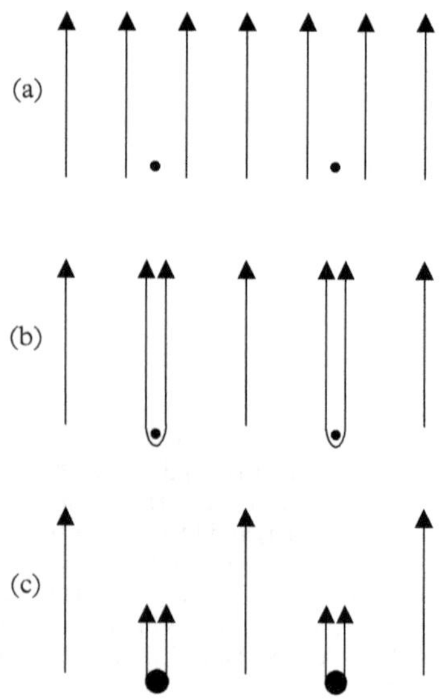

Figure 48.10 *(A) An electron, together with surrounding flux quanta associated with the field, H; (B) the electron "captures" two of the flux lines; (C) the* effective *flux associated with H after the capture of two lines.*

where $R_Q \equiv h/e^2$ and ν is now a rational fraction. The following sequences give a good overall accounting of the Hall plateaus in Fig. 48.9:

$$\nu = \frac{n}{2n+1} = \frac{1}{3}, \frac{2}{5}, \frac{3}{7} \cdots;$$

$$\nu = \frac{n}{2n-1} = \frac{2}{3}, \frac{3}{5}, \frac{4}{7} \cdots;$$

$$\nu = \frac{n}{4n+1} = \frac{1}{5}, \frac{2}{9}, \frac{3}{13} \cdots;$$

$$\nu = \frac{n}{4n-1} = \frac{2}{7}, \frac{3}{13}, \frac{4}{15} \cdots.$$

These and additional sequences can be captured by the following expression:

$$\nu = \frac{n}{2pn \pm 1}, \tag{48.31}$$

where p and n are both integers. We now present a rationale for this expression.

The germ of the idea is summarized in Fig. 48.10. At this point the reader is asked to accept the following three notions: in (A) electrons are in the presence of an external field; in (B) we say that each electron *captures* two of these flux lines to become what is called a *composite fermion*; and in (C) we show that composite fermions see an effective field, H^*, *diminished* by the loss of two flux quanta. If we accept these notions, the *effective* magnetic field, H^*, will be given by

$$H^* = H - 2pn^{(2d)}\phi_0, \tag{48.32}$$

where ϕ_0 and $n^{(2d)}$ continue to be the flux quantum and the two-dimensional number density of the electron gas, with p the integer introduced above; note the effective field can be either *positive or negative*. Recalling the definition of the filling factor in the "conventional" quantum Hall effect given in Eq. (48.18b), $\nu = n^{(2d)}\phi_0/H$, defining an *effective* filling factor for our composite fermions,

$$\nu^* = n^{(2d)}\phi_0/H^*, \tag{48.33}$$

inserting (48.32) for H^*, and eliminating the common factor, $n^{(2d)}\phi_0$, we obtain

$$\nu = \frac{|\nu^*|}{2p\,|\nu^*| \pm 1}. \tag{48.34}$$

The ν^* are filling factors at the effective field, H^*, which can have *either sign*, and are taken to be *integers* for filled levels of composite fermions; from (48.34) the *actual* filling numbers, ν, will then become *fractions*. Taking $\nu^* = n$, Eq. (48.34) is then in agreement with the empirically determined form, (48.31).[19] The fractions with $p = 2$ correspond to two flux quanta attached to the n electrons occupying the first n Landau levels; $p = 4$ involves attaching four flux quanta, and so on. The fractional quantum Hall effect of electrons is thus understood as the integer quantum Hall effect of composite fermions by the process of flux attachment.

Having constructed a physical description for classifying the filling fractions, ν, our task is then to theoretically justify notions (A) through (C). To do this we will need to construct appropriate wave functions that lead to fractionally quantized states, a discussion that is necessarily more formal.

48.2.2 Many-body ground states for non-interacting electrons in the lowest Landau level

The wave function in the symmetric gauge, $\mathbf{A} = (1/2)\,\mathbf{r} \times \mathbf{H}$, for a single electron in the lowest Landau level (LLL), ψ_{l0}, was given in Eq. (48.26). The (unnormalized) many-electron ground state for N *non-interacting* fermions in the LLLs having dimensionless coordinates (scaled by l_H), $z_j = x_j - iy_j$, is an anti-symmetrized product of the form

$$\Psi(z_1, z_2, \cdots z_N) = \sum_P \hat{P}(-1)^P \left(z_1^0 e^{-|z_1|^2/4} z_2^1 e^{-|z_2|^2/4} z_3^2 e^{-|z_3|^2/4} z_4^3 e^{-|z_4|^2/4} \ldots\right)$$

$$= \begin{vmatrix} 1 & 1 & 1 & \cdot & \cdot \\ z_1 & z_2 & z_3 & \cdot & \cdot \\ z_1^2 & z_2^2 & z_3^2 & \cdot & \cdot \\ \cdot & \cdot & \cdot & \cdot & \cdot \\ \cdot & \cdot & \cdot & \cdot & \cdot \end{vmatrix} \exp\left(-\frac{1}{4}\sum_{i=1}^{N} |z_i|^2\right),$$

where $j = 1 \cdots N$ correspond to filling the levels $l = 0, 1, 2, \cdots N$; here, $\hat{P}$ is the permutation operator that interchanges the particle coordinates and in the second

[19] The integer, n, introduced here denotes the *number of filled composite fermion Landau levels* and hence starts from 1; it is to be distinguished from the integer, n, used to denote the energy levels of a harmonic oscillator, which starts from zero.

step we write the antisymmetrized product as a Slater determinant. The determinant can be evaluated using Vandermond's theorem (which we do not prove[20]), with the result,

$$\Psi(z_1, z_2, \cdots z_N) = \exp\left(-\frac{1}{4}\sum_i |z_i|^2\right) \prod_{\substack{j,k=1 \\ j<k}}^{N} (z_j - z_k); \tag{48.35}$$

note that the result satisfies the requirement that the wave function vanishes if any two electrons have the same coordinates and changes sign on interchanging them.

Although not immediately apparent, Eq. (48.35) can be rewritten in the form[21]

$$\Psi = \exp\left[-\frac{N}{4}\left(\left|\frac{1}{N}\sum_{i=1}^{N} z_i\right|\right)^2\right] \prod_{\substack{j,k=1 \\ j<k}}^{N} (z_j - z_k) \exp\left(-\frac{1}{4N}\sum_{i=1}^{N} |z_j - z_k|^2\right). \tag{48.36}$$

When expressed in this way, the factor involving the product is clearly origin independent, while the first factor is interpreted as a *center of mass* and would change with a shift in origin, as expected. With this result we can henceforth work with the simpler form given by Eq. (48.35).

48.2.3 Flux tubes

A flux tube is a hypothetical entity containing one flux quantum in which the magnetic field vanishes everywhere except on an axis, where it is infinite:[22]

$$\mathbf{H}_t(\mathbf{r}) = \phi_0 \delta^{(2)}(\mathbf{r}) \mathbf{n}_z. \tag{48.37}$$

Here, $\mathbf{n}_z$ is a unit vector along z and $\delta^{(2)}(\mathbf{r})$ is a two-dimensional delta function. This equation is confirmed by integrating over the area, $\mathcal{A}$, of an orthogonal plane intersecting the flux tube,

$$\phi = \int \mathbf{H}_t \cdot d\mathcal{A} = \phi_0. \tag{48.38}$$

We can construct the associated vector potential, $\mathbf{A}_t(\mathbf{r})$, with the following argument. Using Stokes' theorem and $\mathbf{H} = \nabla \times \mathbf{A}$, the vector potential must satisfy the requirement

$$\oint \mathbf{A} \cdot dl = 2\pi r \mathbf{A}_t \cdot \mathbf{n}_\theta = \int \nabla \times \mathbf{A} \cdot d\mathcal{A} \\ = \int \mathbf{H} \cdot d\mathcal{A} = \phi_0 \quad ,$$

[20] See, J.K. Jain, *Phys. Rev. Lett* **63**, 199 (1989).

[21] Here the term, $\prod_{j<k}(z_j - z_k) \exp(-(1/4N)\sum_i |z_j - z_k|^2)$, generates multiple copies of the sum over $|z_j|^2$, together with additional sums involving the cross-terms, $z_j z_k^*$; however the term, $\exp\left[-(N/4)\left(\left|(1/N)\sum_{i=1}^{N} z_i\right|\right)^2\right]$, produces the necessary cancellations required to yield Eq. (48.35). See: Elementary theory: the incompressible quantum liquid, R.B. Laughlin in *The Quantum Hall Effect*, 1st edn, Ed. R.E. Prange and S.M. Girvin, Springer-Verlag, NY (1987).

[22] Here we can picture a solenoid with an infinitesimal radius and an infinite current per unit length, such that the total flux in the interior corresponds to one flux quantum.

where $\mathbf{n}_\theta$ is an azimuthal unit vector. Therefore,

$$\mathbf{A}_t(\mathbf{r}) = \frac{\phi_0}{2\pi}\frac{\mathbf{n}_\theta}{r} = \phi_0 \frac{\mathbf{n}_z \times \mathbf{r}}{r^2}. \tag{48.39}$$

This can also be written as[23]

$$\mathbf{A}_t(\mathbf{r}) = \frac{\phi_0}{2\pi}\nabla\theta \tag{48.40a}$$

or in terms of the complex variable, $z = x - iy = |z|e^{-i\theta}$, as

$$\mathbf{A}_t(\mathbf{r}) = i\frac{\phi_0}{2\pi}\nabla\, ln\left(\frac{z}{|z|}\right); \tag{48.40b}$$

we will utilize this form. Note that in the absence of an external field an electron passing outside this flux tube would experience a phase shift by the Aharonov–Bohm effect.[24]

The next step is to adiabatically add this contribution to the single-particle Hamiltonian for an electron in a magnetic field, which then takes the form

$$\hat{H}(\alpha) = \frac{1}{2m_e}\left(\hat{\mathbf{p}} - \frac{e}{c}\mathbf{A} - \frac{e}{c}\mathbf{A}_t(\alpha)\right)^2, \tag{48.41a}$$

where

$$\mathbf{A}_t(\alpha) = \frac{\alpha}{2\pi}\frac{\mathbf{n}_\theta}{r}\phi_0 \tag{48.41b}$$

is a flux tube contribution; here we have included a parameter, α, that will be varied from 0 to ± 1 in order to adiabatically insert the tube. The associated Schrödinger equation is

$$H(\alpha)\psi(\alpha) = \varepsilon(\alpha)\psi(\alpha). \tag{48.42}$$

We leave it as an exercise to show that substituting the form

$$\psi(\alpha) = \exp(-i\alpha\theta)\psi'(\alpha) \tag{48.43}$$

into (48.41a) yields a different Schrödinger equation,

$$H'(\alpha)\psi'(\alpha) = \varepsilon'(\alpha)\psi'(\alpha), \tag{48.44}$$

where

$$\hat{H}'(\alpha) = \frac{1}{2m_e}\left(\hat{\mathbf{p}} - \frac{e}{c}\mathbf{A}\right)^2. \tag{48.45}$$

23 When formulating the vector potential in this way and then calculating the magnetic field we encounter the operation $\nabla \times (\nabla\theta)$; in the present case this does not vanish everywhere since the function θ is singular at the origin.

24 Here electrons traversing opposite sides of the flux tube would encounter different phase shifts; this effect is discussed on most modern texts on quantum mechanics.

Interestingly, the transformation (48.43) has *eliminated* $\mathbf{A}_t$ from the resulting Hamiltonian, $\hat{H}'$. The (unnormalized) α-*dependent* form

$$\psi'(\alpha) = z^{l-\alpha} e^{-|z|^2/4} g \tag{48.46}$$

is a solution to (48.44), but for arbitrary α it does not satisfy the physical requirement of wave function continuity. However, it offers a way to adiabatically insert a flux tube by varying α between 0 and ± 1, and it is valid at the end points where it has the same ground state energy, $\varepsilon' = \hbar\omega_c/2$. But in the process, ψ' is transformed to the *neighboring* eigenfunctions having $l = \pm 1$; i.e., $\psi'(\alpha = \pm 1) = z^{\mp 1}\psi'(\alpha = 0) = z^{l\mp 1}e^{-|z|^2/4}$. The ground state wave function satisfying (48.42) is then

$$\psi(\alpha) = \exp(-i\alpha\theta) z^{l-\alpha} e^{-|z|^2/4}; \tag{48.47}$$

and on evolving α to ± 1 becomes

$$\psi(\alpha = \pm 1) = \exp(\mp i\theta) z^{l\mp 1} e^{-|z|^2/4}. \tag{48.48}$$

Hence, apart from a phase factor, the insertion of a flux line is equivalent to a state, $l \to l \mp 1$, which has the same energy, $\varepsilon = \hbar\omega_c/2$.

48.2.4 Adding flux tubes to the non-interacting many-body ground state wave function

We now use the previous formalism to locate flux tubes at positions, $\mathbf{r}_j$, in anticipation of attaching them to Landau levels. In terms of our complex coordinate, z_j, and taking note of Eqs. (48.40a,b), we can then write the vector potential at position i as

$$\mathbf{A}_t(\mathbf{r}_i) = \frac{\phi_0}{2\pi} \sum_j{}' \boldsymbol{\nabla}_i \theta_{ij}, \tag{48.49}$$

where

$$\theta_{ij} = i\, ln\left(\frac{z_{ij}}{|z_{ij}|}\right) \tag{48.50}$$

and

$$z_{ij} = z_i - z_j \tag{48.51}$$

is the inter-flux tube spacing. Equation (48.49) results in a total field at site, i, of

$$\mathbf{H}_t(\mathbf{r}) = \phi_0 \sum_i{}' \delta^{(2)}(\mathbf{r} - \mathbf{r}_i)\, \mathbf{n}_z. \tag{48.52}$$

Our next task is to construct a mean field Hamiltonian based on our non-interacting composite fermion model. To do this we assume the total magnetic field at the fractional plateaus is given by $H = H^* + 2pn^{(2d)}\phi_0$ (see Eq. (48.32)), which yields the Hamiltonian,

$$\hat{H}_{MF} = \frac{1}{2m}\sum_i \left(\hat{p}_i - \frac{e}{c}\mathbf{A}^*(\mathbf{r}_i) - 2p\frac{e}{c}\mathbf{A}_t(\mathbf{r}_i)\right)^2; \qquad (48.53a)$$

and the associated Schrödinger equation,

$$\hat{H}_{MF}\Psi_{MF} = E\Psi_{MF}. \qquad (48.53b)$$

Here, $\mathbf{A}^*$ is a vector potential that gives the uniform field, $\mathbf{H}^*$, that together with p attached flux tubes per Landau level produces the total field, $\mathbf{H}$.

48.2.5 Trial ground states for the fractional quantum Hall effect

We now construct "trial" wave functions that, when compared with exact wave functions obtained by a "brute force" numerical solution of the true Hamiltonian,

$$\hat{H} = \frac{1}{2m}\sum_i \left(\hat{p}_i - \frac{e}{c}\mathbf{A}(\mathbf{r}_i)\right)^2 + \sum_{i>j}\frac{e^2}{|\mathbf{r}_i - \mathbf{r}_j|}, \qquad (48.54)$$

have been found to show very large overlaps.[25]

We start by considering wave functions having the form

$$\Psi(z_1, z_2, \cdots z_N) = F_S(\{z_i\})\,\Psi_n^{(0)}(\{z_i\}), \qquad (48.55)$$

where $\Psi_n^{(0)}(\{z_i\})$ is a Slater determinant involving all of the fully occupied Landau levels through n, where the $n = 0$ form is given by (48.35), and the overall form is antisymmetric by construction. One role of the factor, $F_S\{z_i\}$, is to *keep the electrons apart* so as to minimize their Coulomb interaction. In choosing this function we will require that it be symmetric on particle interchange in order to respect the overall symmetry.[26] A simple form satisfying these properties is

$$F_S(\{z_i\}, p) = \prod_{\substack{j,k=1\\ j<k}}^{N} (z_j - z_k)^{2p}, \qquad (48.56)$$

where p is an integer. However, this form also has the effect of "mixing in" additional Landau levels, the resulting forms not being the eigenfunctions, $\Psi_n^{(0)}(\{z_i\})$, of the non-interacting many-body Hamiltonian.

[25] J.K. Jain and R.K. Kamilla, Composite fermions: particles of the lowest Landau level, Chapter 1, in *Composite Fermions*, O. Heinonen, Ed., World Scientific, New York (1998).

[26] This part of the wave function is sometimes called a *Jastrow factor*. An analogous factor is commonly incorporated into trial nuclear matter and liquid helium wave functions to account for strong two-body repulsion at short range, causing the wave function to vanish there (R. Jastrow, *Phys. Rev.* **98**, 1479 (1955)).

For the case of $\Psi_0^{(0)}$, corresponding to occupancy of the LLL (as given in Eq. (48.35)), this leads to the wave function,

$$\Psi(z_1, z_2, \cdots z_N) = \prod_{\substack{j,k=1 \\ j<k}}^{N} (z_j - z_k)^{2p+1} \exp\left(-\frac{1}{4}\sum_i |z_i|^2\right). \tag{48.57}$$

If we further restrict to the case $p = 1$, we have

$$\Psi(z_1, z_2, \cdots z_N) = \prod_{\substack{j,k=1 \\ j<k}}^{N} (z_j - z_k)^3 \exp\left(-\frac{1}{4}\sum_i |z_i|^2\right), \tag{48.58}$$

which is the heralded Laughlin wave function,[27] although he used different arguments to construct it. Numerical calculations show that it gives an excellent representation of the $\nu = 1/3$ fractional quantum Hall state; for $p = 2, 3\cdots$, it also describes the $\nu = 1/5, 1/7, \cdots$ states. At the time this was a major advance but the discovery of additional fractions clearly requires a generalization.

We next construct a trial wave function based on the composite fermion approach. Returning to (48.53a,b), we can perform a gauge transformation analogous to that used for the single particle case (see Eq. (48.43)) to remove the vector potential associated with the flux tubes,

$$\begin{aligned} \Psi^{MF}(\{z_i\}) &= \Phi(\{z_i\}, H^*) \exp\left(-2ip\sum_{j<k}\theta_{jk}\right) \\ &= \Phi(\{z_i\}, H^*) \prod_{j<k} \left(\frac{z_j - z_k}{|z_j - z_k|}\right)^{2p}, \end{aligned} \tag{48.59}$$

where $\Phi(\{z_i\}, H^*)$ is the solution to the Schrödinger equation,

$$\frac{1}{2m}\sum_i \left(\mathbf{p}_i - \frac{e}{c}\mathbf{A}^*(\mathbf{r}_i)\right)\Phi(\{z_i\}, H^*) = E\Phi(\{z_i\}, H^*). \tag{48.60}$$

The wave function here, $\Phi_{\pm n}(H^*)$, is a Slater determinant of N particles filling the $n = \nu^*$ lowest Landau levels, *but in a magnetic field*, H^*, with the coordinates measured in units of l_{H^*} The total wave function for a general *fractional* quantum Hall state is then

$$\Psi^{MF}_{n/(2pn\pm1)}(\{z_i\}) = \Phi_{\pm n}(\{z_i\}, H^*) \prod_{j<k} \left(\frac{z_j - z_k}{|z_j - z_k|}\right)^{2p}. \tag{48.61}$$

Since negative n corresponds to negative fields, which in turn corresponds to time-reversed states, we have

$$\Phi_{-n}(H^*) = \left[\Phi_n(H^*)\right]^*. \tag{48.62}$$

It turns out that lower energies are achieved if: (i) the denominators, $|z_j - z_k|$, in (48.61) are removed, which more strongly suppresses the wave function as

[27] R.B. Laughlin, Anomalous quantum Hall effect: an incompressible quantum fluid with fractionally charged excitations, *Phys. Rev. Lett.* **50**, 1395 (1980). Laughlin used different arguments to construct his wave function.

$z_j \to z_k$ (reducing the Coulomb repulsion energy and thereby yielding lower energies), and (ii) $\Phi_{\pm n}$ is evaluated at H rather than H*.[28] The wave function then becomes

$$\Psi_{n/(2pn\pm1)}(\{z_i\}) = \Phi_{\pm n}(\{z_i\}, H) \prod_{j<k} (z_j - z_k)^{2p}. \qquad (48.63)$$

The quantities attached to the electrons are then interpreted as *vortices*, as opposed to flux tubes; by analogy with the superconducting state, this may imply the presence of an overall phase in the many-body ground state wave function. We note that for n = 1, Eq. (48.63) corresponds to the Laughlin wave function. However, with $n = \nu^* > 1$, new fractions are allowed. Comparison of the states, Eq. (48.63) (after projecting them into the lowest Landau level), with exact states of numerical calculations (known for small systems), gives good agreement. Combined with the success in explaining experimental observations and the predictive power of the results, it lends support to the idea that composite fermions with attached flux tubes (or vortices) are, ultimately, the underlying physics of the fractional quantum Hall effect.

48.2.6 The ν = 1/2 state

Interesting behavior occurs at a filling fraction that would correspond to $\nu = 1/2$, as can be seen in Fig. 48.9. Here there is an extended region where the magnetoresistance is large, compared with that at the minimum, but relatively featureless; on the other hand no plateau is observed in the Hall voltage. Figure 48.11 shows an expanded view of this region.[29] We note the following properties: (i) the denominator for a $\nu = 1/2$ state would be even (unlike that for the composite fermion states described previously); (ii) the feature is not accompanied by a minimum in the dissipation; and (iii) the density of features in the neighborhood of this point appears to rapidly increase. Now from Eq. (48.34), we see that for the case, p = 2 and $\nu^* \to \infty$, we will have $\nu \to 1/2$ from both above and below. Here the effective magnetic field for composite fermions *vanishes*, and one can expect a *Fermi sea of composite fermions*. This reasoning provides an explanation for the absence of a FQHE here (just as there is no integral quantum Hall effect for electrons at zero magnetic field).[30]

Note that fractional Hall plateaus are observed at 5/2, 7/2 in the second Landau level, but will not be discussed here.

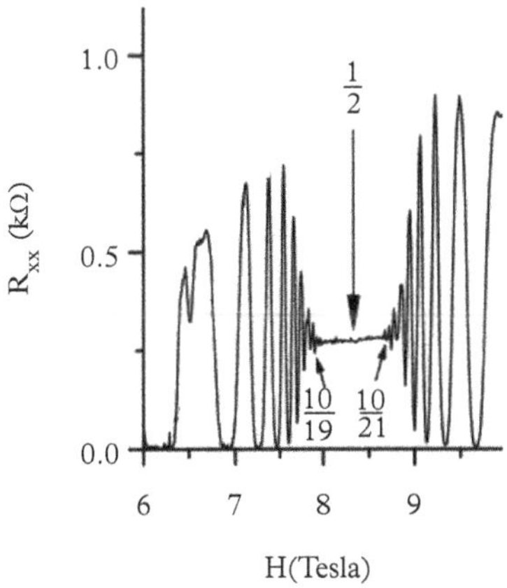

Figure 48.11 *The Hall resistivity in the vicinity of the $\nu = 1/2$ plateau. (From Pan et al (2003).)*

[28] This latter assumption can be justified, but this will not be carried out here; see, J.K. Jain, *Phys. Rev. Lett* **63**, 199 (1989).

[29] Fractional quantum Hall effect of composite fermions, W. Pan, H.L. Stormer, D.C. Tsui, L.N. Pfeiffer, K.W. Baldwin, and K.W. West, *Phys. Rev. Lett.* **90**, 016801 (2003).

[30] J.K. Jain, *Phys. Rev. Lett* **63**, 206 (1989).

ADDITIONAL READING

For a discussion of most aspects of the quantum Hall effect, see: *Composite Fermions*, J.K. Jain, Cambridge University Press, NY (2007).

For older treatments with contributions from a number of the architects of the field, see: *The Quantum Hall Effect*, 1st edition, R.E. Prange and S.M. Girvin,

Springer-Verlag, NY (1987) and *Quantum Hall Effect: A Perspective* A.H. Macdonald, Kluwer (1990).

Problems

48.1. Suppose that an electron moving in the x–y plane is acted on by a magnetic field H parallel to z and an electric field E parallel to y. Show that the trajectory of the electron involves circular orbits the center of which drifts along the x axis with a velocity given by cE/H. Find the time-dependent radius and phase of the electron orbit in terms of suitable initial conditions.

48.2. You are given the Hamiltonian of an electron in a magnetic field in the gauge, $\mathbf{A} = -\frac{1}{2}\mathbf{r} \times \mathbf{B}$, as

$$\hat{H} = \frac{1}{2m}\left[\left(\frac{\hbar}{i}\frac{\partial}{\partial x} + \frac{e}{c}\frac{By}{2}\right)^2 + \left(\frac{\hbar}{i}\frac{\partial}{\partial y} - \frac{e}{c}\frac{Bx}{2}\right)^2\right].$$

(a) Show that

$$\psi(x, y) \propto z^l \exp\left[-\frac{eBzz^*}{4\hbar c}\right]$$

is an eigenstate of this Hamiltonian, where $z \equiv x - iy$ and l is an integer.

(b) What does this state correspond to (hint: what is its energy?)?

(c) What is the normalization constant?

48.3. Verify that the transformation (48.43) leads to the Schrödinger equation (48.44), with the Hamiltonian given by (48.45).

Graphene, Carbon Nanotubes, and Fullerenes

49

Bulk diamond and graphite were discussed in connection with covalent bonding in Section 7.4. But other forms of carbon exist that are of interest due to their usual physical and chemical properties and which have important present and potential applications. Two classes of structures are the carbon nanotubes and the fullerenes, which will be briefly described later in this chapter. A third class of structures is graphene, which we may picture (see Fig. 49.1(a)) as a single cleaved plane of graphite (supported from its perimeter or on some substrate).[1] Specimens containing a few such layers, and with care a single layer, can be prepared experimentally by attaching adhesive tape to the flat surface of a graphite flake and pulling it free. From a conceptual point of view graphene is the simplest of the alternative forms of elemental carbon, with which we will begin our discussion.

49.1 The electronic structure of graphene

Formally, graphene belongs to the symmetry group p3m1 (see plate 14 of Fig. 8.7).[2,3,4] Figure 49.1(b) shows a suitable choice for the underlying two-dimensional oblique Bravais lattice associated with single-layer graphene. The dashed lines connecting nearest-neighbor atoms depict the chemical bonds and show the underlying hexagonal "honeycomb" structure, although the symmetry around any given atom is three-fold. The structure may be viewed as two interpenetrating hexagonal sublattices (each with the space group p6mm, see plate 17 of Fig. 8.7).

The two-dimensional primitive Bravais lattice vectors, **a** and **b** in Fig. 49.1(b), are given by

$$\mathbf{a} = \frac{a}{2}\left(3\mathbf{i} + \sqrt{3}\mathbf{j}\right), \tag{49.1a}$$

$$\mathbf{b} = \frac{a}{2}\left(3\mathbf{i} - \sqrt{3}\mathbf{j}\right), \tag{49.1b}$$

where a is the carbon–carbon spacing, 1.42A. The B atom is located at the point $(3a/2)\mathbf{i}$ in the unit cell.

[1] Two-dimensional atomic crystals, K.S. Novoselov, D. Jiang, F. Schedin, T.J. Booth, V.V. Khotkevich, S.V. Morozov, and A.K. Geim, *Proc. Natl Acad. Sci. U.S.A.* **102**, 10451 (2005).

[2] The 2010 Nobel Prize in Physics was awarded jointly to Andre Geim and Konstantin Novoselov "for groundbreaking experiments regarding the two-dimensional material graphene."

[3] The band structure of graphene was first calculated by P.R. Wallace in: The band theory of graphite, *Phys. Rev.* **71**, 622 (1947).

[4] Our discussion here has utilized some lecture notes prepared by A.J. Leggett.

The Physics of Solids. First Edition. J. B. Ketterson.

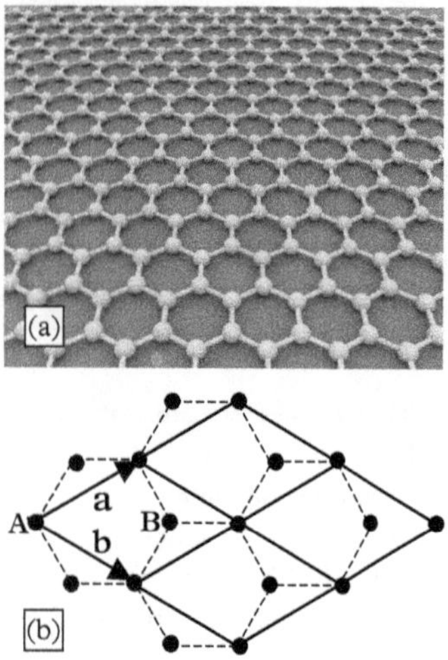

Figure 49.1 *(a) A schematic representation of a two-dimensional graphene sheet. (b) The unit cell where one of the two atoms is chosen as the origin. The dark lines show the Bravais lattice. The large dots indicate the positions of the carbon atoms, while the interconnecting dashed lines represent their directed bonds; note that there are two atoms per unit cell.*

The corresponding reciprocal lattice vectors are

$$a^* = \frac{2\pi}{3a}\left(i + \sqrt{3}j\right) \tag{49.2a}$$

and

$$b^* = \frac{2\pi}{3a}\left(i - \sqrt{3}j\right). \tag{49.2b}$$

The two-dimensional Brillouin zone formed from these two vectors is shown in Fig. 49.2. The points, K and K′, are different and are discussed here.

The three in-plane sp^2 orbitals given in Eqs. (7.30a–c) are primarily responsible for bonding the atoms together into the hexagonal planes making up the graphite structure. The associated bands are filled and, though essential for the structural stability of graphite, they play no role in electron transport; it is the fourth state (7.31d) that governs electron transport. We will calculate $\varepsilon(\mathbf{k})$ for our single graphene layer using the tight-binding method, discussed in Chapter 11. To do this we form the following orbitals:

$$\begin{aligned}\psi(\mathbf{r}) &= \sum_{j=1}^{2}\sum_{\mathbf{R}} a_{j\mathbf{R}}\phi(\mathbf{r} - \mathbf{r}_j - \mathbf{R}) \\ &\equiv \sum_{j=1}^{2}\sum_{\mathbf{R}} a_{j\mathbf{R}}\phi_{j\mathbf{R}}\end{aligned}, \tag{49.3}$$

where $\phi(\mathbf{r})$ is the full (radial and angular) carbon $2p_z$ wave function, the index, j, denotes the positions, $\mathbf{r}_j$, of the two atoms within the unit cell (as shown in

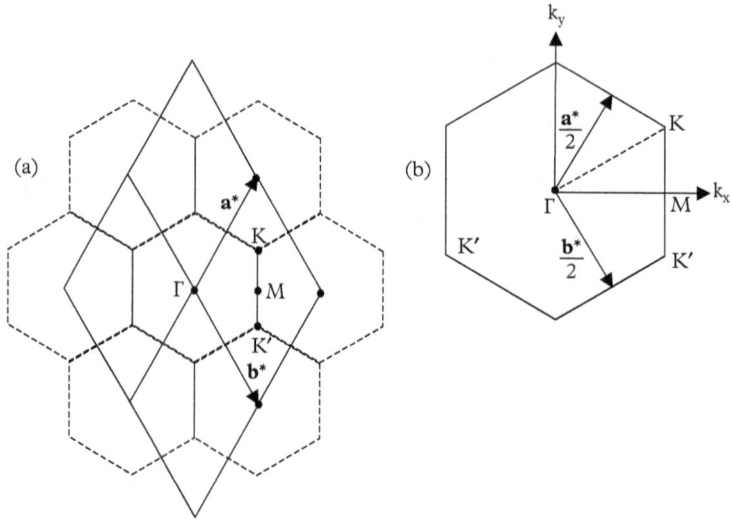

Figure 49.2 *(a) Four unit cells of the two-dimensional graphene reciprocal lattice formed from the vectors,* a* *and* b*. *Shown also are repeated copies of the first Brillouin zone and designated symmetry points. (b) The two-dimensional Brillouin zone for graphene, together with the symmetry points; note,* K *and* K′ *are not equivalent.*

Fig. 49.1(b)), and $\mathbf{R} = n_1\mathbf{a} + n_2\mathbf{b}$ denotes all the real-space lattice vectors. In place of Eq. (11.43), our Hamiltonian now has the form

$$\hat{H} = -\frac{\hbar^2}{2m}\nabla^2 + V(\mathbf{r}), \tag{49.4}$$

where

$$V(\mathbf{r}) = \sum_{j=1}^{2}\sum_{\mathbf{R}} U(\mathbf{r} - \mathbf{r}_j - \mathbf{R}) \tag{49.5}$$

is the full crystal potential energy and $U(\mathbf{r})$ are the atomic potential energies.

The band energies then follow from taking the expectation value of the Hamiltonian (49.4) with respect to the states (49.3), as we did in Eq. (11.45) for a single orbital (in that case an s-state), but now extended to the case of two atoms per unit cell,[5]

$$E = \frac{\sum_{j,\mathbf{R}}\sum_{j',\mathbf{R}'} a^*_{j\mathbf{R}} a_{j'\mathbf{R}'} H_{j\mathbf{R},j'\mathbf{R}'}}{\sum_{j,\mathbf{R}}\sum_{j',\mathbf{R}'} a^*_{j\mathbf{R}} a_{j'\mathbf{R}'} J_{j\mathbf{R},j'\mathbf{R}'}}; \tag{49.6}$$

here we have defined the multi-site "overlap integrals",

$$J_{j\mathbf{R},j'\mathbf{R}'} = \int d^3r \phi^*_{j\mathbf{R}} \phi_{j'\mathbf{R}'}, \tag{49.7}$$

with $J_{j\mathbf{R},j\mathbf{R}} = 1$, and the Hamiltonian matrix elements,

$$H_{j\mathbf{R},j'\mathbf{R}'} = \varepsilon_{2p_z} J_{j\mathbf{R},j'\mathbf{R}'} + h_{j\mathbf{R},j'\mathbf{R}'}; \tag{49.8}$$

where

$$h_{j\mathbf{R},j'\mathbf{R}'} = \int d^3r \phi^*_{j\mathbf{R}} \left[V(\mathbf{r}) - U\left(\mathbf{r} - \mathbf{r}_{j'} - \mathbf{R}'\right)\right] \phi_{j'\mathbf{R}'}, \tag{49.9}$$

and we have used the property $\left(-(\hbar^2/2m)\nabla^2 + U(\mathbf{r})\right)\phi(\mathbf{r}) = \varepsilon_{2p_z}\phi(\mathbf{r})$.

Minimizing (49.5) with respect to the variational parameters, $a^*_{j\mathbf{R}}$, and assuming a form $a_{j\mathbf{R}} = a_j e^{i\mathbf{k}\cdot\mathbf{r}_j} e^{i\mathbf{k}\cdot\mathbf{R}}$, gives the set of equations

$$\sum_{j',\mathbf{R}'} \left[\left(\varepsilon_{2p_z} - \varepsilon_{j'}(\mathbf{k})\right) J_{jj'}(\mathbf{R}') + h_{jj'}(\mathbf{R}')\right] a_{j'} e^{i\mathbf{k}\cdot\mathbf{r}_{j'}} e^{i\mathbf{k}\cdot\mathbf{R}'} = 0, \tag{49.10}$$

where we have used the fact that the periodicity allows us to set $\mathbf{R} = 0$. We formally perform the sum over $\mathbf{R}'$ by writing

$$h_{jj'}(\mathbf{k}) = \sum_{\mathbf{R}'} h_{jj'}(\mathbf{R}') e^{i\mathbf{k}\cdot\mathbf{R}'} \tag{49.11a}$$

[5] Here we are implicitly assuming that due to the formation of strong in-plane bonds, the bands associated with the three sp^2 orbitals are sufficiently split off in energy from that associated with the lone p_z orbital, that they can be ignored in the vicinity of the Fermi level.

and

$$J_{jj'}(\mathbf{k}) = \sum_{R'} J_{jj'}(\mathbf{R}')e^{i\mathbf{k}\cdot\mathbf{R}'}. \qquad (49.11b)$$

Writing $a_{j'}\, e^{i\mathbf{k}\cdot\mathbf{r}_{j'}} = a_{j'}(\mathbf{k})$, Eq. (49.10) then takes the form

$$\sum_{j'} \left[\left(\varepsilon_{2p_z} - \varepsilon_{j'}(\mathbf{k})\right) J_{jj'}(\mathbf{k}) + h_{jj'}(\mathbf{k})\right] a_{j'}(\mathbf{k}) = 0, \qquad (49.12a)$$

where $j' = A, B$. The two energy bands associated with the A an B atoms then follow from setting the 2×2 determinant of the coefficients to zero:

$$\left|\left(\varepsilon_{2p_z} - \varepsilon_{j'}(\mathbf{k})\right) J_{jj'}(\mathbf{k}) + h_{jj'}(\mathbf{k})\right| = 0. \qquad (49.12b)$$

We will restrict ourselves to *nearest-neighbor* interactions. Each carbon atom has three nearest neighbors. The matrix elements between these atoms are identical and we denote them as h. The three overlap integrals are also identical and are denoted as J.

The sums in (49.11a,b) which couple the A atom to the three nearest B atoms, lying in adjacent cells, involve the form

$$g(\mathbf{k}) = e^{i\mathbf{k}\cdot\boldsymbol{\delta}_1} + e^{i\mathbf{k}\cdot\boldsymbol{\delta}_2} + e^{i\mathbf{k}\cdot\boldsymbol{\delta}_3}, \qquad (49.13)$$

where $\boldsymbol{\delta}_i$ are their vector separations, which are given by

$$\boldsymbol{\delta}_1 = \frac{a}{2}\left(\mathbf{i} + \sqrt{3}\mathbf{j}\right), \boldsymbol{\delta}_2 = \frac{a}{2}\left(\mathbf{i} - \sqrt{3}\mathbf{j}\right), \boldsymbol{\delta}_3 = -a\mathbf{i}. \qquad (49.14a–c)$$

Inserting these vectors into (49.13), we obtain

$$g(\mathbf{k}) = e^{-ik_x a} + 2e^{ik_x a/2}\cos\left(\sqrt{3}k_y a/2\right). \qquad (49.15)$$

In place of (49.13), the sum corresponding to the B atom yields the *complex conjugate*, $g^*(\mathbf{k})$. Equation (49.12a) then takes the form

$$\begin{pmatrix} \varepsilon_{2p} - \varepsilon(\mathbf{k}) & \left[h + J(\varepsilon_{2p} - \varepsilon(\mathbf{k}))\right] g(\mathbf{k}) \\ \left[h + J(\varepsilon_{2p} - \varepsilon(\mathbf{k}))\right] g*(\mathbf{k}) & \varepsilon_{2p} - \varepsilon(\mathbf{k}) \end{pmatrix} \begin{pmatrix} a_A(\mathbf{k}) \\ a_B(\mathbf{k}) \end{pmatrix} = 0, \qquad (49.16a)$$

and evaluating the determinant gives

$$\left(\varepsilon_{2p} - \varepsilon(\mathbf{k})\right)^2 + |g(\mathbf{k})|^2\left(h^2 + 2Jh(\varepsilon_{2p} - \varepsilon(\mathbf{k})) + J^2(\varepsilon_{2p} - \varepsilon(\mathbf{k}))^2\right) = 0. \qquad (49.16b)$$

In the limit of small overlap ($J \cong 0$), Eq. (49.17a) becomes

$$\begin{pmatrix} \varepsilon_{2p} - \varepsilon(\mathbf{k}) & h\,g(\mathbf{k}) \\ h\,g*(\mathbf{k}) & \varepsilon_{2p} - \varepsilon(\mathbf{k}) \end{pmatrix} \begin{pmatrix} a_A(\mathbf{k}) \\ a_B(\mathbf{k}) \end{pmatrix} = 0. \qquad (49.17a)$$

The eigenvalues of (49.17a) are then given by[6]

$$\varepsilon(k_x, k_y) = \varepsilon_{2p_z} \pm h\sqrt{|g(\mathbf{k})|^2} \qquad (49.17b)$$

or

$$\varepsilon(k_x, k_y) = \varepsilon_{2p} \pm h\sqrt{1 + 4\cos\left(\frac{3}{2}k_x a\right)\cos\left(\frac{\sqrt{3}}{2}k_y a\right) + 4\cos^2\left(\frac{\sqrt{3}}{2}k_y a\right)}. \qquad (49.17c)$$

We recall that in the first Brillouin zone, shown in Fig. 49.2, we identified the points, K and K′, which have the coordinates

$$\mathbf{k}^{(K)} = (2\pi/3a)\left(\mathbf{i} + \mathbf{j}/\sqrt{3}\right)$$

and

$$\mathbf{k}^{(K')} = (2\pi/3a)\left(\mathbf{i} + \mathbf{j}/\sqrt{3}\right).$$

It is straightforward to verify that

$$g\left(\mathbf{k} = \mathbf{k}^{(K)}\right) = g\left(\mathbf{k} = \mathbf{k}^{(K')}\right) = 0;$$

hence the eigenvalues (49.17b) are degenerate at K and K′.

The – and + signs in (49.17a) correspond to the bonding (π) and antibonding (π^*) bands associated with $2p_z$ orbitals respectively.[7] Figure 49.3(a) shows a plot of $\varepsilon(k_x, k_y)$ in perspective; Fig. 49.3(b) shows cuts through this surface along the symmetry lines, ΓM, ΓK, and KM. Since we have only a single band (the three sp^2 bonding bands already being filled) and there are two atoms per unit cell, the

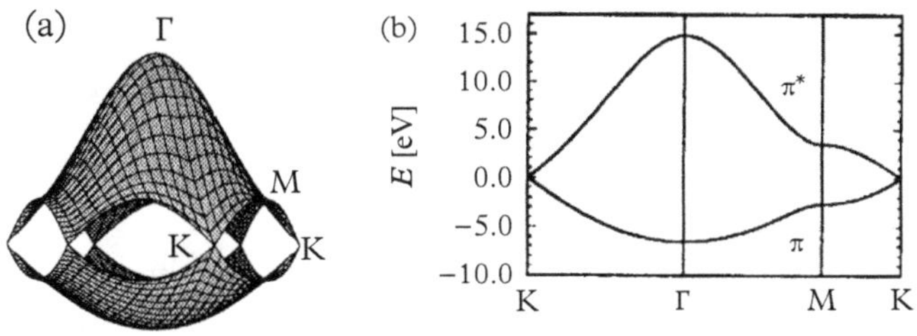

Figure 49.3 *(a) A three-dimensional sketch of the energy bands of graphene; (b) cuts along symmetry lines. Here, π and π^* denote bonding and antibonding bands.*

[6] Since we obtain $\varepsilon_{2p} - \varepsilon(\mathbf{k}) \propto h$, this justifies our neglect of two terms in (6), which are then of order, Jh^2 and h^2J^2; their inclusion does not remove the degeneracy at the points K and K′. If second-nearest-neighbor interactions are included and parameterized by h_{nnn}, Eq.(49.18a) becomes $\varepsilon(k_x, k_y) = \varepsilon_{2p} \pm h\sqrt{3 + f(\mathbf{k})} - h_{nnn}f(\mathbf{k})$, with $f(\mathbf{k}) = 2\cos(\sqrt{3}k_y a) + 4\cos(\sqrt{3}k_y a/2)\cdot\cos(3k_x a/2)$; although the energy is shifted, the degeneracy remains. Note that the addition of second-nearest neighbors leads to different band widths for the higher and lower lying bands, as seen in Fig. 49.3.

[7] Here, the lobes of bonding $2p_z$ orbitals on neighboring atoms have the opposite sign.

π band will be filled and hence the Fermi level passes through K points. We note that the point on the *opposite* side of the zone also corresponds to a K′ point (it is related to the K′ point in Fig. 49.2 by a rotation of −120°. From its definition in (49.15), under the operation, $\mathbf{k} \to -\mathbf{k}$, the function $g(\mathbf{k}) \to g(-\mathbf{k}) = g^*(\mathbf{k})$; if we apply this property to Eq. (49.17a), we obtain $a_B(-\mathbf{k}) = a_A(\mathbf{k})$. Hence the two components of the wave function *switch places* on inversion or, equivalently, on passing from K to K′.

To examine the behavior in the vicinity of the K point, we expand the Hamiltonian matrix in (6) to leading order in the wave vector, $\mathbf{q}$, measured from that point; i.e., $\mathbf{k} = \mathbf{k}^{(K)} + \mathbf{q}$. Carrying out this expansion to leading order yields (see Problem 49.1)

$$g(\mathbf{k}^{(K)} + \mathbf{q}) \cong \frac{3a}{2}\left(\frac{i - \sqrt{3}}{2}\right)(q_x + iq_y) + \mathbf{O}q^2.$$

The complex number in parenthesis, $\exp(5\pi i/6)$, has modulus unity; it corresponds to a phase factor that can be suppressed. We can then write (49.17b) as

$$\varepsilon(q_x, q_y) \cong \varepsilon_{2p} \pm \frac{3ha}{2}\sqrt{q_x^2 + q_y^2}. \tag{49.18}$$

We henceforth measure our energies from ε_{2p} and introduce a velocity, $v = 3ha/2\hbar$, and our linearized expression for the energies in the vicinity of the K point becomes

$$\begin{aligned}\varepsilon(q_x, q_y) &= \pm\hbar v\sqrt{q_x^2 + q_y^2} \\ &= \pm\hbar v q.\end{aligned} \tag{49.19}$$

The two signs refer to two different bands which intersect at the point K. It is important that rather than having a parabolic ε versus k behavior, characteristic of most band extrema, here the behavior is *linear*. Hence, in the vicinity of a K point, an electron moves with a *constant velocity*, $v = \hbar^{-1}\partial\varepsilon/\partial q = 3ha/2\hbar$, *independent of its energy*; experimentally, $v \cong 1.0 \times 10^6$ cm/sec.

Combining these, we have

$$\hat{H}(\mathbf{k}_K + \mathbf{q})\,\vec{\psi}(\mathbf{k}_K + \mathbf{q}) = \varepsilon(\mathbf{q})\vec{\psi}(\mathbf{k}_K + \mathbf{q}), \tag{49.20a}$$

where

$$\hat{H}(\mathbf{k}_K + \mathbf{q}) = \hbar v\begin{pmatrix} 0 & q_x + iq_y \\ q_x - iq_y & 0 \end{pmatrix} = \hbar v\hat{\sigma}\cdot\mathbf{q}, \tag{49.20b}$$

where we have introduced the Pauli matrices in the second step, which operate in a pseudo spin space associated with the two sublattices. Now, a relativistic "electron" with zero rest mass has a similar behavior in the Dirac theory, and the K points are commonly called "Dirac points"; the associated electron and hole bands in the vicinity of K are shown schematically in Fig. 49.4, the structures being referred to as "Dirac cones." Writing $q_x \pm iq_y = qe^{\pm i\theta(q)}$, the eigenfunctions of (49.20) near K immediately follow as

$$\vec{\psi}_\pm(\mathbf{k}_K + \mathbf{q}) = \begin{pmatrix} a_A(\mathbf{k}_K + \mathbf{q}) \\ a_B(\mathbf{k}_K + \mathbf{q}) \end{pmatrix} = \frac{1}{\sqrt{2}} \begin{pmatrix} e^{+i\theta(\mathbf{q})/2} \\ \pm e^{-i\theta(\mathbf{q})/2} \end{pmatrix}, \qquad (49.21a)$$

where the + sign corresponds to positive energies and

$$\theta(\mathbf{q}) = \tan^{-1}\left(\frac{q_x}{q_y}\right). \qquad (49.21b)$$

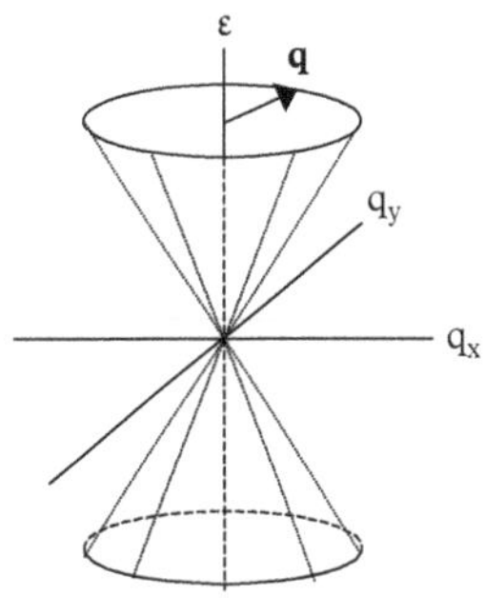

Figure 49.4 *The band structure of graphene in the vicinity of the K points of the Brillouin zone.*

For the point, K′, we have

$$\hat{H}(\mathbf{k}_{K'} + \mathbf{q}) = \hat{H}^*(\mathbf{k}_K + \mathbf{q});$$

hence,

$$\vec{\psi}_\pm(\mathbf{k}'_K + \mathbf{q}) = \vec{\psi}^*_\pm(\mathbf{k}_K + \mathbf{q});$$

i.e., $\theta(\mathbf{q})$ is replaced by $-\theta(\mathbf{q})$ in (49.21a). Note that when $\mathbf{q}$ rotates once around the Dirac point, the phase of $\psi_\pm(\mathbf{q})$ changes by π, thereby requiring two such "windings" to restore the original phase, rather than a single winding, as with conventional rotations; the former is similar to what happens with a spin 1/2 particle. These two cases are then said to have *winding numbers* of 2 and 1 respectively.

Including the spin degeneracy, the number of electrons per unit area at energy, ε, associated with a *single* Dirac point is given by

$$n(\varepsilon) = \frac{2}{(2\pi)^2}\pi q^2(\varepsilon) = \frac{q^2}{2\pi}; \qquad (49.22a)$$

the density of states is then

$$N(\varepsilon) = \frac{dn(\varepsilon)}{d\varepsilon} = \frac{q}{\pi}\frac{dq}{d\varepsilon} = \frac{\varepsilon}{\pi(\hbar v)^2}. \qquad (49.22b)$$

Since there are two such cones in the reciprocal space unit cell, the total density of states is twice this value. The cyclotron effective mass now follows from Eq. (18.28a), as

$$m_c = \frac{1}{2\pi}\frac{\partial A_q}{\partial \varepsilon} = \frac{1}{2\pi}\frac{\partial(\pi q^2(\varepsilon))}{\partial \varepsilon} = \frac{\varepsilon}{v^2}. \qquad (49.23a)$$

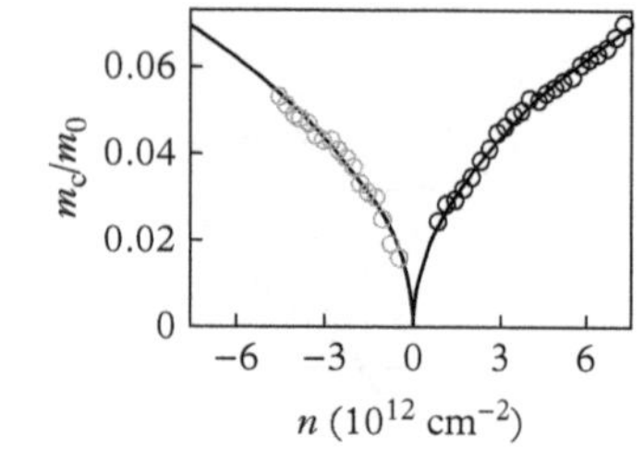

Figure 49.5 *The cyclotron effective mass as a function of the carrier density in graphene. (After Novoselov et al. (2005).)*

When expressed in terms of the number density of electrons below the Fermi energy, $n(\varepsilon_F) = n^{(2d)}$, which can be controlled via a gate electrode, we have

$$m_c = \frac{\hbar}{v}(\pi n^{(2d)})^{1/2}. \tag{49.23b}$$

Note that, defined in this way, $m_c = 0$ at the K point; this behavior is shown in Fig. 49.5.[8]

It turns out that if the layer were perturbed, e.g., by the introduction of a corrugation in the film, Eq. (49.20) would take the form

$$\hat{H} = \begin{pmatrix} +\Delta & \hbar v(q_x + iq_y) \\ \hbar v(q_x - iq_y) & -\Delta \end{pmatrix}, \tag{49.24a}$$

which has the eigenvalues,

$$\varepsilon = \pm\sqrt{\Delta^2 + \hbar^2 v^2 q^2}. \tag{49.24b}$$

We would then have bands split by an energy, Δ, at the K point with the usual locally parabolic behavior. And this happens for arbitrarily small Δ.

If we include the three in-plane sp^2 orbitals in our tight-binding model, Eq. (49.10) will have the form

$$\sum_{j',\alpha',\mathbf{R}'} \left[\left(\varepsilon_{2p_z} - \varepsilon_{j'}(\mathbf{k})\right) J_{j\alpha j'\alpha'}(\mathbf{R}') + h_{j\alpha j'\alpha'}(\mathbf{R}') \right] a_{j'\alpha'} e^{i\mathbf{k}\cdot\mathbf{r}_{j'}} e^{i\mathbf{k}\cdot\mathbf{R}'} = 0, \tag{49.25}$$

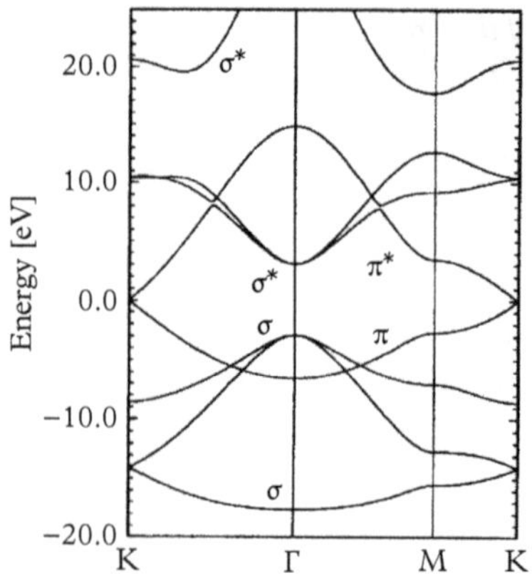

Figure 49.6 *The band structure of graphene, including the six σ bands along with the two π bands.*

where the sum on α runs over all four orbitals; the resulting determinate is now 8×8, but is block diagonal along symmetry lines and at symmetry points. The resulting band structure is shown in Fig. 49.6. Note that along the symmetry axes there is no mixing between the π bands (associated with the $2p_z$ orbitals) and the σ bands involving the three Pauling sp^2 hybrid orbitals (see Eq. (7.30)).

49.2 Graphene multilayers

When attempting to prepare single-layer graphene, one more typically obtains multilayer films. Such films also have interesting properties and here we will discuss the bilayer case, the structure of which is shown in Fig. 49.7. Note that the hexagons, rather than lying directly above each other are displaced such that one of the two atoms in the upper two-dimensional unit lies above an inequivalent one in the lower cell.

Limiting ourselves to the p_z orbitals, the electronic properties of bilayer graphene can be treated using Eq. (49.25), by reinterpreting the orbital index, α, as a layer index; the indices, $\mathbf{R}$ and j, continuing to label the in-plane atom positions. The matrix elements and overlap integrals associated with the A and

[8] Two-dimensional gas of massless Dirac fermions in graphene, K.S. Novoselov, A.K. Geim, S.V. Morozov, D. Jiang, M.I. Katsnelson, I.V. Grigorieva, S.V. Dubonos, and A.A. Firsov, *Nature* **438**, 197 (2005).

$\tilde{B}$ atoms can be ignored since they are adjacent to hexagon centers in the neighboring layer, where there are no atoms. Making the same approximations which yielded (49.24a) from (49.16a), the electronic structure near the points K and K′ of our Schrödinger equation, $\hat{H}\vec{\psi} = \varepsilon\vec{\psi}$, will now be governed by a 4 × 4 form, where

$$\vec{\psi}(\mathbf{k}) = \begin{pmatrix} a_A(\mathbf{k}) \\ a_B(\mathbf{k}) \\ a_{\tilde{A}}(\mathbf{k}) \\ a_{\tilde{B}}(\mathbf{k}) \end{pmatrix} \tag{49.26a}$$

and

$$\hat{H} = \begin{pmatrix} 0 & \hbar v(q_x + iq_y) & 0 & 0 \\ \hbar v(q_x - iq_y) & 0 & h' & 0 \\ 0 & h' & 0 & \hbar v(q_x + iq_y) \\ 0 & 0 & \hbar v(q_x - iq_y) & 0 \end{pmatrix}; \tag{49.26b}$$

Here, $v = 3ha/2\hbar$, where h continues to be the in-plane matrix element and h′ is the inter-layer $\tilde{A} - B$ matrix element. Setting the determinant, $\left|\hat{H} - \varepsilon\hat{1}\right| = 0$, we obtain

$$\left(\varepsilon^2 - \hbar^2 v^2 q^2\right)^2 = \varepsilon^2 h'^2, \tag{49.27}$$

which has four solutions. At the Dirac point where $q = 0$, we have two solutions with $\varepsilon = \pm h'$ and two with $\varepsilon = 0$. For the first pair, the wave functions correspond to the following symmetric and antisymmetric forms:

$$a_{\tilde{A}} = +a_B; \; a_A = a_{\tilde{B}} = 0, \;\; \varepsilon = -h', \tag{49.28a}$$

and

$$a_{\tilde{A}} = -a_B; \; a_A = a_{\tilde{B}} = 0, \;\; \varepsilon = h'; \tag{49.28b}$$

these states would be fully occupied and fully unoccupied respectively in the ground state, and the splitting results from the strong coupling through h′, since the atoms lie one above the other. The zero energy states are associated with the combinations,

$$a_A = \pm\, a_{\tilde{B}}, \;\; a_{\tilde{A}} = a_B = 0, \; \varepsilon = 0. \tag{49.29}$$

Since no inter-layer p_z matrix element couples the $\tilde{A}$ and B atoms, in our approximation their energies at K are unshifted, but near the Dirac point their energies are given by

$$\varepsilon \cong \pm\frac{\hbar^2 v^2 q^2}{h'}; \tag{49.30a}$$

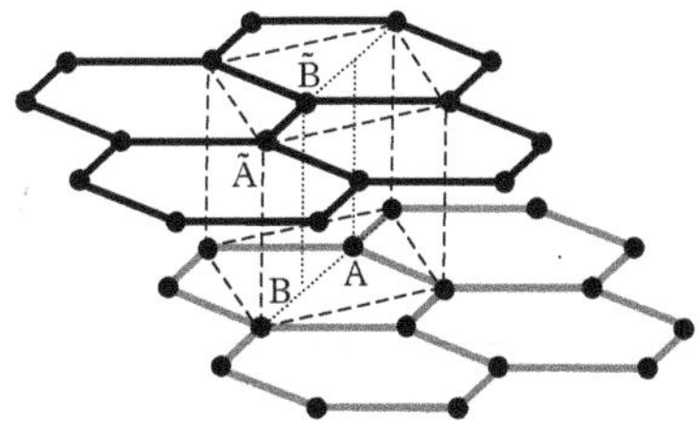

Figure 49.7 *The structure of bilayers of graphene. The two atoms in the 2D unit cell of the upper layer, designated $\tilde{A}$ and $\tilde{B}$, are displaced relative to those in the lower layer, designated A and B, such that $\tilde{A}$ and B are positioned one above the other, while A and $\tilde{B}$ are positioned adjacent to hexagon centers.*

i.e., we have a set of *parabolic* bands which *touch* at q = 0. If a bias is established between the two layers, say via an external gate electrode, this would add equal diagonal elements to the Hamiltonian (49.26b). If the relative shift is ε_0, Eq. (49.30a) will be replaced by

$$\varepsilon \cong \pm\sqrt{\varepsilon_0^2 + \frac{\hbar^4 v^4 q^4}{h'^2}}; \tag{49.30b}$$

i.e., we will have a gap and to leading order an unusual quartic behavior of the two bands.

49.3 The band structure of graphene from ARPES

The electronic structure of graphene has been studied over a wide range of energies using the technique of angle resolved photo-emission spectroscopy

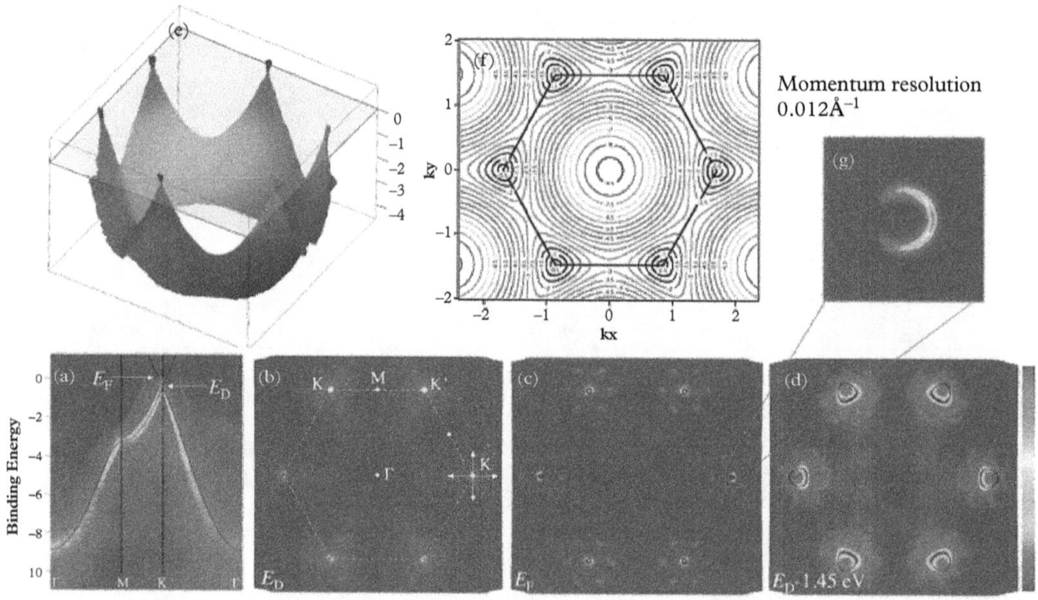

Figure 49.8 *Band structure and constant energy surfaces in graphene. (a) The behavior of ε versus k along the lines, Γ → M, M → K, and K → Γ. (b)–(d) Constant energy cuts for progressively lower energies in the vicinity of the Fermi energy, showing the cone-like evolution of the Fermi line near the K point of the Brillouin zone. Note the constant energy lines become less circular as we proceed away from ε_F. (e) A three-dimensional depiction of the two-dimensional band structure of graphene. (f) The closed lines show the constant energy lines for occupied states in the vicinity of the K point. (g) An expanded view of a constant energy line near K. Note that the line appears not to close. This effect arises from the matrix elements entering the ARPES intensity when the wave vector lies in the neighboring Brillouin zone. (Courtesy of J. McChesney.)*

(ARPES), which was discussed in Section 20.2.1. The graphene layer is prepared by processing the surface of hexagonal SiC, for which the surface normal to the c axis is well matched to graphene.[9] Figure 49.8(a) shows the band structure along the principal symmetry lines, while Fig. 49.8(b)–(d) show the constant energy lines in the vicinity of the Fermi surface. By depositing materials like potassium and calcium on the free surface it has proved possible to dope the underlying graphene, thereby varying the Fermi energy over a wide range of energies.

49.4 Graphene in a magnetic field

Electrons in a magnetic field near the Dirac point behave differently from conventional band electrons, discussed earlier in Sections 18 and 48. From the Onsager–Lifshitz semi-classical quantization condition, Eq. (18.34b), orbit areas near the K points are given by: $A_q = \pi q^2 = 2\pi eH(\nu + \gamma)/c\hbar$ or $q = \sqrt{2eH(\nu + \gamma)/c\hbar}$. Combining this with Eq. (49.19), $\varepsilon(q) = \pm \hbar v q$, yields[10]

$$\varepsilon = \pm v\sqrt{\frac{2e\hbar H(\nu + \gamma)}{c}}; \tag{49.31}$$

note the quantum number ν appears under the square root sign. In the more rigorous treatment given below, we will find that that the factor, γ, associated with the zero point motion vanishes, a prorperty of a so-called "zero mode".

We start by recalling that the effect of a magnetic field is incorporated through the prescription discussed in Section 18.1, which for our two-dimensional electrons has the form $\mathbf{p}_\perp \rightarrow \hat{\mathbf{p}}_\perp - (e/c)\mathbf{A}$, where

$$\hat{\mathbf{p}}_\perp = \frac{\hbar}{i}\nabla_\perp = \frac{\hbar}{i}\left(\frac{\partial}{\partial x}\mathbf{i} + \frac{\partial}{\partial y}\mathbf{k}\right).$$

We start by making the identification,

$$q_x \pm iq_y \rightarrow \frac{1}{i}\left(\frac{\partial}{\partial x} \pm i\frac{\partial}{\partial y}\right) \tag{49.32}$$

in the Hamiltonian (49.20), where x and y denote slowly changing envelope coordinates (on the scale of a lattice constant) and $a_j(\mathbf{q}) \rightarrow \psi_j(x, y)$ in the wave function (49.21a); on incorporating (49.32), using the gauge, $A_x = -Hy$, and seeking a solution of the form $\vec{\psi}(x, y) = e^{iq_x x}\vec{\phi}(y)$, we obtain

$$\hbar v\begin{pmatrix} 0 & \dfrac{\partial}{\partial y} + q_x + \dfrac{eH}{\hbar c}y \\ -\dfrac{\partial}{\partial y} + q_x + \dfrac{eH}{\hbar c}y & 0 \end{pmatrix}\begin{pmatrix}\phi_A(y) \\ \phi_B(y)\end{pmatrix} = \varepsilon\begin{pmatrix}\phi_A(y) \\ \phi_B(y)\end{pmatrix}. \tag{49.33}$$

[9] Photoemission studies of graphene on SiC: growth, interface, and electronic structure, A. Bostwick, K.V. Emtsev, K. Horn, E. Huwald, L. Ley, J.L. McChesney, T. Ohta, J. Riley, E. Rotenberg, F. Speck, and Th. Seyller, *Advances in Solid State*, Springer (2008).

[10] Diamagnetism in graphite, J.W. McClure, *Phys. Rev.* **104**, 666 (1956).

Introducing the magnetic length, l_H (see Eq. (48.5b)),

$$l_H = \sqrt{\frac{\hbar c}{|e|H}}; \qquad (49.34a)$$

and a frequency,

$$\omega_0 = \frac{\sqrt{2}v}{l_H}, \qquad (49.34b)$$

we can rewrite (49.33) as

$$\frac{\omega_0}{\sqrt{2}} \begin{pmatrix} 0 & l_H \frac{\partial}{\partial y} + l_H q_x - \frac{y}{l_H} \\ -l_H \frac{\partial}{\partial y} + l_H q_x - \frac{y}{l_H} & 0 \end{pmatrix} \begin{pmatrix} \phi_A(y) \\ \phi_B(y) \end{pmatrix} = \varepsilon \begin{pmatrix} \phi_A(y) \\ \phi_B(y) \end{pmatrix}, \qquad (49.35)$$

which in terms of the dimensionless variable (with a shifted origin),

$$\xi = \frac{y}{l_H} - l_H q_x, \qquad (49.36)$$

can be written as

$$-\hbar\omega_0 \begin{pmatrix} 0 & \hat{O}^\dagger \\ \hat{O} & 0 \end{pmatrix} \begin{pmatrix} \phi_A(\xi) \\ \phi_B(\xi) \end{pmatrix} = \varepsilon \begin{pmatrix} \phi_A(\xi) \\ \phi_B(\xi) \end{pmatrix}, \qquad (49.37)$$

where we have defined the operators

$$\hat{O} = \frac{1}{\sqrt{2}} \left(\frac{\partial}{\partial \xi} + \xi \right) \qquad (49.38a)$$

and

$$\hat{O}^\dagger = \frac{1}{\sqrt{2}} \left(-\frac{\partial}{\partial \xi} + \xi \right); \qquad (49.38b)$$

these two operators have the same form as the destruction and creation operators that arise in the one-dimensional harmonic oscillator problem and hence obey the commutation relation

$$\left[\hat{O}^\dagger, \hat{O}\right] = 1, \qquad (49.39)$$

with the associated number operator

$$\hat{\nu} = \hat{O}^\dagger \hat{O}, \qquad (49.40)$$

having eigenvalues, ν.

It turns out to be useful to introduce sublattice raising and lowering operators, defined as $\hat{\sigma}_\pm = (1/2)\left[\hat{\sigma}_x \pm i\hat{\sigma}_y\right]$,

$$\hat{\sigma}_+ = \begin{pmatrix} 0 & 1 \\ 0 & 0 \end{pmatrix}; \quad \hat{\sigma}_- = \begin{pmatrix} 0 & 0 \\ 1 & 0 \end{pmatrix}; \qquad (49.41a,b)$$

with the properties,

$$\hat{\sigma}_+ \begin{pmatrix} 1 \\ 0 \end{pmatrix} = 0; \hat{\sigma}_+ \begin{pmatrix} 0 \\ 1 \end{pmatrix} = \begin{pmatrix} 1 \\ 0 \end{pmatrix}; \hat{\sigma}_- \begin{pmatrix} 1 \\ 0 \end{pmatrix} = \begin{pmatrix} 0 \\ 1 \end{pmatrix}; \hat{\sigma}_- \begin{pmatrix} 0 \\ 1 \end{pmatrix} = 0. \tag{49.41c–f}$$

It then follows that (49.37) can be written in the form

$$(\hat{O}^\dagger \hat{\sigma}_+ + \hat{O}\hat{\sigma}_-)\vec{\phi}(\xi) = -\frac{\varepsilon}{\hbar\omega_0}\vec{\phi}(\xi). \tag{49.42}$$

We now note that if there exists a state, $\vec{\phi}_0(\xi)$, where on the left-hand side of (49.42), $\hat{O}^\dagger\, \hat{\sigma}_+\vec{\phi}_0(\xi) = 0$ *and* $\hat{O}\hat{\sigma}_-\vec{\phi}_0(\xi) = 0$, then that state must have $\varepsilon = 0$, a so-called *zero state*;[11] two *inequivalent* conditions satisfying these properties are, $\hat{O}\vec{\phi}_0(\xi) = 0$ and $\hat{\sigma}_+\vec{\phi}_0(\xi) = 0$. The second of these is satisfied by the form

$$\vec{\phi}_0(\xi) = \phi_0(\xi) \begin{pmatrix} 1 \\ 0 \end{pmatrix}, \tag{49.43a}$$

corresponding to a state *localized* on the A sublattice; satisfying the first equation requires

$$\hat{O}\phi_0(\xi) = \frac{1}{\sqrt{2}}\left(\frac{\partial}{\partial\xi} + \xi\right)\phi_0(\xi) = 0, \tag{49.43b}$$

the solution of which corresponds to the ground state of a one-dimensional harmonic oscillator. We recall the rules obeyed by the raising and lowering operators

$$\hat{O}^\dagger \phi_\nu = \sqrt{N+1}\phi_{\nu+1} \tag{49.44a}$$

and

$$\hat{\mathbf{O}}\phi_\nu = \sqrt{N}\phi_{\nu-1}. \tag{49.44b}$$

The general solution to (49.37) is

$$\vec{\phi}_{\nu,\pm}(\xi) = \begin{pmatrix} \phi_\nu(\xi) \\ \pm\phi_{\nu-1}(\xi) \end{pmatrix}, \tag{49.45}$$

where the excited states, $\phi_\nu(\xi)$, can be constructed from the ground state by multiple applications of the rising operator

$$\phi_\nu(\xi) = \frac{1}{\sqrt{\nu!}}\left(\hat{O}^\dagger\right)^\nu \phi_0(\xi). \tag{49.46}$$

That (49.45) is indeed the νth eigenfunction can be verified by directly substituting it into (49.37) and using Eqs. (49.44a,b), which also gives the energies as (McClure, 1956)

$$\varepsilon_{\nu,\pm} = \pm\sqrt{\nu}\,\hbar\omega_0 \tag{49.47a}$$

$$= \pm v\sqrt{\frac{2e\hbar H}{c}\nu}. \tag{49.47b}$$

[11] A zero energy state is unusual in that a zero-point motion usually accompanies the lowest energy state.

Note that these energies are the same as what we anticipated semi-classically in Eq. (49.31), apart from the shift caused by γ, and that it differs from that obtained for "free" electrons in a magnetic field, as discussed in Section 48.1 (see Eq. (48.6)), where $\varepsilon_\nu = (\nu + 1/2)\hbar\omega_c$. Also note that for the magnetic fields typically used, ω_0 is larger than a typical cyclotron frequency, ω_c, of other quantum Hall materials (with parabolic band edges), leading to the observation of Shubnikov–de Hass and quantum Hall phenomena, *even at room temperature.*

To relate (49.47) to what is measured experimentally we equate it to the Fermi energy, $\varepsilon_{\nu,\pm} = \varepsilon_F = \hbar v_F q_F$. Using Eq. (49.22a) for q_F and inserting (49.34) and (49.35) into (49.47), we obtain

$$\nu = \frac{\pi c}{eH} n(\varepsilon_F). \quad (49.48)$$

Hence, the oscillations will be periodic with both the gate voltage and the reciprocal of the magnetic field. Note that because of the presence of the zero state there is no undetermined phase shift, as is present in the quasi-classical Onsager–Lifshitz expression (see Eq. (18.35)).

49.5 The Shubnikov-de Haas and quantum Hall effects in graphene

Figure 49.9 shows the Shubnikov–de Hass (SdH) oscillations of the resistivity, ρ_{xx}, as a function of the magnetic field at a fixed gate voltage.[12] If the correct level number, ν, is assigned to each of the minima of these oscillations and plotted as a function of 1/H, its extrapolation to $\nu = 0 (H \to \infty)$ passes through the origin. This is to be contrasted with the case for the conventional 2D quantum Hall systems where it extrapolates to 1/2, thereby reflecting the difference in the quantization condition given by Eq. (49.47), as opposed to that from (48.6).

Figure 49.10 shows SdH oscillations as a function of the gate voltage at fixed magnetic field; recall that this is done by varying the carrier concentration by sweeping the charge (or equivalently the voltage) on an external electrostatic gate electrode adjacent to the film. Note that the oscillations have a constant period as a function of the gate voltage, as predicted by Eq. (49.48). We leave it as a

[12] Two-dimensional gas of massless Dirac fermions in graphene, K.S. Novoselov, A.K. Geim, S.V. Morozov, D. Jiang, M.I. Katsnelson, I.V. Grigorieva, S.V. Dubonos, and A.A. Firsov, *Nature* **438**, 197 (2005).

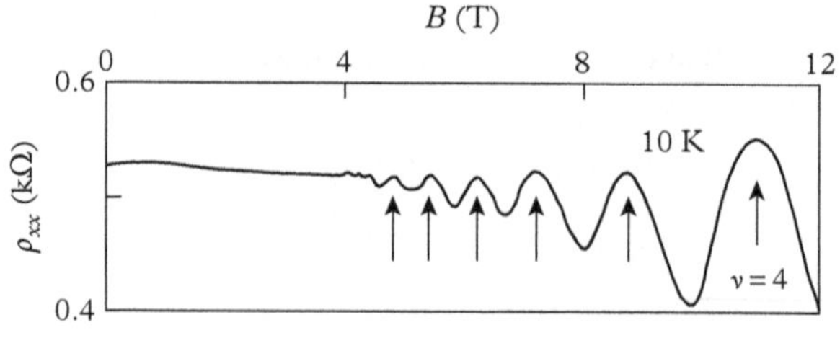

Figure 49.9 *Shubnikov-de Hass oscillations in monolayer graphene. (After Novoselov et al. (2005).)*

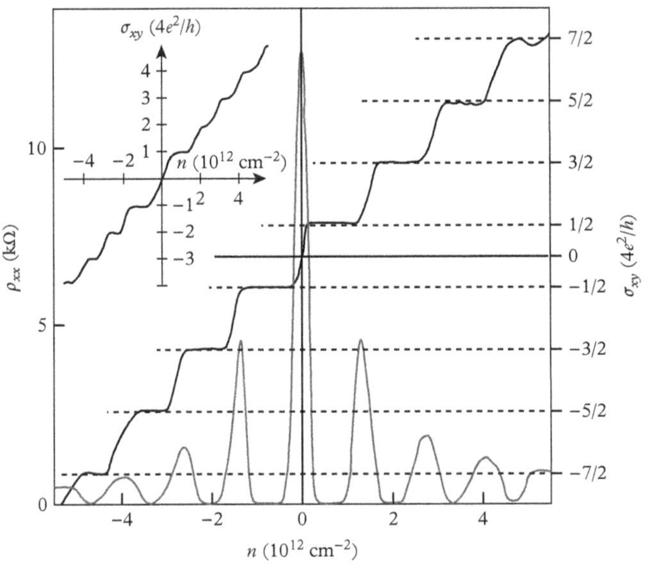

Figure 49.10 *The quantum oscillations in the resistivity, ρ_{xx}, of graphene as a function of carrier density. Also shown is the Hall resistivity, ρ_{xy}; note that the plateaus are now at fractions $1/2, 3/2 \cdots$, of ϕ_0. The inset shows the behavior for bilayer graphene, where the steps are integral multiples of ϕ_0. (After Novoselov et al. (2005).)*

problem to show that transitions between adjacent levels are governed by (49.47), as in a cyclotron resonance experiment, and are consistent with (49.23b).

Figure 49.10 also shows that the Hall resistivity displays the plateaus characteristic of the quantum Hall effect; however, they now occur at the values, $R_H = (\nu + 1/2)\,\phi_0$, again reflecting the difference in the quantization condition. The inset shows the behavior in bilayer graphene, where the steps occur at integral values;[13] here we recall that the band extrema are parabolic, as given by (49.30a,b), where the more conventional quantization applies. In graphene, the QHE features are observed over a much wider temperature range, including room temperature,[14] when compared with typical semiconductor devices, due to the very small cyclotron masses (see Fig. 49.5) and, possibly, a weak electron–phonon interaction arising from the small in-plane compressibility.

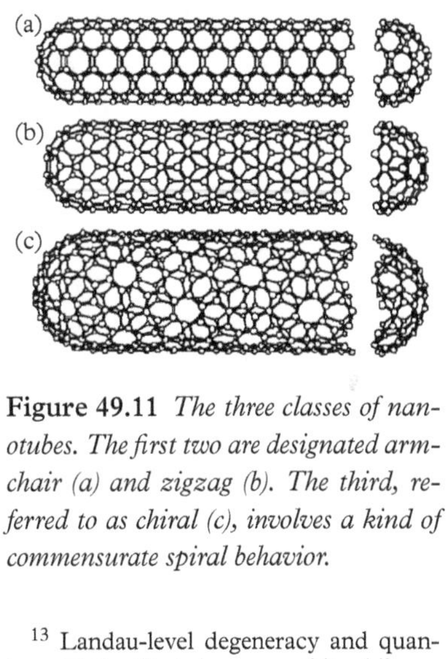

Figure 49.11 *The three classes of nanotubes. The first two are designated armchair (a) and zigzag (b). The third, referred to as chiral (c), involves a kind of commensurate spiral behavior.*

49.6 Single-walled nanotubes

On examining products produced in a carbon arc discharge S. Iijima observed long filaments; structurally, these filaments are "rolled up" graphite. Using chemical vapor deposition methods it is now possible to routinely prepare filaments involving a *single graphene* sheet, so-called single-walled nanotubes. Figure 49.11 shows, schematically some different structures. Two of the structures correspond to joining the two edges of a sheet such that (a) a line joining opposite corners of the hexagons is perpendicular to the tube axis, and (b) the line joining opposite corners is parallel to the tube axis. But there are many other ways to join opposing

[13] Landau-level degeneracy and quantum Hall effect in a graphite bilayer, E. McCann and V.I. Fal'ko, *Phys. Rev. Lett.* **96**, 086805 (2006).

[14] Room-temperature quantum Hall effect in graphene, K.S. Novoselov Z. Jiang, Y. Zhang, S.V. Morozov, H.L. Stormer, U. Zeitler, J.C. Maan, G.S. Boebinger, P. Kim, and A.K. Geim, *Science* **315**, 1379 (2007).

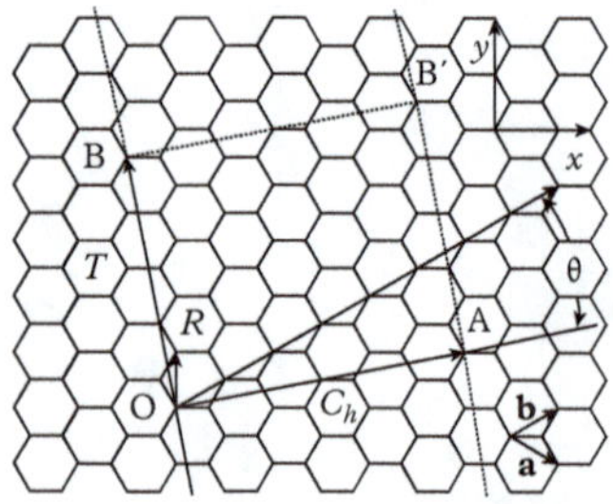

Figure 49.12 *A more general way to roll up a graphene sheet to make a nanotube. Connecting the points O to A and B to B' results in a smooth "atom to atom" match.*

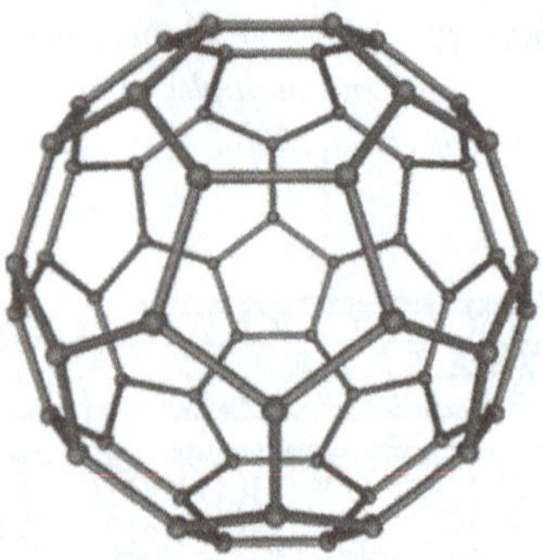

Figure 49.13 *The molecule C60 consisting of 60 carbon atoms at vertices forming interconnected hexagonal and pentagonal faces. It involves six five-fold, ten three-fold, and fifteen two-fold axes, with the over all symmetry of an icosahedron (the group Y_h). Locally the bonding is sp^2-like.*

edges of the sheets since we may slide one edge relative to the other before joining; these are the so-called chiral structure, one of which is shown in (c).

Figure 49.12 shows how one constructs a chiral structure. We connect two points, **O** and **A**, by a vector, C_h, defined in terms of integer multiples of the basic translation vectors of the unit cell, **a** and **b**; i.e.,

$$\mathbf{C_h} = n\mathbf{a} + m\mathbf{b}. \tag{49.49}$$

We then interpret $\mathbf{C_h}$ as the *circumference* of our nanotube, which we take to be constant as we advance along the tube axis, which lies parallel to the lines **OB** and **AB** that are perpendicular to **OA**. The magnitude of the circumference is given by

$$\begin{aligned} |\mathbf{C_h}| &= \sqrt{n^2a^2 + 2nm\mathbf{a}\cdot\mathbf{b} + m^2a^2} \\ &= \sqrt{n^2a^2 + nm + m^2a^2} \end{aligned}, \tag{49.50}$$

where we have used the property, $\mathbf{a}\cdot\mathbf{b} = a^2/2$. The chiral angle, θ, is given by

$$\cos\theta = \frac{\mathbf{C_h}\cdot\mathbf{a}}{|\mathbf{C_h}|\,|\mathbf{a}|} = \frac{2n + m}{2\sqrt{n^2a^2 + nm + m^2a^2}}. \tag{49.51}$$

49.7 Fullerenes

Solids formed from the molecule C_{60}, shown in Fig. 49.13, and related[15] molecules represent a third phase, beyond diamond and graphite, in which elemental carbon can exist. Prior to its discovery,[16] the symmetry point group of the molecule itself, Y_h, was unknown in Nature. The molecules themselves form spontaneously from the vapor phase present in an electrical arc discharge that is maintained between graphite electrodes, the molecules then depositing as a dust on the walls of the associated chamber.[17] The atoms in the molecule, which consists of pentagons and hexagons connected at their edges, lie at the vertices and form sp^2 bonds with their neighbors. Once formed, the molecules will bond with each other through van der Waals force, forming a close packed fcc lattice. At high temperatures, the molecules rotate under thermal activation.

As noted earlier, in Section 46, the solid phases are superconducting.

[15] The 1996 Nobel Prize in Chemistry was awarded to Robert F. Curl Jr., Harold W. Kroto, and Richard E. Smalley "for their discovery of fullerenes."

[16] C_{60} was first prepared in the laboratory in 1985 by R. Smalley, R. Curl, J. Heath, S. Obrien, and H. Kroto, at Rice University. It is reported to be present in interstellar space.

[17] W. Kraatschmer, L.D. Lamb, K. Fostiropoulos, and D.R. Huffman, *Nature* 347, 354 (1990).

ADDITIONAL READING

Graphene is an example of a strongly diamagnetic material. For a discussion, see:

Diamagnetism of graphite, J.W. McClure, *Phys. Rev.* **104**, 666 (1956).

The properties of graphene and nanotubes are discussed in many places. Some references are: The electronic properties of graphene, A.H. Castro Neto, F. Guinea, N.M.R. Peres, K.S. Novoselov, and A.K. Geim, *Rev. Mod. Phys.* **81**, 109 (2009); *Physical Properties of Carbon Nanotubes*, R. Saito, G. Dresselhaus, and M.S. Dresselhaus, Imperial College Press, London (1998); *Science of Fullerenes and Carbon Nanotubes*, M.S. Dresselhaus, G. Dresselhaus, and P.C. Eklund, Elsevier (1996), and *Graphene: Fundamentals, Devices and Applications*, S. Shafranuik, Pan Stanford Publishing, Singapore (2015).

Problems

49.1. Show that for $\mathbf{k} = (2\pi/3a)\left(\mathbf{i} \pm \mathbf{j}/\sqrt{3}\right) + \mathbf{q}$, the function defined in Eq. (49.15) is given to leading order in $\mathbf{q}$ by (49.18).

49.2. Calculate the expression for $\varepsilon(\mathbf{k})$ when second-nearest neighbors are included.

49.3. Verify the forms for the wave functions at the K point in bilayer graphene, as given in Eqs. (49.28) and (49.29).

49.4. Show that the quantities, ω_0 and ω_c, can be related as $\omega_0/\omega_c = 1/2\sqrt{\nu}$.

49.5. Calculate the period of the Shubnikov–de Haas quantum oscillations as a function of the electron density, n_F, and the Fermi energy, ε_F.

Appendix A
The Calculus of Variations

Our primary goal in this appendix is to establish the formalism needed to obtain the equations of motion of a field from the associated Lagrangian, or to obtain an approximate solution for that field, possibly subject to some set of constraints. In this book an example of the former is our derivation of the Ginzburg–Landau equations from a model thermodynamic free energy for a superconductor (see Section 33.1), while the latter include obtaining an approximate solution to the Schrödinger equation for the Bloch states and energy levels in periodic solids in terms of an augmented plane wave basis (see Section 13.1) or a Green's function formulation (see Section 13.2) and the Hartree and Hartree–Fock equations for approximating a many-body system (see Section 15.1). To establish the necessary tools it will be useful to first review the simpler procedures used to obtain the equations of motion in particle mechanics.

A.1 The Euler–Lagrange equations and the mechanics of particles

Assume that we are given a quantity, S, defined by

$$S = \int_{t_1}^{t_2} L(x(t), \dot{x}(t), t)dt, \tag{A.1}$$

with $\dot{x} \equiv dx/dt$. The integrand, L, is a specified function of its three arguments, $x(t)$, $\dot{x}(t)$ and t, but the function, $x = x(t)$, is (initially) not specified. We ask the following question: what function, $x = x(t)$, will result in the quantity, S, being an extremum (usually a minimum). In the case of mechanical systems, L is the Lagrangian, and $x = x(t)$ is some hypothetical one-dimensional path the particle might take, with $\dot{x}$ being the velocity on that path, and S is called the action. It turns out that the *actual path* which the particle follows is the one for which S is a minimum, a property which is referred to as the *principle of least action* or *Hamilton's principle.*

The technique which is used to determine $x(t)$ is called the *calculus of variations.* Let us assume that we are given a function, $x(t)$, which results in S being extremal. Note that S is *not* a function of $x(t)$ in the usual sense, since the effect that $x(t)$ produces is integrated (or averaged) over time (from t_1 to t_2), as we see from Eq. (A.1), and the term *functional* was invented to describe this kind of dependence; we denote such a dependence as $S = S[x(t)]$ (i.e., with $x(t)$ in square brackets).

We now alter $x(t)$ by adding to it a small but arbitrary function, $\delta x(t)$, called the *variation*; i.e., we write

$$x(t) \rightarrow x(t) + \delta x(t). \tag{A.2}$$

If $x(t)$ is a function that yields an extremum of S, then this small change, $\delta x(t)$, should produce no change in S to first order. We now write

$$S[x(t) + \delta x(t)] = S[x(t)] + \delta S \tag{A.3}$$

with the understanding that if $x(t)$ is the function that minimizes S, then δS in (A.3) will vanish. It will turn out that this results in an equation that enables us to find the extremal, $x(t)$.

We may evaluate δS by differentiating under the integral sign with respect to the variation, δx; since varying δx can affect $\delta \dot{x}$, we must differentiate with respect to this variable also, and therefore we have

$$\delta S = \int_{t_1}^{t_2} \left[\frac{\partial L}{\partial x} \delta x + \frac{\partial L}{\partial \dot{x}} \delta \dot{x} \right] dt. \tag{A.4}$$

Writing $\delta \dot{x} = d\delta x/dt$ in the second term and performing an integration by parts yields

$$\delta S = \int_{t_1}^{t_2} \left[\frac{\partial L}{\partial x} - \frac{d}{dt} \frac{\partial L}{\partial \dot{x}} \right] \delta x(t) dt + \frac{\partial L}{\partial \dot{x}} \delta x(t) \Big|_{t_1}^{t_2} . \tag{A.5}$$

Now, if we require that $\delta x(t)$ vanish at t_1 and t_2 then the last term in (A.5) vanishes. Note that this is equivalent to forcing the function, $x(t)$, to go through the two points, $x(t_1)$ and $x(t_2)$. Near the extremum we only require that δx be small. It is, however, arbitrary and therefore the term in square brackets must itself vanish, yielding

$$\frac{d}{dt} \frac{\partial L}{\partial \dot{x}} - \frac{\partial L}{\partial x} = 0. \tag{A.6}$$

Equation (A.6) is referred to as the *Euler–Lagrange equation* and its integration yields the path, $x(t)$, which was our goal.

In one-particle, one-dimensional mechanics, $L = \frac{1}{2} m\dot{x}^2 - V(x)$. The Euler–Lagrange equation is then $m\ddot{x} = -\partial V/\partial x = F$, which, of course, is Newton's Law. Note that in integrating this second-order differential equation we generate two constants of integration; in Newtonian mechanics these are usually chosen to be the initial position and velocity, $x(t_1)$ and $\dot{x}(t_1)$; however, they can just as well be the two points, $x(t_1)$ and $x(t_2)$, introduced earlier.

The above treatment can easily be extended to many particles and dimensions by introducing the Lagrangian, $L = L(x_j, \dot{x}_j, t)$, and the action

$$S = \int dtL(x_j(t), \dot{x}_j(t), t), \tag{A.7}$$

where j runs from 1 to s with s = Nd; here, N is the number of particles and d is the number of dimensions. The Euler–Lagrange equation is then

$$\frac{d}{dt}\frac{\partial L}{\partial \dot{x}_j} - \frac{\partial L}{\partial x_j} = 0; \; j = 1 \cdots s. \tag{A.8}$$

A.2 Extension of the Euler–Lagrange formalism to continuum fields

We now generalize the Lagrangian formalism to the case of a field. We start with a one-component field, denoted by ψ, which is a function of a single space variable, x, in addition to time; i.e., $\psi = \psi(x,t)$. For the case of a field, the Lagrangian of particle mechanics, which is a function of x(t) and $\dot{x}(t)$, is replaced by a *Lagrangian density*, $\mathcal{L}$, which is a function of ψ and its time *and* space derivatives, $\dot{\psi} = \partial\psi/\partial t$ and $\psi' = \partial\psi/\partial x$:

$$\mathcal{L} = \mathcal{L}\left(\psi, \dot{\psi}, \psi'\right) \tag{A.9}$$

and the corresponding action is defined as

$$S = \int_{x_1,t_1}^{x_2,t_2} \mathcal{L}\left(\psi, \dot{\psi}, \psi'\right) dxdt. \tag{A.10}$$

A field may be viewed as the limit of a system in which the number of particles approaches infinity. A good model to keep in mind in the present context is the case of a one-dimensional (1D) solid; we will briefly discuss this model.[1]

Rather than designating each particle's position, $x_j(t)$, it is more convenient to deal with the displacement from their *equilibrium* positions, $\bar{x}_j$, which we write as

$$\psi_j \equiv x_j(t) - \bar{x}_j; \tag{A.11}$$

here it is assumed that the interparticle forces confine the particles locally. We may use the coordinate values, $x = \bar{x}_j$, as a way of replacing the particle number designation by a coordinate designation, $x = \bar{x}_j$, allowing us to write $\psi_j(t) \equiv \psi(\bar{x}_j, t) \rightarrow \psi(x, t)$. Assuming each particle is connected to its neighbor by a spring, a relative displacement between neighboring particles, $\psi_{j+1} - \psi_j$, produces a potential energy, $(k/2)\left(\psi_{j+1} - \psi_j\right)^2$, for the linear (or harmonic) case, where k is the usual spring constant of the harmonic oscillator problem. The kinetic energy of a particle is $(m/2)\dot{\psi}_j^2$. In the continuum approximation these two contributions

[1] For convenience, we will reproduce some of the discussion of Section 16.1, although with different notation.

Table A.1 *Particle versus field designations*

	1D solid	Continuum field
Atom designation or field point	j	x
Atom displacement or field strength	$\psi_j(t)$	$\psi(x,t)$
Kinetic energy or energy density	$\frac{m}{2}\dot{\psi}_j^2$	$\frac{\rho}{2}\dot{\psi}^2$
Elastic energy or energy density	$\frac{k}{2}(\psi_{j+1}-\psi_j)$	$\frac{1}{2}\lambda\psi'^2$

are replaced by $\frac{1}{2}\rho\dot{\psi}^2$ and $\frac{1}{2}\lambda\psi'^2$, where ρ is the linear mass density, $\rho = m/a$, with a being the inter-particle spacing, and $\lambda = ka$ (as we will see below). Table A.1 illustrates the transition from the discrete solid to the continuum elastic field.

The Lagrangian and Lagrangian densities for this problem are

$$\begin{aligned} L &= T - V \\ &= \sum_{j=1}^{N}\left[\frac{m}{2}\dot{\psi}_j^2 - \frac{k}{2}\left(\psi_{j+1}-\psi_j\right)^2\right] \end{aligned} \tag{A.12a}$$

and

$$\mathcal{L} = \frac{\rho}{2}\dot{\psi}^2 - \frac{\lambda}{2}\psi'^2. \tag{A.12b}$$

We see that the Lagrangian and Lagrangian density are related by

$$\begin{aligned} \lim_{N\to\infty} L &= \lim_{N\to\infty}\sum_{N}\left[\frac{m}{2}\dot{\psi}_j^2 - \frac{k}{2}\left(\psi_{j+1}-\psi_j\right)^2\right] \\ &= \int\frac{dx}{a}\left[m\dot{\psi}^2(x,t) - \frac{ka^2}{2}\psi'^2(x,t)\right] \\ &= \int dx\left[\frac{1}{2}\rho\dot{\psi}^2 - \frac{1}{2}\lambda\psi'^2\right] \\ &= \int dx\,\mathcal{L}\left(\psi,\dot{\psi},\psi'\right). \end{aligned} \tag{A.13}$$

The above formalism is ideally suited to a relativistic or covariant treatment of a field since it treats the space and time variables on an equal footing.

Having defined the general structure of the Lagrangian formalism for a (1D) field we must now derive the corresponding Euler–Lagrange equation, which yields an extremal action. We proceed in a manner similar to the particle case by introducing a variation, $\delta\psi = \delta\psi(x,t)$. Differentiating (A.10) under the integral sign yields

$$\delta S = \int dxdt\left[\frac{\partial\mathcal{L}}{\partial\psi}\delta\psi + \frac{\partial\mathcal{L}}{\partial\dot{\psi}}\delta\dot{\psi} + \frac{\partial\mathcal{L}}{\partial\psi'}\delta\psi'\right] = 0. \tag{A.14}$$

Performing an integration by parts on $\delta\dot{\psi}$ and $\delta\psi'$ we have

$$\delta S = \int dxdt\left[\frac{\partial\mathcal{L}}{\partial\psi} - \frac{\partial}{\partial t}\frac{\partial\mathcal{L}}{\partial\dot{\psi}} - \frac{\partial}{\partial x}\frac{\partial\mathcal{L}}{\partial\psi'}\right]\delta\psi + \int_{x_1}^{x_2} dx\frac{\partial\mathcal{L}}{\partial\dot{\psi}}\delta\psi\,\Big|_{t_1}^{t_2} + \int_{t_1}^{t_2} dt\frac{\partial\mathcal{L}}{\partial\psi'}\,\delta\psi|_{x_1}^{x_2}. \tag{A.15}$$

Assuming that $\delta\psi(x,t)$ vanishes at the "end points" (x_1, x_2; t_1, t_2) the last two terms in (A.15) vanish; since $\delta\psi$ is arbitrary the integrand of the first term in (A.15) must vanish, yielding

$$\frac{\partial\mathcal{L}}{\partial\psi} - \frac{\partial}{\partial t}\frac{\partial\mathcal{L}}{\partial\dot{\psi}} - \frac{\partial}{\partial x}\frac{\partial\mathcal{L}}{\partial\psi'} = 0. \tag{A.16}$$

We may extend the above discussion to the case of a multi-component field, ψ_α, and three space variables, x_1, x_2, x_3, in which case (A.16) generalizes to

$$\frac{\partial\mathcal{L}}{\partial\psi_\alpha} - \frac{\partial}{\partial t}\frac{\partial\mathcal{L}}{\left(\dfrac{\partial\psi_\alpha}{\partial t}\right)} - \sum_{i=1}^{3}\frac{\partial}{\partial x_i}\frac{\partial\mathcal{L}}{\partial\left(\dfrac{\partial\psi_\alpha}{\partial x_i}\right)} = 0. \tag{A.17}$$

Equation (A.17) is the three-dimensional Euler–Lagrange equation for a field, ψ_α. For the more limited case of a time-independent field, Eq. (A.17) takes the form

$$\frac{\partial\mathcal{L}}{\partial\psi_\alpha} - \sum_{i=1}^{3}\frac{\partial}{\partial x_i}\frac{\partial\mathcal{L}}{\partial\left(\dfrac{\partial\psi_\alpha}{\partial x_i}\right)} = 0. \tag{A.18}$$

We can view our minimization of the action $S[\psi_\alpha, \nabla\psi_\alpha, \partial\psi_\alpha/\partial t]$ in the Lagrangian formulation of fields as having lead to a generalized "derivative operator" which when applied to the integrand of the action (the Lagrangian) generates the equation of motion the field must satisfy to yield an extremum of the action; it is called a *functional derivative* and is denoted as

$$\frac{\delta}{\delta\psi_\alpha} = \frac{\partial}{\partial\psi_\alpha} - \frac{\partial}{\partial t}\frac{\partial}{\left(\dfrac{\partial\psi_\alpha}{\partial t}\right)} - \sum_{i=1}^{3}\frac{\partial}{\partial x_i}\frac{\partial}{\partial\left(\dfrac{\partial\psi_\alpha}{\partial x_i}\right)}; \tag{A.19a}$$

for the time-independent case the operator is

$$\frac{\delta}{\delta\psi_\alpha} = \frac{\partial}{\partial\psi_\alpha} - \sum_{i=1}^{3}\frac{\partial}{\partial x_i}\frac{\partial}{\partial\left(\dfrac{\partial\psi_\alpha}{\partial x_i}\right)}. \tag{A.19b}$$

Appendix B
The Symmetry of Many–Particle Wave Functions; The Occupation Number Representation

A large part of a typical elementary quantum mechanics course tends to involve a single particle moving in an external potential. Although two-particle problems are discussed in connection with scattering and bound states in the presence of a central potential, they are usually limited to the case of different particles (e.g., an electron and a nucleus). When spin is neglected, the wave function only involves the coordinates of the two particles, x_1, y_1, z_1, and x_2, y_2, z_2. We then find that by constructing coordinates $\mathbf{R} = (\mathbf{r}_1 + \mathbf{r}_2)/2$, describing motion of the center of mass and, $\mathbf{r} = \mathbf{r}_1 - \mathbf{r}_2$, describing motion relative to the center of mass, the wave function can be written as a product

$$\psi(x_1, y_1, z_1; x_2, y_2, z_2) = \Phi(\mathbf{R})\psi(\mathbf{r}), \tag{B.1}$$

provided that the potential only involves the scalar coordinate r. This allows one to treat the problems as one-particle problems.

The helium atom and the hydrogen molecule are "simple" examples of problems where the variables do not separate in the center of mass. Furthermore, they involve two *identical particles*, the two identical electrons; in the case of H_2 (or D_2, the deuterium molecule), we also have two identical nuclei. When *identical* particles are involved special considerations are required, as we now discuss.

B.1 The symmetry of two-particle wave functions

We must first discuss what the word identical means. To focus our discussion we initially restrict ourselves to the case of two identical particles. In classical mechanics we could imagine that the two particles, although having identical physical properties, might start widely separated, but remain *distinguishable* as they approach (and interact with) each other through observing them continuously (so as not to "lose track" of them) at all future times. However, in quantum mechanics the act of observing adds uncertainty, and therefore an experiment which would distinguish them at all future times is not possible.

A way to formally build this required property into the quantum theory of identical particles is to impose a permutation (or, equivalently, particle interchange) symmetry on the associated many-particle wave function. We introduce a *permutation operator*, P_{ij}, which interchanges the ith and the jth particles; for a two-electron wave function, we can then write the interchange operation as

$$P_{12}\psi(\mathbf{r}_1,\mathbf{r}_2) = \psi(\mathbf{r}_2,\mathbf{r}_1). \tag{B.2}$$

We impose the physically reasonable condition that if the particles are identical, the probability density must be unchanged by this operation and therefore,

$$|\psi(\mathbf{r}_1,\mathbf{r}_2)|^2 = |\psi(\mathbf{r}_2,\mathbf{r}_1)|^2. \tag{B.3}$$

From this requirement it follows that the only effect the permutation interchange operation can have is to alter the *phase* of the wave functions; hence,

$$P_{12}\psi(\mathbf{r}_1,\mathbf{r}_2) = e^{i\phi_P}\psi(\mathbf{r}_2,\mathbf{r}_1). \tag{B.4}$$

Applying the permutation operator twice yields

$$P_{12}^2\psi(\mathbf{r}_1,\mathbf{r}_2) = P_{12}e^{i\phi_P}\psi(\mathbf{r}_2,\mathbf{r}_1)) = e^{2i\phi_P}\psi(\mathbf{r}_1,\mathbf{r}_2).$$

Another physically reasonable condition is that two permutations completely restores the original wave function (including its phase)[1] and therefore,

$$e^{2i\phi_P}\psi(\mathbf{r}_1,\mathbf{r}_2) = \psi(\mathbf{r}_1,\mathbf{r}_2); \tag{B.5}$$

thus, $e^{2i\phi_P} = 1,\ 2i\phi_P = 0, 2\pi$ or $\phi_P = 0, \pi$. Therefore,

$$P_{12}\psi(\mathbf{r}_1,\mathbf{r}_2) = \begin{cases} +\psi(\mathbf{r}_2,\mathbf{r}_1); & \text{symmetric case,} \\ -\psi(\mathbf{r}_2,\mathbf{r}_1); & \text{antisymmetric case.} \end{cases} \tag{B.6}$$

Experimentally, the symmetric and antisymmetric cases are found to apply to particles with integer ($s = 0, 1, 2\ldots$) or half integer ($s = 1/2, 3/2\ldots$) spin respectively; the particles are called Bose particles (bosons) and Fermi particles (fermions) respectively.[2]

If our two particles are weakly interacting we may seek an approximate description by writing our two-particle wave function as a product of two one-particle wave functions, having quantum numbers denoted by n and m respectively. In order to satisfy (B.6) these "product" wave functions must have the forms

$$\psi_B(\mathbf{r}_1,\mathbf{r}_2) = \frac{1}{\sqrt{2}}\left[\psi_n(\mathbf{r}_1)\psi_m(\mathbf{r}_2) + \psi_n(\mathbf{r}_2)\psi_m(\mathbf{r}_1)\right] \tag{B.7a}$$

[1] The argument given here follows that of Landau and Lifshitz. It leaves open the question of designing an experiment to perform the permutation without disturbing the system.

[2] The permutation symmetry has a profound effect on the statistical properties of ensembles of particles, particularly when the interaction is weak and the temperature is low. The associated statistics are called Bose–Einstein and Fermi–Dirac and were discovered before the connection with wave function permutation symmetry was identified by Dirac.

and

$$\psi_F(\mathbf{r}_1, \mathbf{r}_2) = \frac{1}{\sqrt{2}}\left[\psi_n(\mathbf{r}_1)\psi_m(\mathbf{r}_2) - \psi_n(\mathbf{r}_2)\psi_m(\mathbf{r}_1)\right], \tag{B.7b}$$

where the subscripts, B and F, identify the Bose and Fermi cases, the factor $(\sqrt{2})^{-1}$ assures normalization, and the subscripts, n and m, denote the (full set of) quantum numbers of the two one-electron states used to construct the product state. For the case, n = m, (B.7b) vanishes identically; this property is called the *Pauli exclusion principle*, which states that: no two fermions may have the same quantum numbers.

In constructing (B.7a,b) we have neglected spin. For simplicity we will assume that our bosons have no spin and that our fermions have spin 1/2, where we may take the latter to be electrons. For the fermions we also neglect any interactions that might couple the spatial coordinates with the spin degrees of freedom (i.e., spin–orbit coupling is neglected). We may then write

$$\psi_F = \phi(\mathbf{r}_1, \mathbf{r}_2)\chi(s_1, s_2). \tag{B.8}$$

We require the *total* wave function, ψ_F, to be antisymmetric under the interchange of *all* coordinates defining the two electrons $(\mathbf{r}_1, s_1)$ and $(\mathbf{r}_2, s_2)$. There are two possibilities: the singlet case, where ϕ is symmetric and χ is antisymmetric and the triplet case, where ϕ is antisymmetric and χ is symmetric. A set of wave functions describing these two possibilities is

$$\begin{aligned}\psi^{\text{singlet}} &= \frac{1}{\sqrt{2}}\left[\psi_n(\mathbf{r}_1)\psi_m(\mathbf{r}_2) + \psi_n(\mathbf{r}_2)\psi_m(\mathbf{r}_1)\right] \\ &\times \frac{1}{\sqrt{2}}\left[\begin{pmatrix}1\\0\end{pmatrix}^{(1)}\begin{pmatrix}0\\1\end{pmatrix}^{(2)} - \begin{pmatrix}0\\1\end{pmatrix}^{(1)}\begin{pmatrix}1\\0\end{pmatrix}^{(2)}\right]\end{aligned} \tag{B.9a}$$

and

$$\begin{aligned}\psi^{\text{triplet}} &= \frac{1}{\sqrt{2}}\left[\psi_n(\mathbf{r}_1)\psi_m(\mathbf{r}_2) - \psi_n(\mathbf{r}_2)\psi_m(\mathbf{r}_1)\right] \\ &\times \left[\begin{matrix}\begin{pmatrix}1\\0\end{pmatrix}^{(1)}\begin{pmatrix}1\\0\end{pmatrix}^{(2)} \\ \frac{1}{\sqrt{2}}\left[\begin{pmatrix}1\\0\end{pmatrix}^{(1)}\begin{pmatrix}0\\1\end{pmatrix}^{(2)} + \begin{pmatrix}0\\1\end{pmatrix}^{(1)}\begin{pmatrix}1\\0\end{pmatrix}^{(2)}\right] \\ \begin{pmatrix}0\\1\end{pmatrix}^{(1)}\begin{pmatrix}0\\1\end{pmatrix}^{(2)}\end{matrix}\right]\end{aligned} \tag{B.9b}$$

where the spin states have been chosen to be eigenfunctions of the total spin angular momentum

B.2 The symmetry of many-particle wave functions

We now extend our discussion to the case of N non-interacting particles. Let ψ_1, $\psi_2, \ldots$ etc. denote the possible one-particle states that N particles *may* occupy. Let p_1, p_2, $\ldots p_N$ be the quantum numbers of the states *actually occupied* by the N particles.[3] Our many-particle wave function for a system of non-interacting particles is made up of products of the form $\psi_{p_1}(\mathbf{r}_1, s_1)\psi_{p_2}(\mathbf{r}_2, s_2)\psi_{p_3}(\mathbf{r}_3, s_3) \ldots \psi_{p_N}(\mathbf{r}_N, s_N)$, where s_i denote the spin projections of the particles, which we continue to take as $\pm 1/2$ for our fermions and zero for our bosons. The interchange of any two particles, i, j, is again denoted by the permutation operator, P_{ij}; thus,[4]

$$\begin{aligned} P_{12}&\left[\psi_{p_1}(\mathbf{r}_1, s_1)\psi_{p_2}(\mathbf{r}_2, s_2)\psi_{p_3}(\mathbf{r}_3, s_2) \ \ldots \ \psi_{p_N}(\mathbf{r}_N, s_N)\right] \\ &= \psi_{p_1}(\mathbf{r}_2, s_2)\psi_{p_2}(\mathbf{r}_1, s_1)\psi_{p_3}(\mathbf{r}_3, s_3) \ \ldots \ \psi_{p_N}(\mathbf{r}_N, s_N). \end{aligned} \tag{B.10}$$

For the Bose case it turns out that we may write the total *symmetrized and normalized* wave function as

$$\Psi_B = \left[\frac{N_1!N_2! \ \ldots}{N!}\right]^{1/2} \sum_P \hat{P}\left[\psi_{p_1}(\mathbf{r}_1)\psi_{p_2}(\mathbf{r}_2)\ldots\psi_{p_N}(\mathbf{r}_N)\right], \tag{B.11}$$

where N is the total number of particles, the sum is over all possible N! permutations of the indices 1....N of these particles, and the N_i are the numbers of particles in each of the occupied states.[5,6]

For the Fermi case the corresponding wave function can be written as

$$\Psi_F = \frac{1}{\sqrt{N!}} \sum_P (-1)^P \hat{P}\, \psi_{p_1}(\mathbf{r}_1, s_1)\psi_{p_2}(\mathbf{r}_2, s_2)\ldots\psi_{p_N}(\mathbf{r}_N, s_N) \tag{B.12a}$$

We again sum over all permutations, except that now we must alternate the sign of each successive permutation. Alternatively, we can write the non-interacting Fermi wave function as

$$\Psi_F = \frac{1}{\sqrt{N!}} \begin{vmatrix} \psi_{p_1}(\mathbf{r}_1, s_1) & \psi_{p_1}(\mathbf{r}_2, s_2) & \ldots & \psi_{p_1}(\mathbf{r}_N, s_N) \\ \psi_{p_2}(\mathbf{r}_1, s_1) & \psi_{p_2}(\mathbf{r}_2, s_2) & \ldots & \psi_{p_2}(\mathbf{r}_N, s_N) \\ & & \vdots & \\ \psi_{p_N}(\mathbf{r}_1, s_1) & \psi_{p_N}(\mathbf{r}_2, s_2) & \ldots & \psi_{p_N}(\mathbf{r}_N, s_N) \end{vmatrix}. \tag{B.12b}$$

The latter is called a *Slater determinant* (after J.C. Slater); it is fully antisymmetric, automatically (i.e., it vanishes if any two p_i are the same), and hence satisfies the Pauli principle.

[3] For example, we might have the string of numbers, $p_1 = 6, p_2 = 8, p_3 = 8, p_4 = 10$. This is interpreted as follows: we have a total of four particles, with the first in state 6, the second and third in state 8, and the fourth in state 10.

[4] In our example in footnote 3, the wave function is $\psi = \psi_6(\mathbf{r}_1, s_1) \psi_8(\mathbf{r}_2, s_2)\psi_8(\mathbf{r}_3, s_3)\psi_{10}(\mathbf{r}_4, s_1)$. Then, $P_{12}\Psi = \psi_6(\mathbf{r}_2, s_2)\psi_8(\mathbf{r}_1, s_1)\psi_8(\mathbf{r}_3, s_3)\psi_{10} (\mathbf{r}_4, s_4)$.

[5] Continuing with our example in footnote 3, we would have $N_1 = 1, N_2 = 2, N_3 = 1$.

[6] If we adopt the usual convention that $0! = 1$, then the N_i in the normalization term can be extended to all basis states, whether occupied or not. However the summation is still restricted to the occupied stases.

B.3 Separation of spatial and spin coordinates in a many-particle system

In non-relativistic quantum mechanics, the Hamiltonian of the system does not contain the spin operators of the individual particles; it must therefore be possible to write the total state function as a product of a function that depends only on spin projections, $\chi(s_1, s_2, s_3, \cdots s_N)$, and one that depends only on spatial coordinates, $\phi(\mathbf{r}_1, \mathbf{r}_2, \mathbf{r}_3, \cdots \mathbf{r}_N)$; i.e., we can write

$$\psi_F = \phi(\mathbf{r}_1, \mathbf{r}_2, \mathbf{r}_3, \cdots \mathbf{r}_N)\chi(s_1, s_2, s_3, \cdots s_N).$$

For identical particles the Hamiltonian must also be symmetric on interchanging the spatial, $\mathbf{r}_i$, coordinates. Now when only two particles are involved, one can always construct functions that are purely even or purely odd with respect to *separately* interchanging the spin and the orbital quantum numbers associated with the two states. This is what has allowed us to write the total wave function of two identical fermions above as a product of spatial and spin state functions that were separately symmetric or antisymmetric, so as to obtain states that *overall* obey the Pauli principle. But for three or more particles we then run into a problem: it is in general not possible to *separately* interchange the spin coordinates and spatial coordinates for an *arbitrary* pair of particles and have the symmetry be globally even or odd for all pairs (although there are special cases where this is possible). But one can construct orbital and spin states that are *even for some pairs and odd for others*. As long as one chooses appropriate products of *complementary* spatial and spin states that *together* satisfy the Pauli principle (odd with respect to the interchange of *any* pair), such states must be admissible. Furthermore, if one is to construct a complete set of states for a given Hamiltonian the inclusion of such states in the overall set is essential. The procedures required to systematically construct appropriately symmetrized states involve the mathematics of the *permutation group* and will not be discussed here. Note that the spatial eigenstates of the Hamiltonian can arise that do not satisfy the Pauli principle when spin is included; such states must be rejected on physical grounds.

B.4 Interactions in many-particle systems

Having described the general form of the wave function for a system of non-interacting Bose and Fermi particles, given by Eqs. (B.11) and (B.12), one usually wants to consider the effect of interactions of the particles with external fields or among themselves using perturbation theory or some other scheme. This generally involves taking matrix elements of various operators entering a Hamiltonian between states of the type (B.11) or (B.12). We now discuss the form that such matrix elements take. We restrict ourselves here to the Bose case; the Fermi case will be considered in Section B.6.

We first consider a quantum mechanical operator that involves only the coordinates of the particles individually. An example is the total kinetic energy operator, $\hat{T} = (1/2m)\sum_{a=1}^{N}\hat{p}_a^2 = -(\hbar^2/2m)\sum_{a=1}^{N}\nabla_a^2$;[7] we call this a *single-particle (or one-particle) operator* and write it formally as

$$\hat{F}^{(1)} = \sum_{a=1}^{N}\hat{f}_a^{(1)} \tag{B.13}$$

where for our example $\hat{f}_a = -(\hbar^2/2m)\nabla_a^2$. Whereas $\hat{f}_a$ is the operator for a single particle, $\hat{F}^{(1)}$ is the operator for all N particles. The matrix element of (B.13) between wave functions of the form (B.11) can be written as

$$\int d^3r_1 d^3r_2 \cdots d^3r_N \psi^* \left(\mathbf{r}_1, \mathbf{r}_2 \cdots \mathbf{r}_N; N_1', N_2' \cdots N_N'\right) \times \hat{F}^{(1)} \psi\left(\mathbf{r}_1, \mathbf{r}_2 \cdots \mathbf{r}_N; N_1, N_2 \cdots N_N\right). \tag{B.14}$$

Here, N_i and N_i', which we will call *occupation numbers*, denote the numbers of particles in each of the single-particle basis states, $\psi_i(\mathbf{r})$, before and after the action of the operator, $\hat{F}^{(1)}$. Since the coordinates are integrated out in evaluating such matrix elements, all that remains is a matrix in *occupation number space*. The result of such a calculation using states of the form (B.11) for that part of the off-diagonal element of $\hat{F}^{(1)}$ involving the basis states, i and k, is given by

$$\langle N_i + 1, N_k - 1 |\, F^{(1)}\, | N_i, N_k\rangle = f_{ik}^{(1)}\sqrt{(N_i + 1)\, N_k}, \tag{B.15}$$

where N_i and N_k are the numbers of particles in states i and k and we define the matrix elements involving the operator, $\hat{f}^{(1)}$, between the basis states, $\psi_i(\mathbf{r})$, as

$$f_{ik}^{(1)} \equiv \int \psi_i^* \hat{f}^{(1)} \psi_k d^3r. \tag{B.16}$$

What we are accustomed to calling "transitions" from states k and i now represent changes in the numbers of particles in states k and i: before the transition we have N_i and N_k particles in states i and k respectively, whereas after the transition we have $N_i - 1$ and $N_k + 1$. The expression obtained for the diagonal elements is given by

$$\langle N_i |\, F^{(1)}\, | N_i\rangle = f_{ii}^{(1)} N_i. \tag{B.17}$$

Note that the particle designations do not appear in (B.15) or (B.17).

[7] We use letters at the beginning of the alphabet to denote individual particles.

B.5 Occupation number representation: Bose statistics

The results, (B.15) and (B.17), may be written in a more elegant form by introducing a new representation called the *occupation number representation.* This involves an (initially strange) kind of state function which we write as $\phi(N_1, N_2 \ldots)$; here, N_1 = number of particles in state 1, N_2 = number of particles in state 2, etc.; these numbers may have any integer values (for this Bose case) *including zero.* Our state function is now simply a column vector, the entries of which are the occupation numbers, N_i, where i refers to the states, $\psi_i(\mathbf{r})$, in the *coordinate representation.* Clearly the basis states to be used for a given problem must be agreed upon, although it is thereafter suppressed.[8] The dimension of this column vector is the same as that of our basis set—if our basis set is denumerably infinite, so is the column vector, $\phi(N_i)$.

For our Bose system, $N_i = 0, 1, 2 \ldots$ and we write $\phi = \phi_B$; ϕ_B is assumed to be normalized and states with different N_i are orthogonal. We next introduce new (Bose) operators, $\hat{a}_i$ and $\hat{a}_i^\dagger$, which act in this occupation number space, and are called *destruction and creation operators* respectively. They are defined to have the properties

$$\hat{a}_i\phi_B(N_1, N_2, \ldots N_i, \ldots) = \sqrt{N_i}\phi_B(N_1, N_2, \ldots N_i - 1, \ldots) \tag{B.18}$$

and

$$\hat{a}_i^\dagger\phi_B(N_1, N_2 \ldots) = \sqrt{N_i + 1}\phi_B(N_1, \ldots N_i + 1, \ldots); \tag{B.19}$$

we then say that $\hat{a}_i$ destroys (removes) one particle from state i and $\hat{a}_i^\dagger$ creates (adds) one particle to state i. Note that

$$\langle N_i - 1\,|\hat{a}_i|\,N_i\rangle^* = \left\langle N_i\,\middle|\hat{a}_i^\dagger\middle|\,N_i - 1\right\rangle = \sqrt{N_i}.$$

From the above we verify that

$$\hat{a}_i^\dagger\hat{a}_i\,|\phi_B\rangle = N_i|\,\phi_B\rangle; \tag{B.20a}$$

and

$$\hat{a}_i\hat{a}_i^\dagger\,|\phi_B\rangle = (N_i + 1)|\,\phi_B\rangle. \tag{B.20b}$$

From these two equations we construct commutation relations for the operators, $\hat{a}_i$ and $\hat{a}_i^\dagger$,

$$\left[\hat{a}_i, \hat{a}_i^\dagger\right] \equiv \left(\hat{a}_i\hat{a}_i^\dagger - \hat{a}_i^\dagger\hat{a}_i\right) = 1. \tag{B.21}$$

[8] When plane waves are used one often writes, $\psi_{\mathbf{k}}(\mathbf{r})$, where the quantum numbers are now numbered by the allowed values of the wave vector, $\mathbf{k}$, in a box of side, L, with periodic boundary conditions.

We verify directly that

$$\left[\hat{a}_i, \hat{a}_k\right] = 0, \tag{B.22a}$$

$$\left[\hat{a}_i^\dagger, \hat{a}_k^\dagger\right] = 0 \tag{B.22b}$$

and

$$\left[\hat{a}_i^\dagger, \hat{a}_k\right] = 0 \; (i \neq k) \tag{B.22c}$$

In terms of the $\hat{a}_i$ operators we may write the *total* one-particle operator, $\hat{F}^{(1)}$, in the occupation number representation as

$$\hat{F}^{(1)} = \sum_{i,k} f_{ik}^{(1)} \hat{a}_i^\dagger \hat{a}_k. \tag{B.23}$$

Note that the term involving the matrix element, $f_{ik}^{(1)}$, in this sum when acting on a state function $\phi(N_i, N_k)$ gives $\sqrt{(N_i + 1)\, N_k} f_{ik}^{(1)} \phi(N_i + 1, N_k - 1)$, in agreement with (B.15).

We also introduce *two-particle operators*, e.g., the total inter-particle Coulomb potential, $\hat{V}^{(2)} = \sum_{a>b}^{N} \hat{V}_{ab}^{(2)} = \sum_{a>b}^{N} e^2/|\mathbf{r}_a - \mathbf{r}_b|$. Formally, we write the total two-particle operator, $\hat{F}^{(2)}$, as

$$\hat{F}^{(2)} = \sum_{a>b}^{N} \hat{f}_{ab}^{(2)}. \tag{B.24}$$

If we take the matrix element of (B.24) with Bose wave functions of the form (B.11) and introduce the operators, $\hat{a}_i$, we find that $\hat{F}^{(2)}$ in the occupation number representation is given by

$$\hat{F}^{(2)} = \frac{1}{2} \sum_{ikl m} f_{ikl m}^{(2)} \hat{a}_i^\dagger \hat{a}_k^\dagger \hat{a}_m \hat{a}_l, \tag{B.25}$$

where

$$f_{iklm}^{(2)} = \langle ik | f^{(2)} | lm \rangle = \int\int \psi_i^*(\mathbf{r}_1) \psi_k^*(\mathbf{r}_2) \hat{f}^{(2)} \psi_l(\mathbf{r}_1) \psi_m(\mathbf{r}_2) d^3 r_1 d^3 r_2. \tag{B.26}$$

An important special case of the above operators is the Hamiltonian

$$\hat{H} = \sum_a \hat{H}_a^{(1)} + \sum_{a>b} U^{(2)}(\mathbf{r}_a, \mathbf{r}_b) + \sum_{a>b>c} U^{(3)}(\mathbf{r}_a, \mathbf{r}_b, \mathbf{r}_c) \; \ldots, \tag{B.27}$$

where

$$\sum_{a=1}^{N} \hat{H}_a^{(1)} = \sum_{a=1}^{N} \left(-\frac{\hbar^2}{2m}\right) \nabla_a^2 + U^{(1)}(\mathbf{r}_a). \tag{B.28}$$

Here, $U_a^{(1)}, U_{ab}^{(2)}, U_{abc}^{(3)}$ are one-particle, two-particle, etc., potentials. In the occupation number representation, $\hat{H}$ is given by

$$\hat{H} = \sum_{ik} H_{ik}^{(1)} \hat{a}_i^\dagger \hat{a}_k + \frac{1}{2} \sum_{ik\ell m} U_{ik\ell m}^{(2)} \hat{a}_i^\dagger \hat{a}_k^\dagger \hat{a}_m \hat{a}_\ell + \dots. \tag{B.29}$$

For non-interacting particles, a so-called one-particle problem,

$$\hat{H} = \sum_{ik} H_{ik}^{(1)} \hat{a}_i^\dagger \hat{a}_k. \tag{B.30}$$

In a representation where $H_{ik}^{(1)}$ is diagonal we may write these elements as $\varepsilon_i \delta_{ik}$; our one-particle Hamiltonian is then written as

$$\hat{H} = \sum_i \varepsilon_i \hat{a}_i^\dagger \hat{a}_i. \tag{B.31}$$

Since $\hat{a}_i^\dagger \hat{a}_i$ yields N_i when operating on ϕ_B, the expectation value of $\hat{H}$, which is the total energy, E, is given by the expression

$$E = \sum_i \varepsilon_i N_i. \tag{B.32}$$

We now introduce the important concept of "$\hat{\psi}$ operators," which are defined by

$$\hat{\psi}(\mathbf{r}) \equiv \sum_i \psi_i(\mathbf{r}) \hat{a}_i \tag{B.33a}$$

and

$$\hat{\psi}^\dagger(\mathbf{r}) \equiv \sum_i \psi_i^*(\mathbf{r}) \hat{a}_i^\dagger. \tag{B.33b}$$

Here, **r** is a *parameter* as opposed to a coordinate (the coordinates of the occupation number representation are the occupation numbers, N_i) and the $\psi_i(\mathbf{r})$ are any complete set of one-particle wave functions appropriate to the space under consideration, i.e., those used to define (B.11). We can rewrite our one- and two-particle operators in terms of $\hat{\psi}$ operators using (B.33), as

$$\hat{F}^{(1)} = \sum_{ik} f_{ik}^{(1)} \hat{a}_i^\dagger \hat{a}_k = \sum_{ik} \left(\int \psi_i^*(\mathbf{r}) \hat{f}^{(1)} \psi_k(\mathbf{r}) d^3r \right) \hat{a}_i^\dagger \hat{a}_k = \int d^3r\, \hat{\psi}^\dagger(\mathbf{r}) \hat{f}^{(1)} \hat{\psi}(\mathbf{r}); \tag{B.34a}$$

similarly,

$$\hat{F}^{(2)} = \frac{1}{2}\int d^3r\, d^3r'\ \hat{\psi}^{\dagger}(\mathbf{r})\hat{\psi}^{\dagger}(\mathbf{r}')\hat{f}^{(2)}\,\hat{\psi}(\mathbf{r}')\hat{\psi}(\mathbf{r}). \qquad \text{(B.34b)}$$

The Hamiltonian can then be rewritten as

$$\begin{aligned}\hat{H} = &\int d^3r\,\hat{\psi}^{\dagger}(\mathbf{r})\left[-\frac{\hbar^2}{2m}\nabla^2 + U^{(1)}(\mathbf{r})\right]\hat{\psi}(\mathbf{r})\\ &+\frac{1}{2}\int d^3r d^3r'\hat{\psi}^{\dagger}(\mathbf{r})\hat{\psi}^{\dagger}(\mathbf{r}')U^{(2)}(\mathbf{r},\mathbf{r}')\hat{\psi}(\mathbf{r}')\hat{\psi}(\mathbf{r})\\ &+ (\text{3body terms}) + \ldots. \end{aligned} \qquad \text{(B.35)}$$

Here, $U^{(1)}(\mathbf{r})$and $U^{(2)}(\mathbf{r},\mathbf{r}')$ are the one- and two-body potentials respectively. An advantage of writing the Hamiltonian in the form (B.35) is that it is *independent of the basis functions*, $\psi_k(\mathbf{r})$, used.

We introduce a *particle density operator* as

$$\hat{\rho}(\mathbf{r}) \equiv \hat{\psi}(\mathbf{r})\hat{\psi}(\mathbf{r}), \qquad \text{(B.36)}$$

and a *number operator*, $\hat{N} \equiv \int d^3r\hat{\rho}(\mathbf{r})$, defined as

$$\hat{N} \equiv \int d^3r\,\hat{\psi}^{\dagger}(\mathbf{r})\hat{\psi}(\mathbf{r}). \qquad \text{(B.37)}$$

Substituting (B.33) into (B.37) and using the orthonormality of the ψ_i, we have

$$\hat{N} = \sum_i \hat{a}_i^{\dagger}\hat{a}_i. \qquad \text{(B.38)}$$

Using the definition (B.33) and the commutation relations for the $\hat{a}_i$ given by (B.21) and (B.22), we obtain the commutation relations for the $\hat{\psi}$ operators,

$$\hat{\psi}(\mathbf{r})\hat{\psi}(\mathbf{r}') - \hat{\psi}(\mathbf{r}')\hat{\psi}(\mathbf{r}) = 0;$$

or

$$\left[\hat{\psi}(\mathbf{r}), \hat{\psi}(\mathbf{r}')\right] = 0 \qquad \text{(B.39a)}$$

and

$$\hat{\psi}(\mathbf{r})\hat{\psi}^{\dagger}(\mathbf{r}') - \hat{\psi}^{\dagger}(\mathbf{r}')\hat{\psi}(\mathbf{r}) = \delta^{(3)}(\mathbf{r}-\mathbf{r}')$$

or

$$\left[\hat{\psi}(\mathbf{r}), \hat{\psi}^{\dagger}(\mathbf{r}')\right] = \delta^{(3)}(\mathbf{r}-\mathbf{r}') \qquad \text{(B.39b)}$$

where we have used the completeness relation $\sum_i \psi_i^*(\mathbf{r})\psi_i(\mathbf{r}') = \delta^{(3)}(\mathbf{r}-\mathbf{r}')$.

To account for the effects of spin, which must be integral in our Bose case, we include a *spin projection coordinate*, α, on all the $\hat{\psi}(\mathbf{r})$ operators; i.e., $\hat{\psi}(\mathbf{r}) \rightarrow \hat{\psi}_\alpha(\mathbf{r})$. The delta functions in the commutators must also be rewritten. Thus (B.39a,b) become

$$\hat{\psi}_\alpha(\mathbf{r}')\hat{\psi}_\beta(\mathbf{r}) - \hat{\psi}_\beta(\mathbf{r})\hat{\psi}_\alpha(\mathbf{r}') = 0 \tag{B.40a}$$

and

$$\hat{\psi}_\beta(\mathbf{r})\hat{\psi}^\dagger_\alpha(\mathbf{r}') - \hat{\psi}^\dagger_\alpha(\mathbf{r}')\hat{\psi}_\beta(\mathbf{r}) = \delta^{(3)}(\mathbf{r}-\mathbf{r}')\delta_{\alpha\beta}. \tag{B.40b}$$

For the spin 1 case, $\alpha = 1, 2$, and 3 to account for the three components of a spin 1 state function.

Some discussions of the occupation number representation simply postulate the commutation relations (B.39a,b) and then go on to derive the commutation relations (B.22a–c).[9] However, this author prefers making the connection to many-particle wave functions in coordinate space and then making the one-to-one association with states in occupation number space. This seems to be much less "magical."

Most many-body calculations in the literature are carried out using the occupation number representation or, as it is sometimes called, the *second quantized representation* (this author prefers the first terminology). The occupation number representation is used extensively because:

(i) it automatically *guarantees* that the relevant states have, at all steps in a calculation, the appropriate symmetry (Bose/symmetric or Fermi/antisymmetric) and

(ii) it allows a convenient description for the appearance or disappearance of excitations (e.g., photon emission or absorption).

B.6 Occupation number representation: Fermi statistics

We now treat the case of Fermi statistics; we again write the state function as a column vector in the occupation number space and denote it as $\phi = \phi_F(N_i)$. However in the Fermi case, the occupation numbers can only be zero or one, to satisfy the Pauli principle. The results turn out to be essentially the same as the Bose case, with the exception that the creation and destruction operators *anticommute*. We establish this immediately below;[10] alternatively the reader can simply proceed to and accept Eqs. (B.48a,b).

To fix the sign of ϕ_F, we: (1) number the basis states ψ_i once and for all, and (2) complete the rows by ordering p_i such that

$$p_1 < p_2 < p_3 \ldots < p_N. \tag{B.41}$$

[9] If we take the functional derivative of Eq. (B.34) with respect to the operator, $\hat{\psi}^\dagger(\mathbf{r})$, we obtain

$$\left[-\frac{\hbar^2}{2m}\nabla^2 + U^{(1)}(\mathbf{r})\right]\hat{\psi}(\mathbf{r}) + \int d^3r'\hat{\psi}^\dagger(\mathbf{r}')U^{(2)}(\mathbf{r},\mathbf{r}')\hat{\psi}(\mathbf{r}')\hat{\psi}(\mathbf{r}) + (3 \text{ body terms}) + \ldots.$$

This form has the appearance of a Schrödinger equation, only the "wave function" is now an operator. Since the Schrödinger equation is sometimes "derived" by regarding the momentum in the kinetic energy part of the Hamiltonian as an operator, which acts on a wave function, now treating the wave function as an operator can be regarded as a *second quantization*; the term second quantization thus emerges. As stated earlier, this author regards the term *occupation number representation* as more transparent.

[10] Here we follow L.D. Landau and E.M. Lifshitz, *Quantum Mechanics: Non-Relativistic Theory*, Pergamon Press (1965).

No two p_i can be the same or the determinant will vanish; this is equivalent to the occupation numbers, N_i, being 0 or 1. As in the Bose case, explicit (but tedious) calculation shows that $F_i^{(1)}$ has non-zero matrix elements in occupation number space between the wave function, ϕ_F, given in (B.12).[11] The off-diagonal elements between states i and k are given by

$$\langle 1_i, 0_k \,|\, F^{(1)} \,|\, 0_i, 1_k \rangle = f_{ik}^{(1)} (-1)^{\Sigma(i+1,k-1)}, \; (i < k) \tag{B.42a}$$

or

$$\langle 1_i, 0_k \,|\, F^{(1)} \,|\, 0_i, 1_k \rangle = f_{ik}^{(1)} (-1)^{\Sigma(k+1,i-1)}, \; (i > k), \tag{B.42b}$$

where 0_i, 1_i signify $N_i = 0$, $N_i = 1$ respectively, and we *define*

$$\Sigma(k, l) \equiv \sum_{n=k}^{n=l} N_n, \tag{B.43}$$

where the sum is defined as zero when $l < k$.

The diagonal elements are $f_{ii}^{(1)} N_i$, as in the Bose case. We define new Fermi destruction and creation operators, $\hat{c}_i$, and $\hat{c}_i^\dagger$ by

$$\langle 0_i \,|\, \hat{c}_i \,|\, 1_i \rangle = \langle 1_i \,|\, \hat{c}_i^\dagger \,|\, 0_i \rangle (-1)^{\Sigma(1,i-1)} \tag{B.44}$$

Multiplying these matrices, we obtain the elements (when i<k),

$$\begin{aligned} \langle 1_i, 0_k \,|\, \hat{c}_i^\dagger \hat{c}_k \,|\, 0_i, 1_k \rangle &= \langle 1_i, 0_k \,|\, \hat{c}_i^\dagger \,|\, 0_i, 0_k \rangle \langle 0_i, 0_k \,|\, \hat{c}_k \,|\, 0_i, 1_k \rangle \\ &= (-1)^{\Sigma(1,i-1)} (-1)^{\Sigma(1,k-1)} \\ &= (-1)^{\Sigma(1,i-1)} (-1)^{\Sigma(1,i-1)+\Sigma(i+1,k-1)}. \end{aligned}$$

Now, since $N_i = 0$, the ith term in the second sum may be omitted and since $(-1)^{2\Sigma(1,i-1)} = 1$, we are left with

$$\langle 1_i, 0_k \,|\, \hat{c}_i^\dagger \hat{c}_k \,|\, 0_i, 1_k \rangle = (-1)^{\Sigma(i+1,k-1)} \tag{B.45}$$

In the case i = k, the matrix, $\hat{c}_i^\dagger \hat{c}_i$, is diagonal, the elements are 1 for $N_i = 1$ and 0 for $N_i = 0$. Thus we can write

$$\hat{c}_i^\dagger \hat{c}_i = \hat{N}_i. \tag{B.46}$$

Comparison of (B.45) with (B.42) yields

$$\hat{F}^{(1)} = \sum_{i,k} f_{ik}^{(1)} \hat{c}_i^\dagger \hat{c}_k,$$

which is the same form as the Bose case, Eq. (B.23).

[11] For the reader who is not interested in the details of the calculations, the results are identical to the Bose case with $\hat{a} \to \hat{c}$, but where the $\hat{c}$ obey the *anti-commutation* relations (B.48); the $\hat{\Psi}$ obey the anti-commutation relations (B.46).

To examine the commutation properties we evaluate (k>i),

$$\begin{aligned}\langle 1_i, 0_k | \hat{c}_k \hat{c}_i^\dagger | 0_i, 1_k\rangle &= \langle 1_i, 0_k | \hat{c}_k | 1_i, 1_k\rangle\langle 1_i, 1_k | \hat{c}_i^\dagger | 0_i, 1_k\rangle \\ &= (-1)^{\Sigma(1,k-1)}(-1)^{\Sigma(1,i-1)} = (-1)^{\Sigma(1,i-1)+\Sigma(i,k-1)}(-1)^{\Sigma(1,i-1)};\end{aligned}$$

in the first sum, $N_i = 1$, so we can rewrite the expression as

$$(1_i, 0_k | \hat{c}_k \hat{c}_i^\dagger | 0_i, 1_k)(-1)^{\Sigma(1,i-1)+\Sigma(i+1,k-1)+1}(-1)^{\Sigma(1,i-1)} = -(-1)^{\Sigma(i+1,k-1)}. \tag{B.47}$$

Comparing (B.45) and (B.47), we have

$$\hat{c}_i^\dagger \hat{c}_k + \hat{c}_k \hat{c}_i^\dagger = 0 \qquad i \neq k.$$

Including the diagonal element by evaluating the operator product with the opposite order of that in (B.46) then yields the relation

$$\hat{c}_i \hat{c}_k^\dagger + \hat{c}_k^\dagger \hat{c}_i = \delta_{ik};$$

i.e., fermion creation and destruction operators *anticommute*,

It is common to introduce the notation, $\hat{c}_i \hat{c}_k^\dagger + \hat{c}_k^\dagger \hat{c}_i = \left[\hat{c}_i, \hat{c}_k^\dagger\right]_+$, which is called an *anticommutator bracket*, and we then write

$$\left[\hat{c}_i, \hat{c}_k^\dagger\right]_+ = \delta_{ik}. \tag{B.48a}$$

Similar calculations show that

$$\hat{c}_i \hat{c}_k + \hat{c}_k \hat{c}_i = \left[\hat{c}_i, \hat{c}_k\right]_+ = 0. \tag{B.48b}$$

Calculations analogous to the Bose case show that for the fermion $\hat{\Psi}$ operators, in place of (B.39a,b), we have

$$\left[\hat{\psi}^\dagger(\mathbf{r}'), \hat{\psi}(\mathbf{r})\right]_+ = \delta^{(3)}(\mathbf{r} - \mathbf{r}') \tag{B.49a}$$

$$\left[\hat{\psi}(\mathbf{r}), \hat{\psi}(\mathbf{r}')\right]_+ = 0. \tag{B.49b}$$

The two-particle operators are written as

$$F^{(2)} = \frac{1}{2}\sum_{ikl m} f_{iklm}^{(2)} \hat{c}_i^\dagger \hat{c}_k^\dagger \hat{c}_m \hat{c}_l \tag{B.50}$$

with $f_{iklm}^{(2)}$ given by (B.26), and the Hamiltonian as

$$\hat{H} = \sum_{ik} H_{ik} \hat{c}_i^\dagger \hat{c}_k + \frac{1}{2}\sum_{iklm} U_{iklm}^{(2)} \hat{c}_i^\dagger \hat{c}_k^\dagger \hat{c}_m \hat{c}_l + \ldots, \tag{B.51}$$

which has the same form as Eq. (B.29). Equation (B.35) for the Hamiltonian in terms of the $\hat{\psi}$ operators also holds for the Fermi case.

To account for the effects of spin we again include a spin projection coordinate, α, on all the $\hat{\psi}(\mathbf{r})$ operators; i.e., $\hat{\psi}(\mathbf{r}) \rightarrow \hat{\psi}_\alpha(\mathbf{r})$. The delta functions in the anticommutators are also rewritten as $\delta^{(3)}(\mathbf{r}-\mathbf{r}')\delta_{\alpha\beta}$. Thus, Eqs. (B.49a,b) become

$$\left[\hat{\psi}_\alpha^\dagger(\mathbf{r}'), \hat{\psi}_\beta(\mathbf{r})\right]_+ = \delta^{(3)}(\mathbf{r}-\mathbf{r}')\delta_{\alpha\beta} \tag{B.52a}$$

and

$$\left[\hat{\psi}_\alpha(\mathbf{r}'), \hat{\psi}_\beta(\mathbf{r})\right]_+ = \left[\hat{\psi}_\alpha^\dagger(\mathbf{r}), \hat{\psi}_\alpha^\dagger(\mathbf{r}')\right]_+ = 0. \tag{B.52b}$$

For the spin 1/2 case, $\alpha = 1, 2$.

Likewise we must include a spin quantum number in the $\hat{c}$ operators. For the case of commonly used plane wave basis states, having wave vector, $\mathbf{k}$, we would write the state creation and destruction operators as $\hat{c}_{\mathbf{k},\nu}^\dagger$ and $\hat{c}_{\mathbf{k},\nu}$, where for the case of a spin 1/2 system, $\nu = 1, 2$ denote the two spin eigenfunctions. The commutation relations (B.48a,b) then become

$$\left[\hat{c}_{\mathbf{k}\nu}, \hat{c}_{\mathbf{k}'\nu'}^\dagger\right]_+ = \delta_{\mathbf{k}\mathbf{k}'}\delta_{\nu\nu'}; \tag{B.53a}$$

and

$$\left[\hat{c}_{\mathbf{k}\nu}, \hat{c}_{\mathbf{k}'\nu'}\right]_+ = \left[\hat{c}_{\mathbf{k}\nu}^\dagger, \hat{c}_{\mathbf{k}'\nu'}^\dagger\right]_+ = 0. \tag{B.53b}$$

B.7 Time-dependent formalism

Up to this point our various occupation number operators have been regarded as time independent. In order to treat time-dependent problems we now allow the $\hat{a}$, $\hat{\psi}$, and related operators to be time dependent; i.e., $\hat{a}_i = \hat{a}_i(t)$ (we will discuss the Bose case, but the resulting equations of motion for the Fermi case are the same, and only the commutation relations employed in carrying out the calculations differ). The Heisenberg representation is used exclusively in the occupation number representation (the state vectors being time-independent columns of numbers). According to the Heisenberg equation of motion $\dot{\hat{a}}(t) = (i/\hbar)\left[\hat{H}, \hat{a}(t)\right]$, where the Hamiltonian is given by (B.29); thus (restricting to two-body interactions) we must evaluate the commutator

$$\dot{\hat{a}}_n = \frac{i}{\hbar}\left[\left(\sum_{ik} H_{ik}^{(1)}\hat{a}_i^\dagger\hat{a}_k + \frac{1}{2}\sum_{iklm} U_{iklm}^{(2)}\hat{a}_i^\dagger\hat{a}_k^\dagger\hat{a}_m\hat{a}_l\right), \hat{a}_n\right]. \tag{B.54}$$

Applying the commutation relations (B.22a–c), we obtain the equation of motion,

$$\dot{\hat{a}}_n = \frac{i}{\hbar}\left(\sum_k H^{(1)}_{nk}\hat{a}_k + \frac{1}{2}\sum_{klm}\left(U^{(2)}_{nklm} + U^{(2)}_{knlm}\right)\hat{a}^\dagger_k\hat{a}_m\hat{a}_l\right). \tag{B.55}$$

For the special case of a one-body Hamiltonian, $\hat{H} = \sum_i \varepsilon_i\hat{a}^\dagger_i\hat{a}_i$, given by Eq. (B.31) (with eigenvalues, ε_i) Eq. (B.55) reduces to

$$\dot{\hat{a}}_n = \frac{i}{\hbar}\varepsilon_n\hat{a}_n \tag{B.56a}$$

or

$$\hat{a}_n(t) = \hat{a}_n(0)e^{i\varepsilon_n t/\hbar}. \tag{B.56b}$$

In a similar manner we may use the Heisenberg equation of motion to calculate the time dependence of $\hat{\psi}(t,\mathbf{r})$ using the Hamiltonian (B.35) (again restricted to two-body interactions),

$$\begin{aligned}\dot{\hat{\psi}}(t,\mathbf{r}) &= \frac{i}{\hbar}\left[\hat{H}, \hat{\psi}(t,\mathbf{r})\right]\\ &= \frac{i}{\hbar}\left[\int d^3r_1\hat{\psi}^\dagger(t,\mathbf{r}_1)\left(-\frac{\hbar^2}{2m}\nabla_1^2 + U^{(1)}(\mathbf{r}_1)\right)\hat{\psi}(t,\mathbf{r}_1)\right.\\ &\quad\left. + \frac{1}{2}\int d^3r_1 d^3r_2\hat{\psi}^\dagger(t,\mathbf{r}_1)\hat{\psi}^\dagger(t,\mathbf{r}_2)U^{(2)}(\mathbf{r}_1,\mathbf{r}_2)\hat{\psi}(t,\mathbf{r}_2)\hat{\psi}(t,\mathbf{r}_1),\ \hat{\psi}(t,\mathbf{r})\right]\end{aligned}$$

Using the commutation relations (B.39a,b), we obtain[12]

$$\begin{aligned}i\hbar\dot{\hat{\psi}}(t,\mathbf{r}) &= \left(-\frac{\hbar^2}{2m}\nabla^2 + U^{(1)}(\mathbf{r})\right)\hat{\psi}(t,\mathbf{r})\\ &\quad + \int d^3r'\hat{\psi}^\dagger(t,\mathbf{r}')U^{(2)}(\mathbf{r},\mathbf{r}')\hat{\psi}(t,\mathbf{r}')\hat{\psi}(t,r);\end{aligned} \tag{B.57a}$$

note that we obtain a *Schrödinger-like* operator equation (but see footnote 9).

B.8 Constructing many-particle state functions from vacuum state functions

Starting from the vacuum Bose state, denoted as $|0\rangle$, one can create states with an arbitrary number of particles in a state, k, simply by operating successively on

[12] It is assumed that all potentials act instantaneously between points $\mathbf{r}$ and $\mathbf{r}'$; hence the $\hat{\psi}$ operators are coupled only for equal times where we may apply our commutation relations. The commutation relation for $\hat{\psi}$ acting at different times cannot be written down in a general form. In relativistic quantum mechanics where an interaction is restricted to propagate at a velocity less than or equal to that of light, the problem becomes much more complex.

$|0\rangle$ with the operator, $\hat{a}_k^\dagger$. In so doing we must account for the factor $\sqrt{N_k + 1}$, which is generated. Thus the form

$$|0, 0, \cdots N_k \cdots\rangle = \frac{1}{\sqrt{N_k!}} \left(\hat{a}_k^\dagger\right)^{N_k} |0\rangle. \tag{B.57b}$$

Proceeding in a similar way we may consecutively add particles to other states:

$$|N_1, N_2, \cdots N_k \cdots\rangle = \frac{1}{\sqrt{N_1!}} \frac{1}{\sqrt{N_2!}} \cdots \frac{1}{\sqrt{N_k!}} \cdots \left(\hat{a}_1^\dagger\right)^{N_1} \left(\hat{a}_2^\dagger\right)^{N_2} \cdots \left(\hat{a}_k^\dagger\right)^{N_k} \cdots |0\rangle. \tag{B.58}$$

For the Fermi system we would write

$$|N_1, N_2, \cdots N_k \cdots\rangle = \left(\hat{c}_1^\dagger\right)^{N_1} \left(\hat{c}_2^\dagger\right)^{N_2} \cdots \left(\hat{c}_k^\dagger\right)^{N_k} \cdots |0\rangle \tag{B.59}$$

Note that in the Fermi case, operating with the same creation operator twice annihilates the vacuum state, and therefore the occupation numbers, N_k, may only be 0 or 1.

B.9 Performing calculations in the occupation number representation

In performing calculations using the occupation number representation, we are often presented with some Hamiltonian for which we wish to find the eigenvalues. If we are using the variational approach, we must calculate the expectation value of our Hamiltonian with some trial state function. Alternatively, if we are doing a perturbation expansion, we must calculate matrix elements of our Hamiltonian between some complete set of many-body basis states, as in conventional single-particle perturbation theory. These state functions are in turn generated by acting with creation operators on the vacuum state, as discussed in Section B.8. Now the Hamiltonian generally contains combinations of creation and destruction operators (often in pairs). The strategy is then to *use the commutation relations to successively commute all destruction operators to the right* (through all operators occurring in the Hamiltonian and through all creation operators used to generate the state functions from the vacuum state function) until, finally, they act on the vacuum state which they destroy, yielding zero. After completing the process of commuting all the destruction operators to the right, the commutation relations "leave behind" delta functions, which give non-zero contributions (to the sums or integrals occurring in the Hamiltonian, see Eq. (B.29) or (B.35), or in generating the state functions); what remains is some c-number function[13] multiplying some state function. If this resulting state function has a *projection* onto the other state function that lies to the left in our matrix element or expectation value (keeping in mind that a state with a well-defined set of occupation numbers is *orthogonal* to a state with any differing occupation numbers), then the resulting c-number is the desired result.

[13] We remind the reader of the following terminology which is regularly used in quantum mechanical discussions. A form that contains operators is called a q-number while an ordinary function is called a c-number.

Problems

B.1. Show that an arbitrary two-particle antisymmetric wave function, $\psi(x_1, x_2)$, in one dimension (two dimensions in configuration space) may be expanded as

$$\psi(x_1, x_2) = \frac{1}{\sqrt{2}} \sum_{n,m} a_{nm} \begin{vmatrix} \psi_n(x_1) & \psi_m(x_1) \\ \psi_n(x_2) & \psi_m(x_2) \end{vmatrix},$$

where $\psi_n(x)$ are a complete set of one-particle wave functions in the same space. Find the expression for a_{nm} and verify that it is correct. Do the same for a three-particle wave function, $\psi(x_1, x_2, x_3)$.

B.2. Assume you are given a many-electron state function in the occupation number representation of the form

$$|\Psi\rangle = \prod_{k} \left(u_k + v_k \hat{c}^{\dagger}_{k\uparrow} \hat{c}^{\dagger}_{-k\downarrow} \right) |0\rangle,$$

where $\hat{c}^{\dagger}_{k\uparrow}$ creates an electron in a state, $+\mathbf{k}$, with spin up, $\hat{c}^{\dagger}_{-k\downarrow}$ creates an electron in a state, $-\mathbf{k}$, with spin down, and $|0\rangle$ is the vacuum state. Using the commutation relations (B.62) and (B.63), find the condition that u_k and v_k must satisfy if the state, Ψ, is to be normalized such that $\langle \Psi | \Psi \rangle = 1$.

B.3. Suppose I wish to create a "Fermi sea" where all states with $|\mathbf{k}| < k_F$ are filled and all states with $|\mathbf{k}| > k_F$ are empty. What form must u_k and v_k in $|\Psi\rangle$ then have?

B.4. Next assume you are given a one-body Hamiltonian of the form

$$\hat{H}_0 = \sum_{k} \xi_k \left(\hat{c}^{\dagger}_{k\uparrow} \hat{c}_{k\uparrow} + \hat{c}^{\dagger}_{-k\downarrow} \hat{c}_{-k\downarrow} \right),$$

where the ξ_k are the energies of states with wave vector, $\mathbf{k}$. Show that the expectation value of $\hat{H}_0$ with Ψ is given by

$$\left\langle \Psi \middle| \hat{H}_0 \middle| \Psi \right\rangle = 2 \sum_{k} \xi_k |v_k|^2.$$

B.5. You are given three non-interacting $s = 1/2$ particles, each of which has spin states, $\begin{pmatrix} 1 \\ 0 \end{pmatrix}$ and $\begin{pmatrix} 0 \\ 1 \end{pmatrix}$.

(a) Assuming that these particles *are not* identical, construct all the spin states that are eigenfunctions of the square of the total spin operator, S^2, and the z axis projection, S_z.

(b) Now assume that the particles *are* identical and are in orbital states, ψ_1, ψ_2, and ψ_3. Write the combined orbital and spin wave states (where the latter continue to be eigenstates of S^2 and S_z) that are permissible for the three cases where: (i) all three orbital states differ; (ii) two of them are the same; and (iii) all three of them are the same.

Index

The manufacturer's authorised representative in the EU for product safety is Oxford University Press España S.A. of el Parque Empresarial San Fernando de Henares, Avenida de Castilla, 2 – 28830 Madrid (www.oup.es/en or product.safety@oup.com). OUP España S.A. also acts as importer into Spain of products made by the manufacturer.

www.ingramcontent.com/pod-product-compliance
Lightning Source LLC
Chambersburg PA
CBHW081415220926
51613CB00010B/211
* 9 7 8 0 1 9 8 7 4 2 9 0 6 *